6th International Conference of Euro Asia Civil Engineering Forum (EACEF 2017)

MATEC Web of Conferences Volume 138 (2017)

Seoul, South Korea
22 - 25 August 2017

Part 1 of 2

Editors:

Jae-Woo Park
Han Ay Lie
Harianto Hardjasaputra
Parama Thayaalan

ISBN: 978-1-5108-5578-6

Printed from e-media with permission by:

Curran Associates, Inc.
57 Morehouse Lane
Red Hook, NY 12571

Some format issues inherent in the e-media version may also appear in this print version.

Printed by Curran Associates, Inc. (2018)

For additional information, please contact EDP Sciences – Web of Conferences
at the address below.

EDP Sciences – Web of Conferences
17, Avenue du Hoggar
Parc d'Activité de Courtabœuf
BP 112
F-91944 Les Ulis Cedex A
France

Phone: +33 (0) 1 69 18 75 75
Fax: +33 (0) 1 69 28 84 91

contact-edps@webofconferences.org

Additional copies of this publication are available from:

Curran Associates, Inc.
57 Morehouse Lane
Red Hook, NY 12571 USA
Phone: 845-758-0400
Fax: 845-758-2633
Email: curran@proceedings.com
Web: www.proceedings.com

TABLE OF CONTENTS

PART 1

PART 2

Preface

The 6th International Conference of Euro Asia Civil Engineering Forum (EACEF 2017) was held at Hanyang University, in August 22nd to 25th, 2017. The conference was jointly organized by Hanyang University, Korean Society of Soil and Groundwater Engineering and Indonesian Universities Consortium. In this conference, the participants from 20 different countries was attended with various research topics in Civil and Environmental Engineering field.

The world pursues to minimize environmental impact on living and industry in these days. Following the flow, EACEF 2017 focused on environment-focused civil engineering. Under the slogan, "green infrastructure for future world", the EACEF 2017 aims to achieve and the promote quality growth of civil engineering.

EACEF 2017 consists of 9 different technical sessions: Sustainable Construction Materials, Structural & Construction Engineering, Concrete Engineering, Geotechnical Engineering, Construction & Safety Management, Hydraulics, Hydrology and Water Engineering, Transportation Engineering, Sanitation & Environmental Engineering, Climate Change, Disaster Management.

It was a great privilege to see that the conferences have become an effective platform to link the researchers, academicians and professional engineers from Europe, Asia and another part of continents. Especially those with a commitment to advanced sustainable development and environmental friendly buildings and infrastructures. Through the EACEF conferences, we can establish long-lasting international collaboration among us in research and even in projects. Also, it is great pleasure that we successfully publish nearly 125 of high-quality papers in MATEC Web of Conference.

I would like to thank our prominent speakers and all participants to the conference, as well as students and colleagues who contributions to the EACEF 2017.

Jae-Woo Park
EACEF 2017 Chair
Professor
Hanyang University, Korea

This work was supported by the Korean Federation of Science and Technology Societies (KOFST) grant funded by the Korean Government.

Statement of Peer review

In submitting conference proceedings to MATEC Web of Conferences, the editors of the proceedings certify to the Publisher that they adhere to the **Policy on Publishing Integrity** of the journal in order to safeguard good scientific practice in publishing.

1. All articles have been subjected to peer review administered by the proceedings editors.
2. Reviews have been conducted by expert referees, who have been requested to provide unbiased and constructive comments aimed, whenever possible, at improving the work.
3. Proceedings editors have taken all reasonable steps to ensure the quality of the materials they publish and their decision to accept or reject a paper for publication has been based only on the merits of the work and the relevance to the journal.

Title, date and place of the conference

The 6th International Conference of Euro Asia Civil Engineering Forum (EACEF 2017)
August 22-25, 2017
Seoul, South Korea

Proceedings editor(s):

Jae-Woo Park, Ph. D.
Han Ay Lie, Ph. D.
Harianto Hardjasaputra, Ph. D
Parama Thayaalan, Ph. D.

Date and editor's signature

November 22, 2017

Embodied Energy Calculation in Mitigating Environmental Impact of Low-Cost Housing Construction

Dewi Larasati ZR[1], *Yuni* Sri Wahyuni[2], *Suhendri*[3], *Sugeng* Triyadi[4]

[1, 3, 4] School of Architecture, Planning and Policy Development, Institute Technology of Bandung, Jalan Ganesha 10, Bandung
[2]Graduate Student of Department of Architecture, Institute Technology of Bandung, Jalan Ganesha 10, Bandung
[1, 2, 3] Green Building Research Center, Department of Architecture, Institute Technology of Bandung, Jalan Ganesha 10, Bandung

Abstract. Reduction of the negative impacts of building construction, or mitigation, should be done in the early phase of a project. Particular method that can be used to reduce the negative impact is Embodied Energy (EE) calculation. This research develops a method to calculate the total EE of materials used in building construction by using the Material Selection Analysis (MSA) based on Indonesia's Construction Work Unit Price Analysis (WUPA). MSA breaks down the types, its unit, and total volume of materials within an object. Using the EE data from Inventory Carbon and Energy (ICE) Bath, in the UK, this research calculate the embodied energy in three small house types in Indonesia: Type 21 M2, Type 36 M2, and Type 45 M2. The result reveals a negative correlation between building's area and EE value. Nonetheless, the EE value is positively related to the volume of material used. Proposed strategies to decrease the value of EE in low cost housing construction manly; a) design approach in improving the efficiency of material usage, b) materials substitution to material with lower value of EE, especially on material on large volume of usage, c) socialization of EE value to stakeholders as consideration in determining the choice of materials used, thus, environmental mitigation is expected to be an important aspect in determining the design decisions and the choice of materials used.

1. Introduction

As a developing Country, Indonesia still has a problem with rapid population growth. Housing development in Indonesia does not follow the population growth, thus create a housing backlog problem. Based on data from Statistics Indonesia 2015 (National Bureau Statistic), housing backlog in Indonesia is about 11 million units. Indonesian Government builds not less than 600-800 thousand houses every year in order to fulfil the needs. The effort is also aided by private sector such as developers, in many Indonesian cities and regions.

Massive housing development clearly makes housing construction in Indonesia as a massive building materials consumer. The development tends to be more into low cost housing than another housing type. This type of housing is affordable to most of Indonesian people, which mostly have low income. Although, the housing unit is small in size, but if it is built in a massive number, there can be a significant material consumption [1].

Further, building material consumption, especially when used in a big number, has something to do with sustainability. Exploitation of natural resources due to building materials production need a special attention. Production processes of building materials use a lot of energy and emit considerable amount of CO^2 to the atmosphere. Octaviana (2007) in Siahaan [2] states that carbon emission of a low cost house relate strongly to three factors such as material source, transport of material to the site, and respiration of labour.

Energy used and carbon emission rate of a material is calculated in the Embodied Energy (EE) Calculation. The calculation is usually done in the phase of preliminary design and design development. The EE value is a useful tool to give an insight about the buildings environmental impact [3]. This paper develops an EE calculation model and how to implement it in housing development in Indonesia. Although the model is developed based on the low cost housing case, it shall be able to use in other type of projects.

2. Methodology

There are four phases in this study: 1) literature review on EE calculation model for low cost housing in Indonesia; 2) developing EE calculation model based on Material Unit Analysis (MUA) by referring to Inventory of Carbon and Energy from the University of Bath and field calculation; 3) comparison of EE value of different types of low cost housing in Indonesia; 4) questionnaire and design approach to proposing strategies in mitigating environmental impacts by reducing EE value.

3. Consideration of Embodied Energy in Selecting Building Materials

Treloar in Dixit [4] mentions that building embodied energy is responsible for 20-50 times of annual operational energy of a building. Another reference states that building materials industry consumes as much as 20% of world's fuel consumption. It will even be higher if fuel used in material transportation is include [5]. Hence, EE calculation is something important to conduct.

All of the phase in building material production require a lot of fuel, and mostly come from fossil fuel. The production process clearly gives impact to the environment in form of carbon emission. Hence, architect should consider the EE value earlier in the design process. Considering EE value earlier can reduce the environmental impact of a building in a lower cost than if it is done later. Reducing EE value of material can be done by selecting low EE materials, low-energy construction method, closer material source, and efficiency in material transportation.

In practice, embodied energy calculation requires a database of materials' EE value. There are many EE value inventories with different level of quality and accuracy. However, Indonesia did not have an inventory of materials' EE until now. Even, there is no standardise EE calculation model. Thus, EE calculation model for Indonesia is important to develop. As a starting, the model can use Inventory of EE materials from other countries.

 DOI: 10.1051/matecconf/201713801001

4. Embodied Energy Calculation Model Development

Haynes [6] says that there are three phases of EE calculation based on the materials production stage. Firstly, cradle to the gate EE calculation, which calculates EE value of materials from natural resources extraction until it becomes a product. This EE calculation approximately counts 59% of the actual total embodied energy of materials. Secondly, EE calculation that measures the materials transportation and it counts for 10% of the total embodied energy. Thirdly, EE calculation that calculates energy used in construction method when the materials applied to a building and it counts for 9%. The third type of EE calculation also measured the recurring embodied energy based on a 'churn rate' during a building life cycle, and counts for 22% of the total embodied energy of materials.

As explained above, EE in the cradle to the gate phase has the most significant value to the total embodied energy. Therefore, this paper focuses on EE value measured in this phase. Embodied energy calculation begins with materials' volume calculation. Every materials used in the building should be calculated, even to the smallest materials like sand, gravel, PC, etc. The materials' volume is calculated based on building components such as foundation, floor, wall, and roof. After determining the materials' volume, the calculation goes through some steps as follows:

- Step 1: Data collection. This step collects all of available data about energy in the materials making process, and then summarise the minimum and maximum EE value of each materials.
- Step 2: EE value selection. The mean EE value data from the first step are then analysed using boundary system cradle to the gate to get the most reliable value.
- Step 3: Eliminating the unnecessary aspects in the EE calculation, in order to get the fine value of embodied energy.

Further, the calculation model is shown by Figure 1 below.

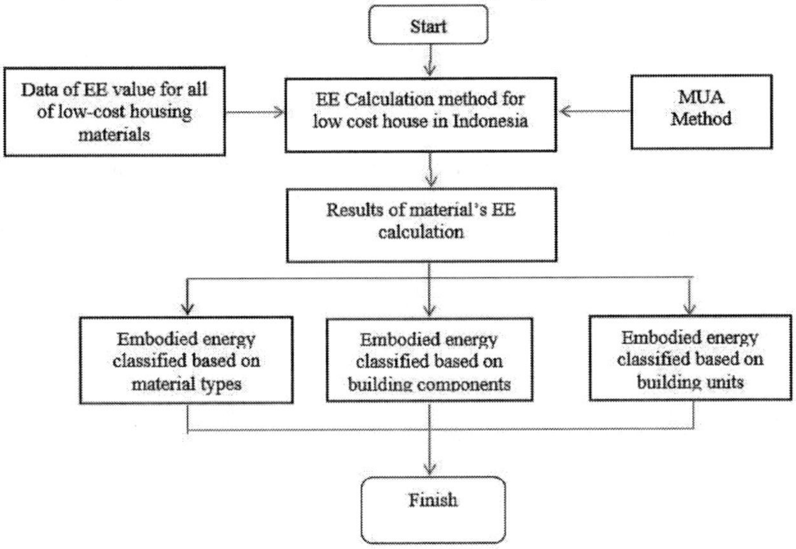

Fig. 1.. Calculation of material's embodied energy value used in this study [7]

4.1 Development of Material Unit Analysis (MUA) and Unit Conversion

Material Unit Analysis (MUA) as a basic tool in the EE calculation is developed in reference to Construction Work Unit Cost or *Analisa Harga Satuan pekerjaan* (the AHSP).

The AHSP is a standard used for bill of quantity and cost estimation of construction projects in Indonesia. AHSP *guideline is issued by the National Standardization Agency (BSN)*. It consists of work components of building, are; 1) materials coefficients, 2) unit, 3) description of work, 4) work volume.

In terms of embodied energy calculation for this study, information used from the AHSP is the materials list, and its coefficients of a construction work. Material's coefficients reveal the volume of materials needed per unit area/volume of a construction work. MUA analysis can extract the all of materials needed in a construction of building components and its measurement unit such as kg, m^2 or m^3.

Data resulted from MUA analysis is then converted into kilogram (kg). This is due to the fact that the embodied energy is calculated in Joule per kilogram (J/kg). So, all of volume units other than kg needs a conversion. The following table (Table 1) explains an example of MUA analysis and unit conversion.

Table 1. MUA Analysis and unit conversion

Code	Coeff.	Unit	Unit Conversion	The Standard Unit	Construction Work's Description
K	70	piece	70 x 3 kg = 210	Kg	Red brick, 5 x 11 x 22
	18,95	Kg	18,95	Kg	Portland Cement
	0,0380	m^3	p = m/V, so m = p x V	Kg	Sand

Table 1 above shows a unit conversion from brick wall construction. The work needs three materials, red brick, Portland cement, and sand, each material has different unit, and thus requires conversion. The coefficient state volume needed for the material in per 1 m^2 brick wall construction. The process is then repeated in all construction works for the low cost house. Based on MUA analysis of general design of low cost housing in Indonesia, there are at least 47 different materials used in low cost house construction. However, there are 8 materials used in a big volume, which the value of EE for each material as displayed in Table 2.

Table 2. 8 materials used in big volume in the low cost housing construction in Indonesia

No	Material Name	EE (MJ/Kg)
1	Wood	2,50
2	Steel	25,00
3	Glazing	15,00
4	Stone	142,40
5	Red brick	2,50
6	Ceramic roof tile	12,00
7	Ceramic	12,00
8	Paint	59,00

4.2 EE Calculation Analysis Based on Building Component

From the 8 (eight) materials, there is only one material comes from sustainable resources (wood). Meanwhile, 7 (seven) other materials are made of non-renewable sources. The result also reveals that red brick and ceramic roof tile have the most energy embodied in them compared to other materials. The amount of EE value on the percentage of material usage in construction of ow cost housing Type 21 M2 can be seen in Figure 2 below.

DOI: 10.1051/matecconf/201713801001

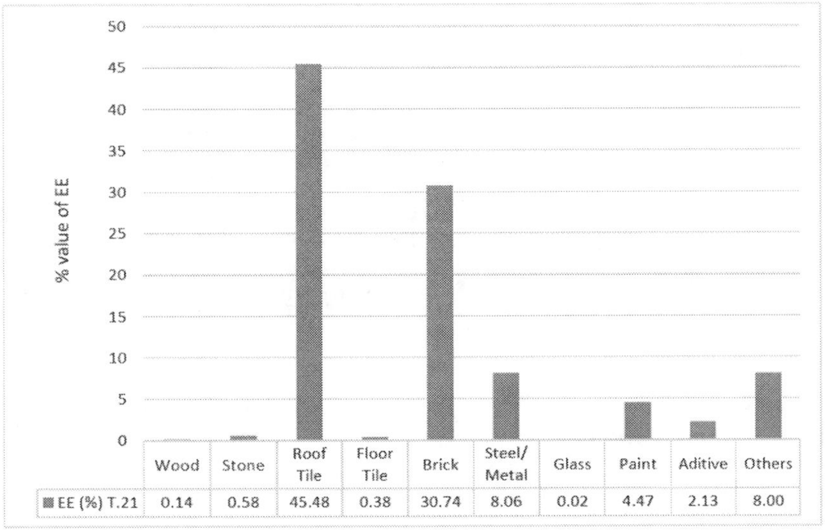

Fig. 2. Value of EE based on percentage of material usage in Type 21 M2 low cost housing construction in Indonesia

Figure 3 below displays EE calculation results in Type 21 M2 low cost house. EE value appears in roof component (46%) and wall (42%). Further, Type 36 M2 low cost housing also shows that roof is the main contributor to the embodied energy of the building (48% and wall is in the second place (38%). Meanwhile, Type 45 M2 calculation result comes with the highest EE value of all, and still with roof as the main contribution (48%) and wall (36%).

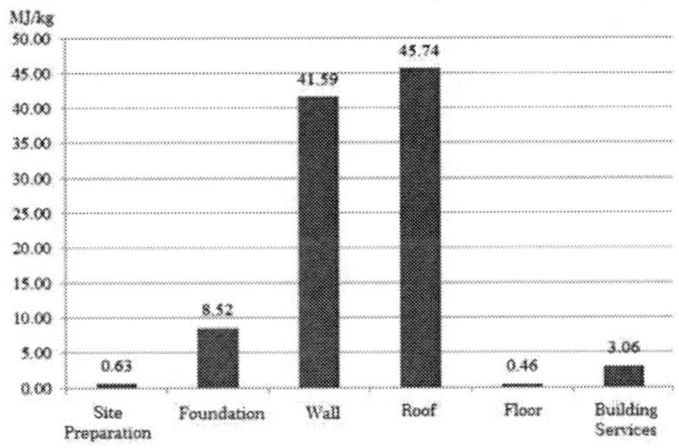

Fig. 3. Comparison of EE value by building components for Type 21 house

4.3 EE Calculation in Comparison in Different Type of Low Cost Houses

Figure 4 below reveals the increasing of total EE value along with the increasing house area. Although, if the comparison is made for two Type 21 house which have total area of 42 m^2 and one Type 45 house, the Type 21 houses have more EE regardless its smaller area.

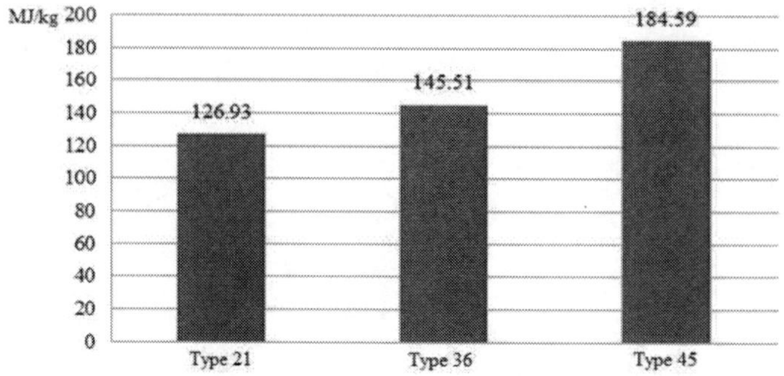

Fig. 4. Comparison of EE value by house types

Furthermore, the comparison of EE value based on building components in Figure 5 shows a same pattern in each component for different type of houses. The figure also shows that roof has the highest embodied energy, followed by wall, and foundation and soil works.

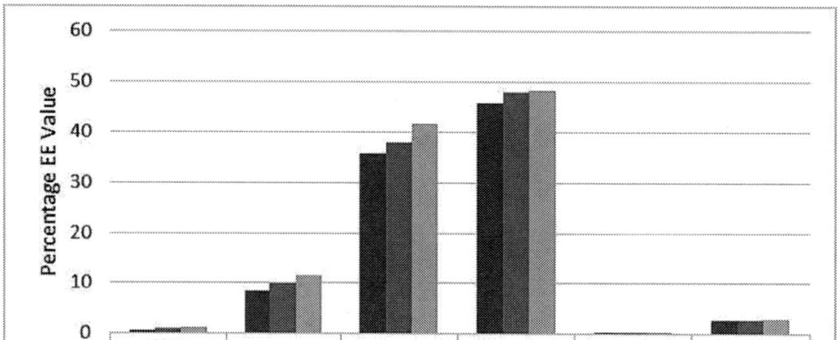

	Site Preparation	Foundation	Wall	Roof	Floor	Utility
■ T.21	0.63	8.52	35.6	45.74	0.45	2.67
■ T.36	1.12	9.91	37.93	47.92	0.46	2.86
■ T.45	1.18	11.53	41.59	48.38	0.47	3.06

Fig. 5. EE value of building components for the low cost house types (MJ/kg)

5. Reducing Embodied Energy of Low Cost Housing in Indonesia

In order to decrease the embodied energy of low cost housing in Indonesia, first thing first is to increase the awareness of the people about the environmental impact of a construction. From that awareness, people with their own motive, will construct a building efficiently.

The results of the questioners to stakeholders regarding the proposed substitute material that have high EE values to material with low value of EE, indicating a good response from stakeholders, including proposal in using sustainable and environmentally friendly materials as long as the material has criteria as follows:

- Easily obtainable during construction,
- The price of material is low and affordable
- The mass of material is light, aesthetic and user friendly

Regarding the strategy for reducing the Value of EE in low cost housing construction, some actions that can lead to embodied energy reduction are mention below:

- Focus on reducing the materials that have large volume such as making the wall area smaller by adding windows or put proportional doors in the wall. Another strategy is by using exposed brick instead of plastered brick wall.
- Focus on the high volume of materials usage, substituting them with alternative materials that have lower embodied energy can be a good practice. This can be done by thoroughly searching for lower embodied energy materials in the inventory.
- Optimisation of low cost housing type. According to the result, Type 21 house is less efficient in material usage than other types. Therefore, Type 21 house should be cut down, and build more type 36 or 45 houses. Also, it is still better to divide the Type 45 house into two 22 m^2 houses than build two Type 21 houses.
- Dissemination of EE calculation model by using simple software to make the stakeholders more aware about the reduction of EE value of their building.
- At the policy maker level, the regulation should be promoted to emphasis the construction of green low cost housing. It can also use a rewards or incentive to encourage the developer to be more sustainable.

Conclusion

Some conclusions can be drawn from the analysis above. Fabricated materials tend to have higher embodied energy compared to materials that extracted directly from natural resources. However, material's volume also drives the embodied energy value of the low cost house types. Thus, no-fabricated materials such as wood or even sand can have a high embodied energy value in the low cost house types. Hence, substituting high volume materials that has high embodied energy with the lower embodied energy one is significant for reducing the total embodied energy value. Government policy is needed in providing incentives for stakeholder who applies materials with low EE values. On the other hand, it is necessary to undertake material dissemination and socialisation of low EE values in facilitating the wider use of environmentally friendly and sustainable building materials by the stakeholder. Further research is needed to explore the possibilities of such alternative materials and the acceptability of them by the low cost housing resident.

References

1. Thormark, C. "A low energy building in a life cycle—its embodied energy, energy need for operation and recycling potential". Build. Environ. **37**, 429–435 (2002).
2. Siahaan, N. "Model Pengendalian Perumahan Sederhana Dalam Sistem Perumahan Berkelanjutan Perkotaan Berbasis Rendah Emisi CO2". (2012).
3. Dixit, M. K., Fernández-Solís, J. L., Lavy, S. & Culp, C. H. "Identification of parameters for embodied energy measurement: A literature review". Energy Build. **42**, 1238–1247 (2010).
4. Dixit, M. K., Culp, C. H. & Fernández-Solís, J. L. "System boundary for embodied energy in buildings: A conceptual model for definition". Renew. Sustain. Energy Rev. **21**, 153–164 (2013).
5. Dixit, M. K., Fernández-solís, J. L., Lavy, S. & Culp, C. H. "Need for an embodied energy measurement protocol for buildings : A review paper. **16**, 3730–3743 (2012).
6. Haynes, R. "Embodied Energy Calculations within Life Cycle Analysis of Residential Buildings Table of Contents : 2010", 1–16 (2013).

7. Wahyuni, Y. S. "Perkembangan Riset Embodied Energy dalam Arsitektur". in Prosiding Seminar Doktor LPPM Universitas Muhammadiyah (UMMIPress, 2015).

Enzyme based soil stabilization for unpaved road construction

Rintu Renjith[1] , *Dilan* Robert[1], *Andrew* Fuller[2], *Sujeeva* Setunge[1], *Brian* O'Donnell[3], and *Robert* Nucifora[2]

[1]Civil Engineering, RMIT University, Australia
[2]Infrastructure Services, Downer Group Ltd, Australia
[3]Center for Pavement Excellence Asia Pacific

Abstract. Enzymes as soil stabilizers have been successfully used in road construction in several countries for the past 30 years. However, research has shown that the successful application of these enzymes is case specific, emphasizing that enzyme performance is dependent on subgrade soil type, condition and the type of enzyme used as the stabilizer. A universal standard or a tool for road engineers to assess the performance of stabilized unbound pavements using well-established enzymes is not available to date. The research aims to produce a validated assessment tool which can be used to predict strength enhancement within a generalized statistical framework. The objective of the present study is to identify new materials for developing the assessment tool which supports enzyme based stabilization, as well as to identify the correct construction sequence for such new materials. A series of characterization tests were conducted on several soil types obtained from proposed construction sites. Having identified the suitable soil type to mix with the enzyme, a trial road construction has been performed to investigate the efficiency of the enzyme stabilization along with the correct construction sequence. The enzyme stabilization has showed significant improvement of the road performance as was evidenced from the test results which were based on site soil obtained before and after stabilization. The research will substantially benefit the road construction industry by not only replacing traditional construction methods with economical/reliable approaches, but also eliminating site specific tests required in current practice of enzyme based road construction.

1 Introduction

Unpaved roads cover more than 56% of the Australian road network [1] and are increasingly being constructed in rural countries due to less construction cost. Major issues of unpaved roads include loss of fines in the form of dust, wasting of natural resources caused by washouts and landfills, detriment water courses, aquifers and ecology and most importantly cause vehicular damage and safety issues. Chemical stabilization using non-traditional additives has been identified as an effective solution for such issues in contrast to the application of traditional stabilizers such as lime, cement or flyash. This is mainly due to environmental and economic viability of traditional stabilizers that can result high emissions of greenhouse gases along with unbearable costs in their applications. Enzyme based soil stabilization is a form of non-traditional chemical stabilization that has been successfully used in road construction for the past 30 years. Despite its use for the last 3 decades, there has been no universal standard or tool that practicing engineers can use to assess the performance of unbound pavement stabilized with enzyme additives. This study aims to produce a validated assessment tool which can be used to predict strength gain within a generalized statistical framework. Within this broad scope, the present study aims to identify a suitable enzyme which can stabilize fine grained soils. Firstly, a series of characterization tests were conducted on several soil types obtained from proposed construction sites. Having identified the suitable soil type to mix with the enzyme, the trial road construction has been performed to investigate the efficiency of the enzyme stabilization along with the correct construction sequence. The enzyme stabilization has showed significant improvement of the road performance as was evidenced from the test results which were based on site soil obtained before and after stabilization.

2 Literature Review

The application of enzymes for road construction dates back to 1981 when it was used to stabilize a half a mile of road in Oklahoma [2]. Since its introduction in the road construction industry, this form of stabilizer is gaining significant exposure. However, the industry is not readily accepting this form of stabilizer due to improper documentation and its unfamiliarity.

There are many reported studies in literature that suggest the advantage of enzyme stabilization for road construction. Scholen [2] reported sites such as Red Maple road of Oucahita National forest that benefitted from enzyme treatment which did not show any signs of damage such as rutting even during the log hauling season when the adjacent untreated sections were prone to severe rutting. Field application of enzymes have also been proved efficient in many cases in nations like Brazil, China, India, Malaysia, South Africa and USA, resulting in increase in load bearing capacity of roads and durability along with significant decrease in road maintenance costs despite harsh weather conditions [3-10]. However, there are several other studies in literature highlighting less efficiency of enzyme based stabilization. Scholen [2] showed that, even though the stabilization has been effective in many sites, non-effective outcomes have also been resulted due to treated layer failure from improper construction techniques. Bergmann [11] identified the failure to attain strength on the harsh conditions such as the dropping of temperature to -3.9 degrees at one tested site. Moloisane and Visser [12] highlighted excessive wet conditions as a plausible reason to the failure and inconsistent results during trial road construction. Therefore, it is evident from the literature that the use of enzymes in road construction needs better understanding by careful characterization of enzyme and soil using laboratory tests prior to apply in the field conditions.

Traditionally, the application of enzymes in the field is followed after a series of testing conducted under laboratory conditions. The purpose of these tests can be various, such as to investigate the response of the proposed trial mix as specified in the manufacturer's specification, find out the optimised mix design or to explore the efficiency of new materials (& design proportions) for applying into road construction. The oldest laboratory tests on enzyme treated soils date back to early 1990's where Khan and Sarker [13] tested the effects of adding enzyme to fly ash stabilised soil. The tests showed significant increase in strength of soil when compared to lime stabilized soil. There are studies that showcased strength improvement range of 136 to 1800 times the original CBR [14]. However, similar to the field applications, few studies highlighted that the treatment of soils with additives can also give non-satisfactory results or minor strength increase. Tingle, Newman [15] treated low plastic clayey soils with enzyme, but only improved the stabilized soil strength by 4-6%. Rauch, Katz [16] also found out that the stabilized samples had not showed satisfactory results due to sample preparation method. In general, the literature suggests that the inefficiency of stabilization can be due to factors such as flaws in sample preparation, inadequate curing, and inappropriate soil type for stabilization [16, 17]

As highlighted above, the efficiency of stabilization can be either positive or negative, depending on many factors such as soil type, soil initial condition, enzyme type, enzyme amount, curing, temperature, etc. As the primary purpose of many studies involved stabilization is to identify the positive spectrum for a given condition, the mechanism of stabilization needs to be explored at the first place. Even though the results of enzyme application have been reported extensively in literature, the mechanism of stabilization process has been rarely identified in majority of studies. Out of very few identified mechanisms, the most accepted hypothesis suggests that the enzyme based stabilization is through the process of organic encapsulation. In this process, the positively charged enzyme particles encapsulates the organic molecules in the negatively charged clay particles which have high affinity for water. With this process, the clay's double layer water as well as the affinity for water decreases. The clay particle is more stable and binds well with other clay particles causing aggregation and reduction of surface area [2]. This hypothesis was further supported by Rauch, Harmon [18] with experiments which showcased the increase in d-spacing, expansion of clay layer, reduction of pore sizing and surface area as shown in Fig.1

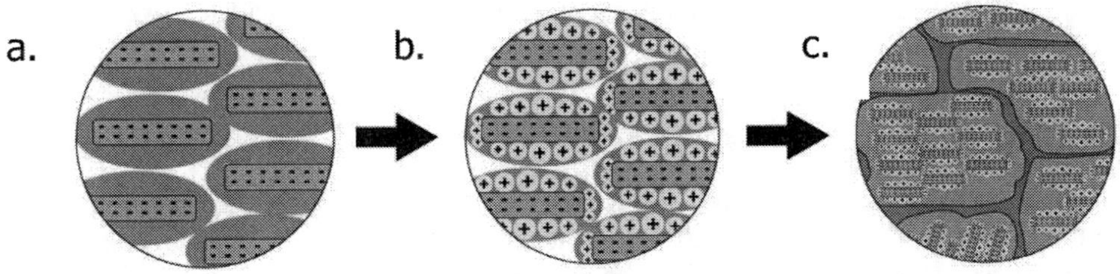

Fig. 1. (a.) Natural clay particle with high affinity for water (b). Organic encapsulation decreasing double layer water and affinity for water (c). The more stable clay sheets bind together with other clay particles causing aggregation and reduction of surface area

3 Methodology

The objectives of the current study were achieved using laboratory experiments, field construction and pavement performance assessment. Firstly, laboratory tests were conducted on the basis of several soils to identify the physical characteristics of the soils obtained from various construction sites in Melbourne. Having identified the suitable soil as per to manufacturer's guideline, the field construction has been performed on the basis of enzyme specification. A mechanistic pavement design has been conducted using CIRCLY to investigate the limit state functionality of the constructed road [1]. The constructed pavement has also been tested using Japanese formula to validate the functionality of the constructed road [1]. This section provides details of lab testing (including properties), field construction stages and the overview/properties of analytical models used for road performance assessment.

3.1 Soil Characteristics

A series of laboratory tests was conducted on different soil types to investigate the suitable soil for soil stabilization. Reference tests such as Gradation tests, Atterberg tests and compaction tests were performed to obtain physical characteristics of soils as per the relevant Australian Standards. Four soil types were tested for this study, which are labelled as S1, S2, S3 and S4 in this manuscript. Results of the reference tests are summarised in Table 1.

Table 1: Reference test results of the soils tested in current study

Soil ID	USCS[1]	% less than 75μ	LL[2]	PL[3]	PI[4]	OMC[5]
S1	SM	13.2	29	13	16	5
S2	SM	4.6	Non-Plastic			6.6
S3	SC	34	21	13	8	6.2
S4	SM-SC	16	28	18	10	6.1
[1]Unified Soil Classification, [2] Liquid Limit, [3] Plastic Limit, [4] Plasticity Index, [5] optimum moisture content						

Based on the current specification of the utilized enzyme, soil types S3 & S4 can be identified as the most suitable soil which works efficiently with the selected enzyme (enzyme specification identifies the suitable soil as having PI between 6-15 for effective stabilization[19]). Therefore, S3 & S4 has been selected for the field test which has been demonstrated in the section 3.3.

3.2 Enzyme used in the study

Eko-Soil was used as the stabilizer in this study. The product, which is a natural, non-toxic and bio-degradable product, is a stabilizing agent that helps in the reduction of voids in the soil layer and in turn increases the load bearing capacity of the soil.

3.3 Trial Road Construction

Two trial road sections of an existing unsealed trial road (Fig. 2) was constructed using enzyme stabilization with the materials S3 and S4 as identified from Table 1. This section summarises the construction stages of the unsealed road construction. Prior to the construction, the trial road subbase was shaped, trimmed and compacted to at least 98% of proctor compaction by measuring the field density of compacted soil using nuclear gauge and manual dry density measurements. Imported materials from the sites having soil type S3 and S4 were added to the pavement at a depth of 150mm at different sections of 300mx5.5m which was selected to be stabilized as the pavements base/wearing layer. Soil samples, before and after adding the enzyme were collected in a sealed plastic bags and delivered to the labs for conducting the mechanical testing (i.e. CBR). The trial road construction was performed on the selected sections of the road (Fig. 3a) in 5 different stages as described below:

DOI: 10.1051/matecconf/201713801002

Fig. 2. Construction site showing trial road sections (Image from Google Earth, 37°06'34.7"S 143°40'09.8"E*)*

Construction Stage 1: Ripping

The compacted pavement was ripped to 150mm depth with the grader, which allowed the moisture levels to be easily adjusted, increase the depth of enzyme penetration and to reduce the possibility of losing enzymes through drainage (Fig. 3b).

Construction Stage 2: Moisture Adjustment

The moisture of the pavement was adjusted via passing the dribble bar from the watercart over the soil several times until it was deemed to be just before optimum moisture content (OMC) as determined by a squeeze test by the enzyme supplier's consultants onsite (Fig. 3c). The controlled water content was based on the selected enzyme percentage (1% by dry weight) and the OMC of the selected soil type.

Construction Stage 3: Dilution of Enzyme and Spreading

After assessing the moisture content of the pavements material, the pre-calculated diluted enzyme was added into the pavement material using the same watercart. The dilution rate used in the current study was 1:500. Once the enzyme was added from the watercart, it was evenly spread over the pavement for uniform bed preparation (Fig. 3d)

Construction Stage 4: Mixing the Enzyme

Once the enzyme was sprayed into the wet soil, soil + water + enzyme were mixed thoroughly using the stabilizer (Fig. 3e).

Construction Stage 5: Compaction, Shaping and Final Moisture Adjustment

Finally, the pavement was compacted by passing the multi-tyre and smooth drum rollers several times over the pavement. Moreover, shaping was performed by the grader until a cross fall of at least 6% was achieved (Fig. 3f).

(a) S4 trial road section before stabilization

(b) Stage 1: Ripping of trial section with grader

 DOI: 10.1051/matecconf/201713801002

(c) Stage 2: Moisture Adjustment with watercart

(d) Stage 3: Soil after dispensing the enzyme

(e) Stage 4: Mixing of soil and enzyme

(f) Stage 5: Finished road section after shaping and compaction

Fig. 3. Construction stages of trial road construction

3.4 CIRCLY and Japan Model

CIRCLY

A mechanistic pavement design has been conducted in this study using CIRCLY to investigate the allowed traffic load which satisfies the strain and rutting limits as specified in the Austroad standards [1]. Results from CIRCLY analysis provide an estimate of the pavement layer depth required to sustain traffic loads. The geometry for the analysis and the input parameters are showed in Fig. 4 and Table 2.

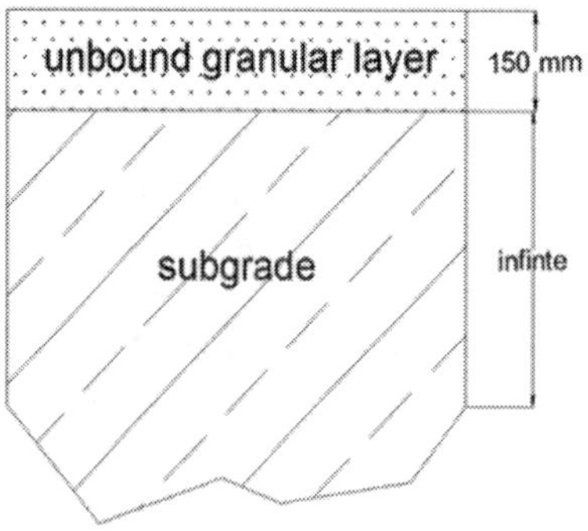

Fig. 4. CIRCLY pavement diagram

 DOI: 10.1051/matecconf/201713801002

Table 2 CIRCLY input parameters

Design Equivalent Standard Axle (DESA)	4 x 10^3 [1]
Project Reliability	85% [2]
Traffic Multiplier	1 [3]
Subgrade Thickness	0.00 [4]

[1] Minor road with 2 lanes, [2] based on Austroads pavement design for unbound pavements, [3] chosen traffic multipliers, [4] 0.00 represents infinite depth

Japanese Model
This model refers to the Japan Road Association formula which determines the equivalent subgrade strength [1]. This model determines the required CBR to be achieved by pavement layers for an average weighted CBR of 5 which is currently being used as a rule of thumb by road contractors from experience. The equivalent subgrade strength is based on the following equation.

$$CBR_m = \left[\frac{\sum_i h_i CBR_i^{0.33}}{\sum_i h_i}\right]^3 \leq 20 \qquad (1)$$

Where CBR_i is the CBR value in layer thickness h_i and $\sum h_i$ is taken up to a depth of 1.0m

4 Results

4.1 Field road performance

A careful monitoring of the moisture and dry density was conducted to ascertain that the construction follows the expected construction standards. Table 3 summarises the monitoring results. It was observed that the construction was able to achieve reasonable moisture ratios, but unable to meet 98% proctor density as specified by Vicroads Code of Practice RC 500.20 [20]. This could be mainly due to the use of multi-tyre roller instead of a vibrating roller.

Table 3 Moisture and density monitoring of the trial road construction

Soil ID	Moisture Ratio (%)	Density Ratio (%)
S3	98.0	92.5
S4	96.5	89.5

CBR tests were performed as per AS1289.6.1.1 [21] on the basis of samples obtained from the road, before and after stabilization. Samples were obtained at two sections of the road where soil type S3 and S4 were utilized. Fig. 5 summarises the test results from the lab tests. As it can be seen, the stabilization has been significantly effective to increase the road strength from prior stabilization to post-stabilization (increase of 69 CBR and 101 CBR respectively).

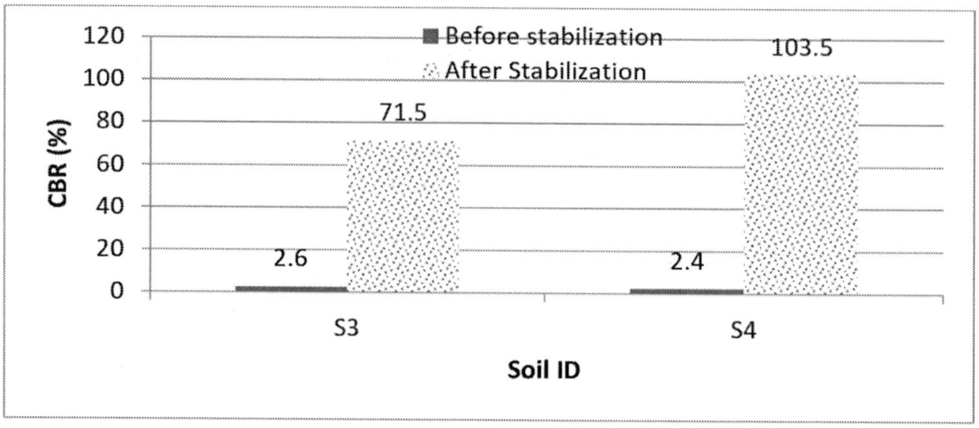

Fig. 5. CBR of the soils obtained from the trial road

4.2 CIRCLY and Japan Model Results

CIRCLY analysis was conducted on the designed pavement to satisfy the cumulative damage factor (CDF<1) of the designed pavement as specified in Austroads [1]. Table 4 summarises the results of the analysis, which shows the minimum depth required for the treated base layer as 350mm and 369mm for S3 and S4 respectively. They revealed substantial reduction of material (22% and 23% reduction for S3 & S4 respectively) for the base layer from prior-stabilization to post-stabilization.

Table 4. Results of CIRCLY analysis

Soil ID	Thickness of layer (before stabilization)	Thickness of layer (after stabilization)	Difference in layer depth
S3	451mm	350mm	101mm
S4	479mm	369mm	110mm

Results also revealed that the required thickness for stabilized layers from the standard mechanistic pavement design tool is substantially higher than what has been adopted (i.e.150mm) in current trial road construction. However, the weighted average CBR for the treated soil obtained from the Japanese model (Table 5) exceeds 5.0 which is currently being used as a rule of thumb by road contractors from experience. Therefore, it can be seen from the current study that the treated soil provides adequate strength to satisfy the constructed road to operate as lightly trafficked unbound road. The road is currently being monitored for its performance under operational traffic loads to verify the effectiveness of the selected enzyme stabilization.

Table 5. Results of Japan Formula

Material	Layer	Depth (m)	CBR per layer	Weighted average CBR
S3(non-treated)	Unbound layer	0.15	2.6	2.57
	subgrade	0.85	2.6	
S3 (treated)	Unbound layer	0.15	71.5	5.63
	subgrade	0.85	2.6	
S4 (non-treated)	Unbound layer	0.15	2.4	2.38

	subgrade	0.85	2.4	
S4 (treated)	Unbound layer	0.15	103.5	6.11
	subgrade	0.85	2.4	

5 Conclusion

The current study presented the efficiency and the correct use of new enzyme in the construction of unsealed road. Several soil types have been analysed to identify the suitable soil to mix with the enzyme in the field application. The enzyme content and dilution mass ratio has been selected on the basis of the current enzyme specification. CBR test results for samples obtained from the constructed road showed a significant increase in the road strength from prior-stabilization to post-stabilization. Though the predictive analysis performed based on standard mechanistic tool suggested increased layer treatment thickness, analysis results based on a practising model revealed that the constructed stabilized road exceeds the minimum specifications required. While this response shows over conservatism of existing standards, the results have potential to identify suitable/economical construction material and the correct construction sequence of unsealed road with the use of the selected material. The research is being continued at present to identify most suitable mix-design for the new enzyme within a generalized statistical modelling framework, incorporating the factors such as soil type, enzyme content, dilution mass ratio and organic content of soil. The results from this research will cater road construction industry for implementing road infrastructure at a reduced cost, but with enhanced road performance.

Authors would like to acknowledge the support from Centre for Pavement Excellence Asia Pacific (CPEAP), Central Goldfield Shire and Downer Ltd for conducting trial road constructions in this research. Authors would also like to thank Mr Malcolm Styles of Engineering Management Styles for sharing valuable resources and expertise during the trial road construction.

References

1. Austroads, *Austroads pavement design guide..* (2001),
2. D.E. Scholen, *Non-standard stabilizers.* (1992).
3. *Ants revolutionize road construction in South America*, in *Toronto Star.*, Toronto Star: Toronto, Ont. p. C5. (1990)
4. A. Hitam, A.Z. Yusof, and O. Samad. *Soil stabilizer for plantation road,*(1999)
5. R. Brazetti and S.R. Murphy. *Objective performance measurement of actual road sites treated with an Organic Soil Stabilizer* (2000).
6. R. Thompson, R. Johnson, and J. Sedgwick, *Road Stabilization Report*, in *Tooele County, Utah..* p. 17. (2002)
7. A. Campbell and D. Jones, *Soil Stabilization in Low-Volume Roads.* T.R.R. Journal. **2204**: p. 172-178. (2011)
8. Y.J. Li, L. Li, and H.C. Dan, *Study on Application of TerraZyme in Road Base Course of Road.* A.M.M.,. **97-98**: p. 1098. (2011)
9. A.R. Shankar, R. Mithanthaya, and B. Lekha. *Laboratory studies on bio-enzyme stabilized lateritic soil as a highway material* (2012).
10. W.S. Guthrie, D.O. Simmons, and D.L. Eggett, *Enzyme Stabilization of Low-Volume Gravel Roads.* T.R.R. Journal,. **2511**: p. 112-120. (2015)
11. R. Bergmann, *Soil stabilizer for use on universally accessible trails.* (2000).

12. R. Moloisane and A.T. Visser, *Evaluation of the strength behaviour of unpaved road material treated with electrochemical-based non-traditional soil stabilisation additives.* J.S.A.I.C.E. **56**(1): p. 28-39. (2014)
13. L.I. Khan and M. Sarker, *Enzyme enhanced stabilization of soil and fly ash.* Fly ash for soil improvement. Asce geotechnical special publication. **36**, (1993.)
14. K.P. Isaac, P. Biju, and A. Veeraragavan. *Soil Stabilization using Bioenzymes for Rural Roads.* (2003)
15. J. Tingle, et al., *Stabilization mechanisms of nontraditional additives.* T.R.R. Journal, (**1989**): p. 59-67 (2007)
16. A.F. Rauch, L.E. Katz, and H.M. Liljestrand, *AN analysis of the mechanisms and efficacy of three liquid chemical soil stabilizers. Research Report 1993-1: VOLUME.* Work,. **1**. (2003)
17. D.J. Robert, R. Renjith, S. Setunge, B. O'Donnell, D. Hewawasmge and E. Tan, *Assessing the stabilizing mechanisms of unbound road pavements,*. (2016)
18. A. Rauch, et al., *Measured Effects of Liquid Soil Stabilizers on Engineering Properties of Clay.* T.R.R. Journal **1787**: p. 33-41. (2002)
19. Manual, E.-S.T. (2015)
20. Vicroads, in *RC 500.20 Code of Practice*. (2014)
21. Standards, A., *AS 1289.6 Methods of testing soils for engineering purposes*, in *6.1.1 Soil strength and consolidation tests - Determination of California Bearing Ratio of a soil.* (2014)

 DOI: 10.1051/matecconf/201713801003

Ductility Of fly ash - slag based reinforced geopolymer concrete elements cured at room temperature.

Mahantesh N.B. [1*], *Amarnath.* K.[2], and *Raghuprasad* B. K.[3]
[1] Alliance University, Bangalore, India
[2,3] The Oxford College of Engineering, Bangalore, India

Abstract Ductility of the flexural element is the main governing property for healthy performance of structural element. Although numerous factors contribute towards the ductility of Reinforced Geopolymer Concrete (RGPC) elements, low calcium based fly ash and GGBS have chemical proportions which make RGPC develop significant ductility along with steel reinforcement – when mixed in an intelligent way satisfying structural and economic conditions. In the present research work influence of low calcium fly ash, GGBS, River sand, M-sand, Steel Grade, manufactured fibres and natural fibres are used to study the ductile behaviour of RGPC sections by load testing 51 under reinforced flexural elements. The study reveals that fly ash - slag based reinforced flexural elements behave in line with OPC based RCC elements. The provisions mentioned in Indian RC designer IS:456-2000 can be used to predict the flexural behaviour of reinforced geopolymer concrete elements. The average flexural ductility of these test specimens observed to lie in between 2 & 3.

Keywords: Ductility, Geopolymer concrete, fly ash, slag, large deflections, flexural behaviour, River sand, Manufactured Sand.

1) INTRODUCTION

Geopolymer Concrete an environmentally friendly concrete having good structural skills is known structural concrete for the last 50 years, but still ordinary Portland Cement Concrete holds the first place in all construction activities. The manufacturing process of geopolymer concrete needs urgent attention to develop a process which suits the requirement of a most consumed private sector of concrete industry like fast moving consumer product ***(FMCG).*** Low calcium based fly ash geopolymer concrete essentially need heat or steam curing which make them unsuitable for cast in situ applications. The cost of curing is the additional factor which has made geopolymer concrete unpopular in most of the construction activities, especially in reinforced concrete structural elements.

*corresponding author: mahanteshb@rediffmail.com

Ambient curing is the most effective remedy to solve this issue. Industrial waste like slag (ground granulated blast furnace slag), when mixed with fly ash, develops strength at ambient temperature without heat or steam curing keeping the final strength same. During the past 10 - 15 years partial replacement of cement by fly ash & GGBS is becoming popular because of economic reasons in these sectors of concrete industry, giving ample evidence and data for a complete replacement.

2) MATERIALS USED AND PROPERTIES.

Fly ash used in this work was collected from Raichur thermal power plant in Karnataka, having sp.gr 2.15, Silicon dioxide (SiO_2) 61.98%, Aluminium oxide (Al_2O_3) 26.06%, calcium oxide(Cao) 3.05% confirming to grade 1 of IS 3812.*Slag – ground granulated blast furnace slag*, was procured from Jindal Steel Plant Bellary-Karnataka, having sp.gr 2.62, Silicon dioxide (SiO_2) 33.88%, Aluminum oxide (Al_2O_3) 18.02%, calcium oxide(Cao) 34.98% confirming to IS 12089.

Manufactured-Sand (M-sand), crushed from granite stone, having Sp.gr 2.45, Fineness Modulus (F.M) 2.70 and *River Sand* of sandstone origin having F.M 2.62 confirming to Zone III of IS 383-1970 are used as fine aggregates and tested as per IS 2386. *Coarse aggregates* of granite origin of sizes 20mm,12.5mm & 4.75mm tested as Per IS2386. These coarse aggregates have water absorption capacity 0.5% by weight at room temperature 16 to 28 degrees Celsius.

Sodium hydroxide of 97% purity and *sodium silicates* with Na_2O=14.7%, SiO_2=29.41%, water = 59.9% by mass are used as Alkaline Activator Solution using ratio of Na_2Sio_3/ NaOH = 2.5. *Sulphonated naphthalene* based superplasticiser, i.e. Conplast SP430 DIS distributed by FOSROC chemicals Bangalore used. Reinforcement is of Fe415 and Fe500 grades.

Four types of fibres used. They are (1) waste fibres produced from workshop lathe machine labelled as ***LMF*** (Lathe machine waste fibre) (2) factory made hook ended steel fibres marked as ***SF2*** of aspect ratio 71from Bekeart Pvt. Ltd, (3) plane steel fibres without hook end labelled as ***SF1*** from Nevatia Steel & Alloys Pvt Ltd (4) Polyester fibres marked as ***PF*** from Recron 3s Pvt Ltd placed sourced from South Indian Private Companies.

Table 1: Constituents for 1m^3 of Geopolymer Concrete

S.No.	Materials	Weight (kg)	Specifications
1	Fly ash	276	70% of total fly ash
2	GGBS (30%)	120	30% of total fly ash
3	20mm to 12mm size CA	451	35% of total CA
4	12mmto 4.75mm CA	451	35% of total CA
5	4.75mm & down sizes	389	30% of total CA
6	River sand	111	20% of total FA
7	M-sand	444	80% of total FA
8	Sodium hydroxide of 8M	45	97% purity
9	Sodium Silicate(Na_2sio_3)	113	Na_2O14.7%,$SiO_2$29.4%
10	Super plasticizer	3.6	SP430DIS (1.5%)
	NOTATION: FA: Fine aggregate, CA: Coarse aggregate		

Table 1 shows the mix proportions used for all reinforced geopolymer concrete specimens listed in Table 2.

3) SPECIMEN PREPARATION

Alkaline Activator Solution (AAS) is prepared 24 hours before mixing of concrete. Molarity of the NaOH solution (SHS) determined by the relation M= 0.25PD. M is the molarity of NaOH solution , P is concentration of sodium , D is the density of SHS for P. Therefore to get 1 liter of SHS of 8, 10 and 12 Molarity (255+745), (306+694) and (354+646) of (Sodium Hydroxide pallets in gms + water in gms) added respectively [4]. Cover blocks of 10mm thick are used for all specimen to provide clear uniform cover to reinforcements. Compaction of geopolymer concrete while filling in formwork containing reinforcement is done by using vibrators. The side from work removed after 24 hours of casting and specimen are left to room temperature curing which varied from 16 degrees Celsius at night & 28 degrees Celsius during peak daytime.

4) SPECIMEN DETAILS

To know the ductile capacity of reinforced flexural elements, different parameters which affect the ductility of components taken into considerations and a careful grouping of different specimens to be tested are listed. In all 41 slabs were tested having four different sizes *1.3m x 0.65m, 1m x1m, 0.8m x0.8m, 0.975m x 0.65m* with aspect ratios 1, 1.5 and 2.0. Two types of loads were applied, i.e. CPL Central Point Load and UDL – Uniformly Distributed Load using monotonically increasing load in 50Mton Loading frame.

Table 2: Specimen Groups

Group Name Fine Agg. - Molarity	No & Element	Size mm L x B x D	Ast- Nx-N_L	fy- fck- & curing days
A1:MS-8M	8-Slabs	1300 x 650 x 75	8mm – 4# & 7#	Fe500-57.87-36
A2:MS-10M	8-Slabs	975 x 650 x 75	8mm – 4#& 6#	Fe500-45.96-26
A3: MS-12M	8-Slabs	800 x 800 x 75	8mm – 5#& 5#	Fe500-41.70-22
B1, B2, B3: MS-8M	12-Slabs	1000 x 1000 x 60	8mm –7# &7#	Fe415-44.20-14
C1: RS-8M	05-Slabs	1000 x 1000 x 60	8mm –11#&11#	Fe415-35-14
D: BM1to BM10 RS-8M	10-Beams	Top :02#-8mm Bot: 03#-12mm 2L-8mm-100c/c	B=150mm D=210mm	Fe415-35-14
Notation: MS- Manufactured sand , RS- River sand : Nx & N_L rebars parallel to Shorter & Longer Sides,Fe500- yield stress 533Mpa & Ultimate stress 587Mpa,Fe415- yield stress 423 Mpa and Ultimate stress 502Mpa,SS- Short sides simply supported & remaining sides free, fck – cube compressive strength Mpa, PF – polyester fibers.PL-plain GPC				

5) LOAD TESTING OF SPECIMENS

Testing of slabs is done by applying crucial point loads and uniformly distributed loads by using 50MTon self-straining loading frame with electrically operated hydraulic jack. The two-point load system at a L/3 distance from both support ends of the beams applied for all the beams. Then the deflections are measured under load points and at the centre of the beam. On the day of testing flexural members - cubes and cylinders were tested.

6) NUMERICAL COMPUTATIONS

Among the most widely used and easily available aggregates, Granite holds the first place having Young's Modulus varying from 10 – 70 GPa - far above the values of sandstone rocks, i.e. 1- 20 GPa. When Manufactured sand of granite origin as fine aggregate mixed with coarse aggregates of granite origin, with fly ash: slag at 70:30 the resulting composite of granite based geopolymer concrete develops better structural properties. The IS:456-2000 describes the relations between compressive strength and flexural strength, modulus of elasticity as $f_{cr}=0.7\sqrt{fck}$ [2] and $E_c= 5000\sqrt{fck}$ after 28 days of curing [9]. These expressions are very closely following for fly ash: GGBS ratio at 70:30. [3]

All slabs and beams are analysed using conventional elastic theory for the applied loads and provided boundary condition using geometrical & material properties like compressive strength, steel strength as listed in Table 1. The failure mechanism of reinforced geopolymer concrete flexural elements follows the conventional OPC based RCC behaviour. The numerical computations are done using IS 456-2000.Until the appearance of the first crack at centre bottom of slab and beam, the composite is linearly elastic. Assessing the strength at this stage, using a full section of concrete results into more moment of inertia gives fewer deflections than measured ones. The composite continues to behave linearly elastic till tensile steel yields. Deflection calculations based on the effective moment of inertia using local code give noticeably matching with actual deflections.

Using 0.67fck as peak stress in compression concrete with parabolic stress blocks gives maximum strain 0.0020 to 0.0025, whereas using 0.85fck with rectangular stress block gives strains around 0.003 to 0. 0035. Failure loads from both peak stresses are acceptably same. The tensile stress of concrete at all stages are neglected. Calculated flexural cracks widths and measured ones are closely matching and are within acceptable limits at service loads. Beyond this stage, the specimen is said to have failed structurally. Further loading on the specimen is treated as post failure stage where significant deflections are observed.

7) FLEXURAL DUCTILITY OF GEOPOLYMER CONCRETE

If the constituent materials of concrete are ductile, then the concrete can be made more ductile by adding tension steel to develop the desired ductility so that the structures respond elastically at low cost. The plain geopolymer concrete under increasing flexure stress develops compression strains from 0.0010 to 0.0045 [5]. While reinforcement steel (Fe415) 0.0045 to 0.015 (yielding to failure). This reflects on strain ductility of 4.5 for GPC and steel 3.33. These values indicate the possible ductility of resulting composite, i.e. RGPC when subjected to pure flexural stresses.

For the known geometrical & material properties - analytically yield load $\mathbf{F_y}$ & ultimate load $\mathbf{F_u}$ is determined. Corresponding deflections Δ_y & Δ_u are read from load Vs deflection curves are drawn by using laboratory measurements. Then Ductility ***(calculated)*** = Δ_u / Δ_y, and Ductile Load D. L= F_u/F_y. Similarly, Strain Hardening Slope for calculated one =SHS cal = Ductile Load (calculated)/Ductility (cal)

Similarly, Ductility***(measured)*** = Δ_{um}/ Δ_y, Δ_{um} is the maximum deflection the component undergone under maximum applied load $\mathbf{F_{um}}$ and Δ_y is the measured deflection at yielding

 DOI: 10.1051/matecconf/201713801003

of steel. Then Ductile Load $D.L_m = F_{um}/F_y$. Similarly, Strain Hardening Slope for measured one =SHS mea= Ductile Load (measured)/Ductility (mea)

Ductility (calculated) & Ductility (measured) represent the minimum and maximum ductility developed by GPC. The Average Displacement Ductility is the average ductility between these two values, having more probability develops under Normal Quality Control.

8) RESULTS AND DISCUSSION

The mix design used is based on 77% of total aggregates to produce 40Mpa after ambient curing at 16 to 28 degrees Celsius for seven days which includes 24 hours of the rest period. Although M-sand and R-sand have nearly same fineness modulus, using Manufactured sand, the mix provided more than 40 Mpa, while using River-sand produced less than 40 Mpa compressive strength. The main reason for the difference in strength may be due to the difference in rock origins of fine aggregates & coarse aggregate, i.e. M- sand being granite origin mixing with coarse granite aggregates produces excellent bonding, whereas River sand being Sandstone origin produced less bonding with coarse granite aggregates. Moreover, the basic compressive strength of granite (100 to 250Mpa) is more than sandstone (20 to 170 Mpa) making granite based concrete stronger.

With steel fibres, the composites develop improved strengths while PF produced a marginal increase in strength. This indicates that all four types of fibres used were able to develop a good bond with GPC mixture of both types of fine aggregate. The increased compressive strengths are attributed to the property of fibres in delaying the failure by increased deflections.

Fig.1 to Fig.3 represent the developed deflections of the first set of slabs ***Group A***. Compared to UDL; Central point loads develop steep & sudden deflections at all the three stages of loading. The first appearance of a crack in tension concrete, yielding of tension steel and development of peak compressive stresses - passed within a short difference of loads because of stress concentration under the load.

Group B slabs: Compared to Fe500 steel, Fe415 steel is more ductile. As seen from Fig 4, Fig 5 & Table 2- it has produced ductility of slabs around 2.81 without fibres. Manufactured steel fibres increased the average ductility to 3.15, while the lathe machine fibres could develop around 2.79 and did not influence much on ductility. The lathe machine fibres are of random sizes in a loop form and thus could not align themselves along the major bending axis. However, it is expected that increased percentage and reducing the sizes by cutting them down to aspect ratio 30 could produce more ductility which is cost effective also.

In Group C: As seen from Fig. 6 & table 2 the average ductility of slabs was limited to 1.64. Using steel fibres slight increase in ductility observed while marginal improvement in ductility was observed using polyester fibres.

In Group D all beams are subjected to two-point load system & beams with steel fibres developed slightly better ductility compared to 4 beams with polyester fibres. As seen from strength tests on cubes and cylinders, river sand of the type used has the less basic compressive strength right from its parent rock, i.e. Sand Stone has less compressive strength than Granite stone. In addition to this - river and does not develop enough bond with fibres and thus develops less ductility compared to M-sand.

As seen from strength tests on cubes and cylinders, river sand of the type used has less basic compressive strength right from its parent rock. That is, Sand Stone has less compressive strength than Granite stone. In addition to this, the river sand does not develop enough bond with fibres and thus develops less ductility compared to M-sand.

 DOI: 10.1051/matecconf/201713801003

The applied load Vs measured deflections of all specimens listed from Fig. 1 to 7. Table 3 shows the list of observed ductilities of all 51 test specimens. These results indicate that M sand based RGPC specimens are more ductile than R-Sand based specimens. With steel fibre, these specimens may further enhance the composite ductility.

Table 4 broadly summarises the ductilities interms of two fine aggregates used. Addition of steel fibres at 1 to 1.5% could increase the ductility of M-Sand and R-Sand based RGPC. The combination M-Sand with Fe415 reinforcement & steel fibres produce maximum ductility of **2.97**. However, R-Sand with Fe415 reinforcement without any fibres produces the lowest ductility of **1.91**.

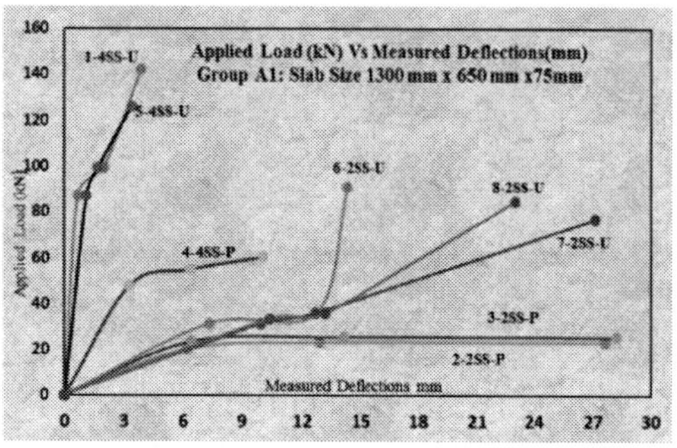

Fig. 1: Applied Load (kN) Vs Measured Deflection (mm) for Group A1- specimens

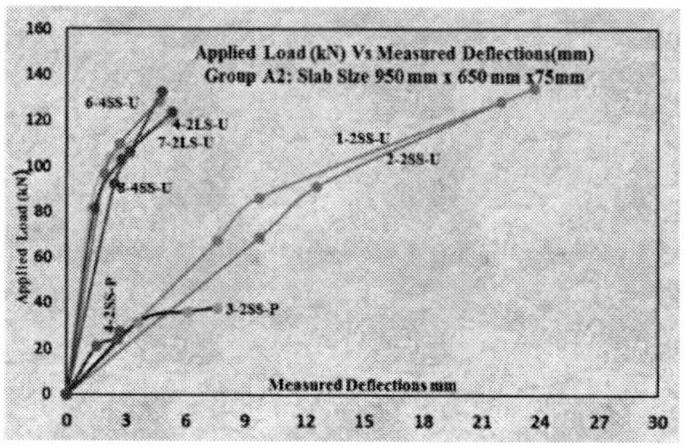

Fig. 2: Load (kN) Vs Measured Deflection (mm) for Group A2- specimens

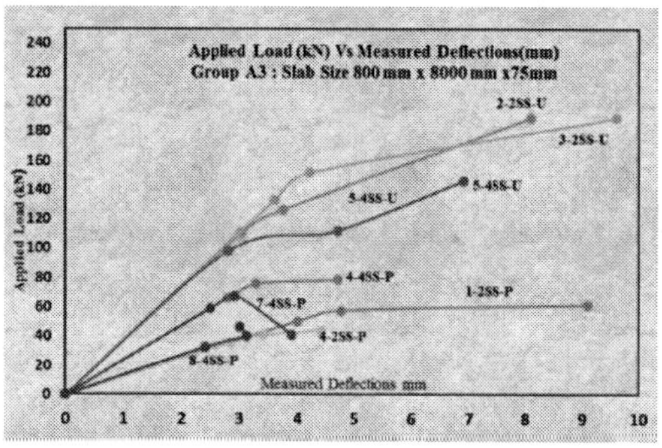

Fig. 3: Load (kN) Vs Measured Deflection (mm) for Group A3- specimens

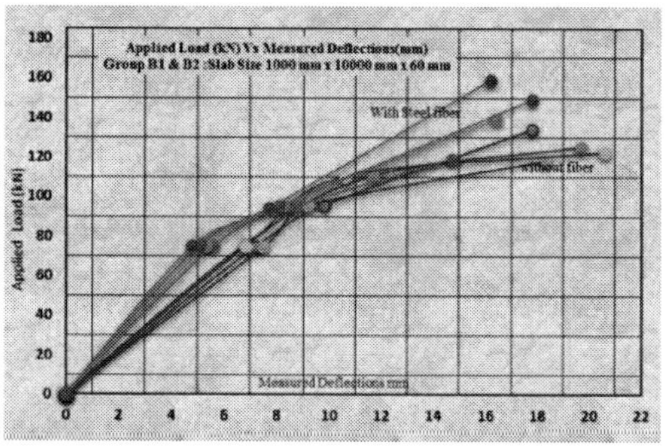

Fig. 4: Load (kN) Vs Measured Deflection (mm) for Group B1 & B2- specimens

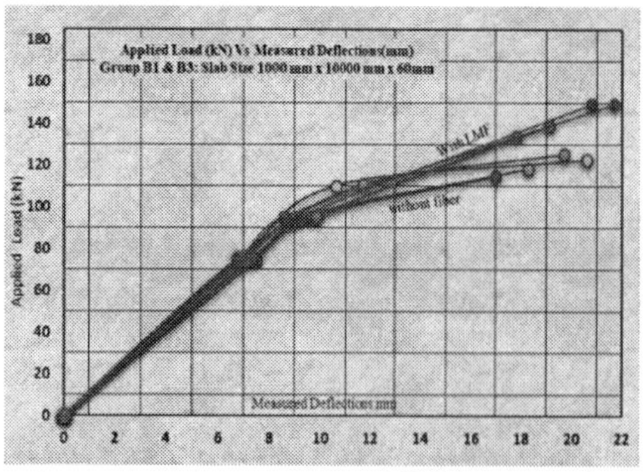

Fig. 5: Load (kN) Vs Measured Deflection (mm) for Group B1-& B3 specimens

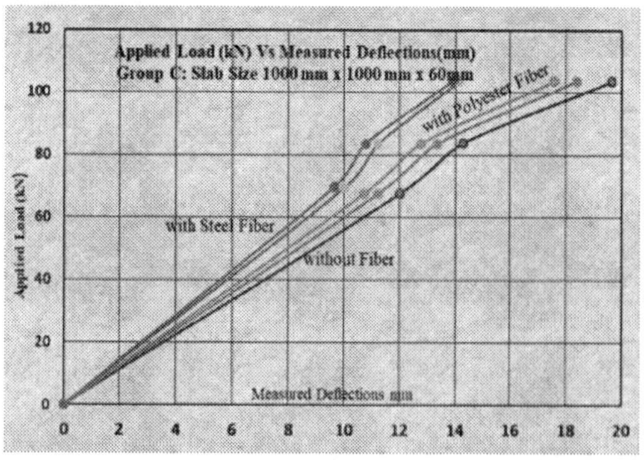

Fig. 6: Load (kN) Vs Measured Deflection (mm) for Group C specimens

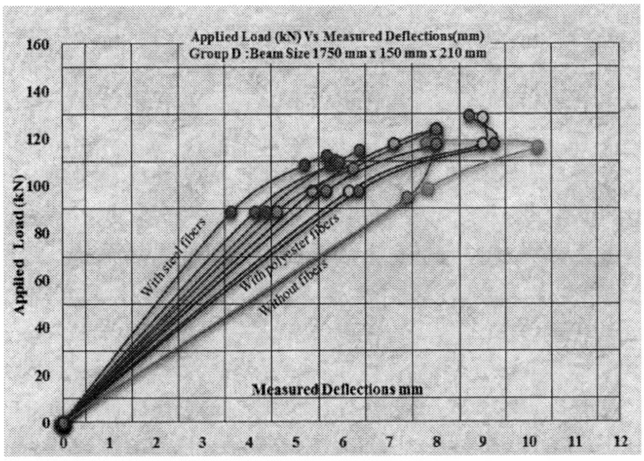

Fig. 7: Load (kN) Vs Measured Deflection (mm) for Group D specimens

Table 3 : Ductility Index of Reinforced Geopolymer Beams & Slab

Sl.No	*Component*	*Fine Agg*	*Fibers*	*fy*	*fck*	*Molarity*	*D.I (Cal)*	*D.L (Cal)*	*D.I (Mea)*	*D.L (Mea)*	*SHS*	*No of Specime*	*Remarks*
1	Reinforced Two Way Slabs	M-Sand	Nil	Fe500	57.87	8M	1.87	1.15	3.46	1.81	0.65	8	
2		M-Sand	"	"	45.9	10M	1.56	1.22	2.74	1.51	0.59	8	
3		M-Sand	"	"	41.7	12M	1.26	1.14	2.02	1.30	0.67	8	
4		M-Sand	Nil	Fe415	44.7	8M	1.47	1.37	2.81	1.67	0.94	4	
5		M-Sand	Steel	"	49.6	8M	1.58	1.26	3.15	1.89	0.80	4	1%,1.5%
6		M-Sand	LMF	"	58.7	8M	1.32	1.26	2.79	1.83	0.96	4	1%, 1.5%
7		R-Sand	Nil	"	35.0	8M	1.19	1.22	1.64	1.54	1.02	1	
8		R-Sand	PF	"	37.6	8M	1.19	1.22	1.64	1.54	1.02	2	0.50%
9		R-Sand	Steel	"	38.1	8M	1.12	1.20	1.68	1.56	1.07	2	1.50%
10	Singly Reinforced Beams	R-Sand	Nil	Fe415	35.00	8M	1.50	1.21	2.17	1.50	0.80	2	
11		R-Sand	Steel	Fe415	38.08	8M	1.46	1.21	2.25	1.58	0.83	4	1.50%
12		R-Sand	PF	Fe415	37.6	8M	1.35	1.22	2.19	1.50	0.92	4	0.50%
13	*Average Value :*						1.41	1.22	2.38	1.60	0.86	51	
Notations used : D.I - Ductility Index, D.L- Ductile load , SHS - Strain Hardneing Slope ,cal-calculated , mea- measured													

Table 4 : Ductility Ranking of RGPC specimens

S.No	F. A	Fy	Fiber	DI_{cal}	DI_{mea}	SHS	D.R
1	M-sand	Fe415	Steel	1.45	2.97	0.88	I
2	M-sand	Fe415	Nil	1.47	2.81	0.94	II
3	M-sand	Fe500	Nil	1.56	2.74	0.64	II
			M-Sand (Average)	**1.49**	**2.84**	**0.82**	
1	R-sand	Fe415	Steel	1.29	1.97	0.95	
2	R-sand	Fe415	PF	1.27	1.92	0.97	III
3	R-sand	Fe415	Nil	1.35	1.91	0.91	
			R-Sand (Average)	**1.30**	**1.93**	**0.94**	

9) CONCLUSIONS

Following conclusions are drawn based on room temperature cured fly ash- slag based reinforced geopolymer concrete flexural elements.

- The flexural behaviour of Reinforced Geopolymer Concrete is similar to Conventional OPC based RCC elements. IS: 456-2000 of Indian RC Design Code can be used to estimate all structural design related output.
- Using M-sand ductility of GPC could be in the range **1.50 to 2.85** and Using River sand it could be in the range **1.30 to 1.95**. The average of these two ductilities of RGPC lies between **1.4 and 2.40.**
- If fine aggregates and coarse aggregates are of the same rock origin, then the ductility of GPC is more with appreciable bond strength.
- Using proper steel grade & fibres at 10 to 15% could increase the strength and ductility of GPC.

REFERENCES

[1] Prakash Desai, K. U. Muthu, A brief review on strength, deflection and cracking of rectangular, skew and circular reinforced concrete slabs, *Indian Institute of Science*, 91-108,(Mar-Apr. 1988).

[2] Djwantoro Hardjito, Studies on Fly Ash-based Geopolymer Concrete, *Thesis Report, Curtain University of Technology*, (Nov.2005).

[3] Radhakrishna et al., Strength Characteristics of Open Air Cured Geopolymer Concrete, *Transactions of the Indian Ceramic Society*, (Feb.2014).

[4] Rajamane et al., Quantities of Sodium hydroxide solids and water to prepare sodium hydroxide solution of given molarity for geopolymer concrete mixes, *ICI technical paper*, (Sept. 2014).

[5] N. Ganesan et al., Development of stress block parameters for geopolymer concrete, *The Indian Concrete Journal*, vol **89**, Issue 9, pp 47-56, (Sept. 2015).

[6] Mahantesh et al., Flexural behaviour of fly ash-slag based reinforced geopolymer concrete flexural elements cured at room temperature, *IJTAM*, ISSN 0973-6085, Volume **12**, Number 4, PP 845-856, (2017).

[7] Mahantesh et al. , Strength and Deformation Behavior of Reinforced Geopolymer Concrete Flexural Elements, *IJCIET*, ISSN 0976-6316, Volume **8**, Issue 5, PP 1108 – 1121, (May 2017).

[8] Robert Park & Thomas Paulay, *Reinforced Concrete Structures*, John Wiley & Sons, INC.UK, (2013).

[9] IS 456-2000, Indian standard code of practice for plain and reinforced concrete for general building construction, *Bureau of Indian Standards* - New Delhi, (2000)

[10] Sumajouw and Rangan, Low Calcium Fly Ash-based Geopolymer Concrete: Reinforced Beams and Columns, *Research Report GC3, Curtin University of Technology*, (2006).

ACKNOWLEDGEMENTS

- The Authors wish to express their grateful feelings and thank the management of **Alliance University** and **The Oxford College of Engineering – Bangalore,** for their kind support while investigating this research work during 2013 - 2016.

- The Authors also wish to thank **EACEF 2017** for their excellent support during the conference at Seoul- South Korea.

The gelsluice, an innovative idea for the present sluice structures

Jacob Gerrit (Jarit) de Gijt[1,2*] and *Bob* Heester[3]

[1]Hydraulic Structures and Flood Risk Section, Delft University of Technology, Delft, the Netherlands
[2]Rotterdam Engineering Department, Rotterdam, the Netherlands
[3]Antea Group, Rotterdam, the Netherlands

Abstract. Sluices have been built worldwide since in 984 when the first sluice was built in China to speed up the operations in the Grand canal. Since that time different types of sluices have been developed world wide both in size and material use and their own shipping and structural characteristics. One of the main drawbacks of the present sluice designs is the waiting time. This problem of the waiting time is in principle overcome by the design of the gelsluice (gellock). This paper describe the development of sluices construction in general and with the emphasis on the characteristics of the innovative gelsluice. A preliminary design of the gelsluice will be discussed and presented. Finally some recommendations are presented for further research to improve the design of the gelsluice.

1 Introduction

Sluices are important civil engineering structures to cope with water level differences and to ease shipping. Since the first sluice was built in China along the Grand canal several types of sluices have been developed. This paper will give a brief overview of the existing types of sluices.

The great disadvantage of most of the present sluice designs is that ships have to wait to pass the sluice . However this is depending of the height difference to overcome a time consuming operation. These disadvantages are overcome with the design of the gelsluice. This paper will discuss the design aspects of gelsluice and will indicate what is necessary to make this type of sluice operational.

2 History of sluice design development

The oldest shipping lock was constructed in China for the Grand Canal in 984. Before that time, the canal was divided in sections by dams to control the water levels and slipways consisted mostly of a row of trees over which the ships were pulled over the dam to the other part of the canal .This caused a considerable of damage to shipping. In Europe, the first lock was introduced in the Netherlands, Vreeswijk in 1358. This construction was a so-called pounded lock with guillotine doors. Recently an old lock with the guillotine doors in

* Corresponding author: j.g.degijt@tudelft.nl

the Rotte River near Rotterdam has been rebuilt. This has been made possible by scanning at location, which revealed enough information for rebuilding. At this moment the biggest lock of the world is in Kallo, Belgium.

The gelsluice idea was first developed in the summer of 2011. That time the idea was made to close the Nieuwe Waterweg, the entrance of the port of Rotterdam by a dam of gel as this might be needed due to sea level rise. The great advantage of this type of sluice is that it has no doors so that creates the possibility that ships can pass the sluice without waiting time. Several options and locations have been considered but finally the selected location was the Nieuwe Waterweg as presented in Figure 1.

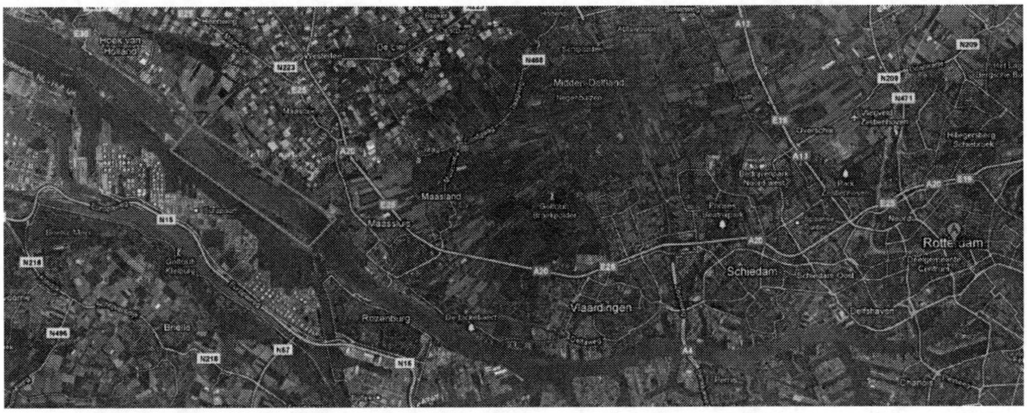

Fig. 1. Location of the gelsluice.

3 The development of gelsluice

The development of the gelsluice after the first idea was made possible by students of Hogeschool of Rotterdam as they investigate different gel compositions and carried out hydraulic testing in the laboratory. Finally the gel which was found the best was a mixture of and gelatine and barit which was added to increase the density of the gel. The physical parameters are presented in Table 1.

The original sketch Figure 2 indicates the principal of the gelsluice which consists of a dam made of mixture of gelatine and barit. The gelsluice consists of a dam made of gelatine and barit with approximately a density of 11 kg/m^3. The mixture has been selected after testing of several other mixtures. From this mixture a dam has been built in a hydraulic flume to test the behaviour of the passing ships and the gel. The original sketch is displayed in Figure 1, which finally involved to the design of the geldam as given in Figure 3.

This idea evolved in the dimensions as presented in Figure 3 which takes into account the results of laboratory testing to find the required properties.

Of course also attention must be paid to the forces on the ship, see Figure 4 when passing the gel as the upward force on the ship in the gel differs from that in water its necessary to have a slope of 1:6 to prevent the development of high stresses in the ships hull.

The ships are pulled through the sluice in the same manner as used in Panama Canal. This is required as the motors are not yet in able to cool the machinery when the ship is in the gelsluice. The gelsluice principally does overcome all the drawbacks of existing sluices. There is no waiting time and hardly any mechanical equipment is required. The gel itself settles and must be maintained and thus a gel suppletion plant is included in the plan. As the gelsluice is impermeable for water the sluice complex should facilitate that the flow of

DOI: 10.1051/matecconf/201713801004

the river is not blocked and that fish passage is possible That result in the design as displayed in Figure 6. The gelsluice also prevent salt water intrusion.

Table 1. Summary of physical parameters of gel.

Parameter	Value	Unit
Specific weight	1100	kg/m^3
Viscosity	0,0605	mPa/s
Kinematic Viscosity	$5{,}5*10^{-5}$	m^2/s
Friction angle	max 1	degrees
Maximum slope	1: 60	1: m
Maximum shearstress	1,04	N/mm^2
Minimal required shear stress	0,13	N/mm^2
Permeability	$< 10^{-10}$	m/s

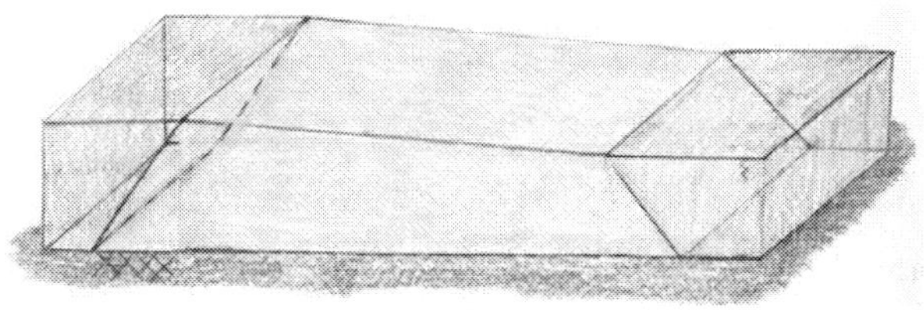

Fig. 2. Original sketch of the gelsluice.

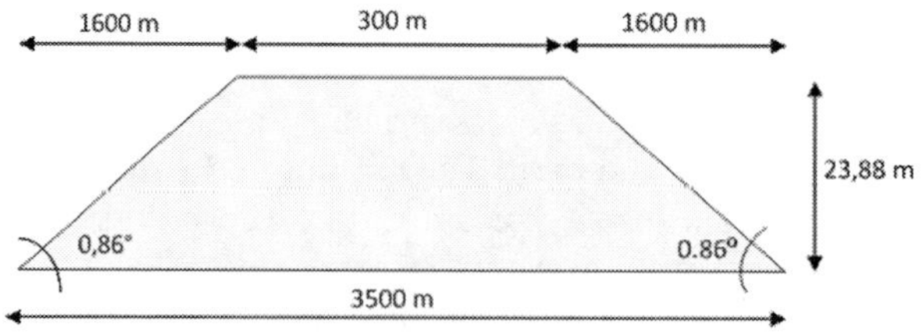

Fig. 3. Shape of geldam.

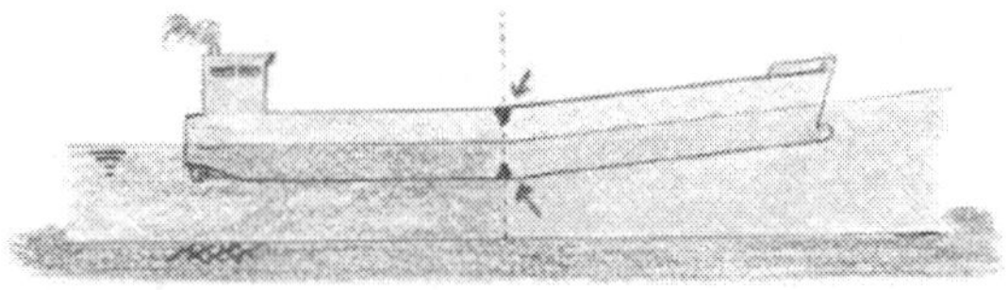

Fig. 4. Indication of influence of density change on the ships hull.

 DOI: 10.1051/matecconf/201713801004

Fig. 5. Ships are pulled through the gelsluice.

Fig. 6. Layout of the gelsluice complex.

4 Costs

In Table 2 a brief cost comparison is presented which indicates that the gelsluice is more expensive then a traditional sluice. However, it is anticipated when the whole life cycle and operation costs are included the gelsluice is more advantigious. The gelsuice shall to be more favourable espicially due the reduction of the waiting time of ships passing the gelsluice.

Table 2. Cost comparison.

Project	Costs (€) in million	% van gelsluice
Maeslantkering	700	54
Nieuwe sluizen Ijmuiden	700	54
Gelsluis	1295	100

5 Conclusions and recommendations

The gelsluice principally does overcome all the drawbacks of existing sluices. However more research is required to improve the gel characteristics. Thereby the following points have to be considered: the behavior of the gel in the long term, the settling, stability behavior of the gel with time, the behavior of ships within gel. Also the cost aspects, the gel production and recycling should be investigated.

This implies that the main research question is the combination of classical hydraulics with the phenomena of hydraulic aspects of pumping fluids to understand and predict the behavior of the gel.

References

1. B. Heester, N. Reijmers, Joolingen van M., *Gelsluis*, B.Sc. Thesis, Hogeschool Rotterdam (2011)

DOI: 10.1051/matecconf/201713801004

2. J.G. de Gijt, *Interview cobouw* (2011)
3. P.H.J.M. van Gelder, N. Doorn, J.G. de Gijt, *Chapter: Responsible innovation in the hydraulic engineering sector* in *New perspectives on responsible innovation,* by Hoven van der D. et al. (2018)

The use of bottom ash for replacing fine aggregate in concrete paving blocks

Antoni[1,*], *Kevin* Klarens[1], *Michael* Indranata[1], *Luthfi* Al Jamali[2], and *Djwantoro* Hardjito[1]

[1]Civil Engineering Department, Petra Christian University, Surabaya, Indonesia
[2]PT Jawa Power - YTL Jawa Timur, Probolinggo East Java, Indonesia

Abstract. Bottom ash that results from coal burning for electrical generation is still much underutilized in Indonesia and it is necessary to increase the usage of this waste. The manufacture of paving blocks using bottom ash, which is normally associated with high water absorption due to its high porosity and carbon content, was examined in this study with the aim to increase the usage of this waste material. The study was done in three phases: (1) the mixture of cement and bottom ash passing sieves of 2 and 5 mm were done with ratios of 1:3, 1:4, and 1:5; from the best proportion, fly ash was used to replace the cement material from 10–80%, and (3) samples with 30% and 50% fly ash replacement ratios were used in combination with 5 mm and 10 mm sieved bottom ash. Compressive strength, water absorption, and abrasion resistance tests were conducted to assess the properties of the resultant paving block. From the result, bottom ash is used to replace natural sand in making paving blocks. By optimizing the particle packing density and using fly ash as a cement replacement, the compressive strength of paving blocks can exceed 40 MPa.

1 Introduction

The production of concrete paving blocks requires natural resources that are difficult to renew. Sand availability is reducing with the more stringent environmental preservation act. In contrast, coal burning waste material from power plants has become an environmental problem. One power plant in East Java uses about 250 tons of coal each hour (2015 figures) and generates combustion waste of 4% and consisting of 25% bottom ash and 75% fly ash. The utilization of fly ash in construction is more common worldwide, reaching 47% of total fly ash produced; however, only 5.28% of bottom ash is used [1,2]. Thus, there is a need to utilize bottom ash to reduce waste deposits in ash ponds and reduce the use of natural sand [3–5].

The effort to utilize bottom ash is constrained by the distinctiveness of its grain shape. A very coarse particle shape facilitates interlocking between particles, causing a decrease in fresh mixed workability when bottom ash is utilized as a fine aggregate. However, bottom ash is a suitable material to substitute sand in a concrete mixture. The use of 100% bottom ash can be applied to concrete where the workability can be ignored, e.g., in the making of

* Corresponding author: djwantoro.h@petra.ac.id

paving concrete [6,7]. Currently, the bottom ash of the power plant is only stockpiled in ash pond in plant location while almost 100% of the fly ash is used by various ready mix plants around East Java to make concrete.

This study tries to maximize the use of bottom ash as a substitute for all sand in producing concrete paving blocks. Maximizing the use of bottom ash was done by sieving separation to evaluate the effect of different particle sizes on the paving block's properties. Bottom ash paving block was made with the aim of achieving the compressive strength, water absorption, and abrasion resistance required by the Indonesian national standard [8]. Fly ash was used in the mixture to reduce cement consumption and increase the workability of the mixture [9]. The water required to obtain a workable mixture also was recorded as a basis for analyzing the resulted concrete.

2 Experimental methods

2.1 Materials

Fly ash and bottom ash are obtained from a power plant in Paiton, East Java, as the experiment was conducted near the power plant location. Results of the XRF test for fly ash and bottom ash are shown in Table 1. The bottom ash used is classified by sieving at 2, 5, and 10 mm to obtain different particle gradations. The cement used was Portland Pozzolan Cement (PPC) from PT. Semen Gresik in a small cement bag. The sand for the control sample is sand passing a 5 mm sieve from Lumajang quarry, East Java. The bulk volume and the specific gravity of each material are shown in Table 2.

Table 1. Oxide compositions for fly ash and bottom ash from the XRF test

Oxide	SiO_2	Al_2O_3	Fe_2O_3	CaO	SO_3	MgO	TiO_2	K_2O	Na_2O	MnO_2	P_2O_5	LOI
Fly ash	34.29	16.62	15.38	18.18	1.63	7.52	0.73	1.35	2.97	0.17	0.25	0.36
Bottom ash	34.39	10.02	18.41	21.16	0.66	9.7	0.65	0.9	0.24	0.22	-	3.54

Table 2. Bulk volume and specific gravity of the material

Material	Bulk Volume kg/m^3	Specific Gravity
Bottom ash (passing 5 mm)	1382	2.26
Bottom ash (passing 2 mm)	1328	2.33
Sand	1598	2.72
Cement	1181	3.05
Fly ash	1276	2.75

2.2 Specimen preparation and test

The study was done in three phases. Phase 1 aims to understand the normal paving block mixture by varying the aggregate to cement ratio and a mixture using two sizes of the bottom ash. Phase 2 started with the best result of phase 1 and used fly ash to replace the cement material from 10–80%. At phase 3, the samples with 30% and 50% fly ash replacement ratio were used in combination with 5 mm and 10 mm sieved bottom ash to increase the density of the matrix. Table 3 shows the detailed mix proportion.

The paving block was made using a semi-automatic mixer and hydraulic press former. The procedure for mixing is as follows. Cement was poured into the mixer and the bottom ash follows in dry condition. If fly ash is used, then fly ash is poured in first followed by cement. Water is added to the mixer sequentially until the desired condition is achieved. The desired mixture is when the wet mixture is clasped and kneaded the mixture remains as a form as shown in Figure 1; that is, the mixture would not crumble after compaction. The mixture then was formed into paving blocks by using hydraulic pressure and vibration. Tests conducted were for compressive strength (7, 14, 28 days), water absorption (28 days), and abrasion resistance (28 days). The targeted properties are "Grade A" based on the Indonesian Standard as shown in Table 4.

Table 3. Mixture proportion for the three phases

No	Code	C	BA	Sand	Description	No	Code	C	FA	BA	Description
Phase 1						**Phase 2**					
1	C1S3	1	-	3	River sand for control (C-S)	1	C9F1-B30-5	0.9	0.1	3	Fly ash replacing cement with 5 mm bottom ash material and 1:3 binder to aggr. ratio (CF-B)
2	C1S4	1	-	4		2	C8F2-B30-5	0.8	0.2	3	
3	C1S5	1	-	5		3	C7F3-B30-5	0.7	0.3	3	
4	C1S6	1	-	6		4	C6F4-B30-5	0.6	0.4	3	
5	C1S7	1	-	7		5	C5F5-B30-5	0.5	0.5	3	
6	C1S9	1	-	9		6	C4F6-B30-5	0.4	0.6	3	
7	C1B3-2	1	3	-	Bottom ash passing 2 mm sieve (C-B)	7	C3F7-B30-5	0.3	0.7	3	
8	C1B4-2	1	4	-		8	C2F8-B30-5	0.2	0.8	3	
9	C1B5-2	1	5	-		**Phase 3**					
10	C1B3-5	1	3	-	bottom ash passing 5 mm sieve (C-B)	1	C7F3-B30K	0.7	0.3	3	Combined size 10 mm and 5 mm 1:1 (CF-BK)
11	C1B4-5	1	4	-		2	C5F5-B30K	0.5	0.5	3	
12	C1B5-5	1	5	-							

Table 4. Classification of paving block grades according to SNI 03-0691-1996 [8]

Grade	Usage	Compressive Strength (MPa)		Abrasion Resistance (mm/minute)		Water Absorption (%)
		Average	Min	Average	Max	
A	Road	40	35	0.090	0.103	3
B	Parking	20	17	0.130	0.149	6
C	Walkway	15	12.5	0.160	0.184	8
D	Garden	10	8.5	0.219	0.251	10

Fig. 1. Water required to make a stiff mixture and the paving specimens produced.

 DOI: 10.1051/matecconf/201713801005

3 Results and discussion

3.1 Water requirement for paving block mixture

The amount of water required for making a stiff mixture was determined for each mix and the water/solid ratio was calculated as the weight of water to combined weight of cementitious and filler. In phase 1, for paving block C-S series, there was no increase in water requirement with the increase of sand content (Figure 2(a)). The water demand was about the same for the mixture when more sand was used in the mixture. The reduction of the water to cement ratio is obvious with the reduction in the cement content as this would influence the compressive strength. When replacing sand with bottom ash aggregate (C-B), the water requirement was nearly doubled that required with the sand specimen. This shows that a change of material changes the water requirement significantly. There was a small increment with the higher aggregate ratio; however, the particle size did not seem to influence the water requirement. The average water to solid ratio of cement and sand was only 0.086 while the use of bottom ash only increased it to 0.182.

The water requirements of phases 2 and 3 are shown in Figure 2(b). When replacing the cement with fly ash (CF-B), there was a reduction of water requirement as replacing more fly ash into the mixture reduces the water requirement because the round particle of fly ash acts as filler in the concrete mix. A significant decrease of water requirement began to be seen in the use of fly ash above 50% of cement replacement. The average water/solid ratio of CF-B is 0.161, showing that replacement by fly ash would benefit the mixture condition.

Phase 3 shows a significant reduction in the water requirement when aggregate gradation is changed (CF-BK), resulting in a water/solid ratio lower than the other mixture that used bottom ash (0.110). Figure 2(b) also shows that with a higher fly ash replacement ratio, the water to cementitious ratio was reduced.

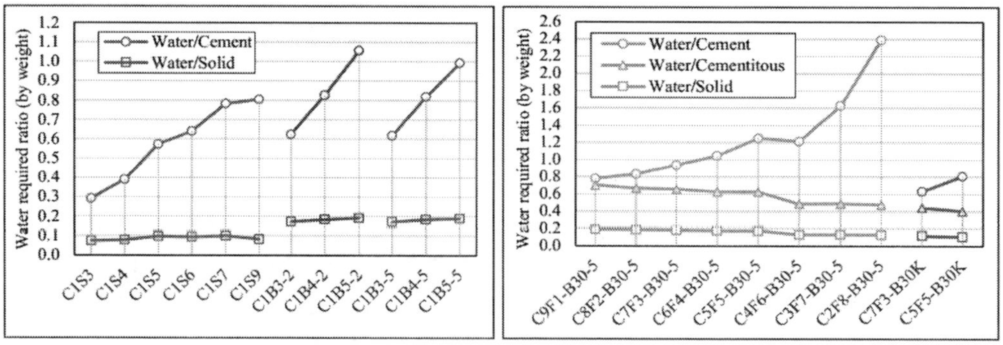

Fig. 2. Water required for making a stiff mixture (a) phase 1 and (b) phases 2 and 3.

3.2 Physical properties of the paving block

3.2.1 Compressive strength

The compressive strength of all paving samples increases with age. Figure 3 shows the compressive strength for all specimens at 7, 14 and 28 days. For the C-S series, the more sand added the lower the strength. Paving made with bottom ash passing a 5 mm sieve has a compressive strength better than that of the 2 mm ash, which is most likely due to less surface area with larger particle size.

For the CF-B series, the cement replacement with fly ash decreased the compressive strength of the paving due to lower hydrate volume. Samples with 20–50% fly ash replacement have about the same compressive strength; however, the strength decreases with a higher replacement ratio.

The compressive strength of paving increased significantly (i.e., above 40 MPa) in all phase 3 samples, where the aggregate was a mixture of 50% bottom ash (CF-BK) passing a 10 mm sieve and 50% passing a 5 mm sieve. This showed that the use of the right combination of bottom ash gradations increases the packing density of the paving blocks and, thus, increases their compressive strength. From the compressive strength, it was found that both phase 3 mixtures could achieve Grade A quality as specified by the standard.

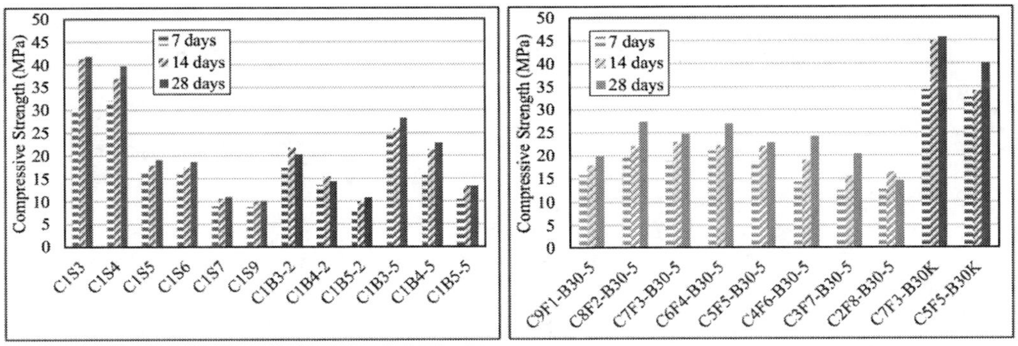

Fig. 3. Compressive strength at 7, 14 and 28 days for (a) phase 1 and (b) phases 2 and 3.

The relationship of the compressive strength water to cementitious ratio is shown in Figure 4(a). The inverse relationship of water to cementitious ratio follows the norm that the lower water to cementitious ratio would have the higher compressive strength. The effort in increasing the compressive strength should focus on reducing the water required by better controlling the gradation of the aggregate. Figure 4(b) shows the relationship of compressive strength and bulk density of the paving block. The CF-BK series had higher bulk volume due to better gradation control, hence, obtaining higher bulk density. The C-S series shows a higher density because the sand's specific gravity is higher than that of the bottom ash.

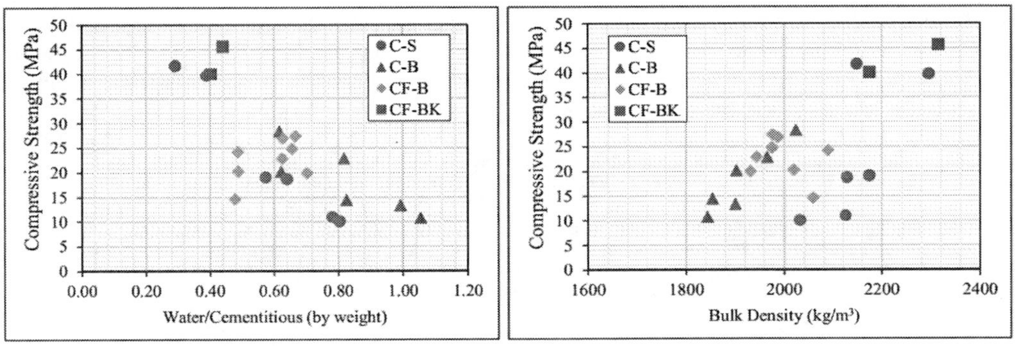

Fig. 4. Relationship of compressive strength of the sample with (a) water to cementitious ratio and (b) bulk density of the sample.

3.2.2 Water absorption

High water absorption adversely affects the durability of paving. The result of water absorbance tests at 28 days can be seen in Figure 5. From phase 1, paving block C-B series with 5 mm bottom ash size has a lower water absorption than the one with 2 mm size. This is due to more uniform particle sizes of bottom ash passing a 2 mm sieve, resulting in more void.

The figure also shows that the use of fly ash as a cement replacement material increases water absorption; however, the higher fly ash content did not have any significant impact on the increment. The CF-BK mixture was shown to be the only mixture with less than 3% water absorption even compared with the sand specimen. The bulk density of the specimen can be an indication of the water absorption rate as higher density results in a lower water absorption ratio.

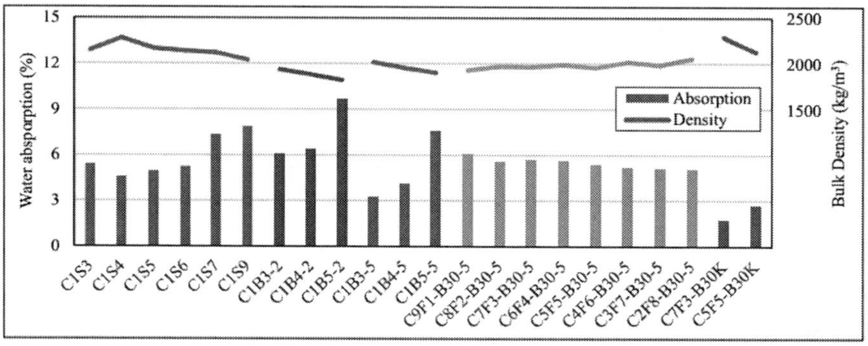

Fig. 5. Water absorption of the paving block sample in relation to its bulk density.

3.2.3 Abrasion resistance

The results from the abrasion resistance tests are shown in Figure 6. All mixtures here perform well under the standard specification of grade A. Mixture C1P3, which uses sand as the aggregate, has the highest abrasion resistance compared with C1B3-5, which has a wear value of 0.0239 mm/min. The use of fly ash as a cement replacement material resulted in reduced abrasion resistance. This means that a high fly ash replacement ratio lowers resistance to abrasion as the strength is also reducing with the increase of replacement ratio. Once again, the gradation combinations of bottom ash aggregate (C7F3-B30K and C5F5-B30K) have the highest abrasion resistance, which is comparable with the control specimen using sand only. Higher compressive strength contributes the abrasion resistance value, but it is also influenced by the type of binder used.

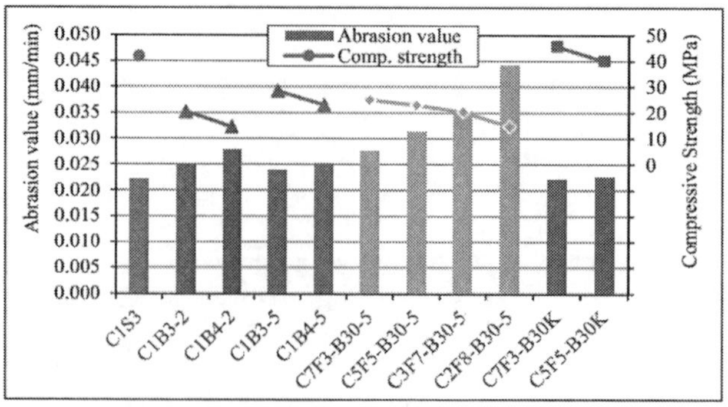

Fig. 6. Abrasion resistance of selected samples and their 28-day compressive strength.

4 Conclusions

From the experimental study, several conclusions can be obtained.

1. Bottom ash can be used as a replacement for 100% sand in making concrete paving blocks. Bottom ash with a size combination of 50% passing sieve 5 mm and 50% passing sieve 10 mm produces optimum mix composition. With 50% fly ash as a cement replacement, the resulting specimens have a compressive strength of 40 MPa at 28 days.
2. The water requirement for making a stiff concrete mixture is more affected by the type of material used than its composition. It is important to use the lowest water requirement so that the mixture has the lowest water to cementitious ratio in order to obtain a higher compressive strength.
3. Using bottom ash increases water requirements compared with the control sand; however, mixing bottom ash passing a 10 mm and 5 mm sieve reduces the water requirements. This indicates that by combining different particle sizes the porosity can be reduced.
4. Water absorption in most of the mixtures was higher than 3%. However, mixing 50% 10 mm and 50% 5 mm bottom ash yields specimens with water absorption below 3%, thus, they can be classified as Grade A, as per target of the paving block quality.
5. Abrasion wear resistance is proportional to the compressive strength of the specimen and the hardness of the aggregate. All paving specimens can achieve a Grade A standard using a combination of bottom ash sizes..
6. From the results, it is suggested that the use of bottom ash can be increased to make high-quality paving blocks and reduce environmental loads due to unused bottom ash deposits. Properties that need to be considered for the use of bottom ash in large quantities include LOI values, size and gradation variations, and the level of porosity of the existing bottom ash.

Acknowledgment to PT. Jawa Power for providing the fly ash, bottom ash and the paving block equipment for the experimental study under the MOU number 002/MoU/FTSP/UKP/2015-YTLJT/LTR/2015/00734.

References

1. H. Kurama & M Kaya, Constr Build Mater, **22,** 1922-1928 (2008)
2. S. Naganathan, A.Y.O. Mohamed & K.N. Mustapha, Constr Build Mater, **96**, 576-580 (2015)
3. S. Mengxiao, W. Qiang & Z. Zhikai, Constr Build Mater, **98** (2015)
4. M. Singh & R. Siddique, J Clean Prod, **91**, 269–278 (2015)
5. N. Chousidis, E. Rakanta, I. Ioannou, & G. Batis, Constr Build Mater, **101** (2015)
6. M. Singh & R. Siddique, J Clean Prod, **112,** 620-630 (2015)
7. H.-K. Kim, Constr Build Mater, **91** (2015)
8. Badan Standarisasi Nasional. SNI 03-0691-1996 Bata Beton (Paving Block) (1996)
9. Antoni, V. Satrya & D. Hardjito, Civ Eng Dim, **17**(1), 38-43 (2015)

Processed bottom ash for replacing fine aggregate in making high-volume fly ash concrete

Antoni[1,*], *Aldi Vincent* Sulistio[1], *Samuel* Wahjudi[1], and *Djwantoro* Hardjito[1]

[1]Civil Engineering Department, Petra Christian University, Surabaya, Indonesia

Abstract. Bottom ash is a coal plant by-product that is abundant and underutilized. There is the potential use of bottom ash as a fine aggregate replacement in concrete mixtures; however, the problems of water absorption and uniformity of quality of the material need to be overcome first. In this study, bottom ash was treated by sieve separation and pounding to smaller particle size for use as a sand substitute. The physical and chemical characteristics of bottom ash were tested after treatment including water absorption, sieve analysis, and fineness modulus. High-volume fly ash (HVFA) mortar specimens were made and the compressive strength and flowability test using bottom ash after treatment are compared with that of the sand specimen. Low water to cementitious ratio was used to ensure higher strength from the cementitious paste and superplasticizer demand was determined for each treatment. The result showed that bottom ash can be used as fine aggregate replacement material. Sieve separation of the bottom ash could produce 75% of the compressive strength compared with the control sand specimen, whereas pounded bottom ash could have up to 96% of the compressive strength of the control specimen. A 28-day compressive strength of 45 MPa was achievable with 100% replacement of fine aggregate with bottom ash.

1 Introduction

The large amount of natural materials used in making concrete has been a concern in the construction industry. In some regions, natural sand has been irresponsibly mined, causing environmental damage. Concern about the large-volume usage of cement also has been raised due to the carbon released during cement production, making concrete a non-environmentally friendly material. A solution is needed to improve the sustainability of concrete from the perspective of raw materials and composition. Concrete developed, manufactured, and used in an environmentally friendly way is called "Green Concrete" [1] and efforts to reduce the use of cement by improving mix composition and replacing natural materials with recycled or waste materials are on-going, with further research needed.

* Corresponding author: antoni@petra.ac.id

Fly ash and bottom ash are coal combustion waste materials from the burning process in power plants. For the use of bottom ash itself, there are still some difficulties such as irregular and rough bottom ash particle shapes [2,3]. In addition, using bottom ash in concrete mixtures reduces concrete strength, increases water absorption, and decreases density [4]. Bottom ash has a larger particle size than fly ash and a rough surface that can interlock easily, which lowers workability of the mixture. In addition, the porosity and water absorption are high, which increases the water requirement of the mixture [5]. Bottom ash can be treated by grinding and sieving to reduce the particle size to reduce water consumption and voids [6]; however, the grinding process requires a lot of energy and cost.

Fly ash material has been used in making high-volume fly ash (HVFA) concrete due to its beneficial effects. Small particle size and the rounded shape of the fly ash particles improve the workability of the paste mix, resulting in a decrease in the water requirement. HVFA concrete has at least 50% fly ash replacing cement, low water content (~130 kg/m^3), and low water to cementitious ratio (w/cm < 0.4) [7,8]. The use of high-volume fly ash in making concrete would reduce the use of cement as well as reduce the water use required for bottom ash mixtures [9,10].

The utilization of bottom ash in Indonesia is still rare due to its irregular, relatively large, and porous particles. In addition, environmental regulation lacks sufficient evidence regarding the use of this waste material. Much of the bottom ash generated by a power plant only accumulates in the ash pond and is then stacked in the fields. Therefore, this research aims to treat and utilize this abundant source of bottom ash as a replacement for sand in a mortar mixture. The HVFA mortar mixture aims to maximize the use of fly ash and bottom ash as well as minimize the use of cement. Early-age and long-term compressive strength and the workability of the mixture are evaluated to ensure a good performance of bottom ash as the aggregate replacement in concrete.

2 Experimental methods

2.1 Materials

The cementitious material used was Portland Pozzolan Cement (PPC) from PT. Semen Gresik and fly ash type C. The fly ash and bottom ash were from PT. Jawa Power at the Paiton power plant complex, East Java. The sand used was from a local quarry in Lumajang, East Java. The XRF test results from fly ash and bottom ash are presented in Table 1. The water used was demineralized water and the superplasticizer (SP) was a polycarboxylate ether-based type.

The bottom ash obtained was designated as raw bottom ash (RBA) and oven-dried before sieving into coarse bottom ash (CBA, with a particle size of 0–5 mm, and fine bottom ash (FBA), with a particle size of 0–2.36 mm. Bottom ash that did not pass the 5 mm sieve was pounded in 5 min increments until all particles passed the 5 mm sieve. These were designated as pounded bottom ash (PBA). The particle shapes of fly ash, sand, and bottom ash are illustrated in Figure 1.

Table 1. Oxide compositions for fly ash and bottom ash from the XRF test

Oxide	SiO_2	Al_2O_3	Fe_2O_3	CaO	MgO	SO_3	TiO_2	K_2O	P_2O_5	MnO_2	Na_2O	LOI
Fly ash	34.29	16.62	15.38	18.18	7.52	1.63	0.73	1.35	0.25	0.17	2.97	0.36
Bottom ash	34.39	10.02	18.41	21.16	9.7	0.66	0.65	0.9	-	0.22	0.24	3.54

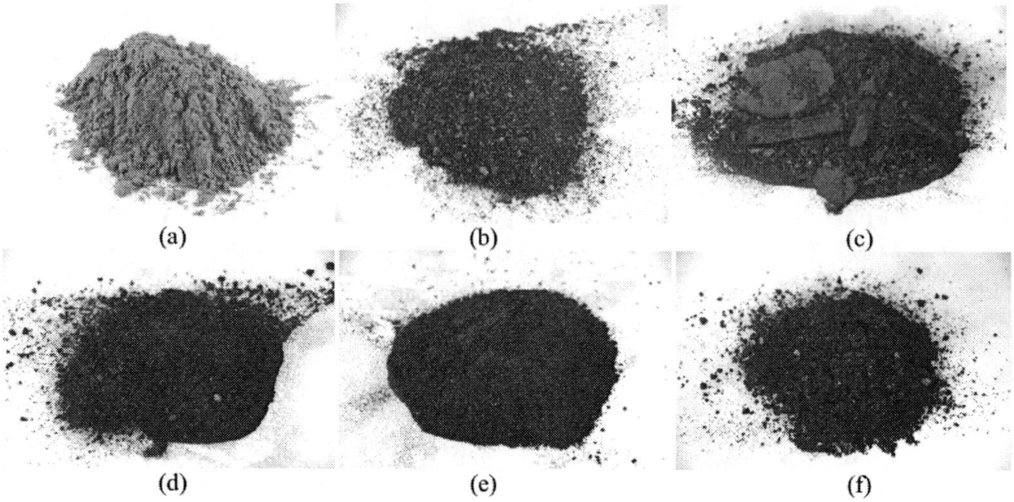

Fig. 1. Photographs of materials used in making the mortar specimens: (a) fly ash, (b) sand, (c) raw bottom ash, (d) coarse bottom ash passing 5 mm sieve, (e) fine bottom ash passing 2 mm sieve, and (f) pounded bottom ash.

2.2 Mixture compositions and variables

For the current experimental study, the water to cementitious ratio (w/cm) was fixed at 0.325, and fly ash was used as cement replacement at level 50%, by mass. Superplasticizer was prepared as required (SP demand) to achieve a flow diameter of 14 ±2 cm. The SP demand is then calculated to the percentage of the cementitious mass. The aggregate to cementitious ratio was determined with the trial mixing with a target of a workable mixture and higher compressive strength. The HVFA mortar was then cast into specimens of 5 cm × 5 cm × 5 cm. The compressive strength test of the mortar was conducted at 3, 7, 14, 28, and 56 days and 1 year, and three replications were conducted for each age testing.

The variable designations are as follow. "CTRL" was for the comparative specimen using Lumajang sand, with a prefix to indicate the bottom ash type: "C" for coarse, "F" for fine, and "P" for pounded. The numbers 100, 080, and 050 denote the mass percentages of the bottom ash replacing control sand. The suffixes (−2, −2.5, and −1.5) show the aggregate to cementitious ratio, in mass.

3 Results and discussions

3.1 Treatment of the bottom ash

Different physical properties of the material were found for different sizes of bottom ash gradation. Water absorption from the saturated surface dry (SSD) condition and bulk volume are displayed in Figure 2(a). Coarse bottom ash had a higher absorption value compared with the fine bottom ash because the voids are reduced as the particles become finer. The bulk volumes of both coarse and fine bottom ash were about the same. Pounded bottom ash has a similar water absorption value with the coarse bottom ash as it has a similar maximum aggregate size, but with a higher bulk volume value, showing it has a higher compaction density. PBA has the largest bulk volume of 1706.67 kg/m^3, which is higher than that of sand 1629.33 kg/m^3. This is because the pounded bottom ash particles have good gradations such that they can pack together well and minimize the air space between particles.

The sieve analysis can be seen in Figure 2(b). It can be analyzed that the fine bottom ash (FBA) has a particle gradation that resembles Lumajang sand. The fineness modulus (FM) of the RBA is highest due to its much larger particle size. The results of the sieve grading analysis were used to calculate the FM of the material. The FM value for sand, CBA, FBA, PBA, and RBA were 2.454, 2.633, 1.991, 3.323, and 3.884, respectively.

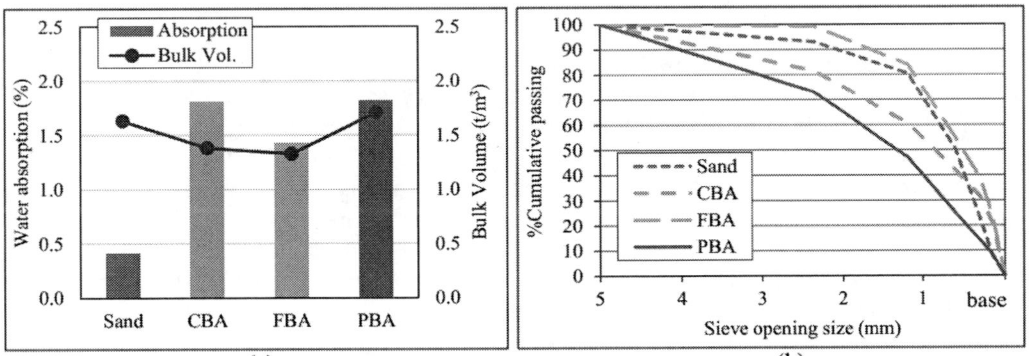

(a) (b)

Fig. 2. Physical properties of the materials: (a) water absorption and the bulk density and (b) gradation of the fine aggregate.

3.2 Behavior of fresh HVFA mortar

Figure 3(a) shows the evaluations of flow diameter and SP demand for each mixed composition. The increased use of bottom ash generally leads to a decrease in flowability; thus, it is necessary to add SP to achieve the target flow diameter. A 50% bottom ash replacement, no additional SP was needed; however, when replacing 80% and 100% sand with bottom ash, further SP was needed.

Mortar specimens with 100% FBA cannot be well made because the SP demand exceeds 2% and still does not reach target workability. Higher than 2% SP was not explored further due to the detrimental effect of SP on the strength and setting time of the mixture as well as economic considerations. An adequate mixture could be made by reducing the ratio of FBA to cementitious to 1.5 by weight. This correlates well with the lower bulk volume that was previously measured.

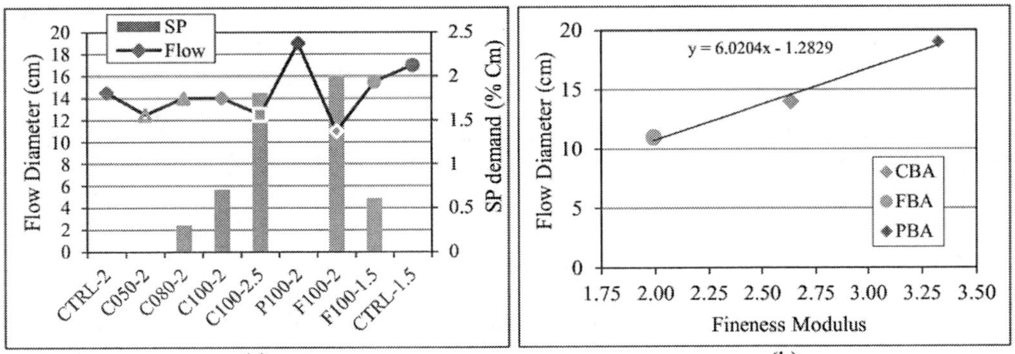

(a) (b)

Fig. 3. (a) Measured flow diameter with superplasticizer demand and (b) relationship of flow diameter and the fineness modulus of the material.

 DOI: 10.1051/matecconf/201713801006

Fresh mortar with PBA has a higher flowability than the control sand due to higher bulk volume of the material. The highest flow diameter was achieved for a mixture using 100% PBA even without additional SP. High flowability corresponds with the high bulk volume value and also correlates with the fineness modulus of the aggregate as shown in Figure 3(b). From the trial, it was found also that higher flow diameter also produced specimens with a smooth surface.

3.3 Properties of hardened HVFA mortar

A compressive strength test was conducted on specimens after curing in water for the specified duration. Figure 4(a) shows the compressive strength of the entire mixture composition. The compressive strength graph shows a corresponding positive increase of strength with time. For the CBA series, C050-2, C080-2, and C100-2 showed that at a higher sand replacement ratio a lower compressive strength was obtained. The 28-day strength of C100-2 was about 75% of the CTRL-2. The composition of C100-2.5 was made to maximize the use of bottom ash and had approximately the same strength with C100-2 at 28 days, but with a smaller diameter flow and more SP demand. However, when considering the later age strength, there was a low increment of strength for C100-2.5 specimen as opposed to still increasing of compressive strength for C100-2 because of the more cementitious material in the specimen.

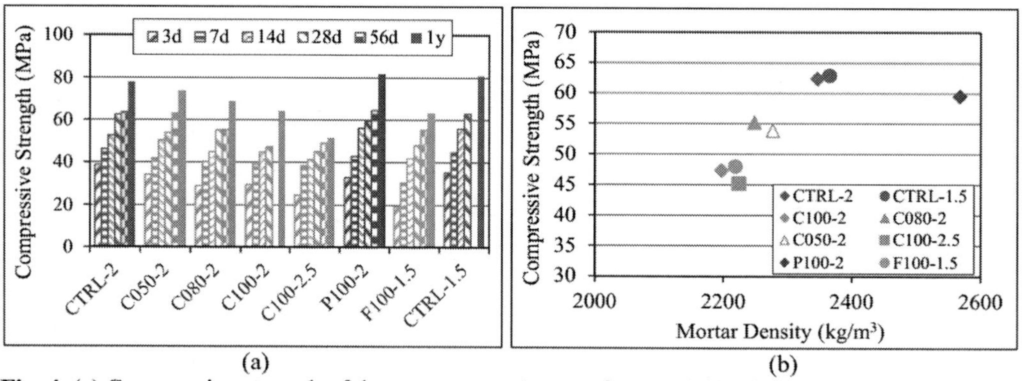

Fig. 4. (a) Compressive strength of the mortar up to 1 year of age and (b) relationship of mortar density and the 28-day compressive strength.

F100-1.5 mixture has lower strength compared with CTRL-1.5, showing that using finer bottom ash could reduce the water absorption of the material; however, it was more beneficial to use coarse bottom ash in the mixture. The compressive strength of the F100-1.5 at 28 days was about 75% of the comparable mortar, CTRL-1.5. There was no compressive strength data for F100-2 due to workability and compactability problems.

Using PBA with coarser particle size, P100-2, had a high compressive strength, almost the same as CTRL-2 mortar. This was because the PBA material has a good gradation so that the bulk volume obtained was optimal.

Long-term compressive strength was conducted at up to one year of mortar age to check whether the bottom ash could have a detrimental reaction in the mortar. All of the specimens had higher strength than that of the younger age. Strength increase was more pronounced for specimens with a more cementitious proportion in the mixture, as the high-volume replacement of the cement with fly ash means a slower and longer reaction rate. The compressive strength of mortar is still increasing, even at later age.

Figure 4(b) shows the correlation of the 28-day compressive strength with the bulk density of each composition specimen. The tendency of higher bulk density to correlate well with the compressive strength was true in this case with the exception of the PBA specimen. It seems that P100-2 could have a higher bulk volume, but with a similar strength as the CTRL-2 specimen.

4 Conclusions

Bottom ash was treated by sieving and pounding to make HVFA mortar. From the study, the following conclusions were obtained.

1. The bulk volume of coarse bottom ash and fine bottom ash is lighter than sand, whereas the bulk volume of pounded bottom ash is 4.75% greater than sand due to a better gradation of the particles such that the gaps between them can be filled. This showed that the gradation of the particle could be one of the main factors in making a good concrete mixture.
2. Water absorption of the bottom ash can be reduced by crushing into a finer particle size; however, the absorption value was higher when compared with the control sand. The use of more bottom ash in the mixture would require higher superplasticizer dosage to obtain a workable mix when using a low water to cementitious ratio.
3. Replacement of fine aggregate using coarse and fine bottom ash at 100% ratio (C100-2 and F100-1.5) resulted in 75% of the control specimen's compressive strength. However, when combining the gradation of the coarse and fine particle (P100-2), the compressive strength obtained was similar to the control specimen. This is due to good particle gradation, which increases the bulk density and, thereby, increases its compressive strength.
4. There is a reduction of fresh mortar flowability with the addition of coarse and fine bottom ash as indicated by the increase of superplasticizer demand. However, when a good gradation of the bottom ash particle was attained, the workability was comparable with the control sand. Therefore, the gradation of particles needs to be controlled to achieve a good workable and strong HVFA mortar.
5. Long-term concrete strength up to one year using bottom ash as the fine aggregate replacement was increased, indicating that there was no long-term detrimental effect on the usage of bottom ash in the concrete mixture.

Acknowledgment to PT. Jawa Power for providing the fly ash and bottom ash for the experimental study under MOU number 002/MoU/FTSP/UKP/2015-YTLJT/LTR/2015/00734 and The Ministry of Research, Technology and Higher Education, Indonesia for providing financial support.

References

1. T. Proske, S. Hainer, M. Rezvani & C. Graubner, Cem. Concr. Res, **51** 38–46 (2013)
2. H. K. Kim. & H. K. Lee, Constr Build Mater **25**(2), 1115–1122 (2011)
3. M. Singh & R. Siddique, J Clean Prod **112,** 620-630 (2015)
4. H. K. Kim. & H. K. Lee, J Mater Civil Eng, **25**(May), 662–665. (2013)
5. L.B. Andrade, J.C. Rocha & M Cheriaf, Constr Build Mater **23**, 609–614 (2009)
6. H. K. Kim, Constr Build Mater **91**, 57–64. (2015)
7. M. Reiner & K. Rens, Pract Period Struct Des Construct, **11**, 58–64. (2006)
8. S. V. Deo, Road Mater Pavement. **16**(3), 707–715. (2015)
9. A. A. Ramezanianpour, *Cement Replacement Materials* (Springer, Berlin, 2014)
10. Antoni ,V. Satrya & D. Hardjito, Civ Eng Dim, **17**(1), 38-43 (2015)

EFFECT OF HIGH TEMPERATURE ON PROPERTIES OF GLASS CONCRETE

Wei Chien Wang[1,*], *Shao Yu* Wang[1], *and Cheng Hsun* Lin[1]

[1] Department of Civil Engineering, Chung Yuan Christian University, Chung Li 32023, Taiwan, R.O.C.

Abstract. In this study, the concrete is mixed with glass fibers which are 0.5%, 1.0% and 1.5% of cement weight and 0.6 cm long. The effect of fiber addition level on the compressive strength, bending strength and residual compressive strength and thermal conductivity after heating of concrete is discussed. The test results show that an appropriate addition of glass fiber to the concrete at room temperature is favorable for the compressive strength. A low addition of relatively short glass fiber has insignificant effect on the bending strength. However, the adverse effect on the bending strength will increase if the addition level is too high. The addition of glass fiber can improve the loss of concrete strength under the effect of high temperature not exceeding 500℃, and a relatively high compressive strength remains under the effect of high temperature not exceeding 800 ℃. The high temperature can reduce the thermal conductivity of glass concrete, and the higher the addition level of glass fiber is, the lower is the thermal conductivity. Keywords: Glass fiber, concrete, heating, thermal conductivity

1 INTRODUCTION

The high temperature resistant fabric made of glass fiber still keeps 50% of strength when it burns at 370℃, and keeps 25% of strength at 538℃, its thermal expansion coefficient is low [1]. The glass fiber has higher compressive strength and elastic coefficient than cement mortar, so that it becomes an ideal reinforcer. However, the alkali resistance is required for cement concrete, if there is ZrO_2, SnO_2 or La_2O_3, the alkali resistance of glass will increase greatly [2]. The glass fiber with ZrO_2 is water resistant, so it is best suited for concrete [3].

The compressive strength of the concrete with glass fiber increases with the volume ratio of fiber, but the strength decreases slowly when a maximum value is reached. As the glass fiber increases, the initial crack flexural strength and ultimate flexural strength are 3 times and 4.9 times of nonfibrous concrete respectively. The flexural strength increases with the glass fiber content, but the strength growth rate does not decrease. The compressive strength is unrelated to the length of glass fiber. Longer glass fiber is more adverse to the concrete slump [4].

* Corresponding author: a654.joy@gmail.com

The concrete is a heat-proof material, but the concrete strength still decreases after burning in the fire. The structure may be damaged if the temperature is too high. This study simulates the effect of an approximate addition of glass fiber to the concrete on the compressive strength, tensile strength and thermal conductivity of concrete in the fire.

2 EXPERIMENTAL PLANNING

The concrete is mixed with glass fibers which are 0.5%, 1.0% and 1.5% of cement weight and 0.60 cm long respectively (mix proportions G5, G10 and G15, control group is C), to discuss the effect of fiber addition level on the compressive strength, bending strength, residual compressive strength and thermal conductivity after heating of concrete. The concrete mix of control group is designed according to ACI method. The slump is 18 cm and the water-cement ratio is 0.5. The additions of water, cement, coarse aggregate and fine aggregate are 218, 436, 720 and 838 kg/m^3 respectively.

In terms of the test for the compressive strength of concrete, the φ15 cm×30 cm concrete cylinder specimen is cured in 23℃±1.7℃ saturated limewater. The compressive strength on Days 7, 14, 21, 28, 49 and 56 is measured respectively. In terms of the test for the bending strength of concrete, a 15cm×15cm×53cm prism concrete is made, cured for 28 days before the thrid-point loading test. In terms of the test for residual compressive strength after heating, the φ15 cm×30 cm concrete cylinder specimen is cured in the 23℃ ±1.7℃ saturated limewater for 28 days, put in the 105±1℃ oven for 3 days, and put in the full electronic high temperature furnace, heated to 440, 500, 580, 800 and 1000 ℃ respectively. According to references, the compressive strength of concrete decreases mainly within one hour under the effect of high temperature [5]. Therefore, there is one hour added after the temperature rises to the set level, it is cooled to the room temperature before compression test. In terms of thermal conduction test, a 15 cm×15 cm×24 cm prism concrete is made, cured for 28 days, the long direction is covered with asbestos, heated in front of the electrothermal high temperature furnace. The temperatures at various points in the long direction of specimen are measured (one point on heating surface, five points inside). The thermal conductivity is calculated by Eq. (1). All data are the mean values of triplicate test results.

$$Q = -KA(\Delta T/\Delta X) \tag{1}$$

Where Q is the heat transfer rate (W), K is the thermal conductivity (W/m℃), A is the heat transfer area (m^2), ΔT is the temperature difference (℃), ΔX is the heat transfer distance (m).

Portland Type I cement is used; The maximum particle size of coarse aggregate is 3/4", the saturated surface dry specific gravity is 2.64, the water absorption is 1.21%, the dry rodding unit weight is 1600 kg/m^3; the fineness modulus of fine aggregate is 2.6, the saturated surface dry specific gravity is 2.63, the water absorption is 2.04%. The filament diameter of the alkali-resistant glass fiber for the test is 14±0.8μm, 2400±240 g/km, tensile strength ≥700 MPa.

3 EXPERIMENTAL PLANNING

The values of test results are shown in Table 1 and Table 2.

3.1 Effect of glass fiber on compressive strength of concrete

The test results are shown in Table 1 and Fig. 1. For the control group concrete (C), the concrete strength increases with the curing age, the compressive strength on Day 28 is 32.3 MPa, the strength on Days 7, 14 and 21 is 73%, 90% and 96% of the strength on Day 28 respectively. The compressive strength on Day 56 increases by only 2%, meaning the strength grows fastest during the first 7 days of curing. The strength grows slowly after 28 days. The strength development trend of the concrete with glass fiber is similar to normal concrete, meaning the addition of glass fiber which is 0.5%–1.5% of cement weight has no significant impact on the time-dependent development trend of concrete strength.

Fig. 1 shows that the 0.5% glass fiber in length of 0.60 cm can increase the concrete strength. However, the compressive strength decreases when the addition is 1.0 and 1.5%. For example, at the curing age of 28 days, the strength is increased by 9.6% when the addition is 0.5%. The strength is reduced by 24.1% and 26.6% respectively when the addition is 1.0% and 1.5%.

The results show that in the environment at room temperature, a low addition of short glass fiber to the concrete is favorable for the strength. The addition shall not be too high, otherwise it is adverse to the strength. The range of appropriate addition of long glass fiber which is favorable for the concrete strength is large, and a high addition has slight adverse effect on the strength.

3.2 Effect of glass fiber on bending strength of concrete

The test results are shown in Fig. 2. The bending strength on Day 28 of control group concrete (C) is 5.10 MPa, which is 15.8% of compressive strength on Day 28.

Fig. 2 shows that the 0.5% glass fiber in length of 0.60 cm can increase the bending strength of concrete. However, the bending strength decreases when the addition is 1.0 and 1.5%. The bending strength is increased by 1.0% when the addition is 0.5%. The bending strength is reduced by 2.0% and 8.6% respectively when the addition is 1.0% and 1.5%.

The results show that in the environment at room temperature, a low addition of short glass fiber to the concrete has insignificant effect on the bending strength. The addition shall not be too high, otherwise its adverse effect on bending strength will grow. When the long glass fiber is added in, the range of appropriate addition level favorable for the compressive strength of concrete is enlarged, and the bending strength is increased significantly, especially when the consumption increases to 1.5%.

Table 1. Average compressive strength and bending strength.

No.	Average compressive strength (MPa) (% of 28 days' strength)						28 days' average bending strength (MPa)
	7 D	14 D	21 D	28 D	49 D	56 D	
C	23.7(73)	29.2(90)	31.0(96)	32.3(100)	32.8(102)	33.1(102)	5.10
G5	22.8(65)	31.4(90)	33.3(94)	35.4(100)	36.8(104)	37.8(107)	5.15
G10	19.6(80)	22.2(90)	23.1(94)	24.5(100)	24.6(100)	26.2(107)	5.00
G15	19.2(81)	20.5(86)	22.3(94)	23.7(100)	24.1(102)	24.7(104)	4.66

DOI: 10.1051/matecconf/201713801007

Table 2. Average compressive strength and thermal conductivity after heating.

No.	Average compressive strength after heating (MPa) (residual strength % compared with strength without heating)				
	440℃	500℃	580℃	800℃	1000℃
C	25.8(80)	25.2(78)	15.7(49)	5.7(18)	5.0(16)
G5	32.2(91)	31.4(89)	20.4(58)	8.5(24)	3.7(11)
G10	22.8(93)	22.1(90)	14.7(60)	7.1(29)	2.7(11)
G15	22.3(94)	21.7(91)	15.5(65)	7.6(32)	3.0(12)
No.	Average thermal conductivity after heating (W/m·K) (compared with 440℃ test result, %)				
	440℃	500℃	580℃	800℃	1000℃
C	2.37	1.70(72)	1.19(50)	0.67(28)	0.45(19)
G5	2.25	1.62(72)	1.13(50)	0.57(25)	0.43(19)
G10	2.09	1.49(71)	0.83(40)	0.52(25)	0.40(19)
G15	1.98	1.33(67)	0.78(39)	0.51(25)	0.38(19)

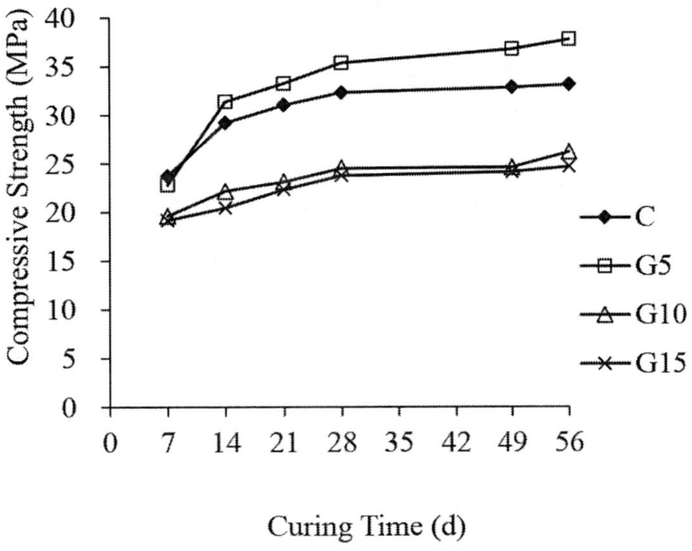

Fig. 1. Compressive strength of glass concrete at different ages.

 DOI: 10.1051/matecconf/201713801007

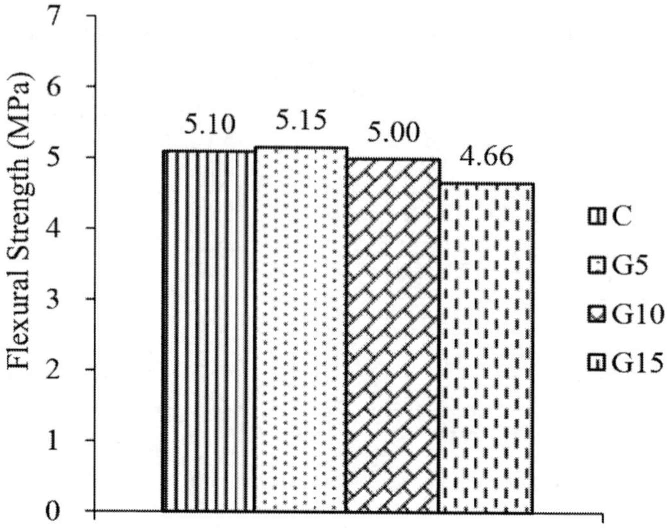

Fig. 2. Glass concrete bending strength comparison.

3.3 Effect of high temperature on compressive strength of glass concrete

The test results are shown in Table 1 and Fig. 3. When the control group concrete (C) is exposed to 440℃, 500℃, 580℃, 800℃ and 1000℃ for one hour, the residual strength is 80%, 78%, 49%, 18% and 16% respectively compared with the specimen free of high temperature. The compressive strength decreases gradually under the effect of high temperature. The compressive strength decreasing amplitude is small before 500℃, it decreases by 0.046% (22%/(500-23)) when the temperature increases by 1℃. The strength is lost because when the temperature is 400 - 500℃, most of bound water of C-S-H colloid has been lost. The compressive strength decreasing amplitude is the largest, about 0.363%/1℃ at 500℃ - 580℃. Because the Ca(OH)2 is decomposed at 500 - 580℃ [6]. The decreasing amplitude is the second largest at 580℃ - 800℃, about 0.141%/1℃. The decreasing amplitude is about 0.010%/1℃ at 800℃ - 1000℃, meaning the concrete strength of control group has been lost severely at 800℃.

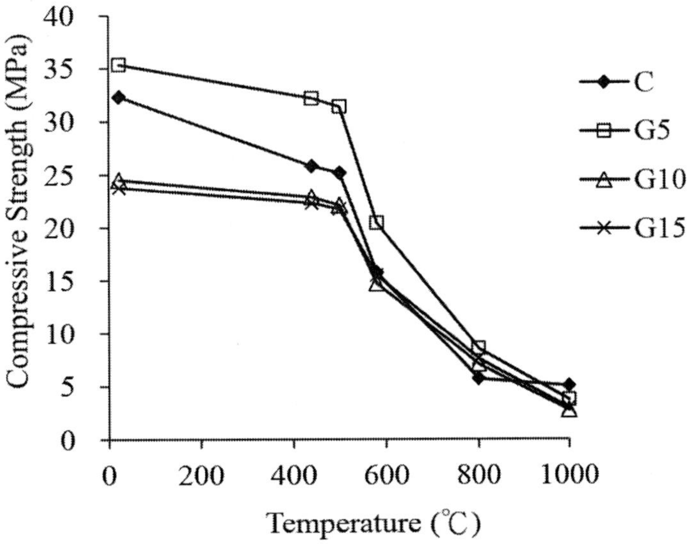

Fig. 3. Effect of high temperature on compressive strength of glass concrete.

Fig. 3 shows when the concrete is mixed with 0.5% short glass fiber (0.60 cm), the compressive strength decreasing amplitude is relatively small before 500℃. It decreases by 0.023% when the temperature increases by 1℃, lower than that of control group. The compressive strength decreasing amplitude is the largest at 500℃ - 580℃, about 0.388%/1℃, slightly different from control group. The decreasing amplitude is the second largest at 580℃ - 800℃, about 0.155%/1℃, slightly different from control group. The decreasing amplitude is about 0.065%/1℃ at 800℃ - 1000℃, higher than control group. The concrete with 1.0% and 1.5% glass fibers has the same trend. In addition, the addition of glass fiber can improve the loss of concrete strength under the effect of high temperature not exceeding 500℃, and the compressive strength is higher than control group under the effect of high temperature not exceeding 800℃. However, when the temperature is 1000℃, the residual strength of the concrete with glass fiber is lower than control group, it may because the glass fiber has lower melting point than aggregate. The loss of strength resulted from high temperature can be reduced by increasing the addition of glass fiber.

3.4 Effect of high temperature on compressive strength of glass concrete

The thermal conductivity of normal concrete is 1.5-3.5 W/m•K. Table 2 shows that after the control group concrete (C) is exposed to 440℃, 500℃, 580℃, 800℃ and 1000℃ for one hour, the thermal conductivity is 2.37, 1.70, 1.19, 0.67 and 0.45 W/m•K respectively. The thermal conductivity decreases gradually as the temperature rises. The test results of 500℃, 580℃, 800℃ and 1000℃ are 72%, 50%, 28% and 19% of 440℃ test result respectively. Fig. 4 shows the decreasing amplitude of thermal conductivity of concrete is the largest at 440℃ - 500℃, followed by 500℃ - 580℃, 580℃ - 800℃, and 800℃ - 1000℃.

Fig. 3 shows the concrete mixed with short glass fiber (0.60 cm), compared with the test results of control group, the thermal conductivity of concrete is reduced under the effect of high temperature, and the thermal conductivity decreases as the addition level of glass fiber increases. However, the trend of effect of high temperature on the thermal conductivity of glass concrete is identical with control group.

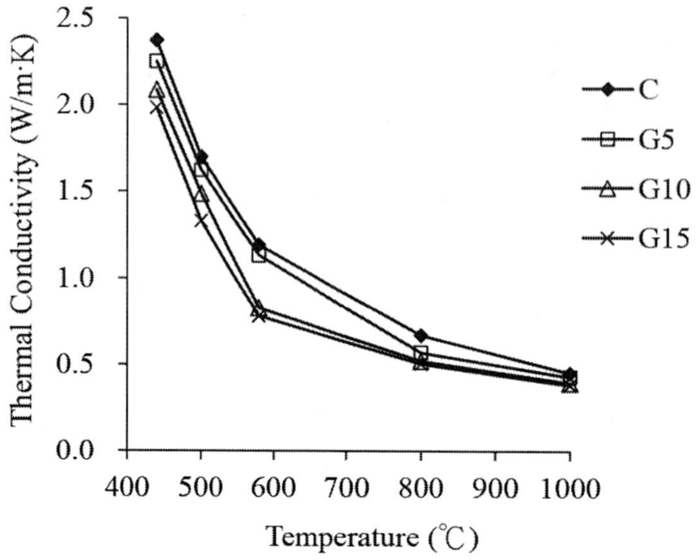

Fig. 4. Effect of high temperature on thermal conductivity of glass concrete.

4 CONCLUSION

The concrete is mixed with glass fibers which are 0.5%, 1.0% and 1.5% of cement weight and 0.60 cm long, to discuss the effect of fiber addition level on the compressive strength, bending strength, residual compressive strength and thermal conductivity after heating of concrete. The conclusions are described below:

(1). In the environment at room temperature, a low addition of short glass fiber to the concrete is favorable for the strength. The addition shall not be too high, otherwise it is adverse to the strength.

(2). In the environment at room temperature, a low addition of short glass fiber to the concrete has insignificant effect on the bending strength. The addition shall not be too high, otherwise its adverse effect on bending strength will grow.

(3). The addition of glass fiber can improve the loss of concrete strength under the effect of high temperature not exceeding 500℃, and the compressive strength is higher than control group under the effect of high temperature not exceeding 800℃. When the temperature is 1000℃, the residual strength of the concrete with glass fiber is lower than control group.

(4). The thermal conductivity of concrete decreases under the effect of high temperature, and the thermal conductivity decreases as the addition level of glass fiber increases. The trend of effect of high temperature on the thermal conductivity of glass concrete is identical with control group.

The authors would like to thank the National Science Council of Taiwan for their financial support of this research under Contract No. MOST 105-2221-E-033 -016 -MY2.

References

1. C.C. Chang, (trans) "Corpora of Glass Fiber and Super POM", Taipei City, Xushi Culture & Education Foundation. (1976)

2. C.C. Chang, "Application of FRP", Taipei City, Xushi Culture & Education Foundation. (1980)
3. C.C. Chang, "Glass Fiber Reinforced Cement Construction Manual", Taipei City, Xushi Culture & Education Foundation. (1987)
4. N. Henry, Jr. Marsh, L. Lawrence, and Jr. Clack, "Glass fiber reinforced concrete base materials", Fiber Reinforced Concrete, ACI Publication SP-44: 147–264. (1974)
5. C.Z. Wolfgang, "Cement chemistry and physics for civil engineers," English Edition. (1980)
6. C.R. Cruze, "Elastic properties of concrete at high temperatures", PAC Research Bulletin. (1971)
7. Y.F. Lin, "Discussion about Fire Performance of Lightweight Concrete Bricks in Mixed Sizes", Master's thesis, National Chung Hsing University. (2004)

Influence of Oil Palm Biomass Waste on Compressive Strength and Chloride Penetration of Mortar

Nor Hasanah Abdul Shukor Lim[1], *Mostafa* Samadi[1], *Mohd Warid* Hussin[1], *Abdul Rahman Mohd* Sam[1], *Nur Farhayu* Ariffin[2], *Mohamed A*. Ismail[3], and *Han Seung* Lee[4], and *Mohd Azreen Mohd* Ariffin[5]

[1]Construction Research Centre (UTM CRC), Institute for Smart Infrastructure and Innovative Construction, Faculty of Civil Engineering, Universiti Teknologi Malaysia, 81310 UTM Johor Bahru, Johor, Malaysia
[2]Faculty of Civil Engineering and Earth Resources, Universiti Malaysia Pahang, Lebuhraya 8 Tun Razak, 26300 Kuantan, Pahang, Malaysia
[3]Civil and Construction Engineering Department, Faculty of Engineering and Science, Curtin University, Sarawak Malaysia, Miri, Sarawak, Malaysia
[4]Department of Architectural Engineering, Hanyang University, Ansan, Republic of Korea
[5]Forensic Engineering Centre, Institute for Smart Infrastructure and Innovative Construction, Faculty of Civil Engineering, Universiti Teknologi Malaysia, 81310 UTM Johor Bahru, Johor, Malaysia

Abstract. The utilization of waste materials which are abundant and cheap, especially from clean resources, has become more pressing than ever. This study investigates the influence of Oil Palm Biomass waste including the Palm Oil Fuel Ash (POFA), Oil Palm Kernel Shell (OPKS) and Oil Palm Fibre (OPF) on the compressive strength and chloride penetration of mortar. The POFA was used as cement replacement up to 80% in nano size. The mass ratio of fine aggregates to binder was 3:1. Test specimens 70×70×70 mm cube were prepared and cured in water for 28 days. Ordinary Portland cement (OPC) mortar was also prepared as control specimen. The specimens were immersed in Sodium Chloride solution up to 18 months. The evaluation was done by visual observation, ultrasonic pulse velocity and mass change before and after exposure. The mortar was then split into two and sprayed with 0.1N Silver Nitrate solution to see the depth of penetration. The test results revealed that biomass mortar showed high resistance to chloride penetration as compared to OPC mortar due to the reduction of cement content in the mixture. Besides, the reactive silica from nano POFA produced more crystalline formation thus, reduced the porosity and crack within the biomass mortar.

1 Introduction

The ordinary Portland cement (OPC) still continues to be the most commonly used in construction field. Many studies have shown that OPC gives poor performance in resistance to extreme climate and chemical condition [1-3]. Moreover, it processes and releases a large amount of green house gasses, i.e. carbon dioxide (CO_2) into the atmosphere. The need towards sustainability and sustainable environment has made the use of pozzolanic material in construction popular. Palm oil fuel ash is generally known as one of the pozzolanic material from natural waste. By 2014, Malaysia has recorded a staggering 5.39 million hectares of oil palm plantations, an increase of 11.0% from the previous 4.85 million hectares in 2010 [4]. Recent research has utilized the use of oil palm biomass waste on the mechanical properties of mortar [5]. As a new material for construction, very little research has been conducted on the durability of biomass mortar particularly with reference to chloride resistance. The durability of mortar is an important requirement for the performance in aggressive environments throughout its design life period.

2.1 Materials

The cement used in this study complies with Portland cement Type I. Palm Oil Fuel Ash (POFA) was obtained from the burning of palm oil shell and husk (in equal volume) from a southern part of Malaysia. The POFA collected were processed until nanoparticles size [6]. The Oil Palm Kernel Shell was used as sand replacement. The Oil Palm Fibre added was 7% by the volume of binder. The mortar was then known as biomass mortar [5].

2.2 Testing Procedures

All mortar specimens were prepared with sand to binder with the ratio of 3:1, whereby the sand was prepared into saturated surface dried condition. The mixing was carried out in a room temperature. The mix proportions are based on weight of materials according to BS EN 998-1:2010 [7]. The test specimens of 70 x 70 x 70 mm cubes were prepared as described in ASTM C109-13 [8]. Chloride ion penetration test was conducted to study the effect of chloride ion to mortar specimens. The test was conducted in accordance with ASTM C1202 'Electrical Indication of Concrete's Ability to Resist Chloride Ion Penetration' [9]. The specimens were immersed in 2.5% sodium chloride solution for 1, 2, 3, 6, 9, 12 and 18 months. The chloride concentration used in this study is similar to the one used by Ariffin (2012) and Hassan (2015) [10,11]. In addition, the ACI Committee 222R-01 (2001) suggested that the concentration of water soluble chloride ion above 0.2% of Portland cement mass can accelerate the corrosion in the environment where moisture and oxygen are available [12]. At the specific test age, the specimens were split into two halves and the cross sections were sprayed with 0.1 silver nitrate solution. The surface of the specimens will changed to white colour as an indication the effect of chloride ions. The changes in mass were determined to monitor the chloride ion resistance during exposure.

3 Results and Discussion

3.1 Compressive Strength

Fig. 1 shows the compressive strength comparison of the OPC and biomass mortars. As the age of water curing increases, the compressive strength of OPC and biomass mortars are increases. This is due to the continuous hydration process and pozzolanic reaction occurred in the mortars. At 1 month curing, the compressive strength of biomass mortar increases by 24% as compared to OPC mortar. This is due to the second production of C-S-H gel from pozzolanic reaction. After 1 year of curing, the OPC and biomass mortars show compressive strength of 33.64 MPa and 47.01 MPa, respectively. Despite the stronger bond produced and filler packing effect, the fibre in biomass mortar help to distribute the load and increase the bond between binder and fine aggregates resulting in an increase in the compressive strength. The compressive strengths obtained at the age of 1 year were approximately 28% higher than the target required design strength of 30 MPa. This continual increase in strength even after 1 year indicates that the biomass mortar does not deteriorate once the OPKS aggregates are encapsulated into the mortar matrix. Similar trends were reported for OPKS as coarse aggregates in the literature [13].

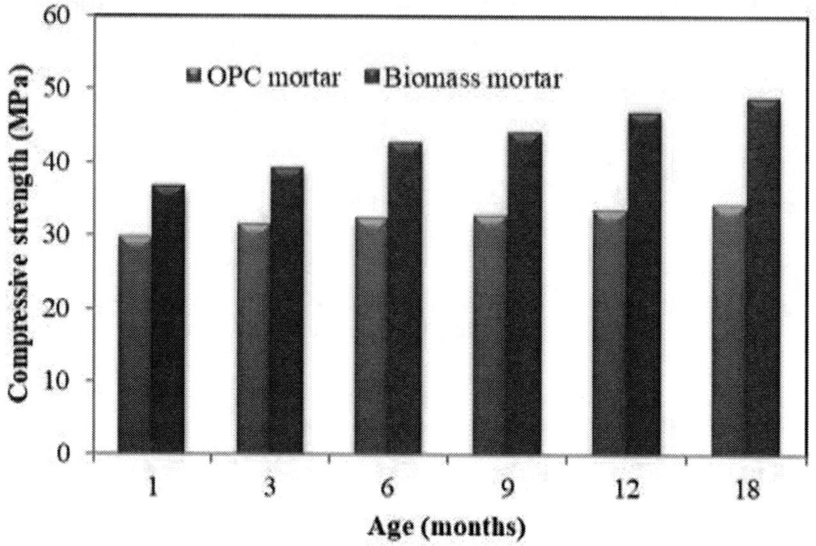

Fig 1 Strength development of OPC and biomass mortars

3.2 Mass Change

Fig. 2 shows the mass change of OPC and biomass mortars after immersion in sodium chloride (NaCl) solution. It can be seen that the mortars had gained some mass when immersed in sodium chloride. The figure shows sudden increase in mass for OPC and biomass mortars to 0.9% and 0.4% from initial mass after 28 days immersion in solution, respectively. Similar results were obtained for immersion after 18 months in which the mass for OPC and biomass mortars gained 4.9% and 2.1% from initial mass, respectively. At the end of immersion, the mass became almost constant and steady. The mass change of OPC and biomass mortars was related to the ability of specimens to absorb the solution. In water absorption test conducted, biomass mortar has lower water absorption compared with OPC mortar. As a result, the mass of OPC mortar has higher percentage of mass change compared with biomass mortar. Similar findings on chloride penetration were also reported by previous research [14,15].

It can be observed that the rate of decrease in penetration was higher in biomass mortar. The permeability resistance of mortar containing NPOFA was about two times that of OPC mortar. This is due to pore refinement within the mortar micro structure by NPOFA, or the transformation of large permeable pores to a smaller pore by the pore refinement of the fine particle and C-S-H gel produced from the hydration process. Therefore, these suggest that the resistance of chloride ion penetration of mortar can be significantly increased with the use of high volume NPOFA replacement. Likewise, high volume NPOFA changes the distribution of cement particle in the matrix, thus producing a decrease in the capillary network as well as reducing the average capillary pore size due to the gradual pozzolanic reaction and cement hydration process. Therefore, it is generally accepted that the incorporation of pozzolanic materials improves the resistance of chloride penetration. This

DOI: 10.1051/matecconf/201713801008

is mainly due to the reduction in the permeability, particularly to chloride ion transportation of the blended cement mortar.

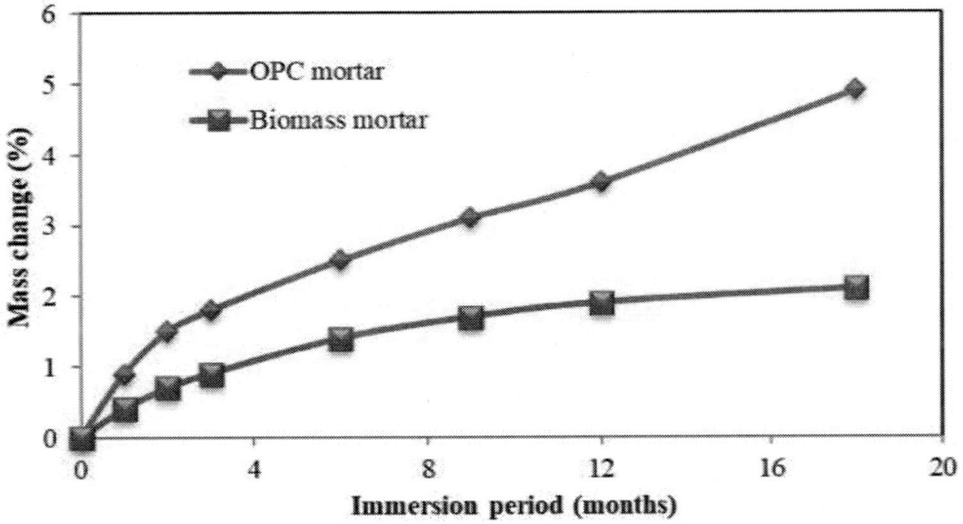

Fig 2 Mass change of OPC and biomass mortars in NaCl

After a certain period of designated immersion time, the specimens were taken out from the solution and tested. The specimen was sprayed with 0.1N silver nitrate solutions to determine the penetration of chloride ions. Fig. 3 shows the effect of chloride on OPC and biomass mortars. The average depth of chloride penetration for the OPC mortar was 15 mm. On the other hand, for biomass mortar the depth was 3 mm. These results show that chloride ions penetrate more in OPC mortar compared with biomass mortar. This is most likely because the biomass mortar contains pozzolanic material that generates more C-S-H gel that makes the mortar denser and less permeable.

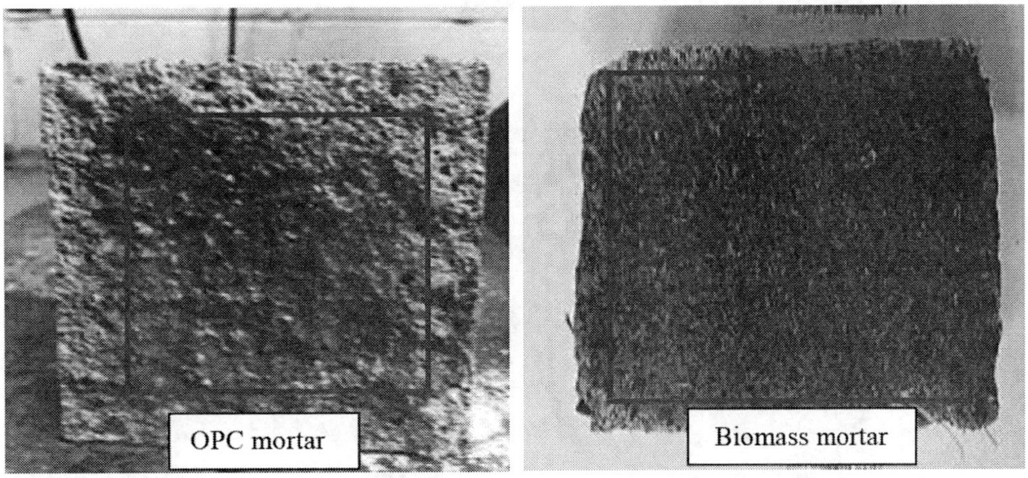

Fig 3 Mortars sprayed with 0.1 silver nitrate

 DOI: 10.1051/matecconf/201713801008

3.3 Microstructure Analysis

The SEM and XRD images of OPC and biomass mortars samples after 18 months of immersion in NaCl solution are shown in Figs. 4 and 5, respectively. It can be seen from the figures, that there are more voids presented inside the OPC mortar compared with the biomass mortar. Fewer voids were observed from the biomass mortar due to the gradual formation of C-S-H gel from the secondary reaction between reactive silica and $Ca(OH)_2$ available from cement hydration. Therefore, higher chloride ions can penetrate into the OPC mortar thus chloride mineral is shown in the EDX analysis. Meanwhile, there is no chloride mineral detected in the biomass mortar due to the dense and less porosity of the mortar. Besides, the NPOFA materials not only generating more C-S-H gel but also act as filler that reduce the size of pore structure of mortar. Furthermore, the rough surface and irregular shape of OPKS used increase the interlocking between the fine aggregates and binder paste. The inclusion of OPF also increases the interlocking by bridging the paste and preventing minor cracks from occurring.

Fig 4 SEM and EDX images of OPC mortar in NaCl

Fig 5 SEM and EDX images of biomass mortar in NaCl

4 Conclusion

The chloride ion penetration test revealed that biomass mortar offers superior resistance to the ion penetration than the OPC mortar. This is attributed to the micro filling ability and pozzolanic characteristic of the binder. Biomass mortar did not show any changes in appearance and dimension when exposed to chloride solution. The depth of chloride penetration for OPC mortar was 15 mm while for the biomass mortar was only 3 mm. Besides, the compressive strength of biomass mortar increases by 24% as compared with OPC mortar at 1 month curing. The compressive strengths obtained at the age of 1 year were approximately 28% higher than the target required design strength of 30 MPa. This continual increase in strength even after 1 year indicates that the biomass mortar does not deteriorate once the OPKS aggregates are encapsulated in the mortar.

The authors are grateful to the Ministry of Higher Education, Malaysia (MOHE) and Research Management Centre (RMC), Universiti Teknologi Malaysia (UTM) for financial support under grant GUP Q.J130000.7322.4B216. The authors are also thankful to the staff of Structures & Materials Laboratory and Faculty of Civil Engineering and Department of Architectural Engineering, Hanyang University for the support throughout the study.

References

1. M. A. R. Bhutta, N. F. Ariffin, M. W. Hussin, and N. S. A. Hasanah, "Sulfate and sulfuric acid resistance of geopolymer mortars using waste blended ash. Jurnal Teknologi", **3**, 1–5 (2013)
2. M. W. Hussin, N. F. Ariffin, M. A. R. Bhutta, and N. H. A. S. Lim, "Study on Dry-Wet Cyclic Resistance of Geopolymer Mortars Using Blended Ash from Agro-Industrial Waste. In Proceeding of Third International Conference on Sustainable Construction Materials and Technologies", August 2013, Kyoto, Japan (2013)
3. J. Hill, E. A. Byars, J. H. Sharp, C. J. Lynsdale, J. C. Cripps, and Q. Zhou, "An experimental study of combined acid and sulfate attack of concrete. Cement and Concrete Composites", **25**(8), 997–1003 (2003)
4. Malaysian Palm Oil Board, (MPOB) (2015, December 1), "Economic and industry development devision. Oil Palm Planted Area", Available at: (http://bepi.mpob.gov.my/index.php/statistics/area.html)
5. N. H. A. S. Lim, M. W. Hussin, A. S. M. Rahman, M. A. R. Bhutta, N. F. Ariffin, N. H. A. Khalid, and M. Samadi, "Properties of mortar containing high volume palm oil biomass waste. Advanced Materials Research", **1113**, 578–585 (2015)
6. N. H. A. S. Lim, M. A. Ismail, H. S. Lee, M. W. Hussin, A. R. M. Sam, and M. Samadi, "The effects of high volume nano palm oil fuel ash on microstructure properties and hydration temperature of mortar. Construction and Building Materials", **93**, 29–34 (2015)
7. British Standard Institution, *Specification for mortar for masonry. Rendering and plastering mortar*. BS EN 998-1:2010, (2010)
8. ASTM C109, 2013, *Standard test method for compressive strength of hydraulic cement mortars*, Annual Book of ASTM Standards, American Society for Testing and Materials (2013)
9. ASTM C1202, Electrical indication of concrete's ability to resist chloride ion penetration, *Annual Book of ASTM Standards, American Society for Testing and Materials* (2012)
10. N. F. Ariffin, "Durability of geopolymer mortars using agro-industrial waste", Thesis, Universiti Teknologi Malaysia (2012)

11. I. O. Hassan, "Self consolidating high performance palm oil fuel ash and pulverised burnt clay blended concrete", Ph.D. Thesis, Universiti Teknologi Malaysia (2015)
12. ACI Committee 222R, *Protection of metals in concrete against corrosion. American Concrete Institute*, (2001)
13. D. C. L. Teo, M. A. Mannan, V. J. Kurian and C. Ganapathy, "Lightweight concrete made from oil palm shell (OPS): Structural bond and durability properties. Building and Environment", **42**, 2614–2621 (2007)
14. D. Mujah, "Compressive strength and chloride resistance of grout containing ground palm oil fuel ash. Journal of Cleaner Production", **112**, 712–722 (2015)
15. P. Chindaprasirt, S. Rukzon, and V. Sirivivatnanon, "Resistance to chloride penetration of blended Portland cement mortar containing palm oil fuel ash, rice husk ash and fly ash. Construction and Building Materials", **22**, 932–938 (2008)

Performance of Lightweight Natural-Fiber Reinforced Concrete

Harianto Hardjasaputra[1], *Gino* Ng[1], *Girum* Urgessa[2], *Gabriella* Lesmana[1], *Steven* Sidharta[1]

[1]Civil Engineering Department, Universitas Pelita Harapan, Tangerang, Indonesia
[2]Civil, Environmental and Infrastructure Engineering, George Mason University, Fairfax, VA, USA

Abstract. Concrete, the most common construction material, has negligible tension capacity. However, a reinforcement material such as natural fibers, can be used to improve the tensile properties of concrete. This paper presents experiments conducted on Super Lightweight Concrete mixed with coconut fibers (SLNFRC). Coconut fibers are regarded as one of the toughest natural fibers to strengthen concrete. Coconut fiber reinforced composites have been considered as a sustainable construction material because the fibers are derived from waste. These wastes, which are available in large quantities in Asia, have to be extracted from the husk of coconut fruits and must pass a mechanical process before being added to a concrete mixture. The Super Lightweight Concrete was made by mixing concrete paste with foam agent that can reduce the overall weight of concrete up to 60% with compressive strength up to 6 MPa. The Super Lightweight Concrete is intended to be used for non-structural walls, as alternative conventional construction materials such as brick walls. The influence of coconut fibers content in increasing the flexural tensile strength of Super Lightweight Concrete was studied in this research. The fiber content studied include 0%, 0.1%, 0.175%, and 0.25% by weight of cement content. Sixteen specimens of SLNFRC mini beams of 60 mm x 60 mm x 300 mm were tested to failure to investigate their flexural strengths. The optimum percent fibers yielding higher tensile strength was found to be 0.175%

Keywords: Super Lightweight Concrete, Natural-fibers, compressive, tensile strength.

1 Introduction

The use of lightweight concrete has increased significantly across countries all over the world. The majority of current research on concrete focuses on high-performance concrete, which is looking for satisfying demanding performance requirements, including durability.

Meanwhile, lightweight concrete can be defined as the type of concrete that has an in-place density of less than 1840 kg/m^3 (116 lb/ft^3) compared to normal-weight concrete with a density in the range of 2240 to 2400 kg/m^3 (140 to 150 lb/ft^3). Lightweight concrete is a versatile material, which consists primarily of a cement-based mortar mixed with low density aggregate (less than 880 kg/m^3) or volume of air up to 40%.

Lightweight concrete has low density that leads to reduced dead load, and lower transport and handling cost. In addition, it also has a good thermal conductivity. However, lightweight concrete has low compression and flexural strength compared to normal weight concrete. The purpose of this paper is to create a lightweight concrete using foaming agent mixed with natural fibers to increase its flexural strength. Based on SNI 03-0349-1989, the density of lightweight concrete has been limited to less than 1200 kg/m^3 and the compression strength has a minimum value of 2.5 MPa.

The compression and flexural strength of concrete depend greatly on the composition and in particular on their density. Even though lightweight concrete discussed in this paper will not be suitable for structural member applications, determining the flexural strength will still be considered. Natural fibers, such as coconut fibers, have been successfully implemented to increase the flexural strength of concrete (Ng et al, 2016). They also act like stirrup reinforcement to prevent premature concrete splitting along the tensile reinforcement and control crack width (Campione, 2013).

Coconut fiber is a waste material that can be easily available in large quantities and is very cheap. In addition, it can improve the flexural strength and form good bond in concrete. Coconut fibers also gives a lot of benefit such as no effect on environment and increase in the strength of concrete (Agrawal, 2014). Coconut fiber has the advantage of stretching beyond its elastic limit without rupturing, as well as having the power to take up a permanent stretch. It also has a unique protection to microbial degradation and salt water (Ali, 2010). Handhinia (2014) found that concrete with 0.25% fibers is the best mix based on crushing load test compared to 1% coconut fiber and regular concrete. It also reported that coconut (coir) fiber may reduce temperature by up to 30% and reduce the weight of the total concrete overall.

Table 1 shows a typical chemical composition of coconut fibers used in this research. Table 2 shows the physical properties of representative coconut fibers.

Table 1. Chemical Composition of Coconut fiber (Alam, 2014)

Lingin	45.84%
Cellulose	43.44%
Hemi-Cellulose	0.25%
Pectin's and Related Compound	3%
Water Soluble	5.25%
Ash	2.22%

Table 2. Physical Properties of Coconut Fiber (Ali, 2010)

Length (in)	6-8
Density (kg/m^3)	670-1370
Tenacity (g/Tex)	10
Breaking Elongation (%)	10-40%
Diameter (mm)	0.1-1

DOI: 10.1051/matecconf/201713801009

Rigidity of Modulus (dyne/cm^2)	1.8924
Swelling in Water (%)	5%
Moisture at 65% RH	10.5%

2 Experimental work

This paper presents an experimental study that investigates the use of coconut fibers in super lightweight concrete. The concrete mix design was selected based on previous super lightweight research that produce concrete with weight less than 1200 kg/m^3 and compression strength up to 2.5 MPa. In this case, 16 specimens of 6 cm x 6 cm x 30 cm concrete beams were made by varying the percentage of coconut fibers by weight of cement. The fiber contents studied are 0%, 0.1%, 0.175%, 0.25%.

2.1 Mix Designs

Figure 1. shows the plot of lightweight concrete's mix design. The optimum mix design was selected based on previous experimental studies about lightweight concrete that produce density of concrete less than 1200 kg/m^3 and compression strength exceed 2.5 MPa. Table 3, Table 4 and Table 5 show the constituents of the concrete used in this research. Foaming agents were used to create the foam to fill up to 40% of final volume. The strength of concrete depends on the quantity of foam that has been used. Adding more foam yields more air in the concrete that will eventually reduce the strength.

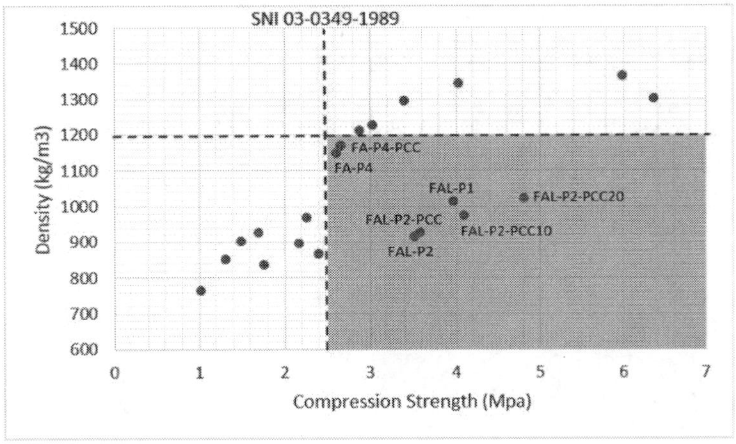

Figure 1. Mix design density vs compression strength of lightweight concrete

Table 3. Constituents of the concrete mix.

Item	Comments
Cement	Ordinary Portland cement type 1
Aggregate	Fine sand with particle size less than 2 mm (sieve #10)
Water	Regular tap water
Foaming agent	ADT Foaming Agent

Table 4. Concrete paste mix proportions of FA-P4.

Material	Mix Design
Cement	402 kg/m^3
Aggregate	777 kg/m^3
Water	181 kg/m^3

Table 5. Foaming agent mix proportions.

Material	Mix Design
Foaming Agent	5 L/m^3
Water	150 L/m^3

2.2 Small Beams Experiments

In this research, sixteen specimens of 6 cm x 6 cm x 30 cm beams were cast with varying percentages of coconut fibers as shown in Table 6 using selected mix designs. In general, procedures for mixing lightweight concrete are similar to those for normal weight concrete. However, the challenging aspect here is mixing regular concrete paste with the foaming agent.

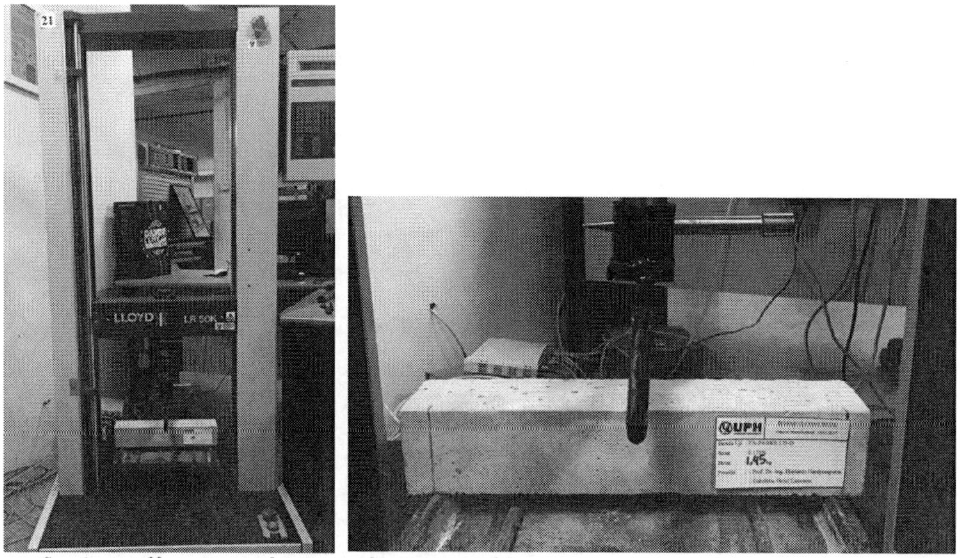

Figure 2. A small concrete beam and a test machine.

The foam concrete was made by mixing two materials, foam and concrete paste. The foaming agent and water are mixed in a mixer until the mixture transforms into foam. The foam was placed into concrete paste slowly until completely mix and becomes foamed concrete. The foamed concrete was placed into mold and the paste was flattened until it is completely distributed. The specimens were stripped from the molds twelve hours after casting and submerged in water for seven days.

Table 6. Percent fibers in the mix design.

Experimental Group	Coconut Fibers (% weight of cement)
1	0%
2	0.1%
3	0.175%
4	0.25%

3 RESULTS

LLYOD type LR press machine with a maximum capacity of 50 KN was used to test all the beams. This machine has transducer technology to measure the load-deformation characteristics. The results were represented by calculating fr (modulus of rupture). Table 7 shows the modulus of rupture of the concrete beams with varying coconut fiber content including the density of the beams.

Table 7. Maximum load capacity of concrete beams with varying coconut fiber content.

Experimental Group	Coconut Fibers (% weight of cement)	Density (kg/m^3)	Load (N)	fr (MPa)
1	0%	1126.54	527.44	0.95
2	0.1%	1136.57	787.42	1.42
3	0.175%	1138.89	927.97	1.68
4	0.25%	1148.15	399.21	0.72

Figure 2 shows the average of modulus of rupture results. It shows that adding coconut fiber until 0.175% clearly increases the flexural strength of the lightweight concrete. However, adding excessive fibers reduces the strength. In fact, adding 0.25% (by the weight of cement) of coconut fiber has same modulus of rupture with normal lightweight concrete. In this case, the effective amount of fiber that can be used is 0.175%, where it is shown to increase the flexural strength up to 76% when compared to regular concrete mix with no coconut fibers.

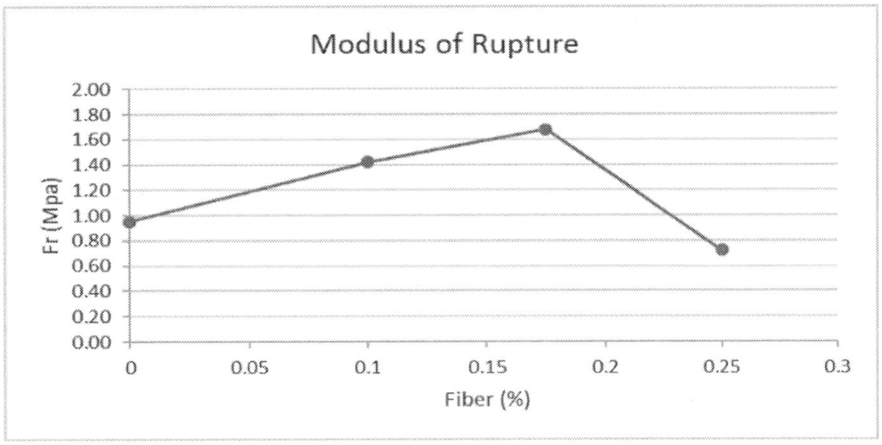

Figure 3. Modulus of rupture of concrete beams with cocunt (coir) fiber reinforcement.

4 CONCLUSION

The results show that the mix design has successfully met code requirements. In addition, it also shows that coconut fibers have a good effect in the mix design. The addition of coconut fibers clearly increased flexural strength and stiffness of the concrete. The optimum percent fibers from this research was found to be 0.175%. However, adding excessive fibers resulted in reduction of the tensile strength due to bonding issue between fibers and concrete. Additional research is needed to determine if there is an optimum value of coconut fiber between 0.175% - 0.25% by weight of cement.

The first Author would like to acknowledge that a part of this research has been done during his research visit at TU Berlin- Germany, granted by Konrad Adenauer Stiftung.

REFERENCES

1. Ali, M., Natural Fibres as Construction Materials, *Journal of Civil Engineering and Construction Technology*. Vol. **3**, No. 3, pp. 80-89 (2012)
2. Agrawal, R.A, Dhase, S.S. and Agrawal, K.S., Coconut Fiber in Concrete to Enhance its Strength and Making Lightweight Concrete, *International Journal of Engineering Research and Development*, Vol. **9**, No. 8, pp. 64-67(2014)
3. Ng, G.P., Urgessa, G.S., Bayleyegn, Y. and Hardjasaputra, H., Mechanical Properties of Coconut Fiber-Reinforced Concete, *Resilient Structures and Sustainable Construction*, Valencia, Spain, (2017)
4. Hardjasaputra, H., Tirtawijaya, J. and Fernandez, D.A., The Application of Natural Fibers in Concrete Mixing to Increase Shear Strength and Performance of Concrete Beam, *Proceeding of The First International Seminar on Sustainable Infrastructure and Built Environment in Developing Countries,* Indonesia, (2009).
5. Hardjasaputra. H, Ng. G, Guntur. P, Widjajakusuma. J, Rahardja. S. Pembuatan Lantai Berbasis Semen Sebagai Bahan Bangunan Hijau Berserat Sabut Kelapa Dengan Teknik Basah dan Teknik Press [Study of cement based flooring (tile) using coconut fiber with wet and press method]. Proceeding of Konferensi Nasional Teknik Sipil 8, ITS-Bandung, Indonesia, (2014)
6. Ramakrishna, G. and Sundararajan, T., "Impact strength of a few natural fibre reinforced cement mortar slabs: A comparative study", Journal of Cement Concrete Composites, Vol. **27**, No. 5, pp. 547-553, (2005)
7. Campione, G. Flexural and Shear Resistance of Steel Fiber-Reinforced Lightweight Concrete Beams, Journal of Structural Engineering, Vol. **140**, (2013).

Factors influencing strength and setting time of fly ash based-geopolymer paste

Adhitya Leonard Wijaya[1,*], *Januarti* Jaya Ekaputri[1,2,*] , and *Triwulan*[1]

[1]ITS Surabaya, Civil Engineering Department, 60111 Surabaya, Indonesia

[2]Konsorsium Riset Geopolimer Indonesia, Concrete Building Material Laboratory, Civil Engineering Department of ITS, Surabaya, 60111, Indonesia

Abstract. In general, fly ash is often used as a binder in geopolymer mixture. In this paper, seven samples of fly ash were collected from five Indonesian power plants. The variation of fly ash may depend upon many factors. However, there is no standardization of fly ash used for geopolymer binder. This experiment was conducted to study and categorize fly ashes both by assessing their physical and chemical characters. All fly ashes were activated with alkali activator to prepare geopolymer paste. Ratio of fly ash to the alkali activator was kept constant at 65 : 35. Compression test was conducted at three, seven, 14, 21, 28, and 56 days to identify the strength development of each sample. The result revealed that the content of CaO in fly ash affected the setting time and early strength development. In three days paste with CaO content more than 20% gain 54% of its 28-day strength. It is caused by both hydration and polymerization reactions. A range of Si to Al ratio contributed to the strength.

1 Introduction

Indonesia has approximately more than 85 coal-fired power plants. These power plants contribute to million tons of fly ash every year. Unfortunately, the utilization of fly ash is still far from optimal. The ashes are just pilled up on the ground without any application. An innovative effort leads to attempt to use fly ash as a sustainable eco-friendly material. Geopolymer concrete is one of the examples of green material, which uses non-Portland cement as a binder. One of the potential Portland-cement replacements is fly ash, which is activated with alkali to form binder.

The problem of application of fly ash in geopolymer concrete is the fact that every fly ash from each power plant has different characteristics. It causes a problem to control gepolymer products. There are many factors influencing fly ash properties. To utilize fly ashes from these power plants, characterization is required to assess its effects on geopolymer mixtures.

The important point to identify fly ash character in geopolymer is its setting time and strength in paste form. Setting time is the most important thing to identify because if the paste set rapidly, it is impossible to apply the paste in concrete. There are many factors that

* Corresponding author: adhityaleonard4@gmail.com and januartije@gmail.com

relate to setting time and compressive strength of geopolymer pastes such as physical properties and chemical properties. [1-3]

Davidovits [4] stated that pH is one of the indicators to determine the character of fly ash. Physical properties such as fineness affected the strength of geopolymer pastes. The Finer fly ash results to higher compressive strength as found by Somna, et al. [5].

From the chemical component Subaer, et al. [6] explained that Si/Al had some impact on the compressive strength of geopolymer, compressive strength increased as the increasing of the ratio of Si to Al but at some point it would decrease. According to Antoni, et al. [7], the higher CaO content made the setting time of the paste become faster. Hardjito [8] explained that high content of CaO make geopolymer experienced both polymerization and hydration reactions that result to higher strength and accelerate the setting time of geopolymer paste.

2 Materials

Seven samples of fly ash from five different coal-fired power plants were being tested. The name of fly ash is arranged in alphabetical order from A to G. Fly ash A and fly ash B were collected from Paiton power plant, C and D were collected from Indramayu power plant, fly ash E from Pacitan power plant, fly ash F from Rembang power plant, and fly ash G was collected from Tanjung Awar-Awar power plant.

These fly ashes were divided into two classes based on ASTM C618 [9]. This categorization was based on the contents of SiO_2, Al_2O_3, and Fe_2O_3 in the fly ash. If the content of SiO_2, Al_2O_3, and Fe_2O_3 are more than 70% then it categorized as F class, and if the content only 50% or more then it classified as C class. As being shown in the Table 1 fly ash A, C, D, E, and F are categorized as F class, while fly ash B and G are classified to C class.

Alkali activator being used was the mixture of sodium hydroxide (NaOH) 8 M and sodium silicate (Na_2SiO_3) BE-58 that consist of 19%, H_2O, 36% SiO_2, and 45% Na_2O. The weight ratio of Na_2SiO_3 to NaOH 8 M that being used was two.

Table 1. Chemical composition of fly ashes (% weight)

NO	OXIDES	A	B	C	D	E	F	G
1	SiO_2	45.82	32.49	41.22	46.67	46.41	37.94	27.04
2	Al_2O_3	24.09	13.43	18.71	24.92	26.58	22.67	18.18
3	Fe_2O_3	11.57	23.59	16.21	11.59	9.80	16.53	17.73
4	TiO_2	1.05	0.78	0.71	0.76	0.79	0.77	0.66
5	CaO	8.03	14.83	12.20	7.07	8.23	11.76	21.08
6	MgO	3.56	10.71	6.03	4.58	3.88	5.78	8.69
7	K_2O	1.66	0.87	0.89	0.69	0.66	0.94	1.04
8	Na_2O	1.20	0.65	1.01	0.97	1.51	1.25	1.96
9	SO_2	0.56	1.03	1.40	0.68	0.63	0.74	2.40
10	Mn_2O_3	0.11	0.44	0.31	0.21	0.17	0.25	0.36
$SiO_2 + Al_2O_3 +$ Fe2O3		81.48	69.51	76.14	83.18	82.79	77.14	62.95
Classification		F	C	F	F	F	F	C

3 Methods

The weight proportion of geopolymer paste mixture used was 65% fly ash and 35% alkali activator. Ratio Na_2SiO_3 to NaOH 8 M by mass was two. This composition was blended to react with fly ash.

Procedure to make the geopolymer paste was by blending the composition using a mixer then put it in the cylinder mold that already prepared. The mold has dimension of 2 cm in diameter and 4 cm in height. All the specimens of paste were kept at moist curing in room temperature until the day of test.

Some series of analysis were conducted for fineness, specific gravity, pH test, and X-Ray Fluorescence (XRF) test. Fineness test of fly ash was performed using a sieve no. 325 (45 μm) based on ASTM C430-96 [10]. The test results were analyzed for each setting time and compressive strength of pastes. Setting time test followed the procedure from ASTM C191 [11]. Setting time is the most important factor to be investigated from the fly ash-paste. This test objective was to understand the time to make the mixture and also determine the suitability of fly ash when being used in geopolymer concrete mixtures. Setting time test was conducted using vicat to calculate the initial and final setting of each fly ash. Initial setting time was calculated as the time elapsed between the initial contact of fly ash and alkali activator and when the penetration was at 25 mm from the surface. Final setting occurred when the penetration of vicat was zero mm. Compressive strength test followed the procedure from ASTM C109 [12]. The compressive strength was tested at three, seven, 14, 21, 28 and 56 days. The strengths were calculated from the average of six identical specimens.

4 Results and Analysis

Setting time of pastes is listed at Table 2. Setting times of paste were greatly influenced by CaO content in the fly ash. Table 2 illustrates fly ash A, D, and E had very long setting time it occurs because CaO content in fly ash A, D, and E was below 10%. Fly ash C and F had a similar setting time to Portland-cement paste. It is shown that CaO content of fly ash C and F is quite high, which is above 10%. Fly ash G set rapidly in 18 minutes. It was affected by its CaO content is more than 20%. It causes fly ash G become impossible to use in the concrete mixture. Figure 1 shows the higher CaO content then setting time of the paste will get faster. It shows that setting time will decrease exponentially as the increase of CaO content. From the Figure 1, it is possible to predict the setting time of geopolymer paste using formula :

$$ST = 4838{,}7.\ e^{-0.262CaO} \tag{1}$$

ST : setting time (minutes)
e : exponential number (2,72)
CaO : CaO content (% fly ash mass)

Diaz, et al. [13] stated that fly ash was suitable as a concrete mixture when the CaO content is between 5% until 15%. Based on data of setting time, fly ashes that applicable for geopolymer mixture are fly ash A and E. Setting time of paste made from fly ash A and E are quite long at about nine hours. In the field, setting time of geopolymer concrete almost half of the geopolymer paste's setting time, for that reason fly ash A and E are the proper choices as the binder of geopolymer.

 DOI: 10.1051/matecconf/201713801010

Table 2. Initial and final setting time of fly ash

Code	A	B	C	D	E	F	G
Initial Setting Time (minutes)	378	72	120	590	377	85	10
Final Setting Time (minutes)	585	129	225	827	530	160	18

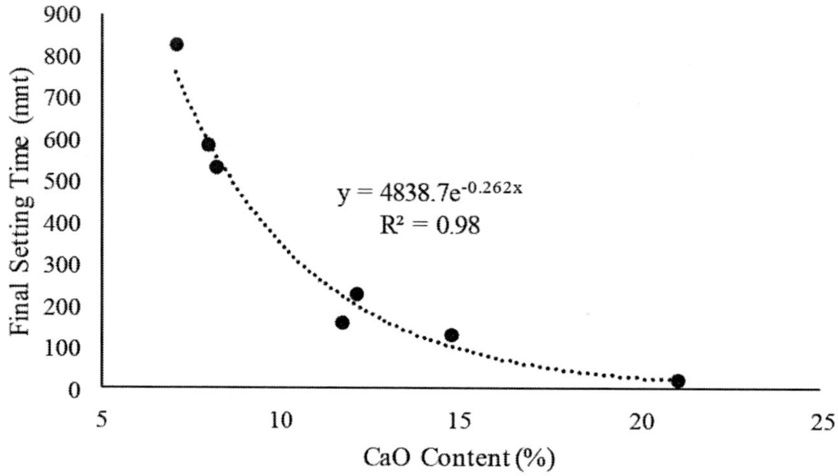

Fig. 1. Relation between CaO content and final setting time

Figure 2 shows the strength of pastes at different days, Nurwidayati, et al. [14] stated that the increase of the age result in higher strength of the paste. The strength of geopolymer paste still kept increasing even after 28 days. Fly ash D gained the highest strength at 28 days 50.30 MPa, while compressive strength of fly ash B is the lowest of all with 32.84 MPa. It shows that even at 28-day fly ash D gain the best strength, but at three days it compressive strength was only 12.26 MPa, while the highest strength at three days was fly ash G with 21.08 MPa. Based on the result of setting time and strength, fly ash A and E are the most applicable fly ash for geopolymer mixtures. Fly ash A and E have high strength and quite long setting time, it is possible to apply them in the field.

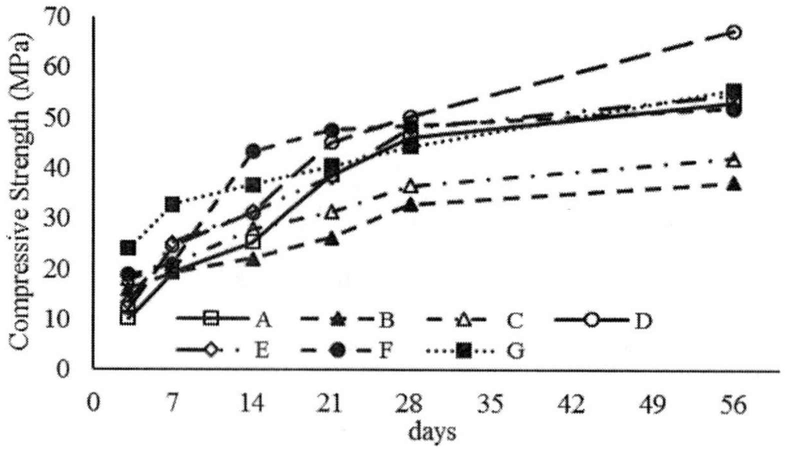

Fig. 2. Strength of geopolymer pastes at different days

DOI: 10.1051/matecconf/201713801010

CaO content in fly ash besides affecting setting time also involved in early strength build-up and pH of fly ash. Table 3 shows pH, strength of pastes at three days, and its percentage of 28 days strength. Higher CaO content will result in higher pH as stated by Antoni, et al. [15], but it only occurred if fly ashes with high CaO content (> 10%) are separated from fly ashes with low CaO (< 10%) as it is shown by Figure 3. Figure 3 shows linear relation between CaO content and pH of fly ashes with CaO content less than 10%, while for fly ashes with CaO content more than 10% the trend is almost linear.

Table 3. pH, and three days strength of pastes

Code	A	B	C	D	E	F	G
pH	11.9	11.5	11.6	11.4	12	11.6	12.1
Three Days Strength (MPa)	10.17	16.08	18.06	12.26	13.40	18.76	24.04
Comparison with 28 days Strength (%)	22%	49%	49%	24%	28%	39%	54%

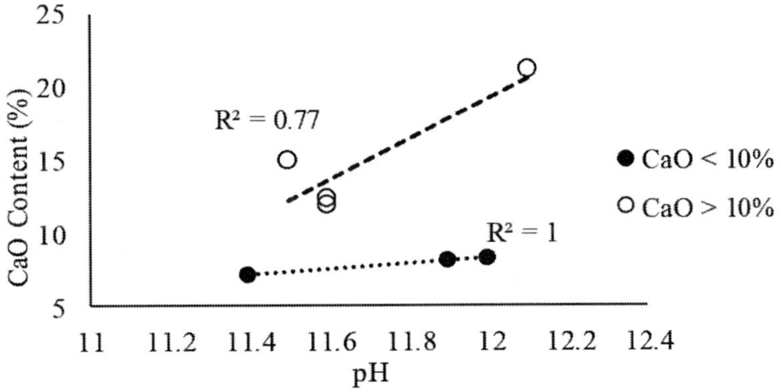

Fig. 3. Relation between pH of fly ash and CaO content

Figure 4 and Figure 5 show the effect of CaO content in early strength of geopolymer pastes. Figure 4 shows the relation of CaO and strength in MPa while Figure 5 shows the relation of CaO and strength in percentage. Both Figure 4 and Figure 5 show that CaO content contributes in early strength development. High CaO content result on a high strength, it as being stated by Antoni, et Al [16]. Risdanareni, et al. [17] explained high CaO content contributed higher strength. It is because of double reaction occurred within the system which are polymerization and hydration reactions. In Figure 4, highest strength occurred at fly ash with highest content CaO. Fly ash G had CaO content of 21.08%, and its strength was 24.04 MPa while fly ash D with the lowest CaO content only had 12.26 MPa. Figure 5 fly ash G with CaO content 21.08% result on 54% of its 28-day strength only in 3 days.

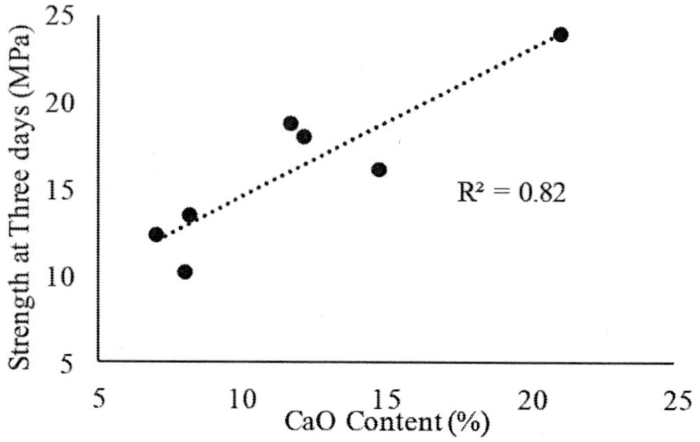

Fig. 4. Effects of CaO content on compressive strength at three days

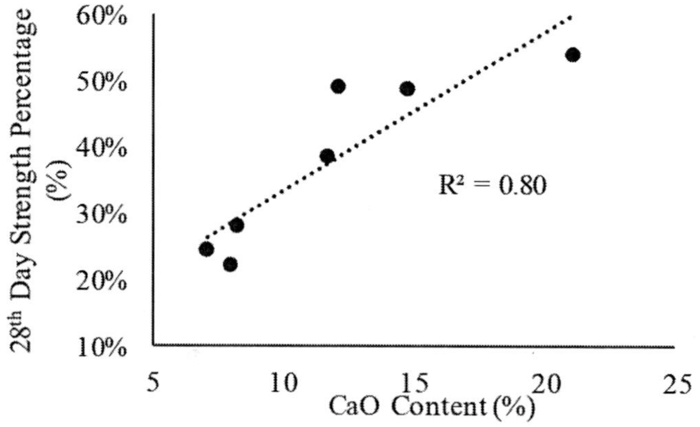

Fig. 5. Effects of CaO content on 28th day strength percentage

Strength of geopolymer paste at 28-day was affected by both physical and chemical properties of fly ash as shown in Table 4. Physical property that dominantly affected the strength is fineness, while chemical property that contributes significantly is ratio Si to Al. The numbers in fineness parameter were indicated percent material retained in the sieve no. 325.

Table 4. Physical and Chemical Properties Affect Geopolymer Pastes Strength

Code	A	B	C	D	E	F	G
Fineness (% retained)	30	41	31	27	30	29	27
Si/Al	2.078	2.883	2.466	2.038	1.902	1.912	1.875
Comp. Strength at 28 days (MPa)	46.06	32.84	36.59	50.30	48.01	48.45	44.39

Figure 6 illustrates that fineness has a significant impact on increasing paste strength. It was as being explained by Soutsos, et al. [18] fineness of fly ash contributes more to the strength of geopolymer compared to the other physical properties.

DOI: 10.1051/matecconf/201713801010

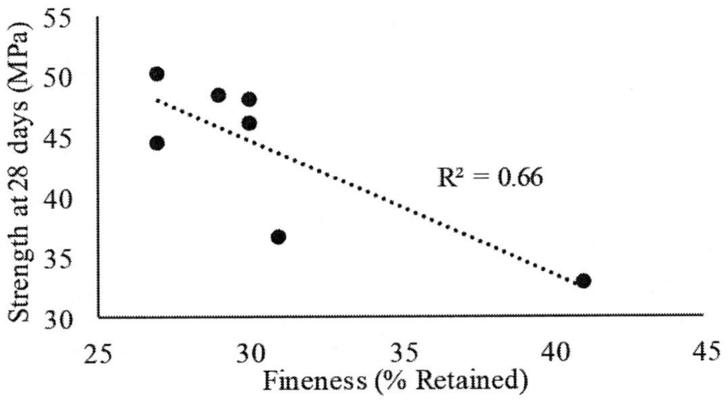

Fig. 6. Relation between Fineness and Compressive Strength at 28 Days

Chemical property contributes to the strength of geopolymer pastes was the ratio of Si/Al. From Figure 7 shows that pastes have high strength when ratios of Si to Al were around two. While the ratios of Si/Al are at the range of 1.875 to 2.078 the compressive strengths are between 44.39 MPa and 50.30 MPa. The strength then keeps decrease as the increase of Si to Al ratio. Asif, et al. [19] explained that the increasing of ratio Si/Al produce higher strength. The optimum strength of geopolymer was occurred when the ratio of Si to Al is equal to two, when the ratio of Si to Al more than two then the strength kept decrease. Asif, et al. conducted the research using various specimens with the ratio of Si to Al from 1.85 to three. Zhou, et al. [20] also stated the optimal condition of fly ash based-geopolymer occurred when the ratio of Si to Al equals to two. Chemical composition has greater effect upon the strength of paste than the physical property of fly ash. Ratio Si to Al contributes more to the strength compares to fineness of fly ash.

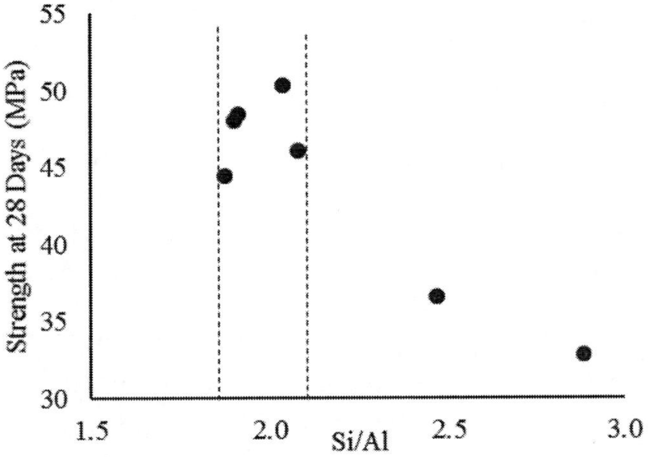

Fig. 7. Relation between Si/Al and Strength at 28 Days

5 Conclusions

1. CaO content of fly ash contributes to faster setting time of geopolymer paste due to double reaction of geopolymerization and hydration process. This double reaction also increased the early strength of geopolymer pastes, especially at three days. Fly ash G with CaO content more than 20% obtained 54% of its 28-day strength.
2. pH of fly ash severely was influenced by CaO content. The pH increased as CaO content increased. However, there should be separation between fly ash with CaO content more than 10% and less than 10% by fly ash mass.
3. Fineness contributes to compressive strength of geopolymer pastes. Finer fly ashes resulted in higher compressive strength.
4. The ratio of Si to Al influences to the strength of geopolymer pastes. Compressive strengths are optimal when the ratios of Si to Al are around two. When this ratio was between 1.875 and 2.078 then the strengths at 28 days were in the range of 44.39 MPa to 50.30 MPa.
5. Chemical composition contributes more to the strength and setting time of paste compares to physical properties of fly ash.
6. Based on setting time and strength, fly ash A and E are the most applicable materials for geopolymer concrete.

References

1. X. Gao, Q. L. Yu, H. J. H Brouwers, *Construction and Building Materials*, **98**, p. 397-406 (2015)
2. S. Alehyen, M. El Achouri, M. Taibi, "Journal of Materials and Environmental Sciences", **8,** issue **5**, 1783-1796 (2017)
3. X. Y. Zhuang, L. Chen, S. Komarneni, C. H. Zhou, D. S. Tong, H. M. Yang, W. H. Yu, H. Wang, "Journal of Cleaner Production", **125**, 253-267 (2016)
4. J. Davidovits, "Geopolymer chemistry and applications 2nd ed. Saint-Quentin (France: Institute Géopolymère)", (2008)
5. K. Somna, C. Jaturapitakkul, P. Kajitvichyanukul, P. Chindaprasirt, *FUEL,* **90**, 2118-2124 (2011)
6. Subaer, A. Haris, Nurhayati, A. Irhamsyah, J. J. Ekaputri, "Materials Science Forum", **841**, 170-17 (2016)
7. Antoni, S, W. Wijaya, D. Hardjito, "Material Science Forum", **841**, 90-97 (2016)
8. D. Hardjito, "Studies on fly ash-based geopolymer concrete (Curtin University of Technology. Perth, Australia)", 2005
9. American Society for Testing and Materials, ASTM C618 (2004)
10. American Society for Testing and Materials, ASTM C430 – 96 (1996)
11. American Society for Testing and Materials, ASTM C191-13 (2013)
12. American Society for Testing and Materials, ASTM C109 / C109M-02 (2002)
13. E. I. Diaz, E. N. Allouche, S. Eklund, *FUEL,* **89**, 992-996 (2010)
14. R. Nurwidayati, M. B. Ulum, J. J. Ekaputri, Triwulan, P. Suprobo, "Material Science Forum", **841**, 118-125 (2016)
15. Antoni, R. Gunawan, D. Hardjito, "Applied Mechanics and Materials", **815**, 153-157 (2015)
16. Antoni, S, W. Wijaya, D. Hardjito, "Material Science Forum", **841**, 98-103 (2016)

17. P. Risdanareni, P. Puspitasari, J. J. Ekaputri, "MATEC Web of Conferences", **97**, (2017)
18. M. Soutsos, A. P. Boyle, R. Vinai, A. Hadjierakleous, S. J. Barnett, "Construction and Building Materials", **110**, 355-368 (2016)
19. A. Asif, Z. Man, M. Azizli, K. Azizi, M.F. Nuruddin, L. Ismail, "Material Science Forum", **803**, 355 (2014)
20. W. Zhou, C. Yan, P. Duan, Y. Liu, Z. Zhang, X. Qui, D. Li, *Materials & Design*, **95**, p. 63-74 (2016)

Flexural strength of walls made of hollow core concrete brick using various notch models as the interlocking device

Nanang Gunawan Wariyatno[1,*], *Yanuar* Haryanto[1], and *Sumiyanto*[1]

[1]Jenderal Soedirman University, Department of Civil Engineering, Jl. Mayjend. Sungkono KM 5, Blater, Purbalingga, 53371, Indonesia

Abstract. The effectiveness of hollow core concrete bricks made of sand and stone ash mixture as alternative substitute materials has been previously studied, considering their weight, load capacity, construction time, and cost. To obtain an optimal notch model as the interlocking device, this further study is conducted. The previous notch model is in the shape of a 2-cm trapezoid and is varied here to consider a 2-cm rectangular notch model and 3- and 3.5-cm trapezoid notch models. The ratio of sand to stone ash in the mixture is 25:75. The wall specimen refers to the previous studies, which investigated specimens with sizes of 90 × 45 cm^2 under out-of-plane loading and 60 × 60 cm^2 under in-plane loading. We found that the 2-cm rectangular notch model is optimal as the interlocking device of the wall made of hollow core concrete bricks, with a flexural strength under out-of-plane loading of 212.29 kN/m^2 and a flexural strength under in-plane loading of 330.92 kN/m^2, representing increases of 14.63 and 175.77%, respectively, compared to the values obtained in previous studies. The failure or collapse initially starts from the material parts and then continues to the connection ones.

1 Introduction

Indonesia is a country with high consumption of resources due to the continuously increasing population and economic growth. The Central Bureau of Statistics [1] reports that the total estimated value of completed constructions based on working types in 2015 reached more than Rp. 635 trillion. According to news recently released by PT Semen Indonesia (Persero) Tbk on 31 October 2016, it is estimated that the national cement consumption in 2017 will reach 68 million tonne, representing a growth of 8.04% compared to the consumption during the previous year of 2016, when it reached only 62.94 million tonne [2]. Red bricks are commonly used for walls of either simple house buildings or reinforced concrete building structures in developing countries and earthquake-prone countries [3], such as Indonesia [4, 5]. However, considering environmental conservation and natural resources, a new brick type is immediately required to replace the red one. Such efforts have been conducted by examining the effectiveness of hollow core concrete bricks

* Corresponding author: nanang_g@yahoo.com

made of sand and stone ash mixture as an alternative substitute material for use in walls by reviewing their weight, load capacity, installation time, and cost. Further studies are required to obtain an optimum notch model of the interlocking device by considering the flexural strength and wall failure mode parameters of hollow core concrete bricks.

An alternative material development for walls has been conducted by Pang *et al.* [6], who investigated a wall with a size of 1,500 × 2,000 × 240 mm^3 made of sand, soil, and lime mixture in a certain ratio with various hollow openings. After the first stage of testing, the wall was coated with reinforcing steel-fibre mortar on both sides and then retested. It is known that increasing the amount of lime in the mixture may improve the wall strength. The hollow opening does not influence the wall-supporting capacity, but the reinforcing steel fibre mortar may effectively improve the wall-supporting capacity. All wall types experience collapse due to losing layers of the reinforcing steel fibre mortar as a result of cohesive failure, in which layers of reinforcing steel fibre mortar suffer only slightly from cracks and damages. This is caused by the damage to the inner wall occurring during the first testing stage.

In response to the need for low-level buildings in rural areas of China, Cao *et al.* [7] developed Recycled Concrete Bricks (RCBs) with a size of 53 × 115 × 260 mm^3 using recycled aggregates obtained from construction waste of buildings destroyed in Beijing, which are further processed to produce fine and coarse aggregate. Monotonic testing was performed against an RCB wall with a size of 780 × 260 × 390 mm^3, while cyclic testing was performed against an RCB wall with a size of 260 × 115 × 53 mm^3 equipped with vertical reinforcing steel. It was found that the RCB wall failure pattern was similar to the red brick wall failure pattern for axial loading pressure. In an RCB wall with vertical reinforcement, the hidden column elements effectively restrict the occurrence of deformations and cracks, so the RCB wall may become more ductile, being supported by a simple and relatively low-cost construction, and may therefore become a good alternative choice for low-level seismic building design in rural areas.

Wariyatno *et al.* [8] conducted a study that aimed to determine the effectiveness of hollow core concrete brick as an alternative wall material substitute due to the weight, load capacity, installation time, and cost. The specimens consisted of two wall types with sizes of 90 × 45 and 60 × 60 cm^2. The results showed that the hollow core concrete brick wall had a weight of 237.61 kg/m^2, which was lighter than the red brick wall, whose weight was 270.45 kg/m^2. However, it was still heavier than the light brick wall, whose weight was only 193.54 kg/m^2. The load capacity of the hollow core concrete brick wall was 4.17 kN, while those of the red brick wall and light brick wall were 5.55 and 2.87 kN respectively. The hollow brick wall had a construction time of 52 min/m^2 with a cost of only Rp. 187.400/m^2, so it could be constructed much faster and more cheaply than the red brick wall, which had a construction time is 182 min/m^2 and a cost of Rp. 302.400/m^2, and the light brick wall, which had a construction time is 140 min/m^2 and a cost of Rp. 279.000/m^2. The effectiveness of the hollow core concrete brick is obviously high, but it still has some weaknesses in the connection part.

2 Methodology

2.1 Equipment and materials

The equipment used in this study included ovens, scales, concrete mixer machines, hollow core concrete brick mould (Figure 1), and a Universal Testing Machine (UTM). The materials used were as follows: sand, stone dust, cement, water, and PVC pipes.

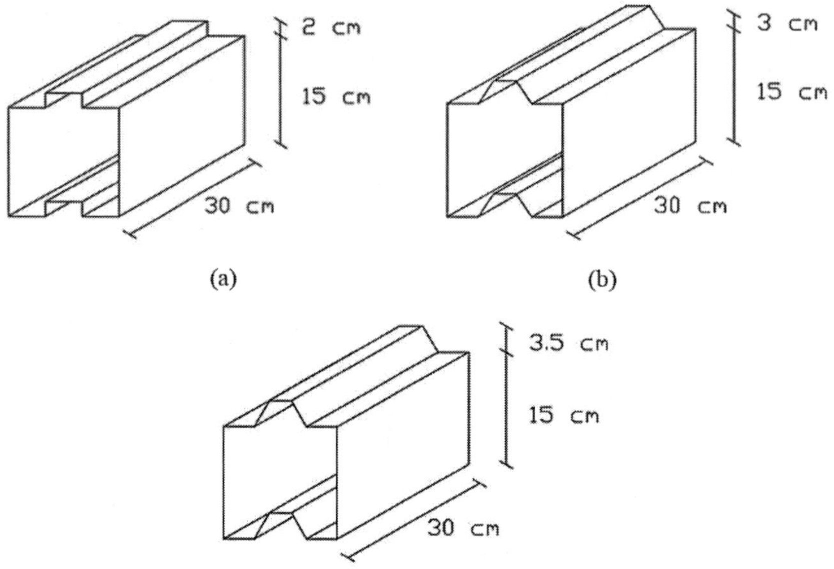

(a) (b) (c)

Fig. 1. Hollow core concrete brick mould; (a) 2-cm square notch, (b) 3-cm trapezoid notch, c) 3.5-cm trapezoid notch

2.2 Specimens

Hollow core concrete brick is made of a sand and cement mixture in a ratio of 6:1. The ratio of sand to stone ash in the mixture is 25:75, resulted from the preliminary tests to obtain the optimum composition. The composition ratio of the mixtures used is presented in Table 1, and the average compressive strength test results of three mortar specimens are presented in Table 2. The wall specimens are made with a size of 90 × 45 cm^2 for out-of-plane loading and 60 × 60 cm^2 for in-plane loading. The specified details are presented in Table 3 and the wall specimen section can be seen in Figure 2.

Table 1. Mixture composition

No.	Number	Ratio	Total weight (g)	Water (g)	Cement (g)	Sand (g)	Stone ash (g)
1.	3	100:0	1343	97.82	177.86	1067.15	0
2.	3	75:25	1380	97.82	177.86	800.36	266.79
3.	3	60:40	1380	97.82	177.86	640.29	426.86
4.	3	50:50	1380	97.82	177.86	533.58	533.58
5.	3	40:60	1380	97.82	177.86	426.86	640.29
6.	3	25:75	1380	97.82	177.86	266.79	800.36

Table 2. Average compressive strength of three mortar specimen.

No	Ratio	Average Compressive Strength (MPa)	
		In 3 days	In 7 days
1.	100:0	2.70	4.90c.
2.	75:25	2.75	6.57
3.	60:40	3.75	7.70
4.	50:50	4.50	8.63
5.	40:60	5.50	8.63
6.	25:75	6.15	11.43

DOI: 10.1051/matecconf/201713801011

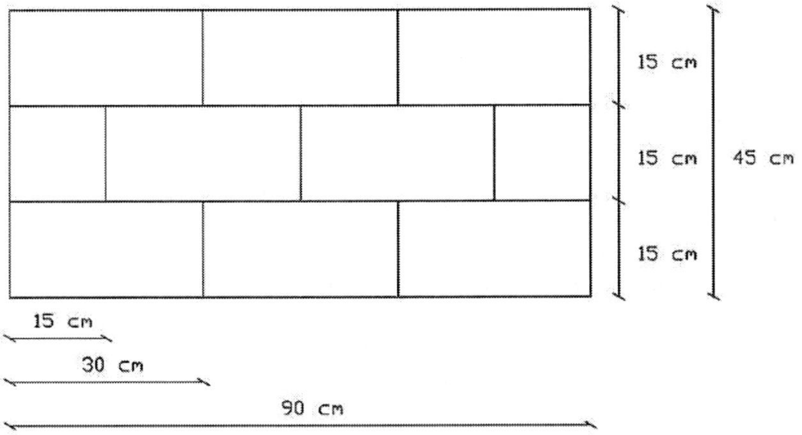

(a)

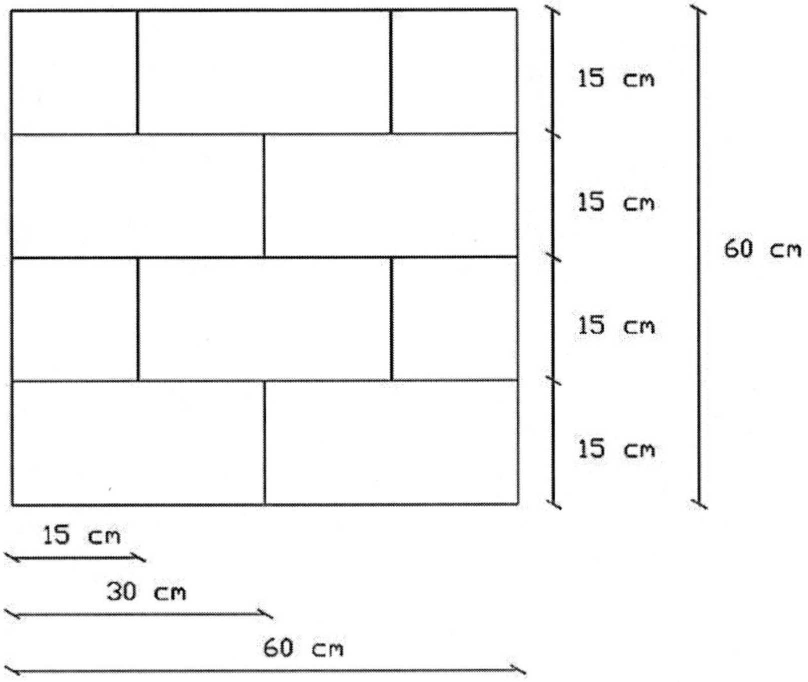

(b)

Fig. 2. Hollow core concrete brick wall sections: (a) for out-of-plane loading, (b) for in-plane loading

Table 3. Details of hollow core concrete brick wall.

No	Size (cm^2)	Loading	Notch model	Notch depth (cm)	Number
1.	90 × 45	Out of plane loading	Square	2	3
2.			Trapezoid	3	3
3.			Trapezoid	3.5	3
4.	60 × 60	In plane loading	Square	2	3
5.			Trapezoid	3	3
6.			Trapezoid	3.5	3

 DOI: 10.1051/matecconf/201713801011

2.3 Experimental Procedures

The experimental procedures refer to the previous study [6]. Hollow core concrete brick is initially made by preparing moulds and a working floor covered with plastic. Next, all materials are weighed and mixed in concrete mixer machines. The mixture is stirred well and then poured mixture into moulds that have been coated with a lubricant and into the middle of which a PVC pipe has been installed. After that, the mixture is allowed to solidify so that there are no air cavities in it. Once the moulds are completely filled up with mixture, the following step is to level out the mould surfaces and hit the surrounding outer part of the mould. The next step is to wait for about 5 minutes before removing the PVC pipes. The process of making the hollow core concrete bricks is illustrated in Figure 3. To obtain hollow core concrete brick wall specimens, three pieces of each brick type are subsequently made by first calculating the requirements for one specimen of hollow core concrete brick. The process of making the specimens of hollow core concrete brick wall can be seen in Figure 4. Measurement of the dimensions and weight of each specimen is then performed and assisting lines are provided to help the researcher define the crack patterns. The flexural strength testing uses two-point loading, as seen in Figure 5.

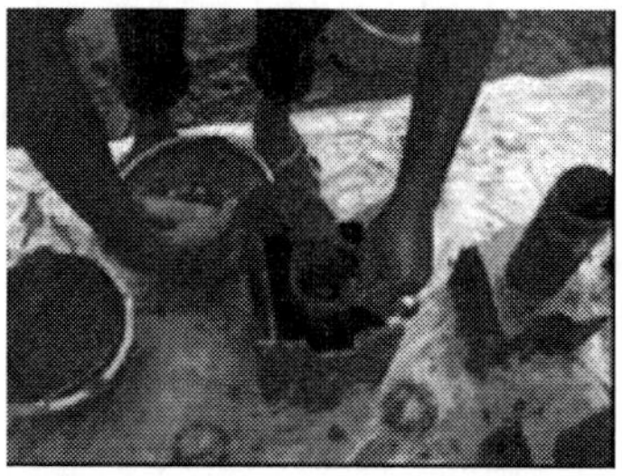
(a)

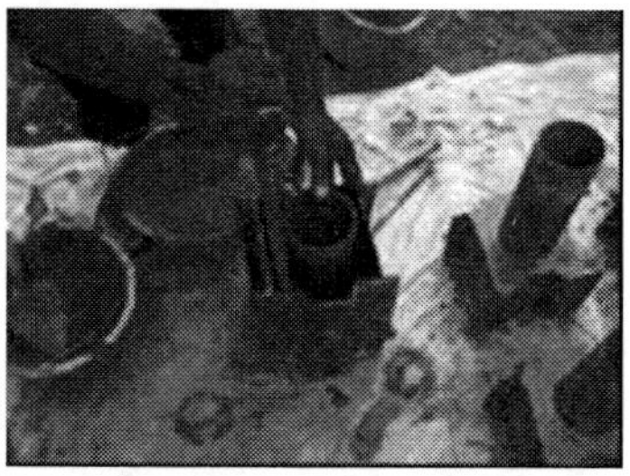
(b)

(c)

(d)

Fig. 3. Hollow core concrete brick making processes [6]: (a) pouring mixture to the mould, (b) solidifying the mixture (c), hitting the surrounding outer part of the mould, (d) hollow core concrete bricks

(a)

(b)

Fig. 4. Process of making specimens of hollow core concrete brick wall: (a) organizing the hollow concrete bricks, (b) hollow core concrete brick walls

Fig. 5. Flexural strength testing under two-point loading

3 Results and discussion

3.1 Flexural strength

The results of the tests conducted show that the notch model as the interlocking device influences the connections of the hollow core concrete brick wall. This is indicated by the increasing flexural strength of the hollow core concrete brick wall in various shapes and the notch depth. The square notch shape has a greater influence on the flexural strength of the hollow core concrete brick wall than the trapezoidal notch shape. Due to the square notch shape, the interlocking device mechanism may become more rigid than the notch depth increase yet still maintain its trapezoidal shape. The average flexural strengths of the three tested hollow core concrete brick walls are presented in Table 4.

Table 4. Average flexural strengths of three tested hollow core concrete brick walls

No	Size (cm^2)	Loading	Notch model	Notch depth (cm)	Flexural strength (kN/m^2)
1.	90 × 45	Out of plane loading	Trapezoid [6]	2	185.19
2.			Square	2	212.29
3.			Trapezoid	2	108.55
4.			Trapezoid	3.5	267.64
5.	60 × 60	In plane loading	Trapezoid [6]	2	120.00
6.			Square	2	330.92
7.			Trapezoid	3	229.63
8.			Trapezoid	3.5	143.79

It is shown that the hollow core concrete brick wall using the 2-cm square notch model has a flexural strength of 212.29 and 330.92 kN/m^2, while in the previous tests conducted by Wariyatno *et al.* [6] on a hollow core concrete brick wall using the 2-cm trapezoid notch model, flexural strengths of 185.19 and 120.00 kN/m^2 were obtained under out-of-plane and in-plane loading, respectively. This shows that there are flexural strength increases of 14.64 and 175.77%, respectively, under out-of-plane and in-plane loading. The comparison of the flexural strengths of the hollow core concrete brick walls with variations in notch shape may be seen in Figure 6.

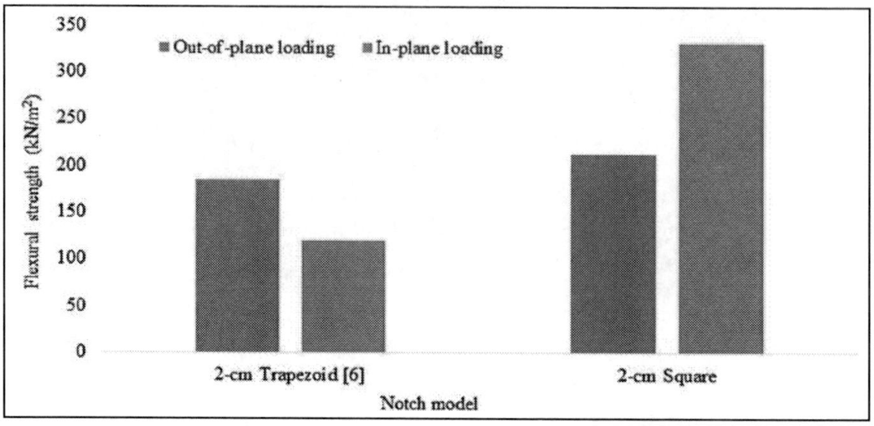

Fig. 6. Flexural strength of hollow core concrete brick wall based on notch shapes

It is also shown that the hollow core concrete brick wall using the 3-cm trapezoidal notch model has flexural strengths of 108.55 and 226.63 kN/m^2 under out-of-plane and in-plane loading, respectively, while the hollow core concrete brick wall using the 3.5-cm trapezoidal notch model has flexural strengths of 267.75 and 143.79 kN/m^2. This shows that the hollow core concrete brick walls using the 3-cm and 3.5-cm trapezoidal notch models have greater flexural strength than the hollow brick wall using the 2-cm trapezoid notch model under in-plane loading. The flexural strength comparison of hollow core concrete brick walls according to variations in notch depth can be seen in Figure 7.

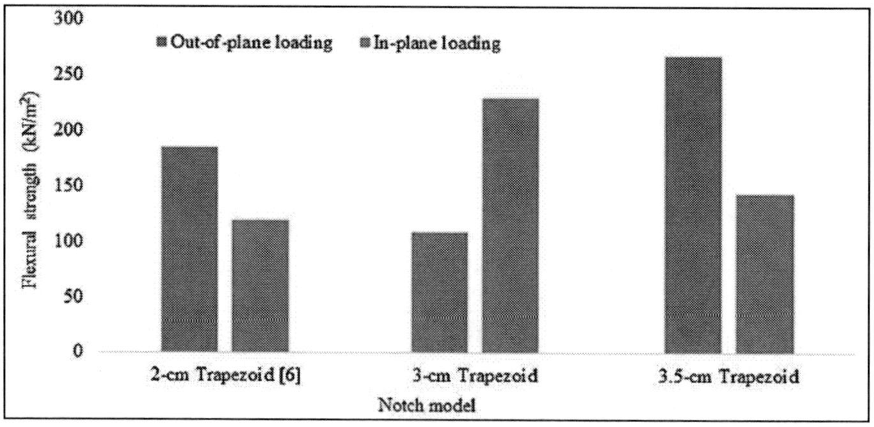

Fig. 7. Flexural strength of hollow core concrete brick walls based on notch depth

3.2 Collapse patterns

In the case of the hollow core concrete brick wall using the 2-cm square notch model under out-of-plane loading, the collapse initially starts from the material parts and then propagates to the connections. Meanwhile, the collapse of the hollow brick core concrete wall using the 3-cm trapezoid notch model initially starts from the connections. As the load increases, cracks on the loading surfaces become wider and propagate to the material parts, eventually resulting in fractures. The hollow core concrete brick wall using the 3.5-cm notch model tends to be more bound since it has the greatest notch depth, but the trapezoidal shape is unable to provide sufficient rigidity. The collapse patterns of the hollow core concrete brick wall under out-of-plane loading can be seen in Figure 8.

 DOI: 10.1051/matecconf/201713801011

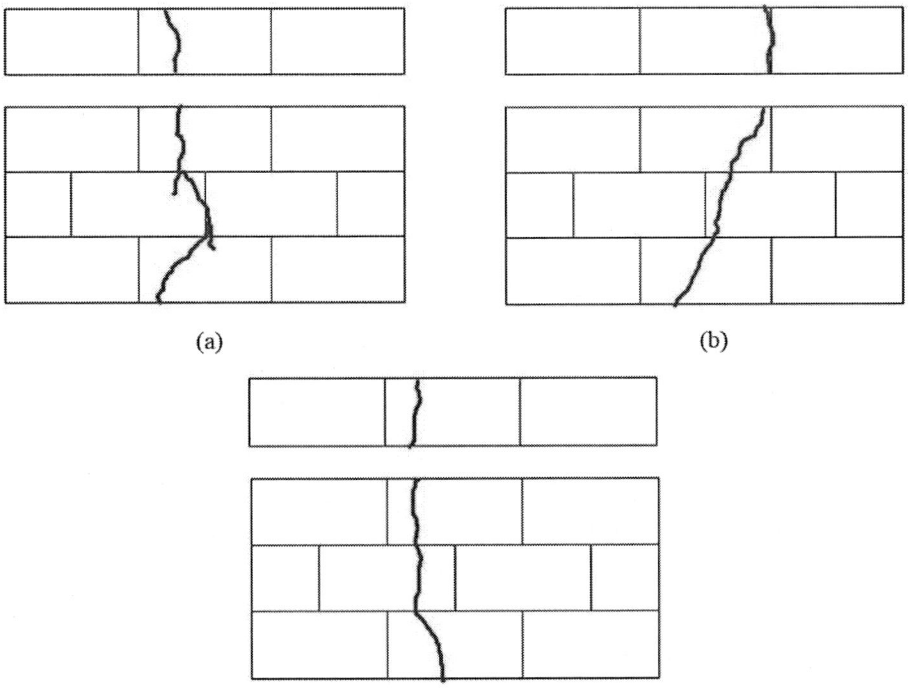

(c)

Fig. 8. The collapse patterns of the hollow core concrete brick wall under out-of-plane loading; (a) 2-cm square notch, (b) 3-cm trapezoidal notch, (c) 3.5-cm trapezoidal notch

The hollow core concrete brick wall using the 2-cm square notch model under in-plane loading initially starts to collapse in the material parts and then the collapse area widens and propagates to the connections. The square notch shape may improve the interlocking mechanism between materials at the connection parts. The collapse of the hollow core concrete brick wall using the 3-cm trapezoid notch model also initially starts in the material. The greater notch depth may influence the bonds between materials, giving a stronger homogenity. Similarly, the hollow brick wall using the 3.5-cm trapezoidal notch model has more stable cracks. The collapse patterns of the hollow core concrete brick wall under in-plane loading can be seen in Figure 9.

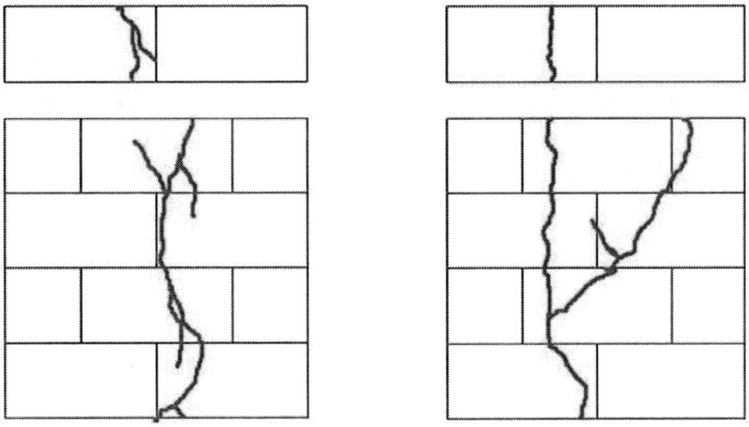

(a) (b)

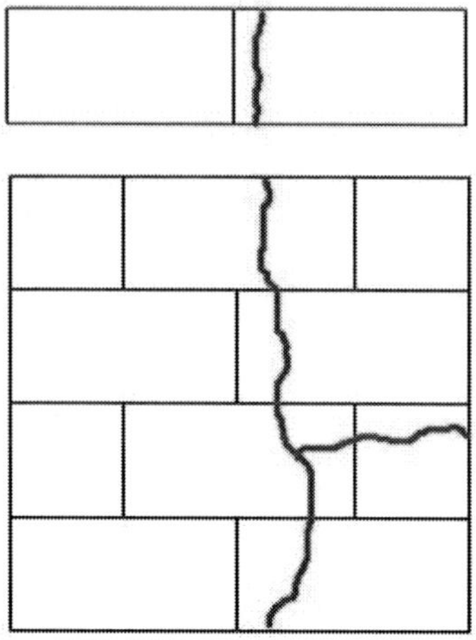

(c)

Fig. 9. The collapse patterns of the hollow core concrete brick wall under in-plane loading: (a) 2-cm square notch, (b) 3-cm trapezoidal notch, (c) 3.5-cm trapezoidal notch.

4 Conclusion

From the results of the study conducted, some conclusions are drawn as follows:

a. The flexural strength of hollow core concrete brick wall using the 2-cm square notch model is increased by 14.63% compared to the one using the 2-cm trapezoidal notch model under the out-of-plane loading.
b. The flexural strength of the hollow core concrete brick wall using the 2-cm square notch model is increased by 175.77% compared to the one using the 2-cm trapezoidal notch model under in-plane loading.
c. The notch model as an interlocking device influences the connections of the hollow core concrete brick wall. The square notch shape provides a greater influence on the flexural strength of the hollow brick wall than the trapezoidal notch shape.

References

1. Badan Pusat Statistik, Nilai Konstruksi Yang Diselesaikan Menurut Jenis Pekerjaan (juta rupiah), 2004–2015, Update 17 October 2016, Available from https://www.bps.go.id/linkTabelStatis/view/id/918, (2016)
2. PT Semen Indonesia (Persero) Tbk, 2017, Konsumsi Semen Naik 8 Persen, Update 31 October 2016 (2016), Available from http://bumn.go.id/semenindonesia/berita/2139
3. J. Tanjung, Maidiawati, J. *Tek. Sip.*, **23**, 99 (2016)
4. Y. Haryanto, G.H. Sudibyo, F.C.A. Effendi, Procedia Eng., **171**, 1025 (2017)
5. Y. Haryanto, B.S. Gan, N.G. Wariyatno, E.W. Indriyari, ARPN JEAS, **12,** 4858 (2017)
6. M. Pang, S. Yang, Y. Zhang, Y., Sustainability, **4**, 2630 (2012)
7. W. Cao, Y. Zhang, H. Dong, Z. Zhou, Q. Qiao, Materials, **7**, 5934 (2014)
8. N.G. Wariyatno, Y. Haryanto, Sumiyanto, J. Il. Tek. Sip., **21**, 136 (2017)

Effect of cantala fiber as textile reinforcement on the flexural behaviour of polymer modified mortar

Edy Purwanto[1] , *Stefanus* Kristiawan[1*] , *Bambang* Santosa[1] and *Pungky* T Istanto[2]

[1]SMARTCrete Research Group, Civil Engineering Department, Universitas Sebelas Maret, Surakarta 57126, Indonesia
[2]Civil Engineering Department, Universitas Sebelas Maret, Surakarta 57126, Indonesia

Abstract. Cantala (Agave cantala), plant of the family Asparagaceae and its fibers, belongs to the leaf fibers group. These fibers are made into coarse twines at about 1.2 mm in diameter. The yarns of the cantala fiber are then interwoven to create textile reinforcements at two meshing size i.e. 5 x 20 mm and 10x20 mm. The textiles are used to reinforce polymer modified mortars to improve flexural properties of the mortars. This research aims to investigate the effect of textile reinforcement made from cantala fiber on the flexural behaviour of polymer modified mortar especially its strain hardening response. The parameters to be observed include strength, stiffness, thoughness, ductility and residual strength. The effect of the meshing size of the textile reinforcement on those properties are assesed. In addition, the efficacy of the cantala fiber textile reinforcement to improve the flexural properties of polymer modified mortar is compared with textile reinforcement made from wire mesh. The results indicate that an inclusion of short fibers could alter the post-peak behaviour of the mortar from brittle to ductile failure with softening response. Futher reinforcement of this mortar with cantala fiber textile would turn the softening response to a strain hardening response. The presence of cantala fiber textile reinforcement has minor effect on the flexural strength and stiffness. However, it significantly improve the thoughness, ductility and residual strength of the mortar. Cantala fiber textile reinforcement with a meshing size of 5x20 mm gives a greater increase on those post-peak properties compared to the the one with a meshing size of 10x20 mm. When comparison is made to the wire mesh, both cantala fiber textile reinforcements show a better properties.

1 Introduction

Concrete has been utilised in the construction of various infrastructures around the globe. The need of this material remains enormous especially in the developing countries where infrastructures development is put into a national priority to drive the economic activities and growth. The versatility of concrete to meet a variety of design requirements is one of

* Corresponding author: s.a.kristiawan @ft.uns.ac.id

the prime reasons to choose this construction material. At the same time, concrete technology and inovation are continuously in progress to meet the current and future issues of infrastructures development.

Sustainability is a major issue of today infrastructures development [1]. The issue challenges the concrete industry to create concretes with minimum exploitation to non-renewable natural resources, less impact to environment, cost-efficient of production, etc. Creation of durable concrete could be a key factor to meet such challenges. Durable concrete promotes a longer service life of the constructed infrastructures with a minimum maintenance throughout their entirely life span. It means durable concrete aiding to reduce the consumption of natural resources for the period of its life span.

It is well known that concrete has a very low tensile strength compared to its compressive strength. Its low tensile strength could be the root of cracking problem even under low imposed load. Cracking may not directly lead to a rupture of the structural concrete but it facilitates the ingress of agggressive agents. The presence of aggressive agents will initiate corrosion of reinforcements which eventually causes a bigger and complicated problems in the future life of the structure. Therefore, concrete industries seek innovative solution by creating concrete which could resist a higher tensile stress. Improvement of tensile properties of concrete has been realised by the development of fiber reinforced concrete (FRC). FRC is characterised by a tensile stress-strain relationship up to failure where this material shows excellent strength, ductility and energy dissipation. Further development of the concrete is the invention of high performance fiber reinforced cement composites (HPFRCC) which do not only exhibit large deformation in tension, but it also provides strain hardening behaviour. The strain hardening behaviour is indicated by the increase of strength after the onset of first cracking while the material exibits large deformation [2-4]. Another cement-based material that has been developed to counter the lack of tensile properties is a textile reinforced composite (TRC). The main difference of this material with FRC or HPFRCC is in the form of fibers inclusion. For FRC or HPFRCC, the fibers are those of short fibers dispersed within the bulk volume of the materials. Meanwhile, for TRC the fibers are woven in the form of textile as a continuos reinforcement of the cement-based material [5].

TRC is a material consisting of two components, namely textile as a reinforcement and concrete (cement-based composite) as a matrix. The type of concrete used in the production of TRC is that of fine-grained aggregate concrete or mortar. Under the increment of tensile loading, matrix will eventually crack at certain level of tensile stress. Once the initiation of cracks occur, all the stresses at the cracks sections are transfered to the textile reinforcement. The post-cracking behaviour of TRC will then be dictated by the strength and the stiffness of the textile reinforcement. High tensile strength of the textile reinforcement will ensure that the reinforcement could carry a higher stress while multiple new cracks are induced in the matrix. The propagation and multiplication of cracks are accountable for the large deformation (ductility) observed at the post-cracking. The propagation of both crack width and crack distribution along the matrix may be controled without compromising the ductility of TRC. This could be achieved by the inclusion of short fibers which are randomly dispersed in the matrix. The short fibers act as crack bridging that minimise the propagation of cracks [6].

Textile reinforcement used in the development of TRC are mostly made of high strength fibers such as basalt, aramid, alkali resistant (AR) glass, carbon (C) and polypara-phenylene-benzo-bisthiazole (PBO). The tensile strength of these fibers are at least 1000 MPa [7-10]. Meanwhile, medium tensile strengths (400-650 MPa) of nylon yarns have been successfully used to create TRC with strain hardening response. The key to achieve such behaviour lies in the minimum amount of reinforcement ratio to ensure the flexural strength is greater than cracking moment [11]. Other relatively low tensile strengths of materials i.e.

polyethylene/PE fibers (240 MPa) and wire mesh (320 MPa) have also been used to produce textile reinforcements [12-13]. Off course high tensile strength is beneficial to obtain high strength-high ductility of TRC as compared to those of fibers with low tensile strength. However, in certain circumstances it may costly effective to use the textile reinforcements made from fibers with tensile strength of 200-400 MPa. Tsesarsky et al [8] tested a TRC with textile reinforcement made from PE fibers. Even though the tensile strength of this TRC only attained 7 MPa, but the material showed ductile behaviour. This TRC has also been applied to strengthen a beam and it gave an increase of flexural load by 170%. Comparing to the strengthened beam by TRC with textile reinforcement made from AR glass and C fibers, the efficiency of PE was 18% i.e. higher than C (13%) but lower than AR glass (33%). Other researchers [12-13] used wire mesh as reinforcement of mortar in a similar manner to that of TRC.

In the authors's point of view, it is necessary to develop cost-efficient TRC for special applications i.e. strengthening low strength and brittle structures such as non-engineered masonry buildings. These buildings particularly exist in developing countries. The materials for manufacturing TRC preferably are available locally. Natural fibers could represent these local resources for the development of TRC. In this current research, natural fibers from the Agave cantala plant are used to manufacture textile reinforcement. The effect of the cantala fiber textile reinforcement on the bending properties of TRC will be investigated. The parameters to be assessed include strength, stiffness, thoughness, ductility and its strain hardening response. The effects of the cantala fiber textile reinforcement on those parameters will be compared with the reinforcement by wire mesh.

2 Materials and method

2.1 Materials

Textile reinforcements used to reinforced polymer modified mortar were made of natural fibers. The natural fibers, extracted from Agave cantala leaves, were supplied by home industry in Kulon Progo, Yogyakarta, Indonesia. These fibers were made into coarse twines at about 1.2 mm in diameter. The yarns of the cantala fibers were then interwoven to create textile reinforcements at two meshing sizes i.e. 5x20 mm and 10x20 mm (see Figure 1). The average tensile strength of the yarns was 150 MPa. The textiles were treated by mean of coating applied on the surface of the yarns using unsaturated polyester resin (UPR). In this way, the fibers were protected from the alkalinity of the mortar. For comparison purpose, a wire mesh having size of 11.2x22.4 mm was also used as reinforcement. The diameter of the wire was 0.8 mm with a tensile strength of 314 MPa.

The textiles were then used to reinforce polymer modified mortars to improve flexural properties of the mortars. The ingredients for making the mortars were cement (C), fine grained sand (S), water (W), polyvinyl acetate (PVA), accelerator (Ac) and superplasticizer (Sp). The S were selected in such they passed through a meshing size of 1.18 mm but retained on a meshing size of 0.5 mm. The proportion of the mortars is given in Table 1. The amount of W shown in the table is a total water consisting of free water and water contained in PVA, Ac and Sp.

Five type of thin beam specimens having size of 10x50x250 mm were made in this research. The first was a mortar specimen (N) with a composition as given in Table 1. The second specimen was a mortar with similar composition of N but with an addition of short cantala fibers (SF) at about 0.45% by weight of mortar. These fibers were added during mixing. The third to fifth specimen were mortars with similar composition of SF but reinforced with three different textiles i.e. cantala fiber textile reinforcement with a meshing size of 5x20 mm TR(0.5x2), cantala fiber textile reinforcement with a mesihing size of 10x20 mm TR(1x2) and wire mesh (WM) with a meshing size of 11.2x22.4 mm. The specimen was prepared as follows: a half layer of mortar (2.5 mm thickness) was laid on the wooden mould and then textile reinforcement was placed to rest on top of this layer. The final layer of mortar was laid to cover the textile reinforcement. The top layer of the mortar was pressed manually to aid in eliminating air within the mortar. Each type of specimens was represented by 5 beams for flexural testing.

(a)

(b)

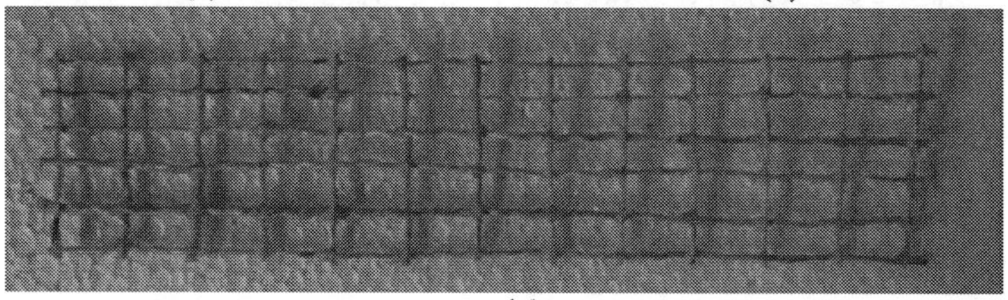

(c)

Fig. 1. Original cantala fibers (a), the yarns (b) and textile reinforcement (c)

Table 1. Proportion of mortars

Spec. ID	**C (kg)**	**S (kg)**	**W (kg)**	**PVA (kg)**	**Ac (kg)**	**Sp (kg)**	**fiber (%)***
N	1	2	0.35	0.2	0.083	0.01	-
SF	1	2	0.35	0.2	0.083	0.01	0.45
TR(1x2), TR(0.5x2), WM	Similar composition to SF but the mortars are reinforced with TR(1x2), TR(0.5x2) and WM, respectively.						

* by weight of mortar

2.2 Method

Flexural testing was carried out on specimens after 28 days of age using Universal Testing Machine (UTM) by mean of third point loading. The span length (support to support) was determined at 210 mm. The load was applied at a displacement rate of 5 mm/min. The load and the corresponding deflection up to failure were recorded automatically by the machine.

3 Results and discussion

3.1 Load-deflection behaviour

Figure 2 shows the load-deflection behaviour up to failure for all types of specimens. The average values are presented in Figure 2.(f). It is interesting to note that at the pre-peak load, all specimens show a similar behaviour i.e. very steep straight curves are observed at the beginning of loading and then at some points between 40-70 N of loading, the curves are shifted slightly to the right indicating an initiation of microcracks have occurred on the specimens. After microcracks arise, the specimens could maintain their proportional load-deflection behaviour with a little lower of stiffness.

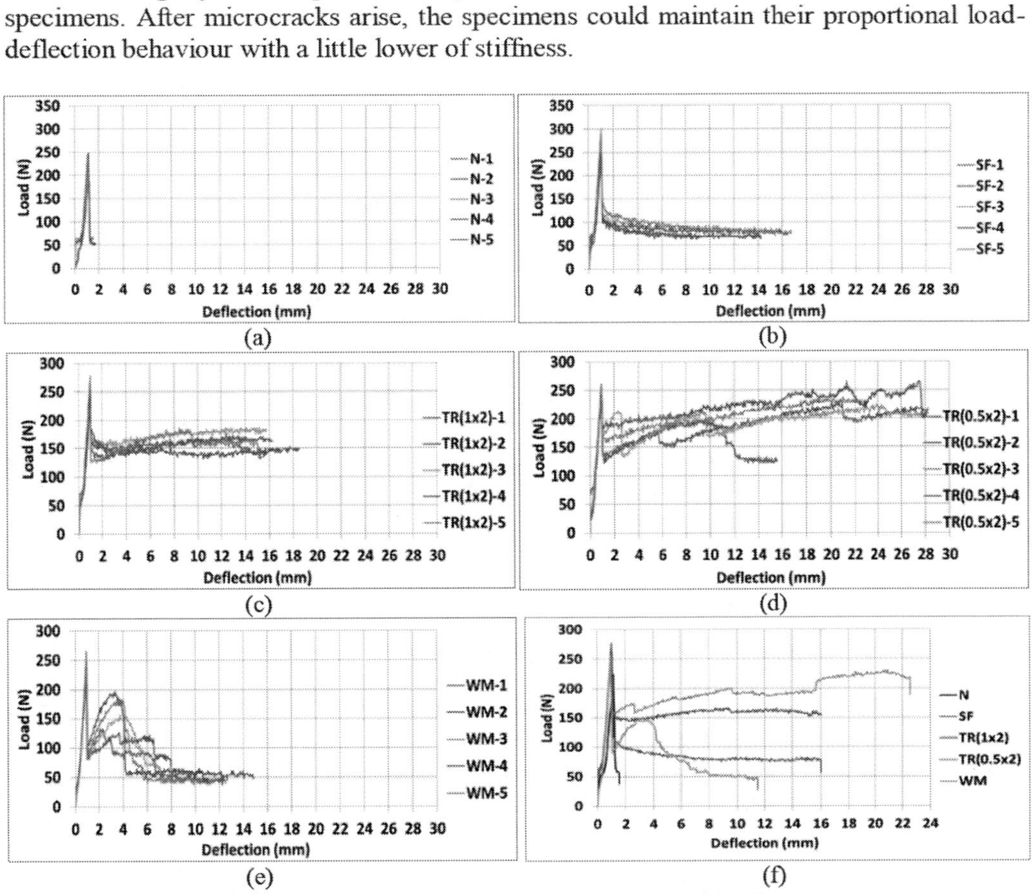

Fig. 2. Load-deflection behaviour of various specimens under flexural loading

Obvious differences are observed at the post-peak load. For normal mortar (N), sudden drop of peak-load leading to failure is observed. This behaviour is typical of brittle material. Inclusion of short fibers in the mortar (SF) significantly transforms from brittle failure to ductile failure. After peak-load drops, a large deformation is observed in SF. A softening response is noted in this material. When the textile reinforcements are used to reinforced the SF mortar, the softening response turns to a strain hardening response. Each type of textile reinforcement provides a unique strain hardening response. For TR(1x2) the hardening level is small even though it gives an increase of residual load more than 200% to that of SF. For TR(0.5x2) the hardening behaviour causes the residual stress approaching the peak-load level. For reinforcement using wire mesh (WM), the strain hardening attain an ultimate value at short deformation level (about 2-3 times of deformation at the first peak-load). After this ultimate value (second peak-load), the material shows softening response.

3.2 Key parameters

This section discusses the key parameters deduced from the load-deflection curves given in Figure 2. The parameters to be examined include ultimate stength, stiffness, toughness, ductility and residual strength with a particular interest to assess the influence of textile reinforcements. These key parameters and their average values are tabulated in Table 2. It is clear that the ultimate strength of the mortar is only slightly affected by cantala fibers both in the form of short fibers and textile reinforcements. Inclusion of cantala short fibers (SF) give an increase of strength at 11.5%. Additional increase of ultimate strength could not be expected when textile reinforcement is introduced. The influence of textile reinforcement on strength is minor compared to its influence on the deformation. The explanation is as follows. Under flexural loading, the mortar specimens undergo tension and compression stresses. Since the property of plain mortar in tension is inferior compared to that in compression, the flexural strength of mortar is controlled by its tension capacity. Inclusion of fibers is expected to improved its tension property with a consequence of increasing flexural strength. The effectivity of the fibers to improve the flexural strength depends on the action zone of the fibers. For thin prism specimen with textile reinforcement sets at the middle section, the textile reinforcement will not effective to improve the flexural strength. However, the induced stresses will be transferred to the textile reinforcement once the mortar develop sufficient macrocracks. These macrocracks occur at peak-load after which causing the sudden drop of the load and activating the tensile action of textile reinforcement. Hence, the influence of textile reinforcement could significantly only be observed at post-peak load especially on the capability of the mortar to undergo large deformation and strain hardening.

The flexural stiffness is a parameter to describe how material will undergo deflection as it is loaded in flexure. The flexural stiffness of the mortars investigated in this study could be identified from the load-deflection relationship curves up to peak load. From Figure 2, it may be reasonably assumed to define the stiffness index as the ratio of the peak-load to its respected deflection. The results show that an addition of cantala short fibers could improve the stiffness index of mortar by 19.44%. If mortar containing short fibers is reinforced with cantala fibers textile reinforcements or wire mesh, the stiffness index is expected to increase futher by no more than 15%. The increase may be attributed to the ability of the fibers and textile reinforcements to some extent restrain the propagation of microcracks.

Table 2. Key paramaters values

Specimen	Strength (N) or (MPa)	Stiffness Index	Toughness Index	Ductility Index	Residual Strength (N)
N	223.78 (9.34)	195.95	14.03	1.29	57.48
SF	249.65 (10.49)	234.20	138.79	11.26	75.48
TR(1x2)	263.09 (11.05)	269.23	253.60	20.83	154.19
TR(0.5x2)	252.31 (10.60)	264.50	435.78	36.11	216.96
WM	255.06 (10.71)	269.94	99.07	9.55	47.22

Note: The stiffness, toughness and ductility index represent relative values. Therefore, no unit is indicated.

Figure 2 suggests that inclusion of cantala short fibers drastically change the deformation of the mortar; from brittle behaviour to ductile behaviour. Further changes in the behaviour are observed when the mortars are reinforced with textile. The changes in the deformation behaviour are best characterised by the toughness and ductility index. The toughness is indicated by the area under the load-deflection curve. This area characterises the capability of the material to absorp energy after peak-load. Meanwhile, the ductility is defined by the ratio of deformation at failure to that at peak-load. This parameter also indicates the capability of material to deform after peak-load before rupture. The ductility could also be quantified by the ratio of toughness at failure to that at peak-load. The latter is adopted in this research. The experimental results show that the toughness index of SF is 138.79. When this mortar is reinforced with TR(1x2), TR(0.5x2) and WM, the toughness index is increased by 83%, 214% and -71%, respectively. An enlargement of the ductility

DOI: 10.1051/matecconf/201713801012

index at about 85%, 221% and -73% could be expected for reinforcement of SF mortar with TR(1x2), TR(0.5x2) and WM, respectively. The increase in the toughness and ductility of SF mortar due to cantala textile reinforcements could be traced from the tensile action of these reinforcements after peak-load. As previouly stated, the stresses transfer to the textile reinforcements could only occur after sufficient macrocracks develop in the mortar. These cracks cause a drop of peak-load and activate the action of textile reinforcements in the post-peak load. The textile reinforcements are accountable for the capability of mortar to undergo large deformation and strain hardening. TR(0.5x2) has a reinforcement ratio twice than that of TR(1x2). Consequently, the toughness and ductility index of TR(0.5x2) much greater than that of TR(1x2). For reinforcement with wire mesh (WM), it is indicated that wire mesh could also turn the mortar from showing a softening response to strain hardening response. However, the material rupture at a lower deformation than that of SF. Hence, the toughness and ductility index of WM is lower than the coressponding SF.

The residual strength of the SF at the end of softening response equals to 75.48 N. The textile reinforcements enhance the residual strength due to the strain hardening response. The highest residual strength is observed on TR(0.5x2) i.e. 217 N. This should be expected since this material have the highest reinforcement ratio. For mortar reinforced with wire mesh, initially the material shows strain hardening response and then follows by softening response. That is why the residual strength of this material (WM) is only 47.22 N i.e. even lower than the SF.

3 Conclusions

This research investigates the influence of cantala fiber textile reinforcement on the flexural characteristics of polymer modified mortar and the main findings are as follow:

- Inclusion of short fibers could alter the post-peak behaviour of the mortar from brittle to ductile failure with softening response. Futher reinforcement of this mortar with cantala fiber textile would turn the softening response to a strain hardening response.
- The strain hardening response of the cantala textile reinforced mortar is influenced by the reinforcement ratio. A higher reinforcement ratio of TR(0.5x2) shows a higher strain hardening response than a lower reinforcement ratio of TR(1x2).
- The strain hardening response of cantala fiber textile reinforcements is different to that of reinforcement using wire mesh (WM). For cantala fiber textile reinforcement the strain hardening is continuous up to failure while for WM the strain hardening is followed by a softening response.
- The reinforcements of the mortars with cantala textile or wire mesh have a minor effect on the flexural strength and stiffness but they signiificantly affect the toughness, ductility and residual strength of the mortars.

This research is financially supported by the University of Sebelas Maret through Maintenance Research Group for the year of 2017.

References

1. K.J.G. Ong, Proc. Eng **171**, 14-21 (2017)
2. S. Gwon, M. Shin, J. ACF **2**(1), 67-80 (2016)
3. T. Simoes, H. Costa, D. Dias-da-Costa, E. Julio, Constr. Build. Mat. **137**, 548-556 (2017)
4. M Singh, A.H. Sheikh, M. S. M. Ali, P. Visintin, M. C. Griffith, Constr. Build. Mat. **138**, 12-25 (2017)

5. M. Mobasher, *Mechanics of fiber and textile reinforced cement composites*, CRC Press, Florida (2012)
6. R. Barhum, V. Mechtcherine, Eng. Frac. Mech. **92**, 56-71 (2012)
7. P. Larrinaga, C. Chastre, H. C. Biscaia, J. T. San-Jose, Mat. Des. **55**, 66-74 (2014)
8. M. Tsesarsky, A. Peled, A. Katz, I. Anteby, Constr. Build. Mat. **44**, 514-523 (2013)
9. L. Ombres, Comp. Struct. **94**, 143-155 (2011)
10. R. Contamine, A. Junes, A. S. Larbi, Constr. Build. Mat. **51**, 405-413 (2014)
11. M. Zargaran, N. K. Attari, S. Alizadeh, P. Teymouri, Constr. Build. Mat. **137**, 459-469 (2017)
12. A. M. El-kholy, H. A. Dahish, Ains Shams Eng. J. **7**(2), 717-728 (2016)
13. H. Zhao, M. N. S. Hadi, Proc. Eng. **14**, 2848-2855 (2011)

DOI: 10.1051/matecconf/201713801013

Optimum concrete compression strength using bio-enzyme

Tony Hartono Bagio[1,*], *Makno* Basoeki[2], *Julistyana* Tistogondo[1], and *Sofyan Ali* Pradana[1]

[1]Narotama University, Surabaya, Indonesia
[2]Engineer of Mirama Consultant, Surabaya, Indonesia

Abstract. To make concrete with high compressive strength and has a certain concrete specifications other than the main concrete materials are also needed concrete mix quality control and other added material is also in line with the current technology of concrete mix that produces concrete with specific characteristics. Addition of bio enzyme on five concrete mixture that will be compared with normal concrete in order to know the optimum level bio-enzyme in concrete to increase the strength of the concrete. Concrete with bio-enzyme 200 ml/m3, 400 ml/m3, 600 ml/m3, 800 ml/m3, 1000 ml/m3 and normal concrete. Refer to the crushing test result, its tends to the mathematical model using 4th degree polynomial regression (least quartic), as represent on the attached data series, which is for the design mix fc' = 25 MPa generate optimum value for 33,98 MPa, on the bio-additive dosage of 509 ml bio enzymes.

1 Background

The quality of the concrete is determined by the materials of concrete preparation. In order to make concrete with high compression strength and has a certain concrete specification cannot be obtained simply by mixing cement Portland cement or other types of coarse aggregate, fine aggregate, and water. But it is necessary to also control the quality of the concrete mix. Improving the quality of the concrete mix will increase the compression strength is generated [6,7].

Mixed concrete technology is currently also developing rapidly, many other added material application technologies to produce concrete with the desired characteristics. One way to improve the quality of the concrete mix is to use the added material, either a chemical such as entraining admixture, reducing water in the form of synthetic or organic. On this paper, we would like to introduce our research on improving concrete quality with green technology using bio enzyme, which is not covered yet in current classification A to F of concrete chemical admixtures used.

[*] Corresponding author: tony@narotama.ac.id

The bio enzyme admixture, is a bio-technology product made of organic/natural materials, denatured proteins, biopolymer surfactant and organic minerals which fermented by beneficial microbes [18]. The quality of the concrete is determined by the materials of concrete preparation. To make concrete with high compression strength and has a certain concrete specification cannot be obtained simply by mixing cement Portland cement or other types of coarse aggregate, fine aggregate, and water. But it is necessary to also control the quality of the concrete mix. Improving the quality of the concrete mix will increase the compression strength is generated.

Mixed concrete technology is currently also developing rapidly, many other added material application technologies to produce concrete with the desired characteristics [1]. One way to improve the quality of the concrete mix is to use the added material, either a chemical such as entraining admixture, reducing water in the form of synthetic or organic such as bio enzyme. The bottom line target in develop and research of this Bio-Enzyme product are:

1. Reduce the cement content, means reduce the CO_2 emission implicate the *eco-friendly impact* [18].
2. Reduce the cement content, means reduce the thermal effect of hydration heat, produced the Low Heat Concrete.
3. Reduce the cement content, means reduce the production cost of concrete [18].
4. Another performances of Bio-Enzyme are a microbe which produce micro fibre, reduce the shrinkage crack event eliminate it, these phenomena show on crack pattern of the samples crushing test [18].

2 Methods

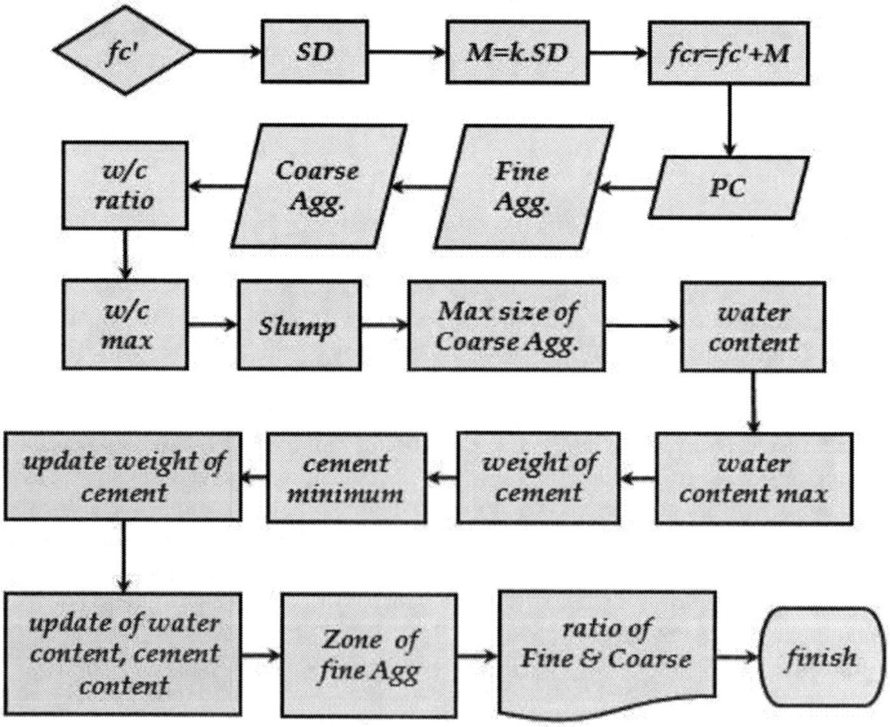

Figure 1. Flow Chart of Mix Design

Mix design of concrete based on DoE (Department of Environment) from British Standard Code's. Adopted to Indonesian's Code [17] to mix design. The method of mix design shows in Figure 1.

Mix design with initial concrete (normal condition) *fc'* is 25 MPa, Portland Cement = 409 kg, water = 225 kg, fine aggregate = 733 kg, coarse aggregate = 1013 kg. Cylinder testing with 6 types of experiments (type A to F). Find the optimum value of the 5 additional variations with different types of polynomial equations to $R^2 = 1$

3 Results

After a physical test [2, 8-16] of the material is used, a concrete mortar plan is prepared using the British Standard 8110 Department of Environment (DoE) [3,4,5] method shown in Table 1.

Table 1. Form of Mix Design

No.	Description	Data	Unit
1	fc' (cylinder 28 days)	25	MPa
2	Standard Deviation	6	MPa
3	Margin	1,64 x SD =9.84	MPa
4	Target Strength (average)	35	MPa
5	Types of Portland Cement	OPC	unitless
6	Types of Coarse Aggregate	Crushed	unitless
7	Types of Fine Aggregate	Natural	unitless
8	free water content ratio	0.55	unitless
9	Water cement maximum	0.6	unitless
10	Slump	150+/-2, (60-180)	mm
11	Max size Aggregate	20	mm
12	Free Water content	225	liter
13	Cement Content	409	kg
14	Cement Content Minimum	275	kg
15	Update Free Water Content	225	liter
16	Zone of Fine Aggregate	Zona II	unitless
17	Ratio of fine / coarse aggregate	43	%
18	Specific gravity of Aggregate	2670	kg/m^3
19	Total fresh Concrete	2380	kg/m^3
20	Total Aggregate content	1746	kg/m^3
21	Fine Aggregate content	733	kg/m^3
22	Coarse Aggregate content	1013	kg/m^3

The proportion of mixture of normal concrete mixture above is also used to make by using bio enzyme. The composition of mix design per m^3 of concrete materials are, cement = 409 kg/m^3, water = 225 kg/m^3, fine aggregate = 733 kg/m^3, coarse aggregate = 1013 kg/m^3, and

DOI: 10.1051/matecconf/201713801013

additive, type A = 0 ml (normal concrete), type B = 200 ml, type C = 400 ml, type D = 600 ml, type E = 800 ml, and type F = 1000 ml.

3.1 Compressive Strength.

After testing and evaluating the average compressive strength of concrete of each type of concrete [2, 8-16], the following recapitulation of concrete compressive strength results, in table 2, and Fig. 2 shows the concrete compressive average strength of each mixture of concrete:

Table 2. Result of compressive strength

Name of Sample	Sample	Weight (gram)	Result (kN)	Result (Mpa)	Average of 3(three) samples
Type A (0 ml)	sample 1	12.82	516.70	29.25	28.58
	sample 2	12.94	501.70	28.40	
	sample 3	12.91	496.00	28.08	
Type B (200 ml)	sample 1	12.65	500.80	28.35	29.47
	sample 2	12.72	523.30	29.63	
	sample 3	12.67	537.40	30.43	
Type C (400 ml)	sample 1	12.93	590.10	33.41	33.58
	sample 2	12.91	597.70	33.84	
	sample 3	12.91	591.60	33.49	
Type D (600 ml)	sample 1	12.87	614.40	34.79	33.79
	sample 2	12.91	594.50	33.66	
	sample 3	12.95	581.60	32.93	
Type E (800 ml)	sample 1	12.97	577.60	32.70	32.99
	sample 2	13.00	568.10	32.16	
	sample 3	12.96	602.60	34.12	
Type F (1000 ml)	sample 1	12.95	582.90	33.00	33.32
	sample 2	13.03	574.00	32.50	
	sample 3	12.98	608.40	34.45	

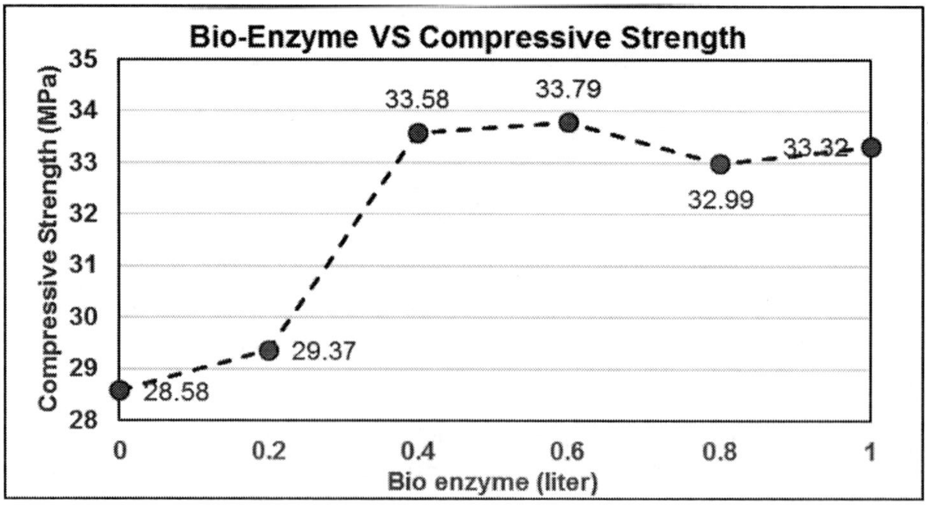

Figure 2. Bio Enzyme Content Vs Compressive Strength

DOI: 10.1051/matecconf/201713801013

3.2 Polynomial Regression Analysis.

The analysis used to find the optimum level of bio enzyme mixture for compressive strength is the fourth degree of polynomial, or least quartic analysis. In this study the free (x) variable is the variation of the bio enzyme's content and the dependent variable ($f(x)$) is the compressive strength of the concrete. The results of least quartic showing in fig. 3.

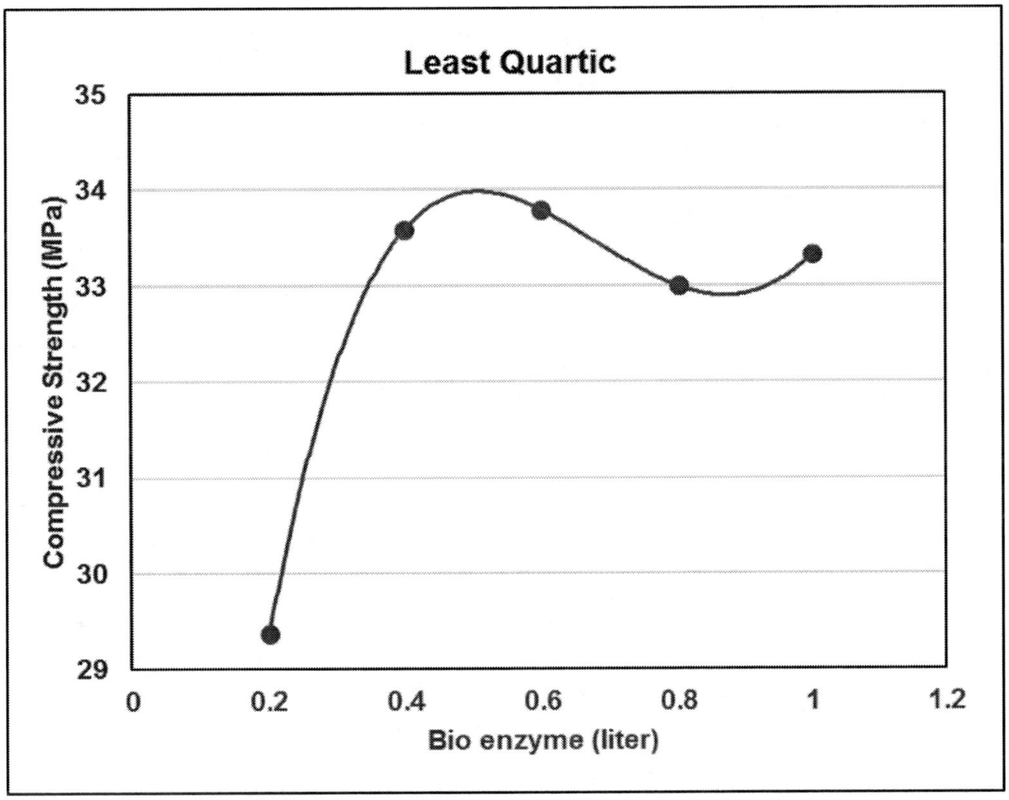

Figure 3. Least quartic / 4th degree of Polynomial

The results of the five concrete model using 4th degree polynomial regression (least quartic),

$$f(x) = -17.9\,x^4 + 95.321\,x^3 - 145.044\,x^2 + 83.007\,x + 17.938 \tag{1}$$

and $R^2 = 1$
derivative of equation (1) is

$$f(x)' = -71.6\,x^3 + 285.96\,x^2 - 290.08\,x + 83.007 \tag{2}$$

x = bio enzyme (liter)
$f(x)$ = compressive strength (MPa)

to find maximum or minimum, $f(x)' = 0$, we get $x_1 = 0.509$, $x_2 = 0.872$, $x_3 = 2.613$, the obtained from the concrete quality normal concrete with $fc' = 25$ MPa, substitution x_1 to equation (1) to find optimum value for 33.98 MPa, and the addition of 509 ml bio enzymes.

4 Conclusion

Based on the discussion that has been put forward in the previous chapter it can be concluded as follows:

1. Concrete B with the addition of bio enzyme 200 ml/m^3 there is increase of 3.1% to normal concrete (concrete A), concrete C addition of bio enzyme 400 ml/m^3 up 17,5% to normal concrete, concrete D addition of bio enzyme 600 ml/m^3 up 18.2% of the normal concrete, the concrete E of the addition of bio enzyme 800 ml/m^3 up 15.4% of normal concrete and decreased the quality of concrete D as much as 2.4%, and F concrete addition of bio enzyme 1000 ml/m^3 3 up 16.6% of normal concrete. It can be seen that D-concrete with the addition of bio-enzyme 600 ml/m^3 is the right mixture of bio-enzyme to produce the compressive strength of concrete among other concrete variations.
2. The maximum x_1 based regression analysis is 0.509 litre or 509 ml, with a maximum compressive strength of 33.98 MPa. It can be concluded that the optimum mixture of bio-enzyme blends to increase the compressive strength of concrete is 509 ml/m^3.

The acknowledgements of this research was funded by competitive grants, from Directorate Research and Community Service - Directorate General of Research and Research Development - Ministry of Research, Technology, and Higher Education Republic of Indonesia, 2017.

References

1. Aprilianti, S., *Analisis Penggunaan Superplasticizer Type Napthalene dan Polycarboxilate*, Tugas Akhir Teknik Sipil, Universitas Muhammadiyah Jakarta (2012)
2. Annual Book of ASTM Standards, Volume 04.02, Concrete and Aggregate (2002)
3. Badan Standarisasi Nasional, *Tata Cara Pembuatan Rencana Campuran Beton Normal*, SNI 03-2834-2000, adopsi SK-SNI T-15 1990-03, Jakarta (2000)
4. Bagio, Tony Hartono, *Handout Perencanaan Campuran Beton Metode DoE British Standart with Microsoft Excel* (2002)
5. Kong FK. and Evans RH., *Reinforced and Prestressed Concrete, 3rd Edition*, Chapman and Hall, London, UK (1994)
6. Mulyono, T., Teknologi Beton, Andi, Yogyakarta (2004)
7. Nawy, Edward G., *Beton Bertulang: Suatu Pendekatan Dasar*, Refika Aditama (1985)
8. SK SNI 03-2834-1992, *Tata Cara Pembuatan Rencana Campuran untuk Beton Normal*, (1992)
9. SK SNI 03-1968-1990, *Metode Pengujian Tentang Analisis Saringan Agregat Halus dan Kasar* (1990)
10. SK SNI 03-1969-1990, *Metode Pengujian Berat Jenis dan Penyerapan Air Agregat Kasar*, (1990)
11. SK SNI 03-1970-1990 *Metode Metode Pengujian Berat Jenis dan Penyerapan Air Agregat Halus* (1990)
12. SK SNI 03-1971-1990, *Metode Pengujian Kadar Air Agregat* (1990)
13. SK SNI 03-1972-1990, *Metode Pengujian Slump Beton* (1990)
14. SK SNI 03-1973-1990, *Metode Pengujian Berat Isi Beton* (1990)
15. SK SNI 03-1974-1990, *Metode Pengujian Kuat Tekan Beton* (1990)
16. SK SNI 03-2493-1991, *Metode Pembuatan dan Perawatan Benda Uji Beton di Laboratorium* (1991)
17. SNI 03-2834-2012, *Tata cara perencanaan ketahanan gempa untuk struktur bangunan gedung dan non gedung* (2012)

18. Basoeki, Makno, *Hands Out Diskusi Panel Peran Enzyme Mikroba dalam Peningkatan Kualitas Beton dan Produksi Beton dengan Reduksi Emisi CO2, Bogor (*2000)

Study of Volcanic-Ash-Impregnated-Bacteria Filler to the Compressive Strength of Concrete

Hendry Anjar Purwanto[1,*], *Ananto* Nugroho[2,*] , and *Ririt* Aprilin S[3,*]

[1,3]Department of Civil Engineering, State University of Jakarta, 13220 Rawamangun, Indonesia
[2]Biomaterial Research and Development Center, Indonesian Institute of Sciences (LIPI), 16911 Cibinong, Indonesia

Abstract. Technology of concrete has been advanced recently to improve the properties of concrete. Utilization of ash as the substitute of cement or filler of concrete also indicates good result in compressive strength concrete in certain amount. Harnessing the bacteria for self-healing concrete has also been successfully performed in order to maintain the destruction of the concrete by itself. Past research indicated that bacteria can be beneficiary to improve the compressive strength, absorption, permeability, and so on. In this paper, *Bacillus Altitudinis* was extracted from the karst (source from the limestone mountain) and bred through liquid media to be impregnated inside the volcanic ash. The impregnated bacteria in volcanic ash were used as the filler to the concrete mixture with the proportion of filler varied from 1%, 1.5% and 2% to the weight of cement. Mechanical properties of the concrete specimen showed that the compressive strength reached its best when the filler proportion was 1.5%. The absorption referred analogously to the increment of compressive strength due to the success of filling the void inside the concrete. Based on compression test to all the concrete specimens, there was enhancement about 18% and the average absorption can be abated around 23%. Future research due to some drawbacks turning up is the need of developing another media for bacteria to live and making sure the exact amount of bacteria impregnated to the media so that the desired advance of concrete compressive strength can be acquired.

Keywords: *Bacillus altitudinis*, volcanic ash, impregnation, compressive strength, concrete

* Corresponding author: hendryanjar@gmail.com (Hendry AP) ; ananto@biomaterial.lipi.go.id (Ananto N) ; ririt-aprilin@unj.ac.id (Ririt AS)

1 Introduction

The approach of biotechnology for increasing the cement-based materials mechanical properties is called bio-mineralization. This process involves living cells which produce mineral from the microbial metabolism under favourable settings of interaction with selective cations [1]. Lately, the use of bacteria as the agent for applying bio-mineralization to the concrete mixture to form the self-healing concrete has been widely investigated. Types of bacteria from the genus of *Bacillus* are one of the most widely used to be lived inside the concrete until the deterioration of the concrete occurred so that the bacteria will work to heal once crack happens.

Methodologies for importing bacteria to the concrete are mentioned some, for instance immobilization through impregnation and encapsulation. Different results of each to the efficiency of bacteria to perform better in mechanical properties or healing process. Immobilization process would significantly increase the compressive strength when the ureolitic activity involved. Impregnation process creates better self-healing particularly for larger cracks if the chemical agent is inserted. Second, encapsulation would be more effective to be applied for healing larger cracks [2].

The major purpose of this research is gaining the information about the effect of the impregnation of bacteria to the volcanic ash to the compressive strength of concrete. Therefore, physical properties and chemical properties test conducted in this paper will be detail discussed in order to expand the analysis.

1.1 Research limitation

Research target for this paper is investigating the mechanical properties of concrete at the maximum age of 28-days. The proportion of concrete mixture is set as 1:2:3 for cement, fine aggregate, and coarse aggregate. Water cement ratio is also fixed to 0.5 with the targeted slump of 100±20 *mm*. Volcanic ash which is originally got from *Kelud* mountain is selected as filler for the concrete and the media for the bacteria to be impregnated. *Bacillus altitudinis* is utilized in this research as got from isolation process of limestone from mountain in Central Java.

The examination of concrete properties was limited to physical properties, mechanical properties, and chemical properties. Density and absorption are the representation for physical properties of concrete, compressive strength is chosen for mechanical properties data, and acid titration to be performed in checking the acid tenacity for the chemical properties. Photo scanning electron microscope (SEM) is performed only for identifying the void size of the volcanic ash to be eligible for *Bacillus altitudinis* to live on. Other limitations are quantification of bacteria and SEM experimentation after the acid titration process is not performed.

2 Material and method

The composition of concrete mixture was taken from the nearest deposit of the research field, whilst the cement used was Portland Cement Type I. The use of volcanic ash as the filler of concrete mixture was selected due to its similarity to cement properties especially the grain size. Filler works best in filling the void inside the concrete so that the failure of concrete can be significantly decreased. As the bacteria is another component tht we want to investigate, then impregnation method is chosen as the method for importing bacteria to the concrete by lodging it to the volcanic ash. The proportion of volcanic-ash-impregnated-bacteria to cement will be discussed in the following.

DOI: 10.1051/matecconf/201713801014

2.1 Concrete compounds

The preliminary test result for fine aggregate and coarse aggregate is provided in table 1, besides the grain size distribution after conducting sieve analysis and mud content test is also proven that the aggregate is eligible to be used for concrete mixture. Qualification of Portland cement is taken from the specified product without any test performed.

Table 1. Experimental result for fine and coarse aggregate

Specimen	Components	Value
Fine Aggregate	Bulk density	1.65
	SSD density	2.65
	Apparent density	1.49
	Absorption	0.6%
Coarse Aggregate	Bulk density	2.54
	SSD density	2.54
	Apparent density	2.63
	Absorption	1.35%

2.2 Filler

Physical characteristic of the volcanic ash especially the grain size is almost the same as cement. Some of the ash also has similarity in chemical properties so that cement can be replaced partially. By performing scanning electron microscope (SEM) experiment, the chemical components of the volcanic ash is served in table 2 below, at the same time the void size can be identified in order to determine the adequacy space for the bacteria to live through impregnation process.

In this research, the proportion of volcanic ash to cement is varied from 0% to 2% with the interval of 0.5%. Before, the impregnation of bacteria to the volcanic ash will be performed. Basically, the utilization of the volcanic ash as filler will improve the compressive strength due to its micro grain size to fulfil the pore between concrete matrix which cannot be accommodated by the fine aggregate and cement.

Table 2. Chemical compounds of *Kelud* volcanic ash

Oxides	Contents (%)
Calcium oxide (CaO)	62.30
Silicon dioxide (SiO_2)	25.40
Aluminium oxide (Al_2O_3)	7.60
Iron (III) oxide (Fe_2O_3)	4.20
Potassium Oxide (K_2O)	0.19
Sodium Oxide (Na_2O)	0.10
Titanium dioxide (TiO_2)	0.21

The dimension of the void is detected in the interval of 50 – 100 μm so that the spore of around 1.2μm *Bacillus altitudinis* is matched to be lodged.

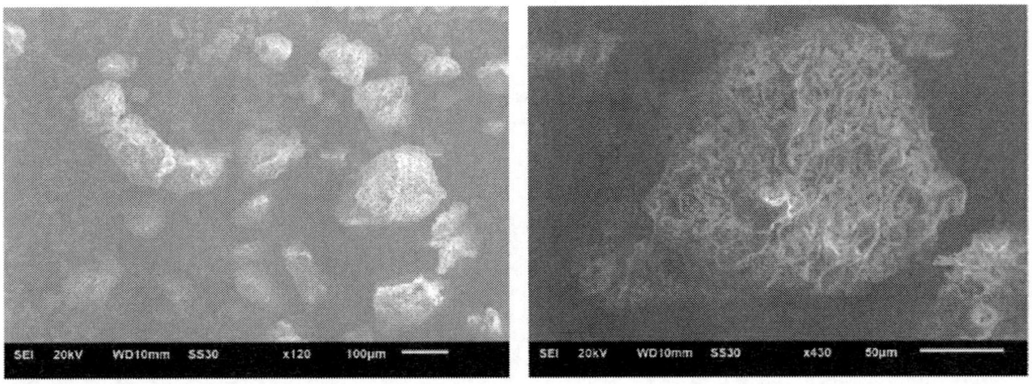

Fig. 1. SEM investigation for *Kelud* volcanic ash

2.3 Bacteria

Type of strain of *Bacillus altitudinis* is 41KF2b^T (= MTCC 7306^T = JCM 13350^T) with a gram positive categorized bacteria. The colonies of *Bacillus altitudinis* on nutrient agar are white, convex with a regular margin and 2-3 *mm* in diameter. Growth occurs between 8 and 45°C at pH 5-8. It tolerates up to 2% NaCl and degrades tyrosine, but tests negative for casein hydrolysis, urease, and phenylalanine deaminase activities, reduction of nitrate to nitrite, utilization of citrate and the Voges-Proskauer test [3].

Culture of bacteria *Bacillus altitudinis* is performed by adding 5 gr/l peptone, 5 gr/l NaCl, and 3 gr/l yeast extract. The bacteria is grown in liquid media using sterilized tube then agitated at 125 rpm for 24 hours [4]. Later, the growths bacteria will be impregnated to the volcanic ash through freeze-dry process.

3 Analysis and discussion

3.1 Physical properties

Water can be seen as the main cause of degradation of building materials, especially concrete [5]. The durability of concrete may utilize the absorption result from total absorption, porosity, surface absorption, and capillary absorption [6]. During the life service of concrete, high porosity compromises the physical-mechanical properties of concrete and its permeability [7].

Absorption and density of concrete will be analyzed to get the physical properties of the volcanic-ash-impregnated-bacteria filler concrete. As shown in Figure 2, the absorption of concrete degrades as the proportion of volcanic ash increases. The maximum of absorption is reached for the 1.5% addition of volcanic ash. The number of pores will be completely blocked by volcanic ash so that the water entrance to the concrete diminishes. When reaching 2% of variation, the absorption rises about 12%.

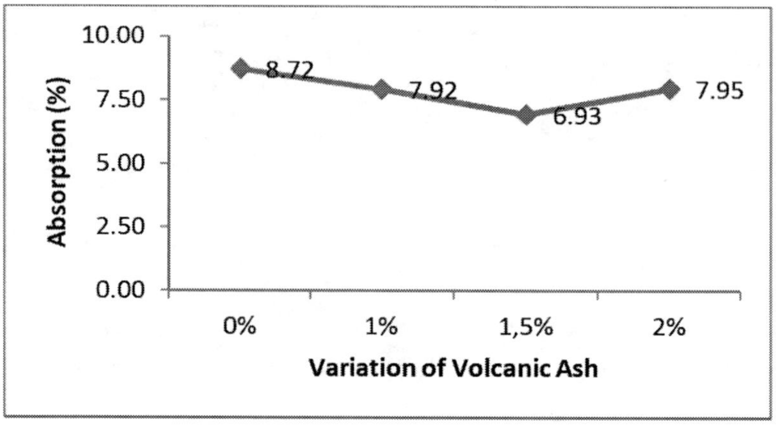

Fig. 2. Absorption of Concrete

Density value can be seen in table 3, where the density of concrete rises along the augment of volcanic-ash-impregnated-bacteria. The increment of 2% from the control concrete specimen is not really significant, but it affects the absorption significantly as the number of volcanic ash added.

Table 3. Density of Concrete

Variation	0%	1%	1.5%	2%
Density	2.56	2.63	2.63	2.61

3.2 Mechanical properties

The slump value of the concrete mixture reaches the target interval so that the workability is good enough during moulding of concrete specimen. The specimens are cylindrical based on ASTM requirements. Compressive strength as the main concern to see the effect of the volcanic-ash-impregnated-bacteria after 28-days is shown in figure 3. Enhancement of the volcanic ash variation indicates the increment of compressive strength; the maximum refers to 1.5% variation. Lesser void in concrete is not only increasing the absorption and density but also compressive strength. Bacteria inside the concrete create the crystal which affect to the compressive strength of concrete [8], [9]. The maximum increment achieves 18% at the 1.5% variation compare to control concrete specimen.

DOI: 10.1051/matecconf/201713801014

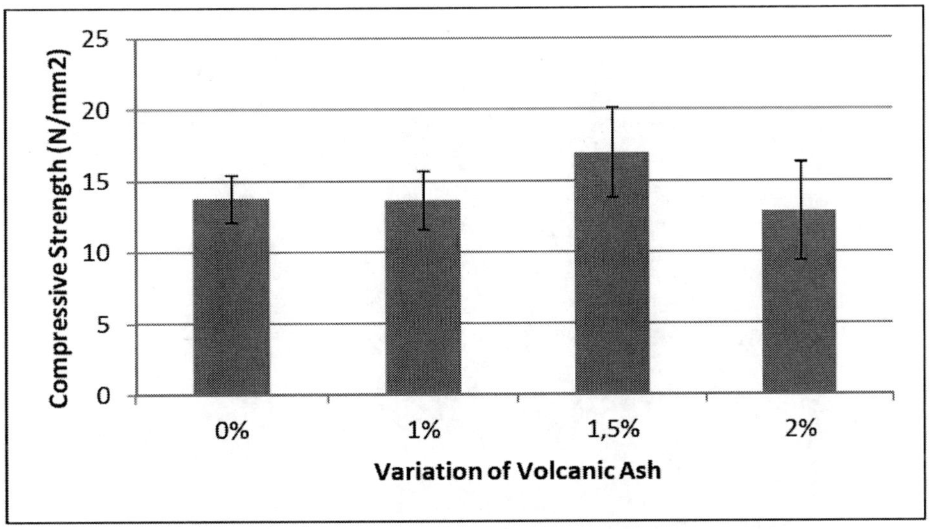

Fig. 3. Compressive strength of volcanic-ash-impregnated-bacteria concrete

The variation of volcanic-ash-impregnated-bacteria inside the concrete results the degradation in density and compressive strength concrete as well as the absorption escalates. Ideally, absorption of bacteria and precipitation of calcite crystals resulted in weight increase of concrete specimen [10], [9]. But in the case of 2% variation concrete specimen, the pore is too dense so that the bacteria have no supply of oxygen to perform ureolytic activity and carbonate biomineralization.

3.3 Chemical properties

Tenacity due to acid performed by using phenolphthalein 1gr/l which is soluble using alcohol content 90% about 50 ml and added by water 100 ml. The result refers to figure 4 that the purple colour in the surface of concrete so that the specimen is acid proof, in other words the condition of concrete specimen is alkali. It also strengthened by the result of the compressive strength of concrete specimen which is constant compare to the usual curing condition.

Fig. 4. Acid titration experiment

4 Conclusion and suggestion

Based on the discussion above, we can conclude and summarize as follows:

a. Utilization of *Bacillus altitudinis* to the concrete specimen has brought enhancement of physical, mechanical, and chemical properties even though the number of addition is very small to the weight of cement.

b. Physical properties of volcanic-ash-impregnated-bacteria filler of concrete specimen indicate analogue result with its mechanical properties. The enhancement of density and compressive is achieved due to calcite crystal formed from the bacteria through ureolytic activity and carbonate bio-mineralization.

c. The maximum satisfying result refers to the volcanic-ash-impregnated-bacteria variation of 1.5% with the 18% compressive strength increment, whilst after 1.5% variation shows degradation.

d. Better detail experiment must be conducted such as quantification of bacteria, SEM investigation after the compression test, and condition of bacteria either live or die. Another type of immobilization process of bacteria to the concrete can also be performed in order to see the effectiveness between them to the physical, mechanical, and chemical properties. Besides, it is also worth to observe the concrete specimen into self-healing activity after it is deteriorated due to compression test.

References

[1] L. S. Wong, "Microbial cementation of ureolytic bacteria from the genus Bacillus: a review of the bacterial application on cement-based for cleaner production," *Journal of Cleaner Production,* pp. 5-17, (2015).

[2] N. Z. Muhammad, A. Shafaghat, A. Keyvanfar, M. Z. A. Majid, S. Ghoshal, S. E. M. Yasouj, A. A. Ganiyu, M. S. Kouchaksaraei, H. Kamyab, M. M. Taheri, M. R. Shirdar

and R. McCaffer, "Tests and methods of evauating the self-healing efficiency of concrete: A review," *Construction and Building Materials,* pp. 1123-1132, (2016).

[3] S. Shivaji, P. Chaturvedi, K. Suresh, G. Reddy, C. Dutt, M. Mainwright, J. Narlikar and P. Bhargava, "Bacillus aerius sp. nov., Bacillus aerophilus sp. nov., Bacillus stratosphericus sp. nov. and Bacillus altitudinis sp. nov., isolated from cryogenic tubes used for collecting air samples from high altitudes," *Systematic and Evolutionary Microbiology,* **vol. 56**, pp. 1465-1473, (2006).

[4] A. Nugroho, I. Satyarno and Subyakto, "Bacteria as self-healing agent in mortar cracks," *J. Eng. Technol. Sci.,* **vol. 47**, no. 3, pp. 279-295, (2015).

[5] A. Fernandez and A. Palomo, "Propiedades y applicaciones de los cementos alcalinos," *Revista Ingenieria de Construction,* **vol. 24**, pp. 15-213, (2009).

[6] N. T. Castellanos, J. T. Agredo and R. M. de Gutierrez, "Evaluation of permeation properties of concrete added with a petrochemical industry waste," in *Calcined Clays for Sustainable Concrete*, Switzerland, (2015).

[7] R. Detwiler and P. Mehta, "Chemical and physical effects of silica fume on the mechanical behavior of concrete," *ACI Mater J.,* **vol. 86**, pp. 609-614, (1989).

[8] S. Krishnapriya, B. D. Venkatesh. and P. A. G., "Isolation and identification of bacteria to improve the strength of concrete," *Microbiological Research,* **vol. 174**, pp. 48-55, (2015).

[9] N. Chahal, R. Siddique and A. Rajor, "Influence of bacteria on the compressive strength, water absorption and rapid chloride permeability of fly ash concrete," *Construction and Building Materials,* **vol. 28**, pp. 351-356, (2012).

[10] R. Siddique and N. K. Chahal, "Effect of ureolytic bacteria on concrete properties," *Construction and Building Materials,* **vol. 25**, pp. 3791-3801, (2011).

MATEC Web of Conferences **138**, 01015 (2017) DOI: 10.1051/matecconf/201713801015
EACEF 2017

Performance Evaluation of Newly Developed Sustainable Blocks for Affordable Housing in Malaysia

Ashwin Narendra Raut[1,2*], *Christy Pathrose Gomez*[2]

[1]Department of Civil Engineering, Lords Institute of Engineering and Technology, India
[2]Department of Construction Management, Universiti Tun Hussien Onn Malaysia, Johor, Malaysia

Abstract. The thermal design of affordable housing in Malaysia could be ineffective and resulting from this, the majority of their occupants are not satisfied with the thermal comfort levels provided. To overcome this issue of thermal performance of building envelope, the development of sustainable blocks has been done. The blocks have been developed using locally available sustainable waste materials. The materials utilized were mainly procured from agro-industrial waste (oil palm industry) and municipal waste (waste glass). The aim of the research was to provide a better performing material for low cost houses replacing the conventional red clay bricks. In this research, the physical, mechanical and thermal performance of the sustainable building blocks were examined and compared with existing conventional building blocks. The experimental findings revealed that the newly developed Sustainable blocks have better thermal performance as compared to conventional burnt clay bricks. However, the physico-mechanical performance complies with various standard practices thus proving to be a sustainable wall material.

1 Introduction

Construction activities are one the major contributors causing environmental pollution. Additionally, it is evident that the construction sector consumes large quantity of ecological resources, resulting in serious negative environmental impacts. Thus, many private and government agencies are focusing on sustainable construction practices to overcome the negative environmental hazards caused by the construction sector. However, the current efforts towards sustainable practices are still found to be lacking. Emphasis on green building concepts for sustainable development is acknowledged to be a key aspect in reducing adverse environmental impact [1].

In the Tenth Malaysian Plan, it is stated that continuous efforts will be put into ensuring citizens of Malaysia of all income groups can have the opportunity to acquire houses that are frequently referred to as "adequate, affordable and of good quality." Currently the issue of housing is found to be more crucial in urban areas, greater emphasis is placed on the low-income group for better urban services and healthy living [2]. The Malaysian government has a requirement that for all private development, 30% of housing units should be of the low-cost housing type. Otherwise, the developer will have to pay the government the cost of building those units, if so agreed. Studies based on the various Malaysia Plans have shown that the government has been continuously encouraging the private sector to develop more low-cost houses and low-medium-cost houses. Currently, there is a huge demand and provision of low-income housing in Malaysia, which is being satisfied mainly by government policy. Although the Malaysian government's commitment is stated to be the provision of quality and affordable housing, the aspect of indoor comfort is noted to be unsatisfactory [3]. Low-cost housing (LCH) is perceived to have a lower standard of housing, and in trying to provide affordable housing, the standard of houses is always being compromised. There is certainly a need for providing a sustainable solution in providing better indoor air comfort levels, which is best done by utilizing sustainable locally available resources (materials).

It is important to improve the thermal performance of the building envelope to improve the indoor comfort levels of the spaces within the building envelope. Building envelopes with better thermal performance retains internal heat, thus maintaining the lower temperature values considering hot climatic conditions outside during the day [4]. The majority of the building envelopes of low-cost housing are covered by walls composed of either clay or cement brick/block assembly. Thus, it is very crucial to have bricks with better thermal performance. A number of methods are used in order to achieve better thermal performance of the building envelope, enhancement of thermal performance can be done by adding pore forming additives in the composition of bricks. The porous additives are mostly light weight in nature. The additives contribute towards the porosity of the final product, which contributes towards a lower rate of heat transfer. Some of the previous researches also focused on the enhancement of thermal performance of building materials especially bricks. Previous research has emphasised on the utilization of contribution of agricultural, municipal and industrial waste to achieve sustainable practices.

Oti et. al. [5] utilized Ground Granulated Blast Furnace Slag (GGBS) as a supplementary cementitious binder along with portland cement and quick lime to develop two varieties of unfired clay bricks. The observed mean thermal values for developed product was found to be 0.2545 W/mK and 0.2612 W/mK using the lime and cement binder specimens respectively. It was concluded that the thermal conductivity values was a function of material density and moisture content. In another research, Khedari et. al. [6] utilized coconut coir fibers to develop soil-cement block to attain the lower values of thermal conductivity. The properties observed were thermal conductivity and compressive strength having values of 0.651 W/mK and 39.55 kg/cm^2. It was concluded that the lower rate of heat transfer in the material was due to the addition of fibers ; however, fiber addition also resulted in lowering of compressive strength.

Turgut [7] studied the performance of bricks developed from combination of glass powder and limestone along with class-C fly ash. The performance of various properties such as water absorption, density, compressive and flexural strengths, and thermal conductivity of the bricks were calculated. Findings showed that the glass powder content enhanced the thermal performance of bricks, due to the fact that thermal conductivity value of glass was found to be approximately 0.96 W/mK. In addition to that the positive rise in the compressive and flexural strength of the bricks containing glass powder were observed.

Binci et. al. [8] studied the fiber reinforced mud bricks for thermal performance and mechanical properties in compliance with ASTM and Turkish Standards. The results showed low thermal conductivity values of 0.2 W/mK due to the addition of straw. Similarly, Alami [9] utilized fibrous materials such as olive husk and straw as a filler in clay bricks. The addition of 40% olive husk and 10% straw showed a thermal conductivity (λ) value of 0.09 W/mK.

Bal et. al. [10] analysed the thermal performance of bricks developed from millet waste mixed with laterite powder. Due to the incorporation of millet waste the thermal performance values reduced from 1.4 W/mK to 0.29 W/mK for dry laterite soil bricks. Whilst Zach et. al. [11] described the use of Polyurethane (PUR) waste of fraction 3-6 mm and Expanded Polystyrene (EPS) waste of fraction 1-4 mm. The PUR and EPS were mixed in 50:50 proportions, but the samples were cured at three different set of temperatures. The thermal conductivity values were found to be around 0.035 W/mK.

Focusing on the use of Rice Husk Ash (RHA), Lertsatitthanakorn et. al. [12] analysed the thermal performance of RHA based sand–cement block as per JIS R 2618 [13]. The comparative results of RHA bricks with commercially available clay bricks showed a significant reduction in thermal conductivity values. It was found that the thermal performance of RHA bricks was 54% better than that of commercial clay bricks. Meukam et. al. [14] performed an experimental study on the influence of addition of a pozzolan namely sawdust in lateritic soil bricks. The partial replacement of sawdust was performed in the Lateritic soil along with cement in 9% wt. proportion. It was found that the developed bricks had thermal conductivity of 0.34 W/mK and 0.36 W/mK by using the flash method and box method respectively.

Wang et. al. [15] analysed the influence of pore structure on the thermal performance of material and its lightweight characteristics. They studied the influence of pores generated from hydration of sewage sludge ash with cement along with metallic alumina powder. Analysis was performed on the pore generated from hydration of sewage sludge ash along with metallic alumina powder. Due to the enanced pores developed from the addition of sewage sludge ash, the thermal conductivity was lower and the values were in between 0.085 to 0.102 W/mK. Having conducted an extensive literature review it was decided that by tackling the problem of improving the thermal performance of the building envelope in low-cost housing with the aim of obtaining a sustainable solutiong, using waste material is the best solution. Hence, this paper describes the evaluation of the performance of bricks and mortar developed from various types of waste generated from agro-industrial as well as municipal activities. The performance of developed prototype was evaluated by comparing its physico-mechanical and thermal properties with the commercially available clay bricks and mortar. Furthermore, the thermal performance was evaluated based on the comparative heat transfer analysis of developed prototype brick wall system with commercially available clay brick wall system.

2 Methodology

2.1 Raw Materials

The materials used for development of bricks were glass powder, palm oil fly ash, oil palm fibers, crusher dust and Lime. The materials for the development of sustainable bricks were obtained from nearby sources to reduce the transportation and energy cost ; the samples used are shown in fig. 1. The waste glass was procured from a nearby municipal dump yard to the university (Universiti Tun Hussein Onn Malaysia, Batu Pahat, Johor) where the research was undertaken. Whereas, the oil palm waste was procured from Kian Hoe Plantations, within a 20 km radius from the university. The primary raw material was mainly sustainable in nature ; whereas, for binding material, lime was used as an alternative to the traditional energy consumptive cement. The X-ray Fluorescence (XRF) test was performed on the raw materials to identify their chemical composition. The test results of chemical analysis is presented in Table 1.

DOI: 10.1051/matecconf/201713801015

(a) Glass Powder (b) Palm Oil Fly Ash (c) Crusher dust

(d) Oil Palm Fibers (e) Lime

Fig 1. Raw materials used for TESH BRICK

Table 1. Chemical composition of glass powder, palm oil fly ash and lime

Chemical Composition	SiO_2 (%)	Al_2O_3 (%)	MgO (%)	Na_2O (%)	CaO (%)	Fe_2O_3 (%)	Cl (%)	P_2O_5 (%)	K_2O (%)	LOI (%)
Glass Powder (GP)	68.89	4.147	2.717	16.938	5.904	0.52	0	0	0.568	0.316
Palm Oil Fly Ash (POFA)	61.663	5.13	4.17	0.49	9.869	5.299	0.162	3.716	8.427	1.074
Lime	3.94	1.37	0.32	0.01	88.7	0.09	-	-	0.02	3.53

2.2 Material Preparation

The prototype for sustainable bricks using locally available waste was developed using laboratory facilities. The newly developed bricks are referred to as Themally Efficient Sustainable Hybrid (TESH) bricks. The raw materials were pre-processed initially to meet the standard requirements for bricks. The glass powder and palm oil fly ash fineness were kept at 90 μm to enhance the pozzolanic reactivity, whilst the crusher dust particle size was maintained below 2.38 mm. The oil palm fibers were cleaned using alkali treatment. The ratios of raw materials were selected based on the Taguchi's method described in a previous research paper by Raut and Gomez [16]. The ratio for development of bricks is given in Table 1. Water content was kept at 0.35 water content to the total weight of constituent materials except for weight of Oil Palm Fibers. The constituent materials were first dry mixed in a mixture machine, and subsequently wet mixed till the mixture attained homogeneity. The homogenous mixture thus obtained was inserted into a mould and further compacted by using a vibrating machine. After 24 hours, sufficiently hardened bricks were demoulded before proceeding with the curing procedure by immersing the specimens in water bath maintained at 20°C.

2.3 Testing Procedure

Various tests were conducted on the developed sustainable blocks as per required standards to validate it as a building material. The tests conducted on developed sustainable blocks were performed to determine and analyse the physical, mechanical and thermal characteristics. Water absorption, bulk density, and porosity tests were performed on the developed samples having size 210×100×100 mm^3 as per the ASTM C20-00 (2015) [17] standard. Also, compression test was conducted as per standard ASTM C67-14 [18] by using compressive testing machine having a capacity of 3000 KN. The flexural strength of bricks were determined using universal testing machine of 10,000 KN capacity as per IS 4860:1968 [19]. Determination of initial rate of suction was done as per BS 3921:1985 [20].

3 Performance of sustainable blocks

3.1 Physico-Mechanical Performances

The physical properties of bricks such as apparent porosity, water absorption, and bulk density are primary properties of building materials. These properties are widely used in the evaluation and comparison of product quality and as part of the criteria for selection and use for construction practices. The physical properties such as bulk density, water absorption, porosity and the initial rate of suction for the material are shown in Table 2. The bulk density observed for developed TESH bricks were comparatively lower that of commercially available clay bricks (refer Table 2). The reduction of bulk density is advantageous due to the fact that light weight bricks will eventually reduce the dead weight of the structure, thus reducing overall steel consumption for reinforcement. Thus, the reduction in dead weight of structure can be helpful in lowering construction cost of the structure. Also, low-density bricks are easy to handle and transport.

From the Table 2, it is evident that values for water absorption and the initial rate of suction are higher for developed TESH bricks compared to conventional fired clay bricks. The higher absorption rate can be attributed to the inclusion of oil palm fibers which enhances the apparent porosity. The high amount of water absorption is a matter of concern considering the durability aspect of materials. However, the developed brick seems to have water absorption within the range provided in IS standard for non-load bearing bricks. However, Malaysian Standard and British Standard does not specify any specific requirement for water absorption of bricks.

Table 2 Comparative analysis of physical performance of bricks

Materials	Bulk Density (kg/m^3)	Water Absorption (%)	Porosity (%)	Initial Rate of Suction (%)
TESH bricks	1467.81	17.67	26.42	2.291
Fired clay bricks	2118	5	-	0.2

Mechanical performance analysis is necessary for understanding the load bearing capacity of structural and non-structural materials. Compressive strength is one of the critical parameters for determining the load bearing capability of the brick/mortar. Common minimum values for compressive strength for non-load bearing, and load bearing fired clay bricks are 3 to 5 Mpa and 5 to 10 MPa respectively. These bricks are in practice part of the brick wall system, and thus the analysis involves both tests on the brick as well as the brick and mortar as a brick wall system. The results of compressive strength tests and flexural strength tests for the TESH bricks that constitute the main part of the wall system are presented in Table 3. The strength of developed TESH bricks is comparatively lesser than the commercially available fired red clay bricks. However, the compressive strength results for TESH bricks were found to be satisfactory as per MS 76: 1972 [21] and BS 3921: 1985[20] for load bearing bricks. This is found to be a non-issue as these bricks are designed for non-load bearing purpose for the low-cost house. The flexural strength results for developed bricks complies with AS/NZS 3700 [22], which recommends that bricks should have a flexural strength of 1 to 2 Mpa. The higher flexural strength of bricks restricts the propagation of cracks thus maintaining good quality.

Table 3 Mechanical properties of the materials used for TESH wall system

Materials	Compressive strength (Mpa)	Flexural Strength (Mpa)
TESH bricks	7.23	1.62
Fired clay bricks	25.65	1.97

The bond strength between bricks and mortar plays a crucial role in mechanical performance of the wall as a whole system. However, the preparation of the fibre reinforced mortar (FRM) is not discussed in this paper, please refer to Raut and Gomez (????) for a complete verification of the performance of the developed FRM. FRM is included in this paper to provide further evidence of the capability of the TESH brick performing as part of the wall system. The wall system should have a perfect bond between its masonry unit and the mortar. This is essential for the masonry wall system to resist the stresses due to different types of loading conditions. To understand the bond between mortar and brick units, shear bond test was performed. The bond between brick units and mortar is mainly attributed to the mechanical inter-locking of cement hydration products into the surface pores of the bricks. The results of shear bond strength for TESH brick masonry system are shown in Table 4. The strength of the bond is found to be comparatively weaker than that of conventional brick mortar bond for clay brick wall system. The shear bond strength mainly depends upon a number of parameters relating to the brick and mortar characteristics. The water content required for bonding of brick and mortar also depends on the initial rate of suction of bricks. If the initial rate of suction is more the bricks might absorb more water from mortar layer, making the mortar layer dry thus lowering the bond. Another important parameter for bond strength is the compressive strength of the masonry units as well as mortar.

Table 4 Result for shear bond strength test

Material	Shear bond Strength (kg/cm^2)
TESH Wall	2.9
Conventional Wall	3.18

3.2 Thermal Performances

The thermal performance of a material is mainly governed by the thermal conductivity, heat capacity, and density. The thermal conductivity for the TESH bricks was compared with the conventional bricks. The developed TESH brick shows lower thermal conductivity values (refer Table 5), which indicates the lowered heat transferred rate. Another important aspect of thermal performance is the materials heat capacity which is the energy required for a material to raise its temperature by 1°c. The heat capacity of material is primarily determined by materials ability to store the heat. The heat capacity of material plays a major role in determining the thermal mass of wall system and also in the dynamic heat transfer process. The heat capacity of TESH bricks were determined by using Digital Scanning Calorimeter (DSC).

The thermal performance of materials for non-steady heat conditions require determination of thermal diffusivity (α) and thermal effusivity (e). Thermal diffusivity of material is helpful in understanding the thermal inertia within the material. Thermal diffusivity is the thermophysical property that defines the speed of heat propagation by conduction during changes of temperature. The higher the thermal diffusivity, the faster the heat propagation. The thermal diffusivity is related to the thermal conductivity, specific heat capacity, and density. The formula for determining the thermal diffusivity (α) is given in Equation 5.1

$$\alpha = \frac{\lambda}{Cp \times \rho} \qquad (1)$$

Where,

α = Thermal Diffusivity (m2/s)
λ = Thermal Conductivity (W/m.K)
Cp = Heat Capacity (J/kg°c)
ρ = Density (kg/m3)

From Table 5, it is evident that the TESH bricks shows lower thermal diffusivity values as compared to the conventional bricks. Thus, it can be concluded that the rate of heat propagation in TESH wall system is slower as compared to the conventional wall system.

Table 5 Thermal properties of the materials used for TESH bricks

Materials	Thermal conductivity (W/m.K)	Heat capacity (J/kg°c)	Thermal diffusivity (m^2/s)	Thermal effusivity ($WK^{-1}m^{-2}s^{1/2}$)	Thermal Resistance (m^2K/W)
TESH bricks	0.3812	1152	2.254×10^{-7}	802.85	0.262
Conventional Brick Layer	0.72	840	4.45×10^{-7}	1079	0.138

Whereas, thermal effusivity (heat penetration coefficient) is the rate at which a material can absorb heat, lower the thermal effusivity better the thermal performance of building material to provide comfort condition. The thermal effusivity depends on the thermal conductivity, specific heat capacity, and density same as that of thermal diffusivity. The formula for determining the thermal effusivity (*e*) is given in Equation 5.2.

$$e = \frac{1}{\lambda \times Cp \times \rho} \quad - \qquad (2)$$

Where,

e = Thermal Effusivity ($WK^{-1}m^{-2}s^{1/2}$)
λ = Thermal Conductivity (W/m.K)
Cp = Heat Capacity (J/kg°c)
ρ = Density (kg/m^3)

From Table 5.6, it is evident that the optimal proportion of fiber reinforced mortar and TESH bricks shows lower thermal effusivity values as compared to the conventional mortar and bricks. Thus, it can also be concluded that TESH wall system has lower heat absorption rate as compared to conventional wall system, which in a key aspect in providing comfort condition within the building envelope.

4 Conclusions

A comprehensive study on the developed TESH bricks using locally waste materials was conducted. The following conclusions were drawn from this research work:

1. The utilization locally available waste material to develop an alternative building material contributes towards the sustainability aspect of a building material.
2. The results fond that the developed bricks are lighter in weight than conventional burnt clay bricks. However, water absorption and porosity was on the higher side.
3. Compressive strength complies with MS 76: 1972 and BS 3921: 1985 for load bearing bricks. Also, flexural strength was also found to be slightly lower than the conventional bricks. The results of shear bond strength showed good bonding with mortar having values in the range of conventional bricks.
4. Thermal conductivity of the TESH bricks was found to be around 50% lower than conventional bricks. Whereas, other thermal properties indicates towards the better thermal performance of TESH bricks over conventional bricks.
5. Overall, the developed TESH bricks are sustainable, light weight, thermally efficient and comply with the standard practices and therefore can be used as an alternative to conventional building material.

References

1. Chan, Albert PC, and Ada PL Chan. "Key performance indicators for measuring construction success." Benchmarking: an international journal **11.2** : 203-221. (2004).
2. Government of Malaysia. Ninth Malaysia Plan (2006–2010). Putrajaya, Malaysia: Economic Planning Unit. (2006).
3. U.R. Abdulazeez, & C.P. Gomez, "Effect of spatial design modification on Malaysian low-cost residential property value: a hedonic price model", J. Teknologi, **76**(1), pp. 367-372. (2016).
4. Raut, A. Narendra, and C. P. Gomez. "Development of thermally efficient fibre-based eco-friendly brick reusing locally available waste materials." Constr. Build. Mater. **133** : 275-284. (2017).
5. J. E. Oti, J. M. Kinuthia, & J. Bai, " Design thermal values for unfired clay bricks". Mater. Design, **31**(1), pp. 104-112. (2010).
6. J. Khedari, P. Watsanasathaporn, & J. Hirunlabh, "Development of fibre-based soil–cement block with low thermal conductivity. Cement Concrete Comp, **27**(1), pp. 111-116. (2005).
7. P. Turgut, "Fly ash block containing limestone and glass powder wastes. KSCE J Civ. Eng. , **17**(6), pp. 1425-1431. (2013).
8. H. Binici, O. Aksogan, M.N. Bodur, E. Akca, & S. Kapur, "Thermal isolation and mechanical properties of fibre reinforced mud bricks as wall materials". Constr. Build. Mater., **21**(4), pp. 901-906. (2007).
9. A.H. Alami, "Mechanical and thermal properties of solid waste-based clay composites utilized as insulating materials". Int. J. of Therm. Env. Eng., **6**(2), pp. 89-94. (2013).
10. H. Bal, Y. Jannot, S. Gaye, & F. Demeurie, "Measurement and modelisation of the thermal conductivity of a wet composite porous medium: Laterite based bricks with millet waste additive". Constr. Build. Mater., **41**, pp. 586-593. (2013).
11. J. Zach, J. Hroudová, M. Sedlamjer, "Possibilities of development of advanced ceramic blocks for external masonry structures of low energy and passive constructions". Proceedings of 3rd International Conference on Sust. Mater. Tech., Kyoto, Japan. (2013).
12. C. Lertsatitthanakorn, S. Atthajariyakul, & S. Soponronnarit, "Techno-economical evaluation of a rice husk ash (RHA) based sand–cement block for reducing solar conduction heat gain to a building". Constr. Build. Mater., **23**(1), pp. 364-369. (2009).
13. Japanese Standard Association. "Testing method for thermal conductivity of insulating fire bricks by hot wire". JIS, 2618, pp. 3-4. (1992).
14. P. Meukam, Y. Jannot, A. Noumowe, & T.C. Kofane,. "Thermo physical characteristics of economical building materials. Constr. Build. Mater., **18**(6), pp. 437-443. (2004).
15. K.S. Wang, J. Chiou, C.H. Chen, & D. Wang, "Lightweight properties and pore structure of foamed material made from sewage sludge ash". Constr. Build. Mater., **19**(8), pp. 627-633. (2005).
16. Raut, A. Narendra, and P.C. Gomez. "Optimization of Mix Design of Thermally

Efficient Blocks Using the Process Parameter Approach." J Mater. Civil Eng. (2016).

17. Standard ASTM. Standard Test Methods for Apparent Porosity, Water Absorption, Apparent Specific Gravity, and Bulk Density of Burned Refractory Brick and Shapes by Boiling Water. West Conshohocken, PA.: ASTM C20-00. (2015).
18. Standard ASTM. Standard Test Methods for Sampling and Testing Brick and Structural Clay Tile. West Conshohocken, PA.: ASTM C67-14. (2014).
19. Indian Sandard. Specification for acid resistant bricks. Bureau of Indian Standard, New Delhi. IS:4860. (1996).
20. British Standard. Specification for clay bricks. United Kingdom. BS 3921. (1985).
21. Malaysian Standard. Specifications for bricks and blocks of fired brickearth. (1972).
22. Australian/ New Zealand Standard. Masonry units and Segmental Pavers -Methods of test Method Determining Initial Rate of Absorption (suction). AS/NZS 4456.17. (1997).

Buoyancy-Driven Ventilation Generated by the Double-Skin Façade of a High-Rise Building in Tropical Climate: Case Study Bandung, Indonesia

Akhlish Diinal Aziiz[1], *Mochamad* Donny Koerniawan[2], *Suhendri*[3]

[1, 2, 3] School of Architecture, Planning and Policy Development, Institut Teknologi Bandung, Jalan Ganesha 10, Bandung

Abstract. High-rise buildings in tropical region is identical to the use of mechanical Air Conditioning in massive scale. Nevertheless, there is an encouragement to high-rise buildings to reduce its energy consumptions, since they consume quite large amount of energy. This challenge can be overcome with various of strategies, one of them, by means of reducing the cooling load of mechanical Air Conditioning in high-rise building. Prospects come from the modern tall building design strategies, for example the use of double-skin façade to give addition of building skin which could provide indoor temperature protection from outside. Double-skin façade system has continued to increase in buildings in a tropical region such as in Indonesia. However, there is another potential of double skin façade, which is the possibility to increase the buoyancy effect in the air gap between the skin and building envelope. The possibility needs to be studied in order to give a proper way in designing double-skin façade of a high-rise building, especially on Bandung-Indonesia tropical climate. This paper explores the potential of double-skin façade in driving the air inside the façade to generate natural ventilation for a high-rise building in Bandung climate condition. Two parameters are used in exploring the buoyancy force, the width of double-skin façade and the temperature of the skin façade. In general, double-skin façade of a high-rise building in tropical climate can generate buoyancy driven ventilation for the building, it relates strongly to the distance between of the double-skin façade and the building envelope.

1. Introduction

Double skin façade system was introduced in Germany in late 18s in factories building in subtropical area. The double skin façade system provided a warmer temperature inside the building when winter and cooler temperature when summer time [1]. This system commonly consists two layers of glasses as a curtain wall to help light intensity come through into the building, that this strategy worked well to decreasing of artificial light consumption, as a main aim is to dynamically respond the temperature outside of the building (Boake, Harrison, Collins, Chatham, & Lee, 2003) [2] [3]. On the previous research, "Variable Analysis on Cause Factors of Glass Material Using in Designing

* Corresponding author: akhlishda@gmail.com

Building" which involved hundreds of respondents also shows that glass material for façade could do the same in Indonesia tropical context [4]. The study informs the tendency of the market in Indonesia to use this material is increasing over the time. However, the fact is that Indonesia has a big amount of sun light through the year. Therefore, the strategy to design double skin façade need to be on certain, research states that the risk of using this system is higher than on the other region [5]

Generally, double skin facade is designed to control indoor temperature from extreme outdoor temperature which could decrease comfort level of user inside. Darmayanty [3] also called this system as envelope that creates some particular space to protect the building. This system also works as curtain wall still to let the natural light come inside, which is one of the reasons this system is often to use in high-rise building [4].

Kraagh (2000) in Poirazis [5] defined double skin façade as system which consists of two skin of building that separated by a gap to reduce heat transfer into the building. Boake (2003) also explaining that this gap has the potential to trap the heat from outside of the building and push out before it could enter the building. The gap between double skin façade system is generally called by air gap, or cavity gap. This part is one of the main parts which define how far your system could control the temperature indoor. Gratia, and De Harde also Lou, Huang (2012) in Barbosa [6] clearly shows that the temperature surface in the system could define the temperature inside the building. Another research done by Iyati [7] in 2012, using Computational Fluid Design (CFD) describe that glass material as façade of the building could be cooled by using air velocity, the more faster the velocity that the surface could get, the more it would be cooler.

This research is trying to get a result by exploring the width of the air gap in Double Skin Façade that could define the temperature of the surface (skin), which that could define the comfort level by also testing the surface temperature of the skin. The model of prototype also created to have the better accurate result, the simulation CFD using ANSYS is added to gain more completed result about how we design air gap in Double Skin Façade system to control temperature inside of the building in specific site.

2. Methodology

This is a quantitative study that utilise computational fluid dynamics to simulate air flow inside the double skin façade. Air flows generated in the space between the double skin façade and the exterior wall of the building is modelled and simulated in ANSYS Fluent. The prototype of the double skin façade is shown by Figure 1. The façade receives heat flux from radiation as much as 200 W/m2. Initial air temperature is 27°C, which the same as surface temperature of the surrounding walls. Inlet is set from the bottom of the model, and the 4 outlets, with 15 cm x 15 cm area, are placed on top of it. Simulation is done for four cases, which characterised by the distance of the secondary skin from the primary façade. The distances are 0.3 m, 0.5 m, 1.0 m, and 1.5 m.

Fig. 1. Prototype of the Double Skin Façade. It designed to adjust the air gap so that could do the test in various width

Along with the simulation, a calculation is also done particularly to the temperature-increase rate inside the double skin façade. From the Eq. 1, a formula to calculate the temperature increase rate that will happen inside the double skin façade for the four cases can be derived. By modifying Eq. 1 as written in Eq. 2, a known variable appears, because Q over t is the heat flux to the space inside the double skin façade. Thus, the temperature-increase rate formula can be derived as Eq. 3. This calculation is later useful in helping the understanding of heat removal rate by the double skin façade strategy.

$$Q = m\,c\,\Delta T \qquad \text{(Eq. 1)}$$

$$\frac{Q}{t} = \frac{m\,c\,\Delta T}{t} \qquad \text{(Eq. 2)}$$

$$\frac{\Delta T}{t} = \frac{Q}{(\rho\,V)\,c\,t}$$

$$\frac{\Delta T}{t} = \frac{1}{\rho\,c\,d}\,\frac{Q}{A\,t}$$

$$\frac{\Delta T}{t} = \frac{\Phi}{\rho\,c\,d} \qquad \text{(Eq. 3)}$$

Q : heat flows to a closed-system, in this case to the space inside the double skin facade (J)
m : mass of the air inside the double skin facade (kg)
c : specific heat capacity of the air (Joule/kg.K)
ΔT : temperature difference, in this case it will always increase because there is only heat gain) (K)
ρ : air density (kg/m3)
V : Volume of the space in the double skin façade (m3), can be substituted by area of the façade (A) times its width (d)
Φ : Heat flux, number of heat flows in a second per unit area (W/m2)
$^{\Delta T}/_{t}$: Temperature-increase rate (K/s)

In terms of simulation results, there are three parameters taken from it, wind velocity, volume flow rate, and temperature. Data for wind velocity are taken at the top outlets and average inside the double skin façade. Meanwhile, for the volume flow rate, the data are recorded from the top outlets. Also, for the temperature data, average temperature inside the double skin façade and the average surface temperature of the primary façade are used in results analysis.

There are four phases in this study: 1) literature review on EE calculation model for low cost housing in Indonesia; 2) developing EE calculation model based on Material Unit Analysis (MUA) by referring to Inventory of Carbon and Energy from the University of Bath and field calculation; 3) comparison of EE value of different types of low cost housing in Indonesia; 4) questionnaire and design approach to proposing strategies in mitigating environmental impacts by reducing EE value.

3. Results and Analysis

Comparison of the contour images of wind velocity inside the double skin façade reveals similar patterns of air movement. Figure 2 also shows that a little difference in the area near the primary skin due to variation in the space width. Arguably the narrower the space,

DOI: 10.1051/matecconf/201713801016

which means the more air movement near the primary façade, will reduce more heat in the primary façade. It means more reduction in indoor air temperature. The average wind speed data in Figure 3 confirm this by showing that the 0.3 m façade has the highest average wind speed among the cases.

However, the temperature contour seems contradictory to the analysis of the wind velocity data. Temperature profiles in Figure 4 show higher value for the narrower width. It is confirmed by temperature data as shown by Figure 5a. If the air movement bring with it heat and reduce surface temperature, trend-line of the temperature is expected to mirroring the pattern of Figure 3 data. But, Figure 5a shows that temperature is lower in the wider double skin façade.

In order to have deeper understanding about the phenomenon, the analysis can use temperature increase rate and heat removal rate. Temperature increase rate is calculated from Eq. 3 and Table 1 sums the calculation results for each case. Meanwhile, the heat removal rate is derived from the data of volume flow rate at the outlets. Table 1 data asserts that heat removal rate is much lower than temperature increase rate. Moreover, a graph shown by Figure 5b displays the difference of the four cases. It indicates that the wider the façade, the smaller the gap between heat gain and heat loss. This explains why a lower facade temperature resulted in the wider façade.

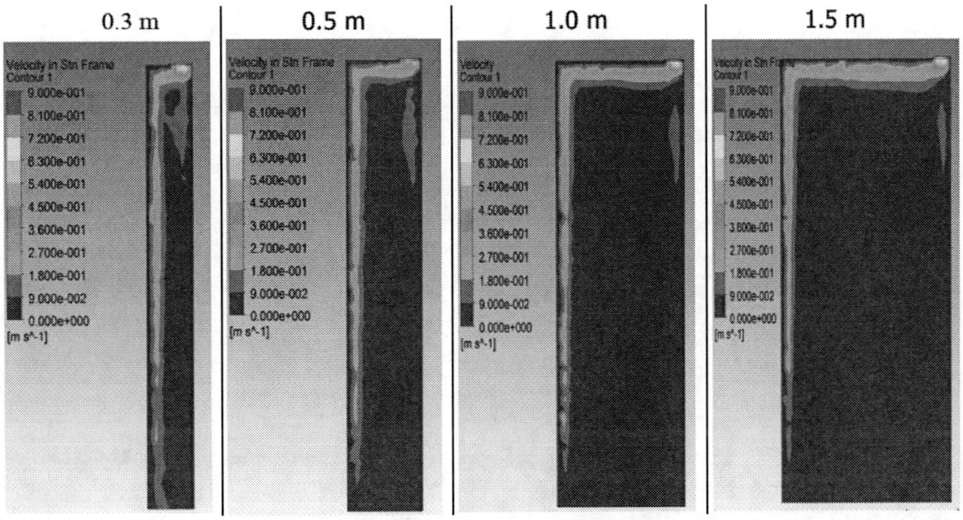

Fig. 2. Various width in the research and the result of velocity each.

Furthermore, the analysis can be extended to the relation between temperature and wind velocity to the width of the double skin façade. Figure 6 below exhibit the relation, by plotting wind velocity and temperature as a function of façade's width. Average wind velocity at the outlets make linear relation to the façade's width, while the average wind velocity inside the double skin façade make a 2nd degree polynomial relation. More, average temperature inside the façade and average surface temperature of the primary skin façade, both have linear relation to the width.

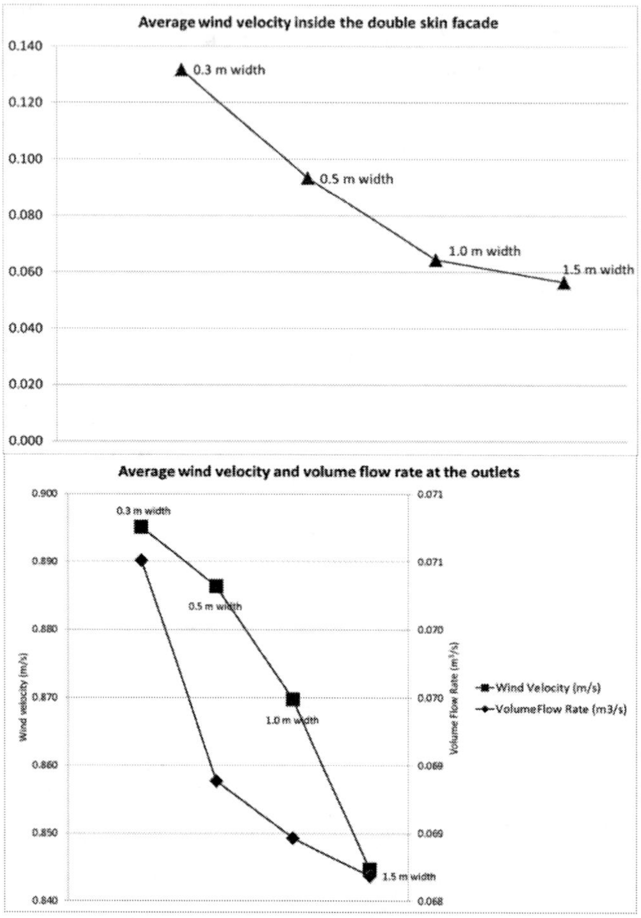

Fig. 3. Diagram that shows the average wind inside air gap DSF and volume rate at the outlet velocity in various width.

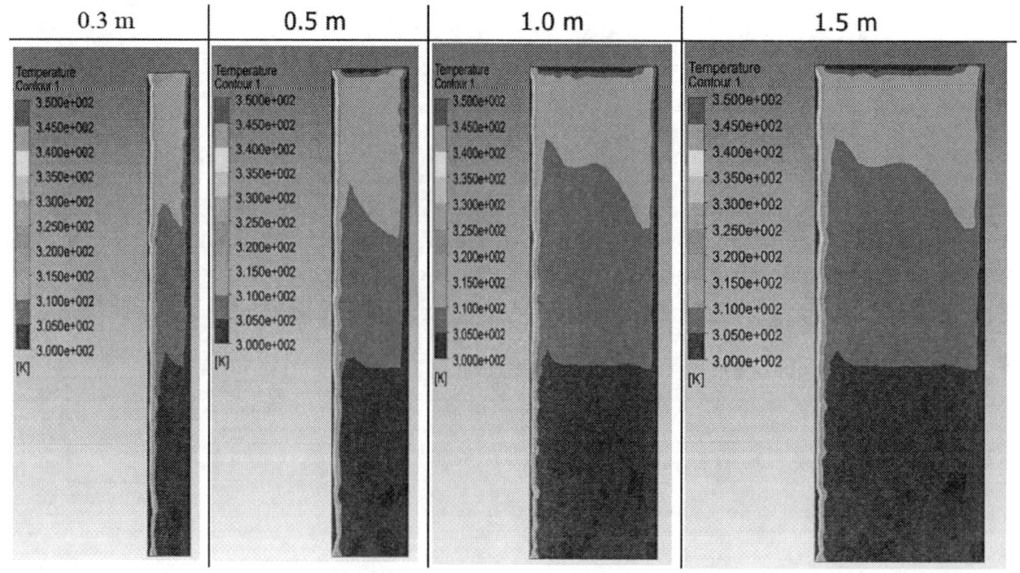

Fig. 4. Contour in various width in air gap of the double skin façade.

Table 1. Effect comparison on setting configuration width in double skin facade

	0.3 m width	0.5 m width	1.0 m width	1.5 m width
Temperature increase rate (K/s)	0.563	0.338	0.169	0.113
Percentage of temperature increase rate	2.09%	1.25%	0.63%	0.42%
Percentage of heat removal rate	0.023%	0.013%	0.006%	0.004%

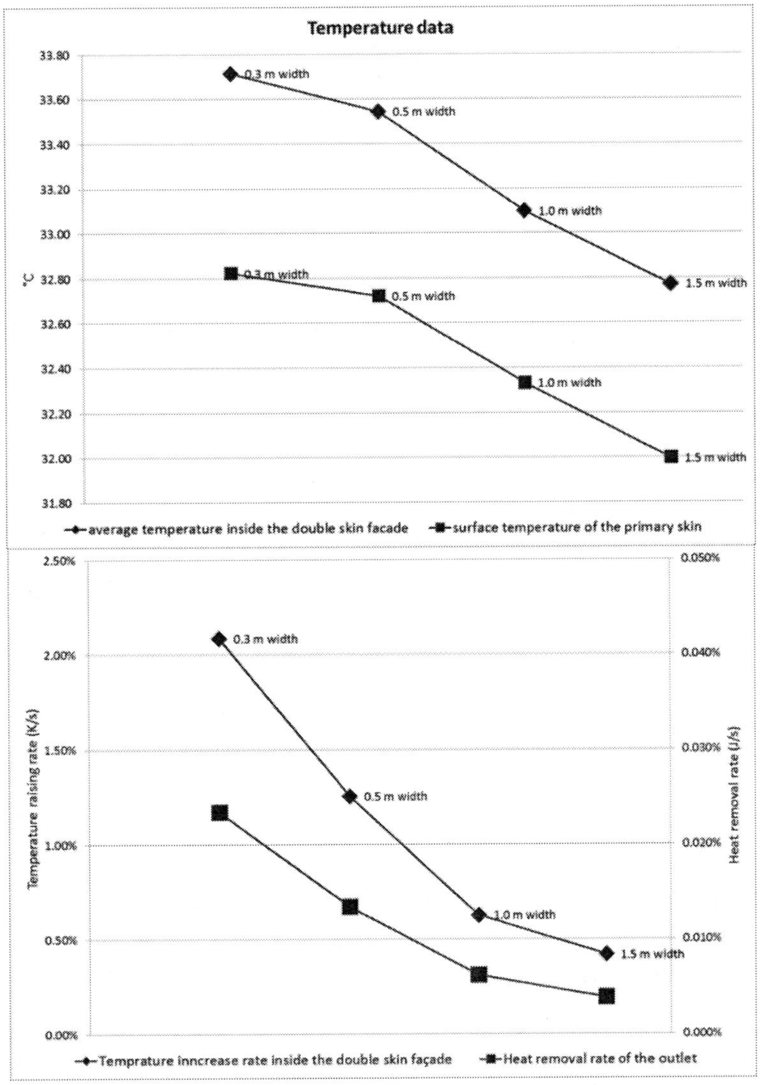

Fig. 5. Various width comparisons in diagram

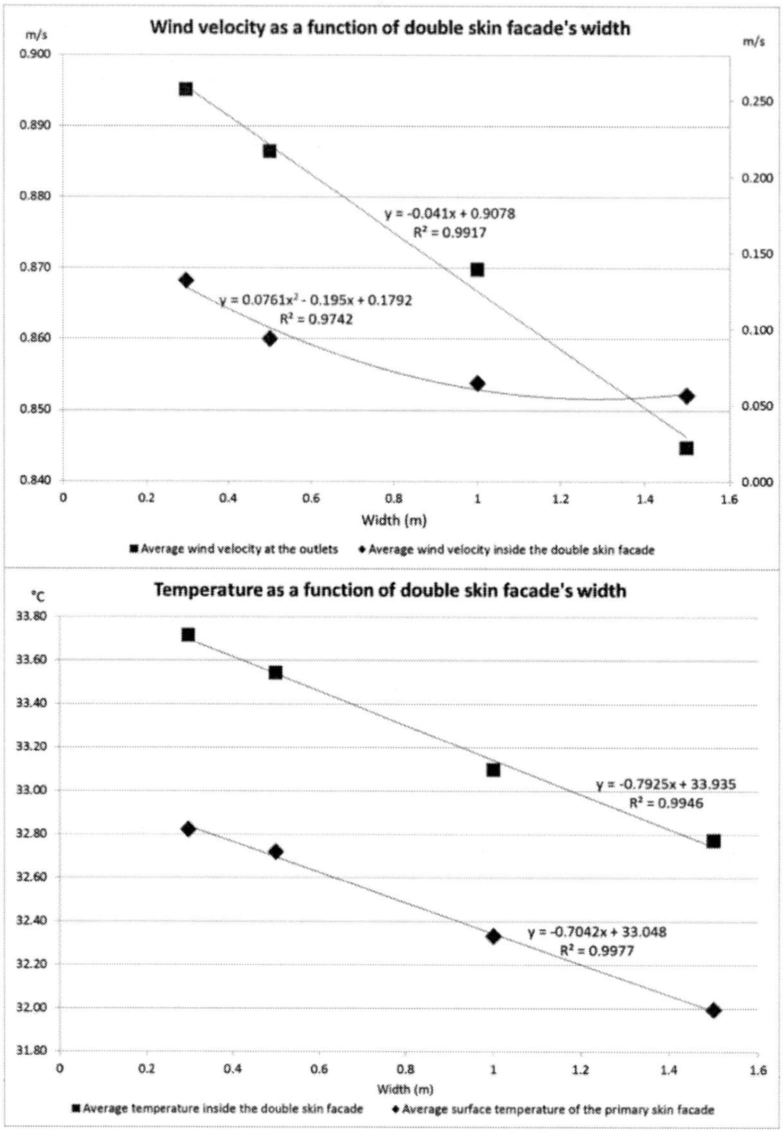

Fig. 6. The diagram shows the result by configure the air gap width in double skin facade

Acknowledgements. This research was funded by Indonesia Endowment Fund for Education (LPDP).

References

1. Baldinelli, G., "Building and Environment", **44 (6),** 1107–1118 (2009)
2. Boake, T. M., Harrison, K., Collins, D., Chatham, A., & Lee, R., "Understanding the General Principles of the Double Skin Façade System", 1–18 (2003)
3. F. Marques and M. Glo., "Energy Building", **40**, 1553–1559 (2008
4. Aziiz, A. D., Kusuma H.E., "11th International Symposium on Architectural Interchanges in Asia", 1-5 (2016)
5. T. M. Boake, K. Harrison, D. Collins, A. Chatham, and R. Lee, "School of Architecture, University of Waterloo", 1–18 (2003)

6. Darmayanty, N., "Institut Teknologi Bandung" (2014)
7. Lee, K., & Lee, D., "Energy Procedia", **78**, 2851–2856 (2015)
8. Poirazis, H., & Kragh, M., "Building Simulation", **11**, 1781–1788 (2009)
9. Barbosa, S., Ip, K., & Southall, R., "Energy and Building", **109**, 397-406 (2015)
10. Iyati, Wasiska, "Institut Teknologi Bandung", (2012)

Experimental Research of Stabilization of Polluted Marine Dredged Sediments By Using Silica Fume

Ernesto Silitonga[1]

[1]Universitas Negeri Medan, Civil Engineering Department, 20221 Medan, Indonesia

Abstract. In order to maintain the harbour activities, big amount of sediments are dredged every year. The traditional solution such as dumping the dredged sediments to the sea/river since its regulated in Europe, is not an option anymore. Therefore the beneficial reuse of polluted marine dredged sediments is urgently needed to realize. This study focused on reuse the dredged sediment in road construction. The main objective of this research is to enhance the physical, mechanical and chemical characteristics of the mix by incorporating binders sediments to fulfil the criteria in their use as material in road construction. The results show that the treatment by hydraulics binders and silica fume could satisfy the needed mechanical characteristics. The main reason of Silica Fume addition is its capability to trap the pollutants so that the pollutants could not escape from the matrix and pollute the environment. After the geotechnical study in laboratory results show as expected, then the study to identify the chemical characteristic realized. To evaluate the environmental impacts of the use material, leaching test are performed. The leaching test was performed to verify the predicted release of pollutants based on total dissolution. And for the final part, the test results show that the polluted marine dredged sediments could be safely used (in term of environmental impact) as a new material in road construction.

1 Introduction

Due to the increase of amount of marine dredged sediments, environmentally reuse of dredged sediment is urgently needed in France. Since the European regulation about the dredged sediment established [1], the port authority is encouraged to find solutions regardful the environmental issues. Available or attainable, there are many constraints on policy implementation need to be designed. However, before deciding if the dredged material is acceptable for a beneficial use, it is necessary to realize serial tests to evaluate the capability of marine dredged sediment in its reutilization domain.

Previous studies with different alternative treatment method to find the domain application for the reuse of marine dredged sediments have been investigated in Europe. This paper concentrate to determine the type and composition of binder to stabilize marine dredged sediments. The ideal percentage of hydraulic binders needed to meet prescribed specifications is very important. Silitonga in his research using dredged sediments from Cherbourg (France) [2], claimed that the ideal amount of hydraulic binder needed to stabilized dredged sediment was between 11%-13%. Behmanesh stabilized dredged sediment of Port du Havre. The amount of hydraulic binder utilized to stabilized dredged sediment was 15%. [3]. Silitonga E., [4] needed 11% of pozzolanic binders (fly ash) to stabilized dredged sediment from Port de Cherbourg (France). Zhibo in his research, stabilized marine dredged sediment from Port of Cherbourg (France) with 15% hydraulics binder. [5].

The primary goal of this research is to investigate the potential reuse of Marine dredged Sediment of Port-en-Bessin as material in road construction. Road construction needs to meet the requirements. In order to fulfil the requirements, the material has to have mechanical characteristics, furthermore if the material polluted. Consequently the addition of binders that capable to fulfil the mechanical requirements and environmental regulation is needed. In this study, the binder utilized to enhance the mechanical characteristics and to prevent the pollutants contaminate environment is Silica fume. Silica fume is by-product of producing certain metals in electric furnaces. Silica-fume, also known by other names such as volatilized silica, micro silica, or condensed silica fume, is a by-product of the induction arc furnaces in the silicon metal industries. From the previous studies [6] and [7], it is well known that the use of Silica Fume enhances the durability of concrete. The Silica Fume addition enhances the compressive strength and reduces the overall permeability of hardened concrete due to its pozzolanic properties, which result in finer hydrated phases (C–S–H gel) and densified microstructure. Raiess-Ghasemi [8] stated in his study that, concretes made with Silica Fume have a finer pore structure, with a lower proportion of fairly coarse capillary pores. For the same composition, these concretes are, therefore, generally denser than pure Portland cement concretes. One reason for this is a lower content of coarse $Ca(OH)2$ crystals in the matrix and the aggregate surface zones of these concretes. Sobolev [9] confirmed in his research that a shift of the pore structure towards finer pores as a result of the pozzolanic reaction is also observed when Silica Fume is used in concrete. The results from several previous studies are the main reason why Silica Fume is being used in this study. Beside to fulfil the criteria needed in Road construction work, by enhancing the mechanical (physical) characteristic, the primary reason is the ability of Silica Fume as a filler and capable to trap the pollutant in to the densify microstructure, so that the pollutant will not break out and contaminate the environment.

2. Materials and methods.

2.1 Materials

This research is realized using marine sediments, were dredged from Port-en-Bessin, France. This harbour is considered as one of the oldest Harbour in France. Port-en-Bessin (PEB) is located in the Northwestern France. The sediments were dredged from four different locations; Avant port exterieur, Avant port interieur, Bassin no.1 and Bassin no. 2. After the dredging process, the marine sediments were taken and stored in experimental area, the experimental area was situated 1,3 km from the Harbour. The experimental area was specially designed and prepared for stabilisation of waste material work. The binder utilized in this study is Cement and Lime and Silica Fume. Cement and lime utilized are typical hydraulics binders used in soil stabilization work. Three different types of Silica Fume (FS) were utilized in this research (FS1, FS2 and FS3). The Silica Fumes utilized are not commercialized yet. In fact, the Silica Fumes used in this study are a prototype of low-cost Silica Fume. The particle size distribution of this Silica Fume is not as fine as Silica Fume that commercialized in the market.

2.2. Index properties.

2.2.1. Particle distribution

In order to identify the particle size of marine sediments, utilized, a laser diffractometer Coulter LS200 was used in this study.

Table 1. Particle size of marine sediment utilized

	PEB A	PEB B	PEB C	PEB D	Lime	Cement	FS1	FS2	FS3
D10 (μm)	2	1.92	1.5	2.2	0.5	0.6	4	18.2	3.2
D50 (μm)	12	11.1	10.6	12.5	11.3	32.6	32.3	153	26,4
D90 (μm)	78.4	66.7	72	75.6	172	730	93	307	84
<2 μm (%)	10.7	10.6	13.8	9	18.3	15.2	7.3	2.3	13.8
2 - 63 μm (%)	77.2	78.8	74.8	78.8	63.1	53.3	70	18.9	68
> 63 μm (%)	12.6	10.6	11.5	12.2	19.7	31.5	22.7	78.8	18.2

The results as shown in Table 1, confirm that the particle distribution of marine dredged sediments that were taken from four different locations (PEAB-A, PEB-B, PEB-C and PEB-D) is almost similar, this means the homogeneity of the marine sediment of PEB. From the point of view of percentage most representative, it shown that marine sediments constituted a large amount particle with size between 2 - 63 μm. The Silica Fume type 3 has finer diameter of particle compared to Silica Fume type 2 (FS2) and Silica Fume type 1 (FS1).

2.2.2. Mineralogical characterization.

In this study, to examine the physico-chemical of dredged sediment of Port en Bessin, the researcher is using Powder X-ray Diffraction (XRD). It is known that with high percentage of smectite can cause problems during the treatment process.

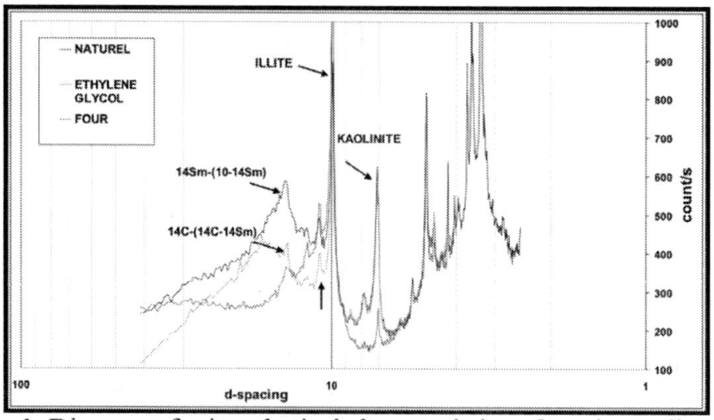

Fig 1. Diagram of mineralogical characteristics of marine sediments

As shown in Fig. 1, it appears that clay sediments of Port en Bessin phases are fairly composed of Smectite, Illlites, and Kaolinite, the presence of Illite and Kaolinite, will not risk importance consequence for the future treatment, however the present of Smectite will cause problem such as swelling. The result shows the percentage of Smectite is 32%, this amount of Smectite considered as a warning of swelling problem in a future and as a reference of a choice of type of lime utilized.

2.2.3. Chemical properties

The marine sediment is considered as a waste material, therefore, it is necessary to assess the environmental risk of the dredged sediment before its land application Leaching Test

DOI: 10.1051/matecconf/201713801017

was performed to identify the chemical properties of three different type of Silica Fume. The results of chemical characteristics are shown in Table 2. The amount of Micropollutants measured from Marine dredged sediment of Port en Bessin was presented in Table 2. In order to verify the level of pollution of the marine sediment, a reference established by European Council 2003/33/EC was used in this study. This reference divided in 3 classes: inert waste, non-hazardous and hazardous waste. These reference values relate to the elements contained in the leachate and not in the raw material

Table 2. Chemical characteristic of Marine Sediments

Pollutants	Port en Bessin	European Council Reference		
		Inert Waste	Non Hazardous Waste	Hazardous Waste
As (mg/kg)	4.9	0,5	2	25
Cd (mg/kg)	1.18	0,04	1	5
Cr (mg/kg)	70.4	2	50	100
Cu (mg/kg)	0.7	0,01	0,2	2
Hg (mg/kg)	0.58	0,01	0,2	2
Pb (mg/kg)	17.3	0,5	10	50
Ni (mg/kg)	19.8	0,4	10	40
Zn (mg/kg)	77.5	4	50	200

Table 3. Chemical characteristic of Silica Fume utilized

Parameters	Silica Fume 1	Silica Fume 2	Silica Fume 3
Colour	Light Gray	Blue	Dark Gray
SiO_2	90 - 92	85-90	90 - 95
Fe_2O_3	1,5 - 2	1,5 - 2	1,5 - 2
AI2O3	1	1	1-1,5
CaO	0,5 - 1	0,5 - 1	0,5 - 1
MgO	1 - 1,5	1 - 1,5	1
Na_2O	0,5 - 1	1 - 1,5	0,8 - 1
K_2O	1 - 1,3	1 - 1,3	1,3 - 1,5
C	0,5 - 1	1-1,5	1 - 1,5
free Si (%)	< 0,2	< 0,4	< 0,4
Free CaO (%)	< 1	< 2	< 1
SO_3 (%)	< 1	< 2	< 1
Cl (%)	< 0,2	< 0,5	< 0,1

The reference values were shown in table 2. According to the European Council Reference, the marine sediments from Port en Bessin content a medium amount of pollutants, as shown in table 2, According to European reference, all amount of micropolluants from 4 different periods are categorized as non-hazardous waste. The differences in chemical analysis between FS1, FS2 and FS3 were shown in table 3. The previous study conducted by Silitonga [4], the result showed that the chemical properties can strongly effect the mechanical properties of the marine sediment stabilized. As shown in table 2, FS3 is characterized by a higher content of SiO_2, followed by FS1 and FS2. Based on this result, we can expect that FS3 will be more reactive than FS2 as a binder. Silitonga [2] on his research noticed that, Silica Fume possesses higher content of SiO_2, Fe_2O_3 or Al_2O3 produces higher resistance on Unconfined Compressive Strength Test. On the other hand FS2 has higher total amount of free CaO, Silitonga [4] on his previous research showed the amount of free CaO has a big role on providing strength on mechanical test such as Unconfined compressive Test and Tensile test

2.3 Methods

2.3.1 Preparations of Samples.

After being dredged, physical characteristics of marine sediment were measured. The measured initial water content is about 169%. The dewatering process was realized, the marine sediment, oven-dried for 5 days at 60°C was pulverized to 2mm sieve size it was initially mixed with determined quantities of Silica Fume, lime and cement as a binder, in a dry state and subsequently mixed with water by a mechanical mixer with a speed of 150 rd/min for a period of approximately 8 minutes (fig 2). After mixing the samples were prepared with the static compaction method, at the optimum moisture content and maximum density determined by Proctor test. Cylindrical specimens (ø = 40mm, h = 80mm) were used for unconfined compressive strength testing. The compressive strength is determined using a 10 kN capacity automatic compression machine according to France regulation NF EN 196-1 (NF, 1995) on a simple speed cross-head moving machine at a speed of 1 mm/s. The samples were pushed out from the mold directly after completion of the compaction and were stored in the curing room until testing.

Fig. 2. Preparation of samples process

2.3.2 Mix design.

In order to identify the influence of the Silica Fume on the strength gained, samples FS1 FS2 and FS3 were set with different type of Silica Fume content. The different formulas in this experiment are given in Table 4. The goal of manufacturing SF2-A, was to compare the influence of using types different Silica Fume in the mixture. The amount of 2-5% of cement was the common amount that normally used in the road construction field, due to this reason. To identify the effect of sand to the mixture thus the amount of Sand utilized on this research varies from 5%-15%.

3. Results

The goal of this study is to find an application for reuse of marine dredged sediments materials as a new material on road construction. Hence, serial tests need to be realized to identify if marine dredged sediment could be utilized for road construction. Serial tests of geotechnical characteristics that are specified in building rules such as Proctor Test,

Unconfined Compressive Strength, Tensile Strength. These tests are aimed to determine either a specific material can or cannot be used as a material in road construction

Table 4. Composition of binders

Name	FS1	FS2	FS3	Lime	Cement	Sand
10-5-3-4FS1	4	-	-	3	5	10%
10-5-3-4FS2	-	4	-	3	5	10%
10-5-3-2FS3	-	-	4	3	5	10%
10-5-3-6FS1	6	-	-	3	5	10%
10-5-3-6FS2	-	6	-	3	5	10%
10-5-3-6FS3		-	6	3	5	10%
10-5-3-0FS	-	-	-	3	5	10%
10-3-3-4FS3	-	-	4	3	3	10%
5-5-3-4FS3	-	-	4	3	5	5%
0-5-3-4FS3	-		4	3	5	0%
CEM1	-	-	-	-	7	10%

3.1 Mechanical Characteristics.

3.1.1 Unconfined Compressive Strength (UCS).

In order to identify the mechanical behavior of mixtures realized, the UCS test was performed. The results in fig 3 show that for curing age 7-14 days, sample mixed with 7% of cement (CEM1) possesses highest UCS strength, this result because of the rapidity of cement hydration to provide strength. Cement hydration initiate as soon as water introduce to the mixture, normally hydration of cement reach its maximal at 14 days or 28 days of curing age. This theory confirmed why UCS value of CEM1 established the highest UCS values (from 7-14 age curing days) and doesn't show any remarkable amelioration UCS values after 14 days curing age. At 28 days the samples with Silica Fume (FS1, FS2 and FS3) show an increase UCS values greater than sample mixed only with 7% cement (CEM1). The hydration of Silica Fume is responsible for this increase. At 28 days, the reaction of normal pozzolans binders will not react perfectly on providing strength, this is because of pozzolans binders need help from cement/ lime hydration, in it self, do not have any cementious properties but when reacts with $Ca(OH)_2$ on hydration of cement produces the gel i.e. calcium-silicate-hydrate (C-S-H) which has good cementious properties. This reaction normally happened after 28 days of curing age for normal pozzolans binders (fly ash), because of high surface area and high content of amorphous silica in silica fume, this highly active pozzolan reacts more quickly than ordinary pozzolans. Effect of different type of Silica Fume (FS1, FS2 and FS3) can be noticed in Figure 3, the UCS values from 7 days to 180 days, the sample with 4% of Silica Fume type 3 (FS3) produces higher strength than samples with FS2 and FS1. This result confirm the theory that binder (especially pozzolanic binder) with finer particle size is more reactive, thus produce higher strength on mechanical properties. As shown before, Silica Fume type 3 (FS3) possesses finer particle size than FS1 and FS2, this is the main reason, why sample with FS3 gained higher strength. In the point of view of chemical properties, FS3 has almost similar content than FS1, we can assume that in this case, particle size distribution plays more important rule to provide strength than chemical properties of Silica Fume. that there isn't any important different between UCS values. At this case we consider that, the chemical properties of Silica Fume used in this study, are not high enough to make any different. According to regulation [4], after the addition of silica fume and hydraulic binders, the UCS values should reach at least 1 MPa at 28 days of curing age. The test confirmed that all samples after stabilization process could be reused as a material in road construction.

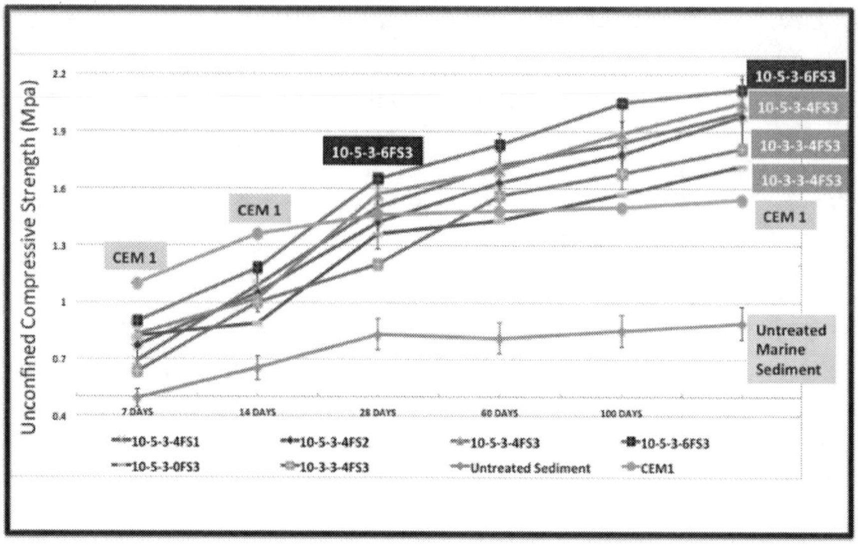

Fig 3. Unconfined Compressive Strength Test

3.1.2 Tensile Strength (TS).

The elastic modulus is obtained using a stress–strain curve of an unconfined compression test. Firstly, the maximum compression strength of the mixture is measured. Then the elastic modulus, E, is measured on another sample. In order to not damage the cylinders, the maximum effort applied corresponds to 30% of the maximum compression strength. The tests are conducted on 360 days aging cylinders. The results of tensile test are presented in fig 4.a, As shown in the figure, only the samples with 6% Silica fume located in class S2 (H and G). According to Scheme of road section and recommended engineering characteristics, the samples with 6% Silica fume in geotechnical view, as material are suitable and secure for sub base and Foundation. Test of California Bearing Ratio (CBR) need to be realized to confirm the selection of road layer. Previous study done by Silitonga [2], Colins [11] and Siham [12] confirmed that dredged marine sediment stabilized with fly ash or hydraulics binders (in geotechnical view) can be used in sub base and foundation layer in road construction base. In this study, the effect of Silica fume addition is remarkable in providing Elastic Modulus and Tensile strength. Despite Samples with 4% Silica fume (A, B, C, D and E) in the same class with sample treated only with 7% cement (CEM 1), yet the Elastic Modulus values are immensely higher than CEM 1. The presents of Silica Fume with its filler effect, due to its fineness make the Silica Fume fit into space and fill the empty space between sediment grains and produces more solid bonds between sediment grain, this is the reason of the enhancement of Elastic Modulus because a stiffer material will have higher Elastic Modulus. The results show that the speciments with 4% of Silica fume (A, B, C, D) and sample treated only with cement (CEM 1) could be used as material only in sub base. In the other hand the samples with 6% of Silica fume (H and G) could be reused in sub base and foundation. The sample without Silica fume (F) is situated within the limit of S0 and S1 classes, it means the reuse for sub base need to be considered.

DOI: 10.1051/matecconf/201713801017

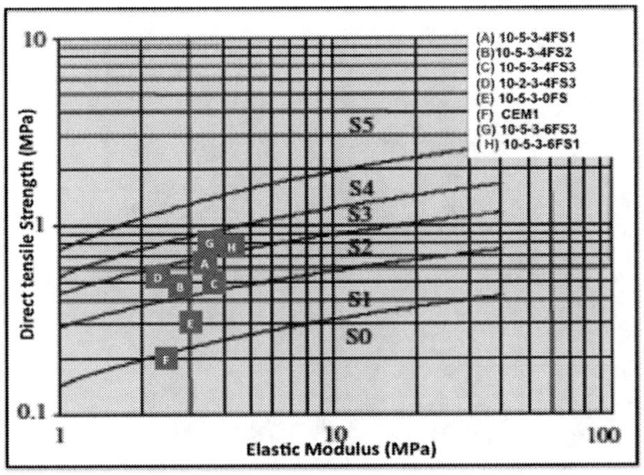

Fig 4.a. Tensile strength,

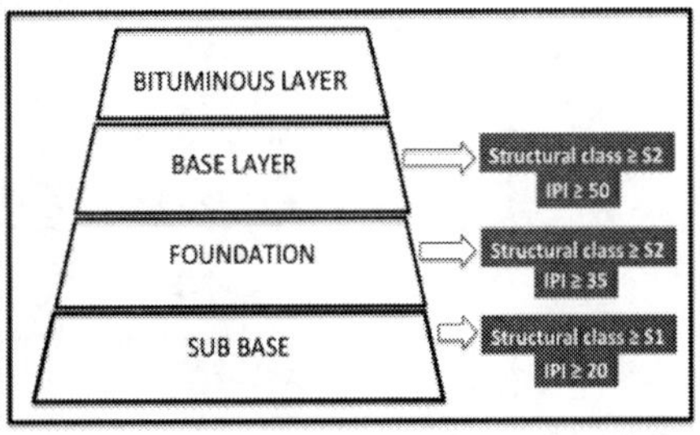

Fig 4.b. Scheme of road Section [4],

3.1.3 Leaching Test

The marine dredged sediment was planning to be reused as a base material in road construction. However, as the land applied dredged sediment is subjected to drying and oxidation, transformations in the chemical forms of heavy metals may affect their mobility and bioavailability may be occurred. The Leaching Test was realized according to French standard NF X 31-210 [13], refers to the solution containing the solubilized elements during the test, which are performed on the analytical characterization. This test is a simulation to identify the mixture reaction to aggressive chemical environment.

In order to classify the dredged sediment of ort en Bessin, the researcher decided to utilized reference values established by European Council No. 2003/33 / EC. The reference used to identify the amount of pollutants in sediment from the Port en Bessin. As shown in fig.5, the amount of Cadmium (cd) is 1.18 mg/kg, according to European reference, the quantity of Cadmium categorized as non-hazardous waste. The leaching test for Cadmium (cd) show the stabilization of marine sediment, with or without Silica fume succeed to reduce the content of Cadmium in the mixture, (fig 5). Effect of Silica Fume in reducing the pollutants content is very dominant in this test. As shown in fig 5, all the samples without Silica fume; B, E, F, J, M, O and P (see fig 5), show unremarkable reduction amount of pollutant (for Cadmium), the samples are still in the same level of pollution as the original amount of pollutants of untreated marine sediment (non-hazardous waste level). Especially for sample utilized only cement (CEM1), show only 5% reduction of pollutant from original amount of pollutants (untreated marine sediment). In the other hand, all the samples mixed with Silica fume, established a major reduction on amount of pollutant. The escalation of percentage of Silica fume (from 4% -6%) produces more reduction of pollutant. The samples with 6% of Silica fume show a very remarkable reduction of pollutant. Sample with 6% of Silica fume established a group with lowest amount of pollutant (H, I, K, L, N and Q). One of the principle physical effect of silica fume is the filler, which because of its fineness can fit into space between sediment grains in the same way that sand fills the space between particles of coarse aggregates and cement grains fill the space between sand grains.

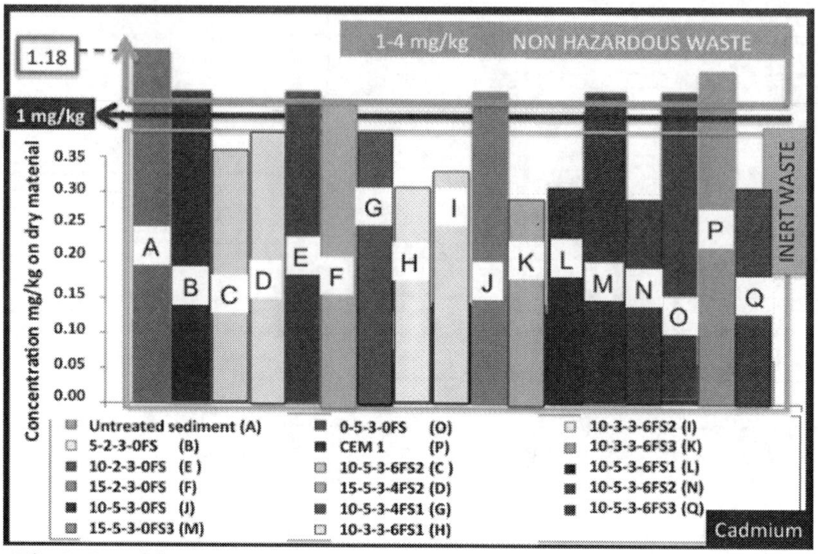

Fig 5. Leaching Test result for marine dredged sediment for Cadmium.

The theory is because of its finesess and after the pozzolans reaction, Silica Fume capable to trap the pollutants so that the pollutans could not escape from the matrix and pollute the environment. This is the main reason Silica Fume is being used in this study. After the results of leaching test presented in fig 5, the theory worked out excatlly as prediction. Except the effect of Silica fume, we can oberserve that the raise of percentage of cement, can enhance the reduction of pollutants. As figured, after addition of cement to 2% (to 5% from 3% Cement), the samples with 5% cement and 6% of Silica fume (C, N and Q) show a remarkble reduction of pollutants (50%) compared to samples with 3% Cement (H, I and K). This results is due to hydration of cement which also produces Ca(OH)2, required for the pozzolanic reaction in Silica Fume and for ion exchange in clays, which accelerates the cementitious and pozzolanic reaction of Silica fume. From the result of leaching test it confirmed that the present of Silica fume in the mixture (with addition of 4%-6%) succeeded to reduce the pollutant concentration up to 50%. According to European reference value for dredged sediment, In the environment point of view, after stabilization process with 4% to 6% of Silica fume, the marine dredged sediment categorized as inert waste, hence harmless to environment and could be reused as a material in road construction.

4. Conclusion

The general intention of this study is to find the domain application for the recycling of marine dredged sediment (contaminated). The reuse of marine dredged sediments as material in road construction purposes requires checking specific geotechnical and chemical criteria, it need to be considered, in order to identify the suitability of marine dredged sediment to be used in road construction. The distribution particle identification of marine dredged sediment from Port en Bessin show the homogeniety of Sediments and that it constituted a large amount particle with silty fraction (2 - 63 μm). The result of leaching test confirmed that, Marine dredged sediment of Port en Bessin content amount of pollutants which is according to European Reference of Dredged sediment, categorized as non-hazardous waste. Several mixtures using marine dredged sediment, sand, cement, lime and Silica fume are realized. In order to identify the effect of Silica fume, cement and sand addition, mixtures are tested with different composition of binders. The experimental results for UCS show that all the samples successfully fulfill the criteria to be used in road construction. The effect of Silica addition identified to provide greater strength in UCS after 28 days of curing ages. Test on tensile strength confirmed that the sample with addition 6% Silica fume conformed to be used as material in sub base and foundation. In the other hand sample with 4% Silica fume only can be used in sub base. Leaching test determine that the present of silica in the specimens (either 4% or 6%) remarkably capable to reduce the amount of pollutant (cadmium) and passed from level non-hazardous waste to categorize as inert waste. Hence the marine sediment with 4%-6% Silica fume can be used in road construction and harmless to the environment.

References

1. OSPAR Commission, Convention for the Protection of the Marine Environment of the North-East Atlantic, *OSPAR Convention,* 33p (1992)
2. E. Silitonga, Valorisation des sédiments marins contaminés par solidification/ stabilisation à base de liants hydrauliques et de fumée de silice. *Thesis doctorat Université de Caen, France*, 160p, (2010)
3. J. Behmanesh, Etude de la durabilité d'un sédiment, traité au ciment et additifs, *Thèse de doctorat de l'Université de Caen*. 214p (2008)
4. E. Silitonga, D. Levacher, S. Mezazigh, Effects of the use of fly ash as a binder on the mechanical behavior of treated sediments, *Environmental Technology,* **vol 30**, p 799-807 (2009)
5. D. Zhibo, Charactérisation, stabilisation et solidification de sédiment marin. *Thèse de doctorat, Université de Caen*, 145p (2008)
6. J. J.Gaitero, I. Campillo, A. Guerrero, reduction of the calcium-leaching rate of cement paste by addition of silica nanoparticles. *Cement Concrete Research,* **vol 38**, p 1112–1118 (2008)
7. Belkowitz JS, Armentrout D. An investigation of nano-silica in the cement hydration process. *Proceeding 2010 concrete sustainability conference, national ready mixed concrete association, USA*, p 1–15. (2010)
8. E. Silitonga, D. Levacher, S. Mezazigh, Utilization of fly ash for stabilization of marine dredged sediment, *Revue Européenne de Génie Civil et Environnement,* **Vol 14** Issue 2, p 253-265, (2010)
9. P.J.P Gleize, A. Müller, et H.R. Roman, (2003). Microstructural investigation of a silica fume –lime mortar, *Cement and Concrete Composites*, **Vol.25**, pp.171-175.
10. European Standard, NF EN 14227-3, Hydraulically bound mixtures. Specifications. Fly ash bound mixtures, *European Committee for Standarization* (2013).
11. D. Colin, Valorisation de sediments fin de dragage en technique routière. *Thesis doctorat, Université de Caen, France*, 181p (2003)
12. K. Siham, B. Fabrice, A. Nor Edine, D. Patrick, Marine dredged sediment as new material source for road construction, *Waste Manaegement* **no.28,** p 919-218 (2008).
13. AFNOR NF X 31-210, Essai de lixiviation relatif a l'entreposage des dechet industriels stabilizes dans des sites de stockage *Normalisation Francaise* (1992).

An Investigation of Damage Factors in Industrial Scale of Light-Weight Bricks Production

Kiki Dwi Wulandari[1,*], and *Januarti* Jaya Ekaputri[1,2]

[1]Civil Engineering Department, Faculty of Civil Engineering and Planning, Institut Teknologi Sepuluh Nopember (ITS), Indonesia
[2]Konsorsium Riset Geopolimer Indonesia (KORIGI), Laboratorium Beton dan Bahan Bangunan, 60111 Kampus ITS Sukolilo Surabaya, Indonesia

Abstract. A damage case of light weight brick's production approximately at 6-7% of total production was found daily in one industry in East Jawa, Indonesia. The physical damage of product always occured. This paper investigates some factors that affect the lost in laboratory analysis. The analysis includes the chemical analysis of raw materials, reactivity of pozzolanic materials, and observation of strength based upon the position of light-weight bricks during autoclaving process. In addition, fly ash is introduced as mineral additive as one of the alternatives to improve the product's quality. It is also concluded that grinding the silica sands particles is the optimum way to improve the quality, but the adding class F of fly ash to five percent in mixture is the most effective solution. Furthermore, maintaining quality of raw materials, curing process, and maintaining the machine will either reduce the product damage that occurred during the fabrication process in industry.

1 Introduction

Light-weight bricks usually called Autoclaved Aerated Concrete (AAC) is normally manufactured from siliceous materials (quartz-rich sand), calcareous materials (lime and cement) and calcium sulphate with traces of aluminium powder as foaming agent to produce porous [1-6]. These components are mixed with water under saturated steam pressure includes the hydrothermal treatment process at high-temperature (typical, 180°-200° C) for several hours [4, 5, 7]. The foaming agent reacts with calcium hydroxide or alkali to form hydrogen gas bubbles, then it contributes to produce high porosity in AAC [1, 6]. A damage case of light weight brick's production approximately at 6-7% of total production was found daily in one industry in East Jawa, Indonesia. The physical damages that were occurred during AAC productions as illustrated in Fig. 1-3.

* Corresponding author: kiki.dwi.wulandari@gmail.com

 DOI: 10.1051/matecconf/201713801018

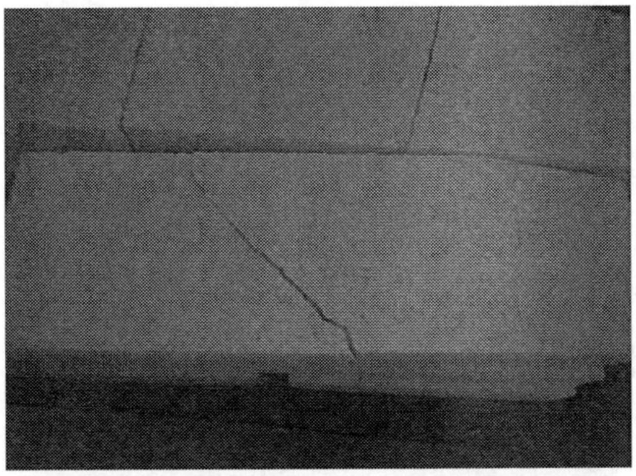

Fig. 1. Crack of AAC.

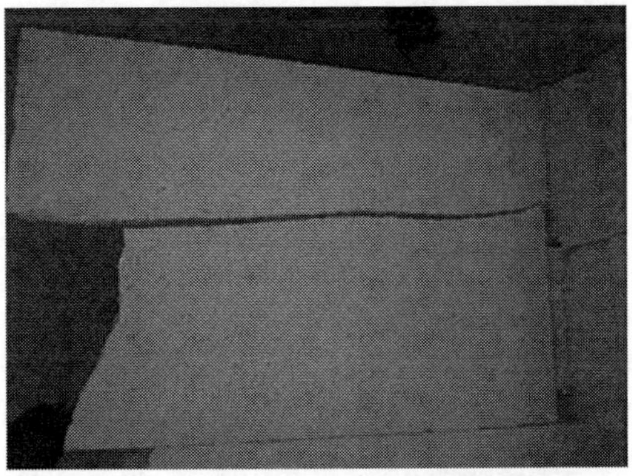

Fig. 2. Unsymmetrical shape of AAC.

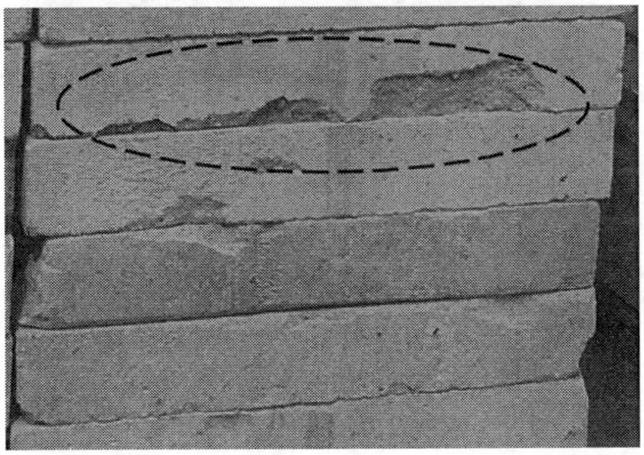

Fig. 3. Edge crack of AAC caused by re-bond after cutting process.

The physical properties of AAC are influenced by its structure, mineralogical composition, and texture. The structure of AAC can be set up by control the addition of foaming agent. Even so, mineralogical composition and texture of AAC, depend on composition and reactivity of raw materials [6]. The influence of mineral composition of raw materials was studied by many authors [1, 2, 3, 4, 6, 8, 9, 10]. This paper investigates some factors that affect the damage case in laboratory analysis. It may involve some further analysis, both mechanical and physical. The analysis includes the chemical analysis of raw materials, reactivity of pozzolanic materials, and observation of strength based upon the position of light-weight bricks during autoclaving process.

2 Materials

Raw materials in this research were obtained from two industries in East Jawa Indonesia. Silica sand SS, gypsum GY, lime CA, and aluminum powder AP, from light-weight brick's industry. The materials were collected in four batches. A class F of Fly Ash was collected from a fertilizer factory in East Jawa, Indonesia. The name of materials was based on their sources and batches.

3 Experimental Test

X-Ray Fluoressence (XRF) and X-Ray Difractometer (XRD) were set up to analyze the chemical composition oxides, and mineral content, respectively in all raw materials. Reactivity test was conducted to analyze the reactivity of pozzolanic materials, based on ASTM C 593-95. The reactivity analyses were applied to silica sands and fly ash. The best silica sand was grinding and compared with fly ash to replace the ordinary silica sand by 5%, 10%, and 15%, then reactivity test were conducted to analyze each other. While the compressive strength were observed based on the distribution of light-weight bricks in autoclave machine were observed at 28 days. Specimens were collected from four autoclaves in different position. They were picked up from upside, center, and below-side (A, B, and C), and closed to the autoclave machine door, center, and backside (D, T, and B), as Fig. 4 below. For each block, there were five specimens (60 cm x 20 cm x 10 cm per specimens). Compressive strength test were applied to two specimens which were chosen separately by the same block notation. It means there were 18 specimens for each autoclave.

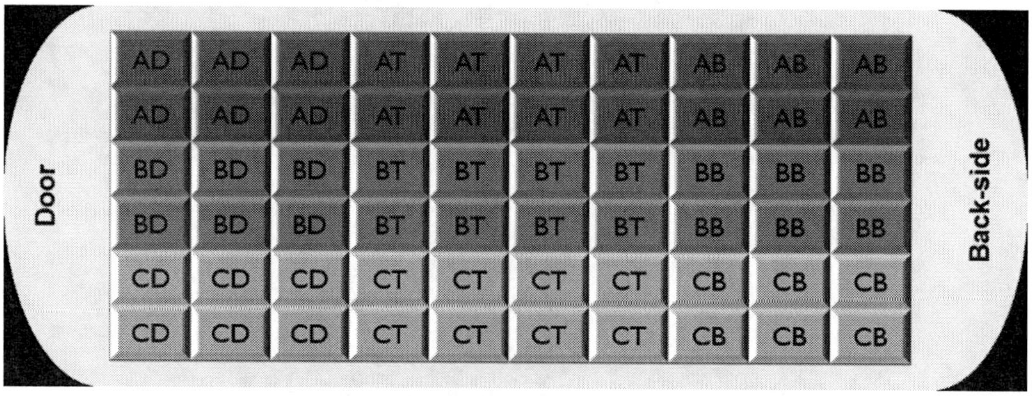

Fig. 4. Identification of brick's location depends on distribution in autoclave machine.

4 Result and Discussion

4.1 XRD Analysis

X-ray diffraction results are shown in Table 1. Silica sands SS consist of quartz mineral are similar to 100%. It was perfect composition, but reactivity of quartz must be compared with reactivity test.

Table 1. Chemical major compositions of silica sand (%) by XRD.

Silica Sand	Quartz (%)
SS1	100
SS2	100
SS3	100
SS4	100

GY3 dan GY4 are mingled with silicon oxide (SiO_2) in the source more than 5% as data on Table 2. All of gypsum materials were bothered by its content. Silicon oxide or quartz were obtained from clay on gypsum sources. Purities of gypsum were bothered by its mineral. Mixtures of AAC were influenced by its condition.

Table 2. Chemical major compositions of gypsum (%) by XRD.

Gypsum	Gypsum (%)	Silicon Oxide (%)
GY1	96	4
GY2	98	2
GY3	88	11
GY4	91	9

The data of XRD analysis implies that the mineral compositions of lime samples are inconsistent. The Table 3 shows the major compositions were obtained by XRD Analysis. Calcite in CA3 and CA4 exist on the large amount, 51% and 72%, respectively. Calcite presents the impurities of samples. Quality of lime was influenced by its mineral and bothered reactivity material while mixing process. The best lime is CA2 with Calcium Oxide at 74% and no content of calcite.

Table 3. Chemical major compositions of lime (%) by XRD.

Lime	Calcium Oxide (%)	Calcite (%)	Calcium Hydroxide (%)
CA1	46	21	32
CA2	74	-	26
CA3	21	51	7
CA4	7	72	14

4.2 XRF Analysis

The chemical composition oxides of raw materials with XRF analysis are listed in Table 4. The number on the material's code indicated different date of material collections. The chemical composition of silica sands SS in different batch and quarry but in the same sources are similar on SiO_2 and Al_2O_3 content. The averages of its content are 97.46% and

1.02%, respectively. According to data on Table 4, the SiO_2 contents are above 97%, but it didn't represent the reactivity of materials clearly.

Table 4. Chemical compositions of silica sand (%) by XRF.

Silica Sand	Al_2O_3	SiO_2	Fe_2O_3	Cr_2O_3	TiO_2	K_2O	CaO	MgO	MnO_2	Na_2O
SS1	1.10	97.25	0.15	0.01	0.15	0.18	0.06	0.10	0.01	0.14
SS2	0.88	97.85	0.12	0.01	0.10	0.12	0.06	0.09	0.01	0.15
SS3	1.15	97.05	0.21	0.01	0.17	0.10	0.16	0.10	0.01	0.09
SS4	0.96	97.67	0.15	0.01	0.03	0.15	0.05	0.10	0.01	0.14

Gypsum in GY1, GY2, GY3, and GY4, show the different on $CaSO_4$ $2H_2O$ content as listed at Table 5. The amount of its contents are 87.34%, 94.22%, 77.49%, and 86.72%, serially. The setting time during the solidification process on light-weight bricks were influenced by its content. The data elucidate the inconsistency of the material's quality. According to the data on Table 5, SiO_2 and Al_2O_3 are present as the composition of all gypsum. Its content indicates the impurities of gypsum and influenced the mixing process.

Table 5. Chemical major compositions of gypsum (%) by XRF.

Gypsum	Al_2O_3	SiO_2	$CaSO_4$ $2H_2O$
GY1	0.64	5.18	87.34
GY2	0.14	2.81	94.22
GY3	0.36	6.69	77.49
GY4	0.16	2.88	86.27

Quality of light-weight bricks were influenced by the usage of lime, which consist of Calcium Oxide (CaO). The CaO content on CA1, CA2, CA3, and CA4 as present at Table 6, are 79.52%, 85.47%, 60.56%, and 61.85%, respectively. The average of its content is 71.85%. CA2 and CA3 have $CaCO_3$ content, which is higher than 50%, but $CaCO_3$ is natural lime and unreactives to cement and silica sand. Calcium which is unreactives to cement and silica sand, have an effect to compressive strength of AAC in early phase. The overdue in solidification process and crack on early time were caused by its condition.

Table 6. Chemical major composition of lime (%) by XRF.

Lime	$Ca(OH)_2$	$CaCO_3$	Total CaO
CA1	60.72	9.81	79.52
CA2	44.09	5.35	85.47
CA3	30.62	53.42	60.56
CA4	37.71	59.42	61.85

Alumunium powder is a foaming agent to form hydrogen gas bubbles in mixture. The chemical composition show the dominant content is Al_2O_3, higher than 99% at all as shown in Table 7. Water content on AP2 and AP3 are 26.67% and 36.72%, but for AP1 is only 0.36%. Alumunium powder is reactive material. The reactiveness of aluminium powder in mixture were influenced by water content and storage humidity related to stock up duration.

Table 7. Chemical major compositions of aluminium powder (%) by XRF.

Alumunium Powder	Al_2O_3
AP1	99.57
AP2	99.49
AP3	99.37

4.3 Reactivity Test

The quality of AAC related to mineralogical composition, and texture were influenced by the composition and reactivity of raw materials. Reactivity tests were applied to pozzolanic materials [11, 12], and were indicated by its compressive strength at 7 days. The result of reactivity of raw material SS is illustrated in Fig.5.

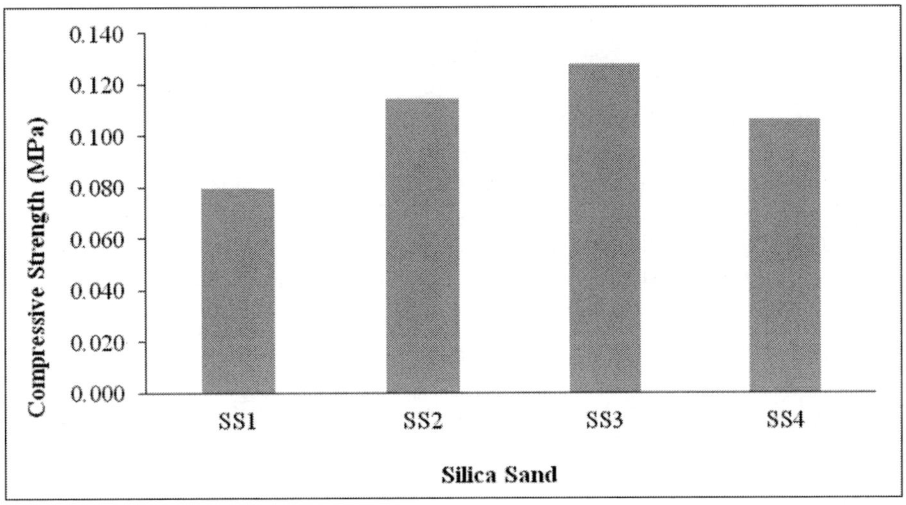

Fig. 5. Compressive strength of silica sand's reactivity.

Generally, Fig.5 described that silica sands as raw material is unreactives. Bonded of silica sand to cement and lime were influenced by its particle size. Coarses particle were indicated that silica sand in AAC composition only as filler and unreactives. Fig.5 shows that SS3 is the best reactive silica sand with 0.128 MPa of compressive strength. By its condition, the improvements were applied. The SS3 was chosen as material in the next experiment. The best silica sand SS3 was grinding to be SSH by 2-4 hours and compared with fly ash FA to replace the ordinary silica sands (OSS) by 5%, 10%, and 15%. Its results are shown in Fig. 6.

DOI: 10.1051/matecconf/201713801018

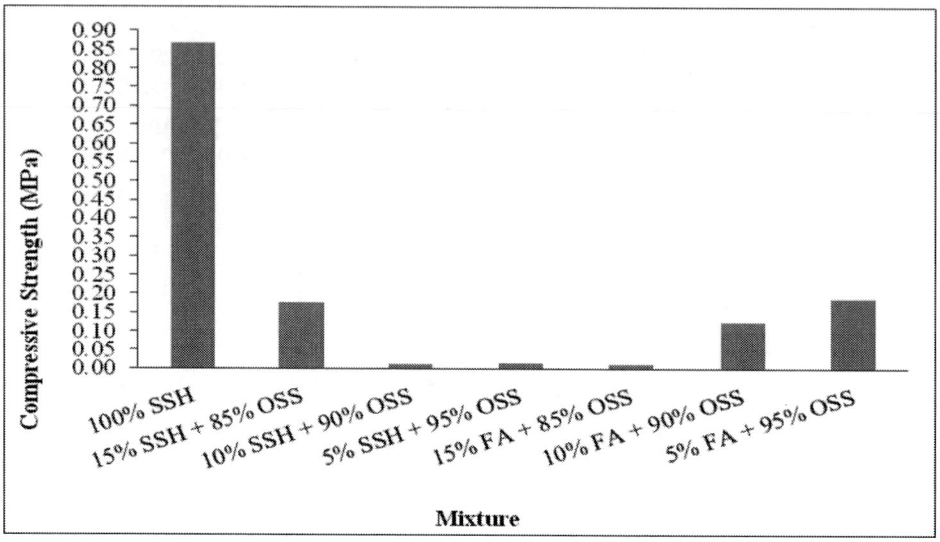

Fig. 6. Compressive strength of recommendation mixture.

From the data above in Fig.6, 100% OSS replaced by SSH shows the best result of compressive strength, and follows by 5% OSS replaced by FA in mixture. Grinding the sands contributes to increase the reactivity of mixture to be more than five times. The data illustrated that the compressive strength increase while the fineness increase. It implies that finer quartz content in silica sands and fly ash, increases the compressive strength of reactivity materials. The finer quartz which has higher surface area could supply larger amounts of dissolved silica. It was same occurred on cement paste partially replaced by fly ash [13] and relationship of fineness and surface area [14]. The utilization of fly ash reduced the amount of silica sands in mixture. The content of silica and alumina in fly ash increased the stiffness of AAC. The utilization of fly ash easier than grinding the silica sand, because it's no need extra machine and time. The fly ash can be used as the alternative siliceous raw materials for the production of AAC [15]. Fly ash were also obtained during AAC productions in industry, but it needs further analysis to determine the class of fly ash.

4.4 Compressive Strength Based on Distribution in Autoclave Machine

The differentiation of compressive strength was influenced by the distribution of specimens in autoclave machine. In this analysis, the strength based on the distribution of light-weight bricks in autoclave machine were observed at 28 days. Samples were collected from four autoclave machines, namely AC1, AC2, AC3, and AC4, in different position as Fig. 4. Before autoclaving process, AAC were prepared as Fig. 7.

Fig. 7. Preparation before autoclaving process.

Generally, the quality related to compressive strength is supported by the maturity of AAC samples. The data were illustrated in pattern as Fig.8-11.

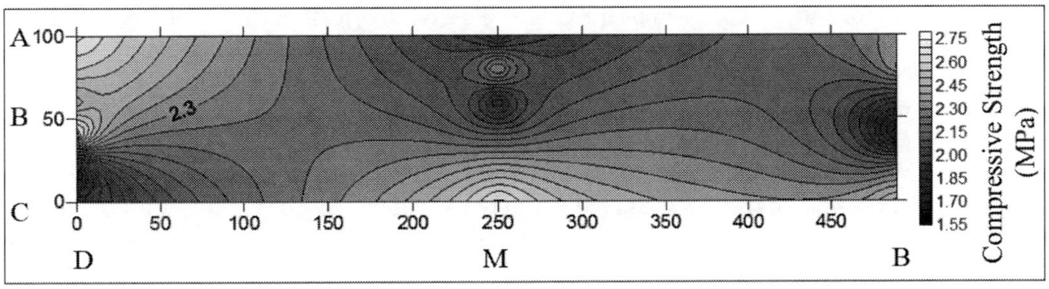

Fig. 8. Contour of compressive strength in autoclave machine AC1.

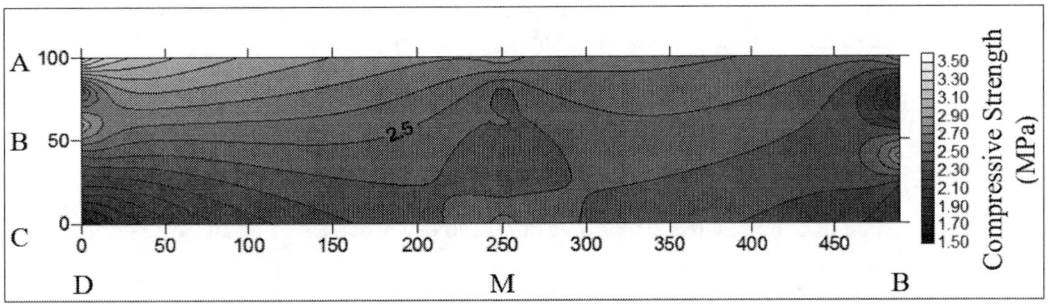

Fig. 9. Contour of compressive strength in autoclave machine AC2.

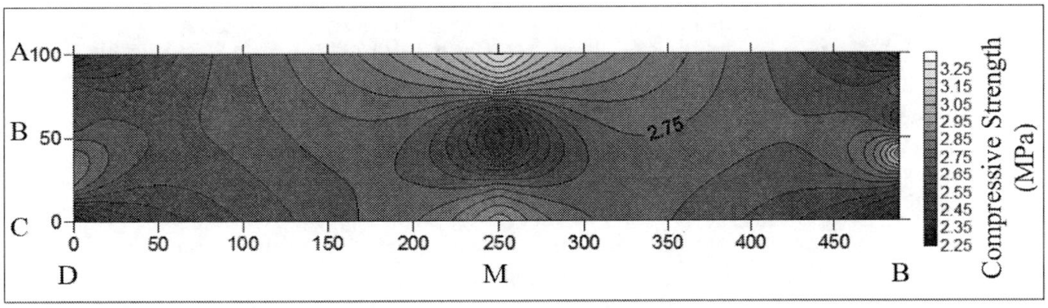

Fig. 10. Contour of compressive strength in autoclave machine AC3.

DOI: 10.1051/matecconf/201713801018

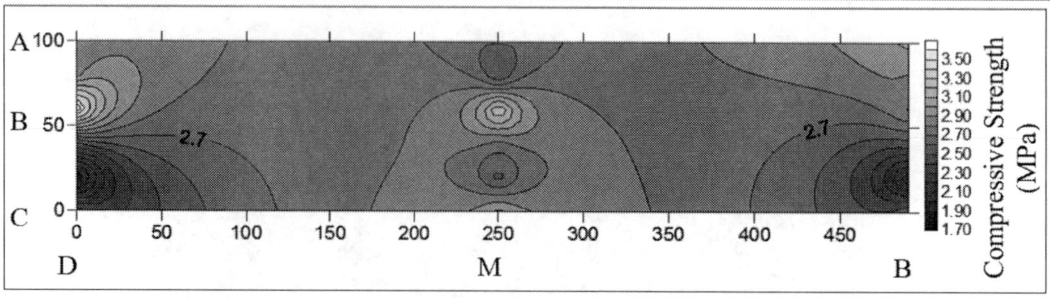

Fig. 11. Contour of compressive strength in autoclave machine AC4.

The strength of AAC samples are similar on AC1, AC2, AC3, and AC4. The averages are 2.22 MPa, 2.41 MPa, 2.65 MPa, and 2.71 MPa, respectively. The highest compressive strength was occurred in AC4, and the lowest was found in AC1. The patterns in Fig. 8-11 indicate variety of AAC quality based on the distribution of samples in autoclave machine during curing process. According to the pattern in Fig.8 and Fig.11, AC1 and AC4 indicate the similar condition. Up-side position contributes the higher compressive strength than back and below-side. Darker pattern represents the smaller strength. Its conditions were found at all autoclave machines, especially closed to the door. It was caused by inconsistency of autoclave machine pressure. The darker patterns in AC1 and AC3 are shown at the center of machine were caused by the composition and maturity of AAC samples. The pattern of compressive strength related with the distribution of samples in autoclave machine represents the pressure quality during the autoclaving process. It needs maintenance process to keep the machine in proper condition. Physical properties of products were caused by its conditions and supported by mineralogical composition and reactivity of raw materials. It was agreed with Hausser et al [6].

5 Conclusion

1. The raw material which is used in this experiment were inconsistent in quality, even they were taken in same source and quarry.
2. Source and storage system affect the moisture of the raw materials. It will cause the difference of material's quality.
3. The compressive strength according to reactivity of materials is more affected by material's fineness and surface area.
4. Replacing ordinary silica sand with 100% finer silica sands in amount deliver the best reactivity and compressive strength. But the extra time to grinding silica sand were needed by 2-4 hours.
5. Replacement of silica sand with 5% FA is the optimum recommendation as it shows the high compressive strength without extra time. Reactivity materials increases the quality of AAC products.
6. The utilization of fly ash as by-product materials in autoclaving process during AAC production reduce the problem with waste, but it need further analysis according to the class of its fly ash.
7. The compressive strength related to the distribution of samples in autoclave machine represents the pressure quality during autoclaving process and the quality of autoclave machine. The maintenance of autoclave machine should do regularly.
8. Physical damage of products were caused by machine conditions and supported by mineralogic composition and reactivity of raw materials. Furthermore, maintaining quality of raw materials, curing process, and maintaining the machine will either reduce the product damage that occurred during the fabrication process in industry.

This research has been supported by Lembaga Penelitian dan Pengabdian Masyarakat ITS (LPPM ITS) and Lembaga Pengelola Dana Pendidikan Republik Indonesia (LPDP RI).

References

1. K. Kunchariyakun, S. Asavapisit, K. Sombatsompop, "Cem. Con. Comp.", **55**, 11 (2015)
2. Y. Song, B. Li, E.H. Yang, Y. Liu, T. Ding, "Cem. Con. Comp.", **56**, 51 (2015)
3. M. Sinica, G.A. Sezeman, D. Mikulskis, M. Kligys, V. Cesnauskas, "Const. Build. Mater.", **50**, 718 (2014)
4. K. Matsui, J. Kikuma, M. Tsunashima, T. Ishikawa, S. Matsuno, A. Ogawa, M. Sato, "Cem. Con. Res.", **41**, 510 (2015)
5. N. Isu, H. Ishida, T. Mitsuda, "Cem. Con. Res.", **25**, 243 (1995)
6. A. Hauser, U. Eggenberger, T. Mumenthaler, "Cem. Con. Res.", **29**, 297 (1999)
7. G. Schober. *ZKG Int.*, **58**, 63 (2005)
8. R. Drochytka, J. Zach, A. Korjenic, J. Hroudová, "En. Build"., **58**, 319 (2012)
9. O. Koronthalyova, "Const. Build. Mater.", **25**, 879 (2011)
10. A. Laukaitis, J. Keriene, D. Mikulskis, M. Sinica, G. Sezemanas, "Const. Build. Mater.", **23**, 3034 (2009)
11. American Society of Testing and Materials C593-95, *Standard Spesification for Fly Ash and Other Pozzolans for Use with Lime, United States*, (2000)
12. JJ. Ekaputri, H. Limantono, Triwulan, T. E. Susanto, M. M. A. B. Abdullah, "Key. Eng. Mater.", **673**, 83 (2016)
13. P. Chindaprasirt, C. Jaturapitakkul, T. Sinsiri, "Cem. Con. Comp.", **27**, 425 (2005)
14. R. Nurwidayati, M.B. Ulum, J.J. Ekaputri, Triwulan, P. Suprobo, "Mater. Sci. Forum", **841**, 118 (2016)
15. N. Huberman, D. Pearlmutter, "En. Build". **40**. 837 (2008)
16. L.N. Halim, J.J. Ekaputri, "Mat. Web. Conf.", **97**, 01002 (2017)

MATEC Web of Conferences **138**, 01019 (2017) DOI: 10.1051/matecconf/201713801019
EACEF 2017

The effect of steam curing on chloride penetration in geopolymer concrete

Januarti Jaya Ekaputri[1, 2,*], *Inne* Syabrina Mutiara[2], *Siti* Nurminarsih[3], *Nguyen* Van Chanh[4], *Koichi* Maekawa[5] and *Davin* H. E. Setiamarga[6*]

[1] Department of Civil Engineering, Institut Teknologi Sepuluh Nopember, Surabaya, East Java 60111, Indonesia
[2] Indonesian Consortium for Geopolymer Research, Institut Teknologi Sepuluh Nopember, Surabaya, East Java 60111, Indonesia
[3] Logistic Engineering Department, Indonesian International University for Cement Studies, Gresik, East Java 61122, Indonesia
[4] Faculty of Civil Engineering, Ho Chi Minh City University of Technology, Ho Chi Minh City, Ho Chi Minh 70000, Vietnam
[5] Department of Civil Engineering, The University of Tokyo, Bunkyo-ku, Tokyo, 113-0033, Japan
[6] Department of Applied Chemistry and Biochemistry, National Institute of Technology, Wakayama College, Gobo City, Wakayama 644-0023, Japan

Abstract. In this paper, we present the result of our study on the effect of steam curing to chloride ion penetration in geopolymer concrete. Class F fly ash was activated using sodium hydroxide (NaOH) and sodium silicate (Na_2SiO_3). The concrete specimens were then steam-cured at 40°C, 60°C, 80°C and room temperature at 24 hours. The treatment was followed by wet curing for 28 days, and then followed by immersion of all specimens in salt water for the durations of 30, 60, and 90 days. Cylindrical specimens were then prepared for compressive strength, chloride ion penetration, pH, and porosity tests. A 16 mm-steel bar was fixed at the center of the specimen concrete blocks (specimen size: 10cm x 10cm x 15cm). Corrosion probability was determined by conducting Half Cell Potential test. Our result showed that increasing the curing temperature to 80°C induced chloride ion penetration into the concrete's effective pores, despite improvements in compressive strength. We also found that chloride ingress on the geopolymer concrete increases commensurately with the increase of the curing temperature. The corrosion potential measurement of geopolymer concrete was higher than OPC concrete even if corrosion was not observed in reinforcing. Based on our result, we suggest that the corrosion categorization for geopolymer concretes needs to be adjusted.

1 Introduction

Seawater greatly affects the strength of concrete for construction, especially on offshore buildings. High salt content gradually decreases the strength of concrete, causing a decrease on the durability of concrete significantly. Such breaches on concrete strength and

* Corresponding author: januartije@gmail.com and davin@wakayama-nct.ac.jp

durability greatly affect the building's service life, and bring safety concerns at the same time. Previous studies have shown that the compromise of concrete strength caused by seawater is due to aggressive penetration of the chloride ions (Cl^-) from the saltwater into the concrete [1]. Further Cl^- ion penetrations deep into concrete pores was then followed by ion exchange reactions between the concrete and chloride ion, causing rebar corrosion and bond-lost between the concrete and the reinforcing materials (e.g. steel rebar). Therefore, in order to reduce the level of chloride ion penetration, concrete mixed with high volume of fly ash is commonly used to build concrete structures in aggressive environments [2, 3].

Previous studies have shown that geopolymer-concrete mix ("Geopolymer Concrete") has a strong resistance to corrosion [4]. Class F of fly ash, which is rich in silica and alumina, is usually used as the main binder in manufacturing of geopolymer concrete. Meanwhile, a liquid alkali solution containing NaOH and Na_2SiO_3 is commonly used to activate the fly ash in the concrete mixtures.

Chloride induced corrosion of the steel rebars in concrete is one of the most concerning durability issues for structures located at aggressive environments. Previously, the corrosion resistance of the reinforcement bar in geopolymer concrete have been studied by some researchers. They set to the effect of alkali solution or additional minerals in the mixtures [5, 6]. It is found that the curing system, especially with steps involving drying, or wetting by immersion in water, influences the water adsorption level of geopolymer concrete. There are some curing methods to improve geopolymer concrete performance. However, only a few number of research have focused on the effect of steam curing on geopolymer concrete to resist chloride ion penetration.

Curing with oven, steam, wet or moist is the most recommended techniques. Differences in temperatures and curing durations may change concrete properties including the characteristics of concrete pores [7, 8]. Past studies have confirmed that steam curing at a temperature range of 60°C-80°C contributes to faster strength development at early age [9] and promotes the reaction of geopolymerization [10, 11]. On the other hand, some studies have suggested that moist curing is more recommended than wet curing [12]. Leaching of alkaline and some compound of alumina-silicates lead to the formation of more pores on the concrete surface. Porous concrete is an indication of a modification of microstructures in concrete, which changes its ability to resist chloride ion.

In this study, we report the result of our detailed investigation into the effect of moist curing using steam on the strength and durability of geopolymer concrete and its rebar, in order to elucidate if such method is effective to prevent the negative effect of chloride ion penetration. We believe our result presented here will provide a very useful insight to solve one of the key durability issues of geopolymer concrete.

2 Materials and method

First, we prepared fresh geopolymer concrete consisted of fly ash, sodium hydroxide (NaOH), sodium silicate (Na_2SiO_3), coarse aggregate, and fine aggregate. To guarantee mixture workability, superplasticizer (SP) from poly-carboxylate material was then applied, which amount was adjusted to be 2% weight-percentages against the weight of fly ash. Na_2SiO_3 composed of 13.7 % Na_2O; 17% SiO_2; 36.4% H_2O. NaOH flake was diluted with distilled water to obtain a 8 molar solution. The ratio of Na_2SiO_3 to NaOH in alkali solution for all mixtures was kept constant at 2.5. Class F fly ash (ASTM C 618) 618 [13] was collected from a power plant in Gresik, Indonesia. The chemical composition of the ash by XRF testing is listed in Table 1.

DOI: 10.1051/matecconf/201713801019

Table 1. Composition of chemical content of fly ash (% by mass).

SiO_2	Al_2O_3	Fe_2O_3	TiO_2	CaO	MgO	Cr_2O_3	K_2O	Na_2O	SO_3	Mn_2O_3
48.47	26.05	12.54	0.92	5.18	2.77	0.02	1.66	0.47	1.05	0.19

The weight ratio of the fly ash-containing binder to the total aggregate is 1:3. In addition, the binder has a mass ratio of 65% fly ash and 35% alkali activator. Geopolymer concrete specimens were cured in a temperature-humidity control room. The chamber was set to vary at 40°C, 60°C, and 80°C for 24 hours. Specimens made of Ordinary Portland Cement (OPC) were prepared as a control system. One day after casting, all specimens were demolded and cured at a specific temperature in the chamber. The specimens were then subjected to wet curing in tap water for 28 days before exposing to the chloride-induced system. The composition of concrete is shown in Table 2.

Table 2. Composition of mixtures (Kg per m^3).

Code	Fine aggregate	Coarse aggregate	Fly ash	OPC	NaOH	Water	Na_2SiO_3	SP	Curing Temperature
GN	720	1080	390	-	60	-	150	7.8	Room temp.
G40	720	1080	390	-	60	-	150	7.8	40°C
G60	720	1080	390	-	60	-	150	7.8	60°C
G80	720	1080	390	-	60	-	150	7.8	80°C
OPC	666	999	-	440	-	220	-	-	Room temp.

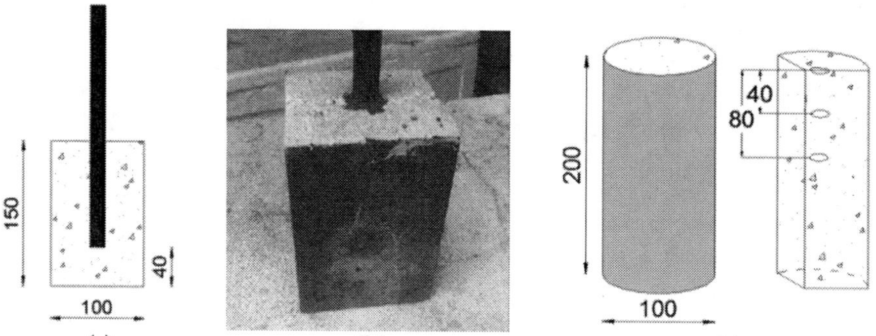

(a) (b) (c)

Fig. 1. (a) size of lollypop specimen in mm, (b) coating at top side, (c) cylindrical specimen.

Three identical lollypop-shape specimens with the dimension of 100 mm wide, 150 mm long, and 40 mm deep (Fig 1(a), (b)) were used as samples in the chloride induction study. Each block of geopolymer concrete specimens has a 16 mm diameter-steel bar set as rebar. The top side of concrete and the bar were coated to prevent direct chloride penetration. The corrosion activity is monitored by measuring the copper/copper sulphate ($Cu/CuSO_4$) half-cell potential according to ASTM C-876 [14]. In each variable, three identical cylinders of 100 mm in diameter and 200 mm in height were prepared for the comparison samples in some tests, such as compressive strength, chloride content, pH and porosity (Fig. 1(c)). Except for the top and bottom side, the surface of all cylinders was covered with a waterproof coating. The specimens were subjected to wet curing in water containing 3.5% NaCl salt for 30, 60 and 90 days. The saline water was replaced with fresh salt solution every week. Tests for chloride penetration depth and pH were targeted at some points in each cylindrical specimen as illustrated in Fig 1 (c).

3 Results and discussion

3.1 Compressive strength

The result of the compressive strength test is presented in Table 3. Because steam curing induces faster concrete maturity, the 28 days specimen G80 showed the highest strength, with the optimal curing temperature at 80°C. Temperature increase is known to increase compressive strength, because it accelerates geopolymerization reaction [15]. This also means that heating causes microstructural properties changes in geopolymer concretes. In our previous study, we reported mechanical properties and elastic modulus changes of geopolymer concrete, due to steam curing, it was found that the mechanical properties change and the elastic modulus on geopolymer concrete due to steam curing. Water evaporation and self-desiccation during the heat curing stage was also reported by Wardhono et al [16]. They found that the Si/Al ratio in the matrix decreased with age indicating an on-going geopolymerization process beyond the 90 day time period. In this condition, continuous gel formation filled up the voids. This is most likely the explaination of the increasing of strength observed in the GN and G40 specimens.

Table 3. Compressive strength of cylindrical specimens.

Age (days)	28	60	90	120
Immersion time (days)		30	60	90
Code	Compressive strength (MPa)			
GN	51.1	64.7	67.5	69.5
G40	63.3	60.2	63.5	70.9
G60	59.3	62.5	68.0	66.2
G80	78.7	78.9	80.5	74.5
OPC	46.6	52.3	60.1	58.1

In Fig. 2, the compressive strength of all specimens at the age of 120 days is actually higher than the strength at 28 days. However, the strengths of specimens GN60 and GN80 increased until the 60th day of immersion, but they decreased gradually until the 90th day of immersion. In addition, the strength of OPC specimen, which was cured in room temperature, shows the same tendency to decrease after the 60th day of immersion. This is due to chloride ion penetration into the concrete pores and precipitation forms as a reaction of chloride ion with calcium mineral constituents, causing a drastic decrease of strength in OPC concrete specimens.

Improved strength of geopolymer specimens when cured in salt water is believed to be related to reduced leaching of alkali activator from the geopolymer concretes to the salt water curing [17]. This mechanism is greater than the effect of intrusion of natrium contained in salt water into the geopolymer specimens. Saline water shield solution for geopolymer most likely help to prevent leaching, and thus explains strength increase when the concrete is subjected to salt water. After casting followed by steam curing, all samples were cured in tap water for 27 days. This might cause alkali leaching of all geopolymer samples. This caused more pores in concrete surface formed generally during wet curing. After 28 days, when all samples were cured in salt water, the leaching is apparently decreased gradually and possibly prevented by excess Na^{+} cations in the surrounding curing system. At later age, decreased strength is associated by formation of inorganic salts in aluminosilicate gel due to ion exchange. Giasuddin et al reported that at ambient temperature, saline water is better for geopolymer curing when compared to freshwater

 DOI: 10.1051/matecconf/201713801019

curing system, because important reactant elements might be leached out from the geopolymer sample to the curing medium in the freshwater system.

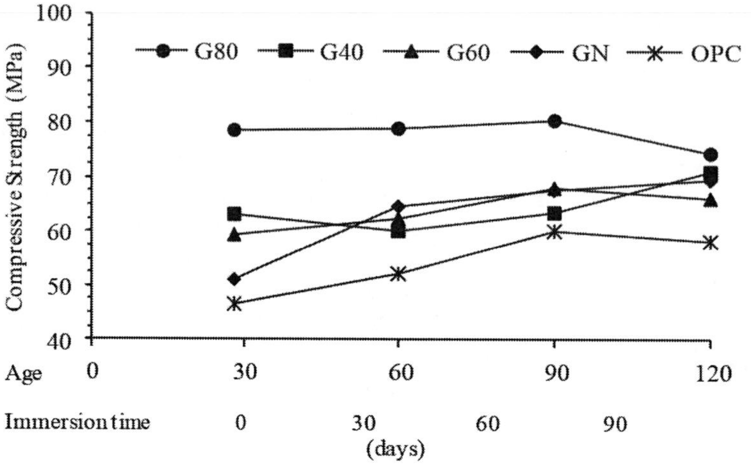

Fig. 2. Compressive strength of specimens curing in salt water.

3.2 Porosity test

Samples for porosity test were selected at the same sample for compressive strength test. Total porosity and closed porosity of all samples are presented in Fig. 3 (a) and Fig 3 (b) respectively. Closed porosity is a percentage of pore space measured by boiling the saturated concrete for five hours. Porosity is the total pores contributed by closed and effective porosity (Fig 3 (c)). However, effective pores have more influence to promote total porosity formation. These results are in good agreement with Kovalchuk et al [18], where the total porosity of 15% was found on geopolymer products subjected to steam curing.

Due to greater reaction occurred in steam curing, all geopolymer samples cured at 40°C, 60°C, and 80°C showed less porosity than specimen GN and OPC at the age of 28 days. Heating mixture contributed not only to the increase of the geopolymer concrete strength, but it also limited the pore size, especially of the effective pores, in the geopolymer paste. During steam curing, geopolymerization process occurred rapidly. When small quantity of water involved in the mixture, which was excluded in the reaction, it was entrapped and filled the closed pores. This is different from reaction occurred in the geopolymer specimen cured at room temperature. Since the reaction is slower, more fly ash particles, aluminosilicate gel, and silicate gel remained un-reacted. Total pores of GN sample is greater than steam cured specimens is associated with leaching of some unreacted constituents to surrounding water from the age of two days to 28 days. Olivia and Nikraz suggested a linear correlation between porosity increase with the decrease of compressive strength [19]. Significant change of porosity is shown by OPC sample in Fig 3(a). This contributes to the decrease of concrete strength immersed for 90 days in salt water.

 DOI: 10.1051/matecconf/201713801019

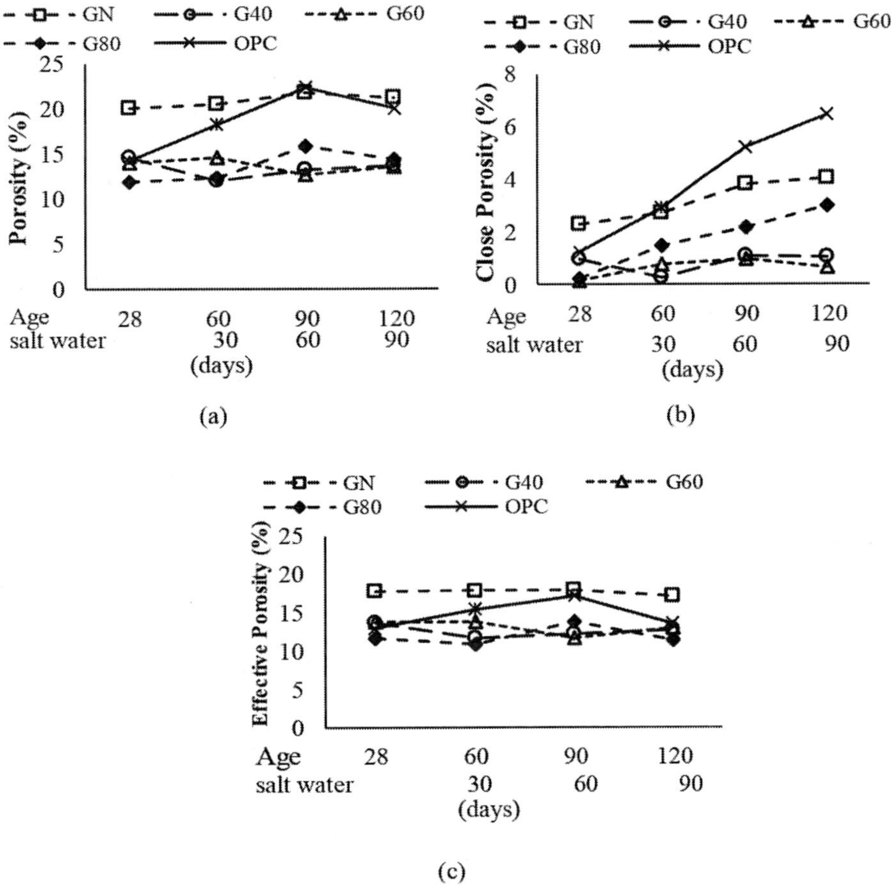

Fig. 3. Porosity test results: (a) Total porosity, (b) closed porosity, and (c) effective porosity.

3.3 Chloride penetration

Fig. 4 presents the chloride content of the specimens at a depth of 0 cm, 4 cm, and 8 cm from concrete surface. The chloride content at concrete surface was quite high, especially at the steamed-curing specimens. The maximum chloride content at G40, G60, and G80 were in a range of 1.2%, 1.5%, and 1.4% respectively. Reduction of chloride content at a depth of 4cm and 8cm was observed but the chloride content is much higher than specimens cured in room temperature. The lowest chloride concentration among the samples was performed by OPC at a depth of 8 cm, which was only 0.04%.

High chloride concentration in the geopolymer matrix could be due to the absence of a chloride binding mechanism or chloride ion exchange in the sodium-aluminosilicate (N–A–S–H) gel system. It is observed at samples G40, G60, and G80. Active chemical reaction in geopolymer concrete exposed to steam curing was observed to happen at early ages (24 hours). This caused early maturity of reactions, causing the subsequent continuous reaction to be very much slower at higher temperature. Smaller size of effective pores formed in the matrix was expected to increase water absorption. This is slightly different from samples where the continuous reaction between alkali activator and un-reacted fly ash particle gradually occurred with the presence of both cation Na^+ as shield solution and ion Cl- ingress to the pores. A similar result was observed by Olivia and Nikraz [20]. When geopolymer concrete was cured at 60°C - 75°C, geopolymer concrete showed better mechanical properties than OPC concrete and lower porosity, but chloride ion concentration was found much higher in geopolymer concrete.

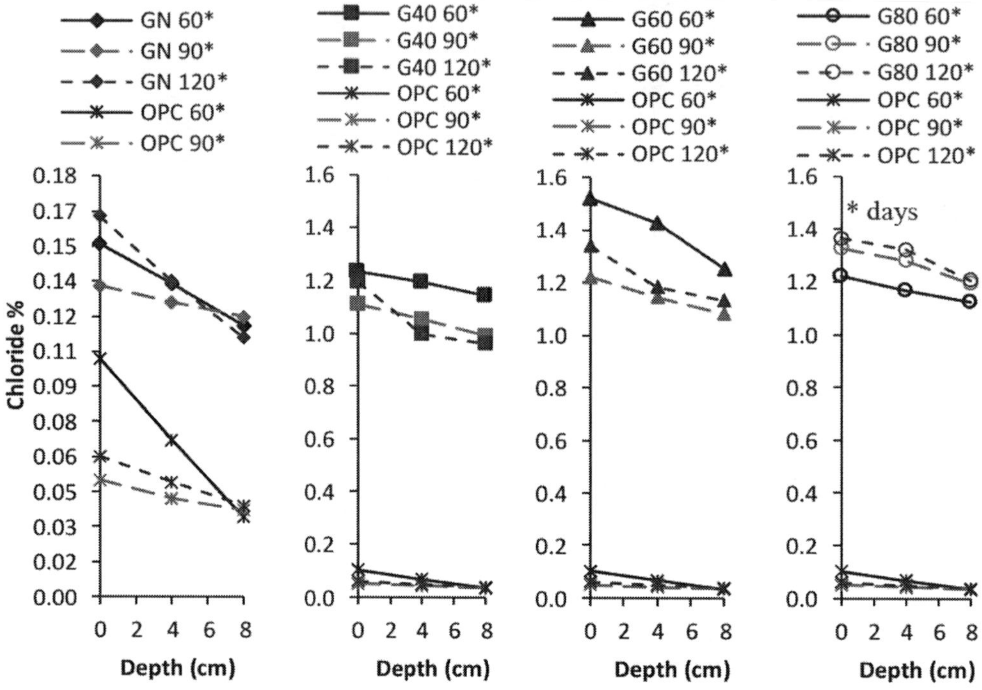

Fig. 4. Chloride ion penetration at several distances from the concrete surface.

3.4 pH in concrete

pH measurement is related to chloride ingress to penetrate concrete pores. Fig. 5 shows that pH of geopolymer specimens is lower than pH of OPC. Besides the fact that geopolymer has lower pH than OPC, water curing caused leaching contributed to concrete pH decreased. Higher pH at the deeper surface of geopolymer specimens subjected to heating shows contradictory results from chloride penetration. When the chloride ingress was found greatly in the samples, the pH is surprisingly similar to that at GN sample. It indicates that chloride ion is not involved during the formation of aluminosilicate gel to lower pH of geopolymer product. A common low pH of geopolymer concrete compared to OPC concrete was also reported by Babaee and Castel [21].

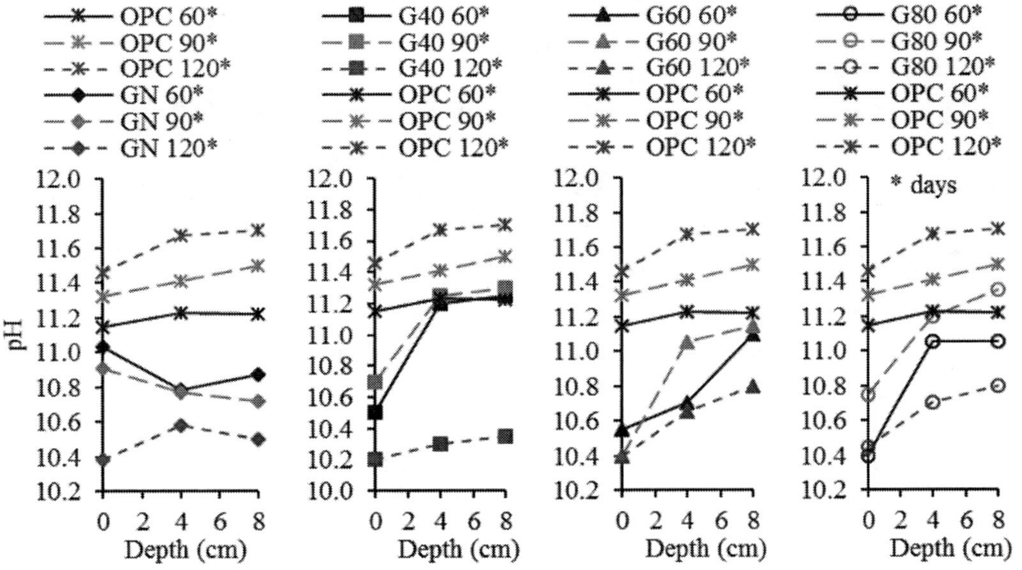

Fig. 5. pH measurement.

3.5 Corrosion potential

Corrosion potential was measured using $Cu/CuSO_4$ electrode mounted on the end of the 16mm diameter reinforcement steel bar of the lolypop test specimen. Measurements are made by connecting the rebars and electrodes to the high impedance digital multi meter.

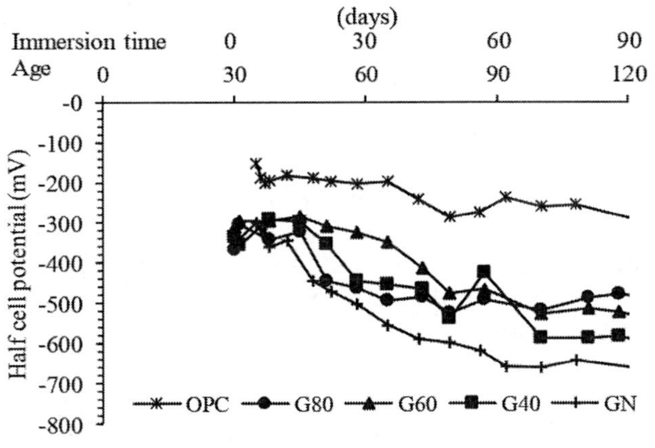

Fig. 6. Half-cell potential measurement.

Fig. 6 indicates that the overall potential of corrosion increases at longer duration of immersion. It is possibly due to the effects of longer exposure time to chloride ions penetration. The half-cell potential of geopolymer concrete at the beginning reached the range between -300 to -500 mV. At the end of immersion period, it was in the range of -470mV to -670mV. All geopolymer concretes showed negative corrosion rate potentials greater than -500 mV, which were more negative than those of the OPC concrete samples. It is apparently caused by the chloride ions penetration. According to ASTM C 876, this potential value is categorized as 90% probability of corrosion. As a result, we can conclude that the lower initial alkalinity of geopolymer concrete certainly leads to more negative corrosion potential values.

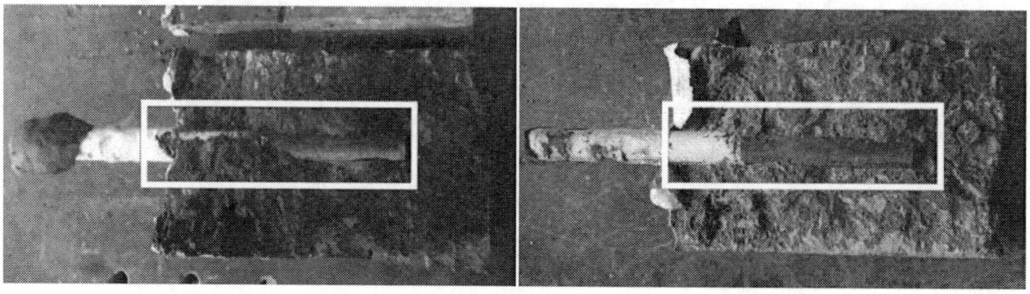

(a) (b)

Fig. 7. After curing in salt water for 90 days (a) G80 specimen, (b) OPC specimen.

Fig. 7 shows the physical condition of reinforcing bars embedded into the geopolymer concrete (a) and OPC concrete (b) after immersing in salt water for 90 days at an ambient temperature. Surprisingly, no corrosion was observed around the steel bar. Although the half-cell potential value is categorized as 90% probability of corrosion in geopolymer specimens, there is no rust on the steel bar in the concrete. This could be considered as one of the evidence that geopolymer concrete strength remains high at the end of immersion time. From this finding, we would like to suggest that the conventional method of the classification system of concrete and rebar corrosions, which was developed based on observations and research on Portland cement-based specimens, needs to be adjusted for geopolymer-based systems.

4 Conclusions

1. Steam curing at 80℃ for 24 hours resulted in the highest compressive strength of geopolymer concrete at 28 days. This is because fly ash reacts slower at room temperature than steam curing system. Steam curing increased the kinetics and reaction rates during the geopolymerization process, causing the development of denser pore system, to produce good mechanical properties
2. After immersion for 90 days in saline solution or at 120 days of age, compressive strength of concrete cured at room temperature increased by 32-36% of its 28-day strength. Strength of geopolymer concrete subjected to higher temperature increased by 12-33% of its 28-day strength. This caused mainly by leaching of Na^+ as the fly ash activator was prevented by salt water, which surrounded the concrete, resulted in continuous reaction of geopolymerization during immersions.
3. Although the compressive strength of geopolymer concrete increased due to steam curing, chloride concentration also increased. The corrosion rate value using half-cell potential for all geopolymer concrete was in the range of -470 to -670 mV, which was categorized as 90% probability in corrosion. However, no rust was found on the reinforcing concrete of geopolymer. The high potential-value is probably due to the low pH of the concrete and the high concentration of chloride in the concrete. Based on this finding, we suggest that different standardized values and settings are needed to measure the corrosion potential of geopolymer concrete.

References

1. J.J. Ekaputri, and Triwulan, "Forensic engineering: From failure to understanding". (2008)

2. T. Cheewaket, C. Jaturapitakkul, and W. Chalee, "Construction and Building Materials", **37**, 693-698 (2012)
3. V. Saraswathy, and H.W. Song, **51 (22)**, 4601-4611 (2006)
4. K. Zerfu, and J.J. Ekaputri, Materials Science Forum, **841**, 162-169 (2016)
5. M. Rostami, and K. Behfarnia, "Construction and Building Materials", **134**, 262-268 (2017)
6. F.U.A. Shaikh, "Advances in concrete construction", **2 (2)**, 109-123 (2014).
7. J. J. Ekaputri, K. Maekawa, and T. Ishida, "Materials Science Forum", **857**, 305-310 (2016)
8. J.J. Ekaputri, S. Junaedi, and Wijaya, Procedia Engineering, **171**, 572-583 (2017)
9. M.M.A. Abdullah, et al., "Advanced Materials Research", **328-330**, 1475-1482 (2011)
10. D. Hardjito, et al., "Civil Engineering Dimension", **6 (2)**, 88-93 (2004)
11. J.J. Ekaputri , et al., "Materials Science Forum", **803**, 63-74 (2014)
12. Halim, L.N., J.J. Ekaputri, and Triwulan, MATEC Web of Conferences, **97** (2017)
13. American Society for Testing and Material, ASTM, *C618 – 03,* (2003)
14. American Society for Testing and Material ASTM, *C876 – 09*, (2015)
15. Triwulan, J.J. Ekaputri, and N.F. Priyanka, MATEC Web of Conferences (2016)
16. A. Wardhono, et al., "Construction and Building Materials", **143**, 272-279 (2017)
17. H. M. Giasuddin, J. G. Sanjayan, and P.G. Ranjith, "Fuel", **107**, 34-39 (2013)
18. G. Kovalchuk, A. Fernández-Jiménez, and A. Palomo, "Fuel", **86 (3)**, 315-322 (2007).
19. M. Olivia, and H.R. Nikraz, ARPN Journal of Engineering and Applied Sciences, **6 (7)** 70-78 (2011)
20. M. Olivia, and H.R. Nikraz. The Concrete Institute of Australia (2011)
21. M. Babaee, and A. Castel, "Cement and Concrete Research", **88**, 96-107 (2016)

MATEC Web of Conferences **138**, 01020 (2017) DOI: 10.1051/matecconf/201713801020
EACEF 2017

The Application of PVA Fiber to Improve the Mechanical Properties of Geopolymer Concrete

Muhammad Lutfi Manfaluthy[1,2,*] and *Januarti Jaya* Ekaputri[1,2]

[1]Institut Teknologi Sepuluh Nopember, Civil Engineering Department, 60111 Surabaya, Indonesia
[2]Konsorsium Riset Geopolimer Indonesia (KORIGI), Concrete and Building Material Laboratory Institut Teknologi Sepuluh Nopember, 60111 Surabaya, Indonesia

Abstract. This paper presents an experimental investigation on the improvement of geopolymer concrete properties through the use of polyvinyl alcohol (PVA) fibers mixed in the fresh concrete. For the purpose of obtaining the optimum mechanical properties, the volume fraction of PVA fibers was varied at 0% ; 0.3% ; 0.5% ; ad 0.8%. All mixtures were cast by mixing fly ash, alkali activator, and aggregates. The activator used in this study was a combination of sodium silicate (Na_2SiO_3) and sodium hydroxide (NaOH). The mechanical properties of geopolymer concrete were obtained from the results of compressive strength, splitting strength, uniaxial tensile strength, elastic modulus, and flexural strength. It is found that the variation of 0.8% PVA fibers resulted in the highest strength for overall test. The utilization of 0.8% PVA fibers also contributed to increasing the direct tensile up to 50%. However, it is noticed that the elastic modulus was more prone to decrease as the fiber content in the mixture increased.

1 Introduction

The application of fly ash in the concrete mixture improved concrete performance. It is because the voids between particles are filled by additional hydrated products [1], in this case, fly ash has exquisite particle size. Geopolymer is formed from chemical reaction rather than hydration reaction like in Portland cement concrete based. Geopolymer concrete is well known to be brittle, several researches established by adding some fibers in geopolymer mixture is one of the solutions to improve mechanical properties and flexural strength. Bhantia and Gupta [2] found that added cellulose based fiber in the concrete mixture, can be used as repair concrete strengthening and reduced the crack width. In [3] added 0.2% kenaf fiber by volume friction, slightly improved the compressive strength as compared the bare lightweight concrete. Likewise, Manfaluthy et al [4] detected by adding abaca fiber in geopolymer mixture improved the compressive strength up to 8% from the non-fiber geopolymer concrete. Based from that reasons, the addition of fiber in the geopolymer mixture can be properly applied in geopolymer concrete. In this paper using Polyvinyl Alcohol (PVA) fiber to improve the mechanical strength varied at 0%; 0.3%; 0.5% and 0.8%

* Corresponding author: Lmanfaluthy@gmail.com

by volume friction. Li [5] was found that the bond-strength of PVA fibers and Portland cement matrices was very high. It resulted that tensile strain-hardening with strain capacity of engineered cementitious concrete (ECC) in excess of 4%.

2 Materials and Test Specimens

2.1 Fly ash

A class F of fly ash was collected from Paiton Coal Fire Power Plant, Probolinggo, Indonesia as raw material. The density of fly ash is 2.6 gram/cm^3 and the chemical composition is provided in Table 1.

Table 1. Chemical composition of fly ash by XRF (% by weight)

Oxides	SiO_2	Al_2O_3	CaO	Fe_2O_3	K_2O	MgO	Na_2O	SO_3
%	43.24	24.43	10.74	11.37	1.12	3.27	2.97	1.13

2.2 Alkali Activator

There are two types of alkali activators were used in this paper, sodium silicate (Na_2SiO_3) and sodium hydroxide (NaOH).The mass ratio of sodium silicate to sodium hydroxide was 2.5 and 14 Molar of sodium hydroxide. The silica to alumina ratio of 1.8 was applied in the binder composition.

2.3 PVA Fiber (Polyvinyl Alcohol Fiber)

PVA fiber has good properties to improve concrete strength. It has high elasticity modulus, durability, tensile strength, and bond-strength. These properties is require for increasing concrete ductility. Properties of PVA fiber are 38 μm in diameter, with eight mm in length, and density of 1.3 gram/cm^3 as shown in Fig 1.. Modulus elasticity of PVA fiber is in the range of 25 - 40 GPa, is higher than natural fiber. The elongation of PVA fiber is 6-10%, and the tensile strength of PVA fiber is in range of 880 - 1600 MPa.

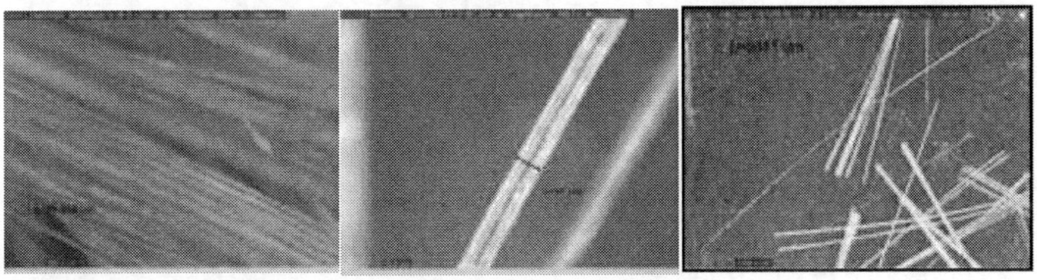

Fig. 1 Properties of PVA fiber which taken with DinoLite camera.

2.4 Aggregates

Aggregates that used in this paper is commonly used in concrete production. Fine aggregate must satisfied ASTM C33 [5]. Coarse aggregate should have maximum size of 20 mm. Both fine and coarse aggregates were prepared in saturated surface dry (SSD) condition.

2.5 Test Specimens

Compressive test to acquire the strength of the specimens was conducted according to ASTM C39 [6]. Splitting test according to ASTM C496 [7]. Both compressive and splitting test specimens were conducted for cylindrical shape with 10cm in diameter and 20cm in height. Dog bone shape specimens were conducted to direct tensile test according to CRD-C 260-01 [8]. Flexural test according ASTM C348 [9] conducted by prism specimen with size 4cm in width, 4cm in height, and 16cm in length. All specimens were cure in ambient room temperature at 28 days.

3 Mix proportions

Specimen mixing proportion is provided in Table 2. All the materials were prepared one day before mixing process. Alkali activator was separated. First, mixed fly ash, sand and gravel. Second,it added 14 M of sodium hydroxide until well-mixed. Third, added sodium silicate in the mixture. Then spread PVA fiber at the end of stage until homogeneously.

Table 2. Mixing proportion (% by volume)

Code	Fly ash	Na_2SiO_3	NaOH (14M)	Sand	Gravel	PVA fiber (by volume fraction)
GC	22%	5.7%	2.3%	28%	42%	-
FGP3	22%	5.7%	2.3%	28%	42%	0.3%
FGP5	22%	5.7%	2.3%	28%	42%	0.5%
FGP8	22%	5.7%	2.3%	28%	42%	0.8%

4 Results and Discussions

4.1 Compressive Strength Test

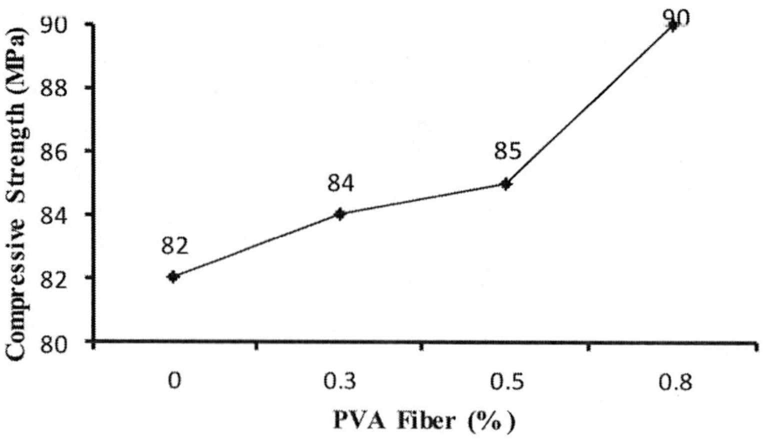

Fig. 2. Contribution of fiber variations to compressive strength.

Fig. 2. shows the addition of PVA fiber in geopolymer concrete increases slightly the compressive strength. It has a good agreement with some previous research [3, 4, 11]. It was found that natural fibers in cementitious material slightly increased the compressive strength. The maximum compressive strength was obtained 90 MPa, by 0.8% of the PVA fiber specimens. Accordingly, Ekaputri [12] found that the addition of PVA fiber in meta kaolin-geopolymer mixture increased the compressive strength. Opposite with Borges et al [13] that the adding of PVA fiber in geopolymer mixture unaffected significantly on the compressive strength. From Ekaputri [12] reported that the addition of PVA fibers as much as 1% on geopolymer paste has a tendency increased the compressive strength. Li et al [14] stated that the optimum of PVA fibers for concrete ECC at 2% when analyzed from the viscosity of the mixture and the distribution of PVA fiber. Generally, the addition of PVA fiber in geopolymer mixture is not significantly contributed in the compressive strength, it caused the properties of PVA fiber was un-resist the compressive load.

4.2 Splitting Test

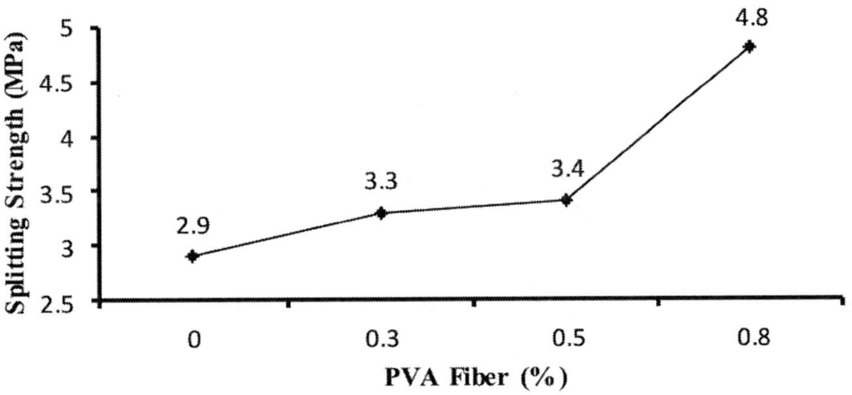

Fig. 3. Splitting strength caused PVA fiber variations.

Fig.3. describes PVA fiber in the mixture increases the splitting strength. The 0.8% of PVA fiber specimen generated splitting strength 4.8 MPa, which is obtained the maximum splitting test. Splitting strength results of fibrous specimens obtained up to 60% of non-fibrous specimens. It was occurred due to the mechanical properties of PVA fiber is more resisted to splitting load, because there was a strong bond between the PVA fiber and the geopolymer matrix. As known that the behavior of geopolymer concrete was very brittle. Adding some fibers in mixture increases the ability of concrete to resist the tensile loads. It aligned with Dehghan [15], due to the brittleness of Portland concrete material, fiber is required to improve the splitting strength up to 25%.

4.3 Direct Tensile Test

Fig. 4. Shows the relation stress and strain from direct tensile test, by adding fiber in geopolymer mixture increased the strength. The specimen contains high volume of fiber is illustrated in Fig. 5.. It indicates that after crack occurs, tensile stress is transferred from concrete matrices to the fiber as it has a great binding capacity with geopolymer matrices. Adding 0.3% of PVA fiber significantly rose up to two times of tensile strength resulted from non-fibrous specimens. Accordingly to Ekaputri [11], the tensile strengths of fibrous

DOI: 10.1051/matecconf/201713801020

specimens were greater than the non-fibrous specimen in a meta-kaolin geopolymer system. Tensile capacity was increased up to 52%. Hossein [16] found that, adding polypropylene carpet fiber increased tensile and flexural strength in OPC concrete. It gained 19.6% of basic OPC concrete.

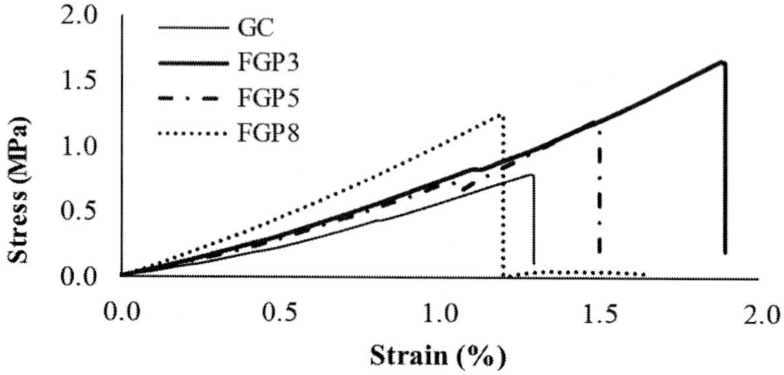

Fig. 4. Stress – strain diagram of direct tensile test.

Fig. 5. Crack pattern in fibrous geopolymer concrete due to tensile load.

4.4 Flexural Test

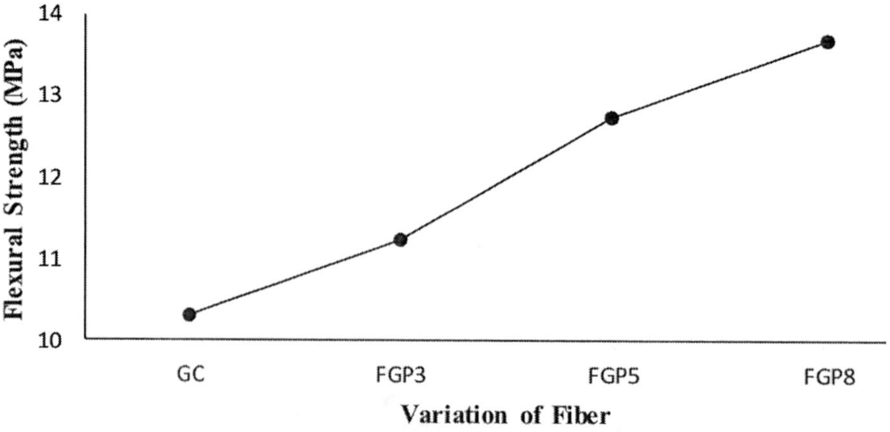

Fig. 6 The effect of fiber adding to flexural strength.

Fig. 6 shows flexural strength of the non-fibrous specimen is gained 10 MPa. The fiber contributed to higher flexural strength as compared to non-fibrous specimen. The flexural strength of fibrous specimen containing 0.3%; 0.5% and 0.8% obtained 11 MPa; 13 MPa and 14 MPa respectively. In general, the application of fibrous specimen containing 0.8% PVA increased the flexural strength up to 30%. Adding some fibers in geopolymer concrete indicated higher strength, it resulted from the compressive, tensile, and flexural strength were increased by adding fibers in the geopolymer mixture. It because after crack occurs in geopolymer matrices, the loads were transferred to PVA fiber. Therefore, fibrous geopolymer concrete accepted higher loads than non-fibrous geopolymer concrete. Bayuaji [17] was found that flexural strength affected by curing system. It suggested oven curing system at 65°C for 24 hours. In the future, combining of PVA fibers adding in the geopolymer mixture and oven curing system must be conducted to produce greater flexural strength.

4.5 Modulus of Elasticity

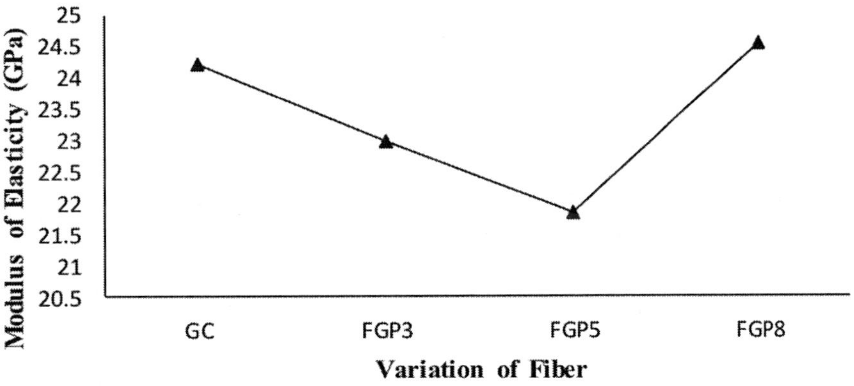

Fig. 7 Modulus of elasticity due to fiber variatons.

From Fig. 7. shows the modulus elasticity test results, it obtained varied results at 14.4 GPa through 24.54 GPa Based Fig. 7 shown that the elasticity modulus of 0.3% and 0.5% of the PVA fiber specimens decreased as compared to 0% specimen. However, addition of 0.8% PVA fiber increased the elasticity modulus. Increasing of elasticity modulus for fibrous specimens caused by the dependent of inter-fiber. It because the fiber covered the micro-cracks and subsequently improved, [18]. It was found that the elasticity modulus of specimens containing 0.4% and 0.5% of polypropylene fiber decreased as compared to non-fibrous specimen. It because the pore formation in the mixtures thus affecting the workability. This is also supported by Karahan and Atis [19], the addition of 0.5% polypropelene fibers in the concrete mixture can increase the modulus of elasticity to 13.3% from the non-fiber concrete mixture.

5 Conclusions

1. The highest strength of the specimens were conducted from 0.8% of PVA fiber specimens. These are, compressive strength increased 9.95% ; splitting strength reached 61.69% ; and direct tensile strength rose 32.78% as compared with non-fibrous specimen. Adding fiber in geopolymer mixture has important role improved the strength of geopolymer concrete.

2. The adding of fiber contributed higher splitting strength and tensile strength. Nevertheless, it showed less contribution to compressive strength. It is because the properties of PVA fiber is only to resist in tensile and flexural load.
3. It is recommended to add 0.5% PVA fiber in geopolymer concrete to improve the mechanical properties of the geopolymer concrete. Whereas, by adding 0.8% PVA fiber improve the strengths of geopolymer concrete, but it is not recommended. It caused by the workability of the mixture very low when added a lot of fiber. Particullary for the fly ash that used in this paper. It contained 10.74% of calcium oxides, which affected to rapidly setting time process for the mixtures.

References

1. Davitdovits, Joseph, *World Resource Review*, **6**(2), pp.263-278 (1994)
2. N. Bhantia, R. Ghupta, *Cement and Concrete Research*, **36**(7), 1263–1267 (2006)
3. M. Manfaluthy, Lutfi, J.E. Ekaputri, Triwulan, *Proceeding of The 6th International Conference of Asian Concrete Federation (ACF 2014)*, 633 – 636 (2014)
4. Manfaluthy, Muhammad Lutfi, Ekaputri, Januarti. Jaya., and Triwulan, *IPTEK Proceeding* (2016)
5. C. Victor, Li., C. Wu, S. Wang, A. Ogawa, T. Saito, "ACI Materials Journal", **99**(5) (2002)
6. ASTM C33, Standard Specification for concrete Aggregates, ASTM (2010)
7. ASTM C39, Standard Test Method for Compressive Strength of Cylindrical ConcreteSpecimens, ASTM Internastional (2013)
8. ASTM C496, Standard Test Method for Splitting Tensile Strength of Cylindrical Specimens, ASTM (2011)
9. CRD-C 260-01, Standard Test Method for Tensile Strength of Hydrulic Cement Mortars (2001)
10. ASTM C348, Standard Test Method for Flexural Strength of Hydraulic Cement Moratrs, ASTM (2013)
11. M. Manfaluthy, Lutfi, J.E. Ekaputri, Triwulan, "Proceeding of The SecondInternational Conference on Sustainable Infrastructure and Built Environment 2013 (SIBE 2013)", 164–171 (2013)
12. Ekaputri et al, *procedia engineering*, **171**, 572 – 583 (2017)
13. H.R. Paulo, Borges, A. Bhutta, L.T. Bavuzo, N. Banthia, *Material and Structures*, **50**, 148 (2017)
14. C. Victor, Li., S. Wang, C. Wu, "ACI Materials Journals", **98**(6) (2001)
15. A. Dehghan, K. Peterson, A. Shvarzman, *Construction and Building Material*, **146**, 238-250 (2017)
16. H. Mohammadhosseini, A.S.M.A. Awal, J.B.M. Yatim, *Construction and Building Material*, **143**, 147-157 (2017)
17. R. Bayuaji, M.F. Nuruddin, S. Francis, J.J. Ekaputri, Triwulan, S. Junaedi, and H. Fansuri, *Materials Science Forum*, **103**, 49-57 (2015)
18. S. Fallah, M. Nematzadeh, *Construction and Building Material*, **132**, 170-187 (2017)
19. O. Karahan, C.D. Atis, *Materials and Design*, 1044-1049 (2011)

MATEC Web of Conferences **138**, 01021 (2017) DOI: 10.1051/matecconf/201713801021
EACEF 2017

Mechanical properties of lightweight aggregate concrete reinforced with soda can waste fibre

Yanuar Haryanto[1,*], *Arnie* Widyaningrum[1], *Gathot* Heri Sudibyo[1], and *Agus* Maryoto[1]

[1]Jenderal Soedirman University, Department of Civil Engineering, Jl. Mayjend. Sungkono KM 5, Blater, Purbalingga, 53371, Indonesia

Abstract. The use of lightweight concrete as building construction material can reduce the building's self-weight and its vulnerability to earthquakes. This paper presents an experimental study on the mechanical properties of lightweight aggregate concrete reinforced with soda can waste fibre. The addition of the fibres may help prevent environmental damage. Twenty-four concrete cylinder specimens with a height of 300 mm and a diameter of 150 mm were cast. The volume fraction of the fibre was varied between 0, 0.3, 0.6, and 0.9%. To facilitate the casting process, superplasticizer was added in a content of 2% by cement weight. We found that the fibrous concrete produced can be categorized as lightweight concrete with a density below 2000 kg/m^3. The mechanical property that experienced the highest improvement, amounting to 61.00%, was the split tensile strength. The optimum fibre volume fraction for better performance in terms of strength is 0.3%.

1 Introduction

Satyarno [1] reported that earthquakes are one of the natural disasters that always come suddenly without any prior warning signs. It cannot be predicted when, where, and with what strength an earthquake will occur. Compared to other natural disasters, earthquakes generally occur in a short period of time of just a matter of seconds, but they can cause damage to building constructions and other infrastructures. Finally, they endanger lives, as shown by the high number of casualties [2, 3]. For example, as reported by GIZ-International Services [4], an earthquake occurred on 4 April 2011, in the early morning at 3:06 am, located about 300 km southwest of Cilacap. The Indonesian National Board for Disaster Management (BNPB) released information about the earthquake on its official website [5]. It was recorded that the earthquake had a magnitude of 7.1 RS at a depth of 10 km. It was found that the damage to houses included 445 collapsed houses, 801 seriously damaged ones, and 1,446 slightly damaged ones. The heaviest damage was found in Bojongsari village, where 102 houses collapsed, 103 were seriously damaged, and 71 were slightly damaged. The number of refugees reached 6043 inhabitants, and the financial losses came to Rp 31.5 billion.

* Corresponding author: yanuar_haryanto@yahoo.com

One of the main causes of the buildings' vulnerability to earthquakes is the use of heavy-weight materials, which can even worsen the performance of the structure [6]. Therefore, lightweight concrete is considered as a solution to reduce the weight of the structure. Lightweight concrete has many and varied applications including multi-storey building frames and floors, bridges, offshore oil platforms, and prestressed or precast elements [7]. It also solves durability problems in buildings and exposed structures. Lightweight concrete has a strength comparable to that of normal weight concrete, but typically 25 to 35% lower, and offers design flexibility and substantial cost savings by providing less dead load, improved seismic structural response, longer spans, better fire ratings, thinner sections, decreased storey height, smaller-size structural members, a need for less reinforcing steel, and cheaper foundation costs. Lightweight concrete precast elements incur lower transportation and placement costs [8]. Many types of aggregates that are classified as lightweight are available, and their properties cover wide ranges. Elastic properties, compressive and tensile strength, time-dependent properties, durability, fire resistance, and other properties of lightweight concrete are dependent on the type of lightweight aggregate utilized in the concrete [9].

Lightweight concrete is defined as concrete with a density of 200–2000 kg/m^3 [10]. Structural lightweight concrete is defined as concrete using lightweight aggregate or a mixture of lightweight coarse aggregate and sand and it has been specified that it should not exceed a maximum density of 1850 kg/m^3 and should meet the requirements regarding compressive strength and tensile strength of lightweight concrete for structural purposes [11, 12]. Generally, lightweight aggregate consists of natural lightweight aggregate and artificial lightweight aggregate. Examples of natural lightweight aggregates include pumice and scoria. Artificial lightweight aggregate is an aggregate created by expanding materials through a heating process, such as slag from the smelting of iron, clay, diatoms, shale, clay stone, perlite, and vermiculite. Akmaluddin [13] argued that because pumice is a natural material of volcanic origin produced by the release of gases during the solidification of lava, it is quite hard in nature and does not deteriorate easily once bound in concrete. Furthermore, he reported that lightweight concrete with pumice aggregate can achieve a compressive strength of 26.6 MPa.

Concrete is the most widely used construction material. However, it has low tensile strength and low energy absorption [14]. An effective method of enhancing the mechanical properties of concrete is by adding a small fraction of short fibres to it [15–17]. Mohammadian and Haghi [18] explained that increasing the amount of fibres in the matrix tends to reduce the compressive strength but provides a substantial increase in flexural toughness and enhances the ability to absorb energy. Furthermore, the reduction in compressive strength of mixtures with fibres may limit their use in some structural applications. With regard to the use of steel fibres, Sivakumar and Sountharajan [19] found that increasing the addition of steel fibres to concrete mixture led to reasonable improvements in tensile strength thanks to the bridging action of the fibres under tension. Heniegal et al. [20] concluded that the addition of 0.5% fibres to concrete increased the indirect tensile strength by about 8–18%. Awal et al. [21] reported that the addition of 0.5 and 1.0% fibres increased the flexural strength by 17 and 1.8% respectively, as well as increasing the split tensile strength significantly and increasing the shrinkage strain slightly compared to normal concrete.

The aim of this work is to present an experimental study of the mechanical properties of lightweight aggregate concrete reinforced by soda can waste fibre. Recent data reveal that over one trillion aluminium soft drink and beer cans have been thrown in the trash rather than the recycling bin [22]. It is estimated that 1,010,000,000,000 (equal to 1 trillion, 10 billion) cans have been wasted since 1972, when the industry started keeping records. Stacked end-to-end, these wasted cans would extend 76 million miles, a distance equivalent to 158 round trips to the moon. A trillion wasted cans weigh in at 17.5 million tons, a quantity of scrap aluminium worth about USD 21 billion at today's market prices. The cumulative environmental damage from the failure to recycle this metal is the real issue. A previous study by Haryanto [23] concluded that the addition of 0.75% aluminium fibres increased the impact resistance of lightweight concrete with artificial lightweight aggregate by 250% for the first crack conditions. Further studies using carpet waste fibres found that the split tensile strength of lightweight aggregate concrete increased up to 75.18% while the flexural strength increased by 7.15% [24, 25]. Utilization of other waste materials to improve the performance of concrete has also been implemented previously [26, 27].

2 Experimental programme

2.1 Material properties

Multipurpose cement was used in this study. A saturated surface dry river sand having a mud content of 3.25% and water absorption of 6.63% was used as a fine aggregate. The fineness modulus and specific gravity of the sand were found to be 3.55 and 2.50 respectively. Artificial lightweight aggregate made from expanded shale [28] with a density of 1050 kg/m^3 having 13.82% water absorption was used as coarse aggregate, as can be seen in Figure 1. The fibres used were obtained from soda can waste (Figure 2) and were 50 mm in length, 2 mm in width, and 0.1 mm in thickness. A control specimen of plain lightweight concrete (PLC) without any soda can waste fibres was also made for comparison. Three fibre volume fractions, V_f, of 0.3, 0.6, and 0.9% were employed in this study. Superplasticizer was added in a content of up to 2% by cement weight to facilitate the casting process.

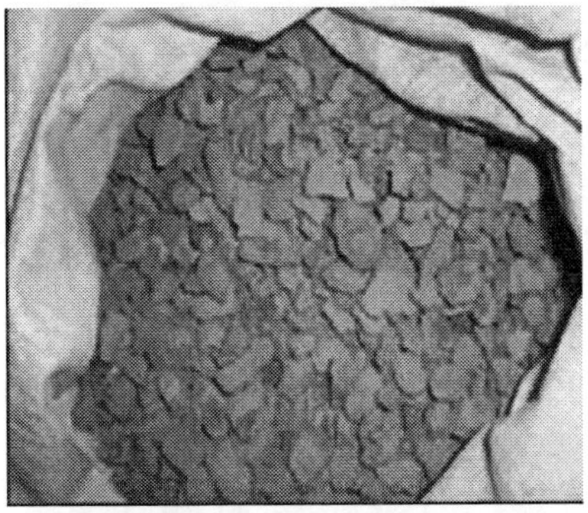

Fig. 1. Artificial lightweight aggregate made from expanded shale. [25]

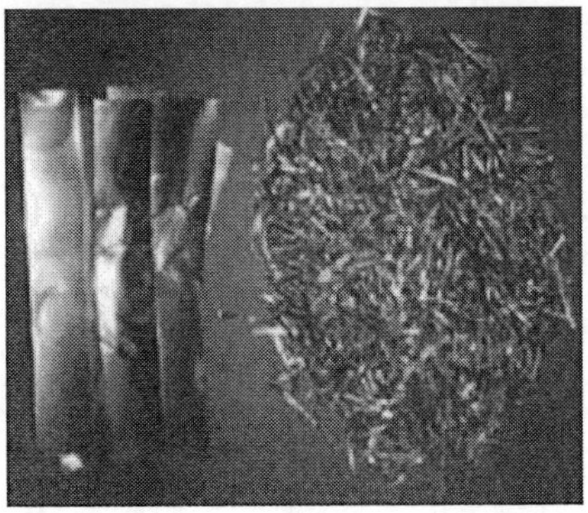

Fig. 2. Soda can waste fibres.

DOI: 10.1051/matecconf/201713801021

2.2 Casting and testing of concrete specimens

The workability of concrete mixes was measured by the slump test in accordance with SNI 1972: 2008 [29]. Compressive strength and split tensile strength tests were conducted on 24 cylinder specimens with a height of 300 mm and diameter of 150 mm at the age of 28 days in accordance with SNI 1974: 2011 and SNI 03-2491-2002, respectively [30, 31]. The modulus of elasticity was measured in accordance with SNI 03-4169-1996 [32]. The compressive strength and modulus of elasticity data were obtained from the same cylinder specimen test simultaneously as presented in Figure 3. Figure 4 shows the concrete cylinder being tested under splitting.

Fig. 3. Compressive strength and modulus of elasticity test. [24]

Fig. 4. Split tensile test. [24]

The compressive strength is determined by Equation 1 while the split tensile strength is determined by Equation 2.

$$f_c' = \frac{P}{A} \tag{1}$$

$$f_t = \frac{2}{\pi}\frac{P}{ld} \tag{2}$$

where:

f_c' = Compressive strength (MPa)

f_t = Split tensile strength (MPa)

A = Area of specimen (mm^2)

P = Maximum load (N)

l = Length of specimen (mm)

d = Diameter of specimen (mm)

3 Results and discussion

3.1 Workability

The fresh concrete properties in terms of workability are presented in Figure 5. As expected, the workability of concrete containing soda can waste fibres in all mixes decreased, with slump values ranging from 1 cm to 4.5 cm. The sturdy fibre–matrix bond and the absence of superplasticizer would have caused this reduction in the slump value of concrete [19]. The addition of superplasticizer affects the slump value significantly because superplasticizer is a water-reducing additive that is effective in improving the workability without having to increase the water content. The range of slump values in all mixes with superplasticizer is 12 to 19 cm.

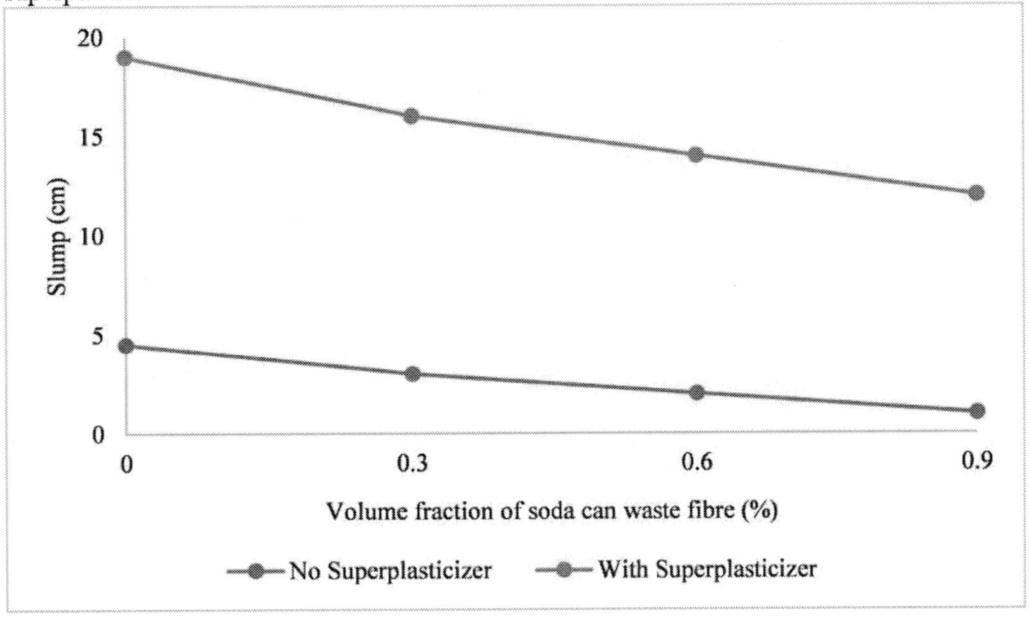

Fig. 5. Workability.

3.2 Density

The use of artificial lightweight aggregate can cause the density of concrete to reach a maximum value of 1940 kg/m^3, so it can be categorized as a lightweight concrete [10]. The addition of soda can waste fibres affects the density of the concrete values. As the addition of the soda can waste fibre increases, the density of concrete increases until it reaches a maximum value. After that, the density of the concrete will drop because the volume fraction of fibre is too high, and as a result the concrete is not solid anymore and contains many pores. The results of the concrete density measurements can be seen in Figure 6.

 DOI: 10.1051/matecconf/201713801021

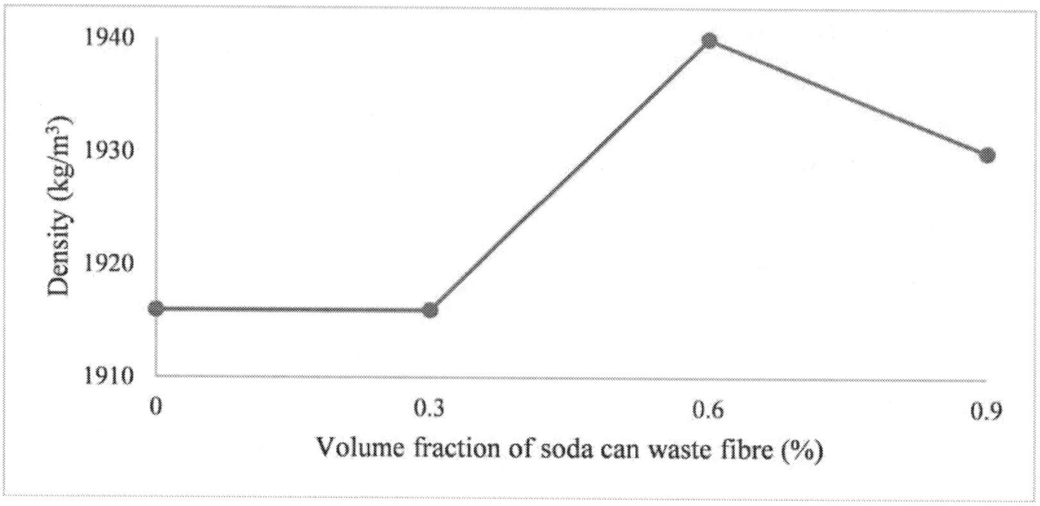

Fig. 6. Density.

3.3 Compressive strength

The test results for concrete at 28 days showed that the compressive strength of lightweight concrete with soda can waste fibre tends to be higher than that of plain lightweight concrete. As can be seen in Figure 7, the maximum compressive strength of 17.36 MPa was found for a 0.3% volume fraction, and this strength value is 37.34% higher than that of plain lightweight concrete. This is caused by the behaviour of the fibre, which can improve the bond between the aggregate and cement-matrix. However, the increase in the volume fraction decreases the workability of the concrete as it tends to become porous, which, in turn, decreases the compressive strength of the lightweight concrete.

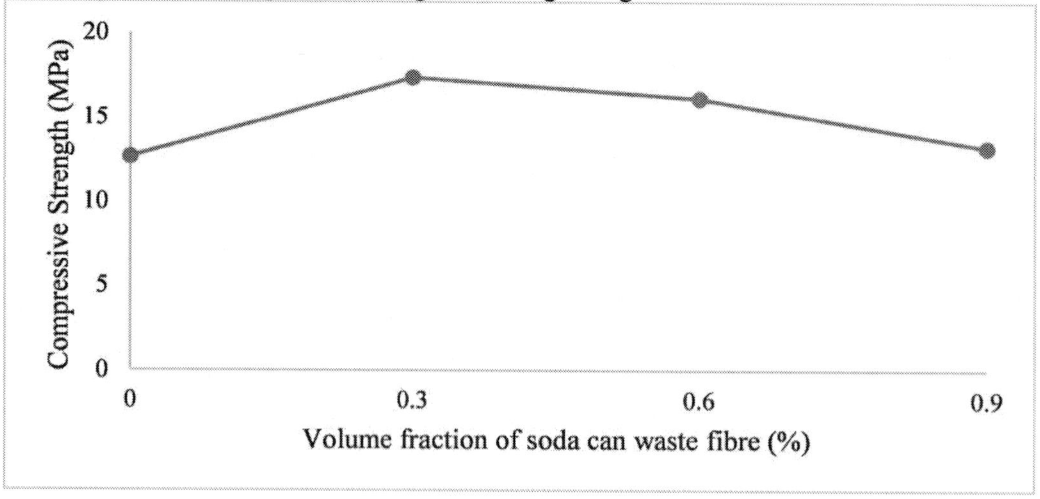

Fig. 7. Compressive strength.

3.4 Modulus of elasticity

The modulus of elasticity is essentially a measurement of the stiffness of a material. The modulus of elasticity of concrete is a key factor for estimating the deformation of buildings and members, as well as a fundamental factor for determining the modular ratio, m, which is used for the design of sections of members subjected to flexure [32]. The modulus of elasticity is often used in sizing reinforced and unreinforced structural members, establishing the quantity of reinforcement, and computing the stress for observed strain values and is especially important in the design of pre-stressed concrete members. So, the modulus of elasticity of concrete is a very important parameter reflecting the ability of concrete to deform elastically. The results obtained indicated that the modulus of elasticity of lightweight concrete increases with the addition of soda can waste fibres to concrete, as shown in Figure 8. The addition of a volume fraction equal to 0.3% gave the maximum value of the modulus of elasticity of lightweight concrete of 14685.50 MPa, representing an increase of 45.50% compared to plain lightweight concrete.

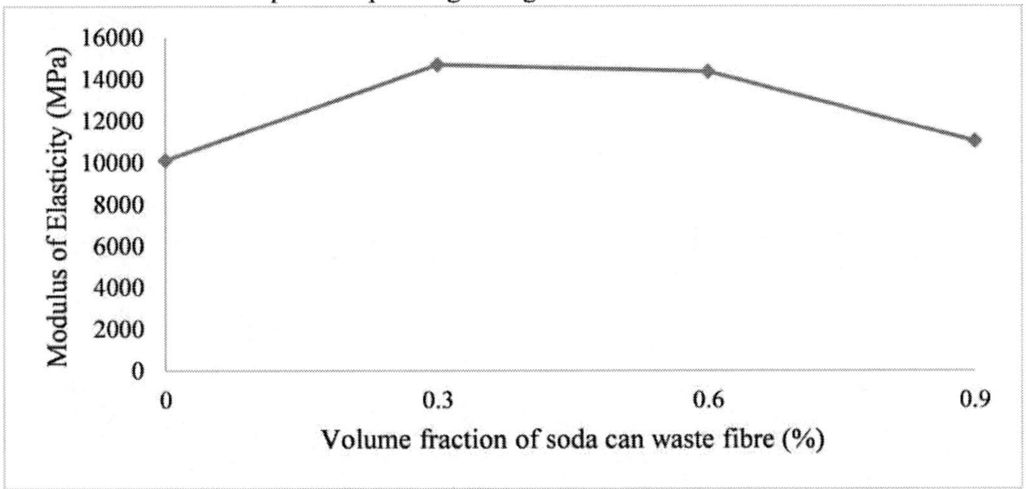

Fig. 8. Modulus of elasticity.

3.5 Split tensile strength

Figure 9 shows that the maximum split tensile strength of 1.98 MPa occurred with the addition of a volume fraction of 0.3%, an increase of 61.5% compared with plain lightweight concrete. In the plain lightweight concrete, the failure occurs suddenly and is not preceded by the early signs. The specimen fracture (Figure 10) was accompanied by a loud explosion. This is very different from the failure of lightweight aggregate concrete reinforced with soda can waste fibre. The specimen did not undergo fracture but only suffered from cracking and remained in one piece (Figure 11). The soda can waste fibres help to bridge and arrest the cracks and restrict the growth of flaws in the matrix, preventing them from becoming enlarged, under stress, into cracks that eventually cause failure.

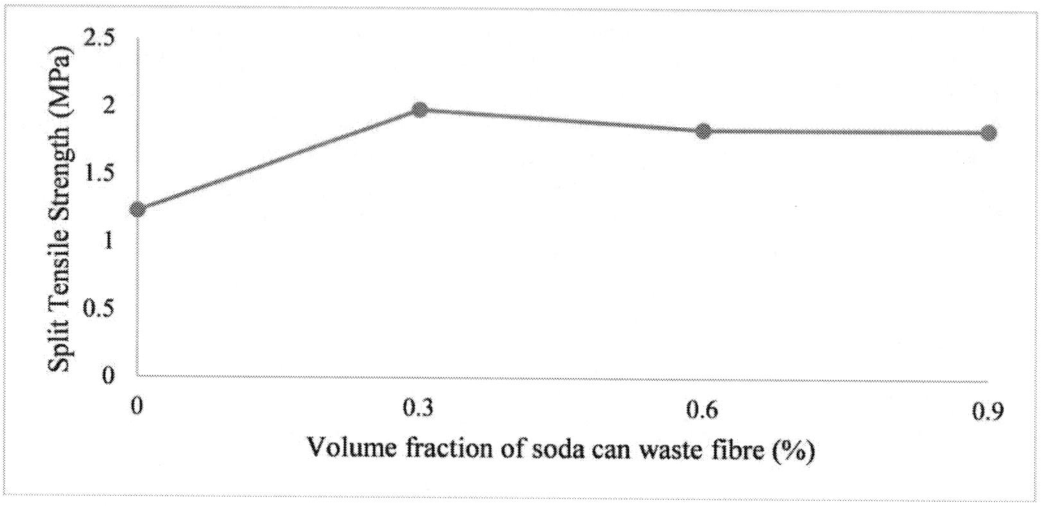

Fig. 9. Split tensile strength.

Fig. 10. Failure of plain lightweight aggregate concrete.

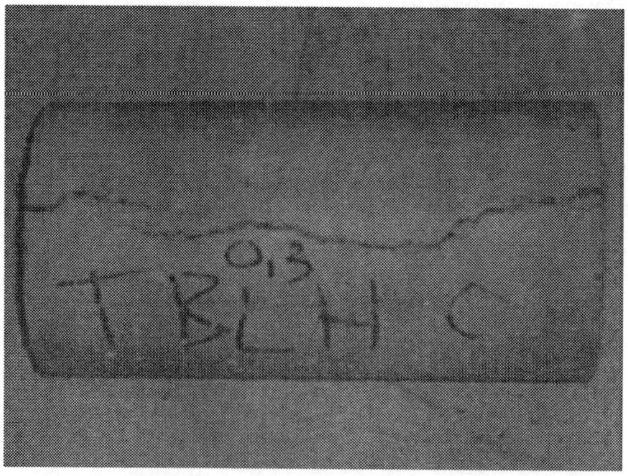

Fig. 11. Failure of lightweight aggregate concrete reinforced with soda can waste fibre.

4 Conclusion

From this experimental study, it can be concluded that the concrete produced can be categorized as a lightweight concrete. The mechanical properties of lightweight concrete are significantly improved by soda can waste fibres. The optimum fibre volume fraction for better performance in terms of strength is 0.3%. The mechanical property that experienced the highest improvement, which amounted to 61.00%, was the split tensile strength. The soda can waste fibres help to bridge and arrest the cracks and restrict the growth of flaws in the matrix, preventing their enlargement, under stress, into cracks, which eventually cause failure.

References

1. I. Satyarno, "Evaluasi Dan Tindakan Pengurangan Kerentanan Bangunan Dalam Rangka Mitigasi Bencana Gempa", Inauguration Speech, Gadjah Mada University (2010)
2. Y. Haryanto, G.H. Sudibyo, F.C.A. Effendi, Procedia Eng. **171**, 1025 (2017)
3. Y. Haryanto, B.S. Gan, N.G. Wariyatno, E.W. Indriyari, ARPN JEAS, **12,** 4858 (2017)
4. GIZ International Services, "Peringatan Tsunami 4 April 2011: Reaksi Lembaga dan Masyarakat terhadap Peringatan Gempa Bumi dan Tsunami di Selatan Jawa" (2011)
5. http://geospasial.bnpb.go.id/wp-content/uploads/2011/04/2011-04-04_gempabumi_cilacap_bnpb.pdf (4 May 2016)
6. T. Boen, "Word Seismic Safety Initiative" (2010)
7. S. Kenneth, P.E. Harmon, "Engineering Properties of Structural Lightweight Concrete", Carolina Stalite Company.
8. "Advantages of Structural Lightweight Aggregate Concrete", Expanded Clay, Shale and Slate Institute, www.escsi.org (4 May 2016).
9. ACI 213R-03, "Guide for Structural Lightweight-Aggregate Concrete", American Concrete Institute (2003).
10. F.H. Rifalldy, "The Study of Toughness and Stiffness of Lightweight Concrete with Aluminum Fiber on Post Fire", Undergraduate Thesis, Department of Civil Engineering, Sebelas Maret University (2010)
11. SNI 03-3449-2002, "Tata Cara Rencana Pembuatan Campuran Beton Ringan Dengan Agregat Ringan", The National Standardization Agency of Indonesia (2002)
12. SNI 03-2461-2002, "Spesifikasi Agreget Ringan untuk Beton Ringan Struktural", The National Standardization Agency of Indonesia (2002)
13. Akmaluddin, J. Eng. Sci., **43** (2011)
14. M. Hsie, C. Tu, P. Song, Mater. Sci. Eng. A, **494**, 153 (2008)
15. Y. Wang, H. Wu, C. Li, Mater. Civ. Eng., **12**, 314 (2000)
16. Z. Suna, Q. Xu, Mater. Sci. Eng. A, **527**, 198 (2009)
17. F. Altun, F. Tanrıöven, T. Dirikgil, Constr. Build. Mater., **44**, 565 (2013)
18. M. Mohammadian, A.K. Haghi, Mater. Plast., **50**, 12 (2013)
19. A. Sivakumar, V.M. Sounthararajan, J. Eng. Technol. Sci., **45,** 257 (2013)
20. A.M. Heniegal, F.F. El Habiby, R.D.A. Hafez, R. J. Eng., **04**, 01 (2014)
21. A.S.M Abdul Awal, H. Mohammadhosseini, M.Z. Hossain, Int. J. GEOMATE, **9**, 1441 (2015)

22. http://www.container-recycling.org/index.php/aluminum-beverage-can-waste-passes-the-qone-trillionq-mark-recycling-rate-drops-to-lowest-point-in-25-years (4 May 2016).
23. Y. Haryanto, Dinarek, **2**, 36 (2006)
24. Y. Haryanto, G.H. Sudibyo, Contribution of Civil Engineering toward Building Sustainable City, 434 (2016)
25. Y. Haryanto, Techno, **17**, 73 (2016)
26. G. Murali, G., C.M.V. Vardhan, R. Prabu, Z.M.S.A. Khan, T.A. Mohamed, T. Suresh, IJERA, **2**, 278 (2012)
27. G. C. Behera, R. K. Behera, Int. Res. J. Eng. Technol. (IRJET), **2** 1942 (2015)
28. Puskim, "Pengembangan Lempung (Shale) untuk Agregat Ringan Buatan (ALWA)" (2002)
29. SNI 1972: 2008, "Cara Uji Slump Beton", The National Standardization Agency of Indonesia (2008)
30. SNI 1974: 2011, "Cara Uji Kuat Tekan Beton dengan Benda Uji Silinder", The National Standardization Agency of Indonesia (2011)
31. SNI 03-2491-2002, "Metode Pengujian Kuat Tarik Belah Beton", The National Standardization Agency of Indonesia (2002)
32. SNI 03-4169-1996, "Metode Pengujian Modulus Etastisitas Statis dan Rasio Poison dengan Kompresor Ekstensometer", The National Standardization Agency of Indonesia (1996)
33. M. Gul, A. Bashir, J.A. Naqash, IJEAT, **3**, 304 (2014)

Mechanical Properties of Geopolymer Concrete Exposed to Combustion

Annisa Rahmadina[1,2,*], and *Januarti Jaya* Ekaputri[1,2,*]

[1]Institut Teknologi Sepuluh Nopember, Civil Engineering Department, 60111 Surabaya, Indonesia
[2]Konsorsium Riset Geopolimer Indonesia, Concrete and Building Material Laboratory Institut Teknologi Sepuluh Nopember, 60111 Surabaya, Indonesia

Abstract. This paper presents the change of material properties, such as decreasing of the compressive strength, splitting tensile strength, and porosity. The main objective of this paper is to analyze the mechanical properties of fly ash-based geopolymer concrete after being exposed to high temperature. The 28th-day test specimens were burned for one hour at specified temperature variation of 200°C, 400°C, 600°C, and 800°C. Ordinary Portland Cement (OPC) concrete was used as a comparison. After burning at 400°C, the compressive strength of geopolymer concrete was surprisingly increased up to 27% of its normal strength. On the other hand, the compressive strength of OPC concrete decreased 67% from its normal strength. The splitting tensile strength of geopolymer concrete also decreased at the range of 50-70%. The porosity of concrete has a sufficient effect to compressive strength and splitting strength. X-Ray Diffraction (XRD) test of geopolymer concrete at temperature 400°C until 600°C showed some minerals change. Geopolymer concrete is proved to have better fire resistance compares to Portland Cement Concrete.

1 Introduction

Fires that lead to the increasing of significant temperature have an impact to the changes of properties of concrete such as mechanical, physical and chemical properties [1]. The high temperature affects the characteristics of concrete such as the decreased at strength, peeling surface on concrete, and the occurrence of cracks.

Geopolymer concrete is known as concrete that uses non-Portland binder in the mixture. As a substitute for cement, Material that has high silica and alumina content such as fly ash is chosen. Fly ash needs alkali activator to create a binder. Commonly used alkali activator is the mixture of sodium hydroxide 8 M to 14 M and sodium silicate with the ratio of 0.4 to 2.5 [2].

Geopolymer concrete posses higher resistance to fire than Ordinary Portland Cement concrete [3]. Geopolymer concrete has little shrinkage. Kong et al. [4] stated that the use of superplasticizer had no significant affect on the performance of geopolymer concrete, especially after being exposed to a high temperature. The Addition of superplasticizer is only to improve the workability of geopolymer concrete. Pan et al. [5] stated that at high

* Corresponding author: annisarahma.jbr@gmail.com and januartije@gmail.com

temperatures chemical properties of geopolymer were more stable, while the OPC concrete was dehydrated. The loss of structural water indicates that Portland Cement-based concrete hydrates and then disintegrate at the increased temperature.

The different temperatures have significant impact to the shrinkage of geopolymer concrete. Geopolymer concrete samples that burned at temperatures below 200°C showed shrinkage at about 6%. At 250°C, the shrinkage decreased to 2%. The conditions of geopolymer concrete tend to be stable at temperature 250°C to800°Cwhile the heating process to 900°C resulted in the greater shrinkage of geopolymer [6].

In the process occurred a cycle of heating and cooling, resulting in a change of phase in the complex physical and chemical basis, it affected the quality of concrete structures and leading to brittle concrete [7]. The increase in temperatures led to the decrease of the compressive strength of geopolymer. The decreasing of the strength was caused by dehydration and dehydroxylation processes at the temperature 400°C to 600°C [7,8]. While heating from room temperature to 800°C, the OPC concrete greatly experienced cracking and spalling [5]. The OPC concrete showed at about 46.5% loss in compressive strength at 600°C. The significant loss of strength in OPC concrete after 400°C was related to the dissociation of calcium hydroxide [9]. At the same temperature, geopolymer concrete samples were more stable than the OPC concrete. Physically, geopolymer concrete is resistant to high temperature up to 1000°C. Cracks in geopolymer concrete were less than in the Portland Cement-based concrete [10]

2 Experimental Program

2.1 Material and mix design

Class F fly ash was used as the binder for geopolymer concrete. The chemical properties of fly ash are shown in Table 1.

Table 1. Chemical composition of fly ash

Chemical	SiO_2	Al_2O_3	Fe_2O_3	TiO_2	CaO	MgO	Cr_2O_3	K_2O	Na_2O	SO_3	Mn_2O_3
Percentage (wt.%)	48,47	26,05	12,54	0,92	5,18	2,77	0,02	1,66	0,47	1,05	0,19

The alkaline liquid in the mixture of geopolymer contains sodium hydroxide of 8 M and sodium silicate. Sodium silicate contains Na_2O, SiO_2, and H_2O by percentage of mass at around 18.5%, 36.4%, and 45.1% respectively. Coarse aggregate and fine aggregates were used in saturated surface dry (SSD) condition. The maximum size of coarse aggregate was 20 mm. All the aggregate requirements for concrete mix design satisfied ASTM C 33 [11].

The material proportion for geopolymer concrete per cubic meter is provided in Table 2. Ordinary Portland Cement (OPC) concrete was used as the control system. The mixing and casting procedures for geopolymer were similar to OPC concrete. All initial strengths of concrete were designed for 56 MPa at 28 days.

Table 2. Mixture proportion of concrete per m^3

Specimen	Materials (kg)							
	Fly ash	Cement	Coarse agregate	Fine agregate	Na_2SiO_3	NaOH	Water	SP 2%
Geopolymer	390	-	1080	720	140	70	-	7.8
OPC	-	539.47	1014.03	621.50	-	-	205	-

DOI: 10.1051/matecconf/201713801022

2.2 Casting and curing of geopolymer concrete specimens

Geopolymer and Ordinary Portland Cement (OPC) concrete were cast into cylinder molds with dimension of 100 mm in diameter and 200 mm in height. The variations of geopolymer, and OPC concrete were shown in Table 3. One day-after casting, geopolymer and OPC concrete were cured in water (wet curing) for 28 days.

Table 3. Hardened concrete specimens test mixture proportion

Specimen	Compressive Strength (Age 28-days)	Sample size	Variation	No. of Samples per Variation
Geopolymer Concrete	56.4 MPa	100mmΦ x 200mm cylinder	GCN; GC200; GC400; GC600; GC800	3/mix
OPC Concrete	56.4 MPa		OPCN; OPC200; OPC400; OPC600; OPC800	3/mix

2.3 Heating methods

At the ages of 28 days, the specimens were heated in the furnace at 200°C, 400°C, 600°C, and 800°C temperatures for one hour. The specimens were heated to a maximum furnace temperature of 800°C and burnt up to 19.5 hours. Detail of the duration of burning process was presented in Figure 1.

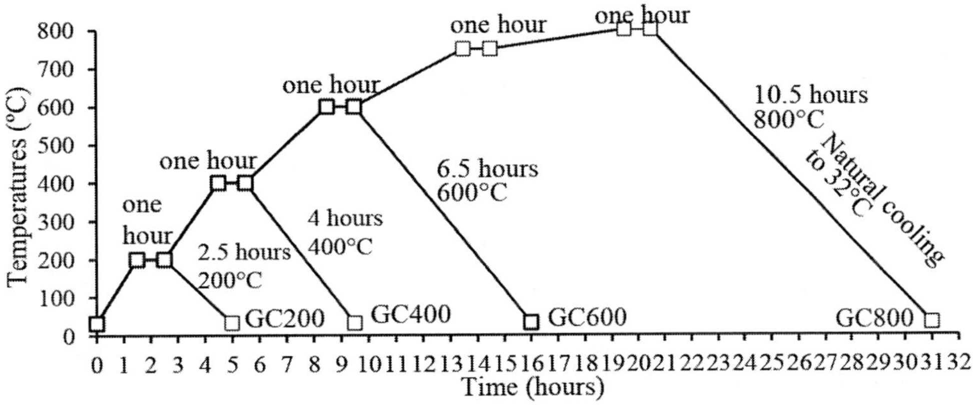

Fig. 1. Elevated temperature-history for specimens exposed to fire.

2.4 Test procedure of geopolymer concrete

After combustion process, all of the specimens were cooled off to normal temperature. The specimens were tested for the compressive strength and splitting tensile strength according to ASTM C 39 and ASTM C 496 [12, 13]. The compressive strength of heated specimens after exposure to the fire then compared to the strength of the unheated specimen. Some fractions of a geopolymer concrete specimen were used for porosity test. X-Ray Diffraction test was also conducted to determine the mineral contents in geopolymer concrete after the combustion.

DOI: 10.1051/matecconf/201713801022

Fig. 2. a. Compressive strength test, b. Splitting tensile test, c. After burning in the furnace.

3 Result and discussion

3.1 Compressive strength of the specimens

The result of the compressive strength of concrete before and after exposure to the fire was shown in Figure 3. At the temperature of 400°C, compressive strength of geopolymer concrete was 14.67% higher compared to the compressive strength of geopolymer at the normal temperature. Geopolymer concrete at 800°C temperatures tends to have a lower compressive strength compared to the compressive strength at the other temperatures. The higher temperatures result in the decreasing of the compressive strength of geopolymer concrete [7], but at specific temperature, the strength of geopolymer increased. It was shown that the strength of the geopolymer concrete was influenced by dehydration and dehydroxylation [8,9].

The compressive strength of OPC concrete was decreased as the increased in temperatures. At the temperature from 600°C to 800°C, the compressive strength of the OPC concrete was decreased significantly. The strength of OPC concretes were 30 MPa and 19 MPa at the temperature of 600°C and 800°C. The strength of geopolymer concrete was 51 MPa and 40 MPa at the same temperature. It was shown that at elevated temperature, the strength of OPC concrete decreased significantly to 56% by its initial compressive strength. This was mainly due to the excessive build-up of vapor pressure that produces large cracks in the matrices of concrete [14].

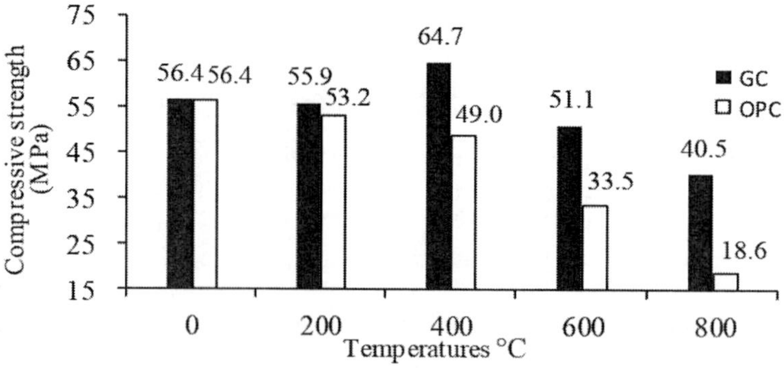

Fig. 3. Compressive strength of geopolymer and OPC concrete at elevated temperature.

3.2 Splitting of the geopolymer concrete

The effects of temperature to the splitting strength of geopolymer concrete were presented in Figure 4. The lowest split strength showed by the specimens exposed to 800°C temperature. Pan [5] stated that the splitting strength of the geopolymer concrete decreases as the heating temperature increases. Figure 5 shows spalling occurs easily at elevated temperatures. Thermal stresses were believed to lead to force spalling of concrete. It is composed of entirely different binder properties [5].

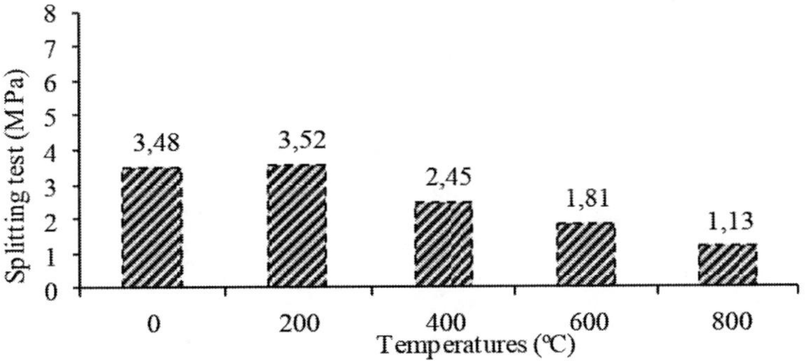

Fig. 4. The effect of elevated temperature to splitting test.

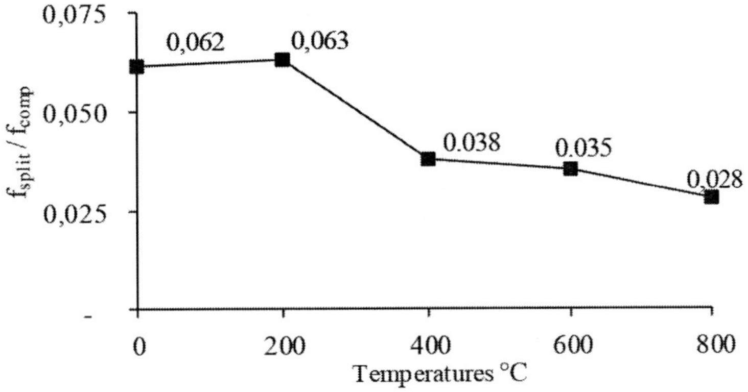

Fig. 5. Ratio of f_{split} to $f_{compressive}$ at different temperature.

3.3 Porosity

Porosity test shows the effect of pores on geopolymer concrete. The test results for the porosity were shown in Figure 6. The highest porosity of geopolymer concrete after exposure to fire occur in GC800 by 19.6% where the unclosed porosity is 14.3%. Meanwhile, the lowest porosity was found in geopolymer concrete GC400 by 15.8% where the open porosity is 6.3%. The pore size after exposure to high temperatures was significantly decreased in fly ash-contained specimens [15]. Large porosity occurs when the water content is high. When the content becomes small then the small porosity occurs [16]. At 400°C, compaction occurs due to mineral changes caused by hydroxylation process. According to Kong et al. [17] high temperature led to change in chemical structure and free dehydration occurs. It resulted in the internal damage of the concrete.

DOI: 10.1051/matecconf/201713801022

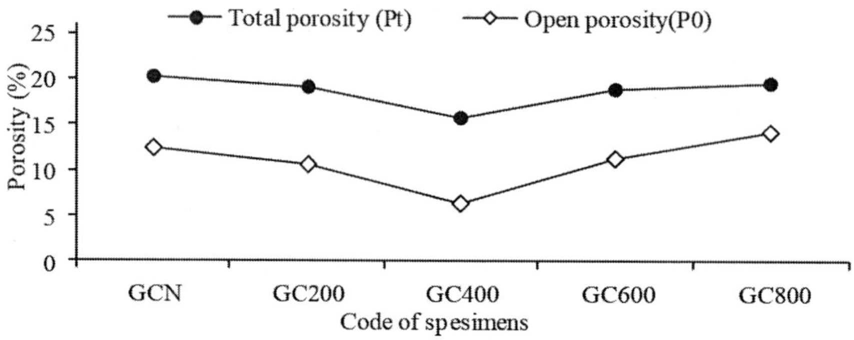

Fig. 6. Total and open porosity of geopolymer concrete

3.4 Residual strength of geopolymer and OPC concrete

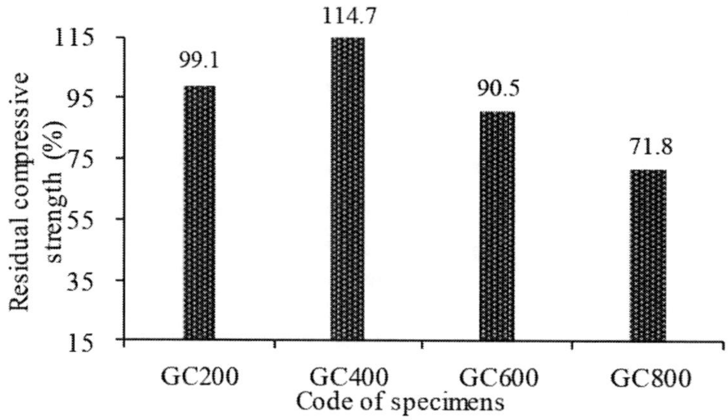

Fig. 7. Residual strength of geopolymer concrete

Figure 7 shows the percentage of residual compressive strength of specimens after exposed to fire to determine the resistance of concrete specimens at high temperatures. At temperature of 400°C, residual strength of geopolymer concrete was increased to 14.7%. At temperature of 250°C to 600°C geopolymer concrete experienced dehydroxylation and resulted to the decreased and increased compressive strength. These phenomenon caused by a cluster of hydroxyl are abundant [8]. The strength of geopolymer concrete decreased was caused by the differential expansion between aggregate and geopolymer. The geopolymer matrix experienced contraction at high-temperature [18].

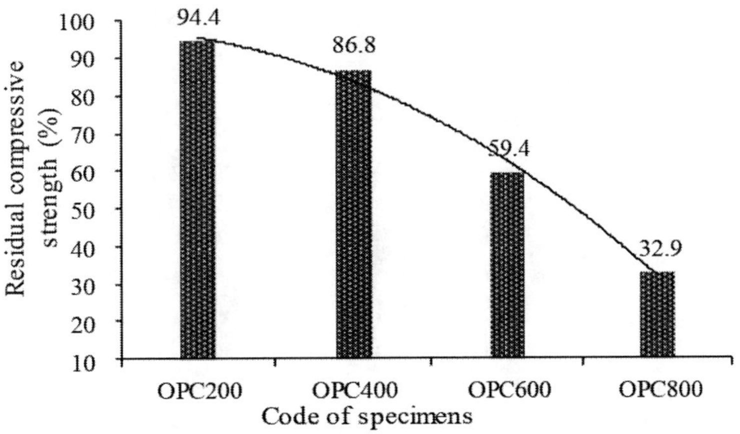

Fig. 8. Residual strength of OPC concrete

Along with the increased of firing temperature compressive strength was decreased. The compressive strength was decreased at the temperature of 400°C, 600°C, 800°C, at about 48.98 MPa, 33.53 MPa, and 18.53 MPa. Residual compressive strengths at each temperature were 86.8%, 59.4%, and 32.9%. The residual strengths of OPC concrete tend to decrease drastically at the temperature of 600°C to 800°C [9]

Figure 7 and figure 8 concluded that the residual strength of geopolymer concrete is higher than OPC concrete. At the maximum temperature of 800°C, the residual strength of OPC concrete was less than 50%. This showed the superior fire endurance ability of fly ash-based geopolymer concrete to the OPC concrete [10]

3.5 Discoloration, spalling and cracking in surface of the geopolymer and OPC concrete

The discoloration on the surface of the geopolymer and Ordinary Portland Cement concrete is shown in figure 9. From the observation of the physical condition before and after exposure to fire, at the normal condition, the color of OPC concrete was changed to be darker than OPC concrete after being burned. At temperature 200°C, OPC concrete began to change its color and the color becomes brighter. The color changed due to the increase in temperature [9]. After the exposure to temperature 800°C, the color appeared to change from grey to pink [19]. Moreover, at temperatures 800°C physical condition of OPC concrete showed a significant change. Crack occurs across the surface of concrete, and the edge of concrete was ruptured.

On the surface of concrete was appeared the limestone. Crack in concrete was caused by a concrete shrinkage process during the heating process [6]. It was the same as that proposed by Sarker et al. [10] after exposed to fire, in geopolymer concrete specimen was found smaller damaged (in this case crack) rather than OPC concrete specimen. After burnt at temperature of 800°C geopolymer, concrete experienced significant discoloration.

DOI: 10.1051/matecconf/201713801022

Fig. 9. Different color of OPC concrete and Geopolymer concrete after heating to fire at 800 °C.

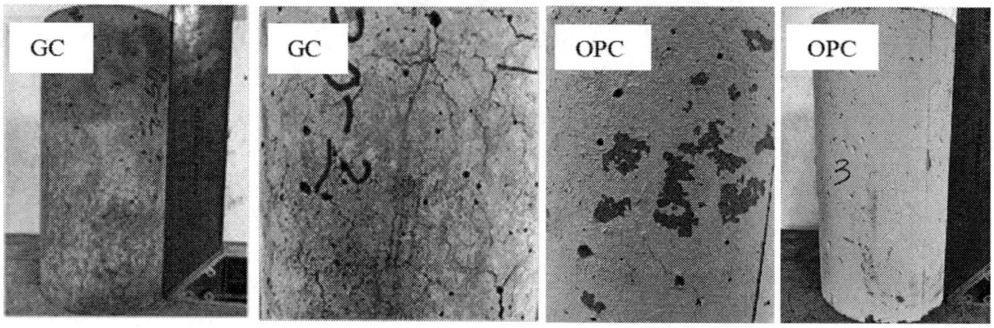

Fig. 10. Discoloration and cracking in surface of geopolymer and OPC concrete at temperature 800°C.

Figure 10 shows cracking patterns across the surface of the concrete. Cracks in the concrete were caused by the shrinkage that occurred after the concrete burnt. Both geopolymer and OPC concrete exploded during fire exposure. At the identical elevated temperature, the result showed that the spalling of geopolymer concrete was less than OPC concrete. The spalling also showed the same result with the result observed by Pan et al. [5]. The geopolymer samples were slightly cracked on concrete surfaces after exposure to fire [19]. The matrix of geopolymer concrete was stable at the elevated temperature.

3.6 X-Ray Diffraction analysis

X-ray diffraction test was carried out to investigate the mineral change in geopolymer concrete during thermal exposure. Mineral changes were presented in Fig. 11.

Figure 11 presents the XRD results of a specimen before and after firing. At 400°C, some minerals such as albite, quartz, rutile were increased significantly. The mineral in geopolymer concrete increased after exposured to high temperature. Quartz and albite were affected by temperatures. Albite is a crystal mineral. Due to the influence of temperature, albite structure was changed. However, the chemical element composition was remained (recrystallizes). At the time of exposure, the amount of albite increased due to the change of mineral from amorph to the crystal. Hematite in the region 35-2θ appeared from the glassy phase of fly ash [19]. At the temperature of 400 to 600°C, zeolites found in alkali-activated fly ash systems was relatively stable. It was characterized by the occurrence of minor dehydration [20].

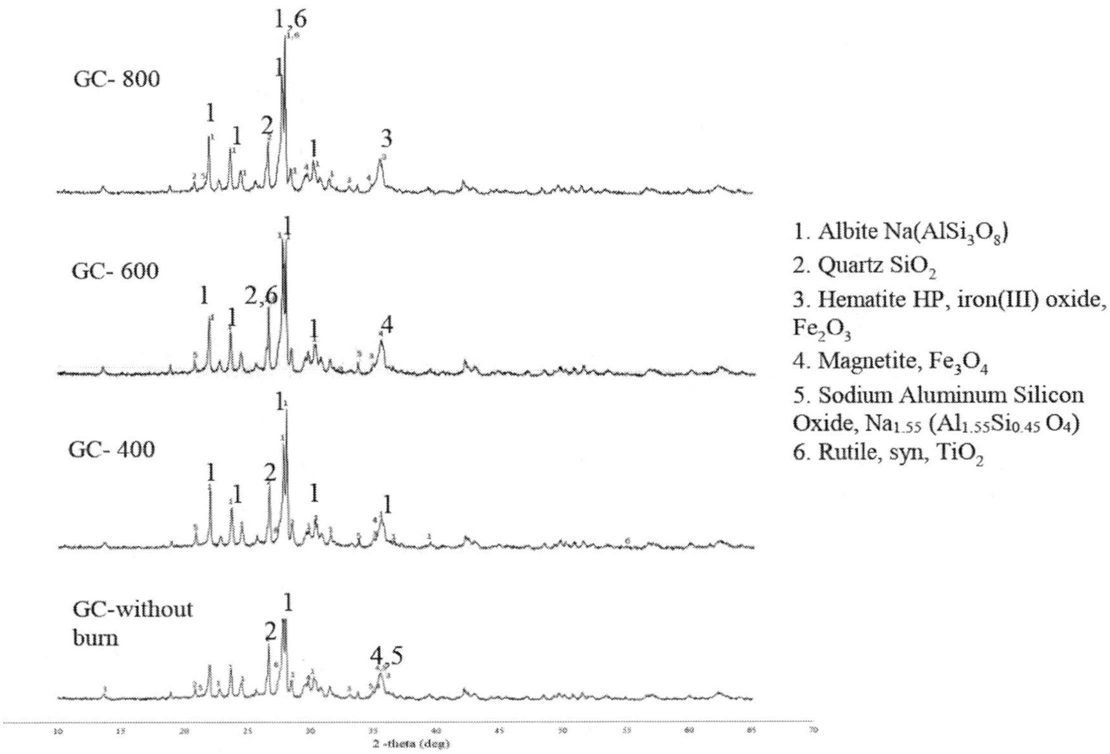

Fig. 11. XRF of geopolymer concrete

4 Conclusions

1. After geopolymer and OPC concrete was burned, both of the compressive strength decreased. But the compressive strength of geopolymer at the temperature of 400°C was increased by 14 to 18%.
2. Geopolymer concrete has good fire resistance due to chemical bonds in geopolymer concrete. After exposed to fire, the amount of some minerals of geopolymer concrete increased at the temperature of 400°C - 800°C.
3. At some temperatures, the increase in compressive strength was indicated by the porosity test results. Compaction occurs at the temperature of 400°C. X-Ray Diffraction test results indicated that some minerals changed. It was caused by dehydroxylation process.
4. Geopolymer concrete has more resistant to elevated temperatures compared to Ordinary Portland Cement (OPC) concrete. The relative residual strength for all geopolymer concrete were greater than OPC concrete after exposure to 200°C, 400°C, 600°C, and 800°C.
5. At the elevated temperature, spalling easily occurred. Spalling occurs due to thermal pressure.

References

1. P.K. Sarker, S. Kelly and Z. Yao, *Material and Design*, **63**, pp.584-592 (2014)
2. D. Hardjito, B.V. Rangan, "Research Report GC 1, Australia: Curtin University of Technology", (2005)

3. F. Skvara, L. Kopecky, J. Nemecek, Z. Bittner, "Microstructure of geopolymer materials based on fly ash, Ceramics-Silikaty", **(4),** 208-215 (2006)
4. D.L.Y. Kong, J.G. Sanjayan, *Cement and Concrete Research*, **40**, 334-339 (2010)
5. Z. Pan, J.G. Sanjayan, D.L.Y. Kong, *Construction and Building Materials*, **36**, 365-372 (2012)
6. Subaer, A.V. Riesen, "Advances in Geopolymer Science and Technology, J. Mater Sci.", **42**, 3117-3123 (2007)
7. O.A. Abdulkareem, M. Al Bakri, H. Kamarudin, I.K. Nizar, A.A. Saif, *Construction and Building Materials*, **50**, pp. 377-387 (2014)
8. L. Vickers, V.A. Riessen, W.D. Rickard, *Springer Briefs in Materials* (2015)
9. F.U.A. Shaikh, V. Vimonsatit, *Fire and Materials* (2014)
10. P.K. Sarker, S. Mcbeth, *Construction and Building Materials*, **90,** 91-98 (2015)
11. American Society for Testing and Materials, ASTM C 33 (2010)
12. American Society for Testing and Materials, ASTM C 39 (2013)
13. American Society for Testing and Materials, ASTM C 496 (2013)
14. I. Turkmen, M.B. Karakoc, F. Kantarci, M.M. Maras, R. Demirboga, *Fire and Materials* (2016)
15. R.Kh. Ibrahim, R. Hamid, M.R. Taha, *Construction and Building Materials* (2012)
16. H. Cheng, K.L. Lin, R. Chi, C.L. Hwang, T.W. Cheng, Y.M. Chang, *Construction anf Building Materials,* **88**, 74-83 (2015)
17. D.L.Y Kong, J.G. Sanjayan, K.S. Crentsil, *Cement and Concrete Research*, **37**, 1583-1589 (2007)
18. D.L.Y. Kong, J.G. Sanjayan, *Cement and Concrete Composites*, **30**, 986-991 (2008)
19. A.M. Rasyad, S.R. Zeeden, *Construction and Building Materials*, **25**, 3098-3107 (2011)
20. A.F. Jimenez, A.G. de la Torre, A. Palomo, G. Lopez-Olmo, M.M. Alonso, M.A.G. Aranda, *Fuel*, **85**, 1960-1969 (2006)

MATEC Web of Conferences **138**, 01023 (2017) DOI: 10.1051/matecconf/201713801023
EACEF 2017

Modelling of chloride penetration in concrete under wet/dry cycle

Sung-In Hong[1] , and *Ki-Yong* Ann[1,*]

[1]Hanyang University, Department of Civil and Environmental Engineering, Ansan, South Korea

Abstract. This present study concerns modelling of chloride penetration in partially saturated concrete. To mimic the intermittent exposure of sea water to concrete, varying environmental conditions for relative humidity and chloride concentration were considered. As for the moisture distribution in concrete, statistical permeability model based on pore size distribution was used to represent influence of material properties on moisture transport. Then, a combined chloride diffusion and convection was modelled in variation of moisture level in concrete. As a result, smaller relative wet duration induces higher rate of chloride penetration due to enhanced moisture permeability from the surface, and also higher concentration gradient near the surface of concrete due to repeated wet/dry cycle. This implies that only diffusion analysis on chloride induced corrosion in concrete structure may underestimate the serviceability in given material performance.

1 Introduction

Chloride penetration in reinforced concrete structure has been regarded as a big problem in the service life design. When chloride ions reach on the steel surface in concrete, a passive film, maintained by a high alkalinity of concrete pore solution, starts to dissolve, thereby leading to corrosion of steel [1]. Corrosion propagation accordingly imposes cover cracking of inner part of concrete, which in turn gives rise to a structure failure in the end of the deterioration process. Thus, predicting the chloride ingress in concrete structure exposed to a saline environment is essentially required to prepare a proper maintenance and thus reduce such high risk of chloride induced corrosion.

For this purpose, many of modelling works have been carried out based on the Fick's law of chloride diffusion. However, most of concrete structures, built in chloride laden environment such as the bridge pier with tidal zone and load pavement with deicer supply, have been experiencing intermittent chloride exposure. Moreover, those conditions, compared with the case for constantly immersed structure in saline water, impose more severe degradation to the concrete durability [2, 3] and difficult to be assessed by using only the conventional diffusion analysis.

This study aims to develop a model for predicting chloride transport in concrete, submitted to intermittent chloride exposure. In this paper, the driving force of chloride ions

* Corresponding auther: kann@hanyang.ac.kr

transporting in partially saturated concrete was defined as diffusion-convection phenomena. To investigate the combined transport mechanism in concrete under wet/dry cycle, moisture distribution, considering a statistical approach on permeability and hysteric effect on water vapour adsorption and desorption, was first addressed, and then chloride diffusion through water filled porosity and chloride convection driven by moisture flow were analysed.

2 Model description

2.1 Moisture distribution

Prior to solving the chloride diffusion-convection problem in concrete, moisture distribution under equal environmental condition with chloride source should be analysed. Under thermodynamic equilibrium in concrete, moisture transport, driven by permeation of both liquid and vapour through concrete pore network, can be described using the one-dimensional mass balance equation such as:

$$(dw/dh)\cdot(dh/dt) = \mathrm{div}[(K_l\cdot\rho_w\cdot R\cdot T)/(M_w\cdot h)\cdot \mathrm{grad}(h) + K_v\cdot P_0\cdot \mathrm{grad}(h)] \quad (1)$$

where, w is the moisture content (kg/m^3), h is the relative humidity, t is the time (s), K_l is the permeability of the liquid water (kg/m·s·Pa), ρ_w is the density of water (kg/m^3), R is the gas constant (J/kg·K), T is the temperature (K), and M_w is the water molecular weight (kg/mol), K_v is the permeability of water vapour (kg/m·s·Pa), and P_0 is the saturation vapour pressure (Pa). According to the statistical permeability model [4], liquid permeability was determined, considering pore size variation from zero size (r = 0) to the critical radius at capillary condensation ($r = r_c$), which is obtained by the Kelvin's law, such as:

$$K_l = \rho_w/(8\cdot\tau^2\cdot\eta)\cdot[\int r_{eq}\cdot\varnothing(r)\cdot dr]^2 \quad (2)$$

where, τ is the tortuosity factor, η is the viscosity of liquid water (Pa·s), r_{eq} is the equivalent pore radius (m) [Chaube et al], r is the pore radius (m), and $\varnothing(r)$ is the cumulative porosity distribution, which in this study was defined by:

$$\varnothing(r) = \varnothing_G\cdot[1 - \exp(-b_G\cdot r)] + \varnothing_S\cdot[1 - \exp(-b_S\cdot r)] + \varnothing_L\cdot[1 - \exp(-b_L\cdot r)] \quad (3)$$

where, $\varnothing_G$, $\varnothing_S$, $\varnothing_L$ are respectively the porosity for gel, small, and large capillary pores, and b_G, b_S, b_L are respectively the distribution parameters for gel, small, and large capillary pores. As for the vapour permeability, the Knudsen law was adopted to consider the water vapour diffusion through unsaturated pore network where only the water content forms on the pore surface by vapour adsorption. Accordingly, the following expression can be used to represent the vapour permeability such that:

$$K_v = D_0\cdot(1 - S)/\{\tau\cdot[1 + l_m/2(r_m - t_a)]\}\cdot[M_w/(R\cdot T)] \quad (4)$$

where, D_0 is the diffusivity of water vapour at free air (m^2/s), S is the degree of saturation, l_m is the mean free path for water molecules (m), r_m is the mean of unsaturated pore sizes (m), t_a is the thickness of water layer on pore wall (m).

To consider the effect of water entrapment on vapour desorption, a simple probabilistic approach [5] was used to determine the degree of saturation at drying condition, S_d, such that:

$$S_d = S_w \cdot [1 - \ln(S_w)] \tag{5}$$

where, S_w is the degree of saturation at wetting condition, which can be derived based on pore size distribution using Eq. 3 such that:

$$S_w = Ø(r)/Ø_t \tag{6}$$

where, $Ø_t$ is the total porosity, sum of gel and capillary pores.

2.2 Chloride transport

Chloride penetration, arose from both diffusion and convection, can be described by a modified diffusion-convection equation, considering the chloride binding, pore saturation, and humidity flux as a solvent velocity such that:

$$(dC_b/dC_f + w_e) \cdot (dC_f/dt) = \text{div}[\, (D_e \cdot w_e) \cdot \text{grad}(C_f) + w_e \cdot C_f \cdot D_h \cdot \text{grad}(h)] \tag{7}$$

where, C_b is the concentration of bound chloride in hydrates (kg/m^3), C_f is the concentration of free chloride ion in pore solution (kg/m^3), D_e is the effective chloride diffusivity (m^2/s), w_e is the water filled porosity, D_h is the moisture diffusivity (m^2/s), which can be obtained by incorporating liquid and vapour transport parameters such that:

$$D_h = [(K_l \cdot \rho_w \cdot R \cdot T)/(M_w \cdot h) + K_v \cdot P_0]/(dw/dh) \tag{8}$$

The binding capacity, dC_b/dC_f [6], was obtained by applying the Langmuir isotherm such that:

$$dC_b/dC_f = \alpha/[(1 + \beta \cdot C_f)^2] \tag{9}$$

where, α and β are the binding parameters of the Langmuir isotherm.

2.3 Boundary condition

The climatic boundary conditions were the relative humidity and chloride concentration. When concrete is submerged by high tidal, 99.5% and 0.5 M respectively for relative humidity as a saturation state and for chloride concentration in sea water were assumed.

Otherwise, 60% for relative humidity and zero concentration for chloride were set as the boundary conditions for drying state. As for the humidity flux at concrete surface, a convective boundary condition based on a mass transfer coefficient was used such as:

$$D_h \cdot \mathrm{grad}(h) = B_h \cdot (h_s - h_{en}) \tag{10}$$

Where, B_h is the mass transfer coefficient for relative humidity (m/s), h_s and h_{en} are the relative humidity at surface of concrete and environment, respectively. By considering the intermittent chloride exposure on the boundary condition, concentration flux of free chloride ions with moisture transfer induced ionic flux [7] was set as:

$$D_e \cdot \mathrm{grad}(C_f) = B_c \cdot (C_s - C_{en}) + C_{en} \cdot D_h \cdot \mathrm{grad}(h) \tag{11}$$

Where, B_c is the transfer coefficient for chloride exposure (m/s), C_s and C_{en} are the chloride concentration at surface of concrete and environment, respectively. Before the tidal zone exposure, concrete was assumed to have zero concentration of chloride and initially saturated state.

As for the numerical procedure for coupled mass transport, the Crank-Nicholson method was used to approximate the non-linear equations with 1 mm for distance increment and 0.1 day for time increment. To enhance the numerical stability on the calculation, the Gauss-Seidel method was used for the iterative solution with a tolerance, 10^{-6}.

3 Result and discussion

3.1 Moisture distribution

The spatial values for rates of moisture flow and resultant water contents through concrete depths are required prior to implement the chloride convection problem.

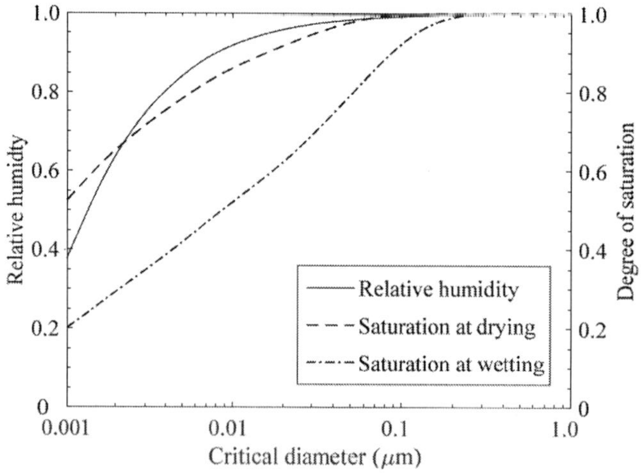

Fig. 1. Relation of relative humidity and degree of saturation under wet/dry conditions

The relation of relative humidity and degree of saturation under wet/dry cycle can be determined using the Eq. 5 and 6. This implies that most of moisture flow occurs through gel and small capillary pores. Large capillary pores above 1.0 μm can be ignored in analyzing the mass transport in unsaturated concrete since such high saturated state can be regarded as the regime of chloride diffusion. The effect of moisture entrapment during desorption process can be indicated by observing the difference between degree of saturations for drying and wetting. Thus, it should be noted that the different moisture properties between wet/dry conditions can occur and thus influence the behavior of chloride transport accordingly.

The moisture related chloride convection is largely dependent on the moisture diffusivity, D_h, which also dependent on the relative humidity, as shown in Fig. 2. As relative humidity increases, the rate of moisture flow decreases below about 20% humidity and then increases over the moisture level. This implies that moisture transport due to liquid permeation is more significant compared with that by vapour phase, which is dominant in low humidity condition. In particular, high water contents over 80% of relative humidity accelerate the moisture flow by Darcy's permeation.

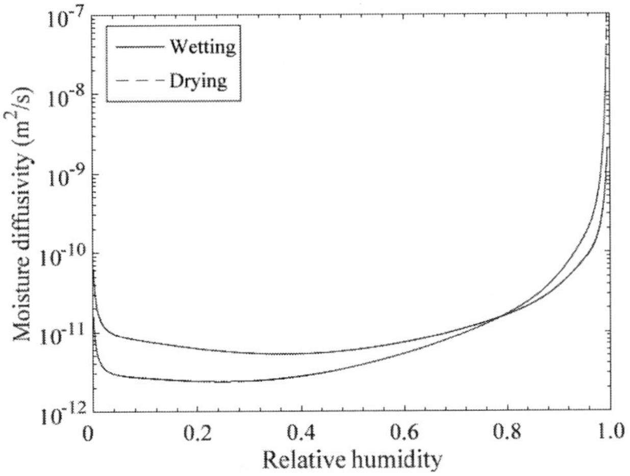

Fig. 2. Moisture diffusion coefficient dependent on relative humidity under wet/dry conditions

3.2 Chloride transport in unsaturated concrete

To assess the influence of drying period on rate of chloride penetration, 12 hours of wetting time per day and three different durations for drying were used to simulate 30 years of chloride concentration distribution, as shown in Fig. 3. As a result, increased dry duration promotes chloride transport under constant chloride diffusivity. This might be attributed to the fact that water supply during wetting cycle enhances the moisture gradient from the surface of concrete, which is more dominant in the case of concrete with low moisture level associated with longer period of drying. Thus, the rate of convective chloride penetration can be enhanced by moisture permeation coming from the wetting phase and propagates into the inner part of concrete.

 DOI: 10.1051/matecconf/201713801023

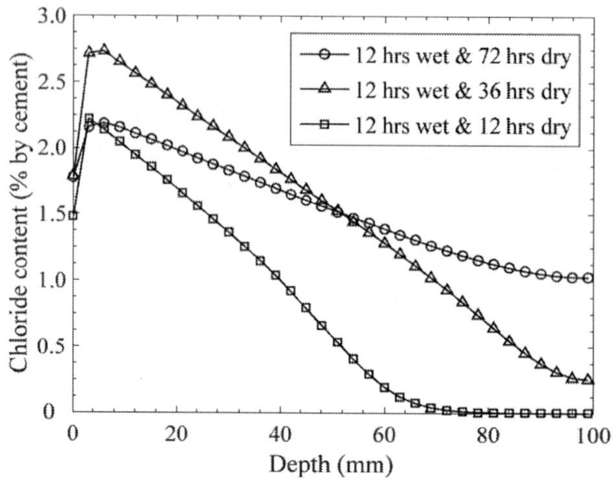

Fig. 3. Chloride concentration profiles varying with dry durations on 30 years of chloride exposure

To assess the effect of convection phenomena on the chloride concentration front, a single diffusion analysis at fully saturated concrete was compared with the case of chloride diffusion-convection, as given in Fig. 4. The surface chloride content at complete saturation shows higher value, accounting for 1.65%, compared with the case of partial saturation, 1.48%, probably due to the limited duration of chloride exposure. Despite the low concentration at surface, peak values, made by the repeated wet and dry, were observed near the surface and increase with time. This can also promote the rate of chloride penetration by enhanced concentration gradient from the surface and thus diffusion process. Also, the moisture flow containing chloride ion moves toward inner concrete depth which is more predominant than flow to the surface. Consequently, the rate of chloride penetration, arose from the coupled action of enhanced diffusion and convection into concrete depth, may surpass the one with diffusion only, irrespective of exposure time. Thus, it is inferred that chloride transport analysis on intermittent mass exposure should encompass the chloride convection, capable of reflecting the effects of thermodynamic equilibrium on water sorption and desorption together with humidity dependent moisture permeability.

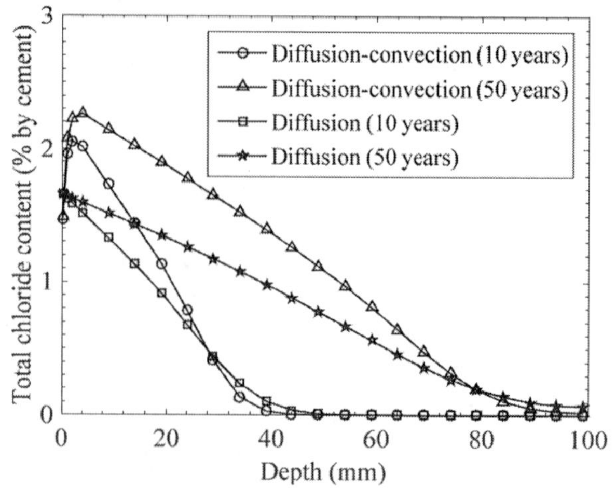

Fig. 4. Chloride concentration profiles in the presence and absence of convection phenomena

4 Conclusions

The present paper focused on investigating the chloride penetration in unsaturated concrete by using the mathematical models in terms of coupled moisture and chloride transport. From this work, the following conclusions were made:

(1) Due to the moisture entrapment, the degree of saturation during the dying cycle shows always higher value, compared with the adsorption process. Moisture diffusivity from the calculation shows high dependence on the liquid permeability, particularly over about 80% humidity condition. Those resultant moisture properties under wet/dry cycle can affect the chloride transport in partially saturated concrete.

(2) From the calculation, the rate of chloride penetration in the presence of moisture flux exceed that with diffusion only, irrespective of chloride exposure time. This was mainly due to (a) enhanced chloride concentration gradient from the surface of concrete by repeated wet/dry durations and (b) moisture permeation from the surface of concrete during wet cycle, which in turn promotes chloride convection velocity toward concrete depth.

Acknowledgements

This research was supported by Basic Science Research Program through the National Research Foundation of Korea (NRF) funded by the Ministry of Science, ICT & Future Planning (No. 2015R1A5A1037548)

References

1. C.L. Page, Nature **256**, 514 (1975)
2. A. Boddy, E. Bentz, M.D.A. Thomas, R.D. Hooton, Cem. Concr. Res. **29**, 6 (1999)
3. Y. Zhang, W.L. Jin, ACI. Mater. J. **108**, 5 (2011)
4. F. A. Dullien, *Porous media : fluid transport and pore structure* (Academic Press, 2009)
5. R. Chaube, T. Kishi, K. Maekawa, *Modelling of concrete performance : Hydration, microstructure and mass transport* (CRC Press, 2005)
6. B. Martın-Pérez, H. Zibara, R.D. Hooton, M.D.A. Thomas, Cem. Concr. Res. **30**, 8 (2000) (2000).
7. A.V. Saetta, R.V. Scotta, R.V. Vitaliani, ACI. Mater. J. **90**, 5 (1993)

MATEC Web of Conferences **138**, 01024 (2017) DOI: 10.1051/matecconf/201713801024
EACEF 2017

Mechanical properties of different bamboo species

Dinie Awalluddin[1], *Mohd Azreen* Mohd Ariffin[1,2*], *Mohd Hanim* Osman[1,2], *Mohd Warid* Hussin[1], *Mohamed* A. Ismail[3], *Han*-Seung Lee[4], and *Nor Hasanah* Abdul Shukor Lim[1].

[1]Faculty of Civil Engineering, Universiti Teknologi Malaysia, 81310 UTM Johor Bahru, Johor, Malaysia.
[2]Forensic Engineering Centre, Institute for Smart Infrastructure Innovative Construction, Faculty of Civil Engineering, Universiti Teknologi Malaysia, 81310 UTM Johor Bahru, Johor, Malaysia
[3]Department of Civil and Construction Engineering, Faculty of Engineering and Science, Curtin University Sarawak, CDT 250, Miri 98009, Sarawak, Malaysia.
[4]Department of Architectural Engineering, Hanyang University, 1271 Sa 3-dong, Sangrok-gu, Ansan 426-791, Korea.

Abstract. Bamboo is a rapid renewable plant that has a fast growth rate as compared to trees, which increases its suitability to be used as a sustainable source for wood industry, especially in construction works. Due to the lack of understanding on bamboo properties, the utilization of bamboo in construction has always been neglected. This paper presents an investigation on the mechanical properties of four species of treated bamboos that are available in Malaysia, which include *Bambusa Vulgaris*, *Dendrocalamus Asper*, *Schizostachyum Grande*, and *Gigantochloa Scortechinii*. A mechanical testing was carried out in various parts along the culm of these bamboo species in order to examine the differences of their compressive strength and tensile strength. The strength development and moisture content of these bamboo species were also monitored at a five-month interval. The results showed that *Bambusa Vulgaris*, *Dendrocalamus Asper*, and *Gigantochloa Scortechinii* possess excellent mechanical properties in compression and tensile strength, which indicate a good quality to be used as a construction material. As bamboo offers promising advantages, thus, it is suitable to be used as a substitute in place of structural timber in construction, which indirectly facilitates the preservation of the global environment.

1 Introduction

Being classified as a grass member of a larger grass family [1], bamboo can be easily found in tropical and some temperate areas of the world.

* Corresponding author: mohdazreen@utm.my

Bamboo has a nature form comprising a cylindrical pole with jointed stem known as a culm. There are over 1500 identified bamboo species in the world [2]. As one of the fastest growing plants, bamboo can reach a full height ranging from 15-30m in a period of two to four months [3]. All bamboo species has a similar anatomy, which consists of nodes, internodes, and diaphragm as shown in Figure 1. Each species can be identified according to their root system, in which there are three known root systems including, sympodial, monopodial, and amphodial [4]. The thickness of a bamboo decreases along the height of the culm, while the fibres density increases from the bamboo culm's inner wall to outer wall.

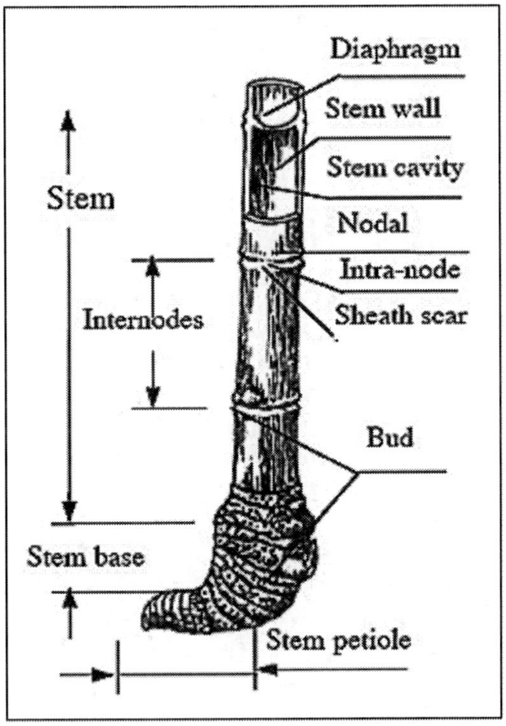

Fig.1 Bamboo anatomy

Bamboo forest has a higher harvest cycle, with up to four times of carbon density per hectare in comparison to spruce forest [5]. This means that bamboo can be harvested in a short period of time, in addition to having a high rate of carbon absorption as compared to timber. Recently, there is a growing interest in using bamboo as a construction material, in place of timber. In China, contractors have utilized bamboo as scaffoldings, while in Bali, bamboo is used to construct a Green School [6], where all structural components of the building is sourced from a local bamboo supplier. In contrast, in other developing country, the utilization of bamboo as a construction material is scarce, even with an abundance of bamboo source available. Lack of study concerning the mechanical properties of bamboo is known to be one of the factors that is associated with a low utilization of bamboo in construction.

The ability and capacity of bamboo to withstand the force applied onto it is very important. The strength and moisture content of the bamboo are among the most important factors, which determine both ability and capacity of full bamboo culm. Bamboo with a moisture content of 15% or lower tends to have good mechanical properties and is less vulnerable to fungus attack [7]. In determining the strength of bamboo, 12% of moisture content in air-dry condition has been regarded as a standard and reference [8]. In terms of durability, it varies according to species and age ranges. It is estimated that an untreated bamboo has a design life of approximately 10-15 years if it is stored appropriately, while a treated bamboo will have a much longer design life [9].

The strength of bamboos also depends on age ranges and species, which determines their suitability to be used in construction when they reach maturity age of around 3-4 years. In maturity, bamboos possess an optimum strength and are suitable to be used for heavy-duty applications. It has also been found that bamboos' compressive strength increases along their height, in addition to increasing compressive strength from the inner part to the outer part of their culm [10]–[13].

The aim of this research is to investigate on the mechanical properties of different species of local bamboos. This research is a part of an experimental work which focuses on the possibility of utilizing bamboo as a construction material.

2 Materials and methods

2.1 Preparation of the bamboo

There are four bamboo species involved in this research, which are sourced from Perak, Malaysia. Those four bamboo species include, *Dendrocalamus Asper*, *Bambusa Vulgaris*, *Schizostachyum Grande*, and *Gigantochloa Scortechinii*. These four bamboo species were selected and left to dry for about three weeks until the colour of the bamboos turned yellowish from their original colour. They were subsequently treated by using boric acid, in order to increase their durability and ability against fungus attack. Figure 2 shows a pile of felled bamboos gathered prior to mechanical testing.

Fig.2 Four bamboo species

2.2 Mechanical testing

All bamboo species were cut into small samples, in which the samples were taken from the internode of the bamboo culm. Three tests have been involved in this research including, compression, tensile, and moisture content, in accordance with ISO 22157 Part 1 [14].

All specimens were cut from the bottom, middle, and top parts of the bamboo and marked with letters B, M, and T, respectively. For compression test and tensile tests, all specimen with free nodes were selected. The outer diameter of the bamboo specimen, D, to be used for compression test was measured from two opposite points of the outer surface. The average diameter was recorded. Average wall thickness, t, of the bamboo culm was measured from four points, separated by a 90° angle around the diameter of the specimen. Compression test was carried out by using a Universal Testing Machine, where the specimens for four bamboo species were tested under a constant rate of 0.01mm/s, as shown in Figure 3. Maximum load for each specimen was recorded by using a data logger.

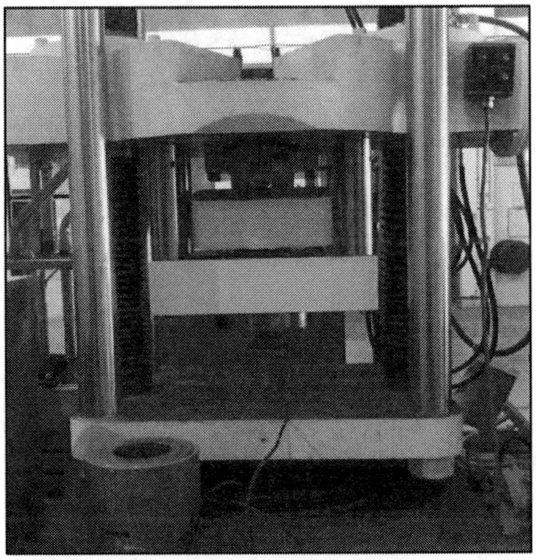

Fig 3. Bamboo specimen under compression test using Universal Testing Machine

Upon completion of compression test, small samples with a dimension of 25mm x 25mm, as shown in Figure 4 were taken and cut immediately from each bamboo species' specimen to undergo moisture content test. The samples were then weighed. Subsequently, they were dried in a hot air drying oven at 103 ± 2°C for 24 hours. In order to quantify the amount of moisture content left after being dried in the hot air drying oven, the samples were weighed again and the weight was recorded.

Fig 4. Specimens for moisture content test

DOI: 10.1051/matecconf/201713801024

Standard tensile strength test as described in ISO 22157 Part 2 [15] directs that specimens should be cut into a 'dogbone' shape. The specimens in this study were cut radially from the culm wall, in order to produce a 'dogbone' shape. All of the specimens were cut and prepared parallel to bamboo's fibre. For testing the specimens' tensile strength, a 0.01mm/s load was applied gradually. The setup of the tension test, which was arranged parallel to bamboo's grain is shown in Figure 5.

Fig 5. Tensile strength test setup

3 Results and discussions

3.1 Compression test and moisture content

Compressive strength and moisture content of bamboo specimens are reported in Table 1. The strength of each bamboo species' specimen was tested at a five-month interval in order to determine the strength development of the bamboo.

Table 1. Compressive strength and moisture content of bamboo

Species	Part	Average Compressive Strength (1st Month) (N/mm²)	Average Moisture Content (%)	Average Compressive Strength (5th Month) (N/mm²)	Average Moisture Content (%)
Dendrocalamus Asper	Top	68.05	20.83	73.65	15.85
	Middle	61.34	20.57	59.84	17.91
	Bottom	60.23	18.32	53.08	18.44
Bambusa Vulgaris	Top	76.52	15.29	78.74	14.01
	Middle	66.09	17.17	78.67	15.10
	Bottom	60.26	21.43	66.43	19.20
Gigantochloa Scortechinii	Top	69.02	16.09	68.62	15.60
	Middle	57.16	20.29	67.11	16.95
	Bottom	48.26	22.38	59.4	18.09
Schizostachyum Grande	Top	30.42	23.36	40.03	16.87
	Middle	28.32	19.67	31.70	17.98
	Bottom	27.05	20.44	25.77	19.63

As reported in Table 1, the average compressive strength of bamboo was observed to be high at the top part of all bamboo species, followed by middle, and lastly, bottom part. The different values signify the different specimen locations where the specimens were cut. Specimens taken near the culm's base were slightly weak and older than the specimens that were taken far from the culm's base. The difference in this property is attributed to the presence of big vascular bundles that are present in bamboos. Vascular bundles are responsible for transporting all the nutrients from culm's base during the life of bamboos. These vascular bundles have indirectly resulted in a lower fibre content at the culm's base as compared the top part of the culm.

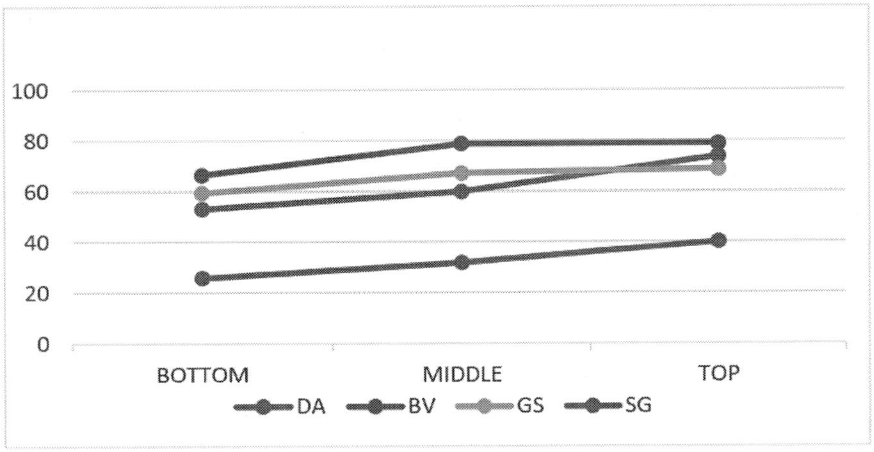

Fig 6. The difference compressive strength at bottom, middle and top.

After a five-month interval, a second compression test was conducted. The result showed a similar trend, whereby the top part of the culm possesses a high compressive strength as compared to middle and bottom parts. In moisture content test, after a five-month interval, the moisture content of bamboo specimens fell. This resulted in a higher compressive strength of the specimens as compared to one month old specimens. The moisture content of the specimens fell gradually, as all the specimens were kept in an air drying storage. As presented in Table 1, the moisture content of all four bamboo species' specimens ranges from 19.63% to the lowest moisture content, 14.01%, which is two percent short from the desired optimum moisture content, 12.00%.

Fig 7. Vertical crack and end bearing failure of specimen after compression test

DOI: 10.1051/matecconf/201713801024

Visual inspection on the specimens after compression test also revealed multiple cracks, which eventually led to splits as individual specimens' sections buckled. These vertical cracks are shown in Figure 7. From this figure, it is the evident that bamboos with air-dried culms experienced brittle failure.

3.2 Tensile test

Tensile strength test parallel to bamboo's grain was carried out in the same condition as compressive strength test, where the specimens used had undergone hot air drying. The result of the tensile test is shown in Table 2. Table 2 shows the tensile strength of *Dendrocalamus Asper* species. The average strength of *Dendrocalamus Asper* species is slightly lower at the middle part of the bamboo, while the strength at the top and bottom parts are almost equal, with recorded readings of 232.80 and 232.31 N/mm^2, respectively.

Table 2: Tensile strength of *Dendracalamus Asper*

Part	Sample	Area (mm^2)	Tensile Strength (N/mm^2)	Average Tensile Strength (N/mm^2)
Top	1T	98.68	262.47	232.80
	2T	91.83	189.49	
	3T	77.91	246.43	
Middle	1M	99.45	275.51	200.75
	2M	94.94	189.59	
	3M	99.89	137.15	
Bottom	1B	119.20	289.43	232.31
	2B	114.504	153.71	
	3B	105.19	253.82	

Table 3 shows the tensile strength of *Bambusa Vulgaris* species, which recorded a strength of above 200 N/mm^2 across all parts of the bamboo considered in this study. The results show that *Bambusa Vulgaris* exhibits an almost similar tensile strength as *Dendrocalamus Asper*, with bamboo's top, middle, and bottom parts recorded readings of, 231.67, 233.98 and 230.63 N/mm^2, respectively.

Table 3. Tensile strength of *Bambusa Vulgaris*

Part	Sample	Area (mm^2)	Tensile Strength (N/mm^2)	Average Tensile Strength (N/mm^2)
Top	1T	56.00	246.71	231.67
	2T	57.12	215.71	
	3T	58.80	232.59	
Middle	1M	60.90	237.32	233.98
	2M	60.90	229.74	
	3M	62.06	234.89	
Bottom	1B	68.48	233.45	230.63
	2B	67.20	210.31	
	3B	65.92	248.13	

For *Gigantochloa Scortechinii*, the recorded readings of tensile strength for top, middle, and bottom parts of bamboo are below 200N/mm^2. The highest strength at the top part of the bamboo recorded a reading of 187.67N/mm^2, while the bottom part recorded a reading of

176.22 N/mm². Meanwhile, middle part of the bamboo recorded a slightly lower reading than the bottom and top parts, with a reading of 144.92 N/mm². However, the difference is not significant. The tensile strength of *Gigantochloa Scortechinii* species is shown in Table 4.

Table 4. Tensile strength of *Gigantochloa Scortechinii*

Part	Sample	Area (mm²)	Tensile Strength (N/mm²)	Average Tensile Strength (N/mm²)
Top	1T	50.00	190.19	187.67
	2T	49.00	188.87	
	3T	49.00	183.95	
Middle	1M	47.00	153.36	144.92
	2M	48.50	125.12	
	3M	50.00	156.27	
Bottom	1B	53.04	151.66	176.22
	2B	53.04	225.07	
	3B	53.04	151.94	

Lastly, Table 5 shows the tensile strength of *Schizostachyum Grande* species. The top part of this species shows the highest tensile strength reading as compared to its middle and bottom parts' readings. However, in comparison to other three bamboo species, *Schizostachyum Grande* shows the lowest recorded readings of tensile strength.

Table 5. Tensile strength of *Schizostachyum Grande*

Part	Sample	Area (mm²)	Tensile Strength (N/mm²)	Average Tensile Strength (N/mm²)
Top	1T	47.43	156.71	149.20
	2T	48.45	164.58	
	3T	48.76	126.32	
Middle	1M	48.88	110.36	114.93
	2M	50.44	114.20	
	3M	48.23	120.22	
Bottom	1B	50.35	97.71	113.01
	2B	49.4	102.34	
	3B	48.76	138.97	

Figure 8 shows failures of specimens taken from tensile test. Most of the specimens experienced similar mode of failure, which comprised of the combination of tension and shear parallel to grain. This mode of failure is different as compared to timber, which has rays and knots that cause it to withstand less stress through the length of stalk.

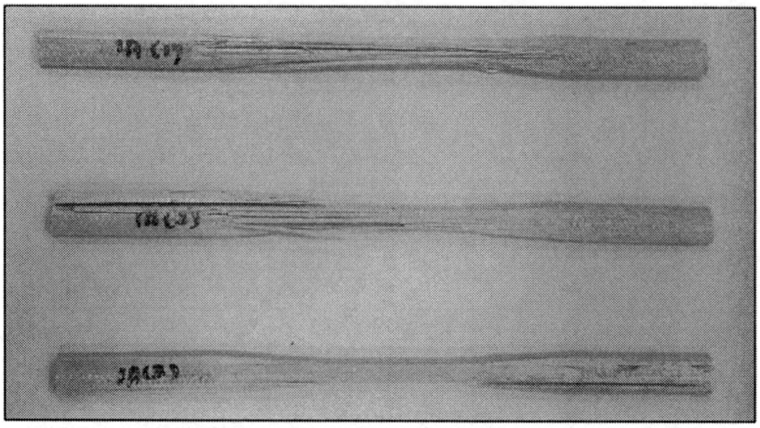

Fig 8. Mode of failure on tensile specimens

Based on previous research findings, bamboos show a good performance in strength properties as compared to other materials such as wood and steel. Table 6 shows the comparison of strength properties of spruce wood, steel, and bamboo [16]. Bamboo possesses higher compressive and tensile strengths when compared with spruce wood, but lower when compared with steel.

Table 6. Comparison of strength properties of bamboo with other materials

Materials	Compressive Strength or Yield Strength (N/mm^2)	Tensile Strength (N/mm^2)	Elastic Modulus (N/mm^2)
Spruce Wood	43	89	11000
Steel	250	410	21000
Treated Bamboo (Bambusa Vulgaris)	78	233	20000

4 Conclusions

The present study indicates that bamboo has a promising potential to be used as a construction material. The findings also indicate that the utilization of bamboo should be practiced widely. In the study, compression, tensile, and moisture content tests were conducted to test four bamboo species' ability and capability. Following a five-month interval, the compressive strength of all bamboo species had increased due to degraded moisture content. In addition, all bamboo species exhibited a good performance in tensile strength test. From the overall result of mechanical testing, *Dendrocalamus Asper* and *Bambusa Vulgaris* species endured the highest compression and possess the highest tensile strength, followed by *Gigantochloa Scortechinii* and *Schizostachyum Grande* species. Thus, the experimental results revealed that bamboo has a good performance in strength properties.

This research was supported by a basic science research program through Research University Grant (RUG) (No.Q.J130000.2522.13H01), with the funding from the Ministry of High Education (MOHE) and Universiti Teknologi Malaysia. This research was also supported by the National Research Foundation (NRF) of Korea, with the funding from the Ministry of Science, ICT, and Future Planning

(No. 2015R1A5A1037548). The authors wish to express their utmost gratitude to these funding entities, which have allowed the research endeavour to be carried out.

References

1. Q. Xu, K. Harries, X. Li, Q. Liu, and J. Gottron, "Mechanical properties of structural bamboo following immersion in water," *Eng. Struct.*, **81**, 230–239, (2014).
2. K. F. Chung and W. K. Yu, "Mechanical properties of structural bamboo for bamboo scaffoldings," *Eng. Struct.*, **24**, no. 4, 429–442, (2002).
3. W. Liese, "Research on bamboo," *Wood Sci. Technol.*, **21**, no. 3, 189–209, (1987).
4. B. Sharma, A. Gatoo, M. Bock, H. Mulligan, and M. Ramage, "Engineered bamboo: state of the art," *Proc. ICE - Constr. Mater.*, no. April, 1–11, (2014).
5. L. Yiping, L. Yanxia, K. Buckingham, G. Henley, and Z. Guomo, "Bamboo and climate change mitigation : a comparative analysis of carbon sequestration," **no 32**, 47, (2010).
6. A. Syeda, B. Shrujal, and J. Kumar, "A case study on bamboo as green building material," **no 2**, 78–82, (2014).
7. M. R. Wakchaure and S. Y. Kute, "Effect of moisture content on physical and mechanical properties of bamboo," *Asian J. Civ. Eng. (Building Housing)*, **13**, **no. 6**, 753–763, (2012).
8. Chinese Standard Agency. Testing methods for physical and mechanical properties of bamboo used in building. JG.T199-2007; (2007).
9. K. Ghavami, "Bamboo: Low cost and energy saving construction materials," *Int. Conf. Mod. Bamboo Struct.*, no. January 2008, (2008).
10. Janssen, J.J.A. 1985. The machanical properties of bamboo. Resent Research in Bamboo. *Proceeding of the International Bamboo Workshop Hangshou, China*. Oct. 6-14, (1985).
11. Kabir, M.F., Bhattacarjee, D.K. & Sattar, M.A. Physical and mechanical properties of four bamboo species. *Bangladesh Journal of Forest Science*, **20**, pp 31-36, (1991).
12. Lavers, C.M. The strength properties of timber. *Forest Product Research Bulletin*. **no. 5**, 62,(1969).
13. Sattar, M.A., Kabir, M.F. & Bhattacharjee, D.K. Effect of age and height position of muli (Melocanna baccifera) and borak (Bambusa balcooa) on the physical and mechanical properties. *Bangladesh Journal Forestry Science*, **19 (1 and 2)**, 29-38, (1990).
14. ISO/TR. 22157-1, "Bamboo - Determination of physical and mechanical properties - Part 1: Requirement," **2004**, 1–8, (2004).
15. ISO/TR 22157-2, "Bamboo - Determination of physical and mechanical properties - Part 2: Laboratory Manual," **2004**, 28, (2004).
16. T. Gutu, "A Study on the Mechanical Strength Properties of Bamboo to Enhance Its Diversification on Its Utilization," **no. 5**, 314–319, (2013).

Lateral Displacement And Shear Lag Effect Of High-Rise Buildings With Diagrid SystemThat Is Constructed Above A Frame

Roslida Abd. Samat[1,*], *Fong Teng* Chua[1] , *Nur Akmal Hayati* Mohd Mustakim[1] , *Fatin Izzaidah* Anuar[1] , *Sariffuddin* Saad[1], and *Suhaimi* Abu Bakar[1]

[1]Faculty of Civil Engineering, Universiti Teknologi Malaysia, 81310 Skudai, Johore, Malaysia

Abstract. Diagrid system has gained a wide acceptance in the design of tall buildings due to its many advantages including its high structural efficiency in resisting both gravity and lateral loads. Most diagrid structures that had been studied have full triangulated members from the ground level to the top of the buildings where comparison in the effectiveness in minimizing the lateral displacement was often made between structures with full diagrid, frame and outrigger system. Nevertheless, no study has been performed on the effectiveness of the diagrid that is constructed above a frame system. The objective of this research is to understand the behavior of the lateral displacement and shear lag effect due to wind load when the diagrid structure is constructed above a frame. Models of sixty storey buildings were analyzed by using Staad.Pro software. The level where the diagrid members started and the spacing of vertical base columns of the frame were altered. The lateral displacement and shear lag effect resembled closely of those of the model of full diagrid when the diagrid was started at level 3, and the vertical base columns were uniformly spaced at 6 metres.

1 Introduction

The struggle to control the lateral displacement of tall buildings has caused the emergence of many tall building systems. The latest tall building system which now has become favourable among today's designer is diagrid system. Diagrid system is composed of many diagonal structures that connect together to form a triangulated shape or can be seen as grid shape. The combination of "diagonal" and "grid" words produces the term "diagrid" which refers to a structural system that is single-thickness in nature and gains its structural integrity through the use of triangulation [1]. Unlike the braced tube structure, diagrid structure has no vertical columns present and thus is considered as the evolution of braced tube structure [2]. Diagrid system that is favoured due to its structural efficiency and aesthetic potential [3,4], is categorized as exterior structure, which is a lateral load resisting system that is located along the building perimeter. Diagrid enables the architect to design a building with some unique elements such as irregular angle and nonlinear shape as

* Corresponding author: roslida@utm.my

portrayed in buildings like Swiss Re (30 St. Mary Axe) in London, Capital Gate in Abu Dhabi and CCTV in Beijing. [5] studied the structural performance of complex-shaped tall buildings such as twisted, tilted and freeform towers that were designed with diagrid system.

The aesthetically unique design of skyscraper that diagrid provides is mainly due to its free of exterior columns [6]. Since diagrid system allows the elimination of almost all the vertical columns, this system is able to provide variety of open floor plans. Diagrid building is an efficient system because it is able to achieve the same displacement as the displacement of buildings that have other tall building systems with significant reduction of the steel volume. [7] reported that the implementation of diagrid system to "The Bow " or Encana Tower that is located in Calgary in Western Canada had managed to reduce approximately 20 percent of the structural steel weight when compared to a building with a conventional moment frame system.

Exterior structural systems including diagrid are designed to control the lateral displacement of tall building. The taller the building is the higher the wind load is as wind speed increases parabolically with height from ground. Damage to non-structural elements such as cracking in reinforced walls, damage to lightweight partitions and impaired operation of windows are common examples of the effect of excessive lateral displacement of tall building [8]. Thus, the control of lateral displacement is a serious issue in the design of tall building. A building should not sway horizontally more than H/500 to consider the design criteria that are strength, serviceability, stability and human comfort [9]. [10] recommends the drift limit to be on the order of 1/600 to 1/400 of the building or story height.

Another important issue in the design of high rise building is shear lag. Shear lag is caused by the lack of shear stiffness that reduces the structural effectiveness of the framed tube. [11] stated that the diagrid tube structure is one of the structural solution to minimize shear lag effect. [12] proved that diagrid structures indeed lower the shear lag compared to frame tube structure. However, the increase of the slope of the braces of diagrid increases the shear lag and becomes more rapid when the angle is higher than 70°. Interestingly, shear lag does not correlate with the lateral displacement of high rise building [12].

Full diagrid structures are buildings that have diagrid system being constructed from the ground floor to the top of the building. Most diagrid buildings in the world are full diagrid structures, for example, Swiss Re (30 St. Mary Axe) in London, CCTV in Beijing, United Steelworkers Building in Pittsburgh, USA and Tornado Tower in Doha, Qatar. There are a few buildings in the world that employ combination of diagrid system and other tall building systems at their perimeter, for example, Lotte World Tower in Seoul, Korea which has its diagrid being constructed only from the 107th floor to the top [13] and Hearst building in New York which has its diagrid system being supported by mega columns and super-diagonals that are constructed from the foundation to the 10th floor [14].

Many studies on the performance of full diagrid structures where the diagrid system begins from the ground floor to the top of building have been conducted. However, there is no systematic research that has been carried out on tall building that has diagrid that are constructed above a frame structure. Thus, the objective of this study is to find the effect of constructing a diagrid system above a frame to the lateral displacement and shear lag of a sixty storey building due to wind load. The parameters of the vertical columns of the frame that transmitted the force from the diagrid system to the foundation were varied to investigate how they influenced both the lateral displacement and shear lag. The parameters that were manipulated are the level of where the diagrid system started, the spacing of the columns of the frame and the arrangement of the columns of the frame.

 DOI: 10.1051/matecconf/201713802001

2 Methodology

The structural analysis software, STAAD.Pro V8i was used to analyse all models statically. All elements in the buildings such as beams, columns and diagrid truss were modelled by using line element where each node has three degrees of freedom which are independent vertical motion, independent horizontal motion and independent rotational motion.

Generally, each model has 60 stories with 4 m storey height and 36 m x 36 m square floor plan as shown in Figure 1. All columns were reinforced concrete columns. The size of the internal column is 1200 mm x 1200 mm, 900 mm x 900 mm and 600 mm x 600 mm at Level 1 to 20, Level 21 to 40 and Level 41 to 60, respectively. Full diagrid model did not have any external columns, while the models with the combination of frame and diagrid systems did have external columns in the frame system part, where the dimensions of those columns were varied such that the total cross sectional area of the columns remained the same for all models.

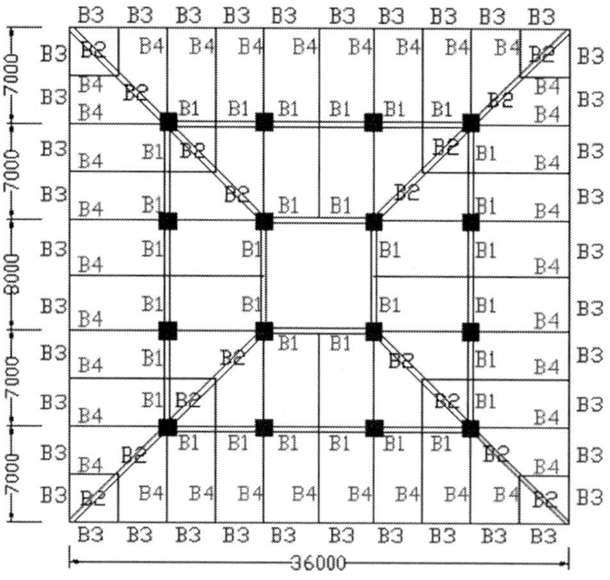

Beam specification :

B1 - 650 mm x 250 mm reinforced concrete beam
B2 - 750 mm x 350 mm reinforced concrete beam
B3 - 700 mm x 700 mm x 25 mm thick steel hollow tube
B4 - UB 406x140x53 steel beam

Fig. 1 Typical plan view of all Building Models.

The angle of the triangulated members was uniform and was approximately about 69 degrees, where one module of diagrid was equalled to two storeys of the building. [15] stated that, 69 degrees is the near optimum angle for 60-story high building with diagrid structure. Furthermore, the intersection between the diagonal members was pin-connection so that axial action merely carries the transverse shear and moment. The diagonal members had pinned supports, while the external and internal columns had fixed-end supports. Models of combination of diagrid and frame systems used reinforced concrete beams of 400mm x 1000 mm as the transfer beams that transmitted the axial forces in the diagrid system to the frame system

Three parameters that were studied in combination of diagrid and frame systems are the height of the frame, the spacing of uniformly spaced exterior base columns and arrangement of exterior base columns. Model 1 is a frame model without any diagrid while Model 2 is a fully diagrid building. For the study of the height of frame of models with combination of frame and diagrid system, the frame system was at the lower levels while diagrid system was at upper levels where the diagrid started at level 1, 3, 5 and 7, for Model 3*h*, 4*h*, 5*h* and 6*h*, respectively, as shown in Figure 2.

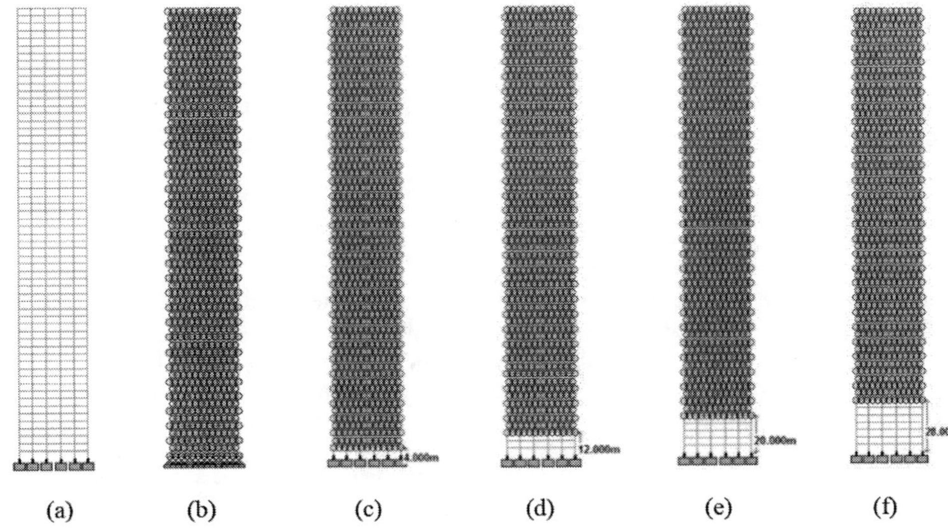

Fig. 2. Side elevation of 60 storey models (a) Model 1: Frame without diagrid, (b) Model 2h: Diagrid started at G level, (c) Model 3*h*: Diagrid started level 1, (d) Model 4*h*: Diagrid started level 3, (e) Model 5*h*: Diagrid started level 5 and (f) Model 6*h*: Diagrid started level 7.

Investigation on parameter 2, which is the spacing of uniformly spaced exterior base columns, was performed by varying the spacing between the base columns to 6m, 7.2m, 9m, 12m and 18m spacing. The height of the frame is 4m and the dimension of the base column building model was varied such that the total cross sectional area of the base columns remained the same for all spacing of the base columns. The third parameter which is the arrangement of the base columns was studied by varying both the spacing and the type of arrangement of the base columns: uniform spacing between the columns (Model 3*r*), varying spacing but symmetrical arrangement of the position of the columns (Model 4*r*, 5*r*, 6*r*, and 7*r*) and varying spacing with non-symmetrical arrangement of the position of the base columns (Model 8*r*, 9*r* and 10*r*).

The wind load was calculated based on the procedure provided in the ASCE7-10. Two wind environments that were considered are Malaysian wind speed which is benign and Hong Kong wind speed which is more aggressive. Determination of the displacement of a building is a serviceability design while the determination of shear lag is a survivability design. Thus, the 3-second gust wind speed for 10 year return period at 10 m high for Malaysian and Hong Kong wind environment which is 28.14 m/s and 42.48 m/s, respectively were applied during the analysis for displacement. Analysis for axial forces which was used to observe shear lag had the 3-second gust wind speed for 50 year return period at 10 m high for Malaysian and Hong Kong wind environment of 33.5 m/s and 57.4 m/s, respectively. The buildings, further, were assumed to be located in an urban area and thus, exposure B as stated in ASCE 7-10 was used.

3 Results and Discussion

Results from the dynamic analysis that was performed on all building models show that the lowest natural frequency of all models exceeded 1 Hz. Thus, the buildings are categorized as stiff buildings, and static analysis is sufficient to determine both lateral displacement and stresses of the models, as in accordance with building code in United States of America, ASCE 7-10 [10].

3.1 Lateral Displacement

The pattern of the lateral displacement is determined under three different conditions : variation of base column height, variation of column spacing and variation of column arrangement.

3.1.1 Variation of base column height

The maximum lateral displacement occurred at the top of the building. For both Malaysian and Hong Kong wind environments, the frame model had the largest lateral displacement while the full diagrid structure had the lowest displacement, where the percentage of difference is 66 percent when compared to the displacement of the frame structure. Increasing the height of the frame reduced the percentage of difference slightly, which is 61, 60, 59 and 57 percent when the diagrid began at Level 1, 3, 5 and 7, respectively.

3.1.2 Variation of column spacing

The larger the base column spacing was, the higher the lateral displacement became, except for base column spacing of 7.2m. Table 1 shows that the lateral displacement for base column spacing of 7.2 m was more than the 9m spacing, and was almost the same as the lateral displacement for 12m base column spacing. This is because the total moment of inertia of the columns for 7.2m column spacing is less than total moment of inertia of the 9m column spacing despite the total cross sectional area of the base columns are kept the same for all column spacing.

The spacing between the base columns did affect the lateral displacement of the diagrid structure where the percentage of difference of the lateral displacement of the structures when compared to the full diagrid structure is 5.3%, 17.4%, 10.1%, 16.2% and 28.7% for the base column spacing of 6m, 7.2m, 9m, 12m and 18m. The lateral displacement is the least for the base columns with the least spacing which is 6m and the highest for base column spacing of 18 m which is the largest spacing.

Table 1. Moment of Inertia and Maximum Lateral Displacement of the models with different Base Column Spacing.

Column Spacing (m)	Moment of Inertia ($x10^{15}$ mm^4)	Maximum Lateral Displacement (mm)
6.0	9.153	141.1
7.2	9.032	157.3
9.0	9.606	147.6
12.0	9.496	155.7
18.0	9.367	172.4

3.1.3 Variation of column arrangement

The combined frame and diagrid system with 6m uniformly spaced base columns had slightly larger lateral displacement when compared to a full diagrid building in the Hong Kong wind environment as shown in Table 2. The total moment of inertia at the ground floor differs for different arrangement of the base columns. The lateral displacement is influenced by the value of moment of inertia and the spacing of the columns near to the corner of the buildings. Model 7*r* which had the smallest moment of inertia and the largest spacing of the columns near to the corner of the building had the largest lateral displacement while Model 3*r* which had one of the largest of moment inertia and small spacing of the column near to the corner of the building had the smallest lateral displacement. Model 8*r* and 9*r* which had the same moment of inertia and the same column spacing at the corner of the building had similar deflection. Model 5*r*, 6*r* and 10*r* had the same moment of inertia, but Model 6 which had the largest column spacing at the corner of the building among the three models had the largest lateral displacement. In contrast, Model 10*r* had the smallest column spacing at the corner when compared to Model 5*r* and 6*r* , and thus had the smallest lateral displacement among the three models.

Table 2. Lateral displacement and total moment of inertia at ground level for different arrangement of the base columns of frame-diagrid models

Model – *(Spacing of Base Column)*	Lateral Deflection (mm)	Moment of inertia (mm^4)
1 - Frame	399.507	
2 - Full Diagarid	134.687	
3*r* - Diagrid-frame *(6.0m-6.0m-6.0m-6.0m-6.0m-6.0m)*	141.777	4.15×10^{12}
4*r* - Diagrid-frame *(4.0m-4.0m-10.0m-10.0m-4.0m-4.0m)*	150.312	4.16×10^{12}
5*r* - Diagrid-frame *(7.0m-11.0m-11.0m-7.0m)*	157.862	2.77×10^{12}
6*r* - Diagrid-frame *(11.0m-7.0m-7.0m-11.0m)*	160.731	2.77×10^{12}
7*r* - Diagrid-frame *(14.0m-8.0m-14.0m)*	174.708	2.08×10^{12}
8*r* - Diagrid-frame *(6.0m-8.0m-10.0m-12.0m)*	152.022	2.43×10^{12}
9*r* - Diagrid-frame *(12.0m-10.0m-8.0m-6.0m)*	151.753	2.43×10^{12}
10*r* - Diagrid-frame *(6.0m-11.0m-13.0m-6.0m)*	150.895	2.77×10^{12}

3.2 Shear Lag Effect

Shear lag effect can be recognized by the uneven distribution of vertical axial forces in the peripheral columns that are located at the flange of the building or by the nonlinear distribution of the axial force in the peripheral columns that are located at the web of the buildings.

3.2.1 Variation of base column height

The stress distribution of the columns that were located on the flange panel at the first storey ($z/H = 0$) of all models, showed positive shear lag effect, where the columns at the edge of flange panel experienced larger axial stresses compared to the middle columns. The axial stress distribution of the columns was almost uniform at fifteenth storey, and became uniform at thirtieth storey and above.

Furthermore, the overturning moment caused by the lateral load formed the tensile axial stress at the windward face and compressive stress at the leeward face. However, the weight of the building balanced the tensile stress that resulted in minimal value of compressive stress at the windward face, while augmented the compressive stress at the leeward face. As a result, the columns at the two corners of leeward face of the building had very large axial load, contrary to the other columns at the two corners of the windward face of the building that resisted minimal amount of axial load.

Shear lag effect was reduced significantly and the value of the stress was lowered when full diagrid building is adopted instead of a frame building. The largest stress at the perimeter column which is located at the corner of the leeward face is reduced by 64 percent that is from 51300 kN to 18700 kN when full diagrid building was used instead of a frame building. The value of stresses that were resisted by the full diagrid building in flange panel and web panel are the least among all models.

However, when the frame is combined with the diagrid, where the diagrid started at the upper level, the effect of the positive shear lag increased significantly. The largest stress at the perimeter columns had the maximum value of 54240 kN at columns at the leeward face for Model 3 where the diagrid started at Level 1. The value of stress in these columns at the leeward face was reduced by 7 percent to 50670 kN, by 10 percent to 48960 kN and by 12 percent to 47890 when the diagrid started at Level 3, Level 5 and Level 7, respectively. In addition, a minor negative shear lag effect where the stress in the columns at the corner was lower than the stress in the columns at the middle, occurred at the fifteenth storey (z/H = 0.25) for all models with combination of diagrid and frame system.

Shear lag between the models can also be compared by using the factor f, which is the ratio between the axial forces at the corner columns to the axial forces at the middle of the panel [12]. A factor more than 1.0 indicates a positive shear lag which occurred at the base of the models; while a factor less than 1.0 indicates negative shear lag which usually occurred at mid-height of the building models.

The use of full diagrid system reduced both the deflection and the shear lag significantly (Table 3). However, when one storey high frame is added below the diagrid system, the shear lag increased tremendously. There was a slight increase of displacement but was still very much lower than the deflection of a frame building. Increasing the height of frame below the diagrid reduced the shear lag, but increased the deflection slightly. The addition of one storey high frame below the diagrid system caused a sudden nonlinear distribution forces in the base columns as these columns carried the loading from the entire fifty-nine storey high of axial forces. As the height of the frame was increased, the forces from the diagrid system above it were redistributed in a better flow to the base column.

Table 3. Deflection and shear lag ratio for models with different height of frame under the diagrid system.

Model	Level where diagrid started	Malaysia Wind Load Case			Hong Kong Wind Load Case		
		Displacement, x-direction (cm)	Shear Lag Ratio, f		Displacement, x-direction (cm)	Shear Lag Ratio, f	
			Base	Mid Height		Base	Mid Height
1	No diagrid	17.47	1.14	0.97	39.87	1.39	0.96
2	Ground	5.87	1.18	0.80	13.42	1.30	0.79
3h	1	6.88	1.64	0.90	15.70	1.70	0.87
4h	3	6.87	1.48	0.90	15.90	1.57	0.87
5h	5	7.24	1.39	0.90	16.53	1.50	0.87
6h	7	7.58	1.33	0.90	17.31	1.46	0.87

3.2.2 Variation of column spacing

The deflection increased slightly when the diagrid was combined with the frame that had uniform spaced columns, the shear lag factor generally decreased to a value close to 1, which indicates an almost uniform stress for all perimeter columns at the base (Table 4). However, when the spacing was 7.2m, both the lateral deflection and the shear lag ratio increased more than when full diagrid system was used. This is the case when the moment inertia is the least compared to other models with different base column spacing and, its perimeter columns being aligned to the internal columns. This alignment disrupts the transmission of the axial force of the columns at the corner to the adjacent perimeter columns and thus causing the columns at the corner to bear large axial force

Table 4. Lateral Deflection and Shear Lag Ratio for Different Building Model.

Type of Model Building	Model Number	Deflection (cm)	Shear Lag Ratio, *f*	
			Base	**Mid-height**
Full Frame	1	39.87	1.390	0.960
Full Diagrid	2	13.42	1.294	0.792
Modified Diagrid with 6m spacing	3-*u*	14.11	1.091	0.875
Modified Diagrid with 7.2m spacing	4-*u*	15.73	1.426	0.872
Modified Diagrid with 9m spacing	5-*u*	14.76	1.032	0.874
Modified Diagrid with 12m spacing	6-*u*	15.57	0.979	0.872
Modified Diagrid with 18m spacing	7-*u*	17.24	0.946	0.868

3.2.3 Variation of column arrangement

Figure 3 shows the distribution of axial stress of the columns at the flange panel for model 3*r*, 4*r*, 5*r* and 6*r*.

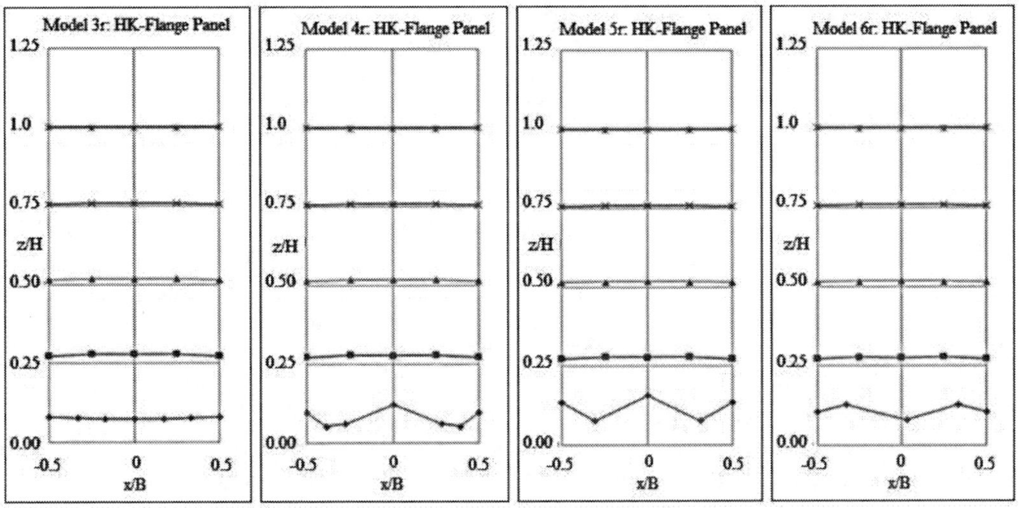

Fig. 3. Shear lag effect on the four models: 3*r*, 4*r*, 5*r* and 6*r*, with combination of diagrid and frame with different base column arrangement in the flange panel of the model in Hong Kong wind environment

Model 3*r* has the most uniform distribution of axial stress at ground level among all the models and has the lowest axial stress among the models that have different arrangement of the base columns. Generally, at mid height of the building where z/H = 0.5, the pattern of axial forces distribution shows the negative shear lag effect for all models regardless the arrangement of the base columns.

4 Conclusion

The use of diagrid system, either full diagrid or combined with frame at lower levels reduced the lateral displacement significantly, when compared to full frame structure. The percentage of reduction of the lateral displacement ranged from 56 percent to 66 percent. The difference of lateral displacement between the models where diagrid started at upper level with fully diagrid model is minor, which is less than 4 cm. Diagrid started at level 3 recorded the least displacement when compared to other models with diagrid started at upper level. Base columns with uniform spacing of 6m had the least displacement when compared with models of combination of diagrid and frame system that had different spacing value and different arrangement of columns. The reduction of the lateral displacement of the models was 64.5 percent, 56.8 percent and 56.3 percent for 6m uniform spacing, 18m uniform spacing and non-uniform spacing of 14m-8m-14m, respectively, when compared to full frame models. It is concluded that lateral displacement increased with increased spacing of the base columns and the non-uniformity of spacing of the base columns.

Shear lag effect was worst at the lowest level of the building where the shear lag ratio is more than one. The two columns at the corner of the building at the leeward face endured the largest axial stresses to resist the overturning moment due to the lateral wind load. The level where the diagrid started affected the shear lag where the worst shear lag occurred when the diagrid started at level 1. However, as the higher the level where the diagrid started was, the lower the shear lag ratio became. Meanwhile, the shear lag ratio at the ground level of the models of diagrid-frame system with uniformly spaced base columns was close to unity regardless the value of the spacing except when the spacing was 7.2 m. The exterior base columns were aligned with the interior columns when the spacing was 7.2m and this reduce the tube action by the exterior columns, and thus, inhibited the axial force from being distributed to the adjacent columns. This phenomena is observed for all models of diagrid-frame combination with different level of where the diagrid started. All of these models had the exterior columns aligned with the interior columns and the shear lag ratio of these models was very much larger compared to other diagrid-frame models where their exterior columns did not align with the interior columns.

Furthermore, the uniform spacing of the base column of models with combination of diagrid and frame system had the least shear lag effect as these models had a uniform distribution of the axial forces among their exterior base columns. In contrast, non-uniform spacing of base column caused the distribution of the axial forces at ground level to become irregular.

As a conclusion, diagrid system combined with frame system decreased the percentage of reduction of the lateral displacement of building when compared to full frame building. However, the value of the decrease of the percentage of reduction of the lateral displacement is very small and thus, the use of the combination of diagrid and frame system is still effective in reducing the lateral displacement of building. Furthermore, the shear lag effect can be reduced by having uniform spacing of the base columns. However, the base columns must not be aligned with the interior columns to avoid the increase of shear lag effect.

Further study on the behaviour of the lateral displacement and shear lag on the combination of diagrid and frame should be done but with the height of the frame increases to 25percent, 50 percent and 75 percent of the height of the building. The same parameters which are the spacing and arrangement of the exterior columns should be studied.

The study was funded by the Ministry of Education Malaysia and Research Management Centre, Universiti Teknologi Malaysia under Research University Grant No. QJ130000.2622.12J25.

References

1. T.M. Boake, *Diagrid Structures: Systems, Connections, Details* (Birkhauser Verlag GmbH, 2014)
2. S.Yadav, V. Garg, Int. j. civ. struct. eng. res, **3**, 394-406 (2015)
3. H. Varsani, N. Pokar, D. Gandhi, IJAREST, *2*, 1–4 (2015)
4. K.S. Moon, M.M. Ali, Archit Sci Rev, **50**, 205-223 (2007)
5. K.S. Moon, Procedia Eng., **14**, 1343-1350 (2011)
6. K.S. Moon, Archit Sci Rev, **51**, pp 239-251 (2008)
7. B.Charnish, T.MCDonnell, *CTBUH 8th World Congress* (2008)
8. L. G. Griffis, ENG J AISC, **First Quarter**, (1993)
9. P. Jayachandran, *National Workshop on High-rise and Tall Buildings* (2009)
10. American Society of Civil Engineer, *ASCE Standard ASCE/SEI 7-10 Minimum Design Loads for Buildings and Other Structures* (2010)
11. B.S.Taranath, *Structural Analysis & Design of Tall Buildings* (1988)
12. J. Leonard, *Investigation of Shear Lag Effect in Highrise Buildings with Diagrid System* (2004)
13. http://www.skyscrapercity.com/showthread.php?t=474259&page=293
14. http://www.fosterandpartners.com/projects/hearst-tower/
15. K.S.Moon, International Journal of Sustainable Building Technology and Urban Development, **2**, 37-42 (2011)

Applications of vertical steel pipe dampers for seismic response reduction of steel moment frames

Junaedi Utomo[1*], *Muslinang* Moestopo[2], *Adang* Surahman[2], and *Dyah* Kusumastuti[2]

[1]Civil Engineering Department, Universitas Atma Jaya Yogyakarta, Yogyakarta, Indonesia.
[2]Faculty of Civil and Environmental Engineering, Bandung Institute of Technology, Bandung, Indonesia.

Abstract. A newly developed vertical steel pipe damper is introduced to improve the seismic performance of steel moment frames. The damper exhibits large lateral stiffness and excellent capability to dissipate energy due to earthquakes. It provides a reliable, compact, inexpensive, and replaceable damper. Improved performance of the structure is verified analitically using a four-story steel moment frame equipped with steel pipe dampers. Vertical steel pipe dampers are placed between any two points where large relative motion exists during earthquake excitation. A nonlinear dynamic analysis of the structure using PERFORM-3D software demonstrated the significant benefit of equipping the structure with steel pipe dampers. All structural components, except the steel pipe dampers, remain elastic during earthquake excitation. Structures properly designed with vertical steel pipe dampers will only require minimum post-earthquake inspection and limited damage. Some practical issues associated with the application of vertical steel pipe dampers to building structure for seismic response reduction are presented in this paper.

1 Introduction

A newly developed vertical steel pipe damper has been introduced to improve the seismic performance of steel moment frames [1]. The damper has large lateral stiffness, small yield displacement and excellent capability to dissipate a tremendous amount energy. The damper is intended to be installed in low to medium rise structures. An improved performance of a four-story steel moment frame with and without steel pipe dampers was verified analytically using north-south components of the El-Centro 1940, Chi-Chi 1999, Fukushima-Hamadori 2011 and Padang 2009 earthquakes accelerograms. The structure was not intended to be designed for a specific site. Therefore no spectral matching, to obtain design ground motion time histories to match a target response spectra of a site, was done. The results of the nonlinear dynamic analysis of the structure demonstrated the significant benefit of equipping the structure with vertical steel pipe dampers. All structural components, except the dampers, remain elastic during four strong earthquake excitations.

* Corresponding author: utomoj@live.com

DOI: 10.1051/matecconf/201713802002

2 Vertical steel pipe damper

Steel pipes in vertical position are sensitive to the diameter (D) to thickness ratio (t) of the pipes. The schedule 80 carbon steel pipe which has small D/t ratio was chosen to avoid buckling in middle part of the pipe. Pipe having D = 114.3 mm, t = 8.6 mm and height (h) = 200 mm was used as the material of the damper. Abebe et al. [2] proposed the ratio of height to diameter of the pipe as √3 so that bending and shear stress yield simultaneously. The results of simple tensile tests to obtain the material properties of the steel pipe and plate are shown in Table 1. Figure 1 shows the results of numerical simulation using ABAQUS [3] for bare pipe fixed at both ends subjected cycles of increased amplitude of lateral displacement by 1xδ_y in each consecutive cycle (δ_y is the yield displacement of the damper). As shown in Figure 1, no local buckling occurred at the middle part of the pipe but local buckling occurred at the ends of the pipe manifesting in unstable hysteretic curve. The specimen of the bare pipe had been tested using the same cyclic loading above, and fracture occurred at the heat affected zone area (HAZ) close to the ends of the pipe. Some kind of strengthening to the bare pipe are required to avoid buckling at ends of the pipe, to relieve the stresses at the ends of the pipe, and to relocate the fracture away from HAZ area.

Figure 2 shows the two improved model of vertical steel pipe dampers. The model of vertical steel pipe dampers shown in Figure 2a had been tested, and fracture occurred at the heat affected zone area (HAZ) close to the ends of the pipe. The details of the strengthening plate outside the pipe were improved so that: (1) buckling at the pipe was eliminated; (2) connection failure at the ends of the pipe was avoided; (3) fractures at HAZ region were avoided; (4) early fractures at points of high intense stress were postponed, and (5) extensive yielding was concentrated in the middle part of the pipe. As shown in Figure 2b, curved and tapered strengthening plate were used to prevent buckling at the ends of the pipe, to concentrate the extensive yielding in the middle part of the pipe and to postpone and to shift fracture locations away from HAZ areas.

Table 1. Material properties of the steel used for numerical simulation

Steel	Modulus of elasticity (MPa)	Yield stress (MPa)	Ultimate stress (MPa)	Breaking strain (%)
Pipe	200.000	330	465	37
Plate	200.000	360	500	25

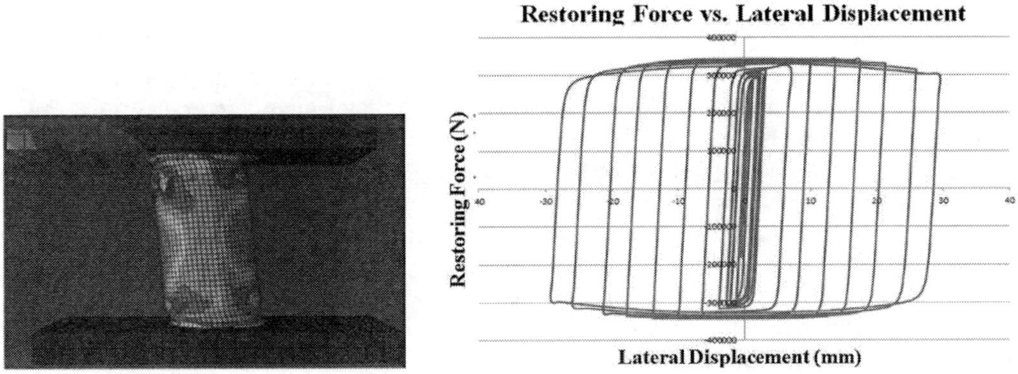

(a) (b)

Fig. 1. Local buckling at the ends of the bare pipe: (a) von Mises stress distribution and (b) Hysteretic curve

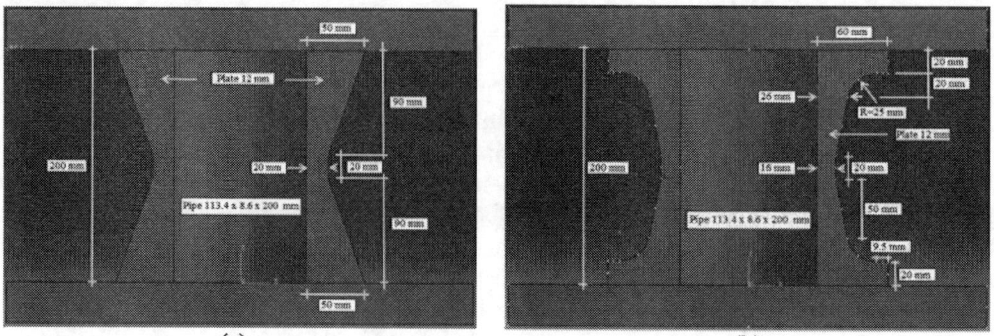

(a) (b)

Fig. 2. Details of the improved model of vertical steel pipe dampers: (a) Strengthened with a pair of trapezoidal shape plates and (b) Strengthened with a pair of curve shape plates

Figure 3 shows the von Mises stress distribution of the improved model shown in Figure 2b. The results of the numerical simulation show: (1) The point of high intense stresses where fractures are expected to happened has been shifted away from HAZ region near the ends of the pipe; (2) Welded connections are placed at low stress areas, and (3) The high stresses at the ends of the pipe are relieved. The improved model of the vertical steel pipe damper has been tested using ATC-24 loading protocol using cycles of increased amplitude of lateral displacement by $1\text{x}\delta_y$ in each of three consecutive cycles.

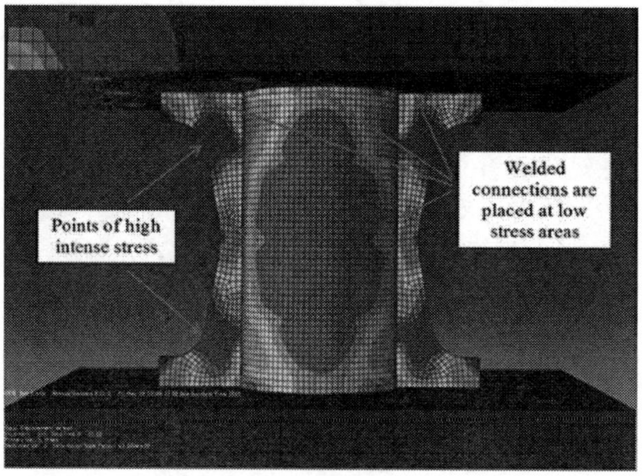

Fig. 3. von Mises stress distribution of the improved model of the vertical steel pipe damper

The stable parts of the hysteretic curves of the test results of the dampers were plotted in Figure 4. Figure 4a shows the plot of two graphs of the bare pipe damper and the vertical pipe dampers strengthened with a pair of trapezoidal shape plates on one Cartesian coordinate system, and Figure 4b shows the same plot of the bare pipe damper and the vertical pipe dampers strengthened with a pair of curve shape plates. The number and distribution of plastic cycles from the test results determined the energy dissipation capacity of the dampers. Table 2 shows the number of plastic cycles and the energy dissipated by each of damper due to the following cyclic displacement loadings: (1) For the bare pipe and the pipe strengthened with a pair of trapezoidal shape plates, the displacement loading amplitude was increased in each consecutive cycle by $1\text{x}\delta_y$, and (2) For the pipe strengthened with a pair of curve shape plates, the displacement amplitude was increased in each three consecutive cycles by $1\text{x}\delta_y$ (ATC-24). The number of plastic cycles from the test results reflects the quality of the detailing of the dampers. As shown in the last row of Table

2, applying smooth geometry configuration to the strengthening plate outside the pipe manifested in the increase capacity of the energy dissipation of the damper significantly.

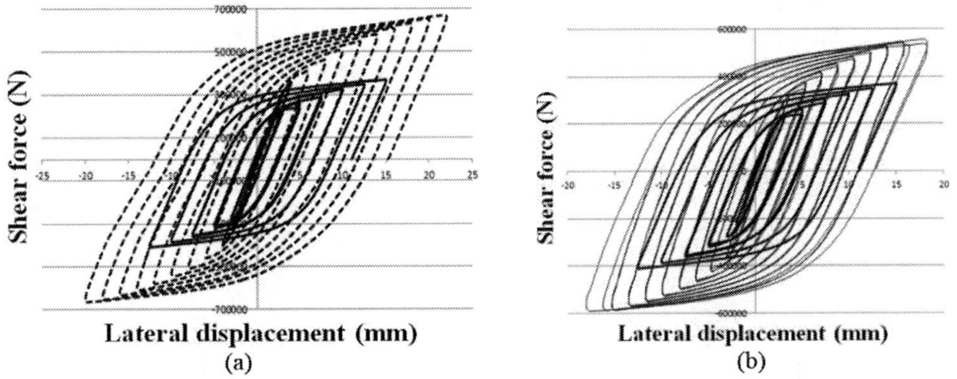

Fig. 4. Plot of two graphs of stable parts of hysteretic curves of vertical steel pipe dampers: (a) Bare pipe vs. pipe strengthened with a pair of trapezoidal shape plates, and (b) Bare pipe vs. pipe strengthened with a pair of curve shape plates

Table 2. Number of plastic cycles and increase capacity of energy dissipation

Vertical steel pipe dampers	Number of Plastic Cycles	Total Dissipated strain energy (N.mm)	Increase capacity of energy dissipation (%)
	7	0.325e+08	-
	8	1.09e+08	236.72
	**29**	**2.45e+08**	**654.08**

3 Application of vertical steel pipe dampers

The vertical steel pipe damper strengthened with a pair curve and tapered plates will be installed in a four-story steel moment frame to reduce the seismic response of the building due to four strong earthquakes. Nonlinear dynamic procedure was used to access the performance capability of the building. Limiting the lateral defection of the damper to 10xδ_y (δ_y = 1.80 mm is the yield lateral displacement of the damper) due to four strong earthquakes was selected as one of the performance objective of the building. A preliminary designed has been carried out to estimate the required number of dampers in each story of the building. A computational model of the building that incorporates the nonlinear load-deformation characteristic of the individual damper was built using PERFORM-3D [4]. The computational model was then subjected to the north-south component of the Chi-Chi 1999, El-Centro 1940, Fukushima-Hamadori 2011, and Padang 2009 ground motion time-histories. The resulting maximum absolute deflection of the dampers was directly compared to the performance objective of the building.

 DOI: 10.1051/matecconf/201713802002

3.1 Steel moment resisting frame

A four-story steel moment resisting frame shown in Figure 5 was used to study the application of the vertical steel pipe damper. A number of end releases were applied so that the lateral load resisting systems in the building are the peripheral frames. One-way slab systems were applied to all floors. The secondary beams were not shown in Figure 5.

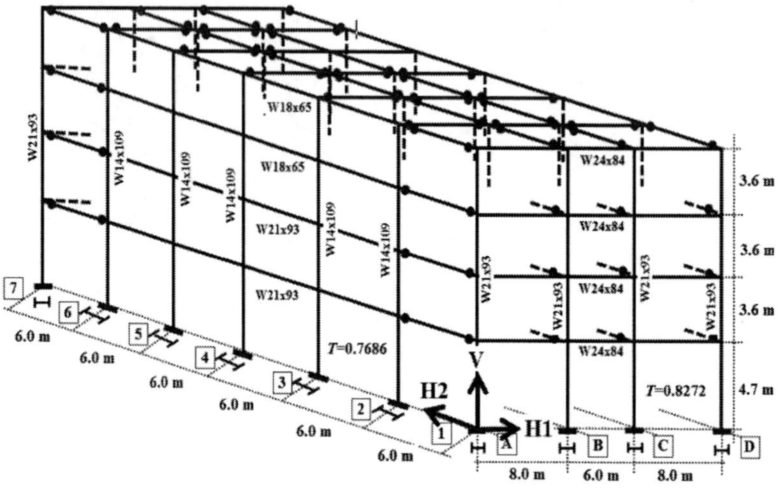

Fig. 5. Four-story steel moment resisting frame

The superimposed specified loads are as follows:
Gravity loading:
- Live load: Roof LL = 1.0 kPa, Floor LL = 2.4 kPa
- Dead load: Floor weight plus topping = 3.6 kPa, Partitions = 1 kPa

Materials:
- Concrete: 24.0 MPa
- Reinforcing steel: ASTM A572 steel, grade 50 f_y = 345 MPa

Elements dimension:
- Exterior columns: W21x93 (along H1), and W14x109 (along H2)
- Interior columns: W12x72 (along H2)
- Exterior beams: W24x84 (along H1), W21x93 and W18x95 (along H2)
- Interior beams: W24x84 (along H1), W21x93 and W18x95 (along H2)

Fundamental periods:
- T = 0.8272 seconds (along H1)
- T = 0.7689 seconds (along H2)

3.2 Modelling dampers as inelastic components

Originally all the columns height were 3.6 m and the four-story steel moment frames had been designed to withstand moderate intensity earthquake. The height of the first floor column was modified to 4.7 m and the doubler plate at panel zones were eliminated to purposely create problems into the four-story steel moment frames. These problems will be eliminated by installing dampers in the peripheral frames of the buildings.

Test results data of the vertical steel pipe damper were used to model shear vs. lateral displacement of the damper. Seismic isolator component in PERFORM-3D [4] was used to model the damper as shown in Figure 6.

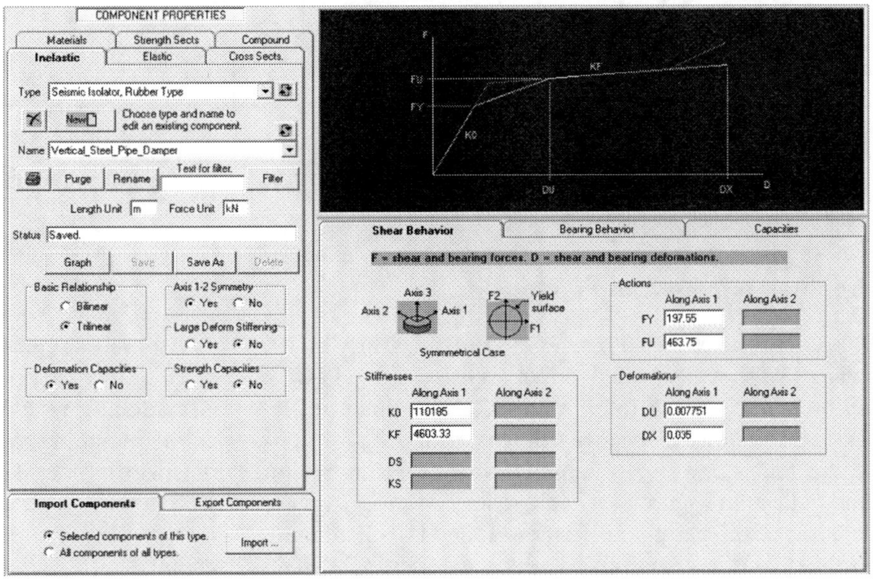

Fig. 6. Trilinear model of the force vs. displacement of the vertical pipe damper

The performance objectives of the steel frames equipped with dampers are:

1. Energy dissipation is concentrated in dampers meaning other components of the steel frames, except the dampers, remain elastic.
2. Inter-story drifts reduction at all levels are significant
3. Absolute lateral deflection of the dampers is less than 10xδ_y.

3.3 Estimating the number of dampers in each story

The required number of dampers in each story had been estimated using energy-based method proposed by Benavent-Climent [5]. The required lateral stiffness and lateral strength of the damper for near-fault ground motion were determined from input energy spectra for moderate-seismicity regions proposed by Benavent-Climent et al. [6]. The configuration of the dampers in the peripheral frames along H1 and H2 are shown in Figure 7. Auxiliary structures in the form of triangular bracings are needed to install dampers between two points where large relative motion exists during earthquake. Auxiliary structures shown in Figure 7 were chosen to minimize the influence of axial forces to the dampers.

A computational model of the four-story frames, that incorporate the nonlinear load-deformation characteristic of the individual damper shown in Figure 6, was built using PERFORM-3D [4]. For assessment of the performance of the building equipped with dampers, the computational model was then subjected to the north-south component of the four strong earthquakes time histories, and the results were evaluated.

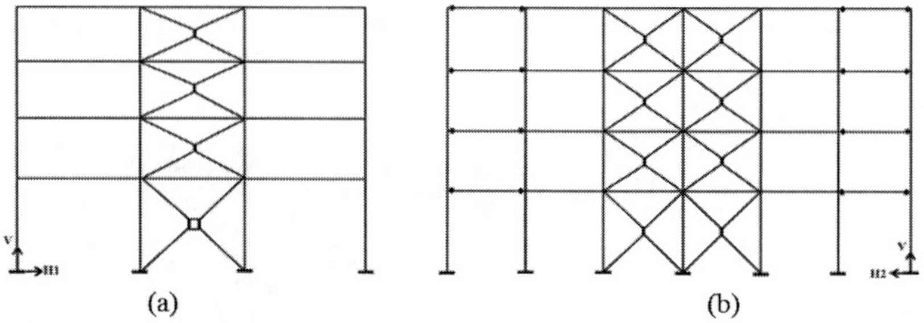

Fig. 7. Configuration of the dampers in peripheral frames: (a) Two peripheral frames along H1, and (b) Two peripheral frames along H2

3.4 Combined acceleration spectrum

The Peak Ground Accelerations (PGAs) of the Chi-Chi 1999, Fukushima-Hamadori 2011 and Padang 2009 were scaled down to El-Centro 1940. Figure 8 shows the combined acceleration spectra of the four earthquake time-histories. Spectral matching would reduce the peaks and valleys in each earthquake time history. No spectral matching was done because the four-story steel moment frame was not designed for specific site. Therefore peaks and valleys in Figure 8 of each acceleration spectra are very obvious.

Two cases of the peripheral frames along H1 considered:

1. Case 0 is the four-story frame without dampers ($T =$ 0.8272 seconds)
2. Case 1 is the four-story frame with dampers ($T =$ 0.4918 seconds)

It is well understood that each point in a response spectra represents the energy content of the earthquake at a certain frequency. For the frame without damper along H1 subjected to Padang earthquake accelerograms, the corresponding point for T=0.8271 seconds lies in a valley. Therefore significant response reduction is not expected for Padang earthquake. Significant response reductions are expected for three other earthquakes.

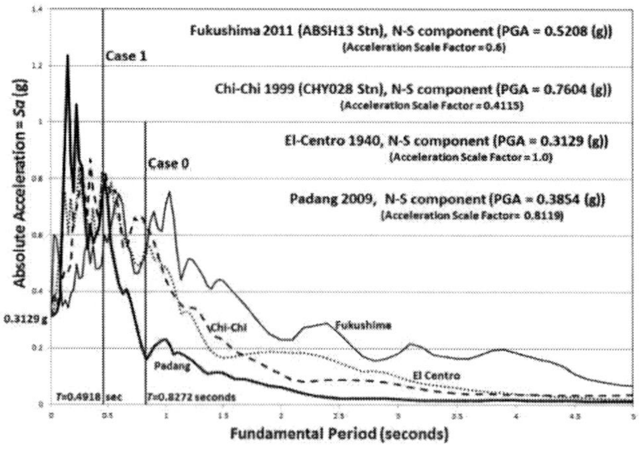

Fig. 8. Combined acceleration spectra

3.5 Inter-story drift reduction

The results of nonlinear dynamic analysis of Case 0 and Case 1 were used to quantify the lateral displacement (drift) and inter-story drift ratio of the four-story steel moment frame. The drift and inter-story drift for Case 0 and Caser 1 are shown in Figure 9 and Figure 10 respectively.

 DOI: 10.1051/matecconf/201713802002

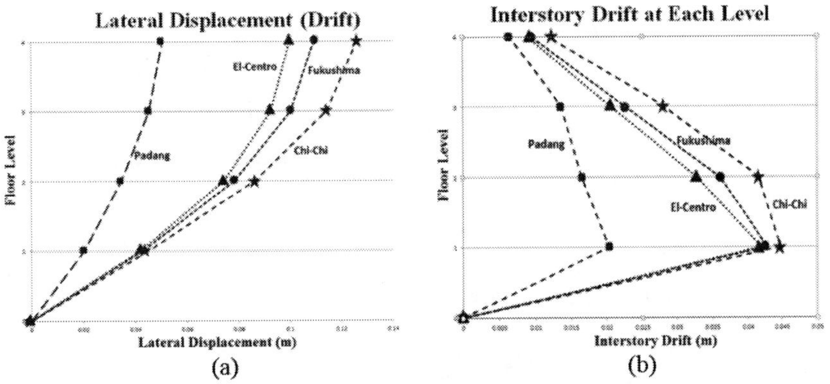

Fig. 9. Drift and inter-story drift for Case 0: (a) Drift, and (b) Inter-story drift

Figure 9a shows the maximum drift at each floor of the four-story frame without damper. The maximum drift due to Padang 2009 earthquake is small as expected (see Figure 8). Figure 9b shows the maximum inter-story drift at each floor of the four-story frame without damper.

Figure 10a shows the maximum drift at each floor of the four-story frame equipped with dampers. Figure 10b shows the maximum inter-story drift at each floor of the four-story frame with dampers installed. Comparing Figure 9 and Figure 10, it can be seen that the lateral displacement and the inter-story drift were reduced significantly due to the present of dampers in the four-story steel moment frame.

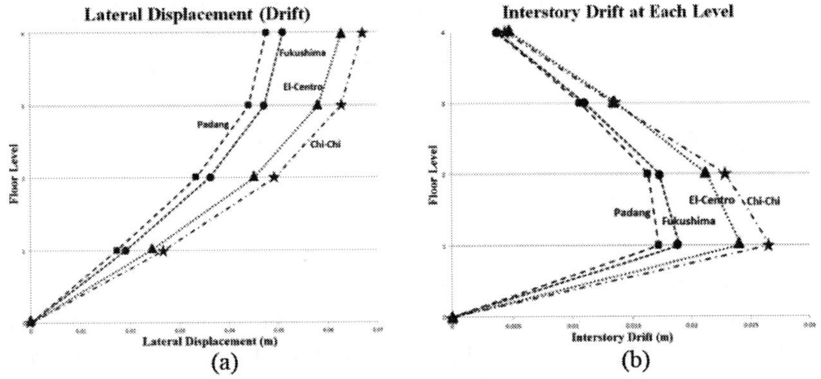

Fig. 10. Drift and inter-story drift for Case 1: (a) Drift, and (b) Inter-story drift

The reduction of inter-story drift at each floor due to the dampers is shown in Figure 11.

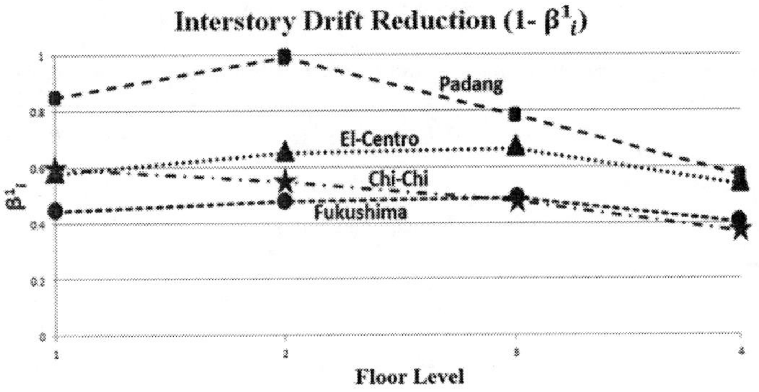

Fig. 11. Inter-story drift reduction

Parameter β^1_i [7] is used to quantify the ratio of maximum drift at each floor for the frame with dampers to the frame without dampers. The inter-story drift reduction is quantified as 1-β^1_i. Except for Padang 2009 earthquake where the inter-story drift reduction is expected to be small, the inter-story drift reduction at each floor due to the other earthquakes is significant. On average the inter-story drift reduction at each floor is about 40%.

3.6 Dissipated inelastic strain energy

Dissipated strain energy quantified at each component of the four-story steel moment frame in nonlinear dynamic analysis is shown in Table 3.

Table 3. Dissipated strain energy at each component of the peripheral frames along H1

	Earthquake			
Group name	**El-Centro - 1940**	**Fukushima - 2011**	**Padang - 2009**	**Chi-Chi - 1999**
Perimeter Columns	0	0	0	0
Perimeter Beams	0	0	0	0
Interior Columns	0	0	0	0
Interior Beams	0	0	0	0
Connection Panel Zones - along H1	0	0	0	0
Connection Panel Zones - along H2	0	0	0	0
Vertical Steel Pipe Dampers	**261.49**	**422.4**	**202.35**	**233.29**
Bracing HSS-H1-1st floor	0	0	0	0
Bracing HSS-H2-1st floor	0	0	0	0
Bracing HSS-H2-other floors	0	0	0	0
Bracing HSS-H1 other floors	0	0	0	0
Total for All Groups	**261.49**	**422.4**	**202.35**	**233.29**

Only the vertical steel pipe dampers dissipate energy. All other components except the dampers remain elastic. The present of the dampers installed at strategic locations in the four-story steel moment frame were able to protect the structure against strong earthquakes. Extensive yielding in dampers is the energy dissipation mechanism. By dissipating strain energy, the dampers control the vibration of the structure during strong earthquakes.

The strain energy dissipated by individual damper is shown in Table 4. The largest value of strain energy dissipated by individual damper located between the 2nd and 3rd floor is due to Fukushima-Hamadori 2011 earthquake. The values of dissipated strain energy of individual damper reflect the damage (the degree of yielding) experienced by the dampers.

Table 4. Dissipated strain energy of individual damper along H1

Earthquake	**Dissipated Energy of Individual Damper (kN.m)**			
	Damper btw 1st and 2nd floor	**Damper btw 2nd and 3rd floor**	**Damper btw 3rd and 4th floor**	**Damper btw 4th floor and roof**
El-Centro	36.40	46.30	11.65	0.0
Fukushima	59.32	**76.71**	15.86	0.0
Padang	26.37	35.18	13.26	0.0
Chi-Chi	34.85	37.38	9.58	0.0

3.7 Maximum absolute shear force and shear displacement

The maximum absolute value of the shear force and the maximum absolute value of the shear displacement shown in Table 5. The maximum absolute shear force is 513.21 kN which is less than the shear-force capacity of the vertical steel pipe damper. The maximum absolute shear displacement is 18.5 mm which is very close the 10xδ_y of the damper (δ_y of the damper is 1.80 mm). Therefore the performance objectives of the structure had been achieved.

Table 5. Maximum absolute shear force and lateral displacement

Earthquake	Maximum Absolute Shear Force (kN)		Maximum Absolute Shear Displacement (m)	
	Damper btw 1st and 2nd floor	Damper btw 2nd and 3rd floor	Damper btw 1rd and 2nd floor	Damper btw 2nd and 3rd floor
El-Centro	502.63	501.88	0.0162	0.0161
Fukushima	480.78	484.38	0.0115	0.0122
Padang	470.01	480.17	0.0099	0.0113
Chi-Chi	**513.21**	508.34	**0.0185**	0.0175

4 Conclusion

The newly developed vertical steel pipe damper was successfully apllied to reduce the seismic response of the moment frame due to four strong earthquakes. The energy dissipation were concentrated in the dampers so that other components of the structure, except the dampers, remain elastic. The performance of the four-story steel moment frame with slender columns at the first floor has been improved significantly by installing dampers at strategic locations. After strong earthquakes, dampers that already experience yielding can be replaced easily.

This research was supported by the Decentralization Research Grant (FTSL PN-1-08-2014 and FTSL PN-1-08-2015) of Directorate General of Higher Education (DIKTI), Ministry of National Education, Indonesia.

References

1. J. Utomo, *Development of Steel Pipe Dampers for Seismic Protection of Buildings*, (In Indonesian), Doctoral Dissertation, Institut Teknologi Bandung, Indonesia (2017).
2. D. Y. Abebe, J. W. Kim, and J. H. Choi, *Hysteresis Characteristics of Circular Pipe Steel Damper Using LYP225*, Steel Innovations Conference 2013, Christchuch, New Zealand, 21-22 February (2013).
3. __________, *ABAQUS user's manual, version 6.11*, Dassault Systems Corp., Providence, RI, USA (2011).
4. G. H. Powel, *User Guide Perform-3D Version 5 – Nonlinear Analysis and Performance Assessment for 3D Structures*, Computer and Structures Inc., California (2011).
5. A. Benavent-Climent, Soil Dynamics and Earthquake Engineering, **31**: 1385-1396 (2011).
6. A. Benavent-Climent, L. G. Pujades and F. Lopez-Almanza, Earthquake Engng Struct. Dyn.; **31**:1151-1172 (2002).
7. C. Tovar, and O. A. Lopez, *Effect of The Position and Number of Dampers on The Seismic Response of Frame Structures*, 13th World Conference of Earthquake Engineering, Vancouver, B.C., Canada (2004).

MATEC Web of Conferences **138**, 02003 (2017) DOI: 10.1051/matecconf/201713802003
EACEF 2017

Flexural Capacity of the Concrete Beams Reinforced by Steel Truss System

Rudy Djamaluddin[1,*], *Pieter Lourens* Frans[2], and *Rita* Irmawati[1]

[1]Hasanuddin University, Civil Department, 90222 Gowa, Indonesia
[2]Hasanuddin University, Doctoral Student of Civil Department, 90222 Gowa, Indonesia

Abstract. Reinforced concrete beam is normally reinforced by steel bars to sustain tensile forces occurring due to flexural response, and vertical steel reinforcement to sustain shear stresses. On the other side, the truss system is a structure composed by triangle. Truss system develops its flexural capacity based on the inter-action of the bar elements. Therefore, introducing of the truss system to the reinforced concrete beams by modifying of the tensile and shear reinforcement may increase the beam flexural capacity. The study on the application of the truss system to the reinforced concrete beams has been conducted. The shear reinforcement was modified to form the triangles that were connected to the tensile reinforcement on the bottom and compressive reinforcement on the top. The steel bar elements were connected by welding to form a truss system. The effect of the buckling of the compression bar element on the truss system may be avoided due to the constrain effect of the surrounding concrete. In order to clarify the effect of the truss system reinforcement to the flexural behaviour, a series of specimens was prepared. The beams specimens have a length of 3.3 m with cross section of the 15 by 20 cm. A normal concrete beam was also prepared as control specimens. Results indicated that the concrete beams reinforced by truss system had flexural capacity higher than the normal beam.

1 Introduction

The constructions using concrete materials are still dominant due to its advantages. The most important characteristic of concrete materials such as workability, low cost and fire resistance as well as its low maintenance cost. The durability has always been a major reason for selecting reinforced concrete as the construction materials for building and other civil engineering infrastructure systems [1]. Due to the mass consuming of concrete materials, the exploration of natural materials such as gravels and sands are increasing to produce a concrete material. Additionally, the cement using in the concrete is almost 90% composed by lime stones that also coming from natural materials.

* Corresponding author: rudy0011@gmail.com

Massive exploration of the natural materials for producing concretes affect to the environment condition and global warning that may cause disasters such as flooding and land-slides. Therefore, the researches focusing on the optimizing of the structural capacity of the reinforced concrete structure elements are one of major research. Nes.L.G et.al [2] studied about structural performance of a concrete sandwich slab element, Pengfei L et.al [3] studied about concrete box girder with corrugated steel webs. The developing of the new concept in structures design to minimize the using of the concrete has been conducted and proposed, such as box type girders, corrugated beams, strengthened beams or column with FRP materials and sandwich structures [4-9].

By increasing the structural capacity while reducing the using of concrete materials will provide advantages on the environmental impact especially in the natural materials exploration. The concrete should be used as efficient as possible. The research on the fields of the concrete efficiency should be conducted intensively in order to save the natural resources.

(a) Roof truss structure

(b) Bridge truss structure

Fig. 1. Truss Structures applied on Roof and Bridge

This paper discussed a study of the integrating of the truss steel structures mechanism into the concrete beams. The truss element develops its flexural capacity based on interaction between truss elements (tension-compression). Stand-alone truss steel structures have been applied on many fields. Due the stability mechanism, the buckling of the truss element of compression strut is an important part that should be considered in design. Fig.1 shows the application of truss system on the bridges and roof structures. Meanwhile, the concrete beam develops its flexural capacity based on the couple moment interaction between tension force on the steel reinforcement and compression force on concrete [10]. The stirrups that commonly exist in a reinforced concrete beams is considered only for shear reinforcement. By modify the stirrups angles to form a series of triangles on the reinforcement system of a concrete beams then the truss mechanism may occur in a concrete beams. The buckling effect of the stirrup reinforcement may be avoided due to the constraint effect of the surrounding concrete. In this concept, the flexural capacity of the concrete beams is the accumulation of two flexural actions, that are due to couple moment of tension and compression forces, and due to the flexural effect of the truss reinforcement system. Fig.2 shows the illustration of the proposed concept. Simply, the triangles were formed by pairing of inclined stirrups. This paper presents the experimental results of the concrete beams reinforced by steel truss system as the reinforcement.

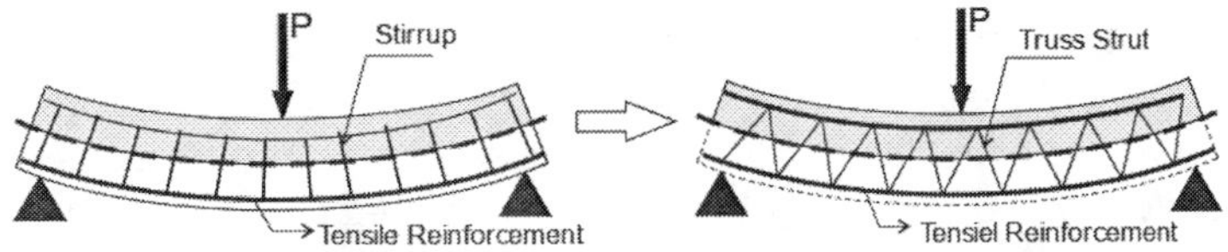

(a) Regular steel reinforced concrete beams (b) Steel truss reinforced concrete beams

Fig. 2. Regular reinforcement system vs steel truss reinforcement

DOI: 10.1051/matecconf/201713802003

2 Specimens

A series of specimen was prepared to clarify the effect of steel truss as reinforcement on the concrete beams. The specimen preparation was divided to the preparation of the truss reinforcements and casting of the concrete beams. The concrete beams specimen dimensions are 3300 mm length with 150 x 200 mm of cross section, respectively. The Detail of specimen is presented in Fig.3. The specimens prepared in this study were three beams for the normal reinforced concrete beams (BN), three beams for truss reinforced concrete beams (BR).

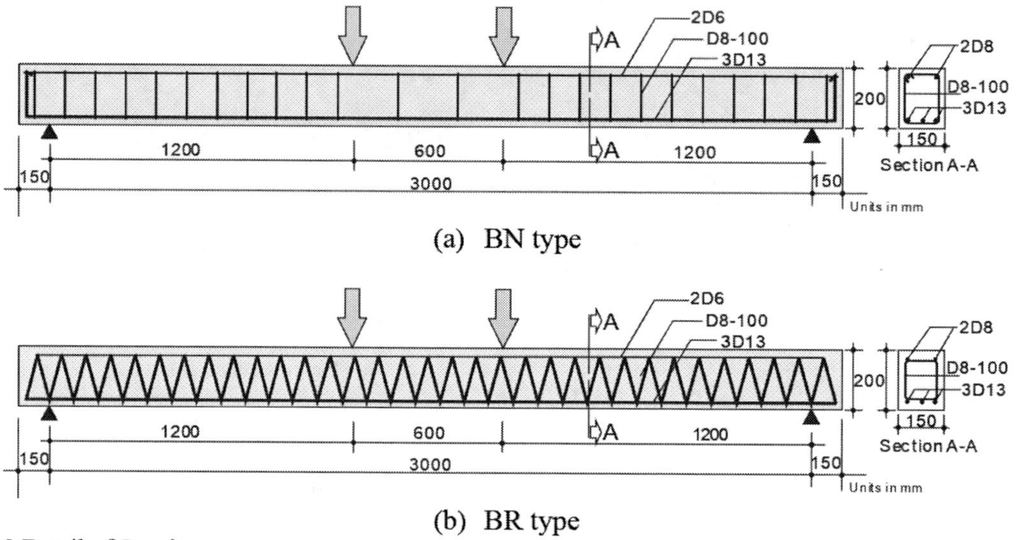

Fig.3 Detail of Specimens

Table 1. Material Properties

Properties	Concrete	Steel Reinforcement
Compressive Strength (MPa)	23	300
Tensile Strength* (MPa)	2.1	300
Elastic Modulus (Gpa)	24	200

**) Yield Strength for Steel*

Specimens BN and BR used three of D13 steel bars as tensile reinforcement and D8 as the shear (vertical) reinforcement. Both BN and BR had two of D6 steel reinforcement at the compression side. For specimen BR, the truss reinforcement was composed by three of D13 steel bar reinforcement for the tension reinforcement, D8 steel bar for diagonal bars, and two of D6 steel on the upper horizontal bars. The space of the diagonal bars on the truss reinforcement was fixed on 100 mm. All connections in the truss reinforcement were done by welding. All beams had the same tensile reinforcement ratio. The strain gauges were patched on tensile reinforcement and, only for BR type, the strain gauges were also patched on truss strut (diagonal bar) at span centre for strain measuring during the loading. The casting of concrete was done using a normal concrete with compression strength of 23MPa. All specimens were cured for 28 days using a wet blanket. Material properties of concrete and steel reinforcement used in this study are presented in Table 1.

3 Test Setup

The specimens were loaded under four point bending test. The supports were prepared to behave as the hinge-roller support. Strain gauges were patched on the concrete surface on the three points at the span centre which were one on the top of beam and two at the concrete web, respectively. Strain gauges patched on the concrete surface as well as on the tension and diagonal bar were then connected to a data logger to measure the strain for further analysis. The specimen was supported by simple support with the span of 3000 mm. Two loading point was applied with the space of 600 mm to the span centre of the beams. Specimen setup is presented on Fig.4. LVDTs were installed on the centre point and both of under loading points to measure the deflection. All data was recorded using a data logger connected to the computer. The load measured using a load cell was applied gradually with the rate of 2 kN per step until first crack of concrete. Further loading, the load was applied with the rate of 5 kN until maximum load up to final failure.

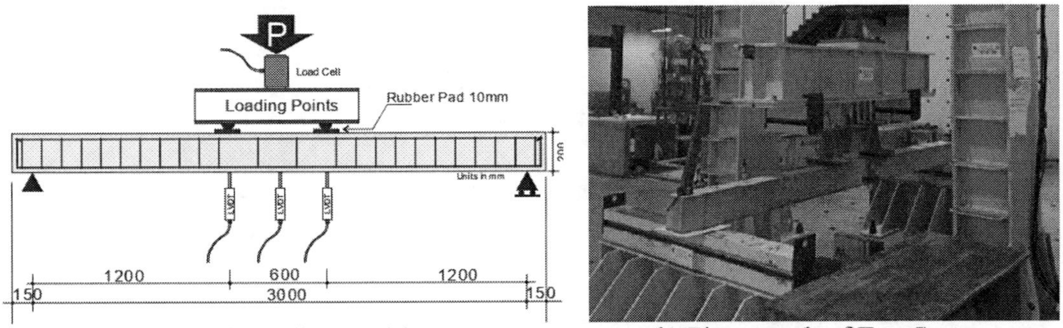

(a) Support and Loading Position (b) Photograph of Test Setup

Fig.4 Detail of Specimens

4 Results and Discussions

Generally, the effect of the truss system on the concrete beams (BR Type) was indicated by the increasing of the flexural capacity as well as the stiffness of the concrete beams compared to the normal steel reinforced concrete beam (BN Type). Detail discussion of the experimental results is presented in the following sections.

4.1 Load-Deflection Relationship

The relationship between the applied load and span centre deflection is presented in Fig.5. At initial stage of loading, all beams were un-cracked beam. On the specimens of BN type as well as BR type, the concrete resisted both compression and tension force. When the applied load reached to the rupture strength of the concrete on specimens, the concrete started to crack. This caused a decreasing of beam flexural stiffness. Once the tension zone of concrete cracked, its tensile force resistance becomes negligible. The tensile force due to external load was primarily carried by steel reinforcement. At pre-crack stage, both specimens had almost same flexural stiffness. However, after first cracking, the specimens BR showed higher stiffness compared to the regular concrete beams. This phenomenon may be understood clearly that on the BR type, the truss reinforcing system provided an additional stiffness in the total stiffness of the concrete beams. The load-deflection curves

DOI: 10.1051/matecconf/201713802003

propagated under different path up to the yielding point of the tensile reinforcement. The average stiffness after first crack of specimen BN was approximately 1.55kN/mm, whereas the stiffness of the specimen BR type was 1.75kN/mm, respectively. The increasing of applied load was followed by the propagation of deflection up to the yielding point of steel reinforcement. On BN type specimen, the yielding of tensile reinforcement occurred when the applied load reached 25.4kN. For BR type specimen, the yielding of tensile reinforcement occurred when the applied load reached 29.0kN. Further loading after yielding of tensile reinforcement caused the decreasing significantly of the beams flexural stiffness on both specimens. As it can be observed that the effect of truss was no longer influencing the beam stiffness. The deflection was propagated without significant increasing of the load capacity. On the BN type specimens, the ultimate load capacity was 26.8kN when the span centre deflection achieved 43.0mm, while on the BR type specimens, the ultimate load capacity was 30.2kN when the deflection equal to 39.5mm, respectively. The final failure of specimens was caused by the concrete crushing on approximately at the span centre. Further steps showed that the beams continuously deflected followed by the decreasing of applied load.

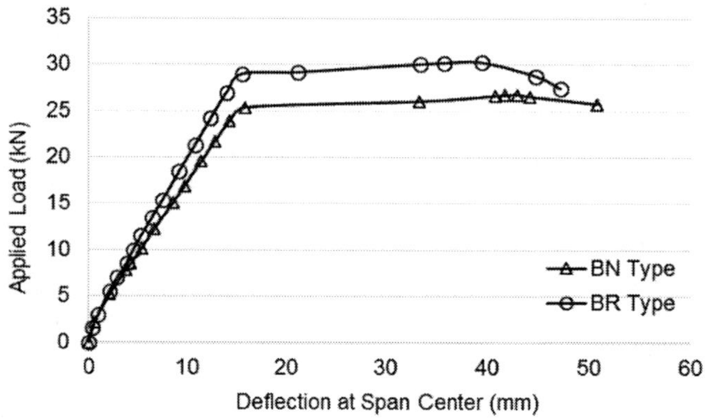

Fig.5 Load-Deflection Relationship

4.2 Flexural Capacity

Table 2 presents the summary of loading steps as well as the deflection accordingly. The first cracking occurred when the applied load reached 5.21 kN on BN type and 5.47 on BR type, respectively. There was no significant effect of the truss reinforcement on the first cracking load. However, the effect of the truss reinforcement was identified when the applied load increased up to the yielding point of the steel reinforcement. The yielding load of the BN type was 14.26 kN and the BR type was 15.57 kN, respectively. The effect of the truss reinforcement was also identified on the flexural stiffness after first crack. The average stiffness after first crack of specimen BN was approximately 1.55kN/mm, whereas the stiffness of the specimen BR type was 1.75kN/mm, respectively

Table 2. Summary of Experimental Results

Type	First Crack		Yield of Steel		Ultimate Capacity		Ratio (BR/BN)	
	Pcr (kN)	Δcr (mm)	Py (kN)	Δy (mm)	Pu (kN)	Δu (mm)	Pu	Δu
BN	5.21	2.21	23.90	14.26	26.77	43.03	1.13	0.92
BR	5.47	2.14	28.97	15.57	30.24	39.47		

DOI: 10.1051/matecconf/201713802003

The ultimate flexural capacity of BR type specimen was higher than the BR type specimens. The ultimate capacity of the BN type was 26.77kN and the BR type was 30.24kN, respectively. The ultimate capacity of both specimens was decided by the crushing of the concrete on compression section. The ultimate load ratio between truss reinforced beams (BR) and the regular beam (BN) was 1.13. This indicated that the effect of truss system was approximately 13% in increasing the ultimate flexural capacity of the reinforced concrete beams.

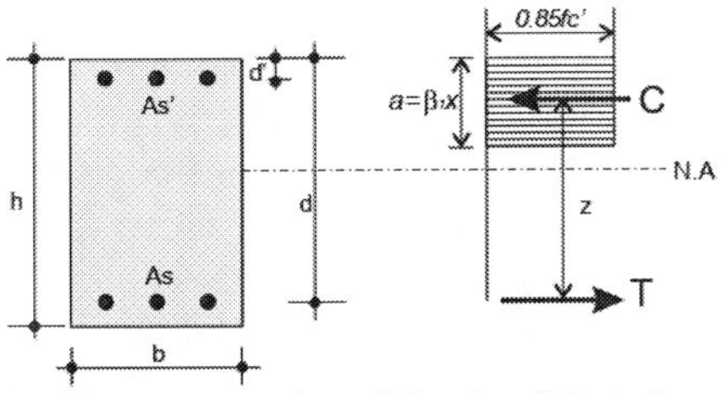

Fig.6 Stress Block Model of the ultimate capacity of the Steel Reinforced Concrete beams

Using the stress model presented in Fig.6, the ultimate capacity of the regular steel reinforced concrete beam may be estimated using the internal couple moment between compression forces on concrete and tensile force on the steel reinforcement [10], as follows :

$$M_r = 0.85f_c' \, b \, a\left(d - \frac{1}{2}a\right) \quad (1)$$

Therefore, simply the total moment M_T of the BR type specimens is the sum of the moment capacity of the RC beams M_r and the truss system M_{ts}.

$$M_T = M_r + M_{ts} \quad (2)$$

The increasing of the moment capacity was the effect of the increasing of the flexural stiffness on the BR type specimens. The increasing of the flexural stiffness was affected by the increasing of the moment of the inertia due to the existing of diagonal bars as on the BR type. By adopting the equation of the moment of inertia of the cracked section of the truss system reinforced concrete reported by Francesco T, et.al [11], then using Fig.7, the moment of inertia of the cracked section of BR type specimens may be expressed by Eq.(3).

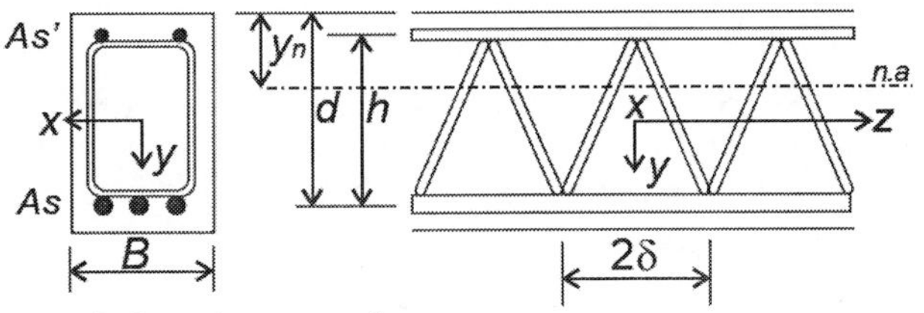

Fig.7 Section of the truss reinforced concrete beam

$$J_{eq} = \frac{By_n^3}{3} + n_E\left(A_s + 2A_{diag}n_z^3\right)(d - y_n)^2 + n_E A_s'(y_n - d')^2 \quad (3)$$

where, B is the width of beam, y_n is the distance from neutral axis ($n.a$) to the upper chord, n_E=Es/Ec, As is the area of the tensile reinforcement (lower steel bars), n_z is the cosine director of the diagonal bars along the z-axis (δ/l_{diag}), l_{diag} is the length of the diagonal bar, A_{diag} is cross section of a single diagonal bar, d is effective depth of beam, As' is the cross section of the upper steel bars and d' is the distance of upper steel bars to the upper chord.

4.3 Failure Mode and Crack Pattern

All specimens were designed to fail under crushing of compression concrete. On BN specimens, the failure of concrete was initiated by the yielding of the steel reinforcement. As the result, the compressive stress of concrete reached the compression strength of the concrete. This failure mode also occurred on the BR type specimens. However, the failure load of the BR type was higher than the normal reinforced concrete beams (BN type). This may be caused by additional flexural capacity of the truss system as the reinforcement in the concrete beam. It should be noted again here that the truss system also affected to the flexural stiffness of the concrete beams. The compression area of concrete was wider on the BR type, therefore the failure load was higher than BN type. The crushing failure of both specimens is presented in Fig.8.

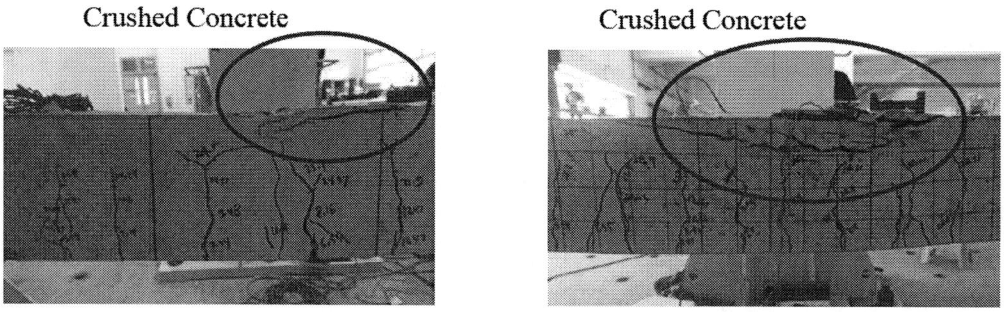

(a) BN type Specimen (b) BR Type Specimen

Fig.8 Failure Mode and Crack Pattern

Fig. 8 presents also the crack patterns on the constant moment region of each specimen types. BN type as well as BR type specimen indicated typical crack pattern of a normal under reinforced concrete beam. Further loading after appearance of the first crack, the other cracks appeared while the existing cracks propagated. The propagation of the cracks moved toward to the compression concrete. The long cracks were concentrated in the constant moment region at span centre. Comparison to the both specimens, it was observed that the BR type specimens had better cracks distributions.

The acknowledgment is presented to the graduate students who involving in this works. The assistance of the Laboratory of structural engineering, Civil Department, Hasanuddin University is also acknowledged.

REFERENCES

[1] Mindess, S., Young, J. F. and Darwin D. Concrete in Special Applications. Chapter 20. Second Edition, Pearson Education Inc. USA, (2003)

[2] Nes, L.G., Overli, J.A. Composite and Hybrids Investigation of Material Parameters and Structural Performance of a Concrete Sandwich Slab Element. fib Symposium PRAQUE 2011, Session 5-6, (2011).

[3] Pengfei Li, Baodong Liu, Bo Li and Mingzhe Yang. The stress retaliated dumping ratio of box girder with corrugated steel web, The open civil engineering journal, **8**, 301-306. (2014).

[4] Jae-ll Sim and Keun-Hyeok Yang. Flexural behavior of reinforced concrete columns strengthened with wire rope and T-Plate units. ACI Structural Journal. V.**106**. No.5. pp.697-705. (2009).

[5] Rudy Djamaluddin, Flexural Behaviour of External Reinforced Concrete Beams. Procedia Engineering **54**, Elsevier, pp.252-260. (2013)

[6] Stevees, C. A., Fleck, N. A. Collapse Mechanism of Sandwich Beams with Composite Faces and A Foam Core. International Journal of Mechanical Science, Elvesier **46**, 30 April 2004, 561-583. (2004).

[7] Despandhe, V. S. and Fleck, N. A. Collapse of Truss Core Sandwich Beams in 3-Point Bending. International of Solid and Structures, Pergamon, **38**, 6275-6305. (2001).

[8] Li, B. and Tran, C. T. N. Determination of Inclination of Strut and Shear Strength Using Variable Angle Truss Model for Shear-Critical. Journal Engineering and Mechanics, Vol. **41**, No.4, 459-477. (2012).

[9] Li, B. and Tran, C. T. N.. Reinforced Concrete Beam Analysis Supplementing Concrete Contribution in Truss Models, Journal Engineering Structures, **30** (11), Elsevier, 13 June 2008, 3258-3294. (2008).

[10] James K.Wight and James G MacGregor. Reinforced concrete mechanics and design. Sixth Edition, Pearson. (2005).

[11] Francesco Trentadue, Erika Mastromarino, Giuseppe Quaranta, Floriana Petrone, Giorgio Monti, Giuseppe Carlo Marano. Bending Stiffness of Truss-Reinforced Steel Concrete Composite Beams, Open Journal of Civil Engineering, Scientific Research, **4**, 285-300. (2014)

Stress distributions of PSC box girder bridge due to creep shrinkage effect

Baskoro Abdi Praja[1]

[1]Civil Eng. Department of Universitas Atma Jaya, 44 Babarsari, Yogyakarta Special District, Indonesia

Abstract. Long-span bridges are usually designed with a service life of more than 20 years. Thus, it is required to consider factors that are affected by time. Creep and shrinkage are parameters to predict time-dependent behavior of a bridge. The purpose of this study is to determine the effects of creep & shrinkage toward total stresess. Lemah Ireng 1 Bridge in Semarang-Solo toll road was used as a case study. The bridge has a total span length of 879 m. It has a service life design of 100 years. Therefore, this creep and shrinkage research is necessary to determine the stress distribution in the box girder bridge until the end of the service life. Computational modelling was conducted to estimate the stress distribution in the structure of the box girder bridge. Several stages of loading on the bridges were applying asphalt leveling, barrier, creep and shrinkage load by using a variation of the long-term duration with intervals of 5, 10, 15, 20, 40, 60, 80 and 100 years. The analysis was using linear static method, Construction Stage Analysis-Time Dependent effects on Midas Civil and only apply non-transient load on the bridge structure. This study is expected to show the impact of the creep and shrinkage effects of the total stresses along the bridge span until the service life of 100 years.

1 Background

Lemah Ireng 1 Bridge (LI1) is a prestressed concrete box girder bridge with a total length of 879 m that uses Free Cantilever Method (FCM) as in Figure 1. In the construction stage, there were two changes of technical conditions. The changes are described as sand material changes density of concrete up to 26 kN/m^3 and changes of moving traveller schedule. Both changes make a difference on the precamber value. The condition allegedly causes the difference between the designed elevation and the actual elevation so that requires asphalt leveling which vary in size. It has a service life design of 100 years, but based on the actual conditions that occured, it is necessary to review the performances and the strength of long-term serviceability. Praja (2016) has performed a deflection analysis for 100 years. This study reviewed the stress distribution along the span at the age of 100 years.

 DOI: 10.1051/matecconf/201713802004

Fig. 1. Construction of PSC box girder bridge using free cantilever method. (Adhi Karya Corporation, 2013)

2 Literature review

Pan et al (2011) analyzed the long-term behavior of a long-span PSC box girder bridge. The study compared the results of stress measurements toward modified creep and shrinkage models. The deviations of stress of the top is 1.7 MPa and the greatest deviation of 2.2 MPa occurs in casting the 18 # segment. The deviation of stress of the bottom slab between calculated values and measured values is around 1.2 MPa and the greatest deviation of 1.75 MPa occurs in casting the 18# segment.

Bao et al (2015) conducted a study on the impact of shrinkage on the box girder bridge. This study shows the effect of shrinkage on the stress along the span. Creep and shrinkage in concrete show a great impact on the internal forces. After 3 years of completion, the results showed that the effects of creep and shrinkage are 1.9 MPa and 0.78 MPa in upper and bottom flange as compressive stress. By considering the shrinkage and creep near the bridge pier, the bridge stress decreased by 15.4% at the maximum; while near the midspan, the stress maximally increased by 7.3%.

The effects of creep and shrinkage can be fatal if neglected; so it becomes the primary object of the study of Lemah Ireng 1 Bridge. By considering the effects of long-term stresses, it is necessary to conduct a research on the performance of Lemah Ireng 1 Bridge.

3 The theoritical calculations of the concrete creep and shrinkage

In normal concrete structures, the continuous dead load causes a creep (Analysis Reference of Midas Civil, 2011). If a unit axial stress $\sigma = 1$ in a concrete type with the age of t_0, then the axial strain result at age t is defined as $J(t,t_0)$.

$$\varepsilon(t) = \varepsilon_i(t_0) + \varepsilon_c(t,t_0) = \sigma.J(t,t_0) \tag{1}$$

$J(t,t_0)$ represents the total that occurs in the unit of stress and is defined as the creep function. Figure 2 shows that the $J(t,t_0)$ creep function can be represented by the initial number of elastic strain and the creep strain as follows:

$$J(t,t_0)=\frac{1}{E(t_0)}+C(t,t_0) \tag{2}$$

$E(t_0)$ represents the modulus of elasticity when the load is applied, and $C(t,t_0)$ represents the resulting deformation creep at the age of t, which is referred to as the specified creep.

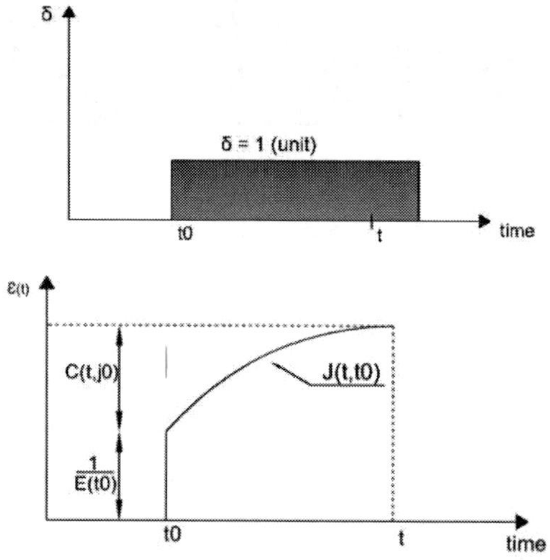

Fig. 2. The definition of the creep and specific creep function (Analysis Reference of Midas Civil,2011)

The creep function $J(t,t_0)$ can also be expressed in a relative comparison to elastic deformation

$$J(t,t0)=\frac{1+\phi(t,t0)}{E(t0)} \tag{3}$$

$\phi(t,t0)$ is defined as the creep coefficient which represents the creep ratio to elastic deformation. The specific creep can also be expressed as follows:

$$\phi(t,t_0)=E(t_0).C(t,t_0) \tag{4}$$

$$C(t,t_0)=\frac{\phi(t,t_0)}{E(t_0)} \tag{5}$$

Midas Civil allows determining the coefficient of creep and shrinkage calculated by the CEB-FIP, ACI, and others or can also be directly determined by the values obtained from the experiment.

Shrinkage is a function of time that is free of pressure in a concrete element. Strain due to shrinkage is usually expressed by time from t_0 to t.

$$\varepsilon_s(t,t_0) = \varepsilon_{s0} \cdot f(t,t_0) \tag{6}$$

Where ε_{s0} represents the shrinkage coefficient at the end time, $f(t,t_0)$ is the time function, t is the observation time and t_0 is the initial shrinkage time.

4 Engineering Background

Lemah Ireng 1 Bridge has a span of 879 m with a configuration of 72 + 115.5 + 120 + 136 + 136 + 120 + 74.5 + 50 m as the main span and 18 + 19 + 18 m as the approach span. The pier height varies from 4 m to 44.5 m. The type of deck on the bridge is cast in situ-prestressed concrete box. The total width of the deck is 2 x 12.6 m. The height of the deck varies from 3 m to 6.5 m as in Figure 3 and 4. The used construction method is Balance Cantilever - Form Traveller.

There were special conditions occured in the construction. First, changes in the material used in the actual condition of the existing type of concrete weight were greater than the deflection analysis,thus the weight itself become larger as the consequences. The concrete compressive strength was increased from the planned 41.5 MPa to 55.51 MPa. Second, there was a change in the sequence of construction methods in the field which is the change of traveller movement due to land acquisition problems as in Figure 5 and 6, so that the precamber was not in accordance with actual conditions that occured.

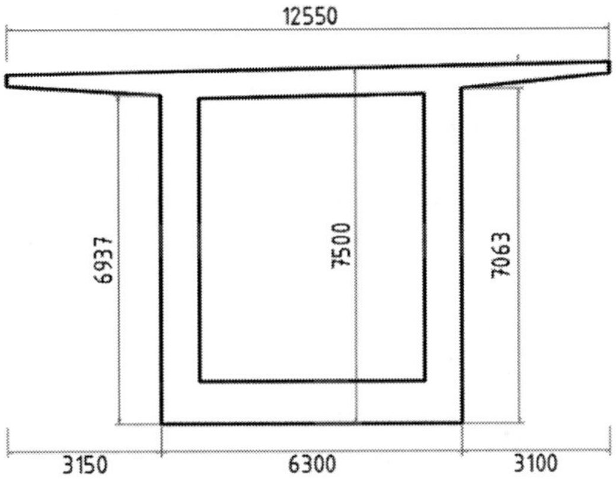

Fig. 3. Girder cross sections at piers 4 in milimeters.

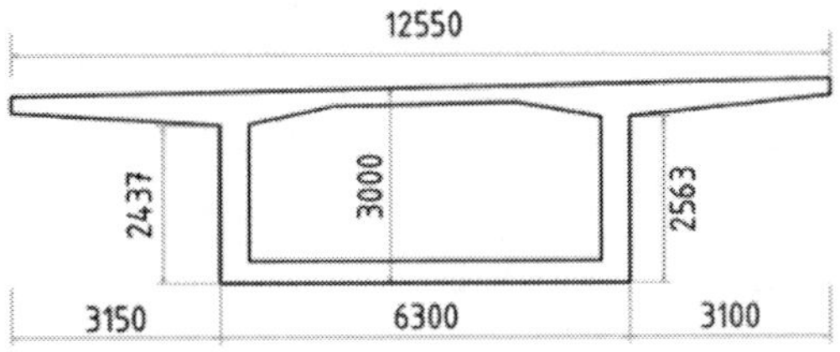

Fig. 4. Girder cross sections at midspan in milimeters.

DOI: 10.1051/matecconf/201713802004

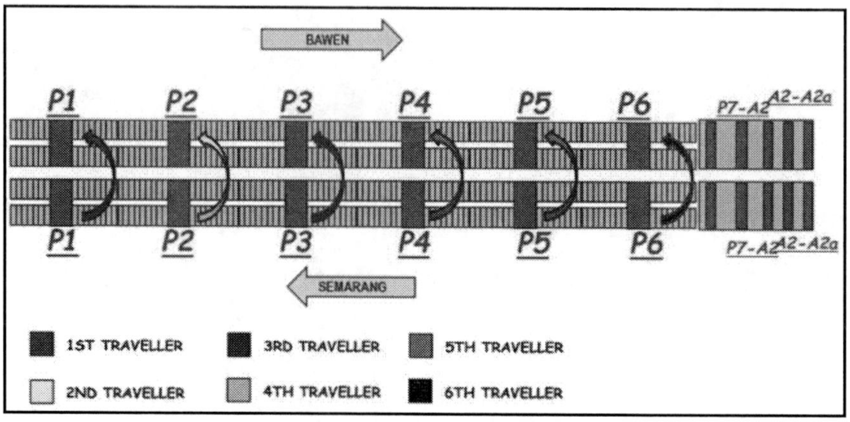

Fig. 5. Plan design of traveller movements. (Adhi Karya Corporation, 2013).

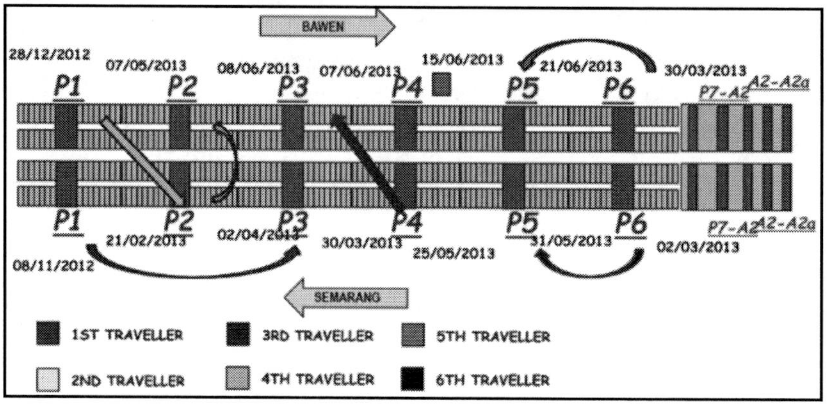

Fig. 6. Actual traveller movements. (Adhi Karya Corporation, 2013).

5 Research methods

The study is based on the analysis of Construction Stage with Civil Midas 2011 Evaluation Licensed. LI1 bridge structure is modeled as a 3D frame element of box girder balanced cantilever method as overall as shown in Figure 7. Some idealizations of the modeling of box girder bridge and its pillars are modeled as frame elements; pedestal structure modeled joints on the bearing pad in abutment; connections with the pillar box girder is modeled with elastic link rigid type; pedestal pillar modeled as fixed.

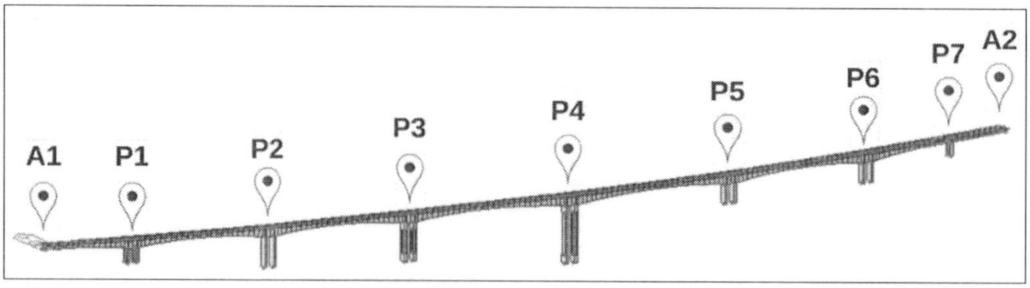

Fig. 7. LI1 Bridge finite element model.

The loading on the construction stage includes the following:

a. Weight of the structure.
b. Prestressed tendons load.
c. Traveller load and wet concrete load. Load of 650 KN traveller form and weight of wet material idealized as a point load in the form of vertical force in the direction of gravity and the moment due to the eccentricity of the vertical force of 0.5Ls (half length of segment).
d. Live Load during construction is idealized as a uniform load of 2 KN/m^2.
e. Additional dead load by inserting a barrier which is modeled as a distributed load.
f. Asphalt load is modeled as a distributed load with a unit weight of 22 KN/m^3.
g. Creep and shrinkage load to100 years.

The analysis of the long-term creep and shrinkage uses the standard of CEB-FIP 1990. The bridge structure was analyzed in a linear static. Construction Stage Analysis in Civil Midas software is used as a media to acknowledge the time-dependent effects. The analysis was carried out based on the condition of the bridge that has been deformed by the construction phase. Praja et al (2016) obtained the span of P4-P5 that has maximum deflection, so this paper only has a review on P4-P5 span. This study compared the design toward actual conditions as in Table 1. In this study, the time span bridge was taken at age 5, 10, 15, 20, 40, 60, 80 and 100 years.

Table 1. Comparison on input of bridge model properties and load.

No.	Subject of differences	Design Model	Actual Model
1	Age of Concrete at the Beginning of Shrinkage	3rd day	1st day
2	Compressive Strength	41.5 Mpa	55.51 Mpa
3	Weight Density of Pier	24 KN / m^3	26 KN / m^3
4	Weight Density of Box Girder	25 KN / m^3	26 KN / m^3
5	Asphalt Levelling	max. 5 cm	max. 20 cm
6	Construction Stage	Design Sequence	Actual Sequence

6 Long-Term Stresses

Long-term loading depends on the effect of time. It changes the internal force, so it can change the stresses on each girder cross section. Creep and shrinkage are time-dependent factors. The diagram shown is a 100-year stress analysis.

In Figure 8, the stress distribution due to creeps and shrinkages on the upper flange shows a tensile – compressive (midspan) - tensile distribution along the span with a maximum tensile stress of 6.615 MPa and a compressive stress of 1.5285 MPa. The effect of shrinkage on 100 years after construction is 26% at most.

DOI: 10.1051/matecconf/201713802004

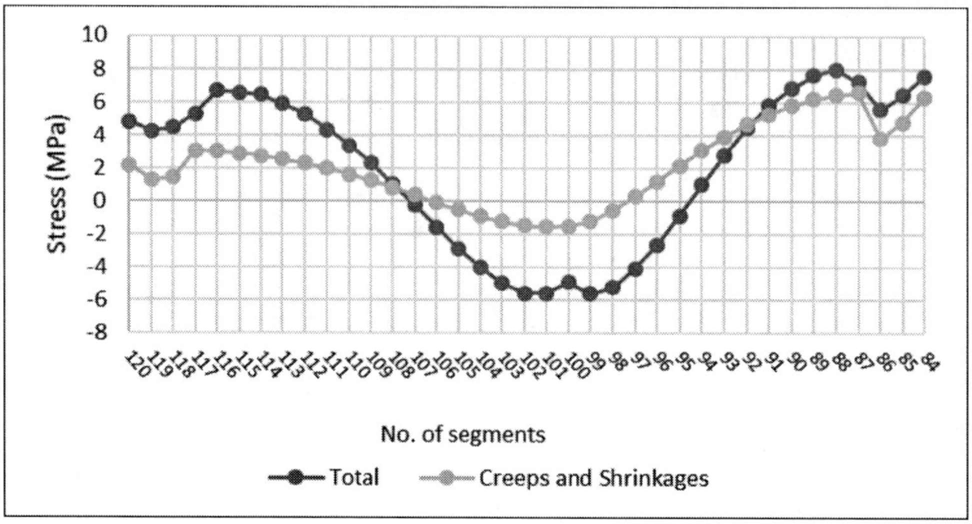

Fig. 8. Stress distribution at upper flange of girder at span P4-P5 (design).

In Figure 9, the stress distribution due to creeps and shrinkages on the bottom side shows a compressive – tensile (midspan) - compressive distribution along the span with a maximum tensile stress of 10.6644 MPa and a compressive stress of 1.1938 MPa. The effect of shrinkage on 100 years after construction is 10% at most.

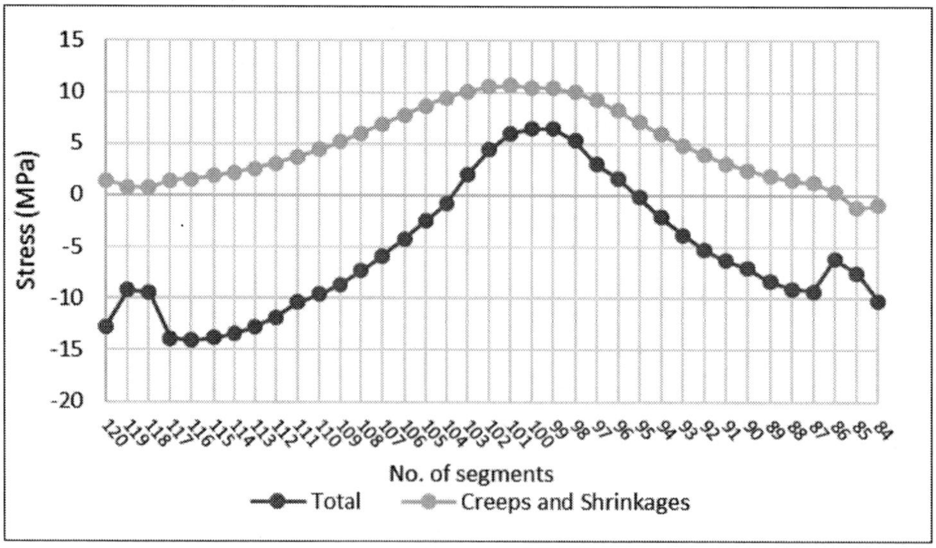

Fig. 9. Stress distribution at bottom side of girder at span P4-P5 (design).

In Figure 10, the stress distribution due to creeps and shrinkages on the upper flange shows a tensile – compressive (midspan) - tensile distribution along the span with a maximum tensile stress of 5.0977 MPa and a compressive stress of 1.4021 MPa. The effect of shrinkage on 100 years after construction is 21% at most.

DOI: 10.1051/matecconf/201713802004

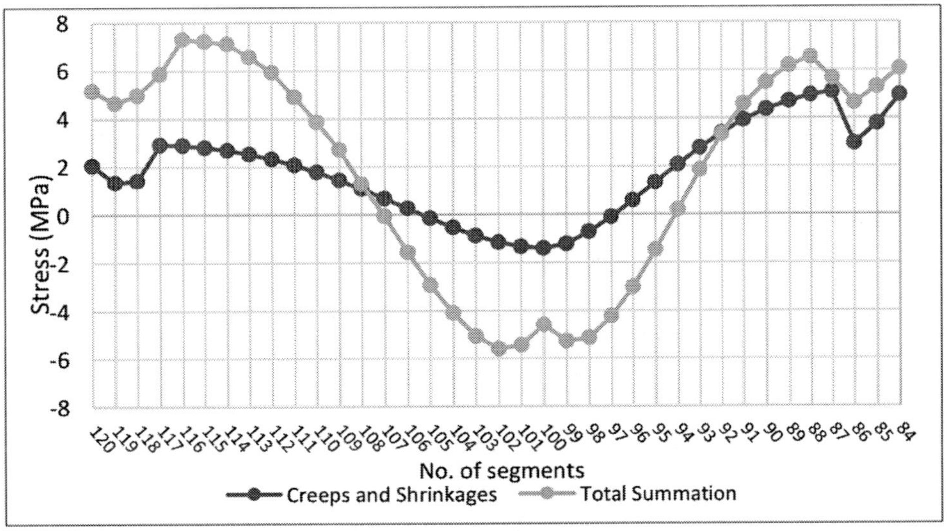

Fig. 10. Stress distribution at upper flange of girder at span P4-P5 (actual).

In Figure 11, the stress distribution due to creeps and shrinkages on the bottom side shows a compressive – tensile (midspan) - compressive distribution along the span with a maximum tensile stress of 8.9655 MPa and a compressive stress of 0.734 MPa. The effect of shrinkage on 100 years after construction is 6 % at most.

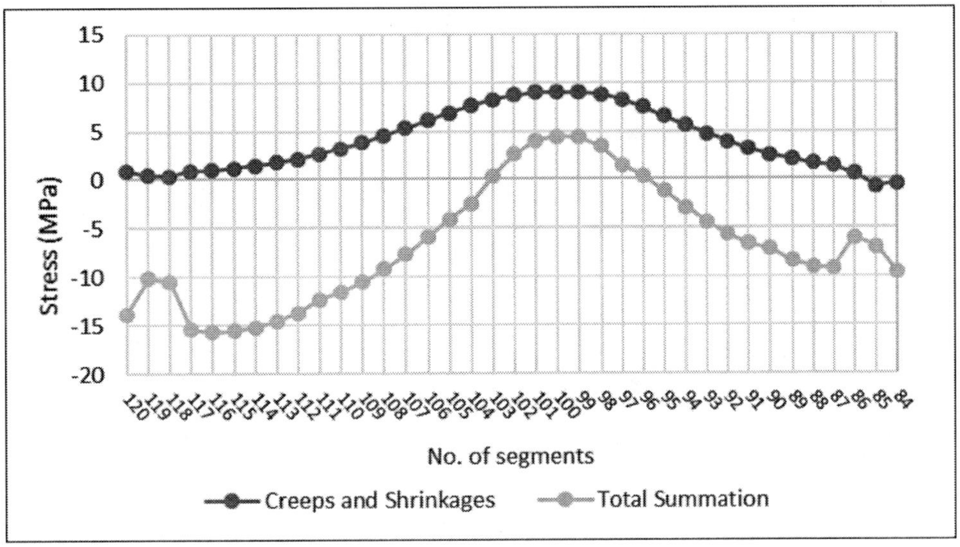

Fig. 11. Stress distribution at bottom side of girder at span P4-P5 (actual).

In Figure 12, the highest deviation of tensile stress between actual and design is in segment 88. Stresses on the segment shows that the tensile stress of the design is greater than the actual tensile stress. In the middle of the span, the compressive stress between the actual tensile stress and the design is quite the same.

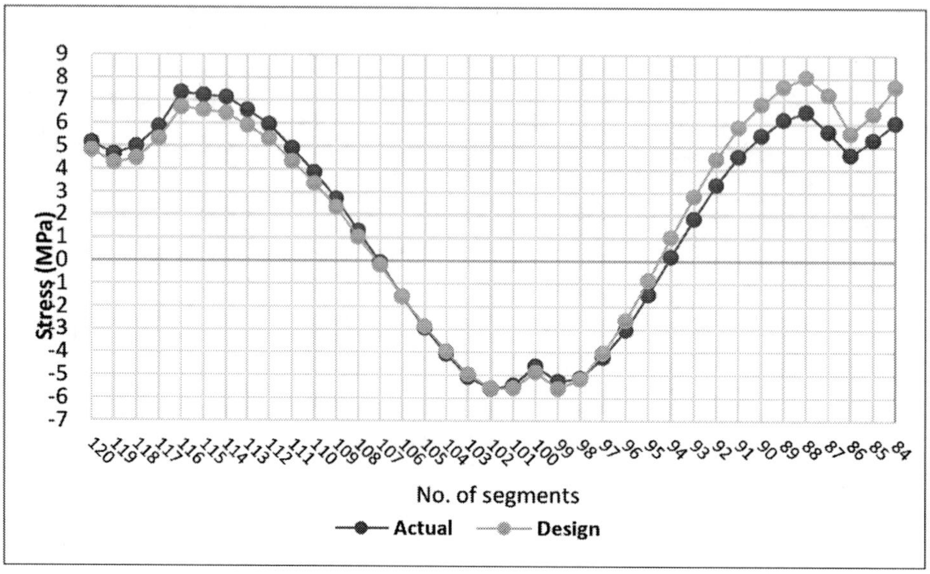

Fig. 12. Total Stress distribution at upper flange of girder at span P4-P5.

In Figure 13, the highest deviation of tensile stress between actual and design is in segment 100. Stresses on the segment show that the tensile stress of the design is greater than the actual tensile stress. Highest deviation of compressive stress occurs on segment 116. Stresses on the segment shows that the compressive stress of the design is smaller than the actual tensile stress.

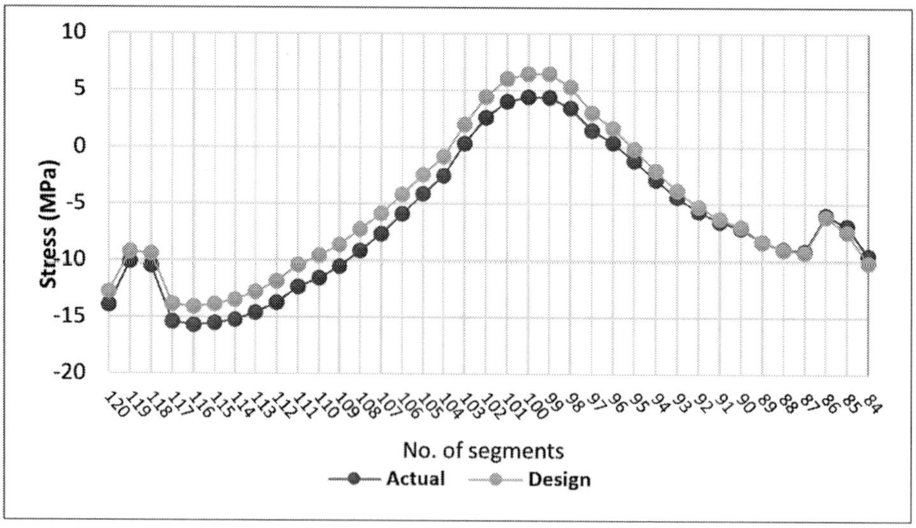

Fig. 13. Total Stress distribution at Bottom side of girder at span P4-P5.

7 Conclusion

Based on the research that has been done, it can be concluded as follows:

a. The actual model has a smaller creep and shrinkage effect than the designed model. The maximum tensile stress due to the effect of creep and shrinkage is 10,644 MPa below the designed model. The maximum compressive stress due to the effect of creep and shrinkage is 1.5285 Mpa.

b. The top section has stress distribution as tensile-compressive(midspan)-tensile. The bottom side of the sections has a stress distribution as compressive-tensile-compressive.

References

1. Adhi Karya Corporation, *Innovation Paper* (2013)
2. Bao, L., Li, G., Yu, L., Zhu, G., "Impact Analysis of Concrete Shrinkage and Creep on a Prestressed Concrete Box Girder Bridge", ICTE (2015)
3. Comité Euro-International du Béton (CEB), *CEB-FIP model code 1990* (1990)
4. Midas Civil, *Analysis Reference, Midas User Support System* (2011)
5. Pan, Z., Fu. C. C., Jiang, Y., "Uncertainty Analysis of Creep and Shrinkage Effects in Long-Span Continuous Rigid Frame of Sutong Bridge, J. Bridge Eng", **16**, 248-258 (2011)
6. Praja, B.A.,Triwiyono, A., *Perilaku Rangkak Susut Terhadap Lendutan Struktur Jembatan Bentang Panjang, DOI: 10.24002/jts.v13i4.934 (In Indonesian)*, (2016)

MATEC Web of Conferences **138**, 02005 (2017) DOI: 10.1051/matecconf/201713802005
EACEF 2017

Fragility function for assessing seismic risk of typical concrete bridge by means of nonlinear static and dynamic analysis

Senot Sangadji[1,*], *Niam A.* Wibowo[1], *Enjels N.* Tropormera[1], *Edy* Purwanto[1], *S. A.* Kristiawan[1]

[1] SMARTQuake, Department of Civil Engineering, Faculty of Engineering, Universitas Sebelas Maret, Sura karta 57126, Indonesia

Abstract. Bridge with concrete pier and deck is a conventional highway infrastructure built and utilized in Indonesia. More than half of these typical bridges have been in service more than 20 years. Evaluating these sort of bridge subject to various intensity of earthquake in its future service life is important. This assessment helps owner to predict damages and economic losses due to earthquake as well as to plan and devise seismic rehabilitation intervention prior to the event. The analysis uses fragility function describing the probability of exceedance of certain damage level due to single predictive demand parameter or earthquake intensity measure, i.e. spectral acceleration on bridge structure. The concrete bridge was modelled deterministically with 3-D computational structural model. Analytical non-linear static and non-linear dynamic pushover (incremental dynamic analysis) was employed to determine the structural response. Fragility function eventually is expressed as series of log normal curves which can be used to assess the seismic risk of the bridge considered.

1 Introduction

Bridge is essential and vital transportation infrastructure for safe and productive mobility of people and goods. In systemic point of view, bridge is actually a complex network of various components like bridge deck, girders, bearings, bents, foundations, abutments, retaining walls, and utility systems. Many of these components are quite vulnerable as affirmed by experience from past earthquakes. Due to minimum bridge structural redundancy, damage to this system ca be catastrophic. As repair of damaged bridge may be quite protractive and rerouting traffic may be notably difficult, the disruption to road networks is fatal. Further it strongly affects the emergency response in earthquake event. In the post-earthquake period, it affects rehabilitation, rebuild and revitalization of business activities of the region affected and neighbouring areas.

* Corresponding author: s.sangadji@ft.uns.ac.id

In many part of the world especially in developing countries, more than half of the existing bridges were developed prior to the modern earthquake resistant state-of-the art principles and practice. Risk assessment of these bridges is of important to ensure its safety as well as to decide rehabilitation intervention prior to the earthquake event. The main purpose of this paper is to demonstrate the development and the use of the fragility function for assessing seismic risk for typical concrete highway bridge. Fragility functions illustrate the probability of exceeding a certain degree of damage given of earthquake intensity parameters (e.g., acceleration spectra) on a typical concrete road bridge structure. In this work, concrete bridge will be modelled as 3D structures. The non-linear static and dynamic analytical method will be used to obtain structural response. The fragility function will be expressed as series of fragility curves that can be used to assess the vulnerability of structures at various earthquake intensities.

2 Fragility functions and methods of its construction

2.1 Fundamental concept of fragility functions

Fragility functions is defined as a functions describing the probability of exceeding different limit states (such as damage or injury levels) given a level of ground shaking. It relates earthquake intensities with the probability of exceeding certain limit states. By means of a functions describing consequences / loss due to ground motion on the damage state, vulnerability functions can be derived from corresponding fragility functions.

Fragility functions is often expressed mathematically by a lognormal probability distribution function and graphically expressed as fragility functions [1, 2]. For infrastructure at risk, fragility functions relates two parameters in the equation; seismic intensity to the exceedance probability of certain damage level as in Eq. 1 [2];

$$P_f(ds \geq ds_i \mid IM) = \phi\left[\frac{1}{\beta_{tot}} ln\left(\frac{IM}{IM_{mi}}\right)\right] \quad (1)$$

where P_f is the conditional probability of being or exceeding a particular damage sate, DS, for earthquake intensity measure, IM, ϕ is the standard cumulative function, IM_{mi} is the median value of earthquake intensity which cause the ith damage state, and β_{tot} is the dispersion representing total standard deviation. Earthquake intensity parameters, IM, mas include peak ground acceleration/velocity/displacement, spectral acceleration, spectral velocity or spectral displacement. In this paper spectral acceleration, S_a is chosen to be the earthquake intensity parameters as easy means to relates with damage level, as expressed in Eq. 2 [3];

$$P(ds \mid S_a) = \phi\left(\frac{1}{\beta_{ds}}\right) ln\left(\frac{S_a}{\overline{S_a}, ds}\right) \quad (2)$$

with β_{ds} is standard deviation for total uncertainty of each damage condition, $\overline{S_a}, ds$ is the spectral acceleration median value when the structure is damaged. The standard deviation value for total uncertainty of any damage condition is determined according to HAZUS-MH MR5[4].

Analytical method has been employed in this paper. The methods is based on the distribution of structural damage obtained from series of numerical simulation of the structure at risk studied. In obtaining structural response the structural model is subjected to seismic input. The inputs were; [a] static as represented by response spectrum and [b] dynamic as represented by site-specific-matched scaled acceleration time-history. 3D computational models were developed based on the existing bridge taking into account trade-off between accuracy of the model nonlinear behaviour and time-efficiency of the simulation. Model with concentrated inelasticity (i.e. plastic hinge) modelling was developed for non-linear static (pushover) procedure with response spectrum input. Meanwhile fibre element model with distributed plasticity was developed for bridge model with time history input.

2.2 Nonlinear static procedures; capacity spectrum methods from pushover

One of the most popular methods to examine structural response and assess its seismic performance is non-linear static or pushover analysis [5 - 8]. Pushover analysis is a non-linear monotonic static analysis performed by providing a lateral static load pattern on the structure model, which is then gradually increased by a scalar multiplier until a target of lateral displacement from a reference point is reached.

The result of a pushover analysis is a capacity curve indicating the relationship of the total base shear force with the reference displacement. In the capacity spectrum method the capacity curve is transformed into a capacity spectrum combined with the demand curve in acceleration-displacement response spectrum format (ADRS) [5 - 8].

The capacity spectrum is a curve that connects spectral acceleration with spectral displacement. To obtain spectral acceleration values, the following formula, Eq.3. [6] is used:

$$S_a = \frac{V_b}{M_1} \tag{3}$$

with S_a is spectral acceleration, V_b is base shear, and M_1 is effective modal mass in fundamental vibration mode. Meanwhile, to obtain the spectral displacement value , the following formula, Eq.4., is used:

$$S_d = \frac{u_N}{T_1 \phi_{N1}} \tag{4}$$

where $T_1 = \frac{\sum_{j=1}^{N} m_j \phi_{j1}}{\sum_{j=1}^{N} m_j \phi_{j1}^{2}}$, with S_d is spectral displacement, u_N is displacement at the reference point, m_j is lumped mass at point j, and ϕ_{j1} is element at point j in fundamental vibration mode.

2.3 Nonlinear dynamic procedures; incremental dynamic analysis (IDA)

Incremental Dynamic Analysis (IDA) is a parametric analysis method that predicts demand and seismic capacity of the structure. In the dynamic analysis of this nonlinear time history the structural model is subjected to one or more earthquake records (accelerograms) multiplied by scale factor (scalar multiplication) for various levels of expected seismic intensity.

The scaling increment level of the quake intensity is chosen in such a way as to 'force' the structure to work across the entire range of behaviours; from the elastic phase to the plastic phase (inelastic) until it finally reaches global dynamic instability (collapses). Due to the dynamic load that works increases in intensity, this analysis can be understood as an equivalence form of static nonlinear analysis or pushover. Even when first developed, this analysis is known as dynamic pushover analysis.

Extensive explanations of IDA's philosophy and methods can be found in Vamvatsikos and Cornell [9 - 11]. They examined IDA's basic formulations and concepts on a 20-storey steel structure with a moment-canard frame system (SRPM).

The IDA input in the structural model is a set of synthetic ground motion matching with the target spectrum response for the bridge location area. The analysis results are the response distribution of the structure expressed in the set of IDA curves. This curve shows the relation of intensity measure (IM) to the structure response (Engineering Demand Parameter [EDP]) .The EDP can be selected such as inter-story drift, drift ratio or peak floor acceleration.

Increment level increments, λ, were set uniformly for all selected artificial accelerograms. Initial intensity starts from 0.2 PGA with increment 0.2 unit up to 2 times PGA. With this increment rule it is expected that the linear elastic structure of the structure can be estimated more thoroughly. Furthermore, the increment of 1 unit, carried out from the intensity of 3 PGA until the structure gives a response assuming that the structure has collapsed. In summary the increment rules are expressed as follows; $\lambda = 0.2, 0.4, 0.6, ..., 1.8, 2. 3, 4, 5, 6 ... \infty$.

2.4 Damage/limit states

The ultimate goal of seismic risk assessment methodology is to assess the performance (or the expected loss) of the infrastructure and all its systems and components when subjected to a seismic hazard. The performance levels of a structures can be defined via threshold of damage (limit states). As damage state defines damage condition of the structure as consequence of earthquake, the limit state delineates the boundary between damage levels. With some consensus on the meaning and content qualitatively, the most common classifications of damages is the following: no damage; slight/minor; moderate; extensive; complete. However, discrete damage scale is usually required for constructing fragility curves. Numerical expression of limit states, either the accumulated consequences of all damages, or the comparison of a demand with a capacity quantity, or the consequence of a mitigation action, may also be defined as performance indicators. The qualitative description of bridge columns (pier) damage level based on HAZUS [4] and the value of limit state described by Kim and Shinozuka (2004) [13] can be seen in **Fig. 1**.

DOI: 10.1051/matecconf/201713802005

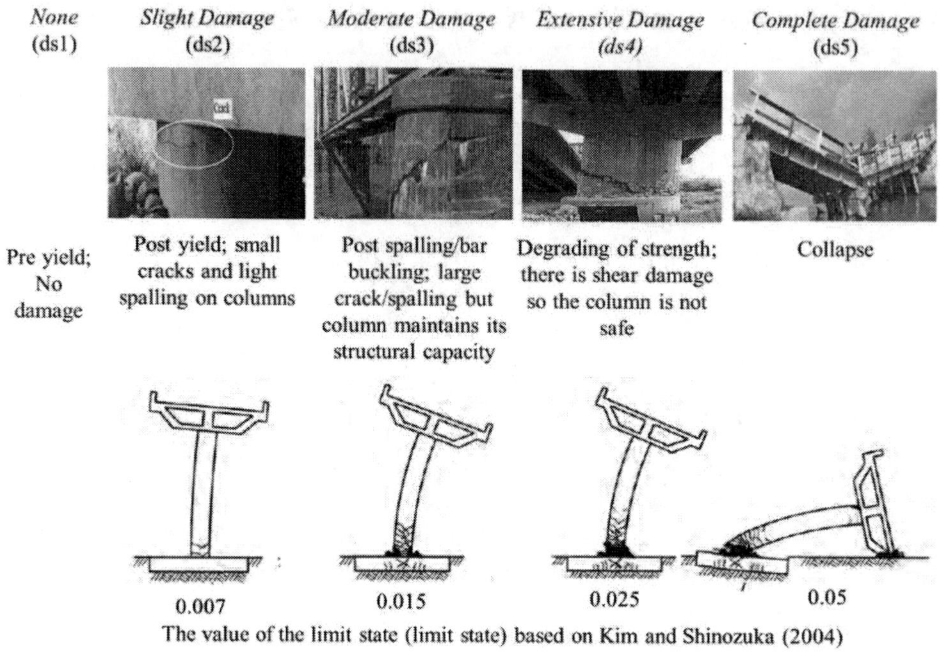

Fig. 1. Damage level based on HAZUS [4] and Kim and Shinozuka (2004) [13].

3 Result and discussion; case study on existing bridge

3.1 Model of the structure

The 3D computational grillage model was built based on the geometry data of Meluang-A bridge, an existing bridge located Bengkulu Province, Indonesia. It has 2 spans of 30 m and 30.1 m concrete girder supported on rectangular pier in top of hard soil as seen in the **Fig. 2**. The pier has cross sectional dimension of 1 × 5 m with the height of 5 m. The lead rubber bearing was modeled as elastic link boundary condition of the model which was connected to the pier head with rigid link. Longitudinal bending moment mode was observed from Eigen problem analysis with fundamental period of 0.96 sec. Fig. 2 also shows conceptually the input vectors of nonlinear static procedure as well as site-specific-matched ground motion records for nonlinear incremental dynamic analysis along with the self-weight and traffic load according to the recent Indonesian Code. Material behavior was assumed to have concentrated inelasticity therefor plastic hinge modelling was applied for push over analysis. Meanwhile distributed plasticity in the pier was modeled with fiber element when incremental dynamic analysis was employed.

Fourteen earthquake acceleration records were chosen from database, PEER *ground motion database* (US) and BRI *strong motion database* (Japan), for response history analysis. These accelerograms (time domain) were transformed into its response spectra (frequency domain). Comparison of these spectra with the *target response spectra* on the interval of 0.2T to 1.5T, where T is fundamental period of the structure revealed the scale factor as expressed by equations [14]:

DOI: 10.1051/matecconf/201713802005

$$SF = \frac{\sum_{i=1}^{n} \bar{A}_i . A_i}{\sum_{i=1}^{n} A_i . A_i} \tag{1}$$

where SF is scale factor; Ā is *spectral acceleration of target response spectrum*, and A is *spectral acceleration of response spectrum* of the source.

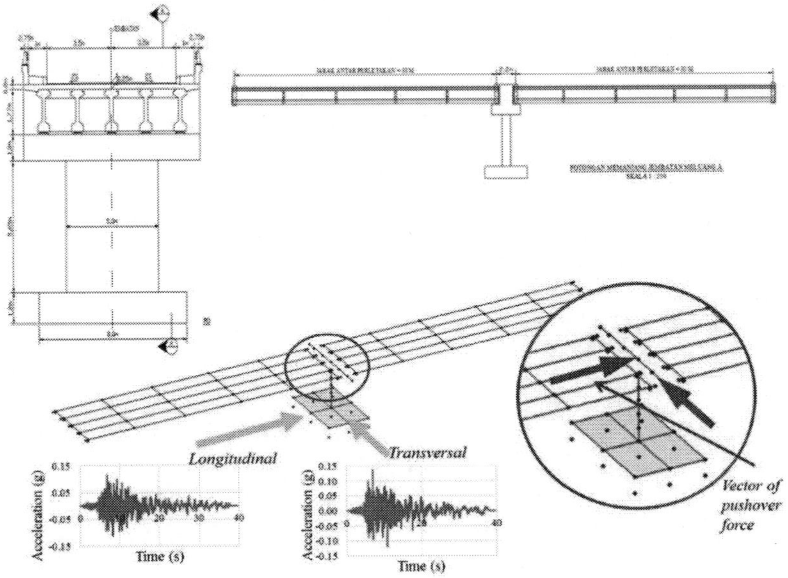

Fig. 2. The existing bridge and 3D computational model

Wavelets algorithm was then employed to spectral matching the earthquake records response spectra to the *target response spectra*. The matched spectrum then was converted into artificial accelerogram (acceleration time history) taking into account stochastic parameters where the results can be seen visually in the Figure 3.

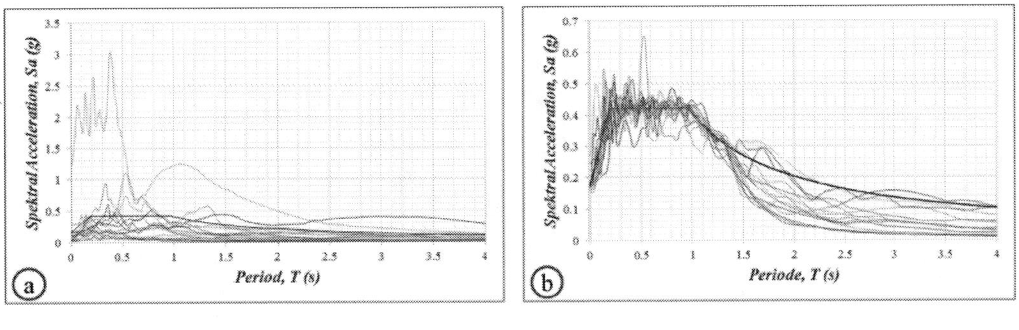

Fig. 3. (a) the unmatched earthquake spectra and (b) matched earthquake spectra with the target response spectra.

DOI: 10.1051/matecconf/201713802005

3.2 Fragility by means of non-linear static procedure.

The results of the pushover analysis and the implementation of the capacity spectrum procedure are presented in the form of spectrum of capacity and spectrum demand combinations, illustrated in Figure 4. Based on the capacity curve of Meluang-A bridge, the structure undergoes a plastic yielding when the column displaced at 0.18 m while base shear force was 12387.2 kN with spectral acceleration of (Sa) 1.28 g. The spectral acceleration value is quite higher compared to the expected peak acceleration value at this bridge location namely 0.4 - 0.5 g based on the earthquake hazard map of SNI 1726: 2012. Meanwhile it is seen that the performance point is in the elastic phase of the structure. These two data lead to the assumption that the Meluang-A bridge column has a high enough capacity to withstand the lateral forces of the earthquake associated with a large enough cross-sectional dimension and a height of only 5 m.

Based on HAZUS criteria, the spectral displacement and its associated spectral acceleration was obtained for each median value from slight to collapse damage level. For instance $S_{d,slight}$ was assumed at first yield at 0.49 m corresponds to its spectral acceleration , $S_{a,slight}$, as seen in the graph Fig. 3. These median value can be seen in the graph Fig. 4.

While it is seen in Figure 4 that the performance point, the point of intersection of capacity spectrum with demand spectrum is in the elastic phase of the structure. These two data lead to the assumption that the Meluang-A bridge column has a high enough capacity to withstand the lateral forces of the earthquake associated with a large enough cross-sectional dimension and a height of only 5 m.

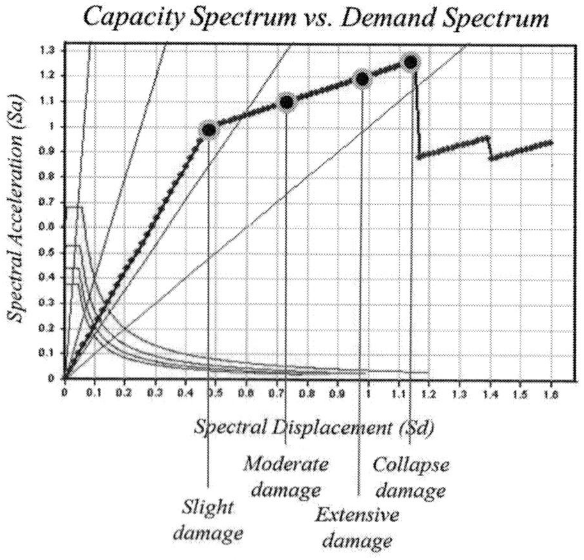

Fig. 4. The capacity spectrum of the Meluang-A bridge with damage criteria

Figure 5 exhibits the fragility curve of the bridge structure based on HAZUS-MH MR5damage criteria. The ordinate of the graph shows the probability of exceedance of certain damage level, from slight to collapse, as a functions of spectral acceleration. In this case fragility function is not directly related to the ground motion intensity measure but indirectly to the structural response parameters as what easily obtained from capacity curve derived from pushover analysis

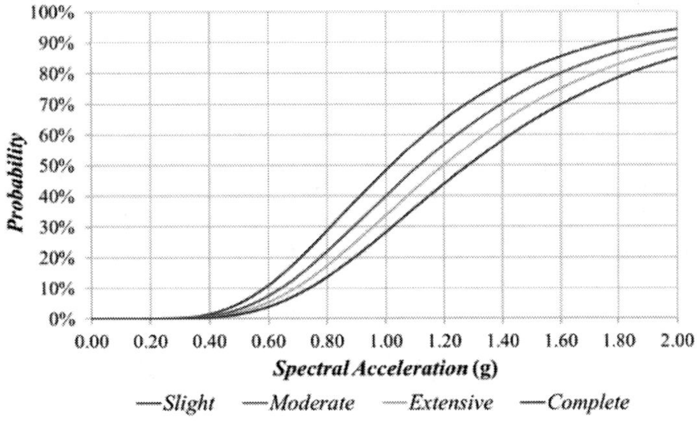

Fig. 5. Fragility curve of the structure with indirect representation by means of pushover analysis

3.3 Fragility by means of nonlinear incremental dynamic analysis.

Figure 6 shows the response of the structure given the applied accelerogram expressed as IDA curves. Generally, the analysis was terminated when 8% drift ratio was achieved. Linear behaviour is shown in the initial phase where all curves mostly coincide up to drift ration about 0.7%. This indicates that in its initial phase the structure tends to behave linear elastic independent of the acclerogram input. Then structure exhibit nonlinear behaviour; hardening and then softening of the structure.

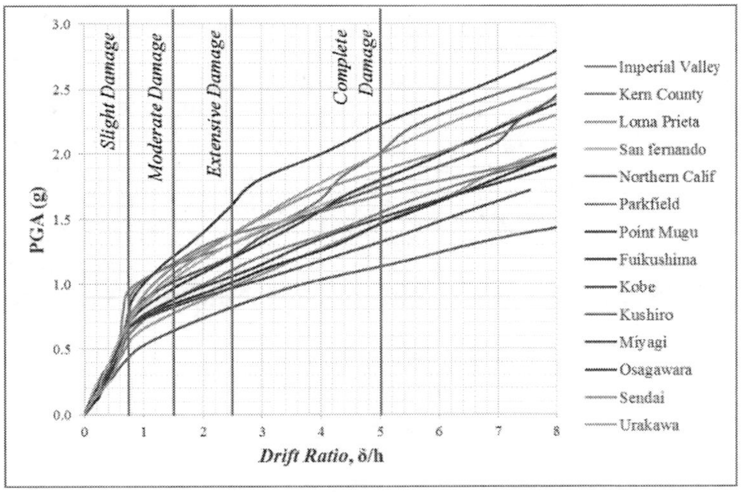

Fig 6. IDA curves from nonlinear time history analysis superposed with damage criteria

The damage criteria was defined based on Kim dan Shinozuka (2004). Drift ratio for the bridge pier is 0.7% for slight damage, 1.5% for moderate damage, 2.5% for extensive damage and 5% for complete damage as seen in the Fig 1. The inherent aleatory and epistemic uncertainties was treated based on the method proposed by Vamvatsikos dan Fragiadakis (2009).

With PGA obtain from from IDA curves and median capacity calculation, Fragility curves was obtained by using equation 2. The results is expressed graphically as cumulative distribution function (CDF) of log-normal distribution as seen in Fig. 7.

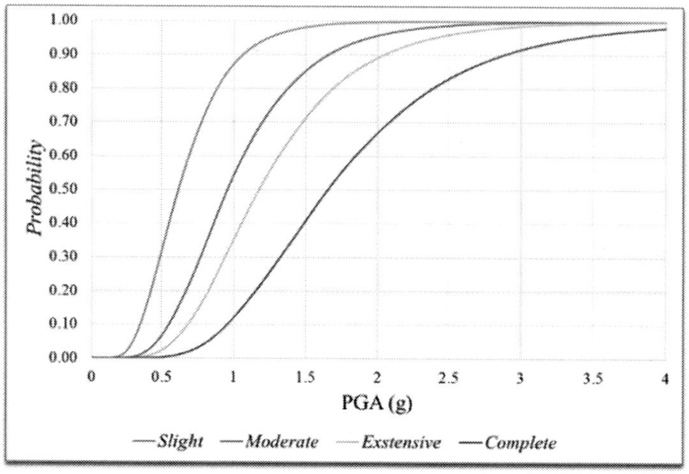

Fig 7. Fragility curve of the structure with direct representation by means of Incremental dynamic analysis

The ordinate of the graph shows the value from 0 to 1 which means zero to hundred percent certainty that some level of damage will be exceeded. The fragility curves is directly related to the expected peak ground acceleration during structures' service life.

With these curves, the evaluation of probability that the bridge structure will be damaged due to variation of earthquake intensity can be done quantitatively and rationally. For example, if the site where the bridge is located experiences earthquake with PGA about 0.4 g there will a chance of 2% the structure experience slight damage. From another perspective, if the 50% probability level is taken by drawing a horizontal line cutting off the fragility curve representing slight damage, then vertical line from the intersection to the PGA show the value of 0.7. This means there is a 50% certainty that the bridge will suffer slight damage as defined by Kim and Shinozuka (2016) in the event of earthquake with PGA 0.7 g.

4 Conclusion

The fragility curve of an existing bridge has been developed using a fragility function equation. The nonlinear response structure is well predicted using a nonlinear static analysis procedure. This procedure is capable of generating a capacity curve transformed into a capacity spectrum. Using the capacity curve spectral acceleration values can be obtained for each damage states according to HAZUS criteria. The indirect representation of fragility curve is then constructed in the form of a cumulative distribution curve as function of structural response parameter. Incremental dynamic analysis predicts the nonlinear response of the structure resulting on the direct representation of the fragility curve as a function of PGA.

References

1. Pitilakis, K., Crowley, Helen, Kaynia, Amir (Eds.). (2014). *SYNER-G: Typology Definition and Fragility Functions for Physical Elements at Seismic Risk*, Springer, The Netherlands.
2. Kaynia, Amir M., (ed)., (2013) *Guidelines for deriving seismic fragility functions of elements at risk: Buildings, lifelines, transportation networks and critical facilities, SYNER-G Reference Report 4*, European Commission, Joint Research Centre.
3. Papailia, A. (2011). *Seismic Fragility Curves for Reinforced Concrete Buildings*. MSc Dissertation, University of Patras.
4. Federal Emergency Management Agency (FEMA). (2010). *HAZUS-MH MR5, technical and user's manual*, Washington, DC.
5. Duan, X. (2008). *A Procedure For Establishing Fragility Functions For Seismic Loss Estimate of Existing Buildings Based On Nonlinear Pushover Analysis*. The 14th World Conference On Earthquake Engineering. China.
6. Chopra, Anil K. and Goel, Rakesh K., (1999) *Capacity-Demand-Diagram Methods for Estimating Seismic Deformation of Inelastic Structures: SDF Systems.*, Report No. PEER-1999/02., Pacific Earthquake Engineering Research Center, University of California, Berkeley.
7. Fajfar, Peter., (1999). *Capacity Spectrum Method based on Inelastic Demand Spectra*. Earthquake Engineering and Structural Dynamic 28, 979 – 993. Wiley.
8. Ger, J., dan Cheng, F.Y. (2012). *Seismic Design Aids for Nonlinear Pushover Analysis of Reinforced Concrete and Steel Bridges*. CRC Press,. Boca Raton,.
9. Vamvatsikos, D. dan Cornell, Allin C. (2002). *Incremental Dynamic Analysis*. Earthquake Engineering and Structural Dynamic 2002; 31:491-514.
10. Vamvatsikos, D. and Cornell, Allin C. (2002.) *Applied Incremental Dynamic Analysis*. 12th European Conference on Earthquake Engineering. London.
11. Vamvatsikos, D. and Fragiadakis, M. (2009). *Incremental Dynamic Analysis for Estimating Seismic Performance Sensitivity and Uncertainty*. Earthquake Engineering and Structural Dynamic 2009; 00:1-16.
12. Mander, J. B. dkk. (2006). *Incremental Dynamic Analysis Applied to Seismic Risk Assessment of Bridges*. University of Canterbury. Christchurch, NZ.
13. Kim, Sang-Hoon, and Shinozuka M. (2004). *Development of fragility curves of bridges retrofitted by column jacketing*. Probab Eng Mech 19(1–2):105–112
14. Kalkan, E. and Chopra, A.K. (2010). *Practical guidelines to select and scale earthquake records for nonlinear response history analysis of structures*, U.S. Geological Survey Open-File Report 2010-1068, 124 p.

DOI: 10.1051/matecconf/201713802006

Development of Acacia Glulam Wood Exterior Beam-to-Column Connection with Angles and Steel Rods System

Djoni Simanta[1], *Bambang Suryoatmono*[1,*], and *Johannes A. Tjondro*[1]

[1]Departmentof Civil Engineering, Parahyangan Catholic University, Indonesia

Abstract. In this paper, two models (M9B1 and M9B2 connection types) of acacia glulam wood exterior beam-to-column connections with angles and steel rod system have been developed. Both models have been tested experimentally under monotonic static and cyclic loadings. The behaviour of each model under static and cyclic loadings, maximum displacement, maximum moment, equivalent viscous damping ratio, stiffness degradation, rotational stiffness and ductility have been observed. From the test results it can be concluded that the behaviour of the connections were influenced by the number of bearing steel rods at the beam, wood quality, and thickness of the angles. The M9B1 connection type has more flexibility than the M9B2 connection type. Three dimensional finite element analyses considering anisotropic plasticity have been conducted for both models. It can be concluded that in M9B1 connection type, the tensile forces due to the moment are received by the combination of tensile steel rods and bearing force in the steel rod and wood around it. Overall performance of M9B2 connection type is better than that of M9B1 connection type because the former has higher initial stiffness and higher rotational ductility than the later.

1 Introduction

In an earthquake resistant timber buildings, one of the important details that needs to be considered is the beam-to-column connections. The connections need to be able to dissipate energy effectively. Moreover, they have to have enough rotation capacity and ductility so that moment redistribution can occur in the building. Several studies have been done to obtain wood beam-to-column connections. A portal quick connection is a semi-rigid moment connection developed by Scheibmair and Queneville [1]. The connection has a concept similar to the usual bolted partial moment connections used in steel structures. The connection consisted of steel rods. Test results of this type of connection showed that failure modes were tension parallel to the grain, shear failure, tension perpendicular to the grain, and bearing failure in the contact area between the beam and the column. Another connection type is dowel moment connection developed by Leitner [2] as seen in Figure 1. It can be seen in the figure that: (a) the initial stiffness and degradation of the stiffness of

* Corresponding author: suryoatm@unpar.ac.id

DOI: 10.1051/matecconf/201713802006

the two models is very different, (b) pinching occurs in experimental cyclic test results, and (c) the plastic behaviour of the material is not modelled sufficiently in the numerical model.

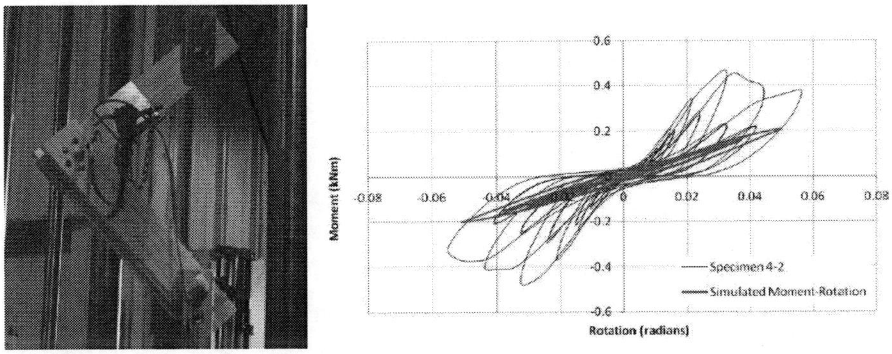

Fig. 1. Moment connection developed by Leitner under cyclic loading (a) Experimental setup, (b) hysteretic loop of experimental test and finite element analysis [2]

In this paper, two models (M9B1 and M9B2 connection types) of acacia glulam wood exterior beam-to-column connections with angles and steel rod system have been developed. What is meant by the exterior connection here does not mean the connection is exposed to the weather, but the connection between the column and the beam exists only on one side of the column. The first model uses one dowel steel rod parallel to the grain and the second one uses two dowel steel rood parallel to the grain. Figure 2 and 3 show the dimensions of M9B1 and M9B2 models, respectively. The presence of steel rod and steel angle in each model is intended to add its ductility. Both models have been tested experimentally and numerically under monotonic static and cyclic loadings. The maximum displacement, maximum moment, equivalent viscous damping ratio, stiffness degradation, rotational stiffness, and ductility have been observed. Simanta et.al [3] discusses the experimental results in more detail. On the other hand, this paper discusses the numerical (finite element) analysis results.

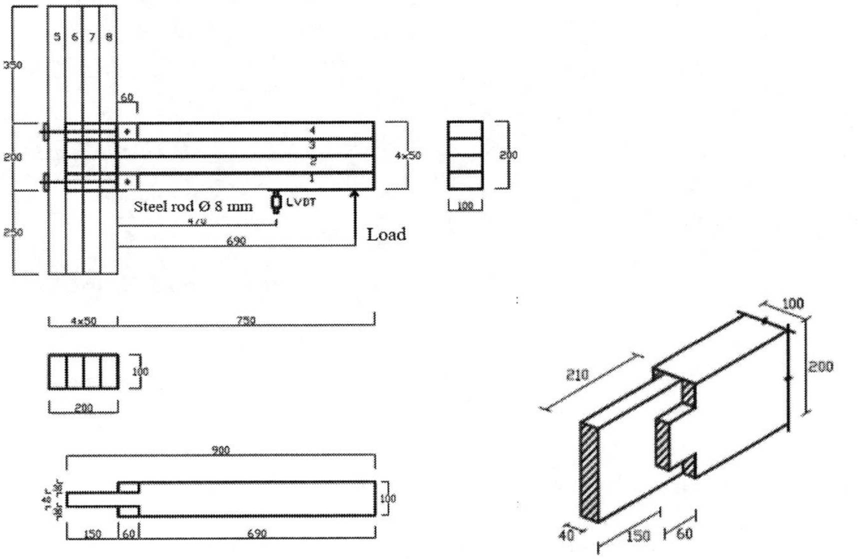

Fig. 2. Dimensions of beam-to-column connection with one steel rod in the beam, type M9B1.

 DOI: 10.1051/matecconf/201713802006

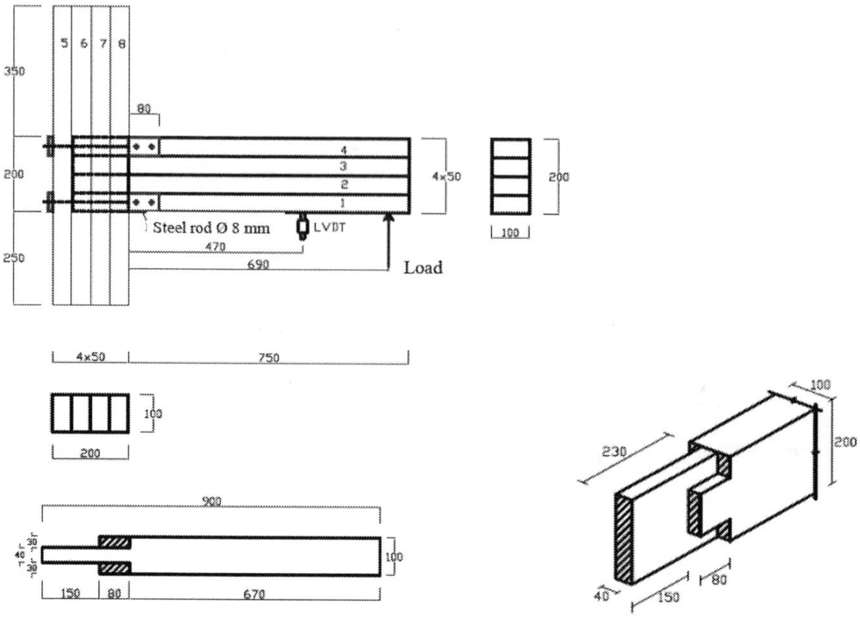

Fig. 3. Dimensions of beam-to-column connection with two steel rods in the beam, type M9B2.

Both types of connection basically use the concept in the steel structure connection, where the bending moment is received by the tensile steel. Then, the tensile force is transferred by a sufficiently rigid steel angle to the lateral steel rods providing a bearing strength to the glulam beam section. The shear force is assumed to be carried by a rigid shear key in the form of a wooden pen that is part of a glulam wood beam that pierces the column. Figure 4 shows the picture of the developed beam-to-column connection.

(a) (b)

Fig. 4. Acacia glulam wood exterior beam-to-column connection with angles and steel rods system, (a) M9B1 connection type, (b) M9B2 connection type [3].

2 Non-linear finite element analysis (FEA)

Experimental tests of wood connections in general require the availability of wood which is quite a lot and requires a considerable cost. Therefore, the use of a nonlinear finite element program capable of simulating the 3D model of the connection would be very useful in the research to complement the experimental test results. The finite element program needed to simulate the behaviour of the beam-to-column connection developed in this paper is the one capable of modelling orthotropic, transverse isotropic and anisotropic plasticity behaviour for wood materials, and various contact behaviour between each part of the connection. Non-linear finite element analysis was performed using ANSYS version 15, or specifically ANSYS Workbench as a CAD system that facilitated the modelling of complex geometric shapes. ANSYS Workbench automatically establishes contact elements and performs meshing according to the geometry of the elements modelled through its designer model. The results of this finite element analysis are compared with experimental results.

The 3D non-linear finite element analysis to study the behaviour of the beam-to-column connection utilizes SOLID45 element with 8 nodes as seen in Figure 5. In this SOLID45 element the plasticity condition is characterized by reversible strain that occurs in a material when a given stress level is achieved by defining the bilinear stress-strain relation as seen in Figure 6.

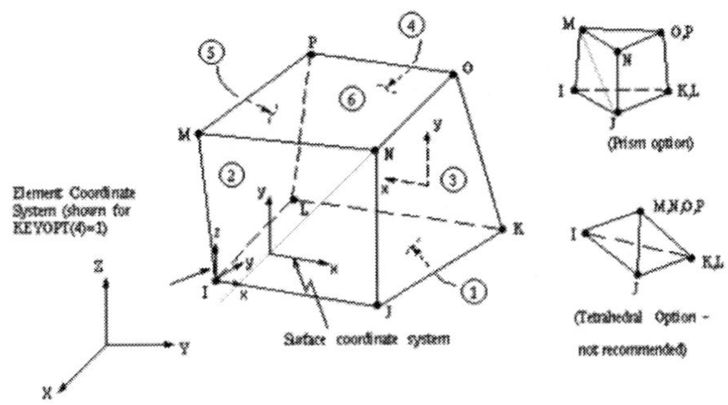

Fig. 5. Element SOLID45 with 8 nodes in ANSYS used in finite element analysis of the beam-to-column connection [4].

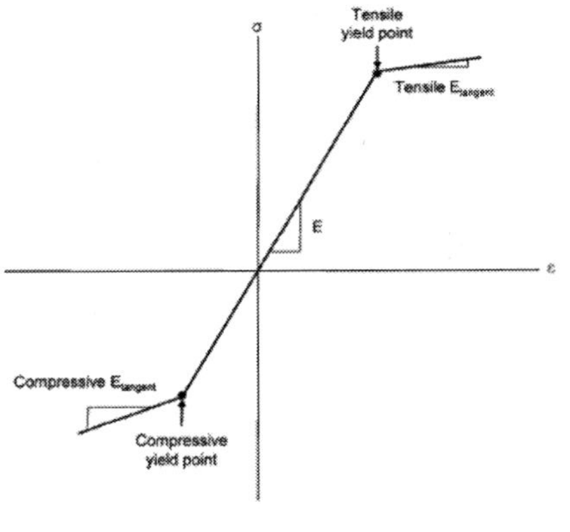

Fig. 6. Bilinear stress-strain relation in anisotropic plastic behaviour model [5]

2.1 Wood mechanical properties

The wood orthotropic properties used as input data for the finite element analysis using ANSYS are expressed in the *L* (longitudinal), *R (*radial*)*, and *T* (tangential) directions as seen in Table 1. The values are obtained from the compression test parallel and perpendicular to the grain according to ASTM D143 [6]. Wood mechanical properties for considering bearing obtained from test results according to ASTM D5764 [7] and used in finite analysis are shown in Table 2. The wood lamina identified as *A, C, D, E, F, G, H* and *I* are the wood lamina used to form glulam columns and beams in the tested beam-to-column connection and modelled in the finite element analysis. To account for the strain hardening of the compressive moduli of elasticity (E_L, E_T, and E_R) and shear moduli (G_{LT}, G_{RT}), all moduli are multiplied by 0.01 [5].

Table 1. Orthotropic properties of wood in finite element analysis.
Moduli and stresses are in MPa.

Lamina	E_L	E_T=E_R	σ_L	σ_T=σ_R	ν_{LT}	ν_{RT}	ν_{TL}	G_{LT}	G_{RT}	γ_{LT}	γ_{RT}	τ_{LT}	τ_{RT}
A	7242.64	459.00	26.54	8.55	0.29	0.43	0.02	850.32	160.02	0.01	0.02	5.88	13.55
C	5734.38	919.58	36.18	8.52	0.30	0.35	0.05	1026.74	339.58	0.00	0.01	4.55	7.90
D	8020.05	511.83	49.09	9.94	0.34	0.26	0.02	933.96	202.37	0.01	0.02	6.78	14.57
E	4569.92	388.46	26.31	7.17	0.20	0.28	0.02	628.74	151.43	0.01	0.01	4.61	9.39
F	9537.34	497.57	46.96	7.25	0.21	0.28	0.01	1038.53	194.36	0.01	0.01	5.29	12.23
G	7676.45	584.13	39.53	14.13	0.32	0.37	0.02	972.09	212.56	0.01	0.02	9.20	19.68
H	7705.64	924.15	44.12	13.86	0.19	0.15	0.02	1253.06	401.63	0.01	0.01	8.15	14.40
I	7846.15	338.23	35.36	6.59	0.31	0.41	0.01	764.82	119.71	0.01	0.02	5.01	12.66

Table 2. Orthotropic properties of bearing wood in finite element analysis.
Moduli and stresses are in MPa.

Wood	E_L	E_T=E_R	σ_L	σ_T=σ_R	ν_{LT}	ν_{RT}	ν_{TL}	G_{LT}	G_{RT}	γ_{LT}	γ_{RT}	τ_{LT}	τ_{RT}
E	370.11	221.38	42.26	19.11	0.20	0.28	0.12	123.59	86.30	0.06	0.07	7.21	8.63
H	609.21	333.79	62.00	27.90	0.19	0.15	0.10	198.03	145.06	0.05	0.06	10.85	12.68

2.2 Steel mechanical properties

In the finite element analysis to obtain the behaviour of beam-to-column connection developed in this paper, the type of elements used to model steel rods and steel angle are SOLID186 and SOLID187. To take into account the plasticity in the analysis, stress-strain curve of steel tensile test results used in the ANSYS data is multi-linear isotropic hardening curve as seen in Figure 7. The curve is used as material property of steel rods. For steel angle, the material is assumed to behave as perfectly plastic isotropic material, with material properties as follows: yield stress F_y = 250 MPa, modulus of elasticity E = 200000 MPa, and strain hardening factor = 0.

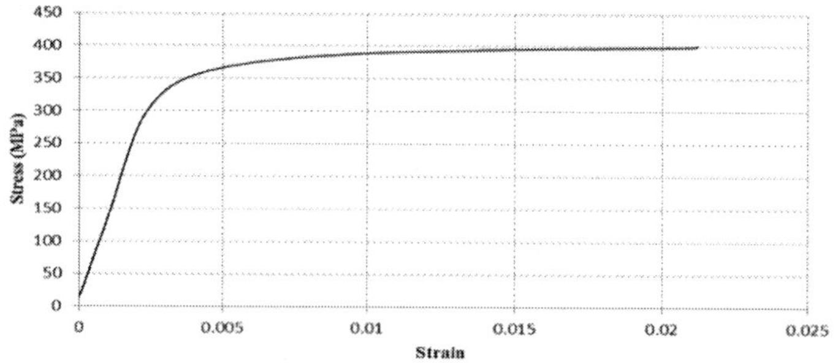

Fig. 7. Multi-linear isotropic hardening stress (MPa) – strain curve for modelling steel rods.

2.3 Modelling the connection for Finite Element Analysis

Finite element mesh to model the beam-to-column connection is generated using ANSYS Workbench. Example of the mesh is shown in Figure 8. In ANSYS Workbench there are some basic options of the contact surface model that can be applied to the finite element model. It is very important to utilize the closest assumption to the reality.

Fully bonded between two surfaces assumes that the contact between both surfaces is perfectly attached so that there is no penetration, no separation and no sliding between both surfaces. Fully bonded is used in modelling contact between each wood lamina in glulam beam and column, between the steel angle and the nut, between the steel bar and the nut, and between the outer lamina of the column and the nut.

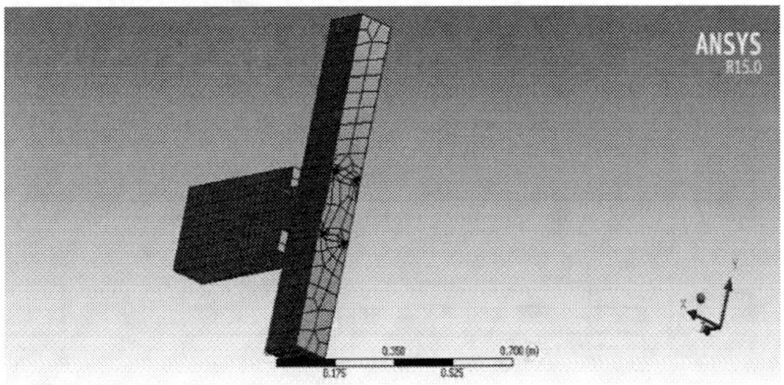

Fig. 8. Example of finite element mesh for studying the behaviour of the beam-to-column connection.

Between the beam pen and the column lamina surrounding it and between the beam lamina and steel angle, it is assumed that no separation exists between the surfaces. This means that one surface can slip without friction along the contact surface but both surfaces cannot separate.

In frictionless model there is no penetration on the contact surface, but both surfaces can slip without friction, and the formation of gap between both surfaces is allowed. Frictionless model is used in the contact surfaces between the column lamina and the steel rod, and in the contact surfaces of the column lamina with the steel angle.

In a situation where contact between surfaces cannot be modelled in ANSYS, it is suggested to use suppressed model. This means that contact between the two surfaces is not activated. Suppressed model is used in contact between steel angle and steel rods, between bearing steel rod and bearing wood in beam, and between steel angle and column.

There are two loading conditions applied to each finite element model, namely static monotonic and cyclic loadings in the form of line load at the beam end. The cyclic load applied to the connection is in accordance to the cyclic load of ISO [8] as seen in Figure 9.

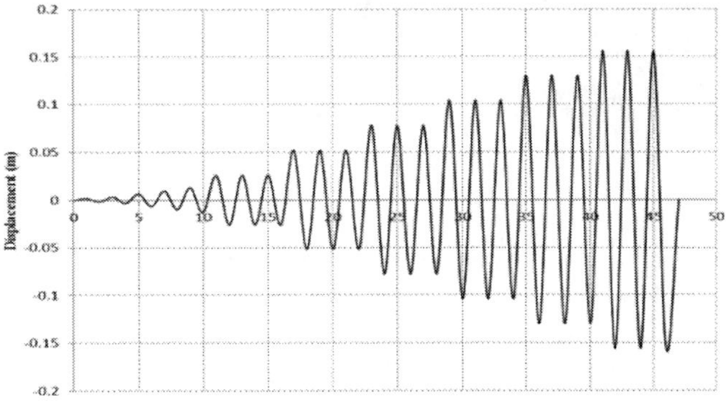

Fig. 9. Cyclic loading of ISO for wood beam-to-column connection [8].

Boundary conditions used in the finite element model are fixed support and frictionless support to model the supports at the top and bottom of the column, respectively.

Nonlinearity in the nonlinear finite element solution occurs because the analysis uses nonlinear material behaviour (plasticity) and contact surfaces between each part of the solid model. Nonlinear analysis solutions generally use iteration. In ANSYS Workbench, this iteration is performed using Newton-Raphson Method.

There are six connection models analysed depending on the lamina in the glulam column and beam as seen in Table 3. Two models are of the M9B1 type that has one bearing steel rod and four models are of the M9B2 type that has two bearing steel rods. As an example, Figure 10 shows all lamina in M9B1_L_1 model. See Table 1 for the material properties of each lamina identified as A, C, D, E, F, G, H, and I. For lamina of beam surrounding the steel rod, the material properties used in the analysis is shown in Table 2.

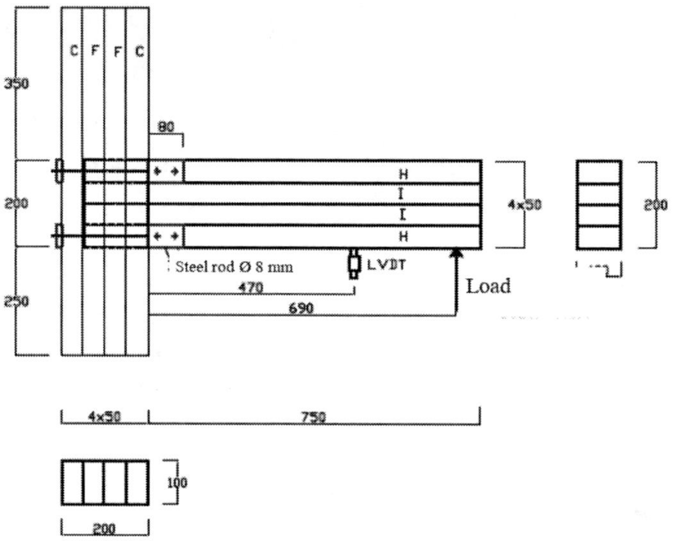

Fig. 10. Lamina in glulam column and beam for finite element model identified as M9B1_L_1.

Table 3. Lamina for glulam column and beam in each model.

Identity*	Column lamina from left to right	Beam lamina from top to bottom
M9B1_L_1	C-F-F-C	H-I-I-H
M9B1_L_2	A-A-G-G	E-D-D-E
M9B2_L_1	C-F-F-C	H-I-I-H
M9B2_L_2	A-A-G-G	E-D-D-E
M9B2_L_3	C-F-F-C	H-I-I-H
M9B2_L_4	A-A-G-G	E-D-D-E

*L stands for lamina

3 Results and discussion

From each finite element analysis of beam-to-column connection, load – displacement curve can be obtained. By using simple statics, the load – displacement curve can be converted into moment – rotation curve. The moment is obtained by multiplying the load by the distance between the face of the column and the load position. The rotation is obtained by dividing the deflection by the distance between the face of the column and the load position. Figure 11 shows load – displacement curve of M9B1_L_1 and M9B1_L_2 obtained from non-linear finite element analysis under monotonic loading condition, as an example. Experimental results [3] for the same models are also shown in the figure. Figure 12 shows the moment – rotation curve for the same models as the ones in Figure 11. In Figure 12, the moment – rotation curves for both models under cyclic loading are also plotted for comparison.

It can be seen from Figures 11, 13, and 14 that for both M9B1 and M9B2 connections, FEA gives higher load (and hence moment) carrying capacity compared to experimental results. On the other hand, experimental results show larger ductility compared to FEA results. Furthermore, from Figure 12, it is seen that the plasticity of M9B1 connection has occurred. However, the cyclic loading results of FEA could not show the expected hysteretic loop as usually happens in a beam-to-column connection. The same phenomenon is also seen in M9B2 connection type. This is partly because the FEA could not model both micro-cracks and macro-cracks that occurred in the experimental tests. Note that the static loading curve in Figure 12 is obtained by assuming an odd function.

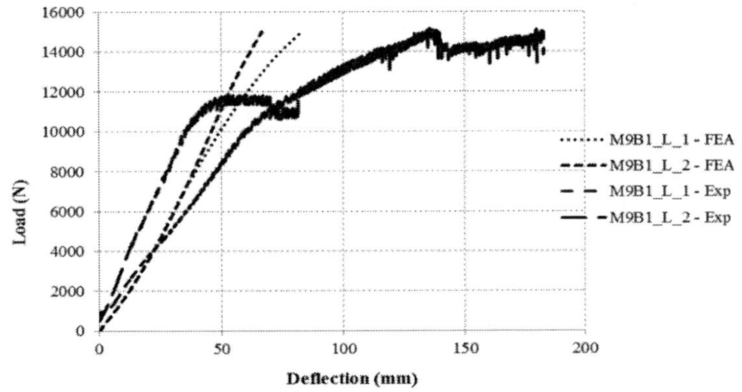

Fig. 11. Load – displacement curves obtained from FEA for M9B1_L_1 and M9B1_L_2 under static monotonic loading. Experimental results [3] are also shown for comparison.

Other properties of the connections developed in this study have been obtained from FEA. Initial elastic rotational stiffness K_e, rotation ductility $\mu_{rotation}$, maximum moment M_{max}, and the deflection at maximum load D_{max} for each connection type are shown in Table 3. Clearly lamina properties have effects on all of these properties. On average, M9B2 connection type has higher initial elastic rotational stiffness and ductility than M9B2 connection type. The maximum moment capacities of both connection types, however, are almost the same.

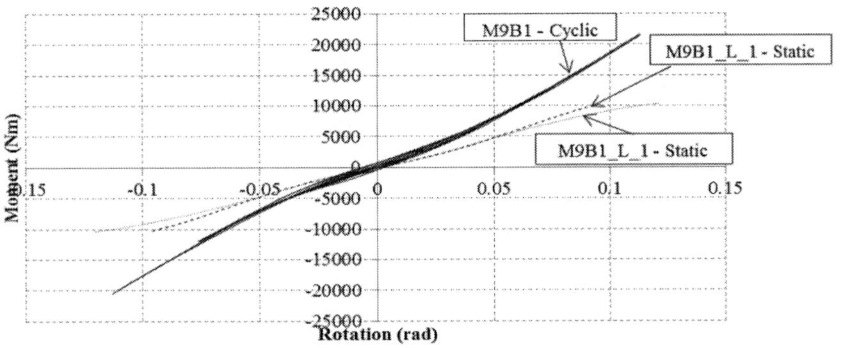

Fig. 12. Moment– rotation curves obtained from FEA for M9B1_L_1 and M9B1_L_2 under static monotonic and cyclic loadings.

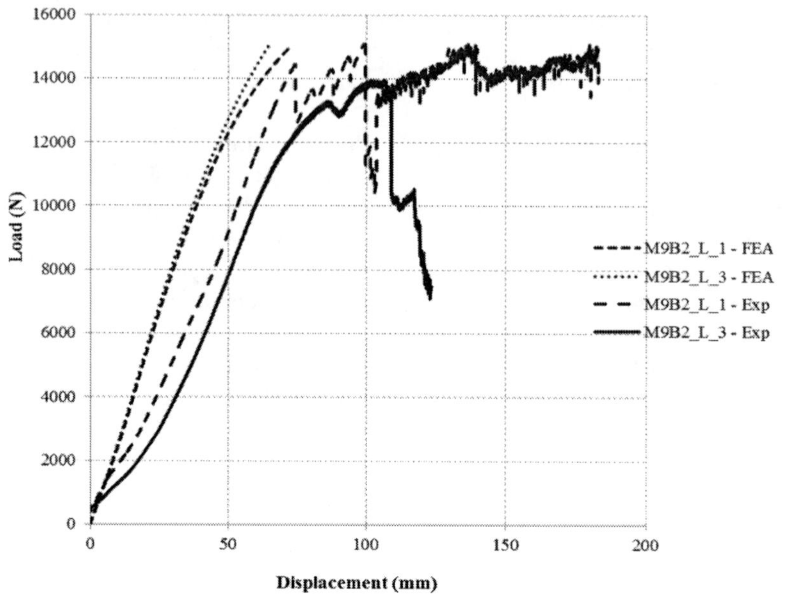

Fig. 13. Load – displacement curves obtained from FEA for M9B2_L_1 and M9B2_L_3 under static monotonic loading. Experimental results [3] are also shown for comparison.

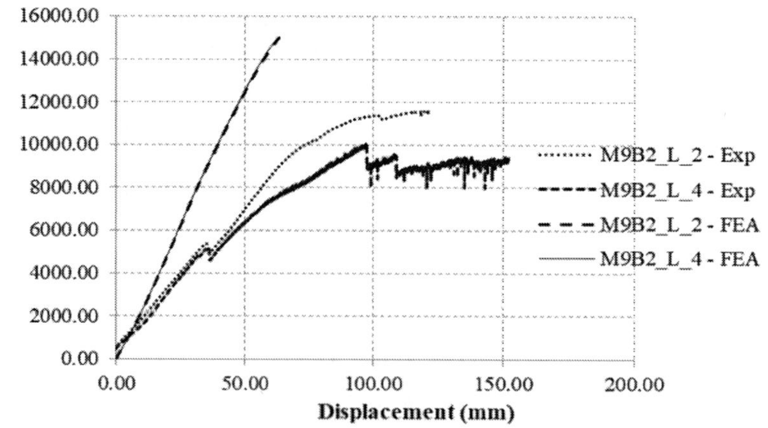

Fig. 14. Load – displacement curves obtained from FEA for M9B2_L_2 and M9B2_L_4 under static monotonic loading. Experimental results [3] are also shown for comparison.

Table 3. Summary of FEA results of M9B1 and M9B2 connection types under static monotonic loading.

Identity	K_e (kNm/rad)	$\mu_{rotation}$	M_{max} (kN-m)	D_{Pmax} (mm)
M9B1_L_1	89.1	1.2	10.35	89.23
M9B1_L_2	110.6	1.0	10.35	66.88
M9B2_L_1	125.0	1.4	10.35	72.4
M9B2_L_2	110.1	1.0	10.43	64.5
M9B2_L_3	125.0	1.6	10.20	64.6
M9B2_L_4	110.0	1.0	10.35	62.8

4 Summary

Two types of beam-to-column glulam connection, namely M9B1 and M9B2, have been developed in this study. It has been shown from non-linear finite element analyses and experimental tests that the combination of longitudinal steel rods, steel angle, and lateral steel bars in both types of connection provide the strength of the connections in resisting moment. The strength is obtained from a combined tensile strength of the rods and the bearing strength parallel to the grain of the glulam beam provided that the steel angle has sufficient flexural rigidity. The M9B2 connection type that has more lateral steel bars than M9B1 is more rigid and more ductile than M9B1 connection type.

The authors wish to thank Parahyangan Catholic University for the financial support of this study and Centre of Research and Development Housing and Settlement for laboratory facilities.

References

1. F. Scheibmair, P. Queneville, "Proceeding of Wood Conference on Timber Engineering (WCTE)",192-201 (2012)
2. E. Leitner, (2011), "A Master of Science Thesis, The Pennsylvania State University, USA", (2011)
3. D. Simanta, B. Suryoatmono, J.A. Tjondro, J. *Mekanika*, **1**, 29-35 (2016)
4. ANSYS, *ANSYS Structural Analysis Guide* (2009)
5. J. P. Hong, "Three Dimensional Nonlinear Finite Element Model for Single and Multiple Dowel-Type Wood Connections", (2007)
6. American Society for Testing and Materials, *ASTM D143-94R07 Standard Test Methods for Small Clear Specimens of Timber* (2009)
7. American Society for Testing and Materials, *ASTM D5764-97a (Reapproved 2002) Standard Test Methods for Evaluating Dowel-Bearing Strength of Wood and Wood-Based Products* (2002)
8. International Standard, *Timber Structures-Joints made with mechanical fasteners-Quasi-static reversed-cyclic test method (ISO 16670)* (2003)

Numerical Modelling of Interior RC Beam-Column Joints for Non-Engineered Buildings Strengthened Using Steel Plates

Yulita Rahmi[1,*], *Ashar* Saputra[2], and *Suprapto* Siswosukarto[2]

[1]Student, Department of Civil and Environmental Engineering, Universitas Gadjah Mada, Yogyakarta, Indonesia
[2]Department of Civil and Environmental Engineering, Universitas Gadjah Mada, Yogyakarta, Indonesia

Abstract. Data examination regarding historical earthquakes in Indonesia has revealed that damages occurred mostly in non-engineered buildings. Many non-engineered buildings have serious design and construction deficiencies in using concrete of inadequate quality and have poor steel reinforcement detailing especially in their beam-column joints. These joints require an effective strengthening technique which is simple to install. This paper presents steel plates to reinforce concrete beam-column joints through numerical studies. The finite element model was developed using ABAQUS. The FEM results were validated by comparisons with previous experimental tests. For this purpose, two RC beam-column joints were analyzed, including original standard beam columns and non-standard steel detailing strengthened with steel plates. The comparison of the results from numerical model indicated that steel plates could increase lateral strength, ductility, energy dissipation and increase the shear strength of the joints.

1 Introduction

An earthquake in Indonesia on a few years ago caused many non-engineered buildings suffered slight or heavy structural damage. Generally, the damage in non-engineered buildings occur in the beam-column connections due to the connection details that are not resistant to the earthquake and low quality of concrete [1]. An effective strengthening technique to non-engineered buildings is simple to be implemented, not require a special skill, low cost, and have a good result. This study developed a numerical analysis of the research that had been conducted by experiment [2]. Several of research have been studied for upgrading beam-column joint.

* Corresponding author: yulita.civil@gmail.com

Enveloped steel plates was used for strengthening RC frame joint. The steel plate was welded to the joint. This research showed that the stiffness, bearing capacity and integrity of the steel-enveloped frame joint are significantly improved, if compared to the primary frame joint [3]. Haunch retrofit solution was studied for retrofitted joints through numerical. This method showed that there is no critical shear occured in RC frame joint, but the haunch retrofit cause an increase in shear [4]. Steel prop and crub were studied for retrofitted reinforced concrete beam-column joints with experimental. This retrofit was used for strengthening of undamaged joints, because the shear stress and number of cracks at the panel zone were reduced. The maximum peak and load were incease 73% [5].

Most of these studies have a good result for beam-column joint, which have a difference technique application. Some of the advantages of the use of steel plate are having a greather tensile strength and easier implementation. Beam-column joints modelled using software ABAQUS 6.11 with two specimens. Numerical approach is expected to provide validate results to the beam-column joint. Therefore, the author perform further analysis of the interior beam-column joint using with steel plates in non-engineered buildings with numerical modeling. The purposes this research are the steel strip plate can be the alternative stiffener which effectively strength, increase the shear strength, and easy in implementation of the beam column joint and comparing the test results experimental that had been studied previously with numerical modeling.

2 Finite element modeling

The numerical study was developed using software ABAQUS 6.11.1. Concrete is modeled as C3D8R element. The rebar material is modeled as truss element. Steel rebars are tensile element on reinforced concrete, which are modeled as T3D2 (Truss, two-dimensional, number of nodes).

Reinforcement elements with concrete is designed as a surface embedded in the bonding relationship. Steel plates material and bolts are modeled as elastic perfectly plastic. Modeling process on steel plates and bolts used C3D8R solid. Contact between the concrete with bolts is defined by tie.

2.1 Specimen details

The FE model analysed the behavior beam-column joint with two specimens. The specimens were used in this study are named as HBK-1NA (normal beam column) and HBK-3A (beam-column joints with non-standars reinforcing strenghtened with steel plate). The beams cross section were 150x150 mm and columns cross section were 150x150 mm. The longitudinal reinforcement for all spesimen was ø10. The transverse reinforcement for all spesimen was ø8. The cross section of steel strip plate was 656.7 mm length and 50 x 3 mm width in the two side of joints. Specimens details that were tested are shown in Figure 1 and Figure 2.

2.2 Constitutive of model material

Material characteristic values are calculated based on experimental results with numerical validation which had been tested by several studies. The calculation of material properties used are described as follows.

 DOI: 10.1051/matecconf/201713802007

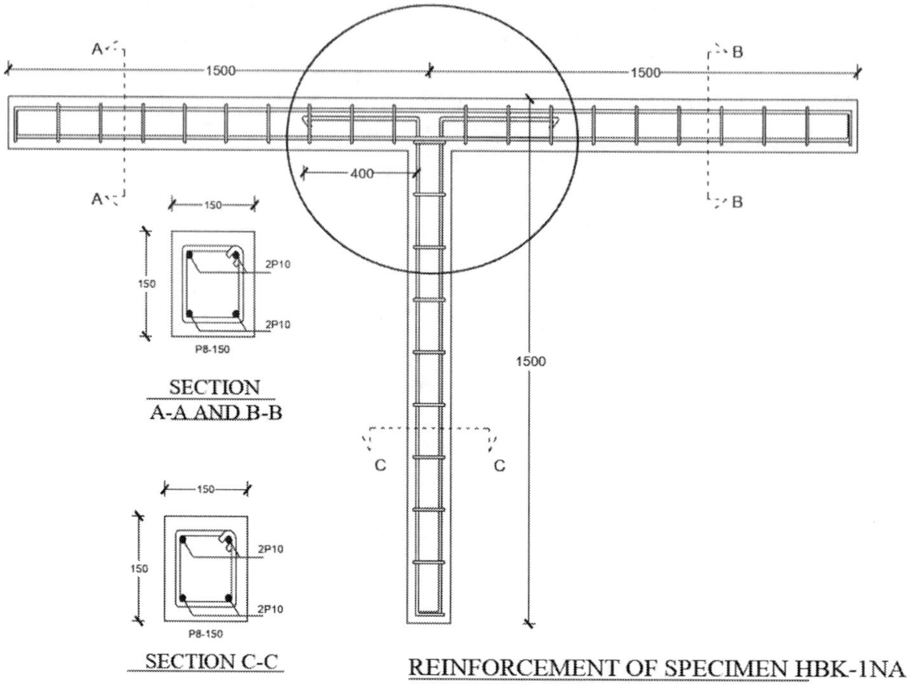

Fig. 1. Specimen HBK-1NA

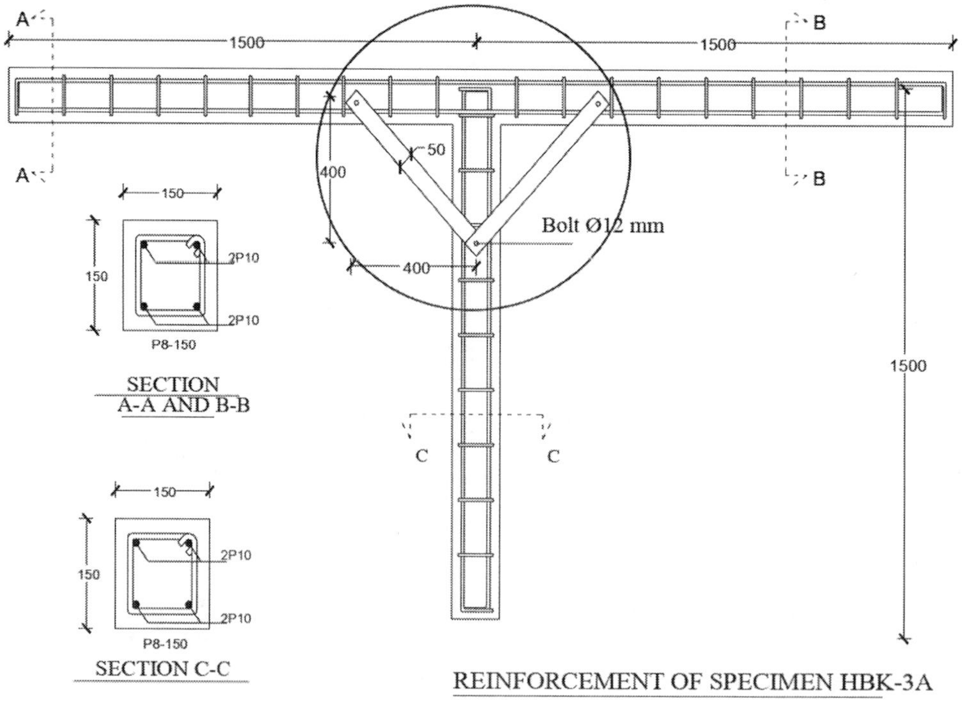

Fig. 2. Specimen HBK-3A

DOI: 10.1051/matecconf/201713802007

2.2.1 Concrete material

Modelling of concrete material uses concrete damage plasticity. Concrete damage plasticity is general capability model for analysing concrete elements under monotonic and cyclic loading [6]. The model of CDP used in ABAQUS is a modification of hypotheses by Drucker-Prager. There are some parameters that are described in the plasticity of concrete namely, K_c parameter, dilatation angle, eccentricity, the ratio of yielding concrete, and eccentricity [7]. The constitutive equation of concrete materials at the compressive conditions use the equation by Todeschini [8]. On tensile condition, material properties are calculated with an equation in CEB-FIP MC 1990 [9]. This equation is used to approximate calculation of the concrete compression stress and tensile stress to beam-column joints using steel plate. The stress-strain behavior of concrete in compression is assumed to be linearly elastic up to initial yield stress (σ_{co}). On the plastic condition, hardening stress that follows the softening until it reaches the ultimate stress (σ_{cu}). Concrete strength (f_{cm}) was used 15 MPa. The concrete stress-strain curve used for compression and tension for all model shown in Figure 3. The constitutive models of concrete materials calculated by the following equation:

$$\sigma_c = \frac{2f_c'\left(\frac{\varepsilon_c}{\varepsilon_c'}\right)}{1+\left(\frac{\varepsilon_c}{\varepsilon_c'}\right)^2} \quad (1)$$

with,

$$\varepsilon_c' = 1{,}71\frac{f_c'}{E_c}$$

where σ_c is compressive stress in concrete, f_c' is compressive strength of concrete, ε_c is strain in concrete, ε_c' is strain corresponding to f_c'.

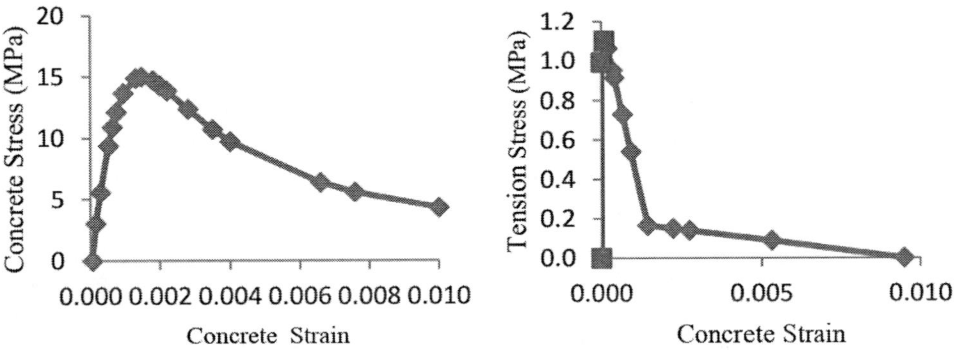

(a) Compression (b) Tension

Fig. 3. Concrete stress-strain curve

The response of concrete for cyclic loading is decrease in elastic stiffness. Elastic stiffeness is an important aspect which is illustrated in Figure 4. ABAQUS uses value factors recovery of stiffness defined as wt = 0 and wc = 1. Wc factor causes the material to regain the full stiffness with no damage. Meanwhile, the value 0 of wc showed no stiffness recovery. However, during the tension loading, stiffness stress was ignored. When the load is removed,

DOI: 10.1051/matecconf/201713802007

the elastic modulus recovery into (1-dt) Eo [10]. Cracks caused material damage defined by dt and dc, which is calculated by the following equation:

$$\bar{\sigma_c} = \frac{\sigma_c}{(1-d_c)} = E_0(\varepsilon_c - \varepsilon_c^{pl}) \qquad (2)$$

Where d_c is damage compression, $\bar{\sigma_c}$ is compressive stress in concrete corresponding to f_c', $\bar{\sigma_c}$ is compressive stress in concrete.

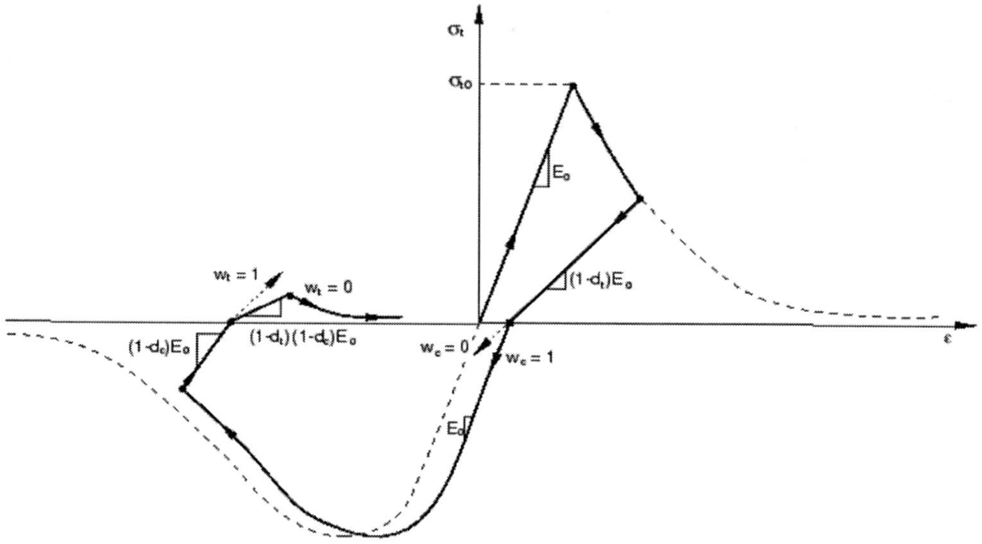

Fig. 4. Cyclic behavior of concrete [10]

In cracking conditions, the strain that occurs is called as the elastic strain $\varepsilon^{in}{}_c$ and plastic strain $\varepsilon^{pl}{}_c$. Inelastic strain can be defined as the total strain which is reduced by elastic strain on the material undamage conditions, with the following equation:

$$\varepsilon^{in}{}_c = \varepsilon_c - \varepsilon^{el}{}_{0c} \qquad (3)$$

with,

$$\varepsilon_{0c}^{el} = \frac{\sigma}{E_o}$$

Where $\varepsilon^{in}{}_c$ is inelastic strain, ε_c is strain in concrete, and ε_{0c}^{el} is elastic strain.

2.2.2 Steel material

The steel reinforcing bars is modeled using a nonlinear isotropic or kinematic hardening. The development of this model consists of two components, namely component that describes the nonlinear kinematic hardening of translation in the yield area between the stress through backstress (α) and component hardening describing isotropic equivalent stress change is defined as the change in the form of plastic in the yield area [10].

The steel material was modelled from the material properties that have been tested by experiment. The material properties are used for the steel rebar, bolt, and steel plate shown in Table 1.

Table 1. Material properties of specimens

$f_{y,main}$ (MPa)	$f_{y,trans}$ (MPa)	$f_{y,bolt}$ (MPa)	$f_{y,steel\ plate}$ (MPa)
545.803	688.816	493.073	452.476

2.3 Meshing, loading and boundary conditions

In this study, beams and columns were meshing with a size of 30 mm for beam-column, 2 mm steel, 1 mm bolts and 2 mm steel plate, for resulting in a stable number of elements. The boundary conditions are set at each end of the beam as hinge. This numerical model using two types of load, axial load of 33.75 kN was applied at the top of the column as a uniform pressure and displacement control given at the bottom of column. The cyclic loading was applied accordance by ACI T1.1-01 [11].

3 Results and discussion

The FE analysis conducted to investigate envelope of the hysteresis load-displacement response, crack pattern, ductility, and energy dissipation in the specimens. The analysis result report are discussed in this section.

3.1 Specimen HBK-1NA

Specimen HBK-1NA is a normal beam-column used as control test specimen against the test specimen HBK-3A. HBK-1NA generated load at peak condition at the positive and negative cycle that is equal to 8.44 kN and 16.31 kN. HBK-1NA stiffness cycle on twelfth cycle to the positive and negative direction were 20.3% and 65.2%. Pattern of collapse on the test specimen HBK-1NA formed as fine cracks that occur at a drift of 3 (0.27%) that is at the interface of the beam-column joint. At drift of 14 (3.75%), yielding in the steel reaching stress about 547.4 MPa and strain equal to 0.0924. Cracks that were occurred on HBK-1NA were dominated by shear cracks in the inside of the beam-column joint. From the pattern of collapse that occurred, it can be concluded that the HBK-1NA has not qualified as strong column weak beam mechanism.

3.2 Specimen HBK-3A

Specimens HBK-3A is a beam-column with the longitudinal reinforcement at the joint does not reach 400mm which strengthened with steel strip plates given at joint of interior beam-column. HBK-3A generated load at peak condition of 10.16 kN on a positive cycle and 15.99 kN on the negative cycle. Stiffness cycle HBK-3A in the twelfth step to the positive and negative direction were 9.4% and 47.8%. Pattern of collapse on the test specimen HBK-3A formed as initial cracks that occur at a drift of 2 (0.36%), that crack in the parts of columns. At drift of 14 (3.76%) yielding in the steel reaching stress about 542.1 MPa at the connection steel strip plate and bolt. Further cracks that occur in the HBK-3A is dominated by shear cracks in the concrete columns and steel strip plate section that began buckling shown in Figure 5. The specimen HBK-3A showed that strengtening using steel plates can increase the shear capacity on beam-column joint, shear cracks at the joint moved to the column which was caused by stress concentration that occurred at the connection between the steel plates and bolts.

Fig. 5. Stress of HBK-3A

3.3 Ductility

Ductility is the ability of a structure and selected structural components to deform beyond elastic limits without excessive strength or stiffeness degradation [12]. Specimen HBK-3 in the positive and negative direction, have a greater ductility 15.38% and 37.78% compared than HBK-1NA. The value of ductility factor result proves that the beam-column joint strenghthened using steel plate (HBK-3A) has a greater ductility in positive and negative cycles.

3.4 Strength degradation

The peak load with displacement on each drift ratio for initial cycle loading shown in Figure 6. Figure 6 shows that in the positive direction, HBK-3A has increased by 0.17% against HBK-1NA. In the negative direction, HBK-3A has a lower lateral load of 0.02% than HBK-1NA. From the results of the three envelope curves shows the HBK-3A generated load and displacement with high average in the positive and negative direction. However, on the negative cycle is lower due to concrete cracking in the connection area.

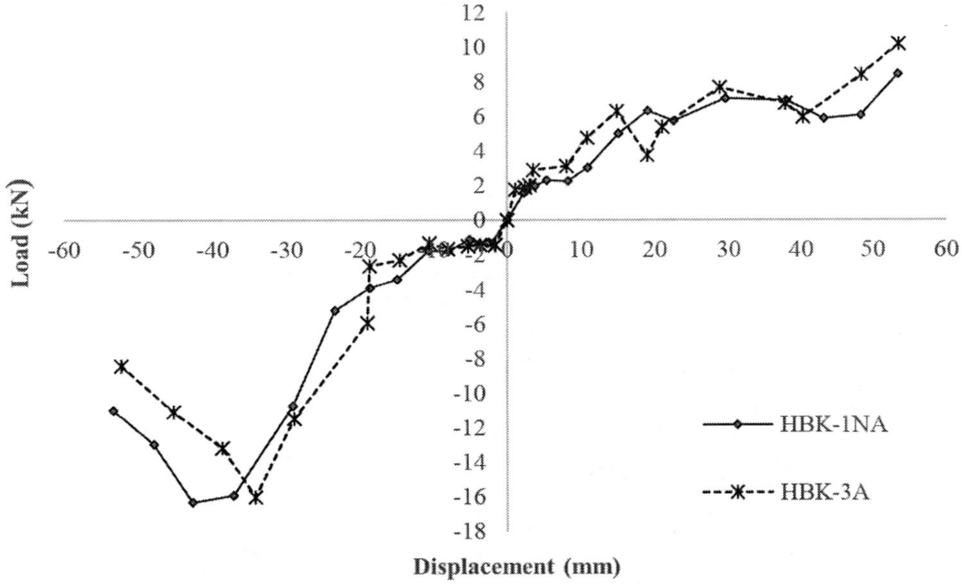

Fig. 6. Comparison of load-displacement of each specimen

3.5 Energy dissipation

The energy dissipation structure describes the ability of the structure to absorb energy due to external loads. The relative energy dissipation ratio (β) is the ratio between the energy dissipated by the actual structure by theoretical calculations. Energy dissipation relative value (β) should not be less than 1/8. Energy dissipation relative to all specimens shown in Table 2.

Table 2. Dissipation energy for all specimens

Specimens	Drift ratio (%)	Ah (kN mm)	Energy dissipation relative (β)
HBK-1NA	3.75%	59.13	0.09
HBK-3A	3.76%	192	0.186

Table 2 shows that the dissipation of energy generated on HBK-1NA was not qualified of the limit values required by ACI TI.1-01 is 0.125. Specimen HBK-3A was qualified in accordance the criteria specified.

3.6 Comparison of experimental and numerical results

The results of numerical analysis in this study compared to experimental results by Pranadita (2016) to observe the validation of the data obtained from the results of a numerical approach. The load and displacement in peak condition resulting in numerical analysis compared with the experimental results are shown the load generated in the specimen HBK-1NA and HBK-3A have differences with the experimental results on the positive and negative cycles. On the positive cycle, HBK-1NA have difference about 11.99% than HBK-1NE and HBK-3A about 9.52% than HBK-3E. It is effected by different of patterns of collapse in the positive and negative cycles, primarily because of the following:

a. Reinforcement on modeling, is considered as a perfect bond with an embedded.

b. The equation on the material properties that are used need to be investigated in order to obtain almost the good results.

3.6.1 Specimens HBK-1NA

The pattern of cracks that occur in the numerical analysis of the test specimen HBK-1NA until the end of the cycle was formed of shear cracks on the joint in the beam-column, crack pattern is similar to the cracking that occurred in the HBK-1NE. The crack of initiation in the numerical results occurred in the third step with a strain of 0.0013, whereas in the experimental initial crack occurred on the sixth step. Comparison of the pattern of cracks that occur on HBK-1NA and HBK-1NE shown in Figure 7 below.

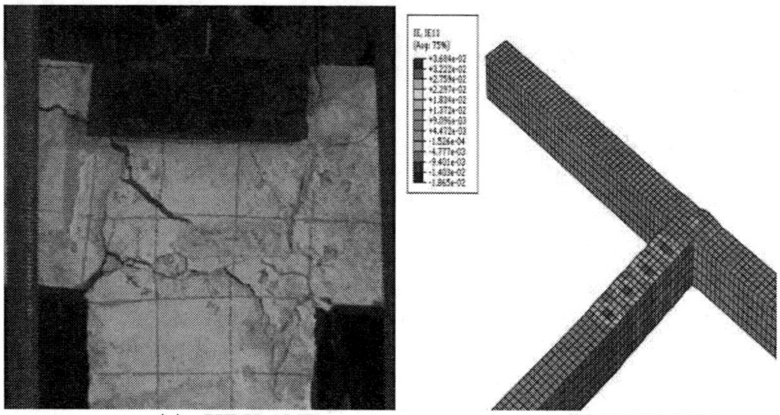

(a) HBK-1NE (b) HBK-1NA

Fig. 7. Comparison of the cracks pattern on HBK-1NE and HBK-1NA

3.6.2 Specimens HBK-3A

The pattern of initial crack at HBK-3A was occurred at the second step, while at the HBK-2E was occurred in the fourth step. HBK-3A cracks at the last cycle were in the form of cracks on the concrete in the columns as shown in Figure 8. On HBK-3E, cracks that were occurred in the form of concrete on the columns experienced which was large and deep, until the main reinforcing steel was exposed (Pranadita, 2016).

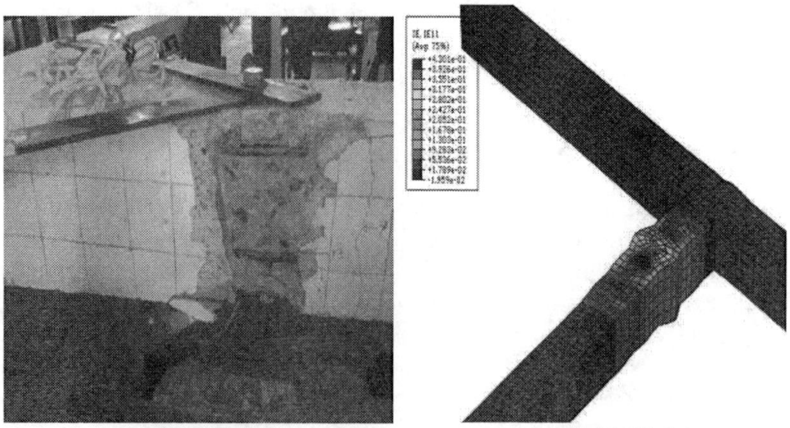

(a) HBK-3E (b) HBK-3A

Fig. 8. Comparison of the cracks pattern on HBK- 3E and HBK-3A

4 Conclusion

The nonlinear material model predicted the behavior of beam-column joint well. Although some differences have been seen amongst the result, but the shear cracks on the joint in the numerical model has good agreement with the experimental results. Based on modeling results, discussion, and analysis of data that has been presented, some conclusions can be obtained as follows:

1. The test specimen HBK-3A generated maximum lateral load in the negative direction and the positive direction to 10.16 kN and 15.99 kN, whereas on HBK-3E to 11.23 kN and 11.58 kN. The resulting differences in numerical results were 9.52% and 38.08% against the experimental results.
2. From the pattern of collapse that occur, HBK-3A can increase the capacity of shear on the join. Crack only occurred in the area of steel plate connection on the column.
3. The relative energy dissipation (β) in the test specimen HBK-1NA and HBK-3A were 0.09 and 0.186. Specimens strengthened using steel plate (HBK-3A) in accordance the criteria required by ACI TI.1-01.
4. In this research, the specimen HBK-1NA and HBK-3A have not qualified as "strong column weak beam" from the mechanism of collapse pattern.
5. The use of steel strip plate on the beam-column connection in non-engineered buildings was more effective, because it did not require a long time and it was an easy installation process.

References

1. T. Boen, *Earthquake Resistant Design of Non-Engineered Buildings In Indonesia* (2001)
2. G. H. Pranadita, *Analisis Perilaku Hubungan Balok-Kolom Interior Pada Struktur Beton Bertulang Non-Engineered Dengan Perkuatan Plat Baja* (Universitas Gadjah Mada, 2016)
3. D.G. Weng, C. Zhang, J.D. Xia, and X.L. Lu, 15 W. C. E. E. (2012)
4. A. Sharma, R. Eligehausen, and J. Hofmann, ACI Struc. J., **111**, 4 (2014)
5. M. K. Sharbatdar, A. Kheyroddin, and E. Emami, J. Els. Const. Build. Mat., **36**, 287-294 (2012)
6. D. A. Pohoryles, J. Melo, and T. Rosetto, Earth. R. Eng. Res. W. (2015)
7. P. Kmiecik and M. Kaminski, Arc. Civ. Mech. Eng., **XI**, 3 (2011)
8. F. Oudah, *Development of Innovative Self-Centering Concrete Beam-Column Connections Reinforced using Shape Memory Alloys* (Department of Civil Engineering University of Calgary, 2014)
9. T. Telford, *CEB-FIP Model Code 1990 Design Code*, (Comite Euro-Du International Betong, 1991)
10. ABAQUS User Manual, *ABAQUS* (2006)
11. ACI, *Acceptance Criteria for Momen Frames Based on Structural Testing T1.1-01*, (American Concrete Institute, 2001)
12. T. Paulay, T. and Priestley, M. J. N, *Seismic Design of Reinforced Concrete and Masonry Buildings* (Wiley and Sons, 1992)

MATEC Web of Conferences **138**, 02008 (2017) DOI: 10.1051/matecconf/201713802008
EACEF 2017

Dynamic Response of AP1000 Nuclear Island Due to Safe Shutdown Earthquake Loading

Buntara S. Gan[1,*], *Dinh Kien* Nguyen[2], and *Ay Lie* Han[3]

[1]Department of Architecture, College of Engineering, Nihon University, Koriyama, Japan
[2]Institute of Mechanics, VAST, 18 Hoang Quoc Viet, Hanoi, Vietnam
[3]Department of Civil Engineering, Diponegoro University, Semarang, Indonesia

Abstract. AP1000 is a standard nuclear power plant developed by Westinghouse and its partners by using an advanced passive safety feature. Among the five principle building structures, namely the nuclear island, turbine building, annex building, diesel generator building and radwaste building, the safety of the nuclear island building is the most concerned. This paper investigates the dynamic response of the nuclear island building of the AP1000 plant subjected to safe shutdown earthquake loadings. A finite element model for the building, which is assumed to be built in a hard-rock base, is developed and its dynamic response is computed with the aid of the commercial finite element package ANSYS. The dynamic characteristics, including the natural frequencies, the vibration modes, and the time histories for displacements, velocities, and accelerations of the building are obtained for two typical safe shutdown earthquakes, El Centro and Kobe earthquakes. The dynamic behavior of the building due to the earthquakes and its safety is examined and highlighted.

1 Introduction

The safety of nuclear power plants in earthquakes is the most concern for engineers and researchers in designing a new plant as well as in evaluating the existing plants [1]. Due to the complexity of the nuclear structure and the seismic loading, it is very difficult to carry out a site test. The dynamic finite element analysis is a preferable tool to predict the dynamic response of the nuclear power structures under the action of seismic loading. Investigations on the dynamic behavior of nuclear power plant by using both the commercial finite element packages and self-developed codes have been extensively reported in the literature.

Tunon-Sanjur et al. [2] considered the soil-structure interaction in the development of the stick and shell finite element models of AP1000 nuclear island for dynamic investigation of the island under seismic loading. Manjuprasal et al. [3] developed a 20-node isoparametric solid finite element for investigating the nonlinear behavior of reinforced concrete containment shell of nuclear plants under seismic loading. It has been shown by the authors that the material nonlinearity which considered in their work has a significant influence on the dynamic response of the structure and the dynamic

* Corresponding author: buntara@arch.ce.nihon-u.ac.jp

displacement increases considerably when the effect of material nonlinearity is taken into account. In [4], Nakamura proposed the methods to transform the frequency-dependent soil impedance to the time domain for predicting the seismic behavior of a deeply embedded nuclear power building. The author then employed the finite element method in combination with the direct integration Newmark method to obtain the vibration characteristics of the building. Continue his work in [4], Nakamura and his co-workers [5] carried out a study on the ultimate seismic response and fragility assessment of a nuclear power building by developing a nonlinear three-dimensional finite element model. Perotti et al. [6] proposed a numerical procedure for evaluating fragility of isolated nuclear buildings under seismic loading. The procedure makes use of the response surface methodology to model the influence of random variables on the dynamic response which can be able to apply to the design of the nuclear power plant reactor. Zhao and Chen [7] used the finite element method to study the dynamic behavior of a nuclear power reinforced concrete containment with or without damping rubber bearing isolators under three-directional seismic loading. Chen et al. [8] employed the ANSYS software in to investigate the dynamic response of isolated and non-isolated nuclear island building under seismic loading. The numerical investigation carried out by the authors showed that the behavior of the isolated structure is very different from that of the non-isolated one, and the natural frequencies and dynamic displacements of the isolated structure are significantly lower than that of its counterpart. A three-dimensional finite element model of the reinforced concrete containment building was developed by Zhai et al. [9] for studying the dynamic behavior of the building under the mainshock-aftershock earthquake sequences. Politopouos et al. [10] modeled the soil domain under the nuclear power plant by the finite elements and Lysmer radiation boundaries for investigation of the effect of foundation embedment on the floor response spectra of the plant. It has been shown by the authors that the base rocking excitation induced by soil–structure interaction, may amplify the non-isolated modes' response. A 3D nonlinear inelastic finite element model was developed and incorporated into the finite element software ABAQUS by Sener et al. [11] for investigating studying the seismic behavior of a pressurized water reactor containment internal structure. Recently, Xu et al. [12] took the effect of water level in water tank into their study of the dynamic response of the AP1000 shield building under seismic loading.

The AP1000 is a standard nuclear power plant developed by Westinghouse and its partners by using an advanced passive safety feature. The system has been approved by some countries and under consideration by the others. The safety system of theAP1000 plant includes the facilities inside and outside of the containment vessel [13]. Understanding the dynamic behavior of AP1000 buildings and their facilities in earthquakes is very important, especially for a country with many earthquakes likes Japan. In this paper, as a first attempt, the dynamic response of AP1000 nuclear island building due to safe shutdown earthquake loading is investigated. With the help of the commercial software ANSYS, a three-dimensional finite element model for the building is developed, and its dynamic response is computed. The dynamic characteristics, including the natural frequencies, time histories for displacements, accelerations and von Mises stresses are obtained for the building under the safe shutdown of 19944 El Centro and 1955 Kobe earthquakes. The dynamic behavior of the building in the earthquakes and its safety are examined and highlighted.

2 Seismic finite element analysis

In the context of finite element method, the equations of motion for dynamic analysis of a structure under seismic loading is of the form [14]

$$\mathbf{M\ddot{D}} + \mathbf{C\dot{D}} + \mathbf{KD} = -\mathbf{MI\ddot{D}}_g(t) \quad (1)$$

where **M**, **C**, **K** are, respectively, the mass, damping and stiffness matrices of the structure; **D** is the vector of unknown relative nodal displacements, $\mathbf{\dot{D}} = \partial\mathbf{D}/\partial t, \mathbf{\ddot{D}} = \partial^2\mathbf{D}/\partial t^2$ are the relative nodal velocities and nodal; $\mathbf{\ddot{D}}_g$ is the vector of ground motion which relates to the vectors of absolute and relative nodal accelerations by

$$\mathbf{\ddot{D}}_g(t) = \mathbf{\ddot{D}}_a - \mathbf{\ddot{D}} \quad (2)$$

with $\mathbf{\ddot{D}}_a$ denotes the vector of absolute nodal accelerations. In Eq. (1), **I** is the influence coefficient vector, having 1 for elements corresponding to degrees of freedom in the direction of the applied ground motion and zero for the other degrees of freedom. The proportional damping is adopted herein. In this regard, the damping matrix **C** is formed as a linear combination of the stiffness and mass matrices as

$$\mathbf{C} = \alpha\mathbf{K} + \beta\mathbf{M} \quad (3)$$

where α and β are, the stiffness and mass proportional damping coefficients, respectively. These damping coefficients can be calculated from the critical damping ratio and the structural natural frequencies as

$$\alpha = 2\xi\frac{\omega_1\omega_2}{\omega_1+\omega_2}, \quad \beta = \frac{2\xi}{\omega_1+\omega_2} \quad (4)$$

where ξ is the damping ratio, depending on the structural material, and in the present work ξ=5% is assumed for both the concrete and steel; ω_1 and ω_2 are natural frequencies of the structure, which are necessary to choose to bound the design spectrum. ω_1 is normally selected as the fundamental frequency, and a value of 30 Hz is chosen for the frequency ω_2 because the spectral contents of seismic design are insignificant above this frequency [15].

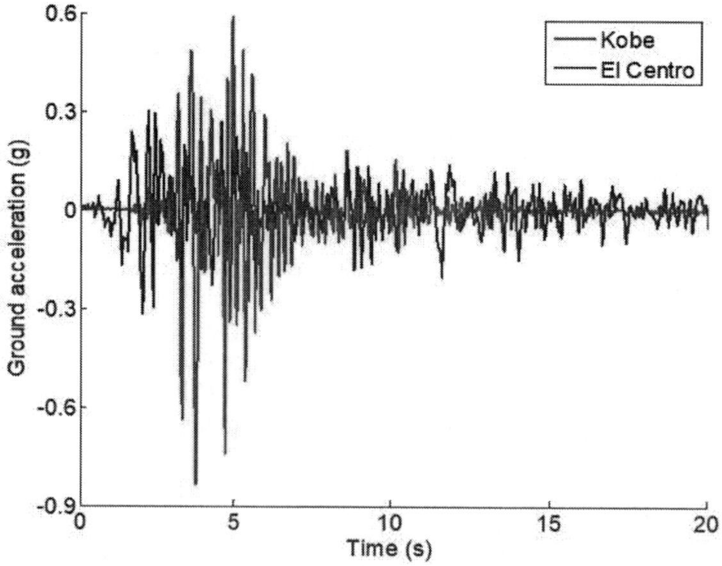

Fig. 1. N-S ATHs of Kobe and El Centro earthquakes.

In order to evaluate the dynamic behavior of the nuclear island building under the safe shutdown earthquakes, the input Acceleration Time Histories (ATH) of the earthquakes are applied at the base of the building. Fig. 1 shows the N-S ATHs of the 1944 El Centro and 1955 Kobe earthquakes, which are plotted from the seismic data of these earthquakes. As seen from the figure, the peak ground motion (PGA) of the Kobe earthquake is more than twice times higher than that of the El Centro earthquake, but the ground acceleration of the Kobe earthquake decays rapidly after 10 seconds.

3 Finite element model

Among the five buildings of the AP1000 nuclear power plant, the nuclear island building contains the steel containment vessel and other important equipment. Therefore the safety of this building is the most important. Fig. 2 shows an imagination of the AP1000 nuclear island building.

Fig. 2. Nuclear island building of the AP1000 [2].

Investigation on the dynamic behavior of the nuclear power plants, as discussed in the introduction section, has been carried out intensively. Various methods for safety evaluation necessary to carry out pre-earthquake and after-earthquake are discussed in [16]. Regarding the AP1000 nuclear island considered herein, due to the complexity of the structure, it is difficult to consider many factors such as the soil-structure interaction, nonlinearity of the material, water levels in the tank and air intakes at the same time. In the present work, the nuclear island building is considered to be built on a hard rock base, and the equipment inside the building is not modeled separately. With the help of the "Geometry" library in ANSYS R15.0 [17], a 3D finite element model for the AP1000 nuclear island building is created and shown in Fig. 3. Both the concrete structures and steel containment vessel are incorporated in the model. The model consists of shell and brick elements, in which the shell elements are used to model the steel containment vessel and the shield building, whereas the remaining parts are modeled by the brick elements. The interface between the upper part of the building and the base mat as well as between the walls of the building and the upper base mat are modeled by the share node option of ANSYS. The convergence of the mesh has been carried out in order to find out an acceptable mesh. Finally, a mesh with a total number of 88349 elements, in which 2192

 DOI: 10.1051/matecconf/201713802008

elements are shared node elements, is decided to employ in the analysis. Table 1 lists the geometric and material data of the nuclear island building which will be used in computation the dynamic response.

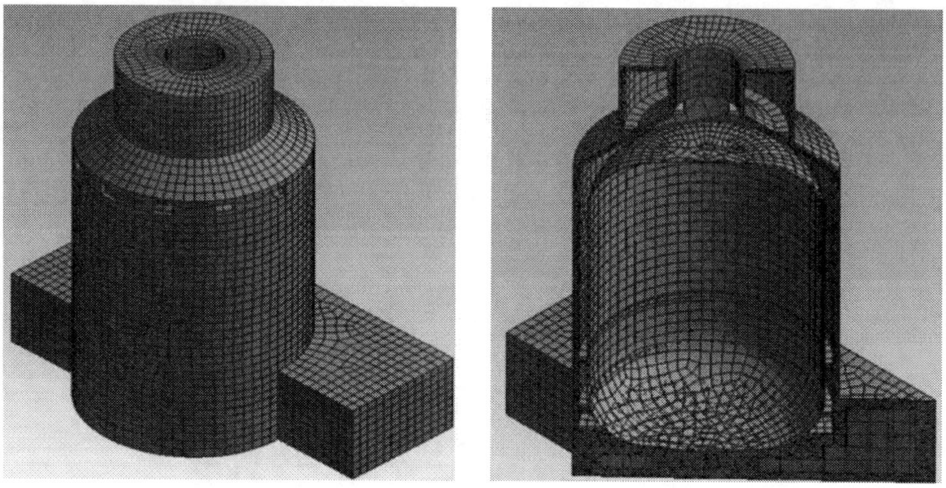

Fig. 3. Finite element model of AP1000 nuclear island building.

Table 1. Geometry and material data of AP1000 nuclear island used in the analysis [8].

Item	Concrete	Steel	Size
Density (kg/ m^3)	2300	7800	
Young's modulus (MPa)	3.35×10^4	2.06×10^5	
Poisson's ratio	0.2	0.3	
Length of nuclear island (m)			77.42
Width of nuclear island (m)			26.67
Height of nuclear island (m)			81.98
Radius of shield building (m)			22.1
Radius of containment vessel (m)			19.8
Wall thickness (m)			0.92
Thickness of containment vessel (m)			0.041
Height of auxiliary building (m)			39.42

4 Results and discussion

4.1 Modal analysis

The results obtained from the modal analysis, namely the first natural frequencies and the corresponding periods of the nuclear island building, are given in Table 2. The first two vibration modes of the building are depicted in Fig. 4, as an example. The first two natural frequencies and the periods of the present work are closed to the values obtained by Chen et al. in Ref. [8] for the case of the non-isolated system. It should be noted that the turbine building, the diesel generator building, the radwaste building in the present work are model as solids, and this can be explained by the difference between the results of the modal analysis obtained herein with that of Ref. [8]

Table 2. The first six natural frequencies and periods.

Mode	Frequency (Hz)	Period (s)
1	3.6219	0.2761
2	3.6294	0.2755
3	3.6322	0.2753
4	3.6781	0.2719
5	3.7547	0.2663
6	3.7553	0.26629

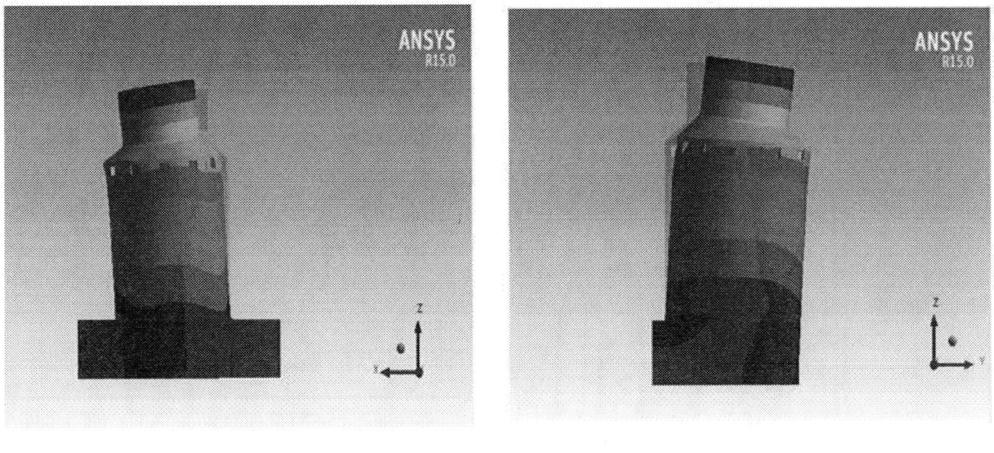

Fig. 4. The first two vibration modes of the nuclear island building

DOI: 10.1051/matecconf/201713802008

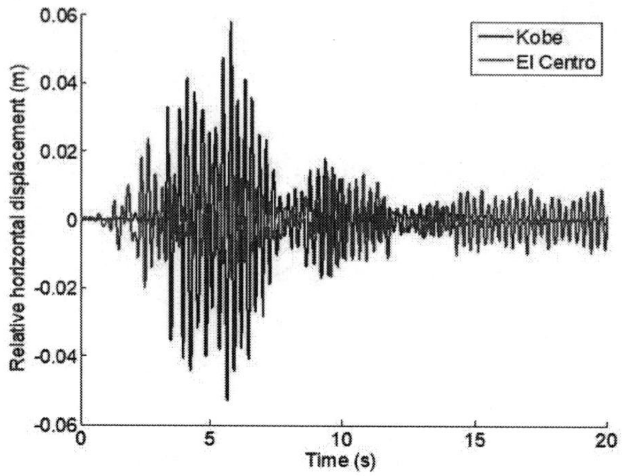

a) Displacement time history

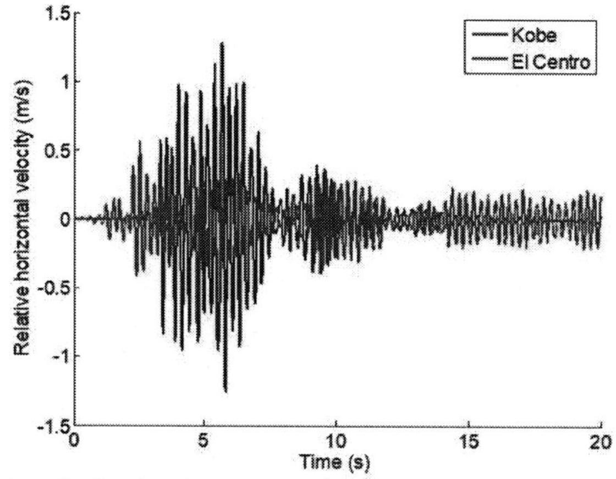

b) Velocity time history

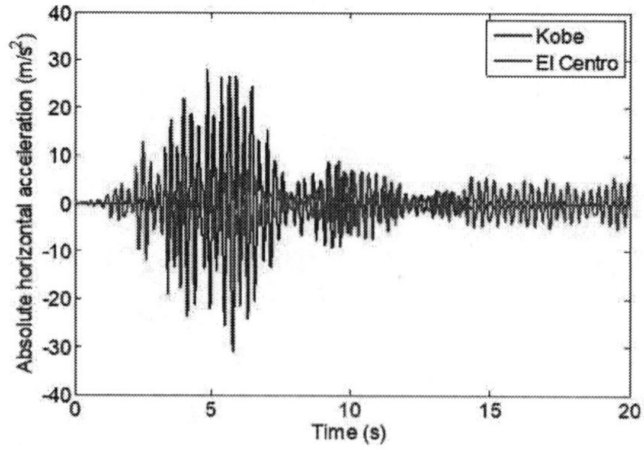

c) Acceleration time history

Fig. 5. Time history of the responses at responses at the top point.

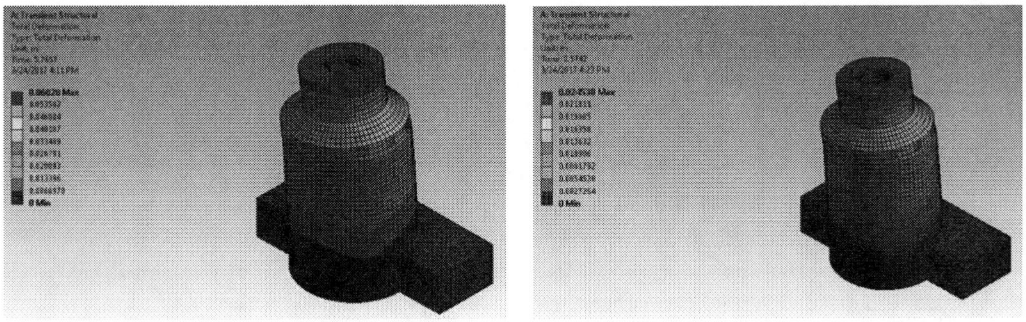

a) Kobe earthquake (t=5.7657s) **b)** El Centro earthquake (t=2.5742s)

Fig. 6. Deformed configuration of the nuclear island under different earthquake loadings corresponding to the maximum displacement

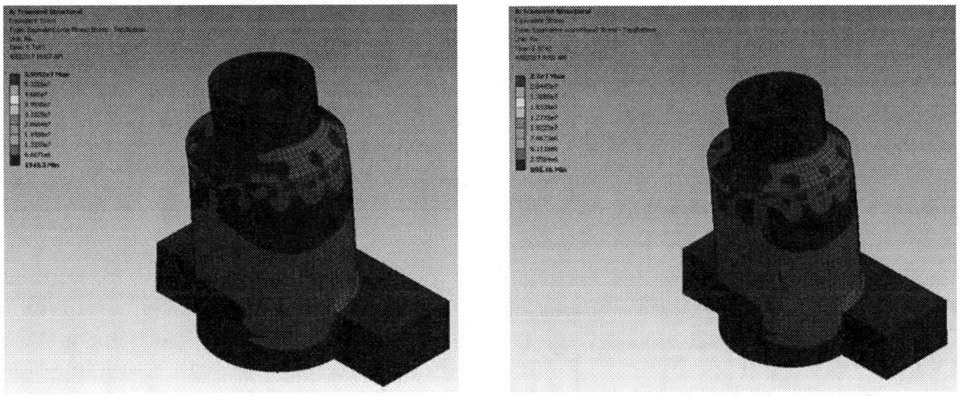

a) Kobe earthquake (t=5.7657s) **b)** El Centro earthquake (t=2.5742s)

Fig. 7. Equivalent von Mises stress distribution of the nuclear island under different earthquake loadings at the time of the maximum displacement

Table 3. The maximum and minimum dynamic response at the top point

Response	**Kobe Earthquake**		**El Centro Earthquake**	
	Maximum	**Minimum**	**Maximum**	**Minimum**
Displacement (m)	$5.7833x10^{-2}$	$-5.2869x10^{-2}$	$2.3512x10^{-2}$	-1.984710^{-2}
Velocity (m/s)	1.2792	-1.2530	0.56181	-0.41577
Acceleration (m/ s^2)	27.9940	-30.9320	12.659	-11.999

4.2 Dynamic response

Fig. 5a show the relative displacement response at the top point of the nuclear island building under the safe shutdown of the Kobe and El Centro earthquakes. The displacement response of the building is in accordance with the ATHs of the two earthquakes depicted in Fig. 1. As seen from the figure, the displacement response at the top point of the structure under the Kobe earthquake is much higher than that of the building under the El Centro earthquake. The response due to the Kobe earthquake is mainly contributed in the first 12 seconds, and it decays rapidly after this period. The displacement response under the El

Centro earthquake is, however still relative large after the 12th second, and even until the 20th second, as seen from the figure. Similar remarks can be drawn for the response of the relative velocity and absolute acceleration at the top point as depicted in Fig. 5b and Fig. 5c, respectively. Both the velocity and acceleration at the point decay of the building in the Kobe earthquake decay rapidly after the 12th second, but they are still relative large until the 20th second when the building is subjected to the El Centro earthquake.

In order to examine the dynamic response of the nuclear island building under the Kobe and El Centro earthquakes in more detail, Figs. 6 and 7 illustrate the deformed configurations and the equivalent von Mises stresses of the building at the times they reach the maximum displacements. The equilibrium von Mises in the figures is defined as follows

$$\sigma_{\mathrm{vM}} = \sqrt{\frac{1}{2}\left[(\sigma_1-\sigma_2)^2+(\sigma_2-\sigma_3)^2+(\sigma_3-\sigma_1)^2\right]} \tag{5}$$

with σ_1, σ_2 and σ_3 are, respectively, the first, the second and the third principle stresses. Figure 6 show the horizontal displacements at the top of the shield building are the largest, and they are smaller in the lower parts of the building. The situation is different for the von Mises stress, and the stress tends to be larger in the areas of the shield building which are near the lower part of the nuclear island. It should be noted that the von Mises stress at some areas of the building under the Kobe earthquake is of the order of 10^7 Pa, which is quite large and it may exceed the yield stress. In this case, a nonlinear analysis is necessary to perform, but it is beyond the scope of this paper. The large dynamic response of the nuclear island building under the Kobe earthquake compares to the El Centro earthquake can be observed again from the figures. Table 3 summarizes the maximum and minimum values of the displacement, velocity and acceleration response of the building in the two earthquakes. One can see from the Table that the critical response values obtained in the Kobe earthquake are more than two times higher than that obtained in the El Centro earthquake. From a safety point of view, the nuclear island building is more dangerous in the Japanese Kobe earthquake than it is in the American El Centro earthquake.

5 Conclusions

The dynamic response of the AP1000 nuclear island building under safe shutdown earthquakes has been investigated in the present paper. With the help of the software ANSYS, a 3D finite element model for the building has been created, and its dynamic response was computed. The dynamic characteristics, including the relative displacements, velocities and the absolute accelerations of the building, assuming to be built on a hard rock base, have been obtained by using the ATHs of the Kobe 1995 and El Centro 1944 earthquakes. The deformed configurations and the von Mises distribution of the building under the earthquakes have also been computed. The obtained numerical results have shown that the nuclear island building displaced larger under the Kobe earthquake than it does under the El Centro earthquake. It has also revealed that the safety of the building under the Japanese earthquake is much more concerned than in the El Centro earthquake. The von Mises stress at some parts of the building when it is subjected to the Kobe earthquake may exceed the yield stress, and more works are necessary to carried out to take the material nonlinearity into consideration in studying the dynamic behavior of the nuclear island building under seismic loading.

The support from Vietnam Academy of Science and Technology (Grant No. VAST01.06/16-17) to Dr. Nguyen Dinh Kien is gratefully acknowledged.

References

1. R.D. Campbell, G.S. Hardy, M.K. Ravindra, J.J. Johnson, A.J. Hoy, Nucl. Eng. Des, 182, 17 (1998)
2. L. Tunon-Sanjur, R.S. Orr, S. Tinic, D.P. Ruiz, Nucl. Eng. Des. **237,** 1474 (2007)
3. M. Manjuprasad, S. Gopalakrishnan, T.V.S.R. Appa Rao, Eng. Struct. **23**, 397 (2001)
4. N. Nakamura, Nucl. Eng. Des. **238**, 1845 (2008)
5. N. Nakamura, S. Akita, T. Suzuki, M. Koba, S. Nakamura, T. Nakano, Nucl. Eng. Des. **240**, 166 (2010)
6. F. Perotti, M. Domaneschi, S. De Grandis, Nucl. Eng. Des. **262**, 189 (2013)
7. C. Zhao, J. Chen, Nucl. Eng. Des. **265**, 484 (2013)
8. I. Chen, C. Zhao, Q. Xu, and C. Yuan, Nucl. Eng. Des. **278**, 117 (2014)
9. C.-H. Zhai, Z. Zheng, S. Li, L.-L. Xie, Soil Dyn. Earthq. Eng. **74**, 46 (2015)
10. I . Politopouos, I. Sergis, F. Wang, Soil Dyn. Earthq. Eng. **78**, 213 (2015)
11. K.C. Senera, A.H. Varma, P.N. Bootha, R. Fujimoto, Nucl. Eng. Des. 295, 804 (2015)
12. Q. Xu, J. Chen, C. Zhang, J. Li, C. Zhao, Nucl. Eng. Tech. **48**, 246 (2016)
13. Lee, D.-S., Liu, M.-L., Hung, T.-C., Tsai, C.-H., Chen, Y.-T., Appl. Therm. Eng. **50**, 207 (2013).
14. A.K. Chopra, *Dynamics of structures, theory and applications to earthquake engineering* (4th edition, Prentice Hall, Boston, 2014).
15. R.D. Cook, D.S. Malkus, M.E. Plesha, *Concepts and applications of finite element analysis* (3rd edition, John Wiley & Sons, New York, 1989).
16. T.J. Katona, Seismic safety analysis and upgrading of operating nuclear power plant, Intech, http://dx.doi.org/10.5772/51368 (2012).
17. Inc. A., ANSYS Release 15.0 (Canonsburg, PA, USA, 2013).

Finite element modelling of composite castellated beam

Richard Frans[1,*], *Herman* Parung[2], *Achmad Bakri* Muhiddin[2], and *Rita* Irmawaty[2]

[1]Doctoral Program, Hasanuddin University, Makassar, Indonesia
[2]Department of Civil Engineering, Hasanuddin University, Makassar, Indonesia

Abstract. Nowadays, castellated beam becomes popular in building structural as beam members. This is due to several advantages of castellated beam such as increased depth without any additional mass, passing the underfloor service ducts without changing of story elevation. However, the presence of holes can develop various local effects such as local buckling, lateral torsional buckling caused by compression force at the flange section of the steel beam. Many studies have investigated the failure mechanism of castellated beam and one technique which can prevent the beam fall into local failure is the use of reinforced concrete slab as lateral support on castellated beam, so called composite castellated beam. Besides of preventing the local failure of castellated beam, the concrete slab can increase the plasticity moment of the composite castellated beam section which can deliver into increasing the ultimate load of the beam. The aim of this numerical studies of composite castellated beam on certain loading condition (monotonic quasi-static loading). ABAQUS was used for finite element modelling purpose and compared with the experimental test for checking the reliability of the model. The result shows that the ultimate load of the composite castellated beam reached 6.24 times than the ultimate load of the solid I beam and 1.2 times compared the composite beam.

1 Introduction

Castellation is one of technique for increasing the bending strength of steel beam and produce a new section property called castellated beam or also known as honey comb beam. Besides increasing the bending capacity of the beam, there are several advantages of using castellated beam such as reducing the total weight of the structures, passing the underfloor services (ducting and plumbing). But, in the other hand, castellated beam have disadvantages caused by holes present of the web section. Local buckling and Vierendeel mechanism sometimes took place. Therefore, composite beam becomes one of solution for avoiding the collapse mechanism. In composite beam, the concrete section will experience compression while tension will occur in the steel profiles. So, local buckling won't happen on the steel profiles caused by there is no compression force on the steel. Besides that, reinforced concrete slab can provide continuous lateral support to beam's flange which can

* Corresponding author: richardfrans.rf@gmail.com

prevent the beam from lateral torsional buckling. The other advantage of using composite on the castellated beam is to increase the plastic moment of the steel beam which automatically increase the bending capacity of the beam. This paper considered a composite castellated beam as a comparison and validation with the composite beam (not castellated) which have been done [1].

2 Castellated beam

Castellated beams are those beams which has opening its web portion. Castellated beams are fabricated by cutting the web of hot rolled steel (HRS) I section into zig-zag pattern and thereafter rejoining it over one another [2]. Fig.1 shows the hexagonal castellated beam geometry section, where, D_o is opening depth, D is overall depth after castellation, t_w is web thickness, t_f is flange thickness, b is flange width, S is gross opening space, e is clearing opening space, α is opening angle.

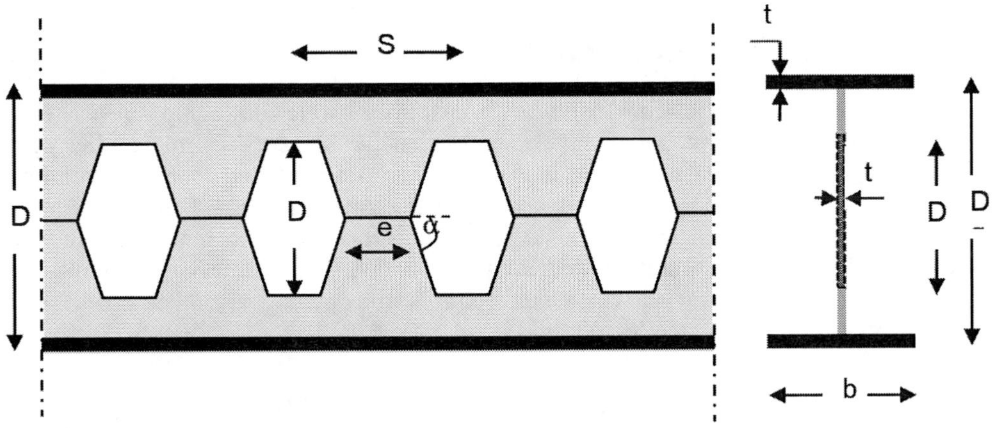

Fig. 1. Geometry of Castellated Beam.

Presence of the hole (opening) makes castellated beam susceptible to buckle. Two of the several failure mechanism which are common are Vierendeel mechanism and local buckling [3]. Vierendeel mechanism is mechanism that cause yield or buckling of the web-post in compression or shear when the other criteria is local buckling which can be occur in three ways [4]:

1. Buckling of the web-flange section (T-section) in compression due to global bending action. For unstiffened openings, this check is require for large opening in regions of high moment.
2. Buckling of the compressive strength.
3. Vertical instability of the sides of the web openings in high shear zones. Plated beams are more susceptible to this form of web instability.

3 Composite beam

Composite beam is a beam which has built with two different material (reinforced concrete slab and steel profile), both of them will work together against the external force. To ensure the composite behavior, the important factor must be noted that the bond between concrete and steel remain unbroken [5], hence the design of the shear connector must be well design. Reinforced concrete slab does not only increase the plastic capacity of the beam but also provide continuous lateral support at beam's flange to prevent lateral torsional buckling then the global instability problem is solved [6]. For local instability, if the composite beam is designed where the plastic neutral axis is lied on the reinforced concrete slab as shown on Fig.2., the local buckling of the steel beam can be neglected caused by there is zero compression force on the steel beam.

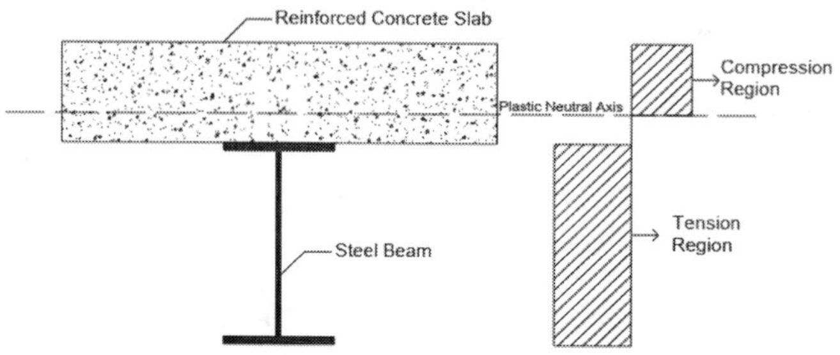

Fig. 2. Plastic Stress Distribution.

4 Composite castellated beam

Because of the advantage of composite beam describes above, the composite castellated beam was conducted in this paper. Fig.3. shows that the compression region lies on the concrete slab, while the tension region is in castellated beam, in this condition, reinforced concrete slab can prevent local buckling and Vierendeel mechanism so that the section can be developed until the plastic moment is reached. But, it must be noted that the shear connector had to design carefully so that composite action be guaranteed. Fig. 3 and Fig. 4 show the section of the composite castellated beam on the opening section and solid section respectively.

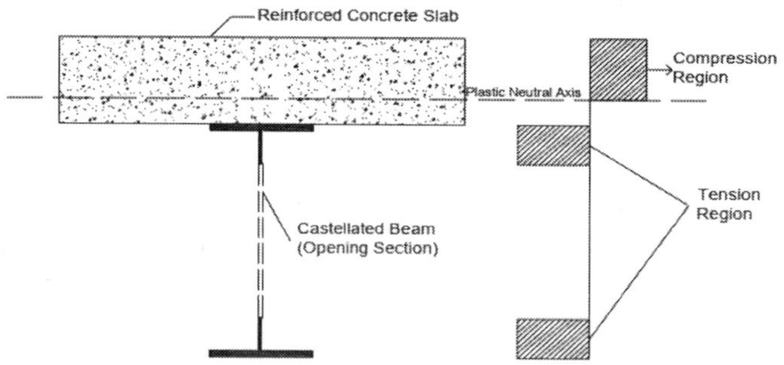

Fig. 3. Composite Castellated Beam (Opening Section).

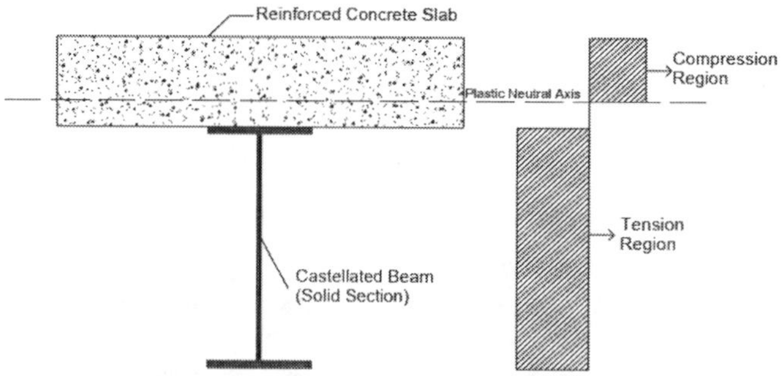

Fig. 4. Composite Castellated Beam (solid section).

DOI: 10.1051/matecconf/201713802009

5 Benchmark problem: composite steel beam

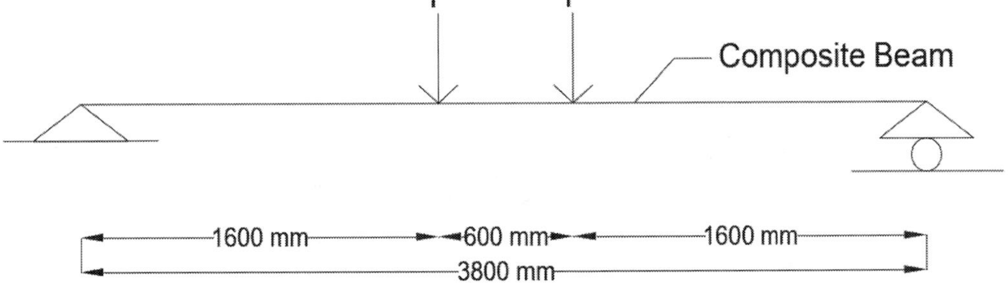

Fig. 5. Simply Supported Composite Beam.

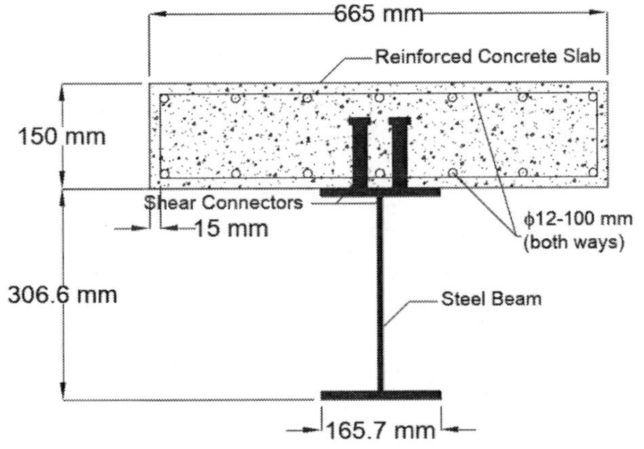

Fig. 6. Section Properties Composite Beam.

Before considered a composite castellated beam, a benchmark problem [1] was taken for test and validation purpose of the numerical modelling. The composite beam are shown on Fig. 5 and Fig. 6 respectively. The material properties and section properties can be seen on Table 1 and Table 2 respectively.

This composite steel beam was modelled using ABAQUS software (Finite Element Package). Eight-node brick elements with reduced integration (C3D8R) was picked for meshing purpose in [1]. But, for simplification of the model cause the cost of computation, the beam element was modelled using doubly curved shell elements 4-node with reduced integration (S4R) [7]. The composite behavior was provided using shear connectors (stud) and modelled using multi point constraint (MPC) option for full composite behavior [8]. This assumption used in this model (Fig. 7) has been validated with the numerical and experimental result in [1]. Monotonic loading (quasi-static) with two point bending has been used for loading condition.

DOI: 10.1051/matecconf/201713802009

Table 1. Material properties of composite beam.

Material Properties		
	Concrete	**Steel**
Density (kg/m^3)	2500	7850
Elastic modulus (MPa)	30000	210000
Poisson's ratio	0.2	0.3
Compressive strength (MPa)	36	
	Hot-Rolled Beam Section	**Reinforcement bars**
Yield Strength (MPa)	300	415
Ultimate strength (MPa)	420	550

Table 2. Section properties of composite beam.

Section Properties	
Steel Beam	
Cross section area (mm^2)	5806
Total depth (mm)	306.6
Flange width (mm)	165.7
Flange thickness (mm)	11.8
Web thickness (mm)	6.7
Reinforced Concrete Slab	
Width (mm)	665
Thickness (mm)	150
Ratio of transverse reinforcement (%)	1.259
Ratio of longitudinal reinforcement (%)	1.259

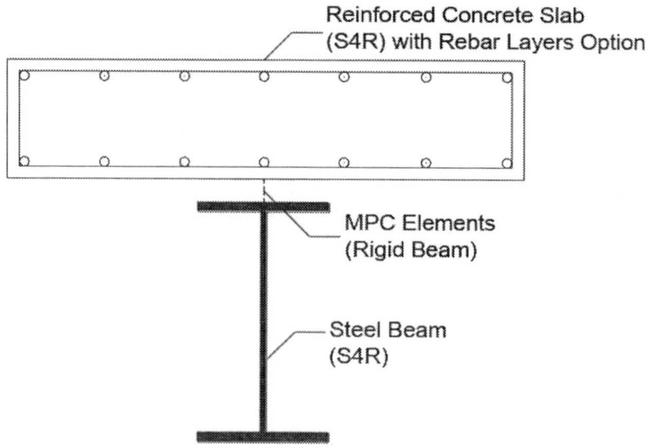

Fig. 7. Shell element assumption used based on [7].

Fig. 8 shows the load-deflection curve of the composite beam. The blue line and red line are experimental result and finite element modelling using FE-Solid (C3D8R) respectively which has been done by [5] while the green line is finite element modelling using FE-Shell (S4R). The yield load which obtained in the experimental result was about 360 kN, this is almost similar with the result of FE-Shell model which has 364,04 kN of yield load. It also can be seen, the ultimate load (peak load) of the experimental result is around 550 kN while the FE-Shell model is 557.96 kN. There is good agreement between the results of FE-shell model with the experimental result and FE-Solid model. So, FE-Shell model can be a good alternative solution for modelling the composite beam with lower cost of computation.

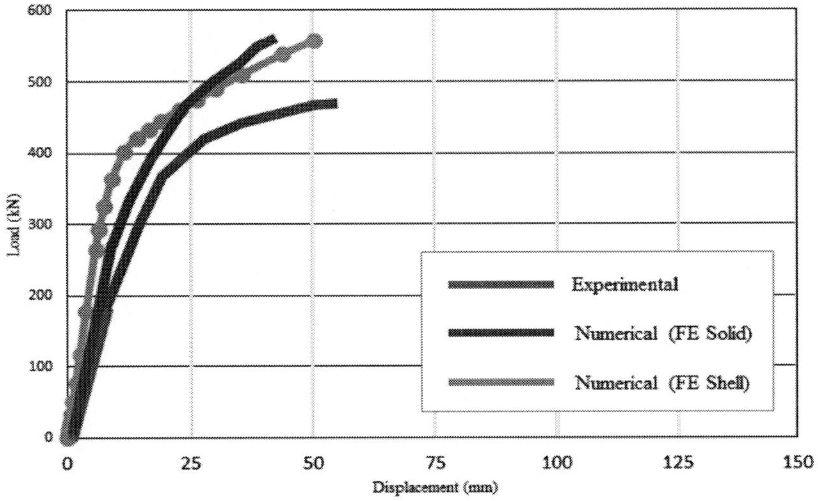

Fig. 8. Load-Deflection Curve for Composite Beam.

6 Application of Composite Castellated Beam

Composite castellated beam which considered were solid I beam with the same section properties as before but after castellation process. Properties of the castellated beam was used cutting pattern according to BS 5950-1:2000 which have an optimum properties for a castellated beam [9]. Fig. 9 shows the cutting pattern for steel beam, this pattern allows reducing of the secondary stress caused by the opening section [10].

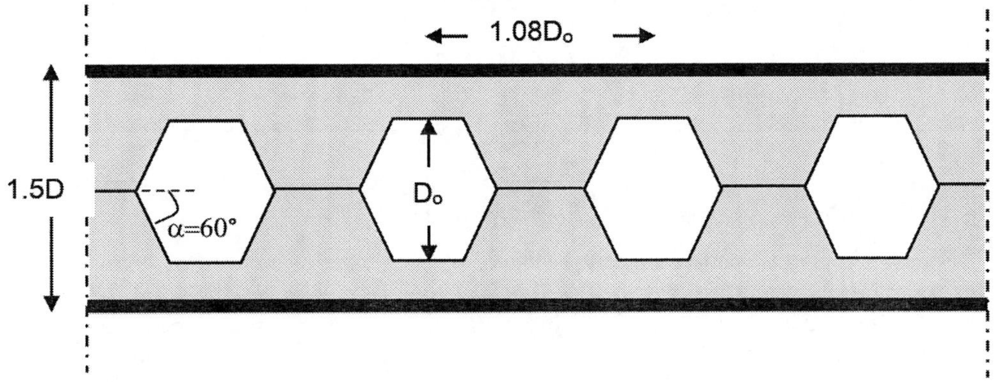

Fig. 9. Cutting pattern according to BS 5950-1:2000.

where, Do is opening height, and optimum angle (α) is 60°.

DOI: 10.1051/matecconf/201713802009

Fig. 10 shows the steel beam before and after castellation process condition. Depth of the beam before castellation process was 306.6 mm and after castellation process the total height of the beam became 459.9 mm. So, the depth increase about 1.5 times from the original beam (solid I beam). The cutting angle was taken 60° with 331.128 mm of the opening space. Thickness of the reinforced concrete slab and the reinforcement bars are still same as before (Fig. 6). Beam's setup is also same as previous.

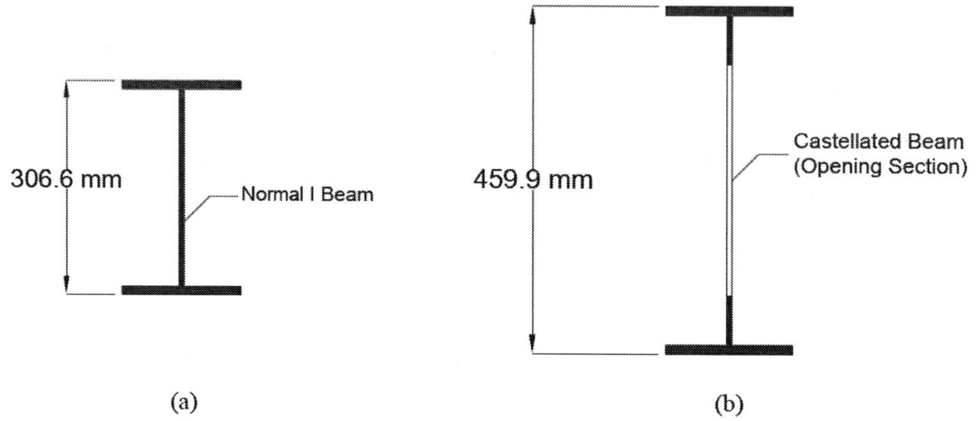

Fig. 10. (a) Normal I Beam, (b) Castellated Beam.

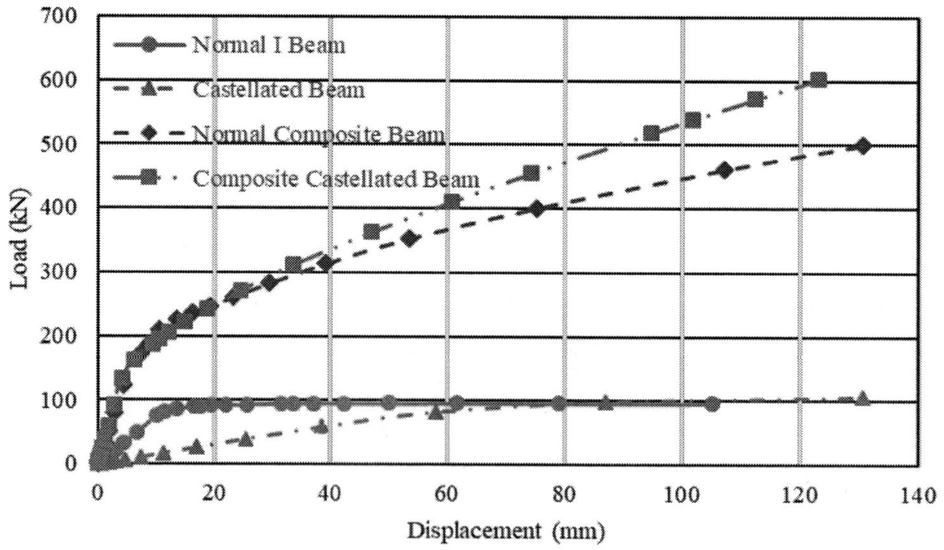

Fig. 11. Load-Displacement Curve for Various Type of Beam.

Fig. 11 shows the load-displacement curve of each type of beam. For comparison, the maximum displacement was taken 130 mm. The blue line indicates the solid I beam while the green line indicates the castellated beam. It can be seen that solid I beam has a higher initial stiffness compared the castellated beam, but after yielding, the load capacity for solid I beam didn't increase significantly while castellated beam did, and at the last, the peak load for the solid I beam and castellated beam at the target displacement were 96.6021 kN and 106.573 kN respectively.

For composite beam (solid I beam + reinforced concrete slab), the maximum load can be reached was 500.389 kN. This is almost five time higher than solid I beam. At the beginning, there is no difference between the composite beam and the composite castellated beam (the yield load is almost similar), but after around 270 kN, the composite castellated beam performed better than composite beam until the target displacement reached. The peak load of composite castellated beam was 603.146 kN. This value is 1.2 times from the composite beam and 6.24 times more than the capacity load of the solid I beam. Table 3 shows the comparison between the peak load which was reached for each type of beam. The peak load of the solid I beam became the reference peak load for comparing with the other type of the beam.

Table 3. Peak load.

Type of Beam	Peak Load (kN)	Percentage (%)
Solid I Beam	96.6021	100
Castellated Beam	106.5730	110.3216
Composite Beam	500.3890	517.9898
Composite Castellated Beam	603.1460	624.3612

7 Conclusions

Finite element which using shell element can be an alternative solution to make a simplification for the solid model. The composite beam which using S4R and MPC constraint (type rigid beam) was success to predict accurately the load-displacement curved compared with the numerical result which using solid element and the experimental result. The result shows good agreement between both of the model. The composite castellated beam shows the increasing of the capacity load significantly, this can be happened cause of the castellation process and the composite reaction between the steel profiles and the reinforced concrete slab. In this paper, the load capacity of the composite castellated beam was almost 6.24 times than solid I beam and 1.2 times compared by composite beam (solid I beam + reinforced concrete slab). Besides that, reinforced concrete slab can be a great lateral support for the beam to prevent lateral torsional buckling effect moreover failure mechanism such as local buckling and Vierendeel mechanism can be avoided cause of the neutral plastic axis is on the reinforced concrete slab. So, the steel beam can develop until the plastic moment reached.

References

1. A.N. Prakash, N.C.K. Anandavalli, C.K. Madheswaran, J. Rajasankar, N. Lakshmanan, "Three dimensional FE model of stud connected steel-concrete composite girders subjected to monotonical loading. International Journal of Mechanics and Applications", **1**, 1-11 (2011)
2. A.M. Jamadar, P.D. Kumbhar, "Parametric study of castellated beam with circular and diamond shaped openings. International Research Journal of Engineering and Technology, India", 2015, **2**, 715-722, (2015)
3. R. Frans, H. Parung, D. Sandy, S. Tonapa, "Numerical modelling of hexagonal castellated beam under monotonic loading", **171**, 781-788, Procedia Engineering (2017)

4. D. Kerdal, D.A. Nethercott, "Failure modes for castellated beams", Journal of Construction Steel Research, **4**, 295-315 (1984)
5. H. Parung, "Composite structures: behavior under earthquake loading", ISBN: 978-602-0924-38-0, Masagena Press, Makassar (2017)
6. W. Dewobroto, "Struktur baja: perilaku, analisis & desain" – AISC 2010 Civil Engineering Faculty, Tangerang (2016)
7. M. A. Gizejowski, W. Salah, "Failure modes of composite beams in the hogging moment region of continuous or semi-continuous systems. Proc. Local Seminar of the Polish Chapter of IASS, Warsaw", 114 – 136 (2007)
8. R.E.S. Ismail, A.S. Fahmy, N.M. Tawfik, "Finite element modelling of partially composite castellated beams. Communications on Applied Electronics (CAE)", **2**, 6-22 (2015)
9. R. Frans, H. Parung, A.B. Muhiddin, R. Irmawaty, "Optimization of short-span hexagonal castellated beam using fast multi swarm optimization", International Seminar on Infrastructure Development (2016)
10. A. Aryatnie, "Studi Parametrik Castellated Beam terhadap Beban Gravitasi dengan Metode Blodgett dan Metode Darwin", Unpublished Thesis at Pelita Harapan University (2015)

Flexural behaviour and punching shear of self-compacting concrete ribbed slab reinforced with steel fibres

Hazrina Ahmad[1] , *Mohd Hisbany* Mohd Hashim[1*], *Afidah* Abu Bakar[1] and *Siti Hawa* Hamzah[1]

[1]Faculty of Civil Engineering, Universiti Teknologi MARA, 40450 Shah Alam, Malaysia

Abstract. This paper investigates the effects of steel fibres as a replacement to the conventional reinforcement under flexural behaviour and punching shear in self-compacting (SCC) ribbed slab reinforced with steel fibres. Four ribbed slabs with similar dimensions of 2.8 m length x 1.2 m width and 0.2m thickness were constructed. Two of the samples were considered as control samples (conventionally reinforced with reinforcement bars and welded mesh) while another two samples were fully reinforced with 1% (80 kg/m^3) volume of steel fibres incorporated to the SCC mix. For the flexural behaviour study, the ribbed slab samples were subjected to two line loads under four point bending. Meanwhile, for the punching shear analysis, the ribbed slab samples were subjected to a point load to simulate loading from the column. The analysis of the experimental results displayed that steel fibres incorporation had been found to effectively delay the first crack occurrence under both flexural and punching shear. The steel fibre replacement has been proven to be able to sustain up to 80% and 73% of the ultimate load resistance for flexural and punching shear, respectively, in comparison to conventionally reinforced ribbed slab structure. The visual observation carried out during the experiment exhibited similar failure mode for both steel fibre reinforced and control samples. This was observed for both flexural and punching shear samples. Overall, it can be concluded that the steel fibres had displayed a promising potential to effectively replace the conventional reinforcements.

1 Introduction

Self-compacting concrete (SCC) has the ability to be properly poured in place, filling the formwork corners and small voids between reinforcement bars by means of its own weight without any compaction. Its rheological property allows SCC to be effectively applied in complex shaped elements and congested reinforcements [1-5]. The flow ability of SCC offers advantage to the construction process, whereas it could minimize the needs of skilled workers to work on concrete finishing. Nevertheless, this could lead to a reduction of costs as well as time needed to construct a structure [6].

* Corresponding author: hisbany@salam.uitm.edu.my

In this research, steel fibres were being introduced to the SCC mix not only to reduce cracks but also as the only reinforcement in the slab structure. Steel fibre reinforced self-compacting concrete (SFRSCC) is a relatively new composite material that combines the benefits of the self-compacting concrete (SCC) technology with the advantages derived from the fibre addition to a brittle cementitious matrix. SFRSCC combines the benefits of SCC at fresh state with high workability as well as its improvement of the properties at hardened state with addition of the steel fibres [7]. Steel fibres has the ability to improve the properties of SCC elements including tensile strength, ductility, toughness, energy absorption capacity [8], [9], fracture toughness [1], [10], [11] as well as cracking resistance [12]. Owing to the positive effects of steel fibres in enhancing the concrete materials properties especially by its contribution against crack propagation and enhancement in ductility, current research is now moving towards investigating the function of the fibres as the reinforcing material to partially or totally replace conventional reinforcements in concrete structures [13]. The aim of this paper is to investigate the flexural and punching shear behaviour of SFRSCC ribbed slab.

2 Materials and mix composition

For all experimental work carried out in this research, similar Grade 30 self-compacting concrete (SCC) composition was used as shown in Table 1. The SCC mix for the sample construction was produced in the concrete plant and delivered to the laboratory. The steel fibres were then manually added to the truck mixer at the end of the mixing process.

Hooked end steel fibres (Stahlcon HE 0.55/35) with the length, l_f = 35 mm, diameter of d = 0.55 mm and aspect ratio (L_f/d_f) of 65 was used in this study. The tensile strength of the steel fibre is 1250 N/mm^2. Hooked end steel fibres is chosen in order to enhance the anchorage between the steel fibre and the surrounding concrete matrix. The hooked end produces greater friction between the fibres and the aggregates in the matrix as compared to straight fibers [7] as well as higher toughness and residual strength [6],[15] as well as greater pull out forces [15]. Similar fibre dosage of 80 kg/m^3, which is equivalent to 1% of the total volume was used for the construction of all slab samples. This volume fraction was chosen based on findings from previous research [16]. The fibres are glued in bundles to facilitate the fibre distribution during concrete mixing and avoid the fibre balling effect [15], [17], [18]. This type of fibres conforms to the requirement of BS EN 14889-1 (2006).

Table 1. SCC Mix Composition.

Cement CEM I 42.5R	Pulverized fly ash (Class F)	Coarse aggregate (10 mm)	Fine aggregate	Water	w/c	Steel fibre content
315 kg/m^3	105 kg/m^3	830 kg/m^3	865 kg/m^3	185 kg/m^3	0.44	80 kg/m^3

3 Size of sample

Four (4) ribbed slab samples were cast in the Heavy Structures Laboratory, Universiti Teknologi MARA (Malaysia). No vibration was applied to the self-compacting concrete mix during casting of the samples. The details of the samples were as shown in Table 2. The dimension of the ribbed slab samples was kept constant with 2.8m (length) x 1.2m (width) with the total slab thickness of 0.2m considering the common size of precast slabs produced in the construction industry. The structural topping for all ribbed slab samples were fixed to 100 mm. The cross-sectional details of the slabs are shown in Fig. 1.

 DOI: 10.1051/matecconf/201713802010

Table 2. Sample details.

Designation	Reinforcement	Testing
CRC(F)	Conventional	Flexural (four point bending)
CRC(P)	Conventional	Punching shear
SFWS(F)	Full steel fibres	Flexural (four point bending)
SFWS(P)	Full steel fibres	Punching shear

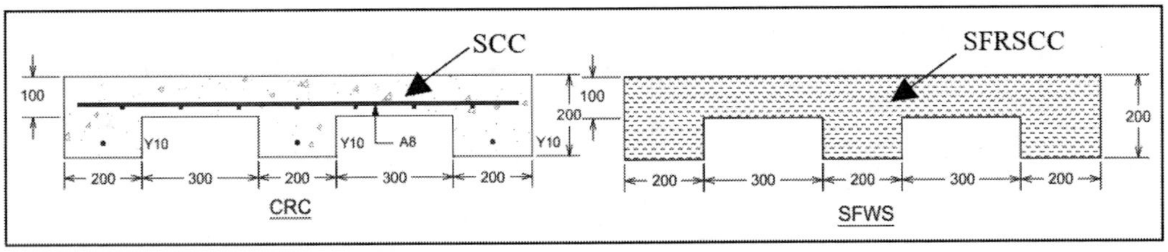

Fig. 1. Cross-sectional details of the ribbed slab sample.

After 28 days of sample curing using the wet gunny sacks, all samples were painted in white to ease the observation of the crack propagation pattern during the commencement of the testing. The measuring instruments involved for the experimental test: (1) linear variable displacement transducers (LVDT) to measure the vertical deflection of the slab samples and (2) Load cell to measure the loads from the actuator. The locations of the LVDT's for the flexural and punching shear test are shown in Fig. 4.

4 Experimental setup

The experimental work for this study consist of two experimental testing; flexural four point bending test and punching shear test. Fig. 2 shows the experimental setup for the flexural test. Four point bending was selected to permit slow gradual failure of the sample. The loading was applied by a hydraulic actuator through longitudinal spreader beam and two transverse loading rollers placed equidistance from the support in accordance to the guidelines given in the BS 12390-5 [20] under static load. The loading rate was set to a constant rate of 0.01 mm/sec in order to be able to meticulously observe the crack pattern and failure mode that takes place during the loading.

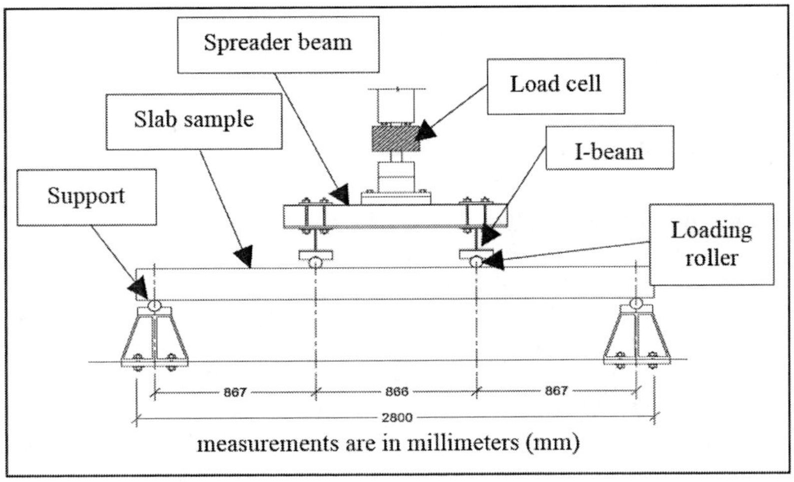

Fig. 2. Flexural testing experimental setup.

DOI: 10.1051/matecconf/201713802010

The experimental setup for the punching shear test is shown in Fig. 3. The punching load was subjected to the centre of the ribbed slab sample using a hydraulic jack controlled by a load cell at a constant rate of 0.01 mm/sec. The cross-sectional dimension of the column applied was 200 x 200 mm to induce sufficient punching shear to the slab sample. The support was fixed to 1.2 x 1.2 m focusing only on the centre of the ribbed slab sample.

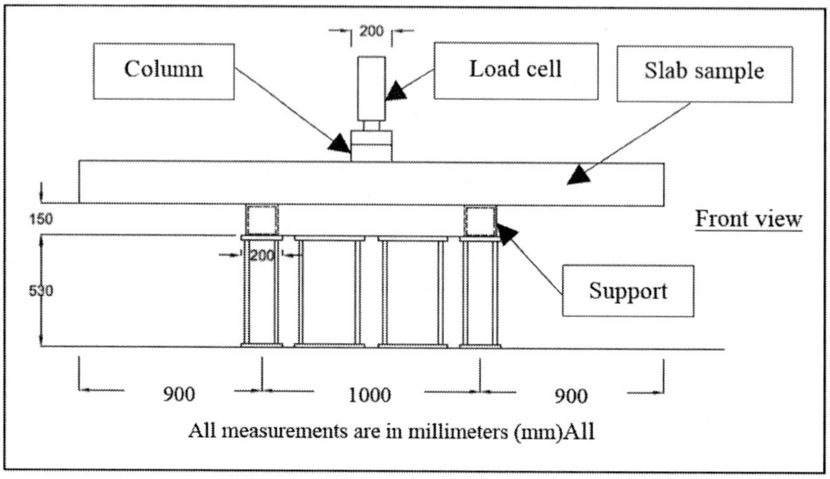

Fig. 3. Punching shear test experimental setup.

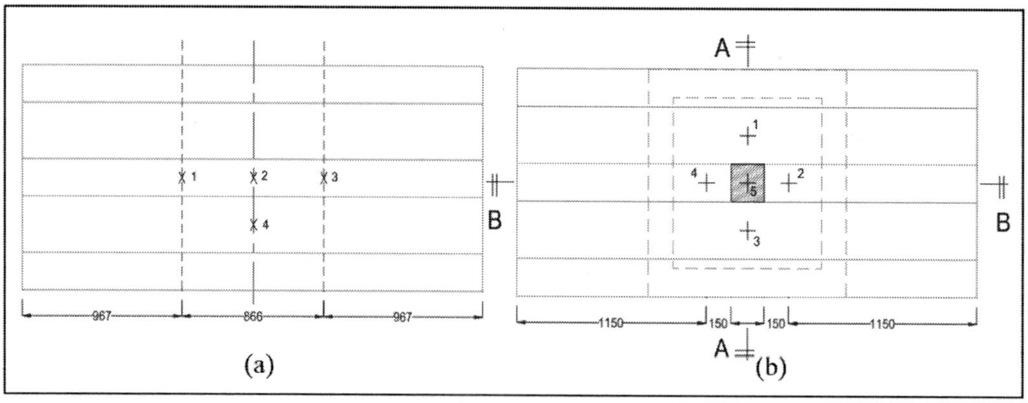

Fig. 4. LVDT locations (a) Flexural test (b) Punching shear test.

5 Flexural testing results and discussion

5.1 First crack and ultimate load

Table 3 presented the results from the flexural four point bending test. The presence of the steel fibres in the self-compacting concrete (SCC) significantly influence the first crack occurrence of the SFWS slab samples. In comparison between the CRC(F) and SFWS(F) samples, it can be observed that the SFWS(F) that is fully reinforced with steel fibres, had experienced crack at later stage and able to resist higher loads before crack is initiated as compared to samples that are conventionally reinforced (CRC(F)). This result proves the ability of the steel fibers to delay first cracks formation by enabling stress transfer across the cracks [21,22]. The values of the loads at first crack occurrence are shown in Table 3. The first crack load for the SFWS(F) sample is 47.67 kN while for the CRC(F) sample is

 DOI: 10.1051/matecconf/201713802010

21.39 kN, with indicated approximately 50% difference. As for the ultimate loading value, the SFWS(F) value was found to be slightly lower than the CRC(F) sample by approximately 20%.

Table 3. Flexural test results.

Sample designation	Topping thickness (mm)	First crack load (kN)	Ultimate load (kN)	Deflection at design load (mm)	Deflection at ultimate load (mm)
CRC(F)	100	21.39	74.93	5.03	63.92
SFWS(F)	100	47.67	60.00	1.27	7.8

5.2 Deflection of the samples

The maximum deflection at the ultimate load was recorded by the LVDT at location L2 which was placed at the middle rib soffit. As specified in the Eurocode2 [23], in general, the deflection of a slab should not exceed span/250 in order to avoid damages to the appearance and general utilities of the structure. Furthermore, in order to avoid damages to non-structural members that is supported by the structure, partition walls for instance, the deflection after construction is limited to span/500. In this study, considering the 2.6 m effective span length of the ribbed slab sample, the allowable deflection is 5.2 mm. Table 3 shows the deflection values recorded at design and ultimate load. At the design load of the ribbed slab sample, which is 31.49 kN (determined from the theoretical calculation), it can be observed that the deflection value for the SFWS(F) sample is 1.27 mm which is lower than CRC(F). This deflection values are acceptable since it is within the limiting value stated in the requirement of the Eurocode2 [23]. These experimental results revealed that both samples has performed accordingly as per designed.

Table 3 also shows deflection values samples at ultimate loading. Comparing these values, it can be observed that the deflection value for the SFWS(F) samples were significantly lower than the CRC(F) sample. Upon reaching the ultimate load, the deflection for the SFWS(F) sample showed only 7.8 mm deflection as compared to the CRC(F) sample that deflected up to 63.92 mm. This result indicates that even though it is not conventionally reinforced, the SFWS samples were still able to sustain the loadings subjected by having the strong interaction between the steel fibres in transferring the stresses effectively even after crack happens. Furthermore, the results also revealed that in terms of deflection at the ultimate load, the SFWS samples has the capability to perform better than the CRC samples. The steel fibres in the concrete mix has functioned effectively in holding the hardened concrete matrix together even after extensive cracking under flexural loading whereas it demonstrates the post-cracking ductility [22].

5.3 Crack pattern and failure mode

Visual observations were being carried out to identify the crack pattern and failure mode of the ribbed slab samples. In general, both types of samples, CRC(F) and SFWS(F) showed almost similar crack pattern at ultimate load, whereas both samples experienced minor cracks in the loading region. Fig. 5 showed the crack pattern of CRC(F) and SFWS(F) samples. The figures showed the side views and the bottom view of the ribbed slab samples at ultimate load.

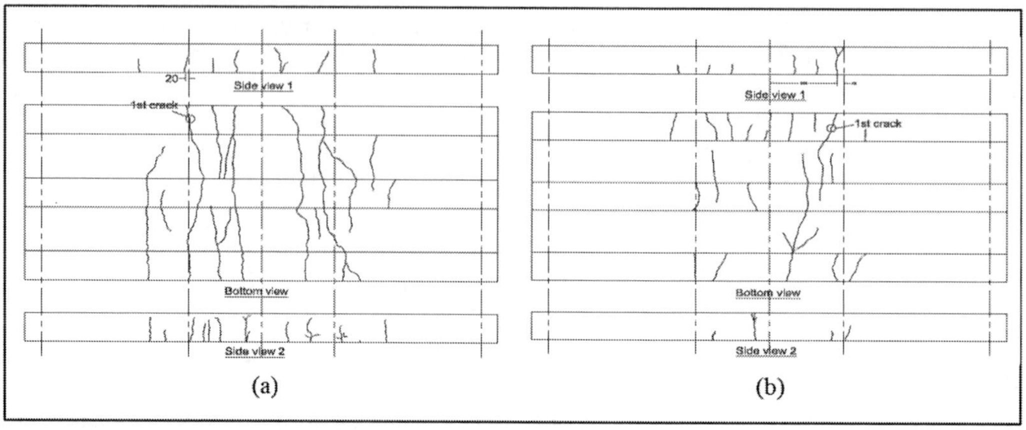

Fig. 5. Crack pattern (a) CRC(F) sample (b) SFWS(F) sample

Based on visual observation, the first crack of the CRC(F) sample occurred exactly under one of the loading line. This small 0.05 mm width crack was detected at the external rib soffit, 433 mm from the centre line. As the loading continues, one major crack formed near the centre line of the slab sample. The flexural behaviour of the CRC(F) slab samples meets the typical behaviour of reinforced concrete structures. The reinforcements in the CRC(F) slab samples was designed to carry the tensile forces transferred by the bonding at the interface of the two materials [24]. The failure mode exhibits gradual yielding of the tension steel reinforcements provided in the rib section.

The crack pattern of the SFWS(F) was almost similar to the CRC(F) samples whereas minor cracks were generated across the centre region of the slab. The first crack was initiated at the external rib, 393 mm from the centre line. As shown in Fig. 5, minor cracks was distributed evenly across the central region of this slab sample. At ultimate load, most of the cracks occurred only on the internal and external rib soffit. Only very small crack lines was observed propagating to the topping of the slab at this stage. A major crack was observed when the slab reached total failure, crossing from near the centre line towards the loading line. Based on the crack pattern and failure mode of the SFWS(F) sample, the explanation that can be drawn is the tensile stress in the sample has been effectively distributed and transferred to the steel fibres that is scattered along the structure. This similar failure mode of the sample as compared to the control sample has proven its ability to sustain flexural load in a similar ductile manner without abrupt failure.

6 Punching shear test results and discussion

6.1 First crack and ultimate load

This section discussed the punching test results. Similarly to the flexural test, observation was been made to the sample to locate and record the first crack load. First crack of the control sample CRC(P) occurred underneath the sample at the column line at 57.06 kN. The crack then propagates to the soffit and continued to the sides of the adjacent external ribs. They were scattered in the punching shear region. The cracks at the top part of the sample occurred only at the perimeter of the column. The punching shear cone can be observed from the middle rib. After the occurrence of the first crack, the sample continued to be loaded until it reached the ultimate shear capacity of 187.13 kN. The crack pattern for this sample is shown in Fig. 6 (a).

Fig. 6 (b) showed the crack pattern for the SFWS(P) sample. The first crack occured at 75 kN load underneath the sample at the middle rib, right below the location of the loading column. The crack then elongates creating a 45° angle towards the soffit and one of the adjacent ribs. The ultimate shear capacity for this sample was 135.68 kN.

In comparison of this two samples, it can be seen that the SFWS(P) that was fully reinforced with steel fibres was able to sustain slightly lower ultimate load than the conventionally reinforced sample (CRC(P)). The difference is approximately 27%. Conversely, viewing at the ability of the samples to sustain loads before the initiation of the first crack, SFWS(P) was found to perform better with almost 31% higher load than CRC(P). This might be due to small spacing between the fibres, that allows it to effectively bridge cracks throughout the whole volume of the concrete structure, thus enhancing the integrity of the slab under the column loading point [25]. Additionally, adequate amount of steel fibres also assist in enhancing the tensile strength of the concrete matrix, resulting an increase in the punching shear force.

Comparing the crack pattern of the two samples, it can be observed that the cracks were scattered around the perimeter of the column and within the supported area. For the control sample, the cracks were more scattered displaying the effective distribution of the stresses across the punching shear region. The pattern of the crack was in combination of punching-flexural mode, whereas cracks occurred at the column face as well as extends radially from the column's corner to corners of the slab's support.

As for the SFWS(P) sample, the crack pattern showed crack lines along the sides of the middle rib supporting the column and extends towards the support. The pattern slightly indicates that the slab underwent flexural failure mode. However, attention should be taken in view of the weak region at the sides of the middle rib.

For the control sample, CRC(P) it can be observed that the concrete cover at the bottom of the slab was about to fall apart due to the brittleness of the concrete matrix. Whereas for the SFWS(P) sample, the cracks occurred on the bottom surface was more intact with smaller width. This is due to the bridging effect of the steel fibres that significantly improve the ductility and the integrity of the slab-column connection [26].

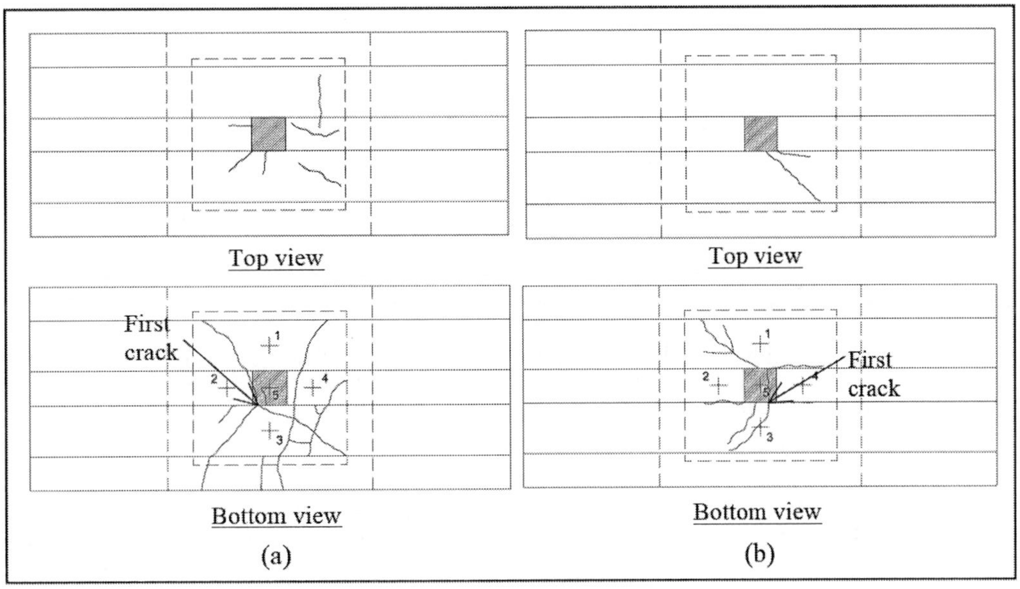

Fig. 6. Punching shear crack pattern (a) CRC(P) (b) SFWS (P).

DOI: 10.1051/matecconf/201713802010

6.2 Deflection

Figures below presents the deflection profile of the ribbed slab sample recorded at the ultimate loading. The comparison of the deflection profile between the CRC(P) and SFWS(P) sample is shown in Fig. 7. The deflection profile were plotted based on the LVDT readings within the support area.

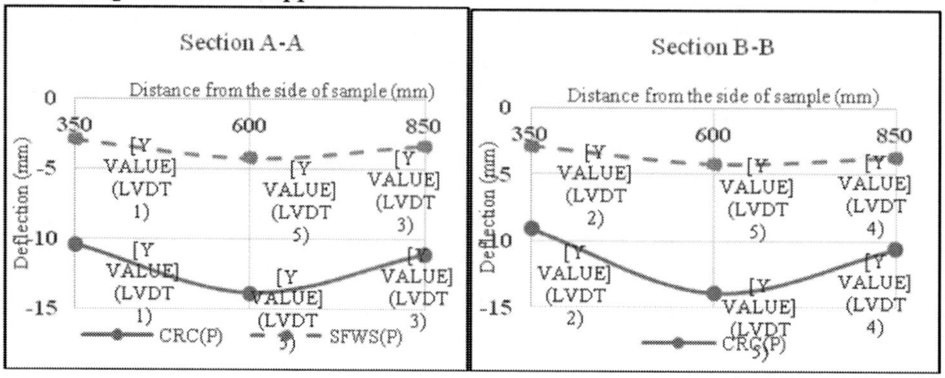

Fig. 7. Deflection profile of the samples at ultimate loading.

Based on the deflection profile, it can be observed that the deflection value for the CRC(P) sample was higher than the SFWS(P) sample. The maximum deflection value for both samples was recorded at the middle point where the load was located. The maximum value for the CRC(P) sample was 13.88 mm while for the SFWS(P) sample the value is 4.25 mm, which is significantly lower. These readings has proven the effectiveness of the steel fibres that is included in the concrete matrix in improving the ductility of the slab-column connections by modifying the failure mode, from pure punching to flexural, that contributes significantly in the ultimate shear capacity [27].

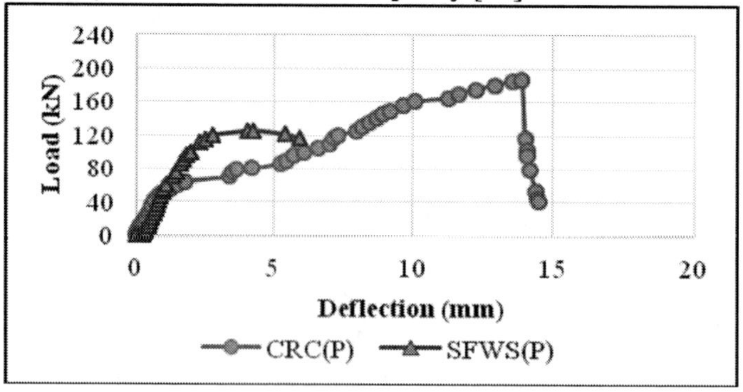

Fig. 8. Load versus displacement of the punching shear samples.

Fig. 8 shows the load versus displacement curve for samples CRC(P) and SFWS(P). Similarly to the flexural behaviour of the ribbed slabs, at the first stage of loading, the graph is approximately linear. Upon crack initiation, the curve tends to move away from the straight line, exhibiting a decrease in the stiffness of the slab. In comparison of the two curves, it can be observed that the deflection value for the CRC(P) sample was higher than SFWS(P) at the same point of load. Beyond the ultimate shear value, the strength of the CRC(P) sample drastically dropped, causing sudden and loss in shear resistance exhibiting that the sample can no longer sustain the punching shear force. On the other hand, for the SFWS(P) sample, upon reaching the ultimate load, the graph gradually descends revealing that the steel fibres still contributes to the stress transfer even after the crack occurrence, showing proof on the enhancement of the structural ductility [27].

7 Conclusion

From the flexural test, it can be concluded that the SFWS sample performed better in terms of its first crack initiation and comparable with regards to its ultimate load. The performance of SFWS is also reflected in the deflection reading where it shows lower deflection in comparison to the CRC(F) sample and the reading was still within the Eurocode2 limit. The crack pattern for both samples were found to be almost similar, experiencing minor cracks with no abrupt, brittle failure, exhibiting ductile behavior.

For the punching shear test, it can be concluded that similarly to the flexural testing results, the SFWS(P) sample was found to perform better in terms of the first crack initiation. However, the ultimate load was found to be 27% lower than conventional sample. In view of the failure mode, the CRC(P) sample underwent punching-flexural mode with brittle cracks on the bottom surface while the SFWS(P) experienced flexural failure mode. The displacement of the SFWS(P) sample was lower as compared to the CRC(P) sample.

As a whole, it can be concluded that the ribbed slab sample (SFWS) has the ability to perform well under both flexural and punching shear. However, special attention should be taken under punching shear since the SFWS(P) sample exhibited failure due to the geometrical aspect of the structure.

Special thanks to laboratory staffs of the Faculty of Civil Engineering, Universiti Teknologi MARA, Malaysia for their help and technical support. Also gratitude to MOSTI for the project funding under E-Science Fund (06-01-01-SF0835) and also the Institute of Research Management & Innovation (IRMI), Universiti Teknologi MARA for their continuous support.

References

1. P. L. Domone, "A review of the hardened mechanical properties of self-compacting concrete," *Cem. Concr. Compos.*, **29**, 1, pp. 1–12 (2007)
2. Y. Fritih, T. Vidal, A. Turatsinze, and G. Pons, "Flexural and shear behavior of steel fiber reinforced SCC beams," *KSCE J. Civ. Eng.*, **17**, 6, pp. 1383–1393 (2013)
3. A. Khaloo, E. R. Molaei, P. Hosseini, and H. Tahsiri, "Mechanical performance of self-compacting concrete reinforced with steel fibers," *Constr. Build. Mater.*, **51**, pp. 179–186 (2014)
4. H. Okamura and M. Ouchi, "Self-Compacting Concrete," *J. Adv. Concr. Technol.*, **1**, 1, pp. 5–15 (2003)
5. L. Ferrara and A. Meda, "Relationships between fibre distribution, workability and the mechanical properties of SFRC applied to precast roof elements," *Mater. Struct.*, **39**, pp. 411–420 (2007)
6. M. Pająk and T. Ponikiewski, "Flexural behavior of self-compacting concrete reinforced with different types of steel fibers", *Constr. Build. Mater.*, **47**, pp. 397–408 (2013)
7. S. Grunewald, *Performance-based design of self-compacting fibre reinforced concrete*. Delft University Press (2004)
8. S. Grunewald and J. C. Walraven, "Transporting fibres as reinforcement in self-compacting concrete," *Heron*, **54**, 2–3, pp. 99–124 (2009)
9. A. Khaloo and M. Afshari, "Flexural behaviour of small steel fibre reinforced concrete slabs," *Cem. Concr. Compos.*, **27**, 1, pp. 141–149 (2005)

10. B. Akcay and M. A. Tasdemir, "Mechanical behaviour and fibre dispersion of hybrid steel fibre reinforced self-compacting concrete," *Constr. Build. Mater.*, **28**, 1, pp. 287–293 (2012)
11. A. S. El-Dieb, "Mechanical, durability and microstructural characteristics of ultra-high-strength self-compacting concrete incorporating steel fibers," *Mater. Des.*, **30**, 10, pp. 4286–4292, (2009)
12. F. Aslani, S. Nejadi, and B. Samali, "Short term bond shear stress and cracking control of reinforced self-compacting concrete one way slabs under flexural loading," *Comput. Concr.*, **13**, 6, pp. 709–737 (2014)
13. J. Michels, D. Waldmann, S. Maas, and A. Zürbes, "Steel fibers as only reinforcement for flat slab construction – Experimental investigation and design," *Constr. Build. Mater.*, **26**, 1, pp. 145–155 (2012)
14. D. V. Soulioti, N. M. Barkoula, A. Paipetis, and T. E. Matikas, "Effects of Fibre Geometry and Volume Fraction on the Flexural Behaviour of Steel - Fibre Reinforced Concrete Effects of Fibre Geometry and Volume Fraction on the Flexural Behaviour of Steel-Fibre Reinforced Concrete," *Strain*, **47**, pp. 535–541 (2011)
15. A. A. Shah and Y. Ribakov, "Recent trends in steel fibered high-strength concrete," *Mater. Des.*, **32**, 8–9, pp. 4122–4151 (2011)
16. N. N. Sarbini, I. S. Ibrahim, and A. A. Saim, "Assessment on the Mechanical Performance of Steel Fibre Reinforced Concrete using Fibres Geometrical Factor," *BEFIB2012-Fibre Reinf. Concr.*, (2012)
17. N. A. Rahman, S. H. Hamzah, and A. Abdulrazzaq, "Effectiveness of SteFib in Composite Structural Member," *J. Eng. Sci. Technol.*, **7**, 5, pp. 552–562 (2012)
18. Z. Tazaly, "Punching Shear Capacity of Fibre Reinforced Concrete Slabs with Conventional Reinforcement," Royal Institute of Technology (KTH) (2011)
19. BS EN 14889-1, "Fibres for concrete: Part 1: Steel fibres - Definitions, specifications and conformity." European Committee for Standardization (CEN), Brussels (2006)
20. BS EN 12390-5, "Testing hardened concrete - Part 5: Flexural strength of test specimens." European Committee for Standardization (CEN), Brussels (2009)
21. R. N. Swamy, "FRC for Sustainable Infrastructure Regeneration and Rehabilitation," in *Fifth International RILEM Symposium on Fibre-Reinforced Concrete (FRC)*, pp. 3–19 (2000)
22. J. Narwal, A. Goel, D. Sharma, D. R. Kapoor, and B. Singh, "An Experimental Investigation on Structural Performance of Steel Fibre Reinforced Concrete Beam," *Int. J. Eng. Adv. Technol.*, **2**, 6, pp. 301–304 (2013)
23. MS EN 1992-1-1, "Eurocode 2: Design of Concrete Structures - Part 1-1: General rules and rules for buildings." Department of Standards Malaysia (2010)
24. B. Mosley, J. Bungey, and R. Hulse, *Reinforced Concrete Design to Eurocode 2*, 7th ed. London: Palgrave Macmillan (2012)
25. L. Nguyen-minh, M. Rov, and T. Tran-quoc, "Punching Shear Capacity of Interior SFRC Slab-Column Connections," *J. Struct. Eng.*, **138**, pp. 613–624 (2012)
26. L. Nguyen-Minh, M. Rovňák, T. Tran-Quoc, and K. Nguyenkim, "Punching Shear Resistance of Steel Fiber Reinforced Concrete Flat Slabs," *Procedia Eng.*, **14**, pp. 1830–1837 (2011)
27. M. H. Harajli, D. Maalouf, and H. Khatib, "Effect of Fibers on the Punching Shear Strength of Slab-Column Connections," *Cem. Concr. Compos.*, **17**, pp. 161–170, (1995).

MATEC Web of Conferences **138**, 02011 (2017)
EACEF 2017
DOI: 10.1051/matecconf/201713802011

The Priority Importance of Economic Motivation Factors Against Risks for Green Building Development in Malaysia

Farid Ezanee Mohamed Ghazali[1], *Rozana* Zakaria[2*], *Eeydzah* Aminudin[3], *Lee* Yong Siang[4], *Gamal* Alqaifi[5], *Darul Nafis* Abas[6], *Nur IzieAdiana* Abidin[7], *Siti Mazzuana* Shamsuddin[8]

[1,4]School of Civil Engineering, Engineering Campus, Universiti Sains Malaysia, 14300 Nibong Tebal, Penang, Malaysia

[2,3,5,6,7]Department of Structures and Materials, Faculty of Civil Engineering, Universiti Teknologi Malaysia, Johor, Malaysia

[8] Faculty of Achitecture, Planning and Surveying, Universiti Teknologi Mara, Shah Alam, Selangor, Malaysia

Abstract. Green building development is an emerging paradigm for the construction industry practice all around the world. The establishement of Green Building rating tool helps to assess the whole life cycle process in planning to operation in a building. The Malaysian construction industry recognizes buildings that have been assessed using established green building tool such Green Building Index, Green RE or My CREST. Eventhough these rating tools provide motivation factors in its criteria and sub-criteria to promote sustainability in Malaysia buildings, there is still a major doubt to developers in terms of risks that may hinder their investments in green buildings. This paper highlights the priority importance of economic motivation factors against risks in the green building development in Malaysia. The data presented in this paper have been mainly derived from responses received through questionnaires completed by building stakeholders involved in green building developments. In order to determine the priority importance of economic motivation factors and risks identified for green building development, the questionnaire outcomes have been thoroughly assessed using the Analytical Hierarchy Process (AHP) method. As a result, lack of government incentive and high capital cost, which classified under green building risks, are the two key factors with highest priority importance that influenced most of the decision making for green building development in Malaysia. The results show green buildings have proliferated as governmental support and incentives with more exampler of higher profit return of investment in enhancing developers preference for green building development.

* Corresponding author: rozana@utm.my

1 Introduction

The green building movement has revolutionized the way the architectural industry envisions and designs buildings for promotion of environmental sustainability, the establishment of green building certification systems globally is one of the most prominent [1]. However, green building faces various types of risks that hinder the penetration of this investment into the industry of sustainable construction. Many studies have highlighted the risks and challenges that face the investment of green buildings in various types of industries [2,3,4,5,6,7,8] Poor management of risk in construction projects may result in the failure to comply with the delivery date and approved budgets, which take place due to lack of appropriate procurement strategy, act of God as well as economy recession on prices of materials. In order to get prepare with the preventive action, contingency sum is needed and it is a provision for every construction project [7]. The identity of each risk also is an important step in risk managing and is by all mean the most difficult especially when it comes to the identification of each source of risk [9]. Thus, the consideration on the influence of each risk factor, will simplify the analysis and management of the risk itself [10]. Among the factors that negatively affect the interest of investors and the public in general regarding the green investment which represent risks factors are a lack of knowledge in sustainable green design and construction in learning institutions, a lack of experience related to the construction of green building in the realms of architecture and designing, the high cost when constructing buildings with green features [11]. The great heterogeneity for developers is the capital cost that contributes in green building development. This has made it challenging to understand the influenced cost implications in improving the performance of green building concept for building occupancy building, the selection of building material and green technology.

2 Risk in Green Building Development

In risk point of view, future cost and availability of green building materials can represent a risk factor to the investment of green building. The cost of green building construction can be increased by the time factor effects due to the unfamiliarity of green investment by the contractors and subcontractors [12]. Negative perception of green building investment in general summarize a cost risk factor of the investment [13]. While technical difficulties represent an internal risk factor in terms of the right way of green building construction due to different environment of practice [14]. Some of the previous research studies clarified that green building cost is higher in terms of material cost, build and maintenance [13,15,16]. [17] identified that some design codes of building do not allow the implementation of green features because the codes are for conventional buildings. In other words, the conventional codes may restrict limited options to the developers such as the choice of design, construction system and new green technology and materials that shall be applied into almost all green building development. According to [11], the difficulty of sourcing for building materials required by developers, the regulatory procedures that inhibit green material production and the fact that there are limited green building examples to emulate The findings clarified that the time frame of the green project can be lengthened due to the lack of expertise and green resources in some communities [18] determined that sustainability measures are not considered by stakeholders and sometimes not even requested by clients while some green features were cancelled in order to achieve others. This statement shows that unreliable sustainable measures with the lack of perception and awareness can be risk factors to the investment of green building. Sustainable measures faces lots of challenges such as cost factors whereby investors may not fund the green project due to the sustainable measures requirements.

Majority of the green building experts believes that the governments can effectively play a major role in promoting sustainable development such as green buildings [19]. At the same time, Governments can enforce the rules and regulations to support green building development. However, government policies and technical standards shall be adopted to satisfy the sustainable measures in order to avoid any issues that may result due to the conflict of conditions of the construction site [18]. [20] highlighted a fact that market-based incentives are one of the most effective and efficient tools to address market failure together with non-market problems to improve the situation for green buildings development. On the other hand, it has been identified that several risks and challenges may hinder the development of green buildings in Asia including Malaysia such as: lack of awareness, Lack of knowledge and training programs on sustainable design, higher cost, special materials, rules and regulation, lack of demonstration, lack of technology and lack of demand [21]. Another study has shown that wrong perception of transition and the expected return on investment represents the highest degree of constraints that may obstruct the development of green buildings [6] including lack of information and the lack of effective management of financial resources [13]. In the same context, lack of financial support, professional experts and implementing of green technologies are considered to be among the key challenges in both public and private sector [22]. Table 1 indicates a summarised ten significant risks that are associated to green building development in Malaysia.

Table 1. Risk Factor in Green Building Development

Item	Risk Factor of Green Building Development
1	High Capital Cost
2	Availability of Experts
3	Lack of Awareness
4	Lack of Enforcement from Government
5	Availability of Green Materials
6	Uncertainty of Future Material Cost
7	Technical Difficulty During Construction
8	Limited Availability of Green Suppliers
9	Lack of quantitative evaluation tools for green performance
10	High Capital Cost

3 Economic Motivation in Green Building Development

Numerous study show that green buildings does performed as more efficient and more sustainable during operations (energy efficiency and water efficiency) as this improvement enhance as its benefits the owner to derive economic value through green marketing. Still, limited rigorous empirical evidence regarding the economics, especially in cost-effectiveness of investments to achieve green building development is being claimed by developer.

Economic motivation factors such organizational, social, environmental and economic are among the key inspiration for the developers to drive forward green building development not only in Malaysia but also other countries around the world. Through identification of economic motivation factors, developer should gain the benefits not only for them but also to the human being and environment.

[23] had simplified ten economic motivation factors for green building development in Malaysia as shown in Table 2.

Table 2. Economic Motivation Factor in Green Building Development [23]

Item	*Risk Factor of Green Building Development*
1	Long Term Value for Money Offset Cost
2	High Market Demand
3	Energy Efficiency Cost Savings
4	Financial Incentive & Tax Exemption
5	Whole Life Cycle Design implies economic benefits
6	Reduce Water Consumption Cost
7	Minimal Fuel/Transportation Cost
8	Low Waste Disposal Cost
9	Higher Return of Investment
10	Minimum Maintenance & Repair Costs

Based on the ambiguity on green building and question above, there is always a question to developer whether to embark and invest on green building development i.e. whether risks associated to green building development always being influenced or emerged as a result of higher initial cost. Therefore, this paper embarked with the analysis to determine the priority importance of economic motivation factors against risks for green building development in Malaysia.

4 Methodology

The quantitative approach is being applied in this study in order to analyze the understanding of practitioners in Malaysia on green building besides determining the influencing factors, which could have a significant impact towards decisions related to green building development. A set of questionnaires had been distributed to various respondents involved in construction and development projects both in Johor Bahru and Kuala Lumpur mainly to the construction practitioners such as contractor, consultant, developer and government servant.

However, the distribution method was not designed to achieve a random or representative sample, and the data were not meant to be subjected to tests of statistical significance. 69 questionnaire survey had been received in return from the involved respondents. The demographic information of the respondents is as follows; 13% (9) of respondents had more than 21 years of working experience, 11.6% (8) have between 16-20 years of working experience, While another 23.2% (16) have between 11-15 years of working experience and another 15.9% (11) have between 6-10 years of working experience. In majority 36.2% (25) had 1-5 years of working experience. For respondent working sector 21.75(15) are from consultants, 24.6% (17) are from contractor, 40.6% (28) are from developers and 13% (9) are from government related construction working sector.

Once the criticality and importance of both key success elements and critical success factors have been determined, the Analytical Hierarchy Process (AHP) approach, which supports any multi-criteria decision-making situations by breaking down problems into relevant hierarchical levels, has been used in the data analysis. [24] regarded the AHP pair-wise comparison matrix as a method that provides more valid results than the Likert-based survey. [25] have generalized the fact that AHP approach allows decision makers to structure complex problems in the form of a hierarchy such as the goal, criteria and alternative compared to other decision-making approaches. Table 3 portrays the fundamental intensity scales developed by [26] (in [27]) for AHP decision-making models. The fundamental concept of pairwise comparison in the AHP is to assess the priority importance of a criterion over another that leads to achieving a hierarchy objective. For example, if the criticality and priority importance of criterion "A" is found to be essential or strong over "B" in pairwise comparison, then an intensity scale of 5 should be awarded to the former. The intensity scale "1" usually represents two activities that have an equal priority importance to the objective when they are being evaluated in the pairwise comparison.

Table 3. Intensity Scale for AHP Pairwise Comparison (Source: [26] (in [27])

Intensity of Importance	Definition	Explanation
1	Equal Importance	Two activities contribute equally to the objective
3	Weak Importance of one over another	Experience and judgement strongly favour one activity over another
5	Essential or strong importance	Experience and judgement strongly favour one activity over another
7	Very strong importance	An activity is strongly favoured and its dominance demonstrated in practice
9	Absolute importance	The evidence favouring one activity over another is of the highest possible order or affirmation
2, 4, 6, 8	Intermediate values between the two adjacent judgements	When compromise is needed

5 Results and Findings

Figure 1 indicates the decision-making hierarchy for factor affecting most to the decision making of Green Building Development which composed of three (3) levels; Goal, Factor and Sub-Factor as a result of cumulative questionnaire-based outcomes generated from a series of interviews carried out with 69 respondents.

L1: Goal

Factors affecting Green Building Development in Malaysia

Green Building Risks

- Lack of Government Incentive
- High Capital Cost
- Availability of Experts
- Lack of Awareness
- Lack of Enforcement from Government
- Availability of Green Materials
- Uncertainty of Future Material Cost
- Technical Difficulty During Construction
- Limited Availability of Green Suppliers
- Lack of quantitative evaluation tools for green performance

Green Building Economic Motivation Factors

- Long Term Value for Money Offset Cost
- High Market Demand
- Energy Efficiency Cost Savings
- Financial Incentive & Tax Exemption
- Whole Life Cycle Design implies economic benefits
- Reduce Water Consumption Cost
- Minimal Fuel/Transportation Cost
- Low Waste Disposal Cost
- Higher Return of Investment
- Minimum Maintenance & Repair Costs

Figure 1. The Hierarchy of Decision Making Factors for Green Building Development

In pairwise comparisons, each of the identified factor and sub-factor are assessed against the hierarchy goal, which revolves around the factor affecting most to the decision making of Green Building Development via the Expert Choice software. The relative importance of each factor and sub-factor is compared with other factor and sub-factor identified within the same decision level with respect to the hierarchy goal, whichever is higher than the assessed factor.

Figure 2 shows pairwise comparisons of factor identified for level 2 in respect to the hierarchy goal. As a result, the Green Building Risks has greater impact compared to Green Building Economic Motivation Factors for the factor affecting most to the decision making of Green Building Development as the scoring 3 leans towards to Green Building Risks. The score 3 given to the Green Building Risks indicates its priority intensity that gives weak importance against Green Building Economic Motivation Factors. Figure 3 shows the overall priority ranking of factor affecting most to the decision making of Green Building Development which the result is composed of all levels in decision-making hierarchy as shown in Figure 1. It is apparent from the research findings that lack of government incentive, followed by high capital cost are two major factors affecting most to the decision making of green building development in Malaysia which fall under green building risk.
Most of the respondents agreed that lack of government incentives to the decision making for green building development. This result can be jusitifed through findings by [28] that supported the contention and consistent of [29] who asserts that the construction industry seems to be lagging in terms of the provision of incentives to developers who meet green ratings, and consultant team members who engage in green design practices on their projects and the lack of incentives for demonstrating best green practice. The provision of incentives will undoubtedly reduce the high up front cost for green projects. [30] support this notion by asserting that incentive provisions are used in construction contracts to reduce contract costs, to minimize contract duration and to maintain acceptable levels in health and safety, productivity, technological progress, innovation, management efficiency, and quality of construction. On the other side, five states in U.S offer limited grants, such as in Kentucky, it's offer the symbolic gestures promoting certification. Besides that, they also offers variety of innovative policy designs and incentives, including property tax reassessment moratoriums, green funds, parking incentives, electric bill discounts and green roof mandates.

Green Building Risks	9	8	7	6	5	4	3	2	1	2	3	4	5	6	7	8	9	Green Building Economic Motivation Factors
Compare the relative importance with respect to: Goal: Green Building Risk Vs Green Building Economic Motivation Factors																		

	Green Building Risks	Green Building Economic Motivation Factors
Green Building Risks		3.0
Green Building Economic Motivation Factors	Incon: 0.00	

Figure 2. Pairwise Comparison on Factor affecting most to the decision making of Green Building Development in Level 2

Besides that, many respondents also concerned that adopting green features into their buildings will involve high capital cost. Compared to conventional buildings, green building projects are often perceived as having higher initial design and constructions costs. The lack of attention to the costs associated with green construction seems to be a global phenomenon. [17] states that one of the major barriers is the need for reliable cost information for green features, without this information, it is difficult for the market to justify the occasionally higher up-front costs for a green development project.

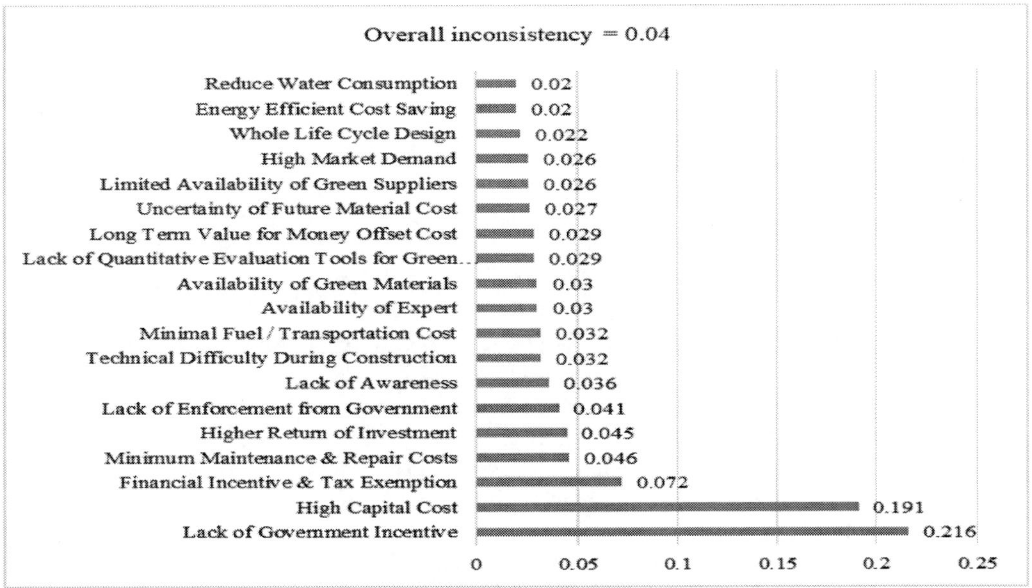

Figure 3. The Overall Priority Ranking of Factor affecting most to the decision making of Green Building Development

[31] reveal that there is still little published data about actual cost premiums for green buildings. For instance, in the United States this information gap is exacerbated by the fact that the USGBC does not require that cost information be included with submissions for LEED certification. Many developers keep cost information proprietary. Accordingly, [32] acknowledges that there are more ways to expand the four pillars of sustainability and one of the ways is to expand Full-Cost Pricing to True-Cost Pricing. Debatably, the True-Cost Pricing goes beyond covering the costs of the infrastructure and includes long-term environmental and community externalities, such as for example, energy savings, green space, and green job creation. While there has been a plethora of research seeking to determine the direct or tangible costs of green building, the indirect or intangible costs remain unexplored in construction. [33] argues that this is because it is difficult, if not impossible to quantify such costs in purely monetary terms. The opinion embedded with [17] emphasizes on the need to look at the indirect costs as well when determining the total upfront or capital costs of green construction.

6 Conclusion

Despite the spur of growth in green buildings across Asia, the shift towards more sustainable goals still lags in Malaysia due to several key challenges. The study of green buildings is further hampered by financial incentives such as lower local government taxes would encourage green building adoption in market. Through financial premiums associated with desired building improvements helps align with costs which green building projects are often perceived as higher capital cost compared to conventional buildings. Capital costs for green building development are often a primary barrier to realizing high performance buildings that can save energy. There is clearly a need to create a coordinated knowledge hub to document and provide the existing and emerging information for green building development especially on the barriers of implementation of government incentive and having high capital cost in this green industry. Understanding the origins and green building risk basically related to green building economic motivation factors which requires an approach of green building practices adopted, and the overall (including uncertified) penetration of green building practices in the built environment.

References

1. Liang, H.-H., Chen, C.-P., Hwang, R.-L., Shih, W.-M., Lo, S.-C., & Liao, H.-Y. Satisfaction Of Occupants Toward Indoor Environment Quality of Certified Green Office Buildings in Taiwan. Building And Environment, **72**, 232-242. (2014).
2. Al Zubaidi, H., & Al Otaibi, S. An Empirical Approach for Identifying Critical Time-Ovrerun Risk Factors in Kuwait's Construction Projects. Journal of Economic and Administrative Sciences, **24(2)**, 35-53. (2008).
3. Hoffman, A. J., & Henn, R. Overcoming the Social and Psychological Barriers to Green Building. Organization & Environment, **21(4)**, 390-419. (2008).
4. Kasai, N., & Jabbour, C. J. C. Barriers to Green Buildings at two Brazilian Engineering Schools. International Journal of Sustainable Built Environment, **3(1)**, 87-95. (2014).
5. Koh, S., Saad, S., Ahmed, A., Kayis, B., & Amornsawadwatana, S. A Review of Techniques for Risk Management in Projects. Benchmarking: An International Journal, **14(1)**, 22-36. (2007).
6. Marker, A. W., Mason, S. G., & Morrow, P. Change Factors Influencing The Diffusion and Adoption of Green Building Practices. Performance Improvement Quarterly. **26(4)**, 5-24. (2014).
7. Mills, A. A Systematic Approach to Risk Management for Construction. Structural survey. **19(5)**, 245-252. (2001).
8. Yang, R. J., & Zou, P. X. Stakeholder-Associated Risks and Their Interactions in Complex Green Building Projects: A Social Network Model. Building and Environment, **73**, 208-222. (2014).
9. Tchankova, L. Risk Identification-Basic Stage in Risk Management.Environmental Management and Health, **13(3)**, 290-297. (2002).
10. Bajaj, D., Oluwoye, J., & Lenard, D. An Analysis Of Contractors' Approaches to Risk Identification in New South Wales, Australia. Construction Management & Economics. **15(4)**, 363-369. (1997).
11. Shafii, F., & Othman, M. Z. Greening Engineering Education for Sustainable Development. *The Conference On Sustainable Building South-East Asia. Johor (Malaysia)*. 416-421. (2007).
12. Griffin, C., Knowles, C., Theodropoulos, C., & Allen, J. Barriers To The Implementation Of Sustainable Structural Materials In Green Buildings. *Paper Presented At The Structures & Architecture: ICSA 2010-1st International Conference on Structures & Architecture, July 21-23 July, 2010 in Guimaraes, Portugal.* (2010).
13. Richardson, G. R., & Lynes, J. K. Institutional Motivations and Barriers to the Construction of Green Buildings on Campus: A Case Study Of The University Of Waterloo, Ontario. International Journal of Sustainability in Higher Education, **8(3)**, 339-354. (2007).
14. Zhang, X., Platten, A., & Shen, L. Green Property Development Practice in China: Costs And Barriers. Building and Environment. **46(11)**, 2153-2160. (2011).
15. Lam, P. T., Chan, E. H., Poon, C., Chau, C., & Chun, K. Factors Affecting the Implementation of Green Specifications in Construction. Journal of Environmental Management, **91(3)**, 654-661 (2010).
16. Shi, Q., Zuo, J., Huang, R., Huang, J., & Pullen, S. Identifying The Critical Factors For Green Construction–An Empirical Study In China. Habitat International, **40**, 1-8 (2013).
17. Choi, C. Removing Market Barriers to Green Development: Principles and Action Projects to Promote Widespread Adoption of Green Development Practices. Journal of

Sustainable Real Estate, **1(1)**, 107-138 (2009).
18. Williams, K., & Dair, C. What is Stopping Sustainable Building in England? Barriers Experienced by Stakeholders in Delivering Sustainable Developments. Sustainable Development-Bradford, **15(3)**, 135 (2007).
19. Samari, M., Godrati, N., Esmaeilifar,R., Olfat, P., Mohd Shafie, M.W. "The Investigation of the Barriers in Developing Green Building in Malaysia", Modern Applied Science, **7**, 2. (2013).
20. Keiner, M. The future of sustainability: Springer. (2006).
21. Shafii, F., Ali, Z. A., and Othman, M. Z. Achieving Sustainable Construction in The Developing Countries Of Southeast Asia. Paper Presented at the Proceedings Of The 6th Asia-Pacific Structural Engineering and Construction Conference (APSEC 2006), Kuala Lumpur, Malaysia, 5-6 (2006).
22. Xiaolu, Z. Investigation of Factors Restraining the Implementation of Green in Mainland China, Organisation, Technology & Management in Construction. **6(3)** (2014).
23. Darul Nafis Abas. The Motivation Factors in the Provision Of Green Building Criteria in Green Building Development. Thesis Universiti Teknologi Malaysia (2016).
24. Fong, P.S.K. and Choi, S.K.Y. Final Contractor Selection Using The Analytical Hierarchy Process., Construction Management and Economics, vol. **18**, pp. 547-557. (2000).
25. Acharya, N.K, Young D.L, and Jung, K.K . Critical Construction Conflicting Factors Identification Using Analytical Hierarchy Process. KSCE Journal of Civil Engineering, vol. **10 (3)**, pp. 165-174 (2006).
26. Saaty, T.L. How To Make A Decision: The Analytic Hierarchy Process, European Journal Of Operational Research, Volume **48**, Issue 1, pp. 9-26 (1990).
27. Mahdi, I.M. and Alreshaid, K. Decision Support System for Selecting The Proper Delivery Method Using Analytical Hierarchy Process (AHP), International Journal of Project Management, Volume **25**,pp. 564-572 (2005).
28. Ndihokubwayo, R., Crafford, G.J. & Buys, F.. Consultant Team Members' Performance Evaluation Against Incentives Towards The Achievement Of Green Building Principles. *Proceedings of 6th Annual SACQSP Research Conference on "Green Vision 20/20", Cape Town, South Africa, 20–21 June 2013* (2013).
29. Milne, N. The Rands and Sense of Green Building: Re-visited. Building the Business Case for Green Commercial buildings in South Africa. Green Building Council, Cape Town, South Africa (2012).
30. Arditi, D. & Yasamis, F. Incentive / Disincentive Contracts: Perceptions of Owners and Contractors, Journal of Construction Engineering and Management. **124 (5)**, 361-373 (1998).
31. Kats, G., & Capital, E. The cost and financial benefits of green buildings: A Report to California's sustainable building task force, developed for the Sustainable Building Task Force, California, USA (2003)
32. Nelson, V.I. Sustainable Infrastructure Management, Coalition for Alternative Wastewater Treatment White Paper (2008)
33. Love, P.E.D. Auditing the Indirect Consequences of Rework In Construction: A Case Based Approach, Managerial Auditing Journal, **17(3)**, 138–146 (2002).

On the effect of drag forces in mooring system restoring forces

Zahid Ullah[1], *Naik* Muhammad[1], *Ji-Hoon* Lim[1], and *Dong-Ho* Choi[2,*]

[1]Ph.D. Candidate, Department of Civil and Environmental Engineering, Hanyang University 222, Wangsimni-ro, Seongdong-gu, Seoul, 04763, Republic of Korea
[2]Professor, Department of Civil and Environmental Engineering, Hanyang University 222, Wangsimni-ro, Seongdong-gu, Seoul, 04763, Republic of Korea

Abstract. Mooring line is a major source of stability and plays a key role in the global response of offshore floating wind turbine. In the current state of the research, a formulation based on the analytical catenary equation is most commonly used for the analysis of mooring lines. However, due to the inability of catenary equations to consider the ocean current drag forces on mooring lines, the effect of drag forces on fairlead restoring forces has not been investigated yet. In this study, we have investigated the influence of drag forces on fairlead forces using discrete catenary formulation for modeling mooring line. The discrete catenary formulation has the ability to incorporate ocean current drag forces. Three types of elements; fully suspended, touchdown and seabed element are formulated to model the suspended, touchdown and seabed portion of a slack mooring line, respectively. The influence of viscous drag on the fairlead restoring forces is demonstrated through the analysis of OC3-Hywind mooring system subjected to ocean currents. It was found that the viscous drag significantly influences the fairlead forces.

1 Introduction

The increasing demand for exploitation of renewable wind energy has dramatically increased the size of the offshore floating wind turbine and pushed the installation of floating wind turbine away from shore in the deeper water. This development of offshore floating wind turbine has resulted in the need for more accurate analysis and simulation tools. Mooring line plays an important role in the stability and the global response of the floating wind turbine. The accurate prediction of mooring loads and fairlead forces is necessary for the safe design of floating wind turbine. Various numerical models have been proposed by various researchers for the analysis of mooring lines. The finite element method has been widely used for analysis of mooring lines [1-2]. However, the high computational cost is a major weakness of the finite element modeling. Due to computational efficiency and easier implementation, the analytical catenary approach has been widely adopted by the various researcher [3-4] for the analysis of mooring lines. However, it is incapable of taking into account the ocean current drag forces. Therefore, the

* Corresponding author: samga@hanyang.ac.kr

effect of ocean current drag forces on the fairlead forces has not been studied. The present is study focused on the influence of drag forces on the mooring line restoring forces.

2 Mathematical formulation of mooring line

An improved quasi-static model based on the discrete catenary formulation is used for the analysis of mooring lines. To model the suspended, touchdown and seabed portion of a slack mooring line three types of discrete catenary elements; suspended, touchdown and seabed are used, respectively, as shown in Fig. 1. The drag forces in vertical and horizontal directions are denoted by f_x and f_y, respectively; F_1 and F_2 denotes the fairlead restoring forces in horizontal and vertical directions, respectively; L_x and L_y represents the horizontal and vertical projections of the mooring line, respectively, and w is the unit self-weight of the mooring line.

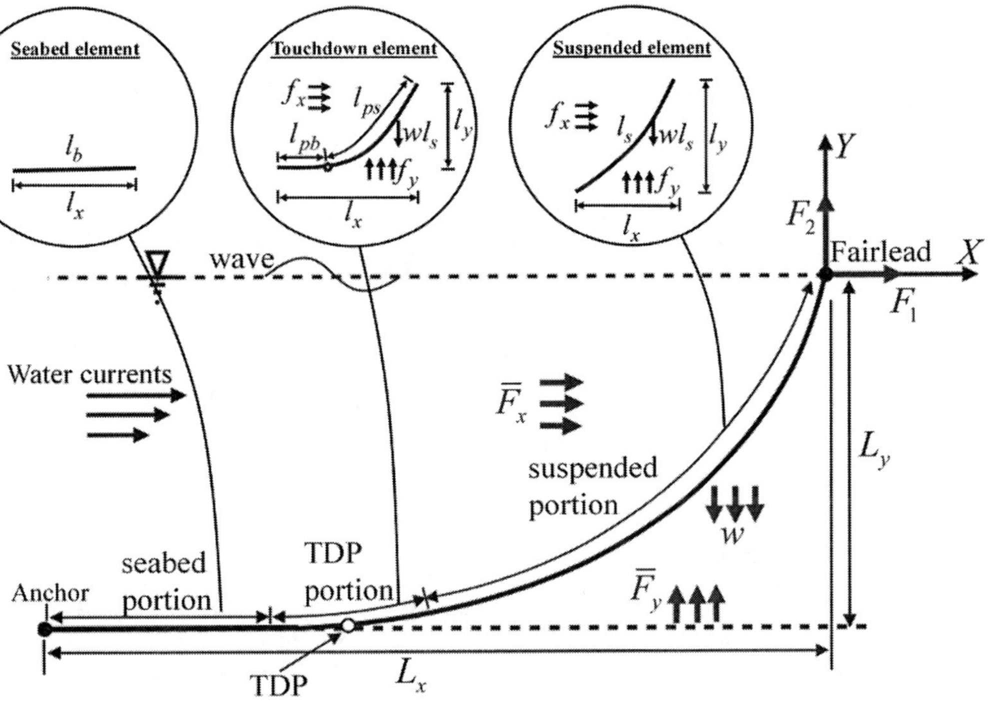

Fig. 1. Mooring line layout and discrete catenary cable elements.

2.1 Suspended element

The suspended portion of mooring line is modeled using suspended element. The horizontal and vertical projection of a suspended element ($i = n-1$) can be determined as:

$$l_x = -\frac{\left(F_1 + \bar{F}_x\right) l_s}{EA} - \frac{\left(F_1 + \bar{F}_x\right) l_s}{\sqrt{\left(F_1 + \bar{F}_x\right)^2 + \left(-F_2 + \bar{F}_y\right)^2}} \tag{1}$$

$$l_y = \frac{\left(-F_2+\bar{F}_y\right)l_s}{EA} + \frac{\left(-F_2+\bar{F}_y\right)l_s}{\sqrt{\left(F_1+\bar{F}_x\right)^2+\left(-F_2+\bar{F}_y\right)^2}} \tag{2}$$

where the l_s represent the unstrained length of the element; $\bar{F}_x = \sum_{i=1}^{n-1} f_x^i l_s^i$ and $\bar{F}_z = \left((n-1)w - \sum_{i=1}^{n-1} f_z^i\right) l_s^i$ are the drag forces on the element in horizontal and vertical directions, respectively and i represent element number ($i = 1$ to N).

2.2 Touchdown element

The touchdown portion of mooring line is modeled using a touchdown element. The touchdown element is partially suspended and partially laying on the seabed. The suspended portion of touchdown element is denoted by l_{ps}, and the seabed portion of touchdown element is denoted by l_{pb}. The horizontal and vertical projections of the touchdown element ($i = n$) can be determined as follow:

$$l_x = -\frac{\left(F_1+\bar{F}_x+f_x^n l_{ps}\right)\left(l_{ps}+l_{pb}\right)}{EA} - \frac{\left(F_1+\bar{F}_x+f_x^n l_{ps}\right)\left(l_{ps}+l_{pb}\right)}{\sqrt{\left(F_1+\bar{F}_x+f_x^n l_{ps}\right)^2+\left(-F_2+\bar{F}_y+\left(w-f_y^n\right)l_{ps}\right)^2}} \tag{3}$$

$$l_y = \frac{\left(-F_2+\bar{F}_y+\left(w-f_y^n\right)l_{ps}\right)l_{ps}}{EA} + \frac{\left(-F_2+\bar{F}_y+\left(w-f_y^n\right)l_{ps}\right)l_{ps}}{\sqrt{\left(F_1+\bar{F}_x+f_x^n l_{ps}\right)^2+\left(-F_2+\bar{F}_y+\left(w-f_y^n\right)l_{ps}\right)^2}} \tag{4}$$

2.3 Seabed element

In the slack configuration of mooring line, a significant portion of mooring line is lying on the seabed. The seabed portion of mooring line is modeled using seabed elements. The unstrained length of a seabed element is denoted by l_b. The projections of a seabed element ($i = n+1$) can be determined as follows:

$$l_x = -\left(1+\frac{\left(F_1+\bar{F}_x+f_x^n l_{ps}\right)}{EA}\right) l_b \tag{5}$$

$$l_y = 0 \tag{6}$$

3 Results and discussions

The influence of drag forces on restoring forces is investigated through the analysis of OC3-Hywind [5] mooring system subjected to an ocean current. The mooring system consists of three mooring lines. The diameter of a mooring line is 0.09 m, mooring line weight in water is 698.094 N/m, and the angle between adjacent mooring lines is 120°. Further details are given in Table 1.

Table 1. OC3-Hywind mooring system properties.

Parameter	Value
Depth to fairlead from mean sea level	70 m
Depth to anchor from mean sea level	320 m
Radius to fairlead from center line of platform	5.2 m
Radius to anchor from center line of platform	853.87 m
Axial stiffness of mooring line	384243000 N

The OC3-Hywind mooring system is subjected to ocean current of velocity 2.5 m/s. The angle θ represents the direction of ocean current with the respect to x-axis. The floating platform is quasi-statically displaced from -24 m to +24 m in a sway direction. The influence of drag forces on the surge, sway and heave restoring forces is demonstrated in Fig. 2 (a-c) for current velocities V = 0, 2.5 m/s, and θ = 0°, 45°. The figure shows that drag forces significantly influence the fairlead forces. The effect of drag forces on the surge and heave restoring forces is more pronounced. The effect of sway restoring forces is insignificant for θ = 0°, however, the effect is increased for θ = 45°. The comparison of results also demonstrates that the net effect of drag forces on total restoring forces depends on the direction of ocean current with respect to the orientation of the mooring lines. For example, the effect of drag forces on heave restoring is insignificant for θ = 0°, But for θ = 45° the influence of drag forces on heave restoring forces increased profoundly.

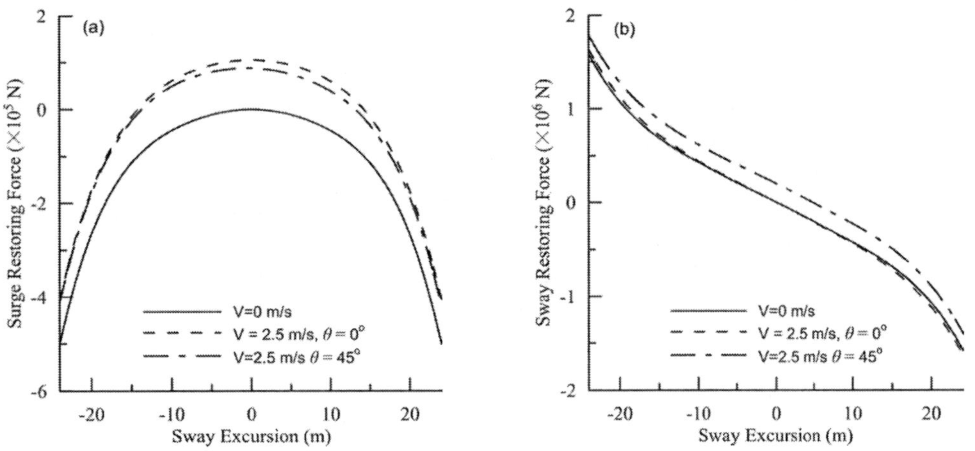

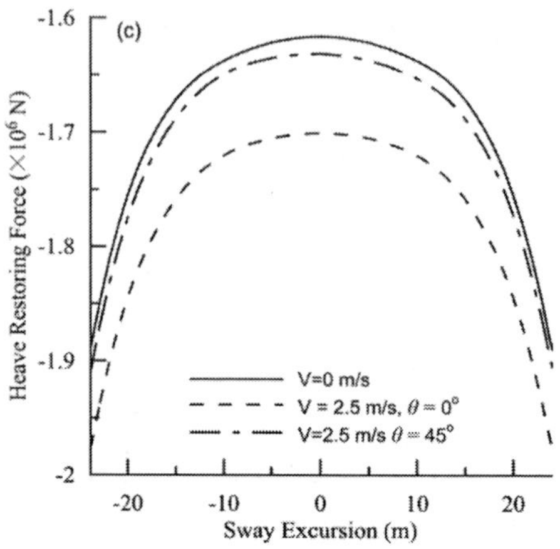

Fig. 2. Restoring forces vs sway displacement (a) surge restoring force, (b) sway restoring force, and (c) heave restoring force.

4 Conclusions

This study investigated the effect of ocean current drag forces on mooring system restoring forces. The mooring system restoring forces play a key role in the stability of floating wind turbine. Mooring lines are modeled using discrete catenary cable element. The effect of drag forces on mooring system restoring forces is investigated through the analysis of OC3-Hywind mooring system. It was found that the drag forces significantly influence the mooring system restoring forces. The effect of viscous drag forces in a specific direction i.e. surge, sway and heave depends on the direction of ocean currents with respect to mooring lines.

This research was supported by the Basic Science Research Program through the National Research Foundation of Korea (NRF), funded by the Ministry of Education, Science and Technology (NRF-2015R1D1A1A09060113). The authors wish to express their gratitude for this financial support.

References

1. N. Haritos, D. He, Modelling the response of cable elements in an ocean environment, Finite elements in analysis and design, **11**, 19-32, (1992)
2. Z. Ran, M. Kim, Nonlinear coupled responses of a tethered spar platform in waves. International Journal of Offshore and Polar Engineering, **7**, 111-118, (1997)
3. R.P. Singh, J. Verma, Evaluation of force-displacement relationship for multicomponent mooring cable by F.E.M., Computers & Structures, **30**, 1079-1089, (1988)
4. JM. Jonkman, Dynamics modeling and loads analysis of an offshore floating wind turbine, ProQuest; (2007)
5. JM. Jonkman, Definition of the floating system for phase IV of OC3, National Renewable Energy Laboratory Golden, CO, USA, (2010)

Investigation of Non-Uniform Rust Distribution and Its Effects on Corrosion Induced Cracking in Reinforced Concrete

Wahyuniarsih Sutrisno[1,*], *Priyo* Suprobo[1] , *Endah* Wahyuni[1] , *Data* Iranata[1]

[1]Department of Civil Engineering of Institut Teknologi Sepuluh Nopember, ITS Sukolilo Campus Surabaya Indonesia

Abstract. Uniform corrosion still widely used by a lot of researchers and engineers to analyze the corrosion induced cracking. However, in practice, corrosion process occurred non-uniformly. The part nearest to the exposed surface is more likely to have faster corrosion initiation compared with other regions. This research is mainly focused on investigating the effect of non-uniform rust distribution to cover cracking in reinforced concrete. An experimental test performed using accelerated corrosion test by using 5% NaCl solution and applied a constant electric current to the concrete samples. The rust distribution and measurement were observed by using a digital microscope. Based on the experimental result, it was found that the rust was distributed in a non-uniform pattern. As a result, the cracks also formed non-uniformly along the perimeter of steel bar. At the last part of this paper, a simulation result of concrete cracking induced by non-uniform corrosion is presented. The result compared with a simulation using uniform corrosion assumption to investigate the damage pattern of each model. The simulation result reveals stress evolution due to rust expansion which leads to concrete cracking. Furthermore, a comparison of stresses induced by non-uniform corrosion and uniform corrosion indicates that non-uniform corrosion could lead to earlier damage to the structure which is specified by the formation and propagation of the crack.

1 Introduction

Reinforcement corrosion induced by chloride ingress is a major durability problem in reinforced concrete structure. Furthermore, this condition can affect the serviceability and finally leads to damage of the structure. The corrosion process starts with depassivation. Depassivation is a condition when a passive layer of the steel is damage. This process followed with the formation of the rust at the interface between steel bar and concrete. The volume of rust will continue to increase as corrosion processes happen continuously. The volume of the rust can exceed 2-6 times the volume of the original steel, and it can generate outward expansive pressure which causes cracks in the concrete. The overall illustration regarding this process shown in Fig. 1.

* Corresponding author: niar1206@gmail.com

DOI: 10.1051/matecconf/201713802013

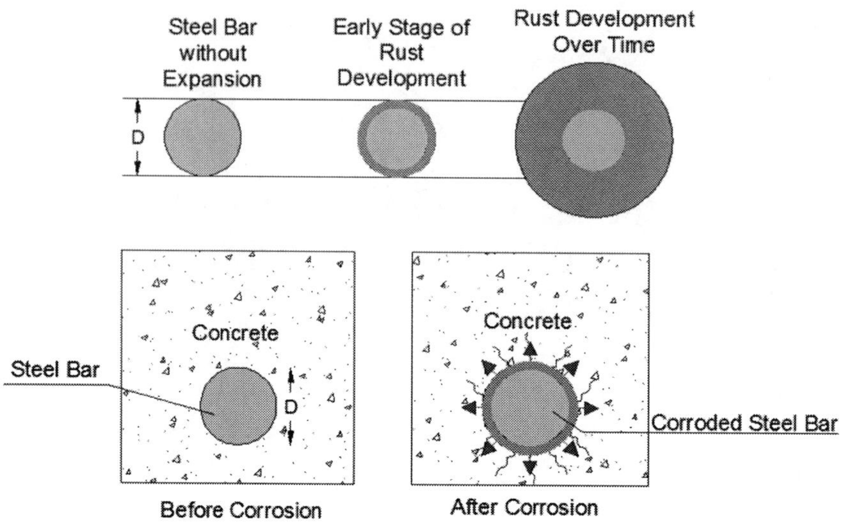

Fig. 1. Illustration of Corroded Steel Bar Induced Concrete Cracking in Reinforced Concrete

Crack on a structural element is undesirable because it can damage the structure by reducing the strength of the structure and increasing concrete permeability. It is also aggravating the bonding between steel and concrete. As the result of concrete cracking development, it can reach the concrete cover and cause failure to the structures. Even though any use of concrete as structural materials will cause a crack, but the crack is still acceptable as long as the width is still within the limits which are not harmful to the user. However, in several conditions, the crack can propagate continuously especially due to environmental influences such as corrosion which can endanger the users of the structure. Therefore, it is imperative to know the damage process of concrete induced by reinforcement corrosion to calculate and predicting the serviceability of the structures.

The total service life of reinforced concrete elements exposed to corrosion is commonly characterized as a two-stage process consisting of an initiation and a propagation period. The crack initiation process is defined as the time taken from corrosion starts until the rust cause crack on the concrete. Furthermore, the propagation period is defined as the time required for a crack to reach specified serviceability or ultimate limit states. Service life has conventionally been taken as equal to the duration of the initiation period, and subsequently, much of the research work has been focused on this area. However, this may be viewed as a conservative failure criterion because the end of the initiation phase only indicates the onset of corrosion and the structure is yet to suffer any effects that may obstruct its functional performance.

Several research have been performed to investigate the corrosion in reinforced concrete and its effects on concrete. Their studies include experimental, analytical and numerical analysis. The numerical and analytical models mainly focused on the development of a predictive model of corrosion induced cracking. The numerical and analytical model consider the realistic failure criteria of the concrete to indicate their structural service life. The end of concrete service life could be shown by depth cracking, cover spalling, and eventually structural collapse. The model usually assumed that the corrosion occurs uniformly around the circumference of the steel bar (Li, Lawanwisut, Zheng, & Kijawatworawet, 2005; Liu & Weyers, 1998). However, this condition did not represent the practical situation. On the real condition, the steel corrosion takes place at locations nearest to the exposed surface of the concrete structure. This situation leads to non-uniform rust distribution which can generate non-uniform expansive pressure and eventually leads to crack to the structures.

This paper provides an experimental investigation of rust distribution in reinforced concrete due to corrosion. The effect of rust distribution to the concrete cracking also discussed in this article. At the end section of this article, numerical simulation of uniform and non-uniform corrosion induced cracking is presented to get realistic comparison related to damage process of the concrete due to corrosion.

2 Corrosion Induced Cracking Process in Reinforced Concrete

It is critical to consider the cracking behavior of concrete cover for the assessment of the remaining strength and service life of concrete structures. Corrosion in reinforced concrete structure due to chloride ingress has a significant contribution to concrete cracking. Corrosion decreases the strength of the concrete by deteriorating the bonding between concrete and steel reinforcement. As a result, the ability of the reinforcement to provide tensile strength and ductility to the concrete decreased [1]. On the steel surfaces, there is a thin passive layer which is used to protect the steel from corrosion. However, when the harmful agent, such as chloride, present in the concrete, the passive layers of the steel can disappear and cause corrosion.

After the depassivation occurs, corrosion of the steel reinforcement is initiated by the reaction between the steel bar with oxygen and water. At the anode reaction, the discrete ions formed and it can react to an insoluble corrosion product. Corrosion reaction of metal are generally written with a simple reaction [2]

$$M \rightarrow M^{2+} + 2e^- \tag{1}$$

The passive layer of the steel will be lost will act as an anode in reaction process of corrosion. The anode reaction equation can be written as follows:

$$Fe \rightarrow Fe^{2+} + 2e^- \tag{2}$$

Sudjono [3] explained that the electrons released from the anode reaction cause the oxygen and water, which is located on the surface of the steel, react. The part of this steel becomes the cathode of reinforcing steel corrosion reaction, and the reaction process can be written as follows:

$$O^2 + 2H_2O + 2e^- \rightarrow 4OH^- \tag{3}$$

Both anode and cathode ions react and generate the corrosion product as the following reaction

$$2Fe + O_2 + 2H_2O \rightarrow 2Fe^{2+} + 4OH^- \tag{4}$$

$$2Fe^{2+} + 4OH^- \rightarrow 2Fe(OH)_2 \tag{5}$$

As the result of the reaction, $Fe(OH)_2$ will be formed as the original form of the corrosion product. Corrosion is a process which will continue if the concentrations of O_2 and H_2O are sufficient in the concrete. However, this corrosion process can take place more quickly when there are catalysts that can speed up the process of formation of the rust like the presence of chloride ions (Cl^-). The Cl^- ions derived from an aggressive environment such as the environment close to the sea water. Chloride essentially has no effect on the formation of rust, but this ion help to destruct the passive layer and accelerate the anode process which resulting speed up the corrosion process. The illustration of corrosion process is shown in Fig 2.

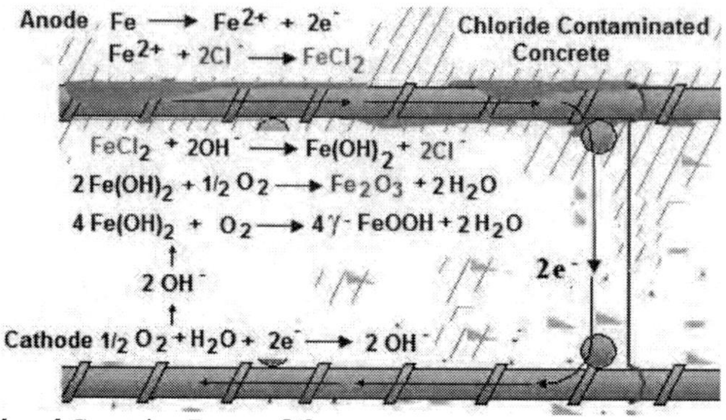

Fig. 2. Chloride Induced Corrosion Process [4]

In general, the occurrence of cracking in concrete due to corrosion on the surface of the concrete and steel covers three stages, which are initiation, expansion and crack stage (Liu and Weeyrs, 1998 in [5]). After the corrosion initiation process, which has been explained previously, the rust start to develop. The development of rust volume can reach two until six times of original steel bar volume. This condition makes the expansive pressure start to present and at one point cause crack on the concrete.

Analysis of cracks which occur in reinforced concrete due to corrosion during this still using the assumption that the corrosion products formed evenly along the perimeter of the reinforcement [2], [6]–[10]. Uniform corrosion assumption illustrate the corrosion products, also known as rust, to be a thick wall which formed around the perimeter of reinforcement and will provide uniform pressure on the expansive concrete.

However, in subsequent studies, it was found that the development of corrosion products (rust) is formed non-uniformly along the perimeter of steel reinforcement [11]–[14]. Research conducted by [12] stated that there the rust was accumulated at the certain parts around the perimeter of the reinforcement, particularly in part nearest to the exposed surface of the concrete which dealing with the outside environment. In this particular area, the aggressive ions such as chloride ions have more quick infiltration process to the concrete which can accelerate the corrosion process.

Crack analysis using the non-uniform distribution of the rust will provide more accurate analysis because by using this assumption, a greater concentrated expansive pressure will be generated on the cover. This condition will lead to an increase in the speed of crack propagation towards the concrete cover so that the analysis results can be obtained more precise and closer to the actual incident.

3 Experimental Setup

The summary of the experimental set up is provided in this section. The sample used in this paper is a concrete cube with dimension 150 x 150 x 150 mm with single reinforcement as shown in Fig. 3. The concrete material used in this experiment is normal concrete with water per cement ratio (w/c) 0.5. Several days before mixing, the coarse and fine aggregate was washed to remove any fine dust which may increase the water demand, and lower the bond strength. The wet aggregate was then placed in a sealed bucket for one day was allowed for absorption. This treatment can stabilize the moisture content of the aggregate and drain the excess water. After one day the aggregate was dried to reach SSD condition.

 DOI: 10.1051/matecconf/201713802013

Prior to mixing, the materials were weighed and batched into sealed buckets at least one day in advance.

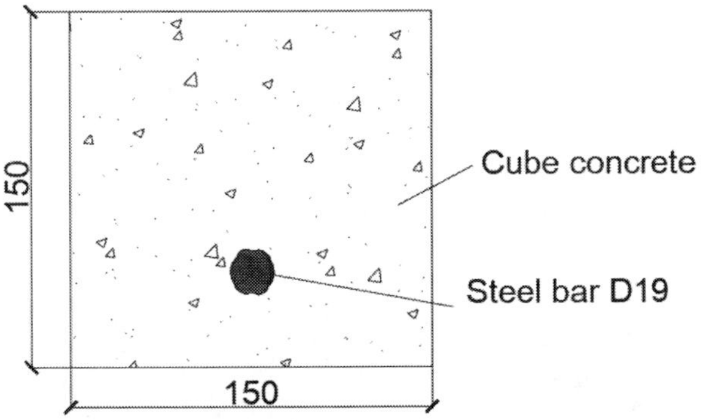

Fig. 3. Cross Section of the Model

The concrete was mixed by a mixer machine, and the special mold was prepared to allow the steel reinforcement embedded in the concrete. The concrete was embedded with single steel reinforcement with diameter 19 mm. The reinforcement was embedded 40 mm from the exposed surface. The concrete was demolded after 24 hours, and it was cured under wet burlap and plastic in the concrete lab environment. After 28 days, the accelerated corrosion test was performed.

The accelerated corrosion test was carried out to investigate the corrosion induced cracking process in reinforced concrete structures. The corrosion process, rust development, crack initiation and propagation were observed using daily visual observation. The accelerated corrosion process was achieved by applying a constant electrical current to the steel reinforcement by a power supply via a current regulator. The current regulator was used in this experiment to maintain constant current over the time. To generate cracks due to corrosion the accelerated corrosion test used approximately 200 $\mu A/cm^2$ of electric current. The concrete specimen placed in the 5% NaCl solution. The accelerated corrosion set up is shown in Fig. 4. The steel bar acts as an anode, and the stainless steel plate which submerged in the NaCl solution acts as the cathode. The pore fluid in the concrete is the electrolyte of the reaction. All of the samples were cut into four identical pieces to observed the rust distribution and crack propagation.

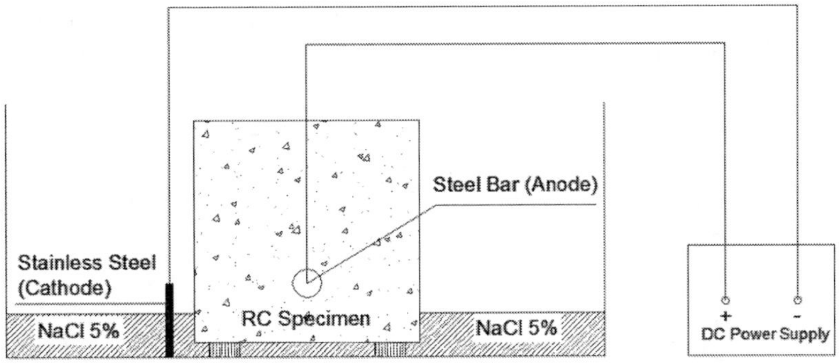

Fig. 4. Accelerated Corrosion Test Set Up

DOI: 10.1051/matecconf/201713802013

4 RESULT AND DISCUSSION

An accelerated corrosion test was performed to investigate the corrosion induced cracking process in reinforced concrete structures. The first corrosion and crack initiation at the concrete surface were observed using daily visual observation. The process of corrosion initiation is indicated by discoloration of the steel reinforcement. This process happens just 5 hours after the accelerated corrosion test started. As time passed, the corrosion starts to developed and cause bleeding of the rust. This condition indicated by discoloration of the NaCl solution as shown in Fig. 5.

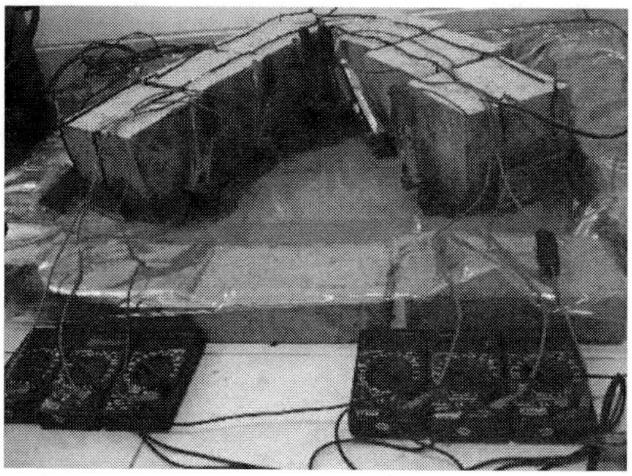

Fig. 5. Discoloration process of NaCl Solution due to Rust Bleeding

The samples were carefully sliced with a concrete cutter into four identical pieces to have a more detailed observation. Each section was then cut sequentially in the cross sectional direction. The rust distribution observed using photographic analysis to get accurate data regarding the thickness of the rust.

As stated at the beginning of this paper, the primary purposed of this paper is to investigate the rust distribution on the circumference of the steel bar and its effect to cracking. Therefore, the thickness of the rust was observed daily and plotted into several graphs as shown in Fig. 6. The perimeter of steel bar is divided into several polar coordinate, and the thickness is measured according to the polar coordinate. Thc non-uniform thickness indicates as symbol Nr on the graph and the polar coordinate (θ) is measured in degree. The closest part of the exposed surface is assume located at 180° and the furthest part located at 0°/360°.

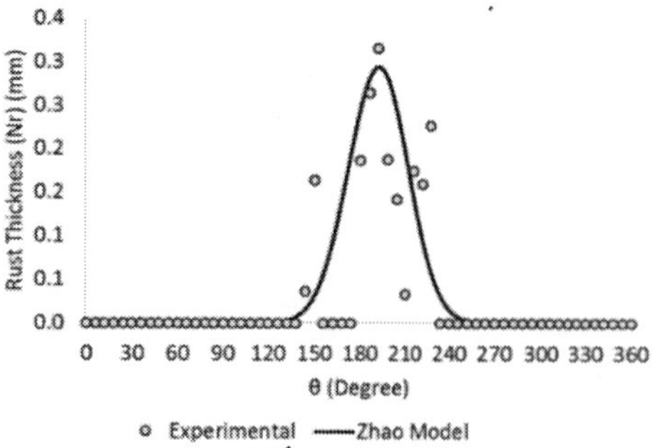

2nd Day

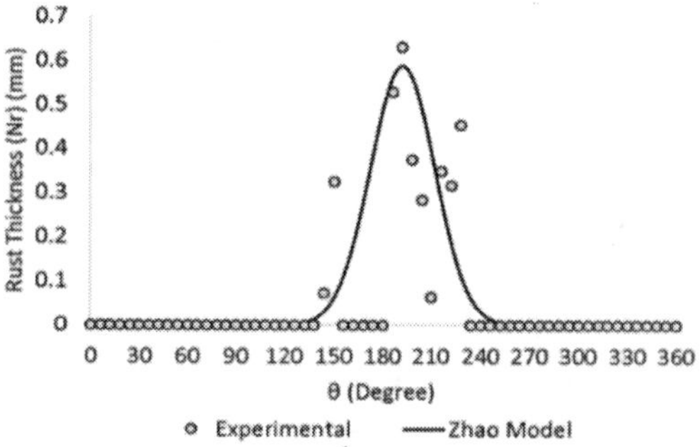

6th Day

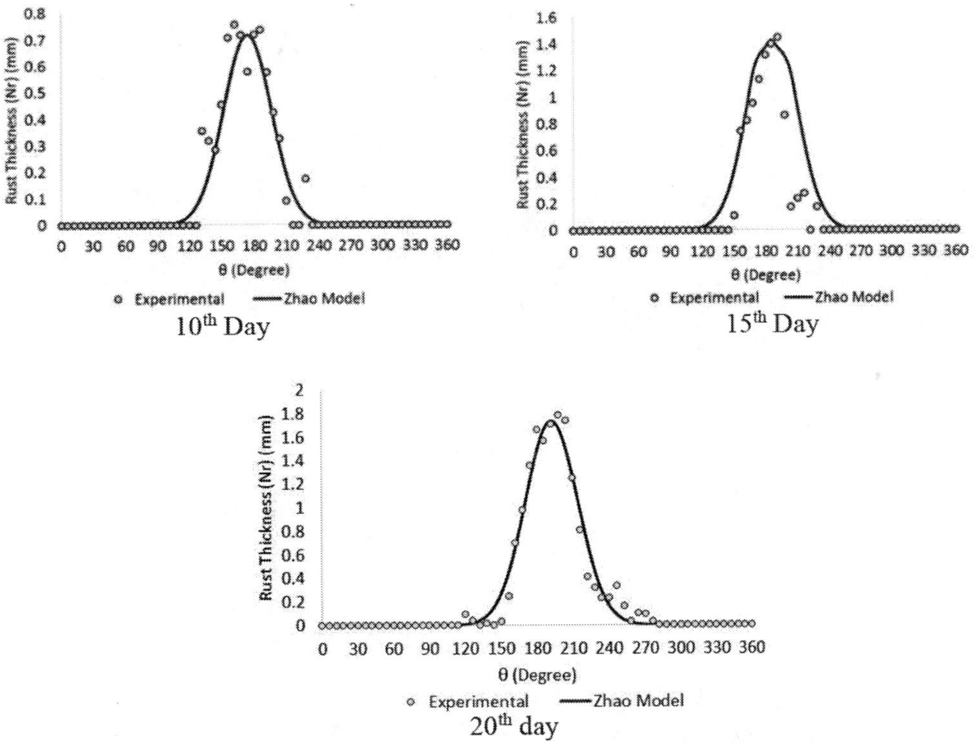

Fig. 6. Rust Distribution Graph from Experimental Result and Its Gaussian Description

In addition to the experimental result, the Gaussian description proposed by Zhao also presented on the Fig. 6. By using the Gaussian function, the non-uniform rust distribution can be idealized. Based on Fig. 6, it can be seen that the thickness of the rust layer is increase as an increase of time. This condition happened because the corrosion process is developed continuously over the time. It is also very clear that the rust distributed non-uniformly along the perimeter of the steel bar. Furthermore, the peak of the curve occurs at the area closest to the exposed surface as expected. This result shows that the area which has the closest distance to the exposed surface is more vulnerable to corrosion because the chloride from outside has faster infiltration to this location than other location.

The curve also showed that the peak of the rust thickness does not always occur at the 180°. The peak of the rust sometimes occurs at 150° or 210°. The reason behind this phenomena is the use of deformed reinforcement. The geometry of the reinforcement affects the distribution of the rust because it has whorl shape which makes it located closer to the exposed surface than the plain part.

The non-uniform rust distribution apparently affects the cracking propagation on the concrete. Due to non-uniform thickness of rust, the non-uniform expansive pressure is generated and crack start to initiate. The cracking development due to non-uniform corrosion is shown in Fig. 7.

Day 4th **Day 6th** **Day 20th**

Fig. 7. Crack propagation due to Non-Uniform Corrosion

The non-uniform crack propagation, as shown in Fig. 7, is developed as the effect of non-uniform rust distribution. The crack development shows that even though there are several cracks is formed in the bottom part of the concrete-steel bar interface, but only single crack manage to propagate into the concrete cover. After the crack occurs, the rust infiltrates into the crack tip as shown in Fig. 8.

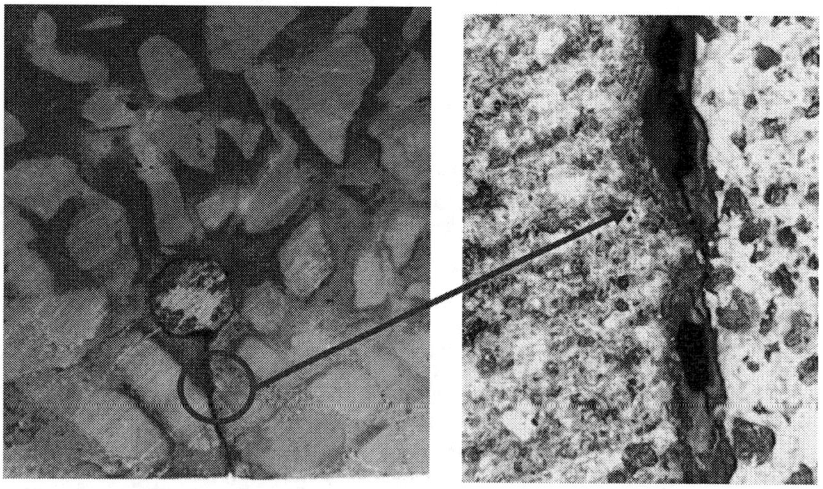

Fig. 8. Rust Distribution at the crack Tip

A finite element simulation is developed to compare the cracking pattern of due to uniform and non-uniform rust distribution. The simulation preformed using the concrete smeared cracking method. The material properties used in this model divided into two categories, elastic and smeared cracking properties. In smeared cracking approach there were three main parameters used to run the analysis such as tension stiffening, failure ratio, and shear retention factor.

The expansive pressure due to uniform and non-uniform corrosion was calculated using the equation proposed by Liu and Weyers [15]. Furthermore, for the non-uniform corrosion, the expansive pressure was calculated by combining the Gaussian model by Zhao and thick walled cylinder based equation by Liu and Wayers [14], [15]. In Liu and Wayers equation, the expansive pressure is affected by the thickness of the corrosion layer.

DOI: 10.1051/matecconf/201713802013

When non-uniform corrosion assumption is used, the thickness of corrosion layer was substituted by non-uniform thickness from the Gaussian function.

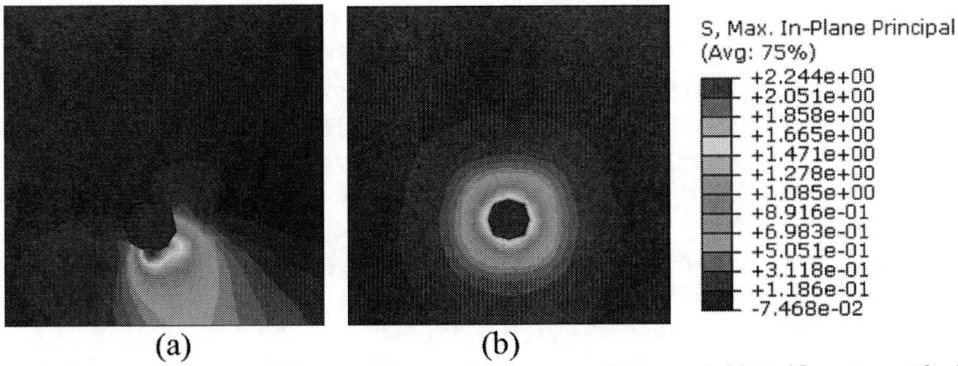

Fig. 9. Finite Element Modeling Result using (a) Non-Uniform and (b) Uniform Rust Distribution Assumption

The Fig. 9 shows stress distribution in the concrete due to expansive internal pressure which reflecting the distribution of the rust. The stress distribution describes the direction of crack propagation in the concrete. Fig. 9a shows that model with non-uniform expansive pressure will lead to a crack which not formed uniformly along the perimeter of the steel bar. This result is contrary to Fig. 9b which indicating crack will form uniformly along the perimeter of the steel bar with uniform expansive pressure. Fig. 9 also illustrate that the stress of the model with non-uniform internal expansive pressure is concentrated in one particular area which indicates the possibility of crack propagation. This result is in good agreement with the experimental result as shown in Fig. 8.

5 CONCLUSION

A series of the experimental test has been performed to investigate the cracking propagation in the concrete due to corrosion. The experimental test it clearly shows that the rust is formed non-uniformly along the perimeter of the steel bar. Even there were several cracks formed in the concrete along the perimeter of the steel bar, there is only one main crack manage to propagate to the exposed surface. The numerical simulation also performed in this paper, and it shows high agreement with the experimental result. The numerical model shows the differences between using uniform and non-uniform rust distribution assumption. The uniform rust assumption produced uniform expansive pressure and stress distribution along the perimeter of the steel bar as expected. However, non-uniform assumption shows that the stress is concentrated in one particular area. This condition could be taken as a representation of cracking propagation direction in the concrete. The non-uniform corrosion assumption can lead concrete to crack faster in the section closer to the exposed surface and vice versa.

6 ACKNOWLEDGEMENT

All of the authors acknowledge the financial support from Directorate General of Higher Education (DIKTI), Ministry of Education and Culture, Republic of Indonesia.

References

[1] C. Fang, K. Lundgren, L. Chen, And C. Zhu, "Corrosion Influence On Bond In Reinforced Concrete," Cem. Concr. Res., Vol. **34**, No. 11, Pp. 2159–2167, Nov. (2004).

[2] Y. Liu, "Modeling The Time-To-Corrosion Cracking Of The Cover Concrete In Chloride Contaminated Reinforced Concrete Structures," Virginia Polytechnic Institute And State University, (1996).

[3] A. S. Sudjono, "Prediksi Waktu Layan Bangunan Beton Terhadap," Civ. Eng. Dimens., Vol. **7**, No. 1, Pp. 6–15, (2007).

[4] X. Shi, N. Xie, K. Fortune, And J. Gong, "Durability Of Steel Reinforced Concrete In Chloride Environments: An Overview," Constr. Build. Mater., Vol. **30**, Pp. 125–138, May (2012).

[5] Y. Zhao, H. Xu, And W. Jin, "Concrete Cracking Process Induced By Steel Corrosion- A Review," No. 2, Pp. 1–10. (2013).

[6] C. Cao, M. M. S. Cheung, And B. Y. B. Chan, "Modelling Of Interaction Between Corrosion-Induced Concrete Cover Crack And Steel Corrosion Rate," Corros. Sci., Vol. **69**, No. 24, Pp. 97–109, (2013).

[7] C. Li, W. Lawanwisut, J. J. Zheng, And W. Kijawatworawet, "Crack Width Due To Corroded Bar In Reinforced Concrete Structures.Pdf," Int. J. Mater. Struct. Reliab., Vol. **3** No 2, No. September, Pp. 87–94, (2005).

[8] Y. Zhao, J. Yu, And W. Jin, "Damage Analysis And Cracking Model Of Reinforced Concrete Structures With Rebar Corrosion," Corros. Sci., Vol. **53**, No. 10, Pp. 3388–3397, Oct. (2011).

[9] Y. Zhao, J. Yu, Y. Wu, And W. Jin, "Critical Thickness Of Rust Layer At Inner And Out Surface Cracking Of Concrete Cover In Reinforced Concrete Structures," Corros. Sci., Vol. **59**, Pp. 316–323, Jun. (2012).

[10] Y. Zhao, J. Yu, B. Hu, And W. Jin, "Crack Shape And Rust Distribution In Corrosion-Induced Cracking Concrete," Corros. Sci., Vol. **55**, Pp. 385–393, Feb. (2012).

[11] C. Cao And M. M. S. Cheung, "Non-Uniform Rust Expansion For Chloride-Induced Pitting Corrosion In Rc Structures," Constr. Build. Mater., Vol. **51**, Pp. 75–81, Jan. (2014).

[12] S. Muthulingam And B. N. Rao, "Non-Uniform Time-To-Corrosion Initiation In Steel Reinforced Concrete Under Chloride Environment," Corros. Sci., Vol. **82**, Pp. 304–315, May (2014).

[13] Y. Zhao, B. Hu, J. Yu, And W. Jin, "Non-Uniform Distribution Of Rust Layer Around Steel Bar In Concrete," Corros. Sci., Vol. **53**, No. 12, Pp. 4300–4308, Dec. (2011).

[14] Y. Zhao, A. Karimi, H. S. Wong, B. Hu, N. R. Buenfeld, And W. Jin, "Comparison Of Uniform And Non-Uniform Corrosion Induced Damage In Reinforced Concrete Based On A Gaussian Description Of The Corrosion Layer," Corros. Sci., Vol. **53**, No. 9, Pp. 2803–2814, Sep.(2011).

[15] Y. Liu And R. Weyers, "Modeling Time To Corrosion Cracking In Chloride Contaminated Reinforced Concrete Structures," Aci Mater. Journals, Vol. **95**, Pp. 675–681, (1998).

MATEC Web of Conferences **138**, 02014 (2017) DOI: 10.1051/matecconf/201713802014
EACEF 2017

Concrete Filled Square Steel Tubular Deep Beam subjected to Bending-shear

Kojiro Uenaka[1,*] and *Hisao* Tsunokake[2]

[1]Kobe City College of Technology, Department of Civil Engineering, 6512194 Kobe, Japan
[2]Osaka City University, Department of Urban Engineering, 5588585 Osaka, Japan

Abstract. This study aims to investigate experimentally mechanical characteristics of S-CFT deep beam under bending-shear through symmetrically three-point loading test method with a constant axial load. Selected test parameters are shear span and the width-to-thickness ratios, which are ranging from 1.0 to 1.5 and 62.5 to 200, respectively. As the results, observed failure modes of the specimens with width-to-thickness ratio being 200 were local buckling of steel tube associated with shear failure of in-filled concrete between two supports. No effect of the constant axial load on bending-shear capacities can be found. The experimental strength of the specimens with shear span ratio between 1.0 and 1.5 can be predicted by the accumulative strength being composed of RC deep beam and web steel plate.

1 Introduction

Concrete filled steel tubular, CFT[1], member consists of hollow circular/square steel tube and filled concrete, as shown in Fig. 1. The CFT member holds good deformability owing to confined strength between the steel tube and filled concrete.

Reviewing the past studies on concrete filled square steel tubular, S-CFT, member, we can first find investigations of S-CFT member under bending-shear and axial load[2-3]. Next, Sakino et al.[4] carried out S-CFT stub column test under cyclic shear force test[4]. Nakahara[5] conducted flexural behavior of S-CFT beam column under two parameters, namely single and double curvature. However, width-to-thickness ratios being less than 50 of square steel tube of S-CFT is mainly discussed in the previously research. The detailed mechanical behavior of S-CFT member with the smaller width-to-thickness ratio is given in Ref. [1].

This study aims to investigate bending-shear characteristics of deep squared CFT(S-CFT) beam, the width-to-thickness ratios of which are 62.5 and 200, through symmetric three-point loading test methods with a constant axial load. The selected testing parameters are the shear-span ratio, width-to-thickness ratio and axial strength. Method to predict bending shear capacity with the large width-to-thickness ratio is given.

* Corresponding author: uenaka@kobe-kosen.ac.jp

DOI: 10.1051/matecconf/201713802014

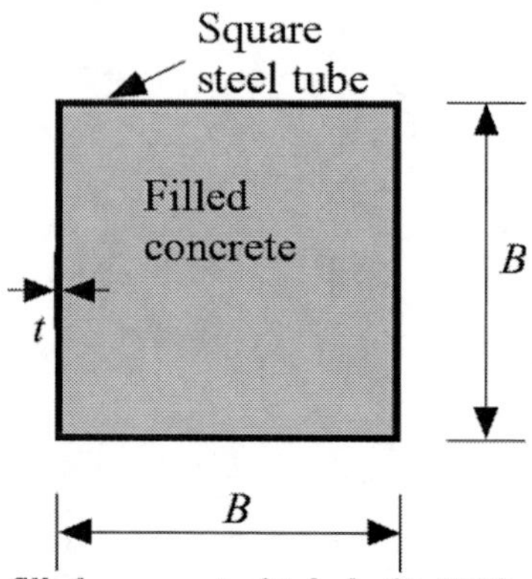

Figure 1. Cross section of concrete filled square steel tubular(S-CFT).

Table 1. List of the specimens.

No.	Tag	Square steel tube: Thick. t(mm)	Width B(mm)	Span length L(mm)	Width-to-thickness ratio B/t	shear-span ratio a/B	Yielding point f_y(MPa)	Fracture point f_u(MPa)	N/N_{cu}	Con-crete f_c' (MPa)
1	t32-s15-n02	3.2	200	600	62.5	1.5	353.0	428.4	0.2	27.5
2	t32-s15-n01	3.2		600	62.5	1.5	353.0	428.4	0.1	27.5
3	t32-s10-n02	3.2		400	62.5	1.0	353.0	428.4	0.2	27.5
4	t32-s10-n01	3.2		400	62.5	1.0	353.0	428.4	0.1	27.5
5	t10-s15-n02	1.0		600	200.0	1.5	216.5	320.0	0.2	27.5
6	t10-s15-n01	1.0		600	200.0	1.5	216.5	320.0	0.1	27.5
7	t10-s10-n02	1.0		400	200.0	1.0	216.5	320.0	0.2	27.5
8	t10-s10-n01	1.0		400	200.0	1.0	216.5	320.0	0.1	27.5

Note : a=shear span length(=L/2), N=axial load, N_{cu}=ultimate axial force

2 Test method

2.1 Test specimens

Table 1 summarizes the detail of the test specimens. The specimens have commonly square steel tube's width B= 200mm. Thicknesses of the steel tubes (t) are 1.0 and 3.2mm. Therefore, width-to-thickness ratios(B / t) are 62.5 and 200, respectively. Two shear span lengths(L=2a) are 400 and 600mm, thus, shear span ratios(a/B) are 1.0 and 1.5.

Axial strength ratio of the S-CFT deep beam is described in Table 1, the N_{cu} in which is calculated from below.

$$N_{cu} = f_c' A_c \tag{1}$$

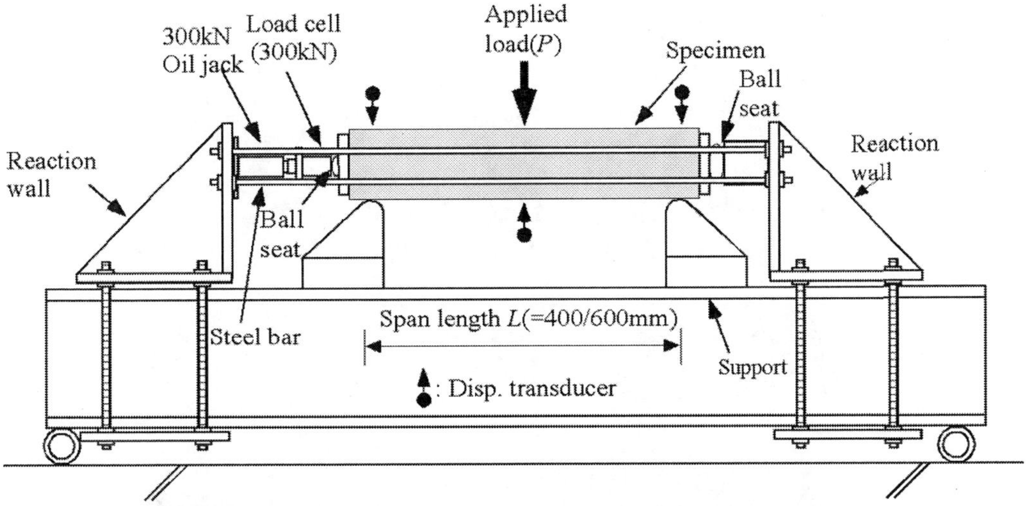

Figure 2. Test apparatus.

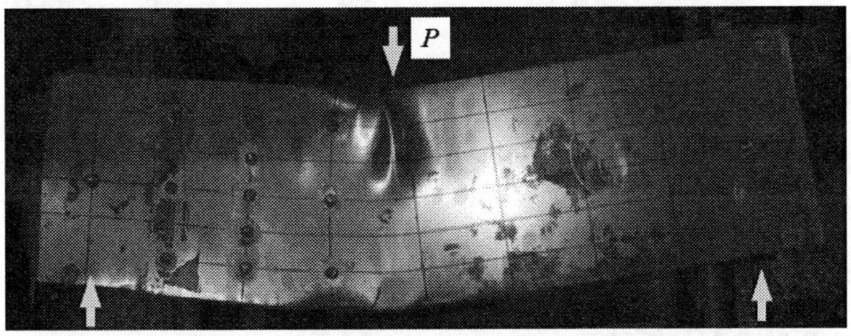

a)Square steel tube

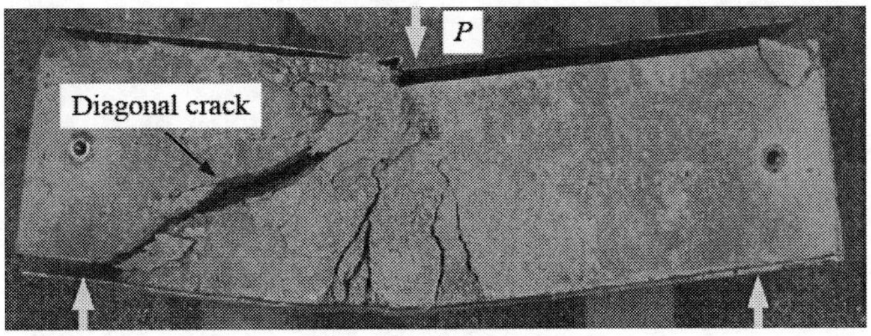

b)In-filled concrete

Figure 3. Failure mode(t10-s15-n02).

where f_c' and A_c are concrete strength and its cross-sectional area.

Figure 2 shows loading method of this investigation. The symmetric three-point loading test method was employed. The constant axial load can be introduced to the specimen by the PC steel bar and two reaction walls.

2.2 Measurements

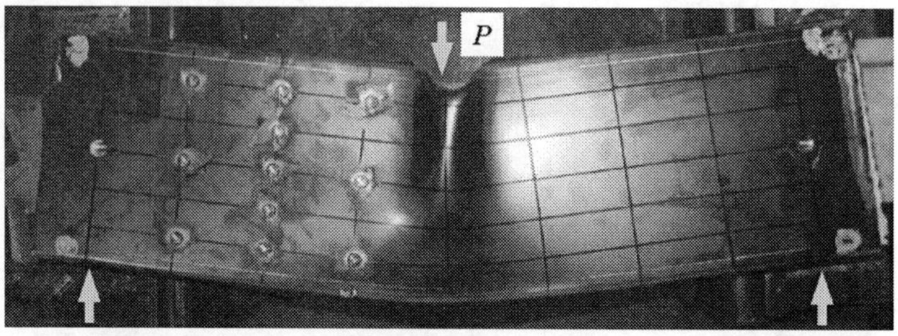

a)Square steel tube

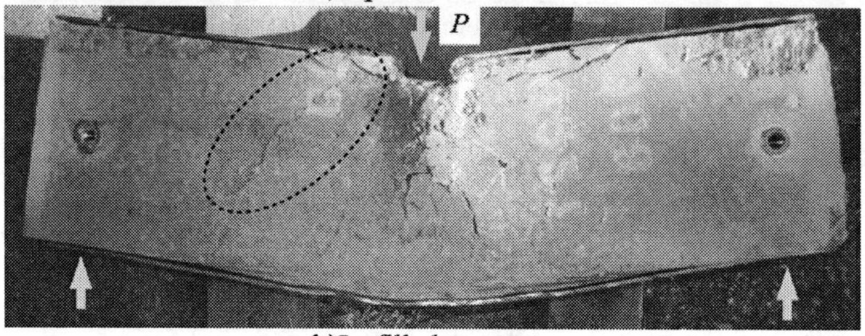

b)In-filled concrete

Figure 4. Failure mode(t32-s15-n01).

Three displacement transducers were arranged under the mid-span of the specimen and upper of the two supports under the constant axial load(N), as shown in Fig. 2. The monotonic bending-shear action was applied to the specimens until fatal failure.

3 Results and discussion

3.1 Failure modes

Figure 3a) and b) show the detailed failure of the specimen with B/t being 200. The local buckling of square steel tube and tensile cracking of steel tube occurred. The large diagonal crack of in-filled concrete can be found clearly.

Figures 4a) and b) show the failure mode of the specimen t32-s15-n01. Local buckling of steel tubes and bending cracking of steel tube can be observed. In addition, compressive failure of in-filled concrete at the loading point occurred as shown in Fig. 4b). Whereas, no large diagonal crack of in-filled concrete appeared in the specimen with B/t being 62.5.

3.2 Load-displacement

Figures 5a), b) provide the relationship between applied load (P) and the angle of rotation of the deep beam(θ), namely, central displacement (δ) divided into shear span length a ($=L/2$). From the figures, it can be observed that applied loads(P) kept maximum the angle θ=0.04. It should be noted that no effect of axial strength on deformability of S-CFT deep beam can be found.

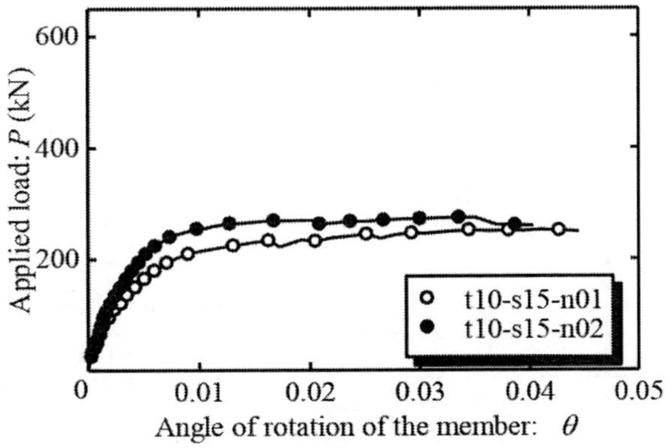

a)t=1.0mm

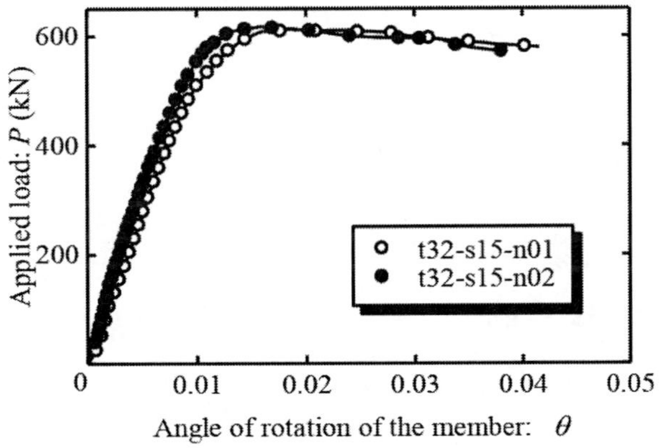

b) t=3.2mm

Figure 5. Applied load and angle of rotation.

3.3 Shear strength

3.3.1 Predicted method

According to JSCE[6], shear strength of RC beam can be expressed as below,

$$V_{cdd} = \beta_d \cdot \beta_n \cdot \beta_p \cdot \beta_a \cdot f_{dd} \cdot D_c \cdot d \tag{2}$$

where,

$f_{dd} = 0.19\sqrt{f_{cd}'}$, $\beta_d = \sqrt[4]{1000/d}$, $\beta_n = 1 + 2M_0 / M_{ud}$, $\beta_p = \dfrac{1+\sqrt{100 p_v}}{2}$,

$\beta_a = \dfrac{5}{1+(a/d)^2}$ and $p_v = \dfrac{t}{d}$

where f_{cd}': concrete strength($f_{cd}'=f_c'$), d: effective depth(B - t/2, mm), D_c: depth of in-filled concrete(B-2t), M_{ud}: pure bending strength and M_o: bending moment obtained from below.

Table 2. Experiment and estimation.

No.	Tag	V_{exp} (kN)	V_{est} (kN)	$\frac{V_{exp}}{V_{est}}$
1	t32-s15-n02	309	313	0.99
2	t32-s15-n01	307	309	0.99
3	t32-s10-n02	365	388	0.94
4	t32-s10-n01	347	377	0.92
5	t10-s15-n02	137	141	0.97
6	t10-s15-n01	126	124	1.02
7	t10-s10-n02	184	211	0.87
8	t10-s10-n01	197	184	1.07

$$M_o = \frac{N}{6}B \quad (3)$$

Next, buckling strength of steel tubes proposed by JSSC[7] is expressed as below,

$$V_{su} = \tau_{cul} \cdot A_w \quad (4)$$

where A_w: cross-sectional area of web and τ_{cul} can be expressed as below,

$$\tau_{cul} = \frac{f_y}{\sqrt{3}} \quad (R_r < 0.60) \quad (5a)$$

$$\tau_{cul} = \left(\frac{0.6}{R_r}\right)^{0.32} \phi_p \cdot \frac{f_y}{\sqrt{3}} \quad (0.60 < R_r) \quad (5b)$$

where ϕ_p is the resistant factor. The parameter of width-to-thickness ratio(R_r) of square steel tube can be calculated from below,

$$R_r = \frac{1}{\pi}\sqrt{\frac{12(1-\nu^2)}{k}} \cdot \sqrt{\frac{f_y}{3E_s}} \cdot \frac{D}{t_w} \quad (6)$$

where E_s and ν are Young's modulus and Poisson's ratio of the square steel tube. The buckling factor k can be obtained from below.

$$k = 5.34 + 4.00(D/a)^2 \quad (a/D > 1) \quad (7a)$$

$$k = 4.00 + 5.34(D/a)^2 \quad (a/D \le 1) \quad (7b)$$

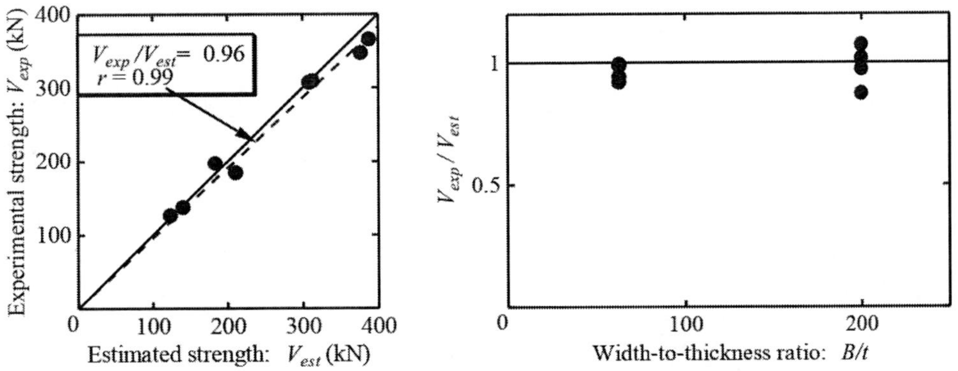

a)Estimations and experiment b) V_{exp} / V_{est} and B/t

Figure 6. Shear strength.

Last, the predicted shear strength V_{est} is sum of shear strength of deep RC beam(V_{cdd}) and buckling strength of V_{su}, namely,

$$V_{est} = V_{cdd} + V_{su} \tag{8}$$

Table 2 compares the experimental and calculated strengths, V_{exp} and V_{est}, respectively.

3.3.2 Comparison of experimental shear strength

Figure 6a) shows the relationship between experimental shear strength V_{exp} and estimated strength V_{est} calculated by Eq. (8). From the figure, The experimental strength V_{exp} is in good agreement with estimated strength V_{est} , namely, with the ratio V_{exp} / V_{est} =0.96 and good correlation factor r =0.99.

The strength ratio(V_{exp} / V_{est}) is plotted against width-to-thickness ratio B/t as shown in Fig. 6b). No effect of the ratio B/t on strength ratio V_{exp} / V_{est} can be found.

4 Concluding remarks

Eight specimens of concrete filled square steel tubular(S-CFT) deep beam subjected to bending-shear under constant axial load were examined. From the results, following remarks can be drawn.

1. Observed failure modes are local buckling of square steel tube with/without shear failure of filled concrete.
2. No effect of the constant axial load ratio on bending-shear deformability on S-CFT deep beam can be found.
3. The ultimate bending-shear capacity of S-CFT deep beam can be correctly predicted by sum of the same method of RC deep beam and steel buckling strength.

References

1. Architectural Institution of Japan: Recommendations for Design and Construction of Concrete Filled Steel Structures, Maruzen(2008)(in Japanese)
2. M. Tomii, K. Sakino, Transactions of the Architectural Institute of Japan, **275**, 55-65(1979)
3. M. Tomii, K. Sakino, *Transactions of the Architectural Institute of Japan*, **280**, 111-120(1979)
4. K. Sakino, H. Ishibashi, *Transactions of the Architectural Institute of Japan*, **353**, pp. 81-91(1985)
5. H. Nakahara, K. Sakino, *Proc. of the 12th World Conference on Earthquake Engineering*, 1923(2000).
6. Japan Society of Civil Engineers: Standard Specifications for Concrete Structures, Design, Maruzen(2007)(in Japanese)
7. Japan Society of Civil Engineers: Design Code for Steel Structures, Part B; Composite Structures, Maruzen(1997)(in Japanese)

Dispute resolution methods in the construction industry sector in the Kingdom of Saudi Arabia

Saad Alshahrani[1,*]

[1] *School of the Built Environment, University of Salford, 4th Floor, Maxwell Building, M5 4WT*

Abstract. The construction industry sector in the Kingdom of Saudi Arabia has undergone significant developments in the economics field, in terms of building tunnels, bridges, railways, airports, universities, etc. It is known that any project can involve disputes between the owner, consultant, and contractor of the project. Therefore, the researcher of this study investigated the methods of dispute resolution used in the economics sector in Saudi Arabia. These methods are negotiation and litigation. The researcher used mixed methods to collect the data. He interviewed 15 experts, academics, and arbitrators and then distributed a questionnaire to 327 experts, academics, and arbitrators. After that, the researcher analysed the gathered data by using both quantitative and qualitative methods. The researcher concluded that litigation is the method commonly used in the construction industry sector to solve disputes, and it is rare to use negotiation method. Also, the researcher found out that alternative dispute resolution methods are rarely used because the public work contract prevents the use of methods other than litigation.

1 Introduction

According to [1], there are three predominant categories of dispute resolution methods in the construction industry, (i) litigation, (ii) arbitration and (iii) alternative dispute resolution (ADR), a process which allows parties to come to a settlement based on their own negotiations, rather than a formal litigious procedure. These methods include negotiating, mediating and arbitrating. According to [2], the increasing cost of litigation has made quicker and less expensive methods of dispute resolution more attractive to parties. Expense not only include the compensation to be paid on settlement, but other financial losses sustained while conducting a construction dispute [3], which could be very costly for the losing party [4]. [5] point out that in order to improve the effectiveness of project organisations in general, it is vital to reduce such non-value-adding costs.

1.1 Litigation

Litigation can be basically described as the system of dispute resolution facilitated by the state judicial system. It is generally a complex and formal process, regulated by a substantial number of rules and procedural requirements that may vary based on the state or county of the judicature [6]. The

*e-mail: saad.888@hotmail.com

parties will find themselves in need of lawyers, solicitors and counsel for assistance in the presentation of their arguments to the courts and the judiciary, for such professionals are generally only those with such rights of audience and expertise in navigating the rules and their compliance. The process and personnel are not renowned for their perceived value for money and time expenditure. The proceedings are open and public, inviting scrutiny and an audience of those members of the public and commercial competition who find interest in the particulars of the dispute [7]. The speed of resolution, due to the complexity of the proceedings and rules, is not conducive to the effective continuation and completion of time important construction projects; research by [8] in India revealed that, on average, it takes between five and fifteen years to reach an adequate resolution in court. A court trial is a considerably more formal process than ADR, slower process and guided by stringent rules of procedure and evidence [9].

1.2 Alternative Dispute Resolution

ADR can be defined as "any method by which conflicts and disputes are resolved privately and other than through litigation in the public courts" [10]. In its various adaptable forms, it is a way of solving disputes without going to trial, and indeed almost 90 per cent of potential court cases are resolved by ADR prior to trial [9]. A considered advantage for the parties is that ADR is a method of considering arguments in a significantly less formal manner than a court case, and less intimidating to the parties involved [9]. Disputes tend to be determined more quickly and at considerably less expense in money and time, whilst still being assessed by experienced professional adjudicators [9]. Where agreed by the parties, the process has potentially the same binding, considered effect as a court verdict, and indeed may be filed at court for enforcement as a final judgement [9]. ADR is evidently has advantages in time and cost savings in a more flexible conciliatory atmosphere [11].

2 Dispute Resolution (Qualitative Method)

2.1 Litigation

Litigation remains a common method in the resolving of disputes in both private and public sectors of Saudi construction enterprises according to respondents. Anecdotal opinion abounds from the research contributors. Interviewee A11, supported by the opinions of Ac1, A12 and A9, asserted "most of construction projects disputes in public and private sectors are solved by litigation. This is because the legal system in the country does not allow using any other methods to solve any disputes in public sector. This is possible in private sector. The reason is cultural awareness". A7 stated "all of the disputes in Saudi construction projects in private as well as in public sectors are resolved by means of litigation and that is a waste of time and money. Some cases last as long as five years and the cost of lawyer and expert reporting is high".

2.1.1 Experts' Reports

Expert reports dealing with particular elements of disputes form a significant part of litigation in Saudi Arabia to assist a judge in his decision-making on matters which are unlikely to be within his own expertise in the construction industry. Ac1 asserts that "in most of the disputes in Saudi construction projects, the judge will ask for assistance from the Consulting Engineering Office (CEO) and engineer experts". A7 confirms this is common practice in court based litigation, and "the expert is usually a representative of the CEO and Engineering Arbitration Centre in the SCE". Interviewee A13 adds: "the Engineering Arbitration Centre in SCE tries to keep in regular, timely communication with the

DOI: 10.1051/matecconf/201713802015

Ministry of Justice to ensure that the forwarded report will be of the quality required, particularly because arbitrators and experts dealing with the opinions therein have high levels of experience and capability under the umbrella of specialised engineering project administration." A11 is however critical of some purported CEO expertise in the area of problem analysis in particular project problems due to lack of their own professional capabilities and conformity with the requirements of a professional body, He adds that the cost of experts' reports is less compared to those of the SCE, suggesting less experience and professionalism in their preparation.

2.2 ADR

In this section different methods of Alternative Dispute Resolution utilised in the Saudi construction projects are considered separately.

2.2.1 Negotiation

In Saudi construction project disputes, negotiation is used in both public and private sectors. A12 notes "that is free of charge and its duration may not exceed a month". It is however apparently rarely used to its full beneficial effect, according to E13. Interviewee A6 added that: "through my long experience, I have not come across negotiation as a means of resolving disputes in private and public sectors but litigation was mainly used as method of resolving disputes". Interviewees believed that resolving disputes by means of negotiation differs from other methods in that does not take long time and the cost is low compared with other methods; A4 adds "if the both sides of a dispute started using negotiation more often, they may no more be in need for other methods". E2 concurs: "negotiation is a developed method and it is based on knowledge and understanding of both parties of disputes. In my perspective, it should be considered as a preliminary method of resolution and it may take only a few hours and will not cost anything at all".

2.2.2 Mediation

Mediation is more broadly used in the private sector in Saudi construction projects rather than in public disputes in which the state or its representatives are involved. It is acknowledged by most respondents as the most effective method of solving disputes in the SCP. E15 states "mediation is a method of much flexibility and it is simple. In my experience, its duration is short and it does not exceed more than 3 months. It costs little. Sometimes it is as little as a tip. It is not used very much. There is no reason for not using it, unless the social culture is preventing that". Ac1 says "mediation is the best alternative disputes resolution, but regrettably arbitrators or experts who try to use mediation feel embarrassed. Since all parties anticipate in the mediation free of charge, this is a negative point for mediation in my point of view." It appears that what is preventing mediation to be used to its full advantage is the lack of utility in public contract projects. The advantages are evident from the anecdotal comments of research respondents. A6, for example, states "mediation is good in solving disputes in Saudi construction projects, and it is the best of all in my perspective, but it is being rarely used. It may take about a month to resolve problems and that is one of its best characteristics". A12 argues "at the moment we need some method other than litigation to solve disputes, such as mediation which is characterised with flexibility, less cost and most importantly, a shorter timescale. Mediation always takes less time". A4 agrees, but suggests it is less used than it should be because of perceptions that there is a "lack of arbitrators and experts' knowledge about this method and its benefits". In the context of Saudi Arabia, interviewee A3 adds "mediation is a good method and with flexibility but is hardly used and that is because of lack of support from Ministry of Justice for its use."

The use of the Dispute Adjudication Board (DAB) as method of alternative dispute resolution is not currently used to its full potential in Saudi construction projects, although eight of the research participants believe it has many potential advantages in this arena. A12, an arbitrator, describes how it would operate; "based on the experiences I had in Saudi Arabia, where I was member of the board and I was involved in this method of handling disputes and from the outset of a project, the dispute arbitration board starts to play its role. The number of its panel members for the hearing of a dispute is uneven, to avoid stalemate. It helps avoid halting a project because all of the disputes will immediately be handled on the site." Interviewee Ac1 stresses speed and flexibility and perhaps the most important positive point of this method is its conduct on site and consequent awareness of dispute details; necessary documentation is also within the dispute arbitration board's reach. The Dispute Arbitration Board has a considerable impact in the quick resolution of problems, but is not currently used effectively in Saudi Arabia.

2.2.3 Arbitration

Arbitration is only used to any significant extent in private sector of Saudi construction projects. Interviewee A11 comments "arbitration is a method that is only used in private sector as the system applied by the government does not allow any other methods but litigation to be used for solving disputes in SCP. As a result, no one generally refers to arbitration to resolve a dispute unless they are semi-state companies or experienced companies or businessmen. Sometimes it is used to save reputation of the business involved in the dispute, a cultural issue to many prominent businessmen who value social status." Interviewee E5 admires the arbitration method and says: "arbitration is an effective method in solving disputes in SCP and is better than litigation. It is less expensive than litigation, and also it takes less time than litigation does. In most of the cases it will not take more than six months". Not all respondents were so effusive in their admiration.

E8 is more critical of arbitration; he complains of high cost to private business, potentially more so than litigation, the cost of which is largely borne by the state. As a relatively formal process compared to other methods of ADR it can be a lengthy process. E15 agrees; "arbitration is within the framework of official litigation and we prefer those alternatives with more flexibility which are far away from being officially imposed and highly expensive measures, which involve longer procedures for both sides of dispute". The time taken by the arbitration process is commented upon by A3; "time is considered as negative point of arbitration since it takes a long time to resolve a dispute in SCP. I should remind you that I took part in arbitration for a case that took three years and this is not shorter than the time litigation takes to resolve a construction project dispute." Interviewee A4 asserts that "one of characteristics of the arbitration method is that it is based on obligation with no executive power. After arbitration has concluded, litigation is still likely to be involved and, due to some prohibition in a country, for example, usury, the judge may reject what has been concluded during the arbitration process. Some domestic or international companies therefore pursue their case to the Gulf Arbitration Centre in Bahrain, or to the Paris Bureau, or Arbitration Centre in London".

Interviewee A9 casts doubt on the character of the arbitration process, asserting "arbitrators should be impartial and fair towards the disputes, but they tend to turn into a lawyer for advice, especially when three or five arbitrators are involved. This is why many companies avoid arbitration and lean towards lawyers since the duty of lawyer is clearly known and the financial expenses are clear from the outset, whereas there is little clarity about arbitrator's wages." On the broader issue of cost, interviewee E10 states: "Arbitration is a costly method, compared to others. It is not easy to take part in it due to the expenses which will be incurred". He adds that there can be a religious or faith element to adjudications, pointing out 'if someone lacks any legal or religious background, he will not be able to occupy a position as an arbitrator; he will only act as an expert and his duty will be limited to expertise reporting, without being allowed to do arbitrating, or take part in a dispute resolution. In my opinion, an expert engineer with long years of experience who has taken part in solving disputes has the ability to take part in any engineering disputes".

The arbitrator (A4) said: There is a live practical example to solve disputes resolution, in terms of construction project needs in Saudi Arabia. A contractor who possesses a big company issued some complaints to an owner. This project consisted of residential buildings with 450 residential units, worth 120 million Saudi Riyal (SAR). The duration of the project was 450 days and the amount claimed by the contractor was 30 million Saudi Riyal (SAR) from the owner. The causes of dispute were numerous and are listed as follows:

- First cause: The owner requested the contractor install extra exit doors for each unit, giving an average of 450 exit doors.
- Second cause: The owner requested the contractor to change the water heater, which was not specified in the original contract.
- Third cause: There was a difference in the real area of the buildings' basement as it was 120 m2 in whilst the design stipulated that it was 140 m2 in real.
- Fourth cause: The contractor asked the owner to extend the time duration for the project, because the owner and consultant delayed approving the materials to be used for the project, and so this impacted on the project timing.
- Fifth cause: The change of consultant during the project by the owner had an impact on the overall timing of the project.
- Sixth cause: A delay by the owner in the time take to provide the contractor with the locations, as there were three separate locations.
- Seventh cause: The differences on the level of the locations with what was specified in the design, as some of the locations were wasteland belonging to the municipality, and so therefore this required extra work from the contractor to level the ground properly.

In return, the owner had some claims from the contractor, and these were as follows: **First claim**: He requested for the implementation of a 10% penalty for delaying compensation based on the contract. **Second claim**: He requested the cost for transferring the remaining project for the rubbish left behind from the contractor, as the owner claims that he hired a subcontractor to accomplish the outstanding work.

Resolving these type of causes of dispute were accomplished through the arbitration method and the final verdict was as follows: These claims and disputes stopped the project for a whole year, and then the disputes were resolved through the arbitration method, which resulted in a ruling in favour of the contractor for the sum of nine millions Saudi riyals (SAR), while the ruling in favour of the owner was for 12 million Saudi riyals (SAR), and a fine for 10% of the contract value.

Figure 1 illustrates that litigation is a means of dispute resolution in Saudi construction projects in both the public and private sector it is the dominant method used by parties in disagreement, especially because most of the large construction projects in the KSA are government sponsored. Simple negotiating is used rarely in both sectors according to research respondents, and mediation used rarely in the private sector, arguably because common practice in dispute resolution is to turn to the traditional court litigation. Arbitration on domestic Saudi based contract disputes is not used in the public sector. Negotiation, mediation, and DAB have clear advantages in terms of speed of resolution and cost which prevent more than necessary disruption to the construction progress; the respondents tended to agree on the characteristics of cost effectiveness, flexibility and time. There was some discrepancy in responses on the time and cost issues of arbitration, with profoundly divergent opinions on expense

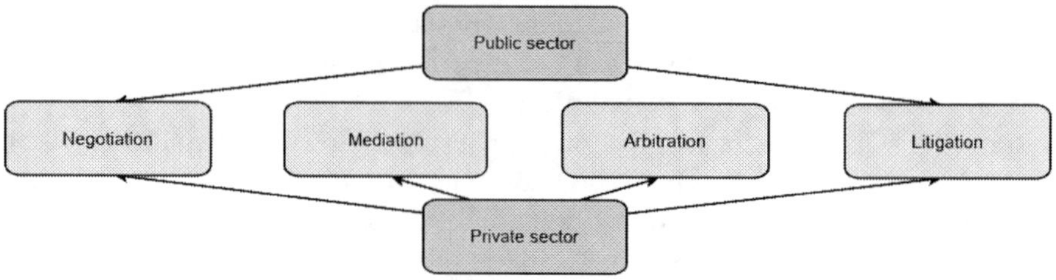

Figure 1. Method of Dispute resolution in Saudi Construction Project.

and length of proceedings reflecting respondents own experiences rather than, perhaps, theoretical perceptions.

3 Dispute Resolution (Quantitive Method)

In this question, a comparison on the approaches used for solving disputes for Saudi Construction Project in the public sector between participants from the public sector, private sector, and those participants who work in both private and public sectors was conducted. Resolutions in public sectors were through litigation and negotiation, according participants from public, private sectors and those who work in both sectors. The figure litigation was over 75%, with that of negotiation in all sectors at less than 40%.

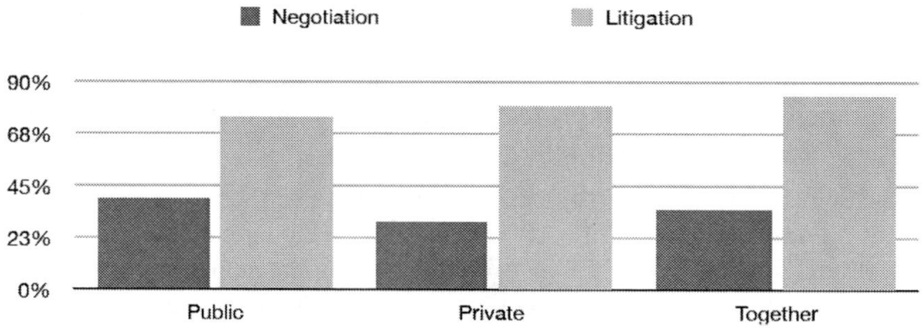

Figure 2. Comparing the percentage of use DRM and the participants who work in the public sector.

Figure 2 illustrates that the approach most commonly used in Saudi Construction Project in the public sector for dispute resolution was litigation which had the highest proportion of 85% according to the participants working in both private and public sectors. Participants working in the private sector considered disputes resolved by litigation to be less than that estimated by those working in both private and public sectors, which stood at 80%. Participants from the private sector believed that the method of disputes resolution used for by litigation in the Saudi construction project in private sector was less, standing at 75%. It is clear that disputes in Saudi construction project in the public sector were mainly resolved by means of negotiation according to participants working in both sectors, with the value standing at 40%. According to participant working both sectors, the method of resolving disputes used in the Saudi construction project public sector was less than what was stated by participants from the public sector, with a figure of 35%. According to participants from the private sector, disputes in the public Saudi Construction Project were resolved through negotiation, standing at 30%.

In this question, a comparison of the approaches used for DRM for Saudi Construction Project in the private sector between participants from the public sector, private sector, and those participants working in both the private and public sectors was conducted. It illustrates that disputes in the public sector were resolved through negotiation, litigation and mediation, according to participants from public, private sectors and those working in both sectors. DRM in the private sector use litigation according to participants working both private and public, having a high value at 86%, while the DRM used in private sector for Saudi Construction Project is through arbitration according to participants, standing at 40% for participants working in both sectors.

Figure 3 illustrates that all of the participants, regardless of the sector, agreed that using litigation as means resolving disputes is the most common, at 86% for participants working in both sectors and it is higher when compared with the figures from those working in private sector and public sector. Participants from the public and private sectors were less than those working in both sectors. Their numbers were similar, at 67% and 66% respectively.

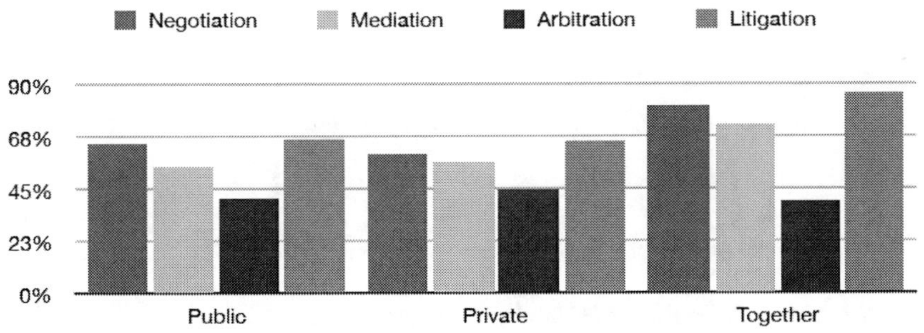

Figure 3. Comparing the percentage of use of DRM and the private sector worker.

Using DRM as a means of negotiation in private sector construction projects is applied mostly after litigation. The figure for this, according to participants who work in both sectors, was 81%. For participants from private sector, it was 65%, very close to the figure from participants working in both sectors. Those working in the private sector in DRM in Saudi Construction Project in the private sector used negotiation the least, with their figure standing at 60%.

The used DRM in private sector Saudi Construction Project was next in third position after litigation and negotiation according to participants. Using mediation, according to those working in both sectors, was more common, at 73%, while participants working in private sector had a different view regarding using mediation, considering it less than did those working in both sectors. The proportion for that was 57%. We can see that, based on the opinions of participants from public sector, mediation has a lower figure, standing at 55%.

The DRM used in the private Saudi Construction Project by means of arbitration was different to the previously used DRM. Previous DRM, according to participants working in both sectors, were higher. While using DRM as a means of arbitration for participants working in private sector was more common, at 45%. Participants working in the public sector were less than those in private sector, at 41%. At the end, we see that the participants working in both sectors together are fewer in number than participants from public sector and in their opinion, DRM by means of mediation was 40

In this question, the time taken to DRM in private sector was compared. As can be easily observed, the same duration was taken for both sectors in DRM. Considering that the four different approaches to DRM are litigation, negotiation, arbitration and mediation, the durations that each approach takes were as follows:

Less than a month, 4-6 months, 7-12 months, 1-3 years and over 4 years. According to the participants, negotiation took less time for DRM resolution and the DRM would not be longer than one month, and the proportion for that is 72%, while DRM by means of litigation took the most time, being over 4 years. The proportion of participants who pointed this out was 33%.

Figure 4 illustrates that dispute resolution method in Saudi Construction Project by means of litigation took the most time. The participants who stated that DRM by means of litigation took 1 to 3 years constituted around 63% of the sample. While the figure for those who stated that it took over 4 years was 33%.

Those who stated that dispute resolution method by means of litigation took about four to 12 months were small in number, not exceeding 1%. Finally, the timing of over 4 months was not mentioned at all and no value was given to it.

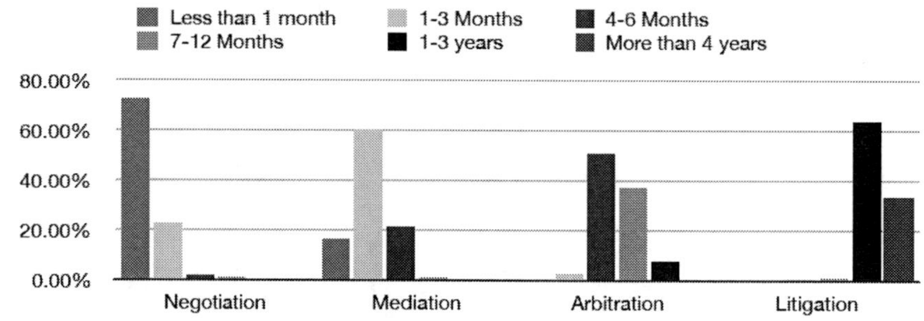

Figure 4. Percentage Time to Duration of Method of Dispute Resolution.

Participants who stated that the dispute resolution method by means of arbitration took four to 6 months were the largest group at 51%, while those who stated that the dispute resolution method by means of mediation took seven to 12 months stood at 37% and came in second position. Next came those participants who said that the dispute resolution method by means of arbitration would take one to three years this group represented a small percentage, at 7%. The smallest group of participants was those who stated that dispute resolution method by means of arbitration would take one to three months, at 3%. Finally, no value was given by the participants with times of less than one month and over four years.

DRM by means of mediation according to participants who was stated that solving disputes by means of mediation takes 1 to 3 months was the most common, at 60%. Those participants who believed that dispute resolution method through mediation took 4 to 6 months were in second position, with 16%. It can be seen that the lowest figure given was very small and relates to those who stated that mediation took seven to 12 months, with a value of 1%. Finally, times of over a year had no value.

Participants who said that dispute resolution method through negotiation took less than a month represented the largest group, at 72%. Meanwhile, we can see that participants who stated that DRM by means of negotiation would take one to three years were in second in position, with a figure of 23%. Those who state that dispute resolution method by means of negotiation will take 4 to 6 months, and its proportion was 2%. But those who stated that the dispute resolution method by means of negotiation would take seven to 12 months were very few, compared to the time that was previously given. At the end, we see some timing that has no value, that being for periods of over one year.

In this question, the cost spent on dispute resolution in the SCE and the level of education of participants are compared.

DOI: 10.1051/matecconf/201713802015

The DRM in the Saudi Construction Project are negotiation, mediation, arbitration and litigation. The levels of education of the participants are bachelor's degree, master's degree, PhD and higher diploma. As can be seen, dispute resolution using litigation has a high cost and, according to participants, the figure for this is over 47%, being the highest of all. The figure given by those with higher diplomas was 100%. Arbitration stood next to litigation in terms of cost. Using litigation for dispute resolution involved great cost and was over 45% for all of the participants. For those with PhD it was the highest, at 64%. Using mediation as means of solving disputes came after arbitration in terms of cost. It was a low based on the participants' beliefs and was not over 33%. For participants with bachelor's degrees it had the lowest cost, at 22%.

Figure 5 illustrates that the method of dispute resolution by means of litigation takes the most expenses and according to participants with higher diplomas, had the highest figure compared to those PhD and master's degrees. The figure for holders of higher diplomas was 100%. It can be seen that participants with higher diplomas believed that litigation had the highest expenses standing in second position, with a figure of 60%. While with a figure of 65%, given by those with PhD considering the cost of litigation in third position. At the end, participants with bachelor degrees stood lowest, with a figure of 47%.

diplomas, master's degrees and bachelor's degrees had the highest cost and a figure of 64%. It can be noticed that the participants with higher diplomas, in terms of cost of arbitration, stand in the second position with a figure of 60%. Participants with master's degrees came in third with 54%. Ultimately, the participants with master's degrees in terms of cost of arbitration represented the lowest figure, at 46%.

Using mediation for dispute resolution was considered a low cost approach by all of the participants and the lowest prices are suggested by those with PhD with the value for the cost not exceeding 10%. It can be observed that figure for the cost using mediation based on those with master's degrees did not exceed 12%. Ultimately, the highest cost of dispute resolution through mediation was the opinion of those with higher diplomas who said that the figure for its cost was 22%.

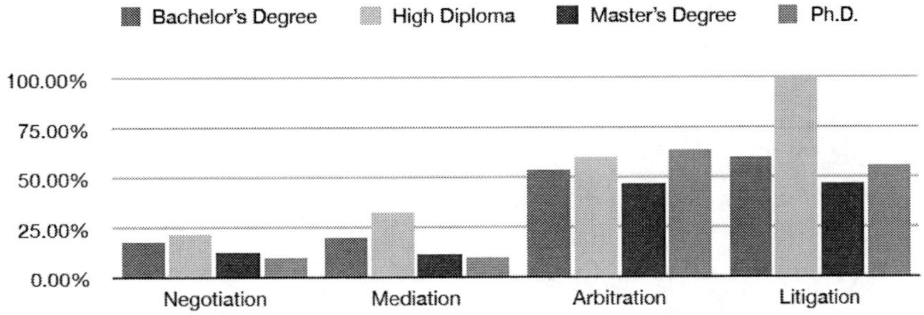

Figure 5. Comparing the percentage the cost of DRM used in the SCE and the level of Education.

Dispute resolution using negotiation, according to participants was very low cost and, in this case, the lowest cost refers to those with master's degrees, with the value not exceeding 20%. Finally, according to participants with higher diplomas, the most costly method of dispute resolution was using negotiation with the figure for the cost of negotiation at 22.

4 Conclusion

This article has considered, with the considerable assistance of experts in the field of Saudi construction projects, the status and value of ADR as an alternative to litigation in the resolution of construction

 DOI: 10.1051/matecconf/201713802015

industry disputes in the Kingdom. There are evident differences in the classification of the nature of disputes, and the factors, which govern the success of resolution. These are essential to ensure the efficacy of the projects and their effective completion. Disputes in Saudi construction projects have obvious impacts on time, cost and quality, which have to be properly managed to avoid costly delay. This article includes the qualitative and quantitative analysis. The aim is to achieve the objectives of research, whereby it is hoped the questionnaire will fill all the gaps in the literature. The questionnaire was distributed to arbitrators, experts and engineers in the public and private construction sector in KSA. With 327 responses being received from participants. . The questionnaire was designed and distributed using Survey Monkey, and the analysis of the questionnaire was performed using figures and tables.

Finally, the dispute resolution method in Saudi construction projects are considered in the context of (i) litigation, and (ii) alternative dispute resolution via negotiation, mediation, DAB and arbitration. The utility of such dispute resolution methods vary between the public sector, which involves the state or its agencies as a party to the contract and tend to seek resolution by negotiation or litigation, and the private sector, which makes use of the wider range of ADR processes.

References

[1] W. Hughes, R. Champion, J. Murdoch, *Construction contracts: law and management* (Routledge, 2015)

[2] P. Love, P. Davis, J. Ellis, S. On Cheung, Engineering, Construction and Architectural Management **17**, 404 (2010)

[3] R.J. Gebken, G.E. Gibson, Journal of professional issues in engineering education and practice **132**, 264 (2006)

[4] M. Li, S. Yu, Y. Zheng, K. Ren, W. Lou, IEEE transactions on parallel and distributed systems **24**, 131 (2013)

[5] S. Cheung, C. Tam, I. Ndekugri, F. Harris, Construction Management & Economics **18**, 281 (2000)

[6] T.J. Kelleher Jr, J.M. Mastin, R.G. Robey, C. Smith, L. Hancock, *Smith, Currie and Hancock's Common Sense Construction Law: A Practical Guide for the Construction Professional* (John Wiley & Sons, 2014)

[7] L. Greenhouse, *Becoming Justice Blackmun: Harry Blackmun's Supreme Court Journey* (Macmillan, 2007)

[8] K. Iyer, N. Chaphalkar, G. Joshi, International Journal of Project Management **26**, 174 (2008)

[9] S. Harms, *Traditional litigation versus alternative dispute resolution (in collection cases)* (2011), `http://www.jdsupra.com/legalnews/traditional-litigation-versus-alternativ-97899/`

[10] K.K. Kovach, *Mediation: Principles and practice* (Thomson West, St Paul, 2004)

[11] V. Chan, S.E. Tett, Pharmacoepidemiology and drug safety **15**, 485 (2006)

Research, Development and Application of High Performance Earthquake Resistant Precast System as Green Construction in Indonesia

Hari Nurjaman[1,*], *Binsar* Hariandja[2], *Gambiro* Suprapto[3], *Lutfi* Faizal[4] and *Haerul* Sitepu[5]

[1]Indonesian Association of Precast and Prestressed Engineer, 13440 Jakarta, Indonesia
[2]Professor of Civil Engineering, Bandung Institute of Technology, School of Civil and Environmental Engineering,40132 Bandung, Indonesia
[3]Wijaya Karya Beton, Ltd., Engineering Department, 17411 Jakarta, Indonesia
[4]Ministry of Public Works and Housing, Puslitbang Perkim, 40393 Bandung, Indonesia
[5]Ministry of Public Works and Housing, Ditjen Penyediaan Perumahan, 12110 Jakarta, Indonesia

Abstract. Sustainable construction is a topic that emerges in the world construction as a response to climate change issue. Building construction stage is a stage in sustainable development. Construction concept that confirm to the concept is referred to as green construction. Precast concrete construction is a construction system that meets green construction criteria, because applies the usage of material and construction method that optimize energy consumption and minimize environment impact during construction. This paper describe the qualitative and quantitative research in precast structures as green construction in Indonesia. Qualitative research was conducted on the assessment criteria of Green Building Council of Indonesia (2012-2014). The quantitative researches are in energy consumption and environmental influence of several construction systems which were conducted at low cost housing construction in Batam (2011-2012). Finally the research (2013-2014) and application (2015-2017) of high performance earthquake resistant precast concrete structure at low cost in Indonesia is described. The result of this research is expected to be a contribution to all construction stakeholders, in order to start to consider optimization of energy consumption in construction, and to become part of sustainable earth movement.

1 Introduction

Sustainable construction is an aspect developed worldwide as a response to climate change. Basic principle of such construction method is the use of energy in integrally optimal fashion starting from design, construction, operation, and demolition. Concrete precast system is a method of construction conforming to green construction, since the system applies efficient working method and material such that energy is used optimal. The paper presents qualitative and quantitative research on concrete precast system as green construction in Indonesia, followed by some applications and conclusions.

*Corresponding author: iappi_ind@yahoo.com

 DOI: 10.1051/matecconf/201713802016

2 Qualitative research

Qualitative research performed on research works by Green Building Council of Indonesia (GBCI) (2012-2014) to inspect how precast construction was elaborated in the determination of green construction category. Generally, there exist seven aspects in building assessment, should that building be designed with sustainable concept, as shown in Table 1 [9].

Table 1. Components of assessment of the building with the concept of sustainable development.

Category	Precast concrete contribute to greenship points
Appropriate Site Development	ASD 6 – Micro climate (Heat island effect – Albedo)
Energy Efficiency and Conservation	EEC P2 – OTTV calculation
Water Conservation	
Material Resources and Cycle	MRC 1 – Building and material reuse MRC 2 – Environmentally processed product MRC 5 – Modular design/prefab material MRC 6 – Regional material
Indoor Health and Comfort	IHC 3 – Chemical pollutant
Building Environment Management	BEM 2 – Pollution of Construcion Activity
Innovation	

Building construction stage is a stage in sustainable development. Construction concept that conforms to this concept is referred to as green construction. Main principle in green construction is the use of 3R (reduce, reuse, recycle) concept in material use. Concrete precast system is a construction system that conforms to this concept. Reduce principle applies to the efficient use of materials and working method. The design of precast system may conserve the use of steel, which has most high emission dan generally available as imported material. The method of construction may conserve the use of formwork as depicted in Fig. 1a, conserve the use of propping as depicted in Fig. 1b, and also may avoid waste in the use of reinfocement and concrete as depicted in Fig. 1c.

The reuse principle mainly applied to the repetitive use of mold as compared to conventional system. Penolin triplex layers as depicted in Fig. 1d can be used 10@15 times compared to the use of only 2@3 times of conventional system. If fiber or steel is used, the mold may be used hundred of times. Recycle principle may be applied by using metal or fiber mold, and on green labeled grouting materials

(a) conventional (b) propping system (c) zero wasted for steel and concrete (d) molding system

Fig. 1. Conventional system and 3R principle of precast system

3 Quantitative research

Quantitative research was performed on the use of energy and its environmental impact on several construction systems of low cost apartments in Batam (2011-2012) [9], and earthquake resistant high performance precast systems with low initial investation in Indonesia (2013-2014) [10-12].

DOI: 10.1051/matecconf/201713802016

3.1 Calculation concept of construction energy and environmental impact

Energy calculation reference of each item of construction work is based on Goedkoop and Oele [14]. Equalization process for comparison of each item is done by LCA (life cycle analysis) method through Simapro 5.0 software to analyze environmental aspect related to a product and its life cycle, starting from retrieval, production, use, maintenance until demolition and bringing the material back to nature. The software calculates input data such as raw material quantities, and produces output to guide safer return to the nature.

3.1.1 Construction system

Rental apartment was made of concrete conctruction, which made through three construction methods, i.e., conventional, partial precast and full precast. Conventional concrete system uses conventional process in view of structural as well as architectural aspects. The construction uses cast in-situ process using timber molding and scaffolding. Architectural wall components are made of bricks or concrete blocks.

Partial precast concrete system uses precast components and architectural components are made of conventional bricks or concrete blocks as shown in Fig. 2a. Full precast concrete system uses precast for both structural and architectural components as shown in Fig. 2b.

(a) partial precast system (b) full precast system

Fig. 2. Variate of precast system

3.1.2 Calculation of construction energy and environmental impact

Construction of rental low cost apartment requires natural resources as construction raw material. The use of natural resources may cost natural damage. Main construction materials with corresponding demands on each construction type may be seen in Table 2.

The use of reinforcement bar in precast system has been significantly decreases from 180 tons in conventional concrete to 77 tons in partial precast. The significant saving on reinforcement is achieved due to the use of prestrssed grid in slab component. Similar condition also occurs on the use of timber, as scaffolding or as molding to 603 m^3. The timber no longer needed for partial as well as full precast concrete. Beside efficiency in the use of construction materials, additional advantage from using partial or full precast concrete is the cost reduction due to shorter construction time, from 8 months to 5-6 months. On the other hand, the total use of labors in precast construction is less then needed for conventional work. However, precast concrete work needs more skilled labors. This is in accordance with that stated by Gibb [1].

The result of the analysis of conventional system of single score in Fig. 3a. shows that steel bar is an element that gives the most impact (8.19 kPt), followed by cement (4.55 kPt). The result single scoring on partial concrete precast system application in Fig. 3b. shows that cement is the element that provides greatest impact (4.7 kPt), folowed by steel reinforcement (3.59 kPt).

Table 2. Building material needs

No	Main component	Conventional	Partially Precast	Full Precast
1	Cement	488 ton	499 ton	617 ton
2	Fine aggregate	3,483 m3	3,354 m3	687 m3
3	Coarse aggregatre	339 m3	398 m3	642 m3
4	Reinforcement bar	180 ton	77 ton	122 ton
5	Mold	wood 41 m3	steel 7.9 ton	steel 9.2 ton
6	Scaffolding	wood 562 m3	steel 7.8 ton	steel 7.8 ton
7	Brick	413,216 nos	413,216 nos	-
8	Skilled labour	20 man	30 man	40 man
9	Unskilled labour	80 man	45 man	15 man
10	Cost (Rp x 1.000 million)	13.657	11.500	11.434
11	No. of stories /area	4 lt/4,600 m2	4 lt/4,600 m2	4 lt/4,600 m2
12	Construction time (month)	8	6	5

*) Precast technology used is from one of the national precaster

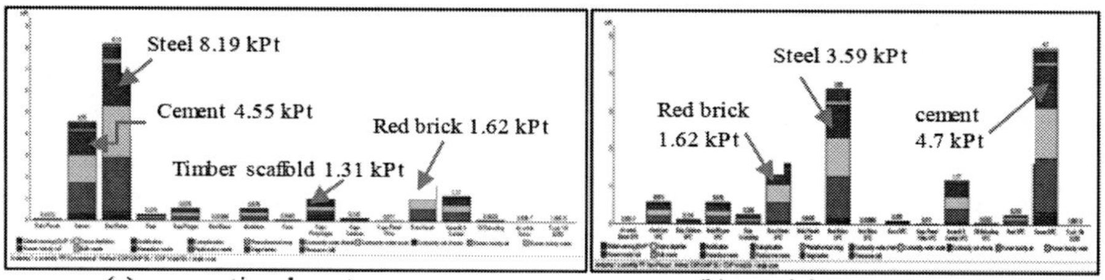

(a) conventional system (b) partial precast system

Fig. 3. LCA single score results

Aspects that differentiate the materials used in the full precast concrete system to partially precast concrete system is a reduction in the use of raw materials for the manufacture of wall plaster, cement, sand, and red brick. The reduction of raw materials of wall was substituted entirely by the use of precast concrete. The result of the analysis of full precast system of single score in Fig. 4a. shows that steel bar is a contributor to the process that could provide the greatest impact (6.81 kPt), followed by cement (3.82 kPt).

The result of a single aggregate score shownin Fig. 4b. Overall environmental impact due to the life cycle of the contributors to the conventional use of concrete is 18.6 kPt. This could be reduced to only 13.8 kPt with full precast concrete and toonly 13.0 kPt with partially precast concrete. While the difference in ability between full precast concrete to partially precast concrete is 6% (0.8 kPt) in reducing the environmental impact.

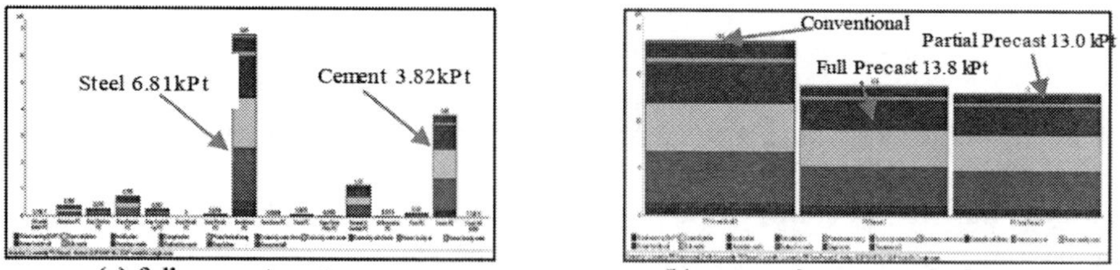

(a) full precast system (b) comparison on each altern

Fig. 4. LCA single score

The main building materials that have the greatest environmental impact are steel bars, cement, aluminium, bricks, ceramics and sanitary, as presented in Table 3. Base on weighting, almost all elements of potential environmental impact of the three dominant (impact catagories), namely: (1) chronic aquatic environmental toxicity (2) acute toxicities of aquatic environments, (3) chronic toxicities of environmental soil. Chronical aquatic environmental poisonous content is caused by materials that contaminate aquatic in long time period. If that contamination is not taken care seriously and correctly, the impact would propagate to accute aquatic poisonous environmental contamination. This is hazardous to human being and living organism that utilize the aquatic.

DOI: 10.1051/matecconf/201713802016

Table 3. Comparison of environmental impact

Construction alternative dan pollutant potential of main building material	Environmental impact catagory (Pt)			Energy (kWh)
	Chronic toxic waters	Acute toxic waters	Chronic toxic soils	
Conventional:				1,253,774.7
-steel bar	5.18×10^8	4.95×10^7	2.53×10^7	
-cement	2.85×10^8		1.39×10^7	
-aluminum		3.68×10^6		
Partially precast:				806,981.9
-cement	2.94×10^8	2.82×10^7	1.43×10^7	
-steel bar	2.27×10^8	-	1.11×10^7	
-red brick		9.97×10^6		
Full precast:				1,008,199.9
-steel bar	4.11×10^8	3.92×10^7		
-cement	2.39×10^8	2.29×10^7	1.97×10^7	
-ceramic & sanitary			3.65×10^6	

3.2 The research of high performance earthquake resistant precast structure

High performance earthquake resistant precast structure is a revolutionary alternative technological solution capable of achieving high-performance (low damage) at low cost. This concept was developed in US-Japan joint research PRESSS Program (1994-2002) and New Zealand in the late 1990s [16], as a response to public demands for the performance of classical ductile design concept that did not comply with their expectation in Loma Prieta (1989) and Northridge (1994) earthquake. The well-known ductile design concept using collapse prevention performance criteria can indeed avoid casualties in a strong intensity earthquake, but the associated structural damage can result in significant business interruption. Furthermore, the post-earthquake repairs can be challenging and costly.

3.2.1 The concept of high performance earthquake resistant precast structure

There are two main components involved within the system,i.e., unbonded post-tensioning that provide re-centering effect, and mild reinforced concrete that provide energy dissipation (dissipater). Unbonded post-tensioning tendon may be applied in beam-column connection, and across vertical connection between precast wall, including walls and foundation [17] as shown in Fig. 5a. Unbonded post-tensioning, while kept elastic, is engineered to prevent residual deformations. Dissipater [17], on the other hand, are used to provide hysteresis damping as shown in Fig. 5b. Dissipaters can be designed as replaceable fuses in the structural system. The experience in Indonesia is that the use of these emerging technologies can be achieved without compromising the architecture of the building.

The difference in seismic response between high performance concept and the classical ductile behaviour may be observed from the test results of beam column joint. Fig. 6a. shows the hysteresis loops and joint damage pattern of classical ductile design that meet the requirements of special moment resisting frame (SMRF) [10]. Hysteresis is described by reasonably "fat" loops that occur in all four quadrants. Inelastic response results from development of plastic hinges, which exhibits structural damage and requires costly repairs. Moreover, such ductile system may experience residual deformations.

In re-centering system, the hysteretic response is essentially described by loops that appear diagonally in two opposite quadrants, as depicted in Fig. 6b. [17]. The beam-column joint in this Fig. exhibits re-centering effect due to the presence of an elastic unbonded post-tension tendon. This concept is also referred to as hybrid concept. The ratio of re-centering and ductile behaviour will produce a spectrum hysteresis hybrid concept known as "flag shape". ACI 550.3-13 recommends prestressing force to be at least 50% of the load in order to conduct a re-centering remaining effective [4].

 DOI: 10.1051/matecconf/201713802016

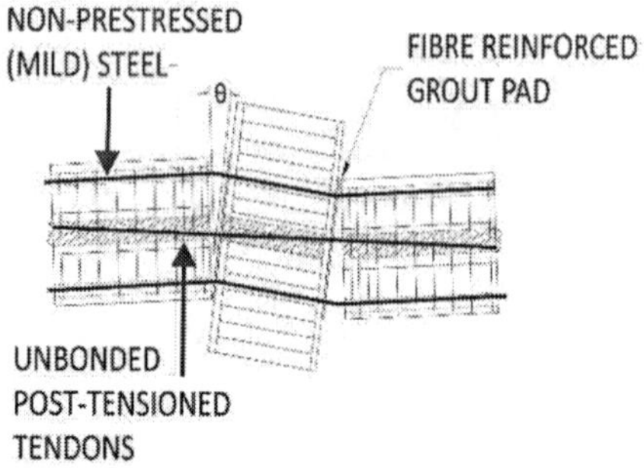

(a) Unbonded Post-tensioning

(b) Dissipater for hysteresis damping

Fig. 5. Two main component of high performance earthquake resistant precast structure [17]

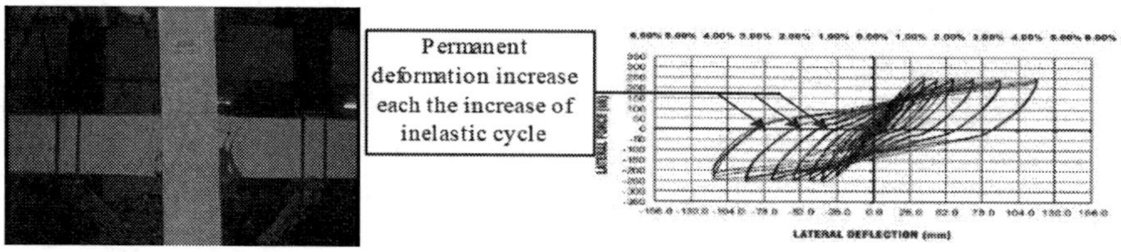

(a) Classical ductile behaviour

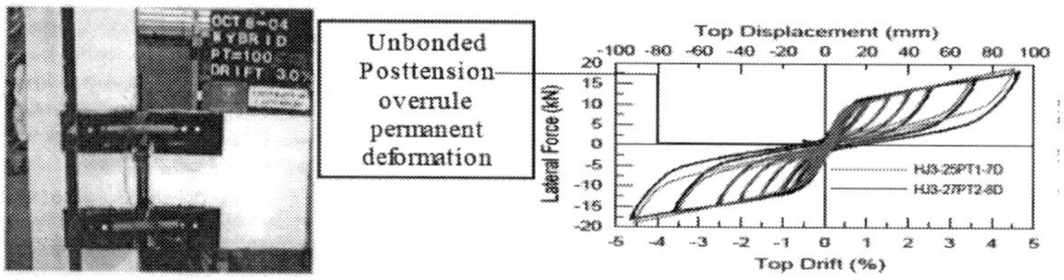

(b) High performance precast

Fig. 6. The difference in seismic response

3.2.2 The development in Indonesia

The development is based on technology and local materials that already exist. The concept of unbonded post tension is relatively well known, so that the material, equipment and construction methods is not difficult to be used as shown in Fig. 7a. The common configuration of a dissipater device consist of the connection of a steel bar using a smaller confined within a metal tube sheet [17]. A local dissipater device was developed successfully in 2014, based on Indonesian method of connecting steel bars, with spiral reinforcement made from plain bars, as shown in Fig. 8b. This spiralis equivalent to metal sheet tube

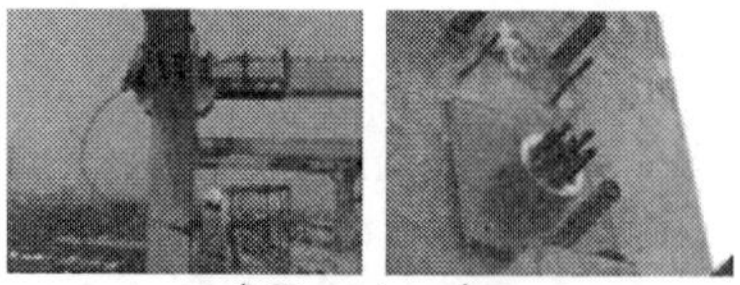

a) Post tension

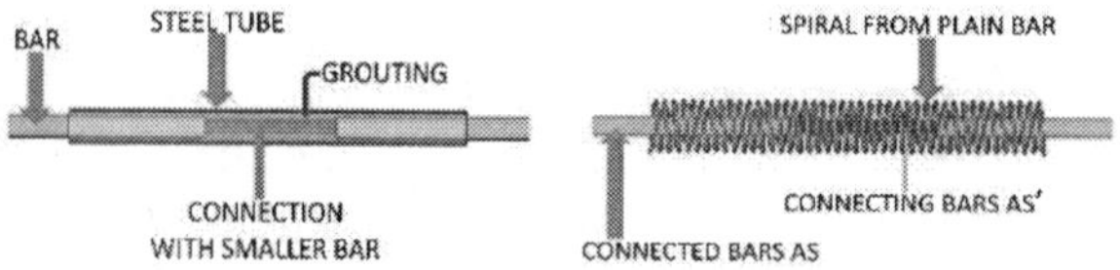

(b) Dissipater development

Fig. 7. High performance earthquake resistant precast system in Indonesia

Dissipater can be mounted externally or internally, as shown in Fig. 8. The advantages of external dissipater is that it can be replaced if damaged, but consequently disturbing the outlook of the system. The development achieved in Indonesia is slipping this tool at the top and bottom of the beam, so it does not disturb the outlook, and still easily replaced if damage.

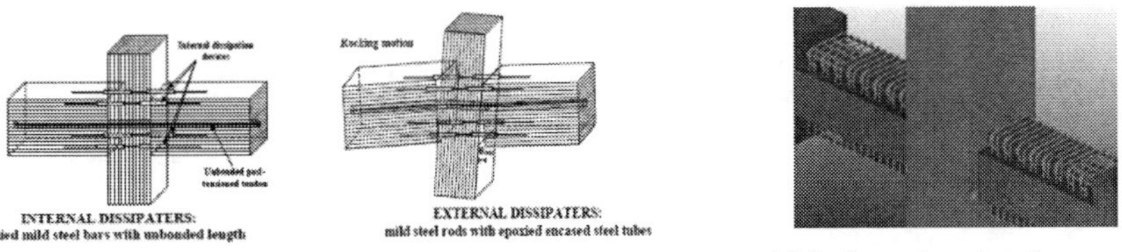

(a) Internal (b) External (c) Indonesian development

Fig. 8. Dissipater location

3.2.3 Laboratory testing

Testing of dissipater development is carried out following ASTM E8 tensile testing procedures [7]. ACI 550.3-13 states that the material must meet ASTM A706 Grade 60 [6]. The ratio of tensile strength to yield strength f_s/f_y is not less than 1.25, in accordance with Article 21.1.5.2 b ACI 318-08 [3]. It directs that the reinforcement ratio between connected bar to connection bar (A_s/A_s') should be between 1 – 1.25, so that yield occurs at connection bar, and the strain hardening phase will not overwhelm yield stress of connected bars. The dissipater test results [10] may physically be seen in Fig. 9.

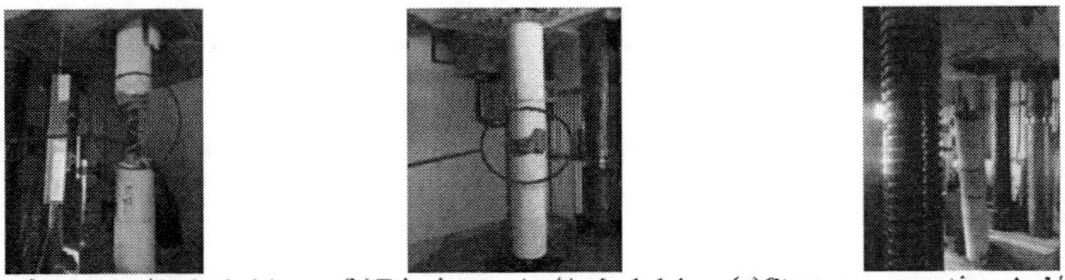

(a) DissipaterAs/As'=1.44 (b)DissipaterAs/As'=1.14 (c)Strong connection As'/As = 1.06

Fig. 9. Dissipater test result

The beam column joint testing was done after some preliminary testing phase. Reinforcement of hybrid beam made according to preliminary test sample with a moment capacity ratio of dissipater bar to the probable moment capacity (M_s/M_{pr}) approaching 0.5, in according with recommendation or article 7.4.2 in ACI 550.3-13 [4]. Connected to connecting bar ratio was A_s/A_s'= 1.14. Joint and column are designed in accordance with article 21.1.3, Article 21.6, and article 21.7 of ACI 318-08 [3]. Loading schemeand acceptance test criteria are based on ACI 374-05[2] as shown in Fig. 10. In drift up to Δ = 3.5%, the specimen must meet three main criteria: strength, energy dissipation and stiffness in order to be categorized as Special Moment Resisting frame (SMRF). Analysis of testing result [12] showed that the beam-column joint meets the criteria of ACI 374-05 [2] as shown in Fig. 11.

(a) Exterior joint (b) Interior joint (c) Loading schemes

Fig. 10. Beam column joint test and loading scheme of ACI 374-05 [2]

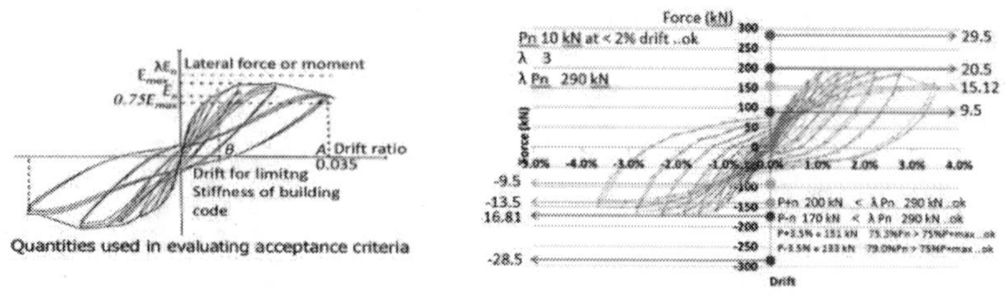

(a) Strength criteria

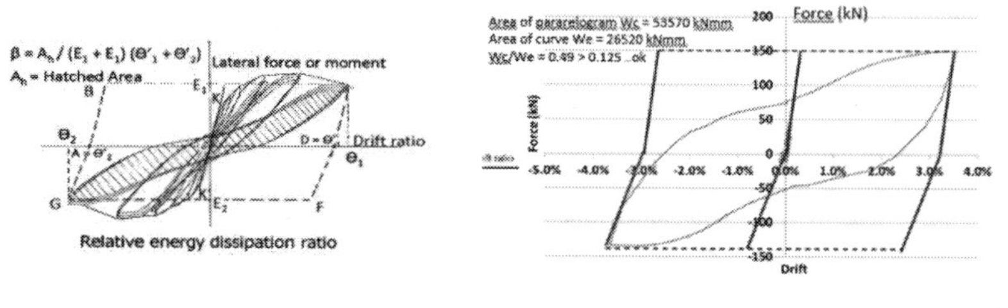

(b) Energy dissipation criteria

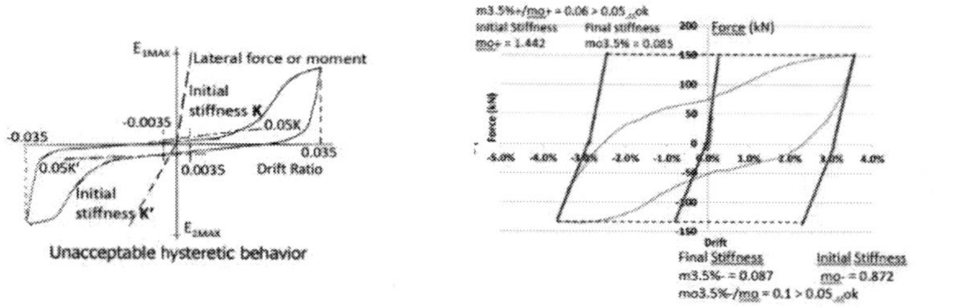

(c) Stiffness criteria

Fig. 11. Beam column join test criteria check by ACI 374-05 [2]

What is more interesting is the failure mechanisms of the test specimens and the performance level at each stage of the test, as shown in Fig. 12. Yield on the connecting bar started at Δ=1%, and began to cause a gap at Δ=1.75 %. This gap opened and closed during cyclic loading, which demonstrates the effectiveness of "re-centering" feature caused by unbonded post tensioning system, as shown in Fig. 13b and 13c. The damage did not occur in other parts significantly until Δ=2.2 %, as shown in Fig. 13d. Δ_a (allowable story drift) on earthquake design (S_{DS}) as required in Table 12.2-1 ASCE/SEI 7-10 [5] is 2% for risk category I or II. So it demonstrates a proven high performance on the structural systems, which the building would return to the original position, with the damaged concentrated in the dissipater components. The retrofitting can be done by replacing dissipater components. The hysteresis loop of hybrid system can be seen in Fig. 13g. This behaviour close to flag shape pattern in proportion of 50:50.

(a) Service Δ = 0.35%

(b) gap in Δ = 1.75%

(c) gap in Δ = 1.75%

(d) Δ = 2.2% equal S_{DS} and Δ_a

(e) Δ = 3.5% equal MCE_R (f) Δ = 5% equal near collapse (g) Flag shape 50: 50 at Δ = 2.2%

Fig. 12. Failure mechanism and specimen

4 Application

The system was applied to a twelve-storey office building in Jakarta (2014); some multi storey low cost housing in Jakarta, Banten and West Java(2015), Riau Island, Riau, North Sumatera, South Sumatera, and West Java (2016);low cost landed house (2015), shop house (2016), and hospital in Jakarta (2017), as shown in Fig. 13.High performance earthquake resistant precast system used in the columns, beams, and hollow core slab or half slab. The application proves resistant and can increase good quality construction with an easy, fast and economical.

(a) 12 story office (b) Low cost housing (c) Hospital

Fig. 13. Application

5 Conclusion

Sustainable construction is a topic that emerges in the world construction as a respond to climate change issue. Precast concrete is a construction system that meets green construction criteria, because the usage of material and construction method that optimize energy consumption and minimize environmental impact during construction are applied.

Precast building in Indonesia started to be applied commonly in government mass rental low-cost housing program (1995). Qualitative research was conducted on the assessment criteria of Green Building Council of Indonesia (2012-2014). The precast concrete building systems is one that meet this rule, that be elaborated in construction material. The quantitative researches were in energy consumption and environmental influence of several construction in Batam (2011-2012), which showed precast system is have less energy consumed and more environmental friendly than conventional system. Finally the research of high performance earthquake resistant precast concrete structure at low cost (2013-2014) had found specific system configuration based on local available material and technology. The system can guaranteed that the building is still functional in its life time even after influenced by major earthquake, and so can save energy consumption in repairment. Based on the successfully application, Indonesian government had set the target application of precast concrete about 30% of construction industry until 2019.

The result of this research is expected to be a contribution to all construction stakeholder, in order to start to consider optimization of energy consumption and environmental impact in construction and to become part of sustainable earth movement.

This research is funded by Pressstekindo Consultant Ltd, and Wijaya Karya Beton Ltd. The writers extend their appreciation to the Indonesian Association of Precast and Prestressed Engineers and to Ministry of Public Work and Housing for providing testing facilities.

References

1. A.G.F. Gibb, *Off-site Fabrication* (John Wiley and Sons, New York,1999)
2. American Concrete Institute, *ACI 374-05 Acceptance Criteria for Moment Frames Based on Structural Testing and Commentary*, 8-9 (2014)
3. American Concrete Institute, *ACI 318-08 Building Code Requirement for Structural Concrete*, 323-324,347-349 (2008)
4. American Concrete Institute, *ACI 550.3-13 Design Specification for Unbonded Post-Tensioned Precast Concrete Special Moment Frame Satisfying ACI 374.1 and Commentary*, 11,19,20 (2013)
5. American Society of Civil Engineers, *ASCE/SEI 7-10 Moment Design Loads for Buildings and Other Structures*, 374,66,97 (2010)
6. American Standard of Testing Material, *ASTM A706/A706M-14 Standard Specification for Deformed and Plain Low-Alloy Steel Bars for Concrete Reinforcement*, (2014)
7. American Standard of Testing Material, *ASTM D : E8-09 Standard Test Methods for Tensioning Testing of Metallic Materials*, (2009)
8. Badan Standar Nasional Indonesia, *SNI 03-1726-2002 Standar Perencanaan Ketahanan Gempa untuk Struktur Gedung*, 21,31 (2002)
9. H.N. Nurjaman, H. Sitepu, H.R. Sidjabat, *Precast Concrete Construction : A Green Construction : Comparison of Construction Energy and Environmental Influence in Low Cost Housing Construction in Batam* (1st International Conference of Sustainable Civil Engineering Structure and Construction Material, Yogyakarta, 2012)
10. H.N. Nurjaman, L. Faisal, H.R. Sidjabat, B. Hariandja, R. Rivky, Y. Put, Elsevier Procedia Eng. J. **95**, 82-86 (2014)
11. H.N. Nurjaman, L. Faisal, H.R. Sidjabat, G. Soeprapto, B. Hariandja, R. Rivky, Y. Put, *Research Development and Application of Precast Systems for Building, with Connection using Unbonded Post-Tension and Local Dissipater Device* (6th International Conference of Asian Concrete Federation, Seoul, 2014)
12. H.N. Nurjaman, L. Faisal, G. Soeprapto, R. Rivky, Y. Put, B. Hariandja, J.I. Restrepo, H.R. Sidjabat, *Research, Design and Application of High Performance Earthquake Resistant Precast Structure in Indonesia* (16th World Conference of Earthquake Engineering, Santiago, 2017)
13. M.J. Scoettler, A. Belleri, D. Zhang, J.I. Restrepo, R.B. Fleischman, PCI. J. **54(1)**, 100 (2009)
14. M. Goedkoop, M. Oele, *Database Manual General Introduction* (Product Ecology Consultant, USA, 2001)
15. R.B. Fleischman, J.I. Restrepo, S. Pampanin, J.R. Maffei, K. Seeber, Earthquake Spectra **30 (1)**, 277-306 (2014)
16. S. Pampanin, *PRESSS Design Handbook* (New Zealand Concrete Society, 2010)
17. S.Pampanin, *Reality-check and Renewed challenges in Earthquake Engineering: Implementation low-damaged System – from theory to practise* (15th World Conference of Earthquake Engineering, Lisboa, 2012)

Application of Impact Vibration Test Method for Bridge Substructure Evaluation

Pham Hoang Kien[1,*]

[1]University of Transport and Communications, Faculty of Civil Engineering, Hanoi, Vietnam

Abstract. The bridge foundations are constructed mainly below ground or water level. The rehabilitation of a foundation is difficult, and the results of evaluation of structural conditions are usually inaccurate. Furthermore, there are many cases that after natural disasters such as earthquakes, typhoons and floods, it is required to rapidly evaluate bridge substructure conditions for a judging decision on being able to open the bridge to traffic or not. Impact vibration test method has been developed to deal with such problems. Impact vibration test method is a non-destructive dynamic test method which allows to evaluate structural conditions of bridge substructures promptly with a good soundness. In this study, firstly, impact vibration test method, which is one of the utilizable non-destructive tests on the scouring of bridge substructure, is presented. Then, application of the test method for substructure evaluation of a bridge in Vietnam is introduced. The soundness and effectiveness of test method is also discussed.

1 Introduction

A bridge substructure is designed to resist vertical settlement resulting from dead and live loads, and lateral movement and rotation caused by temperature change, friction, wind, water, and seismic loads. The bridge substructure and its connection to the footing must resist moments, shear, and compressive forces. The footing must resist lateral, vertical, and rotational movements. The bridge substructure condition rating is governed by the condition of its structure and the dimension and type of footing (shallow, deep, or combined). Deformation and damage of substructure and its foundation will greatly affect traffic service and safety. There are three (3) major phenomena of substructure foundation damage, namely, settlement, inclination and scour. Among these, scour is considered as the most common cause of bridge substructure failure. Scour occurs progressively as supporting material under a footing is removed during flood events and is replaced with material that has little or no bearing capacity.

Accurate information on a bridge substructure is an essential part of implementing a cost effective and safety-conscious bridge management. Substructure foundations are constructed mainly below ground or water level, therefore, visual inspection is very difficult. Dynamic testing is considered to play a key role in bridge substructure performance evaluation and

* Corresponding author: phkien@utc.edu.vn

monitoring system as it can provide more accurate information on the condition, type, and vulnerability of bridge substructures that cannot be obtained by normal visual surveys.

Soundness condition of a bridge substructures can be evaluated on the basis of dynamic settlement, frequency and amplification of vibration excited by passage of vehicle truck or train, in addition to the static measurement of inclination or settlement of the bridge substructure. If the above measuring results exceed the allowable values or show unfavorable time-dependent tendency, the performance of the bridge substructures is suspected. However, the above measuring results generally depend on the weight and passing velocity of the vehicle truck or train; therefore the measuring results vary widely and are insufficient to evaluate the soundness condition of the bridge substructures. Furthermore, there are many cases that after natural disasters such as earthquakes, typhoons and floods, it is required to rapidly evaluate bridge substructure conditions for a judging decision on being able to open the bridge to traffic or not. Impact vibration test method has been developed to deal with such problems. Impact vibration test method is a non-destructive dynamic test method which allows to evaluate structural conditions of bridge substructures promptly with a good soundness. In this study, firstly, impact vibration test method, which is one of the utilizable non-destructive tests on the scour of bridge substructure, is presented. Then, application of the test method for substructure evaluation of a bridge in Vietnam is introduced. The soundness and effectiveness of test method is also discussed.

2 Bridge substructure evaluation by impact vibration test

For bridge substructure inspection and evaluation, the major items to be inspected are: settlement, horizontal movement, inclination and scour. In the past, the item of scour was not included in periodical substructure inspection even if the progress of damage due to scour could be confirmed from inspection results. However, recently, the importance of periodical inspection on scour during/after a heavy rain or flood has been recognized. Damage to substructure by scour is a kind of damage difficult to visually predict. Once it occurs, it often leads to serious accident and disruption of transport. One method developed to evaluate the soundness of bridge pier is using the natural frequency obtained by impact vibration tests. This method uses the tendency for the natural frequency of a pier to decrease as phenomena such as scour deteriorate the soundness of that foundation's pier. The natural frequencies of bridge pier are determined by the mass of superstructures and bridge pier, the stiffness coefficient of bridge pier and the soil characteristics of the ground.
The impact vibration test makes use of the difference of natural frequencies when the scour occurs and riverbed elevation around the bridge pier has changed.

2.1 Procedure of impact vibration test

The impact vibration test procedure is shown in Fig. 1. Firstly, the bridge pier is hit with an iron ball suspended from the superstructure. The response acceleration and velocity are then measured. An iron ball with the weight of around 30kg – 50kg is usually selected for practical reason of workability. To obtain a clear response and reduce noises, iron ball hitting is repeated around five times, because the response acceleration and velocity are quite small with a light weight. Accelerometers can be set at three locations along the bridge pier for the evaluation of vibration modes.

Secondly, the measurement results of impact vibration test are processed and analyzed. The obtained time history response waves are transformed by Fast Fourier transformation to obtain the Fourier Spectrum. Natural frequencies of bridge pier are determined by the result of the Fourier Spectrum and Phase Spectrum.

 DOI: 10.1051/matecconf/201713802017

Fig. 2 gives a comparison between the Fourier Spectrum obtained from impact vibration test and the vibration generator test for the same bridge pier. There is a good agreement between results of two test methods, and it can be said that the Fourier Spectrum obtained from impact vibration test is useful to obtain the natural frequency of bridge piers.

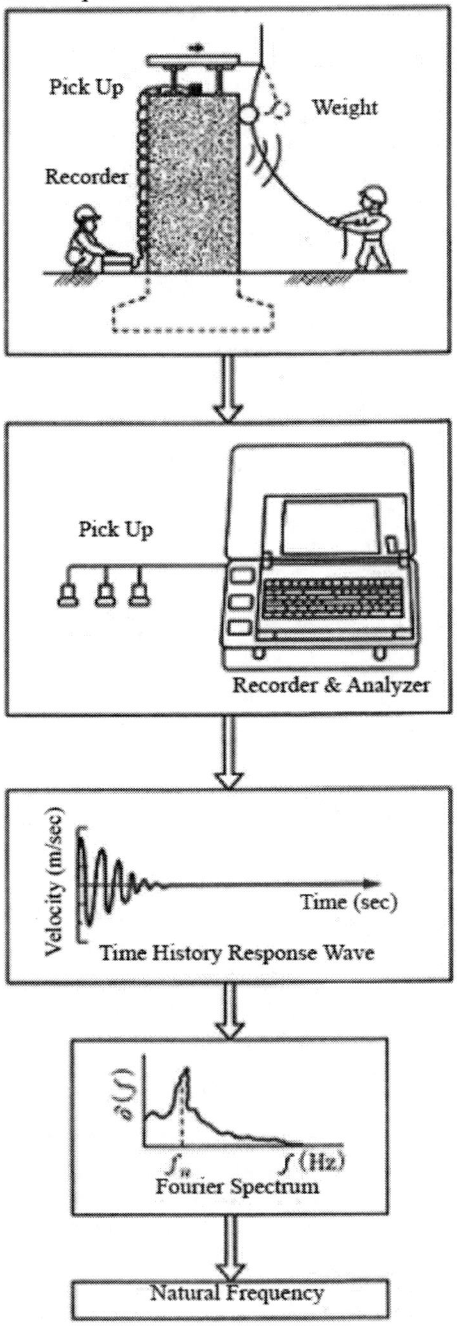

Fig. 1. Procedure of impact vibration test

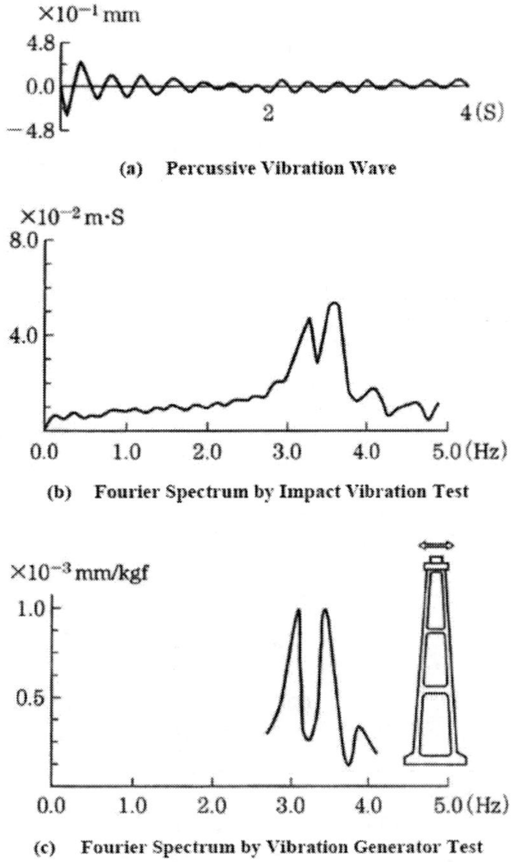

Fig. 2. Comparison between impact vibration test and vibration generator test

DOI: 10.1051/matecconf/201713802017

2.2 Soundness evaluation of bridge pier

The soundness of bridge pier can be evaluated by comparison of natural frequency obtained from impact vibration test to a standard value. The formula for index of soundness which is defined as the ratio of measured natural frequency to the standard value, is given in Equation (1).

$$\alpha = \frac{F_m}{F_s} \tag{1}$$

where α is index of soundness, F_m and F_s are measured natural frequency and standard value of natural frequency, respectively.

An evaluation criteria for the index of soundness is proposed by Railway Technical Research Institute (RTRI), Japan as shown in Table 1.

Table 1. Soundness evaluation criteria

Index of soundness α	Evaluation rank		Countermeasure
$\alpha \leq 0.70$	A	A1	Dangerous state can occur under the action of extreme loads. Other investigation is required and reinforcement should be performed.
$0.70 < \alpha \leq 0.85$		A2	Natural frequency deterioration should be continuously confirmed.
$0.85 < \alpha \leq 1.00$	B		There is almost no problem at present stage
$1.00 < \alpha$	S		The structure is completely sound at present stage

In case where the natural frequency of bridge pier was obtained by measurement immediately after bridge construction, this initial value can be taken as the standard value of natural frequency, and a quantitative evaluation of the foundation's bearing capacity or the degree of deterioration in the pier can be accurately carried out by comparing the standard value with the frequency obtained at the time of evaluation. However, in reality, the initial value is rarely available. Previously measured value of natural frequency can also be taken as standard value.

RTRI (Railway Technical Research Institute, Japan) also proposed an experimental formula to calculate the standard value of natural frequency by the regression analysis on the basis of many field data. RTRI implemented a large number of impact vibration tests on piers which had been judged to be clearly sound as a result of detailed visual inspection, dynamic displacement measurements carried out when trains were passing and so on. Then the accumulated field data were statistically processed with regression analysis considering parameters which are highly sensitive on the natural frequency of bridge pier such as the weight of superstructure and height of the pier. The proposed formula shows a high correlation coefficient, indicating an accurate estimation.

In case where the information of bridge pier can be obtained from design documentations, the standard value of natural frequency can be evaluated by Eigenvalue analysis of bridge pier structure. For this analysis, the reaction forces of the ground are expressed as spring constants. The actual reaction forces are not elastic and they are not modeled as a spring. However, for practical reasons, it may be treated as an elastic spring model because of small deformation.

 DOI: 10.1051/matecconf/201713802017

3 Case history

3.1 Outline of the test

Impact vibration tests were carried out for two piers (pier P4 and P5) of an old bridge in Haiphong city, a port city and the third largest city of Vietnam with a population of over two million people. The Bridge over Lach Tray river on a major urban route connecting city center to suburban areas was constructed in 1980 as shown in Fig. 3. As the old Bridge could not satisfy the increasing demand of traffic volume, it was decided that the bridge superstructure would be destructed and be replaced by a new one. However, two piers (pier P4 and P5) of the old Bridge were decided to be reused. As pier P4 and P5 have been under service with high traffic volume and river discharge for more than 35 years, it is necessary to evaluate their soundness condition to confirm the reuse possibility of these two piers.

Fig. 4 shows the overview of bridge piers after deduction of the old superstructure. Setting work for impact vibration test is shown in Fig. 5. The tests were carried out for both direction of bridge axis and direction perpendicular to bridge axis.

Fig.3. Overview of the Bridge over the Lach Tray river (before deduction of old superstructure)

Fig.4. Overview of bridge piers (after deduction of old superstructure)

Fig.5. Setting work for impact vibration test

3.2 Soundness evaluation of pier P4 and P5

Fig. 6 and Fig. 7 shows a time history of acceleration measured from impact vibration test of pier P4 and P5 subjected to the impact force using 40 kg iron ball. From the measured time history, spectral analysis was carried out to identify the natural frequency. Fourier and phase spectrums of pier P4 and P5 are also shown in Fig. 6 and Fig. 7, respectively.

The standard values of natural frequency of pier P4 and P5 were evaluated by Eigenvalue analysis of the pier structure. Before implementing impact vibration test, geotechnical investigation was carried out and the spring constants which express the interaction between ground and pier foundation piles were calculated based on this geotechnical investigation results. Analytical structural model of bridge pier for Eigenvalue analysis is shown in Fig. 8. Results of soundness evaluation of pier P4 and P5 are shown in Table 2. Based on evaluation results, the judgment of reuse of these two piers was made.

However, for pier P4 in the direction perpendicular to bridge axis, the soundness index α is rather low and it is recommended to continuously confirm natural frequency deterioration in the future.

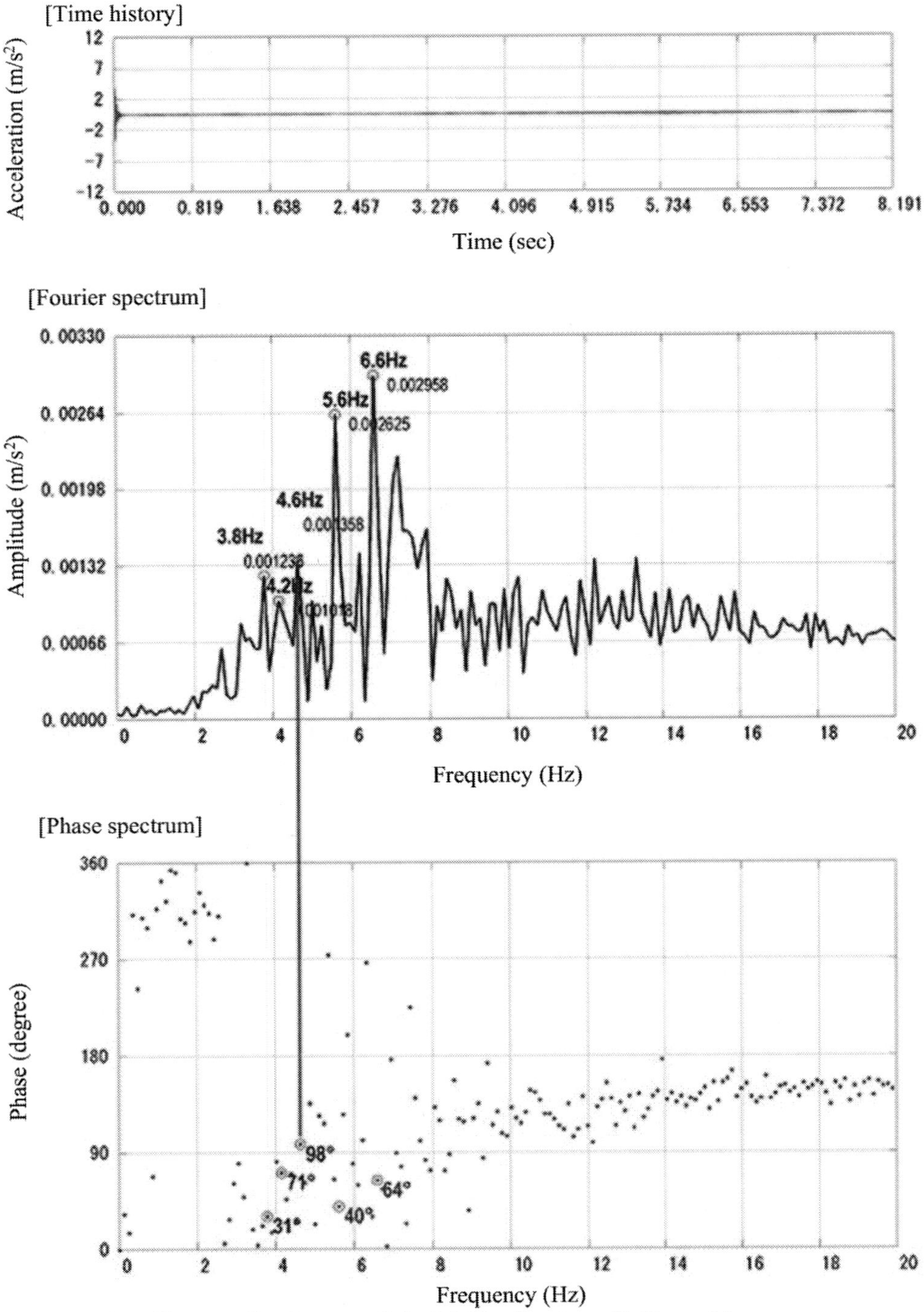

Fig.6. Results of impact vibration test of pier P4 (for direction of bridge axis)

DOI: 10.1051/matecconf/201713802017

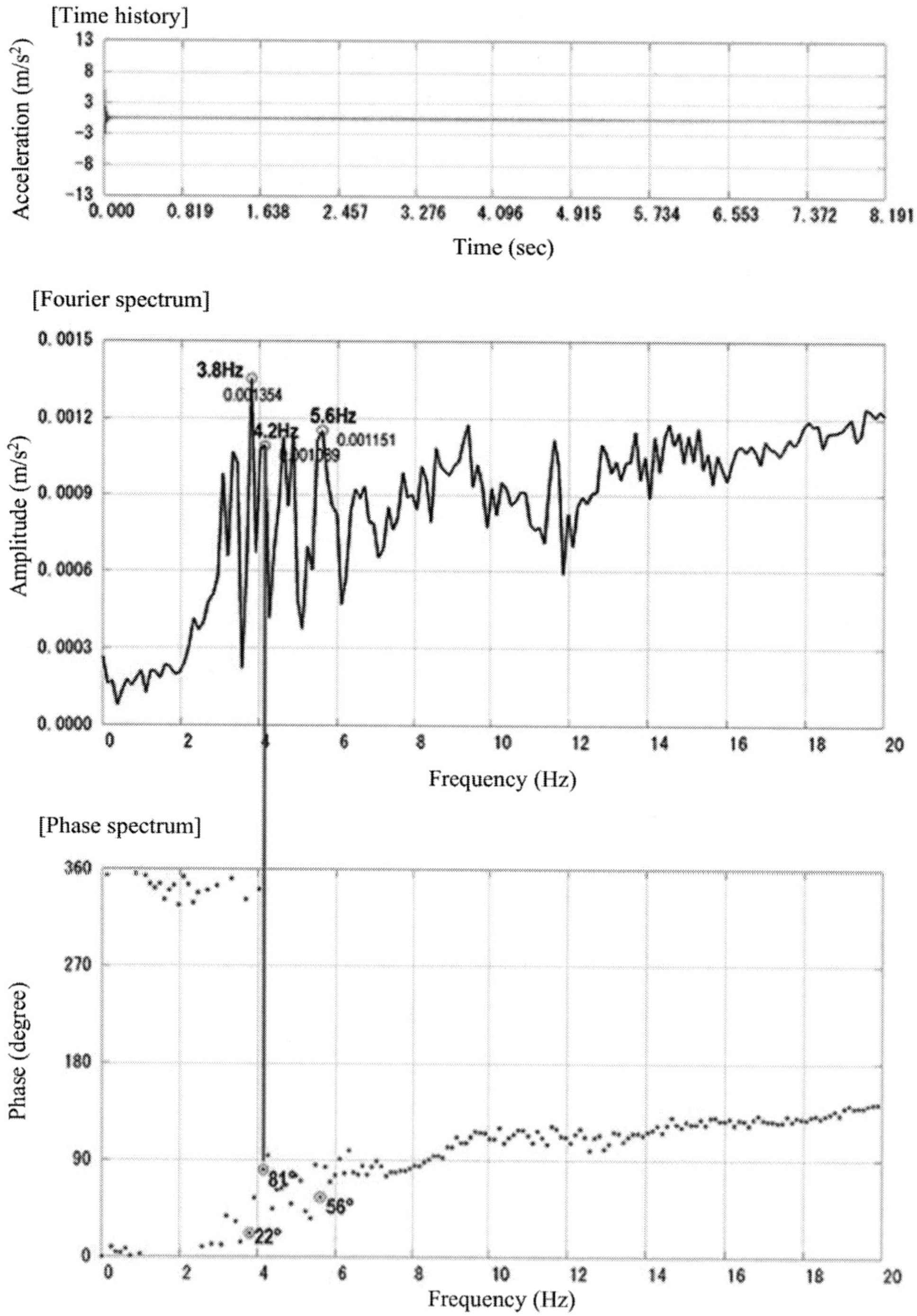

Fig.7. Results of impact vibration test of pier P5 (for direction of bridge axis)

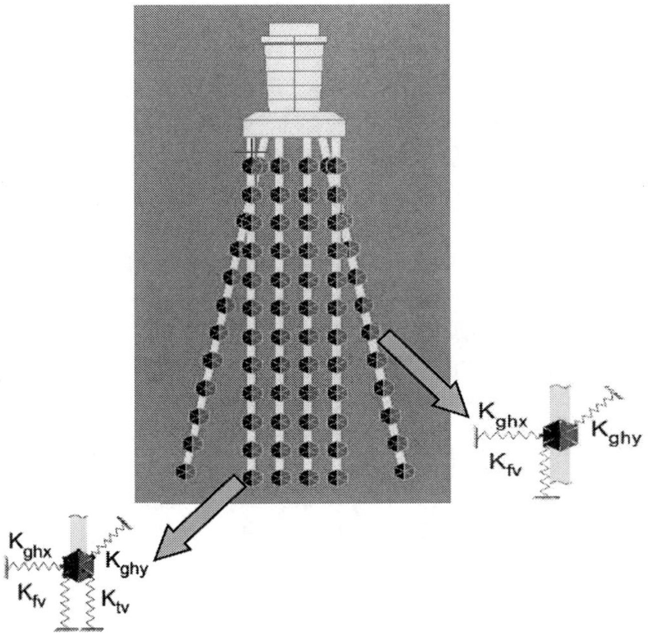

Fig.8. Analytical structural model for Eigenvalue analysis

Table 2. Results of soundness evaluation of pier P4 and P5

		Pier P4		Pier P5	
		Direction perpendicular to the bridge axis	Direction of bridge axis	Direction perpendicular to the bridge axis	Direction of bridge axis
Natural frequency	Measured results	3.9	4.6	4.8	4.2
	Standard value	5.40	5.39	4.27	4.37
Soundness evaluation	Soundness index α	0.72	0.853	1.12	0.96
	Evaluation rank	A2	B	S	B

4 Conclusion

The paper describes a practical technique to evaluate the soundness of bridge substructure by using the impact vibration test. Impact vibration test is one of the utilizable non-destructive dynamic test on the scour of bridge substructure as the test makes use of the difference of natural frequencies when the scour occurs and riverbed elevation around the bridge pier has changed. In order to estimate the measured natural frequency of bridge substructure, a spectrum analysis was conducted with a time history of the free vibration obtained from measurement data. Then the measured natural frequency is compared to a standard value to evaluate the soundness of bridge substructure. Application of the test method for substructure evaluation of two old bridge piers in Vietnam is introduced as a case study. Method to evaluate the standard value of natural frequencies by Eigenvalue analysis of analytical structural model is described in the case study. The application results show that impact vibration test has an advantage of nondestructive, quick and accurate evaluation.

References

1. A. Nishimura, S. Tanamura, RTRI report, *A study on integrity assessment of railway bridge foundation*, **Vol. 3(8)**, 41-49 (1989).

2. Railway Technical Research Institute, Design standard for railway earth structure (in japanese), **Maruzen** (2000).
3. N. Kobayashi, M. Shimamura, JR East technical review (Japanese version), *Development of a scour monitoring method for piers*, **No.3**, 49-52 (2003).
4. T. Sakata, T. Tsuchida, A. Okazaki, *First Japan-Vietnam workshop on urban infrastructure system and transportation*, Life cycle management of bridges and structures, 116-128 (2004).
5. S. Watanabe, M. Samizo, A. Fuchiwaki, T. Sugiyama, RTRI report, *Evaluation of the structural integrity of bridge pier foundation under flood condition by the natural frequency measured with a micro-tremor*, **Vol. 21(1)**, 31-36 (2007).
6. M. Shinoda, H. Haya, S. Murata, *Fourth International Conference on Scour and Erosion 2008*, Nondestructive evaluation of railway bridge substructure by percussion test, 285-290 (2008).
7. P. H. Kien, T. Sakata, A. Nishimura, *40 Years of Cooperation between Vietnam and Japan in Construction Sector: Achievements and Opportunities*, Structural health evaluation for bridge substructures by impact vibration test method, 281-290 (2013).

The Effects of Reduced Beam Section on Castellated Beam

Nini H Aswad[1*], *Herman* Parung[2], *Rita* Irmawaty[3] and *A Arwin* Amiruddin[3]

[1]Doctoral student, Civil Engineering Department, Hasanuddin University, Makassar, Indonesia

[2]Professor, Civil Engineering Department, Hasanuddin University, Makassar, Indonesia

[3]Associated Professor, Civil Engineering Department, Hasanuddin University, Makassar, Indonesia

Abstract. This paper presents reduced beam section (RBS) technique on castellated beam. Aim of this study is to ensure the plastic moment reach far from column face without significant of strength decreasing. Four models of castellated beam with various types of RBS and one model without RBS have been considered. ABAQUS FEA software package was used for modelling purpose. Loading condition, i.e. lateral load as representing earthquake load, was monotonic load using displacement control at the top of the column. The result shows that the maximum load and the maximum displacement with RBS and without RBS do not portray significant difference. The maximum load RBS castellated beam and without RBS were respectively 86.17 kN and 80.63 kN. It is indicated that the RBS on castellated beam might share relatively similar maximum load compared to non RBS castellated beam and the RBS can secure the stress concentration and plastic moment, the location of which is far from the column face. The reduced beam section (RBS) is put on beam-to-column moment connected on castellated beam.

1. Introduction

The various number of research activities has been conducted on the behavior of fully restrained steel connections in light of the 1994 Northridge earthquake, which some has emphasized on the Reduced Beam Section (RBS) moment connection. As regards the castellated beam connection, the RBS is denoted by a circular radius cut in both top and bottom flanges to reduce the flange area and consequently, the plastic moment capacity over a length of the beam near the ends of the beam span.

* Corresponding author : niniaswad@gmail.com

DOI: 10.1051/matecconf/201713802018

This connection was developed to assure greater reliability of steel beam-to-column connections. The joint should have adequate strength and stiffness to resist the internal force induced by the framing members and external force like earthquake and wind loading, and moreover, the castellated model behaved softer therefore dissipate more energy than normal plate beam (Vahid et.al, 2013). The experimental and numerical results observed that cyclic performance of the RBS moment connection was much superior to the connection without RBS (Swati & Gaurang, 2014). In an effort to gain additional insight into the behaviour of the tested specimens, this research investigates castellated beam reduced beam section (RBS) moment connection models by performing simulation on the computer program ABAQUS.

1.1. Material properties and Sections

In this study, profile of the beam is IWF 150 75 with hexagonal castellated beam following the standardized dimension as shown on figure 1.

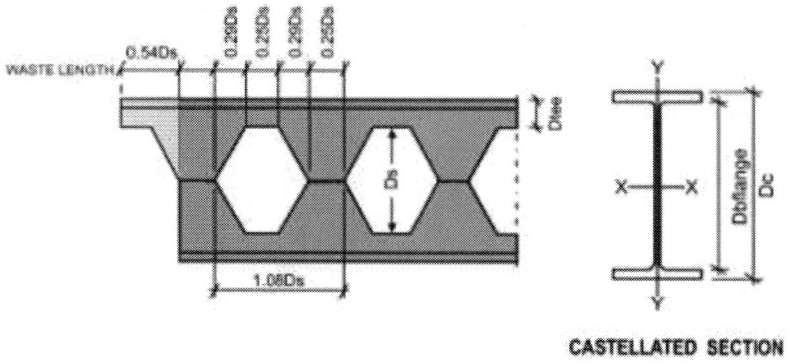

Fig. 1. Standardized ccastellated beam

This castellated beam was modeled in the finite element analysis, together with the column. Following the castellated process, the profiles of the beam for each type are W225x75 and W30x132, while the column profiles are adjusted to be H250x250 and W18x221 for the first two types and the last respectively. The castellated beam has properties 366.17 Mpa and 433.76 Mpa for yield and ultimate strength respectively. On the other hand, the column has 410.11 Mpa and 487.36 Mpa for yield and ultimate strength respectively.

2. Reduced beam section (RBS)

RBS consequently reduces stiffness of the beam and on the contrary, strengthens joint connection. In a RBS moment connection (Figure 2), portions of the beam flanges are selectively trimmed in the region adjacent to the beam-to-column connection. Yielding formation is intended to occur primarily within the reduced section of the beam. RBS connections are prequalified for use in special moment frame (SMF) and intermediate moment frame (IMF) systems within the limits of these provisions.

DOI: 10.1051/matecconf/201713802018

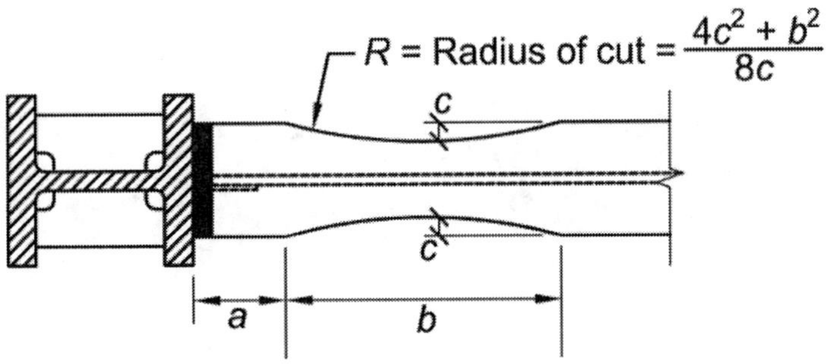

Fig. 2. Reduced beam section connection. (Prequalification limits)

Column sections and RBS dimension *a*, *b* and *c* (Figure 2) subject to the limits:

$$0.5b_{bf} \leq a \leq 0.75b_{bf} \quad (1)$$

$$0.65d \leq b \leq 0.85d \quad (2)$$

$$0.1b_{bf} \leq c \leq 0.25b_{bf} \quad (3)$$

Where b_{bf} is width of beam flange (mm), *a* is distance from face of column to the start of an RBS cut (mm), *b* is length of an RBS cut (mm) and *c* is depth of cut at centre of the reduced beam section (mm) and also *d* is depth of beam (mm). The protected zone shall be as defined for each prequalified connection. Unless specifically indicated in this standard, the protected zone of the beam shall be defined as the area from the face of the column flange to one-half of the beam depth beyond the plastic hinge. The protected zone shall meet the requirements of the AISC seismic provisions.

2.1. Description of model RBS castellated beam

Table 1 shows the description of steel material used for beam and column in each model. The specification of the table are models, spacing hole castellated beam from column, RBS length, cell diameter, cell spacing, and spacing RBS from column.

(a) Web castellated beam

 DOI: 10.1051/matecconf/201713802018

(b) Flange beam

Fig. 3. (a) Web castellated beam and (b) beam flange

Table 1. RBS and castellated beam specimen parameters

Model	Spacing Hole CB from column (mm)	diameter CB (mm)	spacing hole CB (mm)	spacing RBS from column (mm)	RBS Length (mm)
	a	b	c	d	e
WRBS-C	225	125	125	38,5	WRBS
C RBS 1	280	125	125	38,5	280
C RBS 2	310	125	125	38,5	310
C RBS 3	350	125	125	38,5	350

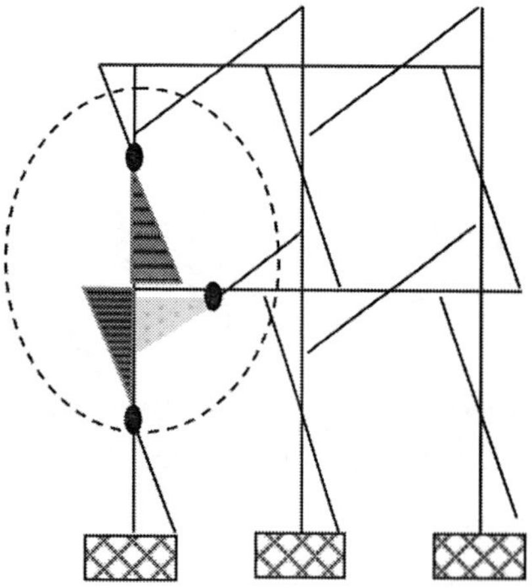

(a) Analysis model

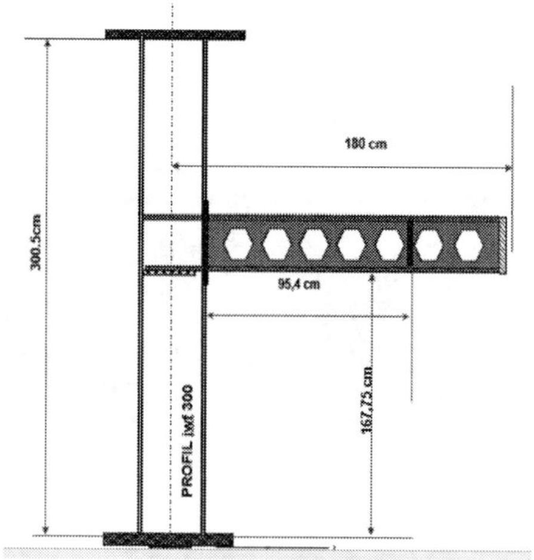

(b) Design of connection

Fig. 4. (a) Analysis model, **(b)** Design of connection

 DOI: 10.1051/matecconf/201713802018

From Figure 4(a), it can be seen the principal design of connection to meet hinge type and Figure 4(b) illustrates the design of connection used in the model.

3. Result of modelling

Based on analysis results, all models of RBS indicated crack on the flange, e.g. as shown on Figure 5. First model, WRBS-C, does not yield high stress concentration on its flange. However, 3 (there) other models, beam with RBS, demonstrate higher stress concentration on their flange section, including at top and bottom flange at the reduced beam section.

Fig. 5. Stress Distribution on RBS Castellated Beam

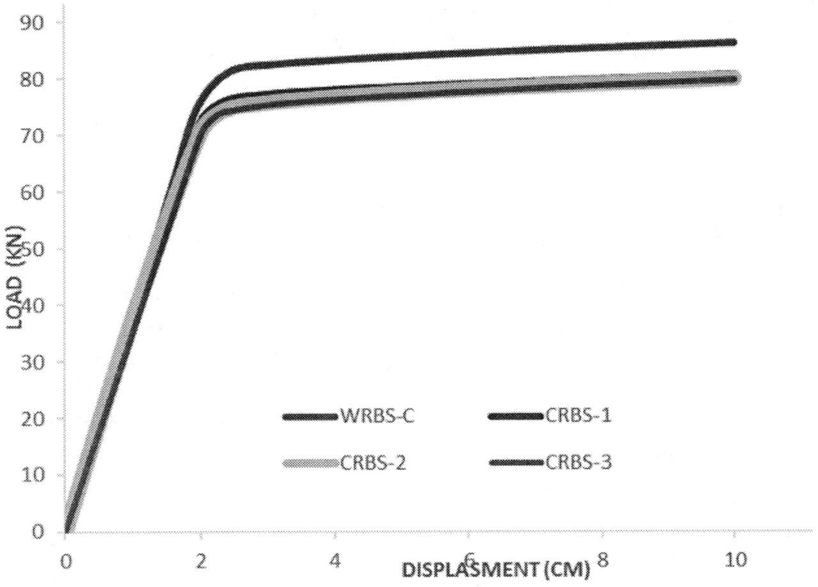

Fig. 6. Load - Displacement curve for monotonic loading

Figure 6 demonstrates the comparison of all models in load-displacement relationship, where maximum displacement was set at 10 cm. The highest maximum load carrying capacity from ABAQUS is WRBSC model i.e. 86.17 kN. Furthermore, the RBS models provided maximum load as follows, i.e. 80.63 kN, 79.85 kN, and79.8 kN for CRBS-1, CRBS-2 and CRBS-3 respectively**.** There is approximately six to seven percent strength difference between non RBS and RBS castellated beam. In addition, at 28 cm from column face (CRBS-1), it performs best result model compared with CRBS-2 and CRBS-3.

4. Conclusion

Four models of castellated beam with one type of non RBS and three various types of RBS were conducted in this study. Beam profiles were modeled based on their standardized size on castellation process. Those sections and material properties were then designed on ABAQUS to obtain numerical analysis result. It can be implied that the difference of the maximum load between RBS and non RBS was not significant, since the RBS location was set as plastic hinge position where with regard to plastic phase in steel material, the yield modulus of both RBS and non RBS share approximately similar value. Therefore, RBS technique on castellated beam provides ability in resulting plastic moment far from the column face without significant loss of strength.

References

1. American Institute of steel Construction *Prequalified Connection For Special and Intermediate Momen Frames for Seismic Aplications (AISC 358-10)* Chicago. (2010)
2. Cheek-Ho Lee and Jae-Hoon Ki, *Seismic Design of Reduced Beam Section (RBS) Steel Moment Connections with Bolted Web Attachment, College of Engineering* (2004)
3. Chambers, J.J., Almudhafer, S. and Stenger F., Effect of Reduced Beam Section Frames Elements on Stiffness of Moment Frames, Journal of Structural Engineering, American Society of Civil Engineers, Vol. **129,** No. 3. (2003)
4. Chen, S.J. and Chao, Y.C., *Effect of Composite Action on Seismic Performance of Steel Moment Connections with Reduced Beam Sections*, Journal of Constructional Steel Research, Elsevier Science Publishers, Vol. **57**. (2001)
5. Chen SJ, Yeh CH, Chu JM. *Ductile steel beam-to-column connections for seismic resistance. Journal of Structural Engineering*, ASCE;**122**(11) : 1292–9 (1996)
6. Mehrdad Memari, Hussam Mahmoud, *Performance of Steel Moment Resisting Frame with RBS Connection Under Fire Loading*, Journal Engineering Structure, Volume **75**, page 126-138, Colorado State University, Fort Collins, CO, USA (2014)
7. Swati, A.K, Gaurang, V., *Study of Steel Moment Connection with and without Reduced Beam Section*, Case Studies in Structural Engineering, Volume **1**, pages 26-31. (2014)

8. Tonapa, S.R., Sandy, D., and Parung, H. "Behaviour of castellated beam-columns subjected to monotonic and cyclic loadings". International Journal of Engineering and Science Applications, pp 147-152. (2015).

9. Vahid, S.Z., Osman, S.A., and Khalim, A.R. "Monotonic and Cyclic Loading Simulation of Structural Steelwork Beam to Column Bolted Connections with Castellated Beam". Journal of Engineering Science and Technology, pp 416-427. (2013).

10. Yousef Ashrafi, Behzad Rafezy and W. Paul Howson, Evaluation *of the Performance of Reduced Beam Section (RBS) Connections in Steel Moment Frames Subjected to Cyclic Loading,* Vol **II** WCE July 1 - 3, London. (2009)

Use of Stress Wave to Evaluate The Repair Quality of Concrete Crack

Keng-Tsang Hsu [1], *Chia-Chi Cheng*[2], *Chih-Hung Chiang*[2], *and Hung-Hua Wang*[3]

[1]Assistant Professor, Department of Construction Engineering, Chaoyang University of Technology, Taiwan
[2] Professor, Department of Construction Engineering, Chaoyang University of Technology, Taiwan
[3] Postgraduate, Department of Construction Engineering, Chaoyang University of Technology, Taiwan

Abstract. The objective of this paper is to develop a nondestructive test technique based on stress wave propagation for evaluating the repair quality of concrete. Concrete structures frequently exist various types of cracks due to inferior environment and excessive loading. There are several crack repairing technique used in concrete construction. Epoxy-injection is the one used mostly in crack repairing. In this paper, a newly developed technique based on the stress wave was used to evaluate the repair quality of concrete crack. Numerical results show that the stress wave propagation of the impact-echo tests is capable of identifying the infill of epoxy-injection in the crack of repaired structures. The new technique under development has a high potential for evaluation of the quality of the repaired cracks.

1 INTRODUCTION

There is an increasing need for structural repairing to extend the service life of deteriorated concrete structures. The cracks of concrete structure are the important indicator for the structure health diagnosis. Because the development of strong epoxy adhesive, epoxy resin injection becomes the most popular method of structural repairing for reinforced concrete members in the repair industry. The repairing method which injects Epoxy resin into the crack of concrete members is relatively simple and cost-effective, increases the strength of existing concrete members, blocks moisture penetration of reinforced concrete members and reduces the chance of corrosion.

The most important factor in determination of the repairing quality is the injection condition of the crack. Epoxy adhesives are used as an injection material to repair cracks but air voids at the crack will reduce the repair strength. To ensure the repairing quality, the techniques able to detect air voids of the crack after concrete members been repaired are needed. In this paper, a newly developed technique based on the Time-of-Flight Diffraction Technique was used to evaluate the repair quality of concrete crack. The research will started from numerical simulation of the stress wave propagation.

2 TIME-OF-FLIGHT DIFFRACTION TECHNIQUE

To measure the depth of surface-opening cracks in concrete, two receivers are used and located on the opposite sides of the crack. Figs. 1(a) shows the instrument arrangement. The first receiver is located on the impact side and has a distance H0 to the impact point. The second receiver is located on the opposite side of the crack. Figs. 1(b) and (c) show typical waveforms recorded at the first and second receivers, respectively. After impact, the R-wave arrival at the first receiver will trigger the monitoring system. If the arrival time, t_1,

 DOI: 10.1051/matecconf/201713802019

and the speed, C_R, of R-wave are known, the time of impact initiation can be calculated as (t_1 - H_0/C_R). The value of t_1 is negative because a pre-trigger mode is used in recording signals to avoid losing important information. The region behind the crack cannot be penetrated by waves generated by impact until diffraction at the crack tip occurs. Thus, the second receiver initially responds to the arrival of the diffracted P-wave. The arrival time, t_2, of the diffracted P-wave can be obtained by identifying the initial disturbance in the waveform recorded at the second receiver. Thus, the shortest travel time for P-wave from the impact point to the second receiver can be obtained as

$$\Delta t = t_2 - (t_1 - H_0/C_R) = t_2 - t_1 + H_0/C_R \quad (1)$$

Knowing the P-wave speed, C_P, in concrete, the shortest travel length of P-wave from the impact point to the second receiver can be calculated as $C_P \times \Delta t$. If the distances from the crack to the impact point and the second receiver are H1 and H2 respectively, the depth, d, of the vertical surface-opening crack can be determined by the following equation:

$$d = \sqrt{\left[\frac{(Cp \times \Delta t)^2 + H1^2 - H2^2}{2 \times Cp \times \Delta t}\right]^2 - H1^2} \quad (2)$$

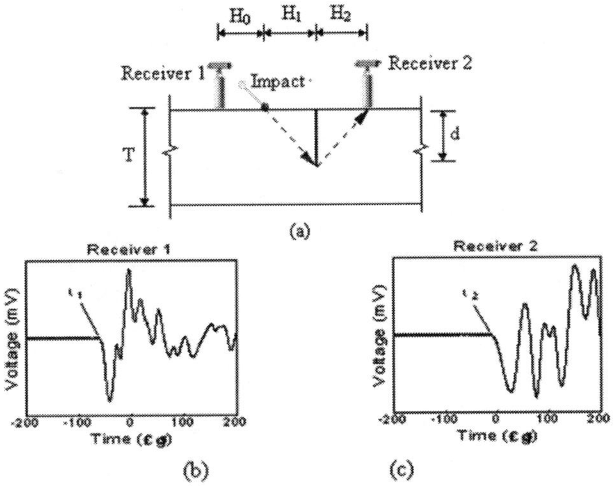

Fig. 1. Test scheme: (a) instrument arrangement; (b) waveform 1; (c) waveform 2.

3 NUMERICAL SIMULATION

In this paper, numerical simulations of the stress wave responses of crack were carried out using the explicit finite-element program LS-DYNA. All cases are simulated using 2-D axi-symmetric, 4-node elements and isotropic models. Concrete is modeled as an isotropic material with a Poisson's ratio 0.2 and a density 2300kg/m^3. A Young's modulus (E) of 33.1 GPa is chosen to provide a P-wave speed of 4000 m/s, the material parameters of Epoxy resin and air used in numerical simulation are also shown in Table 1. The time-history of nodal vertical displacements were recorded every 1.334 μs; and 1024 data were recorded in each simulation.

Table 1. Material parameters of numerical simulation

Material	Wave speed C_P (m/s)	Density (kg/m^3)	Elastic modulus (pa)	Acoustic Impedance Z , (C_P*ρ)
Air	340	1.2	-	408
Epoxy Resin	2400	1141	5.602E9	2738400
Concrete	4000	2300	33.10E9	9200000

3.1 Numerical simulation of the time-of-flight diffraction with variable of different contact time

The correctness of the numerical simulation is carried out in this section. In numerical simulation model, the depth of the crack is 10 cm, the shape of force-time functions are half-sin which is used for introducing stress wave into numerical model as impactor. The distance of impact is setting at 4 cm(H_1) beside the crack. The time-history of nodal vertical displacements were recorded at the distance of 4 cm(H_2) located on the opposite side of the crack. Three force-time functions with different contact times as 10μs, 20μs, and 30μs are chosen to perform the numerical model.

The results of time-history of nodal vertical displacements with different contact time as 10 μs, 20 μs, and 30 μs are shown in Fig.2. The nodal vertical displacements correspond to the arrival of the diffracted P-wave and the arrival time(t_2) of the diffracted P-wave is the same, 55 μs, for different contact time(10μs, 20μs, and 30μs). According the results of numerical simulation, the depth of the crack can be calculated by using Eq.(2) as 9.95 cm. Comparing with 10 cm, the error percentage of numerical simulation results is below 1%, so the numerical simulation was proved to be valid.

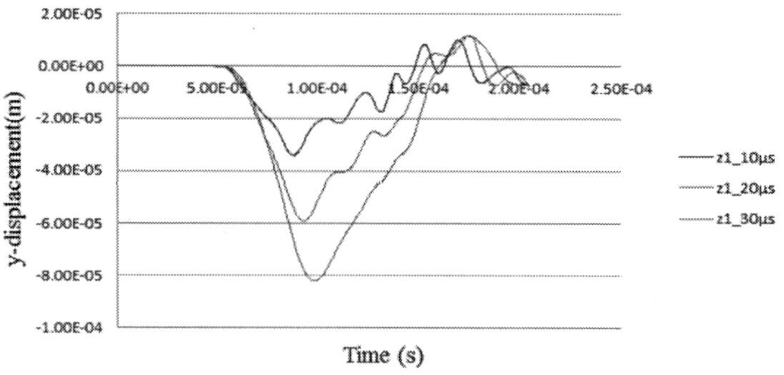

Fig. 2. waveform of time-of-flight diffraction with different contact-time

3.2 Numerical simulation of different injection condition of crack repairing

This section examines the stress wave propagation after the epoxy resin material filled in the crack with different partly filled scenarios. In numerical simulation, the depth of crack is also setting to 10 cm and the contact time of force-time function is 20μs in all cases. The variables of different epoxy resin injection scenarios in numerical simulations are shown in Table 2 and Fig. 3. For all the models, the epoxy is filled from the top surface and the air pack of the crack is underneath the epoxy.

DOI: 10.1051/matecconf/201713802019

Table 2. Variables of different epoxy resin injection scenarios

Code	Length of Epoxy Resin (cm)	Length of Air (cm)	Remark
E0 cm_A10 cm	0	10	Non- epoxy injection
E2 cm_A8 cm	2	8	Partial epoxy injection
E4cm_A6 cm	4	6	Partial epoxy injection
E6 cm_A4 cm	6	4	Partial epoxy injection
E8 cm_A2 cm	8	2	Partial epoxy injection
E10 cm_A0 cm	10	0	Fully epoxy injection
E0 cm_C10 cm	0	-	Non-crack case

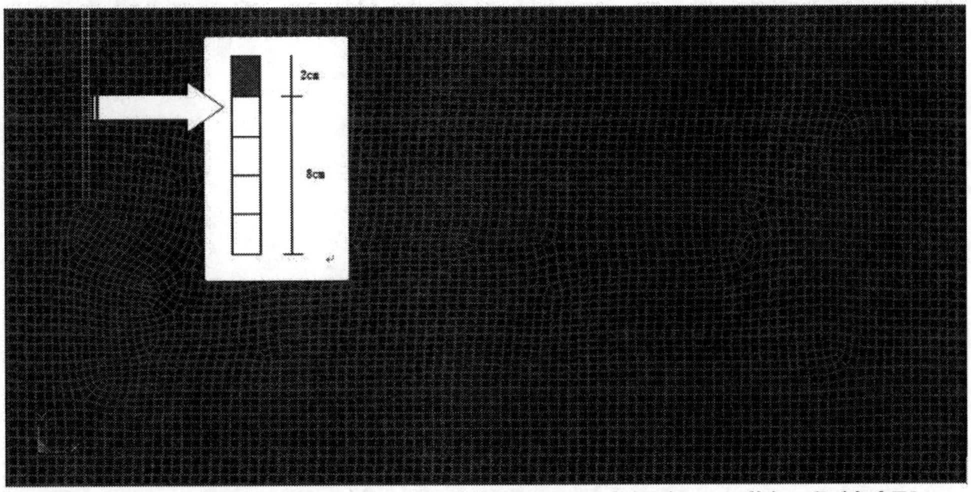

Fig. 3. Numerical model of stress wave with different epoxy injection condition (Labled E2 cm_A8 cm)

The results of time-history of nodal vertical displacements with different condition of epoxy resin injection are shown in Fig.4. The amplitude of R_{max} in time-history of nodal vertical displacement waveform is used as the parameter for investigation as shown in Fig. 5. The time-histories of nodal vertical displacements are varied with different scenarios of epoxy resin injection . Table 3 summarizes the maximum displacement of waveform (R_{max}) for all the cases. The maximum value of nodal vertical displacements (R_{max}) is obtained from the non-crack case. The R_{max} of non-crack case(E0 cm_C10 cm) from the Table 3 can be compared with the fully epoxy injected case(E10 cm_A0 cm), the value of R_{max} attenuation was 11.4%. In addition, the maximum displacement of the R-wave decreases with the decrease of length of epoxy resin injection.. Loss percentage of R_{max} with different epoxy resin injection condition was shown in Fig.6.

Table 3. Maximum displacement of waveform with different epoxy resin injection condition

Code	Maximum displacement of waveform (R_{max}), m	Loss percentage of R_{max} (%) (Base on E0 cm_C10 cm)	Remark
E0 cm_A10 cm	-5.90E-05	72.0	Non- epoxy injection case
E2 cm_A8 cm	-1.09E-04	48.3	Partial epoxy injection case
E4cm_A6 cm	-1.46E-04	30.8	Partial epoxy injection case
E6 cm_A4 cm	-1.76E-04	16.6	Partial epoxy injection case
E8 cm_A2 cm	-1.86E-04	11.8	Partial epoxy injection case
E10 cm_A0 cm	-1.87E-04	11.4	Fully epoxy injection case
E0 cm_C10 cm	-2.11E-04	0.0	Non-crack case

DOI: 10.1051/matecconf/201713802019

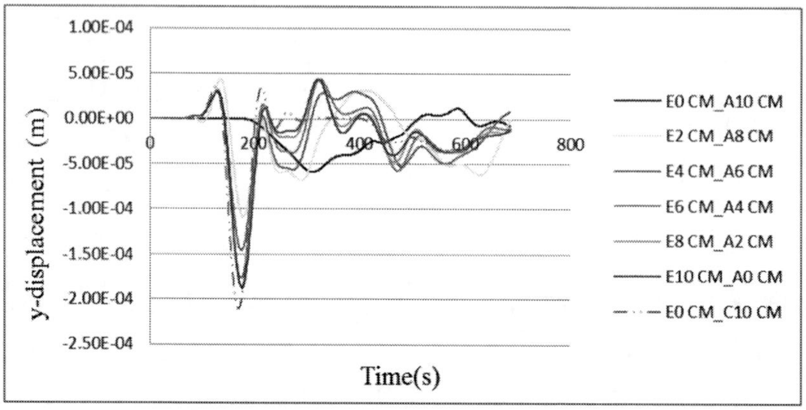

Fig. 4. Time-history of nodal vertical displacements with different epoxy resin injection condition

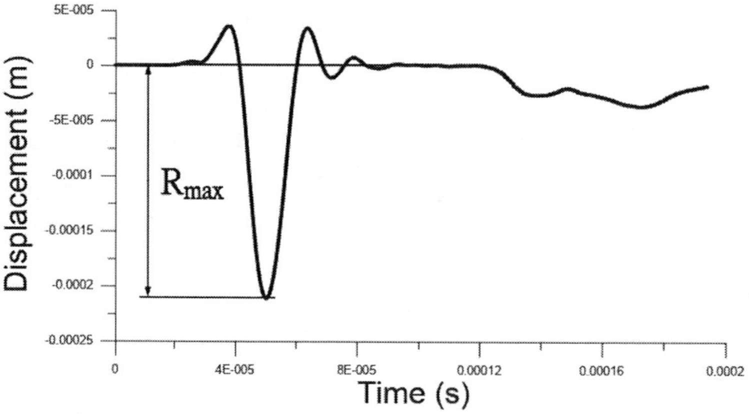

Fig. 5. The length of R_{max} in time-history of nodal vertical displacement waveform

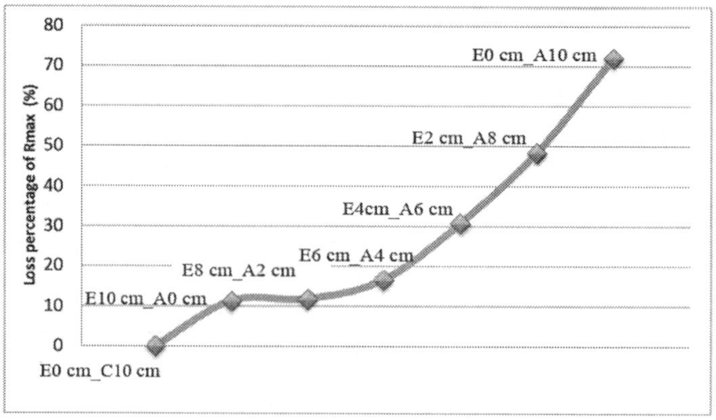

Fig. 6. Loss percentage of R_{max} with different epoxy resin injection condition

4 CONCLUSIONS

Numerical simulation results show that for surface opening crack partially filled by epoxy resin, the maximum displacement of the R-wave can be served as a parameter to evaluate the different injection condition of crack. The maximum displacement of the R-wave decreases with the decrease of the length of epoxy resin in the crack. The technique has a great potential for evaluation of the repair quality of concrete structural cracks injected by epoxy resin.

References

1. Nikolaos Zoidis, Efthymios Tatsis, Christos Vlachopoulos, Anastasios Gotzamanis, Jesper Stærke Clausen, Dimitrios G. Aggelis, Theodore E. Matikas, "Inspection, evaluation and repair monitoring of cracked concrete floor using NDT methods", Construction and Building Materials, Volume 48, November 2013, Pages 1302-1308.
2. Hongqiang Chu, Linhua Jiang, Chuansheng Xiong, Lushen You, Ning Xu, "Use of electrochemical method for repair of concrete cracks", Construction and Building Materials, Volume 73, 30 December 2014, Pages 58-66.
3. D.G. Aggelis, T. Shiotani, "Repair evaluation of concrete cracks using surface and through-transmission wave measurements", Cement and Concrete Composites Volume: 29, Issue: 9, October, 2007, pp. 700-711.
4. Aggelis, D.G., Hadjiyiangou, S., Chai, H.K., Momoki, S., Shiotani, T., "Longitudinal waves for evaluation of large concrete blocks after repair ", NDT and E International Volume: 44, Issue: 1, January, 2011, pp. 61-66.
5. D.G. Aggelis, T. Shiotani, D. Polyzos, "Characterization of surface crack depth and repair evaluation using Rayleigh waves", Cement and Concrete Composites, Volume 31, Issue 1, January 2009, Pages 77-83.
6. Lin, Y., Hsu, K.T., and Cheng, C.C., "Evaluating Bond Quality at Steel/Concrete Interfaces Using the Normalized Impact-Echo Spectrum," Modern Physics Letters B, vol.**11**, no.22, pp.1001-1006,(2008)
7. C. C Cheng , K.C. Pei, and J.H. Wu, "Development of imaging techniques for evaluating the rc plate containing epoxy-repaired delamination," Key Engineering Materials, vols **321-323**,pp.348-351., Aug, (2006).
8. K. T. Hsu, C. C. Cheng, and Y.C. Lin, "Use Impact-echo Method to Evaluate Bond of Reinforced Concrete Subjected to Early-Age Vibration," Journal of Solid Mechanics and Materials Engineering, Vol. **2**, No. 12, Dec, (2008).

 DOI: 10.1051/matecconf/201713802020

Structural damage detection using MAC-Fast Multi Swarm Optimization technique (MAC-FMSO)

Richard Frans[1] , and *Yoyong* Arfiadi[2,*]

[1]Department of Civil Engineering, Universitas Atma Jaya Makassar, Indonesia
[2]Department of Civil Engineering, Universitas Atma Jaya Yogyakarta, Indonesia

Abstract. Damage detection of structures is important to ensure the service of civil engineering infrastructures. Many techniques have been developed for damage detection of structures in the literature. In this paper a new technique, a so called MAC-FMSO technique, is developed by using a combination of Modal Assurance Criterion (MAC) and Fast Multi Swarm Optimization (FMSO) techniques. By using this method, not only the damage location can be assured, but also the loss of strength and loss of stiffness of structures might be quantified. The first stage is to obtain the damaged member's location by using second level of damage detection techniques, such as mode shape curvature method, damage locating vector, or strain energy method. After obtaining the damaged member, the MAC-FMSO technique is used to obtain the loss of stiffness from the candidate of damaged members. The results show that the MAC-FMSO algorithm has the ability to predict the loss of stiffness in the damaged members. In this case, the MAC-FMSO technique can be considered as an efficient and effective tool at the third level of damage detections. This is because this technique successfully measures the loss of strength or loss of stiffness of the damage member or part of structures. At the end of the paper, the MAC-FMSO technique is applied to predict the location of damage and loss of stiffness of three different structures. The first application is a simply supported beam with three different damage scenarios; the second application is a 13-bar plane truss structure with two different scenarios, and the third application is a shear building structure with three different scenarios. The results of simulation show that the MAC-FMSO technique can accurately predict the loss of stiffness of the damaged member as well as the location of the damage.

Keywords: damage identification, loss of stiffness, fast multi swarm optimization, modal assurance criterion, MAC-FMSO

* Corresponding author: yoyong@mail.uajy.ac.id

1 Introduction

Damage detection of structures can be classified into several levels of identification: (a) level 1: detecting the presence of damage, (b) level 2: locating the damaged members, (c) level 3: determining the loss of strength or stiffness reduction of the damaged members, and (d) level 4: predicting the remaining life time. This paper discusses a simple technique on how to obtain the loss of stiffness of damaged members by using the so called MAC-FMSO algorithm. The MAC-FMSO algorithm can be categorized as a third level of damage detection method. This is because the ability of MAC-FMSO to find the damaged member or damaged part of the structure, and determine the stiffness reduction of the damaged members. The MAC-FMSO consists of two stages in order to obtain the loss of stiffness of the damaged member. The first stage is to use second level of damage detection method to obtain the location of the damage member (e.g.: mode-shape curvature method, damage locating vector method, strain energy method, etc.). The second stage is to use modal assurance criterion (MAC) combined with fast multi swarm optimization (FMSO) to get the stiffness reduction of the damaged members.

2 Modal Assurance Criterion (MAC)

One of the techniques that is used to measure the similarity of two mode shapes is modal assurance criterion (MAC). The modal assurance criterion was originated due to the need for quality assurance indicator of the modal vectors obtained from the experimental testing that are estimated from measured frequency response function [1]. The MAC is to provide a statistical measure of consistency between estimates of a modal vector. The resulting MAC is between 0 and 1. If the modal assurance criterion has a value near zero, this means that the modal vectors are not consistent. On the other hand if the modal assurance criterion has a value near unity, this is an indication that the modal vector are consistent [1]. Because of MAC is based on the minimization of the squared error between two vector spaces, therefore the MAC is sensitive to large differences between comparative values and insensitive to small comparative values.

In terms of mode shape of damaged structure (ϕ_d) and mode shape of predicted structural model (ϕ_a), the MAC can be defined as:

$$MAC_j = \frac{\left|\{\phi_a\}_j^T \{\phi_d\}_j\right|^2}{\left(\{\phi_a\}_j^T \{\phi_a\}_j\right)\left(\{\phi_d\}_j^T \{\phi_d\}_j\right)} \quad (1)$$

where $\{\phi_a\}_j$ is mode shape vector of predicted structure for *j*-mode, and $\{\phi_d\}_j$ is mode shape vector of damaged structure for *j*-mode. If the mode shape is identical or similar, the MAC value is 1, otherwise if the mode shape has no correlation, the MAC value is zero.

3 Fast Multi Swarm Optimization (FMSO)

Fast multi swarm optimization (FMSO) that is used in this paper follows the development in [2], from the particle swarm optimization (PSO) technique which was first proposed by [3]. The difference between PSO and FMSO is that the update position and velocity not only depend on equation (2) and (3) but is also influenced by equation (5) as follows:

$$v_{i,j}(t+1) = wv_{i,j}(t) + c_1R_1(p_{best,i,j} - x_{i,j}(t)) + c_2R_2(g_{best,i,j} - x_{i,j}(t)) \quad (2)$$

$$x_{i,j}(t+1) = x_{i,j}(t) + v_{i,j}(t+1) \tag{3}$$

where: $v_{i,j}(t+1)$ is the updated velocity of particle, $x_{i,j}(t+1)$ is the updated location of particle, t represents the iteration-t, $p_{best,i,j}$ is local best location of particle at t-iteration, and $g_{best,i,j}$ is global best location of particle at t-iteration. R_1 and R_2 are random numbers from interval 0-1, c_1 and c_2 are particle acceleration constants. In this paper, $c_1 = c_2 = 2$, and w is positive inertia weight coefficient as a function of ρ_{min} and ρ_{max} as follows:

$$w(t) = \rho_{max} - ((\rho_{max} - \rho_{min})t / maxt) \tag{4}$$

$$x_{i,j}(t+1) = x_{i,j}(t) + (v_{i,j}(t)\, exp(\delta))\delta_i \tag{5}$$

where δ and δ_i denote Cauchy random numbers.

By adding the update equation (5), PSO greatly increases the probability of escaping from the local optimum because the variance of Cauchy distribution is infinite. In this case, Cauchy mutation makes the particles to have a long jump.

In addition, FMSO also uses crossover operation, in order to enhance the capability for searching the optimum solution. For crossover operation, if $rand() < q_c$, the crossover operation is taken as follows:

$$x_{i,j}(t+1) = (1-\alpha) x_{i,j}(t) + \alpha\, p_{best,i,j} \tag{6}$$

$$v_{i,j}(t+1) = rand()\,(p_{best,i,j} - x_{i,j}(t)) \tag{7}$$

where: α is random number with 0-1 interval, and q_c is crossover rate.

4 Fitness and penalty functions

The fitness function is the summations of MAC value from n-modes considered of two different mode shapes, i.e., ϕ_a and ϕ_d. The value of MAC interval is between 0 (no correlations) and 1 (identical). Because in the optimization, we seek the minimum fitness, while the purpose of MAC is to maximize the strong correlation, hence to maximize the function, the fitness should be modified as follow:

$$F = \frac{1}{\sum_{j=1}^{n} MAC_j} \tag{8}$$

where F is the fitness value, MAC is modal assurance criterion of two mode shapes compared, and n is the number of mode shapes to be considered.

In this paper, the first five mode-shapes were considered, hence the fitness value in equation (8) has interval range from 0.2 to ∞. If the fitness equals to 0.2, it represents the two modes are identical; while if the fitness is ∞, this is an indication that no correlation between the two mode-shapes.

The penalty function is activated while one of the particle is less than 0 or larger than 1. This is because particle represents the loss of stiffness, which is in the interval between 0 and 1. Consequently, when this is the case, the fitness value automatically becomes a maximum value that can be accepted by the program. Fig.1 shows the step of MAC-FMSO algorithm.

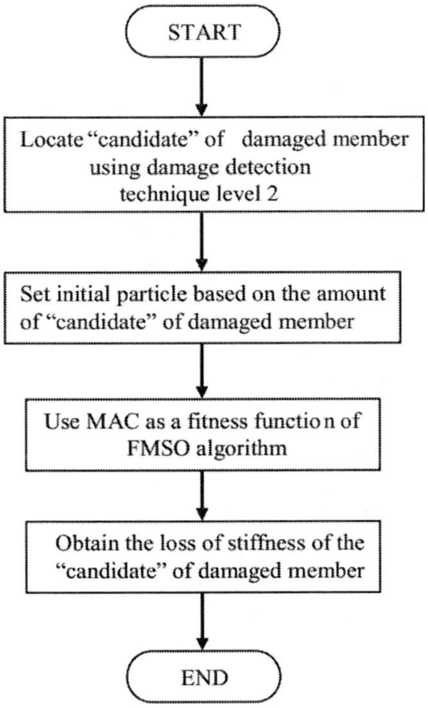

Fig. 1. Procedure of MAC-FMSO for damage detection

5 Case studies

5.1 Simply supported beam model

The first case study is a simple beam as shown in Figure 2. This model is taken from [4] with some modifications in the damage scenarios. The damage scenarios can be seen in Table 1, while the structural properties of the beam are shown in Table 2. In this example, the loss of stiffness is represented by reducing the modulus of elasticity of the damaged member. The structure is divided into 20 parts. The stiffness matrix method is used for structural analysis.

Table 1. Damage scenarios of beam structure

<table>
<tr><th colspan="2">First Scenario (Single Damage)</th><th colspan="2">Second Scenario (Double Damage)</th><th colspan="2">Third Scenario (Triple Damage)</th></tr>
<tr><th>Damaged Member</th><th>Loss of Stiffness</th><th>Damaged Member</th><th>Loss of Stiffness</th><th>Damaged Member</th><th>Loss of Stiffness</th></tr>
<tr><td rowspan="3">10</td><td rowspan="3">50%</td><td>10</td><td>50%</td><td>4</td><td>70%</td></tr>
<tr><td rowspan="2">15</td><td rowspan="2">75%</td><td>10</td><td>50%</td></tr>
<tr><td>15</td><td>75%</td></tr>
</table>

Following Fig.1, the first step is to locate the damaged member, by using level-2 of damage detection technique, to eliminate the healthy member from the "candidate" of the damaged members as the particle in the swarm.

 DOI: 10.1051/matecconf/201713802020

Modal curvature mode shape method [5] was used to locate the damaged member. The mode shape index (MSI) of the first damage scenario is shown in Fig. 3. For this case, the first five natural frequencies and mode shapes were taken to locate the "candidate" of the damaged member (part) using mode shape curvature method (MSC).

Table 2. Section and material properties of beam

Properties	Initial Condition
Width (mm)	20
Height (mm)	200
Area section (mm^2)	4000
Modulus of elasticity (MPa)	200000

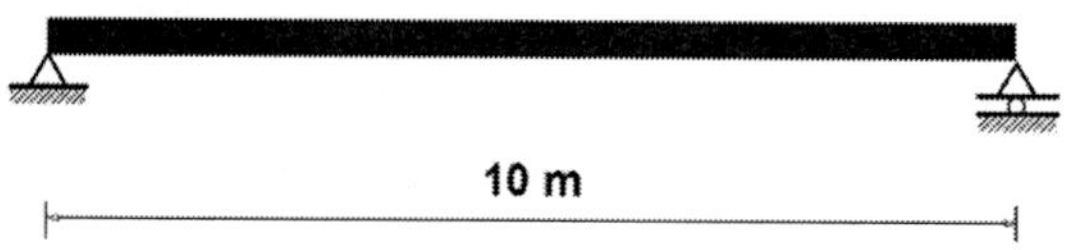

Fig. 2. Simply supported beam

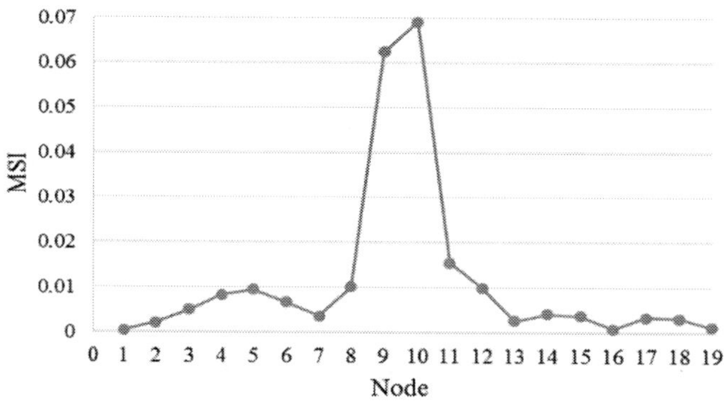

Fig. 3. MSI for first damage scenario

Table 3. Result of stiffness reduction using MAC-FMSO for simply supported beam

First scenario			
"Candidate" of damaged member/part	10		
Stiffness reduction result using MAC-FMSO	50%		
Second scenario			
"Candidate" of damaged member/part	10	15	
Stiffness reduction result using MAC-FMSO	50%	75%	
Third scenario			
"Candidate" of damaged member/part	4	10	15
Stiffness reduction result using MAC-FMSO	70%	50%	75%

After locating the "candidate" of the damaged member, the MAC-FMSO was used to find the loss of stiffness of the "candidate" of damaged members. Table 3 shows the result of MAC-FMSO for each damage scenario.

In order to ensure the convergence, the program was run for four times for each damage scenario, with different lower and upper bounds as shown in Table 4. The parameters of optimization using FMSO are: maximum iteration = 100, number of particles = 10, maximum velocity = 100, crossover rate = 0.8, ρ_{max} = 0.9, and ρ_{min} = 0.4.

Table 4. Lower bound and upper bound limit

Run	Lower Bound	Upper Bound
1	0	1
2	0	100
3	-10	0
4	-10	100

Fig. 4 shows the fitness value of each iteration for first, second, and third scenario, respectively. From Fig. 4 it is clear that MAC-FMSO can accurately predict loss of stiffness of the damaged member/part for all damage scenarios.

5.2 Plane truss model

The second structure considered for MAC-FMSO application is a 13-bar plane truss structure as can be seen in Fig.5. This structure is taken from [4]. Two damage scenarios were considered as follows: (a) a single damage member with loss of stiffness in member-11 is 75% from initial stiffness, and (b) a multiple damage with losses of stiffness in member-8 and member-11 are 50% and 75%, respectively. The stiffness matrix method is used to develop the model.

Fig.6 shows the normalized stress index of first and second damage scenarios by using damage locating vector proposed by [6]. According to [6] member that has normalized stress index value is almost to zero can be considered as a damaged member. From Fig. 6(a) member-11 can be considered as a damaged member, while From Fig. 6(b) and member 8, 11, and 12 might be categorized as a damaged member.

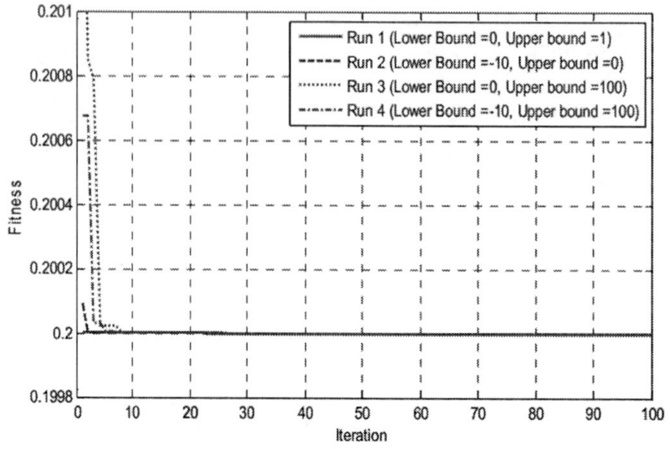

(a)

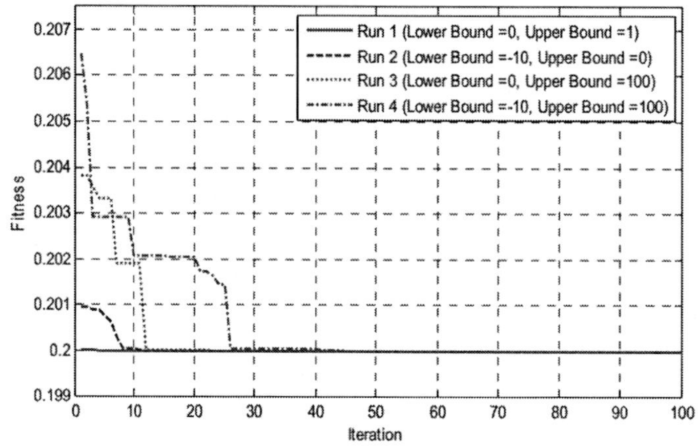

(b)

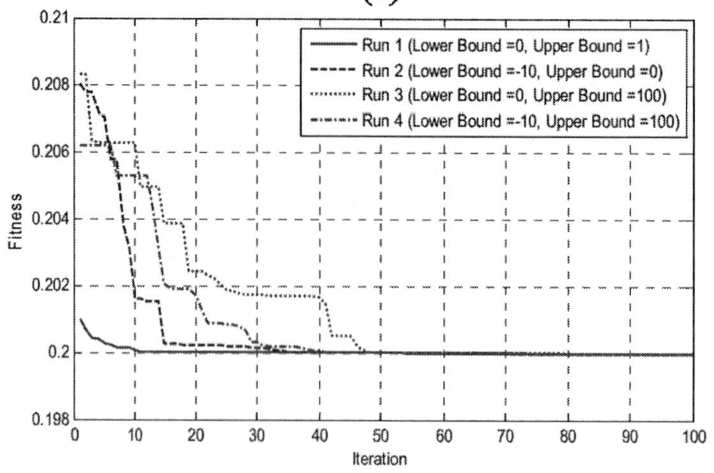

(c)

Fig. 4. Fitness degradation for (a) first scenario, (b) second scenario, and (c) third scenario for simply supported beam

DOI: 10.1051/matecconf/201713802020

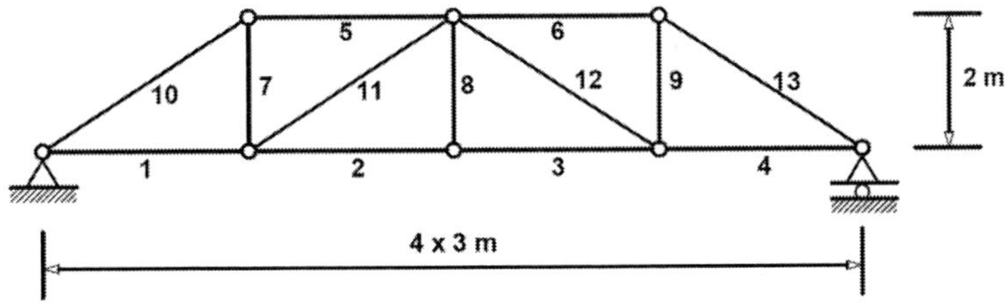

Fig. 5. 13-bar plane truss

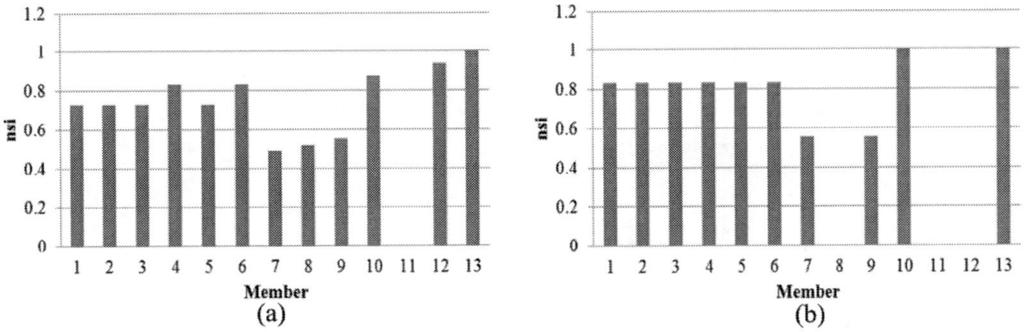

Fig. 6. Normalized stress index (nsi) for (a) first damage scenario, (b) second damage scenario

After having "candidate" of the damaged members, MAC-FMSO was applied to obtain the loss of stiffness of the damaged member candidate. Lower and upper bound candidates, as well as parameters of FMSO that are used for this case are similar to the first case (simply supported beam). In addition, the program was also run four times to ensure the convergence of the result.

Table 5 shows the results of MAC-FMSO. It can be shown that the MAC-FMSO can accurately predict the stiffness degradation for all damage scenarios. The results also shows another advantage of the MAC-FMSO to identify the multiple damage members. As can be seen in Fig. 6(b), member 12 is also identified as a damaged member by using damage locating vector method. However, the results of MAC-FMSO in Table 5 accurately predict that there is no stiffness reduction in member 12.

Table 5. Result of stiffness degradation using MAC-FMSO for plane truss structure

First scenario			
"Candidate" of damaged member/part	11		
Stiffness reduction result using MAC-FMSO	75%		
Second scenario			
"Candidate" of damaged member/part	8	11	12
Stiffness reduction result using MAC-FMSO	50%	75%	0%

 DOI: 10.1051/matecconf/201713802020

5.3 Shear building model

The third example to show the effectiveness of the MAC-FMSO is a shear building model discussed in [6]. Mass and stiffness of each story can be seen in Fig.7. Three different scenarios were considered as follows:

a. a single damage with the loss of stiffness of the sixth story is 60% from initial stiffness,
b. a double damage with the losses of stiffness in the second and sixth story are 50% and 40%, respectively,
c. a triple damage with the losses of stiffness in the second, sixth, and ninth story are 70%, 55%, and 40%, respectively.

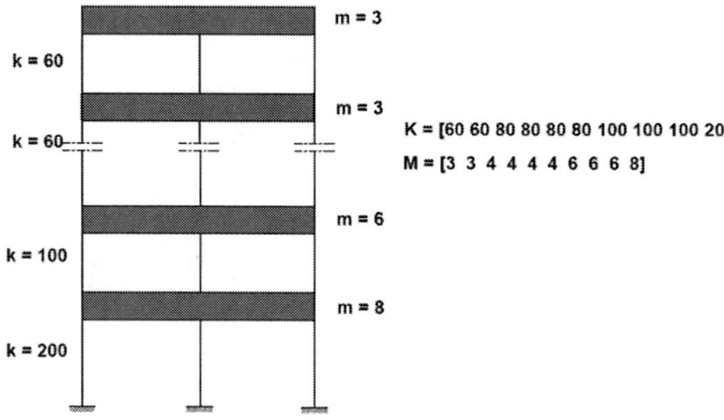

Fig. 7. Shear building properties

The mode shape curvature method was used for this case, and three modes were considered in this simulation. Fig.8 shows the MSI of the second damage scenario. By observing the MSI, it is clear from Fig. 8 that the damage was occurred in the second and sixth stories.

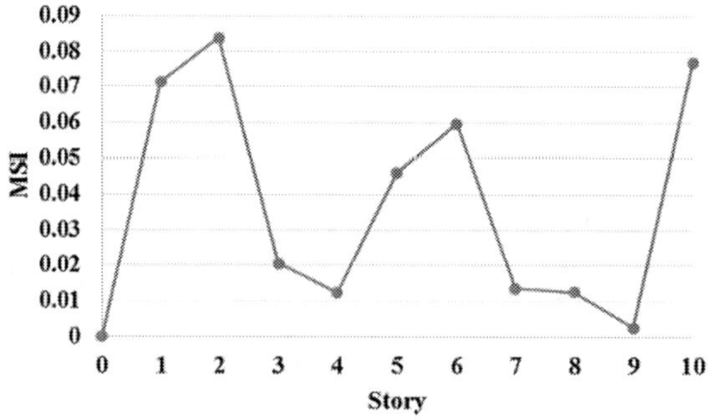

Fig. 8. Mode shape index of shear building for each story

The MAC-FMSO similar to the previous example was used to obtain the loss of stiffness of the structure. The result of simulation is presented in Table 6. From Table 6 it is clear that the MAC-FMSO can accurately predict the loss of stiffness of the structure.

6 Conclusions

The MAC-FMSO method is discussed for identification of structural damage. By using MAC-FMSO, the location of the damage as well as the loss of stiffness can be identified.

The MAC-FMSO consists of two stages. The first stage is finding the damaged member's location by using any of the second level of damage detection techniques, such as mode shape curvature method, damage locating vector, and strain energy method. After obtaining the location of the damaged member, the MAC-FMSO then is used to obtain the loss of stiffness from the "candidate" of the damaged members. The results show that MAC-FMSO has accurately predict the loss of stiffness of the damaged member/part of structures. Therefore, the MAC-FMSO can be considered as a third level of damage detection method, because it can accurately predict the loss of stiffness of the damaged members.

Table 6. Result of story stiffness degradation using MAC-FMSO for shear building case

First scenario			
"Candidate" of damaged story	6		
Stiffness reduction result using MAC-FMSO	60%		
Second scenario			
"Candidate" of damaged story	2	6	
Stiffness reduction result using MAC-FMSO	50%	40%	
Third scenario			
"Candidate" of damaged story	2	6	9
Stiffness reduction result using MAC-FMSO	70%	55%	40%

References

1. R.J. Allemang, DL. Brown, Proc. of the 1st Int. Modal Analysis Conf., 110-116 (1982)
2. Q. Zhang, C. Li, Y. Liu, L. Kang, Proc. of Advances in Computation and Intelligence, Springer-Verlag, Berlin Heidelberg, 344-352 (2007)
3. J. Kennedy, R.C. Eberhart. Proc. of the 1995 IEEE Int. Conf. on Neural Networks. IEEE Service Center, Piscataway, 1942-1948 (1995)
4. R. Frans, Y. Arfiadi, H. Parung, Procedia Engrg., 1263-1271 (2017)
5. A. K. Pandey, M. Biswas, M. M. Samman, J. of Sound and Vibr., **145**, 312-332 (1991)
6. D. Bernal, COST F3, Madrid, Spain, 223-231 (2000)

Cracking behaviour and its effect on the deflection of patched-reinforced concrete beam under flexural loading

Stefanus Kristiawan[1,*], *Agus* Supriyadi[1] , *Senot* Sangadji[1] and *Duwi* Santosa[2]

[1]SMARTCrete Research Group, Civil Engineering Department, Universitas Sebelas Maret, Surakarta 57126, Indonesia
[2]Civil Engineering Department, Universitas Sebelas Maret, Surakarta 57126, Indonesia

Abstract. Under the influence of aggressive environment, structural concrete could exhibit degradation in various forms. One of the symptoms of the degradation could be shown in the form of local delamination of concrete cover. Patching method could be a preference choice to repair this type of degradation to regain its structural performance and durability. This paper presents the structural behaviour of patched-reinforced concrete beam under flexural loading with particular interest on the use of unsaturated polyester resin (UPR) mortar as patch repair material. The repair area is replicated in the tensile zone of the beam. The behaviour investigated in this study is the cracking evolution of the patched – reinforced concrete beam and its effect on the deflection. It is shown that the higher strength of the UPR-mortar leads to a reduction in the intensity of cracks compared to that of un-repaired (normal) reinforced concrete beam. The lesser cracks intensity is beneficial to restore the flexural stiffness of the beam.

1 Introduction

Reinforced concrete (RC) beam is a part of building structural system elements, which is designed to sustain flexural and shear loading. The design approach as specified in various Codes [1, 2] suggests that the RC beam should be designed to fail in a ductile manner. Since the rupture of RC beam due to shear stress is a brittle one [3], consequently the shear capacity of the beam must be intended to be more conservative than the corresponding flexural strength. It means any excessive loads beyond the design load should lead to flexural type of failure. However, a proper design would take into account any kinds of imposed loads throughout its service life and therefore, failure of the beam element should not be expected. A part from the strength, the design of RC beam must also satisfy the serviceability requirements. An example of serviceability requirement is a maximum limit of deflection [4, 5]. The magnitude of deflection is determined by the flexural stiffness of the beam which in turn is influenced by the elastic modulus of concrete and reinforcement,

* Corresponding author: s.a.kristiawan@ft.uns.ac.id

the dimension of the beam, the amount of the reinforcement, creep and shrinkage of concrete. The last two properties contribute to the long-term deflection [6, 7].

The instantaneous deflection of RC beam as a response of loading could be traced from its load-deflection behaviour. Figure 1 illustrates the complete behaviour of RC beam under flexural loading up to failure. The load-deflection curve of RC beam may be divided into three regions [8]: the first region (I) is defined from a zero load to a load causing an initiation of first flexural crack (cracking load). This region could also be viewed as a pre-cracking region where the un-cracked beam behaves fully linear elastic; the second region (II) is counted from the cracking load to a load causing the first yield of the reinforcement (yield load). This region corresponds to post-cracking region where RC beam is cracks at various intensity depending on the load level. The higher the load, the more the intensity of the cracks. The present of cracks certainly affects the rigidity (flexural stiffness *EI*) of the beam. At this region the moment of inertia of the beam lies between the moment of inertia of gross section (I_g) and the moment of inertia of fully cracked section (I_{cr}); finally the third region (III) is counted from the yield load to a failure load where the beam behaves as non-linear elastic material.

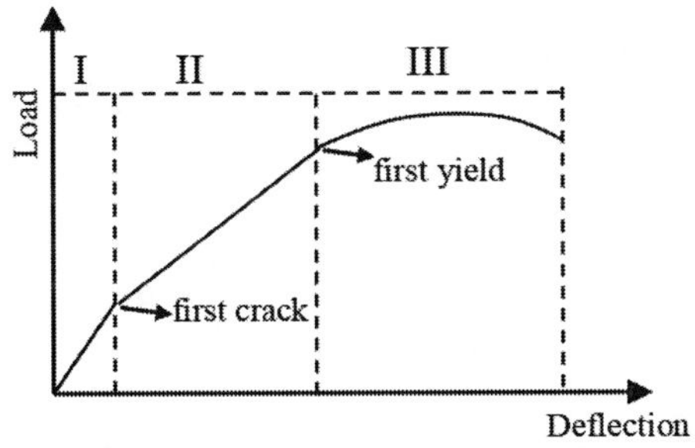

Fig. 1. Load-deflection of RC beam

Generally the service load falls within region II. Hence, the deflection at this load level is of interest. The deflection of the RC beam at service load has to consider the rigidity of the beam at this region. The rigidity of RC beam is a transitional value between rigidity of un-cracked beam and fully cracked beam [9]. This transitional value is represented by the value of effective moment of inertia of the section (I_e) which lies between I_g and I_{cr}. The value of I_e depends upon the extent of cracking caused by applied loads. The following formula is suggested to estimate the I_e [1]:

$$I_e = (\frac{M_{cr}}{M_a})^3 I_g + [1 - (\frac{M_{cr}}{M_a})^3] I_{cr} \qquad (1)$$

where M_{cr} and M_a representing the cracking moment and the maximum moment due to the service load, respectively. M_{cr} could be calculated from:

$$M_{cr} = \frac{f_r I_g}{y_b} \qquad (2)$$

where f_r and y_b is the modulus of rupture of concrete and the distance of the bottom fiber to the neutral axis of the RC beam, respectively. The value of f_r may be predicted from the compressive strength of the concrete f'_c using the following relation:

$$f_r = 0.7\sqrt{f_c^{'}} \tag{3}$$

The moment of inertia of both normal and patched RC beam at un-cracked or cracked section with various amount of reinforcements may be calculated by considering the transformed-area.

The total deflection of RC beam at service load should be maintained within the maximum limit throughout its service life. However, there are circumstances that could alter the magnitude of deflection of the RC beam such as degradation of the beam element. The degradation of RC beam is commonly observed in structures exposed to aggressive enivronments. The degradation could be visually detected in the form of spalling or local delamination of concrete cover. The flexural stiffness of RC beam at this damaged zone will be lower compared to the rest of the beam which in turn altering the magnitude of beam deflection. Any repairs carry out to recover the damaged zone should be expected to resume both the strength and serviceability of the RC beam [10].

Patch repair is a common method to recover the spalling of concrete cover. The method requires a suitable repair material that should satisfy various specifications. They include the requirement of adhesive/bond to the parent concrete, strength, elastic modulus, shrinkage, creep, coefficient of thermal expansion, durability, etc [11]. The requirements are specified to ensure the compatibility of the repair material with the parent concrete in term of dimensional, mechanical and durability. An unsaturated polyester resin (UPR) mortar has been developed by the authors to meet those specifications [12-14]. The performance of RC beams pacthed with this material has been studied with respect to the influence of various patching parameters [15-16]. For the current research, UPR mortar will be investigated with special interest to observe the cracking intensity and its effect on the deflection of patched-RC beam under flexural loading. For this purpose, a comparison of the load-deflection behaviour of patched-RC beam to that of normal RC beam is determined experimentally. Since the load-deflection behaviour of RC beams at region II is influenced by the cracking intensity of the beams, the evolution of cracking of both beams at various load levels within this region are inspected carefully. The influence of the cracking intensity will be further examined to compare the deflection of both beams.

2 Experimental Investigation

For the purpose of this investigation, two type of beams were prepared i.e. normal RC beam and patched-RC beams. Both beams had similar size and reinforcements (see Figure 2). The repair area was replicated at the tensile zone of the RC beam along the maximum moment. The thickness of the repair was specified at 70 mm. The beams were prepared with a concrete where the proportion of which was determined on the basis of average strength of 25 MPa. The beams were cast into wooden moulds and the moulds were striped off on the following day after casting. The beams were then cured with wet burlaps for 28 days and subsequently they were stored in the laboratory environment for about 3 months. A day before testing, one of the RC beams was patched with unsaturated polyester resin (UPR) mortar. The UPR mortar was prepared using the following ingredients and proportion as per 1 m^3: 808 kg of cement, 143 kg of fly ash, 950 kg of sand, 475 kg of UPR and 24 kg of hardener. The hardener was required to initiate the polymerization of UPR. No water was added in the UPR mortar mix. Hence, both cement and fly ash could be considered as filler materials.

The actual properties of concrete, UPR mortar and steel reinforcement are shown in Table 1. The 28 d compressive strength and elastic modulus of concrete were determined on cylinder (150x300 mm) specimens following ASTM C39 and ASTM C469, respectively. The compressive strength of UPR mortar was tested at 1 d of age on cubes of 5x5x5 mm in accordance with ASTM C109. Meanwhile, the elastic modulus of UPR mortar was determined on cylinder (75x150 mm) specimens using similar procedure as that of testing elastic modulus of concrete. The flexural strength of UPR mortar was determined in according to ASTC C580. Finally, the steel reinforcement was tested in tension up to failure using Universal Testing Machine (UTM). All the values given in Table 2 were calculated as an average of at least three results.

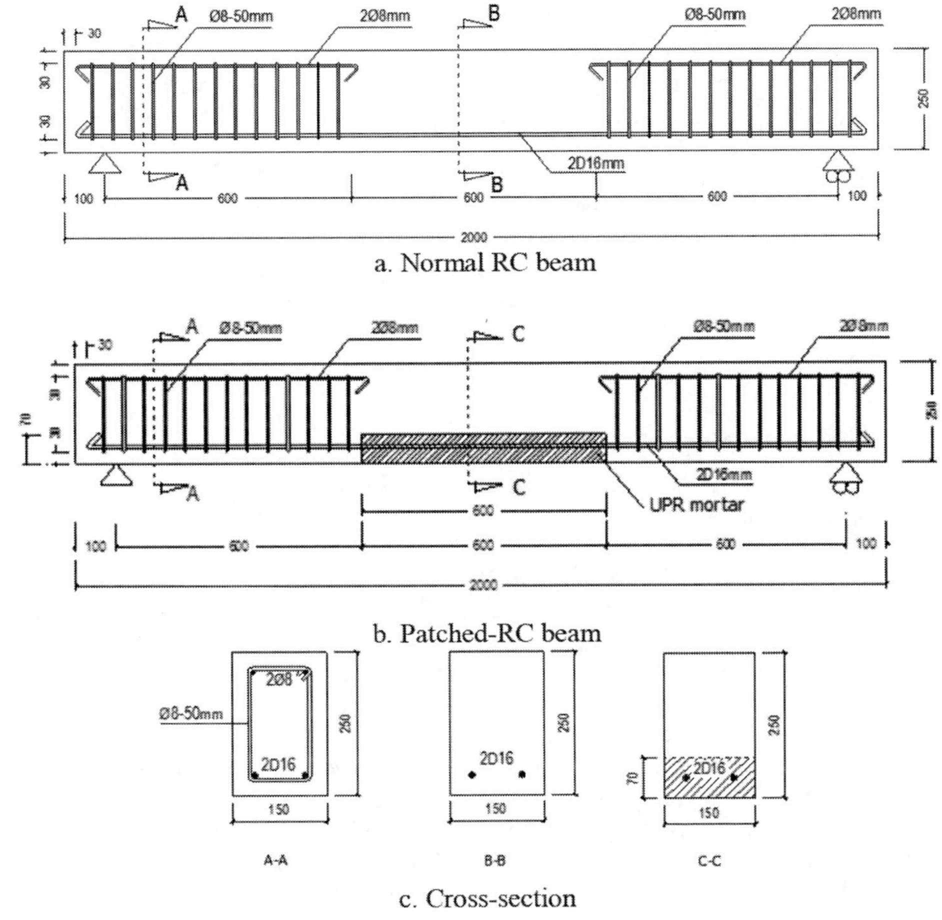

Fig. 2. RC beam specimens with detail of reinforcements

Table 1. Properties of materials

Property	Concrete (MPa)	UPR mortar (MPa)	Steel (MPa)
Compressive strength	30.23	76.33	NA
Flexural strength	3.85*	23	NA
Yield strength	NA	NA	437
Elastic modulus	24225	13557	200000**

* predicted value based on Eq. (3); ** assumed value

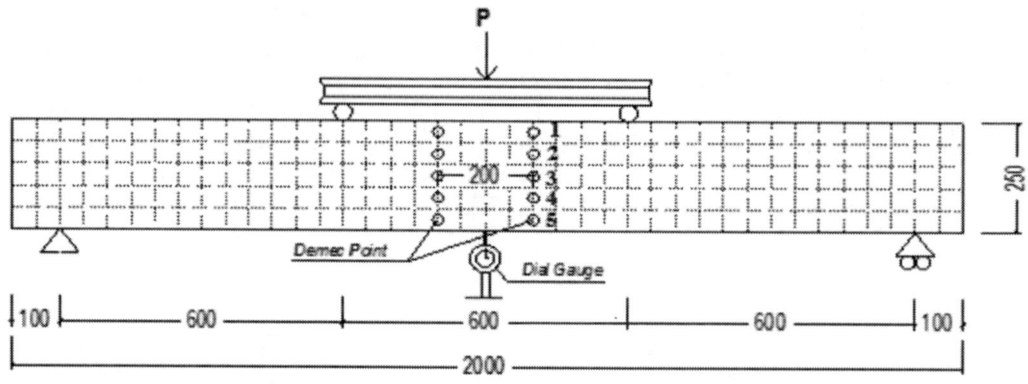

Fig. 3. Schematic loading of RC beam flexural testing

The flexural testing of RC beams were schematically shown in Figure 3. The load P was applied and incrementally increased at 200 kg until the beams were rupture. At each increment of load, the corresponding deflection at the mid-span was measured using dial gauge. The deformations of the beams at both compression and tension zone were checked by measuring the distance between pair of demec points at location shown in Figure 3 using Demec gauge. Visual examination was carried to monitor the evolution of cracks on the beams at each increment of loading.

3 Results and discussion

3.1 Cracking evolution

The evolution of crackings on the normal RC beam and patched-RC beam under increment of flexural loading are illustrated in Figure 4. The first flexural cracks on the normal RC beam appear at a load of 2T. These cracks occur at a zone where the moment is maximum. In contrast to the normal RC beam, the first flexural cracks of the patched-RC beam (induced by load of 2.8 T) appear beyond the maximum moment zone. The present of UPR mortar at this zone could be accountable for this behaviour. At a load of 2.8 T, the calculated maximum tensile stress is only 5.9 MPa. The value of this stress is computed with the moment of inertia of gross section I_g after taking into account the transformed areas. Clearly, the level of this tensile stress is far below the modulus of rupture of UPR mortar (23 MPa). In fact the experimental result shows that the first cracks of UPR mortar occur at a load of 10 T. This level of load induces a maximum tensile stress of 21.24 MPa which is close to the modulus of rupture of UPR mortar. It is also interesting to note that the flexural cracks appearing in the UPR mortar do not propagate to the parent concrete. The cracks is discontinued within the thickness of UPR mortar. In fact the first crack of parent concrete at location above the repair zone occurs at 9 T of loading while no cracks are observed in the UPR mortar at this level of loading.

Figure 4 shows the distinctive cracks development between the normal RC beam and patched-RC beam as the applied load is increased. For the normal RC beam, the cracks are initially within the maximum moment zone when low level of loading is applied; after which the cracks also emerge at both the shear span zones. At high level of loading it seems that the cracks are distributed almost evenly along the span length of the beam. On the other hand, the cracks on the patched-RC beam are initially clustered beyond the repair zone. The cracks only appear at the repair zone (maximum moment) after high loading is applied (at least 10T). It is also clearly shown that the cracking intensity of the normal RC beam is greater than that of the patched-RC beam. Comparing these two cracking intensities, it can be conviniently concluded that the patched-RC beam is better in maintaining the rigidity of the beam at the post-cracking region.

The final cracks patterns at rupture are interesting to be examined. For the normal RC beam, the final flexural cracks reach depths in such that the compression zone of concrete is left to about 50 mm. The mode of failure is initiated by yielding of the reinforcements (shown in Figure 6.a). Any attempts to impose an increase of load at this state will cause a significant deformation of the reinforcements and at some point trigger compression fracture of concrete located just under point loads. For patched-RC beam, the mode of failure is also initiated by yielding of the reinfocements. However, the final cracks display different pattern to that of normal RC beam. The flexural crack of concrete above the repair zone propagates and diverges into two directions toward point loads. These directions of cracks suggest that the propagation of cracks are influenced by the compressive stresses trajectory. The final cracks pattern in the compression zone indicates that an increase of load beyond the post yield limit will ultimately causes concrete to fail predominantly due to high compressive stress at this zone.

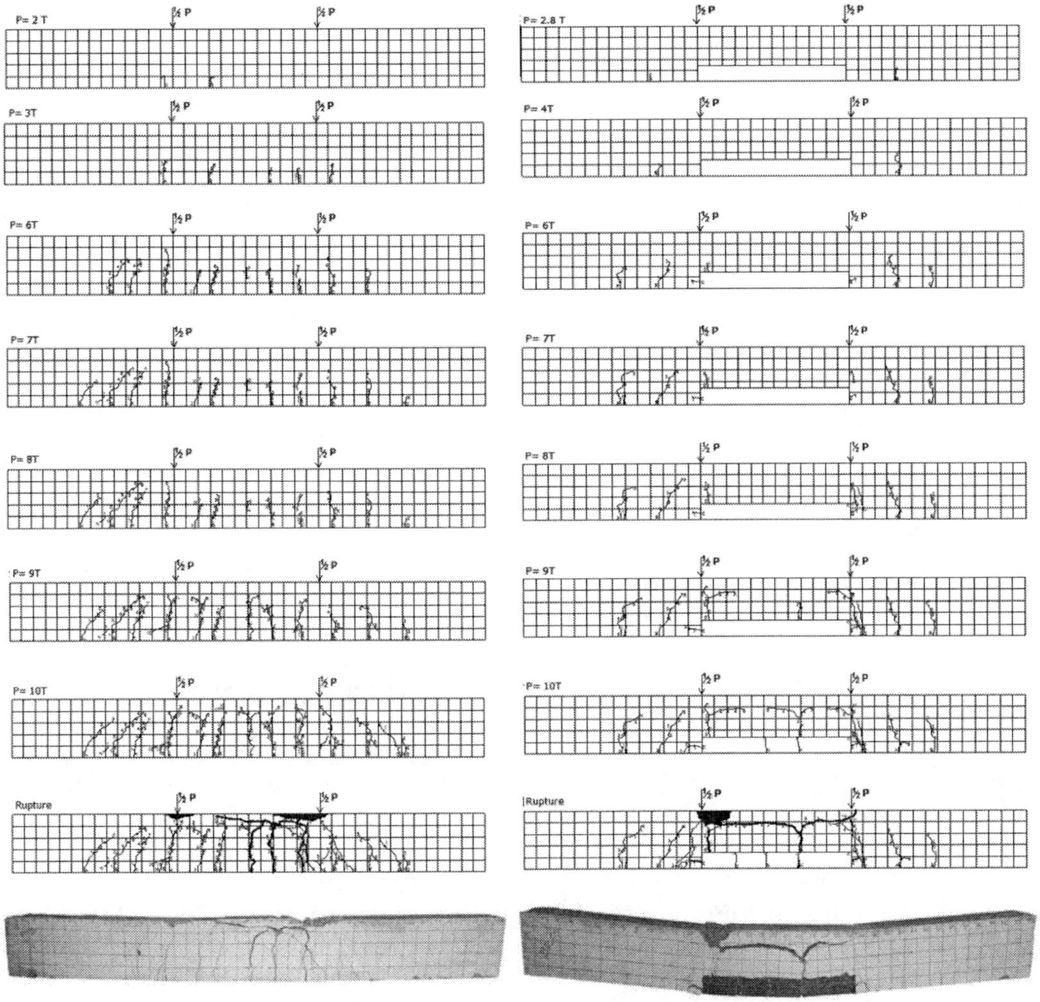

a. Normal RC beam   b. Patched-RC beam

Fig. 4. Evolution of cracking on RC beam under increment of flexural loading

3.2 Deformation of beam at compression and tension zone

RC beam will undergo compressive and tensile deformations at above and below neutral axis, respectively when it is loaded in flexure. The highest deformations will be observed in the middle span of the beam. For this reason the deformations at this location is investigated in this research. The deformations both at compression and tension zone are measured by monitoring the change in the distance between pairs of demec points shown in Figure 3. The measured deformations of the RC beams under increment of flexural loading are given in Figure 5.

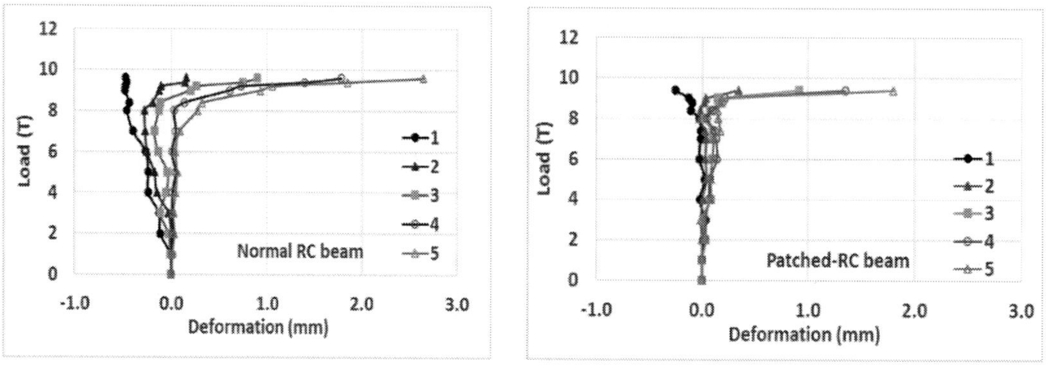

Fig. 5. The deformations of normal (left) and patched (right) RC beam at compression and tension zone

Initially the first three layers designated by number 1,2 and 3 of normal RC beam show compression deformation (negative value) as given in Figure 5.a. These deformations indicate that these layers are within the area of compression zone. The neutral axis of the normal RC beam is lower than a half of the beam height (at layer number 3). The tensile reinforcements in this cross-section is accountable for lowering the neutral axis from its original position of plain concrete beam. The propagation of cracks certainly influence the deformations of the beam. At about 8T of loading, the compressive deformations of layer number 2 and 3 start to change in their directions toward positive values. At this level of loading, the depth of cracks already penetrate the compression zone (see Figure 4.a). Hence, the neutral axis moves upward causing the layer number 2 and 3 turn to be in the tensile zone. Only layer number 1 is still in the compression zone. The 8T of loading also corresponds to yielding of the tensile reinforcements as indicated by large deformations of the tensile zone (see also Figure 6.a).

For patched-RC beam, it appears that only layer number 1 and 2 undergo compressive stresses. The presence of UPR mortar with low elastic modulus (about a half of that concrete as shown in Table 1) is accountable for this characteristic. Considering transformed area, the I_g of this repaired section is lower than that of normal RC beam. The neutral axis of the patched-RC beam is at a higher position compared to that of normal RC beam. Hence, the compression zone of the beam is smaller. At 8T of loading, similar behaviour to that of normal RC beam is observed. At this level of loading, it is marked as the beginning of large deformations in the tensile zone. It is also a starting point where the neutral axis shifts upward causing layer number 2 turns to be in the tensile zone. Comparing to those of normal RC beam, it is obvious that the magnitudes of deformations observed in the patched-RC beam are lower. The behaviour could be related to the cracking pattern of the beams. At 8T of loading, the patched-RC beam shows no cracks at all at the middle span. On the other hand, a number of crack propagations are observed in normal RC beam (see Figure 4). Cracking is a sign that a sufficient magnitude of tensile stresses

DOI: 10.1051/matecconf/201713802021

exhibits in the zone leading to significant deformations which eventually initiate cracks and their propagations. A very high tensile strength of UPR mortar is beneficial to delay the initiation of cracks. Consequently, the deformations at the observed zone of patched-RC beam are retained at low level before cracking.

3.3 Load-deflection behaviour

The deflection of both RC beams under increment of flexural loading is presented in Figure 6. The complete curve determined from the zero loading to the rupture is given in Figure 6.a while the short curve representing load-deflection behaviour from the zero loading to the yield loading is shown in Figure 6.b. The figures suggest that both RC beams generally have a similar behaviour. Initially the deflection is proportional to the applied load. After the normal RC beam exhibits cracks at about 2T of loading, the flexural stiffness of this beam is only slightly reduced. The deflection is a little bit higher at an equal increment of loading than the deflection at the pre-cracking region. The deflection at post-cracking region can still be considered linear with a little lower flexural stiffness compared to that of pre-cracking region. For patched-RC beam the proportional limit of load-deflection is maintained at a higher level of loading compared to that of normal RC beam i.e. at about 5 T before the curve deviates a little downward. It seems that the presence of one little crack at each of the shear span of patched-RC beam do not affect the flexural stiffness of the patched-RC beam. Only cracks appeared at high load level will affect the flexural stiffness of patched-RC beam.

Theoretically, the initial flexural stiffness of the patched-RC beam is lower than the flexural stiffness of normal RC beam. The I_g of patched-RC beam calculated by taking into account the transformed areas equals to 168939220.7 mm^4. This value is smaller than the I_g of normal RC beam which is 209083612.6 mm^4. However, the difference in Ig only occurs on middle span segment where patch repair exists. It may be expected that this difference of I_g would cause the initial deflection of the patched-RC beam is a little higher. The experimental results hardly confirm this. The normal RC beam develops cracks at early loading than the patched-RC beam. Hence, the flexural stiffness of the normal RC beam begins to reduce at early loading. The flexural stiffness of the beam at post-cracking region is calculated on the basis of I_e given in Eq. (1). However, this equation is not appropriate for patched-RC beam which shows difference cracking pattern and intensity. The complexity of the problem is also induced by the alteration of neutral axis position where patched-RC beam deforms differently to the normal RC beam. The cracking pattern and intensity of patched-RC beam make the beam is sufficient to maintain its rigidity up to a very high loading. Consequently, the deflection of this beam at post-cracking region is surpassed by the deflection of the normal RC beam.

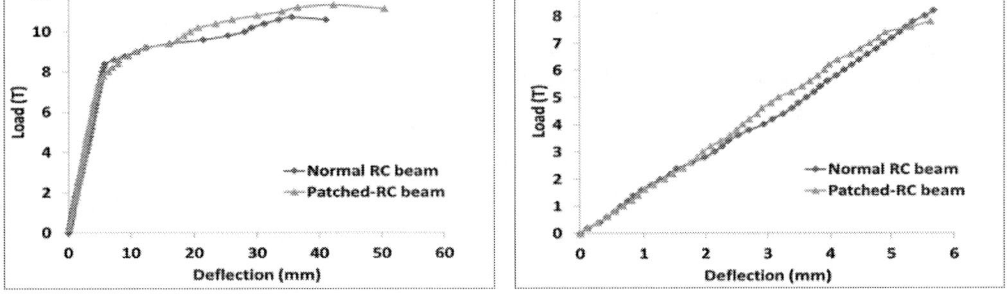

a. Complete curve b. Up to yield load

Fig.6. Load-deflection behaviour

4 Conclusions

UPR mortar has been used to patch a replicated damage in the tensile zone of RC beam. The performance of the patched-RC beam is compared to the normal RC beam. The following conclusions are drawn:

- the presence of UPR mortar could delay the first appearance of flexural cracks
- the evolution of cracks in the patched-RC beam begins and cluster beyond the repaired zone while for the normal RC beam it begins in the maximum moment zone and then distributed evenly along the span length of the beam
- the cracking intensity of patched-RC beam is less than that of normal RC beam
- the presence of UPR mortar could also reduce the magnitude of deformations in the middle span of the beam
- the flexural rigidity of the patched-RC beam is maintained at higher level of loading than that of normal RC beam
- the deflection of the patched-RC beam at post-cracking region is slightly less than that of normal RC beam as a consequence of retaining its rigidity

This research is a part of the research project under the Hibah Kompetensi Scheme which financially supported by DRPM Dirjen Risbang, Kemenristek Dikti, Indonesia.

References

1. ACI Committee 318, *Building Code Requirements for Structural Concrete (ACI 318M-08) and Commentary* (American Concrete Institute, Farmington Hill, 2008)
2. International Federation for Structural Concrete (*fib*), *Model Code for Concrete Structures 2010* (Wilhelm Ernst & Sohn, Berlin, 2013)
3. J.Y. Lee, S.W. Kim and M.Y. Mansour, J. Struct. Eng. **137**, 1017-1029 (2011)
4. M.A. Rashid and M.A. Mansur, ACI Struct. J. **102**, 462-471 (2005)
5. A.S. Desalegne and A.S. Lubell, ACI Struct. J. **109**, 867-877 (2012)
6. W.C. Choi and H.D. Yu, Mat. Des. **51**, 742-750 (2013)
7. A. Abrishambaf, J.A.O. Barros and V.M.C.F. Cunha, Cem. Con. Res. **72**, 21-36 (2015)
8. E.G. Nawy, *Prestressed Concrete: A Fundamental Approach* (Prentice Hall, Upper Saddle River, 2010)
9. J.C. McCormac and R.H. Brown, *Design of Reinforced Concrete* (Wiley, New Jersey, 2014)
10. R. Sahamitmongkol, S. Suwathanangkul, P. Phoothong and Y Kato, Adv. Con. Tech. **6**, 317-336 (2008)
11. International Concrete Repair Institute, *Guide for Selecting and Specifying Materials for Repair of Concrete Surfaces* (ICRI, 1996)
12. S.A. Kristiawan and A.B. Prakoso, App. Mech. Mat. **754-755**, 442-446 (2015)
13. S.A. Kristiawan and A.B. Prakoso, Mat. Sci. For. **857**, 426-430 (2016)
14. S.A. Kristiawan and R.S. Fitrianto, IOP Con. Ser. Mat. Sci. Eng. **176**, 012017 (2017)
15. A. Supriyadi, S.A. Kristiawan and S.B. Raditya, App. Mech. Mat. **754-755**, 457-462 (2015)
16. S.A. Kristiawan, A. Supriyadi and M.K. Muktamirin, Effect of patching thickness on the flexural performance of patched reinforced concrete beams, *Advances in Civil, Architectural, Structural and Constructional Engineering,* Taylor & Francis (2016)

17. S.A. Kristiawan, A. Supriyadi, S. Sangadji and H.B. Wicaksono, K. Eng. Mat. **737**, 441-447 (2017)

MATEC Web of Conferences **138**, 02022 (2017) DOI: 10.1051/matecconf/201713802022
EACEF 2017

Model of Bamboo Strip Notch Reinforced Concrete Beams On The Flexural Capacity

Agus Setiya Budi[1,*], and *Agus* P Rahmadi[2]

[1]Lecturer, Department of Civil Engineering, Universitas Sebelas Maret, Indonesia
[2]Lecturer, Department of Civil Engineering, Universitas Sebelas Maret, Indonesia

Abstract. There are many various species of bamboo trees in Indonesia, one of which is Petung bamboo (Dendrocalamus Asper). Bamboo is one of the potential material as a substitute for steel reinforcement. This paper will discuss the model of bamboo reinforcement to increase the flexural capacity of concrete beams. The use of the notch is expected to increase the bonding effects and minimize slip effects. This study used laboratory experimental method. Bamboo strips are processed and formed with the notches of the form V with the distance between the notch are 2, 3, 4, 5, 6 and 7 cm. Static loading is done with third point loading system (ASTM C78). The test results show that, the average flexural capacity on the specimens of concrete beams with Petung bamboo strip notch as reinforcement on the distance between notch 2, 3, 4, 5, 6 and 7 cm respectively are 0.536, 0.528, 0.432, 0.403, 0.424, 0.366 tm.

1 Introduction

It is a fact that the construction industry is the main consumer of energy and materials in most countries. In consequence of the consumers choosing industrialized products, among other effects, activities are suppressed in rural areas or even in small towns, and renewable materials are wasted and causing permanent pollution. In this sense, it becomes obvious that ecological materials satisfy such fundamental requirements, making use of agricultural by-products such as rice husk, coconut fibres, sisal and bamboo and therefore minimizing energy consumption, conserving non-renewable natural resources, reducing pollution and maintaining a healthy environment [1].

Bamboo is a category of plants that are easy to grow and easy to find especially in Asia. Bamboo has several advantages, one of them is on the side of bamboo fiber which has a high enough tensile strength, especially on the outer bamboo fiber side or on the bamboo skin. Research results by [2], the bamboo is one of the suitable replacements of reinforcing bar in concrete for low cost constructions. Bamboo is natural, cheap, widely available and most importantly strong in both tension and compression.

There are many various species of bamboo trees in Indonesia, one of which is Petung (Dendrocalamus Asper). According to [3] bamboo could be used as a substitute for steel

* Corresponding author: asb0909@yahoo.co.id

reinforcement. This suggests that bamboo can be used to replace mild steel for low cost structural construction.

But the use of bamboo as a substitute for steel directly on the construction is still encountered some obstacles, one of them is the slip effect that occurs between the concrete and bamboo. Therefore, to reduce the slip effect between bamboo and concrete, this paper will discuss the model of bamboo reinforcement to increase the flexural capacity of concrete beams in a simple construction.

2 Material properties

2.1 Bamboo

Just like timber, bamboo is vulnerable to environmental degradation and attacks by insects and moulds. Its durability varies with the type of species, age, conservation condition, treatment and curing. Curing should be initiated when bamboo is being cut in the bamboo grove. There is a strong relation between insect attacks and the levels of starch plus humidity content of bamboo culm. In order to reduce the starch content, bamboo receives a variety of treatments including curing on the spot, immersion, heating or smoke [4].

The bamboos with the following characteristics were selected: 1). Only bamboos showed a pronounced brown color were used to ensure that the plant is matured and at least three years old; 2). Selection of the longest large diameter culms available was made while as far as possible, long straight bamboos without any deformations and cracks were selected. Bamboos used are free of any decay, fungus growth or holes due to white ants; 3). The bamboos were not cut under wet condition because the culms are generally weaker due to increased fiber moisture content [5].

The difficulty associated with durability of raw bamboo is that it absorbs water from the surrounding environment through fine cracks in concrete. This results in swelling of the bamboo. Swelling occurs when there is sufficient time for water to reach the bamboo before the concrete cures. It is then absorbed by the bamboo microstructure, which induces an expansion of the bamboo reinforcement and promotes internal local stress in the concrete surrounding the bamboo. Over a long period of time, the consequences are brittle splitting failure in concrete can result in sudden collapse [6].

In this research using bamboo Petung species (Dendrocalamus Asper), with bamboo above the age of 3 years. Characteristic of this bamboo has a diameter of 20-30 cm, bamboo thickness about 1.5-3 cm and height 10-20 m. Part of culms used is part of a bamboo rod 5 m, which cut 1.5 meters from the base of a tree. This bamboo was obtained from the Banyu Urip village, District Klego, Boyolali, Central Java, Indonesia. The average value of the water content bamboo is 14.67% and the average density is 1.09 g/cm3. The average tensile strength parallel to the fiber (with node of bamboo) is 223.89 MPa.

2.2 Bamboo strip notch model

According to [7], due to remarkable difference between steel shear strength and longitudinal shear strength of bamboo (7.45 MPa for dry and 4.12 MPa for wet bamboo), all of assumptions and calculations are different. In a steel deformed bar, the root connection of the dent (b) regarding to distance between dents (c) is small, see Fig 1.

 DOI: 10.1051/matecconf/201713802022

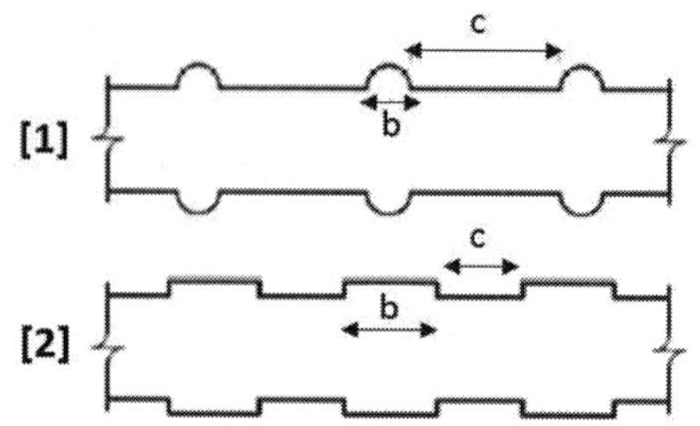

Fig.1. (1) Shape of a steel deformed bar (2) Shape of bamboo corrugated strips.

It can be justified because of remarkable difference between steel and concrete shear strength. As the bamboo longitudinal shear strength and concrete shear strength are close together then the length of dent (b) would be closer to distance between dents (c). Because of proper bamboo workability, it can be easily filed and corrugated in field. By using this method for improving the bond, the necessity of using other chemical coating for increasing the bond will be lessened [7].

See Fig. 2, it was presented the dimension strip notch V model of bamboo reinforcement bar types studied. The selected bamboo strips is then dried for over 15 days, then cut and separated into bamboo strips of 5x20x1650 mm and then notch at any distance of 2, 3, 4, 5, 6 and 7 cm.

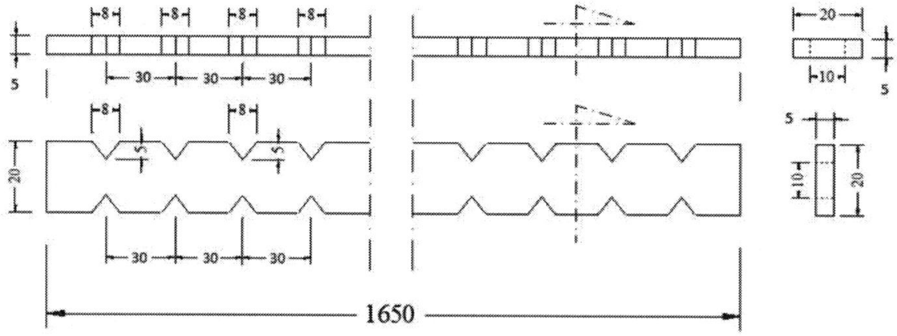

Fig.2. Detail dimension of bamboo strips notch V model (example for distance 3 cm).

Fig.3. Examples of completed bamboo notch V model.

For comparison, in this experiment using a steel round bars Ø8 mm as reinforcement. The tensile test results show that, maximum average tensile strength is 639.91 MPa and yield strength average of steel round bars as reinforcement is 487.87 MPa. The type of cement used is Portland Pozzolan Cement (PPC) which conform with Indonesian National Standard (SNI 15-0302-94).

Based on experimental test results in the laboratory, the value of concrete compressive strength was taken based on the test of 3 cylindrical specimens with the size of 15x30 cm and has 30 days of age. The average value of the compressive strength of the concrete is 18.29 MPa with slump value is 11.5 cm.

3 Testing procedure

The size of the test beam specimen used in this research is 11x15x170 cm. Detail of beam size and reinforcement placement are presented in Fig.4.

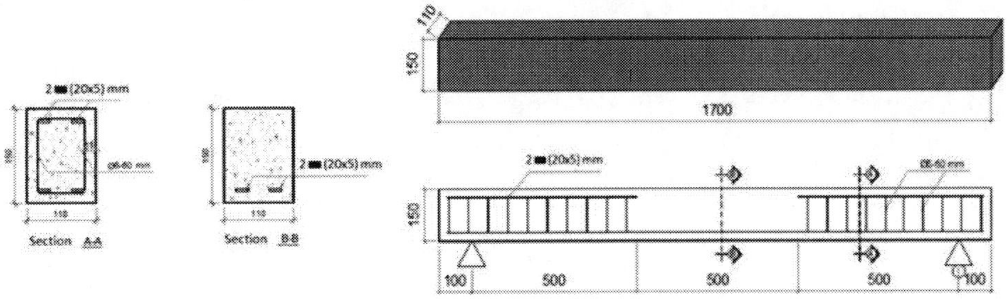

Fig.4. Position of the reinforcement on the cross section of the concrete beam.

The number of concrete blocks used in this experiment is 21 pieces. Which consists of beams for each variation of notch amounted to 6x3 pieces of beam, and 3 pieces of reinforced concrete beams of steel (Ø8mm) as a comparison. Each specimen beams is subjected to static loading in accordance with third point loading system standard (ASTM C78). The loading setting in the flexural concrete beam test experiment can be seen in Fig.5.

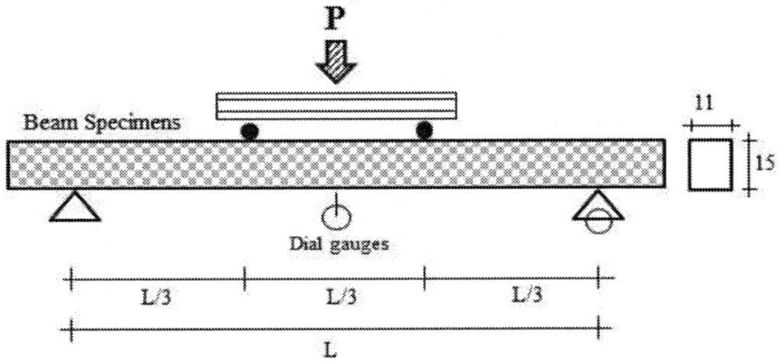

Fig.5. Loading settings on the beam

4 Experimental and analitycal flexural moment

Based on the loading setting, the loading distribution that occurs on the test beam specimen can be seen in Fig.6.

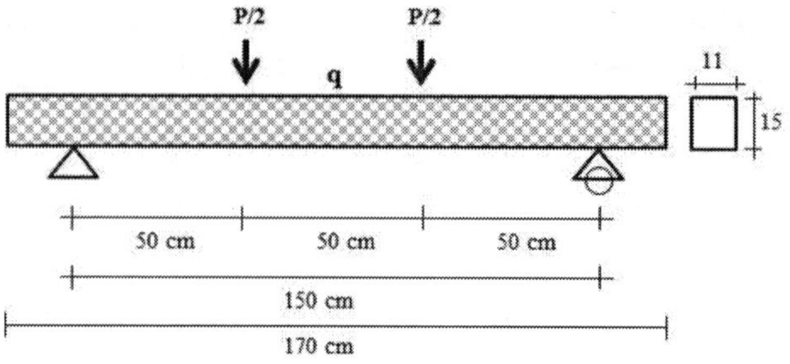

Fig.6. Loading distribution.

The static loading is carried out on the test beam specimen until the maximum load is reached and the beam collapses. The deflection occurring in the dial gauge is recorded at an increased load interval of 50 kg. The experimental flexural moment value can be calculated based on the maximum load data and its own weight of the beam, by the static equilibrium equation of the moment, which is presented in Eq.1.

$$M_{max} = (R_{Av}.L/2)\text{-}(qL^2/(6.228))\text{-}(PL/12) \qquad (1)$$

Based on the data properties of the test beam material and the cross-sectional data, analytical ultimate moment can be calculated using a simplified stress block at the ultimate limit. Stress block diagram on the concrete beam section due to flexural forces presented in Fig.7.

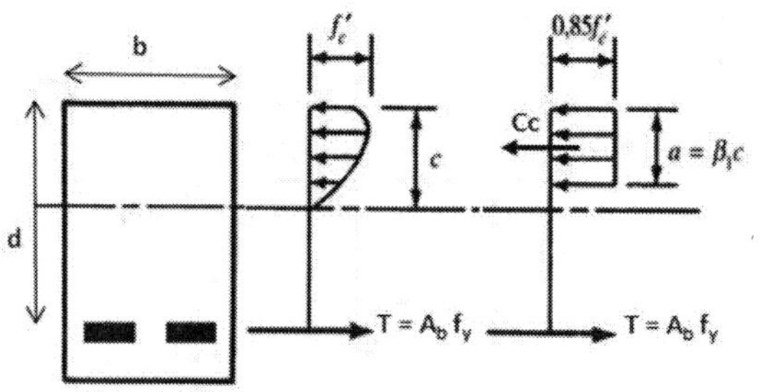

Fig.7. Stress block at the ultimate limit due to flexural forces

Analytical ultimate moment can be calculated based on simplified stress block, with equation below [8].

$$Tensile\ force = Compressive\ force \qquad (2)$$

$$a = A_b f_y / (0.85 f_c'\, b) \qquad (3)$$

$$M_u = (A_s f_y)(d\text{-}a/2) \qquad (4)$$

5 Result and discussions

Based on the results of the test experiments on a number of specimens of concrete beams with bamboo and steel reinforcement, the following is shown in Fig.8 and Fig.9, presented a relationship diagram of load and deflection due to loading until the test specimen collapses.

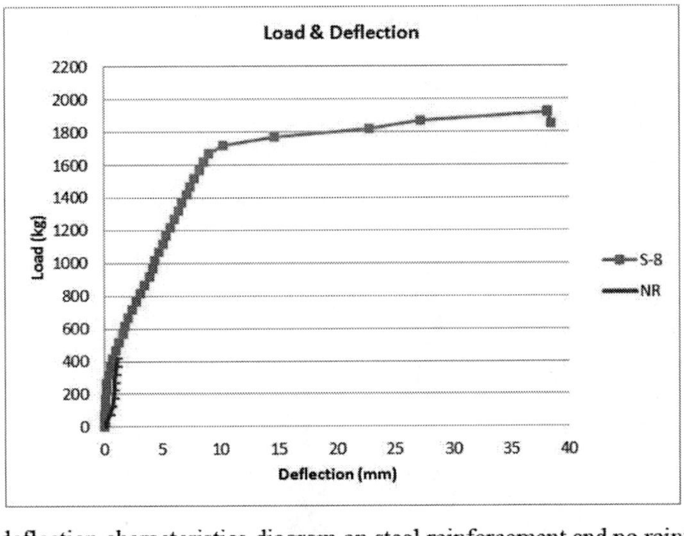

Fig.8. Load and deflection characteristics diagram on steel reinforcement and no reinforcement

Load and deflection characteristics diagram of steel reinforced concrete beam specimens and on bamboo reinforcement specimens, with a spacing between notch of 2, 3, 4, 5, 6, 7 cm, respectively shown in Figure 9.

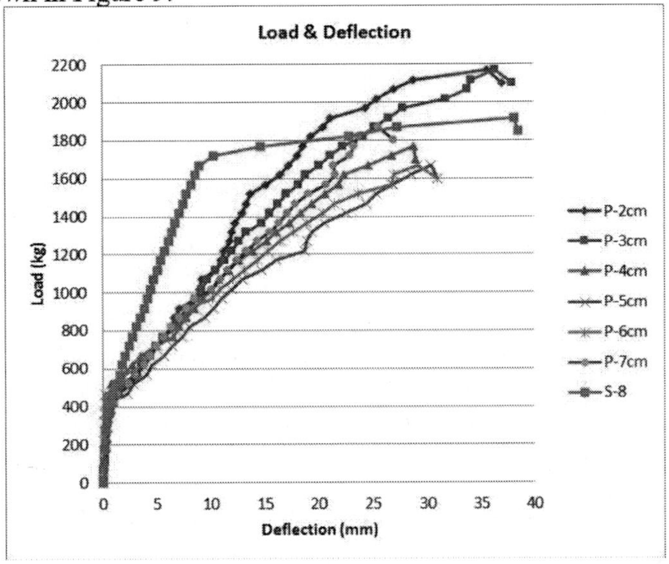

Fig.9. Load and deflection characteristics diagram on bamboo and steel as reinforcement

Note :

P-2cm, etc = concrete beam containing bamboo strips notch at distance 2 cm, etc

S-8 = concrete beam containing steel round bar Ø8 mm

NR = concrete beam no reinforcement

According to the Eq.1, can be calculated the average value of flexural moment experiment for bamboo strip notch at distance 2, 3, 4, 5, 6 and 7 cm respectively are 0.536, 0.528, 0.432, 0.403, 0.424, 0.366 tm. And then, the value of average flexural moment experiment for steel round bar with Ø8mm is 0.516 tm and the value of average flexural moment experiment on concrete beam without reinforcement is 0.11 tm.

According to the Eq.4, can be calculated the analytical ultimate moment for each of the beam specimens. The average value of analytical ultimate moments for steel bar as reinforcement is 0.506 tm and for bamboo strips notch is 0.269 tm.

Below on Fig.10 is presented a diagram the average value of the experimental flexural moment and analytical ultimate moments that occur on concrete beams specimens.

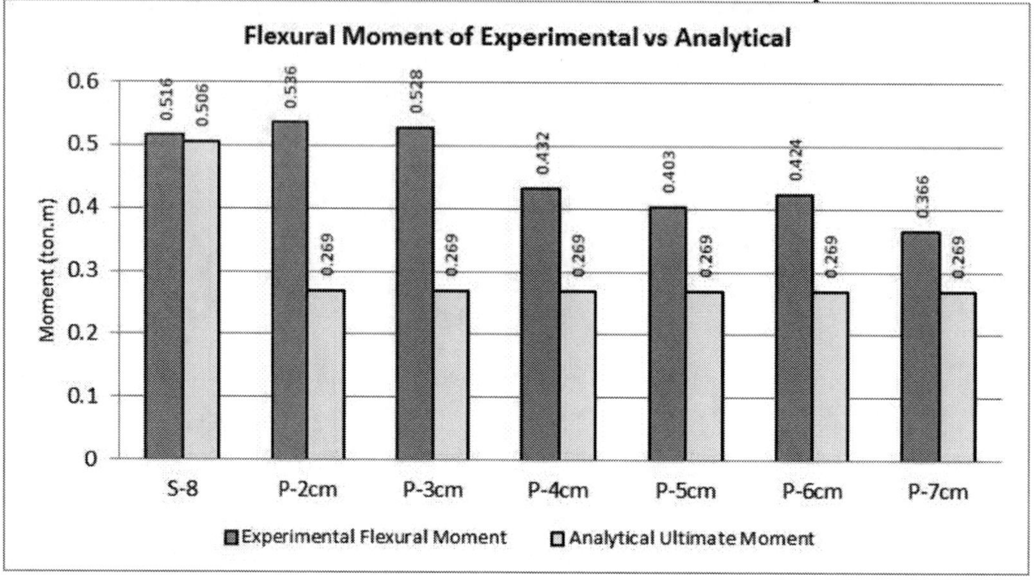

Fig.10. Flexural moment experimental vs analytical ultimate moment

Based on Fig.10, it is shown that the experimental moments tends to increase with the shorter of the distance between the notch of the bamboo strip in the tensile area.

6 Conclusions

Experimental results show that the average flexural capacity on the specimens of concrete beams with Petung bamboo strip notch as reinforcement on the distance between notch 2, 3, 4, 5, 6 and 7 cm respectively are 0.536, 0.528, 0.432, 0.403, 0.424, 0.366 tm. This shows that the notch model V can increase the value of flexural capacity and the smaller the distance between the notches will increase the value of flexural capacity.

The author would like to thank the LPPM UNS for the financial support that has been given to this experimental study.

References

1. K. Ghavami, Cem. & Conc. Comp. Elsev. **27**, 637–649 (2005)
2. M. M. Rahman, M. H. Rashid and M. A. Hossain, IJET-IJENS. **11**, 04 (2011)
3. H. Sakaray, N. V. Togati and I. V. R. Reddy, IJERA. **2**, 077-083 (2012)
4. K. Ghavami, Cem. Conc. Comp. Elsev. **17**, 281-888 (1995)
5. M. A. Salau, I, Adegbite and E. Ikponmwosa, Jour. Sust. Dev. 5, 2, 133–143 (2012)
6. A. Javadian, Const. Build. Mat. Elsev. **122**, 110–117 (2016)
7. A. Azadeh, H. H. Kazemi, KEM. **600**, 69-77 (2014)
8. A.S. Budi, E. Rismunarsi, Sunaryo, AMM. **845**, 208-213 (2016)

Finite element analysis on seismic performance of EWECS composite column

Fauzan[1,*], *Ruddy* Kurniawan[1], and *Zev* Al Jauhari[1,2]

[1]Department of Civil Engineering, Andalas University, 25163 Padang, Indonesia
[2]Department of Civil Engineering, Padang Institute of Technology, 25143 Padang, Indonesia

Abstract. A new hybrid Engineering Wood Encased Concrete-Steel (EWECS) structural system is developed in Japan to solve the limitation of story number for wooden structure. In this structural system, the EWECS column consist of concrete encased steel (CES) core and wood panel as a column cover. An experimental study has been done to investigate the seismic performance of the EWECS column, which is compared with CES column without concrete cover. The main parameter is the presence of wood panel. In order to validate the experimental results, a 3D finite element (FE) model is built by using ANSYS APDL v.14. The FE results indicate that FE EWECS column model has excellent performance, which has a stable spindle-shape hysteresis loops. The presence of wood panel contributes to improve the maximum flexural capacity of the EWECS column about 13%. In addition, the stress distribution on the FE models has good correlation with failure mode of the EWECS column specimens. Generally, the numerical results are able to accurately simulate the seismic behaviour of the EWECS column on test results.

1 Introduction

It has been known that steel reinforced concrete (SRC) structures are typical composite structural systems which have excellent earthquake resistance with high capacities and deformability. However, some disadvantages are found in SRC structures, especially in difficulty to construct both steel section and reinforced concrete. Concrete encased steel (CES) structure consisting of only steel section and concrete is proposed by Kuramoto to overcome these problems [1]. Some experimental studies have been carried out on CES columns subjected to both axial and cyclic loads. The results showed that the CES columns have excellent seismic behaviour.

Based on the SRC and CES ideologies, a new hybrid engineering wood encased concrete-steel (EWECS) structural system is developed for low to high seismic zones. This new composite structural system overcomes the limitation of the number of stories for wooden structures. In this structural system, the EWECS column consist of concrete encased steel (CES) core and wood panel as a column cover. Both economical and structural benefits are realized due to the use of wood panel as column cover. The wood panel acts directly as forming for the composite column during construction, decrease the

* Corresponding author: fauzan@ft.unand.ac.id

labor and materials needed, and consequently reduce the construction cost. From the structural point of view, the wood panel contribute to provide core confinement and resistance to bending moment, shear force, and column buckling [2,3].

An experimental study on the performance of EWECS composite columns subjected to both constant axial and cyclic loads has been done by one of the authors [2] with the main parameter is the presence of wood panel. In this experimental study, the EWECS column is compared with the CES column without concrete cover (the core of EWECS column) in order to investigate the seismic behaviour of EWECS column and the contribution of wood panel on the EWECS column. The performance of both columns is evaluated in terms of hysteresis loops and failure modes.

In order to validate the experimental results, a tridimensional finite element model by using ANSYS APDL v14.0 [4] is developed. This software is a comprehensive general-purpose finite element computer program that contains over 100000 lines of code that very powerful and impressive to be used to solve a variety of problems [5]. ANSYS is capable of performing static analysis that suitable for the current task. The developed model takes into consideration the nonlinear behaviour of material.

2 The geometry of 3D finite element models

All the specimens have a column with 1600 mm height. The column section of EWECS column is 400 mm^2, while that of CES core is 300 mm^2. The EWECS column is covered by a wood panel with a thickness of 45 mm, while its core section is same as that of CES core column, as shown in Fig. 1. Steel encased in each column has a cross section combining two H-section steels of 300x150x6.5x9 mm.

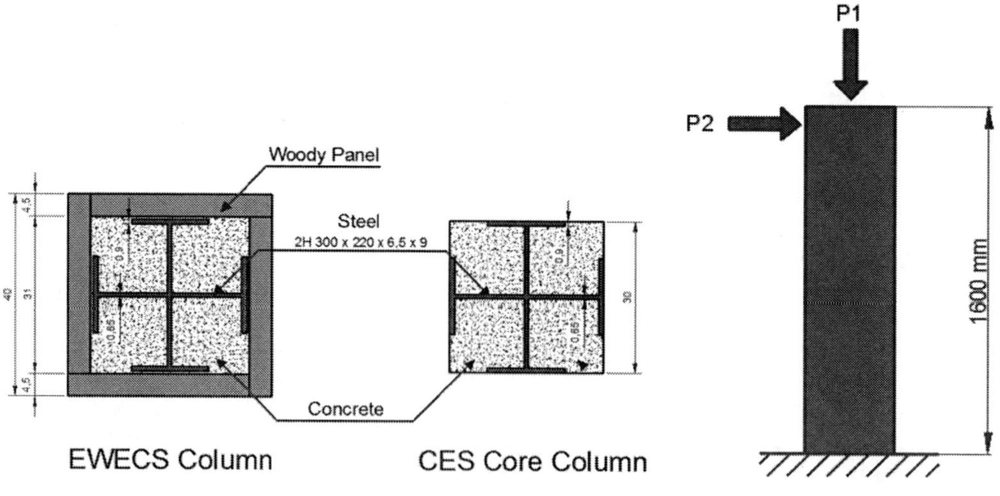

Fig. 1. Details of test specimen of CES core and EWECS composite columns
(P1 = axial constant load, P2 = lateral cyclic load)

The FE models are developed to accurately represent the geometrical configuration and dimensions of the test specimens. The CES core and EWECS composite columns are modelled as illustrated in Fig. 2. The concrete, steel, and wood panels are modelled as block and solid cube with an equivalent length representing the total area of specimens.

DOI: 10.1051/matecconf/201713802023

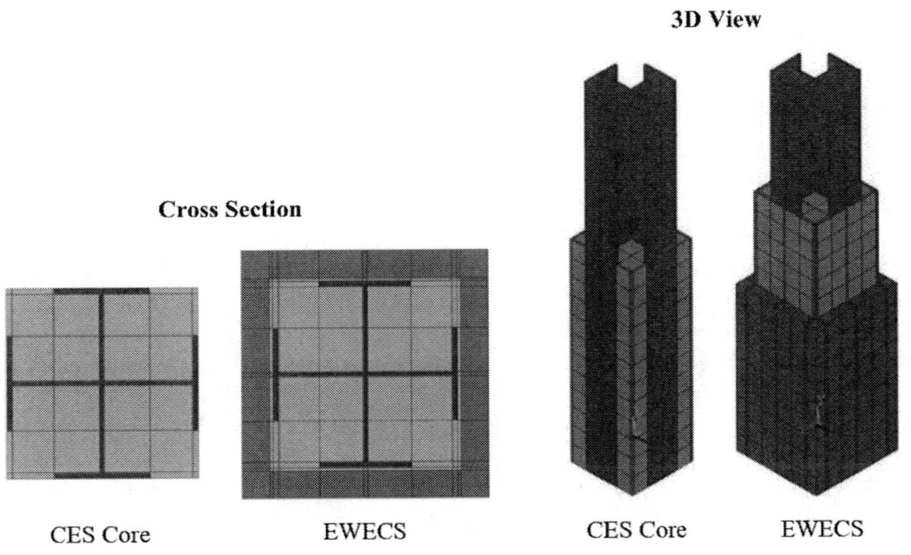

Fig. 2. The FE model of CES core and EWECS composite column models

3 Types of element

Two types of elements are used in the modeling of steel H-section, concrete, and wood panel. ANSYS solid element, SOLID185, is used to modeling steel section and wood panel, while SOLID65 is used to modeling concrete of the EWECS column. The SOLID185 and SOLID65 elements are a 3D hexahedral element defined by eight nodes as shown in Fig. 3. These elements have three translational degrees of freedom at each node in the nodal x, y, and z directions. The SOLID185 element has capability of plastic deformation, hyper elasticity, stress stiffening, creep, large deflection, and large strain. The most important aspect of SOLID65 element is the treatment of nonlinear material properties. The element has the capability of plastic deformation, creep, cracking in three orthogonal directions, and crushing in compression [4].

The FE model in this study is considered perfectly-bonded for material interface between steel and concrete [6], while the unbonded for material interface between wood and concrete that performed by slightly reducing the constitutive model of the materials. This assumption is applied because bonded and unbonded between adjacent materials (wood and concrete) do not much influence for flexural capacities of the columns in the experimental results (the comparison of seismic behaviour between EWECS Column with and without shear studs) [7].

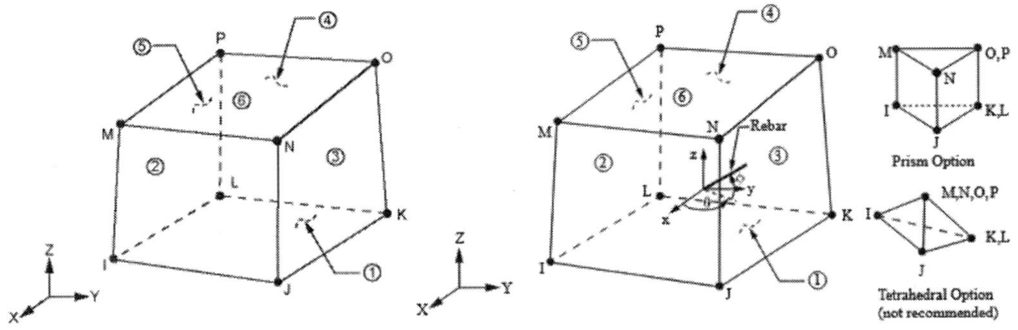

Fig. 3. Solid185 and Solid65 ANSYS Element [4]

 DOI: 10.1051/matecconf/201713802023

4 Material properties, meshing, boundary conditions, and loads

4.1 Material properties

4.1.1 Concrete

The compressive strength of concrete used in both specimens is 35 MPa. A peak concrete strain 0.0025 is used in the analysis. Fig. 4(a) presents the tensile stress–strain curve used for the concrete. The stress-strain relationship in the rising region is designed on the model developed by Saenz [8], which is built in the program. The tensile relaxation (softening) is presented by a sudden reduction of the tensile strength to $0.6f_r$ upon reaching the tensile cracking strain ε_{cr}. After this point, the tensile response decreases linearly to zero stress at a strain $6\varepsilon_{cr}$ as shown in Fig. 4(b).

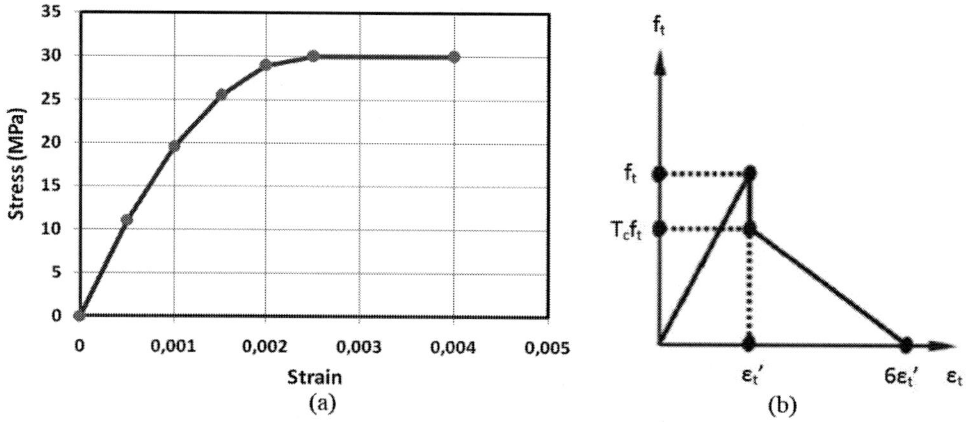

Fig. 4. Idealized (a) compression and (b) tensile stress–strain curve for concrete

Additional concrete material data, such as the shear transfer coefficient β_t for open cracks and β_c for closed cracks are also needed for the concrete constitutive material data table. The shear transfer coefficients β_t and β_c control the amount of shear transfer across the cracks. Typical shear transfer coefficients range from 0.0 to 1.0, with 0.0 represents a smooth crack (complete loss of shear transfer) and 1.0 represents a rough crack (no loss of shear transfer). A shear transfer model developed by Al-Mahaidi [9] is included in the analysis, with value 0.75 and 0.9 is used in the FE model for β_t and β_c, respectively. The fracture criterion of concrete is applied by adopting the five parameter model of William-Warnke [10].

4.1.2 Steel

The yield strength of steel used in each models are 412.5 and 453 MPa for flange and web steel, respectively. The constitutive model of the steel used in this study is a perfectly elastic-plastic criterion [11]. At first, this curve is elastic, then it is assumed to be perfectly plastic (bilinear isotropic model), as shown in Fig. 5(a). This curve is suitable for representing stress-strain characteristics of normal and high quality steel section. Poisson's ratio is assumed 0.3 for this study.

DOI: 10.1051/matecconf/201713802023

4.1.3 Wood Panel

The compressive strength and modulus elasticity of wood are 36.5 MPa and 10500 MPa, respectively. For the design allow the force to be applied in the direction parallel to the annual growth ring of the wood, some existing concrete models built in the program by many researchers might be used in the analysis with some modifications [12]. Stress-strain relationship is modeled by linearly increasing model with slightly reduction (5%), as shown in Fig. 5(b). The fracture criterion of wood is adopted by following the rule of the five parameter model of William-Warnke [10] for concrete with the input of wood material characteristic. The maximum tensile strength of the wood panel during the analysis is taken as 5 MPa. This value is taken into consideration the lower tensile strength in the direction perpendicular to the grain. To account for shear stiffness reduction by shear crack deformation, the shear transfer model for concrete developed by Al-Mahaidi [9] is included by using modified shear transfer coefficient β_c 0.35 for wood.

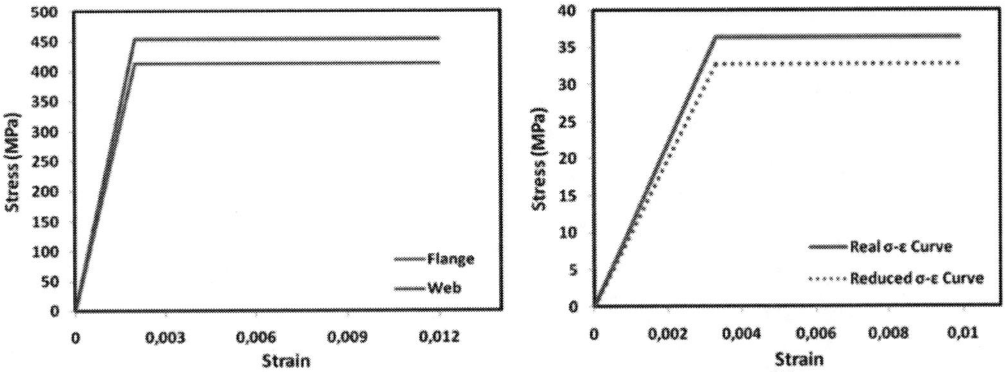

(a) (b)

Fig. 5. Tensile stress–strain curves for (a) steel and (b) wood panel

4.2 Meshing

The mesh density is chosen so that the elements aspect ratio is nearly equal to one. This provided adequate accuracy and fair computational time in modelling the composite columns. The total numbers of element used are 3995 and 5095 elements for CES core and EWECS, respectively. This is an adequate refinement for the constructed the 3D FE models.

4.3 Boundary Conditions

In the FE models, the boundary conditions are as follows: An anchor plate/ stub (700.700.400 mm) is used in the model at the top and bottom of the column. Fixed at the bottom of the column and roller supported at the top of the column as shown in Fig. 6(a). Nodes at the bottom of the column are restrained in all translational degrees of freedom, and the top of the column is restrained in the vertical y direction. The lateral y direction is restrained since the column prevent any displacement in this direction. The nodes at the two edge lines of the column are coupled in the vertical y direction to ensure all the nodes associated with this line move together. The nodes located at the middle face of the column along the direction of the applied displacement are coupled in the horizontal x direction. It should be noted that the boundary conditions are the same for all loading cycles.

 DOI: 10.1051/matecconf/201713802023

4.4 Loads

The loading is applied in the model as follows:

- Constant axial load, approximately 770 kN are applied to the top face columns. This is represented in the FE model by applying a point pressure 1.1 kN on the upper elements (steel plate) of a total nodal area 717.
- The lateral loading history for the model is based on story drift and is the one recommended for use in the experimental program [3]. This is represented in the finite element model by applying the displacement in cycles of loading and unloading at the top edge of the column. The incremental loading cycles are controlled by story drift, R. The lateral load sequence consists of one cycle to each story drift, R 0.5, 1, 1.5, 2, 3 and 4% followed by half cycle to R 5%, as shown in Fig. 6(b). It should be noted that the story drift is equal to the applied lateral displacement at the top of the column face divided by the height of column.

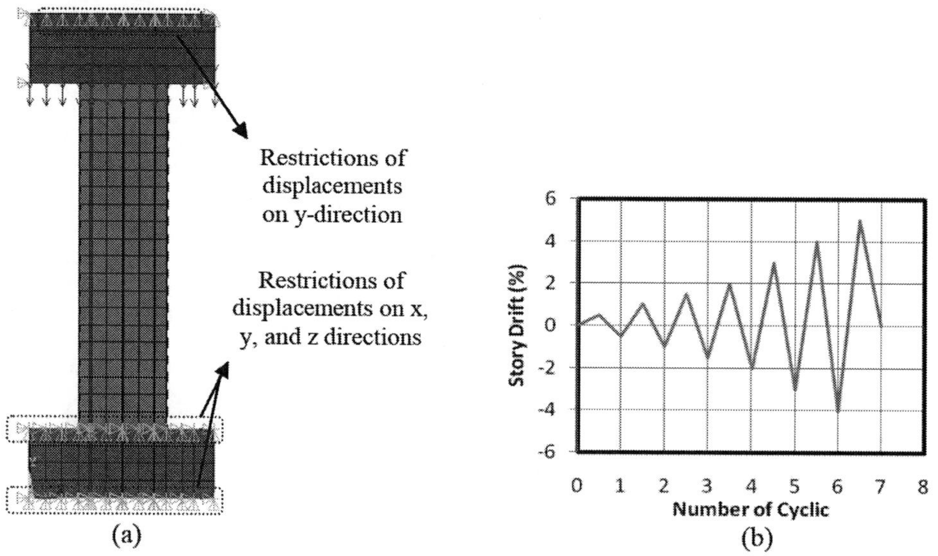

Fig. 6. (a) Boundary conditions, (b) Lateral cyclic load applied in the FE model

The applied cyclic displacements are divided into a series of increments called load steps and load substeps. At the end of each incremental solution, the program adjusts the stiffness matrix of the model before proceeding to the next incremental load step in order to reflect the nonlinear changes in the model's structural stiffness. To predict and control load step size increments, the automatic time stepping option is turned on in this study. The model stiffness is updated in ANSYS by Newton–Raphson equilibrium iterations [13].

5 Result and discussions

5.1 Hysteresis Characteristics

The experimental lateral load-displacement (hysteresis) loops for the CES core and EWECS columns are compared with those obtained from the FE analysis for the entire loading history as shown in Fig. 7. The figure shows that the analytical results for story shear versus story drift of the specimens showed a good agreement with the test results. Both the specimens and models showed ductile and stable spindle-shaped hysteresis loops.

 DOI: 10.1051/matecconf/201713802023

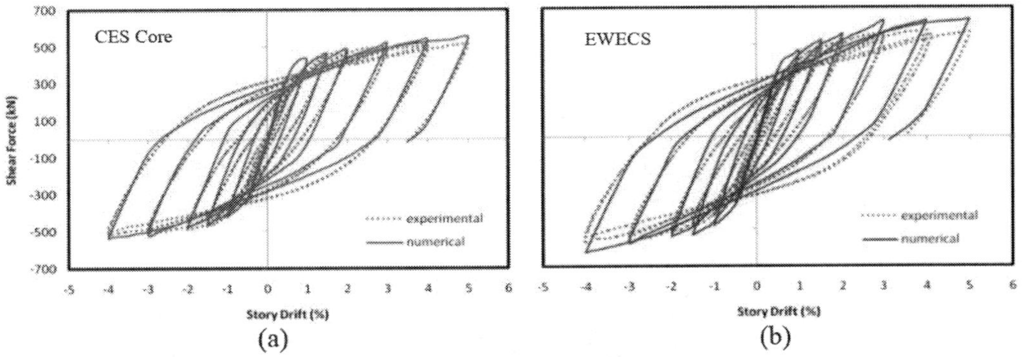

Fig. 7. Comparison between numerical and experimental result of story shear and story drift response of (a) CES and (b) EWECS composite columns

From the FE model, the maximum lateral shear force for CES core column is 541 kN obtained at R 5%. This is approximately 5.4% higher than the results obtained from the experimental (513 kN). It is clear from Fig. 7(a) that the FE and experimental results are almost the same in each stage of cyclic loading. The different percentage of lateral shear force in each stage of loading cycles between the FE analysis and the experimental results is around 5.5%. The maximum shear force for EWECS column is 628 kN at R 5% from the FE analysis, which is 9.1% higher than the experimentally obtained data (573 kN) at R 4%. The FE model behaved less dissipated energy in the last stages of loading cycles than the experimental tested, as seen in Fig. 7(b). The different percentage of lateral shear force in each stage of loading cycles between the FE analysis and the experimental results is around 8.4%. The difference between the FE and experimental results can be attributed to mesh refinement, idealized boundary conditions, material nonlinearity, and the specified coefficient of friction between contact surfaces at the material interface of the column.

5.2 Failure modes

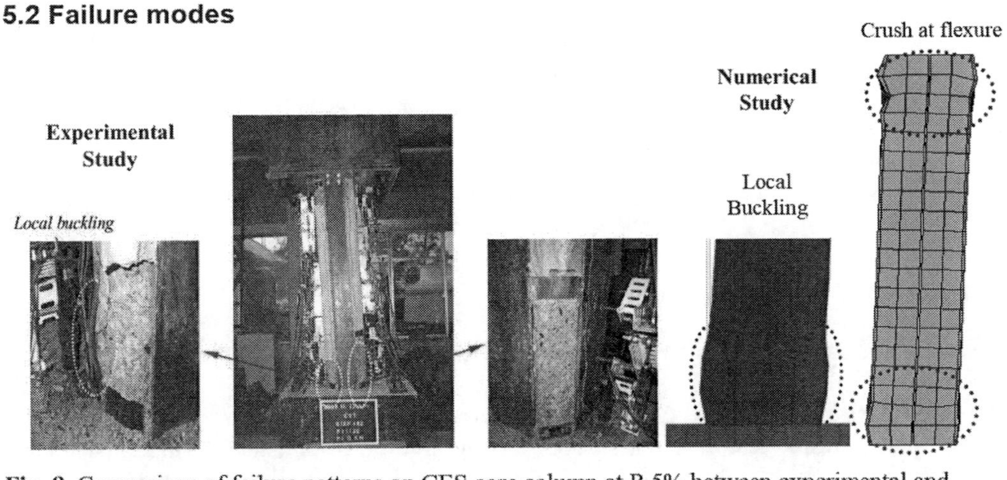

Fig. 8. Comparison of failure patterns on CES core column at R 5% between experimental and numerical studies

Both in the experimental specimen and FE model, flexural cracks in the concrete of CES core appeared to both top and bottom of the column at R 0.5%. Subsequently, the cracks extended at the corners of the column with an increase of the story drift. The crushing of the concrete is found at the corners of both top and bottom parts of the column, and the buckling of the steel flange is observed at the west side in the bottom part of the column, as shown in Fig. 8.

Up to a story drift, R 3%, no damage is observed at the EWECS specimen faces. The wood panel buckled outward, and finally the cracks in the wood panel occurred at the corners of the column at R 5%. Extracting the wood panel after testing, it is observed that the in-filled concrete had crushed in flexure at both top and bottom of the column, and there is no local buckling occurred at the encased steel. This failure mechanism is also observed in the FE model, as seen in the Fig. 9.

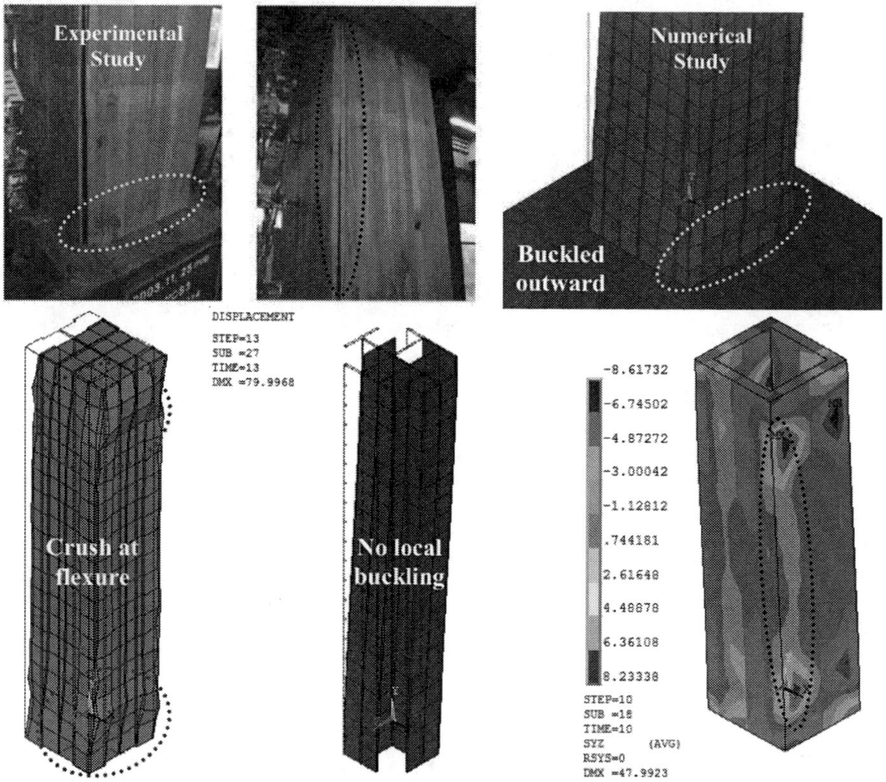

Fig. 9. Comparison of failure patterns on EWECS column at R of 5% between experimental and numerical studies

5.3 Principal Stress Distribution

The stress and strain in the material element are also analysed to validate the FE models. A principal strain 0.002 is reachead in the encased steel of CES and EWECS models at story drift 0.51% and 0.55%, respectively, as indicated that the steel has first yield in red in Fig. 10. The first yield in the corner region both top and bottom of the steel during experimental specimens are at R 0.88% and 0.76%, respectively. These results indicated that the FE model satisfactory portrayed the behaviour of the columns.

Both CES and EWECS models, first crack in the concrete occurs at the story drift 0.4% and 0.55% in the field zone of FE model, indicated by maximum principal stress (tensile) is greater than the tensile strength of concrete (1.8 MPa) as shown inside the oval shape in Fig. 11. These cracks spread on the strut area along the horizontal direction. The principal shear stress in the wood panel of the FE model inside the oval shape in the Fig. 9 shows that cracks occurred on the panel at a shear stress of approximately 8 MPa (maximum shear

 DOI: 10.1051/matecconf/201713802023

force) in the location where the panel is assembled together by using wood glue. The tangential shear strength of the normal wood suggested by Calderoni [14] is average 7.44 MPa. The crack occurring at this location might be attributable to the weak shear strength of the connection using wood glue during construction of the specimens. Cracks also formed at the opposite side of the column, and propagated along the column height.

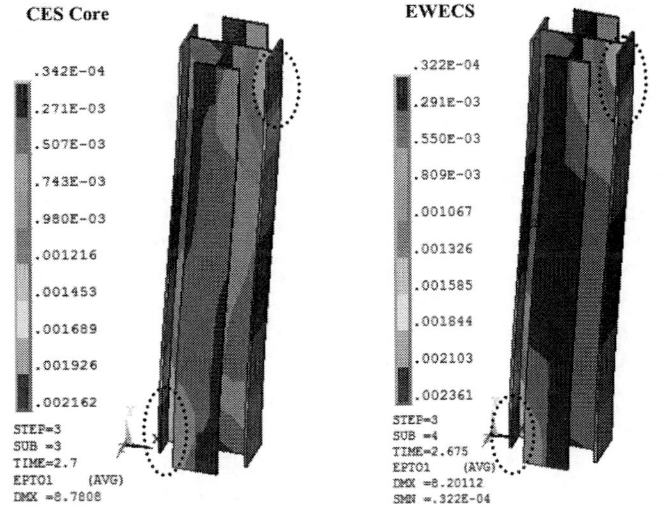

Fig. 10. Principal strain in the encased steel of CES core and EWECS columns at first yielding

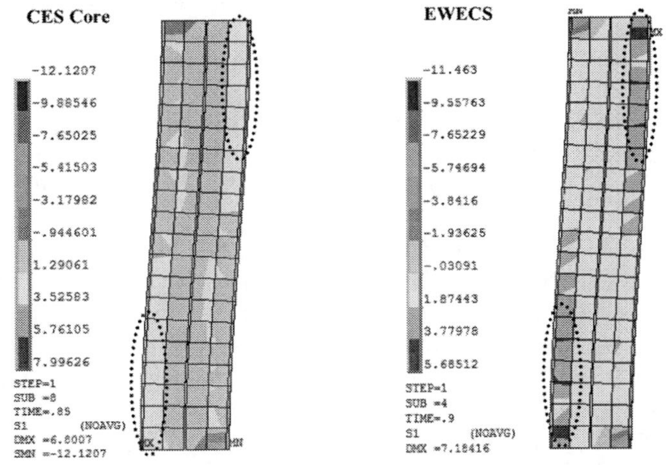

Fig. 11. Principal stress in the concrete of CES core and EWECS columns at first crack

6 Conclusions

Based on the study presented here, the following conclusions can be drawn:

1. The hysteresis loops from the FE analysis have a stable spindle-shape, which correlated fairly well with experimental results. Good correlation exists in all stages of cycling loading. The different percentage of lateral shear force in each stage of

loading cycles between the FEA and the experimental results are around 8.4% and 5.5% for EWECS and CES core columns, respectively.
2. FE analysis results show that the wood panel contributes to flexural capacity of EWECS column until large story drift, R 5%, although the cracks in the wood occured after R 3%. The wood panel contributes to improve the maximum flexural capacity about 13%.
3. The stress distribution on the FE models well represents the failure mode of the CES core and EWECS column specimens.
4. Generally, the numerical results are able to accurately simulate the seismic behaviour of the EWECS composite columns on experimental study.

References

1. H. Kuramoto, H. Takahashi, M. Maeda, Feasibility study on structural performance of concrete encased steel columns using high performance fiber reinforced cementitious composites, *Summaries of Technical Papers of Annual Meeting, Architectural International of Japan (AIJ)*, C-1 pp. 1085–1088 (2000)
2. Fauzan, H. Kuramoto, Y. Shibayama, T. Yamamoto, Structural behaviour of engineering wood encased concrete-steel composite columns, *Proc. of the Japan Concrete Institute (JCI)*, 26(2) pp. 295–300 (2004)
3. H. Kuramoto, Fauzan, Feasibility study on engineering wood encased concrete-steel composite columns, *Proc. 11th Int. Colloquium on Structural and Geotechnical Engineering (11thICSGE)*, Ain Shams Univ., Cairo, Egypt (2005)
4. ANSYS Version 14, User's and Theory Reference Manual, (2010)
5. S. Moaveni, *Finite element analysis: theory and application with ANSYS*, Minnesota State University, Mankato, (1999)
6. A. Leskes, S. Grambicka, Theoretical and experimental studies on composite steel-concrete columns, Concrete and Concrete Structures 2013 Conference, *Proc. Engineering 65 Sciencedirect*, pp. 405–410 (2013)
7. Fauzan, H. Kuramoto, Seismic performance of EWECS composite columns in new hybrid structural system, *Composite Construction in Steel and Concrete VI*, pp. 263–275 (2011)
8. L. P. Saenz, Discussion of equation for the stress-strain curve of concrete, J. American Conc. Inst. (ACI), **61(9)** pp. 1229–1235 (1964)
9. R. S. H. Al-Mahaidi, Nonlinear finite element analysis of reinforced concrete deep members, **79(1)**, Dept. of Structural Engineering, Cornell Univ., NY (1979)
10. K. L. William, E. P. Warnke, Constitutive model for the triaxial behavior of concrete, *Proc. Int. Association for Bridge and Structural Engineering IABSE* 19, Zurich, Switzerland, (1975)
11. J.B.M. Sousa, Caldas R.B., Numerical analysis of composite steel-concrete columns of arbitrary cross section, J. Struct. Eng. ASCE **131(11)** pp. 1721-1730 (2005)
12. H. Kuramoto, B. Li, K. Meas, Fauzan, Experimental and analytical performance evaluation of engineering wood encased concrete-steel beam-column joints, J. Struct. Eng. ASCE **137(8)** pp. 822-833 (2011)
13. R.A. Hawileh, Nonlinear finite element analysis and modeling of a precast hybrid beam-column connection subjected to cyclic loads, J. Appl. Math. Model. Sciencedirect **34** pp. 2562-2583 (2010)
14. C. Calderoni, G. D. Matteis, C. Giubileo, F. M. Mazolani, Flexural and shear behavior of ancient wood beams: experimental and theoretical evaluation, J. Struct. Eng. **28(5)** pp. 729–744 (2006)

MATEC Web of Conferences **138**, 02024 (2017) DOI: 10.1051/matecconf/201713802024
EACEF 2017

Damage Level Prediction of Reinforced Concrete Building Based on Earthquake Time History Using Artificial Neural Network

Reni Suryanita[1,*], *Harnedi* Maizir [2], *Enno* Yuniarto[1], *Muhamad* Zulfakar[1], and *Hendra* Jingga[1]

[1]Department of Civil Engineering, Faculty of Engineering, University of Riau, Pekanbaru, Indonesia
[2]Department of Civil Engineering, Sekolah Tinggi Teknologi Pekanbaru, Indonesia

Abstract. The strong motion earthquake could cause the building damage in case of the building not considered in the earthquake design of the building. The study aims to predict the damage-level of building due to earthquake using Artificial Neural Networks method. The building model is a reinforced concrete building with ten floors and height between floors is 3.6 m. The model building received a load of the earthquake based on nine earthquake time history records. Each time history scaled to 0,5g, 0,75g, and 1,0g. The Artificial Neural Networks are designed in 4 architectural models using the MATLAB program. Model 1 used the displacement, velocity, and acceleration as input and Model 2 used the displacement only as the input. Model 3 used the velocity as input, and Model 4 used the acceleration just as input. The output of the Neural Networks is the damage level of the building with the category of Safe (1), Immediate Occupancy (2), Life Safety (3) or in a condition of Collapse Prevention (4). According to the results, Neural Network models have the prediction rate of the damage level between 85%-95%. Therefore, one of the solutions for analyzing the structural responses and the damage level promptly and efficiently when the earthquake occurred is by using Artificial Neural Network

1 Introduction

Indonesia is one of the countries located in the high seismic region. Some locations in Indonesia were struck by massive ground motion, such as Aceh City (9.3 Ritcher scale) which was followed by a tsunami at December 26th, 2004. At 2005, Nias Island was struck by 8.7 Richter scale earthquake along with tsunami. At May 27th, 2006, a 5.9 Ritcher scale earthquake hit Yogyakarta City which leads to property damage and casualties. At September 30th, 2009 a 7.9 Ritcher scale earthquake hit Padang City with tsunami possibility (1). The design of structure building needs to consider the risk of the loss due to an earthquake.

* Corresponding author: reni.suryanita@eng.unri.ac.id

The risk-targeted ground motions for Indonesia based on the Indonesian Earthquake Resistance Building Code SNI 1726-2012. Seismic loading could significantly impact the structural behavior of building. Therefore it is necessary to investigate the behavior and responses of the building structure under seismic loading.

There are two methods to analyze the responses of a structure due to the earthquake, the response spectrum and time history method (2). Time history analysis is the best used for irregular building structure under seismic loading. The study observed the influence of earthquake loading on the building structure over some duration of time history analysis results. However, the analysis process of non-linear time history taken a long time. To quickly analyze the damage level of building due to the earthquake, an Artificial Intelligent method is applicable. Therefore, this research aims to predict the damage level of the structure under seismic loading using Artificial Neural Network (ANN) technique.

The previous studies stated the Artificial Neural Networks could predict the health of the structure by routinely monitoring (3), (4) and, (5). Artificial Neural Network's algorithm is analogous to human's adaptive neural network which is capable of learning through a set of data with a variable pattern. Moreover, the accuracy of the ANN prediction indicated the accuracy of the provided data in the learning process (6).

The prediction of the behavior of the structure under seismic loading could help the occupants of the building to act prudently when an earthquake occurs. The prediction or output parameter of the ANN model is the performance or damage level of the building structure during an earthquake. Based on FEMA 356 (7), the performance level criteria (Fig.1) are as follows:

1. Fully operational (FO): the building is in fully functional state after the earthquake. The structural have the damage and minor nonstructural damage.
2. Immediate occupancy (IO): the building structure is still safe to functionally operate after the earthquake. There is minor damage with insignificant impact on the occupant of the building. The building could be used directly after the quake.
3. Life safety (LS): the building structure suffers from medium damage which requires rehabilitation. However, the building is still stable and safe for the occupants.
4. Collapse prevention (CP): the building is severely damaged. Nonstructural components collapse and the building is no more useable (operational state).

Fig.1 describes the performance level criteria based on the displacement and force subject to the structure.

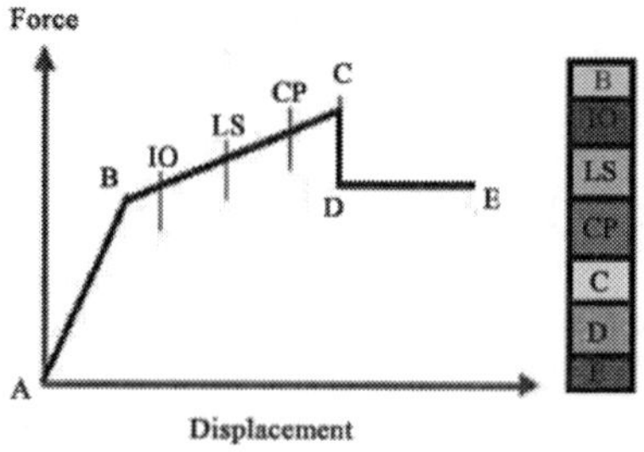

Fig.1. Performance level criteria of building

2. Artificial Neural Networks

The knowledge related to the biological neural network has inspired the development of Artificial Neural Network (ANN) computation method. The data classification, cluster, and non-parametric regression or to simulate the process in the biological neural network need the architecture and operation of ANN. The ANN system comprises of interconnected processing units (neurons) which work simultaneously to provide a solution based on the purpose of the ANN model.

ANN training process is performed to develop the memory and generalization capability of the ANN system. The ANN memorizes how to solve the problem based on the provided learning dataset. The generalization capability means the ANN can recognize the similar pattern (not identical) and provide an output based on the learning phase [8].

The relationship or connection between neurons could determine the quality of the output given by an ANN system (8). The neuron layer stored interconnected neurons. There are three layers in an ANN mode:

1. Input layer, the neurons in the input layer is used to store input data which represents the problem to be solved.
2. Hidden layer, the neurons in hidden layer could not be observed directly and is used to improve the ability of the ANN system to predict a nonlinear relationship between input and output parameters.
3. Output layer, the neurons in the output layer is used to store output data which is the solution provided by the ANN system.

Multilayer perceptron has at least three layers, an input layer, hidden layer, and an output layer. The advantage of multilayer perceptron over single layer ANN is the improved ability to solve complex and nonlinear problems.

There are two criteria used to evaluate the performance and error rate of an ANN system, Correlation Coefficient (R) and Mean Square Error (MSE). The correlation coefficient is the measure of statistic covariance or association between two variables. The ratio between the predicted output value and target (actual) output value indicated R-value. The equation following is the R-value for the ANN model.

$$R = \frac{\sum xy}{\sqrt{\sum x^2 \sum y^2}} \quad (1)$$

where :

x	= X – X'
y	= Y – Y'
X	= observed value (target)
X'	= average of X
Y	= prediction value (output)
Y'	= average of Y

Interprete the strength of relationship between two variables :

$R = 0$	: no correlation
$0 < R \leq 0,25$	: poor correlation
$0,25 < R \leq 0,50$	: medium correlation
$0,50 < R \leq 0,75$	: strong correlation
$0,75 < R \leq 0,99$	: very strong correlation
$R = 1$	: perfect correlation

 DOI: 10.1051/matecconf/201713802024

3. Methodology

3.1. Data of Structure and Loading

The building structure model is the 10-story building with identical floor plan from base to rooftop as shown in Fig. 2. The story height is 3.6 m. The building geometry is symmetrical about the X and Y axis. There are six bays, and each bay spans 5 meters. The rebar material has yield strength, fy = 240 MPa and the ultimate strength, fu = 400 MPa. The dimension of the structural element is as follows:

1. Beam:50 x 70 cm^2 , f_c' = 30 Mpa
2. Column: 80 x 80 cm^2 , f_c' = 30 Mpa
3. Floor-slab:20 cm-thick , f_c' = 30 Mpa
4. Shear wall: 25 cm-thick , f_c' = 30 Mpa

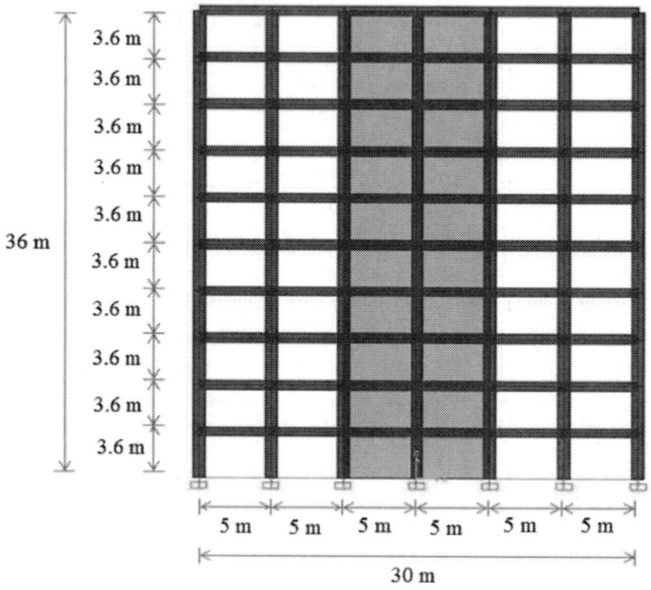

Fig. 2. Ten-story building model

There are nine earthquake-records are used in this study. The list of the earthquake time history is as follows:

1 New Zealand (PGA = 0,255g)
2 San Fransisco (PGA = 0,095g)
3 Cape Mendocino (PGA = 0,15g)
4 El Centro (PGA = 0,319g)
5 Loma Prieta (PGA = 0,12g)
6 San Fernando (PGA = 0,028g)
7 Kocaeli, Turkey (PGA = 0,01g)
8 Nenana (PGA = 0,004g)
9 Denali, Alaska (PGA = 0,008g)

3.2. ANN Training Phase

The damage level of the 10-story building is predicted using Backpropagation multilayer ANN model. The ANN analysis was performed using MATLAB. The ANN testing phase was performed to investigate the prediction accuracy of the trained ANN model with new input data. The testing result will show the capability of the ANN model to predict structure damage level. The ANN model comprises of an input layer with nine neurons, a hidden layer with ten neurons, and an output layer with one neuron. The ANN training was performed by assuming the following training parameters:

1. Number of iteration = 3000 epochs.
2. Target correlation coefficient (R) = 85%.

The level based on structural response parameters during an earthquake. The input parameters are the structural responses: displacement, velocity, and acceleration. Meanwhile, the output neuron has four possible states: Safe, Immediate Occupancy (IO), Life Safety (LS), or Collapse Prevention (CP). Safe is coded with 1, IO with 2, LS with 3, and CP with 4.

Based on the time history analysis results, eight datasets are used for ANN training phase, whereas the remaining one dataset is used for ANN testing phase.

The ANN architecture was varied as tabulated in Table 1. Model 1 has displacement, velocity, and acceleration as input parameters, and damage level as the output parameter. Model 2, 3, and 4 has only one input, that is, displacement, velocity, or acceleration as the input parameter, respectively. The model variation was done to obtain performance comparisons between models with different input parameters.

Table 1. ANN Model

ANN Model	Input Data
Model 1	Displacement, Velocity, Acceleration
Model 2	Displacement
Model 3	Velocity
Model 4	Acceleration

The datasets used for the ANN training phase are 9595 data as shown in Table 2. The source of the data is 8-time history analysis data: New Zealand, San Fransisco, Cape Mendocino, El Centro, Loma Prieta, San Fernando, Kocaeli and Nenana which are scaled to 0.5g, 0.75g, and 1g. The type of the ANN model is Backpropagation ANN. The activation function was Sigmoid Biner. MATLAB program provided two sigmoid functions: Tan-Sigmoid function and Log-Sigmoid function (9).

Table 2. Neural Networks Training Data Set for Displacement as Input

No	Earthquake	Time (Sec)	Input			Output	
			Displacement (m)			Criteria (Target)	
			X direction	Y direction	Z-direction		
1	New Zealand 0,5g	0	0.0000	0.0000	-0.0023	*Safe*	1
2	New Zealand 0,5g	0.05	0.0000	0.0000	-0.0023	*Safe*	1
3	New Zealand 0,5g	0.1	0.0000	0.0000	-0.0023	*Safe*	1
4	New Zealand 0,5g	0.15	0.0000	0.0000	-0.0023	*Safe*	1
5	New Zealand 0,5g	0.2	0.0000	0.0000	-0.0023	*Safe*	1
⋮	⋮	⋮	⋮	⋮	⋮	⋮	⋮
9591	Nenana	29.8	0.0017	0.0007	-0.0023	*IO*	2
9592	Nenana	29.85	0.0050	0.0018	-0.0023	*IO*	2
9593	Nenana	29.9	0.0075	0.0026	-0.0024	*IO*	2
9594	Nenana	29.95	0.0088	0.0030	-0.0024	*IO*	2
9595	Nenana	30	0.0090	0.0030	-0.0024	*IO*	2

4. Results

The Artificial Neural Networks input parameters are structure displacement, velocity, and acceleration, whereas the ANN model predicted the output parameter is the damage level (Safe, IO, LS, CP. Fig 3 shows the calculated R-value 0.94712, which indicates the prediction value closely matches the target value. Therefore it is concluded that the prediction performance of the ANN model is highly accurate and strongly correlated.

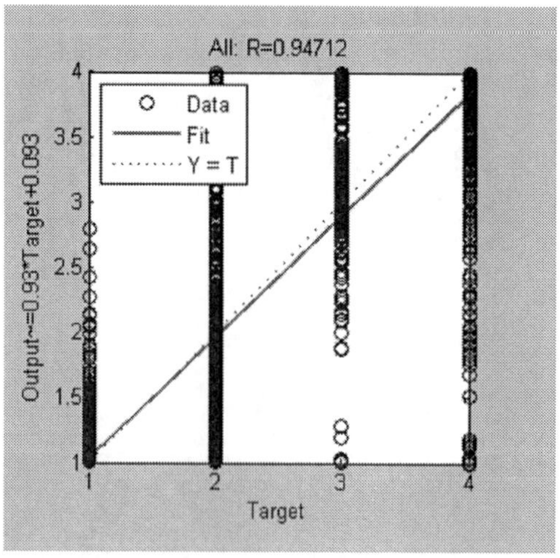

Fig. 3. R Plot for ANN Model 1

The ANN training results for the other ANN models are tabulated in Table 3. Based on Table 3, Model 1 has the best prediction, whereas the worst model is Model 4 with acceleration (X, Y, and Z direction) as the input.

Table 3. ANN performance for all ANN models

NN Model	MSE	R
Model 1	0,11088	0,94712
Model 2	0,10924	0,93841
Model 3	0,23407	0,86543
Model 4	0,33196	0,79339

The results of the ANN testing phase for Model 1 are tabulated in Table 4. The Correlation Regression, R of model 1 is 0.94712 based on 913 data. It means the results are correctly predicted, whereas the remaining 61 data (7.15 %) are wrongly predicted.

Table 4. ANN Model 1 Testing Process

No.	Time History	Time (second)	Target Category		Prediction Result			Status
1	Danieli 0,5g	0	*SAFE*	1	1.019	1	*SAFE*	Match
2	Danieli 0,5g	0.05	*SAFE*	1	1.0175	1	*SAFE*	Match
3	Danieli 0,5g	0.1	*SAFE*	1	1.0245	1	*SAFE*	Match
⋮	⋮	⋮	⋮	⋮	⋮	⋮	⋮	⋮
69	Danieli 0,5g	3.4	*IO*	2	1.7978	2	*IO*	Match
70	Danieli 0,5g	3.45	*IO*	2	2.0011	2	*IO*	Match
71	Danieli 0,5g	3.5	*IO*	2	1.9602	2	*IO*	Match
72	Danieli 0,5g	3.55	*IO*	2	1.4088	1	*SAFE*	***False***
73	Danieli 0,5g	3.6	*IO*	2	1.4343	1	*SAFE*	***False***
⋮	⋮	⋮	⋮	⋮	⋮	⋮	⋮	⋮
85	Danieli 0,5g	4.2	*LS*	3	3.3426	3	*LS*	Match
86	Danieli 0,5g	4.25	*LS*	3	3.8801	4	*CP*	***False***
87	Danieli 0,5g	4.3	*LS*	3	3.8644	4	*CP*	***False***
88	Danieli 0,5g	4.35	*LS*	3	3.2352	3	*LS*	Match
⋮	⋮	⋮	⋮	⋮	⋮	⋮	⋮	⋮
911	Danieli 1g	29.9	*SAFE*	1	1.014	1	*SAFE*	Match
912	Danieli 1g	29.95	*SAFE*	1	1.0239	1	*SAFE*	Match
913	Danieli 1g	30	*SAFE*	1	1.0418	1	*SAFE*	Match

The analog with Model 1, the process was performed for the other ANN models, which results are shown in Table 5.

Table 5. ANN Testing Phase Summary

ANN Model	Prediction		Percentage
	Match	**False**	**Matching**
Model 1	852	61	0,94712
Model 2	841	72	0,93841
Model 3	835	78	0,86543
Model 4	800	113	0,79939

According to Table 5, it can be concluded that Model 1 resulted in the best to predict the desired output with 95% of prediction matching percentage. Model 2 and Model 3 is also conducted above 85% of percentage matching as the target correlation coefficient (R). Meanwhile, the prediction accuracy of Model 4 is only 79 % less than the target correlation coefficient, R. Therefore, the ANN model can predict the damage level of the building structure with the average prediction accuracy more than 90%, which indicates good correlation.

5. Conclusion

According to the results, the ANN training phase by evaluating R-value concluded Model 1 has the best prediction accuracy. The input parameters used in Model 1 are structure displacement, velocity, and acceleration in X, Y, and Z direction. Meanwhile, Model 4 has the worst prediction accuracy with acceleration in X, Y, and Z direction as the only input parameter.

The prediction accuracy of all ANN models based on the ANN testing phase with 913 data sets, 95 % of the data was predicted accurately by ANN Model 1, whereas for Model 2, Model 3, and Model 4 the prediction accuracy was 94%, 86%, and 80%, respectively. The prediction accuracy of the ANN model was above 85%, which indicates that the ANN model is highly accurate as a prediction tool.

The authors would like to thank the Ministry of Research, Technology and Higher Education (Kemenristekdikti), Indonesia, for funding this research through Fundamental Research Scheme 2017.

References

1. BMKG. Earthquake-Query Data [Internet]. BMKG Indonesia. 2017 [cited 2017 Jan 20]. Available from: http://repogempa.bmkg.go.id/index.php?id=101&session_id=zLImgAy7
2. Chopra AK. Dynamics of Structure, Theory and Applications to Earthquake Engineering. Prentice Hall; (2012)
3. Suryanita R, Maizir H, Jingga H. Prediction of Structural Response Based on Ground Acceleration Using Artificial Neural Networks. Int J Technol Eng Stud.

(2017)

4. Suryanita R, Jingga H, Yuniarto E. The Application of Artificial Neural Networks in Predicting Structural Response of Multistory Building in The Region of Sumatra Island. KnE Eng [Internet]. 2016;1(2015):1–6. Available from: http://knepublishing.com/index.php/KnE-Engineering/article/view/526/1619
5. Chakraverty S, Marwala T, Gupta P. Response Prediction of Structure System Subject to Earthquake Motions using Artificial Neural Network. Asian J Civ Eng (Building Housing). Vol 7 (2006)
6. Gharehbaghi K. Artificial neural network for transportation infrastructure systems. MATEC Web Conf. (2016)
7. FEMA356. Prestandard and Commentary for The Seismic Rehabilitation of Buildings. Vol. FEMA 356. Federal Emergency Management Agency; (2000)
8. Rajasekaran S, Pai GAV. Neural Network, Fuzzy logic, and Genetic Algorithms Syntesis and Applications. New Delhi: Prentice Hall of India; (2007).
9. Demuth HB, Beale MH, De Jess O, Hagan MT. Neural network design. Martin Hagan; (2014).

MATEC Web of Conferences **138**, 02025 (2017)
EACEF 2017
DOI: 10.1051/matecconf/201713802025

Factor Analysis on Criteria Affecting Lean Retrofit for Energy Efficient Initiatives in Higher Learning Institution Buildings

Nur IzieAdiana Abidin[1,*], *Rozana* Zakaria[1] , *Eeydzah* Aminuddin[1] , *Abdul Rahim* Abdul Hamid[1], *Vikneswaran* Munikanan[2], *Shaza Rina* Sahamir[3] and *Siti Mazzuana* Shamsuddin[4]

[1]Faculty of Civil Engineering, Universiti Teknologi Malaysia, Johor Bahru, Malaysia
[2]Faculty of Engineering, Universiti Pertahanan Malaysia, Kuala Lumpur, Malaysia
[3,4]Faculty of Architecture, Planning and Surveying, Universiti Teknologi Mara, Shah Alam, Selangor, Malaysia

Abstract. The expansion of Higher Learning Institution (HLI) is a global concerns on energy demand due to campus act like a small city. Intensive mode of operation of a building is correlated to the energy utilization. Improvement in the current energy efficiency is crucial effort to minimize the environmental effect through minimisation of energy in operation by retrofitting and upgrade the existing building system or components to be more efficient. Basically, there are three recommended steps for the improvement known as lean initiatives, green technology and clean energy in response to becoming zero energy solutions for building. The deliberation of this paper is aimed to highlight the criteria affecting in retrofitting of existing building in HLI with lean initiatives in order to achieve energy efficiency and reduction of energy comsumption. Attention is devoted to reviewing the lean energy retrofitting initiatives criteria for daylighting (side lighting), daylighting (skylight) and glazing. The questionnaire survey was employed and distributed to the architects who has an expertise in green building design. Factor analysis was adopted as a method of analysis by using Principal Component with Varimax Rotation. The result is presented through summarizing the sub-criteria according to its importance with a factor loading 0.50 and above. The result found that majority of the criteria developed achieved the significant factor loading value and in accordance with the protocal of analysis. In conclusion the results from analysis of this paper assists the stakeholders in assessing the significant criteria based on the desired lean energy retrofitting initiatives and also provides a huge contribution for future planning improvement in existing buildings to become an energy efficient building.

* Corresponding author: rozana@utm.my

 DOI: 10.1051/matecconf/201713802025

1 Introduction

The critical issue in campus buildings is often associated with the energy efficiency and it indirectly gives an impact to the occupant's comfort and indoor quality [1]. Campus universities have been called as small cities since it has large size, large users and involved with complex activities. The wastage of energy in campus comes from various spaces such as lecture halls, offices, computer rooms, laboratories, street lighting and others. Their activities and operations from the teaching, research, and other services affects the building energy use and to the environment. However, the energy and environmental impact in campus universities able to be reduced with several approaches through technology, organizational and energy optimization initiative [2]. According to [3], the effective implementation to improve the current energy condition and at the same time concern to achieve for sustainability in existing building is through the adoption of energy efficiency. This shall be implemented by replacing the existing mechanical and electrical technology and building system with more energy efficient application which beneficial in reducing the expenditure of energy. [4] stated that, the basic principle with the implementation of energy efficiency is to reduce the operations of energy which including lighting, heating and cooling without denying the comfort and health of users in the space. In addition, the improvement of energy efficiency not only beneficial for environmental protection, but also to the economic especially the operational cost savings. Besides, [5] argued that in order to achieve a great reduction of energy usage, the effective decision in energy plan and management and the technical measures should be under consideration.

2 Literature Review

2.1 Retrofitting in Higher Learning Institution

Retrofitting known as the process to add the existing building systems with new technology, services or functions. The changes are commonly implemented in the structure of interior space in a building which is constructed whether wholly or partially occupied [6]. Basically, the campus energy demand and its building performance are influences from several sources of its existing components or systems such as lighting, building envelope and mechanical system. The improvement for lighting system is the most simple retrofit initiative that could be implemented through integration with daylighting, occupancy sensor installation, convert the existing lighting system with high performance and high lighting efficacy. While higher consideration in building envelope, it able to improve through painting, insulation, shading and glazing system. Meanwhile, for the mechanical system which known as high energy consumers in a building could reduce the usage by adopting the natural ventilation, replace the existing window and installing the variable frequency drives [5].

Table 1 shows the critical summary of energy efficiency initiatives that is commonly implemented in campus building. It consists of several initiatives from lean energy, green technology and clean energy. For the lean energy, the initiative that has been successfully and effectively applied consists of wall, window, roof/atrium and daylighting. Therefore, in order to fulfil the aim of this paper, this information, it assists in finding the lean energy initiatives which are relevant to implement on campus.

Table 1. Summary of Energy Efficiency Initiatives Taken in Campuses [12]

Retrofit Initiatives	Lighting	Ventilation	Refrigeration	Wall	Window	Occupancy sensor	Roof/atrium	Daylight	Solar
a) Indiana State University [7]							√		
b) University of Malaya [8]	√								
c) U.S colleges [9]	√					√	√		
d) College of Desert [10]	√	√							
e) Boston University [11]	√					√		√	
f) Rome Italy University Building [12]	√				√		√	√	
g) UC Berkeley's University [13]		√							
h) University of Mexico [14]	√		√						√
i) Melbourne University [15]	√								
j) University of Seoul Korea [16]	√	√		√	√		√		
k) Curtin University Sarawak [17]	√	√							
l) Guangdong Colleges [18]	√	√				√			
m) Kingsville Texas University [19]	√	√							
n) University of Malaya [20]		√							

Legend:

Lighting	*Solar*	*Window*
Ventilation	*Roof/atrium*	*Occupancy sensor*
Refrigeration	*Wall*	*Daylight*

2.2 Retrofitting Technology

According to [21], there are three steps recommended to be adopted towards the achievement of sustainable building. Firstly, is the utilization of natural resources which known as lean energy, secondly is to take into account on the green technology and lastly is to utilized for clean energy which is the opportunity to have the clean sources towards in achieving zero energy building. The detail explanation of the three steps is discussed in the section below.

2.2.1 Lean Energy, Green Technology and Clean Energy

Lean energy (passive design) is defined as the principals of design that concern with current innovations of technology [21]. The important implementation of lean energy is to less utilize the energy and practice to become energy efficient to reduce the existing demand [22]. The levels of efficiency and cost involve in passive design is different based on the techniques adopted. Therefore, the techniques selected prior to implementation should take into account the function of the building, expected thermal comfort, loss, and gain of energy in the building, potential energy saving, weather, location and environment [23]. Passive design is the strategy that offers an economical and effective approach. Buildings with passive design able to achieve savings over 50% of the total primary energy consumption [24]. There are several passive technologies that could be applied in a building. Firstly, external shading technology is effective to reduce the air-conditioning usage by preventing sunlight admitted into the building during summer but allow it to enter during winter. Secondly, is the natural light utilization by concerning on the window wall ratio at certain orientations in order to reduce the electric light consumption. Thirdly is green roof technology which able to significantly improve the building temperature [25].

Green technology is described as an adoption of technologies that is low carbon as an effort to supply the energy efficiently [21]. Green technology is also an innovation of technology which is developed to protect the environment and reduce the energy by utilizing new technology at the existing electricity equipment [26]. Besides, green technology is to develop and apply systems, equipment and product which has a large contribution to the environment and to conserve the natural resource in order to minimize the impacts from the activities which carried out by humans [27]. It is said that passive technology known as preferred approach and active technology is complementary in green building design, however, it is actually balancing many of the needs to extra indoor performance comfort to a building [25].

Clean energy is the renewable technologies that can be applied as an opportunity to achieve for greater reduction of energy [21]. Clean energy is the renewable energy sources such as wind power, solar photovoltaics and solar hot water which capable to meet the energy demand in the building. Since most of the countries in this global using fossil fuels to generate the electricity and thus indirectly has contributed to the large emission of greenhouse gases and the atmosphere becoming warm. With the clean sources such as wind, sun and water alternative, it provides a large reduction of greenhouse gas emission [28]. Besides, [29] also mentioned that, clean energy approach is the great solution in reducing the greenhouse gas emission because it is capable to minimize the CO_2 emissions which generated from the fossil fuel carbon oxidation.

3 Method

In this paper, lean energy (Passive Design) is only deliberated in its survey and analysis. A questionnaire survey approach was adopted to elicit the important criteria affecting in retrofitting for lean energy that consists of daylighting (side lighting), daylighting (skylight) and glazing. Each of the lean energy retrofitting initiatives has been developed with its own criteria and it is specifically required the respondents to give the degree of importance of each criteria based on the scale given. The establishment of criteria is by reviewing in detail the available information from secondary data such as books, published journal, thesis and online articles. In order to obtain the information pertaining on the criteria affecting in retrofitting of lean energy, a total of 392 questionnaires was sent to the Architects whereby the selections is based on the applicability and reliability of respondents as a designer towards the green building design for HLI. From the total of sample distributed, 85 has successfully returned and response which indicate the response rate of 33.46%.

Then, Factor Analysis was adopted to prioritized and ranked the criteria. Factor Analysis known as a data reduction method by collapsing a large number of variables into several underlying factors and allow to make an arrangement according to its significant value [30].

In order to perform this statistical analysis, the Statistical Package for Social Science (SPSS) was used with the adoption of several protocols. Firstly, assess the Kaiser-Meyer-Olkin (KMO) whereby the requirement for the minimum value is 0.50. Secondly, is Barlett's Test of Sphericity (P) with the minimum value is $P < 0.50$ [31]. Thirdly is to focus on the factor of extraction for which Principal Component Analysis has been selected and it is the default setting of the software. Then, Varimax with orthogonal rotation were chosen as the rotational method. Lastly is the factor loading value with the aim in retaining the factor 0.50 and above. The result obtained were arrange based on the value of importance of the criteria.

4 Findings

4.1 Daylighting (sidelighting)

Daylighting is the passive approach that uses solar energy as the main resources to utilize the natural light in the interior space of a building [32]. Daylighting has become a preferred application in most of architecture design since it offers for uniformity with its illumination distribution and also has the potential to reduce the energy by 20% [33]. In addition, natural light provides the users to have a pleasant environment, comfortability, improves the productivity in comparing to artificial lighting [34]. In daylighting, there are 39 variables were loaded in Factor Analysis that subjected to Principal Component Analysis. From Table 2, Kaiser-Meyer-Olkin(KMO) achieved is 0.872 which shows the adequacy of sampling since it exceeds the recommended value of 0.50, and the Barlett's Test of Sphericity is (P=.000) which indicated as statistical significance.

Table 2. Kaiser-Meyer-Olkin and Barlett's Test of Sphericity for Daylighting (Side Lighting)

KMO and Bartlett's Test		
Kaiser-Meyer-Olkin Measure of Sampling Adequacy.		.872
Bartlett's Test of Sphericity	Approx. Chi-Square	2791.614
	df	741
	Sig.	.000

Then, the Varimax rotation was conducted to retain the factor and to interpret the components into simple structure solution. From 39 variables, 36 variables have achieved the factor loading 0.50 and above. The result as shown in Table 3.

Table 3. Kaiser-Meyer-Olkin and Barlett's Test of Sphericity for (Daylighting-skylights)

Main Criteria	Sub-criteria	Factor loading
Visual Comfort	Cleaning	0.797
	Painting	0.770
	Analyze placement and configuration	0.752
	Locate windows high in a wall	0.662
	Avoid direct beam	0.655
Indoor Quality	Thermal comfort	0.586
	Impact on cooling requirement	0.559
Design	Window shape	0.829
	Size of room	0.748
	Daylight redirecting system	0.708
	Interior design and space planning	0.687
	Incorporated interior and exterior shading	0.658
Environmental	greenhouse gas emission	0.834
Economic Assessment	Maintenance cost	0.834
	Return on investment	0.791
	Construction Costs	0.622
	Simple payback	0.636
Resource Availability	Sun patterns and condition	0.701
	Solar geometry and effective sky angle	0.503

4.2 Daylighting (Skylight)

Skylight is another method to utilize for natural light and being a new trend in modern design architecture especially for commercial building development [35]. Skylight able to improve the quality of lighting in the workspace, improve productivity, space has a variation of appearance and reduce the need of electric lighting during the daytime [36].

The protocol performed in this section is similar from the previous section. For skylights, the sampling adequacy of Kaiser-Meyer-Olkin(KMO) is 0.892 and the statistical significant of Bartlett's Test Sphericity is (P =.000). The result as shown in Table 4.

Table 4. Kaiser-Meyer-Olkin and Barlett's Test of Sphericity for (Daylighting-skylights)

KMO and Bartlett's Test		
Kaiser-Meyer-Olkin Measure of Sampling Adequacy.		.892
Bartlett's Test of Sphericity	Approx. Chi-Square	1446.224
	df	253
	Sig.	.000

Meanwhile, Table 5 shows the summary result achieved after performing the Varimax rotation which contains of factor loading 0.50 and above. It is arranged based on the sequence from most important to least important which performed under 6 main criteria.

Table 5. Summary Result for Daylighting (Daylighting-Skylights)

Main Criteria	Sub-criteria	Factor loading
Design	Integrate with electric lighting system	0.842
	Aim for recommended surface reflectance value	0.782
	Size and shapes of roof lights	0.743
	Determine the daylight factor	0.704
	Building Massing and orientation	0.649
Resource Availability	Solar geometry and sky angle	0.679
	Sun patterns and condition	0.586
Indoor Comfort	Impact on cooling requirement	0.869
	Thermal comfort	0.803
Economic Assessment	Maintenance and repair cost	0.807
	Construction cost	0.783
	Return on investment	0.730
	Simple payback	0.697
	Energy and cost savings	0.563
Environmental	Greenhouse gas emission	0.780
Visual Comfort	Painting	0.533
	Provision for regularly cleaning	0.513

4.3 Glazing

In most of commercial building development, glazing has been widely adopted and become the statement in architectural design. Glazing provides the occupants in a building to harvest the daylight, view to the outside environment and indoor quality [37]. Since window allows to receive for fresh air and connect the occupants to the views outside, it is also easily caused to heat gain and heat loss which then contributed to the building heating and cooling loads. Thus, glazing is the efficient initiative that could be adopted in reducing the energy cost and at the same time ensuring for comfortability, clean and healthier environment [38].

For glazing lean retrofit initiative, there are 32 variables loaded which also performed under Principal Component Analysis. The result from sampling adequacy assessment revealed has exceeded the recommended which is 0.865. Whilst, Bartlett's Test of Sphericity also revealed for statistical significant with (P =.000) . Both of the results can be seen in Table 6.

Table 6. Kaiser-Meyer-Olkin and Barlett's Test of Sphericity (Glazing)

KMO and Bartlett's Test		
Kaiser-Meyer-Olkin Measure of Sampling Adequacy.		.865
Bartlett's Test of Sphericity	Approx. Chi-Square	2119.956
	df	496
	Sig.	.000

Then, based on the Varimax rotation, there are 3 variables require to be eliminated from 32 variables loaded since the factor loading less than 0.50. The summary result of strong factor loading of 0.50 and above is tabulated in Table 7.

Table 7. Summary Result for Glazing

Main Criteria	Sub-criteria	Factor loading
Glazing Selection	Glazing shape	0.845
	Type of insulation	0.834
	Frame types, styles, finishes, colour and durability	0.759
	U-value of glazing properties	0.752
	Visible transmittance and Visible reflectance	0.734
	Spectral Selectivity	0.733
	Glazing sizes	0.725
	Glazing colours	0.717
	Types of glazing	0.708
	Solar heat gain coefficient and Shading Coefficient	0.686
	Ultraviolet Transmittance	0.581
	Sound Transmittance	0.574
	Durability	0.569
	Good safety features	0.564
Design	Ability to comply with technical standards	0.664
	Determine the desired light levels	0.551
	Aesthetic / glass appearance	0.638
	Determine the orientation of glazing	0.574
Occupants Comfort	Health and Indoor air quality	0.797
	Thermal comfort	0.741
	Glare control	0.739
Economic Assessment	Maintenance Costs	0.758
	Energy and cost saving	0.726
	Construction costs	0.679
	Payback Period	0.652
	Return on investment	0.612
Site Assessment	Determine the angle incidence of solar	0.588
	Climate conditions in the location	0.504
Environmental	Greenhouse gas emission	0.541

5. Conclusion

Retrofitting has gained placed by many universities since the government started to promote the effort for sustainability campus due to the realization of rising global energy demands. The campus university building also has always been targeted and being the transformation centers in order to overcome the energy issue for the benefits of current and future generations. Lean energy is an energy efficient initiative that able to significantly contribute to the energy reduction in campus through the utilization of the innovative approach. It is the first step that is recommended to be adopted in order to achieve the aim for sustainable building. Therefore, the development of the lean energy criteria in this paper assists in providing the stakeholders with the information that require being considered during the decision of the retrofit project selected especially in planning and design stage. There are 28 main criteria that consist of visual comfort, indoor quality, design, environmental, economic and others. The findings from Factor Analysis have shown the significant of the criteria based on its factor loading value which provides as a reliable guide for the prioritization process.

References

1. Jomoah, I. M., & Arabia, S. Energy Management in the Buildings of a University Campus in Saudi Arabia – A Case Study. *International Conference on Power Engineering, Energy and Electrical Devices.* 13–17 (2013).
2. Kolokotsa, D., Gobakis, K., Papantoniou, S., Georgatou, C., Kampelis, N., Kalaitzakis, K.,Santamouris, M. Development of a web based energy management system for University Campuses: The CAMP-IT platform. *Energy and Buildings*, **123**, 119–135 (2016).
3. Tan B., Yavuz Y., Otay E.N., Çamlıbel E. Optimal Selection of Energy Efficiency Measures for Energy Sustainability of Existing Buildings, Computers & Operations Research, **66**, 258–271. (2016).
4. Ruparathna, R., Hewage, K., and Sadiq, R. Improving The Energy Efficiency Of The Existing Building Stock: A Critical Review Of Commercial And Institutional Buildings. *Renewable and Sustainable Energy Reviews*, **53**, 1032–1045. (2016).
5. Alshuwaikhat, H,M., & Abubakar, I. An Integrated Approach to Achieving Campus Sustainability: Assessment of the Current Campus Environmental Management Practices. *Journal of Cleaner Production*. **16**, 1777-1785 (2008).
6. Eriksson, R., Nenonen, S., Junghans, A., Nielsen, S. B., & Lindahl, G. Nordic Campus Retrofitting Concepts Scalable Practices. *Procedia Economics and Finance*, 21(15), 329–336 (2015).
7. Appleby P. *Sustainable Retrofit and Facilities Management*. 1st edition. Oxford, United Kingdom. Routledge (2013).
8. Mahlia T.M.I., Razak H.A., Nursahida M.A. Life cycle Cost Analysis and Payback Period of Lighting Retrofit at the University of Malaya, Renewable and Sustainable Energy Reviews, **15 (2)**, 1125–113 (2011).
9. National Grid. Managing Energy Costs in Colleges and Universities. Street Boulder Colorodo: E Source Companies LLC (2003).
10. College of Desert. College of the Desert Campus-wide Lighting Retrofit. Southern California: Edison International (2011).

11. Gevelber M., Choate R., Sheehan K., Vitolo T., Ercolino E., Ricci L., Mchale M. Opportunities for Achieving Significant Energy Reduction in Existing University Buildings: Developing efficient HVAC operations. Boston United States: Boston University (2007).
12. Citterio M., Romeo C. *Retrofit of a University Building in Rome Italy. Energy Conservation in Buildings and Community System Case Study Overview*. Rome Italy: International Energy Agency. (2003).
13. UC Berkeleys., 2012, UC Berkeley's University Hall Wireless HVAC Retrofit Brings the Campus One Step Closer to Its Climate Neutrality, Energy Technology Assistance Program, California, USA.
14. Escobedo A., Briceño S., Juárez H., Castillo D., Imaz M.,Sheinbaum C. Energy Consumption and GHG Emission Scenarios of A University Campus in Mexico Energy for Sustainable Development, **18(1),** 49–57 (2014).
15. Di Stefano J. Energy Efficiency and The Environment: The Potential for Energy Efficient Lighting to Save Energy and Reduce Carbon Dioxide Emissions at Melbourne University, Energy, **25(9)**, 823–839 (2000).
16. Chung M.H., Rhee E.K. Potential Opportunities for Energy Conservation in Existing Buildings on University Campus: A field survey in Korea. *Energy and Buildings*, **78**, 176–182 (2014).
17. Tang F.E. An Energy Consumption Study for a Malaysian, International Journal of Environment, Earth Science and Engineering, **8**, 99–105 (2012).
18. Zhou, X., Yan, J., Zhu, J., & Cai, P. Survey Of Energy Consumption and Energy Conservation Measures for Colleges and Universities In Guangdong Province, Energy and Buildings, **66**, 112–118 (2013).
19. Ayala E., Kocurek J., Wright J., Sijansky L., Chang J., Jr Contreras R. C. *Energy Efficient Method Identification for an Education Building, Architectural EngineeringConference 2011*, California United States. 235–243 (2011).
20. Saidur R., Hasanuzzaman M., Mahlia T.M.I., Rahim N.A., Mohammed H.A. Chillers Energy Consumption, Energy Savings and Emission Analysis in an Institutional Buildings, Energy, **36(8)**, 5233–5238 (2011).
21. Chee Yue, J.L *OTTV Calculation and Energy Simulation Technique for GreenRE Rating System*. Integrated Environmental Solutions (2014).
22. Tymkow,P., Tassou,S., Kolokotroni,M., Jounara,H. *Building Services Design for Energy Efficient Buildings*. USA: Routledge (2008).
23. Goudarzi, H., and Mostafaeipour, A. Energy Saving Evaluation of Passive Systems for Residential Buildings in Hot and Dry Regions. *Renewable and Sustainable Energy Reviews*, **68**, 432–446 (2017).
24. Gong, X., Akashi, Y., and Sumiyoshi, D. Optimization of passive design measures for Residential Buildings in Different Chinese Areas. *Building and Environment*, ***58***, 46–57 (2012).
25. Ma, S. L., Ding, Y., Shen, R. J., and Zhu, N. A Case Study of an Optimization Retrofit of the Heat Supply System in a Campus of Tianjin. *Applied Mechanics and Materials*, 170-**173**, 2670–2674 (2012).
26. Puvanasvaran,A.P., Yop Zain, M.F., Al-Hayali,Z.A., Mukapit,M. Sustainability of Green Technology in Malaysia Industry. *International Conference on Design and Concurrent Engineering*. Universiti Teknikal Melaka (2012).

27. Hassan, F. Application of Green Technology in Malaysia Construction: Have We Got It Right? *International Construction Week & Ecobuild SEA 2014*. Kuala Lumpur (2014).
28. Danby, D., Menter, A., and Faludi, J. *Passive Design Strategies.* Retrieved From:http://sustainabilityworkshop.autodesk.com/buildings/passive-design-Strategies (2011).
29. Wu, M. H., Ng, T. S., and Skitmore, M. R. Sustainable Building Envelope Design by Considering Energy Cost and Occupant Satisfaction. *Energy for Sustainable Development*. **31**, 118–129 (2016).
30. Pallant,J. (2014). *Spss Survival Manual. A step by Step Guide to Data Analysis Using SPSS for Windows (Version 12).* (1st edi). New South Wales: Allen and Unwin.
31. Williams B., Brown T. and Onsman. Exploratory Factor Analysis: A five-step guide for novices, Australasian Journal of Paramedicine, **8(3)** (2012).
32. Syed Fadzil S.F., Byrd H., Energy and Building Control Systems in the Tropics, 1st edition, Universiti Sains Malaysia Pulau Pinang, Malaysia (2012).
33. Zain-Ahmed, A., Sopian, K., Othman, M. Y. H., Sayigh, A. A M., and Surendran, P. N. Daylighting as a Passive Solar Design Strategy in Tropical Buildings: A Case Study of Malaysia. *Energy Conversion and Management*, **43**, 1725–1736 (2002).
34. Lu, J., Birru, D., and Whitehouse, K. Using Simple Light Sensors to Achieve Smart Daylight Harvesting. *Proceedings of the 2nd ACM Workshop on Embedded Sensing Systems for Energy-Efficiency in Building - BuildSys* ',**10**, 73 (2010).
35. Calcagni, B., and Paroncini, M. Daylight Factor Prediction in Atria Building Designs. *Solar Energy*, **76(6)**, 669–682 (2004).
36. Manglani, P., Mchugh, J., Heschong, L., Dee, R., and Group, H. M. *Modular Skylight Systems : Best Practices for Designing Skylights with Suspended Ceilings Potential for Skylight Systems with Suspended Ceilings*. 214–225 (2004).
37. Selkowitz, S., & Lee, E. Integrating Automated Shading and Smart Glazings with Daylight Controls. *International Symposium on Daylighting Buildings*, 10 (2004).
38. Lyons, P., Hockings, B., Reardon, C., Reidy, C., and GramlickHamber, R. *Your Home: Australia's Guide to Environmentally Sustainable Homes- Glazing.* (5th ed). *Australian Government Department of Industry-Glazing* (2013).

The role of cable stiffness in the dynamic behaviours of submerged floating tunnel

Naik Muhammad[1], *Zahid* Ullah[1], *Min-Wo* Park[1], and *Dong-Ho* Choi[2,*]

[1]Ph.D. Candidate, Department of Civil and Environmental Engineering, Hanyang University 222, Wangsimni-ro, Seongdong-gu, Seoul, 04763, Republic of Korea
[2]Professor, Department of Civil and Environmental Engineering, Hanyang University 222, Wangsimni-ro, Seongdong-gu, Seoul, 04763, Republic of Korea

Abstract. Submerged floating tunnel (SFT) is a new solution for the transportation infrastructure through sea straits, fjords, and inland waters and can be a good alternative to long span suspension bridges and immersed tunnels. The mooring cables/anchors are main structural components to provide restoring capacity to the SFT. The time domain dynamic problem of SFT moored by vertical and inclined mooring cables/anchors is formulated. The dynamic analysis of SFT subjected to hydrodynamic and seismic excitations is performed. As the cable stiffness determines the deformation ability of SFT, therefore it becomes crucial to evaluate the effect of mooring cable stiffness on the response of SFT. The displacements and internal forces of SFT clearly specify that the vertical/tension leg mooring cables provide very small stiffness as compared to inclined mooring cables. In order to keep the SFT displacements within an acceptable limit, the effect of cable stiffness should be properly evaluated for practical design of SFT.

1 Introduction

Submerged floating tunnel (SFT) is a novel structural solution for waterway crossings. It floats at a specified depth and is anchored by mooring cables. SFT is subjected to sea waves, currents, tsunamis and earthquakes. The SFT is supposed to be a good alternative for waterway crossing as compared to long span suspension bridges and immersed tunnels especially for deep and wide crossings because the cost of SFT per unit length remains almost constant by increasing length [1]. SFT is supported by the mooring cables, the balance between the net positive buoyancy (Residual buoyancy) and the pre-tension in the mooring cables maintains the structural stability of SFT.

The research efforts for development of the SFT solution began in 1923 from Norway. Since then, many case studies and proposals have been made for different proposed projects, such as Hogsfjord in Norway [2], Funka Bay in Japan [3], Messina Strait in Italy [4], Qindao Lack in China [5], and Mokpo-Jejo SFT in South Korea [6].

Most of the SFT numerical models developed so for, consider the dynamic response of SFT subjected hydrodynamic waves and seismic loadings [5-7]; however, the role of

* Corresponding author : samga@hanyang.ac.kr

mooring cable stiffness in the dynamic response of an SFT is not investigated or evaluated in more detail. The mooring cables are main structural components to provide restoring capacity to SFT. The mooring cable stiffness plays an important role in determination the deformation ability of SFT and therefore needs proper evaluation for practical design considerations.

This study presents the effect of cable stiffness on the dynamic response of submerged floating tunnel for hydrodynamic and seismic loadings. The dynamic problem of the SFT is solved considering the modeling of cables, tunnel, hydrodynamic waves and currents and seismic loadings. The displacements and internal forces of SFT are presented and the effect of cable stiffness on the dynamic behaviors of the SFT have been discussed.

2 Hydrodynamics

The ocean waves and currents are modeled by well-known Airy wave theory, and the wave forces are calculated from the modified Morison's equation, which is given as [9]:

$$\{f(q,t)\}=\frac{1}{2}C_D\rho_w D\{|\dot{w}_\perp \pm U-\dot{q}_\perp|(\dot{w}_\perp \pm U-\dot{q}_\perp)\}+C_M\rho_w\frac{\pi D^2}{4}\{\ddot{w}_\perp\}-C_A\rho_w\frac{\pi D^2}{4}\{\ddot{q}_\perp\} \quad (1)$$

where subscript $\perp$ denotes the perpendicular components with respect to element axis. ρ_w is the density of water, D is the external diameter of SFT, C_D is the drag coefficient, C_M is the inertia coefficient, $C_A = C_M - 1$ is the added mass coefficient, $\dot{w}$ and $\ddot{w}$ are water particle velocity and acceleration, while $\dot{q}$ and $\ddot{q}$ are structural velocity and acceleration respectively. U is the velocity of water currents acting at the centerline of SFT.

3 Equations of motion

The equation of motion for the SFT (Fig. 1) subjected to permanent, hydrodynamic, and seismic loads can be written in matrix form as follows:

$$[M+M_a]\{\ddot{q}\}+[C]\{\dot{q}\}+[K_e+K_m]\{q\}=\{f\}+\{f(q,t)\}-[M+M_a]\{I\}\{\ddot{q}_g\} \quad (2)$$

where $[M]$ is the structural mass matrix of SFT; $[M_a]$ is the added mass matrix; and $[C]$ is the system damping matrix; $[K_e]$ and $[K_m]$ are the elastic stiffness of SFT and the mooring stiffness matrices, respectively; $\ddot{q}$, $\dot{q}$, and q are the vectors representing structural acceleration, velocity, and displacement, respectively; $\{f(q,t)\}$ is the vector of time-dependent hydrodynamic forces; and $\{f\}$ is the vector of the net permanent loads acting on the SFT; $\{\ddot{q}_g\}$ is the vector of ground motions; $\{I\}$ is the influence coefficient vector. The vector $\{q\}$ for an element is given as follows:

$$\{q\}=[u_1 \quad v_1 \quad w_1 \quad \theta_{Y1} \quad \theta_{Z1} \quad u_2 \quad v_2 \quad w_2 \quad \theta_{Y2} \quad \theta_{Z2}]^T \quad (3)$$

where u , v , and w are displacements along the X, Y, and Z directions, respectively, while θ_Y and θ_Z are the rotations about the Y-axis and Z-axis, respectively. The subscripts 1 and 2 represent the first and second node of an element, respectively.

DOI: 10.1051/matecconf/201713802026

4 Numerical example

The dynamic behavior of SFT is simulated for the SFT model shown in Fig. 1 [5,7], which is supposed to be constructed in Qiandao Lake People's Republic of China; same input parameters are used for analyzing the numerical example as those used by [5]. For the dynamic response, the equation of motion (Equation 2) was solved by Newmark average acceleration method for each time step. The hydrodynamic analysis of SFT corresponds to the environmental conditions of Qindao Lake. For the seismic response of SFT, the three-dimensional ground motions of El-Centro (1940) are assumed for the SFT numerical example. The El-Centro ground motions used as input are not shown for simplicity; the ground motion components and peak ground acceleration are shown in Table 1. The ground motion component S90W was applied in longitudinal (X-direction), component S00E is applied in the transverse (Y-direction) and vertical component was applied in the vertical (Z-direction) of SFT, respectively. All three components of ground motion were applied simultaneously. For the dynamic response calculations, 5% damping was assumed for calculating structural-damping matrix using Rayleigh damping model. The simulations were performed for a total time of 300 seconds, for the hydrodynamic case and 50 seconds for the seismic case. The dynamic displacements and internal forces of the SFT numerical model were obtained.

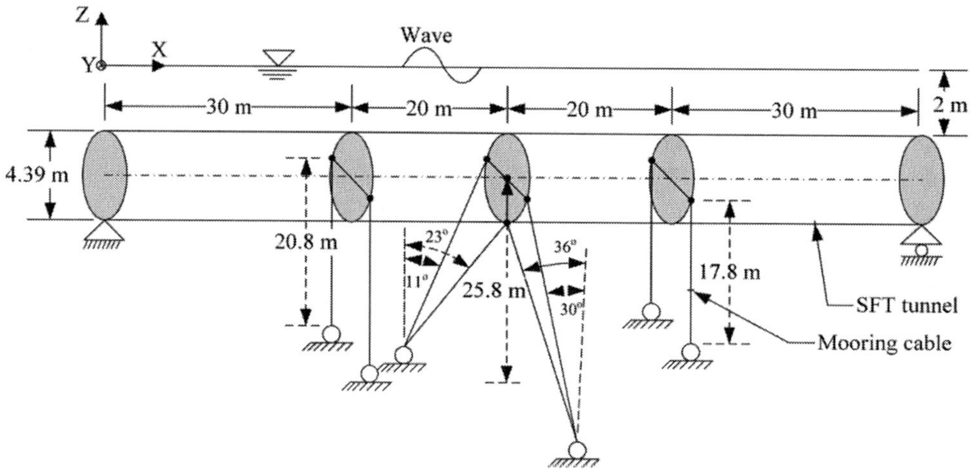

Fig. 1. Structural model of SFT used for numerical simulations

Table 1. Ground motion input (El-Centro 1940).

Component	PGA
S00E	0.35g
S90W	0.21g
Vertical component	0.13g

5 Results and discussions

The displacements and bending moments of SFT at the center of the tunnel for hydrodynamic analysis are shown in Fig. 2. Both transverse and vertical displacements and bending moments show a periodic pattern resembling the wave force model. The transverse bending moment of SFT is very large as compared to the vertical bending moment. The shape of the response curves for hydrodynamic analysis (Fig. 2) shows that the transient motions of SFT are small as compared to the steady state motions. As the SFT is intended to be used for railway transportation, so the maximum deflection criteria ($\Delta_{max} = L/700$=0.143 m) according to Korean railway bridge design specification can be assumed as a benchmark for SFT dynamic deflection. The absolute displacement ($\sqrt{u^2+v^2}$ = 0.0387 m) at the center of the tunnel is within the acceptable limit. The absolute bending moment ($\sqrt{M_Y^2+M_Z^2}$ = 9991.88 kN-m) at the center of the tunnel is within the range of design moment (M_D=70000 kN-m) referred by [5] according to EURO structural codes. The use of w-shape or double mooring is useful in restraining the horizontal displacement of SFT. However, as the wave height in the present case study is one meter and the displacement limit may be reached easily for extreme environmental conditions, as those highlighted by [8].

In order to check the effect of cable stiffness on the dynamic response of SFT, the cable stiffness (i.e. $\beta = E_c A_c = 3.486\times10^8\text{N}$) is varied; where E_c and A_c are the elastic modulus and cross-section area of cable. The effect of cable stiffness on the dynamic response of SFT for hydrodynamic analysis is shown in Fig. 3. As the cable stiffness is increased, the displacements of SFT are reduced. For the present numerical model, both transverse and vertical displacements of SFT can be limited under 0.01 m when the cable stiffness is 4β. When the cable stiffness is less than 4β, SFT behaves as simply supported beam. To keep the SFT displacements within a small range and to avoid the slackness of the mooring cables, an optimized cable stiffness need to be used for the practical design of SFT.

The displacements and bending moments of SFT at the center of the tunnel for seismic analysis are shown in Fig 4. The seismic response curves of SFT show more pronounced transient behaviors. The absolute displacement ($\sqrt{u^2+v^2}$ = 0.138 m) at the center of the tunnel is within the limit of maximum deflection criteria ($\Delta_{max} = L/700$ =0.143 m) according to Korea railway bridge design specification. The absolute bending moment ($\sqrt{M_Y^2+M_Z^2}$ = 43292.6 kN-m) at the center of the tunnel is within the range of design moment (M_D=70000 kN-m) referred by [5] according to EURO structural codes. As both deflections and bending moments are within the range from railway specifications but to avoid the slackness of mooring cables [3] a smaller range of deflections and bending moments need be investigated/fixed for SFT.

The seismic response of SFT is very large as compared to hydrodynamic response. A comparison of maximum absolute displacements and bending moments for hydrodynamic and seismic analysis are shown in Table 2. The response of SFT has an approximate relative difference of 81 % and 32% in transverse and vertical direction for hydrodynamic and seismic analysis.

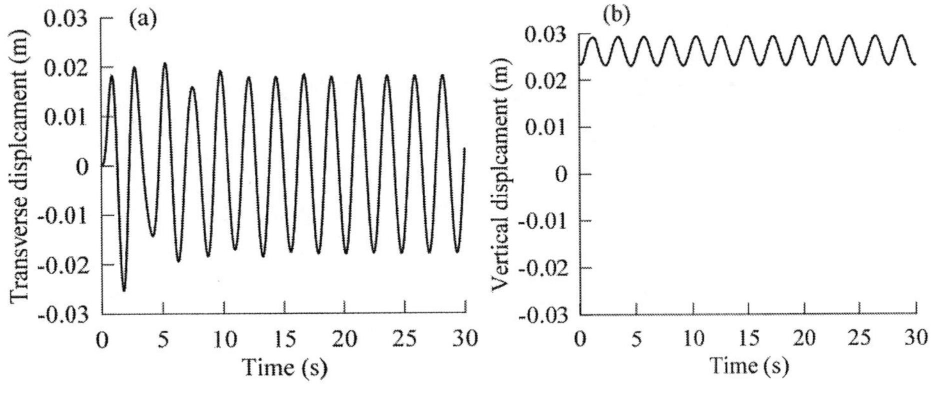

DOI: 10.1051/matecconf/201713802026

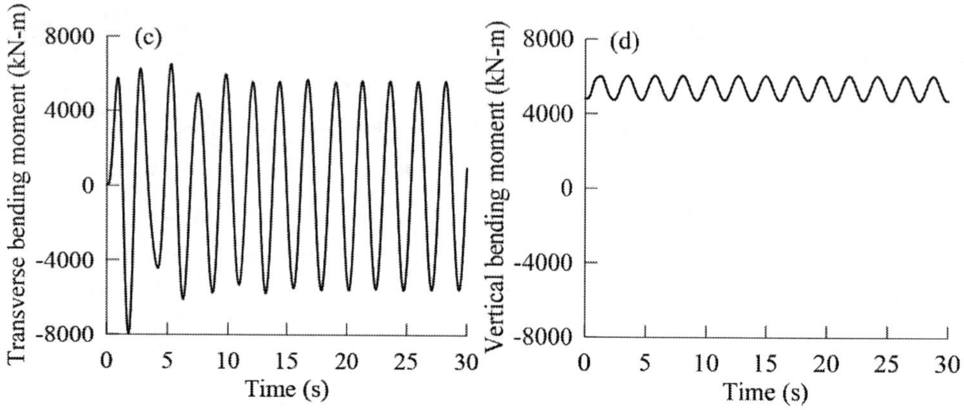

Fig. 2. The displacements and bending moments of SFT at the center of the tunnel for hydrodynamic analysis: (a) transverse displacement, (b) vertical displacement, (c) transverse bending moment, and (d) vertical bending moment.

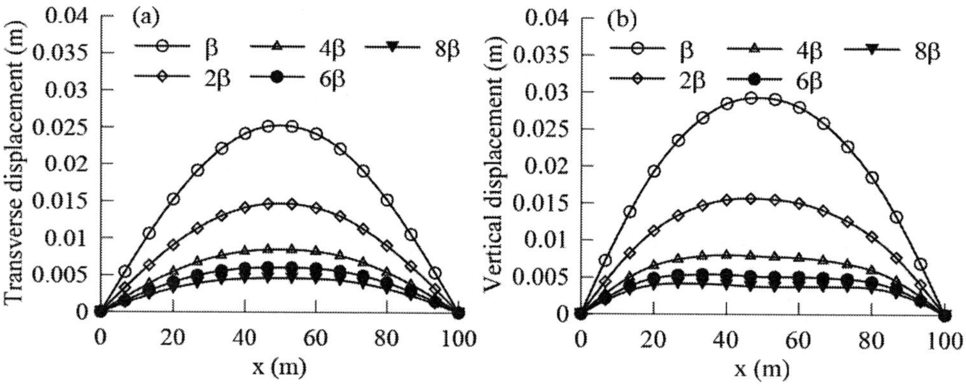

Fig. 3. Effect of cable stiffness on the dynamic response of SFT for hydrodynamic analysis (a) Transverse displacement, and (b) vertical displacement SFT.

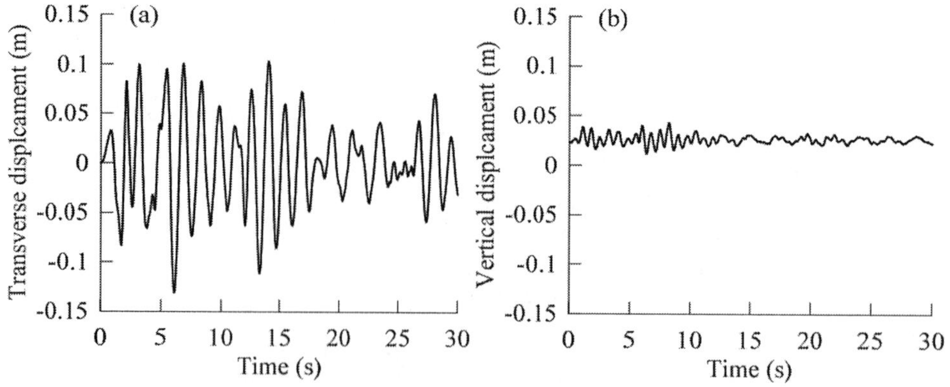

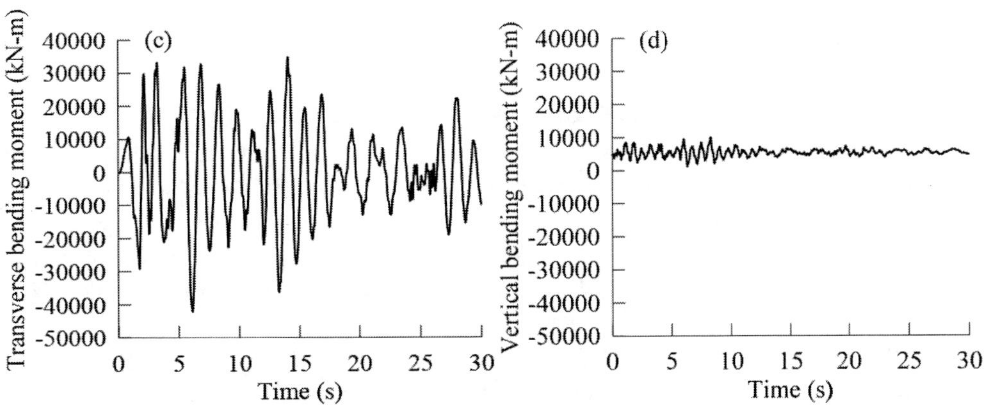

Fig. 4. The displacements and bending moments of SFT at the center of the tunnel for seismic analysis: (a) Transverse displacement, (b) vertical displacement, (c) transverse bending moment, and (d) vertical bending moment.

Table 2. SFT maximum absolute displacements and bending moments at the center of the tunnel.

	Transverse displacement (m)	Vertical displacement (m)	Transverse bending moment (kN-m)	Vertical bending moment (kN-m)
Hydrodynamic	0.0253	0.0293	7961.9884	6036.9302
Seismic	0.1310	0.0428	42112.2048	10040.6098

6 Conclusions

The equations of motion for submerged floating tunnel (SFT) are formulated considering the modeling of structure, cable system and ocean waves and currents. The equations of motion are then solved numerically using Newmark's average acceleration method. The dynamic simulations of SFT are performed for the hydrodynamic and seismic excitations.

Waves and currents are modeled by the Airy wave theory and wave forces are calculated using modified Morison's equation. The displacements and bending moments of SFT for hydrodynamic analysis (environmental conditions of Qindao Lake, Peoples Republic of China) and seismic analysis (El-Centro ground motions, 1940) are presented. The seismic response of SFT is very large compared to hydrodynamic response, due to two reasons (1) hydrodynamic action is small because the wave height of the site is one meter only; for practical design, the SFT need to be checked for a wide range of wave heights (2) SFT is a massive structure.

The effect of cable stiffness on the dynamic behavior of SFT is evaluated. The vertical/tension leg mooring cables are the least effective as compared to inclined mooring cables due to the smaller horizontal stiffness of vertical mooring cables. The transverse bending moments of SFT are found to be larger than vertical bending moments of SFT, and this is attributed to the smaller cable stiffness in the transverse direction. Based on the numerical results of the present study, it is concluded that to keep the SFT displacements within an acceptable range and to avoid the slackness of the mooring cables, an optimized cable stiffness need to be used for the practical design of SFT.

This research was supported by the Basic Science Research Program through the National Research Foundation of Korea (NRF), funded by the Ministry of Education, Science and Technology (NRF-2015R1D1A1A09060113). The authors wish to express their gratitude for this financial support.

References

1. B. Faggiano, R. Landolfo, F. Mazzolani. The SFT: An Innovative Solution for Waterway Strait Crossings, *IABSE Symposium* Report, **90**, 36-42, (2005).
2. L. Skorpa. Innovative Norwegian fjord crossing. How to cross the Hogsfjord, alternative methods. *Proceedings of the 2nd Congress AIOM (Marine and Offshore Engineering Association)*, Naples, Italy, (1989).
3. W. Lu, F. Ge, L.Wang, X.Wu, Y. Hong. On the slack phenomena and snap force in tethers of submerged floating tunnels under wave conditions, Mar. Struct., **24**, 358-376, (2011).
4. R. Bruschi, V. Giardinieri, R. Marazza, T. Merletti. Submerged buoyant anchored tunnels: Technical solution for the fixed link across the strait of Messina Strait crossings, *Strait crossings Rotterdam*: Balkema. (1990).
5. F. Mazzolani, R. Landolfo, B. Faggiano, M. Esposto, F. Perotti, G. Barbella. Structural analyses of the submerged floating tunnel prototype in Qiandao Lake (PR of China), Adv. struct. eng.**11**, 439-454, (2008).
6. J.S. Han, B. Won, W.S. Park, J.H. Ko. Transient response analysis by model order reduction of a Mokpo-Jeju submerged floating tunnel under seismic excitations, Struct. Eng. Mech., **57**, 921-936, (2016).
7. L. Martinelli, G. Barbella, A. Feriani. A numerical procedure for simulating the multi-support seismic response of submerged floating tunnels anchored by cables. Eng. Struct. **33**, 2850-2860, (2011).
8. S.I. Seo, H.S. Mun, J.H. Lee, J.H. Kim. Simplified analysis for estimation of the behavior of a submerged floating tunnel in waves and experimental verification. Marine Struct.**44**:142-158, (2015).
9. S.K. Chakrabarti. *Hydrodynamics of offshore structures*. WIT press: (1987).

Effect on the Load Transferring Properties Fiber Reinforced Steel in Concrete based on Different Mix Ratio

Yoonjung Han[1,*], *Sangkeun* Oh[1] , and *Byoungil* Kim[1]

[1]Seoul National University of Science and Technology, Department of Architectural Engineering, 01811 Seoul, Korea

Abstract. This paper analyses the correlation between the load transfer section and the overall toughness of steel fiber reinforced concrete. The specimens were created with different mixing ratios and different types of steel fiber, and the flexural behaviour properties according to the respective types of steel fiber were analysed. The results showed three types of load – displacement curves, and it was confirmed that the different fiber types affected the load transfer section on the overall toughness.

1 Introduction

The main purpose of steel fiber mixing in steel fiber reinforced concrete is to provide a bridging in the concrete matrix to reduce cracking in the concrete during loading, but they cannot increase the overall strength of the concrete. Following the impact point, the deformation and residual load occur due to the bridging reaction between of the fiber and concrete matrix, and the residual load has an effect on the derivation of toughness of steel fiber reinforced concrete. With the aim of investigating the influence of bridging reaction in the load transfer zone on total toughness, this study conducted an evaluation of the effect of flexural behaviour on steel fiber reinforced concrete, mixed with hooked steel fiber, crimped steel fiber, crimped stainless steel fiber, each having different tensile strength.

2 Materials and Test Procedure

2.1 Test Materials

The material used in this study was grade 1 regular Portland cement mixed with fine and course aggregate produced in Company K. The hooked steel fiber sample was also taken from Company K and the crimped steel fiber and crimped stainless steel fiber were taken from Company S, all respectively used to make the concrete specimens for this study. The aspect ratio of the steel fiber was 70. The fiber types were determined by considering the need to prevent uneven distribution of fiber due to the wall effect, thus the value would

* Corresponding author: yjhan0712@seoultech.ac.kr

 DOI: 10.1051/matecconf/201713803001

have to be smaller than the section of the specimen used in the toughness assessment method divided by 2.5[1]. Table 1 shows the properties and detailed images of the different types of fibers that were used in this study, and the mix design is shown in Table 2.

Table 1. Properties and Details of Steel Fiber

Category	Hooked Type	Crimped Steel Type	Crimped Type
Code	H-SF	C-SF	C-SLF
Material	Steel	Steel	Stainless Steel
Section	Circular	Circular	Circular
Tensile Strength (N/mm^2)	1,300	2,896	1,941
Length(mm)	35	35	35
Diameter	0.5	0.5	0.5
Aspect Ratio	70	70	70
Image	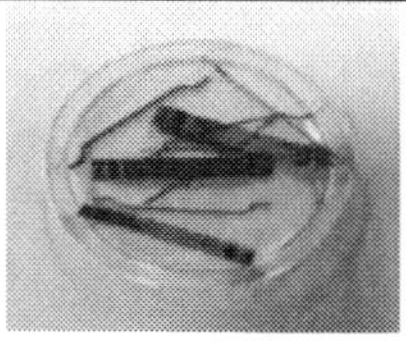	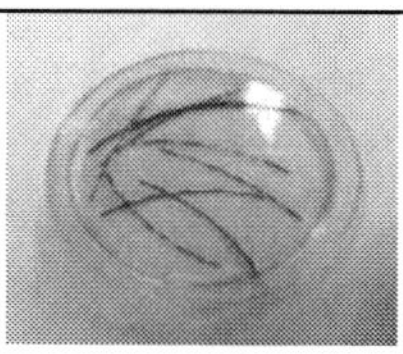	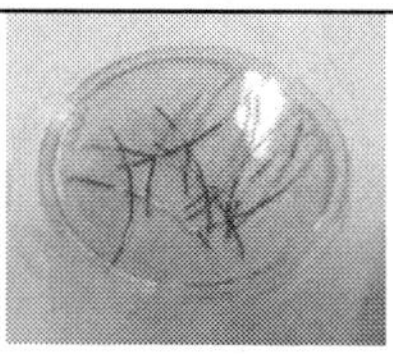

Table 2 Mix design of the Concrete Specimens

<table>
<tr><th rowspan="2">Type</th><th rowspan="2">Fiber Content (%)</th><th rowspan="2">W/C (%)</th><th rowspan="2">S/a (%)</th><th colspan="4">Unit Weight(kg/m^3)</th><th rowspan="2">S.P. (Cx%)</th><th rowspan="2">AE (Cx%)</th></tr>
<tr><th>W</th><th>C</th><th>S</th><th>G</th></tr>
<tr><td>Plain</td><td>0</td><td rowspan="7">45</td><td rowspan="7">51</td><td rowspan="7">200</td><td rowspan="7">440</td><td>778</td><td rowspan="7">748</td><td rowspan="7">1.0</td><td rowspan="7">0.02</td></tr>
<tr><td>H-SF0.5</td><td>0.5</td><td rowspan="6">766</td></tr>
<tr><td>H-SF1.0</td><td>1.0</td></tr>
<tr><td>C-SF0.5</td><td>0.5</td></tr>
<tr><td>C-SF1.0</td><td>1.0</td></tr>
<tr><td>C-SLF0.5</td><td>0.5</td></tr>
<tr><td>C-SLF1.0</td><td>1.0</td></tr>
</table>

2.2 Test Procedure

Three concrete specimens of 100x100x400mm were made to undergo flexural behaviour property testing and toughness assessment. The test was conducted in 4 point bending test based on the instructions outlined in KS F 2566:2014, *Flexural Test Method of Fiber Reinforced Concrete*. The tensile stress load rate was set to 0.06±0.04 MPa (N/mm^2) per second. After the bending test, the toughness of the specimen was derived by calculating the area below the load displacement graph for up to 2.0 mm, which corresponded to 1/150 of the span.

3 Experimental Results

3.1 Flexural Behaviour Properties

The flexural property testing of the OPC specimen showed that its load decreased drastically after reaching the maximum load. However, specimens reinforced with steel fiber showed that its load was being transferred to steel fiber after the crack was formed in the concrete matrix, resulting in a delayed effect of crack diffusion and the residual load maintaining in the matrix due to the fiber bridging effect. The fracture was delayed gradually, showing a decrease in the load after the maximum load or slight increase in the load after maximum load for each of the specimens based on the difference of the shape, quality of material, and content of the fiber.

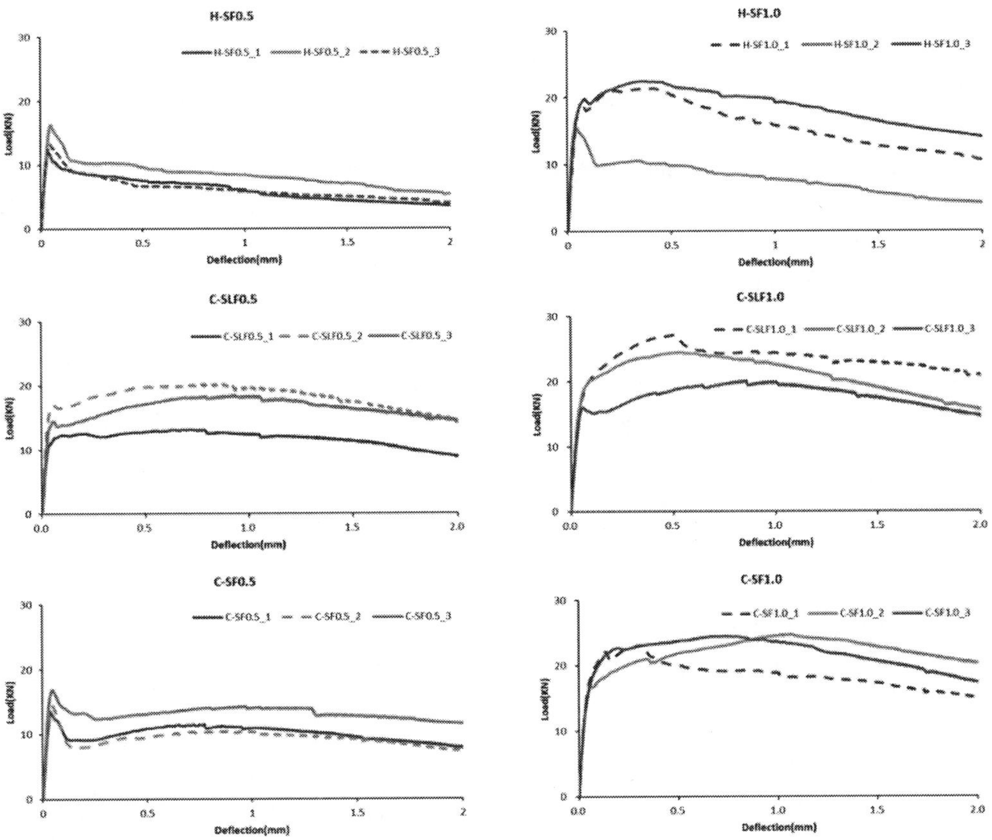

Fig. 1. Flexural Load Displacement Curve of the Specimens Mixed with Different Steel Fibers

3.2 Load Transfer Types by Fiber Mixing Ratio and Material Properties

The load – displacement curve of the SFRC was divided into three cases (Type A, Type B, and Type C). Type A showed that the load was decreased after the peak load and was followed by a more gradual decrease without an increase in the load. Type B showed that the load was decreased slightly after the peak load and was followed by an increase and a subsequent gradual decrease after the second peak load. Lastly, with Type C, the load increased after the peak load without any decrease and then the load gradually decreased and the residual load maintained.

 DOI: 10.1051/matecconf/201713803001

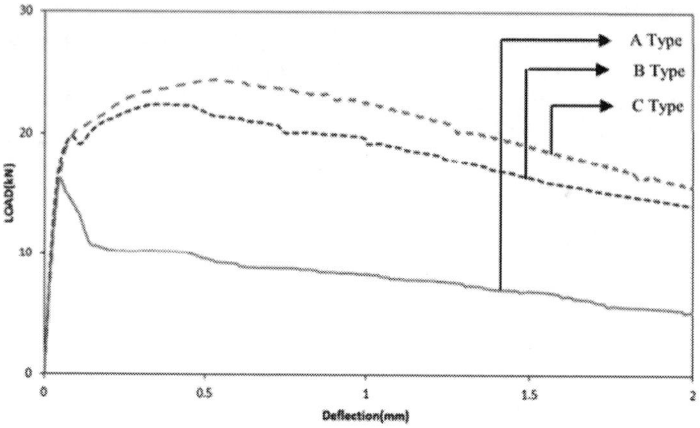

Fig. 2. Load Transfer Properties by different Steel Fiber Type and Mixing Ratio

4 Analysis

4.1 Analysis of the Flexural Behaviour Properties

The area below the load displacement curve is divided into five zones based on the load transfer type for the detailed anlysis of the flexural behavior properties. Table 3 shows a detailed description of the range of each section. Figure 3 shows the diagram of the slope and area by section.

Table 3. Range Variance by Section

Section	Category		Range (d : displacement, mm)	Energy absorption mechanism
	Toughness by section	Slope by section		
1	S1a	G1a	0 ≺d≼ initial crack point	Linear elastic behaviour
	S1b	G1b	initial crack point ≺d≼ first peak load	Progress of fine cracks inside the matrix
2	S2	G2	first peak load ≺d≼ lowest load point	Concrete-fiber load transfer section
3	S3	G3	lowest load point ≺d≼ second peak load	Fiber bridging and pull-out action
4	S4	G4	second peak load ≺d≼ 2.0	

4.2 Toughness Analysis Based on Area

Figure 3 shows the graph representing the toughness of each section. In the case of the first section, the different results were derived despite the same size due to the behavior of the concrete matrix.

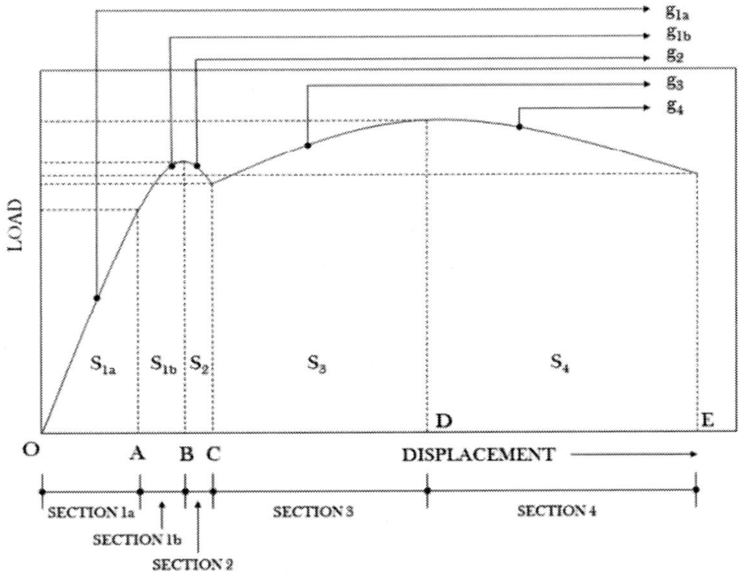

Fig. 3 Area (Toughness) and Slope of the Load Displacement Curve by Section

The second section in Figure 4 shows that the area decreases with the increase of the steel fiber content. It is surmised that the thorough placement of the fiber in the matrix enhances the crack inhibition performance and makes the load transfer rate more efficient.

The third section shows that the energy increases following the increase of the fiber content. This is due to the increased amount of the fiber absorbing the energy and can handle increased amount of strain.

The 4th section shows that the amount of energy absorbed by the fibers was increased as the amount of fiber mixture increased. This is considered to be a natural extension of the fourth section resulting from the decrease of load in the second section and increase in the load of the third section.

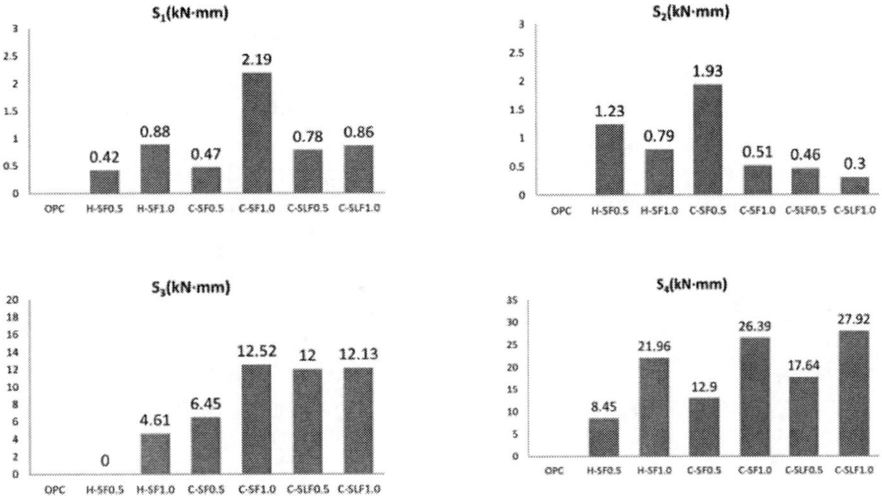

Fig. 4 Toughness by Different Sections

 DOI: 10.1051/matecconf/201713803001

4.3 Equation between the Load Transfer Section and the Toughness

The analysis of the above area shows the area of the load trasfer sesction (S_2) and the area of the load re-increase section(S_3) affect the toughness. In consideration of this factor, the load transfer facter equation was established. And the result of L.T.F. and toughness is shown in the following Table. 4 .

$$L.T.F. = (S_3/S_2)^{1/2} \quad (1)$$

The regression equation of the overall toughness based on the distribution of the LTF was verified by equation (2).

$$T_b = 17.64((S_3/S_2)^{1/2})^{0.45} \quad (2)$$

Table 4. LTF and Toughness

Specimen	Load Transfer Factor (LTF)	Toughness
H-SF 0.5	0	13.86
H-SF 1.0	5.84	28.23
C-SF 0.5	3.34	21.91
C-SF 1.0	24.06	41.61
C-SLF 0.5	26.09	30.88
C-SLF 1.0	40.57	41.22

The analysis results on the degrees of freedom, F-statistic, significance probability, and coefficient of determinant has been calculated accordingly and have been displayed in Table 5. Based on these results, the coefficient of the determinant (R^2) was figured out to be 0.798. The figures derived from the testing indicates that the proposed correlation model can offer a considerable level of accuracy and reliability in determining the appropriate connection between the toughness of the concrete and the load transfer factor.

Table 5. LTF and Coefficient of Determinant (R^2)

Degree of freedom(df)	F-statistic	Significance probability(sigf)	Coefficient of determinant(R^2)	Regression Model
3	11.84	0.041	0.798	$Y=17.64x^{0.45}$

DOI: 10.1051/matecconf/201713803001

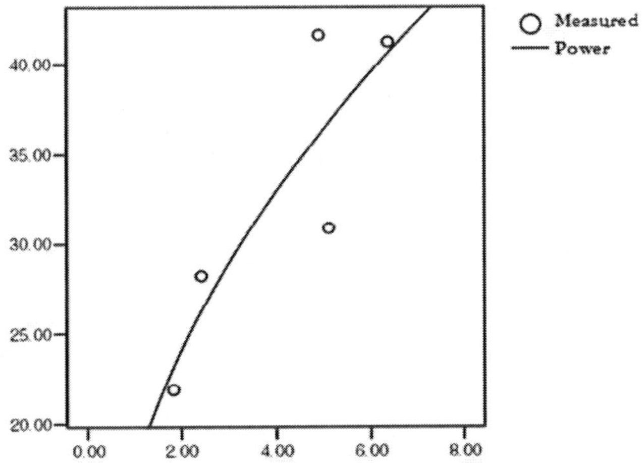

Fig. 5 Regression Model Based on the Load Transfer Factor (LTF) and Toughness

5 Conclusions

In this study, three types of steel fiber were used to investigate the effect of the load transfer performance between concrete matrix and steel fiber of steel fiber reinforced concrete. As a result of section analysis, the area of the load transfer section (second section) and the area after the load transfer section (third section) was shown to affect the overall toughness. Lastly, to determine the correlation between the energy transfer capacity of steel fiber and the overall toughness of the load transfer section, the study developed a regression model (T_b) using the LTF.

This research was supported by a grant (NRF-2016RIC IB2009489) from National Research Foundation of Korea

References

1. ACI Committee, *State – of – the art report in fibre reinforced concrete.* ACI 544 IR, (1982)
2. H. Yoon-jung, *Evaluation of Flexural Behavior Effect in the Load Transfer Zone for Steel Fiber Reinforced Concrete.* Kor. Pp 1-82 (2017)
3. H. Yoon-jung, O. Kyu-hwan, C. Su-young, O. Sang-keun, K. Byoungil. *Fiber Bridge Analysis of Steel Fiber Reinforced Concrete According to Different Fiber Manufacture & Fiber Addition*. Kor. Proceeding of Annual Conference of the Architechtural Institute of Korea. (2016)
4. H. Yoon-jung, M. Seoung-jae, O. Kyu-hwan, O. Sang-keun, K. Byoungil. *Comparison of Load Transferring Property for Steel Fiber Production and Type*. Korea Concrete Institute Conference (2016)
5. H. Yoon-jung, O. Sang-keun, K. Byoungil. *Effect of Load Transfer Section to Toughness for Steel Fiber Reinforced Concrete*. Appl. Sci (In Press)

DOI: 10.1051/matecconf/201713803002

Quick Assessment of the Anomalies in Concrete Structure Using Dispersive Characteristic of Surface wave

Chia-Chi Cheng[1], *Keng-Tsang Hsu*[1], *Chih-Hung Chiang*[1], *Fong-Jhang Ke*[1], and *Hong-Hua Wang*[1]

[1]Department of Construction Engineering, Chaoyang University of Technology, Taichung, Taiwan

Abstract. A waveform generated by a small impact hammer and recorded by single transducer is used to assess different kinds of anomalies in concrete. Using Short-Time Fourier Transform and the reassignment technique, the group slowness spectrogram can be calculated from the waveform. The surface wave response in the diagram of velocity vs. wavelength corresponding to the reassigned spectrogram is used for defect evaluation. Experiments were conduct on a concrete plate specimen containing multiple delaminated cracks, and a honeycomb. The group velocity profiles for test lines passing through different kinds of defects were obtained and compared. It was found the surface wave velocity begins to decrease with the increase of wavelength when the wavelength is about two times of the depth of the defect.

1 Introduction

Plate-like reinforced concrete structures, such as bridge deck, shear wall, or reinforced concrete containing wall of nuclear power station are highly related to structural safety. The internal anomalies may include honeycombs and air voids and delaminated cracks caused by improper casting, corrosion of steel bars, or excessive earthquake load. Although speedy assessment is possible, these anomalies may not be easily identified by high-frequency ground penetration radar (GPR). The electromagnetic waves emitted by GPR can travel through air. The opening of the air gap within 1/4 of the wavelength in the media cannot be distinguishable [1]. For example, for GPR with 2.4 GHz antenna, the distinguishable gap space needs to reach about 100 mm for air gap and 30 mm for gap filled with water. The impact-echo test method (IE) applies transient waves by hitting a steel ball on surface producing stress wave with broad frequency range. As an air gap, less than about 0.1 mm, can produce noticeable reflection of the stress wave from the boundary [2, 3]. The method is more sensitive than GPR in air void detection. The disadvantage of IE is time consuming. SASW (Spectral Analysis of Surface Waves) and MASW (Multichannel Analysis of Surface Waves) methods have also been used for assessing the material properties and thickness of layered structures such as asphalt pavement or concrete plate [4, 5]. The instrumental setups require two receivers for SASW and multiple channels of receivers for MASW placing on the surface. In practice, multiple tapping on the surface by small hammer at different locations may be needed to obtain a dispersive phase velocity profile of surface wave. Using MASW method, not only the velocity profile of surface wave but also the dispersive characters of multiple Lamb wave modes of layered structure can be retrieved [4]. However, the SASW method requires changing the receivers' interval to obtain the wave speed for larger range of wavelength and the MASW method requires multiple measurements or multiple receivers to obtain the accurate phase information. Recently, a device which is conveniently operated at field was invented which combines the impact-echo method (IE) with the ultrasonic surface waves (USV) to evaluate the depth of surface deterioration and defects within concrete. Because the interval between receivers is fixed and the energy of ultrasound is limited, the variation of phase velocity with depth can be evaluated up to 30 cm [6].

The MASW, SASW and USV methods are designed to measure the phase velocity. In present study, the dispersive characteristics of group velocity of surface waves were used for investigation. Tests are performed with one receiver positioned away from the impacting source. The spectrogram of the group velocity obtained by the receiver is calculated by Short-Time Fourier Transform. The resolution of the spectrogram is enhanced by the reassigned technique. [7, 8] The advantage of the technique is the capability on fast investigation of large area, as the method only requires performing single test to obtain multiple lamb wave modal dispersive characteristics.

In the study, experiments were conduct on a plate specimen with delaminated crack, small range crack and honeycomb embedded inside. The group velocity profile of plain concrete, concrete containing different depths of delamination crack, concrete containing honeycombs were obtained and compared.

2 Theoretical Backgrounds

In present study, the technique involving only one test with single impacting source and single receiver is proposed. The signal is analyzed by the Short Time Fourier Transfer (STFT) and reassigning technique to obtain the spectrogram of the group slowness. The dispersive characteristics of group slowness of the test signal can be obtained by two steps. First, STFT was performed on the test signal to construct the preliminary spectrogram. Secondly, the images of dispersion characteristics in the spectrogram are enhanced by Time-Frequency Reassignment technique. For STFT, the signal (s(t)) is separated into a series of overlapping sections. Each section is multiplied by a select window function (h(t)) and transferred into frequency domain by FFT. The STFT, denoted as Sh(ω, t), is defined as Eq. (1).

$$S_h(\omega,t)=\frac{1}{2\pi}\int_{-\infty}^{\infty}e^{-i\omega\tau}s(\tau)h(\tau-t)d\tau \tag{1}$$

The spectrogram of Sh(ω, t) is the energy density spectrum of a STFT, which can be calculated by Eq. (2).

$$E_d(\omega,t)=|S(\omega,t)|^2 \tag{2}$$

In the reassignment method, high-energy corresponding to the dispersive character of a plate can be moved from its original location (ω, t) to a newly reassigned coordinates ($\hat{\omega},\hat{t}$), which is usually at the center of gravity of the high energy band. The reassigned coordinates $\hat{\omega}$ and $\hat{t}$ for a spectrogram is shown in Eqs. (3) and (4) [7].

$$\hat{t} = t - Re(\frac{S_{Th}(x,t,\omega)\cdot\overline{S_h(x,t,\omega)}}{|S_h(x,t,\omega)|^2}) \tag{3}$$

$$\hat{\omega} = \omega + \mathrm{Im}(\frac{S_{Dh}(x,t,\omega)\cdot\overline{S_h(x,t,\omega)}}{|S_h(x,t,\omega)|^2}) \tag{4}$$

Where Sh, SDh, STh are the STFT of the signal x using a normalized window function h(t), $\frac{dh(t)}{dt}$ and t*h(t). The coding for reassigned technique can be found in [9].

3 Experimental designs

A specimen with dimensions 2 m×4.05 m×0.36 m was constructed with some artificial defects buried inside the normal weight concrete. As shown in Fig. 1, artificial cracks made by Styrofoam sheets were placed within concrete at different depths and size, so was a honeycomb block made by mortar bonded coarse aggregates. The honeycomb was wrapped by plastic sheet to prevent cement entering the voids during casting. The layout of defects is shown in Fig. 2. The figure also indicates the positions of the four test lines in present study. The four scenarios under the test lines are summarized by Table 1.

The experiment was conducted with a special impact-hammer applying on one end of the test line. The produced stress waves were recorded by the displacement receiver at the other end of the test line. The distance between the impact-source and the receiver is 0.8 m. In order to find the time origin of the impact, as shown in Figure 3(a), the impact-source is made by a steel half sphere embedded with piezoelectric element. The diameter of the half sphere applied in this study is 5 mm in diameter. For each test, a total of 4096 points was recorded in every 2 μ s. The test setup is shown in Figure 3(b).

Table 1. The scenarios under the test line

Test line Number	Situation beneath test line	Depth of the defect(s)
1	No defect	Thickness = 0.36 m
2	horizontal Styrofoam disk with diameter of 0.5 m	0.19 m
3	horizontal Styrofoam disks with diameters 0.1 m and 0.2 m	0.238 m for 0.2 m dia. defect 0.146 m for 0.1 m dia. defect
4	Honeycomb block	0.127 m

DOI: 10.1051/matecconf/201713803002

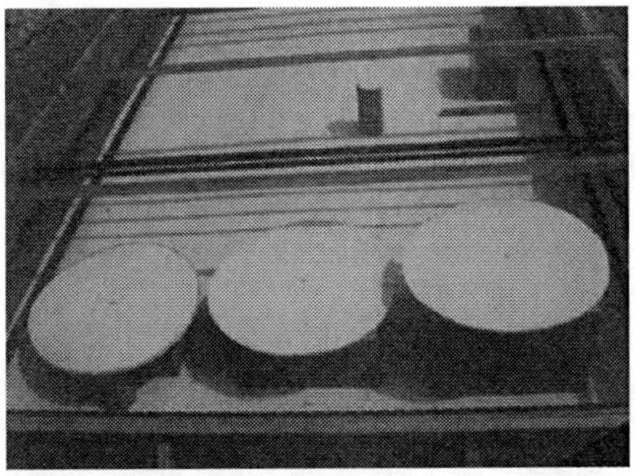

Fig. 1. Photo of defects buried in the specimen.

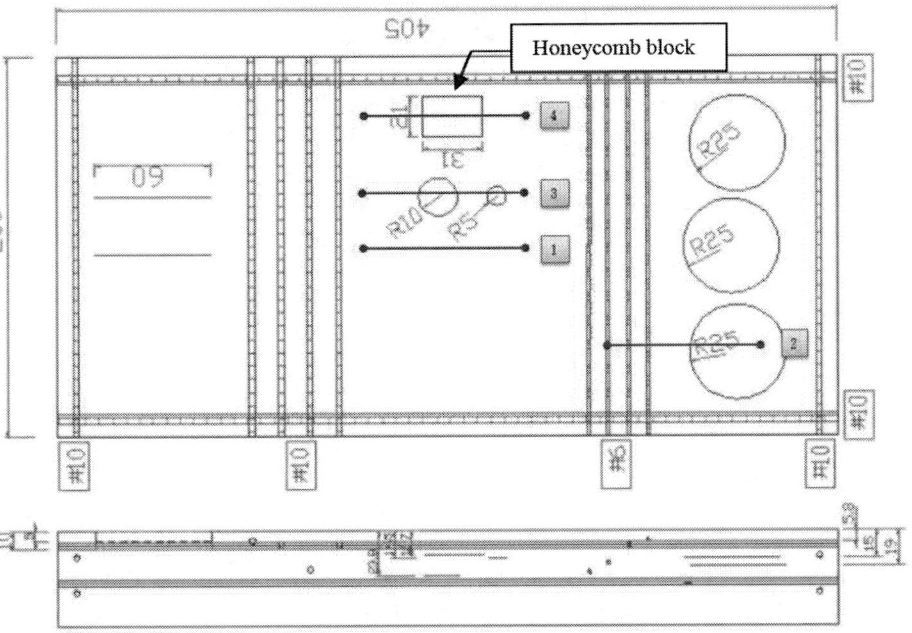

Fig. 2. Illustration of the specimen layout and positions of test lines.

(a) (b)

Fig. 3. Photos of (a) impactor embedded with piezo-electric element, (b) experimental setup.

4 Experimental Results

The algorithms to perform the analysis are as following. First, perform the STFT by chosen a Hanning window with certain window length and window overlap to obtain a

DOI: 10.1051/matecconf/201713803002

spectrogram. Use the reassigned technique to find the center of gravity of the high energy band in the spectrogram. Third, divide the vertical time axis of the spectrogram to the impactor-receiver distance to obtain slowness axis. Fourth, obtain the velocity-wavelength diagram from the slowness spectrogram. Finally, the dominant response with the lowest slowness with respect to frequency in the slowness spectrogram would be shown as the dominant velocity profile with the largest velocity with respect to wavelength. The velocity profile is used for defect assessment.

The results for test lines 1 to 4 are shown in Figure 4 to 7, respectively. Using Fig. 1 as the example, the right figure of Fig. 4(a) shows the waveform obtained from the displacement transducer 0.8 m from the impact hammer. The vertical axis of the waveform is the slowness, which is time divided by source-receiver distance. The spectrogram on the right of Fig. 4(a) is obtained by STFT and reassigned method. In STFT, the window size is 150 points and overlap of the window is 144 points. After reassigned the coordinates of the high energy band to the center of gravity in the spectrogram produced by STFT, the image of the response above -40 dB of the maximum amplitude is calculated and shown in Fig. 4(a). Because the waveform and the spectrogram have the same vertical axis, the frequency contents of the major response in the waveform can be easily identified in the spectrogram. The darker dots means stronger signal. The diagram of velocity vs. wavelength obtained from the data in reassigned spectrogram is shown in Fig. 4(b).

The earliest major response corresponding to the arrival of the Rayleigh wave has noticeable difference for different defect-scenarios. In the reassigned spectrogram, the Rayleigh wave responses are around 0.4 ms/m for frequency range 20-60 kHz for all the test lines. The divergence caused by stress waves traveling around the defects occurs at the frequency range less than 20 kHz. For comparison, the theoretic solution of fundamental asymmetric Lamb wave mode (A0), fundamental symmetric Lamb wave mode (S0), and the second symmetric Lamb wave mode (S1) for a concrete plate with thickness 0.36 m, P-wave speed of 4000 m/s, and poison ratio of 0.2, are also shown in the spectrograms by red, orange, and blue lines.

For the case with no-defect, the Rayleigh wave response shown in Fig. 4(a) generally matches with the theoretical A0 mode at the frequency range larger than about 7 kHz. Below 7 kHz the slowness graduate increases but the slowness remains around 0.4 m/s. For test line 2, which passes through a large delaminated disk at a depth of 0.19 m, the Rayleigh wave response shown in Fig. 5(a) generally matches with the theoretical A0 mode at the frequency range 10-30 kHz and a large increase in slowness can be found below 8 kHz. The slowness is little bit higher than the theoretical solution for frequency higher than 30 kHz. The test line 3 passes through two smaller horizontal Styrofoam disks with diameters 0.1 m and 0.2 m at the depths 0.238 m and 0.146 m, respectively. As shown in Fig. 6(a), the Rayleigh wave response generally matches with the theoretical A0 mode at the frequency range 10-38 kHz and a large increase in slowness can be found below 10 kHz. The slowness is a little bit higher than the theoretical solution for frequency higher than 38 kHz. For test line 4, which passes through a large honeycomb block at a depth of 0.127 m, the Rayleigh wave response shown in Fig. 7(a) generally matches with the theoretical A0 mode at the frequency range 15-20 kHz. For frequency less than 10 kHz the surface wave response seized to exist and a strong response can be found at the slowness of 0.6 ms/m.

The existence and the depth of the defect can be better recognized in the diagrams of velocity vs. wavelength. The velocity profiles of Rayleigh wave can be found as the dominant responses between 2000 m/s and 3000 m/s. For the no-defect case shown in Fig. 4(b), the velocity is around 2500 m/s for wavelength less than 0.2 m, increases to about 2600 m/s for wavelength between 0.2 m and 0.3 m, and remains steady at 2600 m/s for wavelength between 0.3 m and 0.4 m. The velocity is gradually decreased to 2400 m/s for wavelength between 0.4 m and 0.8 m. For test line 2 shown in Fig. 5(b) with delaminated

DOI: 10.1051/matecconf/201713803002

crack at the depth of 0.19 m, again the velocity is around 2500 m/s for wavelength less than 0.2 m and increases to about 2700 m/s for wavelength between 0.2 m and 0.3 m, and the begins to decrease at the wavelength of about 0.38 m. The velocity decreases to 2200 m/s for wavelength equal to 0.8 m. For test line 3 shown in Fig. 6(b) with two cracks at the depths of 0.238 m and 0.146 m, the velocity is around 2500 m/s for wavelength less than 0.2 m and increases to about 2700 m/s for wavelength between 0.2 m and 0.3 m. The velocity begins to decrease at the wavelength of 0.31 m and to the value of 2000 m/s for wavelength reaching 0.5 m. The velocity of the major response decreased to 1500 m/s for wavelength between 0.5 m and 0.8 m as shown by green arrow in Fig. 6(b). For test line 4 shown in Fig. 7(b) with honeycomb block at the depth of 0.127 m, the velocity is around 2500 m/s for wavelength less than 0.18 m. The surface wave velocity is around 2200 m/s and 2500 m/s for wavelengths between 0.18 m and 0.28 m. When the wavelength reaches 0.28 m, the major response near 2200 m/s stopped and reappears at the velocity of 1700 m/s as shown by the green arrow.

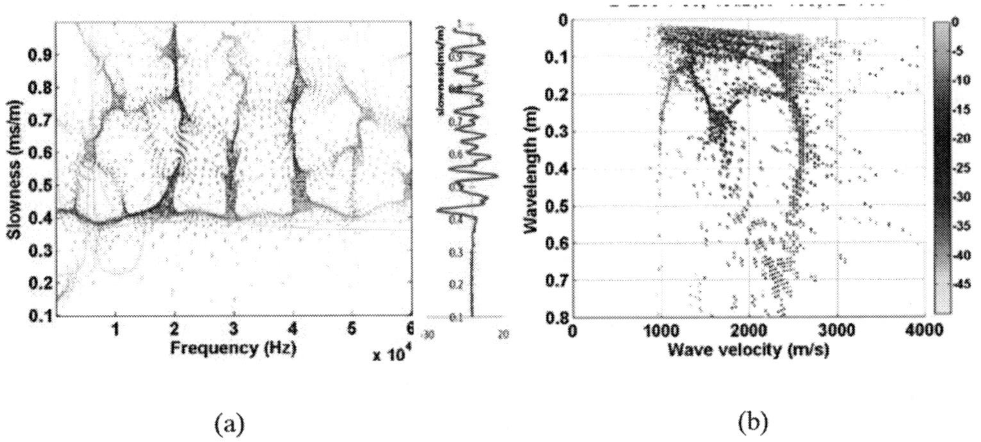

Fig. 4. Test line 1 (a) group slowness spectrogram along with waveform, (b) velocity-wavelength profile.

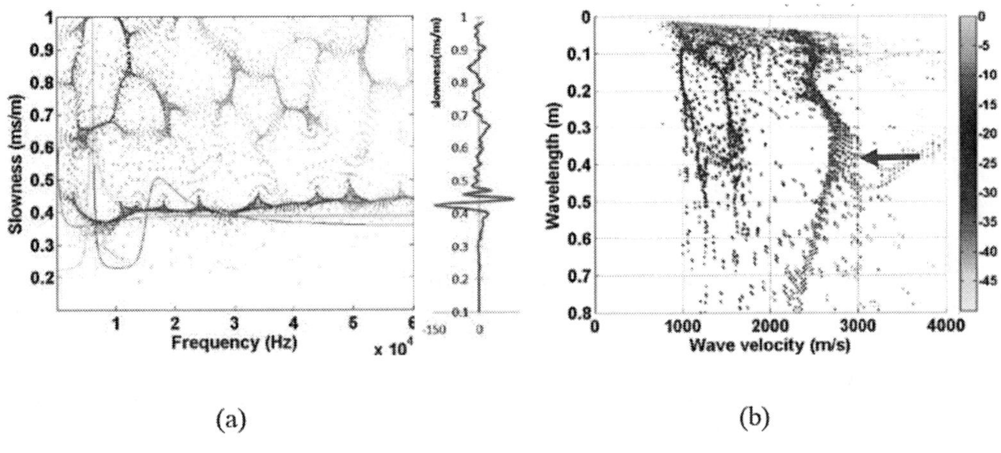

Fig. 5. Test line 2 (a) group slowness spectrogram along with waveform, (b) velocity-wavelength profile.

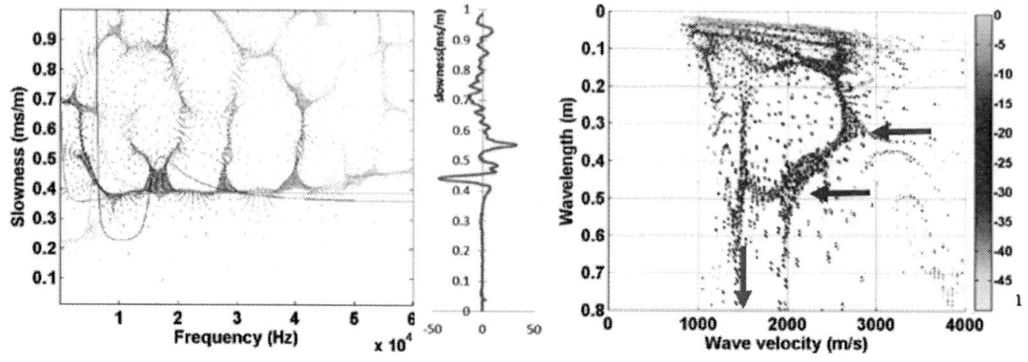

(a) (b)

Fig. 6. Test line 3 (a) group slowness spectrogram along with waveform, (b) velocity-wavelength profile.

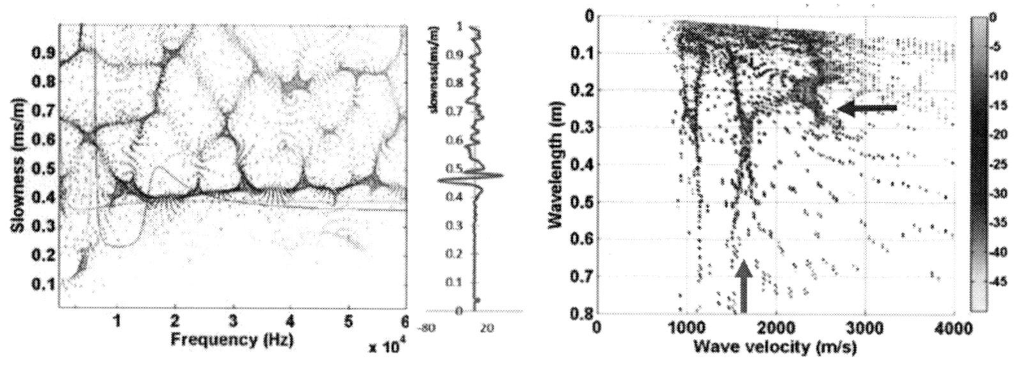

(a) (b)

Fig. 7. Test line 4 (a) group slowness spectrogram along with waveform, (b) velocity-wavelength profile.

5 Conclusions

An algorithm to assess defects within a concrete plate can be developed from the surface wave velocity profiles obtained from present study. If the group velocity remains near 2400 m/s, there should be no defect under the test line. If the group velocity is under 2200 m/s at the wavelength twice the depth of the plate, 0.72 m in present case, it may imply presence of defects under the test line. The group velocity of the surface wave gradually decreases for delaminated crack, and promptly decreases for the existence of honeycombs. As shown in Table 2, the velocity begins to decrease at the wavelength about 2 times of the depth of the defect for all the delaminated cracks and honeycomb. Thus, the depth of the defect under the test line can be estimated from the group velocity profile obtained from one recorded waveform. The impact source-receiver distance of 0.8 m is adequate for investigating defects contained in the 0.36 m-thick concrete plate. As only one test using one receiver is required to investigate the anomalies under a test line of 0.8 m, a quick and cheap assessment of concrete plate-like structure is made possible.

Table 2. Summarize the relationship between defect depth and the wavelength corresponding to decrease of wave velocity.

Test line no.	(1) Depth of defect(s)	(2) Velocity begins to decrease at wavelength	(2)/(1)
2	0.19 m	0.38 m	2
3	0.146 m, 0.238 m	0.31 m, 0.5 m	2.1, 2.2
4	0.127 m	0.28 m	2.2

The authors would like to thank Ministry of Science and Technology, the Republic of China, Taiwan, for financially supporting this research under contract MOST 104-2221-E-324 -022 -MY3.

References

1. Meckel and Nath, *Seismic Stratigraphy-applications to hydrocarbon exploration*, (C. E. Payton, 1977), American Association of Petroleum Geologists, Tulsa.
2. C. Cheng, M. Sansalone, MATER STRUCT, **28** (1995), pp.74-82.
3. C. Cheng, M. Sansalone, MATER STRUCT, **28** (1995), pp.125-132.
4. R.A. Wardany, G. Ballivy, J-L Gallias, K. Saleh, J. Rhazi, ACI MATER J, **104**, 3, (2007).
5. N. Ryden, M. Lowe, J ACOUST SOC AM, **116**, 5 (2004).
6. H. Azari, S. Nazarian, ACI MATER J, **112**, 4 (2015).
7. M. Niethammer, L. J. Jacob, J. Qu, J. Jarzynski, J ACOUST SOC AM, **107**, 5, Pt.1 (2002).
8. S. Fulop, K. Fitz, J ACOUST SOC AM, **119**, 1 (2006)
9. Kelly Fitz, Time-Frequency Analysis, 2008. Site includes code distributed under GNU public license.

MATEC Web of Conferences **138**, 03003 (2017)
EACEF 2017
DOI: 10.1051/matecconf/201713803003

Ultimate Compressive Strength and Its Deformation of Normal and High Strength Concrete Cylinder Confined With External Lateral Pre-Stressing

Harianto Hardjasaputra[1], *Joey* Tirtawijaya[1], *Gino* P. Ng[1], *Selvira* Ayuningtias[1]

[1]Civil Engineering Department, Universitas Pelita Harapan, Tangerang, Indonesia

Abstract. Confining reinforcement for concrete columns have been standardized in building standards ACI 318-11 and SNI 03-2847-2013. The observation of building collapse by earthquake normally caused by malfunction of the building columns as bearing elements. The columns were failed because they did not meet the standard of confining reinforcement. In most cases, external confinement is necessary to strengthen the existing columns of old buildings. This research focused on the methods to strengthen the existing columns. Three types of pre-stressed steel rings were used to confine the concrete cylinders of 10x 20 cm and 10 x 30 cm. Several concrete strengths of 50 MPa, 70 MPa and 100 MPa were selected for constructing 18 specimens of concrete cylinders. The first type of pre-stressed is with zero pre-stressing force represented passive confinement while the second and third types were given different pre-stressing forces on the steel ring represented active confinement. Pre-stressing force on the steel ring will cause radial stress that compress the concrete cylinders. Load test were conducted on concrete cylinders to measure the strength and the relationship of stress strain concrete cylinders which was confined by varied external lateral pre-stressing. The testing results showed that the external lateral prestressing increases the ductility and strength of concrete columns. It proved the method of external lateral pre-stressing can be used effectively for retrofitting and strengthening existing columns in order to meet the requirement of ductility and strength.

1 Introduction

Reinforced concrete columns are widely used for high rise building constructions all over the world. Columns support high compressive forces in tall buildings. However, columns may suffer damage especially due to earthquakes, due to the limited strength and ductility of concrete. The failure of one or more columns will lead to massive building failure. Standard of confinement of reinforced concrete columns is one of the most important aspect in the earthquake resistant buildings design. Lateral confinement by means of individual ties or continuous spiral on reinforced concrete column can increase the strength and performance of the column against axial and lateral forces significantly.

Based on ACI 318M-99 for the first time Indonesian Building Code for Structural Concrete (SNI 2837-2002) has made standard calculation for requirement of RC columns confinement. The total cross sectional of rectangular confinement bar A_{sh} shall not be less than required by these following equations (SNI 2837-2013)

For rectangular hoop reinforcement:

$$A_{sh} = 0,3\left(\frac{sh_c f'_c}{f_{yh}}\right)\left[\left(\frac{A_g}{A_{ch}}\right)-1\right] \quad (1)$$

$$A_{sh} = 0,09\left(\frac{sh_c f'_c}{f_{yh}}\right) \quad (2)$$

For circular hoop reinforcement:

$$\rho_s = 0,12 f'_c / f_{yh} \quad (3)$$

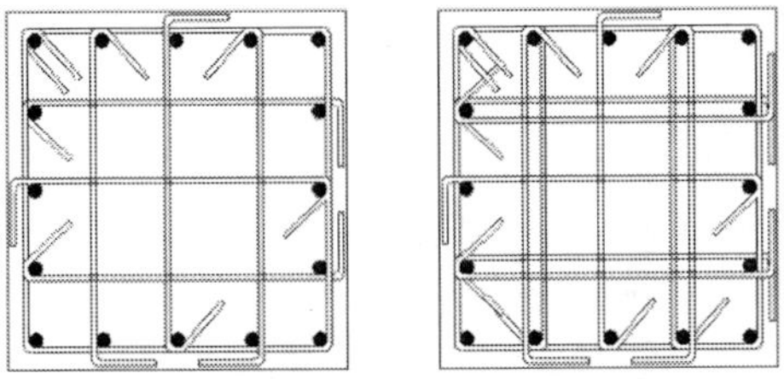

Fig. 1. Typical requirement of lateral ties of RC Columns for concrete strength 40 MPa and 60 MPa based on SNI 2837-2013.

The observation of building collapse caused by earthquake in Indonesia like in Padang (2009) and Yogyakarta (2006) showed that mostly the building columns were failed due to malfunction as the bearing elements. In fact most of the columns had been constructed without sufficient lateral ties to confine the RC columns and secure the ductility. The shortage of confinement provided by lateral ties was the motivation for conducting this research in finding new technique to retrofit the existing columns of the old buildings, constructed before the SNI 2837-2002 was published.

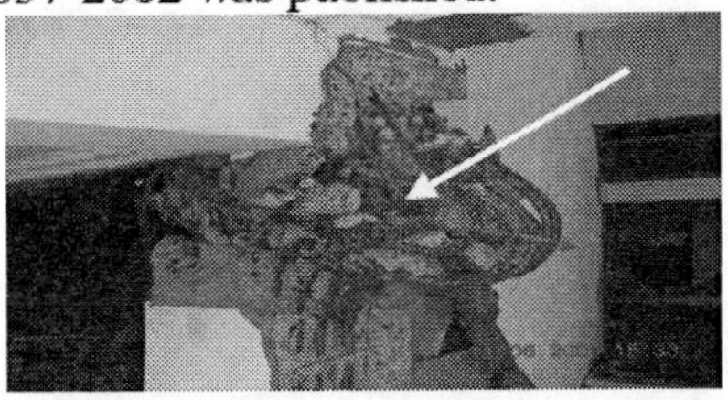

Fig. 2. Column damage due to Yogyakarta Earthquake (2006), because of the shortage of the strength and ductility.

External confinement can be done to strengthen the existing columns by using steel, reinforced concrete and fibre reinforced polymer (FRP) jacketing. With these techniques, the columns will get additional confinement strength. Those techniques have been grouped as additional passive confinement.

This paper reports a new technique of external confinement, called additional active confinement. For circular columns, the additional confinement will be delivered by the pre-

DOI: 10.1051/matecconf/201713803003

stressed force T of the external steel rings, arranged around column perimeter as shows on Fig. 3.

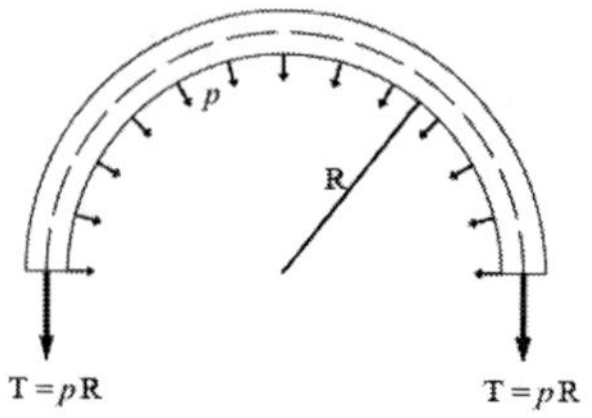

Fig. 3. The pre-stressed force T generates lateral pre-stressed force p at the contact surface of concrete cylinder.

The lateral pre-stressed force p can be calculated with equation 5.

$$T = p.R \tag{4}$$

$$p = \frac{\varepsilon_s . E_s . A_s}{R} \tag{5}$$

where,
ε_s = Strain of steel ring
Es = Modulus of Elasticity of Steel Ring (MPa)
As = Area of steel ring section (mm^2)
R = Radius of cylinder (mm)

The objective of this research is to study the influence of lateral pre-stressed force p to the increasing of strength and ductility of concrete cylinders created from normal and high strength concrete. Table 1 shows the loading test for 18 concrete cylinders of 10x 20 cm and 10 x 30 cm, which laterally reinforced with pre-stressed steel ring.

Table 1. Type of concrete cylinder specimen.

No.	Code	fc' (MPa)	Cylinder Dimension (cm)	Type of Steel Ring
1.	S 10/20-WO	50.2	10/20	WO
2.	S 10/20-W1	50.2	10/20	W1
3.	S 10/20-W2	50.2	10/20	W2
4.	S 10/30-WO	50.2	10/30	WO
5.	S 10/30-W1	50.2	10/30	W1
6.	S 10/30-W2	50.2	10/30	W2
7.	S 10/20-WO	68,11	10/20	WO
8.	S 10/20-W2	68,11	10/20	W1
9.	S 10/20-W2	68,11	10/20	W2
10.	S 10/30-WO	68,11	10/30	WO
11.	S 10/30-W1	68,11	10/30	W1
12.	S 10/30-W2	68,11	10/30	W2
13.	S 10/20-WO	101,8	10/20	WO
14.	S 10/20-W1	101,8	10/20	W1
15.	S 10/20-W2	101,8	10/20	W2
16.	S 10/30-WO	101,8	10/30	WO
17.	S 10/30-W1	101,8	10/30	W1
18.	S 10/30-W2	101,8	10/30	W2

Note:

 DOI: 10.1051/matecconf/201713803003

WO : steel ring non pre-stressed.
W1 : steel ring with moderate pre-stressed force.
W2 : steel ring with high pre-stressed force.

2 Method of research

2.1 Experimental program

Two groups of cylinder 10 cm x 20 cm and 10 cm x 30 cm were selected to produced 18 concrete specimens with concrete strength of 50 MPa, 68 MPa and 100 MPa respectively for each cylinder group. For each group has three types of external steel rings as follows; *Type W0:* steel ring without pre-stressed force, *Type W1*: steel ring with moderate pre-stressed force and *Type W2*: steel ring with high pre-stressed force.

2.2 Design of steel ring

High level grade steel was used for the rings in the test specimens. According to the tension test of the steel plate shows that the yield strength of the external steel ring used in this research is 487 MPa and its Elasticity Modulus is 185.151 N/mm^2. Each steel ring has 25 mm width and 2.4 mm thickness. Figure 4 shows the designs of Steel ring W0, W1 and W2. Steel ring W0 is a type of passive confinement installed to the concrete cylinder and grouted with cement paste to secure the bonding with the surface of concrete cylinder, on the other side, steel ring W1 and W2 can be grouped as type of active confinement. They were equipped with bolt to apply the lateral pre-stressed force p to the concrete cylinder by tightening the bolt.

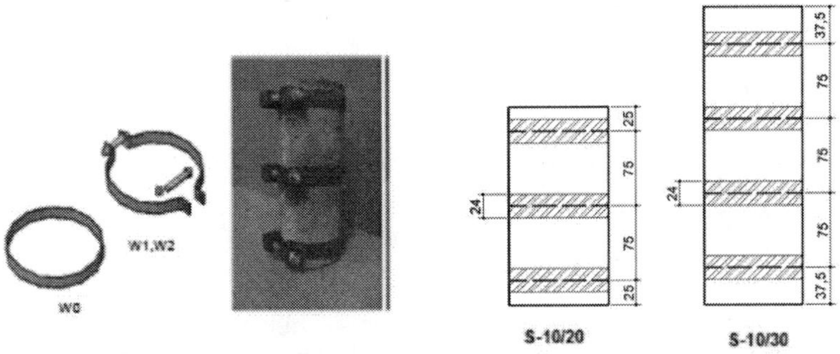

Fig. 4. Steel Ring design and its position at concrete cylinder.

2.3 Concrete properties

Portland Cement of grade 42.5 MPa conforming type 1 of Indonesian Standard (SNI 15-0249: 2004) was used in preparing the test specimen. Locally coarse aggregate was sieved with maximum size of 12.5 mm, while fine aggregate was sieved with maximum size of 0.9 mm. The concrete mixing had been designed for three kinds of concrete strength namely f_c' 50 MPa for normal concrete and f_c' 68 MPa and 101 MPa for high strength concrete.

 DOI: 10.1051/matecconf/201713803003

2.4 Test specimens preparing

Three steel rings were installed on each of concrete cylinder test specimen. In order to secure the same level of lateral prestressed force p all the steel ring W1 and W2 have to be tightened by torque wrench. The steel ring W1 was tightened with 15 Nm force while steel ring W2 was tightened with 20 Nm force. Strain gauge was installed on the steel ring in order to measure the stress level.

Fig. 5. Torque wrench.

Fig. 6. Attaching strain gauge at steel ring.

The lateral pre-stressed force p can be determined with equation 5 based on the measured strain during tightening bolt on each steel ring W1 and W2.

Table 2. Effective lateral pre-stressed force p according to equation 5

No.	Code	Dimension (cm)	Kind of ring	ε_s (10^{-3} ‰)	p (N/mm)
1	S 10/20-WO	10-20	WO	0	0
2	S 10/20-W1	10-20	W1	225.4	49,9
3	S 10/20-W2	10-20	W2	439	97,5
4	S10/30-WO	10-30	WO	0	0
5	S 10/30-W1	10-30	W1	246.5	54.8
6	S 10/30-W2	10-30	W2	413	91.7

Table 1 shows the detail of 18 concrete cylinders of 10 x 20 cm and 10 x 30 cm which reinforced laterally with non and pre-stressed steel ring. Three pieces of steel ring were installed for each 20 cm height concrete cylinder and four pieces for 30 cm height cylinder. There were 3 groups of concrete strength of the specimen test, which consisted of concrete strength of f_c' 50 MPa, 68 MPa and 101 MPa. All test specimens were tested under axial compression until failure by load control testing machine, equipped with LVDT to measure axial deformation.

 DOI: 10.1051/matecconf/201713803003

Fig. 7. Load control testing machine.

2.5 Test result and discussion

Experimental results of axial loading test of 18 concrete cylinders were discussed in the suceeding section. The results are compared in terms of ultimate compressive strength, axial displacement, steel ring deformation and crack propagation.

2.5.1 Ultimate compressive strength

All speciments have been grouped based on their compressive strength. Table 3, 4 and 5 show the loading test result of each concrete cylinder with initial compressive strength of 50.2 MPa, 68 MPa and 101 MPa.

In general all specimens equiped with lateral steel ring W0, W1 and W2 significantly higher ultimate compressive strength than their initial concrete strength. Comparing to the table 1, table 2 and table 3 show that the biggest increment procentage of compressive strength happened for concrete cylinders with f_c' 50.2 MPa. Table 1 shows, the lateral prestressed force p of 97.5 N/mm can increase the compressive strength up to 59.8 %. higher than its initial compressive strength.

Table 3. Concrete Cylinder with compressive strength 50.2 MPa.

No.	**Specimen**	**fc' (MPa)**	**Cylinder Dimension (cm)**	***Lateral pre-stressed p* (N/mm)**	**Ultimate compressive Strength (Mpa)**	**Increasing of compressive strength (%)**
1	S 10/20-WO	50.2	10-20	0	63.7	26.89
2	S 10/20-W1	50.2	10-20	49.9	72.61	44.64
3	S 10/20-W2	50.2	10-20	97.5	80.25	59.8
4	S 10/30-WO	50.2	10-30	0	57.32	14.5
5	S 10/30-W1	50.2	10-30	54.8	68.7	19.8
6	S 10/30-W2	50.2	10-30	91.7	70.06	22.23

Table 4. Concrete Cylinder with compressive strength 68.1 MPa.

No.	Specimen	fc' (MPa)	Cylinder Dimension (cm)	*Lateral pre-stressed p* (N/mm)	Ultimate compressive Strength (Mpa)	Increasing of compressive strength (%)
1	S 10/20-WO	68.1	10-20	0	79.61	16.9
2	S 10/20-W1	68.1	10-20	49.9	**76***	11.6
3	S 10/20-W2	68.1	10-20	97.5	101.9	49.63
4	S 10/30-WO	68.1	10-30	0	73.24	7.55
5	S 10/30-W1	68.1	10-30	54.8	82.8	21.59
6	S 10/30-W2	68.1	10-30	91.7	92.35	35.61

*Failure due to bolt damage.

Table 5. Concrete Cylinder with compressive strength 101 MPa.

No.	Specimen	fc' (MPa)	Cylinder Dimension (cm)	*Lateral pre-stressed p* (N/mm)	Ultimate compressive Strength (Mpa)	Increasing of compressive strength (%)
1	S 10/20-WO	101.8	20-Oct	0	124.2	22
2	S 10/20-W1	101.8	20-Oct	49.9	130	27.7
3	S 10/20-W2	101.8	20-Oct	97.5	**116***	**13.95**
4	S 10/30-WO	101.8	30-Oct	0	114	11.98
5	S 10/30-W1	101.8	30-Oct	54.8	117.2	15.13
6	S 10/30-W2	101.8	30-Oct	91.7	**111.5***	**9.53**

*Failure due to bolt damage.

The resutls show the effectiveness of lateral prestessed force p. Specimens reinforced with steel rings W1 and W2 show higher compressive strength than specimens reinforced with steel rings W0. Fig 8 and Fig 9 show the summary of the influence of steel ring W0, W1 and W2 versus the ultimate compressive strength of each specimen.

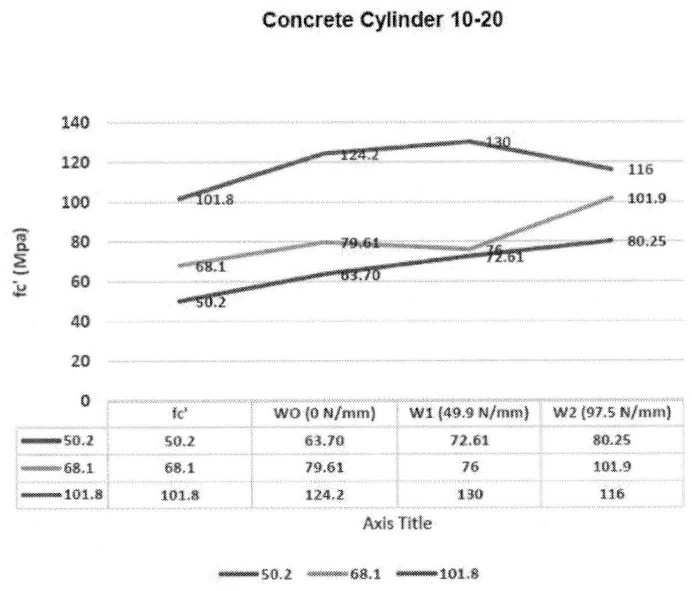

	fc'	WO (0 N/mm)	W1 (49.9 N/mm)	W2 (97.5 N/mm)
50.2	50.2	63.70	72.61	80.25
68.1	68.1	79.61	76	101.9
101.8	101.8	124.2	130	116

Fig. 8. Graphic and table of ultimate concrete strength increment.

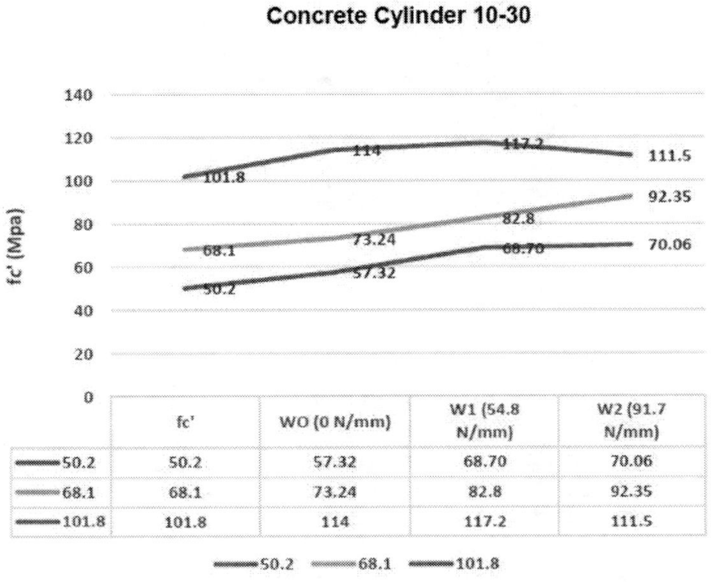

	fc'	WO (0 N/mm)	W1 (54.8 N/mm)	W2 (91.7 N/mm)
50.2	50.2	57.32	68.70	70.06
68.1	68.1	73.24	82.8	92.35
101.8	101.8	114	117.2	111.5

Fig. 9. Graphic and table of ultimate concrete strength increment.

2.5.2 Concrete strain

Alike ultimate compressive strength, all concrete cylinder specimens equiped with lateral steel ring W0, W1 and W2 show higher ultimate axial strain, i.e axial ductility.

 DOI: 10.1051/matecconf/201713803003

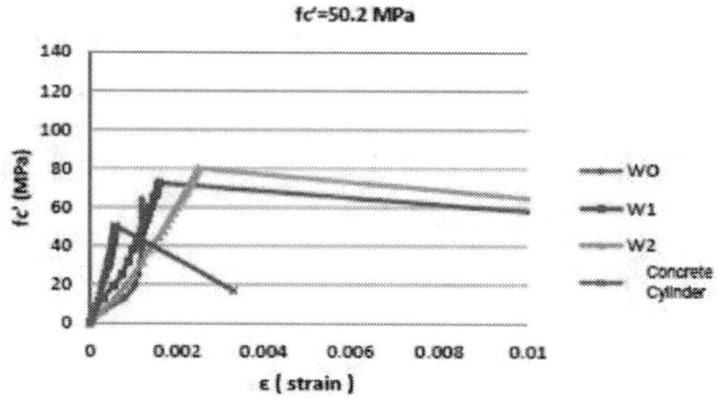

Fig. 10. Stress-Strain of concrete specimens with f_c' 50.2 MPa.

2.5.3 Crack Pattern

Typical failure crack pattern of each concrete cylinder specimen was shown in figure 11 for cylinder height 20 cm and Fig 12 for cylinder height 30 cm. They show that although the ultimate compressive strength was very high, all the specimens could keep their form at the failure. It can be concluded, that lateral steel rings can avoid brittle failure of concrete, especially high strength concrete.

Fig. 11. Crack pattern of concrete cylinder S-10/20.

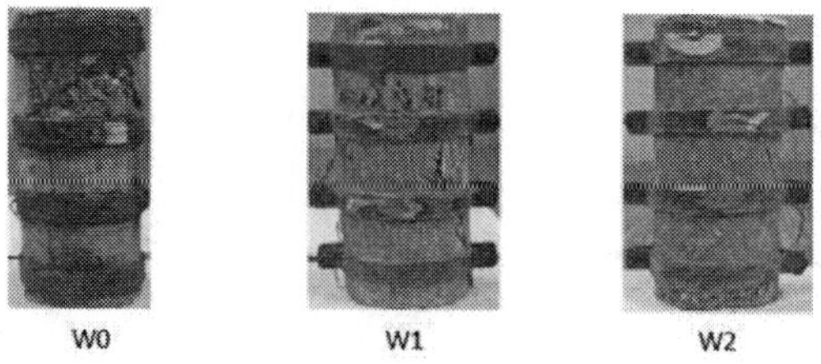

Fig. 12. Crack pattern of concrete cylinder S-10/30.

3 Conclusion

Based on the result of loading test of 18 concrete cylinder specimens in can be concluded that:

1. Confining with Lateral steel ring can improve the ultimate compressive strength effectively including their ultimate axial strain of concrete. The percentage of compressive strength increment for normal concrete is much higher than high strength concrete.
2. Specimens with Lateral steel ring pre-stressed (W1 and W2) have higher compressive strength than specimens with lateral steel ring non pre-stressed (W0). it

shows the effectiveness of lateral pressure force p in improving the strength and ductility.

3. The comparison of steel ring W1 and W2 show the magnitude of lateral pressure force p influences the percentage of strength increment of concrete.
4. Specimens with high strength concrete f_c' 101.8 MPa show lower percentage of compressive strength increment than normal concrete f_c' 50. Therefore, to confining high strength concrete cylinder, high grade steel ring is mandatory.

The first Author would like to acknowledge that some part of this paper has been written during his research visit at Uni-Kassel and TU Berlin in Germany, granted by Konrad Adenauer Stiftung.

References

1. Hardjasaputra, Harianto, T. Joey, S. Ayungtiniyas, "Perilaku Hubungan Tegangan – Regangan Beton Silinder Yang Dikekang *External Lateral Prestress* [Stress Strain Relation of Concrte Cylinder Confined with External Lateral Pre-stressed]", Research Report No.: P-007-FDTP/II/2009, Universitas Pelita Harapan, Indonesia.
2. A.B.M. Amrul Kaish, M. R. Alam, M. Jamil, M. A. Wahed, Ferrocement Jacketing for Restrengthening of Square Reinforced Concrete Column under Cocentric Compressive Load, Procedia Engineering **54**, 720-728, (2013)
3. König, V. Tue and Zink. *Hochleistungbeton.* Germany: Wiley-VCH, 2001 Building Code Requirements for Structural Concrete (ACI 318M-11) Structural Concrete Requirements for Buildings – Indonesian National Standard (SNI 2847-2013)
4. H.E.H. Roy, and M.A. Sozen, 'Ductility of concrete' in 'Flexural Mechanics of Reinforced Concrete', SP-12 (ACI/American Society of Civil Engineers, Detroit, 1965) 213–224.
5. K. Wight MacGregor, *Reinforced Concrete Mechanics and Design,* **4**th ed. Singapore: Prentice Hall, (2008)
6. M.T. Soliman, and C.W. Yu, "The Flexural Stress Strain Relationship of Concrete Confined by Rectangular transverse Reinforcement." *Mag of Concrete Research* , **61**, No. 19, 223 – 238, (1967)
7. J.B. Mander, M.J.N. Priestley, and R.F. Park, "Theoretical stress-strain model for confined concrete", *J. Struct. Engrg., ASCE* **114 (8)**, 1804–1826, (1988)
8. M. Cem. "Retrofitting Existing Concrete Columns by External Prestressing." *U.S.Patent* **6**, 247, 279 B1 (2001)
9. L. Burns. *Design of Prestressed Concrete Structures,* **3**rd ed. New York: John Wiley & Sons, (1982)

Pure rate effect on compressive strength of concrete

Sangho Lee[1], *Kyoung-Min* Kim[1], and *Jae-Yoel* Cho[1*]

[1]Seoul National University, Civil and Environmental Engineering Department, 1 Gwanak-ro, Gwanak-gu, Seoul, Korea

Abstract. Dynamic Increase Factor (DIF) has been used to consider the compressive strength enhancement of concrete at the high and intermediate strain rates. However, DIF formulae suggested until now include the inertia effects as well as the rate effect because the DIF formulae has been assumed as a function of only the strain rate and the inertia effects cannot be avoided in tests at the high and intermediate strain rate region. Therefore, applying the DIF to design or analysis of social infrastructures may be dangerous because the resistance by the inertia effects are considered repetitively. In this study, an apparent DIF formula, which includes the inertia effects, was proposed by introducing terms related with the strain acceleration, which represent the axial and radial inertia effects. Then, a nonlinear regression analysis was conducted to determine the coefficients in the apparent DIF formula with results of Split Hopkinson Pressure Bar (SHPB) tests for concrete. Finally, the DIF formula excluding the axial and radial inertia effects was proposed for compressive strength of concrete at the high and intermediate strain rates.

1 Introduction

When a concrete structure under extreme events like explosion, car, ship, and aircraft crush is analysed or designed, Dynamic Increase Factor (DIF) has been used to consider the compressive strength enhancement of concrete in a constitutive equation. In general, DIF is obtained by conducting Split Hopkinson Pressure Bar (SHPB) test, but the inertia effects cannot be avoided in SHPB test for two reasons. First of all, the material properties of specimen tested are unknown, so a desirable incident wave minimizing the inertia effects cannot be determined. In addition, although the desirable incident wave is known, it is too difficult to obtain the incident wave intended precisely. As a result, test data of SHPB includes the inertia effects as well as the rate effect, so present DIF formulae of concrete compressive strength also include both effects. However, because the inertia effects are already considered in the equation of motion in Finite Element Analysis (FEA) for a structures under extreme loads, the analysis with DIF including the inertia effects, as called as apparent DIF, may give unconservative results by overestimating the dynamic strength of concrete. In this study, an apparent DIF formula, which includes the inertia effects, was

[*] Corresponding author: jycho@snu.ac.kr

proposed by introducing terms related with the strain acceleration, which represents the axial and radial inertia effects, and then, DIF by pure rate effect was proposed.

2 DIF by pure rate effect

2.1 Inertia effects in apparent DIF

The inertia effects in SHPB test were investigated analytically and experimentally, and it is found out that the axial strain acceleration, dimension, and density of a specimen are the key factors influencing the inertia effects [1-4]. Therefore, the inertia effects in SHPB test for concrete can be assumed by Eq. (1). $\Delta f_{inertia}$, $\ddot{\varepsilon}_s$, ρ_s, d_s, l_s, k_2, and k_3 denote the axial and radial inertia effects in SHPB, axial strain acceleration, density, diameter, length of a specimen, and coefficients for each inertia effect respectively.

$$\Delta f_{inertia} = k_2 \rho_s d_s^2 \ddot{\varepsilon}_s + k_3 \rho_s l_s^2 \ddot{\varepsilon}_s \tag{1}$$

And then, apparent DIF can be written such as Eq. (2) by separating the rate effect and the inertia effects, and by substituting Eq. (1) to Eq. (2), the apparent DIF can be assumed such as Eq. (3).

$$\begin{aligned} DIF_{apparent} &= \frac{f_{c,static} + \Delta f_{rate} + \Delta f_{inertia}}{f_{c,static}} \\ &= DIF_{rate} + \frac{\Delta f_{inertia}}{f_{c,static}} \end{aligned} \tag{2}$$

$$DIF_{apparent} = DIF_{rate} + k_2 \frac{\rho_s d_s^2}{f_{c,static}} \ddot{\varepsilon}_s + k_3 \frac{\rho_s l_s^2}{f_{c,static}} \ddot{\varepsilon}_s \tag{3}$$

Meanwhile, the rate effect can be assumed by a power function of loading rate like Eq. (4) in the rate process theory [5-6]. $\dot{\sigma}$, $\dot{\sigma}_0$, and β denote loading rate, reference loading rate, and a constant in accordance with material property. In this study, the rate effect was assumed such as Eq. (5) like fib MC2010 [7]. $\dot{\varepsilon}_{s,static}$, strain rate in quasi-static test, was determined as 10^{-5} s^{-1} by referring ASTM C39/C39M-16b [8].

$$DIF_{rate} = \left(\frac{\dot{\sigma}}{\dot{\sigma}_0} \right)^{\frac{1}{\beta+1}} \tag{4}$$

$$DIF_{rate} = \left(\frac{\dot{\varepsilon}_s}{\dot{\varepsilon}_{s,static}} \right)^{k_1} \tag{5}$$

Therefore, apparent DIF can be expressed such as Eq. (6).

$$DIF_{apparent} = \left(\frac{\dot{\varepsilon}_s}{\dot{\varepsilon}_{s,static}} \right)^{k_1} + k_2 \frac{\rho_s d_s^2}{f_{c,static}} \ddot{\varepsilon}_s + k_3 \frac{\rho_s l_s^2}{f_{c,static}} \ddot{\varepsilon}_s \tag{6}$$

 DOI: 10.1051/matecconf/201713803004

2.2 Concrete SHPB test

2.2.1 Specimen preparation

In order to obtain apparent DIF data for regression of Eq. (6), concrete SHPB tests were conducted with 76.2 mm SHPB of Extreme Performance Testing Center (EPTC) in Seoul National University. Total 72 data for concrete SHPB test was obtained, and the variables of tests are listed on Table 1. I.D. of specimens is defined as Sxx-Dyy-Lzz, and xx, yy, and zz denote static strength from standard cylinder compressive test, diameter, and length of a specimen respectively. Specimens of SHPB test were placed in D50×100 and D75×150 moulds, and were cut and ground on age of 28 days. Table 2 shows concrete mix proportion. Maximum aggregate size is 13 mm for representability of small specimens.

Table 1. SHPB test condition.

Variables	Static strength, MPa	Diameter, mm	L/D ratio	Striker length, mm	Impact vleocity, m/s	Pulse shaper
Values	52, 65, 80	50, 75	0.5, 1	200, 300, 600	12 – 20	Individual design

Table 2. Concrete mix proportion.

Static strength, MPa	G_{max}, mm	W/B, %	S/a, %	Unit weight, kg/m³					
				W	C	SF	S	G	SP
52	13	34	44	176	520	-	715	918	5.2
65	13	36	40	173	480	-	676	1014	4.8
80	13	25	41	160	590	51.3	604	868	6.41

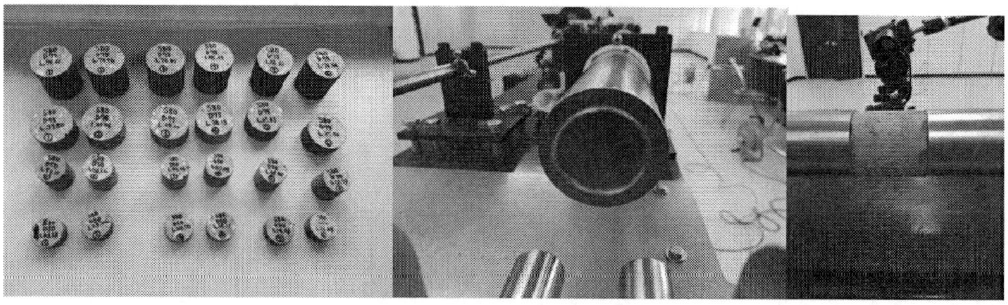

(a) (b) (c)

Fig. 1. (a) 85 MPa specimens of SHPB test; (b) Annular pulse shaper; (c) Positioning specimen.

2.2.2 Test procedure

Annular pulse shapers were used to reduce high frequency wave components due to the radial inertia effect of pulse shaper [9] as shown as Fig. 1. (b), and for each test, each annular pulse shaper was designed by pre-analyses using LS-DYNA. In order to minimize friction effect, the impact faces and interfaces of a pulse shaper and a specimen were lubricated by petroleum jelly as shown as Fig. 1. (b) and (c). Diameter of each bar component is 76.2 mm, and the length of incident and transmitted bars is 5.5 m. The striker length is ranged from 200 mm to 600 mm, and the impact velocity was ranged from 12 m/s to 20 m/s.

2.2.3 Test results

Dynamic equilibrium was checked, and the raw data was transformed to stress and strain by 3 waves post-process. DIF was determined from the static strength of D150×300 standard cylinder specimen. Test results are shown in Fig. 2 and 3.

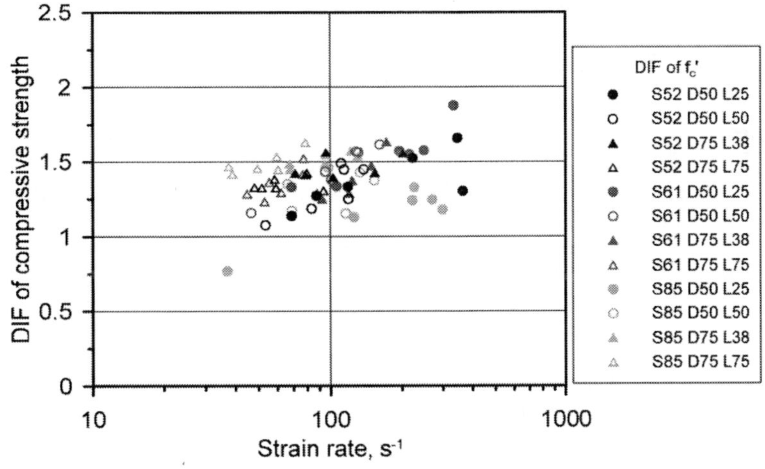

Fig. 2. Strain rate – DIF plot.

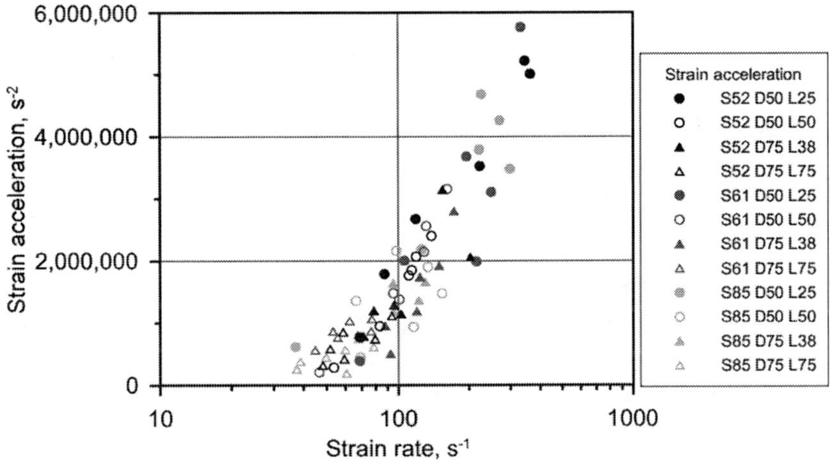

Fig. 3. Strain rate – strain acceleration plot.

2.3 Regression analysis

Nonlinear regression analysis was conducted for Eq. (6) to concrete SHPB test data. The objective function was the least square error such as Eq. (7). It was confirmed that there is the unique minimum point. Apparent DIF and DIF by pure rate effect were obtained such as Eq. (8) and (9) respectively. For dimensionless k_1, k_2, and k_3, units of variables were considered as Table 3.

$$\Pi = \frac{1}{2}\sum_i \left\{ DIF_{test,i} - \left[\left(\frac{\dot{\varepsilon}_{s,i}}{\dot{\varepsilon}_{s,static}} \right)^{k_1} + k_2 \frac{d_{s,i}^{\ 2} \rho_{s,i}}{f_{c,static,i}} \ddot{\varepsilon}_{s,i} + k_3 \frac{l_{s,i}^{\ 2} \rho_{s,i}}{f_{c,static,i}} \ddot{\varepsilon}_{s,i} \right] \right\}^2 \tag{7}$$

 DOI: 10.1051/matecconf/201713803004

$$DIF_{apparent} = \left(\frac{\dot{\varepsilon}_s}{\dot{\varepsilon}_{s,static}}\right)^{0.0147} + 0.3501\frac{\rho_s d_s^{\,2}}{f_{c,static}}\ddot{\varepsilon}_s + 0.4100\frac{\rho_s l_s^{\,2}}{f_{c,static}}\ddot{\varepsilon}_s \tag{8}$$

$$DIF_{rate} = \left(\frac{\dot{\varepsilon}_s}{\dot{\varepsilon}_{s,static}}\right)^{0.0147} \tag{9}$$

Table 3. Unit of variables.

Variables	$\dot{\varepsilon}_s$	$\ddot{\varepsilon}_s$	ρ_s	$f_{c,static}$	d_s	l_s
Units	s^{-1}	s^{-2}	kg/m^3	Pa	m	m

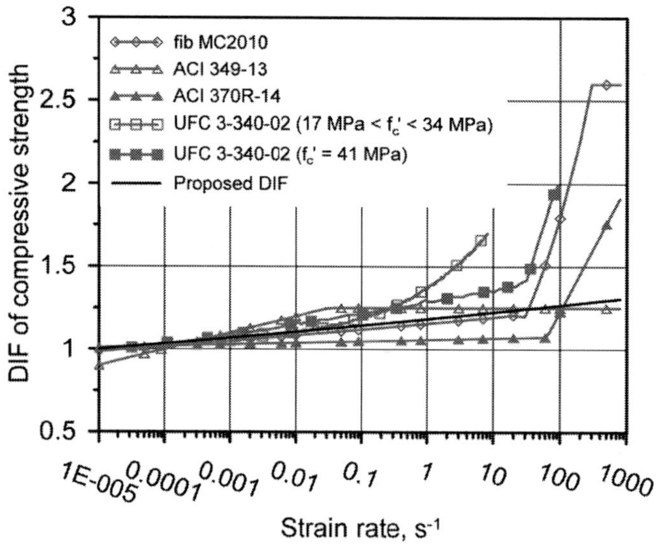

Fig. 4. Proposed and representative DIF. [7, 10-12]

Fig. 4 shows the proposed DIF by pure rate effect. The inertia effects become dominant as strain rate becomes high. First of all, DIF in fib MC2010 below transition strain rate is similar to proposed DIF, and it means that the steep increase of DIF is due to the inertia effects. In the case of DIF in ACI 349-13, the upper limit is just equal to 1.25, so it is similar to proposed DIF at the 10 – 1,000 s^{-1} of strain rate region. For DIF of ACI 370R-14 also, because the transition strain rate is higher than those of the other DIF and the rate effect is underestimated, overestimation of the dynamic strength is started beyond about 100 s^{-1}. Therefore, both DIFs of ACI are conservative relatively. Finally, DIF of UFC 3-340-02 always overestimates the dynamic strength.

3 Conclusion

The inertia effects in apparent DIF by SPHB test were investigated, and DIF by pure rate effect was proposed. And also, following conclusions were derived.

1. DIF by pure rate effect was proposed for application in constitutive equation.
2. DIF of fib MC2010 is unconservative beyond transition strain rate, and this steep increase of DIF in fib MC2010 is due to the inertia effects.

3. DIFs of ACI are conservative relatively, but DIF of ACI 370R-14 can derive unconservative dynamic strength beyond about 100 s^{-1} strain rate.
4. DIF of UFC 3-340-02 is unconservative at all strain rate region.

This research was supported by a grant (17CTAP-C069307-05) from Construction & Transportation Technology Advancement Research Program funded by Ministry of Land, Transport and Maritime Affairs of Korean government and research fund of Doosan Yonkang Foundation.

References

1. E. Davies, S. Hunter, J. Mech. Phys. Solids, **11**, 155-179 (1963)
2. D. Gorham, J. Phys. D: Appl. Phys., **22**, 1888-1893 (1989)
3. M. Forrestal, T. Wright, W. Chen, Int. J. Impact Eng., **34**, 405-411, (2007)
4. M. Zhang, H.J. Wu, Q.M. Li, F.L. Huang, Int. J. Impact Eng., **36**, 1327-1334 (2009)
5. H. Mihashi, F.H. Wittmann, Heron, **25**, 26-28 (1980)
6. J. Ožbolt, A. Sharma, H.W. Reinhardt, Int. J. Solids Struct., **48,** 1534-1543 (2011)
7. fib Commission 8, "Code-type Models for Concrete Behaviour: State-of-the-art Report", **70** (fib Fédération internationale du béton, Lausanne, 2013)
8. ASTM C39/C39M-16b, "Standard Test Method for Compressive Strength of Cylindrical Concrete Specimens" (ASTM International, West Conshohocken, 2016)
9. W.F. Heard, B.E. Martin, X. Nie, T. Slawson, Exp. Mech., **54,** 1343-1354 (2014)
10. ACI Committee 349, "Code Requirements for Nuclear Safety-Related Concrete Structures (ACI 349-13) & Commentary" (American Concrete Institute, Farmington Hills, 2014)
11. ACI Committee 370, "Report for the Design of Concrete Structures for Blast Effects" (American Concrete Institute, Farmington Hills, 2014)
12. Unified Facilities Criteria (UFC) 3-340-02, "Structures to Resist the Effects of Accidental Explosions" (U.S. Department of Defense, Washington D.C., 2008)

An Experimental Study on the Manufacturing Properties of Solid Capsules for Mixing Cementitous Composite Materials for Self-Healing of Crack

Yun-Wang Choi[1], Sung-Rok Oh[1*] and Byung-Keol Choi[1]

[1]Department of Civil Engineering, Semyung University, 27136 65 Semyung-ro, Jecheon-si, Chungcheongbuk-do, South Korea

Abstract. In this study, we fabricated self-healing solid capsules using cement powder, an inorganic-based material that is the same as that of cement composites. Self-healing solid capsules can be divided into two stages(capsule core preparation and capsule wall coating). First, the capsule core mixes the core material and the coagulant to form a core material dough having a proper quality, and then puts into a capsule core manufacturing machine. the next, it was produced through the pore chamber under the pressure of the capsule core making machine. Second, The capsule wall is formed by placing the prepared capsule core into the chamber of the capsule wall coating apparatus and then applying the wall material while the chamber is rotating. The prepared capsules were mixed with cement composites to evaluate fluidity and strength. That is, we evaluated the basic performance changes of cement composite materials with capsules. We will use the results of this paper as the base data to consider the optimal amount of capsule at the level that satisfies the performance required when mixing capsules.

1 Introduction

Smart structures and intelligent materials have emerged as new trends for reducing cracks in concrete, and extensive research is being performed on the self-healing of concrete cracks. Structures capable of self-healing are able to detect cracks in the early stage and heal themselves, thereby saving time and costs involved in maintenance and repair compared to general structures. A major advantage of self-healing is that repairs can be easily carried out, even for structures that are difficult to access physically.[1, 2]

Among the various methods of enabling self-healing, one approach is to encapsulate materials that have self-healing capabilities and introduce them as capsules. Because such capsules can hold a large amount of self-healing materials[3, 4], reactions can be selectively induced at cracks. In one international study, self-healing materials and catalysts were encapsulated in the form of micro-sized capsules[5, 6]. Meanwhile, in Korea, self-healing

* Corresponding author: cgdbs02@nate.com

DOI: 10.1051/matecconf/201713803005

micro-sized capsules capable of reacting with sunlight or moisture without catalysts have been applied in actual construction[7]. Two conclusions can be derived from these case studies at home and abroad: First, the target of self-healing capsules is not the structure's body, but its surface. Second, An organic material is used as the self-healing material.[8, 9].

The number of capsules for self-healing is limited as they are normally utilized as coating to protect the surface, thus making it difficult to heal deep cracks. most of the structure's body is comprised of inorganic composite materials. therefore, the self-healing performance can be enhanced by using inorganic materials that have the same properties instead of organic materials that have different thermal properties. While direct mixing the cementitious composite materials, the thickness of the matrix can be freely adjusted as necessary. The self-healing performance can also be maximized by introducing more self-healing capsules.

This study developed solid capsules that can be directly mixed with cementitious composite materials using inorganic capsules with the same properties as that of cementitious composite materials. The properties of the self-healing solid capsules were also evaluated. This study did not consider the healing effect of self-healing capsules. Instead, it focused on evaluating the effects of capsule mixing on the quality of cementitious composite materials. The results of this study can serve as a valuable reference for developing self-healing capsules and capsule-containing cementitious composite materials.

2 Experiment outline

2.1 Fabrication of self-healing solid capsules (SHSC)

2.1.1 Core material

Self-healing solid capsules (SHSC), which are inorganic self-healing capsules, were fabricated from core materials comprised of cement powder. The cement used in the experiments was ordinary Portland cement. Generally, cement has a mechanism to react by moisture.[10] Therefore, in order to granulate the cement powder, a coagulant having a low water content should be used.

The coagulant used in this study was High-purity methyl alcohol, which facilitates coagulation and is easy to evaporate, was used as the coagulant. The coagulated cement powder was inserted into the capsule fabricating equipment in tempered form to create the core material. Left of Fig. 1 shows a capsule core manufacturing apparatus.

Table 1. Chemical components and physical properties of cement.

Items / Types	CaO (%)	SiO_2 (%)	Al_2O_3 (%)	MgO (%)	Fe_2O_3 (%)	SO_3 (%)	L.O.I (%)	Surface area (cm^2/g)	Density (g/cm^3)
OPC	61.40	21.60	3.40	2.50	3.10	2.50	0.03	3,540	3.15

DOI: 10.1051/matecconf/201713803005

Fig. 1 Manufacture of capsules (capsule core fabricating equipment ; left, Wall material coating equipment ; right)

2.1.2 Wall material

The core material, in which cement is coagulated, can react reactions with the mixing water during the mixing with cementitious composite materials. also it can lost due to the effects of cement composites and mixers. Accordingly, two types of wall materials have been selected to ensure moisture barrier and strength.

The wall materials were liquid rubber (Liquid type Rubber, LR) and an acryl-based material (Poly(methyl methacrylate), PMMA). Methylene chloride (CH_2Cl_2, MC) was used as a dissolving agent to melt the solid PMMA.

The performance of each wall material was assessed by applying the test methods specified in KS L 5105 after standard mortar mixing. Moisture resistance was assessed through the permeability test specified in KS F 4930, and strength was evaluated for samples at day 3 according to KS L 5105. The fabricated core material was inserted into the chamber, and the wall material was lubricated and dried while the chamber rotated at a certain RPM. Right of fig. 1 (b) shows a wall material coating equipment apparatus.

2.2 Quality of SHSC

2.2.1 Slump flow

The slump flow of cementitious composite materials according to mixing of SHSC was assessed using paste samples, fabricated by excluding fine aggregates from the plain mix specified under KS L 5105. SHSC selected 4 levels(0, 5, 10, and 15 mass%) for cement mass.

2.2.2 Compressive strength

The compressive strength of cementitious composite materials according to mixing of SHSC was assessed using paste samples, fabricated by excluding fine aggregates from the plain mix specified under KS L 5105.

3 EXPERIMENTAL RESULTS AND DISCUSSION

3.1 Fabrication of SHSC

3.1.1 Consistency

SHSC is produced by preparing capsule core and coating the wall. If the tempering state is too thin or thick, it is difficult to fabricate the core material. This study utilized OPC as the core material, and applied methyl alcohol at three levels (Step-1: 10, 20, 30 mass%) of the OPC mass to obtain the desired consistency. For quantitative analysis of consistency, a table flow test was performed, and consistency standards were established based on the slump values.

Fig. 2 give the shape of Step-1 samples immediately after tempering and the results of the table flow test. The results show that samples of 10% and 20% consistency were easily coagulated, but the table flow test could not be carried out, as the samples were crushed even by the slightest shock. The 30% samples were too watery and obtained a flow, not a table flow. The core material fabricator creates a core material at a consistency suitable for the chamber, and coagulates the material using pressure created by the chamber lid and mixing blades within the chamber. Therefore samples of 10% and 20%, which are unable to withstand shock, are not suitable since they will be pulverized when the wall material is applied or during drying. Samples of 30% cannot coagulate if they are too watery after the core material is formed. As such, an adequate consistency can be achieved between 20% and 30%. To obtain a suitable consistency, the experiments were repeated with the coagulant at five mixing ratios (Step-2: 21, 22, 23, 24, and 25 mass %).

Fig. 3 show the shape of Step-1 immediately after tempering and the results of the table flow test. Samples of 21% coagulated immediately after mixing, but measurements were difficult due to the exterior collapsing. Samples of 22% and 23% maintained the coagulated form, and slump values were obtained based on the table flow results. Similar to the results for 30% samples, samples of 24% and 25% were too watery to be coagulated, and the flow values were insignificant. These results indicate that an adequate consistency in creating the core material can be achieved by mixing the coagulant at 22-23%.

Fig. 4 show the production and loss of the core material after mixing the coagulant at 22% and 23%. Samples of 22% had a 5% loss in mass, while samples of 23% showed segregation of the coagulant due to pressure within the chamber because of their consistency relatively watery. The latter also had a greater loss in mass with the capsule cores sticking to one another. The production efficiency of the capsule core can be enhanced by mixing the coagulant at 22±0.5% when creating the core material.

 DOI: 10.1051/matecconf/201713803005

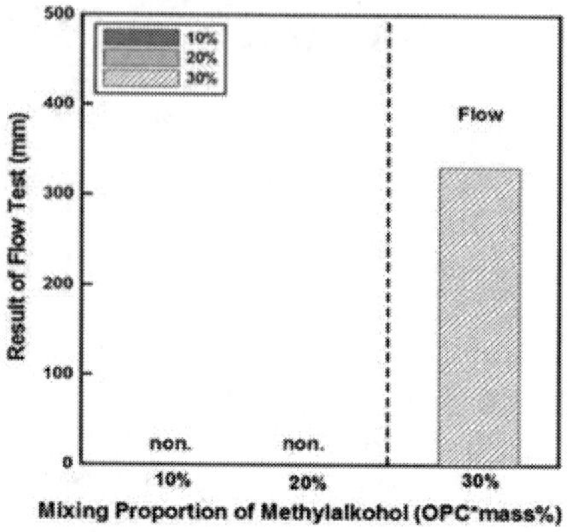

Fig. 2 Result of flow test on group of step-1

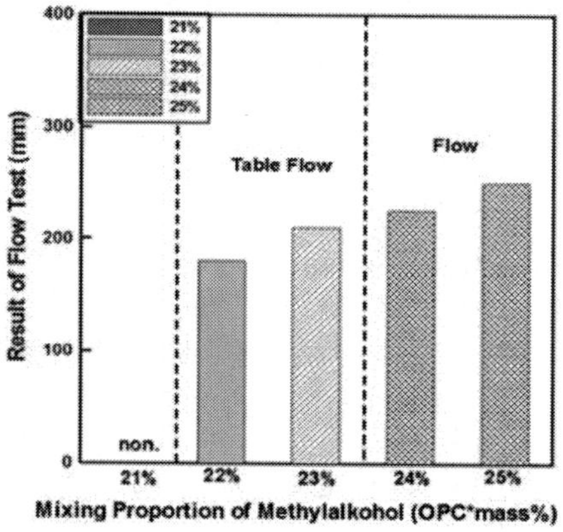

Fig. 3 Result of flow test on group of step-2

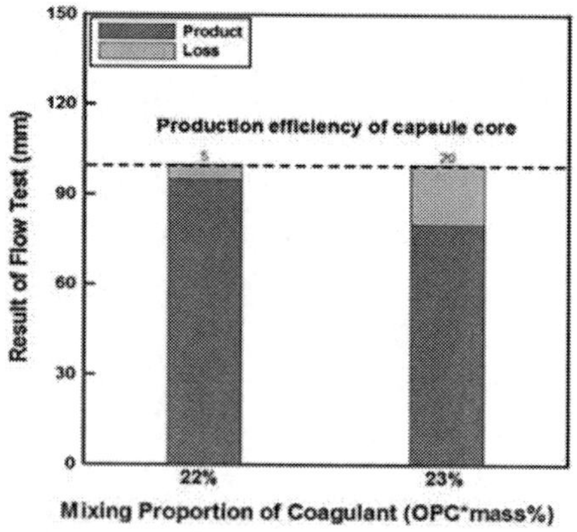

Fig. 4 Production efficiency of capsule core

3.1.2 Properties of wall materials

Since the capsule core has a mechanism that reacts with moisture, moisture must be blocked before the material is destroyed by cracks. so, that it can be possesses activation energy for self-healing. Therefore, the choice of wall material is important. To assess the properties of the wall material, permeability and compressive strength tests were performed after standard mortar mixing.

Fig. 5 and Fig. 6 show the results of the permeability and compressive strength tests. The permeation amounts of LR and PMMA were smaller than the plain sample without any wall material by 88% and 57%, respectively. Compared to PMMA, the permeation amount of LR decreased by 74%. This can be attributed to PMMA being diluted 20 times in the dissolvent, and the less uniform coating of the dissolvent due to the higher viscosity of PMMA than LR. Even if the dissolvent is uniformly coated, the remaining PMMA can form a coating layer after the dissolvent evaporates, resulting in a change in effective content.

Fig. 6 shows that the compressive strength of LR and PMMA is higher than that of the plain sample by approximately 15.8% and 12.7%, respectively. The effects of PMMA dissolution on effective content and coating require further review . However, in this study, LR was selected as the wall material and applied to the capsule core.

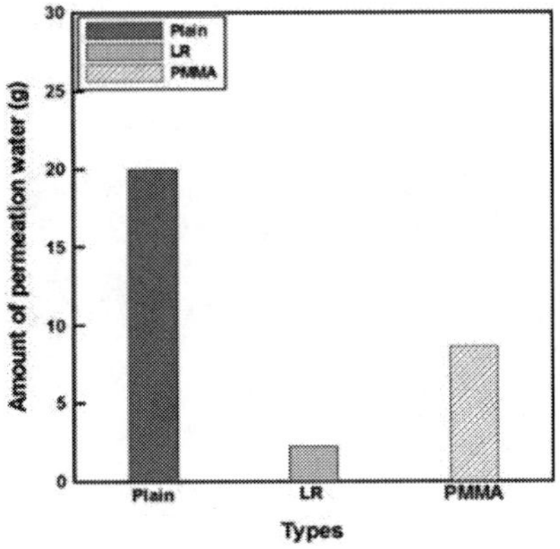

Fig. 5 Amount of permeation water according to wall material

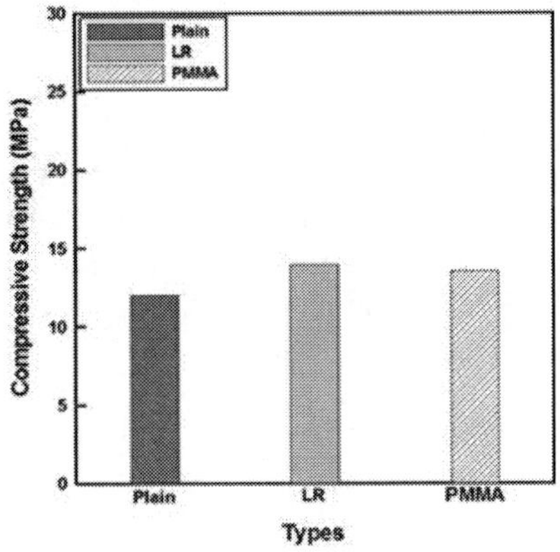

Fig. 6 Compressive strength according to wall material

3.1.3 The shape and size of SHSC

Fig. 7 shows the shape and size of the SHSC. The SHSC is in the form of a spherical shape, and it has a size of more than about 1,000 micrometers. The coating of the wall material was assessed through a qualitative evaluation.

Left of fig. 8 shows a case where the wall material is not coated, and right of Fig. 8 shows a state after the wall material is coated. In the case of the former, since the wall material is not coated, when the capsule core is supplied with water, it is immediately absorbed. In the latter case, since the wall material is coated, even if water is supplied to the capsule core, it is not absorbed.

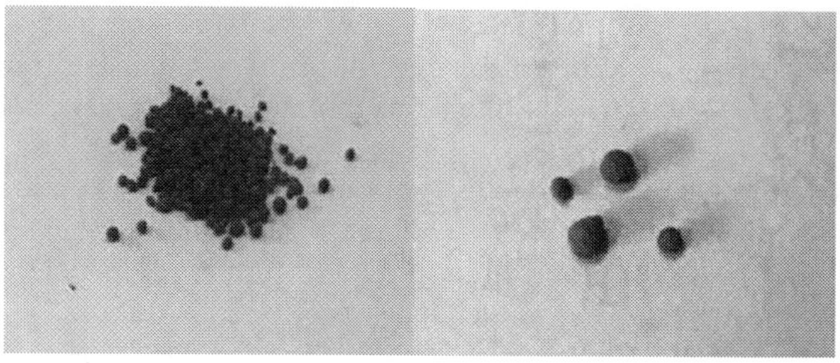

Fig. 7 The shape and size of SHSC

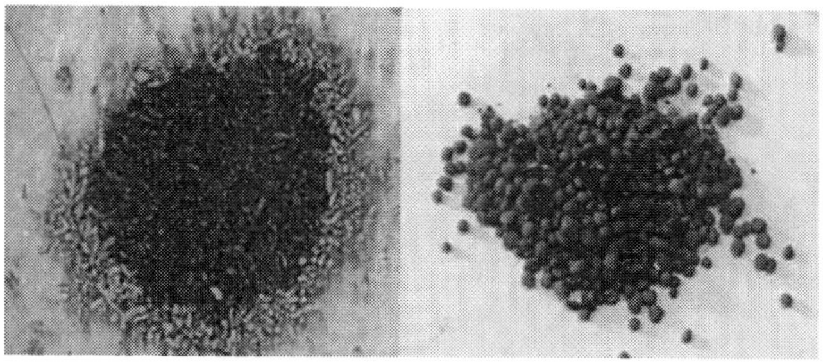

Fig. 8 Qualitative evaluation of wall matarial coating (before coating ; left, after coating ; right)

3.2 Quality of SHSC

3.2.1 Loss rate

Fig. 9 shows the result of SHSC loss evaluation. The SHSC loss was assessed by mixing the cement composites immediately after mixing, and then washing was carried out in a sieve. And it was dried. As can be seen from Fig. 9, the loss was less than 5% for the paste after mixing, and about 10 to 15% for the mortar. When SHSC are mixed with the paste and mortar, they are presumed to be influenced by the mixing blades during mixing. Since mixing with the mortar involves fine aggregates, the rate of loss increases by two to three times compared to the paste. This loss can be compensated by reinforcing the wall material.

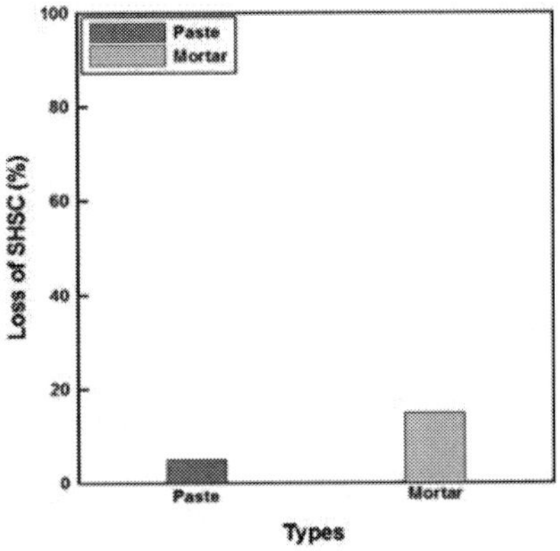

Fig. 9 Loss of SHSC immediately after mixing with cementitious composites

3.2.2 Slump flow

Fig. 10 show the slump flow in relation to the mixing ratio of SHSC. Compared to the plain samples without mixing, the slump flow decreased by 5% with a 5% increase in the mixing ratio. It is considered that SHSC affects the fluidity of cement composites. Further study is needed to examine the possibility of the decrease in slump flow arising from reactions with the mixing water in the case of the wall material having a non-uniform coating.

On the other hand, if the coating is uniform, and assuming that there is no influence on moisture, the SHSC grains are presumed to have caused the decrease in fluidity. The decrease in fluidity with an increase in SHSC, which are needed for the self-healing performance of cementitious composite materials, can be compensated by using binders other than cement or chemical admixtures. Another issue to be considered, although falling outside the scope of this study, is the distribution of SHSC in the cementitious composite materials.

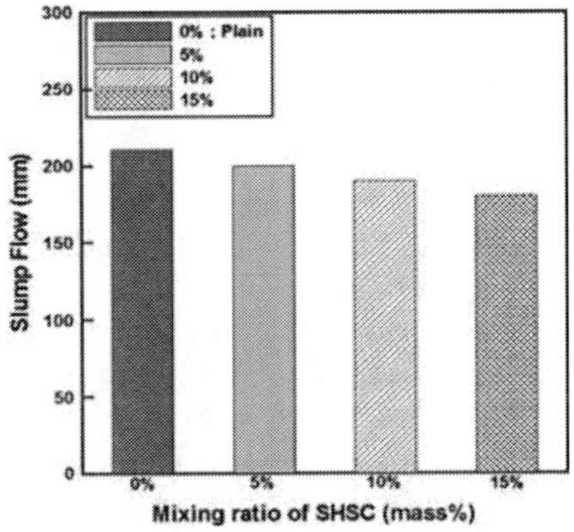

Fig. 10 Slump flow according to mixing ratio of SHSC

3.2.3 Compressive strength

Fig. 11 shows the compressive strength in relation to the mixing ratio of SHSC. Compared to the plain samples without mixing SHSC, the compressive strength decreases

DOI: 10.1051/matecconf/201713803005

proportionately with a 5% increase in the mixing ratio of SHSC. Because the cement powder in coagulated form is distributed in the matrix of cementitious composite materials, the areas in which SHSC are present are relatively weaker, resulting in a smaller compressive strength. Since the weak areas are cross-sections containing SHSC, hydration reactions may be induced at cracks with the leaking of SHSC through cracks, thereby enabling the self-healing of cracked sections.

However, the cracked part can not be healed by the SHSC alone on the entire cracked area., but a significant portion may be restored by adding a function to facilitate swelling or expansion. Considering how strength decreases by at least 15% at a 5% mixing ratio of SHSC, the strength of cementitious composite materials can be improved by adjusting the material proportions, adding additional binders, or using chemical admixtures. The adequate mixing ratio of SHSC should be examined with consideration of the efficiency of compressive strength in relation to self-healing performance.

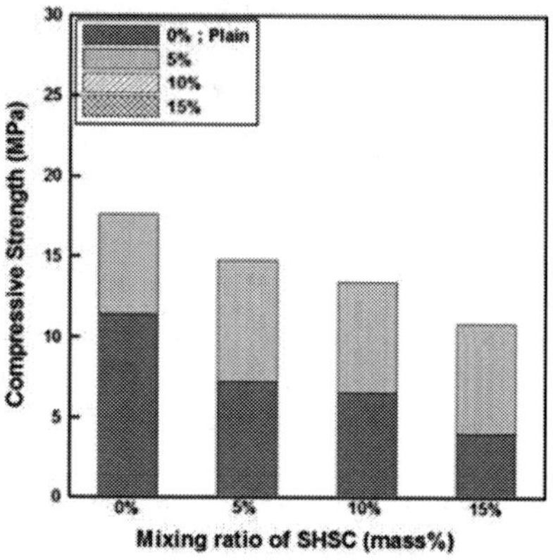

Fig. 11 Compreessive Strength according to mixing ratio of SHSC

4 CONCLUSION

This study developed solid capsules that can be directly mixed with cementitious composite materials using inorganic capsules with the same properties as that of cementitious composite materials. The properties of the self-healing solid capsules were also evaluated. This study focused on evaluating the effects of capsule mixing on the quality of cementitious composite materials. The following conclusions were derived:

1. When cement powder was utilized as the core material, the optimal coagulant mixing ratio for adequate consistency could be obtained. However, the tests have to be repeated to obtain the optimal coagulant mixing ratio if the core material includes a material other than cement powder or consists of at least two materials.
2. Capsule core coating was performed with liquid rubber as the wall material to improve resistance to moisture and strength. However, quantitative analysis on the uniform coating of the wall material is needed, and materials other than liquid rubber must be evaluated.
3. When capsules are mixed with cementitious composite materials, fluidity and compressive strength tend to decrease. It is considered that the quality of the cementitious composites can be improved by adjusting the mixing ratio and improving the chemical mixture to improve the slump and compressive strength.

This research was supported by a grant(17SCIP-B103706-03) from Construction Technology Research Program funded by Ministry of Land, Infrastructure and Transport of Korean government

References

[1] Erik Schlangen and Senot Sangadji, Journal of Procedia Engineering, The 2nd International Conference **54**, 39-57 (2013)

[2] An, E., J. and Shin, M., S., Proceeding of Korea Concrete Institute **26**, No. 2, 477-479 (2014)

[3] Tatyana Nesterova, Kim Dam-Johansena, Journal of Organic Coatings **75**, No. 4, 309-318 (2012)

[4] M. Hunger, A.G. Entrop, I. Mandilaras, H.J.H. Brouwers and M. Founti, Journal of Cement & Concrete Composites **31**, No. 10, 731-743 (2009)

[5] Shazim Ali Memon, H.Z. Cui, Hang Zhang and Feng Xing, Journal of Applied Energy **139**, 43-55 (2014)

[6] Kim, P. S., Jo, C. K., Ju, T. H., Choi, Y. C., and Lee, J. K., proceeding of The Korean Society for Energy, 225-228 (2003)

[7] Song, Y. K., Jo, Y. H., Lim, Y. J., Cho, S. Y., Yu, H. C., Ryu, B. C., Lee, S. I., and Chung C. M., Journal of ACS Applied Materials and Interfaces **5**, No. 4, 1378-1384 (2013)

[8] Xianfeng Wang, Feng Xing, Ming Zhang, Ningxu Han and Zhiwei Qian, Journal of Materials **6**, No. 9, 4064-4081 (2003)

[9] B.J. Blaiszika, N.R. Sottos and S.R. White, Journal of Composites Science and Technology **68**, No. 3, 978-986 (2007)

[10] Korea Concrete Institute, *Concrete Standard Specification* (Korea Concrete Institute, 2009)

 DOI: 10.1051/matecconf/201713803006

Effect of Applying Techniques and Polymer Content on Strength and Drying Shrinkage of Glass Fiber Reinforced Concrete

Ratthanan Ianleng[1,*] ,and *Thatchavee* Leelawat[1]

[1]Department of Civil and Environmental Engineering, Mahidol University, Thailand

Abstract. The purposes of this study were to evaluate compressive strength, flexural strength, and drying shrinkage of Glass Fiber Reinforced Concrete (GFRC) applying different techniques and varying polymer content. Two groups of specimens were classified applying the techniques: sprayed and premixed methods. AR-Glass was used with fiber content of 3 to 4% by volume. GFRC was mixed and applied different techniques with proportions of Styrene Butadiene Rubber (SBR) content of 0%, 3%, 6%, and 9% by weight of cement. Compressive and flexural strength tests were performed at 1 and 28 days. Drying shrinkage tests were measured up to 98 days. The results obtained showed that increasing the SBR content showed a lower compressive strength of GFRC for both sprayed and premixed techniques. In the other hand, 28-day flexural strength results of GFRC for both premixed and sprayed techniques were found to increase with increasing SBR content. The GFRC mixes using sprayed technique exhibited flexural strength higher than the corresponding mixes using premixed technique because of the two-dimensional layer of fiber alignment for the sprayed technique. Increasing the SBR content exhibited the lower drying shrinkage of GFRC. At the age of 98 days, the drying shrinkage of GFRC using 9% SBR content was about 40% lower than that of GFRC using 0% SBR content.

1 Introduction

Glass Fiber-Reinforced Concrete (GFRC) is a cementitious composite material made up of mortar matrix and chopped glass fibers. GFRC is mainly used in exterior building façade panels and architectural precast concrete. The matrix can be made from cement, sand, and additives. The benefits for the use of GFRC are toughness, ductility, and reducing large cracks, when compared to normal reinforced concrete. Generally, alkali resistant glass (AR-Glass) is selected for making GFRC to reduce the degradation of glass fiber due to alkaline content of cement [1-5]. The polymer latex admixture is commonly used to improve the flexural strength, workability, and toughness of GFRC [6]. Two GFRC fabrications were reviewed: namely premixed and sprayed techniques. The sprayed technique, using up to 5% of fiber content, is commonly used to make GFRC. Increasing fiber content could improve

* Corresponding author: ratthanan.ian@gmail.com

flexural strength of GFRC [7]. The production cost of GFRC using sprayed technique is higher than that of GFRC using the premixed technique. Nonetheless, the premixed technique was found to require a higher water demand to obtain the specified workability, when compared to the sprayed technique [8].

However, limited research was found that studied the comparison of engineering performance of GFRC using sprayed and premixed techniques. Therefore, the main aims of the study were to evaluate the effects of applying techniques and polymer content on the compressive strength, flexural strength, and drying shrinkage of GFRC.

2 Experiment

2.1 Materials

White Portland cement conforming to ASTM C150 was used in this research. The specific gravity of the cement was 3.15. Silica sand was selected with the maximum nominal size of 0.84 mm (No.20).

Glass fiber, conforming to ASTM D578, was AR-Glass type obtained from Nippon Electric Glass. The composition of glass fiber was 61% of silicon oxides, 15% of sodium oxides, and 20.8% of zirconia oxides. Specific gravity of AR-glass was 2.74 g/cm^3.

Polymer admixture was a Styrene Butadiene Rubber (SBR). Infrared transmittance of SBR using Fourier Transform Infrared Spectroscopy (FTIR) is shown in Fig 1. Specific gravity of SBR was 1.02 kg/l. The water content of SBR used was 54.5%.

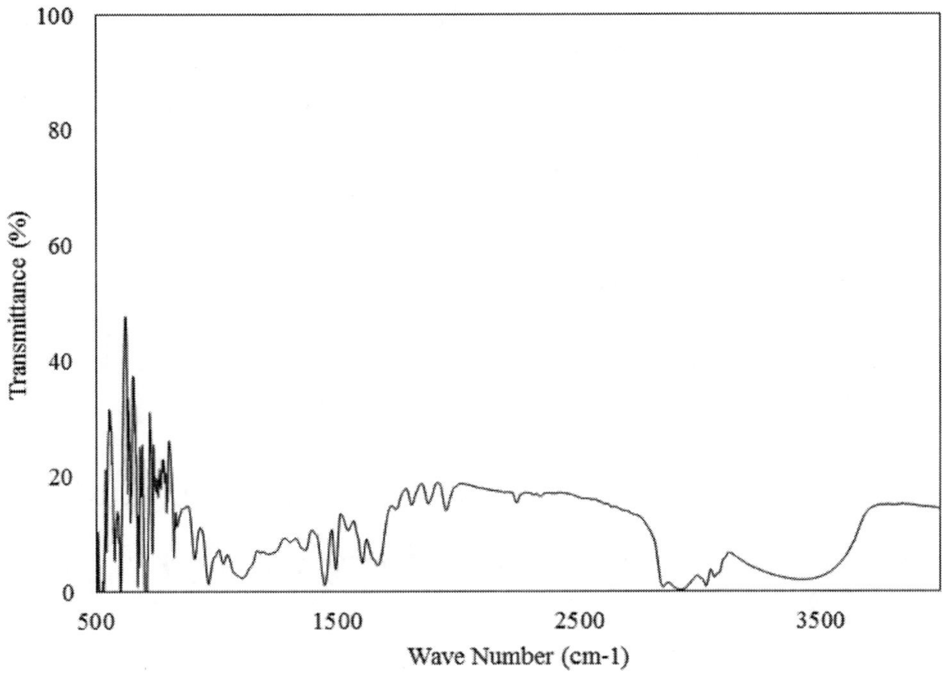

Fig 1. Infrared transmittance of SBR using Fourier Transform Infrared Spectroscopy (FTIR)

DOI: 10.1051/matecconf/201713803006

2.2 Mixing proportions and specimen preparation

The summary of mix proportions for GFRC using sprayed and premixed techniques is given in Table 1. Six cylindrical specimens with the dimension of ø100×200, six prismatic specimens with the dimension of 350×150×25 mm, and three prismatic specimens with the dimension of 300×75×75 mm for each mix were prepared.

The compositions of GFRC using the premixed technique consisted of cement, sand, water, chopped glass fiber, and SBR. The GFRC samples were prepared by combining silica sand and white Portland cement in a mixer for 1 minute. Water and SBR were then added and mixed further for 1 minute. Thereafter, silica sand and white Portland cement, which were properly blended, were added into the mixer and mixed further for 5 minutes. The glass fiber with 3.5% by volume was then added and mixed for 2 minutes. The mixture was cast in the specimens as mentioned previously and vibrated using a vibrating table. It should be noted that the premixed technique could not prepare GFRC mixture without SBR content due to high water demand of glass fiber filaments leading to very low workability of GFRC mixes.

The GFRC specimens using sprayed technique were prepared by mixing silica sand and white Portland cement for 1 minute. Water and SBR were then added and mixed further for 1 minute. Thereafter, blended silica sand and white Portland cement were added into the mixer and mixed further for 5 minutes. The mixture was transferred to the container of the spraying machine and then continuously pumped through the spray nozzle of the spraying machine. The fiber roll was cut to nominal lengths of between 38 and 50 mm and then continuously fed through the mixture at the spray nozzle. The air pressure valve and pumping speed of the machine were adjusted to control the fiber content in the GFRC mixture within the range of 3-4%. The GFRC mixture was then sprayed into the specimens as mentioned previously and consolidated using a tamping rod and vibrating table.

All of the specimens prepared were covered with plastic sheets. After 24 hours, the specimens were demolded and kept in a moist-cured room which was controlled at a temperature of 23°C and a relative humidity of 95% until the age of specimens were 28 days.

Table 1. Mix proportion of GFRC used in this research

<table>
<tr><th>Latex (%)</th><th>White Portland Cement (kg/m³)</th><th>Water (kg/m³)</th><th>Silica Sand (kg/m³)</th><th>SBR (kg/m³)</th><th>W/C</th><th>Fiber content (%)</th></tr>
<tr><td>0*</td><td rowspan="4">930</td><td rowspan="4">372</td><td rowspan="4">930</td><td>0</td><td rowspan="4">0.45</td><td rowspan="4">3-4</td></tr>
<tr><td>3</td><td>28</td></tr>
<tr><td>6</td><td>56</td></tr>
<tr><td>9</td><td>84</td></tr>
</table>

* Only for GFRC using sprayed technique

2.3 Procedure and testing

Flexural strength test, modified from ASTM C1018 [9], was performed at the testing ages of 1 and 28 days for the prismatic specimens with the dimensions of 350×150×25 mm. The thickness of the specimens was adjusted from ASTM standard to be 25 mm. The specimens were placed on the support of the universal testing machine where the specimens were loaded at their mid-point span. The displacement rate of the test was controlled at 1 mm/min using "Merlin" software of the universal testing machine. The compression loads and displacements of the test were monitored. The flexural strength test of three specimens for each testing age was calculated.

Compressive strength test conforming to ASTM C39 [10] was performed at the testing ages of 1 and 28 days for the cylindrical specimens with the dimension of ø100×200 mm. The specimens were loaded at a constant rate of 5. 3 kilo-Newton per second. The compressive strength test of three specimens for each testing age was calculated.

All of the specimens with the dimension of 300×75×75 mm for the drying shrinkage test were kept in the temperature- and humidity-controlled room which the temperature and relative humidity were controlled at 23°C and 50% respectively. The length changes of three prismatic specimens for each GFRC mix were measured starting from 1 day to 98 days. The shrinkage strains of the specimens were calculated.

3 Result and discussion

3.1 Compressive strength of GFRC using sprayed and premixed techniques

Fig 2 shows the compressive strength of GFRC varying the SBR content and using sprayed and premixed techniques. The compressive strength results of GFRC were found to decrease with increasing the SBR content for both premixed and sprayed techniques. The premixed technique exhibited slightly higher compressive strength than the sprayed technique because entrapped air voids were found to form in the cylindrical specimens using sprayed technique more than in the specimens using the premixed technique.

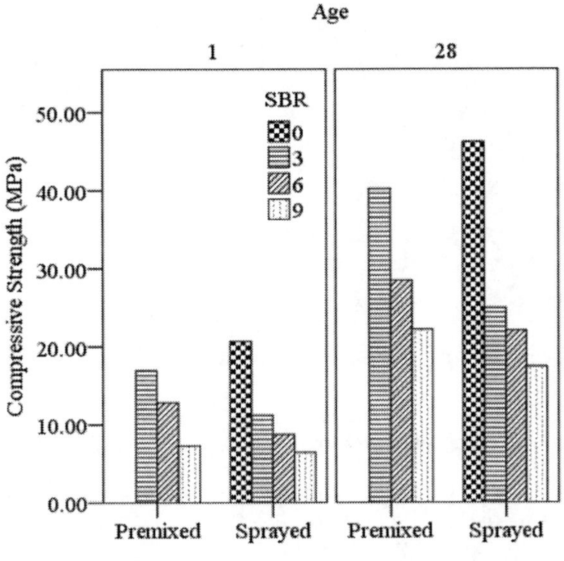

Fig 2. Compressive strength of GFRC using sprayed and premixed techniques

3.2 Flexural strength of GFRC using sprayed and premixed techniques

Fig 3 shows the flexural strength of GFRC varying the SBR content and using sprayed and premixed techniques. The results showed that GFRC specimens containing higher SBR content exhibited 1-day flexural strength lower than those containing lower SBR content. However, the 28-day flexural strength results of GFRC were found to be increased with

 DOI: 10.1051/matecconf/201713803006

increasing SBR content because the polymerization of SBR was developed at the later ages. In addition, the GFRC using the sprayed technique had slightly higher 28-day flexural strength than the corresponding premixed technique because the fiber alignment using sprayed technique was mostly found to be a two-dimensional layer which was different to the premixed technique as shown in Fig 4.

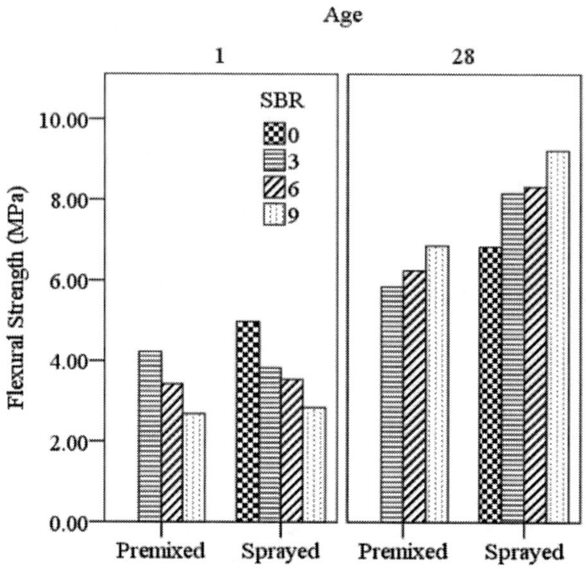

Fig 3. Flexural strength of GFRC using sprayed and premixed techniques

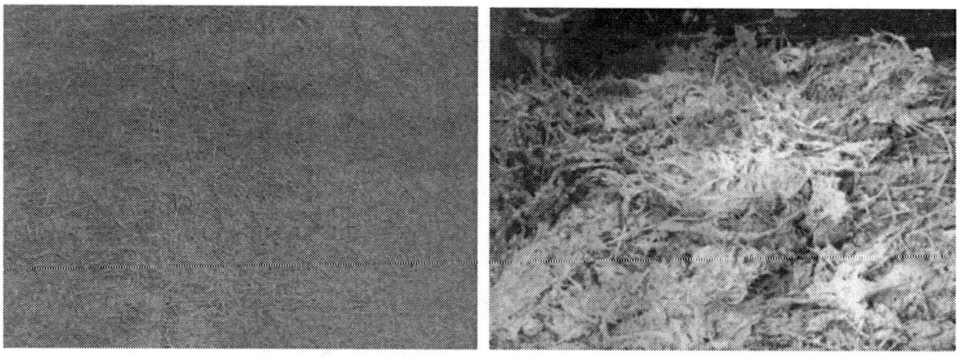

a) b)

Fig 4. Alignment of fiber filaments obtained from a) sprayed technique and b) premixed technique

Fig 5 shows the effect of the premixed and sprayed techniques on flexural and compressive strengths at the testing age of 28 days. The results showed that for a given compressive strength, the sprayed technique exhibited higher flexural strength of GFRC than the premixed technique.

 DOI: 10.1051/matecconf/201713803006

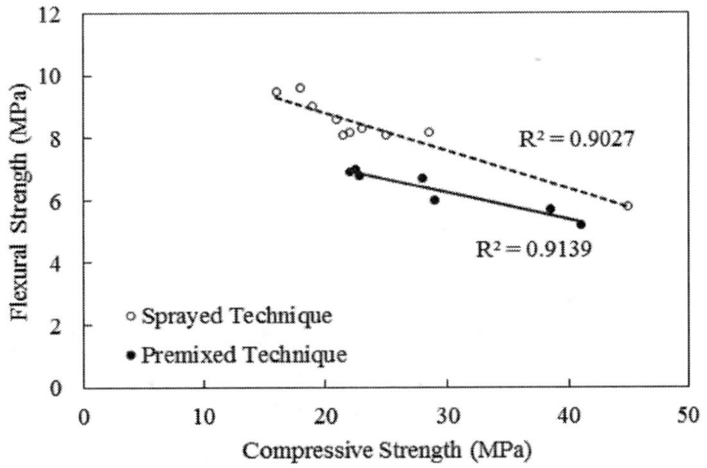

Fig 5. Effect of the premixed and sprayed techniques on flexural and compressive strengths of GFRC

In addition, the toughness value of GFRC using the sprayed technique with 9% SBR content was lower than that of GFRC with 6% SBR content, although the flexural strength of GFRC with 9% SBR content was higher than that of GFCR with 6% SBR content, as shown in Fig 6. The toughness values of GFRC varying with the SBR content and using premixed technique are shown in Fig 7. The toughness values of GFRC were found to increase with increasing SBR content. In addition, GFRC with 9% SBR content using premixed technique was found to exhibit the highest toughness values of the I20 toughness index.

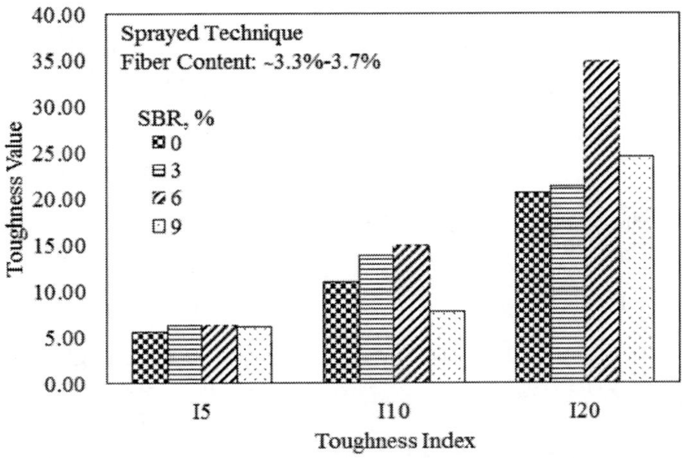

Fig 6. Toughness of GFRC using sprayed techniques and varying SBR content

 DOI: 10.1051/matecconf/201713803006

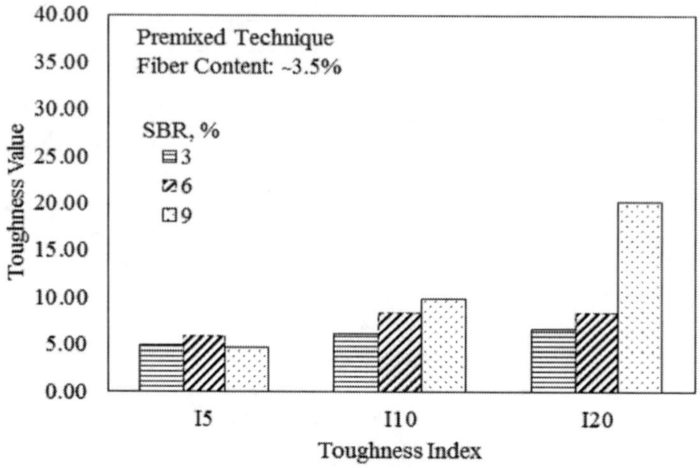

Fig 7. Toughness of GFRC using premixed techniques and varying SBR content

3.3 Effect of polymer content on drying shrinkage of GFRC

Fig 6 shows the drying shrinkage of GFRC varying the SBR contents and using sprayed and premixed techniques. The results showed that the shrinkage strains of GFRC were found to decrease with increases in the SBR content. At the testing age of 98 days, the shrinkage strain of GFRC containing 9% SBR content was about 40% lower than that of GFRC using 0% SBR content. No significant difference in drying shrinkage results between GFRC using premixed and sprayed techniques could be observed.

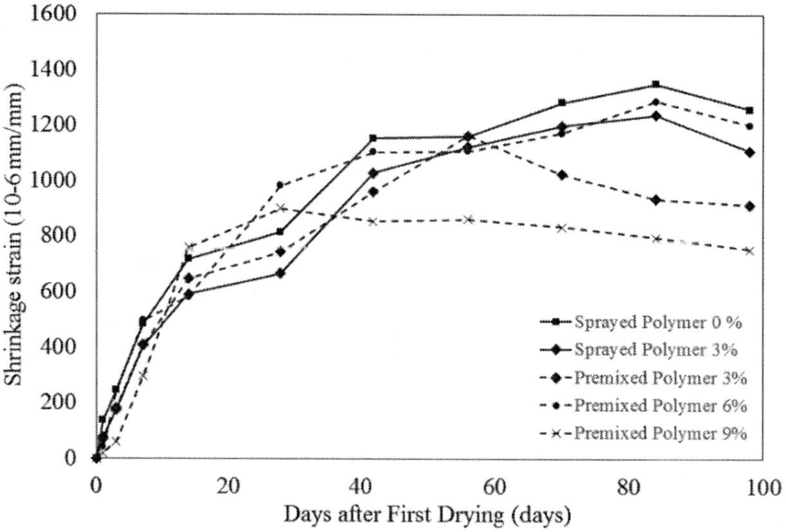

Fig 8. Drying shrinkage of GFRC with and without SBR and applying techniques

4 Conclusions

- The 28-day compressive strength results of GFRC were found to decrease with increases in the SBR content for both premixed and sprayed techniques.
- The 28-day flexural strength results of GFRC were found to increase with increasing the SBR content for both premixed and sprayed techniques.
- The toughness values of GFRC using sprayed technique were higher than corresponding GFRC using premixed technique.
- The shrinkage strains of GFRC were found to decrease with increasing the SBR content.
- The optimum content of SBR latex was in the range of 6% to 9% for GFRC using sprayed technique.

References

1. A. J. Majumdar, R. W. Nurse, *Glass fibre reinforced cement*, Materials Science and Engineering, **15(2)**, 107-127 (1974)
2. D. Hull, *an Introduction to Composite Materials*, Cambridge Solid State Science Series (1981)
3. B. A. Proctor, B. Yale, A. R. Ubbelohde, G. Manfre, *Glass fibres for cement reinforcement*, **294** (1980)
4. A. J. Majumdar, *Properties of GFRC. Fibrous Concrete* (1980)
5. L. J. Larner, K. Speakman, A. J. Majumdar, *Chemical Interactions between Glass Fibers and Cement*, Journal of NonCrystalline Solids, **20**, 43-74 (1976)
6. R. Wang, P. M. Wang, X. G. Li, *Physical and mechanical properties of styrene–butadiene rubber emulsion modified cement mortars*, Cement and Concrete Research, **35(5)**, 900-906 (2005)
7. A. L. Qureshi, A. Ahmed, *an Investigation on Strength Properties of Glass Fiber Reinforced Concrete*, **2(4)**, 2567-2572 (2013)
8. A. Bentur, & S. Mindess, *Fibre Reinforced Cementitious Composites*, **2**, 289-290 (2006)
9. ASTM, *ASTM C1018 Standard Test Method for Flexural Toughness and First-Crack Strength of Fiber-Reinforced Concrete (Using Beam With Third-Point Loading)*, 100 Barr Harbor Drive, PO Box C700, West Conshohocken, PA, United States (2007)
10. ASTM, *ASTM C39 Standard Test Method for Compressive Strength of Cylindrical Concrete Specimens.*, ed. 100 Barr Harbor Drive, PO Box C700, West Conshohocken, PA 19428-2959, United States (2007)

Fiber Orientation Effect on Flexural Response of UHPFRC

Doo-Yeol Yoo[1], *Min Jae* Kim[1], *Soonho* Kim[1], and *Young-Soo* Yoon[2,*]

[1]Department of Architectural Engineering, Hanyang University, Seoul, South Korea
[2]School of Civil, Environmental and Architectural Engineering, Korea University, Seoul, South Korea

Abstract. This study aims to examine the fiber orientation effect on flexural behavior of ultra-high-performance fiber-reinforced concrete (UHPFRC) with various casting methods and loading rates. For this, two different placement methods were adopted for making UHPFRC beams along with two different loading rates (i.e., quasi-static and impact). Test results indicated that the flexural performance of UHPFRC beams at both quasi-static and impact loads was significantly influenced by the placement method, leading to dissimilar fiber orientation. Better fiber orientation, denoting that more fibers are aligned in the direction of tensile load, exhibited better post-cracking flexural behaviors in terms of flexural strength, deflection capacity, and toughness, compared to poor fiber orientation. Strength enhancement in UHPFRC beams was obtained by increasing the loading rates, and their deflection-hardening behaviors were successfully observed.

1 Introduction

Concrete has been considered as one of the most effective and economical construction materials for a long time. However, terrorist attacks against civilians frequently occur in recent years, and particularly, the normal concrete, which is widely used as a construction material worldwide, is known to be vulnerable under extreme loading conditions, such as impacts and blasts, owing to its brittle nature under tension. To overcome this drawback, an ultra-high-performance fiber-reinforced concrete (UHPFRC), which was developed by French engineers in the mid-1990's, is currently considered as the material applicable for structures subjected to those extreme loads, due to its excellent mechanical strength, energy absorption capacity, durability, and fatigue resistance [1].

Accordingly, in this study, to verify the effectiveness of using UHPFRC under extreme loading conditions, the impact resistance of UHPFRC beams was investigated using a drop weight impact test machine. In addition, in order to compare the impact and static behaviors, quasi-static flexural tests were carried out according to the ASTM standard. For evaluating the fiber orientation effect on the quasi-static and impact behaviors of UHPFRC beams, two different casting methods were considered, and an image analysis was carried out to quantitatively estimate the effects of fiber orientation and number of fibers in unit area on the flexural behaviors.

2 Experimental Program

In this study, Type 1 Portland cement, silica fume, silica flour, and silica sand were used as dry components. Coarse aggregate was excluded from the mixture to achive excellent tensile or flexural behaviors. The chemical compositions and physical properties of the used cementitious materials are given in a previous study [1], and the detailed mix proportions are given in Table 1. High-strength short straight steel fibers with a diameter of 0.2 mm and a length of 13 mm were incoporated into the mixture by 2 vol%, similar to the UHPFRC mixture commercially available in North America. A water-to-binder (W/B) ratio of 0.2 was applied, and 2% superplasticizer was added to improve the flowability.

Table 1. Mixture proportions.

W/B (%)	Unit weight (kg/m^3)					
	Water	Cement	Silica fume	Silica flour	Silica sand	SP (%)
20	160.3	788.5	197.1	236.6	867.4	2.0

[Note] W/B = water-to-binder ratio and SP = superplasticizer

To investigate the flexural behaviors of UHPFRC beams with various fiber orientations and loading rates, prismatic beams with a cross section of 100 × 100 mm^2 and a length of 400 mm were fabricated. This satisfied the requirements of ASTM C1609 [2] for width and depth sizes of prismatic beams according to the size of fibers. Two different placement methods were adopted. First, concrete was cast at the end of the mold and allowing it to flow, which is called Case_1, and a large amount of concrete was cast at both ends and the center of the mold, which is called Case_2. The detailed casting methods can be found in a previous paper [3]. In order to accelerate strength development of UHPFRC, steam curing with heat (90 °C) was applied for 2 days after initial 24-hour curing at room temperature.

For the quasi-static loading tests, a four-point load was applied according to the ASTM C1609 [2]. A universal testing machine with a maximum capacity of 25 ton was used, and a loading rate of 0.4 mm/min was applied. The clear span length used was 300 mm. For measuring pure mid-span deflection of the beams, a steel plate with two linear variable differential transformers (LVDTs) was installed at the middle height of the beams. The flexural load was measured from the load cell affixed to the cross head.

A drop-weight impact test machine was also used for evaluating impact resistance of UHPFRC beams. The beam size and clear span length were identical to those used in static flexural tests. An 82-kg hammer was dropped from a height of 600 mm, and thus, a potential energy of 480 Jules was provided. A center point load was applied using a blade load with a striking knife-edge end. The impact load was measured from the strain gauges mounted on the hammer. Since the impact load has an inertial load of the beams, two load cells were installed at supports in order to measure the actual flexural stress applied to the beams. The beams reacted in a very short time under impact loads, so that an accelerometer with a capacity of ± 5000 g was applied to measure the mid-span deflection, which was calculated by double integration of the acceleration measured with time. The data were collected by using a dynamic data logger with a rate of 100 kHz.

3 Experimental Results and Discussion

3.1 Fiber orientation

 DOI: 10.1051/matecconf/201713803007

Fig, 1 shows the pictures of cross section and comparative probability density functions (PDFs) of fiber orientation distribution for the Case_1 and Case_2 beams. The Case_1 beams obviously showed a more left-skewed distribution than the Case_2 beams. The more left-skewed distribution denotes better fiber alignment along the beam length. This is consistent with the findings from the pictures of cross section in the beams that better fiber alignment was observed in the Case_1 beams, compared to their counterpart (Case_2). Based on the image analysis, the number of fibers and coefficient of fiber orientation were calculated. The Case_1 beams exhibited higher number of fibers per unit area (47 ea/cm^2) and higher coefficient of fiber orientation (0.660), as compared to those (37 ea/cm^2 and 0.643) of the Case_2 beams.

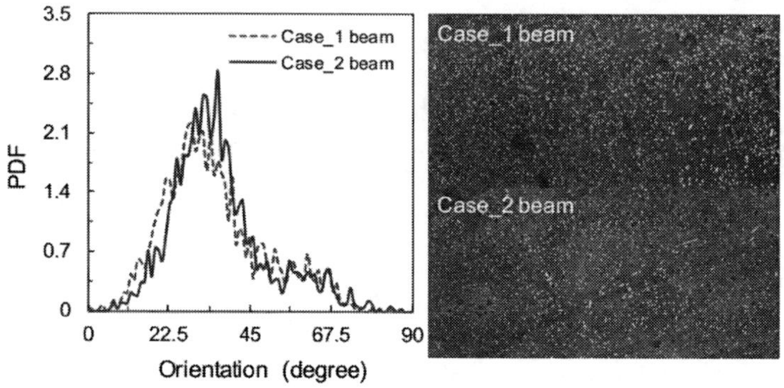

Fig. 1. Fiber orientation distribution

3.2 Flexural behaviors of UHPFRC beams under static and impact loads

The average static flexural stress and deflection curves are shown in Fig. 2. Three prismatic beams were used to obtain the average curve. The flexural stress was calculated by $f = PL/bh^2$, where, P is the applied load, L is the clear span length, b is the width of beam, and h is the height of beam. All of the beams showed a deflection-hardening behavior providing a higher load carrying capacity after the first crack formation. The Case_1 beams provided higher flexural strength and deflection capacity compared to that of Case_2. For instance, the flexural strength of the Case_1 beams was obtained as 36.5 MPa, which is about 37% higher than that of the Case_2 beams. This is caused by the fact that the Case_1 beams had higher numbers of fibers at crack surfaces, due to the better fiber alignment along the beam length, compared to their counterparts. However, the first cracking flexural strength was insignificantly affected by the casting method, because it is more strongly related to the matrix cracking strength rather than the fiber bridging. Due to the better fiber bridging capacity of the Case_1 beams, the higher energy absorption capacity (toughness) was obtained. For example, the toughness of the Case_1 beams at the deflection point of 2 mm, which is $L/150$, was found to be 209 kNmm, approximately 36% higher than that of the Case_2 beams.

The average flexural stress and deflection curves under impact loads are shown in Fig. 3. In this case, three prismatic specimens were also adopted to draw the average curves. Based on a concept of impact behavior of fiber-reinforced concrete (FRC) under flexure [4], the initial peak is related to the first-cracking point in the cement matrix, and the second peak is related to the fiber bridging. It was obvious that the Case_1 beams with the better fiber orientation provided higher first- and post-cracking flexural strengths, compared to those of the Case_2 beams (Fig. 3). In particular, the increased post-cracking flexural strength of the Case_1 beams under impact loading was attributed to the fact that the higher number of fibers more effectively limited the propagation and widening of cracks as compared to their counterparts. This is similar results with the case of quasi-static loading conditions. The first-cracking strength was also improved with the better fiber orientation. This is consistent with the findings from a previous study [5]. In their study, the first-cracking strength of UHPFRC under impact loads increased with the reinforcing index, $V_f(L_f/d_f)$, where V_f is the fiber volume fraction, L_f is the fiber length, and d_f is the fiber diameter. The increased number of fibers in unit area for the Case_1 beams led to an increase in the fiber volume fraction in the reinforcing index. Therefore, a higher first-cracking flexural strength was obtained in the Case_1 beams, as compared with the Case_2 beams.

Due to the better impact resistance of the Case_1 beams, a rebound with decreases in both of the flexural stress and deflection was observed. This means that the Case_1 beams were not completely broken under the impact loads with a potential energy of 480 J. On the contrary, the Case_2 beams were completely failed under the impacts with the identical potential energy of 480 J. Therefore, no rebound was observed.

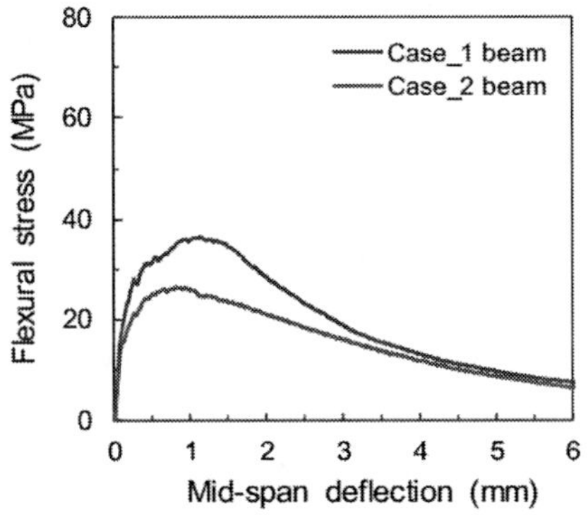

Fig. 2. Quasi-static flexural behaviors

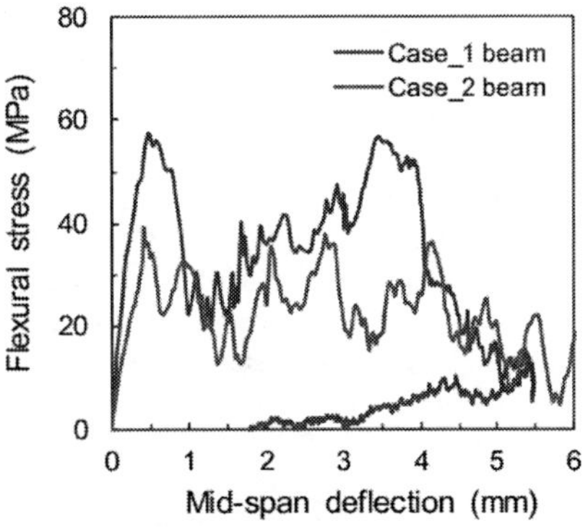

Fig. 3. Impact flexural behaviors

Conclusions

The flexural behaviors of UHPFRC beams under both static and impact loads were investigated with various casting methods. The implications of fiber orientation on the flexural behaviors were also examined based on the image analysis. From the discussions above, the following conclusions were drawn. Better fiber orientation in the longitudinal direction of the beams and a higher number of fibers per unit area were obtained when the concrete was cast at the end of the mold, as compared to pouring a large amount of concrete at both ends and center. Higher static flexural strength, deflection capacity, and toughness were obtained in the beams with better fiber orientation (Case_1). In addition, higher flexural strength and energy absorption capacity were obtained in the beams with better fiber orientation under impact loads. Therefore, it was concluded that providing good fiber alignment in the tensile load direction is effective to improve the flexural performance of UHPFRC under both static and impact loads.

References

1. D.Y. Yoo, H.O. Shin, J.M. Yang, Y.S. Yoon, Compos. Part B-Eng. **58**, 122–133 (2014)
2. ASTM C1609/C1609M, ASTM International, West Conshohocken, PA, 1–9 (2012)
3. D.Y. Yoo, N. Banthia, S.T. Kang, Y.S. Yoon, Compos. Struct, **157**, 62–70 (2016)
4. D.Y. Yoo, Y.S. Yoon, N. Banthia, Cem. Concr. Compos. **64**, 84–92 (2015)
5. S. Pyo, K. Wille, S. El-Tawil, A.E. Naaman, Cem. Concr. Compos. **56**, 15–24 (2015)

MATEC Web of Conferences **138**, 03008 (2017) DOI: 10.1051/matecconf/201713803008
EACEF 2017

Thirty Years Researches on Development for Sustainable Concrete Technology

Jongsung Sim[1] , *Minkwan* Ju[2,*], and *Kihong* Lee[1]

[1]Hanyang University, Department of Civil and Environmental Engineering, 15588 Ansan, Republic of Korea

[2]Yonsei University, Department of Civil and Environmental Engineering, 03722 Seoul, Republic of Korea

Abstract. The enormous amount of concrete production has a serious impact on energy, resources, environment and ecosystem. Therefore, the issue of development of sustainable concrete technology with little impact on the environment is becoming a major issue. In this paper, researches related with sustainable development of concrete are presented in last three decades. FRP has high corrosion resistance and lightweight, thus it can be potential solution for sustainable development of concrete structures as strengthening material or reinforcement instead of steel. Researches and techniques are presented on performance of concrete beam with FRP rebar and enhancing performance of existing concrete structure using FRP strengthening methods. The application of recycled concrete aggregate (RCA) has sometimes been limited in the practice and remained in the low-valued purposes only such as road base materials. In past 10 years, a great improvement in the recycling technique to produce RCA of which quality is close to natural aggregate, hence the applicability and evaluation of RCA are presented in this paper. This paper includes experimental studies for application of waste glass which could decrease CO_2 emission from cement producing. The achievements of these studies are presented in this paper to contribute for sustainable development of concrete infrastructures.

1 INTRODUCTION

Sustainable development means the organizing principle for sustaining finite resources necessary to provide for the needs to the next generations on the earth. It is a process that concretely plans a desirable future state for human societies in which living conditions and resource-use continue to meet human needs without undermining the "integrity, stability and beauty" of natural biotic systems [1]. The concept of sustainable development is strongly originated by the 1987 Brundtland Report. In 1987, the United Nations World Commission on Environment and Development released the report 'Our Common Future', commonly called the Brundtland Report. This report included what is now one of the most widely recognized definitions of sustainable development. "Sustainable development is development that meets the needs of the present without compromising the ability of future

* Corresponding author : j_dean21@naver.com

DOI: 10.1051/matecconf/201713803008

generations to meet their own needs. It contains within it two key concepts: The concept of 'needs', in particular, the essential needs of the world's poor, to which overriding priority should be given; and The idea of limitations imposed by the state of technology and social organization on the environment's ability to meet present and future needs." The significant three pillars of sustainability are shown in Figure 1.

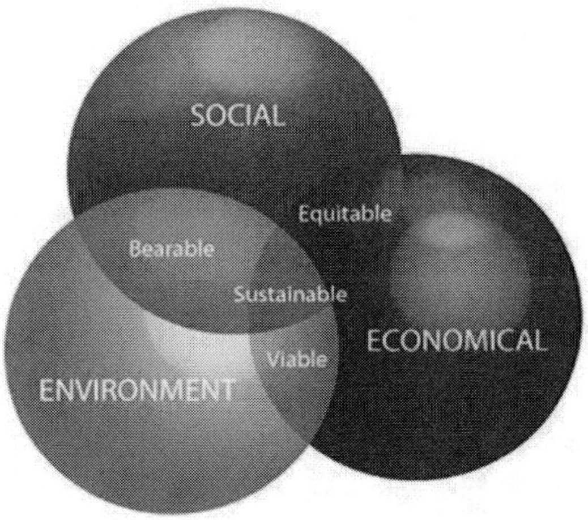

Fig. 1. Significant three pillars of sustainability

2 STRUCTURAL ASSESSMENT RESEARCHES

Since 1987, Prof. Sim as a faculty of Hanyang University, has focused on structural assessment of deteriorated concrete structures [2-4]. Selected research projects are Cheong-Gye bridge, Haeng-Ju bridge, industrial building structures, LNG tank, railway structures, waste water facility etc. In 1995, the collapse of the Sung-Su Bridge (as shown in Figure 2) brought the great change of the paradigm of the structural assessment industry in Korea. Accordingly, the Korea Infrastructure Safety and Technology in government branch was established. As accordance with the advanced structural assessment, Prof. Sim has researched about in-field loading test to the deteriorated bridges (as shown in Figure 3) for evaluating load carrying capacity and systemized the fundamental concept of the artificial decision making technology for the expert system of structural assessment [5].

Fig. 2. Collapse of the Sung Su Bridge in Korea

Fig. 3. In-field loading test for the deteriorated PSC I typed girder bridge [6]

DOI: 10.1051/matecconf/201713803008

3 REPAIR AND REHABILITATION RESEARCH

3.1 Strengthening materials

Deterioration of concrete structure causes degradation of structural performance and durability. Deteriorated concrete structure requires strengthening, rehabilitation, reconstruction, or demolition to guarantee the safety and serviceability of structure. Accordingly, an appropriate strengthening should be strongly employed by using advanced materials such as FRP composite materials. There are various types of fiber reinforced composite materials such as plate, sheet and rod type. Materials made of carbon, glass, aramid, basalt, bamboo as shown in Figure 4.

Fig. 4. Various types of FRP materials for bars and sheets [7-11]

3.2 Strengthening technique of existing concrete structures using FRP materials

Previous strengthening technique using the steel plate causes shortcoming due to its heavy weight and easily corrosive condition against the external environmental attack. Fiber reinforced polymer (FRP) ensures the advantages of high tensile strength, corrosion resistance, and lightweight.

Fig. 5. Externally bonded FRP systems [12, 13]

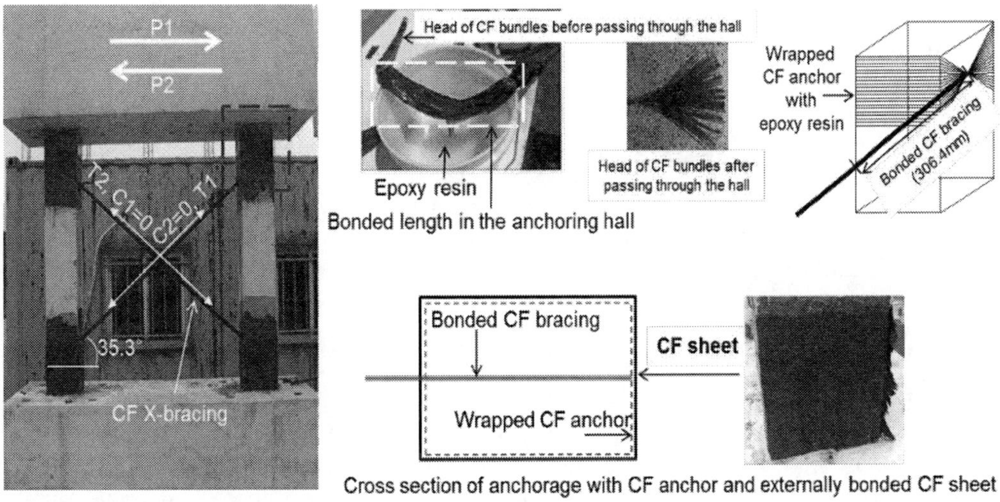

Fig. 6. CF X bracing strengthening system for seismic [17]

The externally bonded reinforcement (EBR) is typical strengthening method using FRP as shown in Figure 5. There are a number of researches about externally strengthening of reinforced concrete structures using FRP for flexural and shear, punching shear, extension of fatigue life, strengthening design methodology [14-16]. Recently, carbon fiber X-bracing strengthening system for seismic as shown in Figure 6 was developed and the structural efficiency was experimentally proved under the earthquake loading condition [17]. Above tremendous experimental results were contributed to establish the design and construction specification using FRP in Korea.

4 INNOVATIVE RESEARCH ON LOW EMMISION OF CARBON DIAOXIDE FOR SUSTAINABLE CONSTRUCTION

4.1 GFRP rebar with surface ribs made of milled glass fibers

Deterioration in reinforcing steel bar in concrete members can lead to other serious structural damage, eventually causing structural failure. The one of the advanced way to prevent corrosion is to use a reinforcing material that does not corrode.

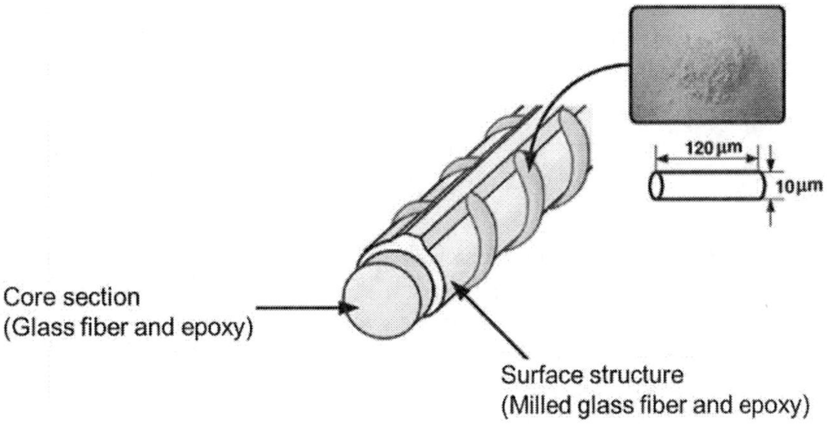

Fig. 7. Newly developed GFRP bar with ribs made of milled glass fiber

DOI: 10.1051/matecconf/201713803008

Many studies have focused on developing new reinforcing materials, such as various types of fibre reinforced polymers (FRPs) as already introduced in Figure 4, to solve the fundamental problems of steel rebar [18]. The milled glass fiber is one that is hammer-milled strands into very small lengths and are usually used to improve thermal properties and surface ribs in the injection molding processes. The surface structure analogous to the ribs of ordinary steel rebar was considered as shown in Figure 7. The spacing and height of a rib were finally determined to be 6 mm and 1.3 mm, respectively [19]. Structural capacity such as strength and serviceability for the GFRP bar reinforced concrete beam was experimentally investigated under flexure [20].

4.2 Smart FRP bar embedded fiber bragg grating (FBG) sensor

This study developed a hybrid fiber-reinforced polymer (FRP) rod for the purpose of new reinforcements with the function of fiber optic sensor (FOS) smart monitoring in concrete structures. FOS provides more stable measurement without the electromagnetic interference (EMI). Besides, the smart FRP bar can provide high tensile strength and better durable property due to non-corrosion reinforcement to the concrete structures. A significant research reported that the hybrid FRP rod could be used for a reinforcement and a smart monitoring in concrete structures [21]. The fabrication process is briefly introduced in Figure 8.

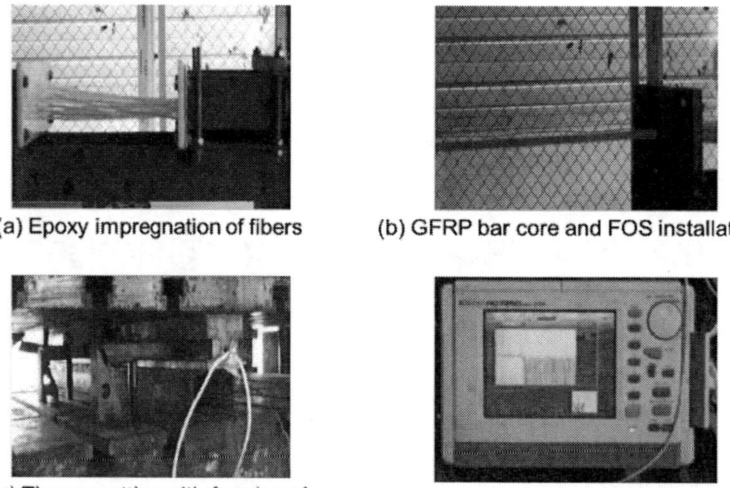

(a) Epoxy impregnation of fibers

(b) GFRP bar core and FOS installation

(c) Thermosetting with forming of fiber ribs

(d) Checking the stability of FOS sensing

Fig. 8. Fabrication process for FOS embedded FRP bar

4.3 Recycled aggregate concrete

Amount of construction waste has been largely increased from the demolition of old structures and re-construction of buildings for improvement of the living standard. Expansion of society warns to reserve of natural aggregate for construction become depleted rapidly. Hence, social and environmental pressures on the construction wastes industry is greatly focusing on ń the recycling of the construction waste. In the past the recycling technique was not satisfactory to produce a sufficient quality RCA for the aggregate. Recently, there has been a great improvement in the recycling technique to produce RCA of which quality is close to natural aggregate as well as economic efficiency [22-24]. The fabrication process is shown in Figure 9.

Fig. 9. Complete production process for recycled concrete aggregate [22]

4.4 Concrete with waste glass powder

Supplementary cementing materials (SCMs) such as fly ashes, slag cement (ground, granulated blast furnace slag), and silica fume, are often added to concrete to make concrete mixtures for more economical, reduce permeability, increase strength, or influence other concrete properties [25-28]. Demands and application of SCMs would increase because of decreasing of CO2 emission from cement producing as shown in Figure 10.

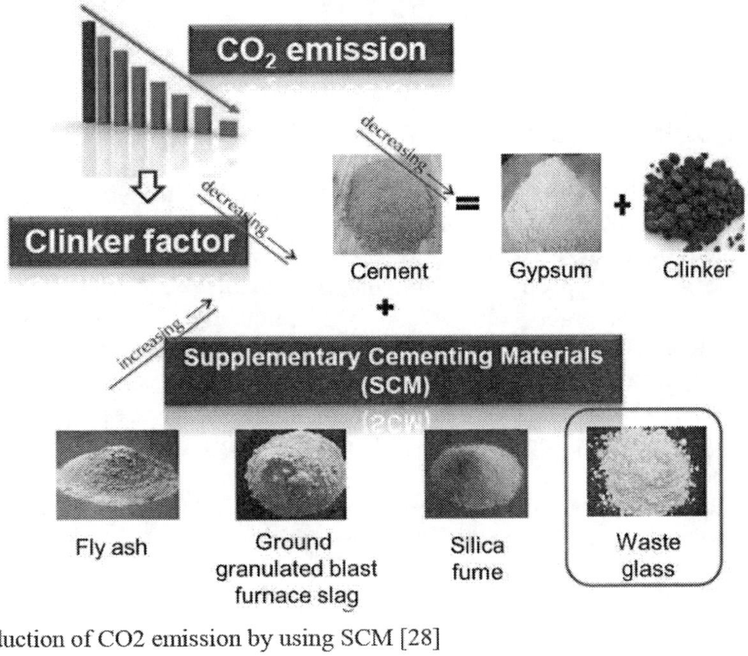

Fig. 10. Reduction of CO2 emission by using SCM [28]

4.5 Precast scour protection block

In order to apply the sustainable concrete technology developed in concrete research structural lavatory (CSRL) to the construction society well, the design and construction specification is required in advance. The publication of the specification for the sustainable concrete technology takes much time, hence, sometimes challengeable approach is needed. The precast product qualified by the experimental and analytical results may be available to rapidly use in the appropriate field of construction. A good practice of this is introduced below. Precast scour protection block made of concrete is an important construction member for the structural safety of footing structures due to consistently dynamic flow around footing soil. One of the significant approach was done to develop high durable and cost effective scour protection block using FRP bar and recycled aggregate concrete as shown in in Figure 11 [29, 30].

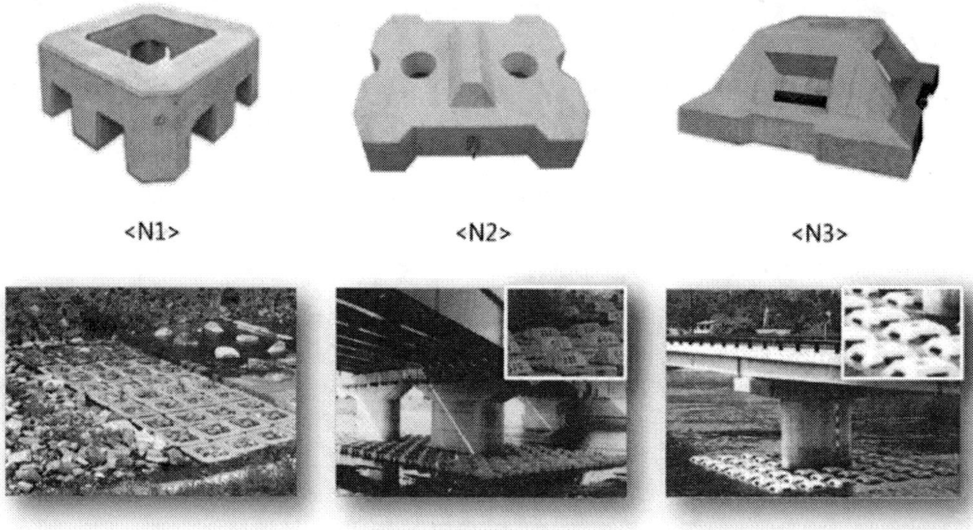

Fig. 11. Prototypes of precast scour protection block made of recycled aggregate concrete [29]

5 DESIGN SPECIFICATIONS

In past 30 years, Prof. Sim have greatly devoted to employ and advance the sustainable concrete technology introduced above in Korea. These technologies are the field of engineering, therefore, they are deserved to be widely used in construction society. For this, first of all, the standard specification or guideline should be developed, reviewed and published. Prof. Sim had largely contributed to publish many of specifications in the field of construction of Korea. Some of representative works are as follows. In 2017, he studied to register the concrete structure design specification of Korea to the international organization for standardization (ISO) [31]. As an editor in chief, he revised the railroad design specification for concrete bridge session [32]. He has participated to revise the highway design specification of Korea [33]. In 2008, design and construction specification of FRP bar reinforced concrete structures was published in order to spread out the fundamental approach of FRP bar to the structural reinforcement in concrete structures [34].

DOI: 10.1051/matecconf/201713803008

6 CONCLUDING REMARKS

In this paper, 30 year research on development for sustainable concrete technology is briefly summarized. Prof. Sim's research has been deeply related with developing concrete structures to preserve the Earth's environment. In summary, there are structural assessment and strengthening method of deteriorated concrete structure, the new conceptual concrete members such as the FRP reinforced or the recycled concrete structures. In last 30 years, the number of published international journals are 38, domestic journals are 25 and scientific conference journals are 314, respectively. Above outputs must truly be correlated with the sustainable concrete technology in 21C.

This research was supported by Basic Science Research Program through the National Research Foundation of Korea (NRF) funded by the Ministry of Education (grant number 2016R1D1A1B03934809).

References

1. World Commission on Environment and Development, Our Common Future, United Nations World Commission on Environment and Development (WCED) (1987)
2. J. Sim and D. Moon, A Proposal for Structural Performance Evaluation of Old Bridge Using Concrete weight Blocks Structural Engineers World Congress (SEWC) 2002, Yokohama, Japan, October, Paper No. T9-1-b-3 (2002)
3. J. Sim and M. Cho, Nonlinear Analysis Technique of R/C Frame Typed Structures Proceedings, Vol. Ⅰ. 4th EASEC Int'l Conference, Seoul, Korea, September, pp. 547-552 (1993)
4. P. Soroushian and J. Sim, Strain Rate Effects on the Dynamic Response of Reinforced Concrete Elements to Impulsive Load Proceedings, The Inter-national Symposium on the Interaction of Dynamic Loads with Protective Structures, Mannheim, Federal Republic of Germany, March, pp. 538-556 (1987)
5. J. Sim, Development of Expert System of Structural Maintenance of Reinforced Concrete Structures, National Research Foundation (1991)
6. M. Ju, D. Moon, G. Kim, J. Sim, Evaluating Rating Factor for Prestressed Concrete Girder Bridges by Nonlinear Finite Element Analysis, Structure and Infrastructure Engineering: Maintenance, Management, Life-Cycle Design and Performance, **11**, 9, pp. 1250-1262 (2014)
7. http://www.toray.com/
8. http://aslanfrp.com/
9. http://www.dostkimya.com/
10. http://www.technobasalt.com/
11. http://www.litrax.com/
12. http://www.structuraltechnologies.com/
13. http://www.luckett-farley.com/
14. J. Sim, G. Kim, C. Park, M. Ju, Shear Strengthening Effects with Varying Types of FRP Materials and Strengthening Methods, ACI SP-230, pp. 1665-1680 (2005)
15. J. Sim and H. Oh, Structural Improvement of Strengthened Deck Panels with Externally Bonded Plates, Cement and Concrete Research, **35**, 7, pp. 1420-1429 (2005)

16. H. Oh, J. Sim, C. Meyer, Fatigue Life of Damaged Bridge Deck Panels Strengthened with Carbon Fiber Sheet, Journal of the American Concrete Institute, **102**, 1, pp. 85-92 (2005)
17. M. Ju, K.S. Lee, J. Sim, H. Kwon, Non-compression X-bracing System using CF Anchors for Seismic Strengthening of RC Structures, Magazine of Concrete Research, **66**, 4, pp. 159-174 (2014)
18. D. Moon, J. Sim, H. Oh, Experimental Characterization of the Bond Performance of a New Type of Glass Fibre-reinforced Polymer Rebar for Application in Concrete Structures, Proceedings of the Institution of Mechanical Engineers Part L, Journal of Materials Design and Applications, pp.113-119 (2007)
19. D. Moon, J. Sim, H. Oh, B. Benmokrane, An Exploratory Study of GFRP Rebar with Ribs Containing Milled Glass Fibers, Composites: Part B, **39**, pp. 882-890 (2008)
20. D. Moon, H. Oh, M. Ju, J. Sim, Experiment and Analysis Study of Beams reinforced with GFRP bar, IABSE Symposium Budapest 2006, Budapest, Hungary, September, pp. 280-281 (2006)
21. J. Sim, D. Moon, H. Oh, C. Park, S. Park, Hybrid FRP Rod For Reinforcement and Smart-Monitoring in Concrete Structure Proceedings of the International Symposium on Bond Behavior of FRP in Structures, Hong Kong, China, December, pp. 401-408 (2005)
22. J. Sim and C. Park, Compressive Strength and Resistance to Chloride Ion Penetration and Carbonation of Recycled Aggregate Concrete with Varying Amount of Fly Ash and Fine Recycled Aggregate, Waste Management, **31**, pp. 2352-2360 (2011)
23. H. Lee, Y. Kim, M. Shahid, C. Park, J. Sim, A Study on the Compressive Strength of Recycled Aggregate Concrete Based on Bonded Mortar Content, The 5th International Conference of Asian Concrete Federation (ACF 2012), Pattaya, Thailand, 24-26 October (2012)
24. C. Park, J. Sim, T. Kang, Y. Kim, H. Lee, W. Choi, Recycling Application of Power Plant Coal Ash for High Performance Shotcrete Material, The 4th International Conference of Asian Concrete Federation (ACF 2010), Taipei, Taiwan, 28 November - 1 December (2010)
25. V.G. Papadakis and S. Tsimas, Supplementary Cementing Materials in Concrete Part I: Efficiency and Design, Cement and Concrete Research, **32**, pp. 1525–1532 (2002)
26. H. Lee, J. Jung, H. Oh, G. Zi, J.W. Shim, J. Sim, Effect of Waste Glass Powder on Concrete Pavement, The 6th International Conference of Asian Concrete Federation (ACF 2014), Seoul, Korea, 21-24 September. (2014)
27. H. Lee, H. Jung, J.W. Shim, H. Oh, J. Sim, An Experimental Study on Durability Characteristics of High Performance Concrete Using Waste Glass, The 4th International fib Congress 2014, Mumbai, India, 10-14 February (2014)
28. J. Sim and K. Lee, Sustainable Concrete Technology, The 5th International Conference of Euro Asia Civil Engineering Forum (EACEF-5), Surabaya, Indonesia, 15-18 September, 2015.
29. J. Sim, Y. Kim, C. Park, Mechanical Properties of Precast Concrete Block Reinforced with GFRP rebars, Engaging Young Engineers and Scientists in FRP Research, Australia-Korea Foundation (AKF), Seoul, Korea, (2013)
30. Y. Kim, J. Sim, C. Park, A. Hanif, Performance Evaluation of Precast Concrete Block Reinforced with GFRP Rebars for Erosion Control, Keynote Lecture, Asia-Pacific Conference on FRP in Structure (APFIS 2012), Hokkaido, Japan, 2-4 February (2012)

31. International Standardization of Concrete Structure Design Specification of Korea, Korea Concrete Institute, 2017.
32. Revision of Railroad Design Specification (Concrete bridge session), Korea Society of Civil Engineer, 2011.
33. Revision of Highway Design Specification, Ministry of Land, Infrastructure and Transport, 2010.
34. Development of Design and Construction Specification of FRP bar reinforced concrete structures, Korea Concrete Institute, 2008.

Benefits Of Using Fiber on Impact Resistance of FRC Slabs

Jin-Young Lee[1], *Tian-Feng* Yuan[1], *Doo-Yeol* Yoo[2] , and *Young-Soo* Yoon[1*]

[1]School of Civil, Environmental and Architectural Engineering, Korea University, Seoul, Korea
[2]Department of Architectural Engineering, Hanyang University, Seoul, Korea

Abstract. The recent increase in terrorist attacks and natural disasters has led to increased requirements for concrete structures to be impact resistant. Concrete normally has excellent impact resistance to such extreme loads in comparison with other construction materials. Nevertheless, existing concrete structures designed without consideration of the impact or blast load with high strain rate are endangered by those unexpected extreme loads. Therefore, the impact resistances of two-way concrete slabs were studied by conducting the multiple impact test. Various fiber reinforcements (Hooked-steel fiber, straight-steel fiber) were considered as variables. In the impact test, the impact load was applied to each specimen by dropping weight on the specimen from a certain height. The applied impact load level was adjusted by changing the height of the weight. Specimens were used for impact loading test, each measuring 1,600mm long, 1,600mm wide, and 105mm thick. All slabs were doubly reinforced (top and bottom layer) and each layer in a specimen had the same reinforcement layout. The performances of the specimens under the impact load were evaluated by comparing their reaction force, deflection, and surface damage. The impact test was stopped when the measured reaction force started to decrease. Based on the impact test, it was observed that the resistance against impact load can be improved by using fiber reinforced concrete.

1 Introduction

The recent increase in terrorist attacks and natural disasters has led to increased requirements for concrete structures to be impact resistant. Concrete normally has excellent impact resistance to such extreme loads in comparison with other construction materials. Nevertheless, existing concrete structures designed without consideration of the impact or blast load with high strain rate are endangered by those unexpected extreme loads. Therefore, the impact resistances of two-way concrete slabs were studied by conducting the multiple impact test. The objective of this study is to evaluate impact damage of structural members and to compare the impact resistance according to the material properties. Before the structural tests, compressive strength and tensile strength were measured through basic material tests. In this study, dynamic impact resistance was evaluated by drop-weight

* Corresponding author: ysyoon@korea.ac.kr

DOI: 10.1051/matecconf/201713803009

impact tests on two-way RC slabs, and the impact resistance was compared by steel fiber contents and concrete strength. In this purpose, two-way RC slabs were tested under drop-weight impact loading condition [1-6].

2 Experimental program

In this experimental study, impact tests were carried out on RC slabs with various strength from 25MPa to 180MPa. The types of fibers were also considered as test variables. Therefore, there are four variables used in this study.

- •NC - Normal Concrete
- •SFRC - Steel Fiber Reinforce Concrete
- •UHPC - Ultar High Performance Concrete
- •HPFRCC. - High Performance Fiber Reinforced Cementitious Composites

2.1 Material properties and mix design

2.1.1 Steel fiber

In Fig. 1 two types of steel fiber were shown. The straight steel fiber was used in HPFRCC, and the hook type steel fiber was used in SFRC in this study. The material properties of fibers were shown in Table 1.

Straight type　　Hooked-end type

Fig. 1 Steel Fiber

Table 1. Material properties of steel fiber reinforcement

Type	Length (mm)	Diameter (mm)	Aspect ratio
Straight	19.5	0.2	95
Hook	35	0.55	65

 DOI: 10.1051/matecconf/201713803009

2.1.2 Mix design

The main variables in this mix design table are w/b and steel fiber content and type. The details of mixture were shown in Table 2.

Table 2. Mix. proportion

Nomen-clature	W/B (%)	Steel Fiber (kg/m³)	Steel Fiber (%)	Unit Weighty (kg/m³)				
				W	C	G	S	SF
HPFRCC 180-F2.0	13.4	156.0	2.0	231.0	734.9	0.0	808.4	73.5
UHPC 180-F0.0	13.4	0.0	0.0	231.0	734.9	0.0	808.4	73.5
SFRC 60MPa-F2.0	35.9	156.0	2.0	163.0	454.0	992.0	707.0	0.0
NC 60MPa-F0.0	35.9	0.0	0.0	163.0	454.0	992.0	707.0	0.0

2.2 Test specimens

As shown in Fig. 2, two-way slabs of 1600 x 1600 x 105mm were tested in this study. there were four variables for this experimental program. Various types of concrete mix were used in this study shown in Table 2. As shown in Fig. 2, SD400-D10 reinforcing bars were used with spacing of 240mm.

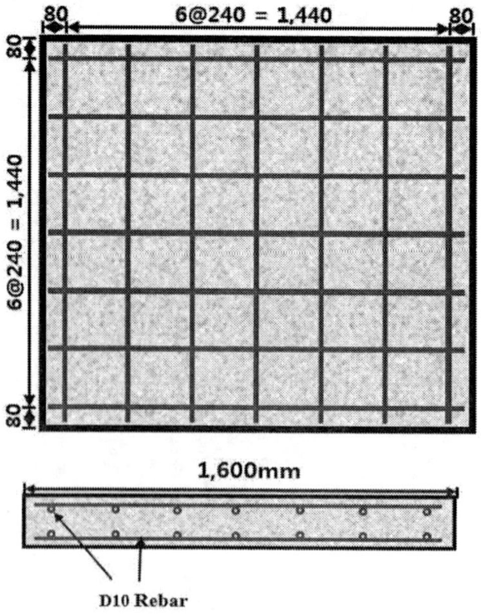

Fig. 2 Details of specimens

DOI: 10.1051/matecconf/201713803009

3 Test results and discussion

3.1 Effect of fiber in FRC slabs

3.1.1 Maximum Reaction and Impact Force at Each Loading Step

In the Drop-weight impact tests, the impact and reaction forces are expressed in Table 3. The number of blows of the test varies depending on the type and mixing ratio of the cement composite. In case of UHPC-180-F0.0, the specimen was failed in 2nd blow in spite of high compressive strength. HPFRCC and SFRC which were strengthened with steel fiber, can resist the 11th and 6th blow, respectively. Therefore, using fiber in structural member subjected to impact load has a lot of benefits in impact resistance.

Table 3. Impact and reaction force

Specimens	Blow No.	H (mm)	Impact force (kN)	Reaction force (kN)
HPFRCC-180-F2.0	1.	500	446.36	119.14
	7.	2000	569.96	369.84
	11.	2000	-	246.23
UHPC-180-F0.0	1.	500	303.62	145.38
	2.	750	172.74	136.26
SFRC-60-F2.0	1.	500	349.24	195.71
	3.	1250	351.10	217.59
	6.	1750	259.08	254.28
NC-60-F0.0	1.	500	303.62	145.38
	2.	750	172.74	136.26

As shown in the Fig. 3, the shape of cracks and punching shear can be shown in the final fracture patterns. Both NC and UHPC showed punching shear. However, the SFRC, HPFRCC slabs reinforced with steel fiber didn't show punching shear.

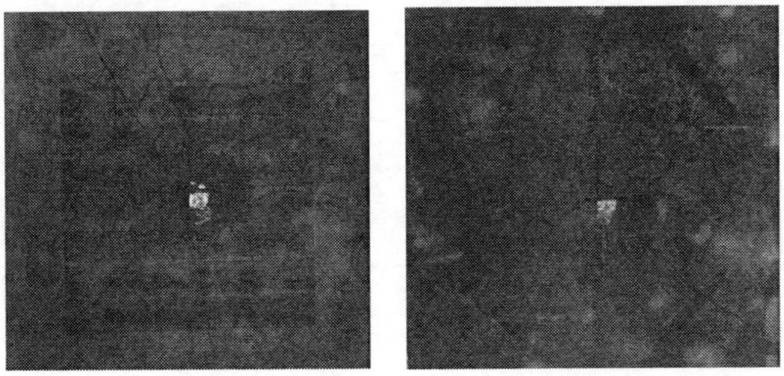

NC-60-F0.0 SFRC-60-F2.0

DOI: 10.1051/matecconf/201713803009

UHPC-180-F0.0

HPFRCC-180-F2.0

Fig. 3 Final fracture pattern of slab

3.1.2 Maximum Crack Width at Each Loading Step

Table. 4. shows the crack width at each loading steps of impact test. In case of NC-60-F0.0 and UHPC-180-F0.0, the crack width could not be measured because the critical shear crack occurred after 1st blow. SFRC-60-F2.0 and UHPFRCC-180-F2.0 failed in 6th and 11th blow. For the comparison, the crack width of SFRC-60-F2.0 and UHPFRCC-180-F2.0 in 5th blow are 5mm and 3mm. In this results, it was confirmed that the strength of matrix can increase the crack resistance and the fiber is the very useful method for resisting crack propagation.

Table 4. Crack Width at Each Loading Step

Loading Step	Maximum Crack Width			
	NC-60-F0.0	SFRC-60 -F2.0	UHPC- 180 -F0.0	UHPFRCC-18 0-F2.0
Step #1. (500mm)	0.6	0.5	0.5	0.2
Step #2. (750mm)	Punching	0.7	Punching	0.35
Step #3. (1000mm)	-	0.9	-	0.8
Step #4. (1250mm)	-	1.4	-	2.0
Step #5. (1500mm)	-	5.0	-	3.0
Step #6. (1750mm)	-	End	-	4.5
Step #7. (2000mm)	-	-	-	6.5
Step #8. (2000mm)	-	-	-	7.5
Step #9. (2000mm)	-	-	-	8
Step #10. (2000mm)	-	-	-	19
Step #11. (2000mm)	-	-	-	End

3.1.3 Maximum Deflection and Residual Deflection at Each Loading Step

Fig. 4 is the graph shows maximum deflection and residual deflection of the specimens. It can be confirmed that the resistance against the impact loading in terms of residual and maximum deflection can be increased by the steel fiber reinforcement. In the both case of NC and UHPC, deflections were decreased significantly.

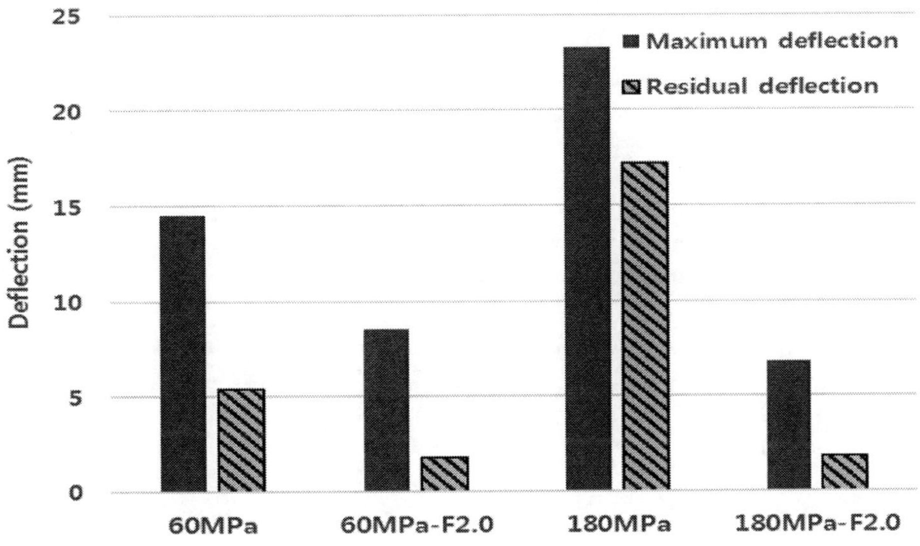

Fig. 4 Maximum Deflection and Residual Deflection (Step #1.)

3.1.4 Total Imparted Energy until Failure

Assuming the kinetic energy can be transferred to the impact energy without any loss, total imparted energy until failure of each specimen can be obtained. Although it varies depending on the loading history, total imparted energy can be a useful index to make a relative comparison between those specimens tested in this program. As shown in Table 5., the same dissipated energy was measured in NC-60-F0.0 and HPFRCC-180-F0.0, which are not contain the steel fiber. On the other hand, the SFRC slab with steel fiber showed 26% increased dissipated energy compared with NC-60-F0.0. And HPFRCC-180-F2.0 showed increase of 73.55% in dissipated energy compared with UHPC-180-F0.0. In this results, it can be confirmed that the reinforcing effect of steel fiber is greater in HPFRCC than in SFRC.

Table 5. Total imparted energy until failure

Specimens	Loading step at failure	Total imparted energy until fa ilure	Failure mode
HPFRCC-180-F2.0	# 11.	82.16kJ	Scabbing
UHPC-180-F0.0	# 2.	6.13kJ	Perforation
SFRC-60-F2.0	# 6.	33.11kJ	Scabbing
NC-60-F0.0	# 2.	6.13kJ	Perforation

3.1.5 Damage Assessment Based on Support Rotation

The value of support rotation was traditionally used as a parameter for evaluating damage in impact and blast tests. This parameter based on the maximum dynamic deflection of structural components, which was measured by using a laser displacement meter and potentiometer.

During the single impact loading test, deflection-time history (see Fig. 5) exhibited by each specimen was recorded and used to calculate support rotation values. And then, corresponding damage level was evaluated according to the design criteria provided by TM 5-1300, the most representative protective design manual.

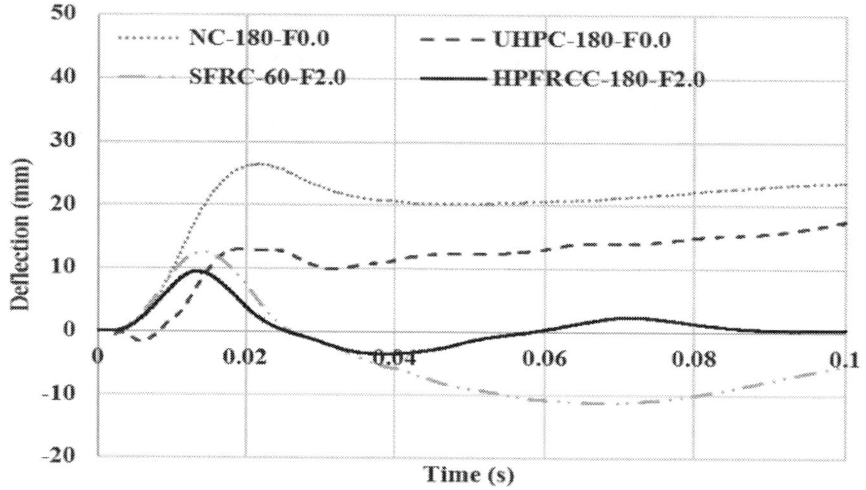

Fig. 5 Deflection-time history under single impact (Step #2.)

Table 5 presents the damage level evaluated using conventional approach based on support rotation. The support rotation values of the slabs without steel fiber were found to be in the severe range of more than 12 degrees. The HPFRCC-180-F2.0 slab was in the light range of 1.8 degrees and the SFRC-60-F2.0 was in the Moderate range of 2.34 degrees. As a results, fiber reinforcement has a lot of benefits on Impact Resistance of FRC Slabs

Table 6. Damage assessment using conventional approach

Specimens	Maximum deflection (mm)	Support rotation (degree)	Damage Level (TM 5-1300)
HPFRCC 180-F2.0	25.10	1.80	Light
HPFRCC 180-F0.0	Punching	-	Severe
SFRC 60-F2.0	32.76	2.34	Moderate
NC 60-F0.0	Punching	-	Severe

4 Concluding remarks

The purpose of this study is to evaluate the impact resistance of FRC slabs strengthened with various fiber. The difference between SFRC and UHPFRCC was also observed by experimental program. The conclusions drawn from this research are steel fiber reinforcement can increase the capacity of RC slabs under impact loading

Experimental assessment of RC slabs under low-velocity impact loading condition was conducted. Fiber reinforcement increased the ductility, shear force, and energy dissipation capacity of the members. The presence of steel fiber also resulted in superior damage tolerance with reduced fragmentation of concrete under low-velocity impact loading condition. In addition, steel fiber reinforcement showed significant enhancement especially in maximum deflection, residual deflection, and vibration reduction.

This work was supported by the National Research Foundation of Korea (NRF) grant funded by the Korea government (MEST) (NRF-2016R1A2B3011392).

References

1. H. Aoude, et al. "Response of steel fiber-reinforced concrete beams with and without stirrups, ACI Strct. J.", **109**, 3, pp.359 (2012)
2. N. Banthia, et al. "Impact behaviour of concrete beams, Mat. & Strct.", **20**, 4, pp.293-302 (1987)
3. V. Bindiganavile, N. Banthia, "Polymer and steel fiber-reinforced cementitious composites under impact loading—Part 2: Flexural toughness" Mat. J., **98**, 1, pp.17-24 (2001)
4. P.H. Bischoff, S.H. Perry, "Compressive behaviour of concrete at high strain rates, Mat. & Strct.", **24**, 6, pp.425-450 (1991)
5. L. Daudeville, Y. Malécot, "Concrete structures under impact, European J. of Env. & Civil Eng.", **15**, 1, pp.101-140 (2011)
6. K. Fujikake, et al. "Impact response of reinforced concrete beam and its analytical evaluation, J. of Strct. E.", **135**, 8, pp.938-950 (2009)

MATEC Web of Conferences **138**, 03010 (2017) DOI: 10.1051/matecconf/201713803010
EACEF 2017

Effect of Glass Fiber on Compressive, Flexural and Spliting Strength of Reactive Powder Concrete

Widodo Kushartomo[1,*], *Richard* Ivan[2]

[1,2]Department of Civil Engineering, Tarumanagara University, Jakarta, Indonesia

Abstract. Reactive Powder Concrete (RPC) is a new technology in high strength concrete industry. Steel fibers are usually used as reinforcements in RPC mixture. In this research, steel fibers in RPC mixture are replaced by fiberglass. The volume of fibers used are 1%, 1.5%, and 2%. Curing process by steam curing in 12 hours. The concrete was tested after 28 days including compressive test, split tensile test, and flexural test. The result shown that optimal value of fiberglass added to mixture was 2% from total volume with compressive strength value at 127.38 MPa, split tensile strength at 9.844 MPa, and flexural strength at 10.36 MPa. Density of fiberglass reinforced concrete at optimum volume is 2224.3029 kg/m^3 it was lower than steel fibers reinforced concrete at optimum volume which has density of 2310.385 kg/m^3.

1 Introduction

The need of high strength concretes is a must in construction industries nowadays. Reactive Powder Concrete (RPC) is one of the high strength concrete technology. RPC was first developed and produced by P. Richard and M. Cheyrezy in the early 1990s as researchers in a laboratory in France, known as Bouygues Laboratory.

RPC is a composite material of cement with high strength and ductility. RPC is a concrete without coarse aggregate. Its microstructure is optimized by the appropriate particle size distributionof the materials contained in a concrete mixture to increase the density of this concrete. RPC consists of very fine aggregates (such as sand, cement, and silica fume), steel fibers, and superplasticizer. The range of the compressive strength of RPC is from 200 to 800 MPa. A high strength could make RPC fragile.Therefore, steel fibers are usually added into the RPC mixture to increase its ductility. The addition of fibers into a concrete can increase its strength and toughness.The direction of the fibers, the length of the fibers, and the tensile strength of the fibers affect concrete strength and toughness [1]. The critical length is the minimum length of the fiber required to retain stress or tension equal to the fiber's strength. The use of shorter or longer fiber beyond its critical length will cause the non-optimum strength and capacity of the fiber to retain stress or tension [1].

*corresponding author: widodo@untar.ac.id

In this research, steel fibers in a RPC mixture will be replaced with fiberglass. Fiberglass is polymer fibers strengthened by glass fibers. Fiberglass is a light and strong material usually used for gliders, boat manufactures, water tanks, cars, etc. Fiberglass was chosen because this material is light but strong against tensile and tension.The tensile strength of fiberglass is higher compared to steel fibers [1-3].

The research of fiberglass use as a reinforcement of a concrete was first conducted in Russia in 1940s and was then introduced to the world's building industry in the early 1970s in the United Kingdom. Glass Fiber Reinforced Concrete(GFRC) is a fiberglass mixture with cement matrix. This mixture maintains the physical and chemical identity of the fiber and even the cement matrix synergistically and reaches a combination of properties which could never been achieved individually. The fibers retain loads while the cement matrix maintains the location of the fiber reinforcement and also protects the fibers from damage caused by the surrounding environment [4-7].

Flexural strength of the fibrous concrete is affected by the volume and ratio aspect of the fibers in a concrete mixture [1]. Because of these factors, we had been conducted experiment in this research by using a different value of fraction of the fiber volume. The length and diameter of steel fibers used were from 11.0 to 13.0 mm and from 0,12to 0,13 mm respectively. The tensile strength of the fibers were 1200 MPa. They were produced by Hebei Yusen Steel Wire Mesh Co.,Ltd. The fiberglass used was produced by PT. Justus Kimia raya conducting continuous roving with a tensile strength of 2400 MPa.

This research was intended to know the difference of compressive, splitting, and the flexural strength of a RPC with steel fibers and fiberglass mixtures. Furthermore, this research was also intended to identify the fraction of the optimum volume of the addition of each materials into the RPC mixture.

2 Research Methods

The materials used in this research are PCC cement, sand with a maximum grain size of 0.6 mm, water, silica fume, superplasticizer *viscocrete-10*, AR-Glass as fiberglass, and steel fibers. The mix designused are as the ones below:

Table 1. Typical RPC composition

Material	Ratio by weight
Cement	1.00
Silica fume	0.25
Sand	1.50
Super plasticizer	0.03
Water	0.20
Fiber fraction volumes (%)	1,0; 1,5; 2

Before making the specimens, a test was done for fine aggregates based on standard examinations. There were 10 cylindrical specimens (7.5 cm x 15 cm) and 2 beams (10 cm x 10 cm x 40 cm) prepared for each kind of fiber and volume fraction. The concrete treatment technique used was the steam method. The examination was done after the concrete has reached 28 days.

The compressive strength examination was done on 5 cylindrical specimens based on ASTM C39 [8]. The splitting strength examination was conducted on 5 cylindrical specimens based on the splitting strength value (F_{ct}) [9], and The flexural strength examination was carried out on the beam specimens based on the flexural strength value (F_r) [10] :

3 Results And Discussions

The addition of fibers into the RPC mixture can increase its strength because the addition of fibers prevent cracks propagation caused by loadings or forces [11,12]. Results of the examination are shown in Table 2. The process of the examination of compressive, splitting, and flexural strength are shown in Figure 1.

The addition of fiberglass and steel fibers in a concrete increases compressive, splitting, and flexural strength of the concrete.The increases those concrette strengths due to addition steel fibers were greater compared to the addition of fiberglass.

We can conclude that the optimum value for the addition of fibers were different between steel fibers and fiberglass. The optimum value for the addition of steel fibers with the volume fraction of 1.5% was 95.89 MPa for its compressive strength are show in Figure 2, 13.11 MPa for its flexural strength are shown in Figure 3 and 12.414 MPa for its splitting strength are shown in Figure 4. The optimum value for fiberglass additoin with the volume fraction of 2% was 90.380 MPa for its compressive strength, 11.020 MPa for its splitting strength, and 10.363 MPa for its flexural strength.

(a) (b) (c)

Figure 1. (a) Compressive Strength Examination, (b) Splitting Strength Examination, (c)Flexural Strength Examination

DOI: 10.1051/matecconf/201713803010

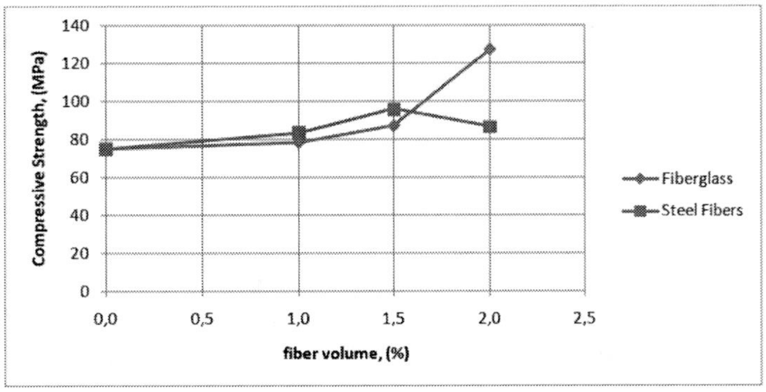

Figure 2. Effect of fibers volume additions on compressive strength of reactive powder concrete

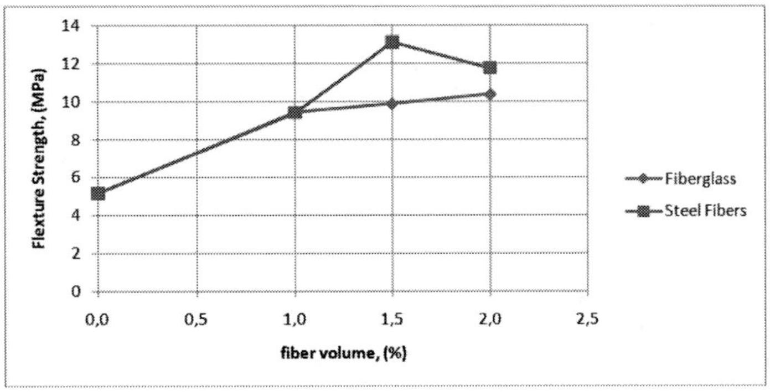

Figure 3. Effect of fibers volume additions on flexture strength of reactive powder concrete.

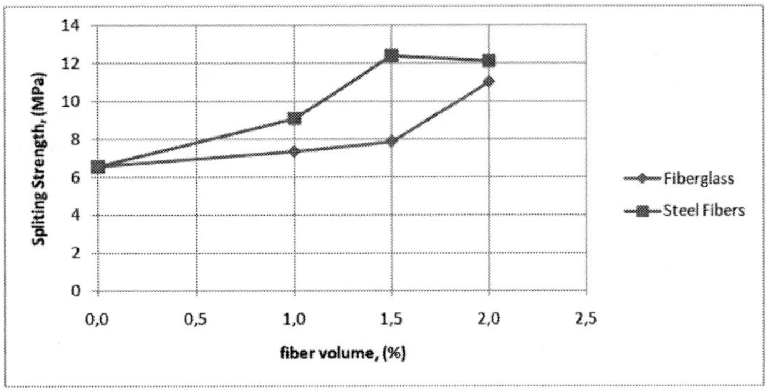

Figure 4. Effect of fibers volume additions on spliting strength of reactive powder concrete.

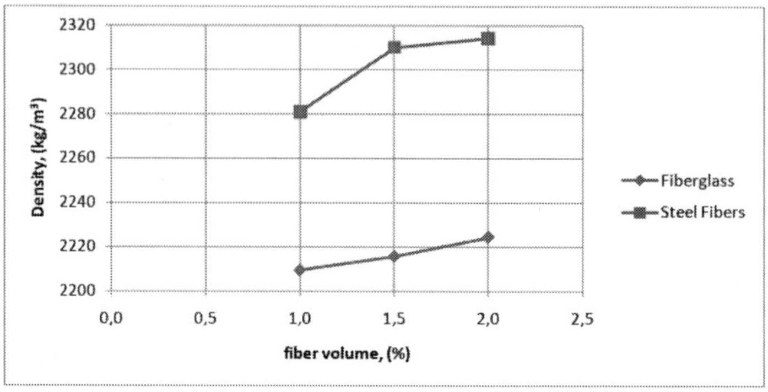

Figure 5. Effect of fibers volume additions on density of reactive powder concrete.

Based on the results of the examination are shown in Figure 2, the changes in the strength of the concrete with steel fibers reinforcement increased after the addition of 1% until the peak which is 1.5% and then decreased when it was added until 2%. The changes in the strength of the concrete with fiberglass reinforcement continuously increased from the addition of 1% until 2%.

The increase of the flexural strength value on Figure 3 and splitting strength value on Figure 4 of the concrete due to the addition of steel fibers was in general higher than the concrete added with fiberglass. However, the overall average density of the concretes for each of the volume fractions with fiberglass is smaller than the one with steel fibers. These results are shown in Figure 5. Moreover, the difference of the mechanical properties value of the concrete, with the addition of fiberglass at the value of 2%, with a lower density,was very small compared to value of mechanical properties the addition of steel fibers at the optimum value which is 1.5%.

The value of flexural and splitting strength of the concrete with the addition of fiberglass was lower compared to the ones with the addition of steel fibers. This was because the ratio aspects which wre too excessive on concretes with fiberglass, where the length of the fiberglass used was between 30-40 mm compared to the diameters of the fibers which is relatively small, about 14 μm per filament. Mean while the ratio aspects on concretes with steel fibers were too large because the length of the fibers wass about 11,0-13,0 mm with a diameter about 0,12-0,13 mm.The addition of fiberglass into the RPC mixture could increase the durability of the concrete against cracks (crack resistance) caused by heat of hydration, creep, and shrinkage.

4 Conclusions

Based on the results of this research, the followings could be concluded:

- The addition of steel fibers and fiberglass into a RPC increases the mechanical property of the concrete.
- The optimum value of the fiberglass added into the RPC was when the volume fraction is 2%.
- The optimum value of steel fibers added into the RPC wass when the volume fraction is 1.5%.
- When the value of volume fractionwas optimum, the flexural and splitting strength of the RPC with steel fibers reinforcement was higher and stronger than the RPC with fiberglass reinforcement. However, the compressive strength of the RPC with fiberglass reinforcement is higher and stronger compared to the RPC with steel fibers reinforcement.
- The density of the RPC with fiberglass reinforcement is overall smaller compared to the RPC with steel fibers reinforcement.

– The use of fiberglass and steel fibers on RPC could prevent cracks caused by heat by hydration, creep, and shrinkage.

References

1. A. Bentur, S. Mindess, , *Fibre reinforced cementitious composites* (Elsevier Sience Publishers LTD, England, 1990)
2. A. Sukumar,E.John,International Journal of Innovative Reasearch in Advanced Engineering,*Fiber addition and its effect on concrete strength*, **1**, 144-149(2014)
3. Q.Zao, J.Yu, G.Geng, J.Jiang, X . Liu, Construction and Building Materials Journal, *Effect of fiber types on creep behavior of concrete*, **105**, 416-422(2016)
4. S. Kakooei, H. Md Akil, M. Jamshidi, J. Rouhi,Construction and Building Materials Journal, *The effects of polypropylene fibers on the properties of reinforced concrete structures*, **27**, 73-77(2012)
5. S. Alsadey, M. Salem, American Journal of Engineering Research,*Influence of polypropylene fiber on strength of concrete*, **5**, pp 223-226(2016).
6. P.Zhang, Q. Li, Composites Part B: Engineering Journal,*Effect of polypropylene fiber on durability of concrete composite containing fly ash and silica fume,***45**, 1587-1594(2013).
7. G.M. SadiqulIslam, S. DasGupta. International Journal of Sustainable Built Environment, *Evaluating plastic shrinkage and permeability of polypropylene fiber reinforced concrete*, **5**, 345-354(2016).
8. ASTM International, *Standard test method for splitting tensile strength of cylindrical concrete specimens*(ASTM C496, 2002).
9. ASTM International, *Standard test method for flexural strength of concrete (Using Simple Beam with Third-Point Loading* (ASTM C78, 2002)
10. ASTM International, *Standard test method for compressive strength of cylindrical concrete specimens*(ASTM C39, 2002).
11. M. A. Mohamed, M. A.Moh, N. W. Akasha, I. Y. I. Elgady, *experimental study on effects of fiberglass and fiber waste in concrete mixes*, International Journal of Engineering Science & Research Technology, 3, 485-493, (2016).
12. B. Graybeal A., F. Baby, *Development of Direct Tension Test Method for Ultra-High-Performance Fiber-Reinforced Concrete,* ACI Materials Journal, 110, 177-186, (2013).

AUTHOR INDEX

AUTHOR INDEX

AUTHOR INDEX

6th International Conference of Euro Asia Civil Engineering Forum (EACEF 2017)

MATEC Web of Conferences Volume 138 (2017)

Seoul, South Korea
22 - 25 August 2017

Part 2 of 2

Editors:

Jae-Woo Park
Han Ay Lie
Harianto Hardjasaputra
Parama Thayaalan

ISBN: 978-1-5108-5578-6

Printed from e-media with permission by:

Curran Associates, Inc.
57 Morehouse Lane
Red Hook, NY 12571

Some format issues inherent in the e-media version may also appear in this print version.

Printed by Curran Associates, Inc. (2018)

For additional information, please contact EDP Sciences – Web of Conferences at the address below.

EDP Sciences – Web of Conferences
17, Avenue du Hoggar
Parc d'Activité de Courtabœuf
BP 112
F-91944 Les Ulis Cedex A
France

Phone: +33 (0) 1 69 18 75 75
Fax: +33 (0) 1 69 28 84 91

contact-edps@webofconferences.org

Additional copies of this publication are available from:

Curran Associates, Inc.
57 Morehouse Lane
Red Hook, NY 12571 USA
Phone: 845-758-0400
Fax: 845-758-2633
Email: curran@proceedings.com
Web: www.proceedings.com

TABLE OF CONTENTS

PART 1

PART 2

MATEC Web of Conferences **138**, 03011 (2017) DOI: 10.1051/matecconf/201713803011
EACEF 2017

Preliminary studies on steel slag as a substitute for coarse aggregate on concrete

Rahmi Karolina*, and *Jannes* Pandiangan

Department of Civil Engineering,University of Sumatera Utara, Medan, Indonesia

Abstract. The development of science and technology in the field of construction that is rapidly increasing, is always followed by the growing community needs for infrastructure facilities, such as buildings, bridges and other construction. One of the key element in that development is concrete. Due to the rapid development of science and technology in the field of construction, it's required a building material which has better advantage than the materials of the existing building. To obtain a better building materials, one alternative is the use of waste as aggregate in concrete mixture. In this study the authors using waste steel waste (steel slag) as a substitute for coarse aggregate. Steel slag used is steel waste from PT. Growth Sumatra Industry. The gravel substitution variations is 0%, 15%, and 25% and the testing was done by the slump test, compressive strength and flexural strength of concrete. From the test results obtained optimum compressive strength variation occurs in 25% substitution of steel slag gravel amounted to 40.481 MPa, whereas for the optimum bending capacity contained in variations of 25% substitution of steel slag gravel amounted to 19.592 N / mm^2. And for optimum slump value obtained on the variation of normal concrete. This shows the workability of the concrete normally higher than the other variation.
***Keywords*:** Steel slag, compressive strength, flexural strength

1 Introduction

Knowledge and technology in construction are developing rapidly and followed by the increase in people's need for facility and infrastructure such as building construction, bridge, and the other constructions. The building itself consists of various components, and one of the main components in construction is concrete.

Concrete is a construction material which consists of fine aggregate, coarse aggregate, cement, and water. There are some factors which have to be heeded in making concrete mixture; they are material, proportion of the mixture, and executing. From these three factors, material is the natural resource which will gradually be used up and cannot be renewed. Therefore, an alternative should found to substitute it and steel slag is considered appropriate to replace it. Slag is the residue of high furnace combustion as the result of steel

*Corresponding author: rachmie_caroline@yahoo.co.id

smelting which produces steel slag which physically resembles. Using steel slag as coarse aggregate will help conserve environment.

In many countries, slag is commonly used as the substitute for aggregate, either as the mixture of concrete or as the material for foundation hardening. In using it, slag is usually considered as aggregate; therefore, physical requirement of aggregate is considered as the same as the physical requirement for aggregate. Since chemical characteristic of slag is far different from the chemical characteristic of natural aggregate, some other components are needed in order that slag can be used as the substitute for standard aggregate; the requirement is preservation BSI, 2007). Some advantages of using steel slag in concrete mixture as follows:

- Heightening the compressive strength of concrete since the increase in compressive strength tends to slow down;
- Increasing the ratio between flexibility and concrete compressive strength;
- Reducing the variation of concrete compressive strength;
- Heightening resistance against sulfate in the sea water;
- Reducing the attack of alkaline silica;
- Reducing hydrated heat and decreasing temperature;
- Heightening preservation due to the influence of volume change;
- Reducing porosity and chloride attack

2 Literature review

2.1 Steel slag

According to Paul Nugraha and Antoni (2007), slag is the residue material from pig iron in which its process uses furnace which fuel comes from blast. Smelting steel begins with eliminating steel polluting ions, such as aluminum, silicon, and phosphor. In order to eliminate the polluting ions, it is necessary to have calcium which is found on the limestone. The mixture of calcium, aluminum, silicon, sand phosphor will shape slag which reacts in the temperature of 1,600°C and form liquid. When this liquid is cooled, crystal will be formed to resemble the form of aggregate. Slag has particle pored granules on its surface and good gradation with different variation in particle shapes.

Slag is defined in ASTM C.989 as "...standard specification for ground granulated blast Furnace Slag for use in concrete and mortar" *ASTMJ, 1995: 494). Slag is a non-metal product, a fine-formed material and the granular of the result of combustion which is cooled by, for example, immersing it into the water.

Some factors for determining cementitious quality in slag are as follows: chemical composition, alkaline concentration and reaction against system, glass content in slag, fineness, and temperature which is caused during the process of hydration.

Steel slag is hard, solid material which contains an amount of free iron so that it gives high density and hardness. The use of steel slag as an concrete aggregate with Portland cement can cause high quality of concrete. Steel slag aggregate has uneven surface texture and has prismatic shape. It also has high volume weight and specific gravity, high friction coefficient, and its water absorption is moderate (until 3%). Steel lag has good character for using aggregate, good abrasion endurance, good characteristics of its strength, and high supporting power.

 DOI: 10.1051/matecconf/201713803011

Fig. 1. Steel slag from PT. Growth Sumatra

Chemical element content in steel slag can be seen in Table 1 below,

Table 1. Chemical element content in steel slag.

No.	Parameter	Unit	Result	Method
1	Lead (Pb)	Mg/kg	26,6	AAS
2	Cadmium (Cd)	Mg/kg	<0,003	AAS
3	Copper (Cu)	Mg/kg	97,5	AAS
4	Chromium (Cr)	Mg/kg	5353	AAS
5	Silver (Ag)	Mg/kg	<0,001	AAS
6	Selenium (Se)	Mg/kg	<0,01	AAS
7	Barium (Ba)	Mg/kg	817	AAS
8	Mercury (Hg)	Mg/kg	0,38	AAS
9	Arsenic (As)	Mg/kg	0,21	AAS

Source: Laboratory of IndustryResearch and Standardization Center Medan 2016

3 Research method

3.1 Flowchart

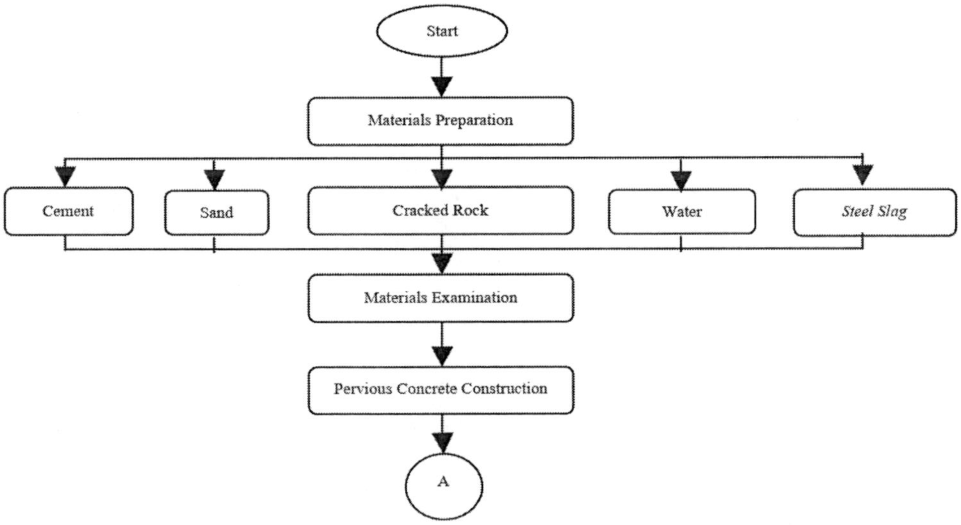

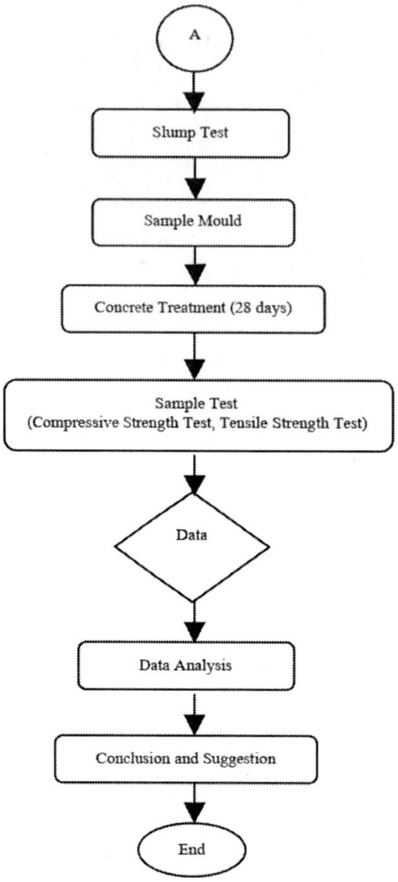

Fig. 2. Research's flowchart

4 Results and discussion

4.1 Slump test

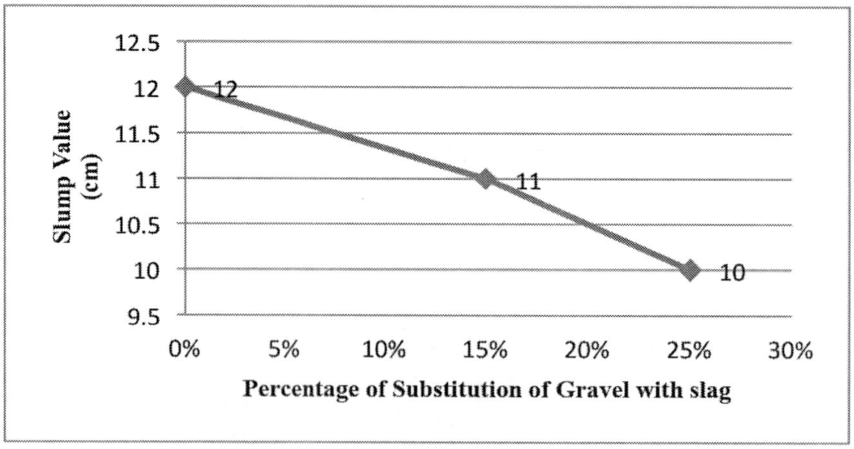

Fig.3. Slump value graph in the percentage of substitution of gravels with steel slag.

 DOI: 10.1051/matecconf/201713803011

4.2 Compressive strength of concrete cylinder

This testing was aimed to find out the value of concrete compressive strength by using various variations of substitution of coarse aggregate with steel slag, compared with normal concrete in which specimens in the form of cylinder which construction and maintenance were carried out in Concrete Laboratory. The testing was done when the concrete was 28 days old, based on SNI 03-6429-2000, using the method of Concrete Compressive Strength Testing.

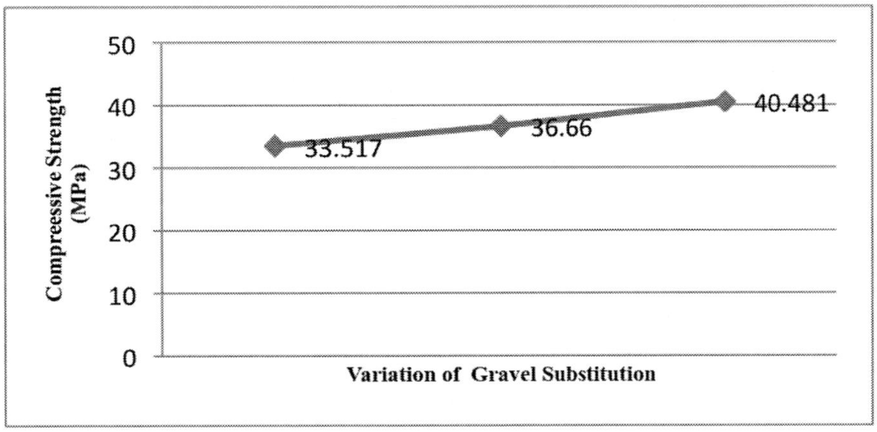

Fig.4. Graph of the result of concrete compressive strength testing

From Figure 4, it was found that there was the increase in the use of steel slag straight proportional to the increase in concrete compressive strength. Therefore, it could be concluded that the increase in the use of steel slag as coarse aggregate could increase concrete compressive strength.

4.3 Reinforced concrete deflection testing

Reinforced concrete deflection testing was done by using Hydraulic Jack and 3 (three) Dial Indicator with the rage of 75 cm respectively,

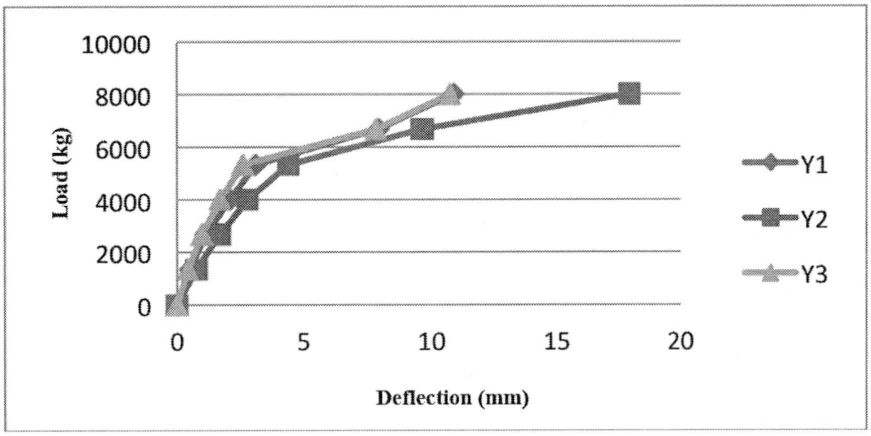

Fig. 5. Graph of correlation of load with reinforced concrete beam deflection (Normal)

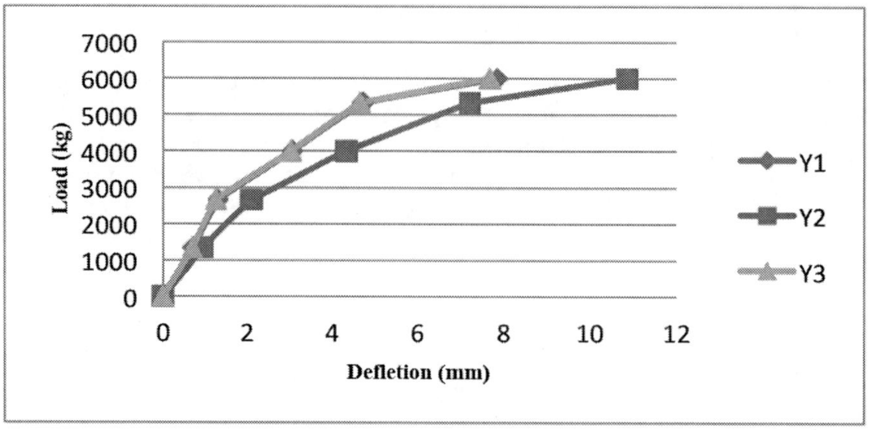

Fig. 6. Graph of correlation of load reinforced concrete beam deflection (substitution of 15% of gravels with slag)

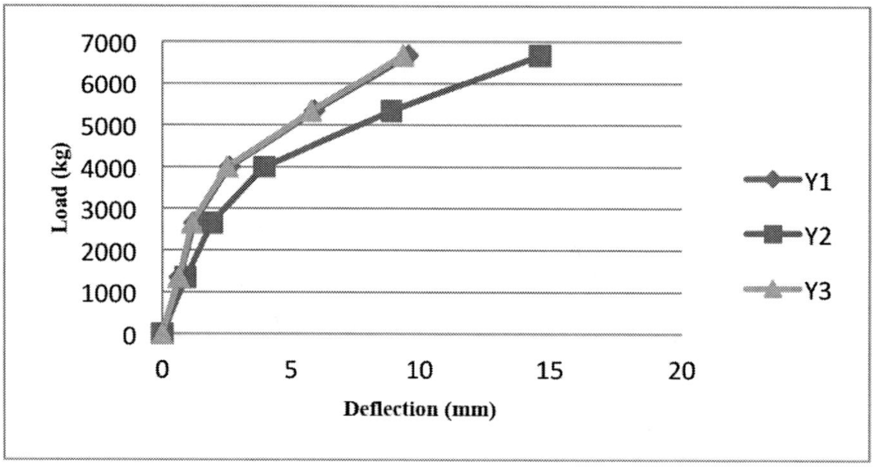

Fig.7. Graph of Correlation of load with reinforced concrete beam deflection (substitution of 25% gravels with slag)

From Figure 5, 6, and 7, it was found that the more substitution of gravels with steel slag was the bigger the load carried by the beam, and the more the deflection occurred. It was the same as the concrete compressive strength, the increase in using steel slag as coarse aggregate could increasere in forced concrete beam deflection.

4.3.1 Pliancy capacity of reinforced concrete

The data of the calculation of pliancy capacity, based on tension for reinforced concrete beam could be seen in Table 2 below,

Table 2. Pliancy capacity of reinforced concrete beam (normal)

Load P (Kg)	εc (× 10^{-4})	εs (× 10^{-4})	Fc (N/mm²)	Fs (N/mm²)	Mn (Nmm) (× 10^6)	Pn (Kg) (× 10^2)	σ (N/mm²)	P/Pn
0	0.00	0.00	0.00	0.00	0.00	0.00	0.00	0.000
1,333	1.54	4.21	4.20	84.14	6.39	12.79	0.88	1.042
2,666	3.41	9.30	9.29	185.99	14.13	28.27	1.96	0.943
3,999	7.02	19.13	19.11	382.60	29.07	58.15	4.03	0.688
5,332	11.72	31.93	31.89	638.56	48.52	97.04	6.73	0.549
5,998.	17.66	48.09	48.04	961.82	73.09	146.1	10.13	0.410
Mean Coefficient								0.606

Table 3. Pliancy capacity of reinforced (substitution of 15% of gravels with slag)

Load P (Kg)	εc (× 10^{-4})	εs (× 10^{-4})	fc (N/mm²)	fs (N/mm²)	Mn (Nmm) (× 10^6)	Pn (Kg) (× 10^2)	σ (N/mm²)	P/Pn
0	0.00	0.00	0.00	0.00	0.00	0.00	0.00	0.000
1,333	1.43	3.90	4.07	77.94	5.94	11.88	0.82	1.122
2,666	3.20	8.72	9.11	174.47	13.30	26.60	1.84	1.002
3,999	6.45	17.58	18.37	351.61	26.80	53.61	3.71	0.746
5,332	14.45	39.37	41.13	787.35	60.04	120.07	8.32	0.444
6,665	23.78	64.79	67.68	1295.72	98.79	197.59	13.70	0.337
Mean Coefficient								0.608

Table 4. Pliancy capacity of reinforced concrete beam (substitution of 25% of gravels with slag)

Load P (Kg)	εc (× 10^{-4})	εs (× 10^{-4})	fc (N/mm²)	fs (N/mm²)	Mn (Nmm) (× 10^6)	Pn (Kg) (× 10^2)	σ (N/mm²)	P/Pn
0	0.00	0.00	0.00	0.00	0.00	0.00	0.00	0.00
1,33	1.24	4.84	3.71	96.86	6.34	12.67	0.82	1.05
2,66	2.81	10.98	8.41	219.54	14.36	28.72	1.86	0.92
3,99	4.63	18.08	13.84	361.60	23.66	47.31	3.06	0.84
5,33	7.28	28.41	21.76	568.22	37.17	74.35	4.80	0.71
6,66	15.96	62.31	47.71	1246.2	81.53	163.06	10.55	0.40
7,99	29.65	115.78	88.65	2315.5	151.48	302.97	19.59	0.26
Mean Coefficient								0.60

From the Table 2, 3, and 4, it was found that the increase in the percentage of gravel substitution with steel slag would increase pliancy capacity of reinforced concrete. Therefore, it could be concluded that steel slag was good to be used in concrete mix.

5 Conclusion

Based on the result of the research, it could be concluded that :

- Compressive strength of normal concrete = 33.517 MPa, compressive strength of concrete with substitution of 15% of gravels with slag = 36.66 MPa (increased 10.18% from compressive strength of normal concrete, and compressive strength of substitution concrete of 25% of gravels with slag = 40.481 MPa (increased 20.78% from compressive strength of normal concrete).
- Deflection in normal reinforced concrete beam at loading P = 5998.5 kg was 10.86 mm, for reinforced concrete beam with 15% of gravels with lag at loading P = 6665 kg was 14.63 mm, while for reinforced concrete beam with substitution of 25% gravels with lag at loading P = 7998 kg was 17.93 mm.
- Pliancy capacity in normal reinforced concrete beam was 10.138 N/mm^2, in substitution beam of 15% of gravels with lag was 13,704 N/mm^2, and in substitution beam of 25% of gravels with lag was 19,592 $N/mm^{2.}$ Thus,usingsteel slag as coarse aggregate could increase pliancy in reinforced concrete.

References

1. R. Aryanti, *Pengujian Lentur Balok Beton Bertulang dengan Menggunakan Modifikasi Alat Uji Tekan* (Universitas Andalas, Padang, 2008), p. .
2. I. Dipohusodo, *Struktur Beton Bertulang* (PT.Gramedia Pustaka Utama, Jakarta, 1999)
3. T. Mulyono, *Teknologi Beton* (Penerbit ANDI, Yogyakarta, 2003), p. .
4. J. C. McCormac, *Desain Beton Bertulang* (PenertbitErlangga, Jakarta, 2004), p. .
5. E. G. Nawy, *Beton Bertulang Suatu Pendekatan Dasar* (Penerbit Refika Aditama, Bandung, 1998), p. .
6. P. Nugraha, *Teknologi Beton* (Penerbit ANDI, Yogyakarta, 2007), p. .
7. Y. Risdianto, *Kajian Kuat Tekan Beton dengan Perbandingan Volume dan Perbandingan Beratuntuk Produksi Beton Massa Menggunakan Agregat Kasar Batu Pecah Merapi* (Universitas Negeri Yogyakarta, Yogyakarta, 2013), p. .
8. D. Setiawan, *Pengaruh Penggunaan Limbah Baja terhadap Kuat Karakteristik Beton* (Universitas Kristen Maranatha, Bandung, 2014), p. .
9. Standar Nasional Indonesia. *Tata Cara Perhitungan Struktur Beton untuk Bangunan Gedung*. SNI 03 – 2847 – 2002.
10. Standar Nasional Indonesia. *Tata Cara Perhitungan Struktur Beton untuk Bangunan Gedung*. SNI 03 – 2847 – 2013.
11. S. Buen, Final project, Universitas Katolik Parahyangan, Bandung, 2013.
12. I. Wikana, *Tinjauan Kuat Lentur Balok Beton Bertulang dengan Lapisan Mutu Beton yang Berbeda* (UKRIM, Yogyakarta, 2007), p. .

DOI: 10.1051/matecconf/201713803012

Strength Properties and Micro-structure of Steel Slag Based Hardened Cementitious Composite with Graphene Oxide

Mao Li[1,*], *Jin-man* Kim[2]

[1]Ph.D student , Department of Architectural Engineering, 1223-24 Cheonan-daero, South Korea
[2]Professor, Ph.D, Department of Architectural Engineering, 1223-24 Cheonan-daero, South Korea

Abstract. For the rapid hardening of concrete, various types of calcium aluminate composite have been used for special purpose as shotcrete or rapid repairing in construction field. However, high cost of calcium aluminates is demerit in the point of expanding usage. In recent years, a new binder for rapid hardening has been introduced as the pulverized ladle furnace slag (RC-LFS) with main components of $C_{12}A_7$ and β-C_2S. Like other traditional cementitious materials, it's use will preserve natural resources and has high environmental advantage with economic efficiency, by transferring the industry byproducts into high value materials. Even though flake state, graphene oxide has high tensile strength and electric conductivity. Then, it can be used to increase strength and thermal properties of concrete and its efficiency depends on the dispersion in the cement matrix. In this paper, we used graphene oxide for increasing the engineering properties in RC-LFS concrete with GGBFS. Test results showed that the 0.05w% of graphene oxide increase the flexural strength to 26% and compressive strength to 16%.

1 INTRODUCTION

Nowadays, many new materials with candidated reinforcing effect for cementitious composite are introduced. Though reinforcing mechanism and reinforcing efficiency are different on case by case, there are special attention to the graphene oxide (GO). GO is mono-layer of 2-hybridized carbon atoms derived from a mixture of carboxyl, hydroxyl, and epoxy functionalities [1,2,3]. It has been reported by Z. Pan and K. Gong et al. that carboxyl acid groups can react with calcium silicate hydrate (C-S-H) to form strong covalent bonds, thereby notably improving mechanical properties of GO-cement composites [4,5].

For high aluminate composite, the properties of graphene oxide reinforced composite haven't been fully documented., This unconventional cementitious material, the binder in this experiment,also been introduced as the pulverized rapid cooling ladle furnace slag (RC-LFS). Ladle furnace slag (LFS) is an industry by-product which generally occupied a large area of the land for store and processing. Thus it has generated a serious environmental

* Corresponding author: jmkim@kongjul.ac.kr

problem for a long time. Moreover, the slow cooled LFS (by natural air) is hard for recycling, since the slowly cooled process allow the generation of crystal structure which lower the reactivity even after pulverized [6]. On the other hand, slow cooled also lead to high amount of f-CaO which will cause the unsoundness problem [6, 7]. But after rapidly cooled by high-pressure air, the material shows remarkable mechanical properties and high early strength after pulverized. More importantly, many studies have been focusing on finding an alternative which can be used as cement replacement materials. Cement replacement materials are the way to utilized untouchable raw materials and should be a break-solution of a green world. Materials that can be used for replacing cement as a binder can come from several sources namely agriculture, industry, marines etc. [8]. The pulverized RC-LFS as an industry by-product has unique properties on reducing environmental problems and such a solution of sustainable development. Owing to its rapid strength development, pulverized RC-LFS was used in structural applications.

Except RC-LFS, ground granulated blast furnace slag (GGBFS) was also involved. GGBFS remains after the steel making process. Application of GGBFS in construction material development is not only cost effective and resource saving but also reduces the energy and green house gas emissions [9].

Since the necessity to document the mechanical properties such as flexural and compressive strength of GO reinforced mortar, those properties were evaluated by series experiment and were presented in this paper.

2 EXPERIMENTAL STUDY

2.1 Materials

Within the scope of this paper, four basic materials such as pulverized RC-LFS, α semi-hydrated gypsum, GGBFS, and GO are used. All those material are getting from Korea market.

The hydration of pulverized RC-LFS process very rapidly in the room temperature since the primary hydraulic material is $C_{12}A_7$. Nowadays, gypsum always been involved in the pulverized RC-LFS based composite to hydrate with $C_{12}A_7$ to form ettringite, thus to prevent from transforming from C_2AH_8 to C_3AH_6 as been illustrated in following equations [6].

$$C_{12}A_7+51H_2O \rightarrow 6C_2AH_8+AH \rightarrow 4C_3AH_6+6AH+18H_2O \quad (1)$$
$$C_{12}A_7+12CaSO_4+137H_2O \rightarrow 4(C_3A\cdot 3CaSO4\cdot 32H_2O)+6AH \quad (2)$$

While the hydration process of $C_{12}A_7$-$CaSO_4\cdot nH_2O$ system is very complicated. First of all, the amount of $CaSO_4\cdot nH_2O$ has to be sufficient to guarantee the stable production of ettringite and can be calculated by using modified Bogue equation [6,7]. Base on the previous studies, chemical retarders are necessary to control the hydration process of C12A7. Since that, citrate acid was involved. GGBFS, as one of the most widely used supplementary cementitious material (SCM) in the construction field, has been documented by many researchers [10, 11,]. The addition of GGBFS to the calcium aluminate cement (CAC) system is reported to change the hydration mechanism of a CAC system. Previous studies have demonstrated that, in CAC–GGBFS blends of suitable ratios, no conversion even at elevated temperatures, even though various types of CAC with different alumina contents were utilized. [12,13,14,15]. The physical and chemical properties of pulverized RC-LFS, gypsum and GGBFS are presented in Table 1.

GO used in this experiment was provided by a Korean Company. To guarantee the good distribution in the cementitious matrix, the aqueous solution was used as presented in Fig.1. The physical properties are shown in Table 2. The package which is provided by the company is 250 ml, and contents 0.75 g solid graphene oxide.

Table 1. The basic physical and chemical properties of LFS powder, gypsum, and GGBFS

Binder	Physical properties		Oxide content(Wt.%)					
	Density (g/cm^3)	Fineness (g/cm^3)	SiO_2	CaO	Al_2O_3	Fe_2O_3	MgO	SO_3
RC-LFS	2.97	6300	10.9	44.5	26.6	4.3	6.6	-
Semi-Gyp	2.72	1100	2.6	40	0.9	0.4	0.3	55.8
GGBFS	2.84	4200	30.3	44.6	13.8	0.5	4.5	4.4

Fig. 1. Graphene oxide aqueous solution.

Table 2. Properties of solid graphene oxide

Carbon (%)	Oxygen (%)	D/G ratio
45-50	40-45	0.99

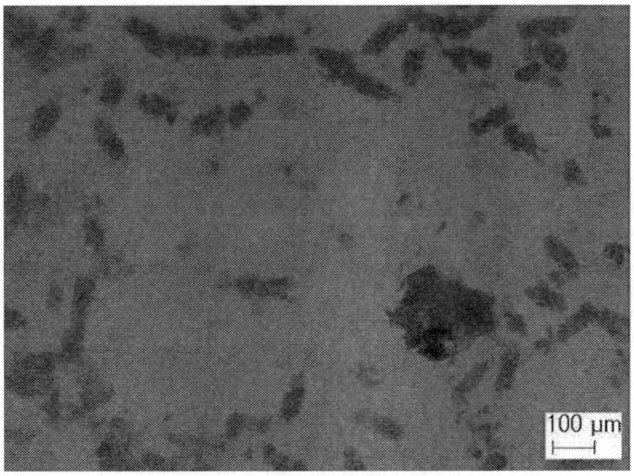

Fig. 2. Graphene oxide aqueous solution under the microscope.

 DOI: 10.1051/matecconf/201713803012

2.2 Experimental plan

The experimental purpose is to illustrate the micro-structure and strength development of RC-LFS base slag cement. Thus the experimental plan is decided and as presented in Table 3. For the fresh composites, the test items include the flow and air content and will follow the ASTM C230 [16] and ASTM C185-08 [17] respectively. For hardened composite, flexural and compressive strength are involved by referencing ASTM C348-14 [18] and ASTM C349 [19].

The fresh concrete was cast in 40*40*160mm prism mold. The curing room for storage

Table 3. Experimental plan

Binder	ID	Dosage of GO	w/b (%)	Test items
Ace*. 100%	A100	0.05 w %	40	-Flow -Air content -Flexural strength -Compressive strength -Microstructure (SEM)
Ace*. 100%+GO	A100GO			
Ace*.75% +GGBFS	A75G25			
Ace*.75% +GGBFS+GO	A75G25GO			
Ace* : pulverized RC-LFS replaced by 25% of α semi-hydrate gypsum				

of the specimens in the mold is maintained at a temperature of 20.0±1.0℃ and relative humidity of 60.0±5.0%. The samples involved in this experiment can be differentiated by different binder material whether containing graphene oxide or not. The control mixtures identified as the mixture without graphene oxide and GGBFS. Other three mixtures contain GGBFS or/and graphene oxide. The different ingredient and ID has been shown in Table 3. The steps of preparing mixtures can be summarized as following:

1. Preparing and measuring binder material.
2. Measuring graphene oxide. The dosage was designed by the solid content of aqueous solution.
3. Modify and recalculate the water amount. Since the graphene oxide solution contains water, adding graphene oxide solution will improve the water-to-binder ratio. So the mixing water will be recalculated.
4. The temperature of all the material will be tested and well keeping in the curing room at 20℃ and 60% relative humidity.
5. Mixing method will be based on the ASTM C305-06 [20]. Graphene oxide solution will be added into the mixtures after mixing water, and continued mixing for additional 1 minute at 285 ± 10 r/min.

3 TEST RESULTS AND DISCUSSION

3.1 Flow, air content of fresh mixtures

As a nanomaterial, GO was documented may have an influence on the fresh properties of the mixtures [21]. Since that, the fresh properties of graphene oxide have been tested by ASTM standard. Fresh cement material always illustrated as yield stress fluids [21, 22], and the fresh properties of the mixtures have directly influence on the hardened composite. Thus, in this experiment, the air content and flow properties of GO composites have been evaluated. The experimental outcome is shown in Fig.3. The GO showed different effects on the different matrix. The tendency of air content and flow showed the similarity. First of all, GGBFS can increase the flow by thinning the RC-LFS. Secondly, the GO can decrease the flow when you compare the mixture A75G25 and A75G25GO. It is because of the large

 DOI: 10.1051/matecconf/201713803012

aspect ratio and nucleation effect [1, 23]. While the increasing flow of the mixture A100GO is because the increasing the air content by the GO.

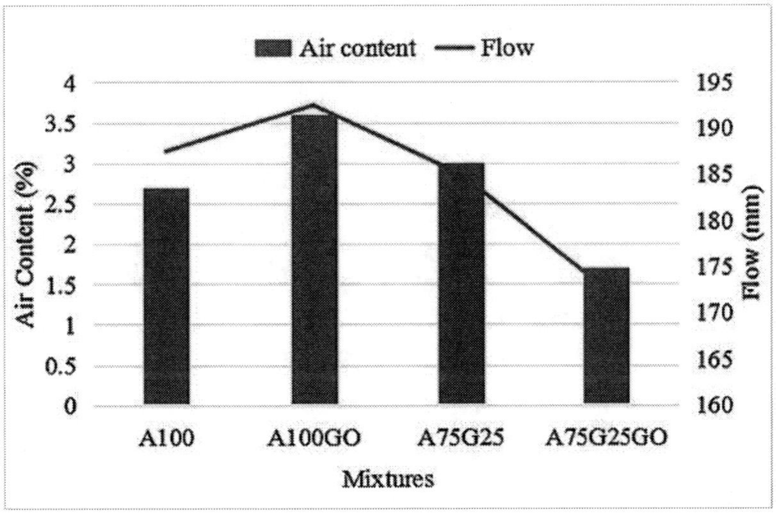

Fig. 3. Air content and flow of GO reinforced composite.

3.2 Flexural, compressive strength and the microstructure

Flexural and compressive strength will be tested after curing for 4 hours, 1 day, 7days and 28 days. The test results as presented in Fig.4.First of all, the binder RC-LFS showed remarkable early strength. The four-hours compressive strength was over 30MPa of mixture A100 and A100GO. Even after replaced by 25 vl.% of GGBFS, the four-hours strength still very closed to 30MPa. After 28 days curing, the compressive of both A100 and A100GO were over 60MPa. The flexural strength also showed high early strength of all four mixtures between 5 to 6 MPa. Secondly, the reinforcing effect of GO showed clearly. From the compare between those four mixtures, the max improvement of compressive strength was over 12.% and the max improvement of flexural strength was over 26%.

The improvement can be summarized as four main reasons. First of all, the function group on the surface of graphene oxide provide high bonding strength between GO and hydrations (such as C–S–H and calcium hydroxide) of cement [1,2,23]. Secondly, as nanomaterial, GO can increase the load-transfer efficiency from cement matrix to the reinforcement [2]. Thirdly, nucleation effect can accelerate the hydration of cement matrix. Moreover, larger regular crystal hydration produce can be generated on the surface of GO and generally cover the GO with curing time. Thus, the concrete with GO shows relatively better engineering performance. Finally, it's also believed that GO nanosheet can provide the crack-bridging effect which has a positive effect on crack resistance [1,2,23]. Base on the previous researches, GO can provide both physical reinforcement and chemical reinforcement at the same time.

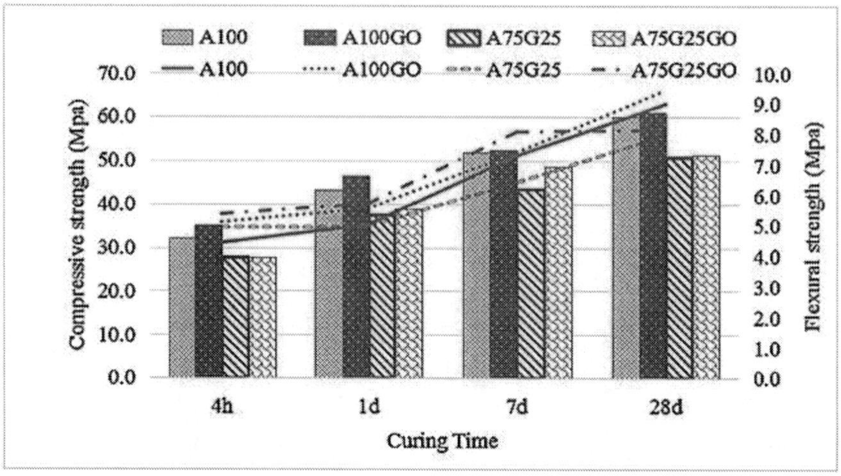

Fig. 4. Flexural and compressive strength.

In order to illustrate the mechanical properties, the microstructure of those mixtures also has been tested by SEM. The test result as shown in Fig.5.

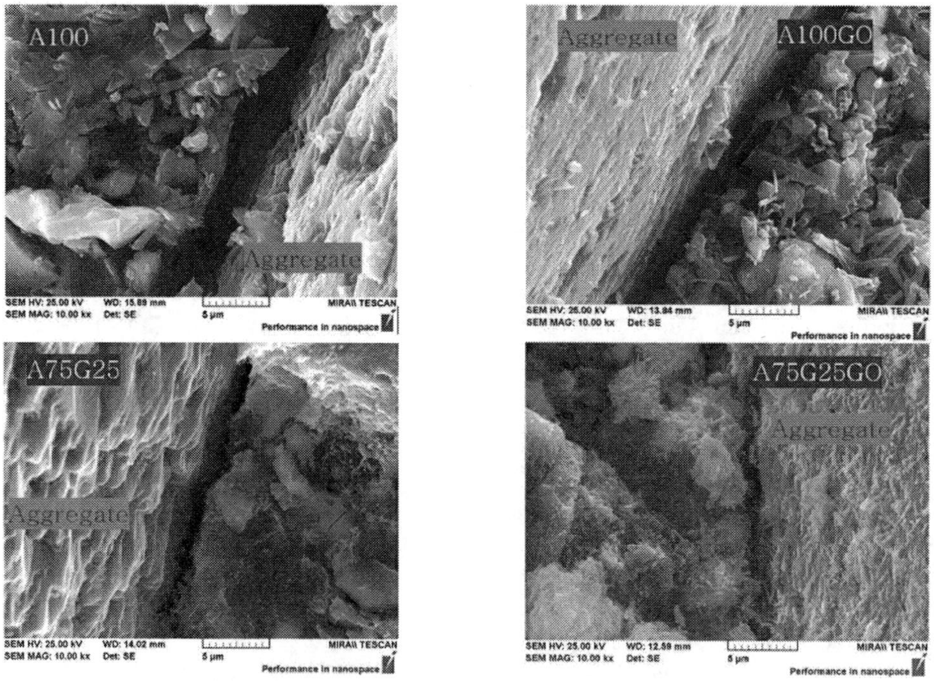

Fig. 5. The microstructure of hardened composite by SEM.

From the test results, the reinforcing effects of GO can also be noticed. The hydration product and the structures showed different. On the other hand, GO is very hard to be located in the matrix due to the nano size and also covered by the hydration productions.

4. CONCLUSIONS

This paper covers the experimental study about the engineering properties of graphene oxide slag based cementitious composites. The experimental outcomes indicate that graphene oxide can reinforce the hydraulic composite by both physical and chemical bounding. As nano sheet and 2D material with controllable functional groups, graphene oxide shows reinforcing effects. Only 0.05w% by the weight of binder can increase the flexural strength up to 26% and flexural strength up to 12 %.

References

1. Wengui Li, Xiangyu Li, Shu Jian Chen, Yan Ming Liu, Wen Hui Duan, Surendra P. Shah, "Effects of graphene oxide on early-age hydration and electrical resistivity of Portland cement paste," Constr Build Mater, Volume **136**, Pp. 506–514, (2017)
2. Y. Cao, J. Zhang, J. Feng, P. Wu, "Compatibilization of immiscible polymer blends using graphene oxide sheets," ACS Nano, **5** (7) , pp. 5920 - 5927,(2011)
3. P.S. Avanish, M. Monika, C. Amita, S.K. Dhawan, "Graphene oxide/ ferrofluid/ cement composites for electromagnetic interference shielding application",
4. Z. Pan, L. He, L. Qiu, A.H. Korayem, G. Li, J.W. Zhu, F. Collins, D. Li, W.H. Duan, C. Wang, "Mechanical properties and microstructure of a graphene oxide-cement composite," Cement Concr. Compos, **58**, pp. 140 - 147,(2015)
5. K. Gong, Z. Pan, A. Korayem, L. Qiu, D. Li, F. Collins, C. Wang, W.H. Duan, "Reinforcing effects of graphene oxide on Portland cement paste," J. Mater. Civ. Eng., **27** (2), p. A4014010,(2015)
6. Jin-Man Kim, Sun-Mi Choi, Dongyeop Han, "Improving the mechanical properties of rapid air cooled ladle furnace slag powder by gypsum," Constr. Build. Mater,**127**, pp. 93-101,(2016)
7. Sunmi Choi, Jin-Man Kim, Dongyeop Han, Ji-Ho Kim, "Hydration properties of ladle furnace slag powder rapid cooled by air," Constr. Build. Mater. **113**,pp.682-690,(2016)
8. Evi Aprianti S, "A huge number of artificial waste material can be supplementary cementitious material (SCM) for concrete production – a review part II," Journal of Cleaner Production, Volume **142**, Part 4, Pages 4178 - 4194,(2017)
9. D.D. Higgins, GGBS and sustainability, Proc. Inst. Civ. Eng. Constr. Mater, **160** (3) , pp. 99 - 101,(2007), http://dx.doi.org/10.1680/coma. 2007.160.3.99
10. Mohamed Elchalakani, Tarek Aly, Emad Abu-Aisheh, "Sustainable concrete with high volume GGBFS to build Masdar City in the UAE," Case Studies in Construction Materials, Vol **1**, Pages 10–24,(2014)
11. Önder Kırca, İ. Özgür Yaman, Mustafa Tokyay, Compressive strength development of calcium aluminate cement–GGBFS blends, Cement and Concrete Composites, Vol **35**, Issue 1,, Pages 163–170, (2013)
12. K.C. Quillin, G. Osborne, A. Majumdar, B. Singh, "Effects of W/C ratio and curing conditions on Strength development in BRECEM concretes," Cem Concr Res, 31, pp. 627 - 632,(2001)
13. A.J. Majumdar, B. Singh, "Properties of some blended high alumina cements," Cem Concr Res, 22, pp. 1101 - 1114,(1992)
14. K.C. Quillin, "Blended high alumina cements", Mater World, 1, pp. 103 - 105.(1993)
15. M. Heikal, M.S. Morsy, M.M. Radwan, "Electrical conductivity and phase composition of calcium aluminate cement containing air-cooled and water-cooled slag at 20, 40, and 60 °C," Cem Concr Res, 35, pp. 1438–1446,(2005)

16. ASTM, ASTM C230/C230M-14 Standard Specification for Flow Table for Use in Tests of Hydraulic Cement, ASTM Int. West Conshohocken
17. ASTM, ASTM C185 - 08 Standard Test Method for Air Content of Hydraulic Cement Mortar, ASTM Int. West Conshohocken
18. ASTM, ASTM C348-14 Standard Test Method for Flexural Strength of Hydraulic-Cement Mortars, ASTM Int. West Conshohocken
19. ASTM, ASTM C349-14 Standard Test Method for Compressive Strength of Hydraulic-Cement Mortars (Using Portions of Prisms Broken in Flexure), ASTM Int. West Conshohocken
20. ASTM, ASTM C305 - 06 Standard Practice for Mechanical Mixing of Hydraulic Cement Pastes and Mortars of Plastic Consistency, ASTM Int. West Conshohocken
21. Qin Wang, Xinyou Cui, Jian Wang, Shiyu Li, Chunxiang Lv, Yichen Dong, "Effect of fly ash on rheological properties of graphene oxide cement paste", Constr. Build. Mater., Volume **138**, Pages 35–44, (1 May 2017)
22. Konstantin Kovler, Nicolas Roussel, "Properties of fresh and hardened concrete", Constr. Build. Mater., Volume **41**, Issue 7, Pages 775 – 792, (July 2011)
23. K.M. Liew, M.F. Kai, L.W. Zhang, "Carbon nanotube reinforced cementitious composites: An overview, Composites Part A: Applied Science and Manufacturing", Volume **91**, Part 1, Pages 301 – 323,(December 2016)

Bamboo Waste as Part of The Aggregate Pavement The Way Green Infrastructure in The Future

Mudjanarko Sri Wiwoho[1,*], *Mayestino* Machicky[1], *Rasidi* Nawir[2], *Indrawan*[1], *and Setiawan* Ikhsan M.[1]

[1]Civil Engineering, Civil Department, Narotama University,Arief Rachman Hakim No. 51, Surabaya, 60117,Indonesia
[2]Civil Engineering, Civil Department, Polinema, Soekarno Hatta No. 9, Malang, 65141 Indonesia

Abstract. Paving block pavement is already widely used in Indonesia. Materials made of concrete paving blocks are easy to make and easy to implement. Currently the manufacture of paving blocks SNI 03-0691-1996 refers to a method that consists of a mixture of cement, sand/gravel: water. Our research uses bamboo pieces of waste material as a substitute for gravel. Comparison of the composition of the bamboo pieces that are used to dry conditions weighing 10 kg, 20 kg and 30 kg. Based on compressive strength testing performed at the age of 7, 14 and 28 days can be known ability paving block receives compressive strength. The research using bamboo fiber material at 28 days had compressive strength 171 kg/cm^2, 190 kg/cm^2, and 199 kg/cm^2. While using bamboo fiber material and fly ash at 28 days have compressive strength 231 kg / cm^2, 176 kg/cm^2, 252 kg/cm^2. Test results meet the quality of type C, D and the use of bamboo waste as aggregate paving can support green infrastructure in the future.

1 Introduction

Infrastructure is one of the most important things in the development effort and one of them is the use of paving material on the sidewalk. The availability of good infrastructure is very helpful for the implementation of economic development goals. The construction of sidewalks is one of the infrastructure to support community activities on foot. The sidewalks are still considered one eye in the development priorities as they are considered as supporting infrastructure and complementary roads. While the priority of development prioritizes the development of motor vehicle modes so that the infrastructure for pedestrians is forgotten. Though the sidewalk is an important pedestrian foot.

The role of the government is related to the provision of access and means for pedestrian protection from the activity of street vendors as it may disrupt the safe, comfortable and humane user community. The integration of pedestrian paths with the layout of the building, the accessibility between environments, and the transportation system in order to manifest well in the development stage for the future. These things

* Corresponding author: sri.wiwoho@narotama.ac.id

certainly need to be handled from the government to pay more attention to the construction of roadside pavement that is more useful according to allotment. So far, paving is used as a sidewalk infrastructure. During this Paving does not penetrate / less response to water so easily damaged. During this paving is made from a mixture of concrete in the form of sand, cement, aggregate and mixed with water. The use of aggregate replacement materials is still rarely used, including the use of waste materials. [10] Sri Wiwoho Mudjanarko, 2016, conducted research on paving using bamboo aggregate (compressive strength test pavement bamboo composite). Obtained research results obtained the largest compressive strength 192.29 with the use of 60% bamboo aggregate and meet the standards of quality SNI B. [6] M. Riang Endarto and M. Heri Zulfiar, 2010, in his research said that The greater percentage of bamboo aggregates in concrete mix , Then the value of compressive strength will be lower. The smallest average compressive strength value is 5.9 MPa that occurs in 100% bamboo aggregate on conventional split aggregate. [3] Bangkit T., T, 2015, conducting preliminary research on paving Pervious concrete with average 28-day compressive strength Pervious concrete with code B1 105,11 kg / cm2 and B2 100,30 kg / cm^2 meet standard paving class D with average weight limit of 100 kg / cm2 for the park road. And has a porosity / drain rate of 247.37 and 226.76 ltr / mnt / m^2.[7] Mokhammad Farid Ma'ruf, 2012, explains that the peak shear strength of bamboo root soil increases with increasing ground root volume ratio. Their relationship can be approached linearly. Although the root-to-ground volume ratio was investigated for only 5%, the peak power peak ratio increased to 55%. (The results show that peak shear strength of bamboo root reinforced soil increase with increasing soil root volume ratio.) Though soil-root volume ratio investigated was only up to 5%, it generated up to 55% additional Peak shear strength). Bamboo has several limitations, namely the fiber is stiff and easily attacked by fungi and insects. From the limitations of the bamboo fiber, the writer wanted to know the extent to which this bamboo fiber is utilized to be an additional material in making paving block. Not only bamboo fiber in this study the author also memamfaatkan fly ash. As we know that fly ash is often utilized in research, especially research in the field of construction. Fly ash is produced from Ash / residual coal combustion which produces a residue called fly ash.

2. Literature Study

2.1 Paving block

Paving block is a building material product of cement used as an alternative cover or hardening of the soil surface. Paving blocks are also known as concrete blocks (concrete blocks) or cone blocks. Based on [9] SNI 03-0691-1996 paving block (concrete brick) is a building material composition prepared from a mixture of portland cement or similar hydrolysis, water and aggregate with or without other materials which does not reduce the quality of concrete bricks. As a cover material and hardening of the soil surface, paving blocks are widely used for various purposes, ranging from simple needs to use that require special specifications. Paving blocks can be used for hardening and beautifying street sidewalks in cities, hardening of streets in residential complexes or residential areas, beautifying gardens, yards and yard, hardening of parking lots, office areas, factories, parks and school grounds, as well as in hotel grounds And restaurants. Paving blocks can even be used in special areas such as container ports, airports, bus terminals and train stations.

 DOI: 10.1051/matecconf/201713803013

Table. 1 Quality Terms of paving blocks [9]

Quality	Used	Compressive strength (kg/cm^2)		Wear resistant (mm/menit)		Average water absorption max
		Rate	Min	Rate	Max	%
A	Ways	400	350	0,0090	0,103	3
B	Parking	200	170	0,1300	0,149	6
C	Walking	150	125	0,1600	0,184	8
D	Garden	100	85	0,2190	0,251	10

2.2 Bamboo

Bamboo is a plant derived from clumps of grasses with cavities and segments dibatangnya. Bamboo easily breed well. Bamboo plants around the world there are more than 80. [1] Agnes Dwi Yanthi Winoto, 2014. In Indonesia there are about 60 species of bamboo, commonly found in open places that are free from puddles. Bamboo is very productive because it has a high growth rate of about 3-10 cm per day. Bamboo plants are used as food, handicraft, and building construction. In addition, bamboo plants are used to improve water catchment. [2] Asih Ardianisa, 2013, Bamboo is a common tree used as one of the building materials and widely available in Indonesia. Approximately 1000 species of bamboo in 80 genera, about 200 species of 20 genera are found in Southeast Asia and in Indonesia there are 154 species of bamboo. Bamboo is an easily cultivated plant. Besides being easy to cultivate, bamboo has a high production amount of about 109.2 tons / ha / year. Bamboo has a very rapid growth with harvest periods ranging from 3-5 years. Bamboo can be utilized as building component such as pole, beam, floor, wall or bulkhead, ceiling, roof, buffer roof, door and others. The formation of this product generally does not use the entire stem section, resulting in a number of waste such as sawdust, shavings, tatal, and the bottom of the stem. Unused or wasted bamboo waste can reach 14.6% - 33.5% or an average of 19.1%. One of the needs of infrastructure materials is a strong paving, cheap and environmentally friendly. Various ways of using bamboo materials as an alternative material and already done research in civil engineering. Bamboo is a clump of grass (perreunial Grass) with woody stems (Woody steams, culus) so in other words the bamboo anatomy is very different from wood. The bamboo network consists of parathyelial cells and vascular groups (rich in reeds). This cluster consists of feathers, thick-walled fibers and pipes. The weakness of bamboo is a relatively short life span due to exposure to environmental changes, so the use of bamboo as a structural element should be capillary shield which is the fibers that give strength to the bamboo. The use of bamboo fiber provides only a small amount of compressive strength, since bamboo has a high wear and tear and shrink rate, and lower structural strength than conventional aggregates. Bamboo Fiber Size made with size 20 mm and diameter 1 mm. The use of bamboo fiber makes it possible to produce paving lighter than paving using conventional aggregates.

2.3 Fly ash

[5] I.Nawaz, (2013). Fly ash is the remaining coal combustion in the form of amorphous fine particles, is an inorganic material formed from the change of mineral materials due to combustion process. From the coal combustion process in the steam generator unit (boiler) will form two types of ash namely: fly ash and bottom ash. The resulting coal ash

 DOI: 10.1051/matecconf/201713803013

composition comprises 10 - 20% bottom ash, while the remainder is about 80 - 90% in the form of fly ash captured by an electric precipitator before being discharged into the air through the chimne. Fly ash is produced from Ash / residual coal combustion which produces a residue called fly ash. Based on the International Journal of Innovative Research in Science, Engineering and Technology Fly ash is considered an amorphous and mineral mixture of Ferro-aluminosilicates. However, the main composition of fly ash depends on geographical factors related to the combustion conditions of coal and the type of ash. The main constituents of Fly ash are oxides, Si, Al, Fe, and Mg which constitute about 95-99% of the total fly ash constituents. [8] Mulyati and Saryeni Maliar, 2015, explained that to improve the quality of paving blocks, such as can be done by using mixed materials that can increase the compressive strength of paving block. Fly ash can be used as a material for the manufacture of artificial aggregates in concrete mixtures, paving block additives, mortar, brick, and lightweight concrete. Currently, fly ash is widely used for concrete mixtures, considering the fly ash contains pozzolan ingredients such as silicate and aluminate and a little calcium

3 METHODOLOGY

This study is an experimental research, which is doing different techniques and types of treatment in each group of research variables. The following research methods are presented in the flowchart below

3.1 Flow chart

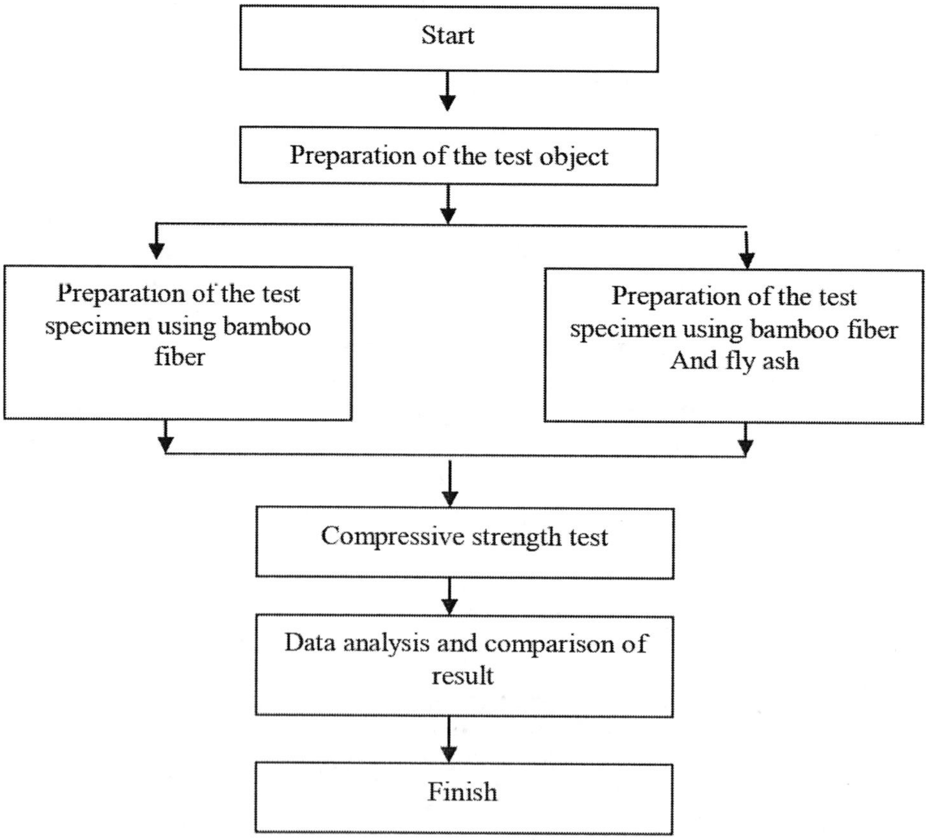

DOI: 10.1051/matecconf/201713803013

3.2 The results of fine aggregate examination (sand)

1. A fine aggregate gradation examination is performed to determine the variation of the aggregate grain distribution and the fine grain modulus number. Aggregates should have varying gradations, so that the pore volume between the aggregates formed becomes small. If the aggregate grains are uniform, the aggregate pore volume formed becomes large. The fine-grained modulus is an index used for the fineness or grain size of aggregate grains.
2. The dry type saturated sand (SSD) was 2.86, while the water absorption from dry conditions was 4.2%.
3. In this study the sand to be used for the mortar is sand under normal circumstances. Water content for sand under SSD condition is 4.2%

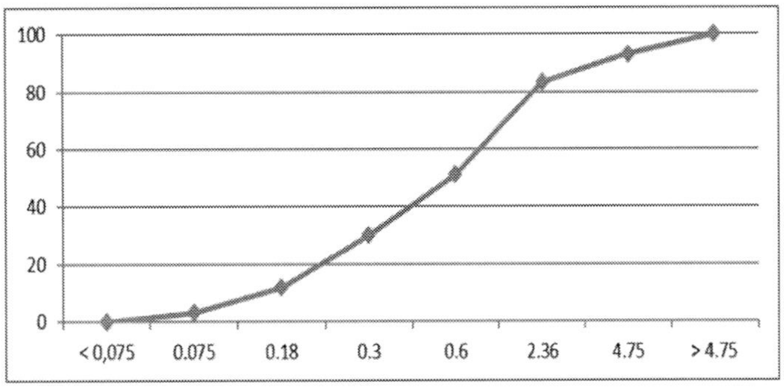

Fig. 1. Test gradation of sand

3.3 Making bamboo fiber

[4] Denny Nurkertamanda and Andi Alvin, 2012, At this stage are the steps to separate bamboo fibers to obtain the desired fiber, the best technique is a slow grinding process, aimed at separating between fiber binding and bamboo fiber. Pengrollan technique is different from beating, if the fiber roll is not interrupted due to beating too hard. If repeated emphasis techniques are applied, sometimes in some parts are missed. So the best technique of fiber separation is by either manual or with the help of mechanical roll. the manufacture of bamboo fiber is done by beating technique so that some bamboo fiber cut off due to beating

Fig. 2. Making bamboo fiber

3.4 Mix Design

Below is a table composition of the material of cement, sand, water and bamboo fiber and fly ash in Table 2 and Table 3.

Table. 2 *Mix Design* paving with Bamboo fiber

Material	Percentage of material			
	33%	**67%**	**100%**	**Unit**
Sand	1200	1200	1200	Kg
Cement	476,9	476,9	476,9	Kg
Water	150	150	150	Kg
Bamboo fiber	10	20	30	Kg

Table. 3 *Mix Design* paving with Bamboo fiber

Material	Percentage of material			
	33%	**67%**	**100%**	**Unit**
Sand	1200	1200	1200	Kg
Cement	476,9	476,9	476,9	Kg
Water	150	150	150	Kg
Bamboo fiber	10	20	30	Kg
Fly ash	10	20	30	Kg

4 RESULTS ANALYSIS AND DISCUSSION

4.1 Effect of Bamboo Fiber to Percentage to Pressure Strength

Bamboo fiber has an effect on strong compressive strength values. In this study with the use of bamboo fiber compressive strength reached the maximum limit at 28 days old with a value of 199 kg / cm2. In this study, the percentage of material density if greater will increase in compressive strength in because bamboo fiber has a high enough density, which affects the compressive strength.

4.2 Effect of Bamboo Fiber and fly ash to Percentage to Pressure Strength

The use of bamboo fiber with fly ash in this penetration obtained value of maximum compressive strength of 252 kg / cm2. In this study, the graph of compressive strength has increased with increasing age of paving

4.3 Compressive strength

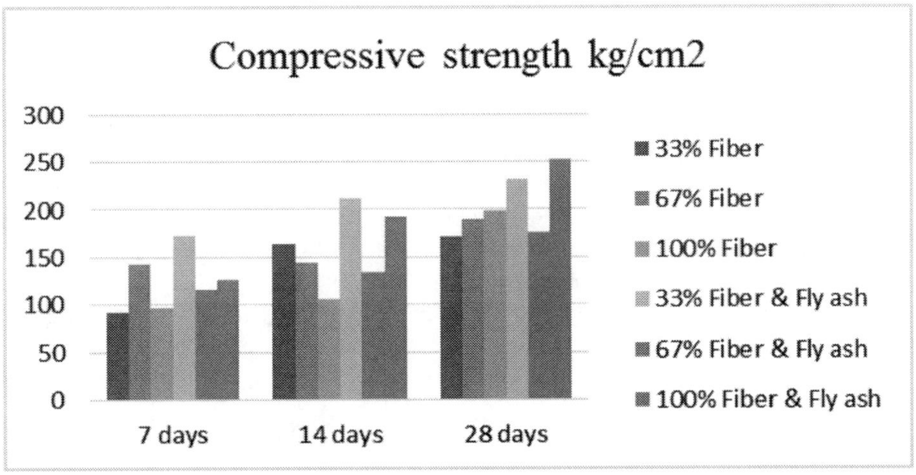

Fig. 3. Percentage Compressive Strength at age 7, 14 and 28 days

Fig. 4. Paving block

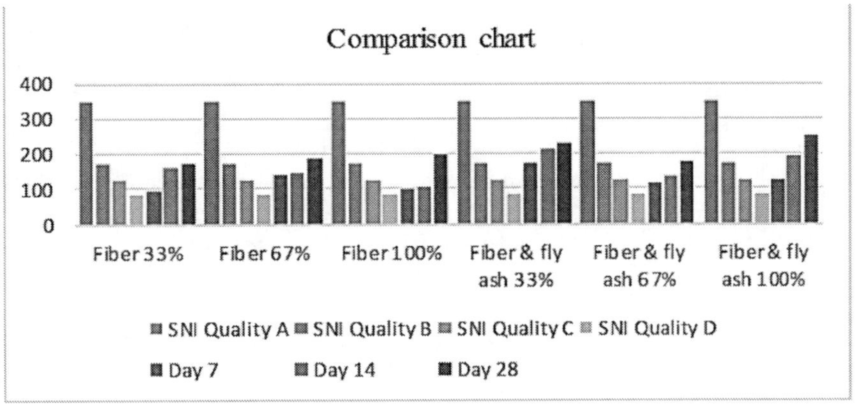

Fig. 5. Comparison of compressive strength with SNI

The following summary:

1. The dose on paving using 100% bamboo and fly ash fibers has the greatest compressive strength value of 252 kg / cm2 at 28 days, where this value is greater. Compared to paving using 100% bamboo fiber. In contrast to the researcher's expectations that the more bamboo used the smaller the resulting compressive strength
2. In paving with bamboo fiber and paving of bamboo fiber with fly ash does not have compressive strength reach 350 kg / cm2 that is value of compressive strength of paving block at quality A, value of highest compressive strength of result of test object is 252 kg / cm2 age 28 day, It is seen that paving blocks do not meet the standards on quality A
3. The value of compressive strength of paving block with class D quality can be achieved at 7 days

4 CONCLUSION

- The greater percentage of bamboo fiber in the paving mixture affects the compressive strength value being greater. The largest compressive strength value is 199 kg / cm2 and the smallest compressive strength value is 92 kg / cm2. Based on SNI the strongest value of the largest paving press meets the quality standard B and the smallest compressive strength paving value meets the quality standard D.
- The greater percentage of bamboo fiber and fly ash in the paving mixture affect the value of compressive strength becomes greater. The largest compressive strength value is 252 kg / cm2 and the smallest compressive strength value is 116 kg / cm2. Based on SNI, the strongest value of the largest paving press meets the quality standard B and the smallest compressive strength paving value meets the quality standard C
- Based from the compressive strength produced by all paving compositions with a mixture of bamboo fiber and bamboo fiber and fly ash, some of them already have compressive strength that meets the SNI 03-0691-1996 standard for use as sidewalks

REFERENCE

1. Winoto, A. D.Y., *Konstruksi bambu untuk bangunan.* Ensikopledia Teknik bangunan. ISBN: 978-602-7526-3 (2014)
2. Ardianisa, A, "Pengaruh Macam Katalis dan Ukuran Partikel Terhadap Sifat Papan Semen Limbah Bambu Petung, Skripsi", UGM, (2013) http://etd.repository.ugm.ac.id/index.php?mod=penelitian_detail&sub=PenelitianDetail&act=view&typ=html&buku_id=64443&obyek_id=4 (2 april 2017)
3. Anoraga, Bangkit T.T., Mudjanarko, S.W., Kurniawan, F., "Desain Perkerasan Jalan Ramah Lingkungan Menggunakan Pervious Concrete Untuk Jalan Setapak Dan Area Parkir", Skripsi-Seminar Nasional Poltek Sriwijaya Palembang (2015)
4. Nurkertamanda, D., Alvin, A., *Desain proses pembentukan serat bambu sebagai bahan dasar produk industri kreatif berbahan dasar serat pada ukm. J@ti Undip,* **VII**, No 3, September (2012)
5. I.Nawaz. *Disposal and Utilozation of fly ash to protect the Environment.* International Journal Of Innovative Research In Science, Engineering and Technology. ISSN: 2319-8753.**2**, Issue 10, October (2013)
6. Endarto, M. R., Zulfiar, M. H.,*Kajian eksperimen kuat tekan beton ringan menggunakan agregat bambu dan bahan tambah beton*, **13**, No. 1, 12-20 (2010)

7. Ma'ruf, M.F., "Shear strength of Apus bamboo root reinforced soil", Ecological Engineering, **14** , pp 84–86 (2012)
8. Mulyati., Maliar, S., *Pengaruh penggunaan Fly ash sebagai pengganti Agregat terhadap kuat tekan paving block.* Jurnal Momentum **17**, No.1, Februari 2015, ISSN : 1693-752X
9. SNI-03-0691-1996, *Tabel syarat mutu paving block,* Badan Standardisasi Nasional (BSN)
10. Mudjanarko, S.W., et all, "Model Paving Pervious Concrete Berkombinasi Agregat Bambu Untuk Pelataran Parkir", Usulan Proposal Hibah Produk Terapan Ristek Dikti Tahun 2016.

DOI: 10.1051/matecconf/201713804001

Overview: Microbial amendment of remediated soils for effective recycling

Soo-Bin Kim[1] and *Jun-Boum* Park[1]

[1]*Department of Civil and Environmental Engineering, Seoul National University, Seoul, Republic of Korea*

Abstract. In recent years, various methods are being considered with appropriate amendments, not with conventional reclamation to recycle deteriorated soils after remediation as agricultural addition, backfilling and construction materials etc. Among these amendments, microbial amendments with effective microorganism(EMs) are known to improve soil qualities such as fertility, strength and toxicity to be recycled into possible utilizations. This study indicates the possibility of recycling the remediated soils by using these EMs most efficiently. Soil samples will be collected from contaminated sites with either heavy metals or petroleum and will be remediated by bench-scale soil washing and thermal desorption. And then the remediated soils will be treated with easily obtainable inocula, substrates (culture media) near our life and they are compared with commercial EM products in terms of the cost and efficiency. Also, after treating with a number of mixing ratios, soil properties of (1) fresh, (2) contaminated, (3) remediated (4) amended soils will be evaluated based on soil quality indicators depending on demands and the optimal mixing ratios which are effective than commercial EM products will be determined. The ratio derived from pre-tests could be applied on the remediated soils with pilot-scale in order to assess suitability for recycling and characterize correlation between soil properties and microbial amendments regarding contaminants and remediation, and furthermore for modelling. In conclusion, application of the established models on recycling remediated soils may help to dispose the remediated soils in future, including environmental and ecological values as well as economical values.

1. Introduction

These days, soils have been significantly contaminated with contaminants such as heavy metals or petroleum any times from everywhere since urbanization and industrialization. Various methods like soil washing and thermal desorption etc. are used to remediate the contaminated soils and also aggressive and subsequent remediation methods are being used to increase remediation efficiency. They remove contaminants from the soils by utilizing physiochemical, biological, and thermal processes and can satisfy soil remediation standards within a limited time. However, such process could affect soil properties and degrade the quality of the soils. Effects of soil washing and thermal desorption, which are the most frequently used remediation methods, on the physiochemical and biological properties of remediated soil were being investigated by many previous studies. After soil washing, particle size, water holding capacity, electrical conductivity (EC), exchangeable cations (potassium, calcium, magnesium), organic matter, cation exchange capacity (CEC), total nitrogen, total microbial number and nutrients (manganese, zinc) bioavailability were decreased. And after thermal desorption, acidity, EC, exchangeable cations, organic matter, CEC, total nitrogen, available phosphate, total microbial number were decreased (Yi et al. 2012, 2013). With the increase of EDTA concentration, the maximum void ratio, soil cohesion, arrangement directionality and the content of illite, albite and montmorillonite decreased while the consolidation coefficient, compression modulus, internal friction angle, plastic limit, elastic limit and content of quartz increased (Wang et al. 2013). Aggressive remediation with high temperature affect the particle size distribution, mass loss, mineralogy and permeability of the soil and it is very likely to affect the dynamic behavior such as infiltration, permeability and shear behavior depending on the sample composition, sand only or sand-clay mixture (Zihms et al. 2013). The range of maximum dry densities under compaction remained reasonably consistent; however, a clear reduction in the optimum moisture content was observed in the thermally treated soils (Khan et al. 2014). Micro CT images have shown that high temperature smouldering remediation makes the grain surfaces significantly smoother and this change may explain changes that have been observed in dynamic soil grain-grain and grain-water interactions such as permeability, cohesiveness, and shear strength of soils (Switzer et al. 2013, 2015). These results show that the remediation processes used to clean contaminated soils affect the soil qualities.

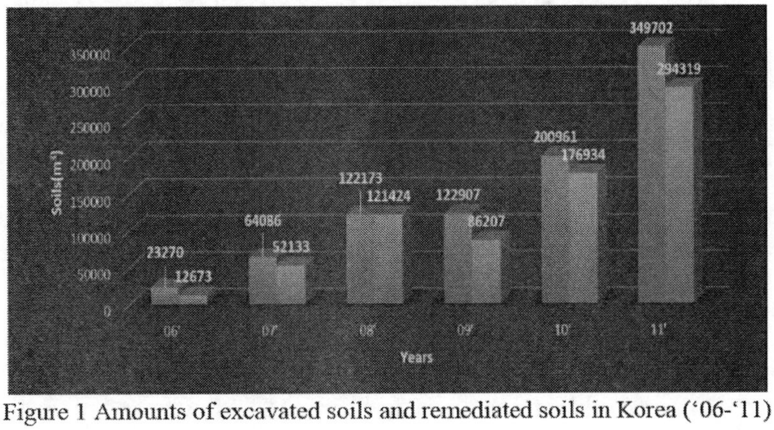

Figure 1 Amounts of excavated soils and remediated soils in Korea ('06-'11)

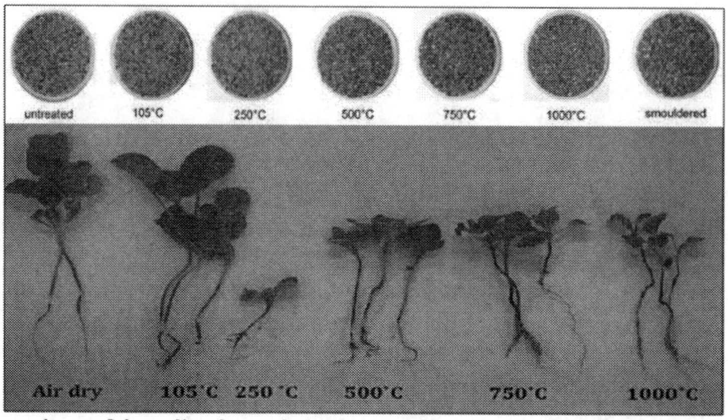

Figure 2 Comparison of the soil color and aboveground length of plants between thermally treated soils with different temperatures and natural soils (Switzer et al. 2015)

Commonly, the most of remediated soils have been used for backfilling and landfilled. Due to high amount of the remediated soils and treatment costs, the possibility to utilize the remediated soils in for example agricultural soil addition, backfilling and construction materials etc. would be desirable. However, there are some functional problems to recycle the remediated soils as diverse usages. The remediated soil could not satisfy the soil criteria for sources of demands and be managed in insufficient regulation. The remediated soils have several problems which are infertility, low strength and residual toxicity depending on demands. For agricultural addition and landscape, it may be inadequate in terms of fertility and toxicity. For construction materials, strength and toxicity. For replantation in roadside cutting-slope, strength and fertility.

In response to the change in soil properties by remediation, the soil material has been amended technologically and ecologically. To utilize the cleaned soil for healthy and more value–added purposes, soil improvement and amendment are needed. Among the amendments, microbiological techniques have potential capacities not to amend the remediated soils, but also to exert multiple functions. The uniqueness of microorganisms and their often unpredictable nature and biosynthetic capabilities, given a specific set of environmental and cultural conditions, have made them solve particularly difficult problems in ecosystem.

For many years, soil microbiologists and microbial ecologists have tended to separate soil microorganism into beneficial and harmful depending on their functions and how they affect soil quality. The microorganisms which are beneficial to soils were suggested by Teruo Higa, University of the Ryukyus, Japan and they were referred to as 'Effective Microorganism (EM)'. The concept of EM means mixed cultures of beneficial and naturally-occurring microorganisms that can be applied as inoculants to increase the microbial diversity of soils. A series of inoculations are made to ensure that the introduced microorganisms continue their dominance over the indigenous populations. Originally, EMs was used for agricultural compost and soil conditioning and these include the suppression of plant pathogens and diseases, conservation of energy in plants, solubilization of soil minerals, soil microbial-ecological balance, and biological nitrogen fixation (Higa et al. 1991). Furthermore, it has been developed as alternative uses for a more sustainable soil management.

And also EMs could be used for bioremediation to remove residual trace contaminants in the remediated soils. TPH concentration before EM treatment was 323.8 mg/kg, whereas TPH concentrations on 2 days after EM treatment was 102.1 mg/kg and on six days after EM treatment TPH was 91.3mg/kg (Lee et al., 2008). But for inorganic toxic compounds such ad heavy metals, microbes are unable to simplify them into harmless compounds, and they should be used according to their specialization for the type of contaminants. Some specific microorganisms are known to reduce the toxicity by different mechanisms such as biosorption, metal-microbe interactions, bioaccumulation, biomineralisation, biotransformation and bioleaching (Dixit et al. 2015).

According to recent geo-biological researches about Microbially Induced Calcium Carbonated Precipitation (MICCP), an alkalophilic soil bacterium with a highly active urea enzyme consumes urea within the microbe, decomposing it into ammonia (NH_3) and carbon dioxide (CO_2). These chemicals diffuse through the cell wall of the gram positive microbe and into the surrounding solution. The reactions spontaneously occur in the presence of water; ammonia is converted to ammonium (NH_4^+) and carbon dioxide equilibrates in a pH-dependent manner with carbonic acid, carbonate and bicarbonate ions. The net increase in pH is due to hydroxyl ions (OH^-) generated from the production of NH_4^+ which exceeds the available Ca^{2+} for calcite precipitation. This provides the alkaline environment and carbonate required for the precipitation of calcite ($CaCO_3$). The negatively charged bacterial cell is attracted to the soil particle surface due to a higher concentration of nutrients adjacent to surfaces in addition to the physicochemical properties of both the bacterial cell and soil particle (Dejong et al. 2010). When diluted commercial EM and EM ceramics are mixed with cement, the compressive strength at 3 days and 7 days was found to be larger by 30 to 50% than that of control (Sato et al. 2000).

Culture-dependent microbes already contribute much to human life, yet the latent potential of vast numbers of uncultured and thus untouched microbes, is enormous (Patil et al. 2014). Indigenous microorganisms are a group of innate microbial consortium that inhabits the soil and the surfaces of all living things. Indigenous microorganisms inhabit the soil with the abilities of biodegradation, bioleaching, biocomposting, nitrogen fixation, and improving soil fertility and they are being used effectively as microbial inoculants that could exist naturally in soil or added as microbial inoculants to soil where they can improve soil quality (Kumar et al. 2015). The use of indigenous inoculum isolated from local soils resulted in a higher arbuscular mycorrhizal fungal diversity in the root of plants growing in the remediated soil with EDTA chelating agent extraction compared to the ones amended with the commercial inoculum (Maček et al. 2016).

Figure 3 Expected demands of the microbially amended soils

2. Materials and Methods

2.1 Materials

2.1.1 Soil

Soil samples will be collected from three different points of soil surface (0-20 cm). The soil samples will be collected from contaminated sites with heavy metals and petroleum respectively. Non-contaminated soil will also be collected from the arable land. The sampling site is located close to an industrial area in Korea and was used for cultivating before.

2.1.2 Microorganism

The commercial EM and EM ceramic will be obtained from the Korean EM Research Organization (Busan, Republic of Korea).

Other beneficial microorganism will be isolated from indigenous solids like cultivated soils, contaminated soils and concrete structure.

2.1.3 Culture media

The commercial molasses will be also obtained from the Korean EM Research Organization.

Chemical agar medium containing nutrient broth and inorganic compounds will be produced by using other researches recipe depending on microorganisms.

Microorganisms are effective only when they are presented with suitable and optimum culture. But, chemical agar medium is expensive to be feasible in real situation. Some organic wastes like food wastes, agricultural by-products and rice water processed with grinding or autoclaving could be used as economical substrates to cultivate microorganism. Addition of an organic compost and liquid swine manure for the removal of soil TPH showed higher efficiency as 84.4% and 92.2% respectively than inorganic nutrients of 80.2% (Kim et al. 2008).

Also, soil solution extracted from indigenous soil by mixing and centrifuging may be effective media to adapt microorganisms to the remediated soil.

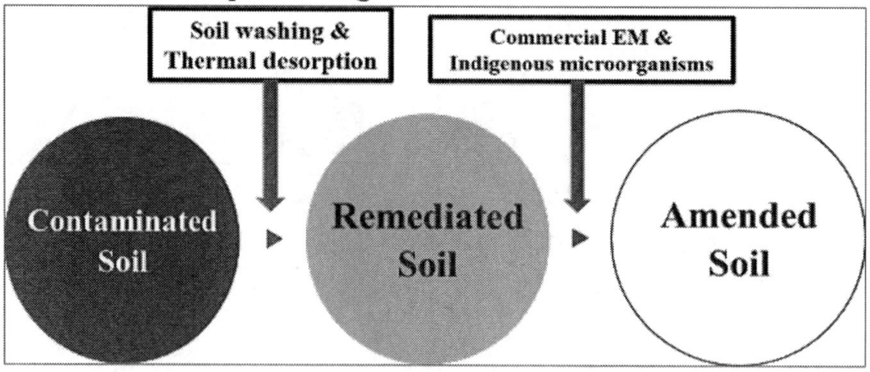

Figure 4 Schematic process of remediation and amendment

2.2 Methods

2.2.1 Soil remediation

2.2.1.1 Soil washing

The soil sample contaminated with heavy metals will be remediated by batch test with EDTA solution as reported by Lei *et al.* The air-died soil samples will be sieved using a sieve (2 mm) to remove large particles, then thoroughly mixed to ensure uniformity. The tubes containing 2 g soil samples and 20ml EDTA (0.075mM) will be shaken at a speed of 280 rpm at room temperature for 3 h and 3 cycle. The suspension will be centrifuged at 5000 rpm for 15 min and the supernatants will be filtered through a 0.45 μm membrane for heavy metal analysis. Chen *et al.* compared the extraction efficiency of several chelating agents for heavy metal contaminated soil, results showed that extraction efficiency of Cd increased with the increasing eluent concentration, extraction efficiency of the same eluent concentration followed the order EDTA> DTPA> NTA> PA> CA> CD> HCl, but only EDTA can achieve the desired leaching effect on Pb and Cd.

2.2.1.2 Thermal desorption

The soil sample contaminated with petroleum will be also remediated by remedy screening test as reported by US. EPA. Among these tests, static tray test full-scale thermal treatment systems. Briefly, An aliquot of 100-500g of fine earth fraction (<2mm) of contaminate soil is heated at 760℃ in a muffle furnace equipped with an electronic temperature controller. The depth of the soil should be kept at a minimum to eliminate temperature and concentration gradients within the soil bed. The time to reach the target temperature should be minimized to a practical laboratory timeframe such as 5 to 10 minutes. Longer time may be required depending on the specific contaminants present in the soil.

2.2.2 Microorganism Isolation and Cultivation

Base on the collection sites, the process of collection and isolation methods are different as they vary from place to place.

For isolation of bacterial strains, 1g of indigenous soils will be homogenized with 9ml of sterile deionized water. Serial dilutions are prepared in 9ml of sterile deionized water and plated on Nutrient Agar (NA). From the NA plates, representative colonies of all different morphologies are chosen at random, purified by sub-culturing and maintained in slants of NA. All culture works will be conducted aseptically.

Cultivation of the microorganism will be conducted under aerobic conditions in a medium containing Nutrient Broth. Inoculated culture will be incubated at 30℃ in an incubator for 120 hr.

2.2.3 Soil amendment

The bacterial isolate $(5 \times 10^7 CFU/soil\ g)$ will be inoculated into 250 ml Erlenmeyer flasks containing diverse culture medium as previously stated and aerobically cultivated at 30℃ by the agitating at the speed of 150 rpm. The cells are harvested from the growth medium at early-stationary phase by centrifugation at

5000 rpm for 10 min or filtering with Whatman No. 1 filter paper. After rinsing in deionized water the cells are again centrifuged or filtered. The cell could be diluted with deionized sterile water prior to use and it can be used as microbial reagent.

Batch incubation will be carried out with 10g of 2-mm sieved remediated soil (oven dried) and different microbial reagents with a number of mixing ratios in 50 ml sterile glass bottle.

Pot incubation will be conducted with spraying the microbial reagents to the remediated soils for 8 weeks with a week interval.

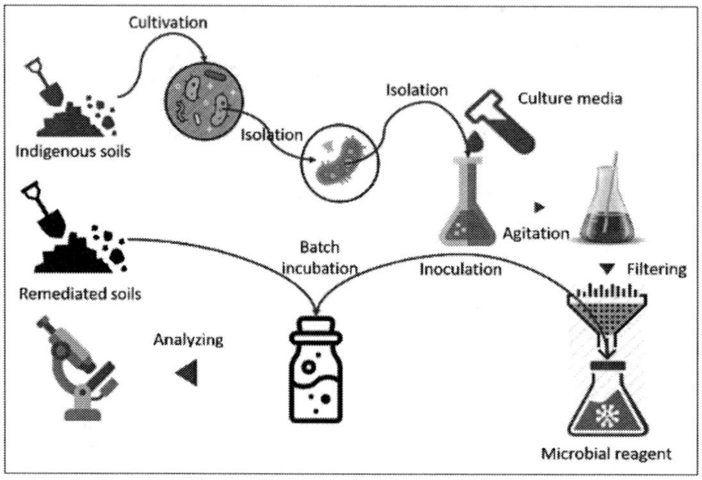

Figure 5 Schematic process of microbial amendment

2.2.4 Soil and Microbes analysis

Soil quality indicators selected by KS | ISO15176 depending on source of demands will be classified into fertility, strength, bioremediation according to physical, chemical, biological, ecological and engineering soil properties.

2.2.4.1 Physical & Engineering properties

Bulk density and water content will be evaluated by USDA method.

Compressive strength will be evaluated with unconfined compression test. Shear strength will be estimated by direct shear test. Plastic index and hydraulic conductivity will be estimated by ASTM D4318-10 and D2434-68.

2.2.4.2 Chemical properties

Soil pH, EC, texture, exchangeable cations (Ca, K, Mg, Na), bulk density and water content will be estimated by USDA methods. CEC will be evaluated by US. EPA method 9081. Total nitrogen and available phosphate will be evaluated by UV/visible spectroscopic method. Organic matter will be estimated by Walkley-Black method.

Heavy metals in the soils will be extracted using sequential extraction procedure and determined by atomic absorption spectrophotometry. Hydrocarbons in the soils will be extracted using a modified standard protocol of determining soil hydrocarbon

DOI: 10.1051/matecconf/201713804001

content in soil according to ISO/DIS GC-method.

2.2.4.3 Biological & Ecological properties

The microbial populations in the microcosm soils will be determined by a most probable number (MPN) procedure.

Five mustard (*B. alva*), and 3 pea (*P. sativum*) will be added separately in triplicate to glass jars containing 20 g of soils re-wetted to 80% WHC. Lids are loosely screwed on to reduce evaporation but allow aeration and the seeds are left to germinate at 22℃, 80% humidity, 16 h full illuminance and 8hr darkness. When >70% seeds in the non-contaminated soil germinated, the number of seeds germinated in all soil samples are recorded; this is after 4, 6 or 7 days exposure for mustard and pea, respectively. Seedling are dried at 105 ℃ for 24 h and their mean dry weight calculated to assess effects (Dawson et al. 2007). Aboveground length of the mustard and pea will be estimated at 8th days after seedling.

2.2.5 Statistical analysis

T-test will be conducted to all soils in process for evaluating effects. The experimental results will be statistically analyzed using SAS program with 0.05 % significance level.

3. Conclusions

This study is aimed to establish a system to enhance effective utilization of the remediated soils reasonably by considering occurrence location, soil properties and soil criteria for source of demands. By doing so, it is intended to develop a soil quality improving device applied to the remediated soil from actual disposal plant, to design an expanded one in field scale, and to derive optimal driving factors.

Ultimately, in this way to a meaningful and significant extent, a correlation of soil properties and microbial amendment will be revealed in bench- and pilot- scale and it could be used to model a function of soil properties and microbial amendments depending on soil demands for effective recycling.

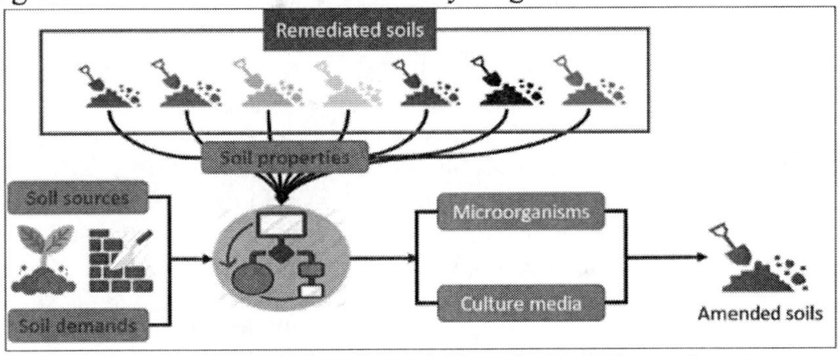

Figure 6 Schematic flow for modelling of microbial amendment

Acknowledgement

This research is supported by the Basic Research Project of the National Research Foundation of Korea (NRF).

References

1. B. L. Kumar and D. V. R. S. Copal, Effective role of indigenous microorganisms for sustainable environment, *Biotech* **5**, 867-876 (2015)
2. C. Switzer, S. Zihms, A Pape, A. Robson and C Knapp, Effects of aggressive remediation on soil properties and function, *EGU General Assembly* **15** (2013)
3. C. Switzer, S. G. Zihms and A. Tarantino, Effects of high temperatures on soil properties: Lessons to share from smouldering remediation experience, *FLAMMA* **6(1)**, 20-22 (2015)
4. G. T. Dejong, B. M. Mortensen, B. C. Martinez and D. C. Nelson, Bio-mediated soil improvement, *Ecological Engineering* **36**, 197-210 (2010)
5. I. Maček, N. Šibanc, M. Kavšček and D. Lestan, Diversity of arbuscular mycorrhizal fungi in metal polluted and EDTA washed garden soils before and after soil revitalization with commercial and indigenous fungal inoculum, *Ecological Engineering* **95**, 330-339 (2016)
6. J. J. C. Dawson, E. J. Godsiffe, I. P. Thompson, T. K. Ralebiso-Senior, K. S. Killham and G. I. Paton, Application of biological indicators to assess recovery of hydrocarbon impacted soils, *Soil Biology & Biochemistry* **39**, 164-177 (2007)
7. Kim GJ, Oh ST, Lee CH, Seo SK, Kang CH and Chang YY, Enhancement of biodegradation rate of petroleum hydrocarbons-contaminated soil with addition of organic composite nutrients and a chemical oxidation, *J. Soil & Groundwater Env.* **13(3)**, 59-66 (2008)
8. Lee EJ, Lee SM, Lee GT, Kim IS and Kim YH, Application of effective microorganisms for bioremediation of crude oil spill in Taean, Korea, *Journal of the Environmental Sciences* **17(7)**, 795-799 (2008)
9. M. Lei, T. Chen, J. Yang, X. Zhou, Y. Wang and Y. Li, Optimized EDTA washing procedure to decontaminate heavy metals from soils in iron and steel work sites, *Asian Journal of Chemistry* **25(1)**, 37-41 (2013)
10. N. Sato, T. Higa, M. Shoya and S. Sugita, Some properties of concrete mixed with effective microorganisms and the on-site investigation of the completed structures, *Hachinohe Institute of Technology, Japan* (2000)
11. P. Wang, J. Li and H. Wang, Engineering properties of heavy metal contaminated soil affected by EDTA washing, *EJGE* **18**, 3909-3918 (2013)
12. R. Dixit, Wasiullah, D. Malaviya, K. Pandiyan, U. B. Singh, A. Sahu, R. Shukla, B. P. Singh, J. P Rai, P. K. Sharma, H. Lade and D. Paul, Bioremediation of heavy metals from soil and aquatic environment: An overview of principles and criteria of fundamental processes, *Sustainability* **7**, 2189-2212 (2015)
13. S. A. Khan, S. J. Tan, E. K. Abdul Rahman and Dk. N. Muneerah, Effect of hydrocarbon contamination and subsequent remedial treatment on the engineering properties of soil, *Engineering and Technology (BICET 2014), 5th International Conference on Brunei* (2014)
14. S. S. Patil, E. M. Adetutu, J. Rochow, J. G. Mitchell and A. S. Ball, Sustainable remediation: electrochemically assisted microbial dechlorination of tetrachloroethene-contaminated groundwater, *Microbial Biotechnology* **7(1)**, 54-63 (2014)
15. S. Zihms, C. Switzer, M. Karstunen and A. Tarantino, Understanding the effects

of high temperature processes on the engineering properties of soils, *Proceedings of the 18th international conference on soil mechanics and geotechnical engineering*, 3427-3430 (2013)
16. T. Higa and G. N. Wididana, The concept and theories of effective microorganisms, In J. F. Parr, S. B. Hornick, and C. E. Whiteman (ed.) *Proceedings of First International Conference on Kyusei Nature Farming*, U. S. Department of Agriculture, Washington, D. C., USA, 118-124 (1991)
17. X. W. Chen, X. Wang, X. Chen, Study on the extraction efficiency of heavy metals by chelates, *Jiangsu Environmental Science and Technology* **18(2)**, 9-10 (2005)
18. Yi YM, Oh CT, Kim GJ, Lee CH and Sung KJ, Changes in the physicochemical properties of soil according to soil remediation methods, *J. Soil & Groundwater Env.* **17(4)**, 36-43 (2012)
19. Yi YM, Kim GJ and Sung KJ, Effect of soil remediation methods on the biological properties of soils, *J. Soil & Groundwater Env.* **18(3)**, 73-81 (2013)

DOI: 10.1051/matecconf/201713804002

Tunnel Portal Construction using Sequential Excavation Method: A Case Study

Rana Muhammad Asad Khan[1*], *Zaka* Emad [2] and *ByungWan* Jo[1]

[1]Department of Civil and Environmental Engineering,
Hanyang University,
Seoul, South Korea.
[2] Department of Mining Engineering,
University of Engineering and Technology,
Lahore, Pakistan.

Abstract. Portal excavation in soft rock is one of the most challenging tasks in the construction of underground facilities. Significant convergence, collapses and surface settlement are usually associated with portal construction. Alluvium is loose unconsolidated material and most often tunnels constructed through tend to destabilize. In this study, portal excavation design has been analyzed using the finite element based computer program known as Phase2. The method of top heading and benching was not a suitable approach, keeping in view the previous experience. Subsequently, a different design of sequential excavation method (SEM) was proposed. Minimum convergence, minimum surface settlement, and machinery constraints were considered to be the vital importance for selection of final design. The results of finite element method (FEM) analysis showed that the finally selected SEM design has a roof convergence of 4 mm for the heading and full face excavation has 25 mm. These values were comparable with the ones obtained from a 5-pin convergence station, installed during the portal excavation (0.5 m inside from the portal). No or very little surface settlement was shown by the numerical model and actual field observation. Consequently, a proper SEM design based on numerical modeling allowed a successful construction of a large portal in alluvial deposits.

Keywords: Tunnel; portal; numerical modeling; finite element method; sequential excavation method

*Corresponding Email: masadkhan87@gmail.com

1. INTRODUCTION

Underground structures have significant importance for the economy of a country. The stability of such structure is of extreme importance because the safety of human lives is directly associated. Excavation of an underground opening always induces some deformation and stresses in the surrounding material. The amount of induced deformations and stresses is highly dependent on the nature of surrounding material as well as the depth of the excavation. The transition between the surface and underground structures are tunnel portals, at this junction the conditions can abruptly change causing several issues of instability as well as the stability of associated structures and adjacent slopes are also attached with tunnel portals. Furthermore, tunnel portal excavation becomes much more challenging when the encountered material is the soft ground of low shear strength such as alluvial deposits. Such material offers great resistance to any excavation as their properties are much closer to those of soil [1-4].

In order to enhance the stability of underground structure during construction and post-construction, always there exist two approaches either to increase tunnel support or to change tunnel excavation sequence. The later one is always given preference because of cost constraints and many other projects factors. Normally, the selection of a suitable excavation sequence can enable safe and stable excavation in any material [5-8]. Among various approaches of excavation, sequential excavation method (SEM) appropriately defines the sequences of excavation by considering the nature of surrounding material and size of structure opening. But the final selection of excavation sequence is highly dependent on the cost, safety, project conditions and project scheduling [9]. In addition to all these, technical factors such as; surrounding material properties, size and shape of the opening, regional geology and many to name more have proven themselves as a handful tools for the selection of an optimum excavation sequence [10].

This study focuses on the selection of an optimum excavation sequence for the safe and stable construction of a large diameter (15.0 m) hydropower tunnel portal in alluvial deposits. This tunnel portal has been constructed at a renowned hydropower project in the North-Eastern areas of Pakistan. At this particular project, large tunnel diameter excavation in the alluvial deposit was a great challenge. Moreover, the occurrences of a number of collapses in a tunnel of 11.0 m diameter excavated in alluvial deposit using heading and benching technique, clearly indicated the insufficiency and incompatibility of heading and benching. Therefore, it was necessary to re-design the excavation sequence for the tunnel portal construction of 11.0 m diameter. Thus, two new sequences of tunnel portal excavation were proposed which were analyzed in 2D using a Finite Element Method (FEM) program Phase 2D 8.0. There exit several numerical approaches such as Discrete Element Method (DEM) [11] and Discontinuous Deformation Analysis (DDA) [12]. FEM has been selected because of its simple modeling and it is easily available. These new excavation designs were proposed by considering project constraints such as cost, pre-defined tunnel support, available machinery and project schedule. Among these designs, one which showed minimum surface and inside portal displacement in FEM analysis to ensure maximum safety was selected and followed exactly in the same pattern at the construction site. Moreover, the post-construction safety was considered during the selection of this excavation sequence. Finally, the tunnel portal was constructed with the selection of an appropriate excavation design. Major contributions of this study are:

1. 2D FEM modelling of tunnel excavation in alluvial deposit of non-uniform nature.
2. Comparison of heading and benching, and sequential excavation sequence methods of tunnel excavation.
3. Selection of an appropriate excavation sequence to ensure maximum safety within the project constraints and pre-defined tunnel supports.

In the present study, Section 2 explains the case study along with the geotechnical properties of lithologies present at the construction site as well it summarizes the properties of encountered alluvial deposit. Moreover, it defines the alternative excavation sequences of tunnel portal. Section 3 describes the conditions of numerical modeling. Section 4 compares the principle stresses, total displacement and yielded elements for heading and benching, and sequential excavation methods. Section 5 emphasizes on the conclusions along with some discussion.

2. CASE STUDY

In recent years, Pakistan has focused on renewable energy resources, among those energy resources hydropower is a paramount energy provider. The case study hydropower project is a renowned under the construction project in the North-Eastern areas of Pakistan. Most of the underground structures at this particular project have been excavated in hard rock, nevertheless, initial 100 m of the headrace tunnel and associated tunnel portal were excavated in soft ground (alluvial deposit). This tunnel portal has a diameter of 15.0 m which reduces to 11.0 m of headrace tunnel. Larger diameter excavation in alluvial deposit imposed potentially instability threats, thus, it was a great challenges to ensure during construction and post-construction safety. The safety was ensured through the selection of an appropriate sequential excavation design. Present study aims at the selection of optimum excavation sequence of the tunnel portal in alluvial deposit to ensure construction and post construction safety. Therefore, a tunnel portal of 15.0 m diameter has been modelled and analyzed in 2D with various excavation designs and sequences.

The bed rocks in studied region lie in the lower western part of Himalayas and belong to Palaeocene to mid-Eocene geological age. Major lithological units are siltstone, sandstone, and mudstone while localized shale composed off fine particles of quartz, silt and clay, is also present up to some extent. Splintery and fractured nature of shales separates it from surrounding mudstone. Sandstone has been classified into two categories: SS1 and SS2 depending on its strength and colour. Sandstone (SS1) is mostly composed of fine to medium size particles of purple grey to greenish grey color, with the dominance of fines and high strength [13]. The bedding contact of this sandstone with contiguous lithologies is very sharp and abrupt. On the other hand, medium to strong sandstone (SS2) is mostly

DOI: 10.1051/matecconf/201713804002

composed off reddish brown sand and silt particles. SS2, siltstone, and mudstone are hard to differentiate because they are pretty much similar in color, only the foremost distinction is the amount of fines, which increases from sandstone to mudstone. Among these three different rocks types, siltstone has an intermediate amount of fines while mudstone can be ranked highest for the utmost quantity of fine particles. Table 1 illustrates the encountered rock mass lithologies along with their geotechnical properties.

Table 1. Lithologies of rock masses and their properties at the study area.

Sr. No.	Lithology	Poison's Ratio	UCS	Young's Modulus	Density	Cohesion	Slake Durability	Tensile Strength
			(MPa)	(GPa)	Kg/m^3	MPa	(%)	(MPa)
1	SS1	0.27	86	58.1	2705	6.02	98.95	9.65
2	SS2	0.25	75	50	2719	5.961	97.9	13.11
3	Mudstone	0.21	42	27.4	2766	5.87	--	--
4	Siltstone	0.31	66	40.6	2766	8.91	--	--
5	Shale	0.3	24.36	22.25	2726	--	--	--

The host material of headrace tunnel portal is an alluvial deposit of late Oligocene to early Miocene geological age. Site and laboratory investigations conducted on this alluvial deposit showed that it is a well compacted to well-cemented material. The main constituents are gravels, clay, and sand. The gravels are well graded and also rounded to well-rounded varying from few centimeter to 20 cm, while, the sand and clay acts as matrix. However, contact between gravels and the filling material is sharp. Gravel or cobble gravel is appropriate terms for this alluvial deposit instead of term conglomerate. Boreholes drilled during preliminary investigations suggested that the only shallow water table and perched on bedrock is present. Therefore, it was a clear indication for dry to damp tunneling in this alluvial deposit. Anyhow, a slightly damped or minor inflow (<1 Liter/min) was observed during the portal construction. Geotechnical investigations classified this material as Well Graded Gravel with Sand and Clay (GW-GC), according to the Unified Soil Classification System (USGS).

Testing in actual field actual field conditions for such material has always been preferred over the scale testing because of the less sample disturbance and determination of accurate properties of materials. Some laboratory scale wet sieve analysis and hand sorting showed 70-75% cobbles or gravels, sand in a range of 5-15% and fines with less than 10 %. Some other grading also indicated 11-12 % of fines. The angle of repose was found in the range of 40-45°, rather than showing it vertical or near vertical. Poison's ratio was found to be 0.25. Before the commencement of portal construction, geotechnical properties such as bulk density, cohesion and friction angle were determined using Mohr-columnb elasto-plastic model. The non-homogenousity of alluvial deposit showed brittle behavior. The regression line with best fitting all points of direct field shear tests gave cohesion (c) and friction angle of 143 kPa and 39°, respectively. Afterward, comprehensive analysis of 80 m high slope, suggested designing tunnel with cohesion=100 kPa and friction angle=38°. Alluvial deposit strength properties have been tabulated in Table 2.

Table 2. Classification and shear strength parameters of alluvial deposit at portal site.

Classification parameters						Shear strength parameters			
Bulk density (t/m^3)	Dry density (t/m3)	Moisture content (%)	Atterberg Limits of fine			Cohesion (kPa)		Friction angle (°)	
			LI	PL	PI	Actual	Design	Actual	Design
2.33	**2.24**	**6.9**	**23**	**16**	**7**	**143**	**100**	**39**	**38**

LI: Liquid limit, PL: Plastic limit, PI: Plasticity index

Barton's Q value system has been applied for the classification of existing sedimentary rock mass at the project site, because of its unsurpassed dealing with all parameters of joints and parameters related to sedimentary rocks. Although, it serves best for sedimentary rocks, it is not only hard but also impossible to classify non uniform alluvial deposit using Barton's Q value classification system.

The main reason behind it is the geotechnical properties of alluvial deposit which are in between those of rocks and soils. Anyhow, support equivalent to Barton Q-5 along with some addition of 4.5 m long grout pipes (dia.= 105 mm) at center-to-center distance of 0.5 m extending towards each side of portal has been considered for computer modelling. Afterwards, these grouted pipes were neglected during the portal construction, as the less stand-up time of alluvial deposit resulted in collapse or even stacking of drilling rod.

The average recommended support for Barton Q5 is fibre reinforced shotcrete (12-15 cm) build up progressively. Three cord lattice girders [ϕ=32mm, ϕ= 22mm (2)], along with the installation of wire mesh (6×100×100 mm). Rock bolts (ϕ= 25 mm, L= 1.2 m @ 1 m) for restricting the longitudinal movement of lattice girders.

During the construction of portal, the excavation sequence has been followed in exactly same manner. For each stage, face advance was not much more than 0.6- to 0.7 m, just to adjust two lattice girders in same excavation. Initially, a thin layer (approx. 4-5 cm) of fiber reinforced shotcrete covered the freshly exposed alluvial deposit surface. This shotcrete layer helped to increase stand-up time of excavation and made it easy for the installation of lattice girders and interconnected wire mesh. The installation of lattice girders and wire mesh has been followed by the installation of rock bolts to fix longitudinal movement of lattice girders. These rock bolts fixed the longitudinal direction of lattice girders. Lastly, shotcrete plain (12-15 cm) progressively build up has been applied to cover the installed lattice girders. The excavation and subsequent support applications for Stage 1 have been shown in Fig. 1 while all other excavation stages followed same support installation sequence.

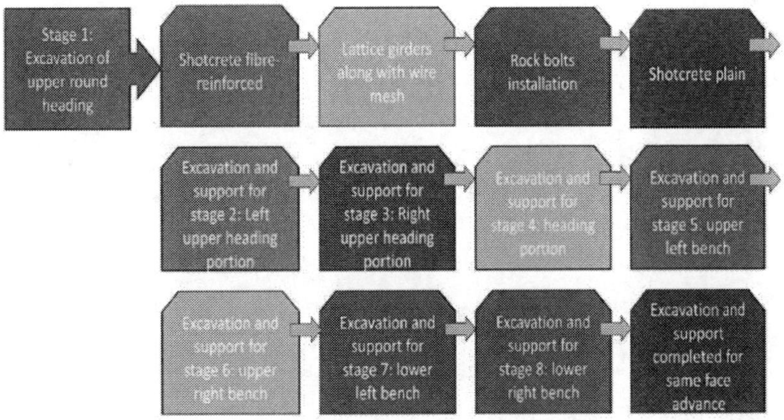

Fig. 1. Stage wise excavation and support sequence of tunnel portal.

The construction sequence of the portal in alluvial deposit was determined with the combination of geotechnical investigation and geological engineering. Fig. 2 represents three different SEM designs. First excavation design Fig. 2 (a) shows the division of complete portal only in heading and benching of height 6.6 m and 8.4 m, respectively. However, by considering the previous experience of slope cutting and tunnel excavation in same material at the same site, it was not possible to excavate portal only in heading and benching. For numerical modeling, another portal excavation design was proposed which divided the entire portal face into four segments, shown in Fig. 2 (b). Third design, divided the complete portal face into eight panels. The method of face excavation in this design was somehow a combination of heading and benching, and sequential excavation method. Fig. 3 (a) describes the sequence of this design. Further, division of portal face was not practically possible due to machinery constraints. Anyhow, it was necessary, to investigate second and third designs in terms of stability and safety, before the execution. Fig. 3 (b) represents the sequence of portal excavation along with activities for third excavation design.

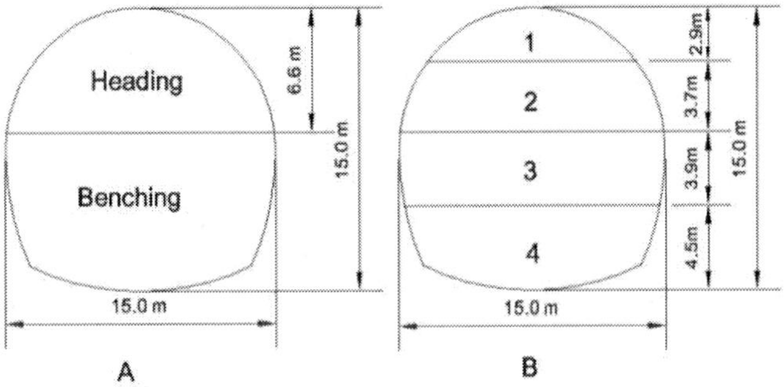

Fig. 2. Excavation designs of 15.0 m diameter tunnel portal using heading and benching method (a) Two stages excavation, (b) four stages excavation design.

 DOI: 10.1051/matecconf/201713804002

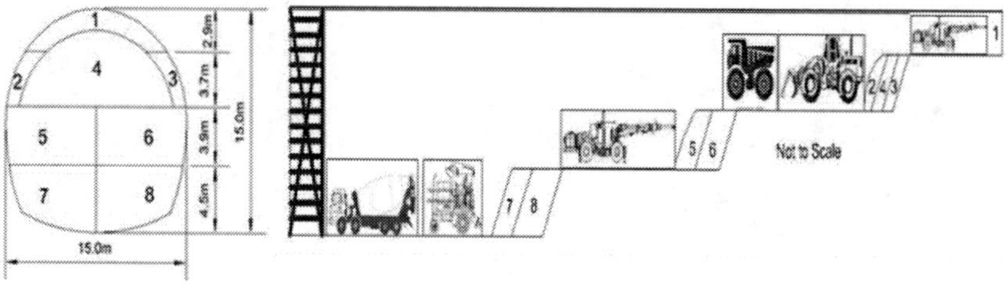

(a) (b)

Fig. 3. Proposed and finally selected sequential excavation design (SEM) for the construction of 15.0 m diameter tunnel portal.

3. NUMERICAL MODELLING

The numerical model of the tunnel was constructed using the well-known finite element code Phase2 by Roc-science. It is a powerful 2D finite element program for soil and rock applications. Phase2 can be used for a wide range of engineering projects and includes excavation design, slope stability, groundwater seepage, probabilistic analysis, consolidation, and dynamic analysis capabilities. The tunnel two dimensional cross-sectional area file was imported in the Phase2 software and tunnel area was constructed. This was followed by construction of model boundaries. The boundaries are kept at a distance of three times the dimensions of the tunnel portal except the top boundary which is free surface. The option of infinite elements was used at the model boundaries which assume that the rock material is continuing in plane. This option enables a user to construct a smaller model while having no or little effects on the model results. In the next stage, material properties were defined for rock material as shown in Table 1 and 2. The in-situ stresses were estimated based on the work of Arjang [14-16]. The estimated stresses are shown in Table 3.

Mesh sensitivity analysis was also performed to optimize the number of elements. This is achieved by constructing a 5.5 m by 5.5 m opening at the center of the model and observing the convergence for different mesh densities. It was observed that some mesh densities yield in to consistent results. Any of these mesh densities can be selected for consistency and precision. In this work mesh density with a total number of 20,500 elements was selected after the mesh sensitivity analysis. The finite element mesh is shown in Fig. 4.

Table 3. In-situ stress parameters.

Parameter	Value
Unit weight, γ (kN/m)	2.71
Depth, H (m)	15
Horizontal to vertical stress ratio, K	1.3
Vertical stress, σ_v (kN)	13.55
Horizontal stress, σ_h (kN)	17.62

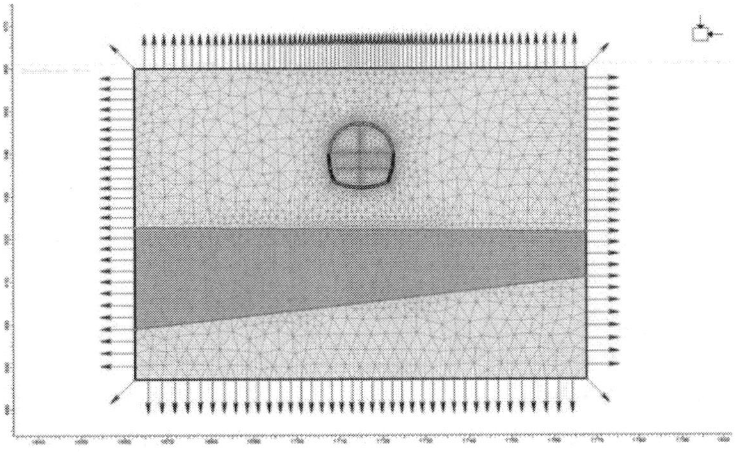

Fig. 4. Numerical model with boundary conditions.

4. RESULTS AND DISCUSSION

Tunnel portal construction methods including top heading and benching, and sequential excavation method, described in Section 3 are compared for stability. The two methods are simulated using the Phase 2 software. The support system including the liners (shotcrete and lattice girder) and rock bolt (dowels) support system employed by the case study project are also simulated in the later part. The results of the numerical model of the two construction methods are compared with respect to major principal stress, total displacement and yielded elements. The results are shown after complete extraction for the heading and benching method and sequential excavation method. The results for two sequences are simulated initially with no support. Later on support system is installed and the results are compared. The results of major principal stress for heading and benching method and sequential excavation method with no supports installed are shown in Fig. 5. As can be seen there is almost no difference in stress magnitude around the tunnel portal, however the yielded elements have decreased significantly. Notably, the maximum stress level near the heading is around 1.25 MPa, which is very low and is actually not enough to provide the clamping force necessary to resist unraveling of blocks. The results of total displacement for the two methods are shown in Fig. 6. The vectors of displacement can also be seen in the Fig. 6. The maximum displacement of 2.5 cm can be observed at the bottom of the tunnel. The crown of the tunnel has moved up to around 0.825 mm. It can be seen that the heading and benching method shows a greater displaced area around the tunnel. The results of two methods are also compared in terms of yielded area around the excavation; Fig. 7 depicts these results. It can be seen that there is a huge yielded area in floor and side walls for simulated heading and benching method. The floor of the excavation is also disturbed. The simulated SEM also shows yielding in the side walls and crown, with relatively lesser yielding in the floor.

 DOI: 10.1051/matecconf/201713804002

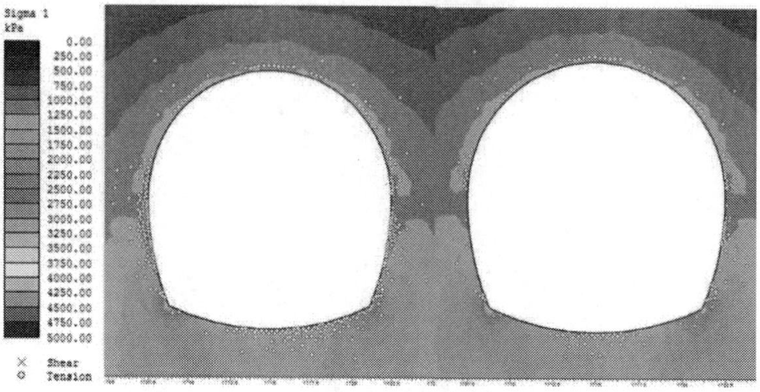

(a) (b)

Fig. 5. Major Principal Stress contours with no support installed, (a) heading and benching, and (b) sequential excavation method.

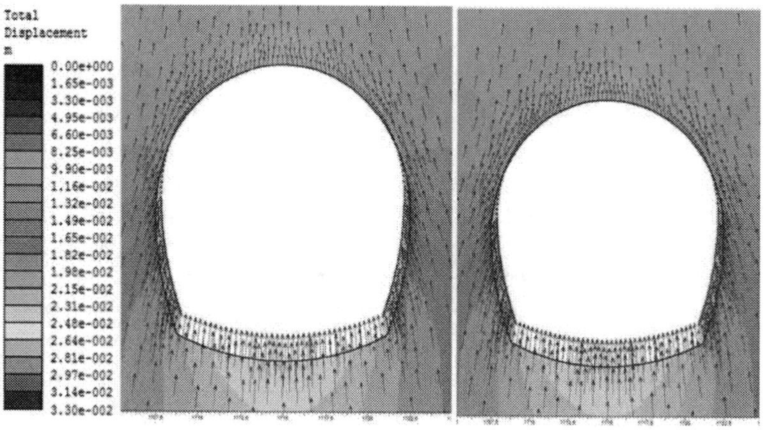

(a) (b)

Fig. 6. Contours of total displacement with vectors, and without tunnel supports, (a) heading and benching, and (b) sequential excavation method.

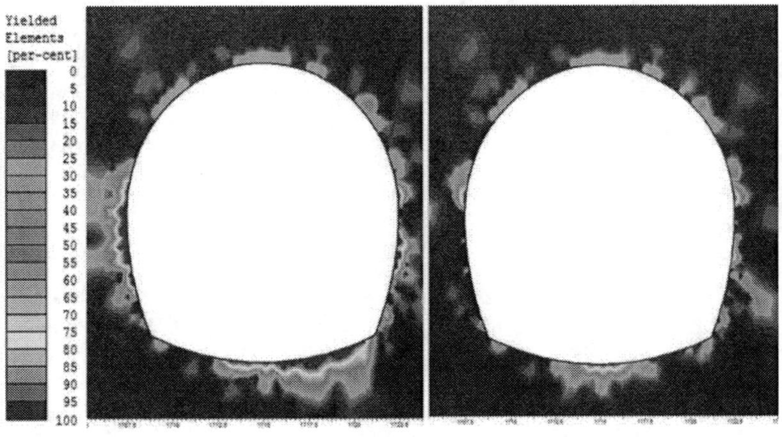

(a) (b)

Fig. 7. Contours of yielded elements when no supports was considered (a) heading and benching, and (b) sequential excavation method.

In the next analysis support system is installed for both the methods while following the excavation and support system installation sequence of case study project. In the first step the excavation is made followed by installation of liners and dowels. The excavation and support installation sequence is followed for both the methods. The results of major principal stress are shown in Fig. 8. It can be seen that the stress concentration around Fig. 8 (a) is higher (1.2 to 3 MPa) than that of Fig. 8 (b) (0.8 MPa to 2.2 MPa). Once again the major principal stress is on the lower side. Similarly the yield zones are greater in number for heading and benching method. Fig. 9 (a) and (b) show the results of total displacement for both the construction methods. It can be observed the total displacement for heading and benching method is around 2.4 cm in floor and 0.5 cm in the crown. The total displacement for the SEM is around 1.8 cm in the floor and 1 cm for the crown.

Fig. 10 represents the contours of yielded volume for the two construction methods simulated with supports installed. As can be seen big yielded volume can be observed for the heading and benching method compared to SEM. No support has been yielded in this analysis. The results found from this analysis can be correlated with the field observations.

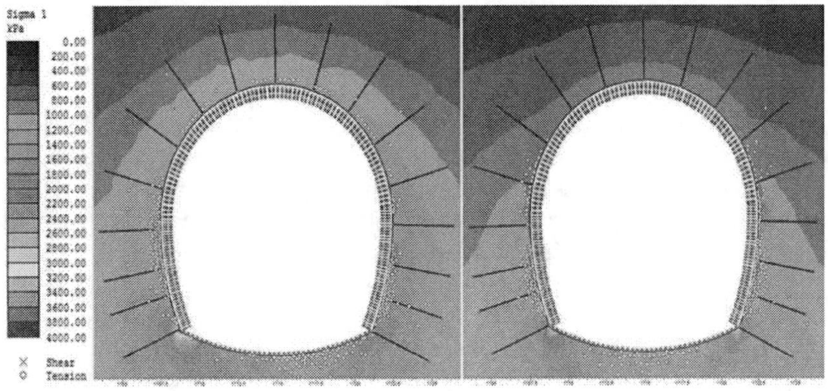

(a) (b)

Fig. 8. Contours of major principal stress with support system installed (a) heading and benching and (b) sequential excavation method.

 DOI: 10.1051/matecconf/201713804002

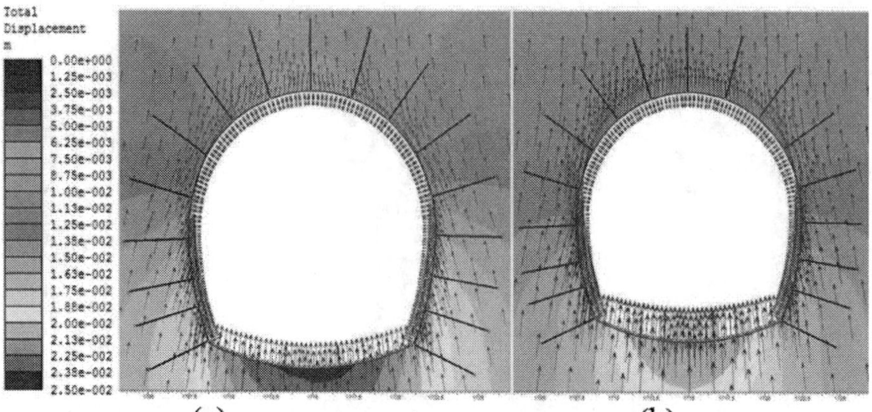

(a) (b)

Fig. 9. Contours of total displacement with support system installed (a) heading and benching, and (b) sequential excavation method.

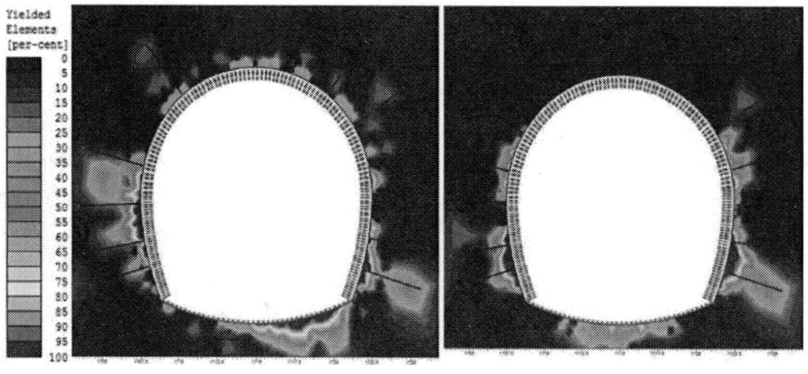

(a) (b)

Fig. 10. Contours of yielded volume with support system installed (a) heading and benching, and (b) sequential excavation method.

The axial stress on all the bolts is calculated by Phase2 software and is shown in Fig. 11. It was found that the maximum stress is 17.5 kPa and 37 kPa for SEM and heading and benching method, respectively. The minimum axial stress in the bolts is 6 kPa for SEM and 5 kPa for heading and benching methods. The radial stress on liners is also computed using Phase2 and is shown in Fig. 12. It was found that the maximum stress is 2 MPa and 1.3 MPa for heading and benching, and SEM respectively. The minimum stress is around 3.75 MPa and 4.30 MPa for heading and benching, and SEM, respectively.

DOI: 10.1051/matecconf/201713804002

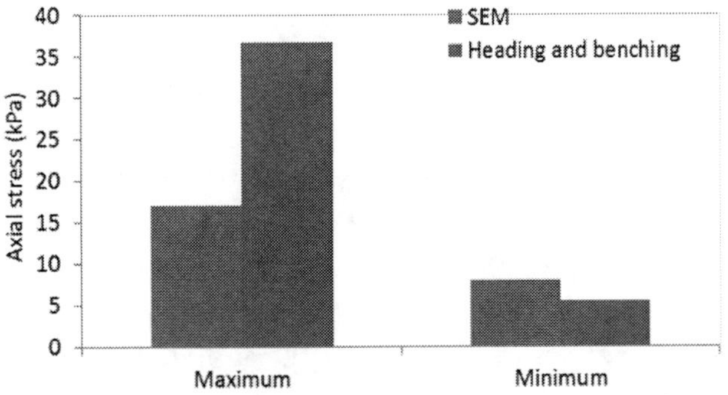

Fig. 11. Axial stress on dowels due to yielding around the tunnel portal.

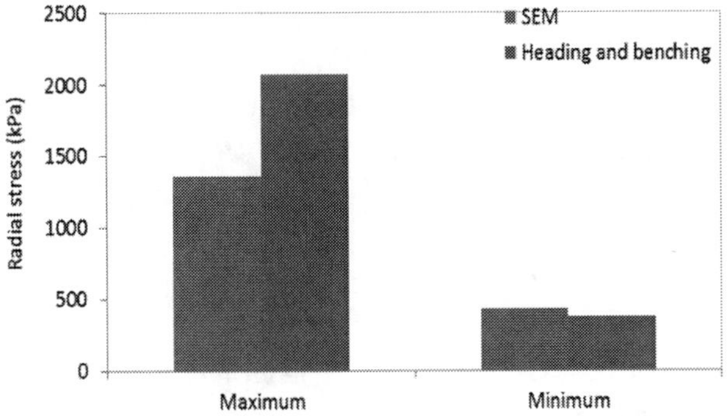

Fig. 12. Radial stress on the combined liners due to yielding around tunnel portal.

The results of the tunnel portal analysis in alluvium show promising results that can also be compared with field observations to achieve meaningful predictions. The limitations of this study include the lack of field data in terms of in-situ stress and a three dimensional numerical modelling code.

5. CONCLUSION

Portal construction in soft rock is one of the most challenging tasks in construction of tunnels. Soft rocks and in-situ stresses underground dictates the stability of excavations. A proper excavation sequence and support installation can mitigate tunnel portal failures in soft rock. Using FEM codes, it is demonstrated that a proper tunnel design and supports can be selected with great success. Two out of three proposed designs are studied with and without the proposed support system using Phase 2. It is found that the sequential excavation design with the proposed support system is more suitable than the heading and benching method. The results are shown in terms of maximum stresses around the tunnel, total displacement and yielded volumes. In the later part the maximum axial stress on bolts and liners are also computed using the numerical models. These results can be calibrated and validated using field measurements to predict ground failures in tunnel portal and design can be altered accordingly.

REFERENCES

1. Lai, K., *Geotechnical properties of colluvial and alluvial deposits in Hong Kong.* The, 2011. **5**: p. 735-744.
2. Zhang, G.-H., Y.-Y. Jiao, and H. Wang, *Outstanding issues in excavation of deep and long rock tunnels: a case study.* Canadian Geotechnical Journal, 2014. **51**(9): p. 984-994.
3. Qiu, S., et al., *Estimation of rockburst wall-rock velocity invoked by slab flexure sources in deep tunnels.* Canadian Geotechnical Journal, 2014. **51**(5): p. 520-539.
4. Zhou, H., et al., *Analysis of mechanical behavior of soft rocks and stability control in deep tunnels.* Journal of Rock Mechanics and Geotechnical Engineering, 2014. **6**(3): p. 219-226.
5. Sharifzadeh, M., R. Daraei, and M.S. Broojerd, *Design of sequential excavation tunneling in weak rocks through findings obtained from displacements based back analysis.* Tunnelling and Underground Space Technology, 2012. **28**: p. 10-17.
6. Koçkar, M.K. and H. Akgün, *Tunnel and portal stability assessment in weak rock*, in *Underground Space Use: Analysis of the Past and Lessons for the Future* E. Solak, Editor. 2005, 2005 Taylor & Francis Group, London, ISBN 04 1537 452 9.
7. Yasitli, N.E., *Numerical modeling of surface settlements at the transition zone excavated by New Austrian Tunneling Method and Umbrella Arch Method in weak rock.* Arabian journal of geosciences, 2013. **6**(7): p. 2699-2708.
8. Zhang, G., Y. Jiao, and H. Wang, *Outstanding issues in excavation of deep and long rock tunnel: a case study.* Canadian Geotechnial Journal, 2014.
9. Hoek, E., *BIG TUNNELS IN BAD ROCK.* Journal of Geotechnical & Geoenvironmental Engineering, 2001. **127**(9): p. 726.
10. Yu, C. and J. Chern, *Expert system for D&B tunnel construction.* Underground Space The 4th Dimension of Metropolises, London, England, 2007.
11. Jing, L. and O. Stephansson, *1-Introduction.* Developments in geotechnical engineering, 2007. **85**: p. 1-21.
12. Wu, J., J. Chen, and D. Lee. *Investigation of the Initiation of the Hsien-du-shan Rock Avalanche in Taiwan Using DDA.* in *48th US Rock Mechanics/Geomechanics Symposium*. 2014. American Rock Mechanics Association.
13. Burg, J., et al., *Infra-arc mantle-crust transition and intra-arc mantle diapirs in the Kohistan Complex (Pakistani Himalaya): petro-structural evidence.* TERRA NOVA-OXFORD-, 1998. **10**: p. 74-80.
14. Arjang, B., *Premining ground stresses at the Bousquet/Dumagami mines, Cadillac, Quebec.* CANMET, Energy, Mines and Resources Canada, Div Report MRL 88-132 (TR), 1988.

15. Sheory, P.R., *A theory for in situ stresses in isotropic and transversely isotropic rock.* Int. J. Rock Mech. Min. Sci. & Geomechanics Abstracts, 1994. **31**(1): p. 23-34.
16. Brown, E.T. and E. Hoek, *Trends in relationships between measured rock in situ stresses and depth.* Int. J. Rock Mech. Min. Sci. & Geomech. Abstracts, 1978. **15**: p. 211-215.

Designing an Ardunio-microprocessor-and-Labview-based system for monitoring wave spread in the ground caused by constructing activities

Nguyen Lan[1*], *Chau Ngoc Bao*[2]

[1] The University of Danang- University of Science and Technology

[2] Center for Science and Technology and Investement Consultancy- DUT.

Abstract. There are many activities such as embankment vibratory rollers, piling, blasting tunnel ... causing wave spread in the ground that affect the surrounding buildings. The high ground wave may damage the surrounding buildings and hence may cause confrontations between the local resident community and the project owners. To determine the radius of vibration of each vibrating source that can damage the neighbouring buildings as the basis for planning, choosing construction technology, and designing damping methods to minimize the risk to the surrounding buildings, It is necessary to measure experimental vibration in the field. This paper describes how to install the Ardunio-circuit-based vibration measuring system , by using labview and shows some experimental results of measuring vibration transmitted to the ground caused by constructing activities in Da Nang, Vietnam. The results prove that the measuring system has a considerably low cost and meets the standards and precision according to Vietnamese ISO 7378 and DIN 4054. The software written on Labview platform is connected to the hardware to collect vibration signals from sensors, analyse measurement data, and then make reports on measurement results in table or chart formats to meet the current standard vibration measuring requirements.

1 Overview of the wave propagation in the soil due to construction activities

The principal wave types that transmit vibratory energy away from a source on or near the ground to a distant receiver include Rayleigh (R-) waves; Shear (S-) waves; Compression (P-) waves. Essentially, these three main wave types can be divided into two categories: body waves, which propagate through the body of the soil; and surface waves, which are transmitted along a surface (usually the upper ground surface) [1].

* Nguyen Lan: nlan@dut.udn.vn

The three wave types produce radically different patterns of motion in soil and rock particles as they pass. Therefore, structures will be deformed differently by each type of wave.

The P-, S- and R- waves travel at different speeds. The P-wave is the fastest, followed by the S-wave, and then the R-wave. Along the surface of the ground, the P- and S-waves decay more rapidly than the R-wave. Therefore, the R-wave is the most significant disturbance along the surface of the ground. The R-waves account for 67% of the total energy, S- waves 26% and P-waves 7% when the exciting force is applied vertically to the propagation direction [2]. Several empirical equations have been proposed that describe vibration attenuation through soil for specific cases. The scaled distance approach, presented as equation (1), is commonly used for blasting and pile driving operations [3].

$$V=k\left(\frac{R}{\sqrt{W}}\right)^{-n} \qquad (1)$$

where: V - PPV at distance R from the source;

W is the energy of source or rated energy of impact hammer;

k and n are parameters found by plotting V versus R on a log-log plot.

Wiss (1981) [4] notes that k and n are unique for each source and soil type combination and this approach cannot be used without detailed site-specific measurements.

Human perception and response to ground vibration varies widely. It depends on individual sensitivity, the frequency, PPV, duration, and on whether or not the event is expected and if so, whether the vibration is expected to cause damage. There are many standards already setting the limits for the vibration which do not affect humans, such as BS 6472-1: 2008 or QCQG 27: 2010 / BTNMT.

Table 1. The maximum value allowed for the vibration acceleration for construction activities (QCQG 27: 2010 / BTNMT) [9].

No.	Area	Time in the day	The level of allowable vibration
1	Special area	6:00 AM –6:00 PM	75 dB
		6:00 PM – 06:00 AM	base level
2	Normal area	06:00 AM – 9:00 PM	75 dB
		9:00 PM – 06:00 AM	base level

Several countries have adopted standards for evaluating the effect of vibration on buildings. A review of international standards identified the following as being the most suitable for providing guidance as to possible building damage from mechanized construction activity:

- ✓ German Standard DIN 4150-3:1999 Structural vibration – part 3: Effects of vibration on structures;
- ✓ British Standard BS 7385-2:1993 Evaluation and measurement for vibration in buildings, part 2. Guide to damage levels from ground-borne vibration;
- ✓ Swiss Standard VSS-SN640-312a:1992 Effects of vibration on construction.

DOI: 10.1051/matecconf/201713804003

Table 2. Vibration guidelines from DIN 4150-3:1999 for assessing effects of vibrations on buildings [6]

Type of structure	Vibration thresholds for structural damage, PPV (mm/s)				
	Short term				Long term
	At foundation			Uppermost floor	Uppermost floor
	0 to 10 Hz	10 to 50 Hz	50 to 100 Hz	All frequencies	All frequencies
Commercial/industrial	20	20 to 40	40 to 50	40	10
Residential	5	5 to 15	15 to 20	15	5
Sensitive/historic	3	3 to 8	8 to 10	8	2.5

Note: When a range of velocities is given, the limit increases linearly over the frequency range.

To support measuring and evaluating the spread of wave in the ground caused by constructing activities so that it is possible to minimize the risk to neighboring buildings, this study designed a low-cost Ardunio-circuit-based vibration measuring system as well wrote a software to collect data and analyze vibration measuring figures. The system has been tested and measured for a number of real projects in the recent periods of time.

2 Designing the system for monitoring ground wave propagation

To design a ground wave propagation system due to the construction work required by TCVN 7378[10] and DIN 4150 [6] and low cost, we use the following hardware: Ardunio circuit and geophone sensor. The LabVIEW environment is used to write a software to connect the sensor to Ardunio and computer, collect vibration signals, and analyse the vibration signals measured. The system block diagram is shown in Figure 1.

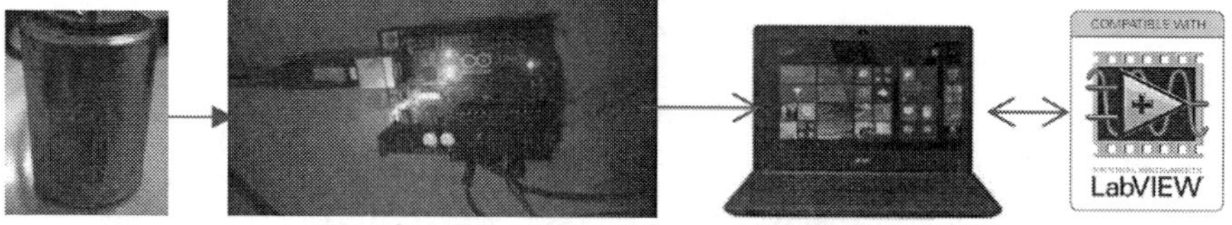

Figure 1. System block diagram for monitoring ground wave propagation

Graphic User Interface of software for collecting and analysing ground vibration signals in figure 2.

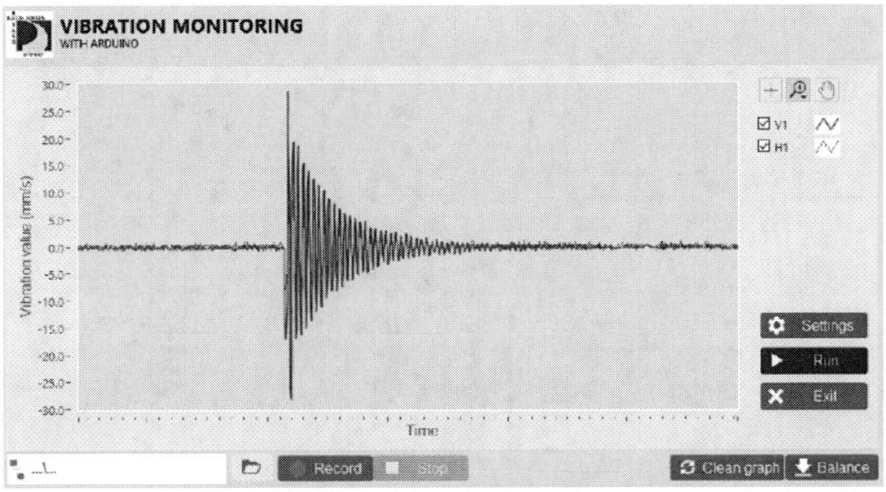

Figure 2. Graphic User Interface of software

Test results of the design system compared with the GPx2 Accelerometer system (USA) on cantilever beam model (Figure 3) show that the free frequency of two systems are similar (10.5 Hz and 11 Hz).

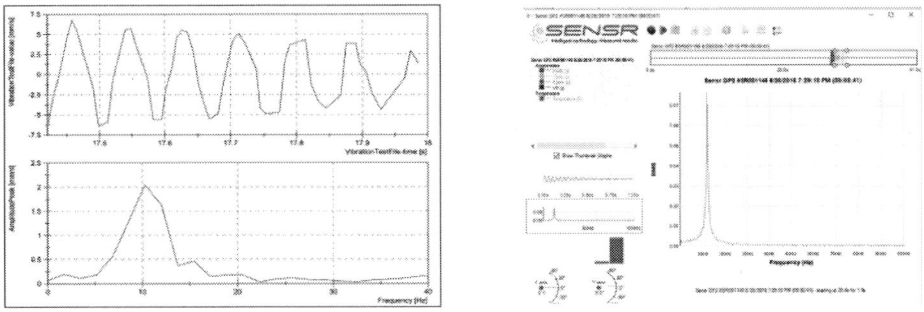

Figure 3. Results of free frequency of cantilever beam model on two systems.

The Arduino circuit cost is about $ 20 compared to other professional dataloggers which cost from $ 500 upwards.

3 Some results of vibration effective radius due to construction activities

Summary results of vibration measurements and determination of the radius affect on the surrounding buildings due to some construction activities in reality are given show in Table 3 [5].

DOI: 10.1051/matecconf/201713804003

Table 3. Synthesis of empirical findings vibration measurement projects by the authors of the paper [5]

No	Name of project / regional geological features	Characteristic vibration source	Empirical relationship V (mm/s) and R (m)	Measured vibration frequency (Hz)	Effective Radius R(m) corresponding to Vgh, mm/s	Date of measurement
1	+ Interchange Hue- Da Nang project-phase 1. Pile driven and train + Small sand and clay powder, light grey, humid conditions (depth <= 2m).	1 pile driven machine, Bauer BG25, Fc=235 KN	V= 1000.Fc0.5/ R2.5	6-29	26 (Vgh=4mm/s)	Aug 15, 2014
		Concreting shaft pile	V = 11000/ R2.5		24(Vgh=4mm/s)	
		Pull out steel casing, W=45 kW	V = 20 . W0.2/ R		11(Vgh=4mm/s)	
		Drop steel casing Q=3,5 T, drop height H=1 m	V = 38.(Q.H)0.2/ R		12(Vgh=4mm/s)	
		Train pass	V = 3,5/ R0.25		2(Vgh=4mm/s)	
2	+ Interchange Hue- Da Nang project-phase 1. Road roller + Small sand and clay	1 vibration roller LiuGon CLG614, high centrifugal forces Fc=27T	V = 1800 . Fc0.5 / R2.5	12	29(Vgh=4mm/s)	Sep 25, 2014

DOI: 10.1051/matecconf/201713804003

	powder, light grey, humid conditions (depth <= 2m).					
3	+ Ring Road south of Da Nang + Farm land, 1m thick clay; Small-medium grained sand grey white	1 vibration roller C250, Fc=246 kN.	V = 0,5. Fc / R0.9	10-50	30 (Vgh=6mm/s).	July 2, 2014
4	+ Expanding Highway 1A, Km 947-987, Quang Nam province + Sand granules medium grey-white	1 vibration roller Dynapac CA25D, Fc= 160 kN	V = 0,7.Fc0.9 / R	15-21	23 (Vgh=3 mm/s).	June 3, 2014
5	+ Extended Nguyen Tat Thanh Street section from Km 2 + 715.04 to Km 4 + 100, Lien Chieu	1 vibration roller Bomag BW 211D-40, Fc= 236 kN	V = 1,2 * Fc / R1.35	27-35	29 (Vgh=3 mm/s).	Oct 9, 2015
		2 vibration roller BW 211D-40,	V = 1,25 * Fc / R1.25	27-35	39.5 (Vgh=3 mm/s)	

DOI: 10.1051/matecconf/201713804003

	District, Danang city. + Granulated sand, white	Fc= 236 kN				
6	+ High Da Nang -Quang Ngai Expressway. Tunnels through the EO mountain (Package 4)	Explosive type P113; weight Ptn=102 kg	V = 4500*Ptn/R2	39-184	391 (Vgh=3 mm/s)	Nov 26, 2015
		Explosive type P113; Explosive weight Ptn=420 kg	V = 4500*Ptn/R2	39-184	794 (Vgh=3 mm/s)	
8	+ Halam Bridge, Da Nang-Quang Ngai Expressway + clay farm land; up ground on the hill.	1 vibration roller Bomag BW 211 D-40; Fc=236 (kN)	V = 0,35 . Fc / R0.9	14-28	33m (Vgh=4 mm/s)	Aug 28, 2014
		1 Vibration hammer Towen 90kW; W is the capacity of vibratory hammers= 90 kW	V = 30 . W0.4 / R	14-28	45m/ vgh=4 mm/s	

9	Construction of wastewater collection system around the Phan Lang lake and sewers for collection and sewage pumping stations along Nguyen Tat Thanh	1 Vibration hammer PCF-350; W is the capacity of vibratory hammers= 45T.	$V = 6,1 * W^{0.99} / R$	29-39	88m (V_{gh}=3mm/s)	Aug 8, 2016
10	Project against the sand sedimentation in the estuary Phu Loc - Danang city.	01 diggers mounted hammer KOMATSU Type DPD350; Fc =300 KN the vibration force of the hammer	$V = 77*Fc/R^4 +1.1$	13.4-25.2	11 m (V_{gh}=3mm/s)	Jan 26, 2016

Note: V (mm/s) = PPVs (Peak Particle Velocity) of ground; R (m)= adversely affect Radius to neighbouring buildings from the vibration source; Vgh(mm/s) = Vibration velocity limits to protect adjacent buildings.

Table 3 shows the radius of influence due to vibration of various construction equipment; it does not only depend on sources of vibration characteristics but also on regional geology conditions.

4 Conclusions

Measurement systems for ground vibration monitoring activities due to construction based on vibration velocity sensor (geophone) and ardunio circuit connected to specialized software built on the platform language labview meet the system requirements of a vibration measuring equipment such as TCVN 7378 or DIN 4150-3. Precision of designed vibration measurement system is equivalent to the imported vibration measurement system as GPX (SENRS-USA) but the cost is smaller.

The application vibration measurement system designed to application possibilities of the system design are very good. The measured data from actual projects can be used to predict the radius of influence of vibration sources from construction activities on neighbouring buildings.

The system can be extended to a wide variety of other sensors for measuring purposes and health monitoring of different structures.

References

1. New Zealand Transport Agency research report 485. *Ground vibration from road construction*, May 2012.
2. Richart, FE, JR Hall and RD Woods. *Vibrations of soils and foundations (*Englewood Cliffs, New Jersey, Prentice-Hall,1970).
3. Svinkin, MR, *Prediction and calculation of construction vibrations (* Proceedings 29th Annual Members Conference Deep Foundations Institute, Dearborn **MI:** 1–22, 1999).
4. Wiss, JF, *Construction vibrations: state-of-the-art. Journal of the Geotechnical Engineering Division* (Proceedings of the American Society of Civil Engineers 107, no. **GT2**, 1981).
5. Center for Science and Technology and Investment Consultancy- the University of Danang. The vibration test reports of projects, 2014-2016.
6. DIN 4150-3, *Structural vibration part 3: effects of vibration on structures*, (Berlin, Deutches Institut fur Normung, 1999).
7. BS 6472-1, *Guide to evaluation of human exposure to vibration in buildings – part1:vibration sources other than blasting (* British Standards Institution, 2008).
8. VSS-SN640-312a, *Effects of vibration on construction (* Zurich, Swiss Consultants for Road Construction Association, 1992).
9. QCVN 27/BTNMT, *National Technical Regulation on Vibration* (2010).
10. TCVN 7378, *Vibration and shock - Vibration of buildings - Limits of vibration levels and method for evaluation* (2004).

Influence of Rice Husk Ash and Clay in Stabilization of Silty Soils Using Cement

*Jack Widjajakusuma** and *Hendo Winata*

Department of Civil Engineering, Universitas Pelita Harapan, 15811 Tangerang, Indonesia.

Abstract. Soil stabilization is needed to enhance the strength of the soil. One popular method of soil stabilization is using cement. Due to the environmental issue, it is a need to reduce the application of cement and/or to replace partially the cement with other environmental-friendly compounds. One of these compounds is rice husk ash (RSA), which is agricultural wastes. The objective of this paper is to study the influence of RSA and clay as partial replacement to cement in soil stabilization of silt soil with high plasticity (MH) using cement. The cement used was ordinary Portland cement, while the RHA was obtained by burning rice husk at temperature of 250°C. The MH soil is stabilized with 4% cement, 4% cement and 3% rice husk ash and 4% cement, 3 % RHA and 3 % clay. The various tests were conducted on the pure and stabilized soils. Results have indicated that application of 4% cement, 3 % RHA and 3 % clay as silt soil stabilization is more favorable in increasing soil strength and reducing brittle behaviour of soil.

1 Introduction

Soils are the most important building materials because soils support all the loads of existing buildings. However, the properties of local soils are often not suitable for the requirements of particular buildings such in the case of many regions in Jakarta, Indonesia. To overcome those problematic soils, one has to enhance their properties. One method to improve the properties of soils is mixing and blending soils with other materials. These materials can be chemical mixers such as cement, rice husk ash, salt, lime, coal ash, asphalt cement, sodium, calcium chloride, paper mill waste and other chemicals [1-3].

The most common and widely used chemical mixer in soil stabilization is cement [1-3]. The reason for applying cement is that the cement is relatively easy to obtain. On the other hand, cement is not a renewable materials and cement production performs very high carbon emissions so that cement production is not environmentally friendly. Other disadvantages are the soil-cement are susceptible to cracks at high temperatures, brittle and corrosive. Furthermore, the price of cement is rapidly increasing due to the sharp increasing price of energy [4]. Therefore, many studies have been conducted to find an alternatives pozzolanas to total or partially replaced cement.

Pozzolona is a siliceous or siliceous and aluminous components which at room temperature and in the presence of water chemically reacts with calcium hydroxide to form cementitious materials.

This study applys rice husk ash (RHA), which contains 80-90% of amorphous silica [5], as alternatives pozzolona to partially replace the cement. RHA is a waste product abundantly available in rice producing countries such as Indonesia (the third-largest rice producing countries). Indonesia produced around 76 million tons of rice in 2015 [6] and will still increase the rice production in order to reach self-sufficiency in rice for Indonesian population.

One big issue in producing rice is the final disposition of rice husks, which are agricultural residues. The residues are huge and can be around 20 % of total rice crop. In many regions in Indonesia, this rice residues are still treated as a waste product. To get rid of the waste, usually the farmers burn rice husks in the fields. Recently, rice husk is started to use as biomass. Burning of rice husks produces RHA, which is around 25 % by weight of rice husk. Due to the low bulk density of RHA, disposition of RHA can be problematic to the land and human health.

The RHA contains large amount of amorphous silica. Amorphous silica has a wide range of applications such as silica gels, silicon chip, cement, catalysts. Due to the low cost and renewable of the RHA, it is worth to study the applications of the silica extracted from the RHA. Furthermore, the results of the study will also help to reduce the amount of RHA waste.

The previous studies with RHA had shown that it had promising potentials of improving the engineering properties of soils and inexpensive ([7-11]). However, RHA cannot be used alone because it has inadequate cementation property, which is required to bind the material to a satisfactory durability [10-11].

* Corresponding author: jack.widjajakusuma@uph.edu

DOI: 10.1051/matecconf/201713804004

Thus, this study focused on investigating the influence of adding RHA and the optimum amount of RHA on improving the engineering properties of silty sands. In order to reduce the brittleness of cementation soils, a small amount of natural clay was added. Clay had been chosen because it contained a lot of alumina and silica and had high plasticity. Thus, the present work study the influence of adding RHA and clay on modifying the engineering properties of silty sands. This work also investigated the effect of mechanical compaction in improving the soil properties.

2 Materials

The materials used in this study are described as follows:

2.1 Soil specifications

For the investigations of this work, typical soil found in West Jakarta such as silty soils. The soil colour was brown. Index properties of investigated soil have been summarized in Table 1. The soil was classified as silt with high plasticity MH in the USCS soil classification system.

Table 1. Summarizes the various index properties of investigated soil

Water Content (%)	**Specific gravity**	**Liquid limit (%)**	**Plastic limit (%)**
53.084	2.631	73.3	53.021

2.2 Admixtures

The admixtures used in this study were cement, RHA and clay. The cement used in this study was ordinary portland cement of type 1. The rice husk was collected from Tangerang. The RHA was obtained by burning the rice husk in a cylindrical furnace at temperature 250˚C. Then, the RHA was grounded with Los Angeles machine and sieved through 0.075 mm aperture before use. The chemical contents of RHA was given in Table 2. The natural clay used in this study was black clay and was collected from Karawang.

Table 2. Summarizes the chemical properties of RHA

SiO_2 [%]	**Fe_2O_3 [%]**	**Al_2O_3 [%]**
95	1.48	2.21

3 Methods of testing

To study the effect of mixing on soil, various laboratory tests were carried out. To study the index properties of soil and its mixtures, particle size distribution and Atterberg limit tests were conducted. To study the engineering properties of soil and its mixtures, California Bearing Ratio (CBR) and triaxial tests were performed. The geotechnical properties of the soil were determined according to ASTM [12].

The laboratory tests carried out first were on the pure soil. Then, the tests were carried out for the soil samples mixed with 4% cement; 4% cement and 3% RHA; 4% cement, 3% RHA and 3% clay. The following abbreviations were used 4% C for mixtures of soil with 4% cement, 4% C 3% RHA for mixtures of soil with 4% cement, 3% RHA, 4% C 3% RHA 3% Clay for mixtures of soil with 4% cement, 3% RHA and 3% clay.

The sample for testing had been prepared in accordance with the requirement of the tests. First, the pulverized soil sample was sieved through the required sieve. The required amount of soil was weighed out for the particular test. The admixture was also sieved through the required sieve and the required amount was weighed out on the weight basis according to the percentage to be added to the soil for the particular test. Then, the soil and admixtures were mixed throughly to ensure homogeneous distribution of the admixtures in the soil. These mixture samples were used for performing the various tests.

4 Results and discussion

4.1 Effect of admixtures on particle size distribution

The range of particle sizes in soils was very large, so the logarithmic scale had been used. Figure 1 shows the particle size curve as percent finer vs. log of particle size. The cement was found to have effect on the grain size of stabilized soil, which can be seen from Fig. 1. The reason was that soil and cement reaction form cementitious compounds in the soil. Consequently, the grain size of the stabilized soils, which contained lesser cement, were lesser modified. This can be clearly seen from Fig. 1. The grain size of stabilized soil with 4% cement was most modified, comparing to other stabilized soils.

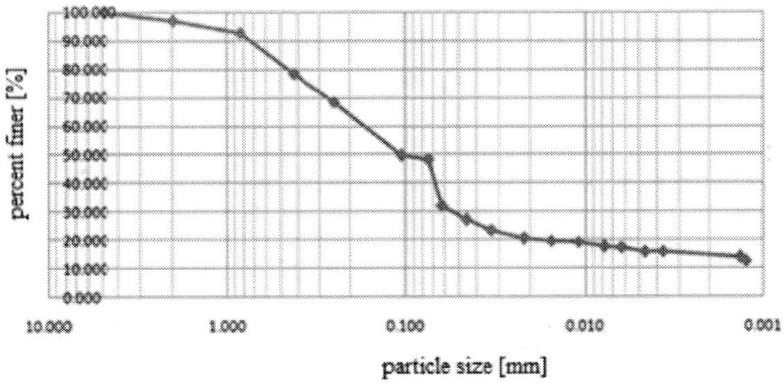

(a) Pure soil

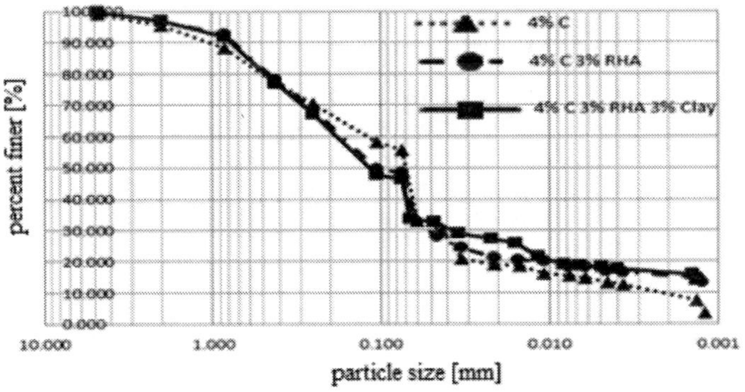

(b). Stabilized soils

Fig. 1. Particle size distribution of (a) pure soil and (b) stabilized soils

4.2 Effect of admixtures on plasticity

The plasticity index (PI) of pure soil was approximately 20.31%. As the silty soil was mixed with cement, cement and RHA, the PI becomes 0 (non-plastic soil). The reason was that soil, cement and RHA reaction produce cementitious soils, which decreased the Atterberg limits to a greater extent [13-14]. Applying the small amount of clay in soil mixtures with cement and RHA decreased the liquid limit (LL) and PI. However, the value of LL and PI of stabilized soil with 4%C 3%RHA 3%Clay were higher than those of stabilized soils with 4% C and 4%C 3%RHA. In general, a reduction of the PI value indicated the improvement of the soil.

Table 3. Summarizes the Atterberg limits

	pure soil [%]	4%C [%]	4%C 3%RHA [%]	4%C 3%RHA 3%Clay [%]
LL	73.33	43.12	35.77	51.43
PL	53.02	-	-	46.03
PI	20.31	-	-	5.40

4.3 CBR tests results

CBR tests, which is common used in the design of base and sub-base material for pavement, had been conducted to evaluate the strength of stabilized soils. In order to study the effect of soaking, the samples had been grouped in three groups. The first, second and third group of samples were soaked in water for 3 days, 7 days and 14 days, respectively. Each group had two series of soil samples. The first series consisted of pure soil, stabilized soils 4% C; 4%C 3%RHA and 4%C 3%RHA 3%Clay, which were compacted with 25 blows. The second series consisted of pure soil, stabilized soils 4% C; 4%C 3%RHA and 4%C 3%RHA 3%Clay, which were compacted with 56 blows.

Table 4 gives the results of CBR tests for pure soil. The soaked CBR values of pure soil decrease with increasing soaking days. The CBR values increased with increasing number of blows. These results had confirmed that water reduced the strength of the soil, meanwhile, the compaction raised the strength of the soil.

Figure 2 shows the results of all CBR tests for the stabilized soils. The CBR values of all samples increased with increasing the soaking days and increasing the number of blows.

 DOI: 10.1051/matecconf/201713804004

The reason for increasing in CBR values of stabilized samples might be because of the gradual formation of cementitious compounds in the soil by the reaction between the cement and RHA with the water.

Table 4. Summarizes CBR values of pure soil

Soaked [days]	Blows	*CBR* (%)
3	25	0.476
7	25	0.458
14	25	0.426
3	56	0.547
7	56	0.469
14	56	0.447

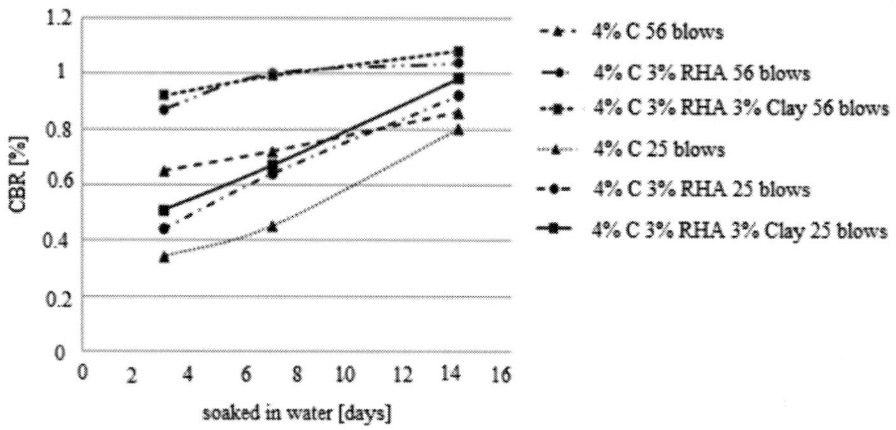

Figure 2. Results of CBR tests

4.4 Triaxial test results

Triaxial tests has been performed on samples without and with admixtures for 3, 7 and 14 days curing. Table 5 shows the development of cohesion intercept c and internal friction angle ϕ. Adding admixtures to the soil caused an increase in cohesion and internal friction angle. The reason was that soil and cement reaction form cementitious compounds in the soil.

Table 5. Summarizes cohesion intercept c and friction angle ϕ

Blows	days	pure soil		4%C		4%C 3%RHA		4%C 3%RHA 3%Clay	
		c [kN/m^2]	ϕ [°]	c [kN/cm^2]	ϕ [°]	c [kN/m^2]	ϕ [°]	c [kN/m^2]	ϕ [°]
25	3	34.4	17.18	30.7	29	49.4	28	49.8	27
	7	30.9	17.12	33.3	29.5	52.2	29	71.	28
	14	28.8	16.73	38.2	30	59.9	29.5	74.5	29
56	3	40.8	17.31	36.4	37	60.9	32	89.6	29
	7	31.4	17.30	60.9	38	69.4	37	92.0	29.5
	14	31.1	16.77	89.6	39	70.6	38.5	101.0	30

The strength of pure soil decreased with increasing soaking days. These results have confirmed that water reduces the strength of the soil. Meanwhile, the strength of stabilized soils increased with increasing soaking days. The reason for increasing of the strength of stabilized soils might be because of the gradual formation of cementitious compounds in the soil by the reaction between the cement, the cement and RHA with the water. The highest soil strength has achieved for the combination of soil, RHA and clay. By increasing the number of blows (compaction) raised the strength of the soil.

5 Conclusions

The results in this paper can be summarized as follows:

- The soft soil was classified as to be silt of high plasticity (MH) in accordance with USCS classified system. Since the soil wass very low, which wass indicated by very low CBR value (below 1%), the soil was needed to be stabilized.
- Compaction, indicating by increasing the blows, had increased the strength of the pure soil and stabilized soils by around 100%.
- All of the stabilized soils after soaking of 14 days had increased the CBR value by around 100 %.
- Treatment with 4% cement had increased the cohesion value of stabilized soil from 28.8 kN/m^2 to 38.2 kN/m^2 (25 blows, 14 soaked days) and from 31.1 kN/m^2 to 89.6 kN/m^2 (56 blows, 14 soaked days). This treatment also increased the value of the internal friction angle from 16.73° to 30° (25 blows, 14 soaked days) and 16.77° to 39°(56 blows, 14 soaked days).
- Treatment with 4% cement 3% RHA had increased the cohesion value of stabilized soil from 28.8 kN/m^2 to 59.9 kN/m^2 (25 blows, 14 soaked days) and from 31.1 kN/m^2 to 70.6 kN/m^2 (56 blows, 14 soaked days). This treatment had also increased the value of the internal friction angle from 16.73° to 29.5° (25 blows, 14 soaked days) and 16.77° to 38.5°(56 blows, 14 soaked days).
- Treatment with 4% cement 3% RHA and 3% Clay had increased the cohesion value of stabilized soil from 28.8 kN/m^2 to 74.5 kN/m^2 (25 blows, 14 soaked days) and from 31.1 kN/m^2 to 101.0 kN/m^2 (56 blows, 14 soaked days). This treatment had also increased the value of the internal friction angle from 16.73° to 29° (25 blows, 14 soaked days) and 16.77° to 30°(56 blows, 14 soaked days).
- Combining the compaction and stabilization of the silt soil had increased the soil strength by around 250 %.
- After the stabilizing, the CBR values were still low, thus, the soils had to be modified.
- Soil stabilization applying RHA in rice-producing countries such as Indonesia is attractive because it reduces the construction cost, especially in the remote area. Furthermore, this method helps to reduce the waste of rice husk and RHA and preserves non-renewable resources such as cement.

The first author sincerely thanks to the University of Pelita Harapan, who provided partially funding for this study. He would also like to recognize DAAD for support during the short visit at Ruhr-University Bochum and some fruitful discussions with Prof. Dr.-Ing. H. Steeb during the stay.

References

1. R.H. Karol, *Chemical Grouting and Soil Stabilization, Revised and Expanded*. CRC Press, Boca Raton (2003).
2. K. Kirsch and A. Bell, *Ground Improvement*, CRC Press, Boca Raton (2013).
3. P.G. Nicholson, *Soil Improvement and Ground Modification Methods*, Elsevier, Amsterdam (2015).
4. A.M. Neville, *Properties of Concrete*, Pearson, Singapore (2000).
5. D.F. Houston, In: *Rice: Chemistry and Technology*, American Association of Cereal Chemists, 301-340 (1972).
6. Badan Pusat Statistik (Statistics Indonesia), Produksi Padi, Jagung, dan Kedelai (Angka Sementara Tahun 2015) (Production of paddy, maize and soybeans (Preliminary Figures 2015)). Jakarta, Indonesia.
7. R. Pode, Renewable and Sustainable Energy Rev, **53**, 1468-1485 (2016).
8. K.M.A. Hossain, J. Mat. in Civil Eng., **23**, 1320-1327 (2011).
9. T.R. Karatai, J.W. Kaluli, C. Kabubo and G. Thiong'o, J. Constr. Eng. Manage. DOI: 10.1061/(ASCE)CO.1943-7862.0001235 (2016)
10. D.G. Nair, K.S. Jagadish and A. Fraaij, Cement and Concrete Res., **36**, 1062-1071 (2006).
11. F.H. Ali, A. Adnan and C.K. Choy, Geotech. And Geological Eng., **10**, 117-134 (1992).
12. A.W. Al-Khafaji and O.B. Andersland, *Geotechnical Engineering and Soil Testing*, Saunders College Publishing, Fort Worth (1992).
13. T.T. Hensley, *Pozzolan stabilized subgrades*, Research Project SPR-I (06) 578, Nebraska Dept. of Roads, Lincoln, NE.
14. M.Y. Fattah, F.H. Rahil and K.Y.H. Al-Soudany, J. Civ. Eng. Urban, **3**, 12-18 (2013)

Systematic Approaches for Signal Processing of Soil Shaking Table Test

Jun Xian Lim[1], *Min Lee Lee*[1,*], *and Yasuo Tanaka*[1]

[1] Department of Civil Engineering, Lee Kong Chian Faculty of Engineering & Science, Universiti Tunku Abdul Rahman, Selangor, Malaysia

Abstract. Selecting an appropriate methodology in signal processing of shaking table test is of fundamental importance for evaluating the soil dynamic parameters. This paper presents the systematic approaches in selecting an appropriate signal processing method for a soil shaking table test. A soil model of 150 mm in diameter and 100 mm in height was tested on a 1g shaking table. The soil model was instrumented with accelerometers to measure the resultant accelerations under different shaking motions. Under normal data processing practices, the acceleration data was adjusted and numerically integrated to obtain the linear displacement data. However, the accuracy and reliability of the computed displacement data and the adopted adjustment technique often raise arguments. In the present study, the acceleration data was subjected to two baseline correction schemes and one method of digital band-pass filtering. The results from the three signal processing methods were compared with the actual displacement data obtained from a direct measurement using laser displacement sensor. Bandpass filtering method was found to be a more reliable data processing method in this particular study. The use of laser displacement sensor can serve as a useful reference when choosing an appropriate adjustment scheme in data processing.

1 Introduction

Ground motion parameters and their characteristics are of importance to seismologists, geologists and earthquake engineers. Among many aspects, residual displacement is essential for investigating the fault rupture after the occurrence of a strong ground motion. The permanent or residual displacement can be caused by plastic deformation of near-surface material or elastic deformation of ground as the result of co-seismic slip on the fault [1]. At the end of each shaking motion, the velocity should return to zero while certain amount of residual displacement can be expected. Unphysical residual displacement will be encountered if the acceleration record is not adjusted accordingly. Under common practices, the interpretation of numerically integrated displacement data from an earthquake event relies upon individual judgement, and hence exposed to numerous uncertainties. In a study of the Chi-Chi Taiwan earthquake, Boore [2] used the technology of Global Positioning System (GPS) to verify his proposed correction schemes on the residual

* Corresponding author: mllee@utar.edu.my

displacement. However, the outcome of that comparison was not convincing due to large discrepancies between the computed and observed values. It may be attributed to the fact that GPS stations were not collocated with the accelerograph stations.

Over the years, various adjustment schemes for processing seismic records have been proposed by numerous authors worldwide [1- 5]. Although there are various correction schemes proposed to recover the actual shaking record, it is generally agreed that it is almost impossible to recover the earthquake record perfectly. Boore and Bommer [1] reviewed the characteristics of a strong-motion record and the effect of noises (short and long-period noise contents) on signals in an earthquake event. They also highlighted the importance of performing baseline correction and selecting a suitable cut-off frequency in digital filtering. Sensitivity analysis was carried out when examining the effect of correction schemes and other parameters on the adjusted results. Boore [2] suggested a simple baseline correction method named as zeroth-order baseline correction. This method requires removal of pre-event mean records from the entire record at the initial stage of signal processing. The subsequent procedure includes identification of obvious changes in velocity baseline. Time instant for a particular change can be identified and followed by subtracting baseline step changes in the acceleration record. After the acceleration record is baseline adjusted, it can be numerically integrated to obtain the velocity and displacement time-series. It was also concluded that the choice of baseline correction does not affect the response spectrum with periods greater than 20s. This finding suggested that most of the structures under seismic activities may not be affected by the choice of baseline correction method as the response spectrums are usually greater than 20s.

In addition, Chiu [4] suggested a "stable" three-step algorithm baseline correction scheme for processing digital strong motion data. In this method, least-square fitting and high-pass filtering of acceleration record was applied. In Japan, Ohsaki [5] proposed a well-known correction method which is fundamentally based on the assumptions that velocity at the end of shaking will become zero whilst certain amount of residual displacement can be expected.

SeismoSignal is one of the popular commercial software to process earthquake strong-motion data. Mollova [6] presented the application of digital filtering using SeismoSignal on an actual earthquake record in Turkey. The effects of various digital filtering models (including Chebyshev, Butterworth, Bessel and Elliptic) were compared and discussed.

Besides using acceleration record to recover the displacement data in an earthquake, similar data processing technique was also applied in other engineering fields. In a study on structural behaviour of pavement, Arraigada and Partl [7] used accelerometers instead of Linear Variable Differential Transformer (LVDT) to measure the acceleration and derived the vertical deformation of pavement from the acceleration data. Slifka [8] used laser displacement sensor as a reference when processing the acceleration data of a moving object.

In light of the non-destructive nature and the ability for accurate measurement of rapid-moving object, laser displacement sensor may serve as a favourable solution for measuring small deformation in soil models. This paper focuses on signal processing of shaking table test on a soil model. Soil displacement is derived from the acceleration data obtained from accelerometers through various adjustment schemes. Laser displacement sensor is used to verify the derived displacement and serves as a reference for recommending a suitable adjustment method for the present shaking table test.

DOI: 10.1051/matecconf/201713804005

2 Instrumentation

2.1 Shaking Table System

Figure 1 shows the shaking table setup used in the present study. A mega-torque motor was used to produce one-dimensional shaking motion on a level platform (2m by 2m) by generating a mechanical torque repeatedly. The shaking table platform was lifted afloat by supplying an air pressure of 2 bars underneath the table platform to minimize the friction between the base and platform during cyclic horizontal movement.

The shaking table was capable of producing frequencies of 0.1 – 20 Hz and horizontal displacements of 0.5-15 mm. Figure 2 shows the relationship between input shaking motions and the resultant peak accelerations. The highest achievable peak acceleration was about 0.55g.

Fig. 1. Shaking table setup.

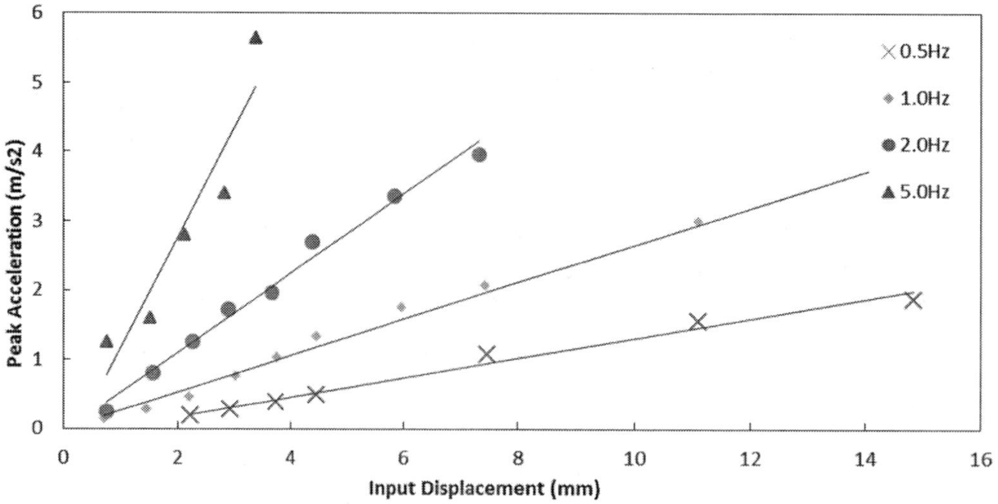

Fig. 2. Relationship between input shaking motions and resultant peak accelerations.

DOI: 10.1051/matecconf/201713804005

2.2 Accelerometer

A pair of accelerometers (Tokyo Sokki Kenkyujo Co., Ltd. (TML) ARH-A model) were attached at the top and base of the soil model as shown in Figure 3. The accelerometer is a strain-gauge type transducer with a measuring range of $10m/s^2$ - $500m/s^2$. The lowest data sampling interval is 1ms. The accelerometer was mounted on the top plate by using adhesive-tape. This mounting method has an advantage of providing electrical insulation between accelerometer and the contact surface [9].

2.3 Laser Displacement Sensor

A laser displacement sensor (Optex CD5 model) was installed near the base of the soil model to provide a direct displacement measurement (Figure 4). The sensor has a measuring range of 85±20mm, with a resolution of 1μm and a sampling interval as low as 100μs.

Fig. 3. Soil model and accelerometers.

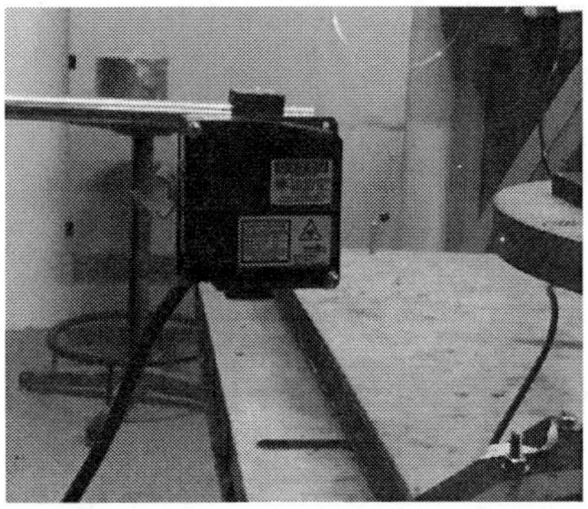

Fig. 4. Laser displacement sensor.

3 Sample Preparation

A remoulded soil sample was prepared for the shaking table test (Figure. 3). Based on the British Soil Classification System, the soil can be classified as Sandy Silt. The soil was compacted to 95% of the maximum dry density ($1664kg/m^3$). The nominal diameter of the sample was 150 mm while the height was 100 mm. The soil sample was confined by a prefabricated rubber membrane and then tightened by a pair of O-ring at top and bottom of the sample. Air confining pressure was supplied into the cylinder chamber and allowed to stabilize for 15 mins.

4 Data Acquisition, Processing & Analysis

Figure 5 shows typical displacement waveforms measured by the laser displacement sensor with shaking motions of 2Hz, 4 Hz and 6 Hz. The shaking table setup was capable of producing a fairly consistent and smooth cyclic movement at different frequencies.

 DOI: 10.1051/matecconf/201713804005

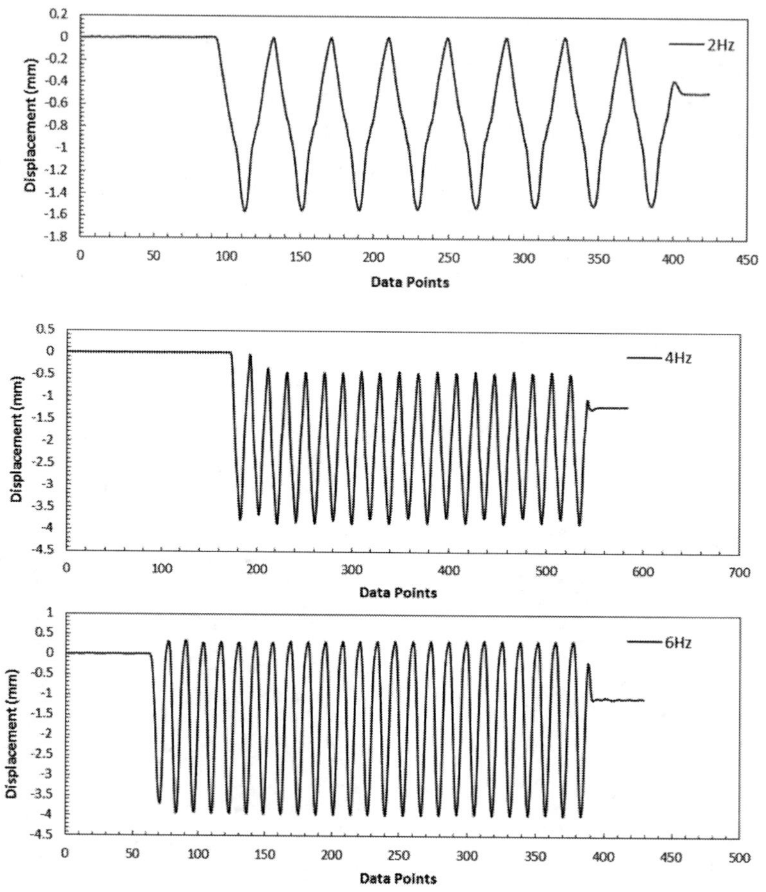

Fig. 5. Displacement waveforms measured using laser displacement sensor.

Figure 6 shows the typical raw acceleration time-series plots obtained from accelerometer measurements. Apparently, an appropriate signal processing approach was required to eliminate the undesirable background noises associated with the acceleration data.

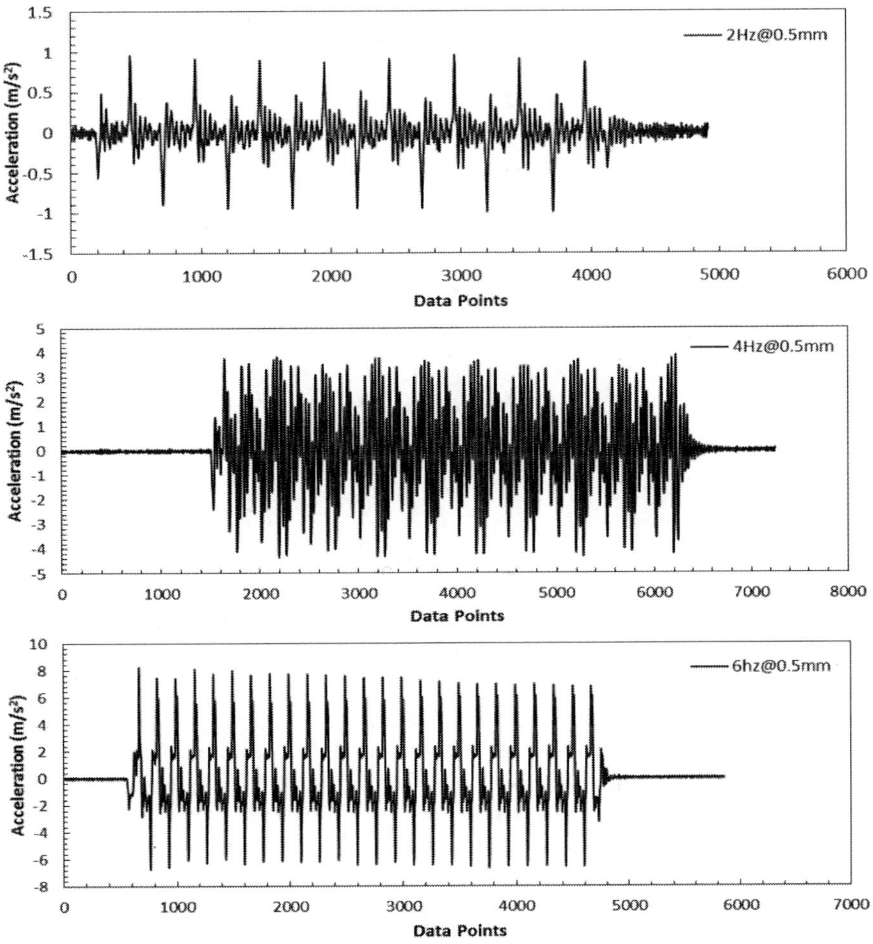

Fig. 6. Acceleration records from accelerometer under different shaking frequencies.

Velocity and displacement time-series can be obtained through numerical integration of the linear-approximated acceleration data points and the assumed initial conditions (i.e. initial velocity and initial displacement). The detailed procedures of the numerical integration may refer to [5, 10].

There is a significant baseline shift in the velocity and displacement profiles, as shown in Figures 7 and 8. The displacement waveform gradually shifted towards the negative sign owing to the negative DC bias in the acceleration time-series record.

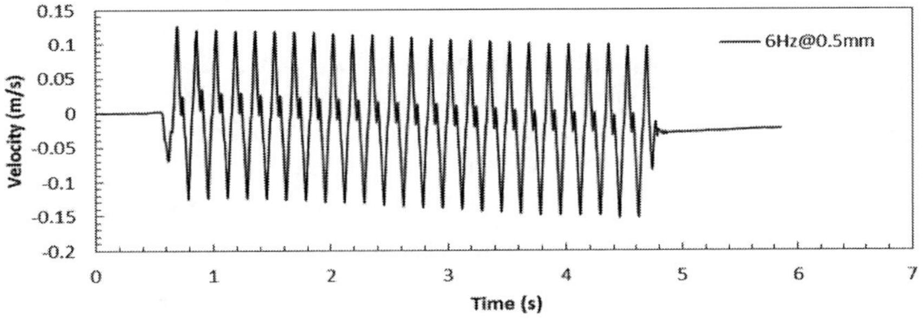

Fig. 7. Derived velocity time-series plot through numerical integration.

If no correction were to be done, the residual displacement at the end of shaking would be unreasonable. As shown in Figure 8, the residual displacement is about 9 cm which was far greater than the input and actual displacement. Therefore, the original shaking record has to be corrected so as to avoid the wavy nature of the displacement waveform. The baseline shift may be attributed to many reasons including ground tilting [11], and contamination of low-frequency noise. Baseline correction is found to be effective in removing low-frequency content and it is actually a method of high-pass filtering with an unknown cut-off frequency [1].

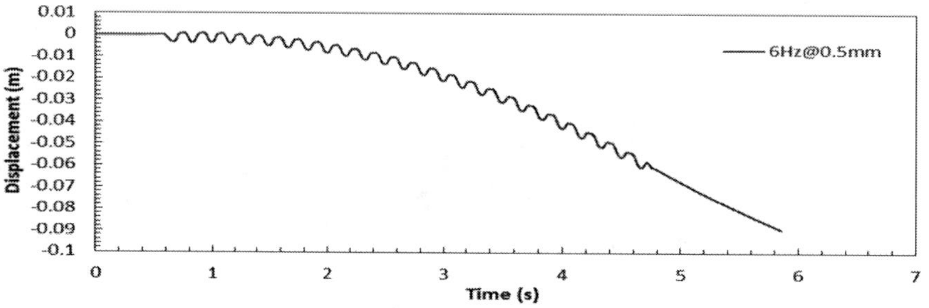

Fig. 8. Derived displacement time-series plot through numerical integration.

The original acceleration record was corrected by attempting three well-established correction schemes, namely Ohsaki's baseline correction, simple quadratic baseline correction, and bandpass filtering methods. The corrected outputs in comparison with the displacement data obtained from the direct laser displacement measurement are shown in Figure 9. It should be noted that the computed displacement traces from accelerometer and the measurement from laser displacement sensor were not oscillated about the same baseline. This is attributed to two standalone datalogger used when capturing the data. The computation of baseline correction by Ohsaki [5] was performed with the aids of a coded programming language, FORTRAN. Apparently, the Ohsaki's baseline correction method renders a movement with a significant residual displacement at the end of shaking. This result was inconsistent with the measurements from the laser displacement sensor (Figure 9).

Through simple quadratic baseline correction method, the problem of baseline drift was remediated significantly as compared to the Ohsaki's baseline correction method. The simple quadratic approach involves subtracting the entire acceleration record from a quadratic least-square fitting line prior to numerical integration. This approach is widely used in commercial processing programs, e.g. SeismoSignal.

Since the baseline-corrected record shows somewhat wavy nature of waveform, it is believed that high-pass or band-pass filtering is capable of removing the low-frequency unwanted signal. Boore and Bommer [1] suggested an approach for selecting a reasonable cut-off frequency in low-pass filtering. In a Fourier Amplitude Spectrum (FAS) plot, shaking record was compared with pre-event or post-event record. Higher cut-off frequency was decided at a frequency where the record to noise ratio is below three. Besides, the frequency response of accelerometer, i.e. 50Hz should be considered when selecting an appropriate cut-off frequency. The relevant comparison is depicted in Figure 10. Nevertheless, the process in choosing a corner frequency in the present study requires a trial-and-error effort.

Band-pass 4th-order Butterworth filtering was subsequently attempted in this study. The filtered results were then compared to that of the laser displacement sensor. Phase difference between the two results can still be observed owing to the filtering process. It was decided to set the amplitude difference between the displacements obtained from filtering method and laser sensor measurement to within 10%, as a criterion when choosing a suitable correction scheme. Slifka [8] reported a similar criterion except that their study involved two types of errors: standard error and peak error. From Figure 9, the displacement waveform computed from the bandpass filtering method does not show any noise effect caused by the low and high-frequency noises. Since the results from bandpass filtering method provided a signal that fulfilled the above-mentioned criterion, it was regarded as the most suitable data processing method in this particular study.

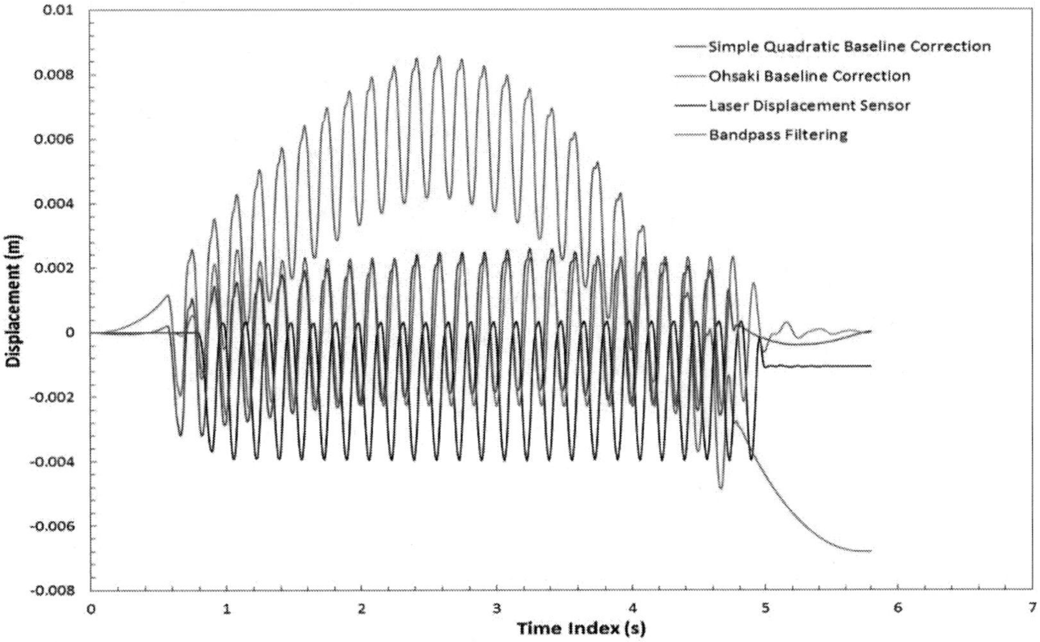

Fig. 9. Comparisons of baseline correction and bandpass filtering methods with laser displacement measurement (6Hz@0.5mm).

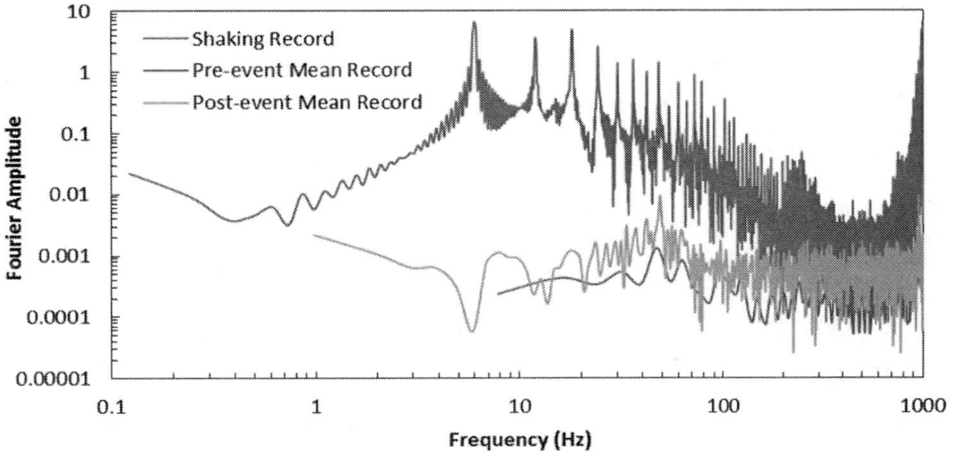

Fig. 10. Comparison of Fourier Amplitude Spectrum for shaking record, pre-event & post-event mean records (6Hz@0.5mm).

Figure 11 shows a typical hysteresis loop obtained upon processing through the bandpass filtering. From the hysteresis loop, secant shear modulus and damping ratio can be easily determined by established methods. Despite of the advantage of using laser displacement sensor as highlighted in the present study, the functions of accelerometer in soil dynamic study should not be overlooked. In many circumstances, acceleration transducer is still required for the measurement of ground response to obtain the synchronized acceleration, velocity and displacement data [12-15].

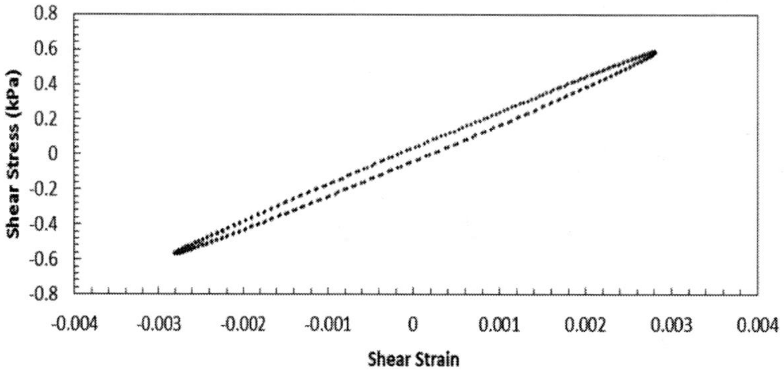

Fig. 11. Typical hysteresis loop for the tested soil model.

5 Conclusion

This study reports the data processing procedures of a shaking table test on a small scaled soil model. Followings are the conclusions that can be drawn from the present study:
(1) A systematic approach of signal processing for a soil shaking table test should include a reliable reference for benchmarking the processed data. Laser displacement sensor can be used as a reliable reference (intuitively similar to the role of a GPS in seismology) when recommending an appropriate adjustment scheme in data processing.
(2) Numerous data processing / filtering methods should be attempted in order to choose the most suitable approach for a particular study.
(3) Digital band-pass filtering is found to be the most suitable method in this particular study. The integrated displacement profile matches reasonably well with the laser displacement measurement.

Acknowledgment

The authors would like to acknowledge the financial supports from the Fundamental Research Grant Scheme (FRGS), Malaysia (Grant No.FRGS/2/2014/TK02/UTAR/02/1).

References

1. D.M. Boore, J.J. Bommer, SOIL DYN EARTHQ ENG., **25**, 93-115 (2005)
2. D.M. Boore, B SEISMOL SOC AM., **92**, 1199-1211 (2001)
3. W.D. Iwan, M.A. Moser, C.Y. Peng, B SEISMOL SOC AM., **75**, 1225-1246 (1985)
4. H.C. Chiu, B SEISMOL SOC AM., **87**, 932-944 (1997)
5. Ohsaki. Y, *Introduction to Spectrum Analysis of Earthquake Motion* (Kajima Shuppankai, 1994)

6. G. Mollova, EURASIP-JASP., **2007**, 1-9 (2006)
7. M. Arraigada, M. Partl, *6th Swiss Transport Research Conference* (2006)
8. L.D. Slifka, *An Accelerometer Based Approach to Measuring Displacement of a Vehicle Body* (Msc Thesis, University of Michigan, 2004)
9. A. Piersol, T. Paez, *Harris' Shock and Vibration Handbook* (Mc GrawHill Education, 2009)
10. G.V. Berg, G.W. Housner, B SEISMOL SOC AM., **51**, 175-189 (1961)
11. V. Graizer, B SEISMOL SOC AM., **96**, 2090-2102 (2006)
12. Kazama. M, Yanagisawa. E, *Eleventh World Conference on Earthquake Engineering*, **367** (1996)
13. Kazama. M, Toyota. H, Towhata. I, Yanagisawa. EJGE., **535/III(34)**, 73-82 (1996)
14. Koga. Y, Matsuo. O, SOILS FOUND., **30**, 162-174 (1990)
15. A.J. Brennan, N.I. Thusyanthan, S.P.G. Madabhushi, J GEOTECH GEOENVIRON., **131**, 1488-1497 (2005)

Stability assessment of enzyme stabilized road embankments

Jaspreet Singh[1], *Dilan* Robert[1*], *Peihan* Wang[1], *Filippo* Giustozzi[1], *Mojtaba* Mahmoodian[1], *Sujeeva* Setunge[1] and *Brian* O'Donnell[2]

[1]Civil Engineering, RMIT University
[2]Center for Pavement Excellence Asia Pacific

Abstract: Road infrastructure is one of the most important factors for a country's development. The vital role that they play in the present economy is reflected in many kilometres of roads worldwide. The external loads due to recurring traffic loads and geohazards can result in road failures with consequences in terms of economic loss to asset managers, public safety and vehicle damage. Due to such adverse effects of road failures, soil stabilization is being widely used as a preventive measure of road failures since recently. However, the assessment (or design) of stabilized road embankments using current standards assume linear material response and two-dimensional geometric idealization. Such assumptions can lead to non-conservative results, leading the road embankments to fail in long-term mainly due to improper assessments in the designs. The current research investigates the assessment of enzyme based stabilized road embankments under external traffic loads using three-dimensional (3-D) finite element (FE) analysis. Firstly, laboratory tests were conducted to characterize the soil/stabilized soil and to calibrate the numerical models. Then, a series of FE analyses were conducted to investigate the performance of stabilized road embankments under external traffic loads. Results showed that the prediction of required pavement thicknesses using elasto-plastic 3-D modelling could be substantially different in comparison to the results obtained from available standard pavement design tools and 2-D idealization of pavement modelling. The study revealed the importance of considering realistic material performance and load applications during the assessment of pavement design in order to maximize the benefits of enzyme stabilization that can be effective in prevention of the road embankment failures.

Keywords: road embankments, enzymes, finite element analysis, traffic loads, pavement design, laboratory tests

* Corresponding author: dilan.robert@rmit.edu.au

DOI: 10.1051/matecconf/201713804006

Introduction

Road infrastructure paves the way for a country's growth and development, particularly in developing countries where road infrastructure has a major influence on economic growth [1]. Road infrastructure stimulates economic activates, improves competitiveness with reduced transportation costs and provides employment opportunities [1]. Failure of this vital infrastructure has detrimental consequences resulting in direct and indirect costs. Direct costs include economic losses due to damages to the road infrastructure whereas indirect costs involve the consequences that the damaged road infrastructure has on society [2]. Within the U.S., annual direct cost exceeds $100 million, whereas annual indirect cost equates to or exceeds direct costs [3]. On one hand, asset managers are faced with economic losses as the road asset that provides services to the public fail and/or become damaged, affecting long term management and maintenance of this vital infrastructure [4]. Further, it can result in fatalities and injuries, in addition to extended road closures due to resulted debris from the road failure [5].

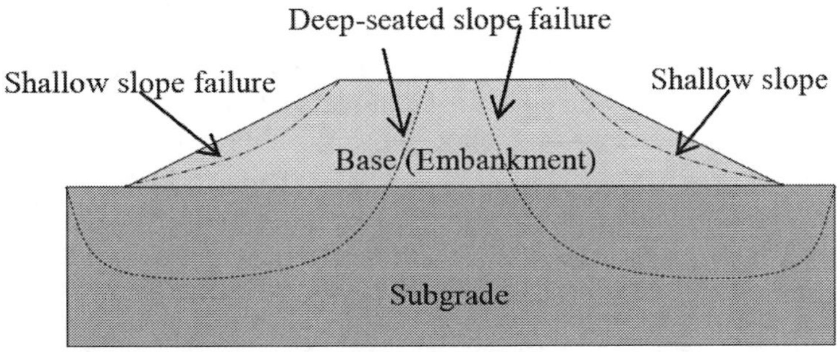

Fig. 1. Potential slope stability failure on a road embankment.

Failures of road embankment can happen when the driving forces exceed the resisting forces from the shear strength of the soil. Driving forces include; external loading from recurring traffic loading and geohazards from rainfall, erosion and earthquakes. Such embankment failures can be identified as a slope failure which can either occur through the embankment as shallow slope failure or deep-seated slope failure [6, 7], as showed in Fig. 1. Deep-seated failures typically occur due to weak foundation soils or after long periods of heavy rainfall, whereas shallow failures commonly occur due to short periods of intense rainfall [7, 8]. The type of failure can be governed by pavement thickness, soil initial condition (embankment and subgrade) and the type of loading applied on the road pavement. However, consequence of failure either due to shallow or deep-seated failure can lead to serviceability concerns and the repair/remediation costs can be substantial. The annual costs of shallow slope failures have been estimated to exceed the repair of deep-seated slope failures [3]. The repair costs of each individual shallow slope failures are low, however due to its common occurrence, the total cumulative costs are extensive [3].

Significant effort has been made in the past to prevent the failure of road embankments through stabilization. Stabilization increases the resisting forces with enhanced shear strength of the soil and reduces the imposed driving forces [9]. Engineers and researchers have adopted numerous stabilization methods for embankments. The current practices adopted by engineers to stabilize shallow slope failures include rebuilding failed slopes, use of geosynthetics, soil nailing and chemical stabilization. However, they do not often provide the most economical stabilization solution. Rebuilding of a failed slope with the failed material is a quick remedial, but is uneconomical in the long term [10]. There are added costs associated with geosynthetic material used in reinforced soil slopes; including; purchasing, transportation and installation [10]. Installation of soil nails can cause displacements in the surrounding wall and its construction and design provisions are expensive [11]. Similarly, the use of lime in embankment stabilization is not the most economical alternative available [12]. On the other hand, enzyme stabilization can be an effective and economic alternative to prevent road embankment failures.

Enzymes are organic catalysts, derived from plants and animals that increase the rate of specific chemical reactions, without being consumed itself. For this reason enzymes are used in low concentrations. Enzymes require mobility to reach the reaction site for it to become activated. Pore fluid in the soil mass provides the required mobility, soil chemistry provides the specific reaction site and time ensures the reaction occurs [13, 14]. Clay soil molecules have a net negative charge, to maintain neutrality; they are attracted to positively charged particles. The negative charge attracts the polarized water molecules, resulting in a water molecules attached and absorbed around the clay soil molecules causing shrinking and swelling of the soil, hence, the properties of clay soil is improved by reducing the amount of absorbed water [14]. Enzyme stabilizers bond with organic molecules, accelerating the reactions between clay soil molecules and the organic cations and accelerate cationic exchange [14]. Consequently, this neutralises the clay molecules negative charge, and thus prevents water absorption and density loss, allowing for soils to become denser [13, 14]. Like other stabilization approaches, enzymes improve and strengthen the soil properties. Some stabilization methods require long cure times or have lengthy stabilization processes and require large quantities, however enzymes provide viable alternative [15]. Enzyme stabilization increases the CBR and durability, reduces the optimum moisture content and plasticity, reduces the soils affinity to water and ensures the recycling of in-situ material [14, 16, 17]. Researchers have found that the unconfined compressive strength (UCS) increases with the use of enzyme soil stabilizers [15, 18]. Enzyme stabilizers have been implemented in the construction of roads to improve the engineering properties of weak soil materials [17]. Enzymes are in liquid concentration form and require dilution; the application process varies between products [16]. At current, there is no model to predict and design stabilized slopes using enzymes. In addition, no standard method exists for the application or testing of enzyme stabilizers in road embankment construction and often manufactures methods are inadequate [15].

Traditionally, the slope stability analysis is expressed as a factor of safety (FOS). Slope stability is determined from limit equilibrium (LE) methods based on assumptions regarding the slip surface [19]. LE method is widely used by engineers due to its level of simplicity; however, the method assumes the FOS is the same along the slip surface, without considering the behaviour of the soil, thus failing to capture the real failure mechanisms of the slope [19, 20]. On the other hand, finite element (FE) analysis can be an effective alternative to overcome these limitations. It provides an understanding of stresses, movements and pore pressures within the slope by using the actual stress-strain characteristics of the soil [20-22]. The improved accuracy and capability of the FE analysis allows slope stability to be monitored and the exact failure mechanism to be obtained [23]. Various slope stability problems have been assessed using numerical modelling [8, 24, 25].

The current approach in the stability assessments of road embankments predominantly assumes 2-D geometric idealizations and linear material behaviour. Unbound pavements are designed using the elastic properties of the material [26]. The practising standard pavement design tools assume material linearity and 2-D geometry/loading simplifications. The performance of unbound pavements subjected to traffic loading have been studied using 2-D elastic FE modelling [27]. Slope stability of embankments have also been determined from 2-D FE models with elasto-plastic behaviour of material [24, 25]. A 2-D model geometry is an over simplification of a 3-D problem. In the 2-D limit equilibrium model, the slip surface is assumed and is infinite and parallel to the slope. Further, traffic loading may not be represented realistically in 2-D modelling as vehicular loads apply

DOI: 10.1051/matecconf/201713804006

pressure on the pavement via patch (i.e. circular) area rather than point or line loading as simulated in 2-D modelling. The approach using 3-D FE modelling can eliminate such simplifications, resulting in a more accurate slope stability analysis and pavement design.

The current research investigates the assessment of enzyme based stabilized road embankments under external traffic loads using three-dimensional (3-D) finite element (FE) analysis. Firstly, laboratory tests were conducted to characterise the soil/stabilized soil and to calibrate the numerical models. Then, a series of FE analyses were conducted to investigate the performance of stabilized road embankments under external traffic loads. Results showed that the prediction of required pavement thicknesses using elasto-plastic 3-D modelling could be substantially different in comparison to the results obtained from available standard pavement design tools and 2-D idealization of pavement modelling. The study revealed the importance of considering realistic material performance and load applications during the assessment of pavement design in order to maximize the benefits of enzyme stabilization that can effectively prevent the road embankment failures.

Methodology

The investigations of the current study are based on laboratory experiments and numerical modelling. Laboratory tests were first conducted on a clay soil obtained from a land excavation site in Melbourne, Australia. Having obtained the soil characteristics of stabilized and non-stabilized soils, a series of FE analysis was conducted to investigate the performance of stabilized road embankment under external traffic loads. The results are compared with standard pavement design predictions to investigate the applicability of current practising tools, which assume material linearity and 2-dimensional approximation of pavement.

Laboratory tests

Laboratory tests, such as hydrometer test, Atterberg test and compaction tests were conducted on the selected clay soil for material characterization (Table 1). A series of direct shear tests was conducted on samples obtained from the stabilized soil mix at various enzyme weights and dilution contents prepared at the optimum moisture and maximum dry density condition. The selected results of the tests are presented in Fig. 2 for enzyme weight of 5% (by dry weight) at the dilution content of 1:500. The soil strength characterization is presented in Table 2.

Table 1. Soil characterization from laboratory tests.

Description	Value
Soil Type	Low plasticity clay
Liquid Limit	28.6
Plastic Limit	20.4
Optimum moisture (OMC)	14.8%
Maximum dry density (MDD)	1.9 g/cm^3
Initial Suction	100 kPa

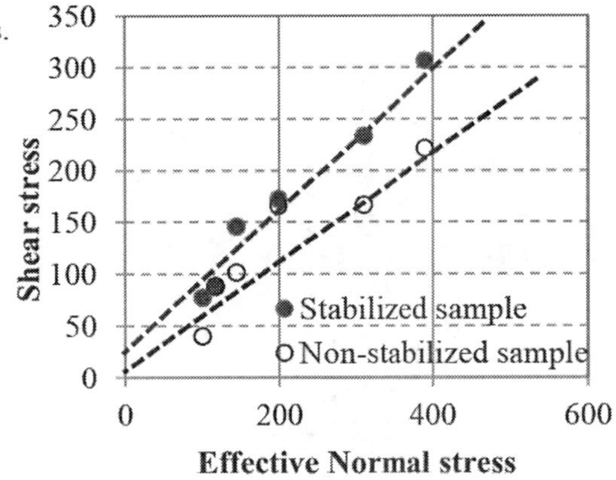

Fig. 2. Direct shear test data for enzyme weight 5% & dilution 1:500

DOI: 10.1051/matecconf/201713804006

Table 2. Soil strength characterization.

	Density (g/cm³)	Elastic Modulus[1] (kPa)	Poisson's ratio	Cohesion[2] (kPa)	Friction angle[2] (degrees)	Dilation angle[2] (degrees)
Stabilised	1.85	150000	0.3	22	34.2	15
Non-Stabilised	1.85	80000	0.3	5	25.6	5

[1.0] From the empirical equation Modulus = 10·CBR based on Austroads (2012)
[2.0] From direct shear test data

Pavement Geometry

In order to select the geometry for numerical modelling, the road embankment was designed according to the Austroads [26] design guidelines, assuming a rural class road with a traffic load of 1.0 x 10^5 ESA, an allowable strain limit of 1.796 x 10^3 and terminal rutting of 20 mm [28]. Allowable strain limit at the subgrade and rutting has been determined based on guidelines [26, 28] and a standard pavement design tool (CIRCLY). For this selected design consideration, the empirical design and mechanistic design (CIRCLY) required a cover of 220 mm and 250 mm, respectively. The roadway design was assumed to be consisting of typical 2-lane rural road with 3.1 m lane width (6.2 m total width). The road embankment is designed with 2:1 slopes, subgrade depth of 5 m and a varying base depth. The selected slope geometry is due to non-recoverability of steeper slopes (>4:1) by traversing cars [29]. Fig. 3 shows the geometry of the derived pavement considered in this study.

FE Model

Three dimensional (3-D) and two dimensional (2-D) finite element analyses were carried out using ABAQUS to investigate the behaviour of road embankments during traffic loading. The pavement and the subgrade soil were represented by 8-noded brick reduced integration elements and 8-noded plane strain reduced integration elements in 3-D & 2-D models respectively. The soil side boundaries of the FE model were assumed to be smooth and are located far (i.e., 2.5 m) from the traffic loads to eliminate any boundary effects. Figure 3 shows the mesh discretization and model dimensions. The applied traffic load on the pavement surface is based on single-axle-dual-tyre (SADT) configuration having total axle load of 80 kN [26].

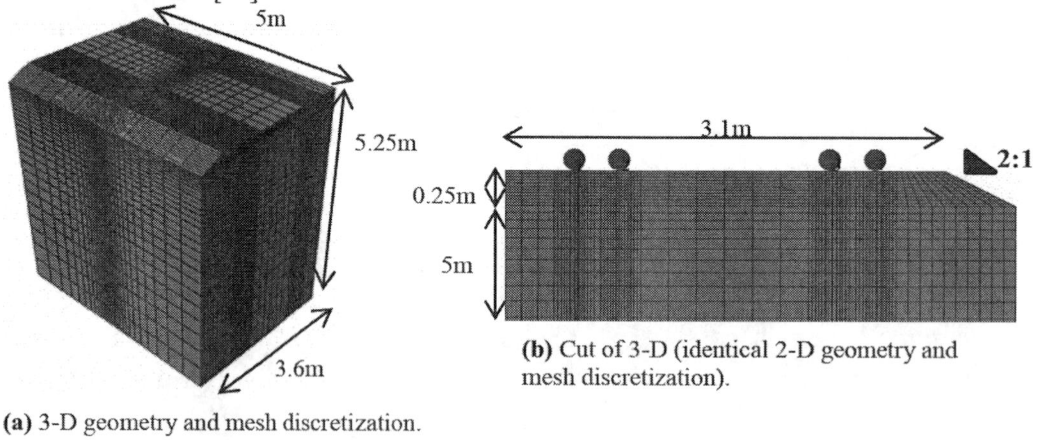

(b) Cut of 3-D (identical 2-D geometry and mesh discretization).

(a) 3-D geometry and mesh discretization.

Fig.3. Geometry and mesh discretization of FE model.

 DOI: 10.1051/matecconf/201713804006

Analysis Plan

Firstly, the analyses were conducted using 3-D & 2-D FE models under traffic loading with linear-elastic material response (Stage 1) for stabilized and non-stabilized pavements. This is to investigate the effect of geometry and loading between 3-D and 2-D (including standard tools) idealizations. Secondly, analyses were performed to study the significance of considering material non-linearity on the selection of stabilized/non-stabilized pavement thicknesses (Stage 2). Finally, a parametric study (with the change in base thicknesses) was conducted to determine the required stabilised base thickness when realistic geometry, loading and material are encountered (Stage 3). In all the analysis, the subgrade depth was kept constant (i.e., 5 m) over the models, while the thickness of the base is varied in order to determine the suitable thickness, as seen in Figure 3

Results/Discussion

The results of the analysis conducted in stage 1 & 2 are summarized in Table 3 & Fig. 4 for investigating the effect of geometry on pavement response for stabilized and non-stabilized materials. The results showed that the 2-D FE modelling of a road embankment does not provide a realistic representation. This is mainly due to the strip tyre pressure loading applied on the pavement rather than a realistic tyre (i.e., patch) loading. Similar to the standard pavement design tool (i.e., CIRCLY) predictions, the 2-D stabilized and non-stabilized pavement models with elastic material idealization satisfies the allowed strain (1.796×10^{-3}) and deformation (20 mm rutting) limits of the pavement as showed in Table 3. However, when non-linearity of the soil is considered within the 2-D modelling framework, pavement fails under both stabilized and non-stabilized conditions of the pavement. This is because of substantial yield induced by 2-D strip loading on the pavement material during traffic loading. The mechanism of failure and shear band formation in the 2-D road embankment is showed in Fig. 4a, which reveals the failure of the road embankment can be due to combined mechanism of shallow and deep soil failures.

3-D FE modelling provides a realistic representation of SADT loading on the road embankment. Similar to what was obtained from 2-D models & CIRCLY analysis, the maximum subgrade strains and ground surface rutting are within allowed limits under elastic stabilized and non-stabilized material conditions (Table 3). However, the road embankment fails when the material was simulated as elasto-plastic for stabilized pavement. Therefore, the pavement, which was accepted under standard pavement design tools and 3-D elastic material idealization, failed due to high deformation of pavement material resulted in soil yielding. This reveals the importance of considering material non-linearity when assessing the stability of stabilized road pavements. The failure mechanism of the 3-D pavement model showed in Fig. 4b (cut from the 3D model across the loading axle direction) which shows the road embankment failure has resulted mainly due to subgrade soil failure.

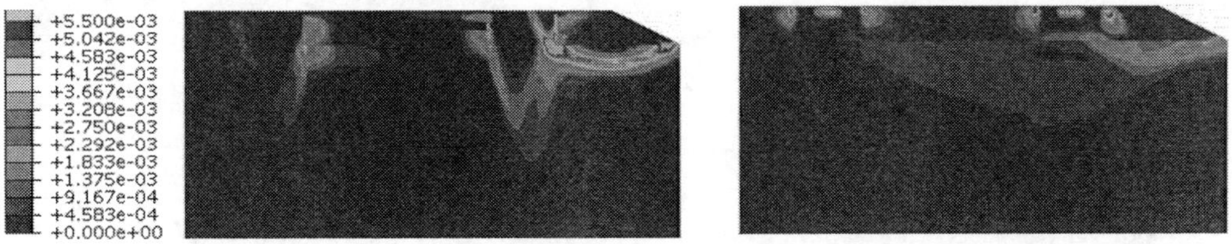

(a). 2-D geometry (30% loading) **(b).** 3-D geometry (100% loading)

Fig.4. Plastic strain and shear band formation in the FE models

 DOI: 10.1051/matecconf/201713804006

Table 3. Strain and rutting in embankment based on different approaches.

Analysis	Dimension	Base Treatment	Pavement Idealisation	Strain at Subgrade, Slope Side (between tyres)	Rutting (mm)
Circly	2-D	Stabilised	Elastic	0.00179	
FE	2-D	Stabilised	Elastic	0.00142	11.8
FE	2-D	Non-Stabilised	Elastic	0.00147	12.6
FE	2-D	Stabilised	Elasto-plastic (with 30.20% loading)	0.00721	5.8
FE	2-D	Non-Stabilised	Elasto-plastic (with 14.45% loading)	0.00148	4.3
FE	3-D	Stabilised	Elastic	0.000527	0.48
FE	3-D	Non-Stabilised	Elastic	0.000569	1.03
FE	3-D	Stabilised	Elasto-plastic	0.00185	0.74
FE	3-D	Non-Stabilised	Elasto-plastic (with 27.30% loading)	0.000690	8.24
Allowable strain: 0.001796, Terminal rutting 20 mm					

The effectiveness of enzyme stabilization can be identified from the maximum strains resulted at the subgrade soil (Fig. 5). Results revealed that when the base is treated with enzyme stabilisation, the strains experienced at the subgrade level are significantly lower than the subgrade strains resulted in non-stabilized road pavement. This can be identified as an improvement of the road performance due to stabilization. i.e. enzyme stabilization has increased the shear strength of soil, enhancing the overall road performance. It can be seen that the subgrade strains getting attenuated towards the slope side due to the loading interaction with the slope. Such response reveals the importance of stabilizing slopes as the road embankment failure is triggered at the slope side of the road.

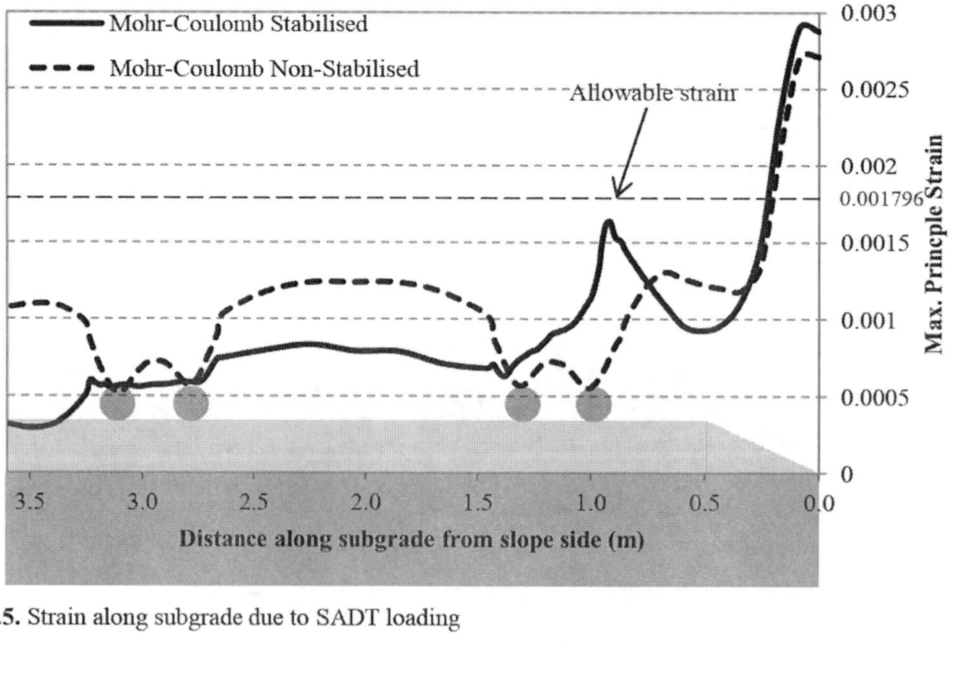

Fig.5. Strain along subgrade due to SADT loading

The results of the parametric study conducted to investigate the adequate stabilized pavement thicknesses is summarised in Fig. 6. It can be seen that the subgrade strain decreases with the increase in treated base (embankment) thickness. The stabilized base thickness required to satisfy the allowed strains in the subgrade (1.796 x 10^3) is identified as 300 mm using the 3-D FE model which incorporates elasto-plastic material response. On the other hand, the standard pavement design tool shows that a base thickness of 250 mm is adequate to satisfy the allowable strains, indicating an under-prediction of 20% for the stabilized layer thickness under the selected design consideration. This highlights the importance of considering the realistic material and geometric idealizations when assessing the stability of stabilized road pavements.

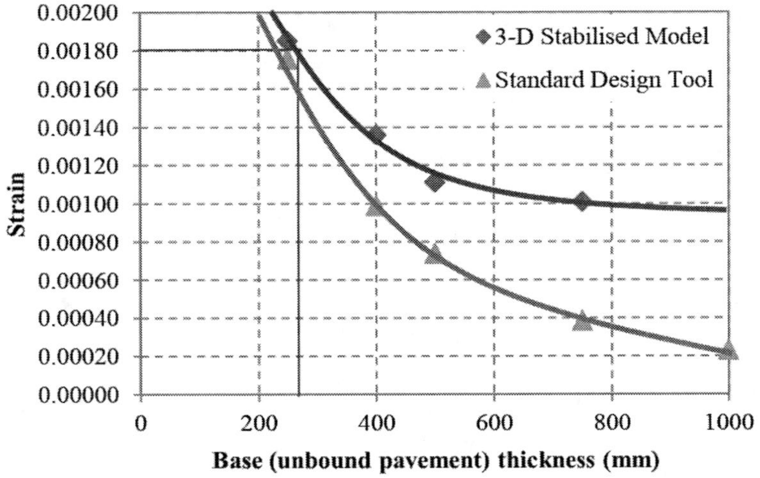

Fig.6. Comparison of strains at various bases thicknesses using 3-D FE analysis (using Mohr-Coulomb plasticity) and standard design tool

Conclusions

The current research investigates the assessment of enzyme based stabilized road embankments under external traffic loads using finite element analysis. Firstly, laboratory tests were conducted to characterise the soil/stabilized soil and to calibrate the numerical models. Then, a series of 2-D & 3-D FE analyses were conducted to investigate the performance of stabilized road embankments under external traffic loads. The results showed that the 2-D FE modelling of a road embankment subjected to traffic loading does not provide a realistic model response due to the strip pressure loading applied on the pavement surface rather than more rational circular patch loading which can be closely simulated using 3-D FE modelling. Results also revealed that, similar to standard design outcomes, the predictions from the numerical models for stabilized road performance are within acceptable deformations when the material behaviour is simulated under elastic idealization. However, the road embankment reaches serviceability failure under the same traffic loading when the stabilized material is simulated using elasto-plastic soil behaviour. The thickness of stabilized base required to satisfy the allowable strains is 20% higher than what is predicted using standard design tools and 2-D/3-D elastic modelling approaches. Thus, the results from the current study implies the importance of considering the realistic material and geometric idealizations when assessing the stability of stabilized road pavements. It is to be noted that the initial moisture (or suction) of the stabilized soil can also influence the overall response of the road embankment subjected to external loading.

The study is being continued at present to investigate the pavement behaviour that can be captured within triple-hardening framework of stabilized soil ('stabilized + unsaturated' cohesion enhancement, 'stabilized + unsaturated' dilation enhancement and strain hardening).

References

1. P. Sahoo and R.K. Dash, *Economic growth in South Asia: role of infrastructure*. Institute of Economic Growth. (2008)
2. L. Masiero and R. Maggi, *Estimation of indirect cost and evaluation of protective measures for infrastructure vulnerability: A case study on the transalpine transport corridor*. Transp. Policy. **20**: p. 13-21. (2012)
3. J. Loehr, T. Fennessey, and J. Bowders, *Stabilization of surficial slides using recycled plastic reinforcement.* Transport. Res. Rec.,(1989): p. 79-87. (2007)
4. S. McNeil and P. Herabat, *Asset Management*, in *The Handbook of Highway Engineering*. CRC Press. (2005)
5. M. Leiba, *Impact of landslides in Australia to December 2011.* Australian Journal of Emergency Management, The. **28**(1): p. 28. (2013)
6. J. Han, et al., *Evaluation of deep-seated slope stability of embankments over deep mixed foundations*, in *GeoSupport 2004: Drilled Shafts, Micropiling, Deep Mixing, Remedial Methods, and Specialty Foundation Systems*. p. 945-954. (2004)
7. K. Briggs, F. Loveridge, and S. Glendinning, *Failures in transport infrastructure embankments.* Eng. Geol. p. 107-117. (2017)
8. F. Cai and K. Ugai, *Numerical analysis of rainfall effects on slope stability.* Int. J. Geomech. **4**(2): p. 69-78. (2004)
9. L.W. Abramson, *Slope stability and stabilization methods*. John Wiley & Sons. (2002)
10. R.W. Day, *Design and repair for surficial slope failures.* Practice Periodical on Structural Design and Construction. **1**(3): p. 83-87. (1996)
11. F. Schlosser and P. Unterreiner, *Soil nailing in France: research and practice.* Transport. Res. Rec. **1330**: p. 72-79. (1991)
12. R.B. Randolph, *Earth materials catalyst stabilization for road bases, road shoulders, unpaved roads, and transportation earthworks.* Transport. Res. Rec.,(1589): p. 58-63. (1997)
13. J. Tingle, et al., *Stabilization mechanisms of nontraditional additives.* Transport. Res. Rec.,(1989): p. 59-67. (2007)
14. M.R. Taha, et al., *Recent Experimental Studies in Soil Stabilization with Bio-Enzymes–A.* Electronic Journal of Geotechnical Engineering. **18**(R): p. 3881-3894. (2013)
15. J. Tingle and R. Santoni, *Stabilization of clay soils with nontraditional additives.* Transport. Res. Rec.,(1819): p. 72-84. (2003)
16. C. Stan and V. Ciobanu, *Using enzymatic emulsions to reinforce road layers.* Bulletin of the Transilvania University of Brasov. Forestry, Wood Industry, Agricultural Food Engineering. Series II. **5**(1): p. 109-114. (2012)
17. V. Saini and P. Vaishnava, *Soil stabilization by using terrazyme.* International Journal of Advances in Engineering & Technology. **8**(4): p. 566-573. (2015)
18. R. Parsons and J. Milburn, *Engineering behavior of stabilized soils.* Transport. Res. Rec.,(1837): p. 20-29. (2003)

19. J.M. Duncan, *State of the art: limit equilibrium and finite-element analysis of slopes.* J. Geotech. Eng-ASCE. **122**(7): p. 577-596. (1996)
20. B. Khadija, et al., *Slope Stability Evaluations by Limit Equilibrium and Finite Element Methods Applied to a Railway in the Moroccan Rif.* Open Journal of Civil Engineering. **2**(1): p.27. (2012)
21. D. Griffiths and P. Lane, *Slope stability analysis by finite elements.* Geotechnique. **49**(3): p. 387-403. (1999)
22. K. Lim, et al., *Slope-stability assessments using finite-element limit-analysis methods.* Int. J. Geomech. **17**(2): p. 06016017. (2017)
23. H. Zheng, D. Liu, and C. Li, *Slope stability analysis based on elasto-plastic finite element method.* Int. J. Numer. Meth. Eng. **64**(14): p. 1871-1888. (2005)
24. S.W. Abusharar, J.-J. Zheng, and B.-G. Chen, *Finite element modeling of the consolidation behavior of multi-column supported road embankment.* Comput. Geotech. **36**(4): p. 676-685. (2009)
25. J.-C. Chai, N. Miura, and S.-L. Shen, *Performance of embankments with and without reinforcement on soft subsoil.* Can. Geotech. J. **39**(4): p. 838-848. (2002)
26. Austroads, *Guide to Pavement Technology* in *Part 2: Pavement Structural Design* Austroads Incorporated: Sydney, Australia. (2012)
27. K. Siripun, P. Jitsangiam, and H. Nikraz, *The design model of unbound granular materials for flexible pavement.(technical paper)(Report).* Aust. Civ. Eng. **10**(1): p. 1. (2012)
28. G. Jameson and K.G. Sharp, *Technical basis of Austroads pavement design guide.* (2004)
29. Austroads, *Guide to Road Design* in *Part 6: Roadside Design, Safety and Barriers* Austroads Incorporated: Sydney, Australia. (2010)

Subsurface Profile Mapping using 3-D Compressive Wave Imaging

Z A M Hazreek[1,2], T B M Faizal[1], M Aziman[1], A T S Azhar[1], S Rosli[1], M I M Ashraf[3] and S M Hafizan[1]

[1] Faculty of Civil and Environmental Engineering, Universiti Tun Hussein Onn Malaysia, 86400 Batu Pahat Johor, MY (E-mail: *hazreek@uthm.edu.my*; *mdfaizal@uthm.edu.my*; *aziman@uthm.edu.my*; *saifulaz@uthm.edu.my*; *hafizanmohd92@gmail.com*)
[2] School of Physics, Universiti Sains Malaysia, 11800 USM Penang, MY (E-mail: *rosli@usm.my*
[3] School of Civil Engineering, Universiti Sains Malaysia, 14300 Nibong Tebal Penang, MY (E-mail: *ceashraf@usm.my*)

Abstract
Geotechnical site investigation related to subsurface profile mapping was commonly performed to provide valuable data for design and construction stage based on conventional drilling techniques. From past experience, drilling techniques particularly using borehole method suffer from limitations related to expensive, time consuming and limited data coverage. Hence, this study performs subsurface profile mapping using 3-D compressive wave imaging in order to minimize those conventional method constraints. Field measurement and data analysis of compressive wave (p-wave, vp) was performed using seismic refraction survey (ABEM Terraloc MK 8, 7 kg of sledgehammer and 24 units of vertical geophone) and OPTIM (SeisOpt@Picker & SeisOpt@2D) software respectively. Then, 3-D compressive wave distribution of subsurface studied was obtained using analysis of SURFER software. Based on 3-D compressive wave image analyzed, it was found that subsurface profile studied consist of three main layers representing top soil (vp = 376 – 600 m/s), weathered material (vp = 900 – 2600 m/s) and bedrock (vp > 3000 m/s). Thickness of each layer was varied from 0 – 2 m (first layer), 2 – 20 m (second layer) and 20 m and over (third layer). Moreover, groundwater (vp = 1400 – 1600 m/s) starts to be detected at 2.0 m depth from ground surface. This study has demonstrated that geotechnical site investigation data related to subsurface profiling was applicable to be obtained using 3-D compressive wave imaging. Furthermore, 3-D compressive wave imaging was performed based on non destructive principle in ground exploration thus consider economic, less time, large data coverage and sustainable to our environment.

Keywords
Site investigation; compressive wave; seismic refraction

INTRODUCTION

The main purpose of the site investigation is to determine sub-surface profile which required the information of soil, rock and groundwater. Generally, site investigation consists of two methods based on conventional and alternative method. Conventional method referred to drilling method such as boreholes, test pit and probes (in-situ test). In the past, several limitations of conventional method were due to time consuming, expensive, limited data coverage and less environmental sustainability. Past researcher has reported that conventional method limitation was due to the expensive, time consuming [1, 2] as well as limited data coverage [3, 4, 5, 6]. Conventional method efficiency may reduce particularly when working on large and difficult sites due to difficulty of equipment mobilization and operational process. Information obtained from drilling methods only provides information on exact drilling point location (1-D) thus promoting to questionable results due to its uncertainties of interpolation approach. Information obtained from drilling method was a single point data (1D) and common interpolation technique between large boreholes spacing can lead to increase the degree of uncertainties of subsurface profile investigated [5, 6]. Cost of site investigation may be increased proportional to the drilling number in order to obtained accurate results. In addition, conventional method requires drilling process which consider as destructive method thus may not be sustainable to our environment during extensive exploration. As reported by [7], problems which may occur during the exploration at difficult sites always related to time and cost consumption due to rugged topography, large coverage area, sophisticated equipment and manpower. According to [8], conventional method particularly invasive technique provide more reliable results but suffering from laborious, time consuming and results reliability turn difficult involving large scale of area. The solutions to these challenges will require multidisciplinary research across the social and physical sciences and engineering [9].

Hence, alternative method related to geophysical method was introduced in order to minimize conventional method constraints. Geophysical method offers the chance to overcome some of the problems inherent in more conventional ground investigation techniques [10]. Geophysical techniques contributes several advantages such as it can be performed fast and low cost and has the ability to cover greater areas more thoroughly [11, 7, 2, 12, 5]. Geophysical studies are measurements of physical quantities (resistance, speed of propagation of sound, density, magnetism, conductivity, etc.) aimed at identifying comprehensively the rock mass structure and lithology characteristics to be used in geotechnical studies, water investigations, etc. [13]. Geophysical method consists of several techniques including seismic, electrical resistivity, ground penetrating radar, gravity and magnetic method. Geophysical methods offer some advantages related to low cost, fast and greater data coverage. As reported by [14], geophysical techniques are non-destructive, non-invasive, relatively fast and cheap.

Geophysical method with particular reference to seismic refraction method was suitable to be adopted in site investigation particularly on characterization purposes. Seismic refraction method was suitable in determination of near surface of geotechnical characterization [15, 16, 17, 18, 19]. The principle of seismic refraction exploration is measuring time taken for a wave to travel from one location to another location. The time taken is a function of elastic modulus of the material through which the wave travels. Layers below the earth surface was investigated using seismic wave (Body wave: Primary & Secondary wave or Surface wave: Rayleigh and Love wave) based on Snell's law. Reflection and refraction of seismic wave will be produced when waves propagate in a medium (geo-materials) follow the elastic characteristic through all directions. The motion of wave's particle will be recorded as a time function at a certain distance. Consequently, layers and structures in the subsurface will be determined. Hence, seismic refraction method was able to determine depth of soil over rock by measuring the first arrival time of wave (Compressive wave: P-wave) at different points along a line. Seismic refraction field measurement can be performed fast with less manpower. Field equipment mobilization and setup was easy due to its portable, lighter weight and can be dismantle into few components. Moreover, seismic refraction method able to provide large data coverage based on compressive wave distribution of 2-D and 3-D imaging. Furthermore, seismic refraction method was performed based on indirect test (non-destructive method) which explore the subsurface profile non-invasively thus consider sustainable to our environment. According to [20], lithology, porosity and interstitial fluids of geo-materials can influenced the success of interpretation of subsurface profile based on the seismic P-wave velocity contrast. Furthermore, the reliability performance of any individual geophysical methods will always depends on fundamental physical constraints, e.g. penetration, resolution, and signal to-noise ratio [6] and [21]. Hence, this study focused on 3-D compressive wave imaging via seismic refraction method to investigate subsurface profile at Parit Seri Mendapat, Batu Pahat, Johor Malaysia.

METHODOLOGY

This study was located at Parit Seri Mendapat, Batu Pahat, Johor Malaysia as shown in Figure 1. Specifically, site study consists of palm oil estate with flat topography. Study area was located at rural areas which consider less noise thus able to produced good quality of seismic data (compressive wave). According to [22], study area was located at boundaries between Quaternary and Triassic period. Common type of materials in Quaternary period consists of clay, silt with minor gravel, while Triassic period consists

DOI: 10.1051/matecconf/201713804007

of sedimentary rocks (Interbedded sandstone, siltstone and shale). Regional geology map of study area was given in Figure 2. Field observation founds that study area consist of sediment material such as sand, silt and clay.

3-D compressive wave imaging was obtained based on seismic refraction method. Three (3) seismic spread lines (total length of each line = 115 m) was performed in parallel direction according to west-east alignment as shown in Figure 1. Basically, seismic refraction method consists of three (3) main components related to source, receiver and record. Source, receiver and record were performed using ABEM Terraloc MK-8 seismograph, 24 channels of 28 Hz vertical geophone, and 7 kg sledgehammer respectively. All spread lines configuration was based on 5 m of geophone spacing, 50 m of offset distance and 10 – 12 m of spread line parallel spacing. In order to obtained adequate seismic wave, seismograph sampling interval (200 μs) and number of sample (4096) was selected based on site condition suitability. Seven (7) numbers of shot points for all spread lines was performed based on 5 in line shot point and 2 offset shot point. 20 to 30 seismic data stacking was performed for all shot point location in order to procedure good seismic wave particularly compressive wave (primary wave, vp). Finally, seismic raw data obtained from field measurement was analyzed using OPTIM and SURFER software to produce 3-D compressive wave imaging.

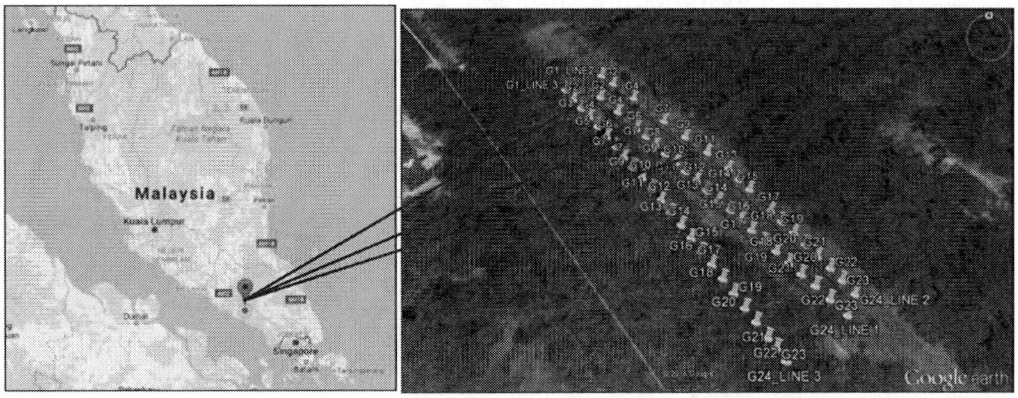

Figure 1. Study area and seismic spread lines performed at site studied.

DOI: 10.1051/matecconf/201713804007

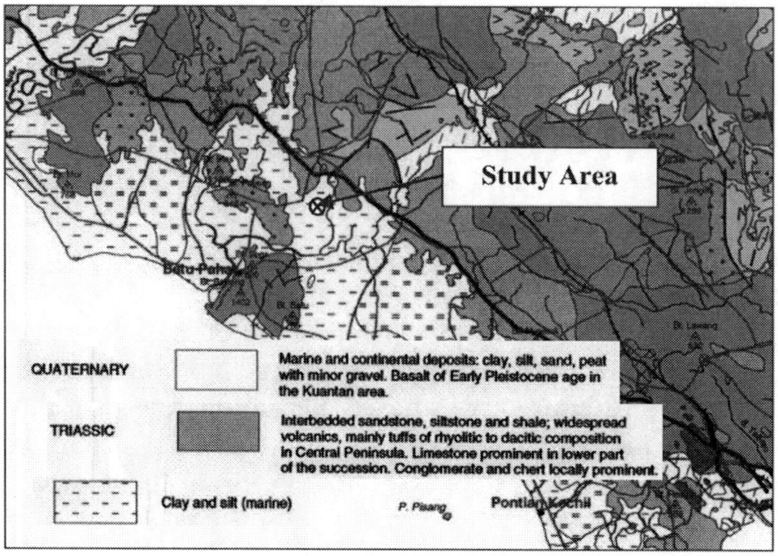

Figure 2. Study area and seismic spread lines performed at site studied [19].

RESULT AND DISCUSSION

3-D compressive wave imaging (3-DCWI) obtained on site studied was given in Fig. 1. According to Figure 3, study area consist of three (3) main layer due to the obvious different contrast of compressive wave (p-wave) representing three (3) types of geo-materials with different characteristics. First layer was found to be thin surface layer with a thickness of 2 m which characterized by low compressive wave (p-wave) value ranging from 376 – 600 m/s. Thickness of second layer was varied from 2 – 20 m depth from ground surface with compressive wave (p-wave) value ranging at 900 – 2600 m/s. Third layer of subsurface profile studied was detected at 20 m depth from ground level based on compressive wave (p-wave) value of 3000 m/s and above. All layers interpretation has been verified based on previous established compressive wave (p-wave) of earth materials. Soil which considers as subsurface overburden materials has p-wave value (vp) of 250 – 600 m/s [23]. Second layer was interpreted as weathered zone which possibly consist of a mixture of highly weathered to moderately weathered zone. Weathered rock may be interpreted by p-wave value of 457 – 3657 m/s [24]. Then, third layer of subsurface profile studied was interpreted as bedrock which possibly derived from slightly weathered to fresh rock. P-wave velocity value of bedrock was 3000 m/s and over [25]. Moreover, groundwater was interpreted occurred at 2.0 m depth and over due to its p-wave velocity value which ranging from 1400 – 1600 m/s. P-wave velocity value for water was varied from 1400 – 1600 m/s [26]. In addition to verification purposes, all geo-materials interpreted from 3-DCWI have

DOI: 10.1051/matecconf/201713804007

shown some good correlation based on geological map as explained previously together with field observation.

3-DCWI was able to present the subsurface profile model in a global perspective which able to contribute for planning related to suitable drilling number and location. 3DCWI was able to obtain image cross-sectional area in wider coverage thus able to compliment drilling results. However, suitable seismic spread line spacing need to be carefully decided in order to obtained best possible of 3-DCWI representing accurate subsurface profile studied. General consideration of suitable spread line spacing was relative to complexity of geological study area and specific target interest (e.g. thickness and layers, boulder, bedrock, etc).

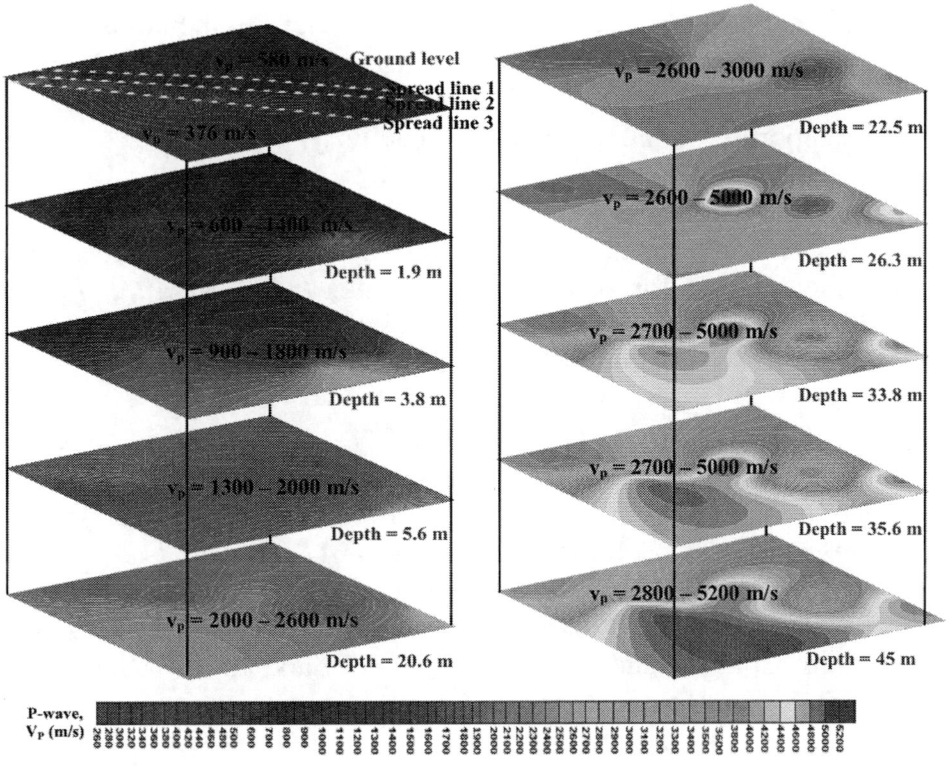

Figure 3. 3-D compressive wave (p-wave, v_p) imaging.

CONCLUSION

Subsurface profile mapping at Parit Seri Mendapat, Batu Pahat, Johor Malaysia was successfully being performed using 3-D compressive wave (p-wave) imaging (3-DCWI). The geometry and geo-materials of subsurface profile studied has been determined by analyzing seismic compressive wave data and the result has shown some good correlation based on previous references. This finding has proved that 3-DCWI technique was applicable as an alternative method in geotechnical site investigation. Moreover, 3-

DCWI via seismic refraction method was able to mapped subsurface profile thus extending surface information observed during the surface physical mapping. The application of 3-DCWI via seismic refraction in conjunction with other records (borehole data, geological information, established resistivity values of geo-materials, etc) was applicable in evaluation of subsurface mapping due to its ability to differentiate various properties of earth materials based on compressive wave analyzed. Geophysical method with particular reference to 3-DCWI is suitable for our sustainable site investigation due to its ability to reduce time, money and compliment others conventional drilling method.

ACKNOWLEDGEMENTS

Thank are due to all research members for their tremendous work and cooperation. Many thanks go to Research and Innovation Fund and IGSP from Vot. U258, Universiti Tun Hussein Onn Malaysia for the sponsor and financial support throughout this research activity.

REFERENCES

Osman, F. I. S. and Osman, S. B. A. S. (2012) Integrating geo-electrical and geotechnical data for soil characterization, *Int. J. of Appl. Phys. and Maths*, **2**.

Liu, C. and Evett J. B. (2008) Soils and Foundation. Pearson Prentice Hall, 2008.

Hazreek, Z. A. M., Aziman, M. Azhar, A. T. S., Chitral, W. D., Fauziah, A. and Saad, R. (2015) The Behaviour of Laboratory Soil Electrical Resistivity Value under Basic Soil Properties Influences, *IOP Conference Series: Earth and Environmental Science*, **23**.

Abidin, M. H. Z., Ahmad, F., Wijeyesekera, D. C., Saad, R. and Baharuddin, M. F. T. (2013) Soil Resistivity Measurements to Predict Moisture Content and Density in Loose and Dense Soil, *Applied Mechanics and Materials*, **353-356**.

Godio, A., Strobbia, C. and De Bacco, G. (2006) Geophysical Characterisation of a Rockslide in an Alpine Region, *Engineering Geology*, **83**.

Mauritsch, H. J. Seiberl, W. Arndt, R. Römer, A. Schneiderbauer, K. and Sendlhofer, G. P. (2000) Geophysical Investigations of Large Landslides in the Carnic Region of Southern Austria, *Engineering Geology*, **56**.

Khatri, R., Shrivastava, V. K. and Chandak, R. (2011) Correlation between Vertical Electric Sounding and Conventional Methods of Geotechnical Site Investigation, *International J. of Adv. Eng. Scie. and Tech*, **4**.

Yamakawa, Y., Kosugi, K. I., Masaoka, N., Sumida, J., Tani, M. and Mizuyama, T. (2012) Combined Geophysical Methods for Detecting Soil Thickness Distribution on a Weathered Granitic Hillslope,

Geomorphology, **145–146**.
Fragaszy, R., Santamarina, J., Amekudzi, A., Assimaki, D., Bachus, R., Burns, S., Cha, M., Cho, G., Cortes, D., Dai, S., Espinoza, D., Garrow, L., Huang, H., Jang, J., Jung, J., Kim, S., Kurtis, K., Lee, C., Pasten, C., Phadnis, H., Rix, G., Shin, H., Torres, M. and Tsouris, C. (2011) Sustainable development and energy geotechnology — Potential roles for geotechnical engineering, *KSCE Journal of Civil Engineering*, **15**.
Clayton, C. R. I. Matthews M. C. and Simons N. E. (1995) Site Investigation. Blackwell Science Ltd, UK, 1995.
Hazreek, Z. A. M., Rosli, S., Chitral, D. C., Fauziah, A., Azhar, A. T. S., Aziman, M. and Ismail, B. (2015) Soil Identification using Field Electrical Resistivity Method, *Journal of Physics: Conference Series*, **622**.
Cosenza, P., Marmet, E., Rejiba, F., Jun Cui, Y., Tabbagh, A. and Charlery, Y. (2006) Correlations between geotechnical and electrical data: A case study at Garchy in France, *Journal of Appl. Geophys*, **60**.
Lukić, I., Barać, D. and D. Zovko (1974) Seismic Refraction Method – Review, *Geoexploration*, **12**.
Kowalczyk, S., Maślakowski, M. and Tucholka, P. (2014) Determination of the correlation between the electrical resistivity of non-cohesive soils and the degree of compaction, *Journal of Appl. Geophys*, **110**.
Nadia, R., Saad, R., Muztaza, N. Saidin, N. A. and Ismail, M M. (2016) Geotechnical parameters study using seismic refraction tomography, *Jurnal Teknologi*, **78**,
Aziman, M., Hazreek, Z. A. M., Azhar, A. T. S. and Haimi D. S. (2016) Compressive and Shear Wave Velocity Profiles using Seismic Refraction Technique, *Journal of Physics: Conference Series*, **210**,
Tajudin, S. A. A., Abidin, M. H. Z., Madun, A. and Zawawi. M. H. (2015) Compressive and Shear Wave Velocity Profiles using Seismic Refraction Technique, *IOP Conference Series: Materials Science and Engineering*, **136**.
Azwin, I. N., Saad, R. and Nordiana, M. (2013) Applying the Seismic Refraction Tomography for Site Characterization, *APCBEE Procedia*, **5**.
Abidin, M. H. Z., Saad, R., Ahmad, F., Wijeyesekera, D. C. and Baharuddin, M. F. T. (2012) Seismic Refraction Investigation on Near Surface Landslides at the Kundasang area in Sabah, Malaysia, *Procedia Engineering*, **50**.
Israil, M. and Pachauri, A. K. (2003) Geophysical characterization of a landslide site in the Himalayan foothill region, *Journal of Asian Earth Sciences*, **22**.
McCann, D. M. and Forster, A. (1990) Reconnaissance geophysical methods in landslide investigations, *Engineering Geology*, **29**.
Malaysia, M. G. D. (1985) Geological Map of Peninsular Malaysia, 8th Ed.

DOI: 10.1051/matecconf/201713804007

Lee, T. S. 2002 Slope Stability and Stabilization Methods: Geologic Site Investigation. John Wiley & Sons, Inc. 2002.

Cornforth, D. H. 2005 Landslides in Practice: Investigation, Analysis and Remedial/Preventative Option in Soils. John Wiley & Sons, Inc. New Jersey 2005.

Blyth F. G. H. and Freitas, M. D. (1974) A Geology for Engineers. Hodder & Stoughton Educational, London, 1974.

Burger, H. R., Sheehan A. F and Jones, C. H. (2006) Introduction to Applied Geophysics Exploring the Shallow Subsurface. W. W. Norton & Company, New York 2006.

MATEC Web of Conferences **138**, 04008 (2017)
EACEF 2017
DOI: 10.1051/matecconf/201713804008

Bearing Capacity Characteristics of Steel Pipe Piles with Wings installed in Soil Cement Column

Pham Hoang Kien[1,*] and *Daisuke* Ito[2]

[1]University of Transport and Communications, Faculty of Civil Engineering, Hanoi, Vietnam
[2]AsahiKasei Construction Materials Corporation, Tokyo, Japan

Abstract. The steel pipe piles with wings installed in soil cement developed in Japan and Vietnam are composite foundation piles made of steel pipes with wings with foundation improvement pillars in soil cement and spiral wings attached to the shaft intermittently. This composite pile has large vertical and horizontal bearing capacities and is a construction method of foundation piles with consideration of environments with little surplus soil, using steel pipes with small diameters. In this study, several full scale loading tests on steel pipe pile with wings installed in soil cement column are presented, and bearing capacity of the composite pile is studied. Especially, combination effect of soil cement column and steel pipe with wings on bearing capacity performance are discussed. Three-dimensional elasto-plastic finite element analysis based on the results of loading tests to help understand this composite pile's mechanical characteristics is also performed.

1 Introduction

The steel pipe piles with wings installed in soil cement (from now referred to as ATT) have been developed in Japan and Vietnam. Up to now, there have been more than 5,000 cases of construction results. ATT is composite foundation piles made of steel pipes with wings with foundation improvement pillars (from now referred to as columns) in soil cement and spiral wings attached to the shaft intermittently. This composite pile provides large vertical and horizontal bearing capacities due to combination effect of soil cement column and steel pipe with spiral wings. ATT is also an environmentally friendly piling construction method with little surplus soil, low vibration and noise, using steel pipes with small diameters.

In Japan and Vietnam, there are many cases of weak ground where pile's support layer is very deep. In these cases, the piles become very long and problems have risen such as the cost for substructure becomes very high in comparison to the cost for superstructure. ATT is a foundation pile construction method developed to solve these kinds of problems. This is because ATT is a pile foundation developed as a friction pile with good cost performance, which does not become fixed to the support layer and displays bearing capacity by pile shaft friction.

* Corresponding author: phkien@utc.edu.vn

In this study, several full scale loading tests of ATT are presented, and bearing capacity of the composite pile is studied. Especially, combination effect of soil cement column and steel pipe with spiral wings on bearing capacity performance are discussed. Three-dimensional elasto-plastic finite element analysis based on the results of loading tests to help understand this composite pile's bearing characteristics is also performed.

2 Standard Specifications of ATT

The standard specifications of ATT's steel pipe with wings and column are shown in Fig.1 and Table 1. It is able to select from two to four types of wing diameter and column diameter for each steel pipe. Fig. 2 shows the toe shape of the steel pipe with wings. The end of the steel pipe is blocked by a steel plate, and there is a drilling blade on the end of the steel plate. Also, the pile head is blocked in the same way and inside of the steel pipe is hollow.

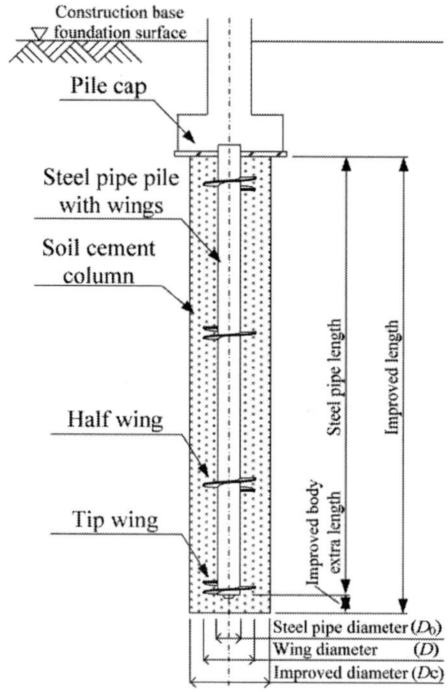

Fig. 1. Schematic drawing of the structure

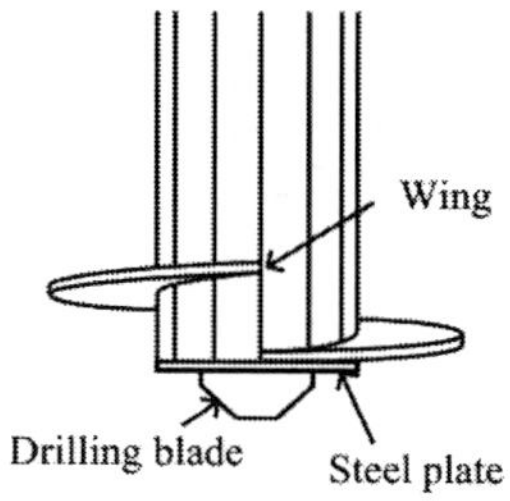

Fig. 2. Toe shape

Table 1. Standard specifications

Wings D (mm)	Column D_c (mm)	Steel pipe D_0 (mm)							
		114	139	165	190	216	267	318	355
250	500	○	○						
300		○	○	○					
350	600	○	○	○	○				
400			○	○	○	○			
450	700			○	○	○	○		
500					○	○	○	○	
600	900					○	○	○	○
700	1000						○	○	○

3 Vertical loading test in Hanoi

DOI: 10.1051/matecconf/201713804008

The site for vertical loading test in Hanoi was situated approximately 15km west from Hanoi's city center. Here, the influence of the existence of steel pipes with wings inserted into the column to the bearing capacity will be examined.

3.1 Soil investigation

Field work was carried out according to the British Standard (BS) 5930:1999 and the American Society for Testing and Materials (ASTM). Investigation results are shown in Fig. 3, Table 2 and Table 3. The ground on which test specimens are placed is sandy silt with clay with SPT N-value of 0-4.

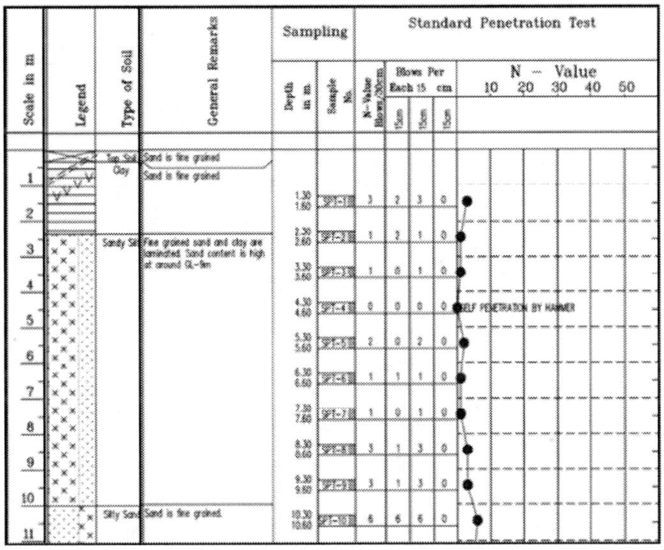

Fig. 3. SPT at Hanoi

Table 2. Index properties and unconfined compression test results

Depth. (GL-m)	Natural Water Content (%)	Wet Density (Mg/m³)	Undisturbed q_u (kPa)	Undisturbed ε_f (%)	Remolded q_u (kPa)	Remolded ε_f (%)	Sensitivity Ratio
2.00 to 2.80	34.0	1.88	51.5	9.2	20.2	19.9	2.5
4.00 to 4.80	45.3	1.71	40.0	5.9	15.3	10.6	2.6
6.00 to 6.80	40.7	1.78	46.7	6.4	13.0	13.5	3.6
9.00 to 9.80	43.1	1.72	51.4	5.6	18.7	10.5	2.7

Table 3. Consolidation test results

Depth (GL-m)	Initial void ratio e_o	Preconsolidation Pressure p_c (kPa)	Average Compression Index *Ccave*	Swell Index *Cr*
4.00 to 4.80	1.621	110	0.56	0.12

3.2 Overview of vertical loading test in Hanoi

Table 4 and Fig. 4 show an overview of the test specimens. N1, N2, and N4 had strain gauges attached to them to measure axial force distribution. Also, N1 and N2's columns were constructed 3m from the pile head. To make each test results easier to compare, for N5, the column's surrounding ground was disrupted by mechanical stirring until GL-3m after column construction. The loading test was conducted after one month of test specimen curing period. The test method of multi-cycle complying with standard of JGS (Japan Geotechnical Society)'s "Method for Vertical Loading Test of Piles and Commentary" was used. As for N6, the aim was to reduce the surface friction resistance until GL-3m, and the column's surrounding ground was drilled until that depth right before the loading test, and then immediately re-buried, and the test was started.

Fig.5 shows the uniaxial compression test results of the core specimen extracted from core boring. Deeper than GL-3m, 200kg/m^3 of cement type solidifying substance was added. Shallower than GL-3m, in the loading tests of N5 and N6, double amount of 400kg/m^3 was added to prevent compression failure.

Table 4. Specifications of test pile

No.	Steel pipe D_0(mm)	Thickness t(mm)	Wings D(mm)	Column D_c(mm)	Depth of Toe (GL-m)	
					Steel pipe	Column
N1	219.1	7.1	500	700	5.7	6.0
N2	168.3	7.3	500	700	5.7	6.0
N4	168.3	7.6	500	—	5.7	—
N5 N6	—	—	—	700	—	6.0

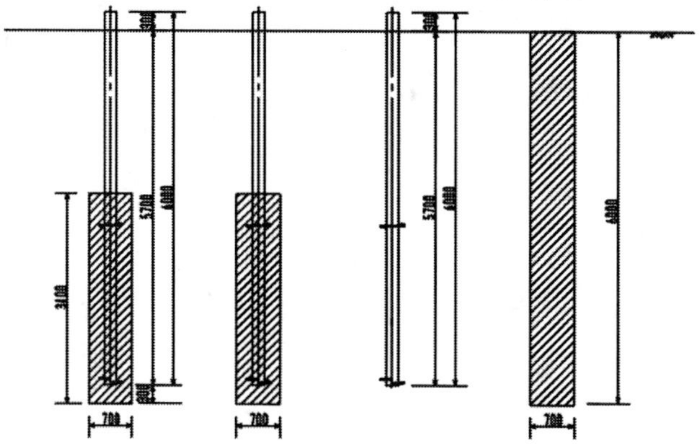

Fig. 4. Detailed diagram of test specimens

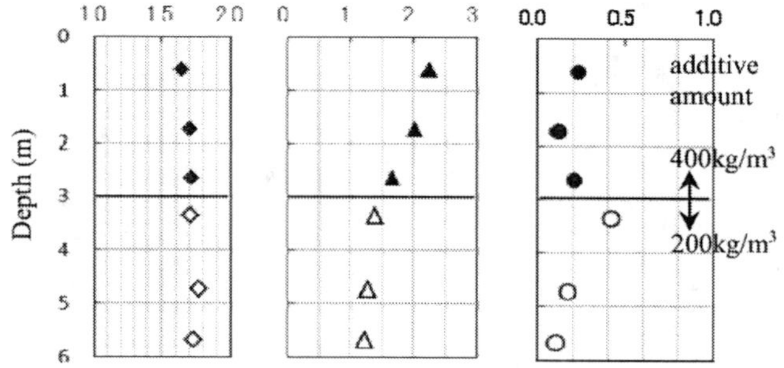

Fig. 5. Uniaxial compression test results

3.3 Test results and discussions

Fig.6 shows the pile head load - pile head settlement curve. (For the curves of N5 and N6's P_0-S_0, continuous measurement results after failure are plotted). The test specimens with steel pipes with wings (N1, N2, and N4) were able to maintain pile head load even at times of large deformation, and no significant decrease in pile head load was seen. On the other hand, test specimens with only a column (N5, N6) had drastically smaller loading capacities after exceeding pile head displacements N5 : S_0=31mm, N6 : S_0=13mm. This can be thought to be caused by compression failure within the column. As can be seen from Fig. 5 that the column's uniaxial compressive strength is approximately 1Mpa within the test interval, and the column cross section is 0.38m^2, therefore, the N6's test maximum load is valid. However, for N5 with the same specifications, it can be inferred that the test maximum load was greater than N6 and the surface friction resistance exhibited even in the shallow section, because although the column's surrounding ground was disturbed until GL-3m that ground recovered to a certain extent during the curing period.

From the test results of N1 and N2, there were no clear differences in load - displacement curve due to differences in steel pipe diameter. This can be explained that the load capacity level was within the steel pipe's elastic deformity range and was small, the steel pipe, wings and column became one and were able to behave together, showing no significant differences in the results of the re-test of the column of the same specifications. Fig. 7 shows the axial force distribution of N2 and N4. The differences in test results are due to the existence and absence of columns. In the interval between the second cross section directly above the wing, and the fourth cross section on the end part, the surface friction degree evaluated from the wing diameter's cylindrical surface was τ_{N2}=88kN/m^2 and τ_{N4}=12.8kN/m^2 respectively, the result being approximately 7 times greater. The judgement of characteristic values was done from the load and displacement gained from the loading test. Table 5 shows the first limit resistance P_{oy} and the second limit resistance P_{ou} for each specimen. It can be seen from the test results of N1, N2, and N4, that the existence of columns has an effect of 4.2-5.3 times at the first limit resistance, and 3.6-3.9 times at the second limit resistance. From the results of N5 and N6, it can be concluded that the variety in column strength and quality greatly influenced the bearing capacity of the single body of the column.

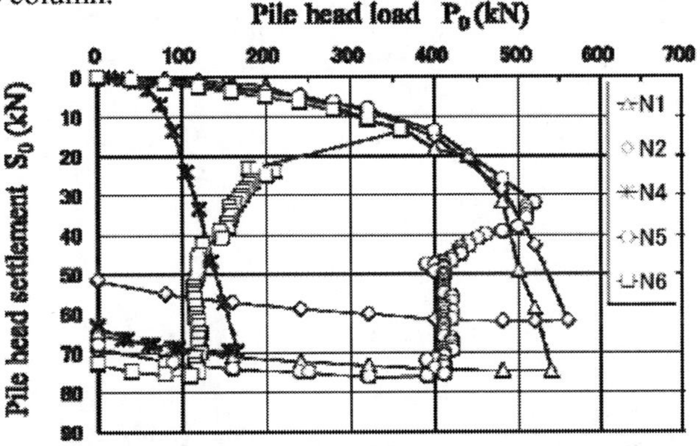

Fig. 6. Comparison of pile head load - pile head displacement

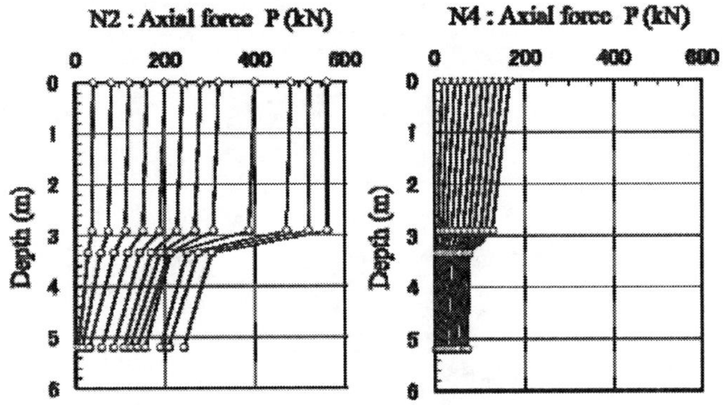

Fig. 7. Axial force distribution (N2, N4)

Table 5. The characteristic values

No.	Steel pipe D_0(mm)	Wings D(mm)	Column D_c(mm)	The first-limit-resistance P_{oy} (kN)	The second-limit-resistance P_{ou} (kN)
N1	219.1	500	700	400	509
N2	168.3	500	700	320	548
N4	168.3	500	—	75	141
N5	—	—	700	320	499
N6			700	270	343

4 Numerical analysis

As shown by the loading test results, due to the combination effect between components of the composite pile, ATT provide greater bearing capacity compared to single body of column and single body of steel pipe with wings. In this section, contributions of column and steel pipe with wings to ATT bearing capacity will be investigated by numerical analysis.

4.1 Analysis cases

The analysis cases are shown in Table 6 and Fig. 8. The loading test carried out in Hanoi is used for finite element modeling of numerical analysis.

Table 6. Analysis cases

	Loading test content	Pile No.	Pile length (m)	Steel pipe diameter D_0(mm)	Steel pipe thickness t(mm)	Wing diameter D(mm)	Column diameter D_c(mm)
Case-A_No.2	ATT	No.2	6.0	168.3	7.3	500	700
Case-B_No.4	Single body of steel pipe with wings	No.4			7.6		
Case-C_No.6	Single body of column	No.6	—	—	—	—	

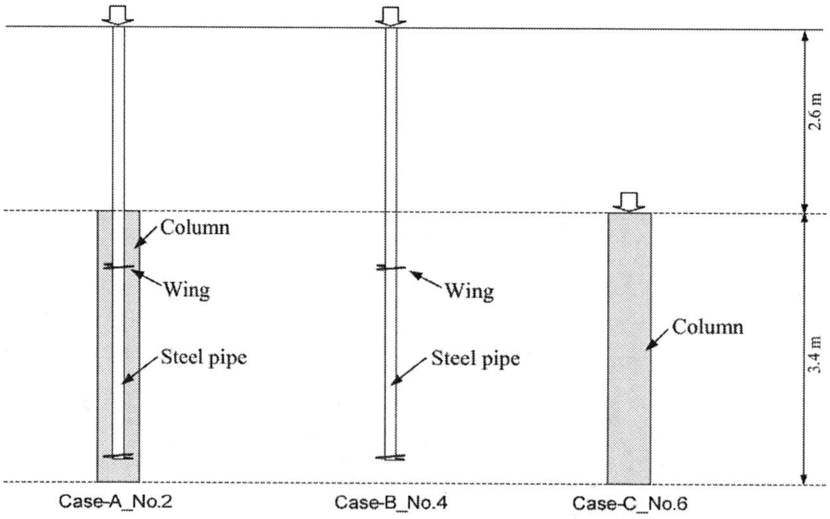

Fig. 8. Analysis cases

4.2 Overview of numerical analysis

 DOI: 10.1051/matecconf/201713804008

Fig. 9 shows finite element modeling of Case-A_No.2. For this case, the number of nodes and elements is 20,727 and 18,400, respectively.

In the numerical analysis, the ground is considered as elasto-plastic body using subloading t_{ij} model. Soil properties of ground layers are shown in Table 7. Parameters of column and steel pipe with wings are shown in Table 8. Density and Young's modulus value of the column are obtained from compression test results of boring core specimens.

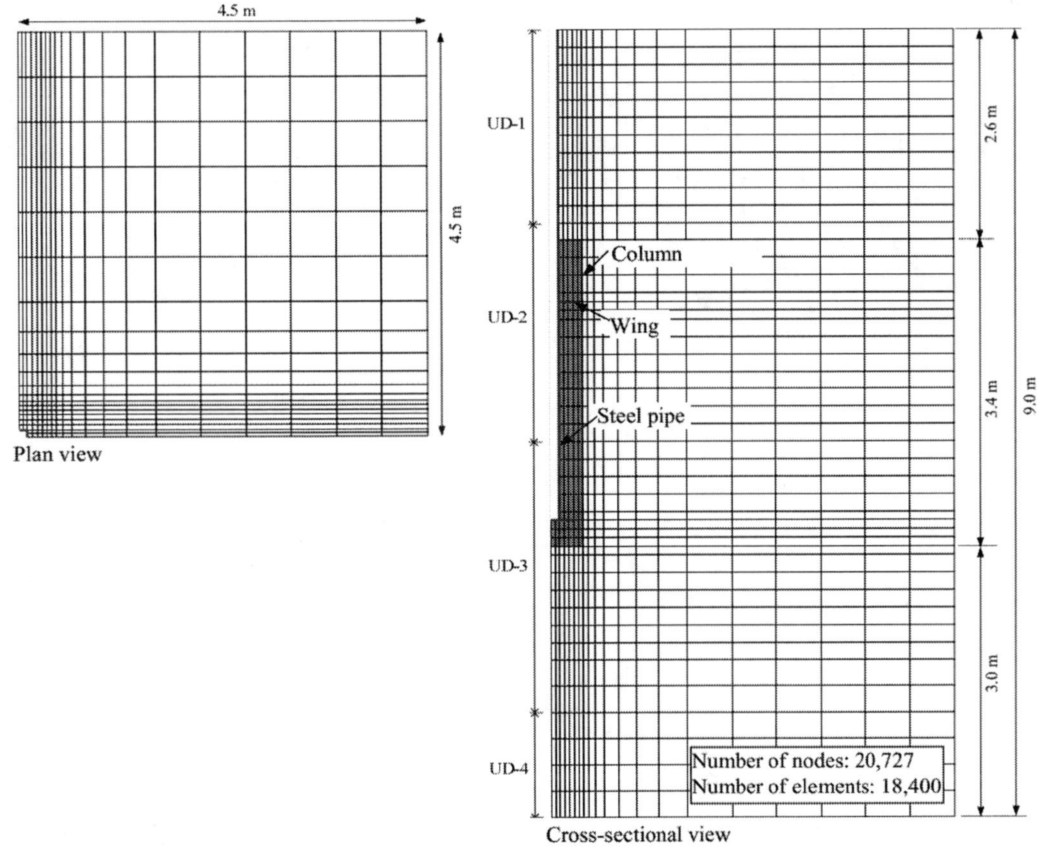

Fig. 9. Finite element modeling (Case-A_No.2)

Table 7. Soil properties of ground layers

Soil properties	Clay	Sandy silt		
	UD-1 (G.L.-0.0～-2.4 m)	UD-2 (G.L.-2.4～-4.9 m)	UD-3 (G.L.-4.9～-7.8 m)	UD-4 (G.L.-7.8～-9.0 m)
Plasticity index	10	19	15	18
Density (10^3kg/m^3)	1.880	1.710	1.780	1.720
Void ratio e_0	0.950	1.320	1.130	1.250
Principal stress ratio at critical state M_f	3.728	3.009	3.310	3.145
Poisson ratio ν	0.300	0.300	0.300	0.300
Coefficient of earth pressure at rest K_0	0.423	0.488	0.464	0.482
β (stress-dilatancy)	1.5	1.5	1.5	1.5
a (ANN) parameter	500	500	500	500
Compression index λ	0.091	0.243	0.117	0.082
Swelling index κ	0.018	0.052	0.020	0.011

DOI: 10.1051/matecconf/201713804008

Table 8. Parameters of column and steel pipe with wings

	Steel pipe and Wings	Column	
		G.L.-2.6∼-4.9 m	G.L.-4.9∼-6.0 m
Density (10^3kg/m^3)	7.850	1.73	1.73
Young's modulus E (N/mm^2)	2.06×10^8	263.7	103.0
Poisson ratio ν_0	0.300	0.298	0.298

4.3 Analytical results and discussions

Relationship between vertical load and pile head settlement in the field experiment and numerical analysis is shown in Fig. 10. Axial force distribution of Case-A_No.2 at the first limit resistance is shown in Fig. 11. It can be seen here that a good accordance is obtained between the experimental and numerical analysis results. Displacement contour plots at the first limit resistance are shown in Fig. 12.

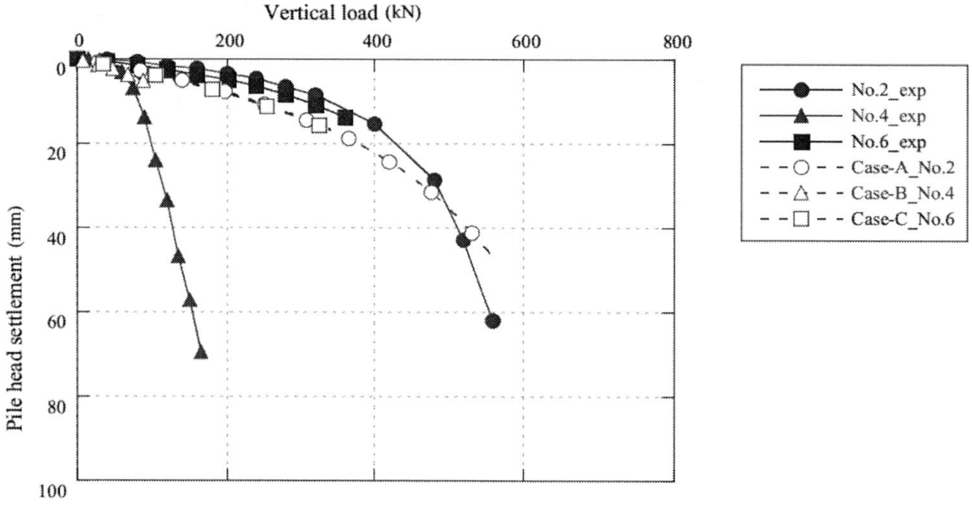

Fig. 10. Relationship between vertical load and pile head settlement

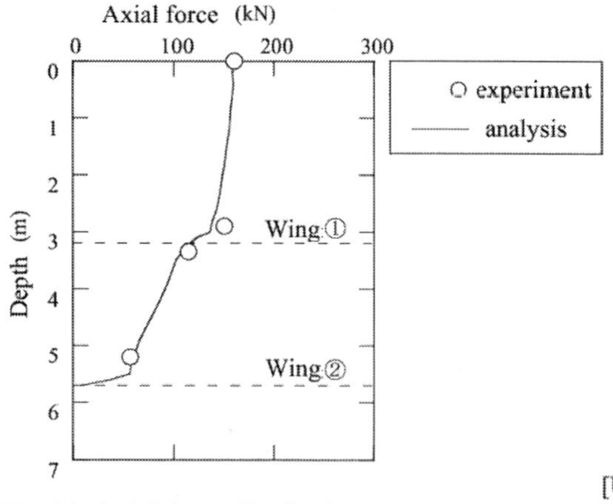

Fig. 11. Axial force distribution
(Case-A_No.2 at the first limit resistance)

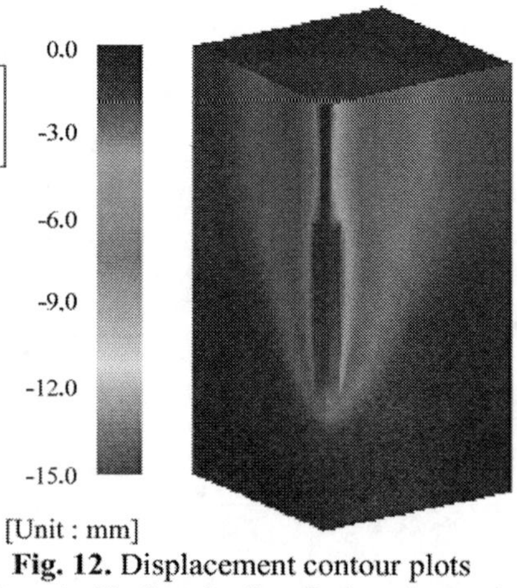

Fig. 12. Displacement contour plots
(Case-A_No.2 at the first limit resistance)

Fig. 13 shows vertical stress distribution of Case-A_No.2 (ATT) and Case-C_No.6 (single body of column) at the second limit resistance. In Fig. 13, the maximum value of vertical stress is taken as 1.0. Enlarged views of test specimen surrounding area are also shown in Fig. 13.

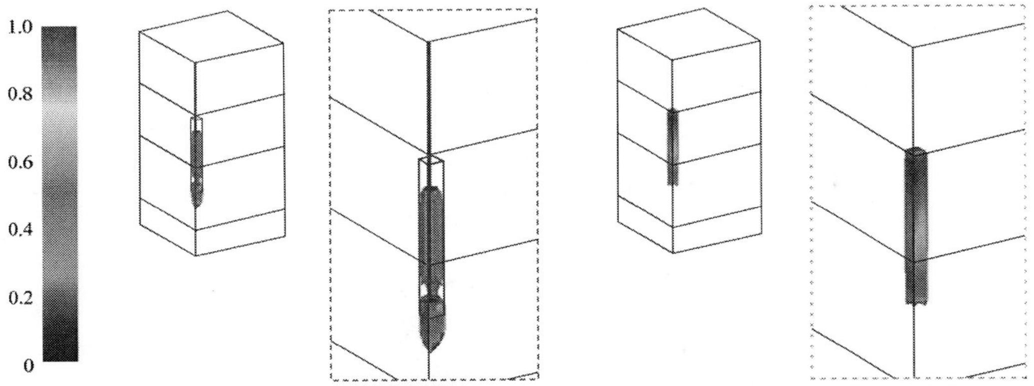

Fig. 13. Vertical stress distribution at the second limit resistance

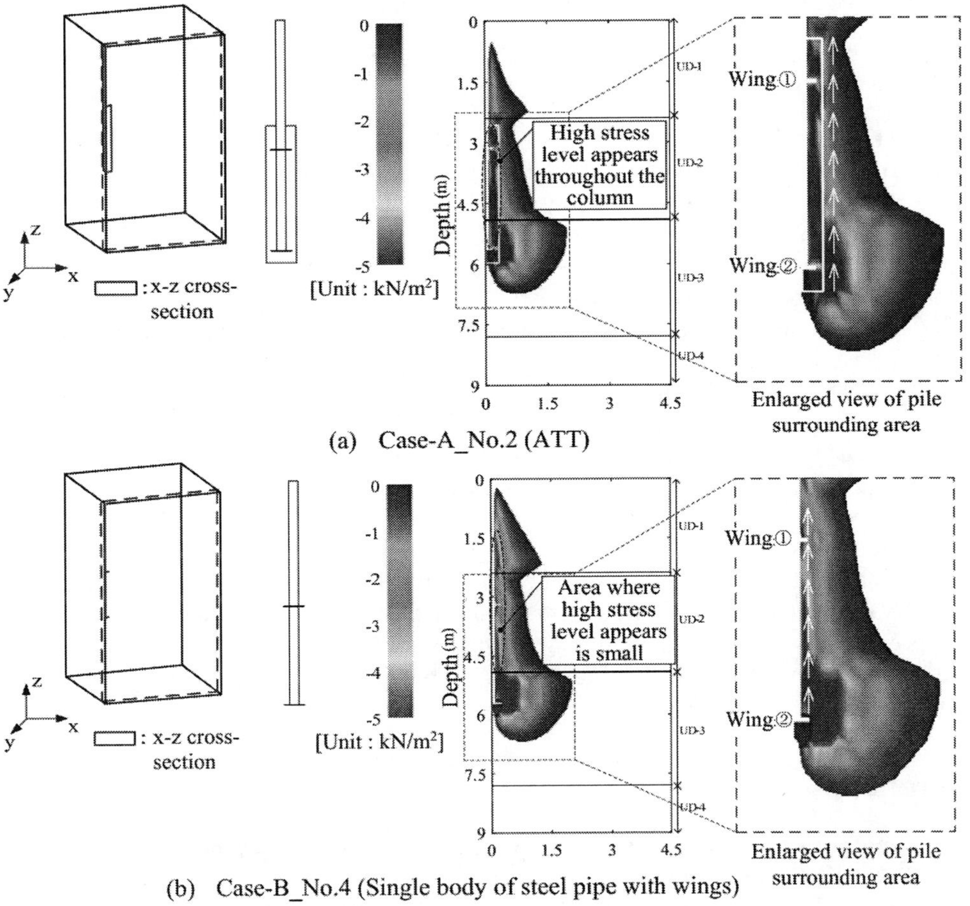

Fig. 14. Contour plots of shear stress τ_{xz} (test loading of 60 kN)

It can be confirmed that for the case of Case-C_No.6 (column single body), stress concentration appeared at upper part of column. With increasing of vertical load during loading test, when the concentration stress becomes larger than column compression strength, compression failure will occur at upper part and the column will reach to its ultimate state. In contrast, for the case of Case-A_No.2 (ATT), high stress level does not appear at column head part and stress distribution in column overall is almost uniform, especially in the part of column under the wings. This means that the existence of steel pipe with wings helps distributing effectively the load acting on pile head to the column and the compression failure which occurs at upper part of column single body can be avoided.

Fig. 14 shows contour plots of shear stress τ_{xz} of Case-A_No.2 (ATT) and Case-B_No.4 (single body of steel pipe with wings). In both two cases, the load acting on steel pipe head is equal to the first limit resistance of Case-B_No.4 with the value of 60 kN. By comparison of shear stress distribution area in these two cases, it can be investigated on how the existence of column effects vertical bearing capacity characteristics. For Case-A_No.2 (ATT), high stress level appears throughout the column in a large area. This means that in the boundary between column and surrounding ground, friction force is exhibited and ATT has received resistance from surrounding ground. In contrast, for the case of Case-B_No.4 (single body of steel pipe with wings), area surrounding steel pipe with wings where high stress level appears is small and stress concentration occurs at area surrounding steel pipe end. It can be concluded that with the existence of column, large friction force with surrounding ground has mobilized and ATT exhibits larger bearing capacity compared to single body of steel pipe with wings.

5 Conclusion

In this study, full scale loading tests of ATT carried out in Hanoi are presented, and bearing capacity of the composite pile is studied. Especially, combination effect of soil cement column and steel pipe with spiral wings on bearing capacity performance are discussed. Three-dimensional elasto-plastic finite element analysis based on the results of loading tests to help understand this composite pile's bearing characteristics is also performed. Below is summary of the knowledge from this study:

1) It was confirmed that due to the combination effect, ATT composite pile can provide a greater vertical bearing capacity compared to the single body of column and single body of steel pipe with wings.
2) The existence of steel pipe with wings helps distributing effectively the load acting on pile head to the column and prevents the occurrence of stress concentration. Consequently, compression failure which occurs at upper part of column single body can be avoided.
3) With the existence of column in ATT, large friction force with surrounding ground has mobilized and ATT exhibits larger bearing capacity compared to single body of steel pipe with wings. It was found out that with the conditions of loading test introduced, approximately 4 times the bearing force was exerted.
4) It was confirmed that ATT composite pile is able to perform the same bearing capacity as in Japan in Vietnam's weak ground.

References

1. Japanese Geotechnical Society, Pile vertical load test method and explanation, **First revised edition**, 23-47 (2002).
2. H. Murakami, D. Ito, E. Mizoguchi, The Foundation Engineering & Equipment, *An application example of ATT Column construction method*, **Vol. 36(10)**, 72-75 (2008).
3. T. Tamai, D. Ito, E. Mizoguchi, Japanese Geotechnical Society Journal, *Study on the in-situ pile loading test and bearing capacity characteristics of steel pipe piles with wings installed in soil cement column*, **Vol. 4(4)**, 273-287 (2009).
4. B.D. Nguyen, H.K. Pham, *2016 International Conference on Sustainability in Civil Engineering (ICSCE 2016)*, A study of bearing capacity of steel pipe pile with wings installed in soil cement column in soft soil of Viet Nam, 283-288 (2016).

Evaluation of the causes of Bukit Antarabangsa 2008 landslide by using fault tree analysis

Danish Kazmi[1], *Sadaf* Qasim[1,*], and *I.S.H* Harahap[2]

[1]Department of Civil Engineering, NED University of Engineering & Technology, Pakistan - 75270
[2]Department of Civil & Environmental Engineering, Universiti Teknologi Petronas, Malaysia - 32610

Abstract. Bukit Antarabangsa 2008 is considered as a deep-seated landslide which occurred in the landslide-prone area of Ulu Klang, Malaysia. The precipitation data obtained from previous studies show that there were comparatively less rainfalls immediately before the landslide which suggests that other potential triggers must be analyzed. This study analyses the causes of Bukit Antarabangsa 2008 landslide by performing fault tree analysis (FTA) which is a logical and diagrammatic method to evaluate the probability of an accident resulting from sequences and combinations of faults. The FTA is performed on four potential landslide contributors of improper shear strength parameters, flawed development plan, erroneous drainage planning & design and, ineffective slope strengthening works in order to determine the causes of the landslide. The analysis shows that the causes of Bukit Antarabangsa 2008 landslide potentially include ineffective slope strengthening works related to monitoring and maintenance which is followed by improper development plan having a probability of failure of 0.194 and 0.15 respectively. These causes actually correlate with the human errors which are often neglected during the slope construction and required to be addressed to increase the safety of the slopes.

1 Introduction

Bukit Antarabangsa is one of the most prominent landslides of Malaysia which occurred on 06 December 2008 in the area of Ulu Klang, Selangor, Malaysia. The history of the landslide area indicates that a number of slope failures have occurred previously and their causes must be thoroughly analyzed in order to take measures for the prevention of slope failures in future.

Bukit Antarabangsa, located in the north of Kuala Lumpur, forms a narrow ridge that extends in the northeast-southwest direction with a maximum elevation of 230 m. The hill is underlain by granite whose characteristics are grey color, coarse grained particles and slightly porphyritic texture. Extensive weathering has transformed the granite into the residual soil (grade VI) and weathered material (grade V). The average thickness of the

* Corresponding author: sadafqasim26@yahoo.com

DOI: 10.1051/matecconf/201713804009

weathered profile is approximately 30 m. The weathered material is sandy and rapidly loses its consistency with increasing amounts of water [1].

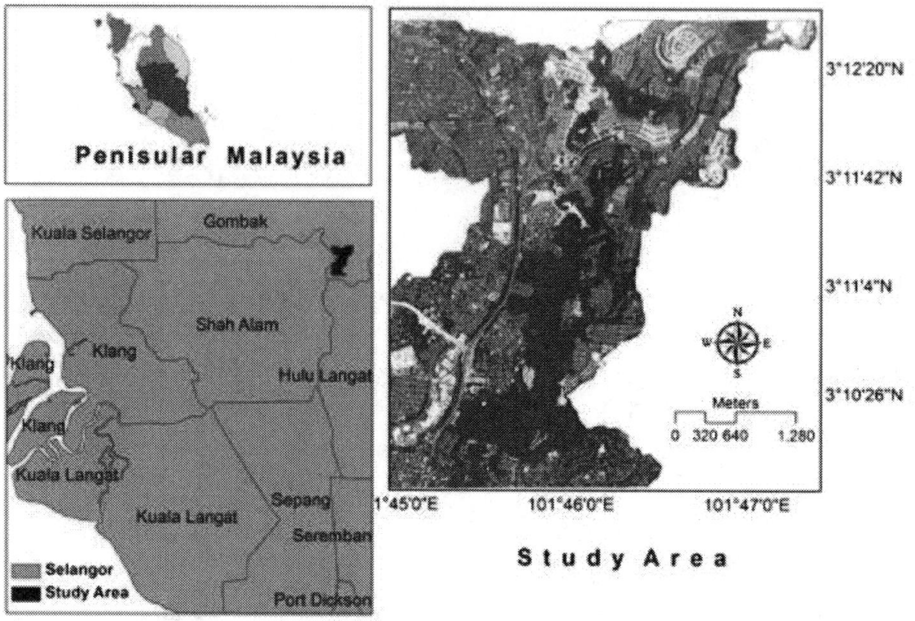

Figure 1. Location of Bukit Antarabangsa Landslide [2]

The Bukit Antarabangsa 2008 is referred to as a deep-seated landslide and several studies have showed that there were multiple causes that potentially triggered the failure. Ng [3] reported that the landslide area had faced a high level of precipitation in November 2008, a month before the landslide. However, no severe landslide happening was reported during this period. From the rainfall study, it is evident that there were no signs of heavy rainfall immediately before the landslide. Therefore, it is necessary to consider the role of other factors which were potentially responsible for the landslide. The potential triggers for the landslide can be related to drainage system, sub-soil conditions, geology, groundwater table, human errors and technical errors.

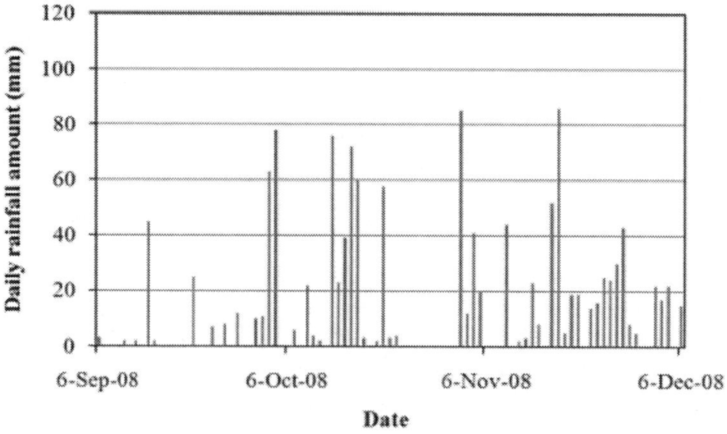

Figure 2. Daily Rainfalls for Bukit Antarabangsa 2008 Landslide [3]

The main causes of Bukit Antarabangsa 2008 landslide as reported by Huat et al., 2012 [4] are:

- Soil creep, due to non-engineered fill on the slope, developed tension cracks over the area.
- The soil creep over the years may have damaged the active water pipe along the abandoned houses and leaked the pipe. The leaks contributed to continuous soil saturation at the lower slope and this, in turn, accelerated creep.
- Prolonged rainfall during the month of October and November resulted in soil saturation, rise in ground water table and increased the rate of creep.
- An increased soil creep further damaged the drainage facilities and widened the existing cracks and opened up new tension cracks

2 Objective of the Study

The objectives of the study are:

- To investigate the causes of the Bukit Antarabangsa landslide in 2008 using fault tree analysis (FTA).
- To evaluate the role of human uncertainties in triggering the landslide.
- To discuss the benefits of using human reliability analysis (HRA) in slope construction.

3 Human reliability analysis

The major shortcoming of simple reliability analysis is that the results obtained from it are conditional as it does not encompass the effect of human uncertainties in the analysis. The utility of applying the human reliability analysis (HRA) is that it takes into account the phenomenon of human uncertainties which provides a more precise estimate of the stability of the slopes.

The core intention of HRA is to accurately weigh up the risks generated from human error and determine ways to mitigate them. A complete HRA process must have the following steps [5]:

a. Recognition and description of human error

b. Quantification of human error probability

c. Analysis of human error modes and effects

d. Design and authentication of protective measures for human error

4 Methodology of the study

This study is based on the application of fault tree analysis (FTA) technique to determine the causes of Bukit Antarabangsa 2008 landslide in terms of four potential landslide contributors of improper shear strength parameters, flawed development plan, erroneous drainage planning & design and, ineffective slope strengthening works.

Fault tree analysis (FTA) is a logical and diagrammatic method to evaluate the probability of an accident resulting from sequences and combinations of faults and failure

DOI: 10.1051/matecconf/201713804009

events [6]. FTA is a powerful and computationally efficient technique to analyze and predict system reliability and safety. Many theoretical advances and practical applications have been achieved in this field to date. FTA is based on Boolean algebra and probability theory and it is consistent with the conventional reliability theory [7].

The subjective probabilities of basic events were determined from the experts of slope engineering division of Malaysian government department, Jabatan Kerja Raya (JKR). This research adopted consensus group method to conclude experts' opinion in relation to the Bukit Antarabangsa 2008 landslide.

5 Results

The fault tree analysis (FTA) is performed on four potential landslide contributors of improper shear strength parameters, flawed development plan, erroneous drainage planning & design and, ineffective slope strengthening works which are given below.

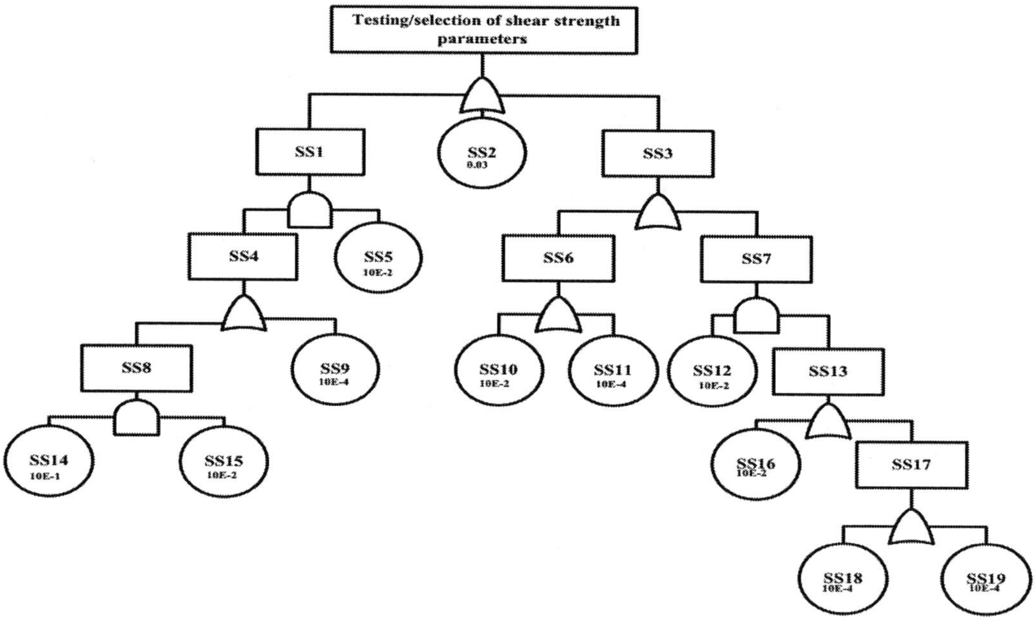

Figure 3. Fault Tree Analysis of Shear Strength Parameters

Table 1. Testing of Shear Strength Parameters and Subjective Probabilities

	Events	Subjective/ Calculated Probabilities		Events	Subjective/ Calculated Probabilities
SS1	Lack in following soil mechanics	0.000011	SS11	No counter check	10E-4
SS2	Improper experimental setup	0.03	SS12	Meet deadlines	10E-2
SS3	Providing results without testing	0.000103	SS13	Staff shortage	0.0102
SS4	Easy follow up of prevailing practices	0.0011	SS14	Poor perception	10E-1

SS5	Overconfidence	0.0011	SS15	Bias records	0.01
SS6	Time period/Pressure	0.000102	SS16	Organizatio nal trend	0.01
SS7	Individuals attitude	0.000001	SS17	Unable to hire related personnel	0.0002
SS8	Misconception about consequences	10E-3	SS18	Out of worth	0.0001
SS9	Deliberately ignored innovations	10E-4	SS19	Lack of intellects	0.0001
SS10	Work load	10E-2			

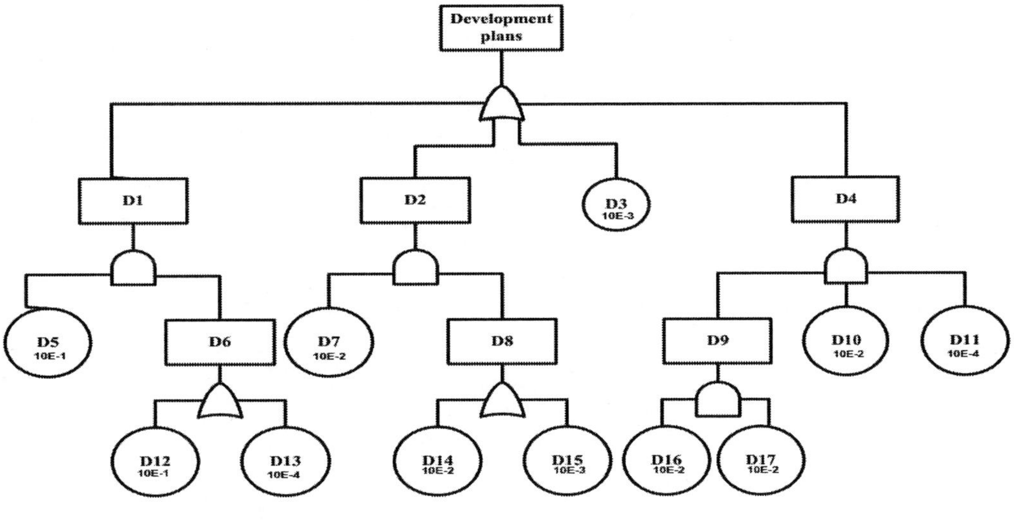

Figure 4. Fault Tree Analysis of Development Plans

Table 2. Development Plans and Subjective Probabilities

	Events	Subjective/ Calculated Probabilities		Events	Subjective/ Calculated Probabilities
D1	Flaws in proposing design strategies	0.01	D10	Nothing for Special Structures like soil nailing, ground anchors	0.01
D2	Consider construction activities less significant	0.00011	D11	No proper check in case of heavy rains	0.0001
D3	No experience of drainage system	0.001	D12	Trend of following thumb rules or patent methods	0.1
D4	Mismatch strategies for monitoring. inspection and maintenance	10E-9	D13	No practicing codes/standards	0.0001

D5	Perception lacking	0.1	D14	No emphasis on sequencing of activities	0.01
D6	Information Shortage	0.1	D15	Lack of specific situation guidelines	0.001
D7	Timing expenditure not estimated	0.01	D16	Least concerned about its consequences	0.01
D8	Not Proposing extreme cases alternatives	0.011	D17	Treat like any other common structure	0.01
D9	Less Significant	0.01			

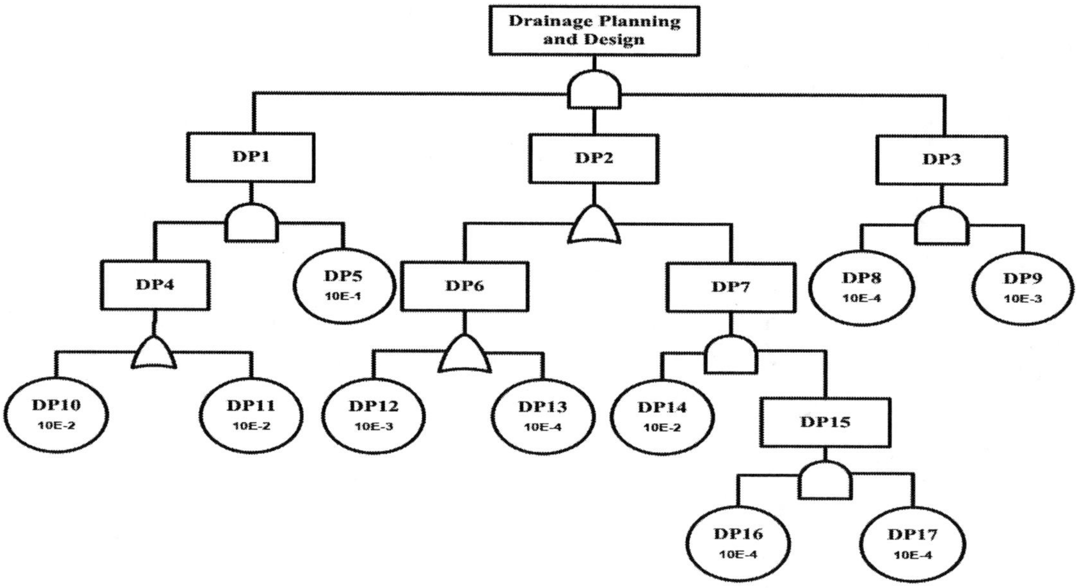

Figure 5. Fault Tree of Drainage Planning & Design

Table 3. Drainage Planning & Design and Subjective Probabilities

	Events	Subjective/ Calculated Probabilities		Events	Subjective/ Calculated Probabilities
DP1	Inaccurate site /topography and soil information	0.002	DP10	Unsuitable outlets proposed	0.01
DP2	Unparsed economic feasibility	10E-4	DP11	Improper layout of drains	0.01
DP3	Estimated inadequate drainage capacity	0.0011	DP12	Consider less preferable	0.001
DP4	Misconception about soil strata/properties	0.02	DP13	Organizational trend	0.0001

DOI: 10.1051/matecconf/201713804009

DP5	No practice to confirm the topography	0.1	DP14	Flaws in geological report	0.01
DP6	Lack of capital	0.0001	DP15	No concept of checking	10E-8
DP7	Lack of resources	0.001	DP16	Time pressure	0.0001
DP8	Unavailability of rainfall data	0.0011	DP17	Work stress	0.0001
DP9	Predicted ground water table	10E-10			

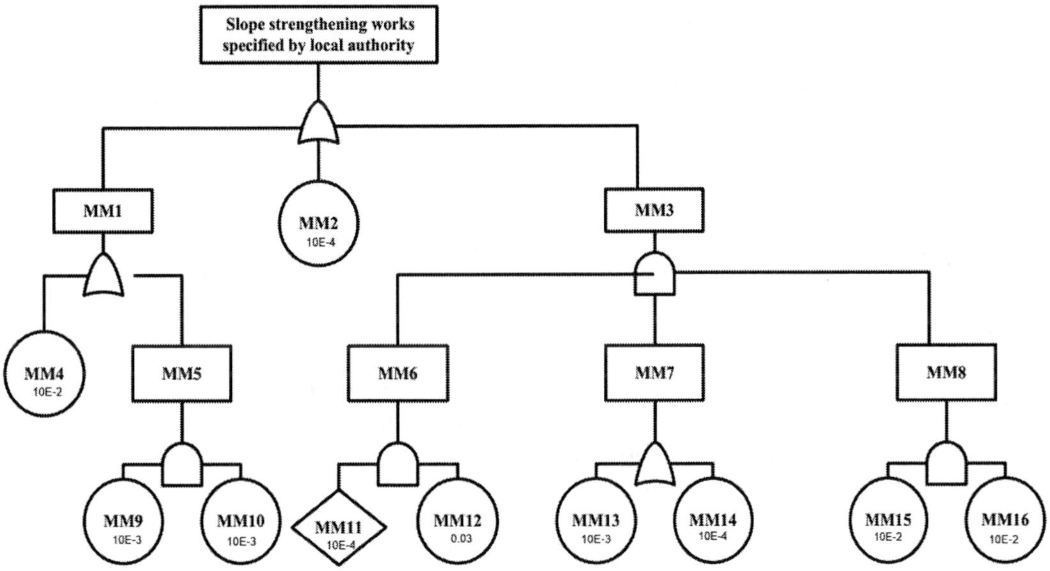

Figure 6. Fault Tree Analysis of Slope Strengthening Works

Table 4. Slope Strengthening Works and Subjective Probabilities

	Events	Subjective/ Calculated Probabilities		Events	Subjective/ Calculated Probabilities
MM1	Biased results through inclinometers	0.01	MM9	Uncelebrated equipment	10E-3
MM2	Maintenance free (assumed)	10E-4	MM10	No QAQC criteria	10E-3
MM3	Water damming effect	0.001	MM11	Unknown cause	10E-4
MM4	Fail to detect in inspection	10E-2	MM12	Improper fixing	0.03
MM5	Faulty deformation data	10E-6	MM13	Rainfall exceeds	10E-3
MM6	Obstruction in drainage	3E-6	MM14	Other outlet diversion	10E-4

MM7	Ground water table rise	0.001	MM15	Consider less preferable	10E-2
MM8	Piezometric pressures inadequacy	10E-4	MM16	Underestimat e its consequences	10E-2

Table 5. Probability of Failure of Potential Contributors

S.no	Symbols	Potential Contributors	Probability of failure using FTA
1	Shear Strength (SS)	Testing and selection of shear strength parameters	0.124
2	Development Plan (D)	Development plan	0.15
3	Drainage Planning and Design (DP)	Drainage planning and design	0.001
4	Monitoring & Maintenance (MM)	Slope strengthening works	0.194

6 Discussion

The analysis shows that the causes of Bukit Antarabangsa 2008 landslide were mainly the human errors committed from design till maintenance phase. It is observed through FTA that the major contributor of landslide was ineffective slope strengthening works related to monitoring and maintenance followed by flawed development plan having a probability of failure of 0.194 and 0.15 respectively. The probability of failure for improper shear strength parameters and erroneous drainage planning & design are found to be 0.124 and 0.001 respectively.

The technique of fault tree analysis (FTA) is fundamentally based on the subjective probabilities which are determined on the basis of respondents' judgment according to their experience and knowledge. The purpose of using FTA is to estimate the likelihood scenario of an accident [8]. According to Silva et al., 2008 [9], geotechnical engineers frequently encounter the challenge of determining the failure probabilities. Previous studies show that the subjective probabilities, such as quantified expert judgment, have been accepted for decades by practitioners and academics. According to Silva et al., 2008 [9], using subjective probability for risk assessment in geoengineering allows engineers to attain 80% of the advantage at 20% of the cost, as recommended by Pareto's principle.

The downside of safety factor approach is that it does not accommodate the influence of uncertainties in the construction which provides a less-accurate estimate about the stability of a slope. On the other side, the reliability analysis technique is comparatively a superior approach as it encompasses the effect of uncertainties particularly in terms of soil parameters. The benefit of utilising human reliability analysis (HRA) technique is that it also considers the effect of human errors during the construction process which is a neglected feature in the previous approaches. The application of HRA provides more accurate estimate about the safety and stability of the slope which also allows the stakeholders to take necessary initiatives if the requirements are not properly satisfied.

7 Conclusion

This study shows that the causes of Bukit Antarabangsa 2008 landslide are mainly human errors which have been identified on the basis of the probability of failure using fault tree analysis. The major contributors of this landslide potentially include ineffective slope strengthening works related to monitoring and maintenance and improper development plan. The rainfall data before the landslide incident suggests that there were no heavy rainfalls immediately before the landslide hence, the rainfall cannot be termed as the sole triggering factor. The findings of the previous studies on Bukit Antarabangsa 2008 landslide also suggests that there were multiple causes that jointly contributed to the slope failure which mainly corresponds to the human errors. Therefore, this study suggests that human uncertainties should be properly addressed during the construction process using human reliability analysis (HRA) techniques to maximize the stability of the slopes.

References

1. Chigira, M., Z. Mohamad, L.C. Sian, and I. Komoo, *Landslides in weathered granitic rocks in Japan and Malaysia.* Bulletin of the Geological Society of Malaysia, **57**, 1-6 (2011).
2. Hassaballa, A.A., O.F. Althuwaynee, and B. Pradhan, *Extraction of soil moisture from RADARSAT-1 and its role in the formation of the 6 December 2008 landslide at Bukit Antarabangsa, Kuala Lumpur.* Arabian Journal of Geosciences, **7**(7), 2831-2840 (2014).
3. Ng, K.Y., *Rainfall-Induced Landslides in Hulu Kelang Area, Malaysia*, 2012, UTAR.
4. Huat, L.T., F. Ali, and A.S. Ibrahim, *An Investigation on One of the Rainfall-Induced Landslides in Malaysia.* Electronic Journal of Geotechnical Engineering, **17**, 435-448 (2012).
5. Kazmi, D., S. Qasim, I. Harahap, and S.B. Azhar, *An extensive study on the quantification of human errors that hampers the performance and construction of structure particularly in geotechnical ambit,* **13**(4), 33-43 (2016).
6. Suresh, P., A. Babar, and V.V. Raj, *Uncertainty in fault tree analysis: A fuzzy approach.* Fuzzy sets and Systems, **83**(2), 135-141 (1996).
7. Huang, H.-Z., X. Tong, and M.J. Zuo, *Posbist fault tree analysis of coherent systems.* Reliability Engineering & System Safety, **84**(2), 141-148 (2004).
8. Kazmi, D., S. Qasim, I. Harahap, and S. Baharom, *Landslide of Highland Towers 1993: a case study of Malaysia.* Innovative Infrastructure Solutions, **2**(1), 21 (2017).
9. Silva, F., T.W. Lambe, and W.A. Marr, *Probability and Risk of Slope Failure.* Journal of Geotechnical and Geoenvironmental Engineering, **134**(12), 1691-1699 (2008).

 DOI: 10.1051/matecconf/201713804010

A Review on the Use of Bamboo for Earthwork Construction

Mokhammad Farid Ma'ruf[1,*]

[1]Civil Engineering Department, Jember University, Jl. Kalimantan 37 Kampus Tegalboto 68121 Jember Indonesia

Abstract. Bamboo becomes a promising engineering material widely utilized for constructions. It is likely a composite material with fibred reinforcement in one direction to make its good flexural strength. Bamboo has been traditionally made use for structural components. Recently, the applications of bamboo for earthwork constructions are reported both experimental and real project. Pile foundations for houses, bamboo pile and nail for slope stability, foundation grid for embankment, and bamboo pile-mattress for railway and reclamation project are among others of the application. This paper discusses the use of bamboo for earthwork construction. Particular discussion is addressed to bamboo pile-mattress that successfully installed for various project in Indonesia. The appropriate field characteristic for bamboo application is specified. The design step and theoretical based are discussed as well.

1 Introduction

Bamboo has a long history to be used as construction material all over the world, especially in tropical and semi tropical countries. It is characterized by its strength and low weight. Bamboo is also easy worked even with very simple tools. Bamboo can be formed to almost all part of traditional house [1], footbridge over small stream, and river bank stabilization (Fig. 1).

Of the author knowledge, research and implementation of bamboo for structural element reinforcement has been reported about five decades ago [2, 3]. Since then many works to do intensively to improve the application of bamboo in structural element [4-11] focusing on bamboo concrete reinforcement and its durability both with and without treatment. The use of bamboo for earthwork construction has a long tradition also as well as for structural construction. Fang and his colleagues reported their work on soil reinforcement for slope stability on highway project [12-14]. However, scarce publications of bamboo application on earthwork construction have been recorded afterward. In the last two decades, the earthwork construction using bamboo has increased significantly. Various applications have been reported either in laboratory scale or field implementations. Bitumen-bamboo mesh composite [15], sheet pile and soil nailing [16, 17], bamboo cell [18, 19], bamboo

* Corresponding author: farid.teknik@unej.ac.id

grid or network [20 - 23], bamboo chips [24,25], bamboo pile [26, 27, 28], bamboo mattress [29,30], bamboo piled - mattress [31,32] are among others.

This paper intends to discuss the use of bamboo on earthwork construction. Various applications both laboratory and field scale are elaborated as well but limited on bamboo as earthwork structural element only. The design step and theoretical based are discussed as well.

a b

c

Fig. 1. Traditional use of bamboo in Indonesia.

2 Mechanical properties of bamboo

Bamboo becomes a promising construction material because of its mechanical properties are comparable with other construction material. The properties strongly depend on bamboo species, growing conditions, and culm position [33]. Bamboo has density about 700 – 800 kg/m^3 in which the fibbers contribute about 60 – 70% of total weight.

Bamboo has very strong tension compare to woods but low to medium modulus of elasticity about 35 – 70 kN/mm2 depends on bamboo diameter and wall thickness of the culm [34]. The high tension is in fibbers direction but more critical in perpendicular one. Since only lignin matrix takes place for bamboo endurance during loading in transverse direction. [35] suggested treating bamboo as structural element instead of material, therefore bending stiffness (EI) is more appropriate than modulus of elasticity. The moment of inertia (I) was calculated as follows:

$$I = \frac{\pi}{64}[D^4 - (D - 2t)^4] \quad (1)$$

 DOI: 10.1051/matecconf/201713804010

D and t are the average external diameter and the average wall thickness of the culm respectively.

3 Earthwork Application of Bamboo

3.1 Bamboo Cell

The use of bamboo for bamboo cell was reported by Hedge and Sitharam [18,19]. The work was on laboratory scale. Fresh bamboo culm was sliced to obtain a thin strip with 25 mm width. The strips were then woven together to form a grid. The grids were fixed together using wire to form a shape which similar in shape and size with commercial geocell. The bamboo cell and bamboo grid were utilized either individually or in combination.

Performance of the reinforcement is represented by ratio between footing settlement (*S*) and footing width (*B*) (**Fig. 2**). The bearing capacity of soil reinforced with bamboo cell and grid increased about 1.2 – 1.5 times than one that reinforced by geocell and geogrid. This high performance may be due to higher tensile strength and surface roughness of bamboo element compare to geosynthetic ones. Bamboo had tensile strength and surface roughness about nine times and 3.5 times respectively compare to that of geosynthetic [19]. However, one should criticize that bamboo strip that being tested for tensile strength was much thicker than one for bamboo cell and bamboo grid.

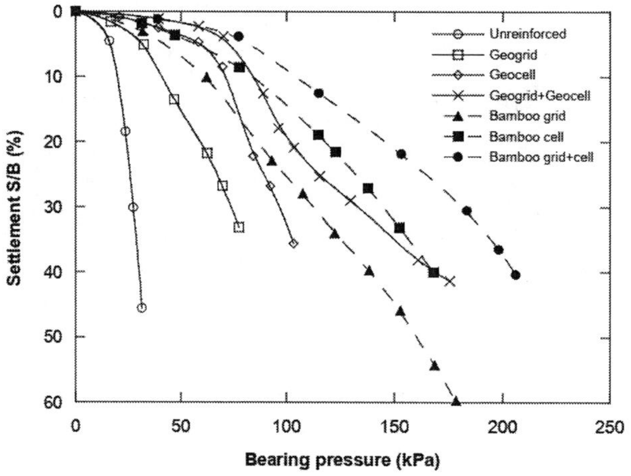

Fig. 2. Bearing pressure – settlement curves for various reinforcement (adapted from [19])

3.2 Bamboo Pile

Bamboo pile can be used as stabilization and reinforcement system instead of being used for direct support of building. [28] suggested to install bamboo pile behind sheet pile wall to reduce active earth pressure or at excavation base in front of sheet pile to increase passive earth pressure . Though the mechanism was unclear, if such assumption is true, then the stability of excavation slope will increase due to the change of the working pressure.

 DOI: 10.1051/matecconf/201713804010

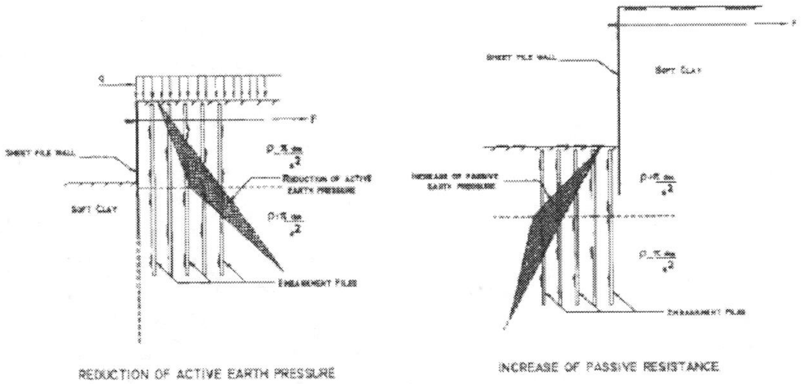

Fig. 3. Stabilization of excavation using bamboo pile (adapted from [27]).

[26, 36] had investigated the behaviour of bamboo pile to support shallow footing in the laboratory scale. Instead of placing the bamboo below the footing, they install the bamboo piles around the footing (Fig. 4). [36] reported that the bearing capacity increased with the pile extend (R) and pile length (L) but decreased with pile spacing (S). However, the bearing capacity converged when the spacing equal to footing width. [36] conducted experiments with only a row of bamboo pile to fence off the footing but with various distance from it. The bearing capacity improved when the bamboo piles closer to the footing. However, it decreased if the pile row was beneath the footing.

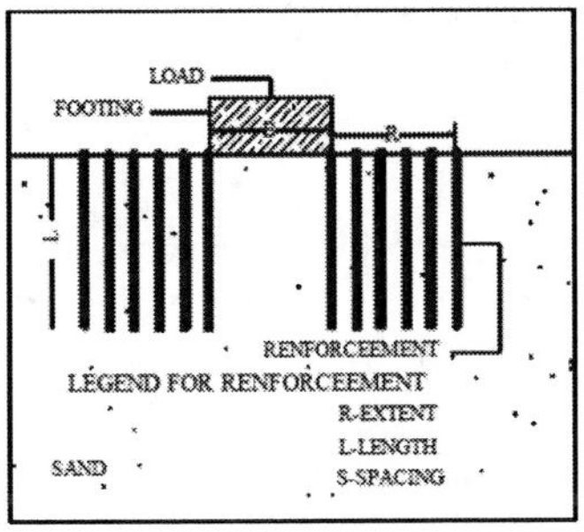

Fig. 4. Bamboo pile plan for footing reinforcement (adapted from [36]).

3.3 Bamboo Soil Nail and Sheet Wall

Bamboo grid that served as sheet wall has a long tradition in remote area of Indonesia (Fig. 1c). However there is no scientific investigation reported. Recent works of Dai and his colleagues [16, 17] combined bamboo soil nails and bamboo pile wall. The bamboo construction was applied to stabilize a 5 m excavation slope on loose silt resting on a deep layer of sludge (Fig. 5). The system consisted of bamboo grid reinforced shotcrete to cover the excavation slope; bamboo soil nails to support the shotcrete; and bamboo piles in a row to serve as sheet wall. Two rows of bamboo piles were installed with 40 cm space in between. The piles were put in through soil boring every 20 cm. Four series of bamboo soil nails were applied and connected to the bamboo grids. The detail installation and calculation method can be found in [16].

a

b

Fig. 5. Combination of bamboo soil nails, bamboo sheet piles, and bamboo reinforced shotcrete for slope stabilization (adapted from [16, 17]).

[16] reported the numerical analysis of bamboo reinforcement system and its monitoring results. The bamboo reinforcement system provided better reinforcement than ones from steel pipe. The bamboo soil nails produced almost 2.8 times of capacity than that of pipe nails. The new bamboo wall brought better stability performance than that of conventional steel pipe both numerically and in real project. The inclusion of bamboo branch during soil nailing had contributed higher soil-structure friction to provide large both soil nail and sheet wall capacity. However, in contrast to conventional soil nailing wall, bamboo soil nails should increase the length of the nail with increasing its depth.

3.4 Bamboo Grids and Mattress

Marto and her colleagues examined the performance of bamboo-geotextile composite (BGC) and high strength geotextile (HSG) for soft clay embankment reinforcement [29, 30]. The bamboo grid was laid down under the geotextile to provide a platform for laying and sewing it and forming a BGC. The reports were rather contradictive since both papers had showed different results. One reported BGC conceded lesser vertical and lateral deformation but another provided the opposite one.

[37] evaluated the bamboo grids and bamboo pile-mattress application for soft soil reinforcement. 4 layers of bamboo grids (Fig. 6) were extended over deep layer of soft soil to support 6.2 m embankment. The grids pattern had a space of 1 m x 1 m and regarded as embankment foundation to increase subgrade stiffness. The numerical analysis provided safety factor of 1.9 for embankment slope stability. The analysis as well as analytical one predicted similar induced consolidation settlement. The prediction was comparable with on-going field monitoring. The reinforcement succeeded to increase the bearing capacity but not to reduce the consolidation settlement.

Bamboo piled-mattress evaluated in [37] was also discussed in detail in [31, 32]. The bamboo mattress was bamboo culm that being set up in parallel with very minimum space in between. The bamboos were tied up with either natural or fibre wire. In case more than a mattress required, the mattresses were set up in zigzag with the adjacent ones. The bamboo pile contained 3 – 5 bamboo culms that were bundled with wire. The top of the pile was fixed to the mattress to form the piled-mattress system. The system served various functions when installed for soft soil reinforcement. The mattress distributed the working load to the soil, change the critical failure surface location, and provide uplift buoyance force. The pile works as friction on to support the mattress.

DOI: 10.1051/matecconf/201713804010

Fig. 6. Bamboo grid for soft soil reinforcement (adapted from [37]).

For design calculation, EI_m of the mattress is easily calculated by multiplying elasticity modulus of bamboo with mattress moment of inertia of the mattress I_m. The inertia is following the equation:

$$I_m = \sum I + \sum A d^2 \tag{2}$$

I is bamboo moment of inertia and can be calculated using (1). *A* is cross area of bamboo. *d* is the distance between bamboo element and the reference point of the mattress. In plane strain modelling, the bamboo pile should be treated as node to node anchor instead of pile beam in which the pile stiffness is replaced by bamboo pile bearing capacity.

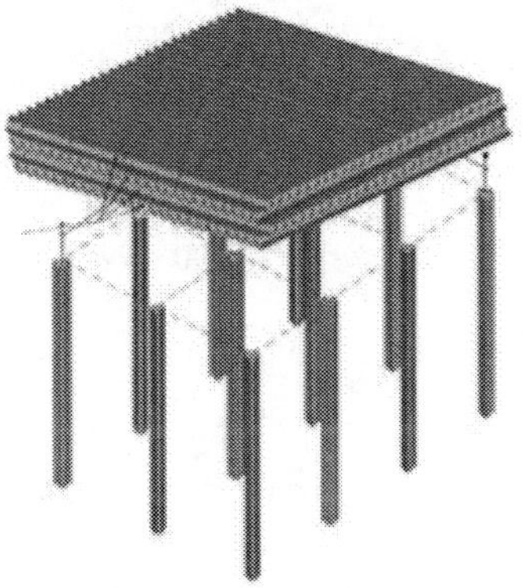

Fig. 7. Bamboo piled-mattress system (adapted from Prof. Masyhur Irsyam presentation)

Following the described procedure, [31, 32, 37] showed that the soil bearing capacity increased significantly to support the embankment. The embankment factor of safety also improved as well. However, the settlement occurred similar to one without reinforcement.

4 Discussions

The most problem of the bamboo is its durability. Without treatment, bamboo durability is less than 2 years. Preservation can make the durability last 4 – 7 years. Soaking bamboo in the flowing water is able to increase the durability. According to Indonesian local belief, it may last more than 10 years. This belief was also reported in other country [38]. The durability restricts bamboo to suit only for temporary structures. Bamboo will serve longer time in case it always in submerged condition. Using bamboo for earth retaining structure can be considered if it is temporary structure [16, 17]. For permanent one, the maintaining schedule and methods should be prepared in advanced to provide sustainable level of services.

Bamboo reinforcement for embankment support can be applied both for temporary or permanent one. The most problem of embankment is excessive strain during embankment process. The soil strength increases with time when consolidation takes place. When the bamboo serviceability reduces, the soil mobilized soil strength alters it.

Bamboo serves its full strength when the culm is in its original condition. Slicing and cutting will reduce bamboo strength significantly. Using bamboo grid or mattress will provide better strength than that of bamboo strips.

5 Conclusions

A review on earthwork construction using bamboo was conducted. Bamboo can be utilized for various earthwork constructions, such as sheet pile, soil nailing, shotcrete reinforcement, grids, mattress and pile. Bamboo reinforcement will increase soil strength to provide adequate stability and bearing capacity. However it cannot reduce settlement. Since durability is a major concern, the use of bamboo for permanent earth structure requires more factors to be considered.

References

1. D. L. Jayanetti, P. R. Follet, *Bamboo in construction* (Trada, UK, 1998)
2. F.E. Brink, P. J. Rush, *Bamboo reinforced concrete construction* (US Naval Civil Engineering Laboratory, California, USA, 1966)
3. K. L. Saucier, E. F. Smith, *Design, Analysis, and Construction of Precast Concrete Elements with Bamboo Reinforcement* (Part 926 of Miscellaneous paper, Waterways Experiment Station U.S, 1967)
4. D. J. Cook, R. P. Pama, R. V. Singh, Mag. Concr. Res. **30**(104), 145 (1978)
5. K. Ghavami, Cem. Concr. Compos. **27**, 637 (2005)
6. H. C. Lima, F. L. Willrich, N. P. Barbosa, M. A. Rosa, B. S. Cunha, Mater. Struct. **41**, 981 (2008)
7. A. Kaware, U. R. Awari, M. R. Wakchaure, IJIRSET **2**(6), 2450 (2013)
8. P. Sharma, K. Dhanwantri, S. Mehta, IJCER **5**(3), 249 (2014)
9. A. Agarwala, B. Nandab, D. Maitya, Const. Build. Mat. **71**, 610 (2014)
10. S. Y. Kute, M. R. Wakchaure, J. Inst. Eng. India Ser. A **94**(4), 235 (2014)
11. A. Javadian, M. Wielopolski, I. F. C. Smith, D. E. Hebel, Consts. Build. Mater. **122**, 110 (2016)
12. H. Y. Fang, H. J. Mehta, Proc. 6th WCEE New Delhi, India, **3**, 25 (1977)
13. H. Y. Fang, Proc. Symp. Soil Reinf. & Stab. Tech. Eng. Prac. Sydney, 191 (1978)
14. H. Y. Fang, T. R. Moore, M. F. Rovi, Proc. 3th CREAAA, Taipei, Taiwan, 413 (1981)
15. D. S. V. Prasad, M. A. Kumar, G. V. R. P. Raju, Global J. Res. Eng. **10**(1), 2 (2010)
16. Z. H. Dai, W. D. Guo, G. X. Zheng, Y. Ou, Y. J. Chen, Int. J. Geomech. **16**(5), 04016012-1 (2016)
17. Z. H. Dai, Y. J. Chen, G. X. Zheng, W. D. Guo, GeoShanghai 2014, 720 (2014)

18. A. Hegde, T. G. Sitharam, J. Mater. Civ. Eng. **27**(9), 04014256-1(2015)
19. A. Hegde, T. G. Sitharam, Int. J. of Geosyn. and Ground Eng. **1**(2), 13 (2015)
20. E. Susila, F. Agrensa, J. Eng. Tech. Sci. **47**(2), 207 (2015)
21. A. Waruwu, J. of Civ. Eng. Res. **4**(3A), 96 (2014)
22. M. Yuan, J. H. Ding, Y. L. Cao, World J. of Eng. Tech. 2, 68 (2014)
23. C. T. Toh, S. K. Chee, C. H. Lee, S. H. Wee, Geotext Geomembranes **13**, 357 (1994)
24. H. Shigematsu, Y. Sakiura, Y. Tanida, Proc. Of CECAR7, - (2016)
25. S. Ismanti, N. Yasufuku, Jap. Geot. Soc. Special Pub. **5**(2) 161 (2017)
26. P. P. Rahardjo, in Ground Improvement Case Histories, 899 (Elsevier, 2005)
27. J. N. Mandal, V. R. Manjunath, Const. Build. Mat. **9(1)**, 35(1995)
28. B.B. Broms, I.H. Wong, Proc. of Third International Geotechnical Seminar on Soil Improvement Methods, Singapore, (1985)
29. A. Marto, B. A. Othman, F.Kasim, I. Bakar, AMM **587**, 77 (2012)
30. A. Marto, B. A. Othman, Proc. of IPCBEE **8**, 129 (2011)
31. M. Irsyam, S. Krisnanto, S. P. M. Wardhani, Proc. of the 2nd Int. Conf. GEDMAR08, Nanjing, China, 165 (2008)
32. M. Irsyam, S. Krisnanto, Civ. Eng. Forum **18**, 667 (2008) in Indonesian
33. J. J. A. Janssen, *Designing and Building with Bamboo: INBAR Technical Report 20.* (International Network for Bamboo and Rattan, Beijing, China, 2000)
34. J. J. A. Janssen, *Mechanical Properties of Bamboo* (Springer Science+Business Media Dordrecht, Netherland, 1991)
35. D. Trujillo, S. Jangra, J.M. Gibson, Proc. of ICE – Struc. And Build. 170:4, 284 (2017)
36. J.N.Jha, A.K Choudhary, K.S Gill, Int. J. of Plasma Sci. & Eng. **03(01)**,167 (2010)
37. E. Susila, F. Agrensa, J. Eng. Technol. Sci.**47(2)**, 207, (2015)
38. N. Adhikary, Proc. of 1st Int. Conf. On Modern Bamboo Structures, Changsa, China, 103 (2007)

MATEC Web of Conferences **138**, 04011 (2017) DOI: 10.1051/matecconf/201713804011
EACEF 2017

The stability of clay using volcanic ash of Mount Sinabung North Sumatera and sugarcane bagasse ash with cbr and uct value

Ika Puji Hastuty[1,*], *Roesyanto*[1] , and *Giovanni* Ramadhany[1]

[1]Departement of Civil Engineering, Universitas Sumatera Utara, Indonesia

Abstract. Soil is the fundamental material that is extremely as the place of establishment of a structure or construction, both building and road constructions. However, not all soil is well used in the constructions field, as there are several types of soil that are problematic in terms of both the soil bearing capacity and deformation. The clay with carrying capacity and low shear strength needs to stabilized in order to meet the technical requirements to be used as sub grade. The add materials that are typically used for soil stabilization are cement, lime or a mixture of two or three of the added materials. In this study, the added material use volcanic ash of Mount Sinabung at North Sumatera and sugarcane bagasse ash. The purpose of this study was to determine an index value of properties as the result of the addition of 4% volcanic ash and variations in content of bagasse ash on clay and then to determine the compressive strength for maximum testing UCT (Unconfined Compression Test) and understand the value of CBR (California Bearing Capacity) as the consequence of the addition of a stabilizing agent, as well as optimum level of addition of bagasse ash. The result showed that the original soil sample has the water content 12.35%, specific gravity of 2.65, liquid limit of 46.73% and plasticity index of 26.44%. The compressive strength value of 1.38 kg/cm^2. Base on the USCS classification, the soil sample including the type CL while base on AASHTO classification, soil samples are include this A-7-6 type. After the soil is stabilized with a wide variety of sugarcane bagasse ash content value obtained the largest unconfined compression test in 4% addition level volcanic ash + 10% sugarcane bagasse ash is equal to 5.1kg/cm^2 and the result California Bearing capacity value on the optimal mix of 4% volcanic ash + 4% sugarcane bagasse ash is equal to 13.91%.

1 Introduction

The entire buildings are closely related to the soil, because the soil can be used as building material and as foundation of a building establish. In each implementation of the development, investigation of the soil is the initial step to conduct in order to determine whether the soil at the construction site meets the requirements of planning, namely stability, deformation and density. Soil stabilization is the mixing of soil with certain

* Corresponding author: ikapuji@gmail.com

substances to improve the technical properties of the soil in order to meet certain requirements. Clay is one type of soil that is frequently stabilized. This is due the clay is very hard in dry condition and more plastic on medium level of water content, but then water content level is high, the clay will be cohesive and very soft, cause large volume change due the influence of water and prompts the soil to expand and swelling in a relative short period of time. This characteristic is the one that becomes the reason of the necessary of the stabilization process, so that it can be corrected in order to increase to carrying capacity of the soil [1, 2, 4]

There are three ways of soil stabilization process, i.e. mechanical, physical and chemical process. The stabilization process utilized in this study is chemical process which was conducted by adding the mixed ingredients in the form Volcanic ash of Mount Sinabung North Sumatera and sugarcane bagasse ash. Sugarcane bagasse ash is derived from the burning of bagasse ash which changes chemically and if left without cleaning that can harm the environment [5]. Volcanic ash is one of the natural substances that are formed in the earth that later becomes volcanic material that is ejected into the air during an eruption. Volcanic ash is not soluble in water, very rough and corrosive.

Based on the function of volcanic ash from Mount Sinabung as the natural material expelled from the volcano, although it can result in damages to the environment, it also has other more profitable usages. The content of SiO_2 is the main constituent element in the forming of cement, therefore the volcanic ash has the pozzolan characteristic. The pozzolan characteristic has the behaviour of binding other minerals that exist in clay so that it hardens in a certain period of time. The content of silica (SiO_2) that is found in volcanic ash is 82.4% and CaO is 5.10%. In sugarcane bagasse ash, the content of SiO2 is 71% and CaO is 3.4%.

The purpose of this study is to understand the physical characteristic of the original soil, knowing the influence of the addition of volcanic ash and sugarcane bagasse ash towards the soil stability and also to discover the optimum mixture which contributes to the value of the Unconfined Compression Test (UCT) and the largest California Bearing Capacity (CBR) value in the clay with is stabilized with volcanic ash and sugarcane bagasse ash.

2 Research Methods

2.1 Preparation work

Finding the literature that are related to the clay that is stabilized, about volcanic ash and sugarcane bagasse ash and the literature regarding the Unconfined Compression Test (UCT) and California Bearing Ratio (CBR).

2.2 Material Preparation

2.2.1 The collection of Soil Sample

The soil sample that is used in this study is from Patumbak, Deli Serdang, North Sumatera. The soil is categorized as the clay with low to medium level of water content.

2.2.2 The procurement of volcanic ash

The volcanic ash that was used is sourced from the eruption of Mount Sinabung. Taken from Desa Tiganderkat Tanah Karo, North Sumatera, Indonesia. The utilized ash is the ash that has passed Sieve No. 200. Based on material test result it is discovered that the chemical composition of the volcanic ash is SiO_2 82.40%, Al_2O_2 is 4.52% and CaO is 5.10%.

2.2.3 The procurement of sugarcane bagasse ash

The sugarcane bagasse ash that was used is the ash obtain from the sugarcane bagasse ash waste that was burned to be ash. Taken from Factory Kuala Madu, Langkat, North Sumatera, Indonesia. The utilized ash is the ash that has passed Sieve No. 200. Based on

 DOI: 10.1051/matecconf/201713804011

material test result it is discovered that the chemical composition of the volcanic ash is SiO_2 71%, Al_2O_2 is 1.90% and CaO is 3.40%.

2.3 Implementation of Testing

The testing was conducted in two parts that are the testing for the original soil and the testing for the soil that has been added with the mixture of stabilizer material. Laboratory analysis at Research Institution of University of Sumatera Utara, laboratory of Soil Mechanic, Department of Civil Engineering, University of Sumatera Utara.

3 Research Results

Table 1. Data of physical properties of the soil, bagasse ash and Volcanic ash

No	Testing	Soil	Bagasse ash	Volcanic ash
1	Water Content	17.89%	-	2.62%
2	Specific Grafity	2.65	2.55	-
3	Liquid Limit	45.49%	Non Plastic	Non Plastic
4	Plastic Limit	15.19%	Non Plastic	Non Plastic
5	Plasticity Index	30.30%	Non Plastic	Non Plastic
6	Sieve Analysis No. 200	52.28%	10.51%	13.40%

DOI: 10.1051/matecconf/201713804011

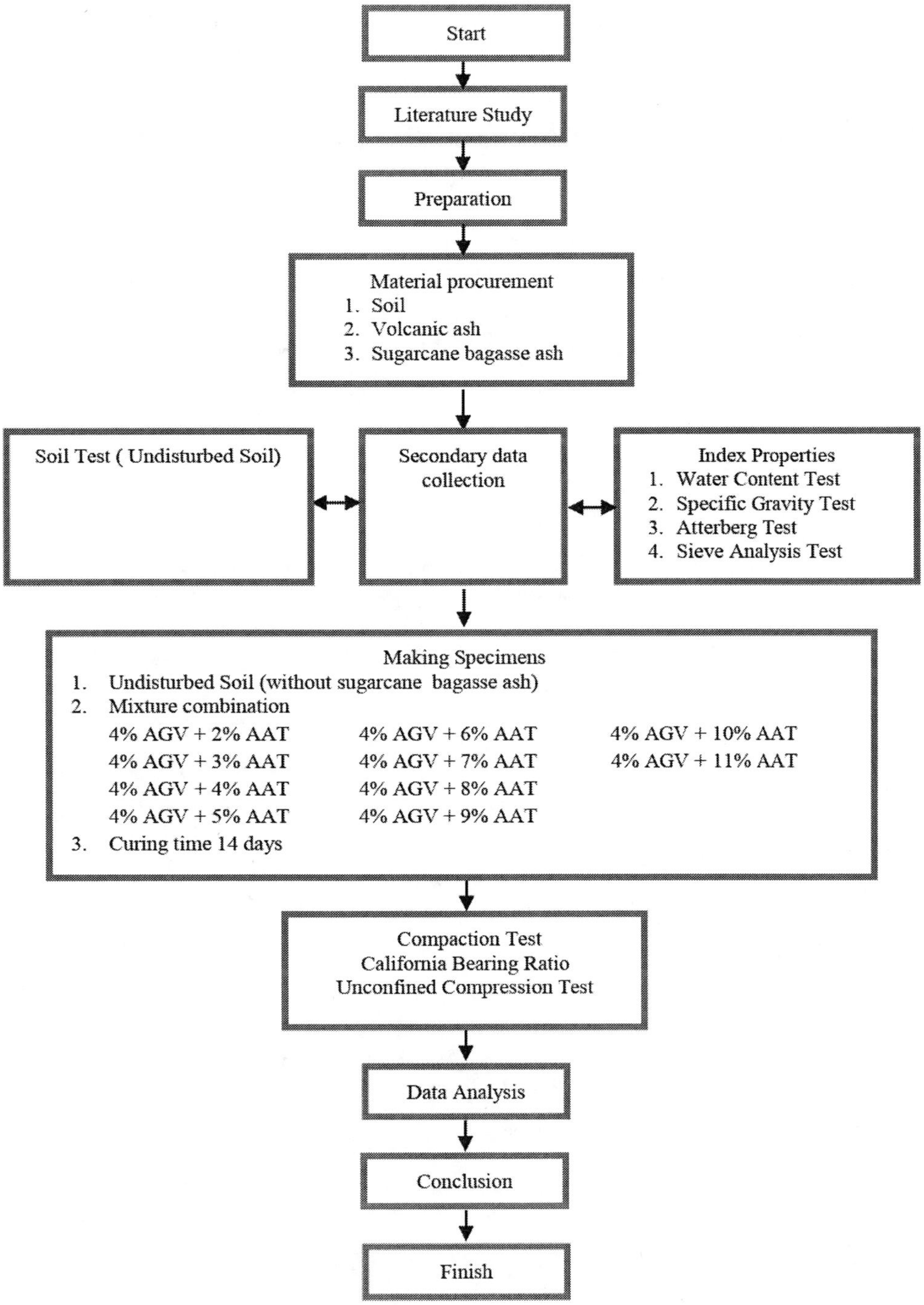

Figure 1. Research Flow Diagram

 DOI: 10.1051/matecconf/201713804011

Base on USCS classification, the sample of the soil is included in CL (Clay – Low Plasticity) type which is inorganic clay with plasticity level of low to medium. Base on the AASHTO classification, the sample of the soil is categorized as A-7-7 type.

3.1 The Test of Physical and Chemical Characteristics of Soil with Stabilizer Material

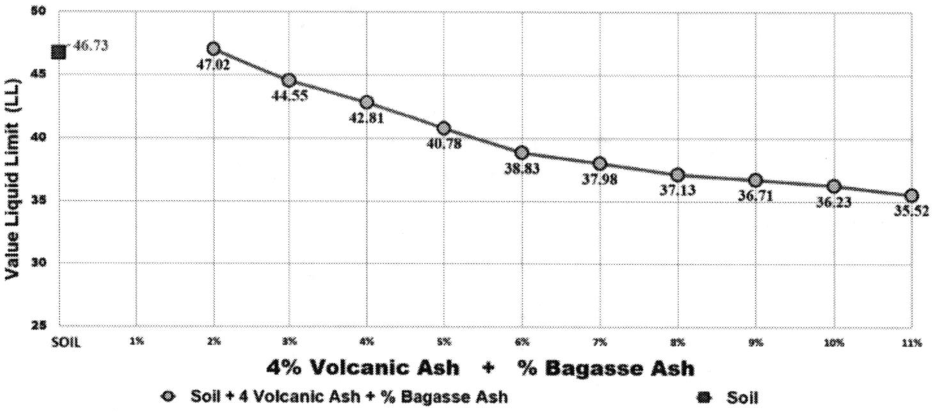

Figure 2. Correlative Graph Between Liquid Limit (LL) and Mixture Variation

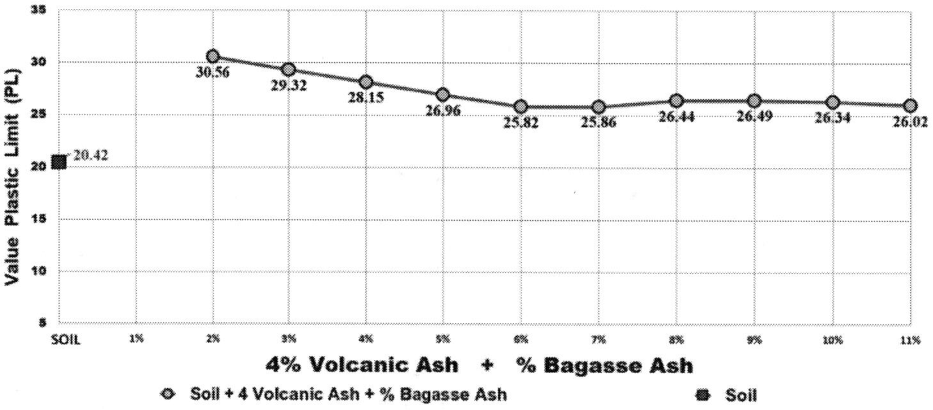

Figure 3. Correlative Graph Between Plastic Limit (PL) and Mixture Variation

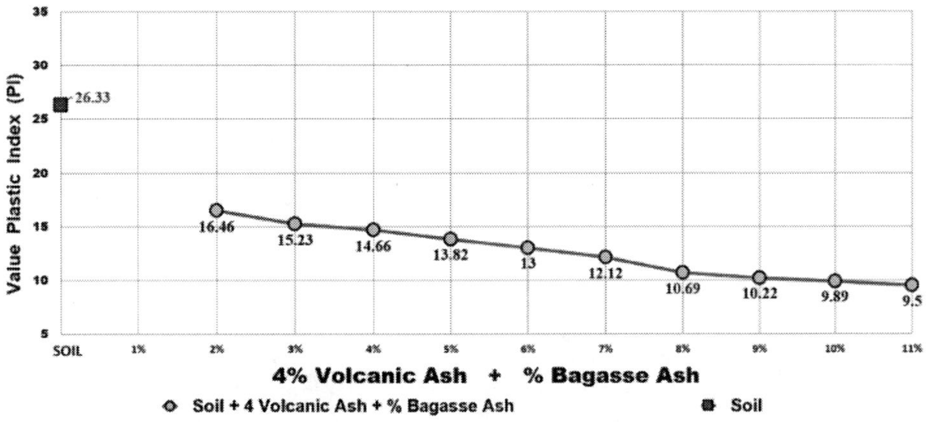

Figure 4. Correlative Graph Between Plasticity Index (PI) Mixture Variation

Figure 2 exhibits that the addition of stabilizer material results in the decrement of Liquid Limit. The higher percentage is the baggase ash, the lower is the liquid limit. Meanwhile, Figure 3 shows the increasing Plastic Limit. On Figure 4, it is the graph of Plasticity Index value of each variation of mixture. The decreasing of Plasticity Index value could reduce the expansion and the shrinkage of the soil.

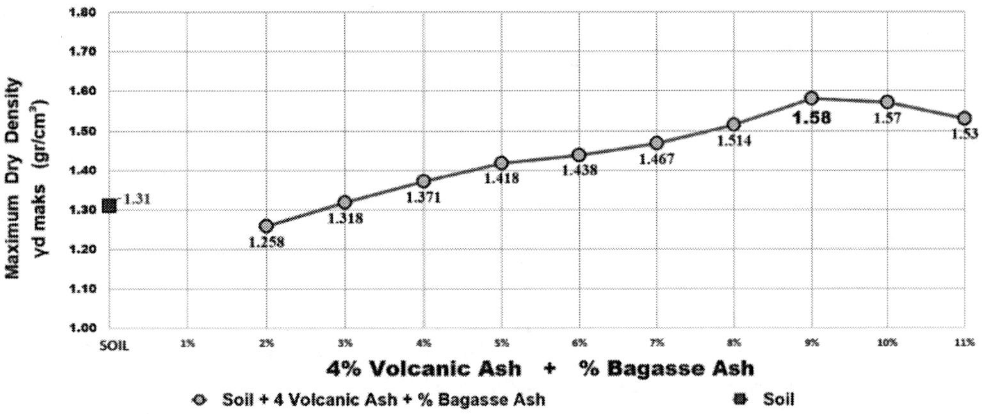

Figure 5. Correlative Graph between maximum Dry Bulk Density of the Soil and Mixture Variation

 DOI: 10.1051/matecconf/201713804011

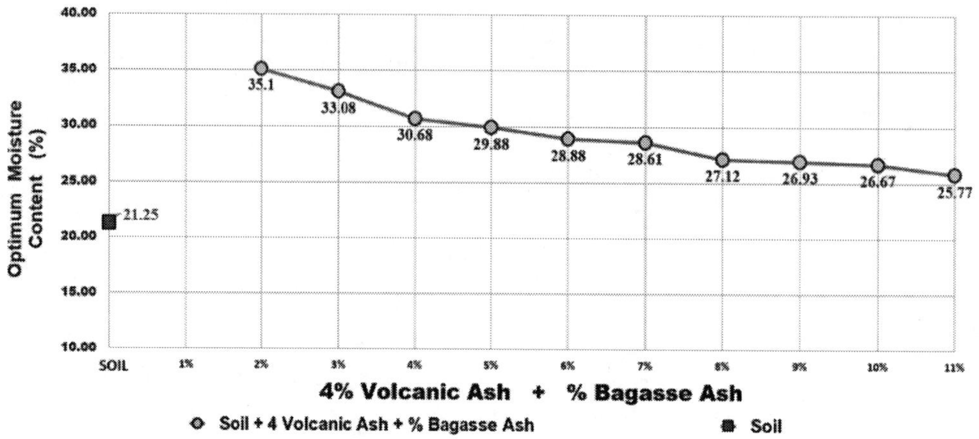

Figure 6. Correlative Graph between Optimum Water Content Level of Soil and Mixture Variation

Figure 7 owed that the Correlative Graph of UCT Value and the Variation of 4% volcanic ash and 2%-11% bagasse ash Addition Percentage. Compressive strength which used 4% volcanic ash and 10% bagasse ash had the highest which was 5.1 kg/cm^2.

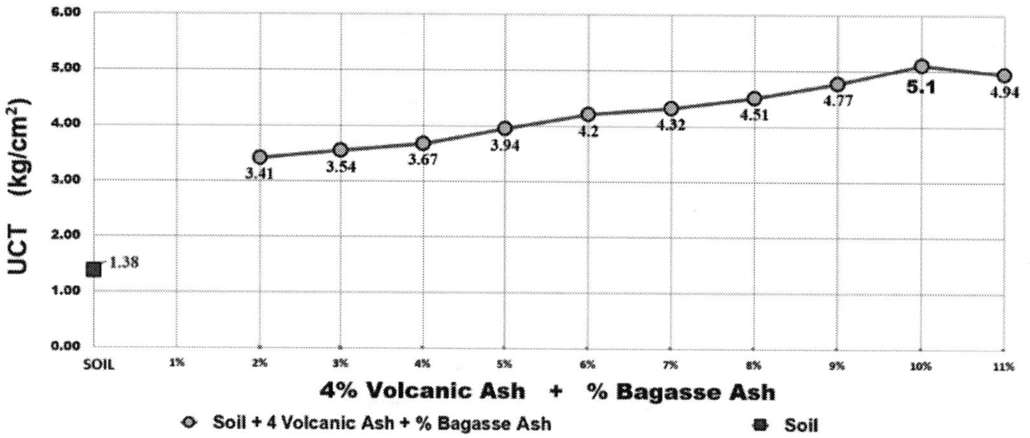

Figure 7. Correlative Graph of UCT Value and the Variation of 4% volcanic ash and 2%-11% bagasse ash Addition Percentage

DOI: 10.1051/matecconf/201713804011

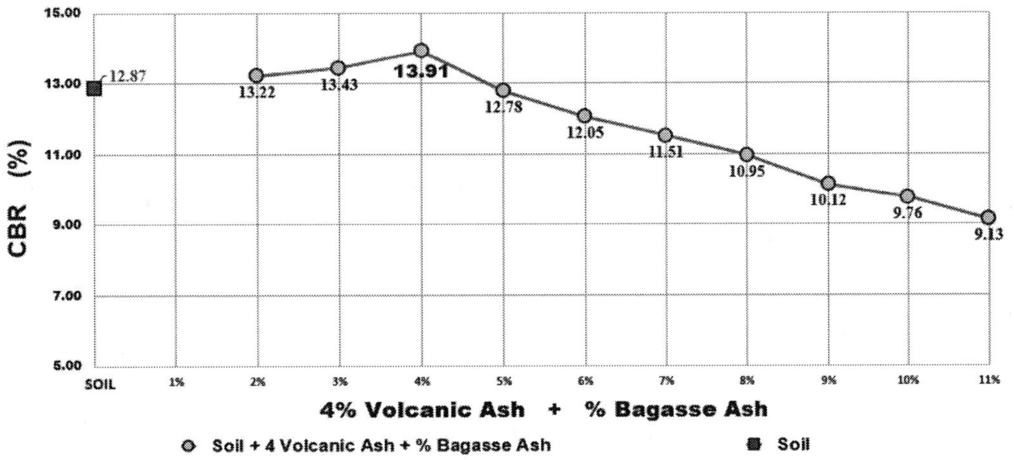

Figure 8. Correlative Graph of CBR Value and the Variation of 4% volcanic ash and 2%-11% bagasse ash Addition Percentage

Meanwhile, the result of the California Bearing Ratio Test is displayed in the graph that is included in Figure 8. The influence of the mixture of baggase ash in the clay towards the strength of the clay can be viewed from the CBR test result in unsoaked condition, with each soil variation that has been mixed with the volcanic ash stabilization material and bagasse ash within 14 days curing time. It can be seen that the increasing of the percentage of baggase ash mixture causes the CBR value to increase to the level of baggase ash percentage of 4% had the highest CBR value is 13.91%

4 Conclusion

Based on the conducted study in influence of the addition of stabilizer material of volcanic ash of Mount Sinabung and sugarcane bagasse ash towards clay with the mixture content that is determined and curing time of 14 days, it can be concluded that the type of soil that is used according to USCS is Clay – Low Plasticity (CL) and according to AASTHO is A-7-6 type. The biggest compressive strength value happens to addition level of 4% volcanic ash + 10% bagasse ash is 5.1 kg/cm^2 and the CBR laboratory test result in the optimal CBR value in the mixing of 4% volcanic ash + 4% bagasse ash with CBR value of 13.91%

Financial support for the research was provide by University of Sumatera, North Sumatera

References

1. Bowles,J.E, *Physical and Geotechnical Properties of Soil*, 2 (1991)
2. Das Sobhan, *Principles of Geotechnical Engineering*, **8** (2012)
3. Hardiyatmo, H, *Mekanika Tanah*, **1** (2002)
4. I.P.Hastuty. I.S.Sembiring,M.I Abidin, The Utilization of Volcanic Ash and High Rusk Ash as Material Stabilization in Clay by Unconfined Compression Test (UCT) and California Bearing Ratio (CBR), IOP Conf Series **180** (2017)

MATEC Web of Conferences **138**, 04012 (2017) DOI: 10.1051/matecconf/201713804012
EACEF 2017

The effectiveness of new austrian tunnelling method (NATM) for hydro power plant construction: lau gunung power plant, north sumatera, indonesia

Rizki Ade Khoir[1,*] , *and Annisa Dewanti Putri* [2]
[1]PT. PP Construction & Investment, Engineering Department of Project, Indonesia
[2]School of Civil Engineering and Architecture, Beijing Jiaotong University, PR. China

Abstract. Hydro Electric Power Plant is a power generating system using gravity fall of water as the main force to move the turbine and generate electricity. The construction purpose of Lau Gunung hydropower (2x7. 5 MW) that is located in Dairi, North Sumatra, is to Supply power to 14,000 house of the surrounding region. The river run-off system, where the water is immediately contained and then flowed through a tunnel considering the discharge flowing river, where it is constant and does not occur in the fluctuating water level. The Lau Gunung river has the minimum flow that can exceed from 15 to 25 m^3/s with a high tunnel dimensions of 4 metres long, 3,9 metres wide and a length of 1,6 kilometres. In terms of the analysis of the time effectiveness of the NATM can be saved because of the continuous work of 24 hours, without any obstacles in which the sub methods used include the drilling & blasting. The tunnel then use the form of steel reinforcement rib and Safety shotcrete lining. The general review may show that using NATM result a tremendous savings, also the use of horse shape conduce small displacement which is effective for the construction.

1 Introduction

Indonesia is famous for its abundant natural resources, one of which located in Dairi, North Sumatra. With a diverse topography that is predominantly mountainous and hilly areas underneath which rivers flow, it is suitable to build power plants, both Micro-hydro power (MHP) to Hydro Electric Power (HEP). One of being constructed is Lau Gunung (2x7.5 MW) Hydropower in Tanah Pinem, North Sumatra Province.

* Corresponding Author : Rizkiadekhoir@gmail.com

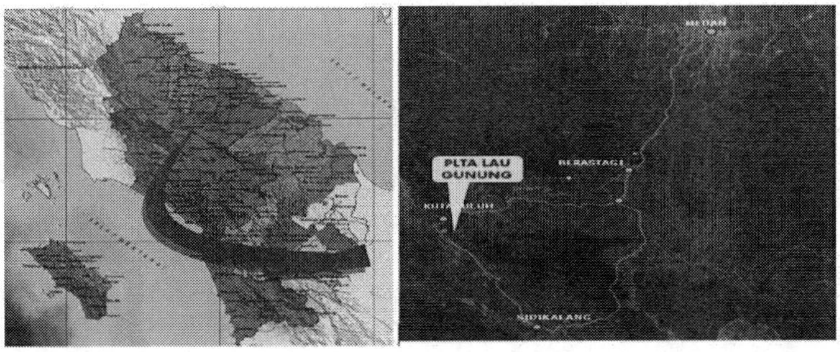

Fig. 1. Construction Location.

The plan design is to apply the river run-off system, which is put on stream flow without dissolution due to constant debit and tend not fluctuate with the minimum discharge of 15 m^3/s and flood discharge of 25 m^3/s. As For the parts of the building itself will include the hydropower, dam, intake, sand trap, tunnel, surge tank, penstock, power house and tailrace [4]. This article will be focusing in the construction of headrace tunnel (aqueduct) which will be using the New Austrian Tunnelling Method (NATM) as an effective and efficient method based on the study. NATM is a system for the manufacture of tunnel by using shotcrete (high pressure sprayed concrete) and the rock bolt as a buffer of the tunnel before concrete lining.

In this case, the geology factor plays a crucial role to decide the kind, shape, cost of a tunnel [7]. Before the tunnel construction started, a deep geotechnical and geological investigation has been conducted including the bore log, SPT (Sand Penetration Test, and incite testing. Therefore, a general review regarding to the quantitative result will be summarized to show the effectiveness of the use in the NATM method for the Lau Gunung Hydropower plant.

2 Research Methodology

Methods of Geological Research were carried out with conventional surface Track system traversing the rivers, cutting the hill, and following the path that already exists. Observations were made on rock outcrops found on the track by petrologist descriptions and measurement of geological elements such as stout and level of rock weathering. Geological mapping review is provided to borrow location search with a close and economic quarry site in accordance with the location of the planned hydropower.

Geological mapping concept, supported also by reference to existing geological map in the study area, is a geological map that already exist in the Study area of the Sidikalang and Sinabang (Sumatera) Regional Geological Map Sheet by DT Aldiss, R. Whandoyo, SA Ghazali and Kusyono. These are the database from Geological Research and Development Centre of Bandung in 1983.

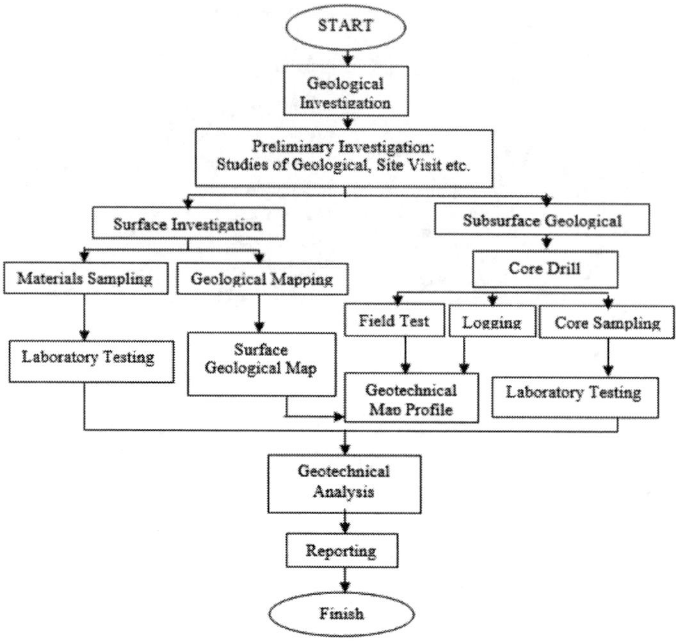

Fig. 2. Geological Research Methodology Scheme

The purpose of the paper will include to find out: (1) the factors considered in the selection of NATM for tunnelling construction, (2) the need for resources, especially man power from start of mobilisation until finishing construction. (3) The tunnel support for construction, (4) the traffic management plan in the tunnel, and (5) the optimize selection of tunnel form by considering the technical aspects.

3 Tunneling Method

Tunnels are extended, flat or only slightly sloping underground cavities with excavated cross-section of over 20 m^2 [6]. There are several methods that are commonly used for tunneling, but the method to be selected is the New Austrian Tunneling Method (NATM). This is based on consideration and analysis by the surrounding environment. The principle of NATM is based on: (1) Mobilization of the rock mass strength, (2) Protection of shotcrete with flexible support, (3) Measurement, (4) Closure invert, (5) Rock mass classification steps, (6) Dynamic design during tunnel construction, (7) Sequential excavated and supported tunnel [5].

Table 1. Advantages and Disadvantages of Different kinds of Methods [1].

Meth od	Advantages	Disadvantages	References
Drill & Blast	Adaptable, flexible, Short mobilization time requirement, Any required shape tunnel cross section is possible, Primary rock support can be installed, Total investment cost is less . Tunnel shape can be changed along the drive length.	Safety of workers is a serious issue , Performance rate of advance excavation is lower Total labor cost is high Involvement of hard and high manual labor. Low level of automation and mechanization of tasks	(Girmscheid and Schexnayder 2002)

TBM	Very high performance and low labor costs, High progress rate, especially in soft ground soil Excellent cost efficiency and high automation level, Continuous operation, Less noise and disturbance to surrounding structures, Best way for constructing deep and long tunnels	Limited flexibility in response to extremes of geologic conditions High investment costs and require high backup systems TBM mobilization take considerable time Fixed circular geometry and tunnel diameter Longer mobilization time and higher capital costs	(Girmscheid and Schexnayder 2002, Abdallah and Marzouk 2013)
Jacked Box	Economical and Better quality control, Time of completion is less Saving in man power & Machinery No involvement of crane & heavy equipment, Less involvement of other Departments	Needs trained staff and skilled supervision Imposition of caution order exists for a longer period No scope of the night working Once the vertical & the lateral alignment of box disturbed it becomes almost impossible to rectify it	(Hiren, Pitroda et al.)
Cut & Cover	Preservation of the environment Safe initiation and completion of highway tunnels Safe work progress in unstable weak ground May applied as sequential construction in case of most adverse geotechnical conditions Cheaper and more practical than other underground tunneling Small risk, relative to other construction techniques	Not suitable for very deep excavations More dust and noise impact may arise Cause interference with traffic and other urban activities	(Sharma 2011, Abdallah and Marzouk 2013, Cut and Cover 2015)

Therefore, based on the table 1, the use of drill and blast method which is the part of the NATM fits the condition of Lau Gunung Hydropower plant in which the mobilization will be the main consideration for the geographical and geological condition.

3.1 NATM Method

Tunneling in those conditions by New Austrian Tunneling Method (NATM) requires close attention on controlling surface subsidence and preventing face collapse [8]. Before the use of NATM method, a wood and steel frame construction were temporary buffered. According to Prof. Singer LV. Rabcewicz from the book of NATM, the weakness for the wood construction is related to the humid condition of woods which the dampness will be very easy to collapse [10]. Even though the steel has a good physical characteristic, working efficiency bow of steel is very dependent From Quality support (Contact steel and rock), Meanwhile, note that the gap result between rock while on time excavation often leads to the decrease of the upper section of the tunnel.

3.2 Construction Sequence

The construction method is the sequence of construction activities in the excavated cross-section and refers to the division of the excavation cross-section into partial sections for excavation, temporary support and lining [6]. The tunnel excavation works include the work of Preparatory, Steel Support Installation, Drilling, Dynamite Charge, Blasting, Ventilating, Mucking, Scaling, Shotcreting, and Rock Bolting. The preparation includes the work of marking and measuring, heating tools, division of labour workers, etc. Next, the installation of steel support following the excavation work. Before the excavation work begins, steel support needs to be designed and fabricated in advance. The shape of steel support usually follows the form of the tunnel lining. The relationship between the steel supports were made of two kinds, wooden beams for compression and iron rods bolted for tension.

Drilling is to make the hole to be filled by dynamite then executed according to a predetermined pattern which in this case is using the full-face method. Next, is a dynamite charge in a borehole using a wooden stick tool with a diameter of 30 millimetres. Dynamite blasting is a process that has been prepared according to the pattern of existing drilling, using Blasting Machine. Ventilating is blowing fresh air from the blower after completion of blasting, to clear the air of smoke and glucose generated by blasting. The material removal work for blasting out the tunnel is using special tools such as the transport wheel loader, dump truck or lorry or conveyor, depending on local conditions. Excavation was done by using a backhoe and dump truck.

Next sequence is Shotcreting which is done after scaling a temporary buffer construction of the tunnel, using a special tool called Robot also Alivia Shotcrete or Placer. The next support is the Rock Bolting. The installation of rock Bolt a temporary buffer construction next to the shotcrete.

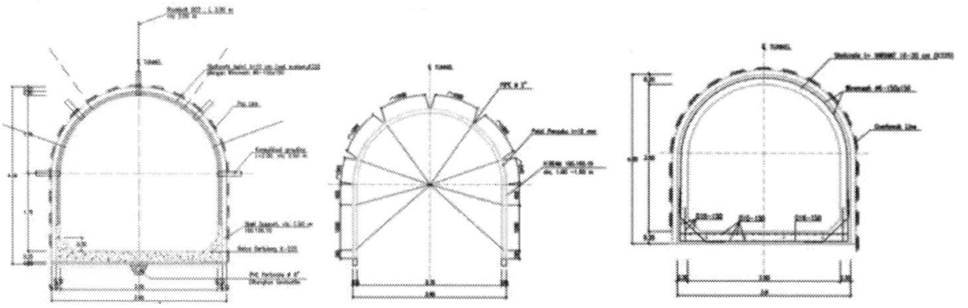

Fig. 3. Cross section: (a) tunnel, (b) tunnel Support, (c) Tunnel Reinforcement.

4 Result Analysis

The analysis of the use of NATM method as tunnelling method will be discussed in this part. Regarding to the technical data, additional data will be used in the field and laboratory test. The data, analysis and design parameters of the design for the carrier channel of the tunnel was using the empirical methods in the rocks related to the aspect of value engineering, man power, shotcrete timing, prevention of upper displacement, tunnel type consideration, and the relation between the effectiveness and efficiency compared to the TBM.

4.1 Value Engineering

The tunnel trail from Station 0 + 00 to Station 1 + 632 will pass through fresh sandstone rocks with a uniaxial compressive strength (UCS) values between 100 to 500 kg/cm^2. Above this rock layer the fragile Toba tuff rocks existed and is calculated by the UCS of 10-100 kg/cm^2. Meanwhile, from Station 1 + 632 to Surge Tank will penetrate the fragile and soft tuba tuff rock [4]. The other consideration of not to choose TBM is that the access to the project site also become an obstacle for the mobilization. The lack of human resources in the operation and maintenance of the TBM and the enormous costs in the form and shape of a horseshoe tunnel into NATM electoral impacts the tunnelling process selection.

In the Borehole of BH-4 (Portal tunnel), the rock condition has many failure with a RQD (Rock Quality Design) of 20-80 %. The top layer shows a weathering moderate level (D) condition. The groundwater surface is at A depth of 8 metres below the surface of the borehole or 13 metres above the surface of the upper tunnel (wet conditions). In the bore hole of TT - 4, the condition of the surface rocks weathered several fractures with a RQD of 10- 20% and with a moderate hardness level (CL-CM). The Surface groundwater is at a depth of 83 metres below the surface of the borehole or 15 metres above the upper surface of the tunnel (wet conditions). For borehole PT 1-2, lane tunnel will pass through the tufa rock conditions which is fragile and easily scattered with a soft hardness level (D). Surface groundwater is at the depth of 83 metres below the surface of the borehole or on the surface of the tunnel (wet conditions).

4.2 Man Power

The man power that are involved in the project implementation from the tunnelling construction mobilisation until demobilisation including the Preparation (mobilization) consist of logistics and heavy equipment Subcontractors. Second, the implementation includes Surveyor, Superintendent, Drilling Operator, Blasting Expert, Concrete and steel Worker, HSEO (Health, Safety and Environmental Officer), Quality Control Officer, and Tunnel Engineer. Meanwhile for the Demobilization will include the logistics and heavy equipment Subcontractors.

4.3 Permanent Shotcrete

After blasting, mucking and scaling, finishing are done before safety shotcrete (t = 10 cm). Safety shotcrete aims to provide security for the next job whether the next job ready to be implemented and checked by engineers to do tunnel reinforcement then shotcrete. In this part, passing bay is created if the tunnel reached a length in half in order to ease the mobilization tool.

4.4 Tunnelling Support

Support measures, both temporary and permanent, are intended to enable, assist and favourably influence the load-bearing contribution of the surrounding rock mass [6]. Tunnelling support is to provide the tunnel from failure. Based on figure 5, the supports are put during or even before excavation, but in some section when the rock mass is stable, support is generally necessary due to the effective cost. In this case, the support is using a H Beam steel with a dimension of 100 mm x100 mm (with a thickness of 10 mm), a bracing pipe (Ø= 2 inches), and supported by rock bolts. The rock bolts can be stressed with about 100 to 150 kN but their action is also due to an additional dowel effect. [6].

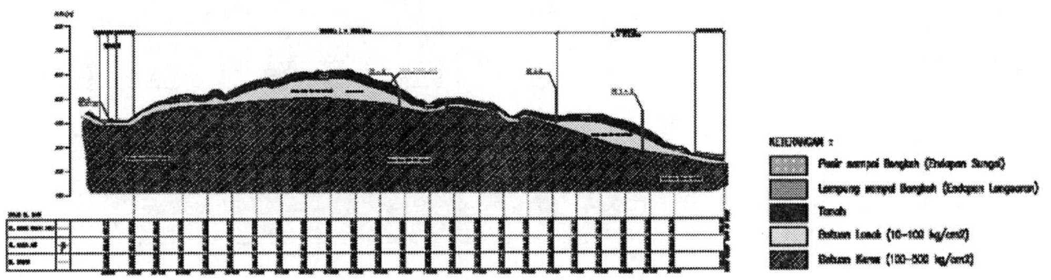

Fig. 4. Geotechnical Condition of Tunnel.

Table 2. Rock Mass Rating (RMR) System of Tunnel Route [2].

A. CLASSIFICATION PARAMETERS AND THEIR RATINGS							
	Parameter		Range of values				
1	Strength of intact rock material	Point-load strength index	>10 MPa	4 - 10 MPa	2 - 4 MPa	1 - 2 MPa	For this low range - uniaxial compressive test is preferred
		Uniaxial comp. strength	>250 MPa	100 - 250 MPa	50 - 100 MPa	25 - 50 MPa	5 - 25 MPa / 1 - 5 MPa / < 1 MPa
	Rating		15	12	7	4	2 / 1 / 0
2	Drill core Quality *RQD*		90% - 100%	75% - 90%	50% - 75%	25% - 50%	< 25%
	Rating		20	17	13	8	3
3	Spacing of discontinuities		> 2 m	0.6 - 2 . m	200 - 600 mm	60 - 200 mm	< 60 mm
	Rating		20	15	10	8	5
4	Condition of discontinuities (See E)		Very rough surfaces Not continuous No separation Unweathered wall rock	Slightly rough surfaces Separation < 1 mm Slightly weathered walls	Slightly rough surfaces Separation < 1 mm Highly weathered walls	Slickensided surfaces or Gouge < 5 mm thick or Separation 1-5 mm Continuous	Soft gouge >5 mm thick or Separation > 5 mm Continuous
	Rating		30	25	20	10	0
5	Ground water	Inflow per 10 m tunnel length (l/m)	None	< 10	10 - 25	25 - 125	> 125
		(Joint water press)/(Major principal σ)	0	< 0.1	0.1, - 0.2	0.2 - 0.5	> 0.5
		General conditions	Completely dry	Damp	Wet	Dripping	Flowing
	Rating		15	10	7	4	0
B. RATING ADJUSTMENT FOR DISCONTINUITY ORIENTATIONS (See F)							
Strike and dip orientations			Very favourable	Favourable	Fair	Unfavourable	Very Unfavourable
Ratings	Tunnels & mines		0	-2	-5	-10	-12
	Foundations		0	-2	-7	-15	-25
	Slopes		0	-5	-25	-50	
C. ROCK MASS CLASSES DETERMINED FROM TOTAL RATINGS							
Rating			100 ← 81	80 ← 61	60 ← 41	40 ← 21	< 21
Class number			I	II	III	IV	V
Description			Very good rock	Good rock	Fair rock	Poor rock	Very poor rock
D. MEANING OF ROCK CLASSES							
Class number			I	II	III	IV	V
Average stand-up time			20 yrs for 15 m span	1 year for 10 m span	1 week for 5 m span	10 hrs for 2.5 m span	30 min for 1 m span
Cohesion of rock mass (kPa)			> 400	300 - 400	200 - 300	100 - 200	< 100
Friction angle of rock mass (deg)			> 45	35 - 45	25 - 35	15 - 25	< 15
E. GUIDELINES FOR CLASSIFICATION OF DISCONTINUITY conditions							
Discontinuity length (persistence) Rating			< 1 m 6	1 - 3 m 4	3 - 10 m 2	10 - 20 m 1	> 20 m 0
Separation (aperture) Rating			None 6	< 0.1 mm 5	0.1 - 1.0 mm 4	1 - 5 mm 1	> 5 mm 0
Roughness Rating			Very rough 6	Rough 5	Slightly rough 3	Smooth 1	Slickensided 0
Infilling (gouge) Rating			None 6	Hard filling < 5 mm 4	Hard filling > 5 mm 2	Soft filling < 5 mm 2	Soft filling > 5 mm 0
Weathering Ratings			Unweathered 6	Slightly weathered 5	Moderately weathered 3	Highly weathered 1	Decomposed 0

F. EFFECT OF DISCONTINUITY STRIKE AND DIP ORIENTATION IN TUNNELLING**			
Strike perpendicular to tunnel axis		Strike parallel to tunnel axis	
Drive with dip - Dip 45 - 90°	Drive with dip - Dip 20 - 45°	Dip 45 - 90°	Dip 20 - 45°
Very favourable	Favourable	Very favourable	Fair
Drive against dip - Dip 45-90°	Drive against dip - Dip 20-45°	Dip 0-20 - Irrespective of strike°	
Fair	Unfavourable	Fair	

Table 3. Rock Mass Condition Data

No	Item	Sta.0+000 to 0+100 (Bore Hole BH-4)		Sta.0+100 to 0+200 (Borehole TBIN1+TBIN2)		Sta.0+200 to 0+300	
		Value	Rating	Value	Rating	Value	Rating
1	Uniaxial Compressive Strength (qu) I1	365,92 kg/cm2	4.00	365,92 kg/cm2	4.00	365,92 kg/cm2	4.00
2	Rock Quality Designation (RQD) I2	50-60 %	13.00	40-50%	8.00	40-50%	8.00
3	Spacing of Joints I3	30-800 mm	8.00	10-1000mm	8.00	40-1000 mm	10.00
4	Condition of Joints I4	Slightly rough surface, seperation<1mm, Slightly Weathered wall	25.00	Slightly rough surface, seperation<1mm, Slightly Weathered wall	25.00	Very Rough Surface, Not continous, No Separation, Unweathered Wall rock	30.00
5	Groundwater Condition I5	Damp	10.00	Damp	10.00	Damp	10.00
6	Strike and Dip Orientation I6	Fair	-5.00	Unfavourable	-10.00	Fair	-5.00
	I=I1+I2+I3+I4+I5+I6		55.00		45.00		57,00

Based on the table, the rock condition and Uniaxial Compressive Strength at Sta 0+000 – 0+100 and 0+200 -) +300 to be estimated with no drilling and laboratory test. The figure

DOI: 10.1051/matecconf/201713804012

of rock mass class compared to the theory of Rock mass classification, shows that the rock mass class is categorized as Rock Mass Class III which is classified as Sand Rock condition.

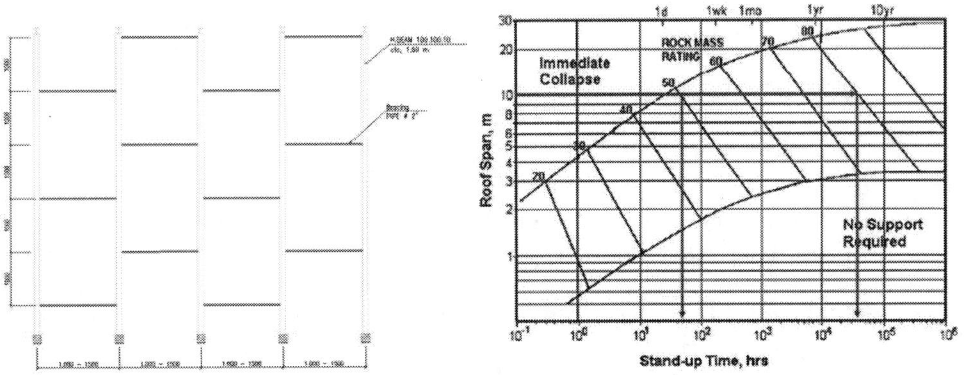

Fig. 5. (a) Supports are put if the rock condition is unstable, (b) Support Condition which is related to the Value of RMR.

The figure 6 indicates the support with a span length of 10 metres. Assuming the value of RMR (Rock Mass Rating) obtained from the result and relation of drilling test, the time known to resist the load will remain for 2 days and 4 years. The higher the RMR value, then the lesser the support can be or even without support. Besides that, the support can be given for more duration. For the two days duration may stand for two days, then shotcrete must be done. Meanwhile, the four years may last for four years, then shotcrete should also be done.

4.5 Horseshoe Tunnel Type

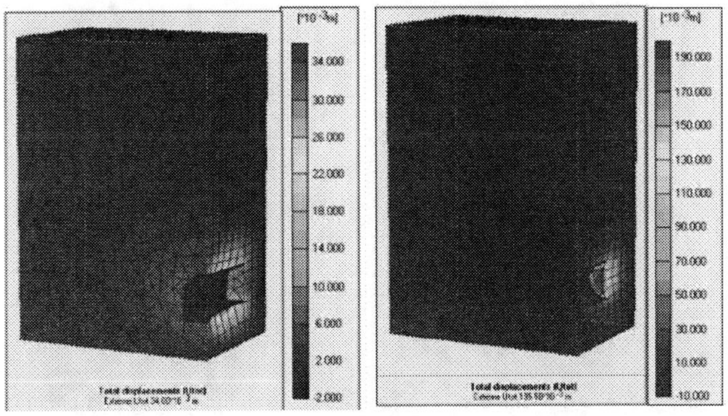

Fig. 6. Total Displacement for (a) Horseshoe and (b) Circle shaped tunnel. Abaqus, 2016.

The shape and size of the design cross-section derive firstly from the purpose of the tunnel (rail tunnel, road tunnel, sewer, water tunnel or pressure runnel for a hydropower station) and thus the required clearance gauge [7]. In this case, the use is for water tunnel for a hydropower station. The analysis has been done by numerical analysis using the finite element method based on Abaqus, a computer program used for the modelling. The result shows the maximum of total displacement of 34,0 x 10^{-3} meters for the modelling of the horse shaped tunnel. Meanwhile, the maximum total displacement of the circular shape tunnel result is 195,59 x 10^{-3} meters. This means that the model generated Horse hoe shaped displacement value due to the deformation are smaller than the circle shaped

 DOI: 10.1051/matecconf/201713804012

tunnel. Beside that the cross-section shape is also influenced by the selected construction process and the associated machinery; for example a tunnel boring machine can only bore a circular section [7].

4.6 A Comparison to TBM

The tunnel boring machine (TBM) is usually used for large size tunnel with consistent rock condition. The process of extracting the drilling machines are continuously used, being fitted with equipment that threw out the results of excavation at the same speed. Thus the drilling machine can run continuously. When the tunnel pass through loose soil layers, then the drilling machine needs to be equipped with a shield, so the progress cannot match the speed when through consistent rocks. Meanwhile for the soft soil, the use of the drilling machine will cause troubles. Because, the drilling machine may change its position (the soil does not hold the load of heavy drilling machine), which will compound the control of the tunnel. In such cases in the soft soil, the use of grouting before bypassed by the drilling Machine will be a must.

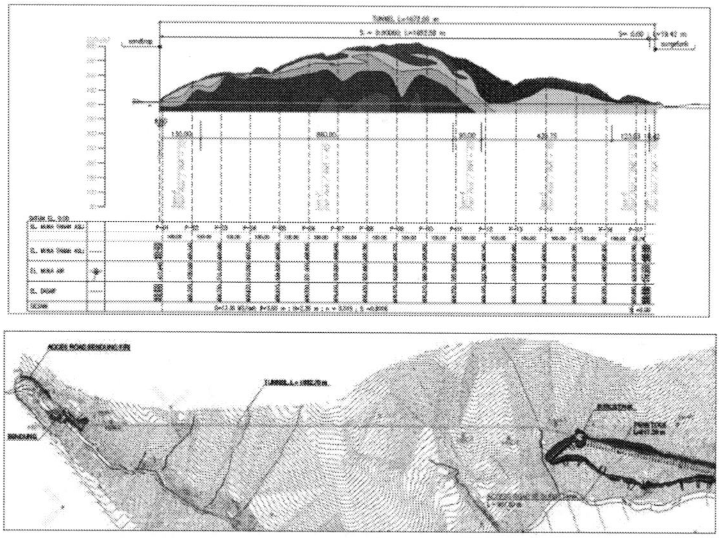

Fig. 7. Contour/geotechnical Relief of Lau Gunung Region, Dairi, North Sumatera. [4]

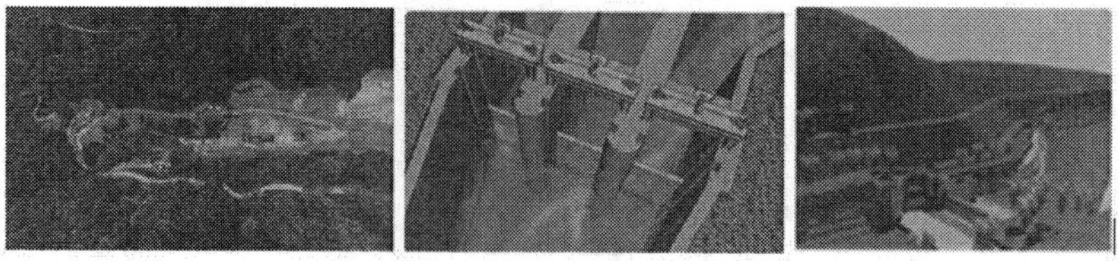

Fig. 8. Site condition and Modelling of Lau Gunung Hydropwoer Plant.

5 Conclusion

The method of NATM consisting of preparatory work, surveying, drilling, charging, blasting, ventilating, mucking, scaling, shotcreting, and rock bolting. The function of the tunnel is made to drain the water that has been free of sediment to be streamed to the surge tank. Technical geological factors and geotechnical conditions are a major cause chosen

NATM. Horseshoe shape displacement value is smaller than the circle. The construction of with NATM method is an indispensable engineering geological surveys regular to get the result of strength of the rock surveyed. Meanwhile, the cost consideration, especially the aspect of mobilization, implementation and maintenance, will be the factors considered in the method chosen.

A special Acknowledgments goes to PT. Inpola Meka Energi in welcoming to conduct a preliminary study to one of the project owned. Besides that, special thanks to PT. PP (Persero), tbk, Construction & Investment for providing data and substantial information about the Lau Gunung Hydropower plant. Additionally, acknowledgement for Beijing Jiaotong University for support related to the paper and conference.

References

1. Arshad, R.A. Abdullah, *A Review on Selection of Tunnelling Method and Parameters Effecting Ground Settlement*. Mal. Eng. Vo. **21** Bund 14. (2016)
2. Bieniawski Z.T., *Engineering rock mass classifications*. (John Wiley & Sons, New York, 1989)
3. PT. PP, *Buku Referensi untuk Kontraktor PP*. (Gramedia Pustaka Utama Press, Jakarta, 2003)
4. PT. IME, *Buku Geology Report of Proyek PLTA Lau Gunung*. (Jakarta, 2016)
5. L.C Vydrova. *Comparison of tunneling methods NATM and Adeco RS*. (2015).
6. B. Maidl, M. Thewes, U. Maidl. *Handbook of Tunneling Engineering: Volume I: Structures and Methods.* (Ernst & Sohn, Berlin, 2013)
7. B. Maidl, M. Thewes, U. Maidl *Handbook of Tunneling Engineering: Volume II: Basics and Additional Engineering Services for Design and Construction.* (Ernst & Sohn. Berlin, 2014)
8. K. Haruyama, S. Teramoto, K. Taira. *Construction of large cross-section double-tier Metropolitan. Inter-city Highway (Ken-O-Do) Ome Tunnel by NATM*. Tun. & Und. Space Tech. **20**. 111–119. (2005)
9. The International Tunneling Association (ITA). *Guidelines for the design of tunnels*. Tun. & Und. Space Tech., Vo. **3**, pp. 237-249. https://www.ita-aites.org/media/k2/attachments/public/Tust_Vol_3_3_237-249.pdf (1988)
10. M.A. Wiweko. *Pembuatan terowongan (Tunneling)*. www.pusdiklat-minerba.esdm.go.id/index.../335-pembuatan-terowongan-tunneling. (2015)

Bridge Deterioration Prediction Model Based On Hybrid Markov-System Dynamic

Jojok Widodo Soetjipto[1,2,*], *Tri Joko* Wahyu Adi[1], and *Nadjadji* Anwar[1]

[1]Department of Civil Engineering, Institut Teknologi Sepuluh Nopember, 60111 Surabaya, Indonesia
[2]Department of Civil Engineering, The University of Jember, 68111 Jember, Indonesia

Abstract. Instantaneous bridge failure tends to increase in Indonesia. To mitigate this condition, Indonesia's Bridge Management System (I-BMS) has been applied to continuously monitor the condition of bridges. However, I-BMS only implements visual inspection for maintenance priority of the bridge structure component instead of bridge structure system. This paper proposes a new bridge failure prediction model based on hybrid Markov-System Dynamic (MSD). System dynamic is used to represent the correlation among bridge structure components while Markov chain is used to calculate temporal probability of the bridge failure. Around 235 data of bridges in Indonesia were collected from Directorate of Bridge the Ministry of Public Works and Housing for calculating transition probability of the model. To validate the model, a medium span concrete bridge was used as a case study. The result shows that the proposed model can accurately predict the bridge condition. Besides predicting the probability of the bridge failure, this model can also be used as an early warning system for bridge monitoring activity.

1 Introduction

Indonesia has 89,000 bridges with a total length of 1,050 km connecting city/regency, provincial, and national roads [1]. The number of bridges will predictably increase as they are still needed to connect separated areas. Most areas in Indonesia are separated by rivers, valleys, seas, etc. As the areas mutually need each other, bridges are needed to connect them.

According to the 2012 Statistical Information Book of the Ministry of Public Works, 32.49% of bridges on the national roads were damaged and 1.5% collapsed [1]. The causes of the damages and collapses include overloading, foundation damage, structural damage, bridge floor damage, etc.

Many studies have been conducted to early detect bridge damage or failure. FMEA (failure modes and effect analysis) is used to identify risks on scouring protector design of the pier and the abutment, and to determine a risk priority number in calculating risks and potentials as well as the mitigating scenarios [2]. FTA (Fault Tree Analysis) is also often employed to determine the causative factors and incident estimations of bridge failure [3].

* Corresponding author: jojok.teknik@unej.ac.id

 DOI: 10.1051/matecconf/201713805001

SHM is a brige damage prediction method which is currently developing. The method is commonly employed for monitoring bridge structural behavior due to natural disasters, temperature, bridge loading, etc [4].

Such studies still have some drawbacks, which include the bridge elements and components assumed to have independent behavior. Yet, the behavior and damage of bridges are complex, influenced by the relationships between bridge elements and components. The damage of one element might influence other elements and the other way around. Therefore, research on the behavioral relationships between bridge elements/components to detect elements/components predominantly affecting the condition of a bridge is needed. In this research, joint probability method was used for identifying the most dominant relationship between bridge components. On the other hand, Markov Chain was used for predicting the probability of bridge condition [5]. The two methods were integrated with System Dynamic (SD). SD is a tool which has the ability to create a simulation in a complex system. Furthermore, system dynamic has often been used to analyze bridge damage and its monitoring [6].

2 Model Constructions

For developing a model, sufficient data and proposed model are required. Such data are used for making and validating the model. Meanwhile, the proposed model should be able to solve the calculation analysis more easily and applicable for further analysis, for example optimation, sensitivity, simulation, etc.

2.1 Data Collections

In Indonesia, Manual Book for Bridge Inspection published by Directorate General of Bina Marga, the Ministry of Public Works, is used as the main reference for bridge monitoring. The book is used for Bridge Management System (BMS) aimed at assessing and evaluating bridge condition. The bridge structure is classified into five levels: bridge, upper and sub structure of bridge, foundation and pier, pile and bridge wall, and small elements individually. The element assessment system consists of five factors: structure, damage, volume, function, and influence. Meanwhile, the value given is 1 if there is a damage, and 0 if there is no damage or the damage has not taken place. The condition value of bridge components is between 0 (no damage on the bridge) and 5 (bridge cannot be used) [7].

In this research, data were obtained from 235 bridges which have the same construction, that is reinforced-concrete girder bridge. The data were obtained from Directorate of Bridge, Directorate General of Bina Marga, the Ministry of Public Works and Housing. The data were from visual inspections based on BMS in East Java from 2010 to 2016. For modelling, data from 2010 to 2014 were used as a model, while data from 2015 to 2016 were used for calibration/testing.

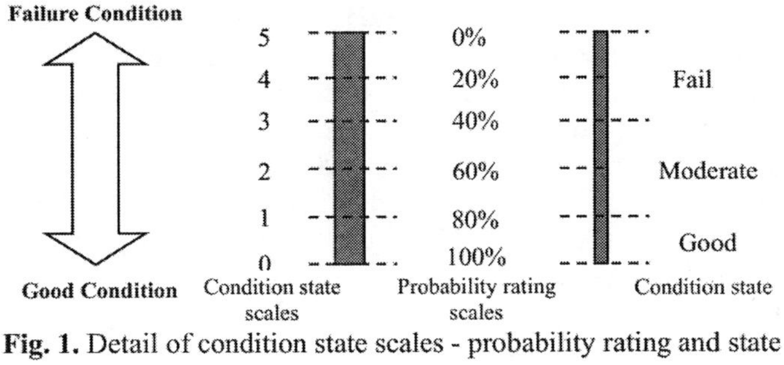

Fig. 1. Detail of condition state scales - probability rating and state

The condition value of BMS can be employed for predicting the long-term condition of bridge elements [8]. The employed method used integrates the Markov Chain method (if the data are available) and BPM (Backward Prediction Model) as generating missing years' condition data with regression [9]. For the analysis, the condition value (0-5) on BMS was changed into probability rating and condition categories (Good (G), Moderate (M), and Fail (F)) on components and bridge (see Fig. 1).

Bridge condition is influenced by the condition of bridge components (abutment/pier, beam/girder and deck) and the relationships between such components. Bridge construction can collapse due to construction materials, such as fatique and deterioration. For predicting the deterioration of concrete, ANN (Artificial Neural Network) model can be employed [10], while FRM (Fatique reliability model) can be used for predicting fatique [11]. The relationships between components can also cause bridge construction collapse. This is shown by data of bridges which collapsed due to abutment/pier conditions although the beam and deck bridge were in a good condition, vice versa. Therefore, research on the relationship behavior of the components is needed.

BMS data recording was applied to every element or bridge component, and the relationship pattern between the elements was not known. To identify the pattern of relationship between the component and bridge statistically, categorical data analysis was used. The analysis was able to evaluate whether the data were independent or dependent. Meanwhile, joint probability theory can be used to determine the strongest relationship probability between variables in a particular event. A more thorough analysis can be learned from the reference [12].

Generally, categorical data and joint probability analyses used chi-square test with the following steps: (i) making the matrix of relationship between variables (equation 1); (ii) making the matrix of probability variable (equation 2); (iii) making estimated-expected frequencies matrix (equation 3 and 4); (iv) calculating the varian value (equation 5); (v) conducting statistical test using chi-square.

$$\theta_{ij} = \text{Observation on category row i dan column j} \tag{1}$$

$$p_{ij} = \frac{\theta_{ij}}{n} \tag{2}$$

$$E_{ij} = n\, p_{i*}\, p_{*j} = \frac{n_i m_j}{n} \tag{3}$$

$$p_{i*} = \frac{n_i}{n} \quad \text{and} \quad p_{*j} = \frac{m_j}{n}\ ;\ i = 1, 2, 3, \ldots, 12 \text{ and } j = 1, 2, 3, \ldots, 12 \tag{4}$$

$$W = \sum_{\substack{i=1 \\ j=1}}^{12} \frac{(\theta_{ij} - E_{ij})^2}{E_{ij}} \tag{5}$$

From BMS data (Table 1) and calculation (equation (1)-(5)), a critical value of W = 11,256.14 was obtained. Meanwhile, according to chi-square, the distribution table for $\chi^2_{(121;0.05)}$ = 140.17. Thus, it can be concluded that the condition of the bridge and its components (Br: bridge, A: abutment, B: beam and D: deck) are dependent.

The probability of an event given one or more events can employ the conditional probability. The conditional probability can be calculated using bayes theorem. Calculating the probability of bridge condition given the condition of bridge components can be done using equation (6) and (7). The conditional probability of bridge given all of bridge components can be seen in Table 2. The most dominant bridge conditional probability occured if all bridge components are in good condition.

$$P(Y|X) = \frac{P(Y,X)}{P(X)} \tag{6}$$

 DOI: 10.1051/matecconf/201713805001

$$P(Y_1, Y_2, Y_3, \ldots Y_n) = P(Y_1)\, P(Y_2 \mid Y_1)\, P(Y_3 \mid Y_1, Y_2) \ldots P(Y_n \mid Y_1, Y_2, \ldots Y_{n-1}) \quad (7)$$

Table 1. Probability condition of components of bridge based on BMS data

		Bridge **(Br)**			Abutment **(A)**			Beam **(B)**			Deck **(D)**			Prob. row
		G	M	F	G	M	F	G	M	F	G	M	F	p_{i*}
Br	G	0.0899	0.0000	0.0000	0.0899	0.0000	0.0000	0.0899	0.0000	0.0000	0.0331	0.0568	0.0000	0.2697
	M	0.0000	0.0203	0.0000	0.0047	0.0156	0.0000	0.0142	0.0061	0.0000	0.0091	0.0109	0.0003	0.0610
	F	0.0000	0.0000	0.0009	0.0001	0.0001	0.0007	0.0002	0.0004	0.0003	0.0001	0.0005	0.0003	0.0026
A	G	0.0899	0.0047	0.0001	0.0948	0.0000	0.0000	0.0920	0.0026	0.0001	0.0349	0.0596	0.0003	0.2843
	M	0.0000	0.0156	0.0001	0.0000	0.0157	0.0000	0.0121	0.0036	0.0000	0.0074	0.0082	0.0001	0.0470
	F	0.0000	0.0000	0.0007	0.0000	0.0000	0.0007	0.0002	0.0003	0.0001	0.0000	0.0004	0.0003	0.0020
B	G	0.0899	0.0142	0.0002	0.0920	0.0121	0.0002	0.1043	0.0000	0.0000	0.0399	0.0640	0.0004	0.3129
	M	0.0000	0.0061	0.0004	0.0026	0.0036	0.0003	0.0000	0.0066	0.0000	0.0023	0.0041	0.0001	0.0197
	F	0.0000	0.0000	0.0003	0.0001	0.0000	0.0001	0.0000	0.0000	0.0003	0.0000	0.0002	0.0001	0.0008
D	G	0.0331	0.0091	0.0001	0.0349	0.0074	0.0000	0.0399	0.0023	0.0000	0.0422	0.0000	0.0000	0.1266
	M	0.0568	0.0109	0.0005	0.0596	0.0082	0.0004	0.0640	0.0041	0.0002	0.0000	0.0683	0.0000	0.2049
	F	0.0000	0.0003	0.0003	0.0003	0.0001	0.0003	0.0004	0.0001	0.0001	0.0000	0.0000	0.0006	0.0018
Prob. Colum	p_{*j}	0.2697	0.0610	0.0026	0.2843	0.0470	0.0020	0.3129	0.0197	0.0008	0.1266	0.2049	0.0018	1.0000

Table 2. Conditional probability of bridge

No	Conditional Probability	Condition component	Probability
1	P (Br | A,B,D)	Biggest probability	0.0853
2	P (Br | A,B,D)	Good	0.0853
3	P (Br | A,B,D)	Moderate	0.0348
4	P (Br | A,B,D)	Fail	0.0012

Dominant conditional probability between the components toward the bridge and between the bridge components can be determined by making joint probability between 2 variables. Thus, matrix in Table 1 was rearranged by reducing the variables, from 4 variables to 2 variables. The most dominant conditional probability value on the bridge and bridge components can be calculated using equation (6) and (7). The most dominant conditional probability can be seen in Table 3.

Table 3. Order of bridge and its' component relationship.

No	Conditional Probability	Probability
1	P (Br=G | A=G)	0.2023
2	P (Br=G | B=G)	0.2023
3	P (Br=G | D=M)	0.1278
4	P (A=G | B=G)	0.2070
5	P (B=G | D=M)	0.1439
6	P (A=G | D=M)	0.1342

As seen in Table 3, the strongest interrelationship between the bridge and components occurred on abument and beam (each 0.2023), while the deck had a smaller relationship, that is 0.1278. The relationship between components having the strongest interrelationship occured on the abument/pier with bean/girder, that is 0.2070. The relationship between beam/girder with deck is 0.1439, while the relationship between abument/pier with deck has the smallest value, that is 0.1342.

2.2 Markov-System Dynamic (MSD) Modelling

This study used Markov model for predicting bridge condition based on the current condition randomly [13]. This theory can be applied to predict future damages of bridges based on current conditions. The process of Xt can be said to have the Markov property if it fulfills equation (8). Markov is widely used because it can be applied to almost all fields of research. However, this model has some drawback; it requires repeated and longer calculation; it is more difficult to understand because it does not produce graphics automatically; furthermore, repeated calculations are also needed to determine steady conditions. Besides that, modelling is also required to generate an easier calculation method which can be applied to further calculations, such as optimation, simulation, sensitivity analysis, etc. Therefore, this research employs hybrid markov-system dynamic (MSD).

$$P(X_{t-1}=j \mid X_t=i) = P(X_1=j \mid X_0=i) \text{ ; for } t = 0, 1, 2, \ldots, n. \tag{8}$$

To build Markov-System Dynamic Model have several steps which are:

Step 1: Constructing Conditional Probability Table (CPT)

Data from the observation of bridge and the components are defined into 3 states: Good (G), Moderate (M), and Fail (F) (see Fig.1). The observation data used for making the model are data from 2010 to 2014, turned into CPT matrix using equation (9). Meanwhile, for calibrating the model, observation data from 2015 to 2016 are used.

$$P_{ij}^{(\upsilon)} = \begin{bmatrix} P_{00} & P_{o1} & \cdots & P_{0j} \\ P_{10} & P_{11} & \cdots & P_{1j} \\ \vdots & \vdots & \vdots & \vdots \\ P_{i0} & P_{i1} & \cdots & P_{ij} \end{bmatrix} \tag{9}$$

Step 2: Building the MSD model

For building the MSD model, causa-loop diagram (see Fig. 2a) and stock-flow diagram (see Fig 2b) are needed based on the behavior of Markov System (see Fig 2).

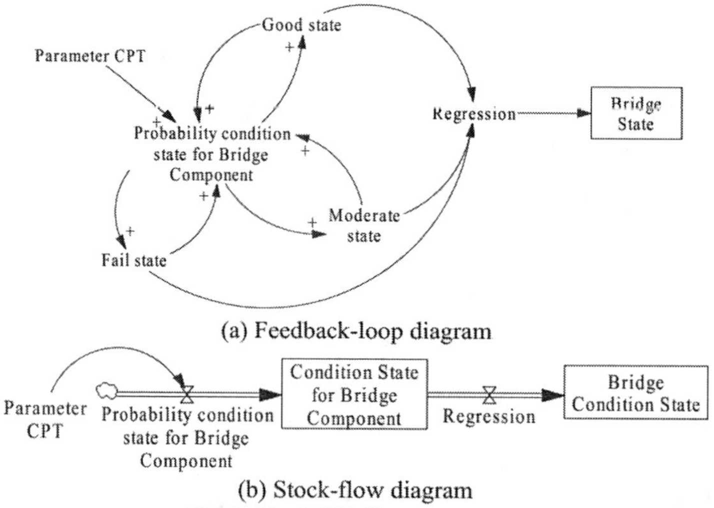

(a) Feedback-loop diagram

(b) Stock-flow diagram

Fig. 2. The MSD diagram concept

The basic concept of MSD model can be constructed based on Fig 2. CPT is a parameter for transition "probability condition state of bridge component" which has 3 states: G, M, and F. CPT becomes an input for stock on "condition state for bridge component" which also has 3 states: G (Good), M (Medium), and F (Fail). For predicting the bridge condition value based on the bridge component condition (abutment/pier, beam/girder, and deck), an approach using multiple linear regression equation is needed. Due to the limited features of Vensim PLE version, the multiple linear regression was done manually using matrix method (see step 7th in the following MSD simulation)

Step 3: MSD simulation

After the MSD model has been constructed, the next step was making the simulation of the MSD model by using CPT at step 1. The simulation used system dynamic software for running the model. The following is the alogarithm used:

Step1: *calculating CPT by using equation (8) and (9) with a time limit of T in the system determined 100 years*

Step 2: *Initially set PG(t) =1 and PM(t) = PF(t)= 0*

Step 3: *A conditional loop was formed if t < T*

Step 4: *From one process to the next one, t will increase as much as dt where* $t = t_0 + dt$.

Step 5: *Condition state was calculated based on the following equations:*

$$PG(t) = PG(t-dt) + CPT*dt \quad (10)$$

$$PM(t) = PM(t-dt) + CPT*dt \quad (11)$$

$$PF(t) = PF(t-dt) + CPT*dt \quad (12)$$

Step 6: *The output of MSD on "Condition State for Bridge Component" can be presented in the forms of numbers and graphics*

Step 7: *The Multiple Linear Regression was calculated using matrix method with equations (13)-(18). Data Y is the output MSD bridge condition, while* X_{1i}, X_{2i}, X_{3i} *is the output MSD bridge component with i = 1 until steady state on output MSD.*

$$Y = \beta_0 + \beta_1 X_{1i} + \beta_2 X_{2i} + \beta_3 X_{3i} + \ldots + \beta_p X_{pi} + \varepsilon_i;\ i = 1, 2, 3, \ldots, n \quad (13)$$

$$[Y] = \beta [X] + e \quad (14)$$

$$\beta = [X'X]-1 [X'Y] \quad (15)$$

$$R^2 = 1 - \frac{\sum ei^2}{\sum yi^2} \quad (16)$$

$$\sum ei^2 = e'e \quad (17)$$

$$\sum yi^2 = Y'Y \quad (18)$$

This regression equation can be used as an equation on flow "regression".

Step 8: *Output of MSD and regression prediction "Bridge condition state" can be presented in the forms of numbers and graphics.*

Step 4: MSD scenario

For observing the bridge condition behavior, a scenario by changing the CPT behavior of bridge component can be employed.

3 Results and Discussions

MSD model was used to predict the bridge condition and its components in 3 states: probability of Good, Moderate and Fail. The MSD simulation result can be seen in Fig 3. The bridge probability for good, moderate and fail are 0.84, 0.14 and 0.02 respectively. On average, the steady condition was achieved on year 26th. Based on each component of bridge: (i) the probability of good condition of abutment, beam and deck is 0.78, 0.73 and 0.82 respectively; (ii) the probability of moderate of abutment, beam and deck is 0.21, 0.24, 0.17 respectively; and (iii) the probability of fail condition of abutment, beam and deck is 0.01, 0.03, 0.01 respectively.

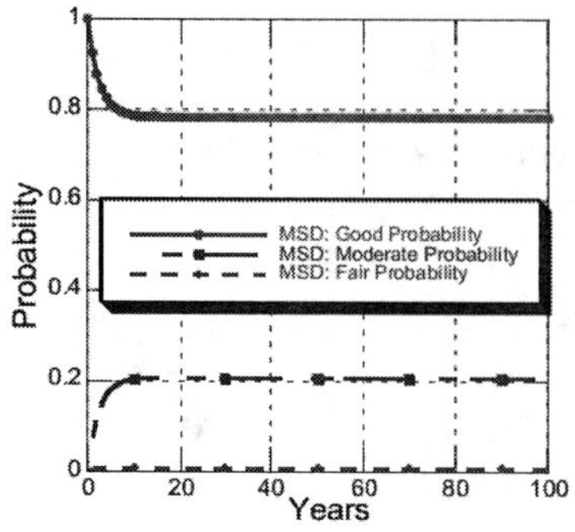

(a) MSD Abutment

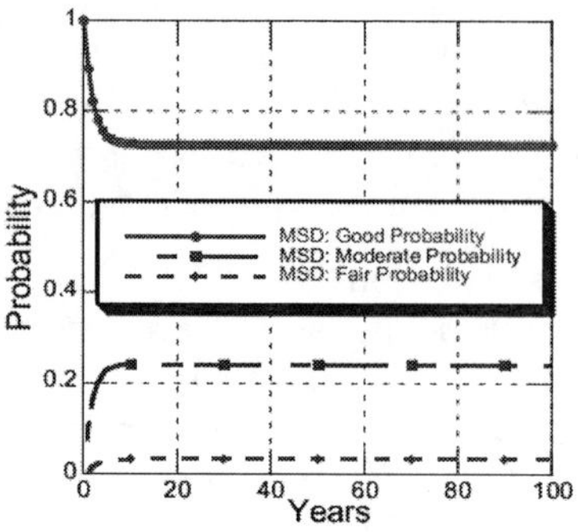

(b) MSD beam

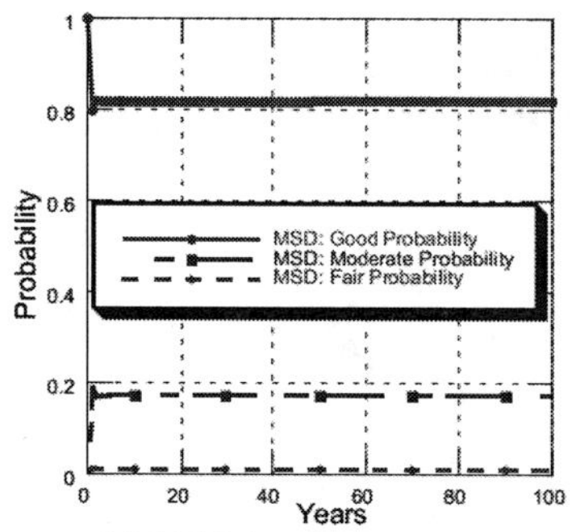

(c) MSD Deck

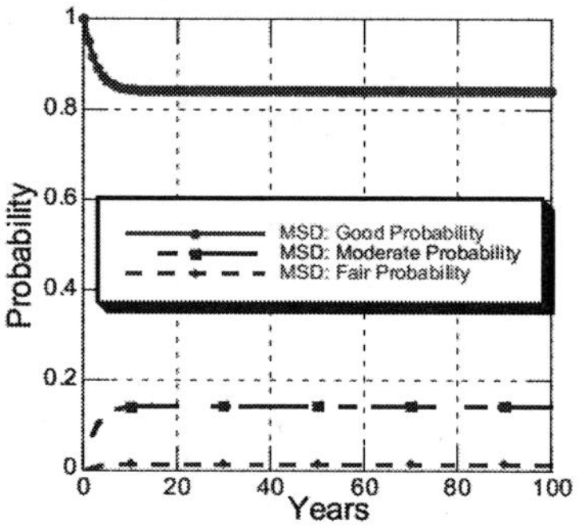

(d) MSD Bridge

Fig. 3. MSD Bridge and it's components based on model

The output of MSD model for each bridge component was used as data for making multiple linear regression for predicting the overall bridge condition. The bridge regression equation can be seen on equation 19 – 21.

Probability of good: $Y=0.4003+0.4357A+0.2458B-0.0818D;\ R^2=0.9902$ (19)
Probability of moderate: $Y=0.0000-0.1766\,A+0.7174B-0.0797D;\ R^2=0.8163$ (20)
Probability of fail: $Y=0.0000+0.4775A+0.0390B-0.2880D;\ R^2=0.3568$ (21)

For validating the MSD model, the model should be run through data testing. The data testing employed BMS observation data from 2015 to 2016. The data were also calculated using the regression equations and the values were compared. The value obtained from MSD and regression is presented in Fig 4. As the Figure indicates, the two values has similar trends and values, although the still contain errors.

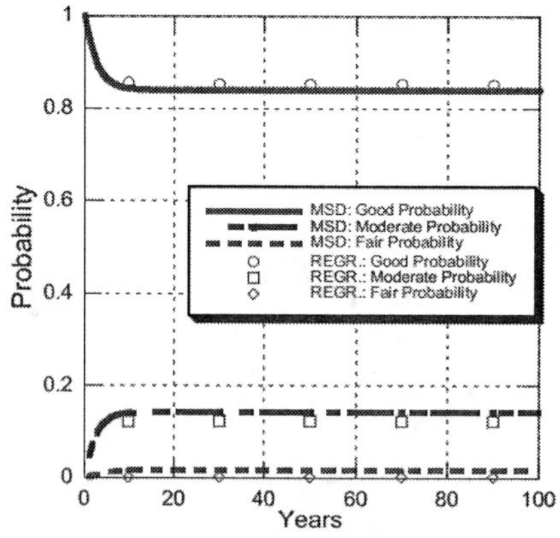

Fig. 4. Comparison of MSD and regression bridge and the component probability prediction

To observe the bridge condition behavior due to the change of CPT bridge components, a model scenario can be used. The applied scenario can be seen in Table 4. The scenario was designed by changing one variable only so that the outputs can be compared and the behavior can be identified.

Table 4. The scenarios of MSD model

No	Scenario	Code	CPT values of each bridge components	No	Scenario	Code	CPT values of each bridge components
1	Scenario 1	A1B1D1	$\begin{matrix} G \\ M \\ F \end{matrix}\begin{pmatrix} 1 & 0 & 0 \\ 0 & 1 & 0 \\ 0 & 0 & 1 \end{pmatrix}$	3	Scenario 3	A3B3D3	$\begin{matrix} G \\ M \\ F \end{matrix}\begin{pmatrix} 0 & 0.5 & 0.5 \\ 0 & 1 & 0 \\ 0 & 0 & 1 \end{pmatrix}$
2	Scenario 2	A2B2D2	$\begin{matrix} G \\ M \\ F \end{matrix}\begin{pmatrix} 0.5 & 0.5 & 0 \\ 0 & 1 & 0 \\ 0 & 0 & 1 \end{pmatrix}$	4	Scenario 4	A4B4D4	$\begin{matrix} G \\ M \\ F \end{matrix}\begin{pmatrix} 0.5 & 0.25 & 0.25 \\ 0 & 1 & 0 \\ 0 & 0 & 1 \end{pmatrix}$

In the first scenario, MSD model was given CPT values (G-G=1, M-M=1, F-F=1 and the others =0) producing identity matrix. Scenario 1st generated bridge condition value for Good = 1 and Moderate = Fail = 0. Thus, the behavior of bridge is only determined by the initial condition of bridge components, which are Good = 1 and Moderate = Fail = 0 (see Fig 5a). This state was used as a reference for making the next scenario in which the CPT change was only given in a good condition (moderate and fail were fixed). Thus, it can be concluded that the bridge behavior is only affected by good condition. Therefore, the bridge behavior can be identified clearly if good is tranformed into moderate and fail in all bridge components, which were not affected by transformed the moderate or fail into other conditions.

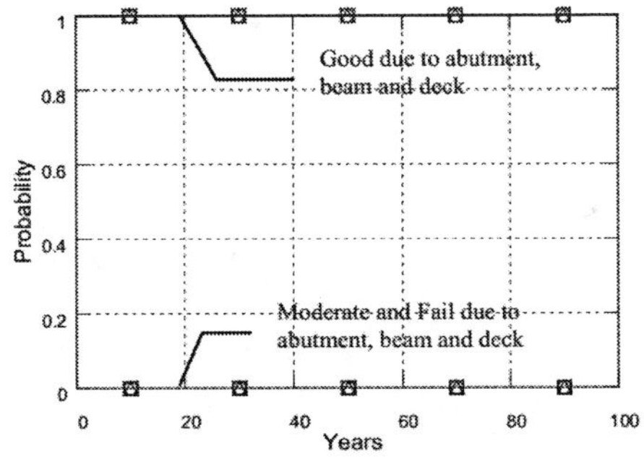

(a) Probability of bridge condition in A1B1D1

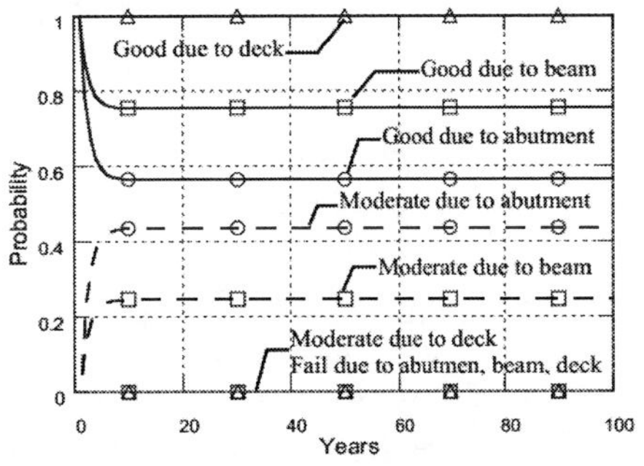

(b) Probability of bridge condition in A2B2C2

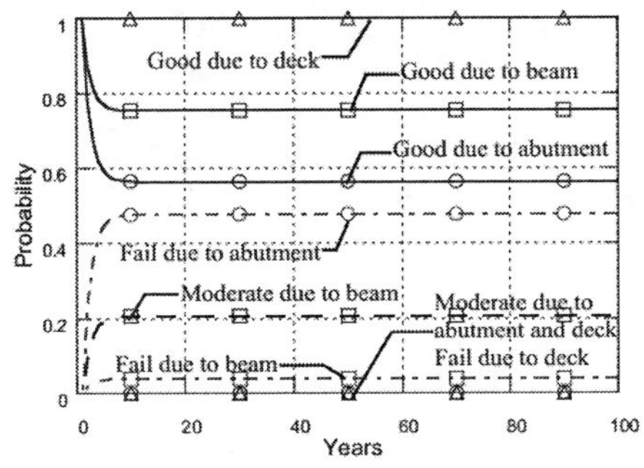

(c) Probability of bridge condition in A3B3D3

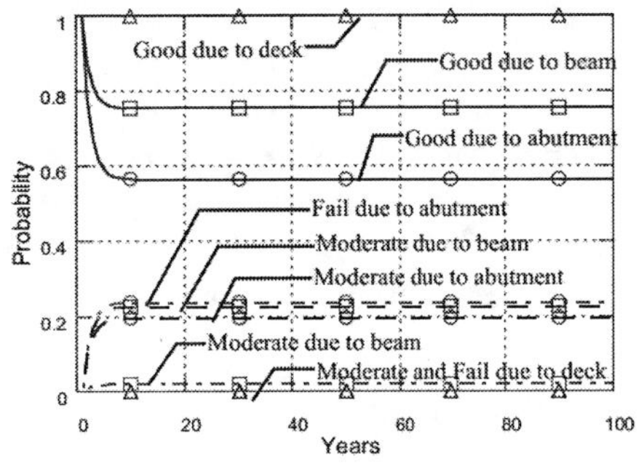

(d) Probability of bridge condition in A4B4C4

Fig.5. Probability of bridge condition in variance scenario

The second scenario was designed to observe the bridge behaviors if all bridge components in good condition were transformed into 50% to moderate condition. The bridge behavior indicates that the good probability condition decreased from 1 to 0.56 (due to abutment), 0.75 (due to beam), and deck still had the value of 1. The bridge moderate probability condition which previously had 0 value increased due to the transition, which is 0.44 (due to abutmen), 0.25 (due to beam), and for deck still according to the initial value. Meanwhile, the condition probability of bridge fail had the value of 0 for all bridge component effect. The result of scenario 2 can be seen in Fig 5b.

The next scenario was made by transforming all good condition of the bridge components to 50% to moderate and 50% to fail. The result obtained had a trend similar to scenario 2 although they had different values. There was a decrease of good from the initial condition (=1) to 0.56 (due to abutment) and 0.75 (due to beam). The moderate probability changed from 0 to 0.21 (due to beam) and remained the same for abutment and deck. The fail probability of the bridge changed from 0 to 0.48 (due to abutment) and 0.04 (due to beam). The change of CPT on the deck had almost no significant effect on the bridge behavior. The graphic of MSD scenario 3 can be seen in Fig 5c.

To observe the effect of the condition value transition from good to moderate and fail simultaneously, scenario 4 was made. This scenario generated graphic presented in Fig 5d. The transition generated: (i) the decrease of good condition probability on the bridge from 1 to 0.56 (due to abutment) and 0.75 (due to beam); (ii) the increase of moderate condition probability on the bridge from 0 to 0.20 (due to abutment), 0.23 (due to beam); (iii) the increase of fail condition probability on the bridge from 0 to 0.24 (due to abutment), 0.02 (due to beam); and (iv) no effect on the bridge condition due to transition change on the deck.

Based on the result of MSD model, it can be concluded that bridge behavior is significantly influenced by the condition of bridge components, where the most influencing component is abutment, followed by beam and deck. This results align with the BMS data analyzed using the joint probability theory.

4 Conclusion

The proposed model, MSD, can be used to predict the condition of the bridge accurately. Simulation result shows that the condition of bridge and its components are dependent; thus, the bridge condition is strongly influenced by its components. The most dominant relationship is between the abutment and beam. MSD model can predict bridge failure in 3 state probability (good, moderate and fail).

MSD model can also be used as a reference for determining bridge condition probability, so that it can be used as a guide for the maintenance and operation of bridge. Besides predicting the probability of the bridge failure, this model can also be used as an early warning system for bridge monitoring activity.

This model, however, still has some drawbacks. The CPT calculation and multiple linear regression are still conducted manually. This is due to the limited features of Vensim PLE version used. This model also requires lengthy calculations if more scenarios are given. This is because CPT calculation is still done manually. Therefore, this model needs to be developed further to address the drawbacks.

Acknowledgement

The author would like to thank the Directorate of Bridge, Directorate General of Bina Marga, the Ministry of Public Works and Housing for providing data, manuals, and other needed information for this study.

References

1. P. Work, *Book Statistical Information*. Jakarta: Ministry of Public Works (2013)
2. F. U. Ashraf, P. A. Johnson, and M. Asce, J. Bridg. Eng., **2**, 1–10 (2015)
3. W. Pang, D. Ph, M. Asce, and K. Dey, J. Infrastruct. Syst., **19**, 326–334 (2013)
4. F. N. Catbas, M. Susoy, and D. M. Frangopol, Eng. Struct., **30**, 9, 2347–2359 (2008)
5. S. Ranjith, S. Setunge, R. Gravina, and S. Venkatesan, J. Perform. Constr. Facil., **27**, 319–325 (2013)
6. J. W. Soetjipto, T. J. W. Adi, and N. Anwar, Int. J. Eng. Technol. Innov., **6**, 4 (2016)
7. P. Work, *Bridge Inspection Guide*. Jakarta: Public Work (1993)
8. S. Hasan, S. Setunge, D. W. Law, and Y. Koay, Int. J. Eng and Tech., **7**, 1, 40-44 (2015)
9. G. Bu, J. Lee, H. Guan, M. Blumenstein, and Y. Loo, J. Perform. Constr. Facil., **28**, 330–340 (2014)
10. Y. Huang, J. Perform. Constr. Facil., **24**, 597–602 (2010)
11. K. Kwon and D. M. Frangopol, Probabilistic Eng. Mech., **26**, 3, 471–480 (2011)
12. A. Agresti, *An Introduction to Categorical Data Analysis*, London, UK: A John Wiley & Sons, Inc. (2007)
13. S. Leu and T. J. W Adi, Eng. Appl. Artif. Intell., **24**, 4, 658–665 (2011)

MATEC Web of Conferences 138, 05002 (2017) DOI: 10.1051/matecconf/201713805002
EACEF 2017

A framework of initiatives for successful application of life cycle costing (LCC) in industrialised building system (IBS) in Malaysian construction industry

Siti Mazzuana Shamsuddin[1*], *Rozana* Zakaria[2], *Norfashiha* Hashim[1], *Norazian* Mohamad Yusuwan[1], *Shaza Rina* Sahamir[1], and *Nur IzieAdiana* Abidin[2]

[1]Faculty of Architecture Planning and Surveying, Universiti Teknologi MARA, Shah Alam, Selangor
[2]Faculty of Civil Engineering, Universiti Teknologi Malaysia, Skudai, Johor

Abstract. Industrialised Building System (IBS) contributes a great shift in sustainable construction and benefits the society. Previous studies have proven in general that one of the hindrances in promoting sustainable in IBS, is a high cost for any pre-cast material specifically thus effected the overall cost. The introduction of Life Cycle Costing (LCC) leads into providing a better and comprehensive cost estimation, including projecting actual cost to operate the building, hence providing a better baseline for decision making. The lacking application of LCC in IBS is still in trivial impact, therefore, this paper presents a framework which produced by expressing the successful initiatives of LCC and IBS in Malaysia construction industry. The framework developed based on findings from the extensive literature reviews, 164 responded questionnaires, and 19 expert opinions, which has three sections: Strategy Development, System Development and the last part is Decision Level. Aspects of objectivity, practicality, reliability, completeness that were likely to be implemented in the Malaysian construction industry were used. There are significant opinions on the usefulness and completeness of the proposed framework in providing a comprehensive cost estimates which helps much in deciding to carry out IBS or remain with a traditional construction method.

1 Introduction

Industrialised Building System (IBS) contributes efficient construction system in which it expedites construction time, produced in quality workmanship and reduce labour resources. Thus, this contributes to sustainable development whereby the reduction of construction waste is manageable, therefore, indirectly preserve natural resources and environment for the society. Previous literature proved that one of the hindrances in promoting sustainable in IBS was the stigma of high cost for any pre-cast material specifically, the overall cost of development. While construction industry nowadays, needs to respond to the greater

* Corresponding author: mazzuana5962@gmail.com

demand for social, economic and environmental improvements of the industry, the price or cost to adhere such project, somehow is a burden to the contractor [1]. If sustainable design and construction are approached holistically, using integrated design with modern materials and systems, it is entirely possible to create a built environment that is both sustainable and economically viable[2]. There are many terms used to refer to the ultimate cost or total cost of building ownership. Even though the terms used are different, the Life Cycle Costing (LCC) is preferred as it is widely used in the practice and better known to represent the total cost related to the building ownership throughout the study life. In a layman term, LCC is best known for generating the total discounted cost of building ownership throughout the building's life[3]. A key indicator of the validity and usefulness of LCC should surely be the level of practical application and implementation within the construction and engineering industries. LCC had gained favour in the construction industry, but the practical implementation of it is decelerating [4]. There are several processes involves in building construction life cycle, starts from design stage until demolition stage. Whereas, the construction sector involved with the operation and it needs to be maintained until the building reached its disposal as well as life expectancy for higher life cycle costing (LCC) [1].

2 Life Cycle Costing In Industrialised Building System

In IBS projects, in common, production cost is the main cost factor considered in selecting the most appropriate construction method. Normally, the cost is set to the minimum, which is not necessarily improving the lifetime performance of the buildings [1]. However, the sustainably designed building can reduce the higher production cost which generally high consumption of the energy during factory production, and additional energy during the construction phase, furthermore, more than half of total energy consumed in operating and maintenance. Increasing interest in LCC provides better understanding [2] for the design process of IBS buildings, which cost factors are considered in building life cycle.

LCC adds all the costs of alternatives over their life period and enables an evaluation on a common basis for the period of interest. This enables decisions on acquisition, maintenance, refurbishment or disposal to be made in the light of full cost implications. In IBS, since the stigma of initial investment and the production cost of the project is always on a high note, LCC analysis will slowly diminish the negative perception of the cost of IBS project. In IBS the high production or material cost can be countered back in another cost such as labour and maintenance cost. LCC can be used as a decision-making tool as its clearly shows options in deciding whether to build or not.

A set of questionnaire was distributed to 164 IBS practitioner across Malaysia in order to gain their insight pertaining to the real factors influencing the construction cost of IBS projects. From the analysis, the most important factor contributing to the construction cost of IBS project in Malaysia is repeatability and standardization of the elements. Followed by the suitability of IBS in any design and type of project, accuracy, and consistency of the design and specification throughout of the project and urgency of the task completion. These factors are considered as *'Project-IBS characteristics'*. While another three factors which include quantity ordered, government incentive and assistance and supply stability of the building materials, are related to economic and marketability. In IBS projects, the production cost is the main cost factor that mostly considered while selecting the most appropriate construction method.

2.1 Integrating cost estimates and life cycle costing

A preliminary research using questionnaire survey was conducted among 64 Quantity Surveyors throughout Malaysia to identify the most widespread or preferred Incidence-In-Use (IIU) of cost estimation model being employed by the Malaysian Quantity Surveyors for projects using IBS. Traditional types of cost model where, in general, still the most widely used. The newer, only the value management, resource- based model, and life-cycle cost model costs being in general use. The construction industry really needs a paradigm shift towards an innovative and dynamic cost model [5].

From the study, it can be summarised that the continued and overpowering utilization of the conventional sorts of cost model at the expense of the newer non-traditional types have somewhat stigmatised Brandon's call for worldview change[6] [7]. Here are finding the reasons why the traditional cost estimation modelling is still being widely used; (i) lack of familiarity with the newer techniques, (ii) time constraint, (iii) absence of data and learning, (iv) questions whether these strategies are replicable to different tasks, (v) most development undertakings are not sufficiently vast to warrant the utilization of these methods or examination into them, (vi) they require the accessibility of sound information to guarantee certainty, (vii) the degree of sophistication is seen as too superfluous for an average project, (viii) the dominant part of dangers are contractual or development related, and (ix) a genuinely subjective reason such that they can be managed better on the premise of individual experience or from past contracts attempted by the company [8].

As a dynamic and unique industry, the construction industry is constantly improvising its methods in managing resources, building materials, mechanical and electrical and Information Technology system, business processes, procurement methods and managements techniques. Since building frameworks are perplexing and incorporate a vastly different sort of parts, the capacity of the frameworks to ceaselessly perform their obliged capacities is of essential [9]. Therefore, it is natural to develop an integrated costing method to keep up with other changes. An integrated cost method will able to provide a logical method for accurate determination of a true cost for the project [10].

3 The Framework of Success Initiatives Between LCC and IBS

The framework consists of three (3) sections: Strategy Development (Barriers of success and Solution which furthermore broken down into sub-parts consists of costing success elements and failure reduction initiatives.), System Level (Breakdown from success elements, failure reduction activities and strategies in calculation solution), and the last part is Decision Level (breakdown from failure reduction activities and strategies in calculation of cost).

DOI: 10.1051/matecconf/201713805002

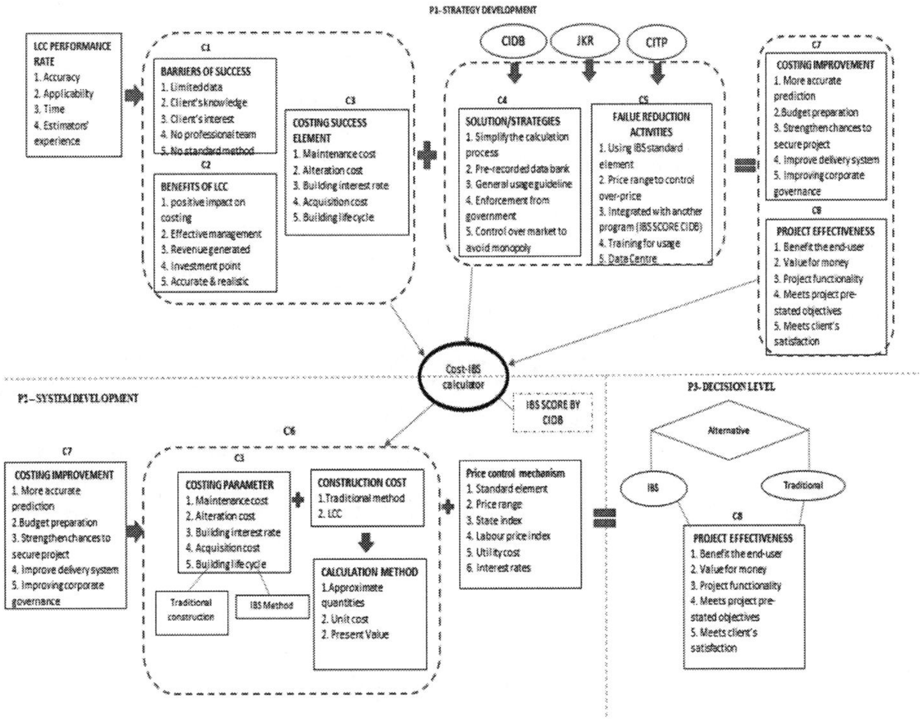

Figure 1: A framework for Initiatives for Successful Application of LCC for IBS in Malaysian construction industry

In the first section, called Strategy Development. In part one (1) of this section. It shows the LCC Performance Rates, focuses on the costing performance. The performance rates are accuracy/reliability, applicability, time and estimators' knowledge. From the analysis, it shows that most of the respondents highlight that accuracy of the estimates determine the project performance while the applicability is measured in term of the usage in a variety of project type. The second subsection concentrate on the barriers of success, benefits, and the costing critical elements. The first components are Barriers to Success (C1) refers to the highlighted barriers that exhibit LCC being used widely in Malaysian construction industry. The second component focus on the benefits (C2) offered when adopting LCC in the costing exercise. While the last component is Costing Success Element (C3). The costing success elements highlight on the criteria to achieve a reliable cost estimate. In the third subsection, it concentrates on the Solution or Strategies (C4) to achieve higher accuracy in cost estimating exercise. Based on the questionnaire survey and further semi-structured interviews conducted, with support from the literature review, five (5) strategies were found to be important including simplified process, pre-recorded data bank, general usage guideline, enforcement and education & awareness training. This component leads to the birth of the next level initiatives which is the introduction of COST_IBS CALCULATOR, a comprehensive application combining all the strategies and the costing success element (C3). The fourth subsection is focusing on the Failure Reduction Activities (C5), which then produce a Costing Improvement parameters (C7) as well as Project Effectiveness (C8). In this subsection, it concentrates on the steps that can be taken into consideration to ensure less failure in cost estimating works. The steps listed are obtained as results of the questionnaire and semi-structured interviews.

The second section concentrates on the details extracted from Costing Success Element (C3) and Failure Reduction Activities (C5). This section called System Development. This level is extracted from C3 contains the parameters to calculate the construction cost, maintenance cost, alteration cost, building interest rate, acquisition cost and the building life cycle (C6). This is done by providing a base of the calculation in a range form. As an example, failure reduction activities no 1 is the selection of price to control price manipulation. Therefore, to ensure the price is in control, range selection is adopted. Price for each item will be in a form of range (the range is obtained from the document analysis, ranging from a minimum to the maximum price per item). The last section of the framework is Decision Level. The component is labelled as P3- Decision level. It contains the project effectiveness parameters which benefit the end user, value for money, project functionality, meet pre-stated objectives and meet client's satisfaction.

The framework proposed will give benefit to Government agencies, in preparing their estimates in the feasibility stage. The outcome gives the alternative to proceed with the investment or not. For contractors, the framework will help to reduce the likelihood of over estimates, hence reducing their chance in winning the bid. For a consultant, the framework benefits them in term of preparing a sound and reliable comprehensive cost estimates which are more realistic and incorporating future cost which leads to a better decision making.

4 Validations of Framework

The expert validation survey was conducted to the above described framework to express as whether this proposed framework leads for successful initiatives of LCC and IBS in Malaysian construction industry. Validation is an official determination that a framework is acceptable with respect to a set of explicit standards for a specific purpose [11]. As for this research, an expert panel opinion validation is utilised as suggested by other researchers [12-16]. The expert validation is obtained using a questionnaire by means of a structured interview. The interview was conducted with ten (10) respondents that involved closely with IBS and have experience in IBS project cost estimating regardless the estimating method. Table 1, portray the expert respondents profile. This includes three (3) Government agencies, four (4) consultants and three (3) contractors, based on the recommendation by [17] that the targeted panels are from different background. They were selected based on two criteria: (1) they were directly involved in the previous data collection (questionnaire and structured interviews) and (2) have pertinent experiences dealing with the construction industry and IBS project cost estimating. The validation aspects used are appropriateness to be implemented in Malaysian construction industry, objectivity, practicality, reliability, completeness to cover all aspects needed in basic cost estimating activities, applicability to a simple project and applicability to a complex project.

4.1 Design of the validation questionnaire survey

To validate the framework developed from this research study, a questionnaire method was adopted. This survey was similar to [18] and [19], wherein their study, a questionnaire study was adopted to validate the success factors model which was developed. The parameters include are objectivity, practicality, reliability, completeness to cover all aspects needed in basic cost estimating activities, applicability to a simple project and applicability to a complex project. The overview respondents were first given the motivation behind the survey, some foundation data, direction on the activity, clarification on the model and the achievement activities of LCC and IBS structure created from this exploration. Toward the finish of activity, the respondents were solicited to rate their degree from fulfilment for each of the seven approval viewpoints as indicated by a scoring scale from 1 to 5, where 1 spoke to "poor" and 5 spoke to "excellent". A frequency analysis by mean score is utilised in determining the degree of agreement from the expert point of view.

Table 1. Profile of validation respondents

Respondent	Type of respondent	Organizations	Respondents experience	Involved with previous data collection
V1	Head of R & D Center	Government agencies	25	Yes
V2	Director	Government agencies	33	Yes
V3	Project Manager	Contractors	20	No
V4	Senior Quantity Surveyor	Contractors	25	Yes
V5	Director	QS Consultant	20	Yes
V6	Managing Director	Contractors & Manufacturer	35	Yes
V7	Senior Quantity Surveyor	QS Consultant	21	No
V8	Director	QS Consultant	18	No
V9	Director	Government agencies	32	No
V10	Senior Quantity Surveyor	QS Consultant	15	No

4.2 Results of the validation

Table 2 shows the results obtained from the validation exercise which was conducted with the respondents mentioned in the previous section. The respondents were solicited to rate their degree from fulfillment for each of the five approval viewpoints as indicated by a scoring scale from 1 to 5, where 1 spoke to "poor" and 5 spoke to "excellent". A score above 3 would represent the satisfactory performance of that aspect. The result shows that the highest mean comes from “Degree of completeness (covers all important aspects for cost estimates and LCC)”, which the mean value of 4.70. And the lowest is “Degree of applicability to a complex project”, with the mean value of 3.30.

Table 2. Results of the validation survey

Validation aspect	Respondents										Mean
	1	2	3	4	5	6	7	8	9	10	
Degree of appropriateness	4	4	5	4	4	4	4	4	3	4	4.00
Degree of objectivity	3	3	3	4	5	4	4	4	3	3	3.60
Degree of practicality	4	4	4	4	4	4	4	4	4	4	4.00
Degree of reliability	4	5	4	4	5	4	4	4	4	3	4.10
Degree of completeness (covers all important aspect for cost estimates and LCC)	5	5	5	5	5	5	4	4	5	4	4.70
Degree of applicability to a simple project	4	4	4	4	4	4	4	4	4	4	4.00
Degree of applicability to a complex project	3	3	3	4	3	4	3	4	3	3	3.30

Respondent V3 highlighted that for a complex project, the researcher could expand the range of material used not only from IBS Standard Catalogue from Construction Industry Development Board Malaysia (CIDB). Respondent V1 suggested that the system should be created in a very simple form, where no complex parameters key in needed. He also suggests to includes the inflation rate into one of the parameters. Therefore, it can be construed that the newly proposed framework for successful initiatives of LCC and IBS in Malaysian construction industry was validated to it appropriateness that fit objective, reliable, complete and applicable to be utilised for a simple and complex project.

5 CONCLUSION

This paper discussed the important of LCC that should be adopted in IBS project. A framework for initiatives that has to be carried out in decision-making for cost accountability is proposed and the later lead to the development of LCC calculator for IBS components. An expert validation survey also conducted to foresee the suitability of the framework. From the result of validation resulted in the newly developed framework for a successful initiative of LCC and IBS in Malaysia construction industry are mainly satisfactory and highly present the degree of completeness of life cycle cost accountability which shown the suitability for its application in a simple and complex project.

This work was supported in part by the BESTARI Grant (600-IRMI/MyRA 5/3/BESTARI (037/2017), Universiti Teknologi MARA, Shah Alam, Selangor.

References

[1] S. M. Shamsuddin and R. Zakaria, "Integrating Life Cycle Costing (LCC) in Industrialised Building Sytem (IBS) in Malaysian Construction Industry," presented at the International Conference on Innovation in Business and Strategy, Universiti Teknologi Malaysia, Kuala Lumpur, 2016.
[2] F. Olubodun, J. Kangwa, A. Oladapo, and J. Thompson, "An appraisal of the level of application of life cycle costing within the construction industry in the UK," *Structural Survey,* **28**, 11, 2010.
[3] M. F. A. K. A. Rashid, *Investigation on Reliability and Validity of Life Cycle Cost (LCC) Data Inputs of Building*. IIUM, Malaysia: IIUM Press, 2016.
[4] D. J. O. Ferry and R. Flanagan, "Life cycle costing - a radical approacg," Construction Industry Research and Information Association London, UK, London1991.
[5] M. R. Nor Azizah and A. Zainal Abidin, "Integratig Life cycle costing model: A literature review," in Conference Proceedings, Management in Construction Researchers Association 9th Annual Conference and Meeting Shah Alam, Selangor, 2010.
[6] P. S. Brandon, *Building Cost Research–Need For A Paradigm Shift?. In Brandon, P.S. (ed.). Building Cost Techniques: New Directions.* London: : E&FN Spon, 1982.
[7] A. B. Nor Azmi, "Exploring the types of construction cost modelling for IBS projects in Malaysia," in International Conference on Industrialised, Integrated, Intelligent Construction, London, UK, 2008.
[8] S. M. Shamsuddin, R. Zakaria, S. F. Mohamed, A. L. Saleh, M. Z. Abd Majid, C. Utomo, *et al.*, "Developing Methodology for cradle to grave cost estimating for IBS in Malaysia," *Jurnal Teknologi (Sciences & Engineering),* **77**, 37-42, 2015.
[9] K. Hunter, H. Subashini, and J. Kelly, "A Whole Life Costing Input Tool for Surveyors in UK Local Government," *Structural Survey,* **23**, 346-358, 2005
[10] O. Mohamed, F. Mohd Nor, S. B. Abdul Karim, and K. M.Y., "The Practice Of Life Cycle Costing (LCC) In The Malaysian Construction Industry- Application During Design Stages Management," in Conference Proceeding in Construction and Researchers Association (MICRA) Meetings and Conference, 28-29 August, Shah Alam, Selangor, Malaysia, 2007.
[11] R. G. Sargent, "Verifying and validating simulations models," in Winter Simulation Conference, USA, 1996.
[12] D. E. Tigelaar, D. Dolmans, I. Wolfhagen, and C. Vlueten, "The development and validation of a framework for teaching competencies in higher education," *Higher education,* **48**, 253-268, 2004.
[13] K. Hyrkas , K. Appleqvist-Schmidlechner, and L. Oska, "Validating instrument for clinical supervision using an expert panel," *International journal of nursing studies,* **40**, 619-625, 2003.
[14] S. Lauesen and O. Vinter, "Preventing requirement defects : An experimeny in process improvement," *Requirement engineering,* **6**, 37-50, 2001.
[15] T. Rosqvist, M. Koskela, and H. Harju, "Software quality evaluation based on expert judgement " *Software quality journal,* **11**, 39-55, 2003.
[16] R. Subramoniam, D. Huisingh, R. Chinnam, and S. Subramoniam, "Remanufacturing decision making framework (RDMF): Research validating using the analytical hierarchichal process," *Journal of Cleaner Production,* **40**, 212-220, 2013.
[17] S. Beecham, T. Hall, C. Britton, M. Cottee, and A. Rainer, "Using an expert panel to validate a requirement process improvement model," *The journal of system and software,* **76**, 251-275, 2005.

[18] E. Cheung, "Developing a best practice framework for implementing PPP projects in Hong Kong," Doctor of Philosophy, School of urban developement, Queensland University of Technology, 2009.

[19] M. Y. Norazian, "Developing a framework for a successful extension of time (EoT) for Malaysian construction industry," PhD in Specialism in Built Environment, Post Graduate Study, Universiti Teknologi MARA, 2016.

Improving construction claim management in Malaysian construction industry

Nor Azmi Bakhary, Hamimah Adnan, Azmi Ibrahim

Universiti Teknologi MARA Shah Alam, Malaysia

Abstract. The construction industry in Malaysia is large and it involves multidisciplinary participants. There is high frequency of disputes arising from construction contract claims. Even with the most expert understanding of construction contract clauses and the most equitable risk-allocation management, claims will continue to present problems if they are poorly managed in practice. The significantly increasing number of construction claims indicates the need for the implementation of an effective construction claim management. This research aims to improve the process of claim in construction industry. This research explores the problems related to the process of claim from contractors' and consultants' point of view by studying common procedure of claim management. The findings highlight on the three key issues. There are claim identification, claim notification and claim documentation. These issues are the most troublesome and contribute to the significant problems in construction claim process. It also appears that a standard and transparent procedure should be put in place whereby contractors would be able to follow to properly prepare the claim. In addition, staff awareness, training in how to document and submit a well-supported claim, and negotiations with clients are becoming a necessity.

Introduction

The construction sector is one of the main economic engine sectors supporting the Malaysian economy. The expansion of this sector has led to the revival of the construction profession, and the promotion and encouragement of new investments. The sector has played a crucial role in extending job opportunities for the Malaysian labour force. Its expansion has generated many jobs for skilled, semiskilled and unskilled workers (Fadhlin, 2004). It is central as being the catalyst for development and a major indicator and determinant of domestic performance in the economy. Construction claims are considered by many project participants to be one of the most troublesome and unpleasant events of a project [1]. The high competition coupled with the sluggish global economy has forced contractors to bid projects with minimum profits to stay in business. In addition to their multiparty nature, projects are becoming more complex and risky. This has placed an added burden on contractors to construct increasingly sophisticated and risky projects with less resources and profits. Under these circumstances, it is not surprising that the number of claims within the construction industry continues to increase [2].

Construction projects can be described as temporary multi-organisations, made up of a large number of different people from different backgrounds, based on different geographical

 DOI: 10.1051/matecconf/201713805003

locations making their contributions at a different time [3]. These people are interdependent but unfortunately, because of their conflicting interests, are often highly insensitive to the needs of others around them. Moreover, construction is not a single, vertically or horizontally organised industry dominated by a small number of national or multinational corporations. It is a large, fragmented, decentralised industry characterised by thousands of small and medium sized companies and material suppliers. In addition to that, thousands of architects, engineers, bankers, insurance brokers, lawyers, accountants, public administrators, government inspectors, and contracting personnel play a key role [4].

Construction activity is a complex process involving many disciplines with differing skills. Therefore, problems are bound to arise in undertaking complex projects, such as a construction project, due to the widely differing values and goals among the project participants [5]. Some of these problems have the potential to escalate to become significant disputes. In addition, construction, like many other industries in a free-enterprise system, has sizeable risk built into its profit structure [6]. Although all the parties in a construction contract start with the best of intentions to get the work both complete satisfactorily in the agreed time and at the least expense to owner, whilst ensuring that the general contractor and all other specialist contractors and suppliers make a reasonable profit. Somewhere between the beginning and the end, disagreement, disputes, disruption and delay arise which can destroy the best of intentions. Therefore, the construction industry has a poor reputation for coping with risks, failing to meet deadlines and cost targets. Hence, the clients, the contractors, the public and others have suffered as a result. Construction claims have such high impact on construction project's cost and time that an effective claim management system in any construction organisation deserves serious attention [7].

Claim

A claim as a right given to the party who deserves a request for compensation for damages incurred by the other party. Claim is a request by a construction contractor for compensation over and above the agreed-upon contract amount for additional work or damages supposedly resulting from events that were not included in the initial contract [8]. In any construction project, significant additional costs can be experienced by the contractor, the owner, or both, due to the actions of the other party or parties involved. Disputes over the right to a compensation as well as over the amount of time and/or money to be given often necessitate a resort to litigation, arbitration, or other forms of dispute resolution methods for settlement [9].

A construction claim arises when a party to a construction contract believes that in some way, by act or omission, the other party has not fulfilled its part of the bargain [6]. To put it in other words, a claim arises when one party to the contract has suffered a detriment for which that party should be compensated by the other party. Therefore, a construction claim is an assertion of and a demand for compensation by way of evidence produced and arguments advanced by a party in support of its case [10].

Many public agencies have the perception that "claims" are attempts by contractors to pursue exaggerated demands for payments, compelling the government agency to defend itself accordingly [4]. To most contractors, the term "claim" evokes a negative reaction because it signifies the necessity to participate in a frustrating process in order to gain the relief to which one is contractually and equitably entitled usually after suffering an economic loss [11]. Although public and private entities arrive at their negative view of claims from entirely different perspectives, those perceptions are intensified by an ever-increasing resentment of the consistency of the process of dispute resolution which is customarily utilized in public contract claim resolution [8].

Result

Claim Identification

It is best to avoid construction claims from the beginning of a project and resolve claims quickly and efficiently once they arise. Identification of any changes needs to be carried out early and distribution of the information to all stakeholders is a necessity. This will help any decision-making process by all parties. Early identification of potential claims is necessary to successfully prevail on claim disputes arising during construction, particularly those involving additional work and compensation. Proper claim management begins with identification of a claim. The contractor must be able to recognised and identify a claim situation goes together with notification. Most public and private contracts contain clauses requiring notification of differing site condition, changes and delay within a stated period before equitable adjustment can be pursued. It is very important to identify the source of claim. Typical sources of disputes and claims are worth noting.

A successful claim process needs people with "claims consciousness," familiarity and awareness of potential claim situations. Although the identification of claim is automatic with the direction of change orders from the client, the contractor must be able to recognize and identify a claim situation when it first develops, not after it has become a controversy. Prompt identification and notification are very important to fulfil the contractual requirements. The contractor cannot wait too long to take actions. All rights of the contractors to claim can be lost. It may also be practical to look for assistance from expert consultants at an early stage of the claims process. A project manager must take a proactive role in the early identification of potential claims and disputes. He is the best person to evaluate the progress of the work and identify any developing problems on the project. Besides that, management personnel need to have a basic understanding of the terms and conditions of the contract documents. The failure to identify a claim until the end of the project may expose the contractor to contractual defenses. Therefore, project managers must be vigilant in identifying potential claims during construction.

The project staff also must have a good working knowledge of the contract documents. Familiarity with relevant technical and general terms is essential for project personnel to recognised their contract rights and duties. Communication ideas and problems is essential for efficient and effective claim management. The project staff must be knowledgeable and aware of actual conditions that differ from those expected such as soil types, interference of structures, other contractors, traffic, weather and labour are likely to be sources of changed condition. Project staffs also need to have a working familiarity with the legal concepts and right that will affect the outcome of potential claim situation.

Claim Notification

Identification of claim must be followed by notification. It is frequent for construction contracts to require a party to give timely notice of claims to the other. Most standard form contracts provide some mechanism and contain clauses explaining the process of giving such notices and the likely consequences that will arise if notices are not given as specific in the contract. The obvious purpose of notification clauses in a contract is to ensure that both parties have on record the dates and facts that initiate a claims situation, and to protect their respective rights. Notice clauses are useful because they allow the owner of the construction project more time to react to claim problems.

The procedures and reporting systems adopted by any contractors to notify claims need to be systematic. The system should ensure that all claims are notified to owner as soon as possible after the project manager become aware of any complaint or dispute, written or oral, alleging or suggesting any deficiency in the provision of the services. On the other hand, the owner

DOI: 10.1051/matecconf/201713805003

need to make it clear what the contractor has to do if it hits problems such as delay and needs to extend the completion date or claim additional money. Most of the construction contracts require written notice for changes, differing site condition, extra work or other events which may affect the contractor's time and cost performance. A notice clause should be precise and unambiguous about what is necessary and the contractors should fulfil the conditions of the notice clause.

Prompt notification is very important to avoid the possibility of breaching contract policy conditions by late notification. Late notification is always arising from misconceived view that problem will eventually resolve itself, embarrassment of the individual and/or the contractor concerned, concern for the personal consequences of the individual involved and fear that the facts will be made public with the attendant risk to the contractor's reputation.

Claim Documentation

A complete and strong claim document is important in presenting a claim and to resolve any disputes. Contractors must comply with the contract requirements such as the notice and deadlines for submitting any documentation supporting the claim. To have the strongest documentation, the information should be contemporaneously, documenting and closing out the work as it is performed. Information should be consistent and understood. The process of preparing and presenting the claim document is very important because it requires the contractor to refine and produce the claim from beginning to end. The claim must be supported with all the required documents to explain the dispute in a simple, complete and comprehensive approach. This documentation should include the method and date of correction, with approval by the involved parties. The writing style should be clear, precise and easy to be understood. Detail pricing the claim and supporting the damage calculations is very important. Charts, graphs, drawings, and photographs are very helpful and should be included into the claim document. Systematic photos or video of completed work, testing conducted or other quality control activities provide unquestionable visual evidence of the actual conditions.

Documents also should include specifications, special condition, specific instruction as well as the contractor's calculation and bid preparation documents are generally acceptable evidence as to what was intended by a construction contract. Make sure a project diary is kept along with diaries for key personnel and should record the weather, manpower, visitors and contractors on site, key deliveries and any notable event such as problematic or hidden site conditions or events that may cause delay or affect productivity. Well maintain an as-planned schedule and regularly update it with an as-built schedule are vital. Record all key events, especially ones that may lead to a claim, and specifically record when the event occurred, what it was, who noticed it, the projected impact it may have on cost and time, whether notice was given and to whom and response to notice. Contractor should ensure that a fully signed written agreement is in place before commencing work on a project. While oral agreements are generally enforceable, written ones are easier to prove.

The contractor should regularly update the record set with the goal of capturing all changes as they are approved and with sufficient detail so that there is less remembering necessary when the final record set is compiled. Even before a claim is identified, it is important that the contractor keep good records of its activities on the project. Most documentation problems stem from, three basic conditions:

- Getting the proper amount and type of information.
- Getting this information to the appropriate individual.
- Delivering it while it is timely.

When these considerations are addressed properly, those responsible for problem resolution will be able to respond more effectively as project challenges arise. Beyond maintaining good records of daily activities and of the project in general, the contractor must be both proactive and reactive in developing claim documentation. With respect to proactive documentation, the contractor should take immediate steps to create a record of potential claims as they arise. It is essential that the contractor respond promptly in writing to any communication concerning alleged facts or positions that may impact the viability of its claim. Where the owner, general contractor, construction manager or other relevant project participant disputes the claim or indicates that the problem has been addressed, the contractor must respond in a timely manner.

Conclusion

It was found that by having a proper claim management system, contractors can utilize it to improve their claim management process. This is to guarantee that they will have proper management setup to justify, quantify, and present claims for events under the control of the owner or his agents then chances of prolonged disputes are reduced. The results in this study also can be used to assist contractors in adjusting to a business environment that demands measuring business processes to form a basis for continuous improvement. A need for an overall step-by-step procedure for claims analysis and administration is very crucial for achieving proper resolutions and for preventing claims from developing into disputes.

References

1. Winch, G. M. (2010). Managing construction project – 2nd edition. Sussex: Wiley-Blackwell.
2. Zaneldin, E.K. (2006). Construction claims in United Arab Emirates: Types, causes and frequency. International Journal of Project Management, 24(5): 453– 459.
3. Lucko, G. & Rojas, E. (2010). Research Validation: Challenges and opportunities in the construction domain. Journal of construction engineering and management. 136(1), 127-13
4. Ishan, Z., Mustaffa, N.E. & Jaafar, M.S. (2012). A revisit on the current practice of dispute resolution and ADR in the Malaysian construction industry. Journal of Design and Built Environment, Vol.10, June.
5. Thomas-Mobley, L., Oberle, R., and Kangari, R. (2007)." Design and Construction Challenges of a Federal Laboratory Building: A Case Study." J. Archit. Eng., 13(4), 224–229.
6. Tadayon, M., Jaafa, M. & Nasri, E., 2012. An Assessment of Risk Identification in Large Construction Projects in Iran. Journal of Construction in Developing Countries, Volume Supp. 1, pp. 57-69.
7. Chovichien V. & Tochaiwat K. (2006). Information System for Managing Employer's Construction Claims. Technology and Innovation for Sustainable Development Conference (TISD2006).
8. Assaf S.A., Al-Hejji S. (2006). Causes of delay in large construction projects. International Journal of Project Management, 24(4), 349–357.

9. Abdul-Malak, M., El-Saadi, M., and Abou-Zeid, M. (2002). Process Model for Administrating Construction Claims. Journal of Management Engineering. 18(2), 84–94.
10. Kululanga, G., Kuotcha, W., McCaffer, R., and Edum-Fotwe, F. (2001). Construction Contractors' Claim Process Framework. Journal of Construction Engineering and Management. 127(4), 309–314.
11. Ibrahim, M. and Madon, Z. (2006). Understanding contract document. Proceedings of the 6th Asia-Pacific Structural Engineering and Construction Conference. C12-C18.

Construction Worker Fatigue Prediction Model Based on System Dynamic

Tri Joko Wahyu Adi[1,*], *Lila* Ayu Ratnawinanda[1]
[1]Department of Civil Engineering, Institut Teknologi Sepuluh Nopember, Indonesia

Abstract. Construction accident can be caused by internal and external factors such as worker fatigue and unsafe project environment. Tight schedule of construction project forcing construction worker to work overtime in long period. This situation leads to worker fatigue. This paper proposes a model to predict construction worker fatigue based on system dynamic (SD). System dynamic is used to represent correlation among internal and external factors and to simulate level of worker fatigue. To validate the model, 93 construction workers whom worked in a high rise building construction projects, were used as case study. The result shows that excessive workload, working elevation and age, are the main factors lead to construction worker fatigue. Simulation result also shows that these factors can increase worker fatigue level to 21.2% times compared to normal condition. Beside predicting worker fatigue level this model can also be used as early warning system to prevent construction worker accident

Keyword: Construction Safety; Worker Fatigue; Prediction model; System dynamic

1 Introduction

The construction industry has a very important role in the development of a country. The unique characteristics and the dynamic nature of the construction industry lead to a dangerous condition and prone to accidents. Construction industry becomes the riskiest industries compared to other industry, because construction industry is based on the workers as its main power source. The use of man power become one of the reason, why construction industry suffered occupational accident more than other industries. As seen in Fig. 1, the number and rate of construction worker accidents in 2013 is very high. In Indonesia, the death rate due to accidents in the in 2015 increased by 4% compared to 2014 [1]. The problem of occupational safety and health are often ignored. Unsafe behaviour of workers is the main cause of 88% of accidents in workplace in construction industry, 10% due to unsafe conditions, and 2% are unpreventable [2]. The unsafe workplace conditions significantly determined the types of accidents and severity of injuries, while safe work equipment was essential to the complexity of the construction activities. The unsafe condition of the equipment refers to hazards that are not detected in the security apparatus.

* Corresponding author: trijokowahyuadi@gmail.com

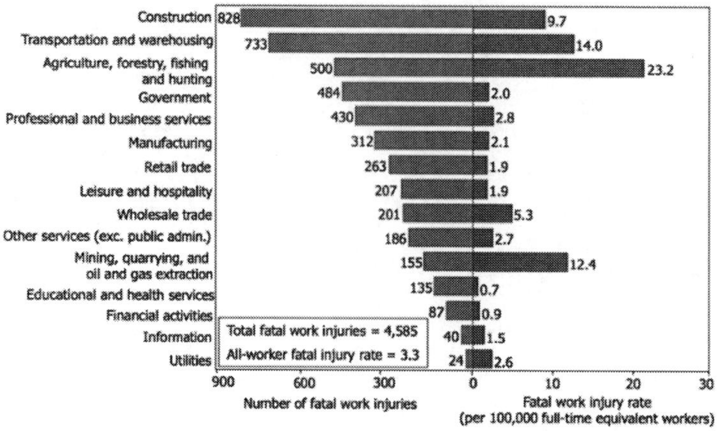

Fig. 1. Fatal occupational accident among industries

Construction accidents can negatively impact the project, including the disruption of work schedules, cost overruns, psychological effects for workers and company reputation. Tight schedule of the project forces the construction worker to work in a high pressure condition. These situations can trigger stress and fatigue to construction worker. Excessive fatigue level will trigger the occurrence of human error and unsafe act which lead to worker accidents [3,4,5,6]. This paper aims to propose worker fatigue prediction model based on system dynamic. Beside predicting worker fatigue level this model can also be used as early warning system to prevent construction worker accident

1.1 Factors causing work fatigue

Construction accidents are caused by three main factors, namely: *unsafe acts*, *unsafe conditions* and *unsafe equipment* [3,4,5,6]. Factor of human error can lead to unsafe acts. According to statistics, 80% of the incidence of human error, 50% of them are caused by fatigue [4]. Fatigue is the body's protective mechanisms to avoid further damage. Fatigue is a condition characterized by feelings of fatigue, decreased alertness, and decreased worker productivity. According to Oesman & Simanjuntak [7], fatigue can be caused by internal factors and external factors. Internal factors are the characteristics of workers, such as gender, age, weight, height, nutritional status and lifestyle; while external factors include workload and working conditions [2,7,8,9,10], revealed that the excess load factors greatly affect worker fatigue. 32% of workers who lift heavy items between 75-100 kg will experience back pain.

Age increases susceptibility to respiratory disease. Respiratory problems are also triggered by an unhealthy lifestyle, such as smoking and alcohol consumption. Empirical data also show that age will affect psychomotor ability and cause a decrease in muscle strength. Productive age with maximum muscle strength is between 25-50 years. At age 45, the strength of the worker remaining only 70% -80% [4,7]. At every 0.4 °C increase in body temperature, an increase in pulse rate 5 times/ minute. So that the effect on worker fatigue is 1/0.4x5 = 12.5 times/minute. (12.5/60) x 100% = 20.8%. Based on interview with occupational health and safety expert in construction project, alcohol and cigarettes also affect the fatigue of worker by 10% [8].

According to Maurits, as noted in the Limbong's paper [11] external factors that causing fatigue associated with the work environment; such as weather, wind pressure, noise, and sun exposure. While fatigue associated with work stress, include: long working hours (overtime), rest periods and improper work shift arrangements [8,12]. Working at height makes the job needs more oxygen. Lack of oxygen made unstable working conditions and reduced vision. As a result, could endanger workers. Workers falling from a height, such as the roof and scaffolding, is the largest contributor of accidents in the construction industry [13]. Rest will eliminate fatigue. Rest done at break time, will return the stamina of workers by 70% [14]. In normal working hours, the lighting depends on the elevation of the workplace. According to Chang, sun Chuang and Hsu, changes in floor elevation (6 to 10 floors) will have an effect of lighting degradation of 3%. [8]. High ambient temperature and humidity can cause the heart to pump more blood. As the result the pulse rate increases 4-10 times per minute. If it is taken an average of 7.5 times / min at a temperature of 29 °C it gives an effect of 12.5%

1.2 Method for measuring work fatigue

Prediction of fatigue, is done by measuring several indicators of fatigue. In general, the methods for measuring fatigue is divided into six groups; they are: 1. measurement based on the quality and performance of workers, 2. measurement based on subjective impressions, 3. measurements using electroencephalography (EEG), 4. measurement based on subjective observation of eye blink frequency, 5. measurement of psychomotor, and 6. mental testing. Measurement of work fatigue is conducted before work, during work and after work activities. Based on these tests, levels of fatigue and sources of work fatigue can be inferred [2].

Measurement of fatigue can also be done with the direct measurement method (EMG). EMG uses indicators of fatigue such as changes in heart rate, changes in body temperature, changes in breathing rate, changes in visual ability and changes in muscle strength, to predict level of fatigue [8,15]. Measuring the level of worker fatigue, can interfere the activities of the workers. Workers feel disturbed by the equipment installed, resulting in bias result due to the perceived inconvenience of workers. However, by doing objective measurement, realistic and real-time results can be obtained [8,15, 16,17,18,19,20,21,22,23].

2 Model Construction

2.1 Causal Loop Diagram (CLD)

This paper proposes construction worker fatigue prediction model based on system dynamic (SD). System dynamic is a computer-aided approach to understand the nonlinear behavior of complex systems over time. More detail information of SD can be read in [24]. In this model, system dynamic is used to represent correlation among variable and to simulate the fatigue level of construction worker. 93 construction workers from 3 high-rise building construction project in Indonesia were used as respondents. To validate the model a high rise office building project in Surabaya-Indonesia is used as case study. Causal Loop Diagram which represent the correlation among variables is derived from in-depth literature review and validated by occupational health and safety expert. In CLD, variables are divided into internal and external factors. Internal factors consist of: worker characteristic such as age, lifestyle, height, weight of the worker and physical condition such as body temperature, breathing, pulse and vision. External factors consist of: project condition such as workload, work elevation, equipment noise, weather condition, wind velocity, sun exposure, ambient temperature, and worker schedule such as work shift arrangement, over time and recess. Based on internal and external factors, a Causal Loop Diagram which represent the relationship among variables and stock flow diagram are presented in Fig. 2 and Fig. 3 respectively.

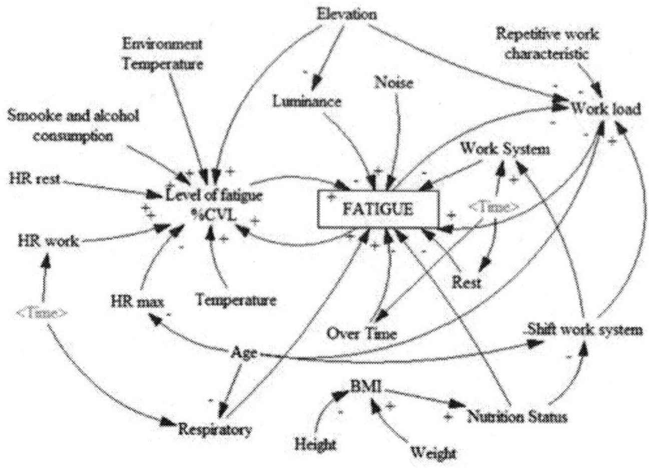

Fig. 2. Causal Loop Diagram for construction worker fatigue model

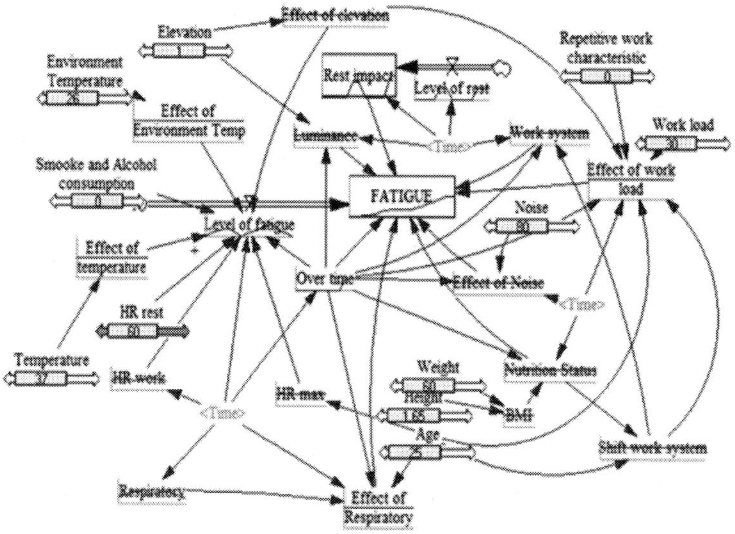

Fig. 3. Stock flow diagram of Construction worker fatigue model

Fatigue level of the worker is predicted using Percentage of Cardio Vascular Load (%CVL) indicator. The classification of the CVL condition can be seen in Table 1. In SD simulation, Total fatigue level of construction workers is formulated in equation 1.

Table 1. Classification of % CVL related to worker fatigue

% CVL	Classification of CVL
<30%	No Fatigue
30% - 60%	Need to be recharged
60% - 80%	Allowed to work in short period
80% - 100%	Need special treatment
> 100%	Not allowed to work

$$Total\ \%CVL = \%CVL\ internal\ factor\ effect + \%CVL\ external\ factors\ effects \qquad (1)$$

DOI: 10.1051/matecconf/201713805004

3 Model Implementation

3.1 Data collection

To collect data, internal and external sensor are used. Sensor node A (Internal/Body sensor) are used for collecting real time worker data such as: heart beat data, body temperature, respiration and working elevation. While sensor node B (External/Environmental sensor) are used to collect real time environmental information such as, noise, and luminance. The mechanism of the sensor is depicted in Fig. 4.

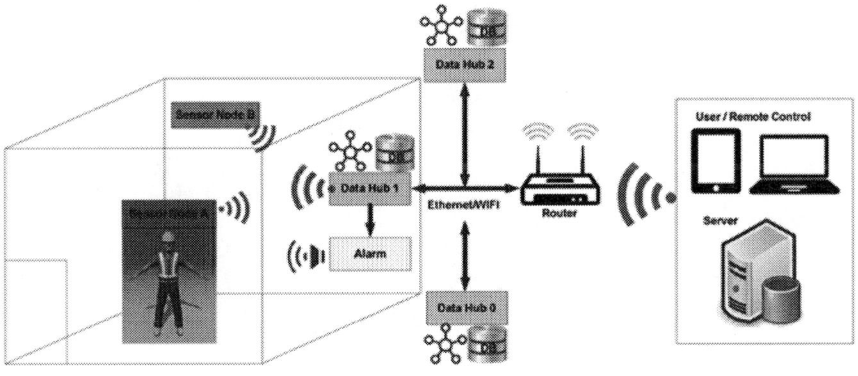

Fig. 4. Sensor system Network

80% of 93 construction workers ware used to build model equation. Input data from body sensors and environmental sensors are collected by Data Hub 1. These information is then stored as database in Data Hub 2. Further process is done by calculating level of worker fatigue, using SD model proposed in Figure 3 and Equation 1. Final result is transmitted to alarm when level of fatigue higher than 60%.

3.2 Model validation

To validate the proposed model, 20 construction workers from a high rise building project are used as case study. Validation is done by comparing worker fatigue prediction and observation result from sensor data in the normal condition. Data were taken every hour from construction worker and converted to %CVL. As seen in Fig. 5, the proposed model provides quite good prediction of % CVL. The average error is 12.6%.

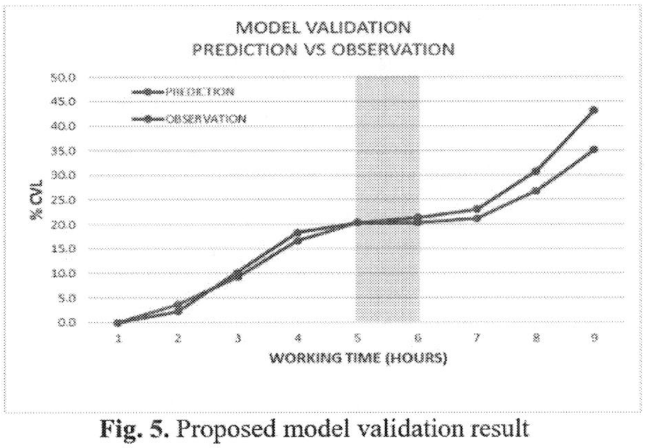

Fig. 5. Proposed model validation result

DOI: 10.1051/matecconf/201713805004

3.3 Simulation scenarios

To implement the proposed model, 2 simulation scenarios are designed to see the impact of internal and external factor to the construction worker fatigue level. First scenario is designed for comparing worker fatigue in normal load condition (load < 40 kg), versus worker fatigue with excessive load (varies load between 40-75 kg). The second scenario is designed by increasing value of internal and external factors such as age (>40 years old) and working elevation (> 6thfloor). In both scenarios, every worker is given break for 1 hour. The scenarios result can be seen in figure 6 to 7 respectively.

In Fig.5, it can be seen that after 2 hours the fatigue level of construction worker with excessive load increase significantly. The average of fatigue level increase 64.3%.

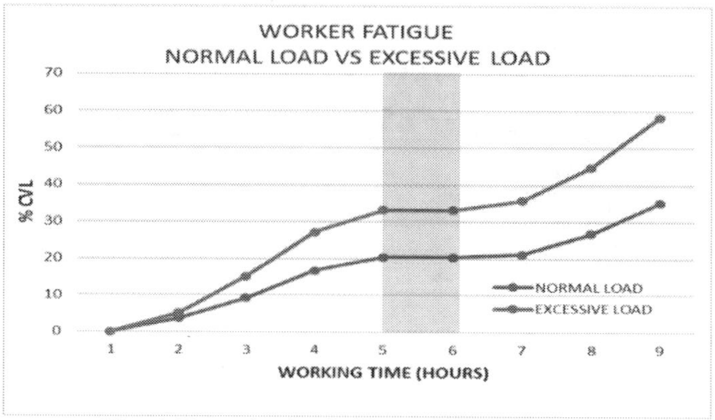

Fig. 6. Excessive load effect to worker fatigue

Fig. 7 shows the change in worker fatigue rate compared to normal conditions. Model simulations with changes in internal factors (age, nutrition status, body temperature and smoke and alcohol consumption) showed increase in fatigue rate 39.5%. Based on the scenario on the changes of external factors (work system, work load, noise, luminance, environment temperature, over time and elevation), worker fatigue level increase 60.7%. While the effect of simultaneous changes between internal and external factors the fatigue rate increase 74.2%. It can be concluded that the more complex the changes in internal and external variables, fatigue level increased significantly.

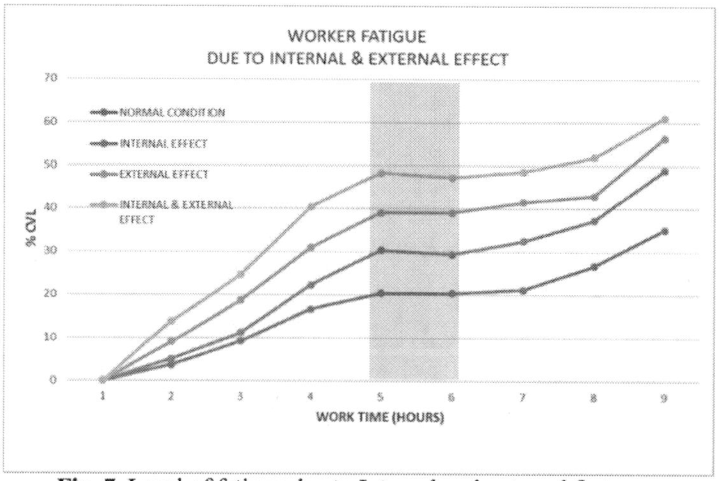

Fig. 7. Level of fatigue due to Internal and external factors

4 Conclusion

This model is under development as part of construction occupational health and safety monitoring system. The proposed model presents the relationship among internal and external variables that cause fatigue of construction worker. To predict the level of fatigue, percentage of Cardio Vascular Load (%CVL) is used as main indicator. Simulation result shows that excessive work load, working elevation, and age are the factors that most influencing construction worker fatigue. It also shows that the effect of external factors is higher compared to internal factor.

References

[1] Anto, A. Erma, S. J. Industrial Eng. **7 (1)**, pp. 83-90, (2010)
[2] Cahyani, W. D. J., *J. Knowledge and Tech.*, (2010)
[3] Cheng, C.W. Lin, C.C. Leu, S.-S. *Safety Science* ***48***, 436-444., (2010)
[4] Fadel, M. Muis M. Russeng, S. S., Master Thesis, unhas.ac.id, (2014)
[5] Han, S. Lee S. Mora, F. P. *J. Comput Civil Eng. 27 (6)*, 635-644. (2013)
[6] Haslam, R., Hide, S., Gibb, A., Pavitt, T., Atkinson, S., & Duff, A. *Appl. Ergon.* ***36 (4)***, 401-415., (2005)
[7] Oesman, T. I. Simanjuntak, R. A. *Proc. 11th Nat'l Conf. of Indonesian Ergon. Society*, (pp. 268-276), (2011)
[8] Chang, F. L. Sun, Y. M. Chuang, K. H. Hsu, D. J. *Appl Ergon*, **1(6)**. (2008)
[9] Javaid, N., Faisal, S., Khan, Z., Nayab, D., & Zahid, M. *IEEE 8th Int'l Conf. on Broadband and Wireless Comp.,* Compeigne, France, (2013)
[10] Irma, M., Russeng, S. S., & Wahyuni, A. Master thesis, Hasanudin university, (2014).
[11] Limbong, N. Y. Josephus, J. Kawatu, P. A. Master thesis, Sam Ratulangi university, (2015)
[12] Hallowell, M. R.,Professional Safety, (2010)
[13] Goh, Y. M. Sa'adon, N. F. *J Constr Eng Manag.* **141 (16)**., (2015)
[14] Indonesian labor law, Indonesian Labor Law, article 13, (2013)
[15] Wang, D. Dai, F. Ning, X. J. Constr Eng Manag. **141(6)**, (2015)
[16] Fang, D. Jiang, Z. Mingzong, Z. Han, W. *Saf Sci 73*, 80-91., (2015)
[17] Lin, Y. H. Ho, S. C. Lai, C. Y. *Proc. 19th Congress of the IEA*, (pp. 1-6) Melbourne., (2015)
[18] Mocci, F. Serra, A. Corrias, G. *Occup Environ Med* ***58***, 267-271., (2001)
[19] O'Neill, C. Panuwatwanich, K. *Proceedings of the 4th International Conference on Eng. Proj., and Prod. Manag. (EPPM, 2013)*, (pp. 993-1005)., (2013)
[20] Fang, Y. C. Dzeng, R. J. *Creative Construction Conf. 2014, CC2014* (pp. 147-156). Procedia Engineering, (2014)
[21] Hsu, D. Sun, Y. M. Chuang, K., Juang Y. Chang, F. *Saf Sci* ***46***, 833-843. (2008)
[22] Isaac, S. Edrei, T., *Auto. in Constr.* ***63***, 66-78. (2016)
[23] Li, H. Chan, G. Huang, T. Skitmore, M. Tao T. Y. Luo, E., et al., *Automat Constr.* ***55***, 58-65., (2015)
[24] Katsuhiko, O. Pearson Education Limited, (2014)

 DOI: 10.1051/matecconf/201713805005

Model of Construction Waste Management Using AMOS-SEM for Indonesian Infrastructure Projects

Elizar[1,*], *Suripin*[2] , and *Mochamad Agung Wibowo*[2]

[1]Engineering Department, Islamic University of Riau, Pekanbaru, 28284, Indonesia
[2]Engineering Department, Diponegoro University, Semarang, 50275, Indonesia

Abstract. Construction waste is being produced by the infrastructure project. Thus, waste minimization is an important area to implementation of the construction waste management in Indonesia. This paper aims to build model of construction waste management and to determine the significantly of factors that can improved construction waste management for Indonesian infrastructure projects. This study identified 20 factors to improve the construction waste management that grouped into 5 groups, they are policy, asset, technology, human resources and knowledge. Methods of research carried out by site observation and discussion with contractors using questionnaire instrument and focusing to contractors engaged in various type infrastructure projects in Indonesia with amount of 383 respondents. The 5 groups of factors were developed into AMOS-SEM model to determine significant level in contributing to construction waste management. Data was analyzed uses the Structural Equation Modeling (SEM) techniques with the aid of Analysis of Moment Structures (AMOS) statistical package to overcome these factors, where a family of statistical models that seek to explain the relationships of covariant among multiple variables were formulated by combining a measurement model with structural model into a statistical test. The model will be recommandation to improve construction waste management for Indonesian infrastructure projects future.

1 Introduction

The construction industry is one of the important in Indonesia because it provides various infrastructure that serves to support various economic activities of the community. Crucial issues facing the Indonesian nation today is the condition of inadequate infrastructure, both quantity and quality. One of the infrastructure that describes this phenomenon is the road infrastructure [1]. These condition depend on the project management system implemented by the contractor and owner of the infrastructure project. The low commitment of the contractor to the contract resulted in a significant delay. Such facts can be identified as construction waste that has an impact of the cost, time and quality

*elizar2011@gmail.com

of the infrastructure project management system. In the activities of many construction projects are faced with various problems such as low productivity [2], lack of supply capacity [3], low technology innovation [4]. These problems either directly or indirectly result in significant waste in the construction process [5]. Base on the chain of analytical processes of waste management and waste control business, the identification of seven factors of success criteria for waste management, they are waste management regulations, low waste technology, waste management system, awareness of waste management, reducing design change, development of science waste and waste management practise [6]. The concept of waste management and identification methods helps the management team to control and manage waste on construction project [7]. In this paper discusses construction waste management model which aims as a recommendation to reduce waste waste during the process of implementation of infrastructure projects. The relationship of component witch to support of construction waste management are analysed using Analysis of Moment Structures (AMOS) based on the concept of Structural Equation Model (SEM). Conclusions based on the model results are presented. Last, this paper presents other significant construction waste management of research that could potentially benefit from AMOS-SEM modelling techniques.

1.1 Construction Waste Management

Construction waste management is effort to minimize construction waste on the construction phase from the project feasibility until the project maintenance. This can be achieved by educating stakeholder and the benefits of implementing construction waste management system for effectiveness of waste reduction [8]. Success factor of construction waste management system can be done by among others; waste management regulation, waste management system, awareness of waste management, low-waste building technologies, fewer design changes, research and development [6]. Approaches that consider the identification of the application of methodologies depend on the type of problem at the stage of the process. Knowledge management methodology is a approach using the quality and quantity of available information. The methodology divides knowledge about a process into five main classes which contribute in different levels of waste minimization; crucial knowledge is the one responsible for the initial description of the process, prior knowledge refers to information obtained previously about objects, actors and actions needed to perform, subordinate knowledge is the one without which, the performance of the process might be unsatisfactory or unsuccessful, contextual knowledge refer to the contribution of any situation affecting the perception of the actors or performance of the objects and causal knowledge is identified in the form of relationships between object, actors and variables [9]. The factors of commitment of the parties involved in the implementation of waste management system [10]. Likewise with adequate working environment, reliable cooperation, practice and health workforce can improve the construction waste management system [11]. The success of construction waste management is also influence by regulatory factors, systems, technology, and awareness on waste management [6]. Technology identification and right technology is also influence on construction waste management [7]. Furthermore, development of construction waste management needed to build of contractor commitment [12], on the other side, Waste management plan has the most important effects on the assessment waste of equipment, labour and time [13]. With good communication and documentation system, construction waste can be identified and reduced [14].

DOI: 10.1051/matecconf/201713805005

1.2 Structural Equation Model (SEM)

Structural Equation Model (SEM) is a statistical technique to test and estimate causal relationships by integrating factor and path analysis. SEM can examine a series of dependence relationships simultaneously. It is particularly useful in testing theories that contain multiple equations involving dependence relationships. SEM is the development of General Linier Model (GLM) with multiple regression as main part. SEM is more reliable, illustrative and robust than regression techniques when modelling interaction, nonlinearity, measurement errors, error term correlations and correlation among multiple independent variables [15]. Although SEM models can be tested in different ways, all structural equation model are distinguished by three characteristics [16]:

1. Estimation of multiple and interrelated dependence relationships
2. An ability to represent unobserved concept in these relationships and account for measurement.
3. Defining a model to explain the entire set of relationships Structural Equation Modelling can examine a series of dependence relationships simultaneously.

SEM provides a useful way in which to determine whether observed data concur with a priori hypotheses on the structure of incentives [17]. Figure 1 illustrates a simple SEM model incorporating both the measurement and structural relationships of two constructs with four indicators each.

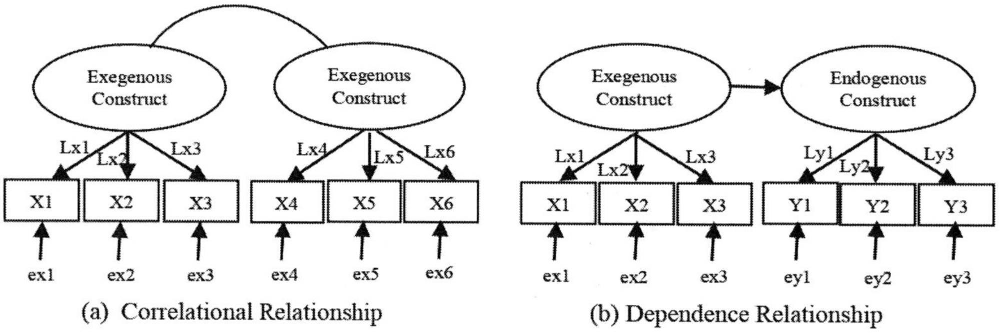

Fig. 1. Visual representation measurement and structural model relationships in a Simple SEM Model [16]

In Figure 1a, there is a correlational relationship between the two constructs, indicated by the curved arrow. The indicators (four on each construct) are labeled *X1* to *X8*. Figure 1b depicts a dependence relationship between the exogenous and endogenous construct [16].

1.3 Analysis of Moment Structures (AMOS)

Analysis of Moment Structures (AMOS) is statistical package to overcome these issues, where a family of statistical models that seek to explain the relationships among multiple variables were formulated by combining a measurement model with structural model into a simultaneous statistical test [17]. AMOS is included in Covariance Based Equation Modelling (CB-SEM) analysis. Commonly, CB-SEM aims to estimate the theoretical structural model and examine the clausal relationship between constructs and measure model feasibility and confirm according to empirical data [18]. The model can be said fit if it meets three criteria of standard measurement. The criteria in the Goodness of Fit (GOF) measurement are absolute fit measures, incremental fit indices and parsimony fit indices, Table 1 shows the GOF characteristic [16].

Table 1. Model fit measurement indices [16]

Characteristic	Good of Fit Measures	Cut-off Values Based
Absolute fit indices	CMIN/DF	< 5
	GFI	0 to 1
	RMSEA	0.05 to 0.08
Incremental fit indices	AGFI	0 to 1
	TLI	0 to 1
	NFI	0 to 1
Parsimonious fit indices	PRATIO	0 to 1
	PNFI	0 to 1
	PGFI	0 to 1

Table 1 shows the model fit measurement indices. Absolute fit indices is a direct measures of how well the model specified by researcher reproduces the observed data and analysed by CMIN/DF is the value of chi square divided by degree of freedom, Goodness of Fit Index (GFI) was an early attempt to produce a fit statistic that was less, Root Mean Square Error of Approximation (RMSEA) one of the most widely used measures that attempts to correct for the tendency of the chi square. Incremental fit indices, the most common baseline model is referred to as a null model and analysed by Adjusted Goodness of Fit (AGFI) is the development of a GFI to the degree of freedom ratio for a proposed model with a degree of freedom for the null model to as a null model, Tucker Lewis Index (TLI) is actually a comparison of the normed chi-square values for the null and specified model, Normed Fit Index (NFI) is comparison between model proposed with null model. Parsimonious fit indices is improved either by a better fit or by a simpler model and analysed by parsimony ratio (PRATIO) is calculated as the ratio of degrees of freedom used by a model to the total degrees of freedom available, Parsimony Normed Fit Index (PNFI) is a modification of NFI, Parsimonious Goodness of Fit Index (PGFI) is a modification of GFI [16].

2 Research Method

The research method was done by observations and interviews with the contractors who involved in infrastructure project using the questionnaires. Variables and indicators was collected based on observation and literature review from selected study have been done in past nearly 10 years. Although these studies are not definitely similar about the purpose and method of survey, the comparison is useful for understanding the problems of construction management in infrastructure project due to waste problems in selected studies as shown in Table 2.

Table 2. Comparison between selected studies

Study	Construction Waste Management Variables				
	1	2	3	4	5
Indonesia, this study [19]	Policy	Asset	Technology	Human Resources	Knowledge
Vietnam [14]	waste management ability	conduct training	audit and provide	sequence activities	punishmen ts
China [11]	physical working environment	operative	safety	practitioners	long-term health
China [6]	waste management system	low waste technology	fewer design change	research	practice
Korea [10]	manpower	material and equipment	construction method	management practice	industry policy
UK [9]	crucial knowledge	prior knowledge	subordinate knowledge	contextual knowledge	causal knowledge

Table 2 shows the comparison construction waste management variables. In this study, each variables have indicator that to support construction waste management. The indicator is shown as in Table 3.

Table 3. Variables and indicators of construction waste management

Variable	Indicator
Policy	construction waste management rule
	construction waste management certificate
	construction waste management incentive
	reward and sanctions
Asset	construction waste management operational
	HR asset : creative, initiative, innovative
	Social asset : care of environment
	Equipment asset : management system
Technology	Construction waste management information
	Data base of waste construction
	Low waste of construction method
	equipment innovative
Human Resource	Degree of study
	Skill Certificate
	Commitment
	Analysis capability
Knowledge	waste construction of knowledge
	Construction waste management of knowledge
	Reduce, reuse and recycling of knowledge
	Improve skill of knowledge

DOI: 10.1051/matecconf/201713805005

Table 3 shows type of variable and indicator for construction waste management which uses in questionnaire. The respondents are requested to select using liker scale of option between 1 to 5 terms of relative importance. A score of "1" represents "not influence" to improve construction waste management whereas a score of "5" represents "very large influence" to support of construction waste management. The data was collected from 383 respondents and estimated the correlation model constructed through the Analysis of Moment Structures (AMOS) version 22.0 for windows.

3 Result and Discussion

This study create analyses measurement of construction waste management model aims to determine the effect of relationships between indicators. The model of construction waste management serves as a recommendation for improving the infrastructures project in Indonesia. Construction waste management model have five components, they are policy, asset, technology, human resources and knowledge. Correlation between components as shown in Figure 2.

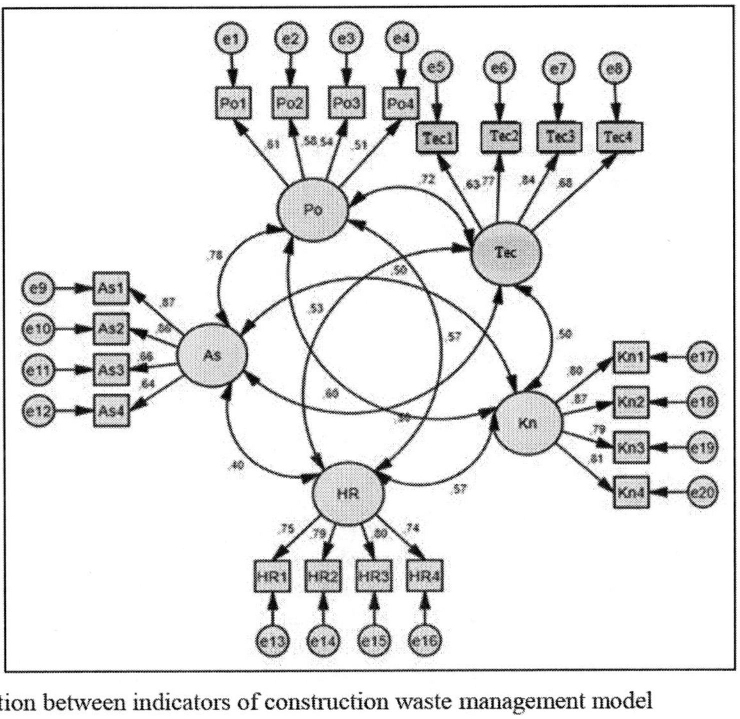

Fig. 2. Corrélation between indicators of construction waste management model

Base on figure 2 shows corrélation between indicators, the result is corrélation of policy and asset with path coefficient value of 0.78 ; policy and technology with path coefficient value of 0.72 ; policy and knowledge with path coefficient value of 0.60 ; policy and human resources with path coefficient value 0.57 ; asset and technology with path coefficient value of 0.50 ; asset and human resources with path coefficient value of 0.40 ; asset and knowledge with path coefficient value of 0.50 ; technology and knowledge with path coefficient value of 0.50, human resources and knowledge with path coefficient value 0.57. The coefficient result show value is equal to greater 0.05. It means having a significant correlation except the correlation between asset and human resources is below 0.4. The result of the Goodness of Fit characteristic as shows in Table 4.

Table 4. Measurement model fit indices

Measurement	Value	Cut-off values based	Conclusion
CMIN/DF	2.524	< 5	*Fit*
GFI	0.911	0 to 1	*Fit*
RMSEA	0.063	0.05 to 0.08	*Fit*
AGFI	0.883	0 to 1	*Fit*
TLI	0.921	0 to 1	*Fit*
NFI	0.895	0 to 1	*Fit*
PRATIO	0.842	0 to 1	*Fit*
PNFI	0.754	0 to 1	*Fit*
PCFI	0.786	0 to 1	*Fit*

In Table 4 shows the results analysis of construction waste management model base on the Goodness of Fit characteristic is fit.

According to [6] that the success of construction waste management is also influenced by regulatory factors, management system, technology, awareness on waste management, waste management research and practice. So it can be said that construction waste management is an important part in strategy to reduce construction waste both physical and non-physical.

The result of this study can be said in accordance with [6] study but need to be strengthened with assets that support construction waste management system. So that with construction waste management model can give the final result of efficient infrastructure project.

4 Conclusion

The result show that outcome the model is fit with CMIN/DF value of 2.524, GFI value of 0.911, AGFI value of 0.833, TLI value of 0.921, CFI value of 0.933 and RMSEA value of 0.063. The significant level to contributing construction waste management from 5 groups were covariant corrélation of policy and asset with path coefficient value of 0.78, policy and technology with path coefficient value of 0.72, policy and knowledge with path coefficient value of 0.60, policy and human resources with path coefficient value of 0.57, asset and technology with path coefficient value of 0.50, asset and human resources with path coefficient value of 0.40, asset and knowledge with path coefficient value of 0.50, technology and human resources with path coefficient value of 0.53, technology and knowledge with path coefficient value of 0.50, human resources and knowledge with path coefficient value of 0.57. The results of the model show that relationship between policy and asset has the highest correlation to support of construction waste management system. The model will be useful to improve construction waste management for minimize construction waste on infrastructures projects in Indonesia.

Aknowledgements

The authors are very grateful to the Program of Overseas Seminar Aid, Directorate General of Strengthening for Research and Development, The Ministry of Research, Technology, and Higher Education Republic of Indonesia. The authors are also grateful to Islamic University of Riau and the respondents of various infrastructure projects in Indonesia for their help with the quetionnaire.

References

1. M. Abduh, *Konstruksi Indonesia* (Kementerian Pekerjaan Umum, 2012)
2. C. Ho, P.M. Nguyen, M.H. Shu, Information and Management Science, **18**, 403-426 (2007)
3. T.V. Luu, S.Y. Kim, H.L. Cao and Y.M. Park, Construction Management and Economics, **26**, 373-386 (2008)
4. T.N.Q. Nguyen, P.A. Neck and T.H. Nguyen, International Business Research Journal, **2**, 3-16 (2009)
5. H.D. Khanh and S.Y. Kim, Construction Management Journal, **18**, 867-874 (2014)
6. W. Lu and H. Yuan, Resources Conservation and Recycling, **55**, 201-208 (2010)
7. J. Yang and D. Mitchell, *The Construction, Building and Real Estate Research Conference of the Royal Institution of Chartered Surveyor* (2010)
8. O.F. Kofoworola and S.H. Gheewala, Waste Management Journal, **29**, 731-738 (2009)
9. A.P.R. Cardoba, P.N. Sharratt and J.A.A. Sanchez, Process Safety and Environment Protection Journal, **86**, 375-388 (2008)
10. H.S. Cha, J. Kim and J.Y. Han, Construction Engineering and Management, **135**, 647-656 (2009)
11. H. Yuan, Waste Management Journal, **32**, 1218-1228 (2012)
12. T. Ninmann, *Waste Management: Plan for it Before, During and After Construction* (2011)
13. O.J. Oladiran, Architectural Engineering and Design Management, **5**, 165-176 (2009)
14. F.Y.Y. Ling and D.S.A. Nguyen, Built Environment Project and Asset Journal, **3**, 141-156 (2013)
15. Jogiyanto, *Konsep dan Aplikasi Structural Equation Modeling Berbasis Varian Dalam Penelitian Bisnis* (2011)
16. Jr.J.F. Hair, W.C. Black, B.J. Babin and R.E. Anderson, *Multivariate Data Analysis* (2014)
17. U.K.J. Mudalige, J.M.M. Udugama and S.M.M. Ikram, Applied Statistics, **13**, 15-37 (2012)
18. H. Latan, *Aplikasi Analisis Data Statistik Untuk Ilmu Sosial Sains dengan IBM SPSS* (2014)
19. Elizar, M.A, Wibowo and P. Koestalam, Procedia Engineering, **125**, 46-52 (2015)

Characteristics of small medium contractor entrepreneurs: a study of Indonesia

*Harijanto Setiawan**, *Ferianto Raharjo*, and *Koesmargono*

Atma Jaya Yogyakarta University, Civil Engineering Department, 55281 Yogyakarta, Indonesia

Abstract. The success of entrepreneurship as an important driving force for business success cannot be separated from the role of entrepreneurs. Therefore study about entrepreneurs in contractor business in Indonesia is considered important to be carried out. This study is aimed at identifying the characteristics of contractor entrepreneurs who owned small medium contractors in Indonesia. The study is conducted based on eight characteristics of entrepreneurs that have been identified from the literatures: risk taking, innovativeness, need for achievement, internal locus of control, self-confident, tolerance of ambiguity, need for autonomy and proactiveness. The study adopts qualitative research approach in which the data are collected from interviews with the owner of small medium contractors in Indonesia. Then the data are categorized by a coding process using NVivo 10 software and analysed using thematic analysis with inductive approach. The study found several specific characteristics of Indonesian small medium contractor entrepreneurs that are related to the eight characteristics of entrepreneurs but under the circumtances of contractor business. After all characteristics were examined further, they are mainly linked to projects, clients and business. This finding is in accordance with the nature of contractors as project based companies.

Keywords: Entrepreneurship, Contractors, SMEs, Indonesia

1 Introduction

The construction industry has been considered as an industry that drives a country's social and economic establishment [1, 2, 3]. This role is increasingly important in many rapidly developing countries such as Indonesia.

Several types of companies such as contractor, sub-contractor, consultant, and supplier are involved in construction Industry. Among all type of companies, contractors running their activities in many unique ways. Contractors are project-based firm (PBFs) that run their business based on projects that are unique and designed based on client's demands.

In running their business, contractors must deal with a risky situation, required to be innovative and face high competition [4, 5]. This business situation also happens in

* Corresponding author: haris@mail.uajy.ac.id

Indonesia, therefore appropriate business strategy needs to be developed in order to achieve success in business.

Over the last few decades, entrepreneurship became very popular, phenomenon and widely discussed among people around the world [6, 7]. Entrepreneurship has been considered as a driving force for business success in many business areas [8]. Considering the reliability of entrepreneurship to support business success and the need of contractors to implement the right strategy to achieve business success, it was established that entrepreneurship can be adopted by contractors as a right strategy.

However the implementation of entrepreneurship cannot be separated from the role of people called entrepreneurs. Entrepreneurs have been defined as persons who run their own business [9, 10, 11, 12, 13, 14, 15]. In order to be success in running their own business, entrepreneurs are required to be innovative, creative and independent [10, 11, 13, 14, 16].

Although the role of entrepreneurship to support business success in several fields has been recognized, there is very little study directed towards entrepreneurship as well as entrepreneurs in construction. Therefore this study will focus on bridging this gap in construction management research by exploring the characteristics of entrepreneurs in small medium contractors. The study is carried out in Indonesia, therefore the particular circumstances of the construction industry in Indonesia will characterise this study.

In order to achieve the aim of this study, the specific research objectives have been established as follows: (1) to explore characteristics of entrepreneurs from literatures in a systematic way and (2) to identify the characteristics of small medium contractor entrepreneurs in Indonesia.

2 Literature Review

2.1 Characteristics of Entrepreneurs

Over the years various characteristics of entrepreneurs have been widely studied. Entrepreneurs are distinguished from owner-managers; because not all owner-managers can be considered as entrepreneurs [17]. In running the business, an owner-managers tend to focus on surviving, while entrepreneurs have a desire to grow in addition to survive. Therefore owner-managers only have survival instinct while entrepreneurs have both survival and growth instinct. Furthermore, the characteristics that are considered as growth instinct are: opportunistic, willingness to take greater risks, innovative, self-confident, proactive and decisive with high energy, able to live with greater uncertainty, self-motivated, having vision and flair.

Six psychological characteristics of entrepreneurs which are need for achievement, internal locus of control, propensity to take risk, tolerance of ambiguity, self-confidence and innovativeness have been investigated [18]. These six characteristics became frequently cited in the previous similar studies. In addition, autonomy is included because it is found as one of entrepreneur's needs and this need is higher than people in general [19]. Another study used three of these characteristics: need for achievement, risk taking propensity and internal locus of control to link psychological traits of entrepreneur and business performance [20].

Later on the personality characteristics of entrepreneurs such as embracing risk taking, innovativeness, need for achievement, internal locus of control, proactive personality, stress tolerant, and with passion for work have been described in the previous study [21]. These characteristics, together with human capital and environment factors, can support the active performance of the entrepreneur to achieve success.

After reviewing all of these characteristics, it was found that some characteristics have similar meanings but are identified by using different terms or words, and some characteristics can be considered as a part of other characteristics, Finally this study considers eight characteristics that distinguish people who have entrepreneurial tendencies from those who do not. The eight characteristics are: risk taking, innovativeness, need for achievement, self-confident, internal locus of control, tolerance of ambiguity, need for autonomy and proactiveness. All of these characteristics are explored further in the following section.

Risk taking

Risk taking refers to a tendency to take an opportunity which required decision making under uncertain conditions [18, 22]. In particular it was emphasized that an entrepreneur's risk taking must be distinguished from the behaviour of gambler [22]. The entrepreneur takes a risk under controllable circumstances, therefore the risk taking behaviour of entrepreneurs is considered as calculated risk taking, which involves not only luck but skill as well. It was found higher performance entrepreneurs have a higher propensity to take risk taking action compared to lower level performers [20].

Innovativeness

Entrepreneur's ability to create innovation has been considered as one of two most important characteristics of entrepreneurs; the second major characteristic is being opportunistic [17]. However, innovation is needed in order to create and to exploit opportunity. Innovativeness is defined as creating and implementing new ideas such as new product, new service, new system or new strategy in order to achieve success of the company [21]. He considered innovativeness is related to self-starting, although they are different. Innovation is an important aspect of self-starting because self-starting is contrary to imitating what has already been done by others.

Need for achievement

Need for achievement is related to the ambition of individuals to achieve their goal rapidly and perfectly. This ambition motives the individual to defeat all constraints that prevented the goals from being reached [19]. Particularly entrepreneurs are characterized by high need of achievement [22]. It was found that the higher the need of achievement, the higher the entrepreneurs' performance [20]. However the achievement levels of peoples may vary depending on the individual and an entrepreneurs' achievement is usually associated with money [17].

Internal locus of control

Locus of control is related to an individual's beliefs about who or what controls the actions and achievement of his or her life. This characteristic is distinguished into internal and external locus of control. People with an internal locus of control tend to assume that his or her behaviour is under his or her control; they take responsibility for their actions. Those people with an external locus of control tend to assume that his or her achievement is influenced by external factors and so that individual has little or no control over the outcome of their actions. The entrepreneurs are characterised by having an internal, rather than external, locus of control personality, as do many successful people in a range of professions [18, 19]. Internal locus of control makes people feel competent to achieve their expected goal or goals [22]. Furthermore internal locus of control was found positively related to positive entrepreneurial performance [20].

Self-confident

Uncertainty is the nature of entrepreneur's circumstances; In order to deal with uncertainty, people need to be self-confident [17]. Self-confidence is defined as having the conviction of his or her own ability to be able to achieve expected goals [18, 19]. Furthermore self-confidence directs people to act properly and effectively in order to be successful [19].

Tolerance of ambiguity

An ambiguous situation exists when the information about that situation is not fully or sufficiently available [18]. This situation usually occurs in new and complex situations of high risk and high uncertainty. In many cases, an entrepreneur will need to make a deal in an ambiguous situation. Entrepreneurs who characterised by having a high tolerance of ambiguity, consider this situation as a challenge, not a barrier, and will responded positively in such a situation. The limited information available will be fully utilized to achieve optimal results [18, 19].

The need for autonomy

The need for autonomy has a variety of meanings, depending on the context of discussion; however, when it is associated with the personal need for autonomy, the most suitable definition is personal preference to work and to make decisions independently, free from the influence from outsiders [23]. Autonomy leads entrepreneurs to have new and great ideas, therefore business ideas will emerge [24]. Entrepreneurs need independence in order to control the idea, the goal, the business, to be able to work differently, to reach the potential, etc. [17].

Proactiveness.

Entrepreneurs must be able to seize opportunities quickly and accurately, they should not just wait for 'good luck' [17]. Entrepreneurs should not only be determined and able to seek opportunities but also to anticipate problems that will arise from those opportunities and then to find ways to overcome those problems [22].

3 Research Method

Two research methods have been implemented in this study. The study was started by comprehensive review on several literatures to get the characteristics of entrepreneurs in general. Then it was followed by exploratory approach to identify specific characteristics of entrepreneurs in small medium contractors under specific circumstances of Indonesia. This approach was appropriate for this study due to a lack of previous research in entrepreneurs in the area that is particularly related to contractors in Indonesia. The study was started by collected data and then followed by data analysis as explained in the following sections.

3.1 Data collection

Data for this study was collected through face to face semi-structured interviews with the owner of small medium contractors in Indonesia. The owners of contractors are considered as an appropriate source of data because they are the ultimate decision maker that intensively involved in daily activities of contractors' business. The choice of interviewees constitutes judgemental sampling which determines potential participants by considering their capacity to provide proper and appropriate information related to the investigated issues [25].

Prior to the interview, an interview guide was sent to the participants in order to give them an idea about the interview. The interview guide consists of several questions about the entrepreneur's characteristics that were explored based on several references. During the interviews, communication varied depending on the flow of the conversation therefore the format and sequence of questions did not always expressly follow those outlined on the interview guide. Extra questions were asked to follow up on issues which appeared to be important and relevant to the topic of the interview. The interviewer controlled the interviewee to make sure the focus of discussion was maintained. The interviews lasted between 30 to 60 minutes and each was audio-recorded.

3.2 Data analysis

The data collected was analysed using thematic analysis 'bottom up' approach that is data driven. In this approach, themes are identified mainly based on the data [26]. Data driven approach was chosen because of a lack of previous research in the area of entrepreneurs in construction. Then identified themes are directed to the issues that relate to the characteristics of contractor entrepreneurs.

The following processes of data analysis were carried out to identify the characteristics of contractor entrepreneurs.

- Audio-recording of each interview was heard carefully and continuously in order to enable the researcher to be familiar with the data and then initial ideas were extracted from the data.
- Then these audio-recordings were heard carefully but part by part, the sentences, that indicated any characteristics of entrepreneurs, were written-down
- The next step is coding process that was started manually. During manual coding, any sentences, that indicated the same issue, were coded under one node that is still nameless.
- The next coding process was done using NVivo 10 software. In this stage, codes were refined and re-categorized into the appropriate nodes that are considered as important issues related to the characteristics of entrepreneurs.
- Coding process using NVivo 10 software was carried out continuously in three rounds to check whether the issues work in relation to the entire data set or not.
- Finally important issues that explained the characteristics of small medium contractor entrepreneurs were identified then the names and the definitions of the theme is decided based on the sentences that were coded under each node.

4 Research Findings

4.1 Background of small medium contractor entrepreneurs

The study has interviewed 8 owners of small medium contractors. On average their companies employ full-time staffs ranging from 10 to 20 persons in the main office, whereas if calculated together with the staff and workers in the field projects, they have totally about 200 employees. They are not only owners but also are actively involved in running business and project activities.

Based on the interviews, this study found few similar conditions in all interviewees that are considered to be influencing characteristics.

- Starting their business with very limited capital.
  None of them get any support from their families or friends in order to start their business therefore they have to fight themselves from scratch without facilities from anyone.
- Starting working as a staff of contractors
  Most of them start their career in contractors as a staff. Among them, a few worked for contractors only in a short period, while the rest works in contractors for a long period and got the position of Project Managers.
- Committed to their career step.
  All of them always start working in the same industry, such as general contractors or subcontractors for foundation.
- Peoples who are strong in their religions.
  All of them believe that all of their achievements cannot be separated from the blessing of God.

4.2 Entrepreneurial characteristics of small medium contractor entrepreneurs

The data analysis found seven out of eight characteristics of entrepreneurs have been implemented by small medium contractor entrepreneurs. The only factor that is not implemented is the need of autonomy. Autonomy is not considered as an important issue for contractor entrepreneurs because as the owner, they have had the freedom to run the business in their own way. The characteristics specifically attached to small medium contractor entrepreneurs are presented in Table 1. Each characteristic is clarified by short explanation and examples that raised from interviews.

In addition to the characteristics presented in Table 1, the following characteristics were found during interviews with the entrepreneurs in small medium contractors in Indonesia.

- Integrity to build clients' trust which is embodied as honest, having moral principles and keeping promises to clients
- Passion fot working for the project because they consider working in construction projects as a hobby.
- Survival spirit in order to deal with the unstable condition of construction industry in Indonesia. For example they are able to survive during the monetary crisis in 1998.

5 Discussions

After all characteristics of entrepreneurs in small medium contractors in Indonesia examined further, they are mainly linked to projects and clients. This finding is in accordance with the nature of contractors' business. In order to be successful in business, contractors face two main challenges, those are to win the competition to get the project and to deliver the project successfully [27]. These two challenges are mainly related to clients and projects.

In addition to manage projects, contractors also need to manage their business, however project and business have different characteristics. Business processes involve repetitive activities while projects usually involve temporary and unique activities [28].

Table 1. Specific characteristics of small medium contractors entrepreneurs

General characteristics	Characteristics of small medium contractor entrepreneurs and their explanations and examples
Risk Taking	Risk taking in dealing with clients Bold actions to make a deal with clients because of a risk that come from clients, such as late payment, design change, unreasonable requests, etc.
	Risk taking to complete project on time, on budget and on quality Bold actions to complete project on time, on budget and on quality because they should be committed before the project begins.
Need for Achievement	Ambition to complete the projects quickly and flawlessly This spirit supports client's satisfaction and trustworthy making them to be a loyal client and then they will bring in a repeat order.
	Ambition to expand the scope of business This ambition is a growth instinct that was implemented through several activities such as: • investment on new equipment in order to get more projects • running business diversification such as property developer
Internal Locus of Control	Awareness that the success of business depending on hard work The business sustainability is the result of hard work in order to complete the projects properly and to satisfy clients' needs.
Self-confidence	Self-confidence to compete for projects In some occasions, contractors must compete with bigger contractors therefore self-confident is needed.
	Self-confidence to make a deal with clients To be confident is needed in order to acquire equal position with the client as service provider and user
Tolerance of Ambiguity	Tolerance to the projects under uncertain condition During construction stage, contractor entrepreneurs face ambiguous situation in technical, political, social and economic aspects, therefore they must have high tolerance to this situasion and anticipate the way out to solve the problems.
	Tolerance to the needs of clients Clients who involve intensively to the construction process, sometimes have unexpected needs, such as short project duration and design changes. In order to deal with these needs, contractor entrepreneurs have to be tolerant with ambiguous situations by devising appropriate strategies
Proactiveness	Proactive to get the projects This proactive action is manifested into several activities such as • proactively find information about new projects and then approaching the owners. • carrying-out marketing activities such as preparing web-site, sending company profile to the potential clients
	Find opportunities in a new market Contractor entrepreneurs should be responsive to the opportunities that available in the market, for example when infrastructure projects are booming, they expand the market from private to public clients
Innovativeness (Instead of innovative, creative was found more suitable)	Meet client's demands and solve clients' problems creatively Client's demands and problems are unpredictable, therefore contractors entrepreneurs need to be creative to meet client's demands and to solve client's problems
	Manage the business creatively and effectively The contractor entrepreneurs need to consider the following strategies: • managing projects' cash flow and resources allocation, • implementing profit sharing system with project teams in order to increase the loyalty and the performance of project teams.

Considering their different characteristics, instead of managing the projects, contractors also have a challenge to manage their business. Therefore the findings of this study also shows the characteristics of small medium contractor entrepreneurs that linked to how to develop their business, such as: proactive to expand the company, creative to meet client's demands and to solve clients' problems, creative to manage the business effectively.

Regarding the findings of this study that has been discussed above, the characteristics of small medium contractor entrepreneurs were directed to get projects, to serve clients and to manage business. Figure 1 shows a clear picture about the findings of this study.

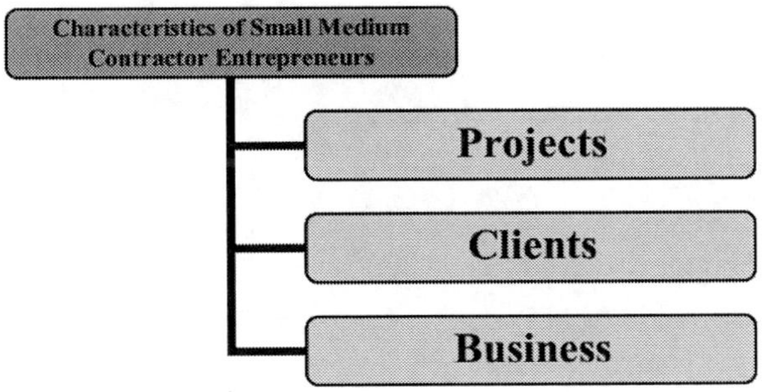

Fig. 1. The findings of this study

6 Conclusions and Recommendations

The literature review of the characteristics of entrepreneurs has found eight characteristics that distinguish entrepreneurs from peoples who do not have entrepreneurial tendencies. The eight characteristics are: risk taking, innovativeness, need for achievement, self-confident, internal locus of control, tolerance of ambiguity, need for autonomy and proactiveness. After obtaining these characteristics, this study conducted research to investigate the characteristics of small medium contractor entrepreneurs.

Thirteen characteristics were identified by this study, these 13 characteristics can be classified into the characteristics that have been adopted in this study, except need of autonomy. In addition, this study also found another three characteristics such as: integrity, passion and survival instinct, therefore overall this study found 16 characteristics of small medium contractor entrepreneurs.

These findings were explored through the experiences of small medium contractors entrepreneurs in Indonesia, therefore they were influenced by the particular circumstances of each contractor's business environment in Indonesia. For this reason these characteristics need modifications before they can be adopted by contractors entrepreneurs in other countries.

During the study, several areas have been identified to be investigated further in the future. The potential research topics that can be studied further are listed below.

- To investigate the characteristics of entrepreneurs in a different country or in a different project based industry
- To develop a comprehensive questionnaire as a tool to assess characteristics of contractor entrepreneurs
- To assess the entrepreneurial characteristics of contractor entrepreneurs in Indonesia, or another country, by assessing a large number of contractors entrepreneurs based on the entrepreneurial characteristics found in this study.

Acknowledgement

We thank the Ministry of Research, Technology and Higher Education of the Republic of Indonesia that has funded this study through the Scheme of Fundamental Research.

References

1. J. M. W. Wong, S. T. Ng, A. P. C. Chan, Habitat Int J, **34**, 256 (2010)
2. G. M. Winch, *Managing construction projects : An information processing approach* (2010)
3. D. W. Halpin, R. W. Woodhead, *Construction management* (2011)
4. D. Dvir, A. Sadeh, A. Malach-Pines, PMJ, **37, 5**, 36 (2006)
5. J. Schaufelberger, *Construction business management* (2009)
6. S. Yalcin, H. Kapu, JDE, **13, 2**, 185 (2008)
7. A. S. Spencer, B. A. Kirchhoff, C. White, ISBJ., **26, 1**, 9 (2008)
8. J. Wiklund, D. Shepherd, SMJ. **24, 13**, 1307 (2003)
9. S. R. H. Brockhaus, AMJ., **23, 3**, 509 (1980)
10. V. Luchsinger, D. Ray Bagby, SAM AMJ. **52, 3**, 10 (1987)
11. R. F. Hebert, A. N. Link, SBE, **1, 1**, 39 (1989)
12. P. Sharma, J. J. Chrisman, ET&P, **23, 3**, 11 (1999)
13. E. P. Lazear, JOLE, **23, 4**, 649 (2005)
14. B. Thompson, *Entrepreneurs : Talent, temperament, technique* (2004)
15. R. A. Baron, SEJ, **1, 1-2**, 167 (2007)
16. W. B. Gartner, AJOSB, **12, 4**, 11 (1988)
17. P. Burns, *Corporate entrepreneurship: Building an entrepreneurial organisation* (2005)
18. H. C. Koh, JMP, **11, 3**, 12 (1996)
19. K. Tajeddini, S. L. Mueller, JIE, **7, 1**, 1 (2009)
20. S. Sidek, F. A. Zainol, IBM, **2, 1**, 170 (2011)
21. M. Frese, FTE, **5, 6**, 437 (2009)
22. D.C. McClelland, *The achieving society* (1961)
23. M. Van Gelderen, P. Jansen, S. Jonges, SCALES-paper N200315 (2003)
24. G. T. Lumpkin, G. G. Dess, AMR, **21, 1**, 135 (1996)
25. C. Quinlan, *Business research methods* (2011)
26. V. Braun, V. Clarke, QRP, **3, 2**, 77 (2006)
27. P. Volpe, P. J. Volpe, *Construction business management* (Wiley, Canada, 1991)
28. D. M. Gann, A. J. Salter, RP, **29**, 955 (2000)

DOI: 10.1051/matecconf/201713805007

Using BIM as a Tool to Teach Construction Safety

Biemo W. Soemardi[1,*], and *Ray G.* Erwin[2]

[1]Faculty of Civil and Environmental Engineering, Bandung Institute of Technology, Indonesia
[2]Graduate Student, Civil Engineering, Bandung Institute of Technology, Indonesia

Abstract. The advancement of computer technology allows the construction education to exploit the various teaching strategies. Unlike other engineering subjects, teaching construction subjects often requires media beyond traditional class room environment that would enable the creation of field-like construction atmosphere. Construction visualization allows students to exercise and experience various field settings that are too difficult to deliver using traditional teaching approach. This paper aims at describing an attempt to develop and implement building information modelling (BIM) as a platform for teaching construction safety. Using BIM technology students can learn various aspects of safety in construction, which are typically hidden from project documents. Visualization features in BIM are utilized by students to help them identifying potential hazard, unsafe acts, and unsafe conditions in a simulated construction project. The result shows that students are more aware and capable to identify any potential hazard and related safety regulations of construction plan and design pertinent to specific site conditions than that if they were only equipped with a set of construction drawings and specifications.

1 Introduction

Its apparent that teaching engineering subjects often requires learning media beyond traditional classroom environment that would enable the creation of field-like environment. One of the challenges in teaching construction safety is to raise students' awareness of the importance of creating safe construction workplace by avoiding or minimizing potential accident. While theory and knowledge can be taught through traditional learning method, it is believed that forming correct attitude toward work safety and accident can effectively be done through the creation of an atmosphere in which students can as much as possible be experiencing what are perceived as unsafe acts or conditions. Within the context of construction safety, the creation of positive safety culture rests on the attitude of individuals engaged in the construction, which is the compound reflection of personal behaviour, knowledge, and environment/situation factors [1].

* Corresponding author: b_soemardi@si.itb.ac.id

Construction safety is commonly taught using texts and manuals on safety issues, such as those from OSHA [2], ILO [3] and many other sources. Using this approach students are expected to learn and understand various aspects of occupational safety and health based on the teaching material provided. Yet, the questions remain; how effective is this learning method, will then the students capable of identifying and adequately responding to safety issues in the construction field? To cope with this, instructors can bring students to the construction field to get direct exposure of field conditions, and/or alternatively examine construction documents. In most cases, instructors used the latter approach to teach students to identify potential hazards in a construction project. However, as the complexity of the construction project increases, the use of 2D or 3D construction drawings will no longer sufficient, and the needs for alternative learning approach becomes apparent.

2 BIM and Construction Safety

The advancement of computer technology has enabled construction educators to bring the actual construction field environment closer to the classroom. Visualization technology allows students to exercise and experience various field settings that are too difficult to deliver using traditional teaching approach. BIM technology is deemed as a tool to help students to understand safety issues in construction project by means of enhanced visualization of the construction process. With all of its rich visual features, BIM technology can be regarded as a potential learning tool to assist students to understand safety issues in construction project by means of enhanced visualization of the construction process.

2.1 Visualization and the construction hazard

The objective of construction safety is to provide work environment that enable the construction process to proceed without or with minimum hazard. In general, this objective can be achieved through the elimination or avoidance of unsafe condition and unsafe acts that may cause hazard. Construction hazard is natural characteristics of an object that has the potential to disrupt a process and/or cause harm or injury to people. Construction risk-related accident is defined as the probability of disruption of process and/or harm/injury to people, which can be determined by the level of exposure to hazard and the magnitude of disruption of a construction process or injury to person. Risk of construction accident can be controlled through the following steps: 1) identification of construction hazard, 2) assessment of risk hazard, 3) control of accident risk, and 4) review of risk control implementation.

Common practices of construction hazard identification are typically done by reviewing accident reports and planning the construction process while looking at possible occurrence of unavoidable hazards and accidents. These practices involve review of design/construction drawings, trying to reveal any visual information that may lead to identification of hazards.

However, with the ever increasingly more complex design and construction process, such review of drawings has become cumbersome and less effective. Traditional 2D and 3D drawings, as well as the technical specifications, can only capable of providing limited visual information on geometry, dimension, and perhaps some information on the quality of an object under review. Paper-based drawings are also not flexible enough to allow quick changes as well as inability to describe a process. Therefore, paper-based 2D 3D drawings can only provide limited opportunities to identify potential hazard in construction, whereas other potential hazards can only be identified by means of other method.

2.2 BIM and visualization of construction process

In the last decades, the application of Building Information Modelling (BIM) in building design and construction planning have grown rapidly due in part to the advancement of computer technology and acceptance of its benefits by the construction industry. BIM's powerful visualization feature has provided the construction practitioners with the opportunity to explore the richness of information derived from a design object, far beyond the capability of a traditional 3D modelling. BIM technology allow design engineers and construction planners to exercise various design and construction scenarios.

In the area of construction safety, the applications and researches on BIM has also been widely recognized, focusing on the planning and design phases to improve safety during construction by connecting construction safety issues to construction planning [4], [5], and [6]. BIM-based modelling and 4D simulation allow design engineers to exercise various design alternatives that best address safety issues, as well as determining whether any proposed design change is integrated into final construction plans. BIM can even include automated safety rule-checking algorithms for identifying potential hazards [7] or integrating BIM with other technology such as location tracking and augmented reality as part of integrated construction safety management and visualization system [8].

Using BIM technology design engineers and construction planners can develop and review various scenarios of construction process within different site conditions and layouts. Based on the visual information provided by BIM they then can identify potential hazards and to propose solutions.

3 Methodology

This study aims at improving students' ability to improve students understanding on the importance of safety in construction worksite. Specifically, this study was designed to enhance students' ability to better identify potential hazards in a construction project by means of BIM visualization of construction process. In this study, a hypothetical construction project was used to be examined the students. The object being studied is a simple building structure located in a sloping site surrounding by some trees and other objects. Prior taking this assignment students must have already acquired basic knowledge of construction engineering and blueprint reading. First, the students were to identify as many potential hazards as possible based on a set of typical paper-based construction drawings, afterward they were assigned with the same task but equipped with the digital BIM version of the design. After reviewing the project, identified hazards were compared to determine any improvement in the ability to identify potential construction hazards.

The methodology to identify potential construction hazards using BIM is as follow:

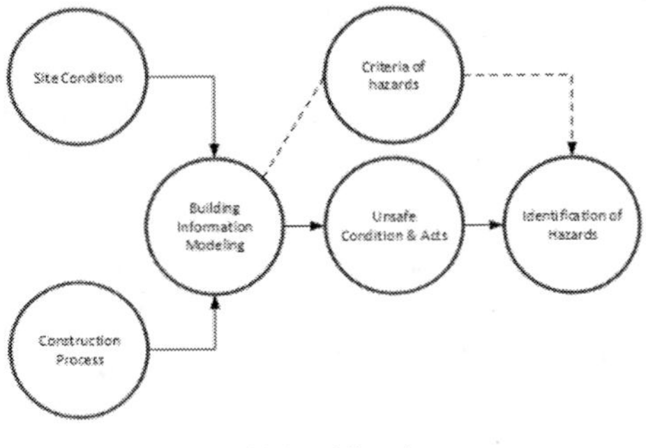

Fig. 1. Framework for BIM construction hazard identification.

 DOI: 10.1051/matecconf/201713805007

Basic information on site condition is developed from 3D design drawings. Additional information is latter added into based on site observation, such as existing structures, utility lines or other physical obstacles. At the same time, identification of construction process is developed based on knowledge and understanding of traditional construction methods. These two components are then stored in BIM databases as the basis for analysing the potential construction hazards.

To gain better understanding of the construction site, students are allowed to observe the construction site. On the next step, the students are asked to develop criteria and define the conditions that make construction accident possible based on types of hazard in OSHA - Job Hazard Analysis [9]. From 25 types of potential hazards, the assignment was to identify as many construction accidents as possible based on only four types of hazard, which are fall (slip, trip), excavation (collapse), struck by (mass acceleration), and moving objects. The types of hazard were limited to the top four most frequent accidents. For each type of hazard, having basic knowledge of construction safety and the site condition, students were asked to make what-if-analysis of the construction process and site conditions based on their interpretation and understanding of project documents (drawings) and BIM model. This what-is-analysis aimed to determine possible existence unsafe conditions and unsafe acts.

4 Analysis of BIM for Hazard Identification

One of many approaches to control accident is by carefully planning the design that focuses on the creation of safe work environment based on the principle of prevention through design (PTD). BIM allows to bring knowledge of construction site into the design [10]. In this case, design is described as visual representation of a process toward the realization of object. PTD principles ensure that accident can prevented before the real construction process is executed, by ways of providing prior information on where and what would be the un-safe conditions and/or un-safe acts happen, so that prevention actions can be planned.

4.1 Visual parameter for hazard identification

To facilitate that design that can be utilized to prevent accident and to assist identifying potential location of hazards, as set of visual parameters must be set.

Table 1. Visual characteristics of prevention through design

BIM visual parameter	description	application
writing and symbols	number, words, sentences, sign	label, type, property, symbol
dimension and geometry	line, distance, volume, elevation, height, depth	to observe working/equipment zones, distance between object, collision of two or more objects, clear zone, workspaces, opening/holes
animation	imitation of movement of objects	to represent walkthrough (emergency route), work progress
colour	visual impressions that are reflection of light from objects	to provide indicators/location of hazards
shading	colour intensity on the surface of objects exposed to light	to locate area of hazards due to lack of insufficient visibility
texture	surface condition of objects (rough, smooth, reflective, etc.)	to identify surface of objects with potential hazards

All those visual characteristics are facilitated in BIM features, whereas traditional drawings are mostly limited to the first two, which in turn limits the possible identification of construction hazards.

4.2 Hazard potential analysis using BIM

3D models with high detail and complex shading and colour provide better representation of object under study, in contrast to what can be derived from paper-based drawings. The situation surrounding the object are better represented that enables students to experience the near-actual complexity of the actual site condition. The size and positions of objects (trees, fence, etc.) give more information on the intricate condition of the site under observation that lead to better chance to identify potential hazards.

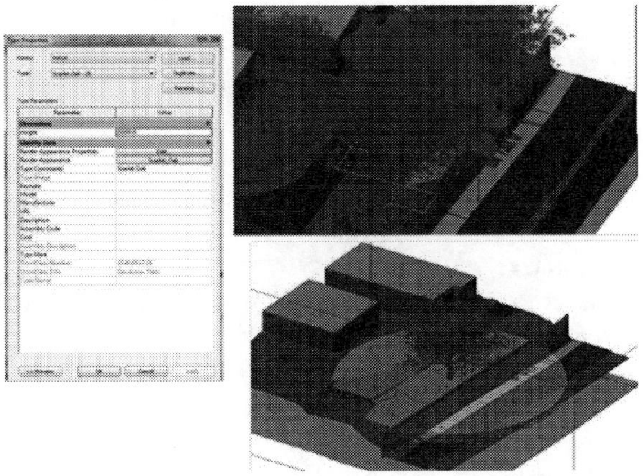

Fig. 2. Construction site situation.

BIM's 3D geometric representation gives the students the opportunity to better recognize the characteristics of elevation and slope of the site than that of lines of contour in traditional drawings. This information helps them to identify potential hazard stems from elevation differences, such as fall and slippage. In addition, other visual parameters such as symbol/sign and texture of objects also enhance the quality of information about the object and the adjacent constraints. Physical and mechanical property of soil, for instance, can be used by the students to determine whether the soil where an object is located has the potential to cause an accident, such as collapse.

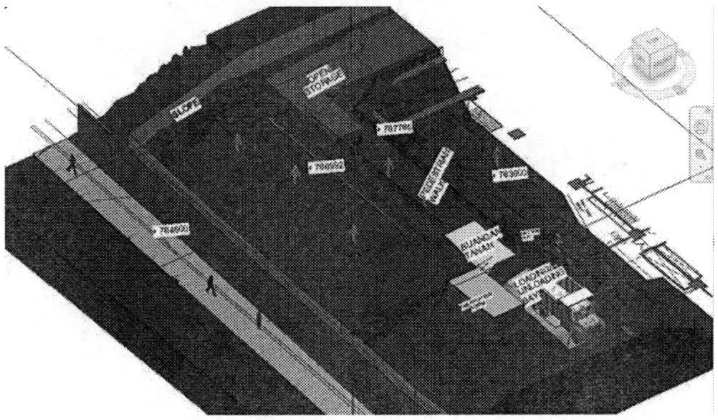

Fig. 3. Animation of construction traffic.

Animation is an important feature of BIM that enables students to exercise various scenarios of construction process and examine the consequences with regards to the potential existence of construction hazards. As an example, students designed flow of traffic in construction sites, where they can examine potential accident involving vehicles and workers, as well damages to properties. Here collision detector feature of BIM can be used to accurately position objects within spaces.

4.3 Hazard identification using BIM

3D capability of BIM can provide the representation of geometric form of topography of the construction site and other objects. By combining this geometric information with other parameters (such as property of soil and other objects) students can determine whether a particular location is potential to become an unsafe condition that may lead to an accident, such as falling from higher elevation. On the other side, BIM's animation feature can be used to exercise the possible emergence of unsafe acts at a certain site condition.

The construction hazard in the form of struck by objects, which primarily in the form of falling objects can be identified by BIM technology. In addition to the potential location of unsafe condition, the severity of hazard can be further determined by examining the information about the height of fall, weight and other characteristic of the falling objects, such types of object (liquid, solid metal, fine grains), physical property (sharp edges, rough surface texture etc.), and any other property. This information can be stored and later retrieved from BIM database.

Excavation hazards are commonly resulted from unsafe conditions and/or unsafe acts around excavation area. The potential and severity of excavation hazards is influenced by geometry and dimension of excavation as well as the property of excavated material (density, friction coefficient, water content, etc.). BIM technology allow students to identify the potential hazard by examining the height of excavation and information about the characteristics and mechanical property of material, which can be retrieved from BIM database.

Students can also BIM visualization to identify potential hazard from movement of objects. Hazard from movement objects can be identified using the animation feature of BIM. In addition to utilizing collision detector feature to determine safe perimeter, the use of colour and geometry parameters in BIM can be used to design path for moving objects and safe zones. Students can utilize different colour to assign different degrees of safe zone associated with different kind of moving objects.

BIM allows student explore conditions that are no available in traditional approach. to exercise the consequence of a construction action. For example, inappropriate installation of construction ladder has the potential of falling. Such condition can only be experienced by students through the use BIM technology, while to who rely on traditional paper-based drawings and specification will find difficulty to visualize.

Students can also study the effect for natural light to the construction site. Site clarity will influence the ability to visually identify potential hazard. Solar study feature in BIM technology helps students to identify site potential hazards because of lack of illumination. As the sun moves during the construction progress, the students can experience different clarity of site visibility.

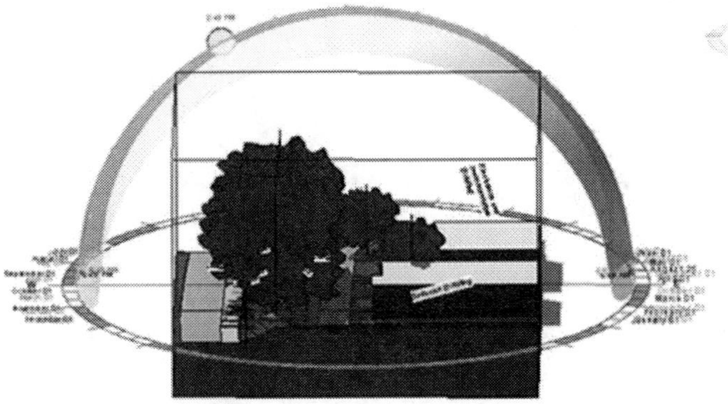

Fig. 4. Solar study of construction site.

In summary, by using BIM technology students were able to identify 64 potential construction hazards. This is significantly higher than number of potential hazards that can be identified using traditional paper-based construction drawings.

Table 2. Potential construction hazards

Hazard	frequency
Fall (trip/slip)	28
Falling objects	21
Movement of objects	9
Excavation	6

The significant improvement in identifying hazard by using BIM technology as opposed to using traditional design drawings are hazards related to fall, falling object and movement of objects. Students are able to identify more fall and falling object hazards because they are provided with more detailed and comprehensive information about the characteristics of the objects. Animation, which is not available in traditional design drawings clearly has the advantages for identifying hazards related to movement objects. Although students can make prediction of potential hazard resulting from the movement of objects in a space using traditional design drawings, they were not sure about the correctness and accuracy.

5 Conclusion

It is clear that visualization plays important role in understanding object under certain conditions. BIM allows student to experience the sensation of depth and other physical site condition that are absent in the traditional paper-based drawings. The higher the quality and accuracy of the information, the better and more comprehensive understanding of the students toward the objects under study. More features in visual representation enables students to explore their imagination and creativity. However, there are conditions that must be met before allowing students to explore the potential of BIM. BIM technology is expensive and students must be better prepared and equipped with knowledge and principles of construction practices as well as understanding the principles of construction safety. Such conditions are necessary to prevent students from excessively exploiting the

DOI: 10.1051/matecconf/201713805007

capability of BIM without clear understanding the principles of prevention through design to address construction safety issues.

References

1. R. M. Choudhry, D. Fang, and S. Mohamed, Developing a Model of Construction Safety Culture, Technical Notes, J. of Mgt. in Eng., Vol **23** Issue 4, pages 207 (2007)
2. OSHA, *Recommended Practices for Safety & Health Programs in Construction*, (2016)
3. ILO, *Training package in occupational safety and health for the construction industry*, (2010)
4. M. Weinstein, J. Gambatese, and S. Hecker, Can Design Improve Construction Safety? Assessing the Impact of a Collaborative Safety-in-Design Process, J. of Const. Eng. and Mgt., Volume **131** Issue 10, pages 1125-1134 (2005)
5. Sulankivi, K. Kähkönen, K. Mäkelä, and T. Kiviniemi, M., 4D-BIM for Construction Safety Planning, *Proceedings W099 - Special Track 18th CIB World Building Congress*, Salford, United Kingdom, pages 117-128, (2010)
6. W. Zhoua, J. Whytea, and R. Sacks, Construction safety and digital design: A review, Automation in Construction, Volume **22**, Pages 102–111, (2012)
7. S. Zhanga, K. Sulankivib, M. Kiviniemib, I. Romoc, C. M. Eastmand, and Jochen Teizere, BIM-based fall hazard identification and prevention in construction safety planning, Safety Science, Volume **72**, Pages 31–45 (2015)
8. Chan-Sik Park, and Hyeon-Jin Kim, A framework for construction safety management and visualization system, Automation in Construction, Volume **33**, Pages 95–103 (2013)
9. OSHA, *Job Hazard Analysis, 3017, revised edition* (2002)
10. A. Ganah and J. A John, Integrating Building Information Modeling and Health and Safety for Onsite Construction, J. Safety and Health at Work, Volume 6 Issues 1, pages 39-45 (2015)

Implementation of life cycle costing for a commercial building: case of a residential apartment at Yogyakarta

Peter F Kaming[1]

[1]Department of Civil Engineering, Universitas Atma Jaya Yogyakarta, Indonesia.

Abstract. Analysis of a design process is very important in controlling the initial costs and future costs in possession of an investment project such as commercial building. Therefore, it should be wise to perform a life cycle cost analysis to determine the cost of any category contained in future cost of the building. The analysis also provide information to see how much the total cost incurred by a development project from initial to the future cost by implementing BS ISO 15686 part 5: 2008, regarding life cycle costing. The purpose of this study is to identify the cost proportion and make long-term plans of a commercial building in term of its life cycle costing from a case of a residential apartment in Yogyakarta, Indonesia. Results of the study show that there are three groups that make up the life cycle cost: the cost of development of the building, the operating costs, and the cost of maintenance and replacement. For a long-term plan the life cycle cost for 25 years the percentage obtained as follows, initial development cost of 42%, operational costs 39%, maintenance and replacement costs 19%. The results would also make comparison with other existing commercial buildings.

Keywords: *Life cycle costing; service life; residential apartment; BS ISO 15686-5:2008; Yogyakarta*

1 Introduction

In most developing countries, the government has attempted to develop in all aspects of economy, social, politics and technology to support their programs. For running the programs, the countries need infrastructure and buildings as to facilitate all the program. When developing new building and infrastructure, they mostly did not consider the operating and maintenance of the facilities, this was due to lack of knowledge of life cycle costing (LCC) and encountered with limited funds.

Lack of operation and maintenance strategies to maintain building system performance leads to increased operating and maintenance cost and less healthy buildings. The first step toward improved practices that take advantage of potential operating savings is to identify the O&M practices routinely performed in buildings. Understanding LCC and service life of building component has two major benefits. First, baseline for service life is the benchmark from which to estimate cost for O&M practices. And second, service life baseline practices can be used as a guide to direct the long terms O& M cost estimate for the assets.

Indonesia as a developing country in the world today is promoting development in all fields. In implementing directed development, planned, integrated and construction strategy, it is expected construction to be able to deliver efficient and effective results and environmentally accepted. This refers to the constructions sustainable concept, where a development can work together between the concept of responsible development, has the spirit of maintenance, and accountability.

Integrated development has been regulated Building Acts No. 28/2002 [1], Water Resources Acts 74/2002 [2], Highway Acts 38/2004 [3] on the road construction, and Spatial Planning Acts No.26 / 2007 [4], which aims at balancing between buildings and its environment, as well as be a legal umbrella in harmonizing the utilization of space. Once, a top Public Utility Officer explained that "the future of development policy should be able to boost the quality of the environment, no exception public works infrastructure must meet the characteristics of balance and equality, holds a long-term and systematically saving". The policy the development of which was to apply the concept of construction sustainable, maintain and promote an increase in the percentage of green space against other cultivated areas, maintaining conservation areas, especially urban region, as well as realizing eco-city improve supervision in the aspect of environment control construction implementation.

In line with the concept of sustainable constructions, costs incurred in the development process is based on the calculation of life cycle cost, which is an integrated process in decision making, planning and control, procurement, operational, security, and the final value assets, see Barringer and Weber [5].

The purpose of the life cycle cost is to manage the process repeated from the planning to the destruction or replacement of assets, to manage the lifecycle cost (long-term) of the short-term savings, to ensure the appropriate consistent quality service of the designed building, to improve sustainability and lowers the risk of failure and maximize the potential and advantages of the provision of services, in order to minimize the associated costs throughout the life of the building itself, see Langdon [6].

Analysis of the life cycle cost of a design process that is important in controlling the cost of initial and future costs in possession of an investment project. Therefore, it is necessary to do a life cycle cost analysis study to determine the cost of any category contained in university building construction project in the university, and also see how much the total costs incurred by a project development a building ranging from the design phase to the technical life of the building.

In planning life cycle cost, information of the service life of component used such as equipment, and building materials are needed. This is something interesting because whenever service life assessment is only done on a building in overall activities. Determine the service life calculation aims at facilitating the maintenance and replacement of components of the building materials that have overdue the limit of its service life.

The objectives of the study were: 1). to identify service life of component of the material a building; and 2). to provide a long-term plan of 25-year life cycle cost a university building on the Universitas Atma Jaya Yogyakarta.

The building is located in the complex of UAJY Yogyakarta. This development is a tradition separate design and construct project. Contract system used was a lump sum method. This university building plans take as long as 1 year. The building is planned to have 5 floors and functioned for offices, class rooms, auditorium.

2. Literature Review

2.1 Definition of Life Cycle Costing (LCC)

There is some understanding of the LCC, according to some experts, including as follows:
1. According Asworth [7], the cost of the life cycle of a building or structure covers total costs associated ranging from stage the beginning until the end of the demolition phase.
2. According to Barringer and Weber [5], LCC is a modeling concepts cost calculation of its early to demolition of an asset of a project as a tool to take a decision on a study analysis and calculation of the total existing costs during its life cycle.
3. According Pujawan [8], the cost of the life cycle (Life Cycle Cost) of aitem is the sum of all expenses relating to the item since designed to unused. In other words, the cost of the building is the cost over the life of the plan building. Therefore, Life Cycle Cost can be formulated as follows:

LCC = Initial Cost + Cost + Cost of Care and Use of replacement wherein, the initial cost is the cost of the planning and execution of the building, the cost of usage is the cost incurred during the building operations, and maintenance and replacement costs are costs for maintenance and replacement the constituent components of the building during the design life of the building.

The main usefulness of Life Cycle Cost is at the time of the evaluation of alternative solutions to certain design problems, for example, an option may be available for the roof of a new project. It is necessary to review not only the initial cost, but also the cost of maintenance and repair, service life plan, appearance, and things that might affect the value as a result from the available options. Although aspects of appearance is a consideration aesthetics, and so it is subjective, but it cannot be ignored in the overall evaluation of these alternatives. Thus, Life Cycle Cost is a combination of calculation and wisdom.

Application Life Cycle Cost for major projects in the industry construction causes utilization of buildings and structures form can really different. However, a problem arises in practice because although the initial construction costs are relatively clear and unexpected at the design stage not so for the usage fee. It required calculation thorough review of the costs incurred from construction to the demolition of the building.

2.2 Life Cycle Cost Plan

Life Cycle Cost Plan is a plan for spending the proposal of a construction project over the life of the project. On execution of development, ranging from ideas, feasibility studies, planning, implementation, to maintenance and dismantling operations require miscellaneous costs, which are grouped into several components, namely: 1) Cost of construction: a) direct costs; b) indirect costs. 2). Operational costs: a) cleaning costs; b) The cost of utilities; c) cost of administration. 3).The cost of maintenance and replacement. 4).The final value of the building.

2.3 LCC Previous Studies

The following were reviews of previous studies regarding LCC that mostly carried by Indonesia researchers and mostly in the context building projects.

2.3.1 LCC for Teluk Bayur Port [9]

Title of research compiled by alzahri is the application of life cycle costs the construction of the pier (pier case studies CPO Teluk Bayur Port). This Research aimed at identifying factors, and variable life cycle cost in planning the development of the construction dock crude palm oil (CPO) on Teluk Bayur port fields, and determine the amount of the life cycle cost. The methodology used in this study is, first conduct a study literature, both perform model selection using a variable life cycle cost Izzati method, the third conduct field surveys, the fourth calculating life cycle cost, fifth conclusion. The conclusion of this study were: 1) Conceptual Model was selected Izzati model life cycle costing. Variables used in the planning model CPO pier construction that acquisition costs, operating costs and maintenance, and residual value. 2) The total life cycle cost CPO dock bay Bayur is Rp. 163,196,675,124. 3) In the 9th year have experienced payback period. 4) The application of life cycle cost for construction of the pier CPO Gulf Bayur in the year to 30, earned a profit of Rp. 48,640,121,124.

2.3.2 LCC Mixed Office & Residential Building at Manado [10]

Title of research compiled by Grace Priscilla Kamagi is Analysis life cycle cost of the building (case study: building projects Rukan (Mixed Office and Residential) Bahu Mall Manado. In this study the LCC was applied for 20 years. The methods used is the conduct literature and field studies. The second method used was mutual support to achieve the ultimate goal of research. It also made use of the Internet to search for information relating with research. This study used a life cycle analysis of the actual cost, namely life cycle cost calculation is based on building materials as per specifications cost planning of 9 buildings Rukan Block-N Bahu Mall Manado. To facilitate calculation then there were some cost data needed to be benchmarked. Data cost were: budget planning (RAB) and the list of unit prices of materials and wages. Analysis jobs were reviewed for the work of walls, floors and roofs.The conclusion of this study were: 1) By using the basic calculation of the project life cycle cost of development of Rukan Bahu Mall Manado, for work items walls, floors, and roofs, it was obtained a total cost = Rp1,722,634,337. 2) Based on the calculation of the life cycle cost of the construction project Rukan Bahu Mall Manado, for work items, walls, floors, and roof, then the result is the breakdown as follows: cost of capital construction (beginning) of Rp.574.598.000 (33.36%), cost for maintenance of the wall of Rp. 722,742,750 (41.96%), cost for floor maintenance Rp. 1,059,551 (0.06%), cost for roof maintenance Rp. 418,818,353 (24.31%), and costs of dismantling Rp. 5,415,681 (0.31%).

2.3.3 LCC Campus Center Building Institute Building (ITB) [11]

Title of research conducted Kawtharazlanshah Koento namely “Estimates life cycle cost (initiation of the concept of green building) campus center ITB”. Life-cycle cost calculations performed included some aspects of cost: initial cost, the cost of electricity, clean water costs, maintenance costs, costs maintenance, demolition costs, and resale value. Calculations were performed taking into account the choices made after the service life of the building exhausted. Building service life was assumed to be up to 30 years. The first option was destroyed building after passing the maid, and therefore costs demolition taken into account in the calculation of life-cycle cost. The second option was the building was sold after passing the maid, so the calculation of life-cycle taking into account the cost resale value ITB Campus Center Building. The conclusion of this research was, most cost aspects the calculation of life-cycle cost was the initial cost of building Campus Center ITB which reached Rp. 16,627,638,200. The cost reached 83.93% of the overall total life-cycle cost at the first option and 80.31% in the second option. For an annual cost incurred, costs of maintenance was an aspect that had the largest percentage, reaching Rp.2.536.221.568. The cost reached 12.8% of the total life-cycle cost at the first option and 12.25% in the second option.

2.3.4 LCC Diamond Building at Malaysia [12]

Based on the analysis conduct by Firsani and Utomo (2012) the study with the titled "Life Cycle Cost Analysis on Diamond Building Green at Malaysia "could be concluded as follows: 1) Cost category contained on Diamond Project Building, namely the Initial Cost,

Energy Cost, Operational and Maintenance Cost, and Replacement Cost. 2) Using the Present Worth Method on the level of interest rate (i) = 6% and the period of analysis (n) = 10 years. Total cost of living (Life Cycle Cost) from Diamond Building Malaysia were: Rp. 759.290.649.000, with the details as follows: a) Initial Cost of Rp 572.000.000.000, Energy Cost Rp. 5.599.000.000, Operational and Maintenance Rp 182.000.000.000, and Replacement Cost of Rp.201.000.000. b) In a sensitivity analysis performed on large Life Cycle Cost value of the rate changes, and interest rates range ± 30%, it can be seen that changes in interest rates are not sensitive to change Life Cycle Cost. c) At LCC modeling, it can be seen that the percentage each cost category against the total cost if no taking into account the residual value, i.e. Initial Cost of 75.306%, amounting to 0.737% Energy Cost, Operational and Maintenance Cost of 23.932%, and Replacement Cost of 0.026%, while if taking into account the residual value, the percentage of each transformed into cost categories, namely Initial Cost were 62.44%, 0.61% Energy Cost, Operational and Maintenance Cost of 19.84%, Replacement Cost of 0.02%, and Residual Cost was 17.09%. Modelling of LCC could also ben note that with or without taking into account the residual value, the largest percentage of the LCC on Diamond Building Project was the initial cost.

2.3.5 LCC Hostel Building at Kediri, Eastern Jawa, Indonesia [13]

The study regarding LCC had carried out in Kediri at a hostel building with the following components: 1) There were three groups of estimation of LCC for the hostel building, namely development cost Rp. 4.290.000.000 (46%), operational cost 2.360.412.125 (26%), and maintenance and replacement cost Rp.2.179.307.000 (28%). 2) In operational, the biggest cost go to administration Rp.56.000.000 (61%), followed by utility Rp.22.116.485 (22%), and cleaning Rp.16.300.000 (17%). 3) For maintenance and replacement, the biggest was ME Rp.987.685.000 (38%), sanitary Rp.625.000.000 (24%), wall Rp.431.328.000 (16%), ceiling Rp.197.500.000 (7,6%), accessory Rp.83.500.000 (3,3%), floor Rp.202.947.000 (7,8)% and roof Rp.62.940.000 (2,7%).

2.3.6 LCC for a University Building [14]

The purpose of this study is to identify the cost proportion and make long-term plans of a university building in term of life cycle costing in the area of Univeristy of Atma Jaya Yogyakarta (UAJY). Results of the study show that there are three groups that make up the life cycle cost: the cost of development of the building, the operating costs, and the cost of maintenance and replacement. For a long-term plan the life cycle cost for 25 years the percentage obtained as follows, initial development cost of Rp. 10,301,450,000 (43%), operational costs Rp. 11,085,000,000 (46%), maintenance and replacement costs Rp. 2.660.570.080 (11%).

3. Research Method

3.1 Data Collection and Analysis

Data collection methods used in this study includes following stages. 1) Preparation: stages of preparation is done is to formulate the research problem, research objectives, and explore data to the consultant's office to run smoothly. 2) Collect data: steps taken in this phase are: a). The field survey to see if there are projects eligible to be used as a research location and conduct the permitting process to the performer, or the owner of the project. b). send questionnaires to the respondents, to collect data for supporting the research. 3) Data Analysis: after the data collected lifecycle cost calculation using MS-Excel. 4) Discussion: step taken is to discuss the results of a study of life cycle cost calculation results for the conclusion. 5) Conclusions and Recommendations: steps being taken is to give conclusions and suggestions on the results obtained in this study.

DOI: 10.1051/matecconf/201713805008

3.2. The Building Under Studied

The apartment building is situated at Babarsari region, in Yogyakarta. The building consists of 8 floors, 203 units with a total of floor area of 16,719 m2. It was built in the land of 4000 m2 in the year of 2013, but the building cost was recalculated using cost data of 2016. It was then estimated life cycle cost of 25 years using history data obtained from the office for building maintenance of the apartment.

3.3 Cost Model for LCC

The model estimate was adopted the example from ISO 15686 part 5 for LCC [15].

3.3.1 Construction Cost.

All costs of initial construction work on the building: 1) Building Works Costs. Cost of building works and external works (Presented in the SFCA element cost structure); 2) Other Construction Related Costs. All costs payable by the client in connection with the building or constructed asset.

3.3.2 Maintenance Costs

All costs of replacement, maintenance, repair and adaptation of the constructed asset (Presented in the SFCA element cost structure. They consists of 1) Major Replacement Costs. It includes scheduled replacement and major components. This will form the detailed asset life cycle replacement program. 2) Minor Replacement and Repairs Costs. Scheduled replacement of parts and scheduled maintenance and repairs to components and associated making good and minor redecorations including planned preventative maintenance. 3) Unscheduled Replacement, Repairs and Maintenance Costs. Allowance for unforeseen maintenance arising from early failure, inappropriate use etc. 4) Redecorations Scheduled redecorations..5) Refurbishment and Adaptation Costs Scheduled refurbishment and adaptation during the period of analysis. Exclude refurbishment and adaptation carried out as part of the initial construction, which should be included under the construction costs.

3.3.3 Operation Costs

All costs of operating the building or facility arising from the building itself rather than from its occupancy, excluding maintenance costs. It includes periodic: routine and specialist cleaning; windows and external surfaces cleaning windows, curtain walling, glazed screens, cladding, sun screening etc.; internal cleaning; and external works cleaning. Part of the operation cost is fuel consumption. Fuel Show cost of different fuel separately e.g. 1) Gas, 2) Electricity. 3) Fuel oil. 4) Solid fuel (define). 5) Other Include any income from the sale of energy.

3.3.4 Administrative Costs

User support costs related to the operation of the building or facility. Property Management All costs involved in managing the operation and maintenance of the building. Includes: 1) Supervisory staff e.g. building supervisor, maintenance manager, facilities manager etc. 2) Professional staff e.g. architects, engineers, surveyors etc. 3) Clerical staff and administrative staff. 4) General and regulatory surveys commissioned by or on behalf of the client

4. Result of Analysis and Discussion

4.1 Data Analysis

Distributing questionnaires were conducted to determine service life of components of building materials used in the construction of the university building. From 15 people who responded to the questionnaire obtained an average service life and testimony of building

DOI: 10.1051/matecconf/201713805008

materials used in the construction of buildings. Table 1 below will explain the average service life and testimony of building materials.

Table 1 Service life for Material Components, Source: Kaming & Mardiansyah, [13] and other primier data.

Cost Group	Component Material	Service Life (Year)	Material Brand and Features	Reference/ Interview)
ME		30	Sony/Samsung/Krisbow	supplier
	Water Heater	15	Ariston/Paloma/Rinnai	supplier
	AC	15	LG/Panasonic/Sharp	consultant
Roof Structure	Steel roof frame	30	Prima Truss	supplier
	Roof clay tile	20	Sokka/Mutiara/KIA	consultant
	Gutter	15	Steel/Zinc	consultant
Wall	Wall paint	8	Taka/Dulux/Movilex/Nippon	supplier
	Aluminium paint	12	Taka/Dulux/Movilex/Nippon	supplier
	Wood paint	15	Taka/Dulux/Movilex/Nippon	supplier
	Frame aluminium	15	Lokal/Aluplus/Alurre	consultant
	Door aluminium	15	Lokal/Aluplus/Alurre	consultant
	Timber frame	30	Jati/Merbau/Bengkirai	consultant
	Timber door	30	Jati/Merbau/Bengkirai	consultant
	Timber windor	30	Jati/Merbau/Bengkirai	consultant
	Door and window glass	20	Glass 3 mm	consultant
Floor	Granit	20	Roman/Platinum	supplier
	Ceramic	20	Asia Tile/ Roman/Platinum	supplier
Ceilimg	Gypsum	15	Local	consultant
	GRC	18	KalsiBoard	consultant
Sanitary	Wastafel	8	Toto/America Standard	consultant
	Stainless steel sink	12	Enchanting	consultant
	Water crane	3	Enchanting	consultant
	Water closet	15	Toto/America Standard	consultant
	Shower	8	Enchanting	consultant
	Floor drain	8	Enchanting	consultant
	PVC pipe	30	Waving	consultant
Accessory	Slot door & window	10	Solid /Paloma/Dorma	supplier
	Handle door & window	10	Solid /Paloma/Dorma	supplier
	Hinge door and window	20	Solid /Paloma/Dorma	supplier
ME	Lamp standard	5	Philips/Krisbow/GE/Osram	supplier
	Lamp down light	5	Philips/Krisbow/GE/Osram	supplier
	Wall Stop contact	20	Broco/Jung/Panasonic	consultant
	Saclar single	20	Broco/Jung/Panasonic	consultant
	Sacla multiple	20	Broco/Jung/Panasonic	consultant
	Fitting	20	Broco/Jung/Panasonic	consultant
	Lift	30	Hyundai/Otis/Schneider	consultant
	Fire alarm	30	Siemens	supplier
	Telephon	15	Telkom/Panasonic/Samsung	consultant

4.2 Cost of Initial Construction for the Apartment

The initial cost comprised of: preparation, excavation, concrete, reinforcement, bore piles, etc .In the study, the initial construction cost was calculated using 2015 cost data. The total initial cost was estimated Rp.104.421.000.000. with the floor area of 13.331 m2, the unit cost was over Rp.7.833.000 /m2.

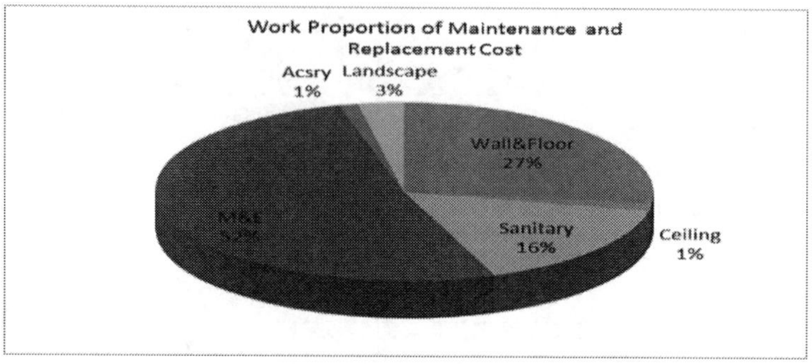

Fig. 1. Proportional of Estimated Maintenance & Replacement Cost over 25 years.

4.3 Maintenance and Replacement over 25 Years.

Maintenance and replacement cost were calculated based on period of service maintenance and service life data obtained from Table 1. Total maintenance cost was about Rp. 47.540.000.000 over 25 years. The biggest proportion went to component of the cost of electrical and mechanical equipment and devices (51.62%); floor and wall maintenance including repainting was 26.5%; and sanitary work was almost 16%, the rest was for others. See Figure1.

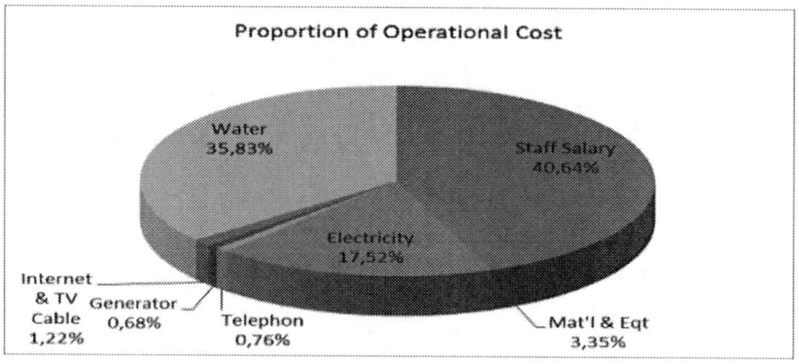

Fig. 2. Propotional of Estimated Operational Cost of the Building over 25 years.

4.4 Operating Cost over 25 Years

The operational cost was estimated with a total of Rp. 98.545.000.000 over 25 years with a biggest proportion went to administration cost of 40.64%, water 35.83%, electrical 17.52%, the rest was for other. See Figure 2.

4.5 Proportional LCC for the Apartment.

The proportional of overall construction cost, operational cost and maintenance and replacement cost were showed in Figure 3. The construction cost was the biggest proportion with 41.68%, followed by operational cost of 39.34%, and the maintenance and replacement cost of 18.89%.

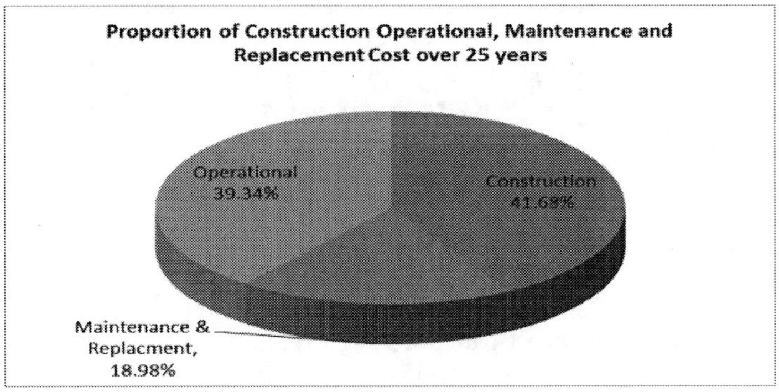

Fig. 3. Proportional of Total Cost of the Building over 25 years.

4.6 Comparison with Previous Studies

Since the LCC model of previous studies [9-12] were varied in terms of periods of LCC, and methods, they were difficult to compare. The comparison were made for other previous studies [13,14] because they were applied the same period of 25 years of LCC, and utilized the same method. Table 2 shows the comparison for average proportional of three type of building with the initial construction cost (43.67%), operational cost (37.00%), and maintenance and replacement costs (19.33%).

Table 2. Comparison LCC Three Studies.

Project	**Initial Construction Cost (%)**	**Operating Cost (%)**	**Maintenance & Replacement Cost (%)**
Hostel [13]	46	26	28
University [14]	43	46	11
Apartment *	42	39	19
Average	43.67	37.00	19.33

*This study.

5. Conclusions and Recomendation

The study regarding LCC has been carried out in Yogyakarta at a university building with the following cost components. 1) There were three groups of estimation of LCC for the university building, namely development cost Rp. 104.421.000.000 (41.68), operational cost Rp.98.545.000.000 (39.84%), and maintenance and replacement cost Rp. 47.540.000.000 (18.98%). 2) On maintenance and replacement cost, the biggest proportion went to maintenance and replacement of a) electrical and mechanical divices (52%), b) floor and wall ceramic tile of almost 27%; followed by c) sanitary replacement of 16%; the rest was for other works. 3). On operational, the biggest cost went to administration cost of 40.64%, water cost of 35.83%, electricity cost of 17.52%, and the rest for other expenses.

This study provided the following recommendation. The service life used in this study was been gathered from stakeholder involved in the suppliers, consultants, and academics, involved in the construction industry, as well as the maintenance officers from the university had some drawbacks in term of its quality and quantity. Perhaps the similar information should be gathered for accurate based the empirical research methodology.

The cost items used in this study was subjective and mostly be relied on the experience of the individual professional and available of service life of the components. The cost could also be adjusted when it could be reduced based on the limited resource available. In future, it was recommendated method of the Davis Langdon and ISO 15686 part 5 be applied in whole life cycle costing (WLCC) [16]. Future research should be emphasize at service life for building and infrastructure components for local condition adopting factor methods of ISO 15686 part 1, and create specific WLCC research model for more type of building and civil engineering projects in developing country such as Indonesia.

Acknowledgement

This research is internally funded by Universitas Atma Jaya Yogyakarta (UAJY). The author was grateful to the Research and Service Boards of UAJY, building officers, supplier, consultant who participated in this study. The author also thankful to reviewers for their commend and critics for improvement the quality of the paper.

References

[1] National Acts of Indonesia, number 28 year of 2002 regarding Building Development (2002).
[2] National Acts of Indonesia, number 74 year of 2002 regarding Water Resource (2002).
[3] National Acts of Indonesia, number 38 year of 2004 regarding Highway Development (2004).
[4] National Acts of Indonesia, number 26 year of 2007 regarding Spatial (2007).
[5] P. H Barringer, & D.Weber. "Life Cycle Cost Tutorial", *Fifth International Conference on Process Plant Reliability and Hydrocarbon Processing*, Gulf Publishing Company., Texas. (1996).
[6] D. Langdon. Life Cycle Cost (LCC) as a construction to sustainable construction: a common methodology, Darvis Langdon Management Consulting (2007).
[7] A. Asworth. Cost Planning for Building (in Bahasa), PT. Gramedia Pustaka Utama, Jakarta. (1994).
[8] I.N Pujawan, Engineering Economics, (in Bahasa) Guna Widya, Surabaya. (2004).
[9] Alzahri. Implementation of Life Cycle Cost in Developing a Dock: Case of the Dock for CPO at Teluk Bayur Padang. (in Bahasa) (2013).
[10] G.P Kamagi. Analysis of Life Cycle Cost for a Building: case of mixed office and residential building of Bahu Mall, at Manado (in Bahasa). *Jurnal Sipil Statik*, **1**(8) 549-556. (2013).
[11] K.A Koento, Estimating Life Cycle Cost: Initial development of the concept for green building development for Campus Center ITB. (in Bahasa). (2012).
[12] I. Firsani, & C. Utomo. Analisis LCC pada Green Building Diamond Building Malaysia, Jurnal Teknik ITS Vol **1**(34-39). (2012).
[13] P.F Kaming.& J. Mardiansyah, Implementation of Life Cycle Costing: A Case of Hostel Building in Kediri, Eastern Jawa, Indonesia, Proceeding *ICRMCE*, Surakarta, UNS. (2015).
[14] P.F Kaming Implementation of Life Cycle Costing for a University Building, Indonesian Journal of Life Cycle Assessment and Sustainability, Vol.**1** (1). (2016).
[15] BSI. BS ISO 15686-1:2008 Building and Construction Assets – Service Life Planning – Part 1: Service Life Methodology, BSI-BCIS. (2008).
[16] BSI. BS ISO 15686-5:2008 Building and Construction Assets – Service Life Planning – Part 5: Life Cycle Costing, BSI-BCIS. (2008).

MATEC Web of Conferences **138**, 06001 (2017) DOI: 10.1051/matecconf/201713806001
EACEF 2017

Stability analysis quay wall at the Amazonehaven, port of Rotterdam

Nadevah K.N. Mourillon[1] , *Jarit* de Gijt[2], *Klaas Jan* Bakker[3], *Henk* Brassinga[4], *Erik* Broos[5]

[1] Structural Engineer, Hydraulic & Civil Engineering dept., Municipality of Rotterdam, The Netherlands. [2] Ass. Professor Hydraulic Engineering Dept., University of Delft, Senior consultant, Engineering Dept., Municipality of Rotterdam, The Netherlands. [3] WAD 43 bv Consulting Engineers, The Netherlands. [4] Geotechnical consultant, Delft, The Netherlands. [5] Port of Rotterdam, The Netherlands.

Abstract. The stability of any structure is an important aspect in civil engineering. This aspect is the subject of the researched quay wall at the Amazonehaven, port of Rotterdam. The quay wall with a relieving platform structure had in various section, over the entire 900m length of the quay, large deformations at the toe of the combined wall. The purpose was to analyse and quantify the influence of the deformed combined wall on the stability of the quay wall, its service lifetime. To obtain a better insight into the concept of stability, analytical methods based on the Blum theory, beam on elastic foundation method and finite element method using Plaxis 3D were applied and compared. The finite element method, Plaxis 3D, proved to be a better method to investigate the quay wall. Plaxis 3D takes into account the 3-dimensional effects of the quay wall and considers the actual soil behaviour during calculation which is a sophisticated manner of modelling a quay wall. A calibration model (which is the actual designed quay wall) and a series of models with various penetration depth of the combined wall are modelled. Also, a safety analysis of the soil parameters were applied to the various models.

1. Introduction

In the 1980's, it was of great importance to the port of Rotterdam, while competing with other major loading ports in countries such as Australia, the United States of America, Canada, South America and South Africa, to be able to facilitate vessels greater than 100 000 dwt. With this in mind, the E.M.O. deep-sea terminal (Fig. 1) was constructed for the throughput of iron and coal which was realized in 1990. This 900 m long deep-sea quay terminal is located at the Maasvlakte I at the Amazonehaven. In Fig. 1 the cross section of the quay wall is illustrated.

Fig. 1 Overview of the 'Europees Massagoed Overslagbedrijf' (E.M.O.) at the Amazonehaven (Left); Cross section of the quay wall at the Amazonehaven [10] (Right)

The widening at the south side of the Amazonehaven harbour basin must be realized due to the rapid growing of the container vessels mooring in the Amazonehaven. During the demolition of the old quay wall much damage are discovered at the tip of the open tubular and intermediate sheet piles. This was an unexpected discovery which according to the recorded events that took place during its service lifetime was not known. The damages to the open tubular piles consist of folding of the pile toe, completely closed pile toe, ovalisation at the pile toe or other damages. In Fig. 2 typical damages to the tubular piles are illustrated.

Fig. 2 Various types of damages to the tubular piles [10]

A rough estimation of the damages to the open tubular piles is to about 15% to 20% of the piles. In the case of the intermediate sheet piles roughly 50% seems to be damaged. All the damages are at the toe of the sheet pile because of the interlock openings that occurred during construction. These unexpected damages lead to the investigation of the structural stability mechanisms of the deep-sea quay wall. In this paper the stability of the quay wall at the Amazonehaven for its service lifetime will be discussed.

2. Methods

Various design calculation methods are available for the design of a quay wall. These methods are:

2.1 Blum method – Classic (analytical) method

This design calculation method is used to determine the strength of the combined wall of the quay wall. The results of the analytical calculations with Blum, no specific answers were found on the stability of the quay wall. The reason for this is that with this method based on the load distribution only the penetration depth can be determined. The corresponding moment and shear force can also be derived. However, the actual earth pressure redistribution caused by arching of the soil behind the combined wall is not taken into account in this method. Furthermore the various construction stages are not taken into account in this analytical method.

2.2 D-Sheet Piling – Beam on elastic foundation method

The calculation method for beam on elastic foundation (D-Sheet Piling) is more complex than the previous method. The calculations with D-sheet piling software are restricted to determining sheet pile dimensions. The interface relation between soil and structural elements will be assumed linear in the elastic area and the springs. The behaviour of the soil is represented both horizontally and vertically and mutually uncoupled.

The inclined combined wall for the quay wall is considered vertical when modelling in D-sheet piling software. The inclination effects of the combined wall which are favourable, are not considered. The eccentricity of the axial load reduces the maximum bending moment, meanwhile second order moment effects will be unfavourable to the stability of the combined wall. That is why, the second order effects will not be taken into account.

2.3 Plaxis –Finite Element Method

In favour to previous methods, the finite element method (Plaxis 3D) is based on a model in which the soil behaviour and the contained construction elements are integrated. This method is done based on the stress equilibrium and deformation of soil and bending behaviour of the construction elements are described by a linked system consisting of partial and ordinary differential equations.

The advantage of this method is the displacement of the soil directly around objects, in this case the soil against the combined wall can be calculated. For that reason, this method should be applied in case that objects nearby are sensitive to the deformations of the soil during the construction of the wall. The effect of the construction on the object can then be calculated. This method is also applied for three-dimensional problems when the distribution of the earth pressures over the primary members and intermediate piles in a combined wall have to be taken into account.

3. Results and Discussion

A static analysis under the serviceability loading conditions has been preliminary performed. The relationship of the stability of the quay wall is determined by using Blum method, D-Sheet Piling and Plaxis 3D.

DOI: 10.1051/matecconf/201713806001

3.1 Modelling

For the design of the quay wall three methods were used. The first two methods, Blum and D-sheet Pilling, were the least reliable for determining the stability of the quay wall. With these methods an estimation is obtained of the penetration depth, bending moment, shear forces and displacement.

The quay wall is the input model in Plaxis 3D. This allows for a more accurate quay wall modelling, specifically oblique installed combined walls and the analysis and evaluation of plastic deformation are also possible. In order to understand the deformation behaviour of the quay wall, specifically the combined wall, a calibration model (Fig. 3) is set up. This model is based on the original designed quay wall without surface loads. This model can be compared to the results from the other five models (Fig. 4, 5 and 6) based on the deformation of the foundation elements and relieving platform structure. The reason for this is the stability of the quay wall model is not only dependent on the deformation of the combined wall but also on the displacement of the foundation elements and relieving platform structure as well.

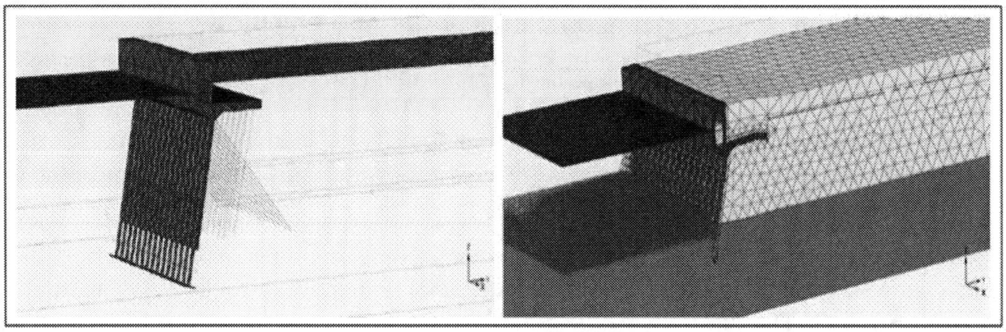

Fig. 3 Deformed mesh of the calibration model (Left) and schematization quay wall model (Right) [1]

3.2 Five quay wall model simulations

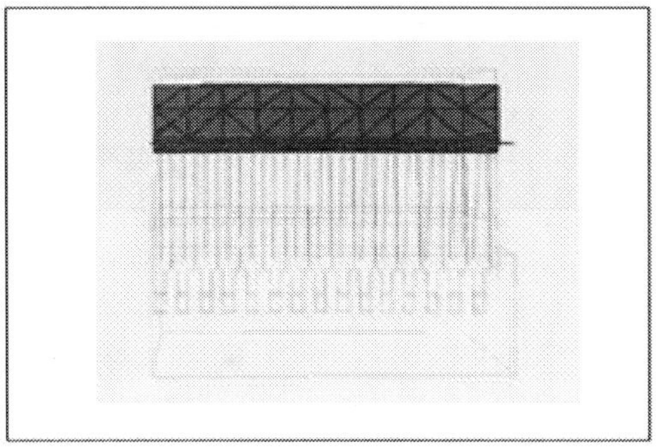

Fig. 4 Quay wall without combined wall

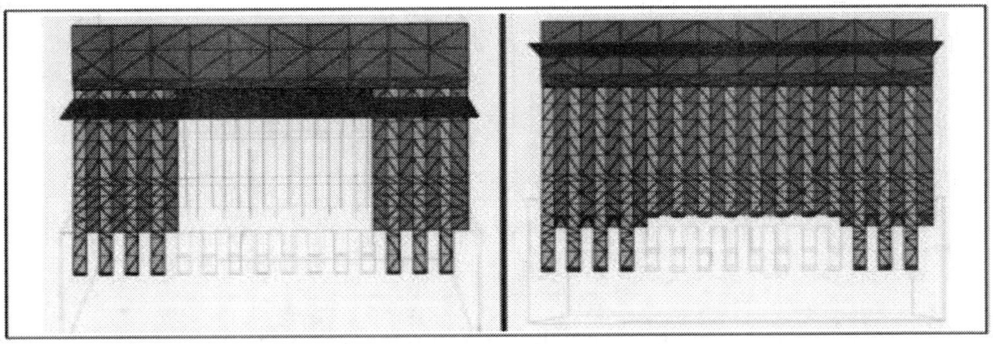

Fig. 5 Quay wall simulated on two supported ends – Model 2A (Left) and Model 2B (Right)

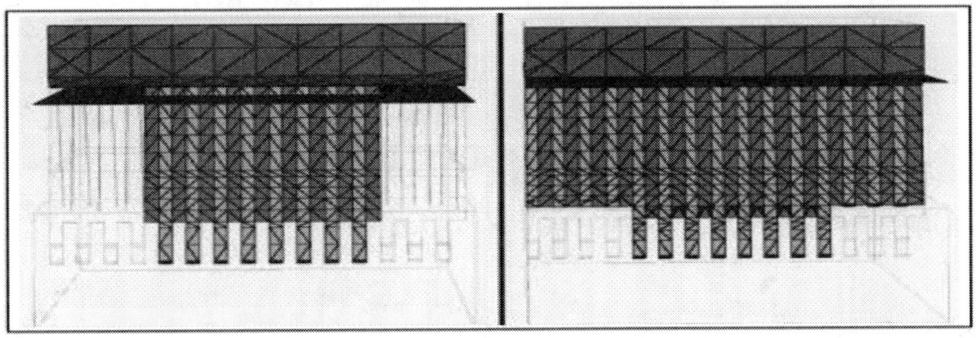

Fig. 6 Quay wall simulated on one supported end – Model 3A (Left) and Model 3B (Right)

From the results, comparison between the various quay wall models and the calibration model can be concluded that Model 2B and calibration model can be further investigated.

4. Results and discussion

In this section, the results from D-sheet and Plaxis 3D are presented. In the Fig. 7 below the results from D-Sheet is analysed for both combined wall models (Uniformed and trapezium distributed surcharges). The relation between the minimum penetration depth and bending moment are presented with different surcharges (0, 200 and 450 kPa) with a construction depth of NAP-25,50m.

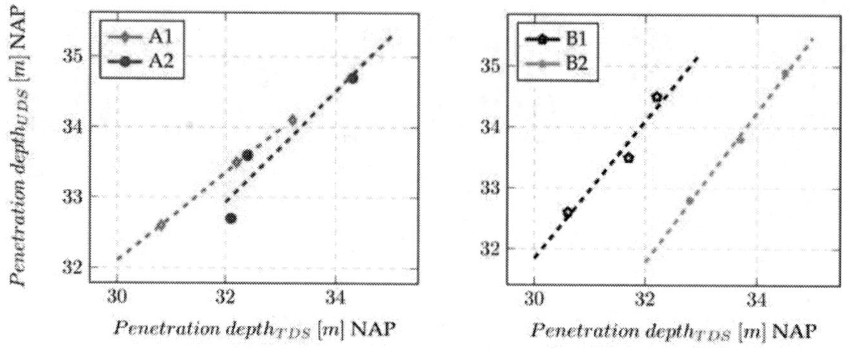

Fig. 7 The relation between the minimum penetration depths for a Trapezium Distributed Surcharges (TDS) and Uniformed Distributed Surcharges (UDS) for various load cases [1]

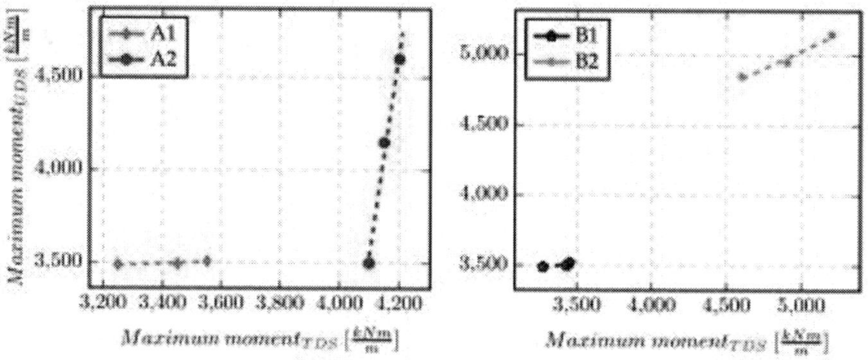

Fig. 8 The relation between the minimum penetration depths for a Trapezium Distributed Surcharges (TDS) and Uniformed Distributed Surcharges (UDS) for various load cases [1]

From the comparison between the two models it can be concluded that by modelling the distribution of the surface loads as a trapezium or uniformly the difference of minimum penetrated depth of the combined wall is considerably small. Furthermore, the field bending moment increases as the surface load increases which leads to a freely imposed bending moment. When the combined wall is shorter than the calculated minimum penetrated depth the quay wall model becomes unstable. With a 2-dimensional software such as D-sheet piling it is apparent that the (in) stability of the quay wall is not investigated for the entire length of a quay section of 45 meters. It is only a cross section per meter that is calculated.

Therefore, to implement the 3-dimensional effects, a 3-dimensional method would be very helpful in analysing the actual stability of the entire quay wall. The three dimensional finite element models of the quay wall is presented.

DOI: 10.1051/matecconf/201713806001

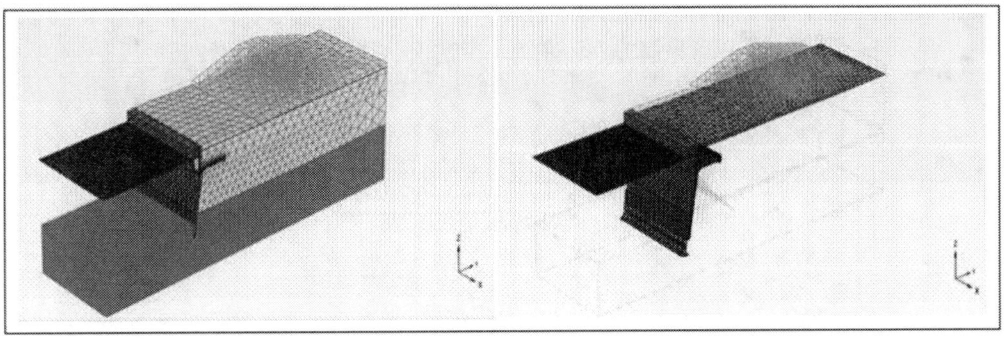

Fig. 9 Deformed mesh calibration model with designed surcharge (Left) and deformed mesh calibration model with designed surcharge without soil (Right)

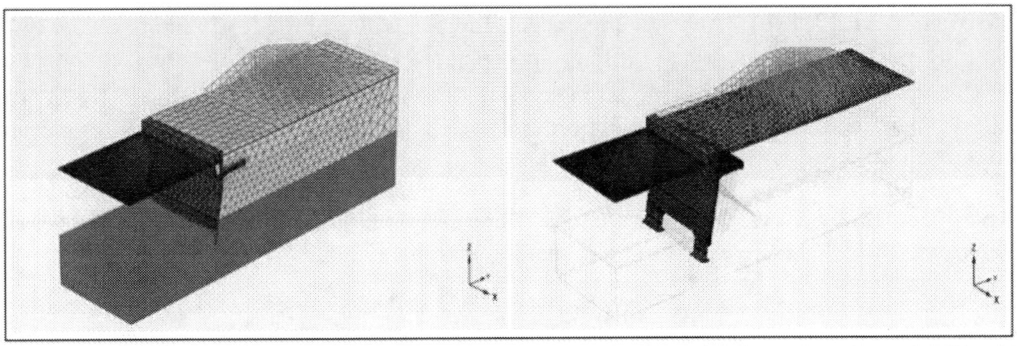

Fig. 10 Deformed mesh model 2B with designed surcharge (Left) and deformed mesh model 2B with designed surcharge without soil (Right)

The results of the previous models of the quay wall are compared to the calibration model and model 2B based on the structural stability.

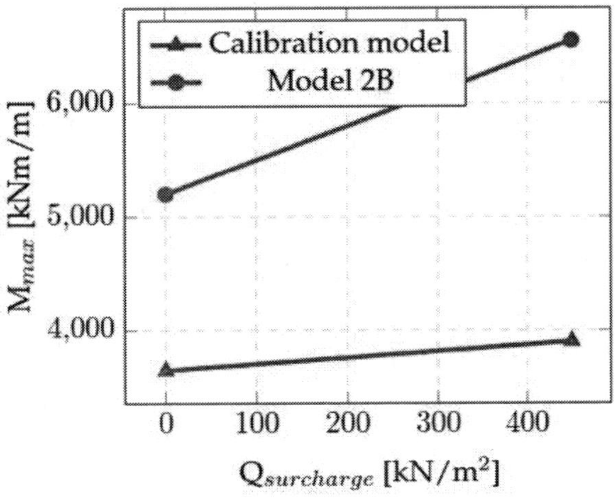

Fig. 11 Comparison maximum moment between calibration model and model 2B

In Fig. 10 a comparison between the two models is illustrated. As it can be observed from the graph the maximum bending moments in model 2B are higher than in the calibration model. The reason for this is that model 2B has an opening at the centre of the combined wall. Because of this opening the redistribution of the moment will be higher at the sides than the centre of the combined wall where the opening is situated.

Table 1 Safety comparison between calibration model and model 2B

	Calibration model with surcharge	**Model 2B with surcharge**
Unity Check	0.738	1.436
Safety margin	1.355	0.696
Safety factor	1.494	1.473

The safety analysis results in a safety factor which can be obtained by reducing the strength parameters incrementally. Starting from un-factored values that are inputted when creating the soil layers of the models ϕ_{avail} and c_{avail}, until equilibrium can no longer be achieved in the calculations [7, 8, 9]. The results represents the behaviour of the loads acting on the quay wall, in this case the self-weight, at the defined construction phase. The safety factor value (S.F.-value) for the calibration model is equal to **1.494** and for model 2B this is **1.473**. The safety factor value implies the reduction factor that is imposed on the friction angle and cohesion until the quay wall becomes unstable.

Apart from the S.F.-value a unity check and safety margin of these two models were determined which can be expressed in a mathematical equation:

$$\sigma = \frac{N}{A} + \frac{M}{W} \leq \sigma_{yield} \tag{1}$$

$$\frac{\sigma}{\sigma_{yield}} < F.S.Value \tag{2}$$

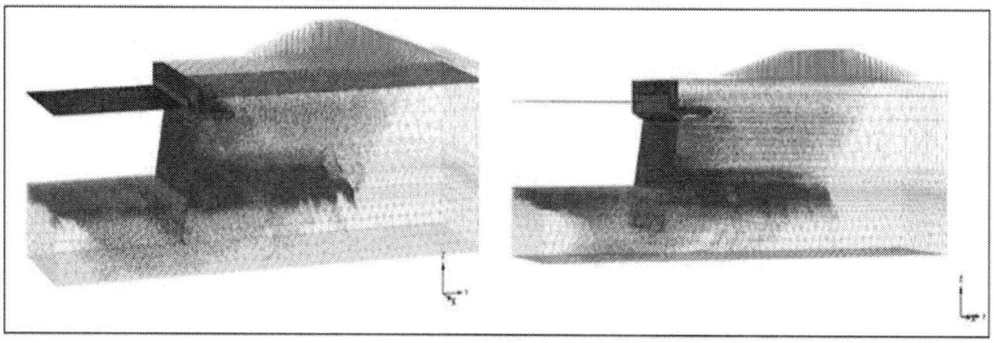

Fig. 12 Plastic points calibration model (Left) and plastic points model 2B (Right)

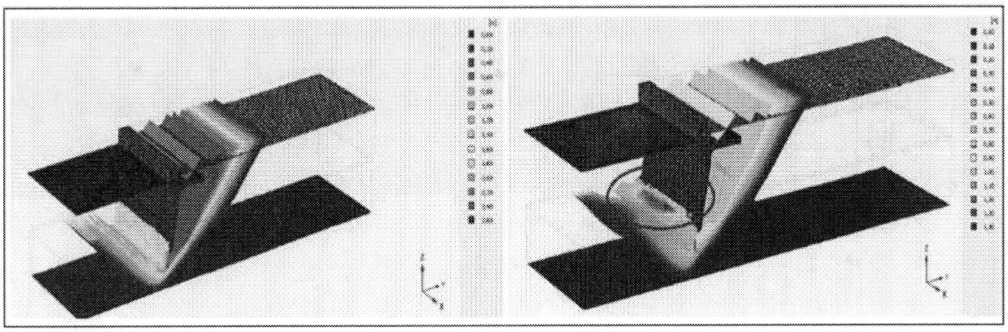

Fig. 13 Phase displacement calibration model (Left) and phase displacement model 2B (Right)

In Fig. 11 a comparison is shown for both model which shows that underneath the combined wall higher stresses occur and these are very interesting to analyse. This figure shows for both model the potential of numerical methods when investigating failure mechanisms soil-structure interaction should be taken into account. The structural elements, specifically the combined wall, will yield due to the effective horizontal soil pressure from the active side. The effective horizontal soil pressure on the active side underneath the relieving platform structure, which can be recognized by the red points, becomes larger than the effective horizontal soil pressure on the passive side.

However, the phase displacements of the models (Fig. 12) a difference can be observed at the passive side. The contour lines are different at the passive side in front of the combined wall. In Fig. 12 at the passive side the phase displacement is larger and at the surface level of the harbour basin. This is because of the opening being two meters below the construction depth, N.A.P. -25.50m. According to the φ/c –reduction calculation for Model 2B and the calibration model the calculated deformations are not realistic, but it can illustrate how the quay wall model will behave.

5. Conclusions and Recommendations

- The actual quay wall shows no instability.
- In Plaxis 3D, Model 2B will become unstable when exposed to unfavourable conditions. The deformation behaviour of the combined wall will be too large due to higher loads than anticipated. However, the safety factor value of model 2B comes close to the safety coefficient of 1.5 which was used on the quay wall during the design process. Therefore, it can be concluded that even though the quay wall was never fully loaded it was strong enough to retain the loads it was exposed to during its service lifetime. Also the superstructure was strong enough and capable to retain all the excess loads without collapsing although the combined walls never reached its designed penetration depth.
- The results shows how the quay wall behaved during its service lifetime. And it also illustrates both models will fail due to steel yielding and subsequently failure of the passive soil resistance.
- The influence of the interaction between deep and hard soil layers and open tubular piles during pile driving should be investigated.

DOI: 10.1051/matecconf/201713806001

References

1. N.K.N. Mourillon, J.G. de Gijt, K.J. Bakker, T. Vellinga, P. Taneja, S.H.J. van Es, E.J. Broos, *Stability analysis quay structure at the Amazonehaven port of Rotterdam* (2015)
2. J.G. de Gijt , M.L. Broeken, *Quay walls, Second edition SBRCURnet Publication 211E* (2014)
3. J.G. de Gijt, A.A. Roubos, D. Grotegoed, *SBRCUR 186 Binnenstedelijke kademuren.* (2014)
4. J. Grabe, *Sheet piling handbook design* (2008)
5. CUR Bouw & Infra, *Damwandconstructies. SBRCURnet Publication CUR 166* (2012)
6. C.N. van Schaik, J.G. de Gijt, *Diepwater terminal Amazonehaven zuidzijde* (1989)
7. Plaxis, *Reference Manual* (2013)
8. Plaxis, *Material Models Manual* (2013)
9. R.B.J. Brinkgreve, H.L. Bakker, *Non-linear finite element method analysis of safety factors. In Proc. 7th Int. Conf. on Comp. Methods and Advances in Geomechanics. Cairns, Australia* (1991)
10. R. Nejad, J.G. de Gijt, J. Verlaan, H. Pacejka, P. Middendorp, S.H.J. van Es, *Research into pile toe failure in Amazonehaven* (to be published)

Characteristic of macroinvertebrates abundance in cascade-pond system at Universitas Indonesia Campus, West Java, Indonesia

Muthiah Fadillah Prinasti[1,*], *Dwita* Sutjiningsih[1] , and *Evi* Anggraheni[1]

[1]Civil Engineering Department, Universitas Indonesia, 16424 Depok, Indonesia

Abstract. Catchment area of cascade-pond system at Universitas Indonesia (UI) campus is an area with high population density. The high population density has led to increasing impervious cover since it has to provide the housing needs. Increasing of impervious cover affected stream quality in the catchment area. The state of aquatic system quality can be assessed among others using biotic index and observed macroinvertebrates distribution. The management of UI campus plans to reduce the negative effect of decreasing the aquatic system quality by conducting audit the health state of cascade-pond system. The objective of this study is to examine the characteristics of macroinvertebrates abundance at cascade-pond system at UI campus. Macroinvertebrates are commonly used as indicator of pollution level at aquatic system. Sampling is conducted using biomonitoring method, and is carried out at cascade-pond system, which consisted of six ponds, by two times repetitions. The identification of any individual macroinvertebrates caught during the sampling is conducted in the laboratory. The characteristics of macroinvertebrates were assessed based on Average Score Per Taxon (ASPT) Index dan Family Biotic Index (FBI).The results indicates that there is a tendency of improving diversity of macroinvertebrate in the downstream direction of the cascade-pond system.

1 Introduction

In order to become a green campus, UI built aquatic ecosystem covering approximately 30 hectares. It system consists of six ponds, which serves various purposes including as landscape element as well as natural laboratory for both academic staffs and students. Along with the population growth, its catchment area is becoming a region which a high population density. The population density has led to increasing impervious cover caused by the growing need for a place to stay. Increasing impervious cover can affect water quality in the catchment area [1]. After more than 20 years of service, the presence of contaminant in the cascade-pond is visible. Water contamination can occur due to human activities that eventually becomes runoff into water bodies [2].

* Corresponding author: muthiah.fadillah@ui.ac.id

DOI: 10.1051/matecconf/201713806002

Biomonitoring is the systematic observation of organism (macroinvertebrates) and their response to environmental quality. Macroinvertebrates are widely used to determine the parameters of water quality because it has a sensitivity to each type of pollutant and has a quick reaction, do not have ability to migrate if there are alterations in the conditions of the waters, is arrested for visible and separated into several types [3].

The study aims to identify the abundance of macroinvertebrates found in the cascade-pond system at Universitas Indonesia, Depok and to know the characteristics of macroinvertebrates by identifying the by Average Score Per Taxon (ASPT) and the Family Biotic Index (FBI). Most of biomonitoring were conducted in lentic waters like a river. This study is conducted in lotic waters by applying biomonitoring method in cascade pond system, UI Depok.

2 Research Method

2.1 Reserach Site

The research site is the cascade-pond system at Universitas Indonesia, Depok, West Java (Fig.1). The samples were taken in each inlet, middle and outlet of each pond. In this study, Rawa Besar Pond and Pladen Pond not included to futher analysis, because we the sampling conducted only within the campus area.

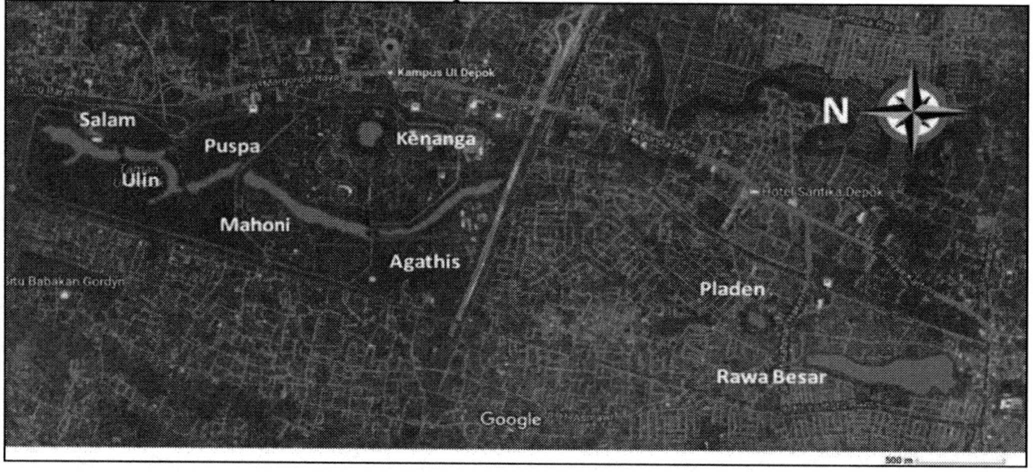

Fig. 1. Location of the study [4]

2.2 Sampling Method

Sampling macroinvertebrates needed some intruments and tools such as biomonitoring nets, sieve to separate the mud, glass bottle as a place to put samples, sample cards macroinvertebrates, and pH meters. Materials used during the macroinvertebrates sampling is 70% alcohol and formaldehide 4% to preserve macroinvertebrates. Macroinvertebrates sampling were carried out in each of the determined sample points. Macroinvertebrate sampling per sample point approximately 30 macroinvertebrates sought. The samples were preserved in the sample box for identification purpose. The identitification using microscope in laboratory [5].

2.3 Analysis of Macroinvertebrates

In the determination of water quality based on the biotic index, there are several indices that are commonly used, such as, Family Biotic Index (FBI), EPT Index, Biological Monitoring Working Party (BMWP), and Average Score Per taxon (ASPT). Each of these indices has different calculation methods, but still based on the diversity of the biota to be found.

2.3.1 Average Score Per Taxon

ASPT index itself is a development of BMWP. Both biotic index was indices were developed in the United Kingdom. ASPT index can be calculated using the formula [6]:

$$ASPT\ Index = \frac{\Sigma(BMWP\ score \times n)}{\Sigma n} \quad (1)$$

Where:

BMWP score : scores of taxa found
n : number of macroinvertebrates found during the sampling

The score in BMWP index ranges from 1 to 10. The higher figure score of BMWP, the more vulnerable or sensitive to contaminants. The results of the ASPT index that shows the quality of water is as follows:

Table 1. ASPT Score [3]

ASPT Value	Water Quality Assessment
>6	Clean Water
5-6	Doubtful quality
4-5	Probable moderate pollution
<4	Probable severe pollution

2.3.2 Family Biotic Index

Family Biotic Index (FBI) is an index that is based on the level of taxa resistance to contamination. The score ranges from 0 to 10, the higher the score the more resistant to pollutant [7]. The following is the FBI formula:

$$FBI = \frac{\Sigma x \cdot t}{\Sigma n} \quad (2)$$

Where:

FBI : Family Biotic Index
x : Macroinvertebrates found
t : Scores against pollution tolerant taxa
n : number of macroinvertebrates found during the sampling

The value of water quality index based on the Family Biotic Index are as follows:

Table 2. Family Biotic Index Score [7]

Family Biotic Index	Water Quality	Degree of Organic Pollution
0.00-3.75	Excellent	Organic Pollution Unlikely
3.76-4.25	Very Good	Possible slight organic pollution
4.26-5.00	Good	some organic pollution probable
5.01-5.75	Fair	Fairly substansial pollution likely
5.76-6.50	Fairly Poor	Substansial pollution likely
6.51-7.25	Poor	Very substansial pollution likely
7.26-10.00	Very Poor	Severe organic pollution likely

2.4 The Influence of Riparian Buffer in Macroinvertebrata Abundance

In Webster's New Universal Unabridged Dictionary, riparian is located on the bank of a natural watercourse (such as a river) or sometimes of a lake or a tidewater. Riparian zone are a combination of several landscapes, communities and environments that serve to understand the organization, diversity, and associated community dynamics in the aquatic ecosystem [8].

Many of the functions of the riparian zone are ecological functions. On the ecological function of the riparian zone serves as a sources of nourishment, as nutrient filters, and habitat [9]. As nutrient filters riparian zone serves as a biological buffer. Riparian forests are especially important sites for biotic accumulations of nutrients because transpiration may be quite high, increasing the mass flow of nutrient solutes toward root systems, and because morphological and physiological adaptations of the many flood-tolerant species facilitate nutrient uptake under low-oxygen condition [9]. Other than as a biological buffer, also serves as a Physical buffer. Which filters out sediments carried by runoff [10].

As a habitat, the riparian zone affects the abundance of macroinvertebrates. Riparian buffer zones could significantly reduce the impacts of deforestation on tropical streams. Macroinvertebrates are sensitive animals to existing pollution. Riparian buffer appeared to significantly reduce the effects of environmental degradation on benthic communities, as macroinvertebrate diversity [11].

3 Result and Discussion

Based on the sampling result, the abundance of macroinvertebrates found in each pond from upstream to downstream, namely, Kenanga, Agathis, Mahoni, Puspa, Ulin, and Salam are as follows:

Table 3. Macroinvertebrates found in cascade-pond.

Makroinvertebrates	Amount macroinvertebrates in each cascade-pond system					
	Kenanga	Agathis	Mahoni	Puspa	Ulin	Salam
Filopalodina javanica	3	1	63	24	20	13
Melanoides tuberculata	39	2	13	10	1	1
Tarebia granifera	0	0	8	1	0	0
Gyraulus sp.	0	19	0	0	0	1
Indoplanoubis	0	1	0	0	0	0
Lymnea Rubignosa	0	3	0	0	0	0
Pilla scutata	3	1	21	0	12	17
Pomacea canaliculata	3	1	4	0	1	0
Parathelphusa convexa	0	3	1	2	3	10
Parathelphusidae	0	0	6	5	0	22
Gyrinius sp.	0	47	0	0	0	13
Hydrophillus larva	0	1	0	0	0	0
Hydrophorus sp.	0	1	0	0	0	0
Corixa sp.	0	3	0	1	1	3
Notonecta sp.	0	1	0	0	0	4
Ilyocoris cimicoides	0	0	0	0	2	2
Gerris sp.	0	2	0	0	30	7
Sialis larva	0	1	0	0	0	0
Aeshna sp.	0	1	0	1	0	0
Gomphus sp.	0	1	0	2	0	0
Nereiridae	0	1	0	0	0	0
Tubificidae	0	5	1	0	0	0
Total	48	95	117	46	70	93

After processing by macroinvertebrates found ASPT value and FBI obtained as follows:

Table 4. ASPT and FBI Value

Score	Cascade-Pond System					
	Kenanga	Agathis	Mahoni	Puspa	Ulin	Salam
ASPT	3.19	4.27	4.77	5.39	4.98	5.08
FBI	6.98	5.43	6.41	5.93	5.73	5.84

ASPT values of Kenanga pond amounted to 3.19 indicating that is heavily polluted. Similarly, base on Family Biotic Index with a value of 6.98. The macroinvertebrates are found largely a group of gastropods. If in a body of water gastropods are dominant, then the waters classified as polluted. The riparian of Kenanga has been converted into riprap walls. If a body of water has undergone development, it will reduce the diversity of macroinvertebrates.

The ASPT value of Agathis pond is 4.27, which indicates the waters are moderately polluted. The same conclusion when viewed from Family Biotic Index with a value of 5.43. If seen from the distribution of macroinvertebrates in Agathis, the types of macroinvertebrates are quite vary, as including *Oligochaeta, Gastropods, Crustacean*, and *Insecta*. *Oligochaetta* and *Gastropods* are indicators of pollution. The most probable reason that Agathis has a better FBI score is the condition of its riparian, which is still in natural condition.

Mahoni pond has two inlets, from Agathis and from outside the campus, while its wall is already constructed by using riprap. The ASPT value is equal to 4.77, which indicated that Mahoni is moderately polluted. Similarly by Family Biotic Index the value of 6.41 indicates a rather heavyly polluted. The dominant macroinvertebrates found are gastropods, almost 93% are snails. The gastropod signifies the polluted waters. Similarly, if seen from the riparian condition of Mahoni, which has undergone development, will influence the spread of macroinvertebrates.

Puspa pond located downstream of Mahoni. The conditions of the riparian are still natural, surrounded by city forests. The ASPT value of Puspa is 5.39, which indicates lightly polluted conditions. However, if assessed based on Family Biotic Index, Puspa has a value of 5.93 indicating lightly heavy polluted waters. Distribution of macroinvertebrates consists of *Gastropods, Crustaceans, Insecta*, and Dragonfly larvae.

Ulin pond has three inlets. The first comes from Puspa, the second, and the third inlets come from outside the campus. Based on the ASPT, Ulin has a value of 4.98 which means Ulin is moderately polluted. A Family Biotic Index value of 5.73, indicates that the waters is moderately polluted. Distribution of macroinvertebrates at Ulin consists of *Insecta, Gastropods*, and *Crustaceans*. Although the riparian of Ulin still surrounded by city forest, the inlet side wall is already constructed by using riprap.

Salam pond has two inlets, first comes from the Ulin and the second comes from outside the campus. Based on ASPT value of 5.08, Salam is lightly polluted. While based on Family Biotic Index value of 5.84, meaning that Salam is polluted rather bad. Macroinvertebrates found in Salam is of *Crustaceans, Gastropods*, and *Insecta*. Riparian zone of Salam is city forest, while at the side of the wall has been constructed by using riprap.

The overall distribution of macroinvertebrates in the cascade-pond system is presentedin the following figures:

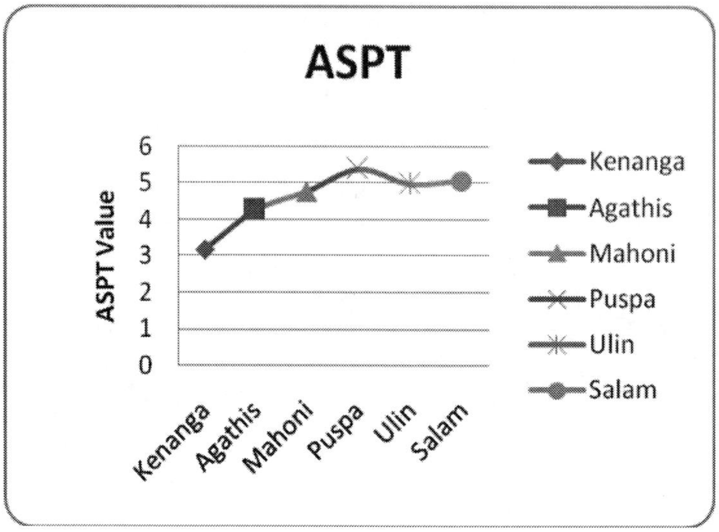

Fig. 2. ASPT Index of Cascade-pond System

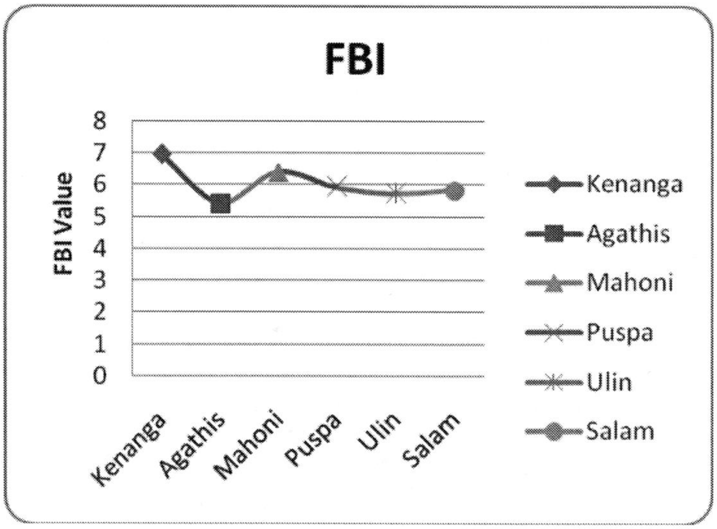

Fig. 3. Family Biotic Index of Cascade-pond System

Considering the cascade-pond system as a whole, in the downstream direction, in general, the distribution of macroinvertebrates and water quality is getting better. However, valuation by the ASPT index, cascade-pond Puspa have the most excellent value than the others. When viewed as a cascade-pond system, Salam, the most downstream pond should be better than Puspa. The most probable reasons are the riparian of Puspa is still natural, while part of the Salam wall is already constructed by using riprap.

Although in general Salam has better macroinvertebrates distribution, Agathis has the best FBI score. Family Biotic Index is essentially based on the level of taxa resistance to contamination. In Agathis encountered *Gomphus Sp* and *Aeschna Sp*, which is the larvae of a dragonfly. They are susceptible to pollution. If one of these macroinvertebrates is found, meaning that Agathis is still under natural condition.

4 Conclusion

Based on the stages and research methods undertaken, it can be concluded as follow:

1. Based on ASPT index in general cascade-pond system in Universitas Indonesia is in medium contaminated condition. Except in the Kenanga, which is in a heavy polluted state. Puspa and Salam are lightly contaminated.
2. Based on the Family Biotic Index the condition of Universitas Indonesia's cascade-pond is slightly polluted. Except cascade-pond Kenanga, which is heavily polluted. Whereas, Agathis and Ulin are in medium contaminated condition.
3. In general the spread of macroinvertebrate abundance is increasing in the downstream direction of the cascade-pond system, and also is the diversity
4. Based on the ASPT Index, Puspa is the best, while based on FBI, Agathis is the best. The difference is influenced by the condition of the riparian zone of each pond. The abundance of macroinvertebrate is also affected by riparian zone.

Acknowledgements are addressed to Universitas Indonesia for giving financial support through the PITTA 2017 funding scheme. A special thanks to Drs. Wisnu Wardhana, M.Si as Freshwater Ecology Lecturer to help us identify macroivertebrates and 16 Surveyors which consist of college students and staffs of Civil Engineering Department Universitas Indonesia in conducting field survey.

References

1. Center for Watershed Protection , *Watershed Protection Research Monograph No.1: Impacts of Impervious Cover On Aquatic Systems* (Center for Watershed Protection, Ellicot City, 2003).
2. J. Sartor, G. Boyd, F. Agardy, J. Wat. Poll. Cont. Fed **46**, 458 (1974)
3. S. Mandaville, *Benthic Macroinvertebrates in Freshwaters Taxa Tolerance Values, Metrics, and Protocols* (Soil & Water Conservation Society of Metro Halifax, Nova Scotia, Canada, 2002)
4. D. Sutjiningsih, Int. Jour. Tech **1**, 145 (2017)
5. W. Wardhana. *Metoda Prakiraan Dampak dan Pengelolannya pada Komponen Biota Akuatik* (Universitas Indonesia, Jakarta, 2006)
6. W. Wardhana. *Perubahan Lingkungan Perairan dan Pengaruhnya Terhadap Biota Akuatik* (Universitas Indonesia, Jakarta, 1999)
7. F. Haurer, G. Lamberty, *Methods in Stream Ecology: Second Edition* (Elsevier, Academic Press, California, 2007)
8. S. Gregory, F. Swanson, W. Mckee, K. Cummins, J. Bio. Sci **41**, 540 (1991)
9. R. Naiman, H. Decamps, Annu. Rev. Ecol. Syst **28**, 621 (1997)
10. T. Dillaha, R. Reneaus, S. Mostaghimi, D. Lee, Trans. Am. Soc. Agric. Eng **32**, 513 (1989)
11. C. Lorion, B. Kennedy, J. Fresh. Bio **54**, 165 (2009)

MATEC Web of Conferences **138**, 06003 (2017) DOI: 10.1051/matecconf/201713806003
EACEF 2017

Green infrastructure in oil rich states; Abu Dhabi Emirate-Alain as a case study

Mohamed El Amrousi[1,*], and *Mohamed* Elhakeem[2]

[1]Department of Architecture, Abu Dhabi University, 59911, United Arab Emirates
[2]Department of Civil Engineering, Abu Dhabi University 59911, United Arab Emirates

Abstract. The Abu Dhabi 2030 Vision highlights the importance of Sustainable Design and Green Infrastructure inside the UAE and internationally. Future development plans in Abu Dhabi and Alain developed by the Urban Planning Council-UPC manifest emerging interest in Green Infrastructure and Environmental planning. The development of several landscape projects that integrate water elements such as Jahili Park, Al-Ain Al-Fayada and Sheikh Zayed Desert Learning Centre cement the 2030 vision. In order to understand the future of projects with water canals in Desert Cities it is essential to examine the Alain Oasis and its irrigation system known as *falaj*, that is similar to historic aqueducts and antecedents in the Islamic world such as Qasr Alhambra in Granada/Spain that also had a complex canalization system. This paper studies emerging projects with water features in Alain, and the water flow system in the *falaj* canalization system in the Oasis by conducting a hydrologic study to quantify the amounts of water needed to irrigate these green areas and simulate the flow condition in the *falaj* complex channel system using the 2D-hydrodynamic Finite Element Surface Water Modeling System (FESWMS). This paper aims at giving projection of the amounts of water needed sustainable green infrastructure and will be helpful for the decision makers if they want to repeat this model in the future.

1. Introduction

The United Arab Emirates has developed its landscape in a desert region to have the largest number of fruit-bearing palm date trees in the Arab World, estimated at more than 16 million trees. Abu Dhabi's hinterland city of Alain takes its name from the Arabic translation of the word Spring or Oasis because it includes vast fresh water reserves in its underground rock foundation. The Oasis in Alain has a *falaj* canalization system that distributes water through its networks via gravitational slope similar to Roman aqueducts, yet it is not covered therefore a percentage of its water evaporates while irrigating palm tree plantations. Alain's complex *falaj* canalization system is supplied by ground water via submersible pumps between rock formations in Buraimi region, a shared borderland region between the United Arab Emirates and Oman. The Falaj Alaini and *falaj* Daoudi are the main water supply systems for the Palm Oasis located in the Desert city of Alain, hence they represent a valuable reserve water resource of symbolic and cultural value for the community. In the Arab world fruit bearing gardens "*bustan*" have a prominent position in history because of their association with paradisiacal metaphors in the Quran and have been described in texts, and depicted in miniature paintings and mosaic panels. Alain city includes many fruit bearing gardens interwoven within its urban fabric. Following the unification of the United Arab Emirates in 1971 urban development of Desert cities such as Alain aimed to attract tribes to the city through offering them large land plots to construct houses for extended families and establishing a national university in 1976 titled the United

* Mohamed El Amrousi: Mohamed.amrousi@adu.ac.ae

Arab Emirates University (UAEU). Decrees by the founder of the modern Nation Sheikh Zayed bin Sultan Al Nahyan aimed to develop Alain into a Garden City in a challenging desert environment interpreted the planning of the city as a modern city with the Oasis in the center of the downtown area and retained a few traditional adobe forts and residences while the rest of the city expanded horizontally to include modern buildings designed to respect the socio-cultural and environmental criteria of the region such as UAE University Campus, Tawam Hospital, Oasis Hospital, and the Sheikh Zayed Desert Learning Centre are all landmarks that support the green infrastructure of Alain. Alain horizontal growth supplemented by palm tree farms was supported by a building code that sets the limitation of building heights to 20 meters i.e. Ground floor + 4 floors [1].

2. Green Infrastructure and the Modern City

The role of the Oasis in the centre of Alain city was crucial in the development of the green infrastructure of the city as a whole and its conservation from unplanned urban sprawl, other Oasis in the Alain such as Al-Qattara Oasis followed a similar plan with its traditional market, Arts and Crafts centre manifest a consistent urban strategy that follows the Abu Dhabi 2030 Vision for the development of green infrastructure in Alain and sustainable buildings in Abu Dhabi. A significant section in the 2030 vision emphasized the importance of sustainable development through guidelines known as 'Estidama', a building program for innovative green standards, other legislations such decrees banning the destruction of trees in favour of urban sprawl supported the growth of Alain as the Garden City of the UAE. The city has a population of 627,000 and aims to grow to 1 million by 2030 which may create a challenge on its water resources, while Alain's urban growth is warranted as part of the development of cities in the UAE, Abu Dhabi's Urban Planning Council (UPC) 2007, aims to improve the quality of life in the city through projects that employ water features such as al-Jahili Park (Figure 1a) and contextualized green infrastructure, therefore management of ground and surface water resources is crucial because historically many cities have been deserted as a result of dilapidation of water resources. The UPC Plans exhibit support for the cohesive development of the Oasis and its integration in the modern city, through increased pedestrian access, shaded walkways and extensive public landscapes. Palm trees Oasis gardens and parks and other open spaces in cities offer attractive settings for outdoor recreation that are important for residents' satisfaction with their living environment [2].

Traditional Arab urban enclaves and cities have overtime responded to the harsh Desert climatic conditions and scarce water resources by developing dense urban fabrics, with narrow shaded pedestrian pathways, and adobe courtyard houses with thick walls to protect its inhabitants from the heat. Small windows, and niches were both functional and aesthetic and protected socio-cultural values of Islamic and tribal communities. This is best manifested by the Nizwa Fort and its surrounding buildings that intertwine with the palm trees. The conservation of a traditional building fabric is evident in Nizwa and provides a more picturesque perception to the visitor as a reflection of environmental/regional architecture, however, it also limits the urban expansion of the city. In Abu Dhabi and Alain development of green infrastructure in conjunction with modern transportation means reflect a different approach than that of Nizwa, it is a green infrastructure plan that addresses planning requirements of the modern city hence it is less picturesque than the image of the traditional Arab city. Planners are therefore challenged to balance between tradition and modernity in the same manner that 19th Century planners were as exemplified by Ebenezer Howard in planning the *Garden Cities of To-morrow* and Frederick Olmsted's Central Park designs. The creation of green infrastructure and healthy cities warrant community education regarding the importance of green infrastructure is part of the

 DOI: 10.1051/matecconf/201713806003

community's day-to-day activities and highlighting the significance of creating a healthy city that would preserve resources for future generations.

In Alain the new Sheikh Zayed Desert Learning Centre sets another example of the development of green infrastructure in Abu Dhabi. The building which revisits many of the traditional design elements such as narrow windows on the exterior walls and central courtyard reflects Abu Dhabi's initiatives for innovation and reinterpretation of the past. The massive fort like building UAE's first sustainable building to be awarded the 5 Pearl rating from Estidama. The Sheikh Zayed Desert Learning Centre reduces dependency on mechanical air-conditioning through its form and use of solar power. In addition to its narrow lozenge shaped energy efficient windows, and sand textured walls that emulate its Desert environment. The Centre vital for community education and has created around it vast gardens with Palm tree plantations and a reconstructed *falaj* system (Figure 1b).

Fig. 1a & b Al-Jahili Park and Sheikh Zayed Desert Learning Centre-Alain

3. Developing Alain's urban landscape historical precedents and the living city

In Abu Dhabi's many projects are developing along water canals such as the new Central Park that includes several kilometres of water bodies along which residential, leisure and entertainment projects are constructed. The main difference between urban developments along seawater canals in Abu Dhabi versus development of projects in Alain is that in first there is abundance in sea water while in the later it is essential to sustain fresh water resources through appropriate management, in the simulation model created in this paper we address the areas water flow in the *falaj* canalization system in Alain and the highlighting the critical areas, such a study can be expanded on a larger scale to include the study of water flow in the emerging project in Alain Al-Fayda Canal villas in order to sustain them in the future. President studies are also important for understanding the challenges that may face Alain's green infrastructure and urban expansion along water canals. One of the good examples is New York's Central Park that was designed and engineered by Frederick Olmsted, today it sets an example on how water bodies, lakes and canals can reduce the heat island effect and create public space of inclusion in the city. Frederick Olmsted in planning the Central Park in New York advocated for creation of natural landscapes and green areas in cities. At the time such a public park would promote democratic values and improved social life. Even though Olmsted adhered to a naturalistic aesthetic, his draining design, road-making, planting, brush-cutting, and digging artificial ponds were all part of the green infrastructure warranted to create such a project [3]. In the Abu Dhabi Central Park which will be developed by Sorouh Investments, water canals are planned to be a dominant site attraction, with 4 kilometres of waterfront/canal side pedestrian Abu Dhabi is thereby constructing an engineered environment as a green Oasis in the modern city. In Alain the opening of Jahili Park located in the downtown area next to Jahili Fort and the recreated Falaj in Sheikh Zayed Desert Learning Centre further cement Abu Dhabi's interest in the development of green infrastructure (Figure 2a & 2b). The open garden includes diverse forms of trees and water elements that cascade from geometric basins on several levels. The Park was constructed next to Jahili Fort one of the UAE's historic landmarks and thereby accentuates its visual identity and attracts visitors to it. The introduction of new landscape areas in the city juxtapose modernity and tradition since reconstructing the past as in the case of Nizwa was not park of the planning scheme on the contrary green infrastructure within a contemporary context aimed to complement the city via the creation of lungs in the Downtown area that reduce pollution resulting from exhaust of cars. Similar to Sheikh Zayed learning Centre Jahili Park was planned to increase community awareness of the importance of plants and green spaces and their role in the protection of environment. Here, green infrastructure interventions such as vegetated walls, in addition to the preservation of palm trees and parks mitigate extreme urban heat islands from rising temperature associated with climate change especially in a hot dry climate present in the UAE and the region. Ebenezer Howard in his book the Garden Cities of To-morrow published in 1898, highlights the problems of London as a metropolis that was in need of resolutions for its transformation into a liveable city capable of producing health societies [4]. He proposed a concentric circles diagram that reduces the density of the city centre and sets within its centre a large garden surrounded by leisure pavilions, he also thereby caps the population growth of his Garden Cities by creating a series of small cities that would be interconnected by a railway as public transport. Howard attempts to offer communities a combination of the positive aspects of Cities and Countryside, in that he acknowledges the importance of urban landscape as manifested in his diagrammatic plans that are still valid today. Ideas such as integration of nature into cities translate into green infrastructure and constitute the fundamentals of sustainable urbanism. Howard comprehended the effect of the metropolis in attracting people such as jobs opportunities and leisure facilities and therefore he attempted to combine services of the city and the advantages of the suburban lifestyle such as better communal interaction and larger family oriented living spaces. Setting strict building codes and building height limits for the Downtown area in Alain is a major step to preserve cultural identity of the Downtown area from unplanned urban sprawl and protect the Gold market and the Oasis. Development plans of the existing area of the Downtown and the Oasis display awareness of the importance of landscape and visual connectivity between the Gold Market and the Oasis through the integration of more public spaces development of clusters that take into consideration the sociability and the UAE's cultural values [5]. Similar to Ebenezer Howard's Garden City ideology to connect between the Country Sides with its landscape to Cities via public transport, Alain's development plans exhibit connectivity through green corridors that connect between the edges of the Oasis and the Downtown area. Furthermore, through the construction of a new cultural complex, will enhance the presence of the Oasis similar to the visual connectivity between Jahili Park and Al-Jahili Fort. The concept of transforming urban centres and large metropolises from heterotopias to spaces of inclusion and healthy cities has constantly been a challenge to urban planners and landscape designers.

DOI: 10.1051/matecconf/201713806003

Fig. 2a & b Al-Jahli Park, and *Falaj* Garden in Sheikh Zayed Learning Center

The creation of gardens and landscapes in association with urban settings have always been an essential part of planning for better cities, the Shalimar Gardens in Lahore designed in the 17th century employed water elements as an essential part of their composition. Viewing pavilions constructed amidst water ponds and along them further consolidates the importance of water as an essential element in landscape. In Arab world two of the best known examples to us are Madinat al-Zahra on the outskirts of Cordoba and Qasr Alhambra in Granada/Spain. Both cities although several centuries apart can be viewed as examples of Islamic Garden cities that incorporated fruit bearing gardens, open views of landscapes, pavilions and water fountains. Water features in Islamic gardens had many functions from agricultural to aesthetic for example mirroring the formal elements of the irrigated agricultural landscape. The pleasure garden displays the most important formal elements of the irrigated, cultivated landscape: the grid of garden beds where plants are organized and tended, shallow channels for the transport of water, and pools for water storage. They were supported by a complex hydraulic system of *falaj* supplied by aqueducts. In Madinat al-Zahra landscape was designed based on geometry and symmetry the governed the relationship between buildings and open spaces. This is manifested in reception halls, administrative and government offices, gardens, barracks, residences, and baths. The buildings of Madinat al-Zahra were built on large, stepped terraces were formulated carefully to direct viewers in shaded pavilions to water pools and landscape [6]. In Qasr Alhambra in Cordoba water was used extensively in the exterior landscape and within the bathhouses of palatial complexes. The canalization system allowed water to be elevated to the top (without an electromechanical system) then channelled through the canalization systems with multiple divisions that allowed it to be used in the fountains of the Palace, its canals and then cascaded to the fruit bearing Gardens of the Generalife, via a system of cisterns that made the development of Alhambra gardens and agricultural landscapes around it in the city below it possible. As outlined by Nasser Rabbat the effect achieved by the manipulation of water was certainly intentional as manifested by the Fountain of the Lions were water canals form links between the Gardens and the rest of the palatial city [7]. In Alhambra water was used as an essential resource and also as a method of climate controlling and as an aesthetic element, here Roman aqueducts must be acknowledged as fine examples of water resource management, the role of repair and maintenance must have been critical for a proper operation of the aqueduct system as well as the quality of the waters [8].

DOI: 10.1051/matecconf/201713806003

4. Alain's Oasis and the *falaj*, application of the Water Modeling System

Downtown Alain has the largest Oasis area in and urban setting in the UAE (Figure 3). The preservation of the Oasis and its *falaj* water system is essential because of declining groundwater its replacement by artificial recharge of groundwater from TSE and desalinated irrigation water which is now widely used to keep the garden city of Al Ain green. The *falaj* in the Oasis is largely supplemented from pumped groundwater from support wells and piped desalinated water from the relatively new Qidfa desalination plant in Al Fujairah [9].

Fig. 3 Alain Oasis *falaj* irrigation system

Due to the complex landscape geometry of of Falaj Al Ain (Figure 4) irrigation system (canals), the 2D Finite Element Surface Water Modeling System (FESWMS) is used to investigate the flow pattern in the cannels. FESWMS is part of the commercially available Surface water Modeling System (SMS) package version 12.2 Graphical Interface that combines a series of hydrodynamic/ sediment codes [10].

FESWMS solves the differential forms of the continuity and the momentum equations in the stream wise and transverse directions using the Galerkin method of weighted residuals, providing water depth and depth-averaged velocity magnitude in x and y directions at each node in the grid [11]. The governing equations are written in the conservative form hence, the momentum is conserved along the streamline and the model is capable of capturing shock effects [12]. The conservative form was chosen because of its robustness in solving critical and transcritical flow fields under both low and high flow conditions by allowing dry-elements to exist within the computational mesh [12].

FESWMS solves the following equations simultaneously:

$$\frac{\partial z_w}{\partial t} + \frac{\partial q_1}{\partial x} + \frac{\partial q_2}{\partial y} - q_m = 0 \qquad (1)$$

$$\frac{\partial q_1}{\partial t}+\frac{\partial}{\partial x}\left(\frac{q_1^2}{d}+\frac{1}{2}gd^2\right)+\frac{\partial}{\partial y}\left(\frac{q_1q_2}{d}\right)+gd\frac{dz_b}{\partial x}$$
$$+gn^2\frac{q_1\sqrt{q_1^2+q_2^2}}{d^{7/3}}\sqrt{1+\left(\frac{\partial z_b}{\partial x}\right)^2+\left(\frac{\partial z_b}{\partial y}\right)^2} \quad (2)$$
$$-2d\varepsilon_{xx}\frac{\partial^2\overline{u}}{\partial x^2}-\varepsilon_{xy}\frac{\partial}{\partial y}\left(\frac{\partial\overline{u}}{\partial x}+\frac{\partial\overline{v}}{\partial y}\right)=0$$

$$\frac{\partial q_2}{\partial t}+\frac{\partial}{\partial y}\left(\frac{q_2^2}{d}+\frac{1}{2}gd^2\right)+\frac{\partial}{\partial x}\left(\frac{q_1q_2}{d}\right)+gd\frac{dz_b}{\partial y}$$
$$+gn^2\frac{q_2\sqrt{q_1^2+q_2^2}}{d^{7/3}}\sqrt{1+\left(\frac{\partial z_b}{\partial x}\right)^2+\left(\frac{\partial z_b}{\partial y}\right)^2} \quad (3)$$
$$-2d\varepsilon_{yy}\frac{\partial^2\overline{v}}{\partial y^2}-\varepsilon_{yx}\frac{\partial}{\partial x}\left(\frac{\partial\overline{u}}{\partial x}+\frac{\partial\overline{v}}{\partial y}\right)=0$$

Eq. (1) is the continuity and Eqs. (2) and (3) are the momentum equations in x and y directions, respectively. In the equations, t is time (s), d is water depth (m), ρ is water density (kg/m^3), g is the acceleration due to gravity (m/s^2), n is Manning's coefficient of roughness, z_w and z_b are water surface elevation and bed elevation above certain datum (m), q_1 and q_2 are the unit discharge fluxes (m^2/s) defined as $\overline{u}_d$ and $\overline{v}_d$, respectively, $\overline{u}$ and $\overline{v}$ (m/s) are the depth-averaged velocities of an element in the stream wise and transverse directions, respectively, q_m is the resultant inflow or outflow from that element (m/s), v_{xx} and v_{yy} are the normal components of the eddy viscosity (m^2/s) in the x and y directions, respectively, and v_{xy} and v_{yx} are the shear components of the eddy viscosity (m^2/s) applied to the $x-y$ plane.

FESWMS inputs are the Manning's coefficient of roughness n, and the eddy viscosity v. Model inputs must be assigned correctly to represent the physical processes occurring in the modeled reach and to produce accurate model predictions. Manning's n is an empirical coefficient that accounts for the total flow resistance caused by flow interaction with the boundary [11]. FESWMS utilizes Manning's n to account for momentum loss due to bed-shear, which may vary significantly in a stream reach in accordance to bed-bathymetry and roughness. The second input variable used by FESWMS is the eddy viscosity v. Eddy viscosity accounts for flow resistance due to the internal shear stresses, or the Reynolds' stresses of the fluid incorporating the added energy dissipation due to turbulence in the flow [13]. Therefore, eddy viscosity is not a physical property of the fluid, but rather a turbulent characteristic of the flow. For isotropic flows in prismatic channels, a single eddy viscosity value is sufficient to describe the turbulence flow characteristics within the modeled reach. Although FESWMS possesses the capability of spatially distributing both Manning's n and eddy viscosity in the computational mesh, it is common to use averaged values for Manning's n and eddy viscosity. In prismatic channels under uniform flow conditions, an average eddy viscosity value can be assigned to the study reach as a function of depth and bed slope [14]. Figure 4 shows the model simulation for velocity and the flow direction in the irrigation system of Falaj Al Ain. The irrigation system consists of one main canal contributing to three other canals (branches) with 19 outlets that are controlled with sluice gates. The canals are made from concrete of rectangular cross-sectional area with a width of 1.0 m and height of 0.5 m. The Manning's roughness coefficient and eddy viscosity were estimated as 0.025 and 0.05 m^2/s, respectively. The simulation is made under the extreme condition when all the gates are open. Recorded flow depth at the outlets was 0.4 m and the flow rate at the inlet was 0.25 m^3/s. The simulations show that the velocity is the range of 0.1 to 0.5 m/s with higher values in the main canal and lower values in the branches. Recommended minimum velocity in rigid boundary channels should be in the range of 0.2 m/s such that sediment is not deposited, aquatic growth is inhibited, and sulphide formation does not occur, while recommended maximum velocity should be in the range of 1.0 m/s to avoid high turbulence in the flow [15]. From the model simulation, it can be seen that the velocities are within the accepted range recommended for rigid boundary channels.

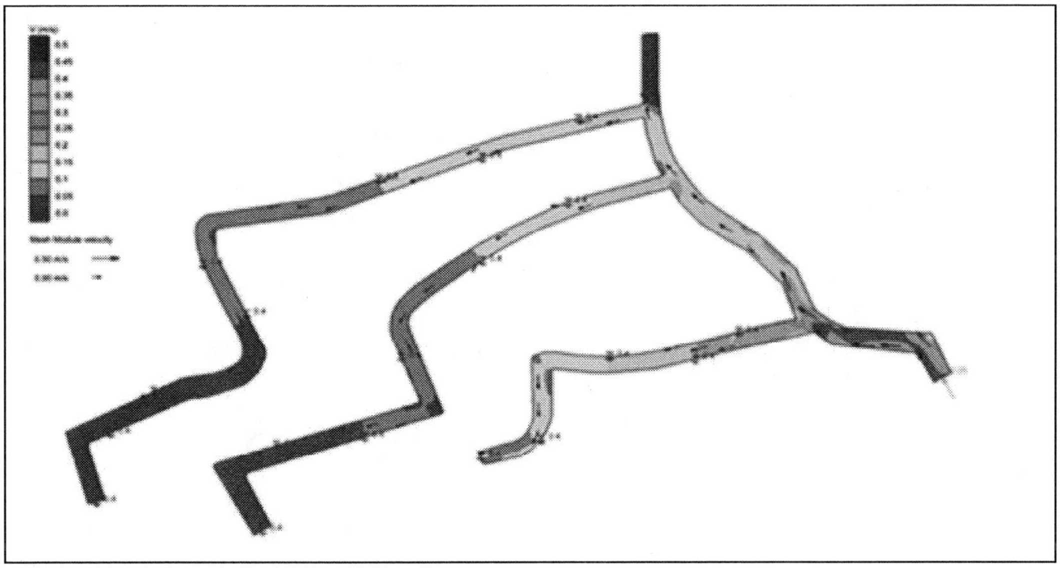

Fig. 4 Simulation of the flow velocity in Alain Oasis falaj irrigation system

Conclusion

The use of water elements in the urban fabric of the city is essential to create green infrastructure that reduces the heat island effect and creates a health city. Alain is known as the Garden city of the UAE, and its future development through a continuous network of green corridors that emanate from the Oasis to connect the major green areas of the city reinforces Abu Dhabi's plans to create better living spaces for its communities. Linking parks, buildings with courtyards and public outdoor spaces to a continuous network of shaded pedestrian paths pedestrian connectivity to the Downtown area and the Oasis. Water management for irrigation has always been a challenge in Desert cities, today the emergency of community education centres such as the Zayed Desert Learning Centre provide cultural services in addition to education and recreation that further explain to the community how hydraulics and water canalization systems such as the *falaj* work. Positioning of buildings and their integration within landscape is part of the Abu Dhabi 2030 Vision and it further accentuates how architecture can be used in shaping water to create better views of the city, this is exemplified by the investment in the new Abu Dhabi Central Park. While landscape proposals demonstrate how the new development could be integrated with the existing cityscape, it is essential to study precincts especially that many of the Islamic gardens such as those of Qasr Alhambra required consistent maintenance to retain their vitality while many others have been abandoned and perished as a result of water supply shortages. In a city like Abu Dhabi which is emerging in a rapidly homogenizing world, the degree of success of sustaining its green infrastructure needs to be constantly monitored in order to ensure its presence for future generations.

References

1. Abu Dhabi Urban Planning Council, *Abu Dhabi Vision 2030* (2007)
2. O. H. Caspersen, C.C. Konijnendijk, A.S. Olafsson, J. Geografisk Tidsskrift **106,** 2 (2006)
3. T.S. Eisenman, J. of Planning History **12,** 4 (2013)
4. E. Howard, *Garden Cities of To-Morrow* (London, 1902)
5. M.A. Benedict, E. McMahon, *Green Infrastructure: Linking Landscapes and Communities* (Washington, DC: Island Press, 2006)
6. D. F. Ruggles, F., Humanities Research Group **12,** (2003)
7. N. Rabbat, Journal of the Islamic Environmental Design Research Centre **2,** (1985)
8. H. Chanson, *Hydraulic Engineering and Roman Aqueducts: Modern Perspectives,* JIA **4,** (2015)
9. M. Brook, H. Al Houqani, *Current status of Aflaj in the Al Ain area United Arab Emirates* (2006)
10. D. Froelich, *User's Manual for FESWMS, Release 3* (2002)
11. D.M. Hicks, P.D. Mason, *Roughness Characteristics of New Zealand Rivers* (New Zealand DSIR Marine and Freshwater Resources Survey, Wellington, NZ, 1991)
12. H.M. Chaudhry, *Open Channel Flow* (Springer, New York, 2008)
13. A.N. Papanicolaou, R. Hilldale, J. Eng. Mech. **128**, 9 (2002)
14. D.R. Richards, *Proc. 1990 National Conf. Hydraulics Engineering* (ASCE, Reston, VA., 1990)
15. L.R. Mays, *Hydraulic Design Handbook* (McGraw-Hill, New York, 1999)

Enhancing community role in catchment area of cascade-pond system at the campus of Universitas Indonesia, Depok, West Java, Indonesia

Caya Maitri[1,*], *Dwita* Sutjiningsih[1] , and *Evi* Anggraheni[1]

[1]Civil Engineering Department, Universitas Indonesia, 16424 Depok, Indonesia

Abstract. The population in the catchment area of Universitas Indonesia (UI) campus cascade-pond system is growing rapidly. The rapid growth of population increases the activities intensity in the catchment area, so that a better water resource management is urgently needed. The Dublin Principles in Integrated Water Resource Management (IWRM) among others says that water development and management should be based on a participatory approach, which involves users, planners and policymakers at all levels. The Watershed Stewardship Program is one of the Eight-Tool for Watershed Protection developed by Center for Watershed Protection. Community as the main role holder, besides as beneficiary, is also as planner and executor of the program. The study aims to facilitate community in designing the activities towards water-friendly neighborhood, in the framework of community-potential-based watershed management, to increase the health of UI aquatic system and its catchment area. A combination of Participatory Action Research and Appreciative Inquiry method is applied. The required data and information is collected through field survey and focus group discussion (FGD). Field survey is conducted by interviewing random population inside the catchment area to get the behavior pattern overview of the community. FGD aims to unite the stakeholder and discuss their knowledge and potencies.

1 Introduction

In this globalization era, especially in Indonesia as a developing country, there are many things that are being developed in every sector. Development is closely related to construction project. Construction project has its own advantages and disadvantages. The advantages of the construction project are definitely what the purpose of the construction. Besides, it decreases the total area of the pervious cover of land. Watershed is an area of land which is bounded by topographic separation and shore, and it is influenced by the activity of the area [1]. Watershed is divisible into a number of catchment areas. It eases community to maintain their own little catchment area by their own ability and potencies. Regencies became the first rank of land cover in the catchment area of cascade-pond system at the campus of UI. The outlet of cascade pond system at campus of UI is mid-Ciliwung watershed. The catchment area of cascade-pond system at campus of UI includes 6 villages (Kukusan, Beji, East Beji, Depok, Depok Jaya, and Kemiri Muka) and a little

* Corresponding author: caya.maitri@ui.ac.id

part of Pondok Cina and Pancoran Mas village. There are 6 cascade-ponds in UI Campus (Kenanga, Agathis, Mahoni, Puspa, Ulin, Salam).

Center for Watershed Protection define eight tools of watershed protection in developing areas, one of them is Watershed Stewardship Program [2]. One of principles applied in Integrated Water Resources Management (IWRM) is Dublin Principles. One of these principles is that Water resources development and management should be based on a participatory approach, involving all relevant stakeholders [3,4]. Those statements implied that there should be the nearest stakeholders to participate in protecting and managing the watershed. That place is best filled by its own community, especially in this study case, because the community takes the most intervention in the catchment. In this study, Appreciative Inquiry (AI) is applied to strengthen the community role. AI is about the coevolutionary search for the best in people, their organizations, and the relevant world around them. Four stages in AI: **Discovery**-mobilizing a whole system inquiry into the positive change core; **Dream**-creating a clear result-oriented vision in relation to discovered potential and in relation to questions of higher purpose; **Design**-creating possibility proportions of the ideal organization, an organization design which people feel is capable of magnifying or eclipsing the positive core and realizing the articulated new dream; and **Destiny**—strengthening the affirmative capability of the whole system enabling it to build hope and momentum around a deep purpose and creating processes for learning, adjustment, and improvisation like a jazz group over time [5]. This study is a Participatory Action Research (PAR). PAR means an action research that needs objects participation. The use of AI changes the role of object, in this study is community, become partner. PAR is generally thought to involve a spiral of self-reflective cycles of Planning-Acting and Observing-Reflecting-Replanning-Acting and Observing again- and so on [6].

Participation potency of community in the catchment of cascade-pond system at UI campus is actually good. The awareness of environment cleanliness is a very basic thing to be acculturated in community. This can be started from their neighborhood, sub-village, village, to sub-district government. In Beji sub-district, which parts of them included in the catchment of cascade-pond system at UI Campus, always conduct a Jumsih (Clean Friday) event every Friday to keep the environment clean so that they can maximize their service to the community. This event also gives motivation to the community to actively participating in keeping their environment clean.

The use of Appreciative Inquiry (AI) Approach in Participatory Action Research (PAR) to enhance community role of contributing in catchment area management is not applied yet, especially in the catchment area of cascade-pond system at the campus of Universitas Indonesia.

2 Methodology

This study is a Participatory Action Research (PAR) using Appreciative Inquiry (AI) approach. The method of data collecting is by surveying and discussing. Collected data from survey then extracted to get the pattern of community existing participation and potency. The output of FGD is to unite the stakeholders and discuss their knowledge and potencies.

 DOI: 10.1051/matecconf/201713806004

2.1 Research Site

This study takes place in the catchment of cascade-pond system at the campus of Universitas Indonesia. Following picture shows the catchment area and the name of the village.

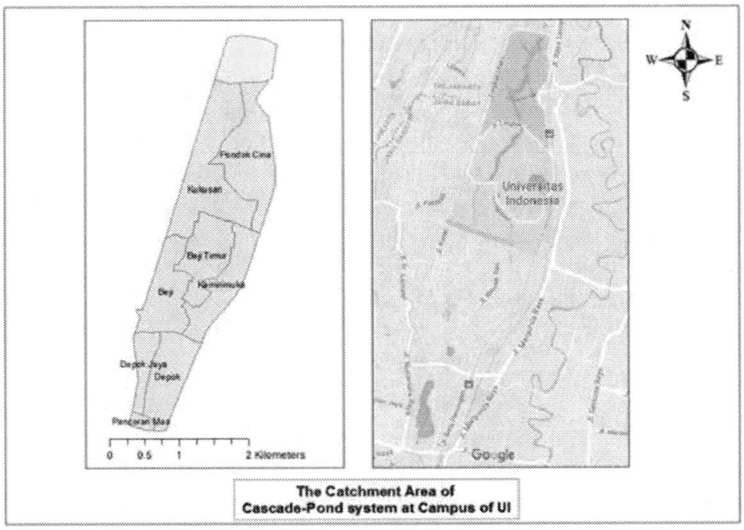

Fig. 1. The catchment area of cascade-pond system at the Campus of Universitas Indonesia.

2.2 Survey

There are 3 main parts of surveying, those are preparing, surveying, and post-surveying. Preparing the survey is about making the questionnaire, determining the location, and the respondents of survey and also briefing the surveyors. The questionnaire is divided into three parts; those are respondents' basic info, existing conditions, and respondents' participation potencies. Respondents' basic info contains name, address, gender, age, and latest formal education. Existing conditions contains their sanitation [7] and their existing event aiming for clean environment. Respondents' participation potencies contains if they will be participating while there's any environment protection event or not and good things they will do to their unused spaces around their house. The surveyors are divided into 7 teams which will survey different villages. The villages we chose to be surveyed are Beji, East Beji, Kemiri Muka, Pondok Cina, Depok Jaya, Kukusan, and Depok. Pancoran Mas village is not surveyed because the part which included in the catchment area is too small and consist of trading areas only. This survey doesn't plot any smaller area, so the surveyors can survey to anyone in anywhere bounded by their village area. The term post-surveying is about data recapitulating and extracts it to get the pattern of community potencies and existing participations. Extracted data show the relationship between gender, age ranges, last formal education and the existing condition and participation potencies.

2.3 Focus group Discussion (FGD)

The FGD had been attended by any stakeholder of the catchment of cascade-pond system at UI campus protection effort. They came from National Government, Local Government, Non-Government Organization, University, and Community Delegation. FGD started by panel discussion by the speakers from National Government to give the general information about catchment area health assessment. FGD continue by dividing the attendants into two main groups, decision makers and beneficiaries, those group were divided again into two groups so that we have four focus groups. After dividing the attendants into four groups then the discussion started in those groups discussing about the topic which had been made before. The materials of the discussion are their knowledge about catchment area health quality, the factors influence catchment area health quality, the stakeholders who had responsibility to catchment area quality, the effort to achieve better catchment area quality, and the way how to monitor the catchment area quality. After collecting the opinion of the attendants by discussing in four focus groups, then the attendants will reunite into two groups; Decision Makers and Beneficiaries to bring their opinion into one agreed opinion for those groups. The second discussion mastered by facilitators and helped by reporter and time keeper. After having agreed opinion then the attendants reunited in the plenary discussion. The plenary discussion aims to get the agreement of the discussion results. The result is their knowledge and awareness of catchment area protection. It is expected to trigger the community to enhance their participation in protecting the catchment area.

3 Result and Discussion

The result of surveying is the pattern of community's existing condition and potencies compared by their gender, age range, and latest formal education. The existing condition is about the way they drain off their grey water and black water, the way they treat their solid waste, and the existing event aiming clean environment. The potencies are about their willingness to use their unused space around their house and their willingness to actively participating in any environmental protection event. The result of FGD is their knowledge and awareness of catchment area protection.

3.1 Survey

Collected data from survey then extracted to see the percentage of total respondent's respond and the relationship between respondent's respond and their latest formal education, gender, and age range. Survey results then divided into three parts; Existing condition, Existing Participation, and Potencies.

3.1.1 Existing Condition

Existing condition contains the way they dispose waste; grey water, black water, and solid waste. Following figures show the examples of the way the dispose waste.

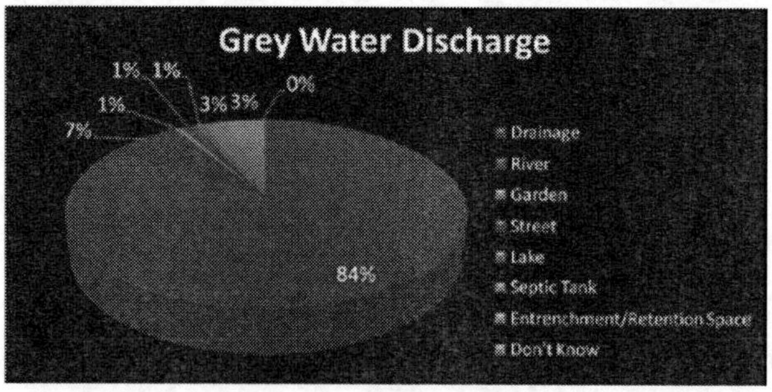

Fig. 2. Respondent's respond about the way they drain their grey water off.

 DOI: 10.1051/matecconf/201713806004

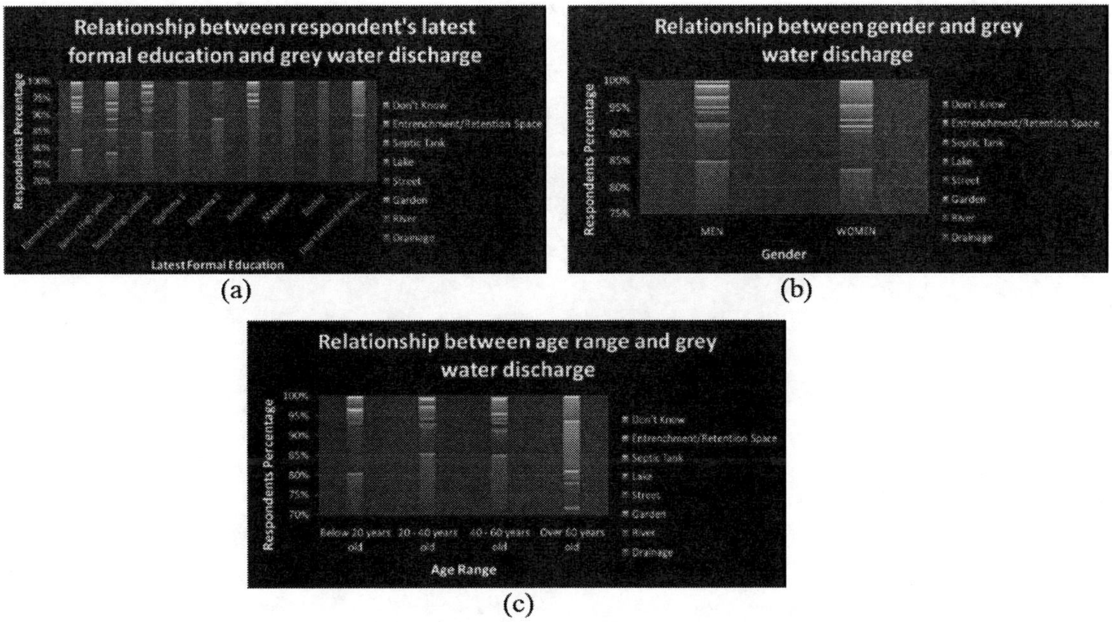

Fig. 3. Relationship between grey water discharge and respondent's (a) latest formal education; (b) gender; and (c) age range.

Grey water discharge management dominated by direct discharge to drainage with the total percentage of 84%. This result shows that grey water discharge management is the same as rainfall runoff discharge system. There was no significant differences when they are classified by their gender, latest formal education, or age. The existing condition result shows the same result that gender, latest formal education and age do not change any significant differences. For black water discharge, most of member of the community already know that it is better to drain the black water off to the septic tank. Besides, there still 3% of respondent who discharge their black water to the river without any treatment. The treatment of solid waste is dominated by paying the garbage collector to dispose it to the landfill but 10% of respondent still treat the garbage by burning it and 4% honestly say that they throw the garbage to the river.

3.1.2 Existing Participation

Existing participation is their participation in past event aiming clean and health environment, for example cleaning the drainage up, planting plants, restoring the street, etc. following figures show their participation in those event.

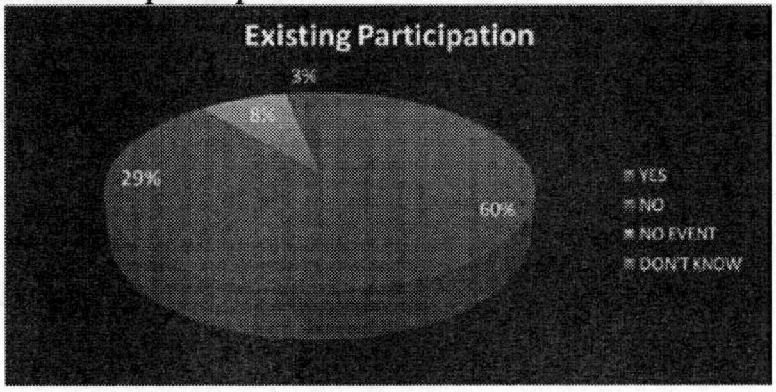

Fig. 4. Total Respondent's existing participation

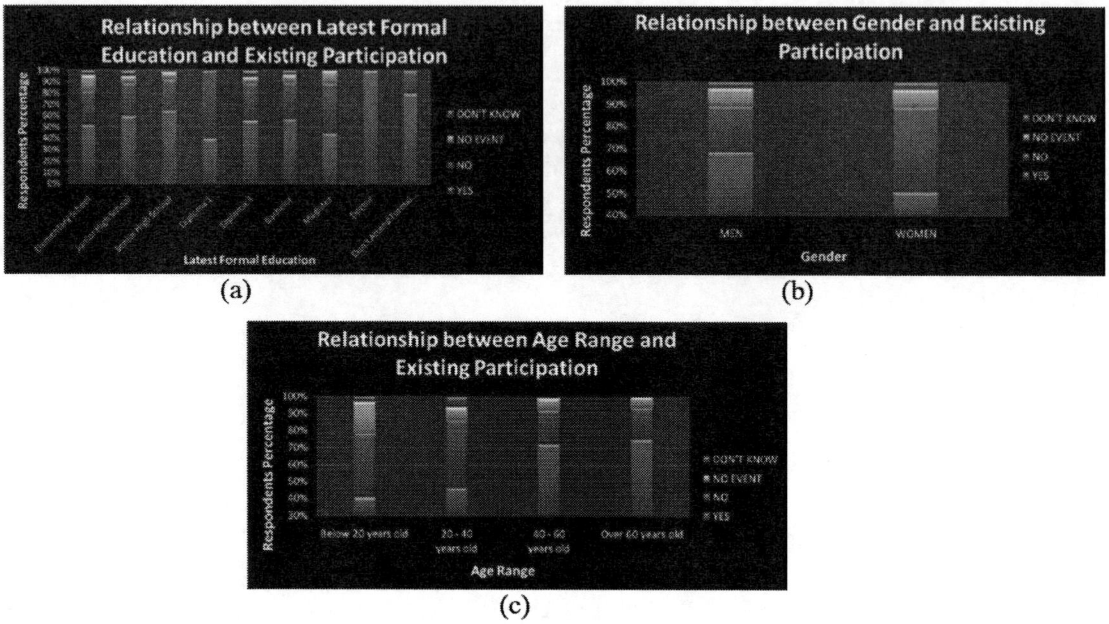

Fig. 5. Relationship between existing participation and respondent's (a) latest formal education; (b) gender; and (c) age range.

The figures show that the member of the community mostly participating in their event aiming health environment although there were 29% of the respondents who do not participating in the event and it was not a few numbers. The relationship graphs show no relation between respondent's latest formal education and their participation. The relationship between gender and their participation shows that men have the higher willingness to participate in the event because women usually do not take part in the activity but they participate in preparing the lunches and drinks. The relationship between age range and their participation show that the older their age the more they are willing to take a part in achieving better environment.

3.1.3 Potencies

The Potencies is their willingness to actively taking part in any event aiming clean and health environment/watershed and their willingness of functioning their unused space around their house in effort to achieve a better environment/watershed. Following figures show their willingness to actively take a part of future program.

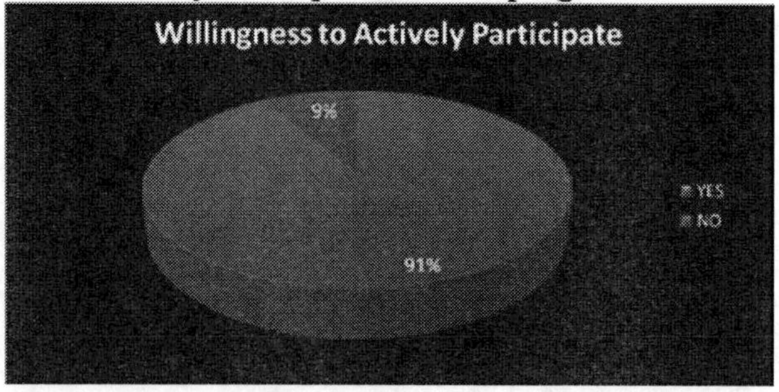

Fig. 6. Respondent's willingness to actively participating in future program.

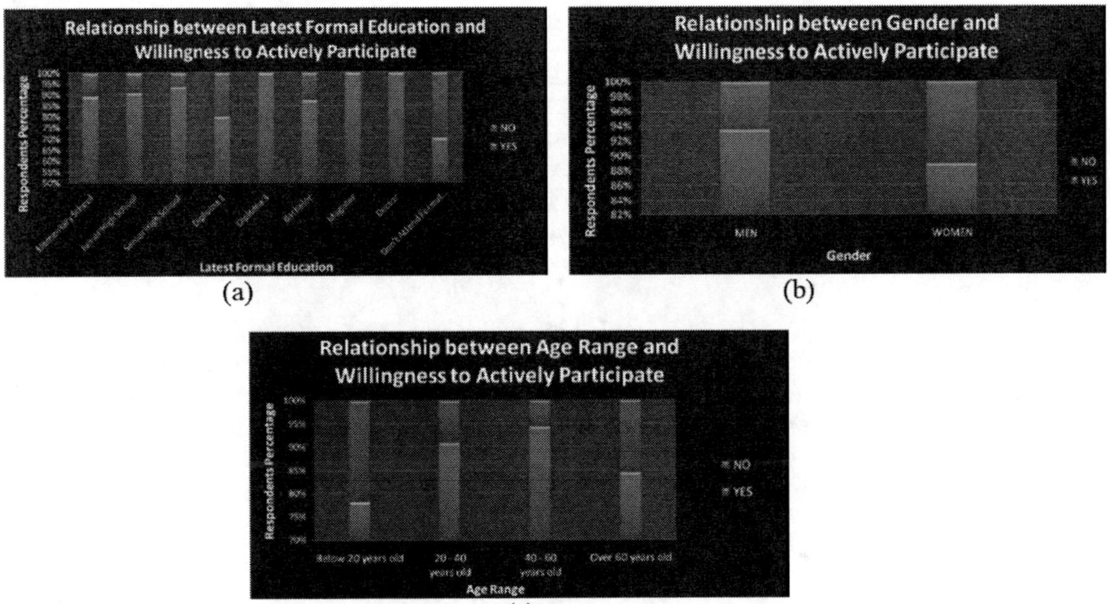

Fig. 7. Relationship between their willingness to actively take a part of future program and respondent's (a) latest formal education; (b) gender; and (c) age range.

Respondents mostly have willingness to actively participating in future program aiming clean and health environment but there is still 9% of respondents say that they do not want to actively participate. The relationship graphs show that the highest percentage of respondent who do not want to actively participate is the groups of respondents who do not attending formal education. If it compared by gender, women has the higher percentage in refusing actively participating in future program. It also shows that the older age range, the higher the percentage of respondent's willingness to actively participate in future program, but there is decreasing percentage for the age range above 60 years old. Following figure shows

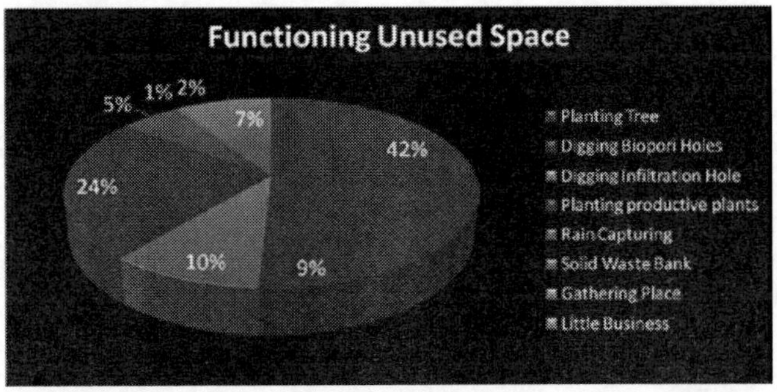

Fig. 8. Respondent's willingness to functioning their unused space around their house.

Based on the pie chart above, we can conclude that mostly the member of community want to functioning their unused space around their house by planting trees and productive plants because it is easy. They are not preferred to do another effort because they do not know the way it works. Embarrassingly, 7% of respondents still prefer functioning their unused space by building small business.

3.2 Focus Group Discussion (FGD)

The first topic for discussion is about the knowledge of water and catchment area health quality and its changes possible consequences. Decision Makers team said that water health quality is when the water body is free from wastes and the consequence of its changes is human health quality. Beneficiaries team said that water body health quality is the indicator of the condition of water is good or not, it can be used to determine whether the water can be used or not in the future. Water quality changes influences water ecosystem, water resource for human or any living creature consumption, and human health.

The second topic of the discussion is about the importance of letting the water body and catchment area quality good and the factors of the quality changes. Decision Makers said that it is important to let the water body quality good and the factor of quality changes is wastes. Beneficiaries team said that it is important to let the water body quality good in order to balance the ecosystem and because it relevant to human health quality. They argue that the factors of the quality changes are wastes, water treatment, and community awareness of water quality.

The third topic is those responsible to let the environment/water body quality good. Decision Makers team said that the responsibility should be held by the government, community, and any other parties who doing activity in the catchment area. The other team states the same answer.

The forth topic is the possible effort aiming enhancing water quality they can do, the easiest effort that can be a recommendation to enhance water quality and how to do it. Decision Maker team opinion is law enforcement, criminal penalty, evaluation and monitoring in any stages, and enhancing every stakeholder awareness. They state that the easiest effort that recommended to be applied is enhancing every stakeholder's awareness by educating them from sub-neighborhood scale to higher scale. Beneficiaries team opinion is to do "Gerakan RT Ramah Air" (Sub-Neighborhood Water Friendly Movement). This movement contains counseling, solid waste treatment, partnership, non-structural approach (communication, education, rainfall and another gauges), and structural approach (urban master plan).

The last topic is about monitoring water health quality. Decision Maker team state that monitoring is needed to make sure the water quality good and do not suffer community's comfort and quality. The parties who will monitor is the government around water body, Local Department of Highways and Water Resources, and other authorized stakeholder by periodically evaluating and monitoring water quality and following up the result. Beneficiaries team state that monitoring is needed to ensure the quality is good and later the data become public database. Monitoring will be held by action monitoring.

4 Conclusion

The existing solid waste and black water treatment condition of the catchment area of cascade-pond system at UI campus is mostly good but there also several percent of respondent state they still have a bad habit in managing those wastes, it causes bad water quality. More than 50% of respondents said that they had been participated in their past event aiming clean and health environment, while 29% did not participated, 8% said that there was no event aiming clean and health environment, and 3% even do not know whether there was an event aiming clean and health environment or not. For future event, 91% of respondents agree to actively take part in it. In determining good things to do in their unused space around their house, they mostly choose planting trees and productive plants because it is easy to apply. Through FGD, it can be concluded that actually both Decision Maker and Beneficiaries have a basic knowledge about water health quality, how to make it better, and who have responsibility to it although somehow their perception about those things still different and need a further discussion to equalize their perception. The further step of this study will be choose one neighborhood which directly flow their runoff to the cascade-pond at UI campus, build a partnership which contain every stakeholder in catchment area of the cascade-pond system at UI campus, discuss what they will do to enhance their role in managing health quality of catchment area of the cascade-pond system at the campus of Universitas Indonesia, and implement the agreed activity/ program.

Acknowledgements are addressed to Universitas Indonesia for giving financial support through the PITTA 2017 funding scheme, Head of Ciliwung-Cisadane River Basin Management board, Sahroel Polontalo as a head of Ciliwung- Depok Community for giving advice and suggestion, 14 Surveyors which consist of college students and staffs of Civil Engineering Department Universitas Indonesia in conducting field survey and data input, and the community in catchment area of the cascade-pond system at the campus of Universitas Indonesia as the respondents in this study.

References

1. Ministry of Public Works and Housing, *Peraturan Menteri no. 04/PRT/M/2015* (Ministry of Public Works and Housing, Jakarta, 2015)
2. H. Kwon, R. Winer, T. Schueler, *Eight Tools of Watershed Protection in Developing Area* http://www.epa.gov/watertrain (2016)
3. P. Borchardt, *Introduction to Integrated Water Resource Management* http://www.iwrm-education.de/#!borchardt2 (2012)
4. P. Zaag, H. Savenije, *Principles of Integrated Water Resource Management* (UNESCO-IHE, Delft, 2014)
5. D. Cooperider, D. Whitney, *A Positive Revolution in Change: Appreciative Inquiry* (Berret-Koehler Publishers, Inc., California, 2005)
6. S. Kemmis, R. McTaggart, *Participatory Action Research: Communicative Action and the Public Sphere* (Sage Publication Ltd, California, 2007)
7. N. Lies, *Laporan Pemutakhiran Studi EHRA* (Pokja AMPL/sanitasi Kota Depok, Depok, 2015)

MATEC Web of Conferences **138**, 06005 (2017) DOI: 10.1051/matecconf/201713806005
EACEF 2017

Assessment of hydropower potential in small karst catchments: the case of the Rocche Plateau, Central Italy

Maurizio Leopardi[1], and *Anna Rita* Scorzini[1,2,*]

[1] University of L'Aquila - Department of Civil, Environmental and Architectural Engineering, L'Aquila, Italy

[2] Politecnico di Milano - Department of Civil and Environmental Engineering, Milano, Italy

Abstract. Estimation of flow duration characteristics is key in assessing hydropower potential in natural catchments. However, such analysis is not usually straightforward, especially in ungauged sites and/or in complex catchment areas. In this study we evaluate the feasibility of revamping of a small hydroelectric power plant, located in a karst plateau in central Italy, by assessing the hydropower potential of its feeding surface and subsurface stream network. A thorough analysis of runoff processes occurring in the examined area is carried out in order to corroborate regionalization studies based on measured specific flows in neighboring homogeneous basins. The results show an appreciable availability of water resources to be exploited for hydropower purposes.

1 Introduction

The assessment of flow regime in natural streams is essential for a proper management of water resources. Flow duration curves have proven to be useful for e.g. water supply and hydropower design studies, where it is important to define the variability of flows in natural catchments. The most reliable means for the determination of a flow duration curve is to use actual stream flow data measured at the point of interest. However, regionalization techniques from a gauged to an ungauged location are more often applied in practice, although with a certain degree of uncertainty, since flow records are not always available.

This paper analyzes, by means of a case study in Central Italy, the importance of performing detailed analyses of runoff generation and propagation processes as a support to the results of regionalization techniques for a comprehensive assessment of the hydrologic regime and hydropower potential in complex catchments.

2 Study area overview

The Stiffe spring flows out from a fracture in the limestone mass near the homonymous village of Stiffe, L'Aquila (Italy). This water, originating from several waterfalls, reaches

* Corresponding author: annarita.scorzini@polimi.it

the downstream valley, where, after being channeled, it is used for industrial purposes. This spring deserves a particular mention in relation to the peculiar orography and geology of the upstream catchment of the Rocche Plateau, which, as shown by Perrone [1], would have become a lake of about 10 km^2 if water had not breached through the fractures in the limestone layers, creating a characteristic sinkhole system in the valley. Indeed, the endorheic Rocche Plateau is a closed drainage basin, with no direct outflow to external water bodies, which converges to a sinkholes area in Saline Camp that connects it, through a karst system with a 600 m drop, to the Stiffe spring (Figure 1).

3 Previous surveys and studies for the determination of the mean annual flow

In this section we report the results of existing studies and surveys, with the first ones dated back to the late XIX century, aimed at the determination of the mean annual flow for the Rocche Plateau catchment.

3.1 First flow measurements in Stiffe spring (1898-1899)

The first isolated flow measurements recorded in Stiffe spring took place at the end of the XIX century, as follows: June, 2 1898: Q_{St}=122 l/s; August, 28 1898: Q_{St}=80 l/s; November, 18 1898: Q_{St}=227 l/s; August, 19 1899: Q_{St}=68 l/s.

3.2 Surveys and studies conducted by the Mezzogiorno Development Fund (1983)

In a study conducted in the early 1980s, the Mezzogiorno Development Fund [2] estimated the following annual flows in Stiffe spring: maximum flow $Q_{St,max}$=900 l/s, minimum flow $Q_{St,min}$=10 l/s and mean flow Q_{St}=200 l/s.

Differential flow measurements indicated that the runoff generated in the sinkholes area of the Plateau contributed to about 70% of the flow registered in Stiffe spring. The remaining part was considered to be attributable to the widely diffused conduits dug out into the limestone rock mass, with one of the main entrance in the Pozzo Caldaio sinkhole.

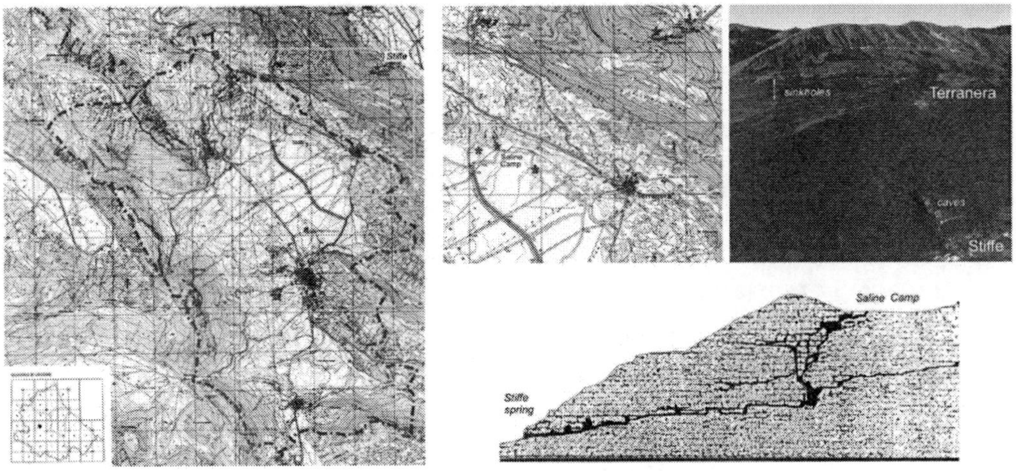

Fig. 1. Overview of the Rocche Plateau and the karst system in Saline Camp.

3.3 Regionalization study led by ENEL and University of L'Aquila (1987)

The study [3] was focused on the identification of potential locations for hydropower generation in mountain areas of regional basins between Aso and Trigno Rivers. After a preliminary phase, only few sites were considered for further detailed geological, hydrogeological and hydraulic investigations, with the aim of assessing their hydropower potential. For Stiffe, which was one of these selected locations, the mean annual producible power was estimated by means of "artificial" flow duration curves. The procedure consisted in the reconstruction of the mean annual isohyets and in the assessment of annual flow based on runoff coefficients calculated from water balances studies performed in regional stream gauge stations (river basins located between Chienti and Trigno Rivers). Long-term time series of mean annual flow measured in 61 gauged stations and geomorphologic features of the related drainage areas (extension, mean altitude, effective drainage area) were the main input data. A power-law relationship of specific flow to drainage area was found by means of the method of least squares, as follows:

$$q = a \cdot A^{-b} = 42.5 \cdot A^{-0.26} \quad (1)$$

where q is the specific flow [l/s km^2] and A the drainage area.

Based on the geological characteristics of the endorheic Rocche basin, it was possible to consider about 50% of the total drainage area (A=52 km^2) as non-contributing to the flow of Stiffe spring. Therefore, with an effective area of about 30 km^2, the calculated specific flow was q=17.5 l/s km^2 and then the mean flow expected at the sinkholes and, consequently, at Stiffe spring was estimated to be Q_{sink}=526 l/s and Q_{St}=750 l/s, respectively (assuming that about 70% of the flow observed in Stiffe may be attributable to the flow entering the upstream sinkholes, according to [2]).

3.4 Regionalization study for ENEL-VALOREN EU Project (1992)

In this project [4], given the peculiarity of the area, the number of stream gauge stations used for the regionalization of the specific flow was restricted to 6 sites (Table 1) located in the neighboring Aterno, Tasso and Sagittario river basins, which were considered to be part of the same homogeneous hydrological region. With this limited dataset, the regional relationship obtained for the specific flow was the following:

$$q = a \cdot A^{-b} = 27.15 \cdot A^{-0.2291} \quad (2)$$

Adopting the exponent of the function as an homogeneity index, Eq. (2) could be referred to the average specific flow expected in the nearest stations of Treponti, L'Aquila and Molina (Table 1), as shown in Eq. (3):

$$q = 26.77 \cdot A^{-0.2291} \ (Treponti); \ q = 36.77 \cdot A^{-0.2291} \ (L'Aquila); \ q = 21.72 \cdot A^{-0.2291} \ (Molina) \quad (3)$$

Table 1. Mean annual specific flows measured in the stations of the Aterno River basin.

River	*Station*	*Drainage area [km²]*	*Specific flow [l/s km²]*	*Permeable part [%]*
Aterno	Treponti	114	9.04	34
Aterno	L'Aquila	531	7.31	46
Aterno	Molina	1303	4.20	60
Tasso	Scanno	80	8.62	97
Sagittario	Villalago	108	12.54	89
Sagittario	Capo Canale	599	11.60	93

In this case, assuming for the scale factor the minimum value (21.72) between the three calculated in Eq. (3), the specific mean annual flow for the Rocche basin was estimated to be q=10 l/s km^2, i.e. an expected mean annual flow at the sinkholes Q_{sink}=300 l/s and at Stiffe spring Q_{St}=420 l/s.

4 Analysis of runoff processes occurring in the Rocche Plateau

Analysis of runoff processes occurring in the examined basin is carried out herein in order to corroborate the results shown in the previous section (and specifically, the relationship existing between Q_{sink} and Q_{St}) in order to use recent flow measurements to derive a flow duration curve for the study area.

Depending on actual data availability and accuracy requirements of the results, the runoff processes can be modeled at different levels of detail, from practical formulas to complex representations of runoff generation and propagation in the catchment.

4.1 Estimation of runoff using Iszkowski formula

In the literature many formulas exist for relating the expected streamflow to the pluviometric regime of the tributary catchment. An example is the simple Iszkowski formula [5] which relates the streamflow Q to the mean annual rainfall depth h:

$$Q = \lambda \cdot m \cdot h \cdot A = \psi \cdot h \cdot A \tag{4}$$

where λ and m can be expressed by a coefficient of proportionality, i.e. the runoff coefficient ψ, as a function of the ratio of impervious drainage area A_{imp} to the total drainage area A, as follows:

$$\psi = 0.315 \cdot \left(\frac{A_{imp}}{A} \right)^{0.393} \tag{5}$$

Considering the rain gauge station of Rocca di Mezzo, with a registered mean annual rainfall of about 1000 mm, the application of Eq. (4) and (5) for the Rocche Plateau catchment led to an estimated annual flow Q_{sink}=389 l/s.

4.2 Estimation of runoff using SCS-CN method

With the aim of calculating a probable high-frequency runoff in the study area, the SCS-CN method [6] was used for estimating an expected low return period flow of about 1.5÷2 years. The SCS-CN model combines infiltration losses with surface storage by Eq. (6):

$$Q_p = \frac{(P - I_a)^2}{P - I_a + S} \tag{6}$$

where Q_p [mm] is the accumulated runoff, P [mm] is the rainfall depth of given return period, I_a [mm] is the initial abstraction, which includes surface storage, interception and infiltration prior to runoff (if $P<I_a$, $Q_p=0$), and S [mm] is the potential maximum retention, given by:

$$S = 254 \cdot \left(\frac{100}{CN} - 1 \right) \tag{7}$$

where the Curve Number, CN [dimensionless, ranging from 0 to 100], is derived from data reflecting land cover, hydrologic soil group and antecedent soil moisture condition (AMC).

DOI: 10.1051/matecconf/201713806005

Therefore, when $P>I_a$, the runoff generated by a rainfall event of given return period (1.5÷2 years, in this case) can be estimated as:

$$Q = \frac{Q_p \cdot A}{t_c} \tag{8}$$

where t_c is the time of concentration of the basin.

4.2.1 CN parameter estimation

Area-weighted *CN* for the examined catchment was calculated using a GIS procedure described in Leopardi and Scorzini [7]. This consists in combining information from the regional land cover map and the hydrologic soil group map, which is produced by assigning a particular hydrologic soil group (A, B, C or D) depending on the soil's minimum infiltration rate. Subsequently, to create a CN map, the hydrologic soil group field from the soil theme and the land use field from the land cover theme are selected for intersection. This operation generates a new polygon shapefile indicating the merged soil hydrologic group and land cover themes. Appropriate *CN* values according to [6] are assigned to each polygon of this new map under the hypothesis of average moisture condition (AMC II).

Data extracted with this procedure were then used to compute the area-weighted *CN* for the Rocche catchment, CN=67. Therefore, by substituting this value in Eq. (7), we obtained S=125.1 mm and an initial abstraction I_a=$0.2\ S$=25 mm, calculated according to the USDA expression [6].

4.2.2 Estimation of time of concentration

Several definitions and methods to compute the time of concentration are in use in the literature. According to [8], it can be defined as the time taken by a water drop to move itself, within an hydraulic path, from the most distant point in the basin until the outlet point. In other words, it can be expressed as a travel time, including surface t_s and channel flow t_{ch}:

$$t_c = t_s + t_{ch} \tag{9}$$

The estimation of t_s is usually affected by large uncertainties related to the complex storage phenomena occurring in natural catchments. Nevertheless, the so-called "velocity method" can be a hydraulically sound way for determining the time of concentration [9]. This method assumes that this parameter can be expressed as the sum of travel times for reach segments along the hydraulically most distant flow path (with segment lengths L_i and average flow velocities V_i):

$$t_c = \sum_i \frac{L_i}{V_i} \tag{10}$$

During field surveys in the study area on February-April 2016, we identified along the main reach in Saline Camp some characteristic cross-sections which were used for flow velocity measurements. Figure 2 shows the localization of the cross-sections and the identification of the different reach segments, with measured average minimum and maximum flow velocities. Based on data of Figure 2, we estimated, by means of Eq. (10), a time of concentration t_c for the examined area ranging from 3.3 to 3.4 hours.

In addition, different practical formulas developed by other authors were applied to the case study for comparison purposes. These included Giandotti [10] and Kirpich [11] formulas (Eq. 11 and 12), which were found coherent with our experimental result:

DOI: 10.1051/matecconf/201713806005

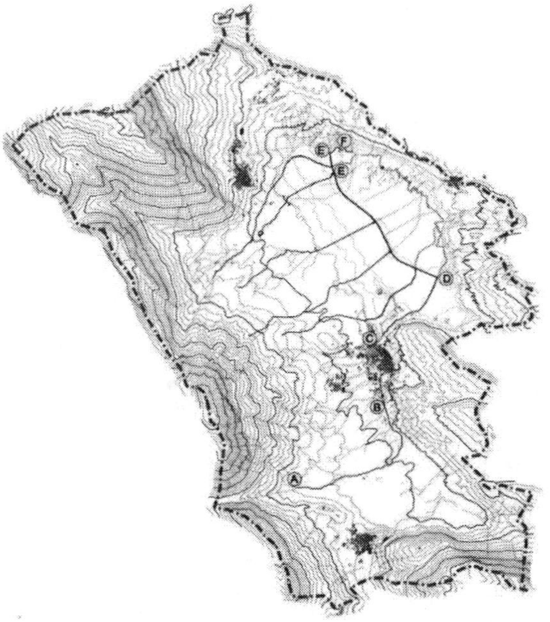

Reach	L_i	Δq	i_i	V_{min}	V_{max}
	[m]	[m]	[m/m]	[m/s]	[m/s]
A - B	3.129	65,5	0,0209	3,0	3,5
B-C	1.142	5,5	0,0048	2,5	3,0
C-D	1.664	13,0	0,0078	2,0	2,5
D-E	2.859	2,0	0,0007	0,2	0,2
E-E'	265	1,0	0,0038	0,5	0,7
E'-F	410	4,5	0,0110	2,0	2,5

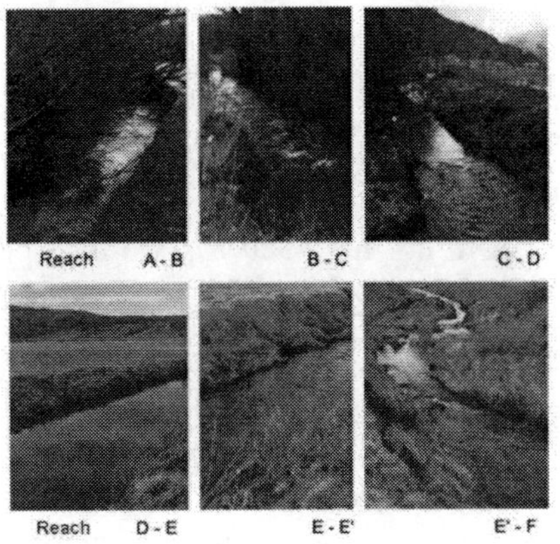

Fig. 2. Identification of the reach segments used for flow velocity measurement in Saline Camp.

$$t_c = \frac{4\sqrt{A} + 1.5L}{0.8\sqrt{H_0}} = 3.9 \; hours \qquad (11)$$

$$t_c = 0.000325 \cdot L^{0.77} \cdot s^{-0.385} = 3.5 \; hours \qquad (12)$$

with L=total length of the main channel, H_0=difference between the mean basin elevation and the outlet elevation and s=mean slope.

4.2.3 Expected flow determination

The expected flow in Saline Camp sinkholes area was then determined by means of Eq. (8) using the parameters defined above and an accumulated runoff P of about 27÷29 mm calculated for a 1.5÷2 years return period rainfall event with a duration equal to the time of concentration of the basin. The resulting Q_{sink} ranged around 340÷360 l/s.

4.3 Recent in situ flow measurements in the Rocche Plateau and Stiffe spring

The results shown in the previous section were integrated with a streamflow measurement campaign (February-April 2016) in the Rocche Plateau and specifically in Saline Camp sinkholes area and in the spring in Stiffe cave.

First, we checked the actual functioning of the sinkholes system, since during the 1960s land reclamation and drainage works were carried out to solve the problem of frequent water stagnation phenomena occurring in the Plateau. Consequently, all the minor streams flowing in the Plateau towards Pozzo Caldaio sinkhole (now fed only by some remaining minor streams) were drained into a main channel which ended in Saline Camp. Here, during our surveys, five sinkholes were found to be active.

In addition, stream flow measurements were performed on February 23, March 15 and April 6 2016 in the three control cross-sections shown in Figure 4, on Rio Gamberale (XS1), Rio Caporitorto (XS2) and upstream from the sinkholes area (XS3). Cross-sections

DOI: 10.1051/matecconf/201713806005

1 and 2 were identified by two broad crested weirs (Figure 3), whose discharge rating curve was given by:

$$Q_{1,2} = 0.385 \cdot L_w \cdot h \cdot \sqrt{2g \cdot h} \tag{13}$$

where 0.385 is the weir coefficient, L_w is the weir length and h is the head above weir crest. Section 3 was located upstream from sinkhole 1 (Figure 3) in Saline Camp, at the end of a straight 120 m long reach segment. Here, we installed a trapezoidal Cipolletti weir, whose discharge rating curve was expressed by Eq. (14):

$$Q_3 = Q_{sink} = 0.415 \cdot L_w \cdot h \cdot \sqrt{2g \cdot h} \tag{14}$$

Average flows measured in the three control sections during the different field surveys are reported in Table 2.

During these surveys we also performed parallel flow measurements by means of an electronic hydrometer located on a weir in the Stiffe cave, in order to validate the aforementioned relationship found by the Mezzogiorno Development Fund [2] between the flow observed in Stiffe spring and in Saline Camp. Spring discharges registered in Stiffe Q_{St} (averaged over a 5-days period for each survey) are shown in Table 2, together with the related flows observed in Saline Camp, Q_{sink}. The calculated ratio Q_{sink}/Q_{St} confirmed the results obtained by the Mezzogiorno Development Fund [2], i.e. water flowing to the sinkholes area in the Rocche Plateau contributes to about 60÷70% of the flow discharging out from Stiffe spring.

Table 2. Recent field flow measurements in the Rocche Plateau: in the sinkholes area of Saline Camp and Stiffe spring.

Survey	*Measured flow [l/s]*				Q_{sink}/Q_{St}
	Q_1	Q_2	$Q_3 = Q_{sink}$	Q_{St}	
Feb, 23 2016	250	138	400	639	0.63
Mar, 15 2016	290	156	460	724	0.63
Apr, 6 2016	144	82	234	335	0.70

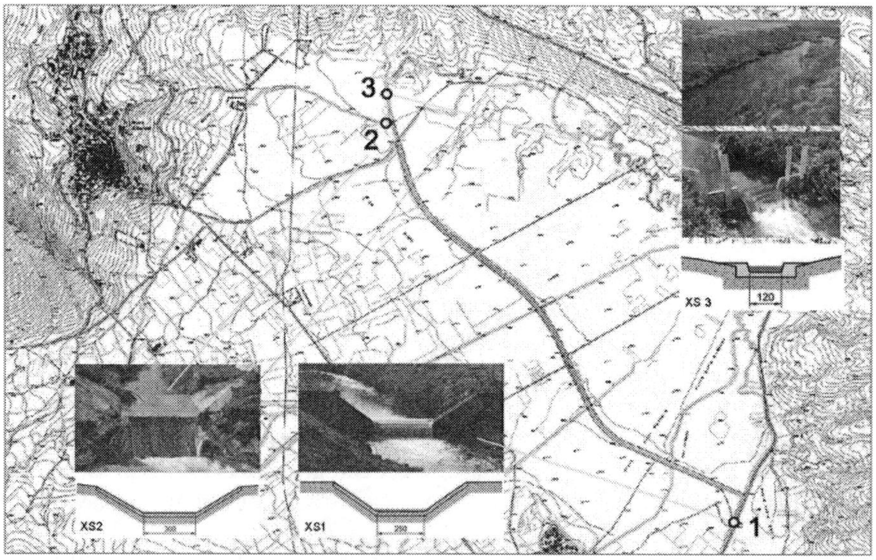

Fig. 3. Localization of the control sections used for flow measurement in Saline Camp.

DOI: 10.1051/matecconf/201713806005

5 Assessment of hydropower potential and conclusions

In order to estimate the availability of water usable for electricity generation, i.e. to assess the hydropower potential of the study area, it was essential to derive a flow duration curve for Stiffe spring. This could be easily accomplished by using flow records at the gauge station in Stiffe cave. However, such time series was limited to the periods 1995,1999, 2003÷2008 and 2014÷2015; in addition, the years 2003, 2006, 2007 and 2014 were excluded due the large amount of missing data, resulting in an available 6-years long time series.

It is clear that a flow duration curve based on a short record would be an unreliable tool for decision making. Nevertheless, in this case we could take advantage, for validation purposes, of the results obtained from the analysis of runoff processes and from the verified relationship existing between the flow generated in the Rocche Plateau and the discharge flowing out from Stiffe spring. Therefore, we used the available flow records to obtain information on flow characteristics in the examined location. Figure 4a reports the flow records for the different years, together with that of the mean hydrological year.

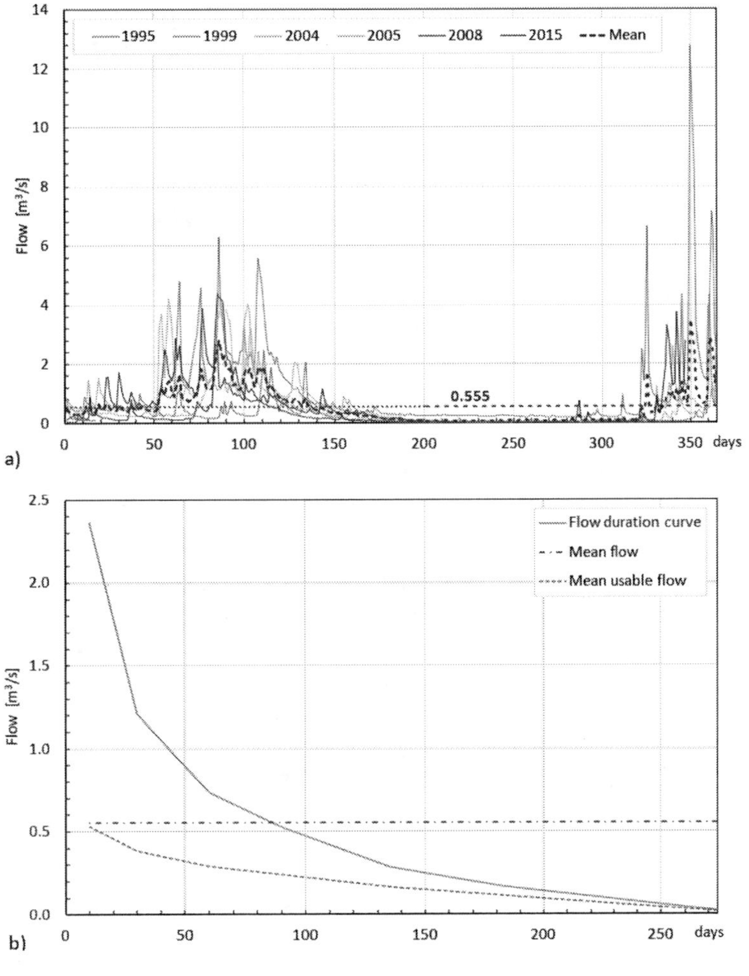

Fig. 4. a) Flow time series measured in Stiffe spring; b) Flow duration curve and mean usable flow curve derived from flow measurements in Stiffe spring.

Despite the limited sample and the instable flow pattern, it was possible to recognize a well defined hydrological regime, characterized by significant flow values in spring (March, April) and fall (November, December) and a mean annual flow for the mean hydrological year equal to Q_{St}=555 l/s.

In conclusion, by comparing this value to the calculated runoff expected in the sinkholes area of Saline Camp, Q_{sink}=340÷390 l/s (depending on the selected approach) and by verifying the ratio $Q_{sink}/Q_{St} \approx 0.7$, it was possible to prove the reliability of the result obtained by using the short flow time series recorded in Stiffe.

Therefore, based on these data, the flow duration curve and mean usable flow duration curve shown in Figure 4b were finally determined and used as fundamental input data for the design of the hydroelectric power plant in Stiffe.

References

1. E. Perrone, *La cascata di Stiffe nel territorio di San Demetrio nei Vestini*. Ed. Pertilia, L'Aquila (1905)
2. Mezzogiorno Development Fund, *Progetti speciali per gli schemi idrici del Mezzogiorno: Idrogeologia dell'Italia centro-meridionale. Quaderni della Cassa per il Mezzogiorno*, Roma (1983)
3. ENEL and University of L'Aquila, *Indagine per l'individuazione di impianti idroelettrici di limitata potenza nell'area montana dei bacini dall'Aso al Trigno*. Final Report (1987)
4. ENEL Programma VALOREN, *Studi di fattibilità degli impianti idroelettrici sul Fiume Vella, Fiume Orfento e Rio di Stiffe*. Final Report (1992)
5. R. Iszkowski. Beitrag zur Ermittlung der Niedrigst-, Normal- und Höchstwassermengen auf Grund charakteristischer Merkmale der Flussgebiete. Zeitschrift des Oesterr.Arch. und Ingenieurvereins, 69-98 (1886)
6. USDA, *Urban hydrology for small watersheds*. Technical Release 55, USDA, Washington, DC, USA (1986)
7. M. Leopardi, A.R. Scorzini, iForest-Biogeosciences and Forestry, **8**, 302-307 (2015)
8. V.T. Chow, D.R. Maidment, L.W. Mays, *Applied Hydrology*. McGraw-Hill. New York, USA (1988)
9. USDA, *National Engineering Handbook. Part 630, Hydrology* (2010)
10. M. Giandotti, Istituto Poligrafico dello Stato, **8**, 107-117 (1934)
11. Z.P. Kirpich, Civ. Eng., **10**(6), 362 (1940)

Lowering the Beneluxtunnel

Tomas Weeda[1,*], *Jarit de* Gijt[2,3], *Klaas Jan* Bakker[2], *René* Braam[2], *Eric* Broos[4], and *Bas* Jonkman[2]

[1]City of Amsterdam, Engineering Department, PO Box 12693, 1100 AR Amsterdam, The Netherlands
[2]Delft University of Technology, Hydraulic Engineering Department, PO Box 5048, 2600 GA Delft, The Netherlands
[3]City of Rotterdam, Engineering Department, PO Box 6575, 3002 AN Rotterdam, The Netherlands
[4]Port of Rotterdam, PO Box 6622, 3002 AP Rotterdam, The Netherlands

Abstract. As a result of the persistent increase of container vessel dimensions, future problems regarding navigation draught in the presence of tunnels are becoming more likely to occur. Hence, possible solutions to this problem have been investigated for the Beneluxtunnel. Several design options have been elaborated to determine the technical and economic feasibility of a possible lowering of the tunnel. Important subjects involved are the consequences of increasing the slopes, the cross-sectional concrete capacity, the use of joint rotations, the construction methods and the costs. This initial exploration of the subject shows lowering the Beneluxtunnel seems to be possible and is expected to be economically attractive. However, further research into certain boundary conditions and risky aspects of construction is required to ascertain this statement.

1 Problem

The minimum depth of tunnels crossing waterways is related to the expected maximum draught of navigation during the operational phase of the tunnel. Due to the large and unexpected increase in containership capacity as of 2005, the minimum depth estimates for tunnels built before this period were rather shallow. Consequently, tunnels crossing seaport access channels could become obstacles in the near future.

Rotterdam, Amsterdam, Antwerp and Hamburg are amongst the cities of which port navigation channels are tunnelled and problems regarding the draught of the latest generation of container vessels could occur. Hence, the municipality and the port authorities of Rotterdam have decided to determine a solution strategy to this problem.

2 The Beneluxtunnel

The Beneluxtunnel lies underneath the river Nieuwe Maas on the western side of Rotterdam and therefore crosses the access channel of the so-called "city ports". The connection consists of 2 tubes built in 1967 and in 2002, and houses a 2x4 highway, a 2x1 metro and a passage for bicycles and pedestrians.

* Corresponding author: t.weeda@amsterdam.nl

 DOI: 10.1051/matecconf/201713806006

Fig. 1 indicates the draught and the beam of the largest container vessels at the time of construction of the tunnels and now, relative to the depth of the tunnel tubes. It shows the depth related problems if a container ship of the latest generation would attempt to cross the Beneluxtunnel. Currently however, this is only a potential scenario as these types of ships do not (yet) visit the city ports because they are being served elsewhere in the port. But given the ongoing growth of container vessel dimensions and the expected unlikeliness of the port being expanded further into the sea in the coming years, the probability of the Beneluxtunnel becoming an obstacle for navigation is significant and preparing for this scenario by determining solution strategies to this potential problem appears to be sensible.

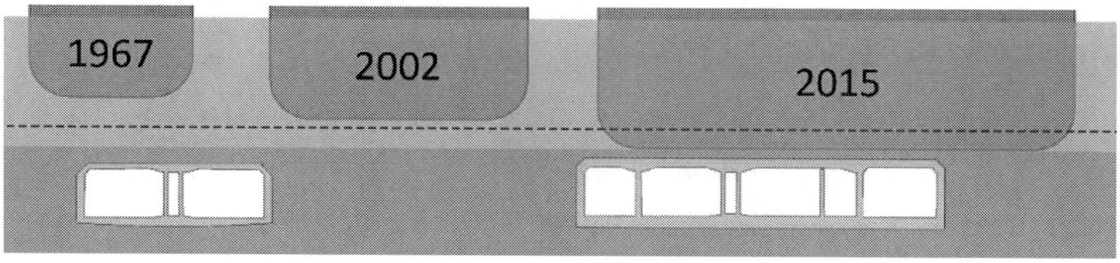

Fig. 1. This image indicates the draught and beam of the largest container vessels at the time of construction of both Beneluxtunnels and now, relative to the depth and width of the tunnel tubes (lowest point roof: NAP - 16.5 m) and the depth of the navigation channel for two-way traffic (red dotted line, NAP - 14.0 m) at low tide (NAP - 0.5 m).

3 Immersed tunnels

Both tubes of the Beneluxtunnel are built using the immersion method. This method for constructing tunnels has originated in the United States in 1910 with the construction of the Michigan Central Railway tunnel, which was built by immersing steel tubes onto the bottom of the waterway and connecting them to each other. This method was further developed in the Netherlands by introducing the reinforced concrete rectangular cross-section, resulting in the construction of the Maastunnel in 1942. Since the 1960's this method has been often applied for the construction of tunnels underneath waterways.

An immersed tunnel consists of two abutments and a number of immersed elements. The abutments are constructed on site and are thoroughly founded with tension-piles/anchors to withstand the upward pressure of the groundwater. The elements are prefabricated elsewhere and are sailed to the construction site when the abutments are finished, where they are immersed into a trench between the abutments, which is then backfilled and covered with bottom protection.

The 1st Beneluxtunnel consists of 8 elements of 8 x 24 x 93 m and the 2nd Beneluxtunnel consists of 6 elements of 8.5 x 45 x 140 m. The abutments are built independent from each other with a spacing of about 30 m and contain amongst others an underwater concrete floor with tension piles and sheet pile sidewalls with anchors.

4 Research

As far as known by researchers the possibility of lowering an existing immersed tunnel has never been investigated. Hence, effort has been made to maintain a broad scope and to use as little fixed variables as possible. By determining the most promising solution strategies as generally as possible, this study can act as a basis for further research in this yet unexplored field of research.

This method is amongst others applied for determining the maximum slope of the road by varying it from the current 4.5 % up to a maximum of 7 %. With a larger slope, the available draught can be increased without having to apply major adjustments to the abutments. To allow fair comparison, the associated lowering of the speed has also been included in the evaluation.

This strategy contributes to objective value determination and therefore to the determination of the economic feasibility. In addition, also certain aspects concerning structural safety and constructability have been examined, aiming to determine the technical feasibility. At this stage, it is presumed that for lowering to be allowed, the structural safety of the tunnel may not be negatively influenced by any of the adjustments.

5 Design options

An initial limited deepening of the navigation channel could be achieved by replacing the bottom protection at the critical points near the edges of the navigation channel with a thinner layer of protection. However, as it turns out this measure has already been applied to allow for the depth of the current channel. Hence, the height of the tunnel roof should be decreased at the critical points to allow for additional lowering, for which possibilities have been examined in this study.

For the additional lowering of the immersed part, mainly method of construction is of importance, for which two main alternatives can be distinguished:

I_1 The elements can be disconnected and re-floated – which could be interpreted as the inversed immersion process – for them to be re-immersed and re-connected later on a lowered bed.

I_2 One could also try to lower the elements while they remain connected by using the (limited) freedom of rotation in the joints. To allow lowering, also the soil must be removed from underneath the elements and restored afterwards by underflowing.

For the abutments on the land side of the tunnel, the following design options can be distinguished:

A_1 One could decide not to adapt the abutments, which would have the lowest impact on the costs but lowering would need to occur in the immersed part only and is therefore limited.

A_2 The transition points between the abutments and the immersed part could be adapted using the space provided by the presence of dewatering cellars. Consequently, rotation and limited translation of the connecting element could occur without having to alter the water- and soil retaining functions of the abutment.

A_3 If more depth is required, the abutments will have to be drastically adjusted. The side walls could most likely be reused but the underwater concrete floor and the tension piles will have to be replaced. This is expected to be a very difficult and costly operation but it is not regarded impossible at this stage.

The longitudinal profile of the tunnel is mainly determined by the chosen land part alternative and the maximum allowed slope. An example of possible longitudinal profiles and the associated dimensions of the enlarged navigation channel is shown in Fig. 2. In this example, the transition point has been adapted on one side (A_2) and on the other side the abutment is kept unaltered (A_1).

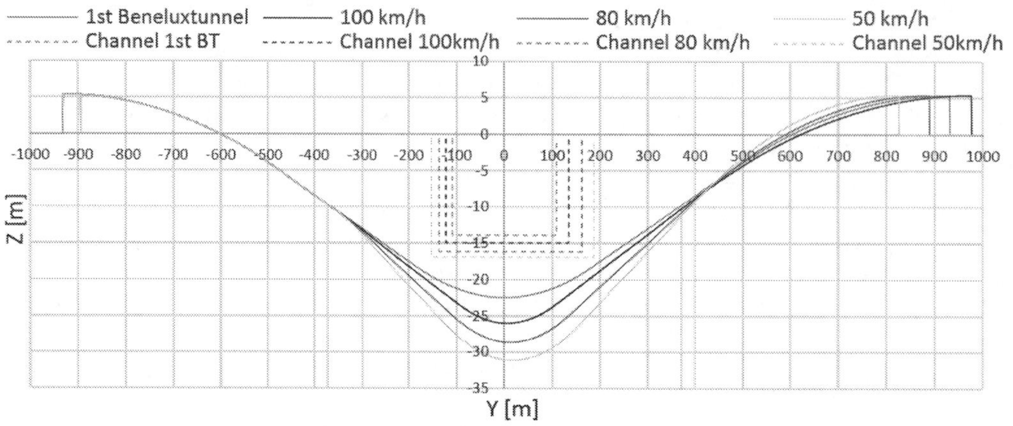

Fig. 2. This graph gives for the 1st Beneluxtunnel, the current and the possible longitudinal profiles for different maximum speeds and the associated maximum slopes, if the transition point has been adapted on one side (A_2) and on the other side the abutment is kept unaltered (A_1). Also, the associated cross-sectional profiles of the navigation channel are displayed.

For this analysis, the relations between navigation, the navigation channel and the tunnel have been determined using [1]. Because navigation has not only increased in draught but also very strongly in beam – as indicated in Fig. 1 – also the width of the required navigation channel increases very strongly. Consequently, the critical points in the longitudinal profile of the tunnel moves towards the shores if more depth is required.

This analysis gives a first impression of the possible design options. However, the structural safety of the immersed part may impose additional constraints to these design options. Crucial aspects to take into account are the cross-sectional strength of the elements and the consequences of rotations in the joints.

6 Cross-sectional strength

An important condition when lowering a tunnel element is that its strength needs to be sufficient to resist the occurring increase in hydrostatic pressure. To determine the maximum possible amount of lowering, this strength is compared with the forces occurring when the hydrostatic pressure is increased. Only the cross-section of the tunnel has been analysed as al forces are transferred in this direction. Because of available data, only the 1st Beneluxtunnel has been examined.

The results of the analysis are displayed in Fig. 3 as the maximum possible lowering for different load factors. In all scenarios, material factors have been applied as prescribed in the Eurocode. As the situation is based on the maximum possible water level – based on the height of the dikes – increased with an extra 2 meters for waves, a load factor of 1 could be regarded sufficient.

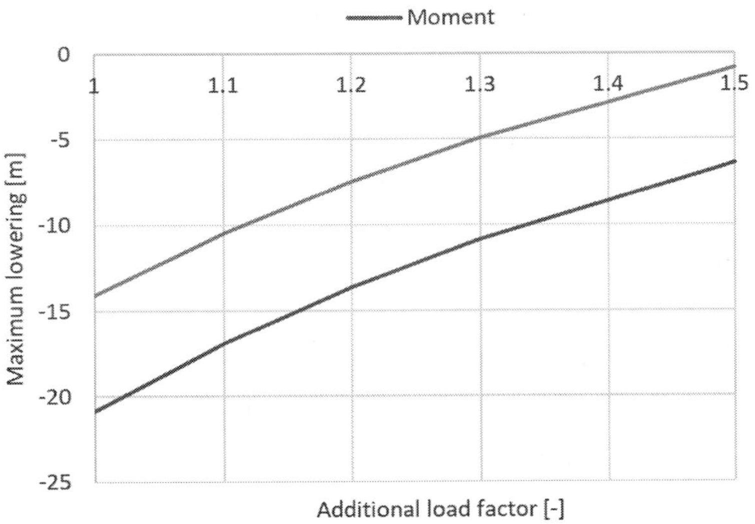

Fig. 3. This graph shows the maximum possible lowering for the deepest elements of the 1st Beneluxtunnel for different load factors, based on the maximum allowed moments and shear forces.

Based on the calculations performed in the study, the cross-section of the 1st Beneluxtunnel has sufficient strength to allow a significant amount of lowering. Depending on the safety philosophy, 5 to 14 meters of lowering would be possible for the deepest elements. The shallower elements are dimensioned less strongly but also require less lowering.

This residual strength is particularly expected to be a consequence of the applied safety philosophy during the design of the elements, when a safety factor of 1.5 has been applied on the maximum considered hydrostatic pressure, resulting in a corresponding water level being much higher than the height of the dikes. In this study however, the design water level has been determined taking the probability of exceedance of the water level into account, resulting in a lower design value of the hydrostatic pressure than based on previous criteria. If a similar residual strength is found for the 2nd Beneluxtunnel is given the 25 years between the construction of the tunnels not certain.

7 Joint rotations

With the existing methods for constructing immersed tunnels it is it is customary to cast the elements in segments with a length of about 15 to 20 meters. This prevents for temperature related cracks to develop during the hardening process and for loads occurring as a result of differential settlements to become unmanageable. Water tightness is provided by mounting rubber-metal profiles (W9ui) in the joints between the segments. Between the elements themselves GINA-profiles are mounted, which provide the primary seal during the immersion process. OMEGA-profiles later provide the final watertight seal. To prevent transverse displacements of the segments relative to each other, both the segment joints and the immersion joints contain a collar structure acting as a shear key.

The rotational capacity of these joints could be used to allow adjustment in the longitudinal profile of the tunnel. Small rotations in many joints can provide significant lowering of the tunnel. This method is especially interesting because the immersion joints in the 1st Beneluxtunnel have been permanently casted meaning the elements are securely fixed against each other. For the 2nd Beneluxtunnel, both joint types could be used. In this article however, only the segment joint will be discussed.

It is important for the strength and the water tightness of the joints not to be compromised. The segment joints are designed to allow rotation and if one examines the functioning of the collar structure in more detail, it turns out that regarding the strength, there is very little difference between the situations in which an opening of a few millimetres occurs or a few centimetres. The pull-side of this situation is shown in Fig 4. This figure implies that with an enlarged opening, the probability of shear failure of the collar system would increase. However, for openings up to a few centimetres, the functioning of the reinforcement is practically the same. The W9Ui profiles turn out to be governing with a maximum elongation of 22 mm. [2]

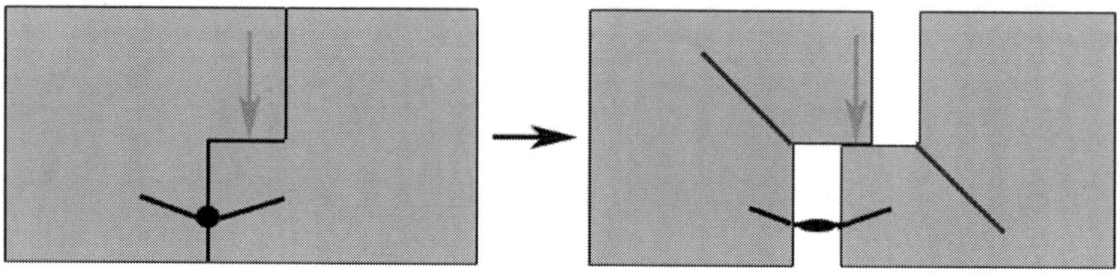

Fig. 4. This figure shows the collar on the pull side of the segment joint, before and after rotation has occurred. The yellow arrow indicates the working line of the shear force. The red lines indicate possible fractures. The horizontal part of the collar has a length of about 25 cm.

To determine the maximum magnitude of the initial rotation, one has to estimate the differential settlements that are expected to occur. This however, is very hard to predict. Partially because limited measuring data is available, but also because after lowering, new foundation is required. Moreover, would the additional joint rotations caused by settlement lead to damage of the profiles, repair is principally possible as is described in [3]. A probabilistic approach could clarify this issue, but for this study it is chosen to estimate the maximum initial rotation as half of the maximum elongation, which is 11 mm or 1.6 x 10-3 rad.

8 Results

Now that the possible profiles like the example in Fig. 2, and the technical boundary conditions are known, the scope of the solutions can be determined. Fig. 5 shows for the 1st Beneluxtunnel, for different combinations regarding the abutments and for different maximum speeds that are related to certain slopes, the maximum possible lowering of the navigation channel. The dotted line indicates the depth of the Maeslantkering its sill, what could presumably be regarded being the maximum required depth increase. Also, a distinction is made between rotation in only the immersion- or the segment joints – providing an almost equal amount of lowering – or in both joint types.

The technically possible lowering is clearly less than if one would only look at the abutments and the slopes, which is mainly caused by the restricted rotation capacity. The maximum depth regarding the strength of the cross section is only reached in case of large lowering of the abutments, as the adjacent elements are dimensioned less strongly and the relative increase of hydrostatic pressure is larger.

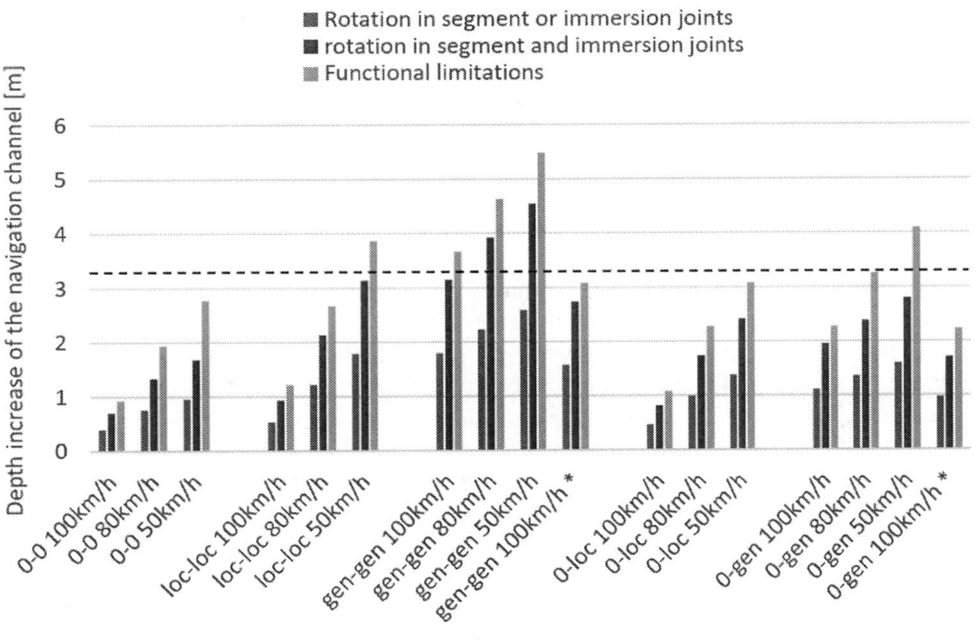

Fig. 5. This graph shows the maximum depth for different combinations of abutment adjustments and for different maximum speeds. 0 = no adjustments (A_1). Loc = locally adapt the transition point (A_2), gen = general lowering of the abutment (A_3). The two options with the * have a slope of 4.5 %, like in the existing tunnel. The dotted line indicates the depth of the Maeslantkering at NAP - 17.0 m.

9 Construction

For all design alternatives, the extent to which problems concerning the constructability are to be expected is examined. On the one hand, this is performed to demonstrate the (technical) feasibility, but it also serves a purpose as a base for cost estimates. The results will be briefly described here:

For the immersed part:

I_1 To be able to re-float the elements, they must be excavated first. Also, the weight of the elements must be reduced by removing some of the ballast and external pre-tensioning must be installed to keep the segments together. Next, the elements must be disconnected. To be able to do so, bulkheads must be placed and the collars and OMEGA-profiles need to be removed from the immersion joints. To remove the elements, also the closure joint must be disconnected, which is expected to be very complicated but we believe it could be accomplished. Instead of re-floating the elements, they could be hoisted up to the surface which would allow for better control during this process. The elements can then be transported to a construction site where they are temporarily stored and revised. When the abutments are adjusted to the increased depth, the elements can be re-immersed.

I_2 Alternatively, the elements could be lowered without re-floating them. To allow this, the soil would have to be removed from underneath the elements with specialized dredging equipment. The axial compression – present in immersed tunnels due to their construction method – will allow limited spans to exist without the occurrence of tensile stress in concrete or rotations in the joints. Consequently, spans of 20-30 meters are expected to be possible. To undermine the entire tunnel however, external pre-tensioning and temporary supports will be required. Both in the pre-tensioning and in the supports, mechanical components must be implemented to allow and control the

 DOI: 10.1051/matecconf/201713806006

lowering process. Also, at least one joint must be opened to allow horizontal displacement of the elements when the tunnel is lowered. This joint will later be closed similar to a closure joint. Once the tunnel has reached is desired depth, the elements must be underflown again to restore their foundation.

For the abutments:

A_1 No adjustments requires no constructional effort.

A_2 To locally adjust the transition point, it must be accessible, which requires for the adjacent element to be removed so a water retaining screen can be placed. The space provided by the dewatering cellar could be utilized to reconstruct a new transition point at a lower level, possibly at an increased angle.

A_3 General lowering could be achieved by wet reconstruction. First, additional anchors must be inserted in the side walls. The adjacent element must be removed and a water retaining screen must be placed which will allow for the entire land part can be inundated. The underwater concrete floor and all structures above have to be removed. New tension piles and vertical anchors must be placed. Finally, a new underwater concrete floor can be constructed on the desired depth, the water can be removed and all structures above can be rebuilt.

Apart from the construction methods of the alternatives themselves, also the consequences of their combinations are important. For instance, to adjust the land part space is required, making it necessary for at least one element to be re-floated in most scenarios. Also access to the elements during construction is an important advantage, making the adjustment of only one of the land parts much more attractive. This will however require for the navigation channel to be moved towards one of the shores.

Also, some differences between the 1st and the 2nd Beneluxtunnel exist. Important is for instance the fact that in for the 1st Beneluxtunnel, the immersion joints are filled with concrete, which highly complicates the re-floating capabilities of this tunnel its elements. Also the foundation is different. Underneath the 1st Beneluxtunnel lie temporary foundation tiles that need to be removed separately. The 2nd Beneluxtunnel is founded directly on gravel which could be removed using the right dredging equipment.

Based on the described methods, constructing the design alternatives is expected to be possible. However, additional research is required to increase the certainty of this statement. To prove constructability, it is required to perform a full safety analysis in which is checked whether the strength, stiffness and stability of all structural components satisfy the limit states during all phases of construction. This was not possible during the limited time of this study. Instead, only the structural components and construction phases that are expected to be most critical are examined. Hence, at this stage a large risk on unexpected setbacks exists.

10 Evaluation

Based on the construction methods, the costs of the different design alternatives are estimated. Also some other aspects are expressed in money for the comparison to be as fair as possible. For instance, the economic damage of traffic hinder during construction and as a consequence of lowering the maximum speed is estimated. The costs of building a new tunnel and demolishing existing one has also been estimated.

This information is used to determine what combination of design alternatives at which desired depth is the most economically attractive lowering method. The results are given in table 1, in which the total costs of the project are shown, including traffic costs. The effects of possible restrictions on combinations and the differences between the 1st and the 2nd Beneluxtunnel have been neglected.

Table 1. This table shows the most economically attractive combinations of design alternatives per for different lowering ranges. The costs include the entire project (reconstructing both tunnels, including traffic costs).

Lowering range	Abutments	Immersed part	Costs (millions)	Slope
0.0 - 0.9 m	Local adjustments on both sides	Remain immersed	€ 295	5 %
0.9 - 1.9 m	General lowering on one side	Remain immersed	€ 415	5 %
1.9 - 3.2 m	General lowering on both sides	Remain immersed	€ 500	5 %
3.2 - 3.6 m	General lowering on both sides	Re-float	€ 665	5 %
> 3.6 m	New immersed tunnel		€ 730	5 %

This cost analysis shows that for all desired depths up to the depth of the Maeslantkering, lowering the existing Beneluxtunnel is economically more attractive than building a new one. Also, increasing the slope with more than 5 % is not desired. The lowered construction costs do not outweigh the economic damage of lowering the speed. However, due to recent changes in design regulations, the slope can be slightly increased from 4.5 % to 5.0 %. [4]

When interpreting these results one should be aware of the large uncertainties in these estimates as they strongly depend on construction methods that are currently not sufficiently detailed to be able to exclude large unforeseen costs.

11 Conclusions

This preliminary study shows that lowering the Beneluxtunnel down to the depth of the Maeslantkering is expected to be technically possible as well as economically attractive. Furthermore, for this depth range can be concluded that:

- The 1st Beneluxtunnel has sufficient excess strength in order to allow lowering.
- Joint rotations can be utilized to adapt the longitudinal profile of the tunnel.
- For the abutments, the preferred solution depends strongly on the desired depth. General lowering is by far the least attractive solution, but is inevitable from a desired depth of more than 1 m.
- For the immersed part, lowering the elements while they remain immersed preferred. To create space however, it is desired to re-float at least one element. Hence, a combination of both methods is preferred.
- The large expected costs of lowering the maximum speed makes this an unattractive option.

It is recommended to perform additional research on certain boundary conditions and construction methods which could only be treated partially during this study and could therefore be defined as large risks during this stage. Important subjects are:

- Determining the excess strength of the 2nd Beneluxtunnel.
- Determining the maximum allowed joint rotations taking into account the effect of deformations.
- The removal of structural components (including the closure joint) and the effects of this on the tunnel.
- The stabilisation of the tunnel elements during excavation and during the lowering process.
- The replacement of the underwater concrete floor and the tension piles of the abutments.

References

1. PIANC, *Harbour Approach Channels Design Guidelines*, PIANC Secrétariat Général, Brussel (2014)
2. Trelleborg, *Waterstops,* Trelleborg Ridderkerk B.V, Ridderkerk (2010)
3. L. Leeuw, *Lekkage in Tunnels: Dilatatievoegen en beton*, Rijkswaterstaat Bouwdienst (2008)
4. Rijkswaterstaat, *NOA: Nieuwe Ontwerprichtlijn Autosnelwegen,* Rijkswaterstaat Adviesdienst Verkeer en Vervoer, Rotterdam (2007)
5. Weeda, T.R. (2015). *Feasibility study of lowering the Beneluxtunnel. Master thesis,* Technische Universiteit Delft, 6 oktober 2015
6. Schols, I. (2012). *Segmentvoegcapaciteit van de Kiltunnel.* Master thesis, Technische Universiteit Delft, mei 2012
7. Vervuurt, A.H.J.M., Suma, A.B., Bakker, K.J., Hendrix, B & Schols, I. (2011). *Vervormingen van afgezonken tunnels in Nederland*. InfraQuest, Delft, 21 december 2011
8. Eysink, W.D., Luth, H.R., Vroeg, J.H. de & Wijhe, H.J. van (1995). *Tweede Beneluxtunnel: Nautisch, hydraulisch en morfologisch onderzoek.* Waterloopkundig laboratorium, Delft, april 1995
9. Rijkswaterstaat (2005). SATO: *Specifieke Aspecten TunnelOntwerp. Ontwerphandleiding*, Rijkswaterstaat Bouwdienst, Utrecht

Placement of BMPs in urban catchment area using SUSTAIN model: case study at Universitas Indonesia Campus, Depok, West Java, Indonesia

Tina Kesuma Warganda[1,*] , and *Dwita* Sutjiningsih[1]

[1]Civil Engineering Department, Universitas Indonesia, 16424 Depok, Indonesia

Abstract. Aquatic system quality degradation by urban runoff owing to the effect of increasing impervious cover due to urbanization is becoming a serious issue. Based on 2015 data, with total imperviousness more than 60% in aquatic system catchment area of Universitas Indonesia (UI) Campus, the aquatic system quality is in poor condition. Best Management Practices (BMPs) as the practical form of Low Impact Development (LID) concept are proposed to be implemented in the catchment area. The performance of BMPs implementation can be quantitatively assessed using System for Urban Stormwater Treatment and Analysis Integration (SUSTAIN) model developed by US Environmental Protection Agency. As a decision support system, SUSTAIN model as the extension of ArcGIS can be applied for placement of BMPs at strategic locations in urban watersheds. This study aims to find the suitable location of BMPs placement in nine sub-catchments of UI Campus aquatic system catchment area. The suitability criteria for placing structural BMPs comprise nine database layers including slope, soil type, urban land use, percent impervious, land ownership, roads, water table depth, stream location, and drainage area. The study shows that there are ten types of BMPs that suitable to be implemented in UI campus catchment area, while three types of BMPs are not suitable. Based on the validation analysis, the applicable result from SUSTAIN model is 18 of 20, or has 90 percent of trust.

1 Introduction

A decrease in the quality of aquatic systems due to rainwater runoff in urban areas as a result of the increasing extent of impervious cover (IC) due to urbanization has become a very serious problem. The development of an urban area is always accompanied by an increase in population, activities and land needs, both for economic activities and settlements to meet the human needs. The conditions of land constraints result in urban activity intervention on land that should function as conservation areas and green open spaces. As a result, the number of impervious cover (IC) is increased, infiltration areas are

* Corresponding author: tina_kesuma@yahoo.com

getting narrower resulting in increased surface flow and erosion that ultimately lead to the degradation of water quality [3, 5, 7 - 9, 12, 15].

The relationship between IC and water quality mentioned that the quality of an aquatic system of a catchment area can be indicated based on the percentage of its impervious cover from total CA [3, 5, 7 - 9, 15]. Referring to that theory, based on 2015 data, with a total impervious cover of 60% [8] within the Catchment Area (CA) of the Universitas Indonesia (UI) [2, 8], Depok, its CA aquatic system is in a range of impacted and heavily damaged condition [8, 15].

Refering to the understanding of the relationship between IC with the water quality degradation mentioned, a Low Impact Development (LID) approach is highly required for site development of catchment area [10]. LID is intended to preserve or reinstate the natural, pre-developed condition of an urban site to maximize on-site stormwater controls [10, 13]. As components of LID, structural Best Management Practices (BMPs) are designed to prevent, control or treat stormwater discharges. LID/BMPs integrate small-scale measures (e.g., bioretention or rain garden, porous pavement, grassed swale, green roof, rain barrel or cistern, vegetated filterstrip, infiltration basin, constructed wetland, etc.) scattered throughout the urban site [10].

One of the most important thing in determining the successful application of LID in a CA is in determining the suitable site or location for placing the BMPs, in order for the BMPs to function optimally [4, 10, 11, 14]. In a catchment area with a complex variations of land use, such as urban areas with high density of buildings, and still developing like CA in Universitas Indonesia campus, the decision to choose the suitable location becomes more complicated compared to areas with lower building density. Based on these conditions, this research aiming to answer the challenge by performing suitability analysis using public domain tools that are developed specifically to finally be able to produce a water-friendly planning in the Universitas Indonesia Depok catchment area using SUSTAIN model.

Some research on LID has been done before especially focused on the runoff quantity as the decrease of flood discharge and the effectiveness of each type of LID practice separately. However, no specific research has been found which discussed specifically about the site selection to determine the specific type of BMPs to be installed in a CA, especially in UI campus CA. In addition to the scope of the study, the novelty of this research is on the public domain tools used, namely System for Urban Stormwater Treatment and Analysis Integration (SUSTAIN), which is a decision-making tool developed by US Environmental Protection Agency (US EPA) since 2003 [14]. SUSTAIN is a relatively new tool in the analysis of rainwater runoff management, especially in Indonesia.

2 Method

2.1 Framework and Tools of Study

This study is conducted in 2 (two) main steps, which are; (1) Suitability Analysis, to determine the type of BMPs to be applied/placed in certain location within the UI Campus CA, and (2) Validation Analysis to ensure the result from the suitability analysis which will be cross checked with the actual condition in the field. The location of this study case is in catchment area of UI Campus in Depok as one of the example of urban catchment area.

The tool used for suitability analysis is BMP Siting Tools (BST) module on the SUSTAIN program which is an extension of the ArcGIS software. BMP siting tool is used to assist users in selecting suitable locations for different types of LID (BMP) techniques.

DOI: 10.1051/matecconf/201713806007

This tool is implemented using ESRI ArcGIS 9.3 and 10.1 and Spatial Analyst extension [4, 6, 11, 14]. This study is using ArcGIS 10.1 for its analysis process.

Validation of SUSTAIN model analysis is performed as the conformity how accurate is the result from BST suitability analysis in accordance with actual conditions in the site. Total twenty points are set as the checking point, whereas the actual condition situation pictures is taken by direct observing and using the information from Google street view as one of the Google maps feature.

2.2 Study Area Description

UI Campus in Depok is located at the boundary of Jakarta Special Capital Region and Depok City, West Java Province, Indonesia which geographically located between $6^0$20'54"N - $6^0$23'19"S and 106°49'E - 106°49'48"E. The campus area was designed with better landscaping to preserve the topography as well as the water bodies. There are total 8 ponds identified within the UI campus CA, namely Kenanga, Agathis, Mahoni, Puspa, Ulin, and Salam, which existed inside the campus area, and pond of Pladen and Rawa Besar which existed outside of the campus area.

With total catchment area around 580 Hectares, where around 305 Hectares (the northern part) is located inside the campus area, and the rest (the southern part) is located outside campus area [8], the difference of land use condition inside and outside the UI campus in its catchment area is contrasting. As can be seen in Figure 1, whilst most of the land use inside of the campus area are covered with the vegetation, the outside part of the campus is dominated by improperly planned and high density residential areas [8].

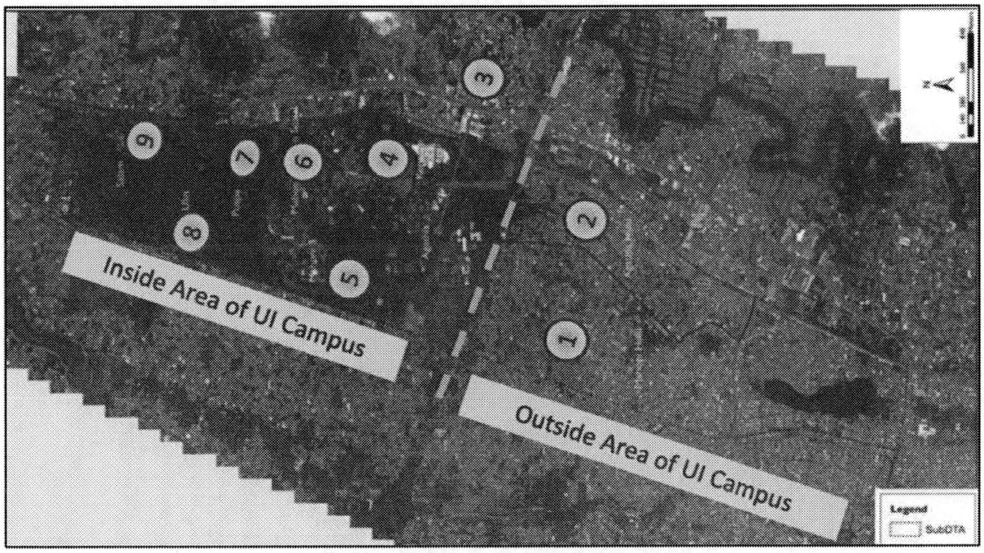

Fig. 1. Sub Catchment Area of Universitas Indonesia Campus, Depok

UI campus CA is divided to 9 (nine) sub catchment area which are (1) Sub CA Mahoni Hulu, (2) Sub CA Agathis Hulu, (3) Sub CA Agathis, (4) Sub CA Kenanga, (5) Sub CA Mahoni 3, (6) Sub CA Mahoni Rumpun Sosial, (7) Sub CA Puspa, (8) Sub CA Ulin, and (9) Sub CA Salam (see Figure 2). The division is done based on the topographical border and the location of the ponds in the CA. This sub CA partition and numbering will be used in the analysis process to get more detailed results especially in the suitability of the BMPs.

2.3 Data Requirement

Data required for this research using BST in SUSTAIN mainly is the thematic map (in raster and shape format) for data layer in ArcGIS. There are nine base data layers required including; slope, soil type, urban land use, percent impervious, land ownership, roads, water table depth, stream location, and drainage area [14], as seen in Table 1.

Table 1. Data Requirement for BMP Siting Tools, SUSTAIN

GIS layer	Format	Description	Data Used
DEM	Raster file	The DEM is used to calculate the drainage slope and drainage areas that are used to identify the suitable locations for BMPs.	Topography Map of Depok City, *Badan Perencanaan dan Pembangunan Daerah*, 2012, scale 1 : 25.000
Land Use	Raster file	The land use grid (e.g., NLCD land cover) is used to eliminate the unsuitable areas for BMPs.	- Land Use Map of Depok City, *Badan Informasi Geospasial* 2011, scale 1 : 25.000 - Google Earth, Quickbird, 2015 - Bing Satellite, Microsoft, 2016
Percent Impervious	Raster file	The percent impervious grid is used to identify the suitable locations for BMPs for the given suitability criteria.	
Urban Land Use	Shape file	The urban land use data contain the boundaries for the buildings and the impervious areas needed to identify suitable locations for LID elements.	
Road	Shape file	The road layer is used to identify suitable locations for some BMPs that must be placed in a specific road buffer area.	
Stream	Shape file	The stream layer is used to define a buffer so that certain BMP types can be placed outside the buffer to minimize the impact on streams.	
Soil	Shape file	The soil data contain the soil properties such as hydrological soil group, which are used to identify suitable locations for BMPs.	Soil Map of Depok City, *Pusat Penelitian Tanah*, Bogor, 2011, scale 1 : 25.000
Groundwater Table Depth	Shape file	The groundwater table depth layer is used to identify suitable locations for the infiltration BMPs; derived from monitoring data.	
Land Ownership	Shape file	A parcel layer is used to identify the locations on the public or private land.	

Source: [14]

Every each of the data listed are the final format which will be used as one layer. The layer of Percent Impervious, Urban Land Use, and Land Ownership are identified and created from basic data of land use map with the scale of 1 : 25.000, which then digitize referring to Google Earth, Quickbird and Bing Satelite, Microsoft to get the layer map with the scale of 1 : 10.000 as the result.

GIS data input for BST in SUSTAIN required a specific format, type and field type, which is very sensitive for error. Making sure that the data is made properly as required is

DOI: 10.1051/matecconf/201713806007

really important in this case. The other highlight is that GIS layers with different coordinate systems will cause errors during the spatial analysis. It is important to project all GIS data into the preferred coordinate system of the study area before starting the project. To project the GIS data from one coordinate system to another, use the Project utility under Data Management Tools of ArcToolbox [14].

Table 2. GIS data format requirement for BMP suitability analysis

GIS layer	Data type	Field name	Field type	Field value/description
DEM	Raster	VALUE	Floating	ESRI grid
Land Use	Raster	VALUE	Integer	ESRI grid
Land Use Lookup	Table	LUCODE	Integer	Should be same as in the VALUE field in Land Use raster layer
		LUNAME	String	Description about the land use type
		SUITABLE	Short	1 or 0 (1 = suitable; 0 = unsuitable)
% Impervious	Raster	VALUE	Integer	0 – 100
Soil	Shape	MUKEY	String	Hydrological soil classification (STATSGO or SSURGO or local dataset)
Soil Lookup	Table	MUKEY	String	Should be same as in the MUKEY field
		HYDGRP	String	A or B or C or D
Urban Land Use	Shape	LU_DESC	String	Buildings (for Green Roof BMP type and defining the buffer around the buildings) or Roadways or ParkingLots (for Porous Pavement BMP type)
Road	Shape	N/A	N/A	N/A
Stream	Shape	N/A	N/A	N/A
Groundwater Table Depth	Shape	GWDEP_FT	Double	Depth to groundwater table
Land Ownership	Shape	OWNERSHIP	String	Public or Private

Source: [14]

2.4 Suitability Parameter

Suitability analysis performed by setting the criteria for all nine parameters which considered to be appropriate for each type of BMPs. The criteria used in this research mainly is based on the default criteria derived from the BMP Siting Tools User Guide, which refers to two of US EPA reports in 2004, which is shown in Table 3 [14].

These criteria can be adjusted to our preference based on local knowledge and situation, though the parameter of column (5), (7), (8), and (9) have a limited option, whereas we need to select the option which already been offered (e.g for building buffer some of the options are <30, 30 – 50, 50 – 100, or >100), and can not be filled with a random number nor alphabet. As for the other columns, the limitation is off and can be adjusted with any number needed [14].

Based on the local situation in research site, some modification was made for some of BMPs suitability criteria. The default criteria made by US EPA in the BST guide book mostly based on the site condition in the United State of America, whereas having a lower density of urban area compared to Indonesia. This modification crossed checked by referring to the consideration of BMPs guidelines [1]. The changes made are;

- Porous pavement; suitability criteria for slope = <5%
- Rain barrel; suitability criteria for building buffer = 50 – 100 ft
- Vegetated filterstrip; suitability criteria for road buffer = 50 – 100 ft

Table 3. Default Criteria for each Parameter in BMP Siting Tool, SUSTAIN

BMP Type	Drain-age Area (acre)	Drain-age Slope (%)	Imper-vious (%)	HSG	Water table depth (ft)	Road Buffer (ft)	Stream Buffer (ft)	Build-ing Buffer (ft)
Bioretention (Br)	< 2	< 5%	> 0%	A–D	> 2	< 100	> 100	--
Cistern (Cs)	--	--	--	--	--	--	--	< 30
Constructed Wetland (Cw)	> 25	< 15%	> 0%	A–D	> 4	--	> 100	--
Dry Pond (Dp)	> 10	< 15%	> 0%	A–D	> 4	--	> 100	--
Grassed Swale (Gs)	< 5	< 4%	> 0%	A–D	> 2	< 100	--	--
Green Roof (Gr)	--	--	--	--	--	--	--	v
Infiltration Basin (Ibs)	< 10	< 15%	> 0%	A–B	> 4	--	> 100	--
Infiltration Trench (Itr)	< 5	< 15%	> 0%	A–B	> 4	--	> 100	--
Porous Pavement (Pp)	< 3	< 1%	> 0%	A–B	> 2	--	--	--
Rain Barrel (Rb)	--	--	--	--	--	--	--	< 30
Sand Filter non-surface (Sfns)	< 2	< 10%	> 0%	A–D	> 2	--	> 100	--
Sand Filter surface (Sfs)	< 10	< 10%	> 0%	A–D	> 2	--	> 100	--
Vegetated Filterstrip (Vfs)	--	< 10%	> 0%	A–D	> 2	< 100	--	--
Wet Pond (Wp)	> 25	< 15%	> 0%	A–D	> 4	--	> 100	--

Source: [14]

The assumption used in this study is for the water table depth data, due to limited data availability, which is 4 m (13,12 ft). The geology formation for the CA is Qav formation (Alluvial Fan) which consist of bedded fine tuff, sandy tuff, inter bedded with conglomeratic tuff, thus the water table depth is assumed to be the same for all of the CA. Furthermore, the study area is relatively a small area and flat.

2.5 Operating BMP Siting Tools

The suitability analysis operated by BMP Siting Tools (BST) in SUSTAIN is done in three steps. The data specified are entered simultaneously in the Data Management tab of the BST module is the first step. Then, the next step is to select the type of BMPs to be analyzed, by moving the wanted BMP type on the left box to the right box. These 2 steps shown in Figure 2. Total fourteen of BMPs type are provided in BST. In this study, all types of BMPs listed are included as an alternative, nonetheless, the Cistern is opted out as it is considered has the similar function with rain barrel but requires a bigger area.

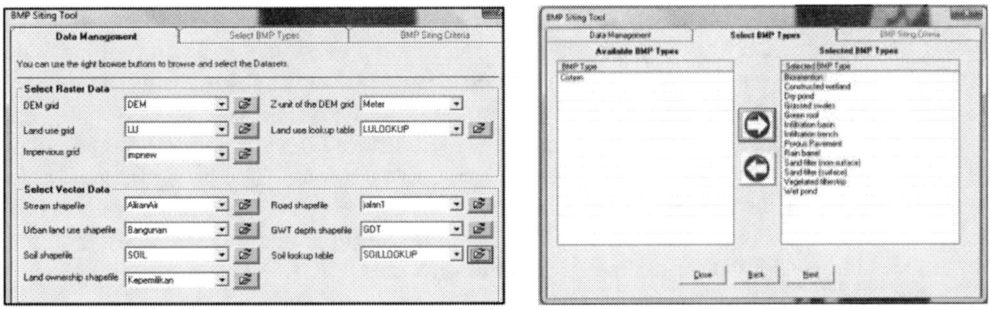

Fig. 2. Step 1 Filling Data Management (left); and Step 2 Selecting BMP Types (right), BMP Siting Tool, SUSTAIN [14]

After selecting the BMPs type, the next step needed is to establish the siting criteria by entering the parameter values that have been set as in Table 3. Select the BMPs type one by one, then fill in the criteria. This step is done repeatedly for each type of BMPs that will be assessed. Figure 3 shows the example of setting the criteria for Bioretention.

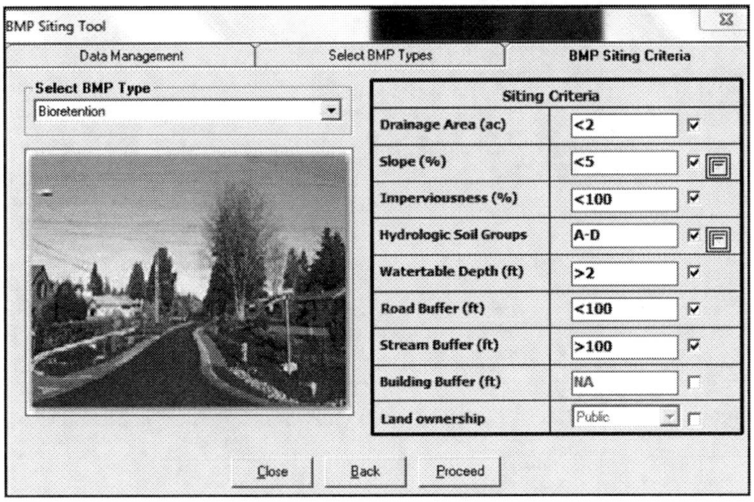

Fig. 3. BMP Siting Criteria, BMP Siting Tool, SUSTAIN [14]

Make sure to click the proceed button only after finishing all the criteria for all type of BMPs. The output from BMP Siting Tool Analysis is a spatial map that marks areas in the CA that meet the standards or that meet the specified criteria required.

3 Result and Discussion

3.1 Distribution of BMPs

The BST executed to perform BMPs suitability analysis is performed simultaneously at one time for each selected BMPs. Suitability analysis using SUSTAIN BST shows that there are 10 types of BMPs that are suitable to be placed in UI Depok catchment area, namely; (1) Bioretention; (2) Green Roof; (3) Porous Pavement; (4) Vegetated Filterstrip; (5) Grassed Swale, (6) Rain Barrel, (7) Infiltration Trench, (8) Infiltration Basin, (9) Sand Filter surface, and (10) Sand filter non-surface.

The output of BST is the layers showing the location for each one of BMPs type and a layer of composite map showing all of the combination of suitable BMP types at each location. The distribution of location and type of BMPs of this suitability analysis is presented in Figure 4. The result of this suitability analysis also shows that green infrastructure in the upstream area still has the potential to be developed, although it is not continuous (still segmented) as situated in a dense residential area. The green line from Situ Rawa Besar located on the upstream of CA can be connected to the downstream (Lake Agathis on UI campus or Situ Pladen) by following the middle channel of CA, by applying Bioretention, Grassed Swale and Vegetated Filterstrip types

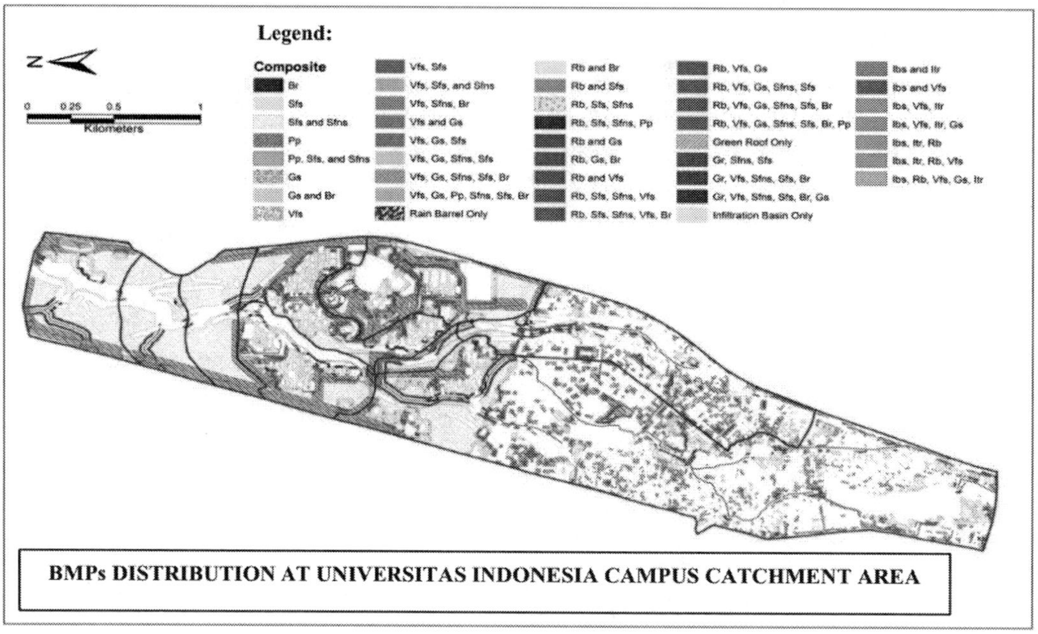

Fig. 4. Composite Result of BMPs Distribution in Universitas Indonesia Depok Catchment Area

3.2 BMPs Suitability and Validation

Analysis of the suitability of BMP types to the locations within the UI Depok CA illustrates that certain types of BMP have a higher suitability compared to other types of BMPs in terms of the number of appropriate Sub CA (see Table 4).

Types of BMPs of constructed wetland, dry pond and wet pond are not suitable to be installed in any sub CA of UI Depok. From the simulation process, the cause of the non-suitable BMPs is found in drainage area and the water table depth parameter, whereas the type of non-suitable BMPs requires a large area, like dry pond and constructed wetland.

The validation step is performed by determining 20 observation points as sample, scattered and representing all nine sub CA. The result from suitability analysis by SUSTAIN model are then validated with actual conditions in the field, thus the validation result will come as applicable or not applicable. From total 20 observation points, 18 points are applicable, while 1 point is not applicable and 1 other point is still applicable but not maximum. Table 5 shows some of the example of the validation process and findings from total 20 observation points.

DOI: 10.1051/matecconf/201713806007

Table 4. BMPs Suitability for Each Sub Catchment Area

	Sub Catchment Area Code Number	1	2	3	4	5	6	7	8	9
BMPs type	Bioretention (Br)	√	√	√	√	√	√	√	√	√
	Constructed Wetland (Cw)	×	×	×	×	×	×	×	×	×
	Dry Pond (Dp)	×	×	×	×	×	×	×	×	×
	Grassed Swale (Gs)	√	√	√	√	√	√	√	√	√
	Green Roof (Gr)	√	×	×	×	×	×	×	×	×
	Infiltration Basin (Ibs)	√	√	√	√	√	√	√	√	√
	Infiltration Trench (Itr)	√	√	√	√	√	√	√	√	√
	Porous Pavement (Pp)	×	×	×	√	√	√	×	×	×
	Rain Barrel (Rb)	√	√	×	√	√	√	×	√	√
	Sand Filter (non-surface) (Sfns)	√	√	√	√	√	√	√	√	√
	Sand Filter (surface) (Sfs)	√	√	√	√	√	√	√	√	√
	Vegetated Filterstrip (Vfs)	√	√	√	√	√	√	√	√	√
	Wet Pond (Wp)	×	×	×	×	×	×	×	×	×

Note : (√) Suitable; (×) Not Suitable

Table 5. Sample of Validation of BMPs Suitability

Point No	Sub CA Number Code	BMPS's Suitability Result	Actual Condition	Picture	Validation
1	1	Sfs, Sfns	Building, grassy yard		Applicable
2	2	Br, Vfs, Sfs, Gs	Vegetation		Applicable
3	4	Sfs, Pp	Grassed Cover, Open Space		Applicable
4	6	Br, Vfs, Sfs, Gs	Building, road, vegetation		**Applicable, not Maximum**
5	7	Br, Vfs	Vegetation and Building		**Not Applicable**

The validation for point 5, BST model shows that the area is suitable for Bioretention and/or Vegetated Filterstrip. The actual condition for this area is the parking yard with a full pavement, which is not applicable for those type of BMPs. In this case, SUSTAIN may identify the area as an open space caused by the scale of the map or data input.

4 Conclusion

Suitability analysis using SUSTAIN BST shows that there are 10 types of BMPs that are suitable to be placed in UI Depok catchment area. The type of BMPs which are not suitable, including constructed wetland, wet pond and dry pond, mainly caused by the limited availability of land. As the type of suitable BMPs are still dominant in this case, this study shows that LID concept can be applied in UI campus CA with high possibilities.

DOI: 10.1051/matecconf/201713806007

Based on the validation analysis, the applicable result from SUSTAIN model is 18 of 20, or has 90% level of trust. Thus, this model is proper to be used and applied for a suitability analysis for BMPs placement. For further study, using the large scale of map as data input is highly recommended for a better result.

Acknowledgment

The author would like to thank Universitas Indonesia for the financial support for this study through the PITTA 2017 funding scheme under grant no. 857/UN2.R3.1/HKP.05.00/2017, managed by the Directorate for Research and Public Service, Universitas Indonesia. Special thanks to Dr. Rer. Nat. Eko Kusratmoko and Chandra Pradana from Department of Geography, Universitas Indonesia for providing the geospatial data, the member of Hydraulic, Hydrology and River Engineering Laboratory for the support accordingly, and to Wisnu Tri Winarendro for the troubleshooting related to SUSTAIN software during the process of this study.

References

1. Applied Ecological Services, Inc, *Alternative Stormwater Best Management Practices Guidelines* (Public Works and Utilities Department Watershed Management Division City Of Lincoln, Eudora, 2006)
2. Badan Perencanaan, Pengembangan dan Pengendalian Universitas Indonesia (BP3UI), *Persiapan Audit Kesehatan Perairan dan DTA Kampus UI Depok* (BP3UI, Depok, 2015)
3. Center for Watershed Protection, *Watershed Protection Research Monograph No.1: Impacts of Impervious Cover On Aquatic Systems* (Center for Watershed Protection, Ellicott City, 2003).
4. C. Chen, M. Sheng, C. Chang, S. Kang, J. Lin, Jour. Water. **6**, 3575 (2014)
5. C. Arnold, C. Gibbons, Jour. Amr. Plan. Ass. **62**, 243 (1996)
6. C. Viriyadhika, I. Kustiwan, J. Prc. Wil. Kot. **3**, 239 (2014)
7. D. Theobald, S. Goetz, J. Norman, P. Jantz, Jour. Hydrol. Eng. **14**, 362 (2009)
8. D. Sutjiningsih, Int. Jour. Tech. **1**, 145 (2017)
9. E. Brabec, S. Schulte, P. Richards, Jour. Plan. Lit. **16**, 499 (2002)
10. J. Lee, A. Selvakumar, K. Alvi, J. Riverson, J. Zhen, L. Shoemaker, F. Lai, Jour. Env. Mod. Soft. **37**, 6 (2012)
11. L. Shoemaker, J. Riverson, K. Alvi, J. Zhen, R. Murphy, Tetra Tech, Inc., *Report on Enhanced Framework (SUSTAIN) and Field Applications for Placement of BMPs in Urban Watershed* (U.S Environmental Protection Agency, Cincinnati, 2011)
12. P. Flinker, *The Need to Reduce Impervious Cover to Prevent Flooding and Protect Water Quality* (Dodson Associates, Ltd., Ashfield, 2010)
13. S. Muthukrishnan, B. Madge, A. Selvakumar, R. Field, D. Sullivan, *The Use of Best Management Practices (BMPs) in Urban Watershed* (U.S Environmental Protection Agency, Cincinnati, 2004)
14. Tetra Tech, Inc, *BMP Siting Tool: Step by Step Guide, ArcGIS 10.1 Service Pack 1* (Tetra Tech, Inc, Fairfax, 2013)
15. T. Schueler, L. McNeal, K. Cappiella, Jour. Hydrol. Eng. **14**, 309 (2009)

MATEC Web of Conferences **138**, 06008 (2017) DOI: 10.1051/matecconf/201713806008
EACEF 2017

Cascade-pond System Health Assessment Based on Macroinvertebrate Indices and Its Relationship with Impervious Cover and Aquatic Buffer Zone in Urbanized Catchments

Faris Zulkarnain[1,2*], *Dwita* Sutjiningsih[1], and *Yasman*[3]

[1]Department of Civil Engineering, Faculty of Engineering, Universitas Indonesia, Indonesia
[2]Center for Applied Geography Research, Universitas Indonesia, Indonesia
[3]Department of Biology, Faculty of Mathematics & Natural Science, Universitas Indonesia, Indonesia

Abstract. A cascade-pond system consists of six ponds located at Universitas Indonesia Campus, Depok. Its catchment area is dominated by high density urban area with moderate to high imperviousness. Some of riparian buffers surrounds six ponds are also occupied by high imperviousness that may lead some ecohydrological problems i.e. water quality degradation, declining freshwater biodiversity and food web changes. The aim of this study is assessing the current state of cascade-pond system health. The assessment of macroinvertebrate indices is based on SingScore that have been developed by Public Utilities Board of Singapore for macroinvertebrate biotic index. Impervious cover data is obtained from high-resolution imageries and processed using ArcGIS 10.5. Qualitative statistics methods, Chi-squared test describes the relationship of macroinvertebrate indices with catchment area imperviousness and aquatic buffer zone. The health assessment based on macroinvertebrates indices shows that the lower ponds are relatively healthier than the upper one. There is also any significant relationship between macroinvertebrate indices with impervious cover based on chi square test and cross tabulation analysis.

1 Introduction

A cascade-pond system consists of six ponds located at Universitas Indonesia (UI) campus, Depok. Its catchment namely UI catchment, as a part of Ci Liwung watershed is dominated by urbanized area with high imperviousness. Previous studies found that UI catchment has varied imperviousness, from 22% to 76.9% [1]. According to Center for Watershed Protection, current state of imperviousness in UI catchment is classified from impacted to heavily damaged for its stream health [2]. High imperviousness would lead some

* Corresponding author: faris.zulkarnain@sci.ui.ac.id

ecohydrological problems for 6 ponds at UI catchment i.e. water quality degradation, aquatic habitat destruction, declining freshwater biodiversity, etc [1,3,4,5].

Macroinvertebrate indices believed as one of bioindicator tools to assess the current state of freshwater environment in catchment scale [6,7,8,9,10]. There are many advantages of macroinvertebrate as bioindicator i.e. their relatively easy to collect and identify, their low mobility, their high sensitivity for several environmental parameter changes, found in almost freshwater environment both lakes or stream, and their numerical abundance [11,12,13]. Macroinvertebrate also has important roles for aquatic ecosystem. They act as first consumers in aquatic food webs and play a role in the process of recycling nutrients [14,15,16].

Macroinvertebrates indices is being used widely in the world, from subtropics to tropics region. In the Southeast Asia region, Public Utilities (PUB) of Singapore developed The Singscore as macroinvertebrate biotic indices to assess the aquatic environment health [17]. This study is using SingScore to calculate the macroinvertebrate indices, assumes that geographical, meteorology, climatology, and indigenous macroinvertebrates are quite similar between Singapore and Indonesia.

Macroinvertebrate indices would be varied spatially. Previous studies show that there is any relationship between macroinvertebrate indices and impervious cover. If the percentage of impervious cover is high, then the macroinvertebrate indices score would be low [18,19,20]. Besides impervious cover, aquatic buffer zone is also influencing macroinvertebrate indices [21]. High score indices would be associated with high riparian cover buffer zone. Reducing riparian buffer zone would altered the macroinvertebrate indices [22,23]. On the other hands, impervious cover area buffer zone would affect macroinvertebrate indices badly.

The objectives of this research are applying the macroinvertebrate indices that developed by Public Utilities Board of Singapore namely SingScore and determining the relationship between macroinvertebrate indices and many environmental variables, which the most significant environmental variables that affecting the macroinvertebrate indices. Many research has been conducted to determine the relationship between macroinvertebrate indices but mostly in lentic ecosystem. Until now, there is only a few research about macroinvertebrate indices in pond or lotic ecosystem particularly cascade-pond system.

2 Materials and Methods

2.1 Study Site

This study was conducted at UI catchment area that located at the boundary of Jakarta Special Capital Region and Depok City, West Java, Indonesia. UI catchment consists of 6 catchments. (see Figure 1). The upper part of UI catchment area which is A1 and A2 sub-catchment has the highest percentage of impervious cover than the lower part (see Table 1). Land use in the lower part of UI catchment area are dominantly vegetation cover (see Figure 1). The total area of UI catchment area is approximately 580 hectares where around 305 hectares is located inside the UI Campus and 275 hectares is located at the southern of UI Campus. The UI Campus was well designed with a suitable landscape to maintain the natural condition of topography, water bodies, natural vegetation, and indigenous organism. On the other hand, the outside of the campus is dominated by unwell planned, high density, and fast growing residential area [1].

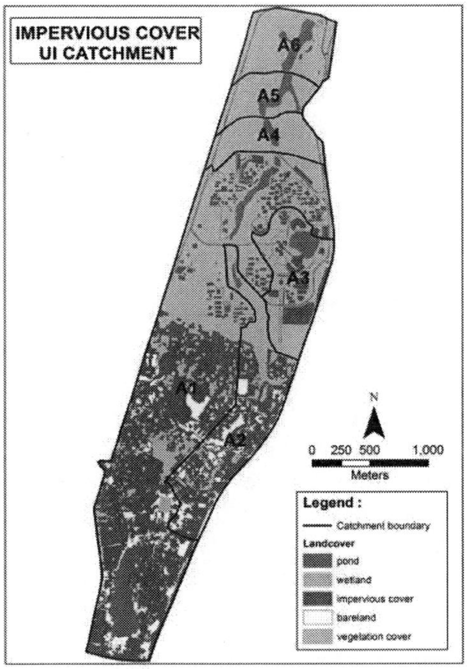

Fig. 1. Impervious cover map of Universitas Indonesia catchment

Four of six catchments in Universitas Indonesia are classified as non-supporting for its stream health while the others classified as heavily damaged and impacted [2]. Each catchment has one pond except A1 catchment. A1 catchment has two ponds namely Rawa Besar pond and Kenanga pond. In this study, Rawa Besar pond is not included for further analysis because it is not included as the cascade-pond system.

UI catchments cascade-pond system has 6 stormwater ponds, known for its acronym KAMPUS (Kenanga-Agatis-Mahoni-Puspa-Ulin-Salam). All of this ponds was built in 1990's. The function of cascade-pond system at UI campus is built firstly for capturing and storing storm water runoff temporarily or for some periods of time. Cascade-pond system is also known as the "end-of-pipe" term to maintain water quality and/or quantity impacts from upper part of catchment [24].

Table 1. Subcatchment Imperviousness.

Catchment Code	Pond (Point of Interest)	Imperviousness	Classification
A1	Mahoni	44.20%	Non-supporting
A2	Agathis	76.90%	Heavily damaged
A3	Kenanga	22.40%	Impacted
A4	Puspa	41.50%	Non-supporting
A5	Ulin	39.30%	Non-supporting
A6	Salam	36.20%	Non-supporting

2.2 Sampling methods

Field sampling was carried out at six ponds. All ponds were divided into three stations which is inlet, middle, and outlet. All macroinvertebrates samples were taken from January to March 2017 within two replicate during the rainy season using a standard 1 mm mesh hand net. All samples were randomly caught from each substrate habitat i.e. riprap, clay, vegetation, and mud. After being caught, samples preserved with 4% formaldehyde, then identified using microscope.

2.3 Macroinvertebrate Indices

Macroinvertebrates were identified to family level. Identification of macroinvertebrates based on [17]. After the process of identification, macroinvertebrates were calculated using SingScore that developed by Public Utilities Board of Singapore. SingScore gives varied tolerance score for each family level of macroinvertebrates. The range score is from 1 to 10 that the smaller score indicates the most tolerant family and vice versa. The SingScore then were classified from 1-10 tolerance score into three classification: 1-3 is tolerant, 4-6 is moderate, and 7-10 is sensitive (see Table 3). The tolerant classification indicates that the pond has poor water quality while the sensitive classification indicates has good water quality [17]. The overall tolerance score of Singscore is shown below:

$$Singscore = \frac{\sum_{i=1}^{i=S} a_i}{S} \times 20 \tag{1}$$

Where S = the total number of taxa in the sample and a_i is telorance value for the ith taxon. The classification of macroinvertebrate indices for water quality assessment are as follows:

Table 2. SingScore Classification [26]

SingScore	Likely Water Quality
0-79	Poor
80-99	Fair
100-119	Good
>120	Excellent

2.4 Impervious Cover and Aquatic Buffer Zone Estimation

Estimation of catchment area imperviousness and aquatic buffer zone were generated by applying GIS technology [1]. This study using ArcGIS 10.5 software and digitized the impervious cover feature (rooftops, asphalt road, concrete road) from high resolution imagery by visual interpretation method. The classification of impervious cover is based on Center for Watershed Protection [2]. Aquatic buffer zone is divided into three categories i.e.: high density forest, forest, and built-up area. The distance of aquatic buffer zone is 50 meters from ponds.

Qualitatively statistics method, chi squared test and cross tabulation options was used to investigate the relationship between macroinvertebrate indices with impervious cover and aquatic buffer zone. Macroinvertebrate indices becomes dependent variable and impervious cover and aquatic buffer zone acts as independent variables. Chi squared test was executed using IBM® SPSS® statistics version 22.

 DOI: 10.1051/matecconf/201713806008

3 Results and Discussions

3.1 Macroinvertebrate Indices at 5 Ponds

The percentage of macroinvertebrate indices that found in cascade-pond system of UI catchment is shown below:

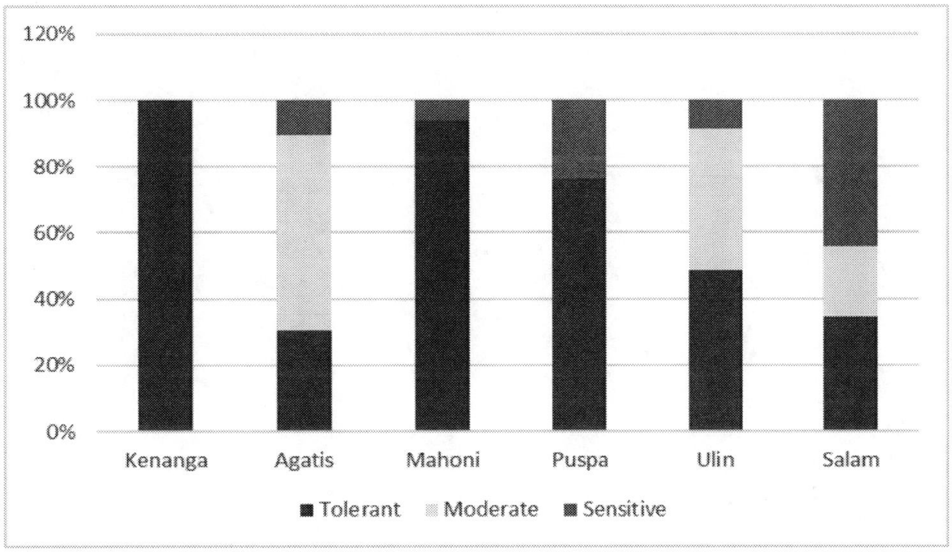

Fig. 2. Percentage of Macroinvertebrates

Water quality in Kenanga pond, the upper of cascade-pond system is poor with overall score 26 because 100% of macroinvertebrates that have been collected is classified as tolerant. The other upper cascade-pond system, Agatis pond is better than Kenanga pond because the moderate macroinvertebrates was dominating rather than tolerant macroinvertebrates. It was also found low percentage of sensitive macroinvertebrates at Agatis pond. The overall macroinvertebrates indices score of Agatis is 101, classified as good water quality. Mahoni pond that located after Agatis pond, classified as poor with high percentage of tolerant macroinvertebrates and very low percentage of sensitive macroinvertebrates. The overall score for Mahoni pond is 58, classified as poor water quality. The middle part of cascade pond system, Puspa pond is also quite similar with Mahoni pond although the overall score is slightly increasing after Mahoni pond but still classified as poor water quality. The lower pond, Ulin pond is relatively better than Mahoni pond and Puspa pond which located in the upper of Ulin pond.

Table 3. Macroinvertebrates Indices Score

Score	Cascade-Pond System					
	Kenanga	Agathis	Mahoni	Puspa	Ulin	Salam
Singscore	26	101	58	75	86	113
Likely Water Quality	Poor	Good	Poor	Poor	Fair	Good

 DOI: 10.1051/matecconf/201713806008

The "end-of-pipe" of cascade-pond system which is Salam pond has the best water quality condition based on macroinvertebrate indices. The percentage of sensitive macroinvertebrate is dominant, it makes the overall score of Salam pond is the highest (see Table 3).

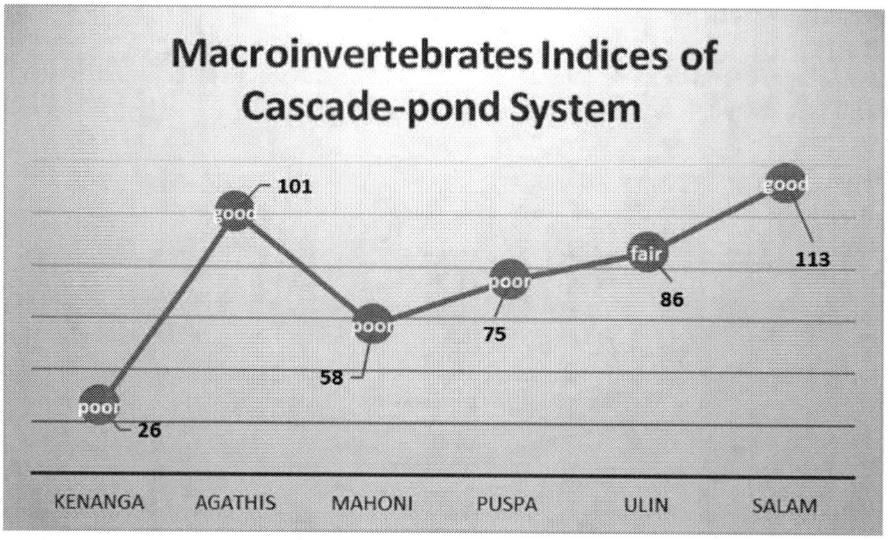

Fig. 3. Macroinvertebrates Indices of Cascade-pond System

Viewing all of the pond into a whole cascade-pond system, the distribution of macroinvertebrates indices is getting better and healthier from upstream to downstream but there is unique evidence in the upper ponds. Kenanga pond and Agatis pond that both located in the upper of catchment has big different scores and water quality classification. This different score between Kenanga pond and Agatis pond might be influenced by impervious cover [18,19,20,22,23].

3.2 Impervious Cover and Aquatic Buffer Zone

Table 4 shows the impervious cover and aquatic buffer zones in each cascade-pond system catchment area.

Table 4. Impervious cover and aquatic buffer zone [1]

Pond	Aquatic Buffer Zone	Imperviousness	Classification
Kenanga	Low riparian cover (<90%)	22.40%	Impacted
Agatis	Moderate riparian cover (90-95%)	76.90%	Heavily damaged
Mahoni	Moderate riparian cover (90-95%)	44.20%	Non-supporting
Puspa	Good riparian cover (>95%)	41.50%	Non-supporting
Ulin	Good riparian cover (>95%)	39.30%	Non-supporting
Salam	Good riparian cover (>95%)	36.20%	Non-supporting

 DOI: 10.1051/matecconf/201713806008

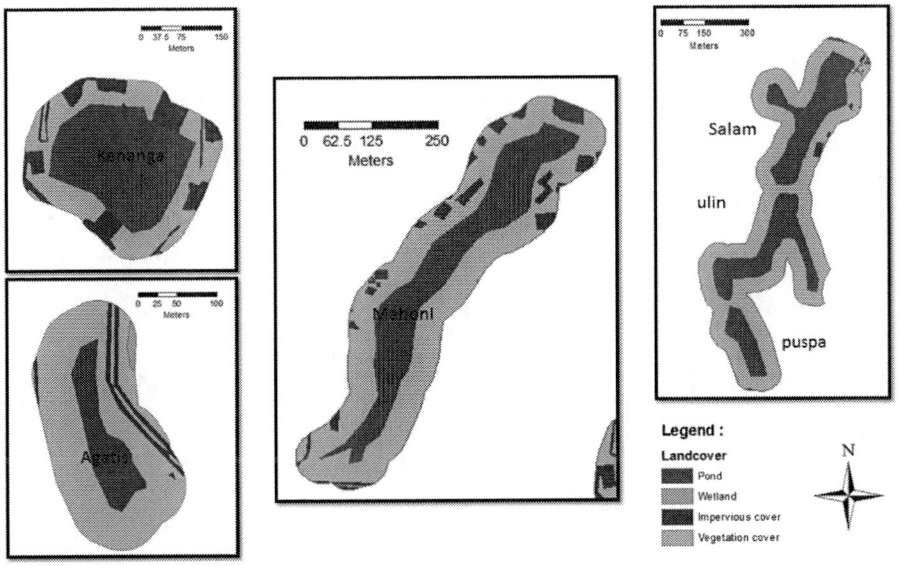

Fig. 4. Riparian buffer of Cascade-pond System

Most of cascade-pond system has moderate to good riparian cover except Kenanga (see Figure 4). Riparian cover has many ecological functions that support aquatic ecosystem. Losing riparian cover would impacted the macroinvertebrate indices. Some studies before shows that converting riparian cover into impervious cover may reducing macroinvertebrate indices. Riparian zone altered the macroinvertebrate composition particularly at low riparian cover zones [22,23].

3.3 Relationship Between Macroinvertebrate Indices with Impervious Cover and Aquatic Buffer Zone

Cross tabulation analysis shows that there is no specific pattern of relationship between macroinvertebrate indices and impervious cover. Only catchment that classified "non-supporting" catchment reach the expected count for each tolerance categories (see Table 5). The Pearson Chi-square score is 112.096. Chi squared test calculation indicates that there is any relationship between macroinvertebrate and impervious cover because the score is greater than 9.49 for 0.05 probability with degree of freedom 4. The asymptotic significance (2-sided) also shows that there is any relationship because the value is 0.000 which is smaller than 0.05 probability. Cross tabulation analysis for macroinvertebrate indices and aquatic buffer zone shows some specific pattern that may indicating the relationship between them (see Table 6). 56.2% of macroinvertebrates was found at pond which has low riparian cover. On the other hands, 51% of moderate macroinvertebrate was found at pond that moderate riparian cover and 61.8% of sensitive macroinvertebrate was found at good riparian cover catchment. The Pearson Chi-square score is 142.096. Chi squared test calculation indicates that there is any relationship between macroinvertebrate and aquatic buffer zone because the score is greater than 9.49 for 0.05 probability with degree of freedom 4. The asymptotic significance (2-sided) also shows that there is any relationship because the value is 0.000 which is smaller than 0.05 probability.

Table 5. Crosstab option between macroinvertebrates indices and impervious cover

			IC			Total
			Heavily damaged	non-supporting	Impacted	
Tolerance	tolerant	Count	27	221	49	297
		Expected Count	55.6	210.7	30.6	297.0
		% within tolerance	9.1%	74.4%	16.5%	100.0%
	moderate	Count	52	50	0	102
		Expected Count	19.1	72.4	10.5	102.0
		% within Tolerance	51.0%	49.0%	0.0%	100.0%
	sensitive	Count	10	66	0	76
		Expected Count	14.2	53.9	7.8	76.0
		% within Tolerance	13.2%	86.8%	0.0%	100.0%
Total		Count	89	337	49	475
		Expected Count	89.0	337.0	49.0	475.0
		% within Tolerance	18.7%	70.9%	10.3%	100.0%

Table 6. Crosstab option between macroinvertebrates indices and aquatic buffer zones

			Aquatic buffer zone			Total
			Low	Moderate	Good	
Tolerance	tolerant	Count	167	62	68	297
		Expected Count	108.8	85.0	103.2	297.0
		% within tolerance	56.2%	20.9%	22.9%	100.0%
	moderate	Count	0	52	50	102
		Expected Count	37.4	29.2	35.4	102.0
		% within Tolerance	0.0%	51.0%	49.0%	100.0%
	sensitive	Count	7	22	47	76
		Expected Count	27.8	21.8	26.4	76.0
		% within Tolerance	9.2%	28.9%	61.8%	100.0%
Total		Count	174	136	165	475
		Expected Count	174.0	136.0	165.0	475.0
		% within Tolerance	36.6%	28.6%	34.7%	100.0%

4 Conclusion

The term of "end-of-pipe" at cascade-pond system at UI catchment affecting the macroinvertebrate indices. the distribution of macroinvertebrates indices is relatively getting better and healthier from upstream to downstream. Macroinvertebrate indices at cascade-pond system at UI catchment is also affected by aquatic buffer zone and impervious cover. Aquatic buffer zone has the most significant relationship than impervious cover because the Chi-squared value is greater.

Acknowledgment

The author would like to thank the financial support provided by Universitas Indonesia through the PITTA 2017 funding scheme under grant no. 857/UN2.R3.1/HKP.05.00/2017, managed by the Directorate for Research and Public Service, Universitas Indonesia and member of Hydraulics, Hydrology, and River Engineering, Department of Civil Engineering, Universitas Indonesia for helping the field survey. Special thanks to Drs. Wisnu Wardhana, M.Si. from Department of Biology, Universitas Indonesia for helping the identification of macroinvertebrates.

References

1. D. Sutjiningsih. Int. Jour. Tech. **1**, 145 (2017)
2. T. R. Schueler, L. Fraley-McNeal, K. Cappiella. J. Hydrol. Eng. **14**, 309 (2009)
3. H. Kim, H. Jeong, J. Leon, S. Bae. Water.**8**, 111 (2016)
4. T. J. Mackintosh, J. A. Davis, R. M. Thompson. Sc. Tot. Env. **536**, 527 (2015)
5. R. Schiff, G. Benoit. J. Amr. Wat. Res. Ass. **43**, 712 (2007)
6. M. N. Damanik-Ambarita, K. Lock, P. Boets, G. Everaert, T. H. T. Nguyen, M. A. E. Forio, P. L. S. Musonge, N. Suhareva, E. Bennetsen, D. Landuyt, L. Dominguez-Granda, P. L. M. Goethals. Iimnologica. **57**, 27 (2016)
7. J. E. Helson, D. D. Williams. Ecol. Ind. **29**, 167 (2013)
8. J. Xia, Y. Zhang, C. Zhao, S. E. Bunn. J. Hydrol. Eng. **19** (2014)
9. C. J. Goodnight. Trans. Amr. Micr. Soc. **92**, 1 (1973)
10. E. Mylinski, W. Ginsburg. J. Amr. Wat. Wor. Ass. **69**, 538 (1977)
11. L. Li, B. Zheng, L. Liu. Proc. Env. Sci. **2**, 1510 (2010)
12. B. A. Markert, A. M. Breure, H. G. Zechmeister, *Bioindicators & Biomonitors* (Elsevier, Oxford, 2003)
13. T. Alemneh, A. Ambelu., S. Bahrndorff, S. T. Mereta, C. Pertoldi, B. J. Zaitchik. Ecol. Ind. **73**, 452 (2017)
14. W. J. Van de Bund, W. Goedkoop, R. K. Johnson. J. Nor. Amr. Benth. Soc. **41**, 532 (1994)
15. J. B. Wallace, J. R. Webster. Ann. Rev. Ento. **41**, 115 (1996)

16. A. P. Covich, M. A. Palmer, T. A. Crowl. Bioscience. **49**, 119 (1999)
17. T. J. Blakely, H. S. Eikass, J. S. Harding. Raff. Bul. Zoo. **62,** 540 (2014)
18. Y. Zhang, D. Dudgeon, D. Cheng, W. Thoe, L. Fox, Z. Wang, J. H. W. Lee. Hydrobiol. **652**, 71 (2010)
19. K. J. Collier, B. L. Clements. Hydrobiol. **664**, 35 (2011)
20. B. Wang, D. Liu, S. Liu, Y. Zhang, D. Lu, L. Wang. Hydrobiol. **680**, 39 (2012)
21. C. M. Lorion, B. P. Kennedy. Fresh. Biol. **54**, 165 (2009)
22. A. B. Moraes, A. E. Wilhelm, T. Boelter, C. Stenert, U. W. Schulz, L. Maltchik. Environ. Monit. Assess. **186,** 7063 (2014)
23. J. P. Duehr, M. J. Siepker, C. L. Pierce, T. M. Isenhart. Jour. Iowa. Acad. Sci. **1**, 49 (2006)
24. D. Sutjiningsih, E. Anggraheni. The 12th International Conference on Quality in Research (2011)

Analysis of Spring Development and Gravity Flow System to Capture Water for Local Communities

Cita Adiningrum[*]

Department of Civil Engineering, Atma Jaya Yogyakarta University, Indonesia

Abstract. Springs as water sources are relatively inexpensive but highly susceptible to contamination since they are fed by shallow groundwater. Proper spring development helps protect the water from contamination. This study presents an analysis and design of spring development including the type of broncaptering/collecting wall, the dimension for the spring box and the conduction line. In addition, a guideline on "Springwater Construction" published by the Ministry of Public Works has been used in this design. A concentrated spring in Wates, Magelang, Central Java is used as a case study. The design calls for the collection of water from a spring using sets of broncaptering and a spring box, then piping it by gravity a distance of 5.1 kilometers to Van Lith Senior High School. Analysis was done using a manual calculation, which is subsequently compared to the result of HYDROFLO 3 software. Results show that the spring with a flow rate of 0.12 litre/s (manual) and 0.17 litre/s (software) will be collected into a 5 m^3 volume of spring box. The spring box with a +543 m water surface elevation is being supplied to Van Lith +384 m ground elevation using a uniform PVC pipelines with a ¾ inch of diameter.

1 Introduction

Capturing water from springs is a simple yet inexpensive process that has been used since ancient times. Inhabitants of rural mountain villages or underprivileged areas are primarily concerned about the development of springs because of the possibility to obtain pure water from neary sources with easy access at no cost.

In a case study on a concentrated spring in Wates, Magelang, Central Java, the spring is located in the private land of a villager. To clarify the ownership of the chosen spring, some negotiations about an ownership change have been done between a villager and Van Lith Senior High School. So, the ownership was transferred from a privately-owned to a school-owned spring.

Prior to the use of spring water, the school had got the majority of their water from shallow wells, especially for domestic purposes. For other uses, such as to operate a swimming pool, the water was collected from drainage canals. Because of continuous contamination that leads to water quality deterioration, an access to cleaner water sources from the springs was needed.

[*] cita_adiningrum@staff.uajy.ac.id

A range of studies have been conducted on the spring development and the design of a spring collection system. A study on water supply planning in Taratara village, West Tomohon, North Sulawesi was done by capturing springwater using broncaptering and spring catchment and then distributing the water to public hydrants [6]. In this study, the pipe distribution system was modeled using the EPANET2 software. Another study was done on a pond construction to sustain springs for community use and this has been successfully tested in Kavrepalanchok District, east of Kathmandu, Nepal [9].

This research was conducted to design broncaptering or collecting wall and spring box, as well as to determine the pipelines route and the flow that simulated in two different ways, manual and using HYDROFLO 3 software tool. These two simulations will be used in the comparison of the maximum discharge that occurs in the pipelines, as well as the pipe's type and the pipe diameter with regard to the total pressure drop during internal flow.

2 Literature Review

2.1 Definition and Types of Springs

A spring is a type of groundwater that emerges naturally to the land surface which may flow by gravity from a water table aquifer or by pressure from an artesian aquifer [3]. Springs can be categorised into two types, namely gravity spring and artesian spring. Gravity springs result where the land surface intersects the water table by some cracks or fractures and flows horizontally out of the ground [1]. On the other hand, an artesian spring occurs when water is trapped between impervious layers and is forced to the surface under pressure [4].

A gravity spring can occur at a distinct point as a concentrated spring [3]. It is unhidden and often found along hillsides where groundwater is forced through openings in fracture bedrock. The other type, seepage spring occurs when shallow groundwater seeps from the ground over a large area and has no defined discharge point.

2.2 Spring Development Considerations

2.2.1 Quantity and Quality of Springwater

When considering using a spring as a source of water, it is important to ensure that the flow rate of the spring is reliable during all seasons of the year. The flow rate should be monitored over a year, but it is most critical to have measurements during late summer and fall when the groundwater level at its lowest condition. Water quality is also important to consider because springs may be at risk of contamination from surface run off which infiltrate to the ground.

2.2.2 Methods of Spring Development

According to the Directorate General of Human Settlements, Ministry of Public Works, there are three categorizations of spring development [2]. Firstly, based on the spring type, springs can be divided into four types, namely Type 1A (concentrated artesian spring type), Type 1B (seepage artesian spring type), Type 1C (vertical artesian spring type), and Type 1D (gravity contact spring type). The second category is due to the size of spring box, which is one of the main components of spring development, namely Type IIA (volume 4 m^3), Type IIB (volume 8 m^3), and Type IIC (volume 10 m^3). The third category is based on how the springwater will be conveyed, whether by gravity flow or by a pumped system.

DOI: 10.1051/matecconf/201713806009

2.2.3 The Spring Catchment

The catchment area can be divided into five components as shown in Fig. 1. The intake area includes the area from which the spring is supplied (by infiltration and percolation of rainwater) to the spot where the water comes to the surface. the other components are the catchment, the supply pipe, and the spring chamber or spring box. The typical spring catchment consist of three elements, namely a broncaptering (collecting wall), a permeable construction behind the broncaptering (either in the form of a filter package or as a perforated pipe), and a catchment cover [4].

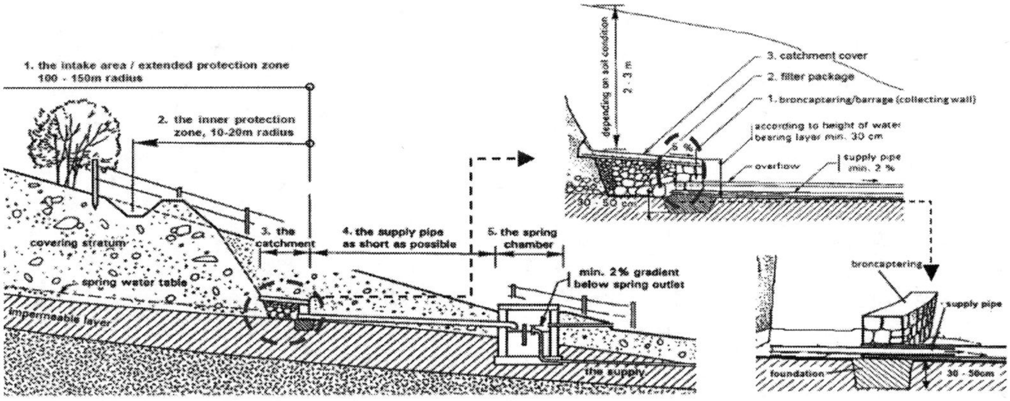

Fig. 1. The five components catchment area of springs (Source: Meuli and Wehrle, 2001)

2.2.4 Spring Box Design

The main components of a spring box consist of six parts, namely inlet pipe from broncaptering, overflow pipe and delivering pipe, cleanout drain, manhole, box valve (operation room), and surface drainage diversion ditch [3, 4, 10]. Table 1 shows the recommended size of water basins for a range of minimum spring yields with coresponding to total users. Water demand is assumed 30-60 litres/day and flowing in 8-12 hours a day.

Table 1. Size of springwater basins

Users/population	**Discharge < 0.5 lt/sec**	**Discharge 0.5-0.8 lt/sec**	**Discharge 0.7-0.8 lt/sec**	**Discharge > 0.8 lt/sec**
200 – 300	5 m^3	2 m^3	2 m^3	2 m^3
300 – 500	10 m^3	10 m^3	5 m^3	2 m^3

2.3 Pipeline Design

2.3.1 Energies of a Flowing Fluid

A fluid particle situated a distance z above datum (measured in unit meter, m) possesses a potential energy. They have energy due to its pressure above datum expressed as the pressure head, $h = p/\gamma$ (m). A body of mass m when flowing possesses a kinetic energy, $KE = ½mV^2$ and for a unit weight it can be written as $V^2/2g$ (m). In real fluid, there is an additional force acting because of fluid friction, namely shear stress (τ). It occurs over the boundary of the element causing a loss of energy (h_f). So, the total energy equation for steady flow of incompressible fluids in a real fluid can be written as [5].

$$z_1 + \frac{p_1}{\gamma} + \frac{V_1^2}{2g} = z_2 + \frac{p_2}{\gamma} + \frac{V_2^2}{2g} + h_f \quad (1)$$

2.3.2 Hydraulic Grade Line and Energy Line

According to Bernoulli's theorem, the quantity $z + p/\gamma$ refers to the piezometric head, because it shows the level to which liquid will rise in piezometer tube, whose connecting end in a plane parallel to the flow. The line drawn through the liquid surfaces in that tube is called the Hydraulic Grade Line (HGL). The total energy head is measured by a pitot tube which is a small tube with its open end in the flow pointing upstream and will intercept the kinetic energy of the flow. From the piezometer and pitot tube reading, the vertical distance between it indicates the velocity head or $V^2/2g$ [5].

3 Methodology

3.1 Study Area

The study was conducted in the area of Wates, Dukun District, Magelang Regency, Central Java which is geographically located at 7° 33' 50.36" LS and 110° 20' 07.76" BT. Wates is one of the villages surrounding Mount Merapi which has contoured or uneven ground. Two locations with three discharge point of springs were discovered in an overgrown area and categorized as fracture/tabular-gravity springs.

3.2 Data Collection

In this study, several field measurements were conducted in order to get necessary data for analysis. A topographic survey was conducted to identify the condition near the springs and trace pipeline route. A discharge measurement for both springs was also conducted on 29th April 2016. According to the head of facilities and infrastructure of Van Lith Senior High School, the springwater was quite good and safe. It was physically clear with no odor and taste,. It also has a low level of dissolved material/metal (Fe^{2+}) to become a source of water supply. As a result, the water quality testing was not carried out in this study.

3.3 Broncaptering and Spring Box Design

A water system to capture the springs consists of broncaptering or collecting wall to hold the water emerges from land surface and a spring box as a temporary storage. The design is based on the "Springwater Construction" guidelines from the Ministry of Public Works.

3.4 Pipeline Flow Analysis: HYDROFLO 3

A pipeline system was built to convey the water from distant springs. Pipelines were installed in series from the spring box to Van Lith Senior High School. Because the location of spring is higher than the school, the flow in pipelines is influenced by gravity. The flow in the pipes was simulated in two different ways: manual and using HYDROFLO 3.

HYDROFLO 3.0.0.4 Academic version from Tahoe Design Software is a fluid conveyance system design tool. In this analysis, the type of fluid conveyance system is categorized to open source and discharge systems driven by gravity.

3.4.1 Supply Element Theory

Supply elements are points of source and discharge at the beginning and end of the main line. In this study, the lines start with open tanks elements and end with discharge to free (open to atmosphere) pressure [7].

3.4.2 Pipe Theory

The Hazen-Williams equation was used for modeling water systems under turbulent flow conditions. A Hazen-Williams friction factor C is determined by empirical test methods.

$$h_f = 0.002083L\left(\frac{100}{C}\right)^{1.85} x\frac{gpm^{1.85}}{d^{4.8655}} \tag{2}$$

where h_f = friction head loss (feet of liquid); L = pipe length (feet); C = Hazen-Williams friction factor; gpm = flow of liquid (US gallons per minute); d = inside diameter (inches)

Minor friction losses through fittings (which include inlets and exits) in water systems were modeled by multiplying this K value by the velocity head, $V^2/2g$. The variations of K value highly depend on the local changes in the pipes, such as an enlargements/contractions, nozzles/orifices, and control valves.

4 Results and Discussions

4.1 Topography Conditions

Fig.2 shows the topographic conditions of the springs and the surrounding which are traced and recorded by GPS. It also shows the proposed pipeline route which spans 5.1 kilometer. This route passes a bark garden and a paddy field, and crosses a river.

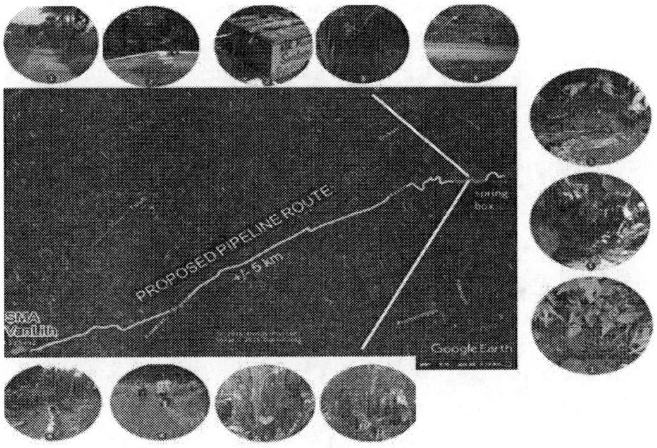

Fig. 2. The proposed pipeline route and the situation in springs (contour map by Google Earth)

4.2 Broncaptering Design

Based on investigations, two sources of springs were discovered and they can categorized as gravity springs. The water emerges from one defined discharge in the earth's surface by fractures or gaps in rock as a concentrated flow. Because these two springs are separated by

 DOI: 10.1051/matecconf/201713806009

some distance and different ground levels, the capturing process of both springs will be done using two broncapterings as illustrated in Fig. 3 [8]. According to the Directorate General of Human Settlements, Ministry of Public Works, the springs can be categorized as a concentrated springs (Type 1A) [2]. The detailed design of the spring collection system, the spring box, and the pipe arrangement are shown in Fig. 4.

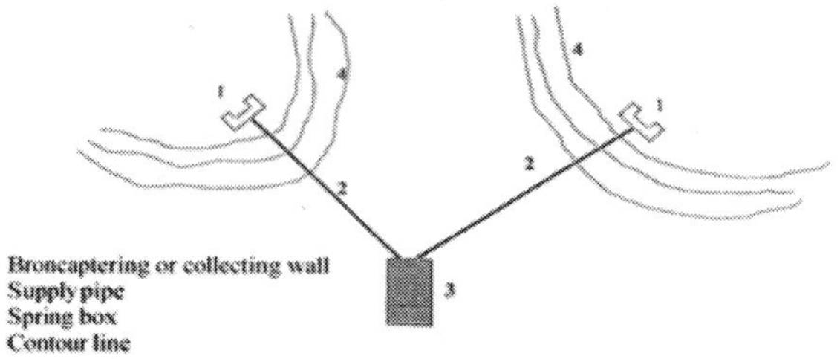

Fig. 3. A set of springwater construction with more than one sources (Source: Lowa, 2016)

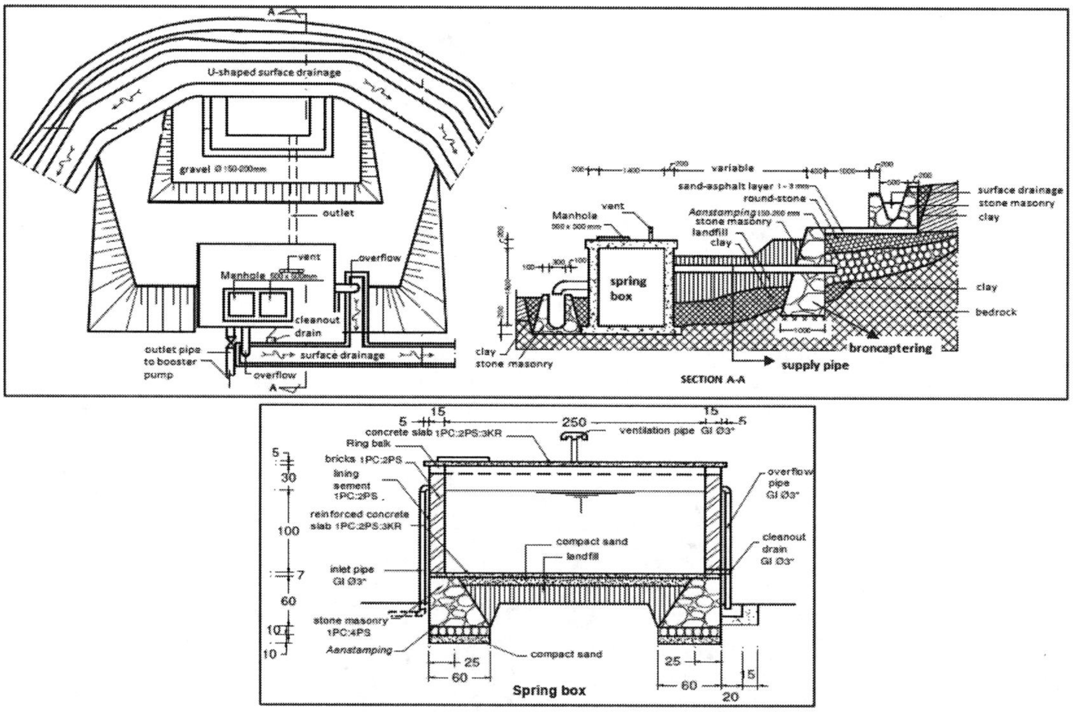

Fig. 4. A plan view and the details of the spring collecting system Type 1A for a concentrated spring (Source: Directorate General of Human Settlements, Ministry of Public Works, 1996)

4.3 Spring Box Design

The first spring has a flow rate of 0.04 litre/s, while the second spring has a double rate of 0.08 litre/s. To store the water for both springs, a total flow rate of 0.12 litre/s must be considered. In Table 1, the size of spring box was determined by two parameters: the discharge of the springwater and the number of population. It was assumed that the total population was about 200-300 people, because the spring development was done primarily to operate a swimming pool. Based on the requirements with a discharge of less than 0.5 litre/s and a population of 200-300 people, the recommended size of the spring box is 5 m^3 of volume or with a detail dimensions of length 2 m, width 2 m, and height 1.25 m.

4.4 Pipeline Flow Calculation

The first spring is located in +543 m of elevation, while the second spring is located higher at elevation +545 m. So that the spring box was put in the same elevation of the first spring to avoid flow change in another direction if the spring box is higher than the spring's location. The water surface elevation at the spring box is +543 m and the ground elevation in Van Lith Senior High School is +384 m. The total static head is therefore 159 m.

The distribution pipelines were arranged in series with uniform cross sections of pipes. The losses that were taken into account were the head loss due to friction (major losses) and the losses at the entrance as the spring box enters the pipe, as well as the losses by the bends. The last two losses were considered as minor losses. A manual calculation was done with an initial assumption of the pipe diameter until it meets the total head loss is less than the available total head and the remainder pressure head is not negative. The pipelines calculation was divided into several segments as tabulated in Table 2. It was divided into 13 pipes in series where the first pipe was located in elevation +542 m.

Table 2. The trial result of pipe diameter (manual calculation)

pipe	elevation (m)		angle of bends	angle of α	length of pipe (m)	diameter of pipe (m)	major losses hf (m)	minor losses		HGL (m of liquid)	∑ head loss hL
	begin	end						kb, koef of bends	hb (m)		
1	542	540			77	0.019	1.317				
			125	55				0.305	0.003	541.667	1.333
2	540	539			67	0.019	1.146				
			95	85				0.800	0.007	540.514	2.486
3	539	534			115.5	0.019	1.975				
			140	40				0.140	0.001	538.537	4.463
4	534	529			106	0.019	1.813				
			138	42				0.162	0.001	536.723	6.277
5	529	528			53	0.019	0.906				
			142	38				0.131	0.001	535.816	7.184
6	528	518			163.5	0.019	2.796				
			157	23				0.064	0.001	533.019	9.981
7	518	516			200	0.019	3.420				
			124	56				0.316	0.003	529.596	13.404
8	516	509			172.5	0.019	2.950				
			136	44				0.184	0.002	526.645	16.355
9	509	505			112	0.019	1.915				
			148	32				0.104	0.001	524.728	18.272
10	505	485			586	0.019	10.021				
			-	0				0.000	0.000	514.707	28.293
11	485	458			737	0.019	12.603				
			168	12				0.030	0.000	502.103	40.897
12	458	422			1265	0.019	21.633				
			162	18				0.045	0.000	480.470	62.530
13	422	384			1405	0.019	24.027				
								0	0	456.443	86.557

total length of pipe	5059.5	m
total head loss	86.557	m
the remainder of the pressure head	72.443	m
Total head	159	m
CEK :	feasible	

According to Bernoulli's theorem, if the velocity head is constant as a uniform diameter, the drop in the hydraulic grade line (HGL) between any two points is equal to the loss of head in it. Table 2 shows the optimum diameter of pipe of 0.75 inch so that the total head loss (86.557 m) does not exceed the available total static head (159 m). The total static head is obtained from the difference between elevation of spring box water surface (+543 m) and the pool line (+384 m). Additionally, the remainder of the pressure head at the end of the system is more than zero (72.443 m) so the water will be able to flow without any blockage.

A pipe flow simulation using HYDROFLO was done to verify the result from the first method (manual calculation) and to provide more detailed results including the pressure head graphics. The first step is set up the plot plan of the proposed system as depicted in Fig. 5.

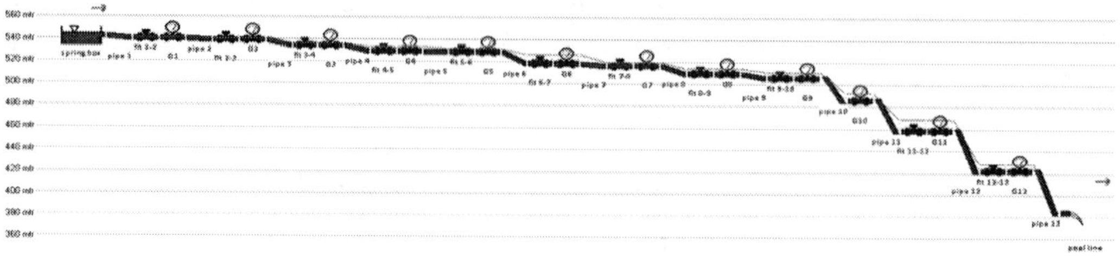

Fig. 5. Plot of main line in HYDROFLO 3

In the HYDROFLO simulation, the primary use of a gauge element is to obtain a gauge pressure reading at a particular point. The gauge was placed immediately after a fitting (See Fig. 5). According to the 1st attempt simulation result with 0.75 inch of uniform pipe's diameter, a negative pressure was found at gauge G2 (-5.40 kPa). It was probably due to the size of pipe 2 and fitting 2-3. A fitting in this system accommodates the flow through the bends in particular a pipe elbow. The fitting 2-3 is a 85° pipe elbow which is the largest bend angle in the pipelines route that located between pipe 2 and 3.

A smaller diameter will cause a higher velocity and a higher head loss. On the other hands, the effect of the flow through a pipe elbow is a large increase in pressure losses as compared to the straight pipe of equivalent length. As such, the location G2 displays a negative pressure reading. The HGL at gauge G2 is under the pipe, so the flow will be blocked. Increasing the pipe diameter is one of the solutions to deal with negative pressure, so the diameter of pipe 2 and fitting 2-3 have been changed from 0.75 inch to 1.00 inch. The further results show that the pressure at gauge G2 becomes 9.77 kPa (see Table 3).

Table 3. The gauge pressure reading comparison for the 1st attempt and 2nd attempt simulations

Gauges: ID		Elevation (mtrs)	Demand (L/s)	1st attempt Pressure (Kpa)	1st attempt HGL (mtrs)	2nd attempt Pressure (Kpa)	2nd attempt HGL (mtrs)
G1	at the beginning of pipe 2	540.00	0.00	5.45	540.56	5.21	540.54
G2	at the beginning of pipe 3	539.00	0.00	-5.40	538.44	9.77	540.01
G3	at the beginning of pipe 4	534.00	0.00	7.87	534.81	22.69	536.34
G4	at the beginning of pipe 5	529.00	0.00	24.03	531.48	38.52	532.97
G5	at the beginning of pipe 6	528.00	0.00	17.56	529.81	31.89	531.29
G6	at the beginning of pipe 7	518.00	0.00	64.69	524.67	78.51	526.10
G7	at the beginning of pipe 8	516.00	0.00	23.13	518.39	36.32	519.75
G8	at the beginning of pipe 9	509.00	0.00	38.42	512.96	51.08	514.27
G9	at the beginning of pipe 10	505.00	0.00	43.07	509.44	55.39	510.71
G10	at the beginning of pipe 11	485.00	0.00	58.48	491.03	68.99	492.12
G11	at the beginning of pipe 12	458.00	0.00	95.76	467.88	104.00	468.73
G12	at the beginning of pipe 13	422.00	0.00	59.50	428.14	63.83	428.59

The first simulation was completed using uniform diameter of PVC pipelines, but there was a negative pressure. The second simulation attempt (after changing diameter) in Table 4 gives information about the changing HGL (m) in sequence of elements. There is no residual head left at the end of the segment and the optimum flow that may occur is 0.17 litre/s, which is more than the flow rate of the springs (0.12 litre/s).

Table 4. The report elements in sequence after adjusting the diameter of pipe and fitting

Open Source/Discharge System

Source Elev:	
Source Pressure:	
Source HGL:	543.00 mtr
Discharge Elev:	384.00 mtr
Discharge Pressure:	0.00 Kpa
Discharge HGL:	384.00 mtr
Total Static Head:	-159.00 mtr

System Information

Pipe Head Loss Eq.: Hazen-Williams
Flow Tolerance: 0.00063 ltr/s
Atmospheric Press: 101.00 Kpa
Total Elements: 38 (approx)
Total Branches: 0 (used in solution)

Liquid Properties

Water,Saturated at 80 deg F
Temperature: 26.70 C
Specific Gravity: 0.99
Viscosity: 1.13E-06 m2/s
Vapor Pressure: 3.52 Kpa (abs)
Specific Heat: 1.00 Cal/grC

Element Data by type...

Supplys:

ID	Type	Pressure (Kpa)	Elev (mtrs)	Flow (L/s)	In/Out	HGL (mtrs)
spring box	reservoir/tank (open)	0.00	543.00	0.17	out	543.00
	Inlet/Exit: pipe inlet-inward projec, Elev=542.00 mtrs, dia=0.750 in, hl=0.13 Kpa					
	one storing system for two springwater					
pool line	free discharge	0.00	384.00	0.17	in	384.00
	the distributed line to the pool					

Elements in sequence

ID	Type	Desc	Pressure (mtlq)	Elev (mtrs)	Head Change (mtlq)	HGL (mtlq)
Start of Main Line - source side...						
spring box	Supply	reservoir/tank (open)	0.00	543.00		543.00
		Inlet/Exit: Type=pipe inlet-inward projecting			-0.01	542.97
		one storing system for two springwater				
pipe 1	Pipe	PVC Sched 40			-2.44	540.54
fit 1-2	Fitt	Custom Fitting			-0.01	540.54
		55 degree of bend				
G1	Gauge	at the beginning of pipe 2	0.54	540.00		540.54
		at the beginning of pipe 2				
pipe 2	Pipe	PVC Sched 40			-0.52	540.01
fit 2-3	Fitt	Custom Fitting			0.00	540.01
		85 degree of bend				
G2	Gauge	at the beginning of pipe 3	1.01	539.00		540.01
		at the beginning of pipe 3				
pipe 3	Pipe	PVC Sched 40			-3.67	536.34
fit 3-4	Fitt	Custom Fitting			0.00	536.34
		40 degree of bend				
G3	Gauge	at the beginning of pipe 4	2.34	534.00		536.34
		at the beginning of pipe 4				
pipe 4	Pipe	PVC Sched 40			-3.36	532.98
fit 4-5	Fitt	Custom Fitting			0.00	532.97
		42 degree of bend				
G4	Gauge	at the beginning of pipe 5	3.97	529.00		532.97
		at the beginning of pipe 5				
pipe 5	Pipe	PVC Sched 40			-1.68	531.29
fit 5-6	Fitt	Custom Fitting			0.00	531.29
		38 degree of bend				
G5	Gauge	at the beginning of pipe 6	3.29	528.00		531.29
		at the beginning of pipe 6				
pipe 6	Pipe	PVC Sched 40			-5.19	526.10
fit 6-7	Fitt	Custom Fitting			0.00	526.10
		23 degree of bend				
G6	Gauge	at the beginning of pipe 7	8.10	518.00		526.10
		at the beginning of pipe 7				
pipe 7	Pipe	PVC Sched 40			-6.35	519.75
fit 7-8	Fitt	Custom Fitting			-0.01	519.75
		56 degree of bend				
G7	Gauge	at the beginning of pipe 8	3.75	516.00		519.75
		at the beginning of pipe 8				
pipe 7	Pipe	PVC Sched 40			-6.35	519.75
fit 7-8	Fitt	Custom Fitting			-0.01	519.75
		56 degree of bend				
G7	Gauge	at the beginning of pipe 8	3.75	516.00		519.75
		at the beginning of pipe 8				
pipe 8	Pipe	PVC Sched 40			-5.47	514.27
fit 8-9	Fitt	Custom Fitting			0.00	514.27
		44 degree of bend				
G8	Gauge	at the beginning of pipe 9	5.27	509.00		514.27
		at the beginning of pipe 9				
pipe 9	Pipe	PVC Sched 40			-3.55	510.72
fit 9-10	Fitt	Custom Fitting			0.00	510.71
		32 degree of bend				
G9	Gauge	at the beginning of pipe 10	5.71	505.00		510.71
		at the beginning of pipe 10				
pipe 10	Pipe	PVC Sched 40			-18.60	492.12
G10	Gauge	at the beginning of pipe 11	7.12	485.00		492.12
		at the beginning of pipe 11				
pipe 11	Pipe	PVC Sched 40			-23.39	468.73
fit 11-12	Fitt	Custom Fitting			0.00	468.73
		12 degree of bend				
G11	Gauge	at the beginning of pipe 12	10.73	458.00		468.73
		at the beginning of pipe 12				
pipe 12	Pipe	PVC Sched 40			-40.14	428.59
fit 12-13	Fitt	Custom Fitting			0.00	428.59
		18 degree of bend				
G12	Gauge	at the beginning of pipe 13	6.59	422.00		428.59
		at the beginning of pipe 13				
pipe 13	Pipe	PVC Sched 40			-44.59	384.00
pool line	Supply	free discharge	0.00	384.00		384.00
		the distributed line to the pool				

 DOI: 10.1051/matecconf/201713806009

5 Conclusions

From the calculation and analysis, it can be concluded that: (1) the type of spring development can be categorized as Type 1A, which is a concentrated spring; (2) the recommended size of the spring box is 5 m^3 with a length of 2.00 m, a width of 2.00 m, and a height of 1.25 m; (3) from the manual calculation, the diameter of the PVC pipe should be 0.75 or ¾ inch to convey the flow of 0.12 litre/s, while the system that modeled by HYDROFLO will convey more with a flow rate of 0.17 litre/s; (4) from the first HYDROFLO simulation, with the same diameter (0.75 inch) and material (PVC with Hazen-William coefficient, $C_{HW} = 130$), a negative pressure was read at gauge G2; (5) the negative pressure is caused by both the size of pipe 2 (too small) and the condition of the bend between pipe 2 and 3 (too large of the bend angle) that lead to enlarged pressure losses; (6) a proposed solution to overcome the negative pressure is by increasing the diameter of the pipe and fitting.

References

1. U.S. Department of Agriculture, *Springs and Wells*, chapter **12**, eDirectives of NRCS, (1983)

2. Directorate General of Human Settlements, Ministry of Public Works, *A Practical Guide of Spring Development and Construction*, module **2.1** (1996)

3. G.D. Jennings, *Water Quality and Waste Management: Protecting Water Supply Springs*, North Carolina Cooperative Extension Service, AG 473-15, (1996)

4. C. Meuli, K. Wehrle, *Spring Catchment – Series of Manuals on Drinking Water Supply*, SKAT, Vol.**4**, (2001)

5. E.J.Finnemore, J.B. Franzini, *Fluid Mechanics with Engineering Applications*, (2009)

6. M.C. Hajia, A. Binilang, E.M. Wuisan, *Planning of Water Supply System in Taratara Village West Tomohon*, TEKNO, Vol.**13**, No.**64**, (2015)

7. Taho Design Software, *HYDROFLO User Guide*, (2015)

8. A. Lowa, Water *Supply Pipe Networks by Gravity Flow System*, (2016)

9. B. Sharma, S. Nepal, D. Gyawali, G.S. Pokharel, S.M. Wahid, A. Mukherji, S. Acharya, A.B. Shrestha, *Springs, Storage Towers, and Water Conservation in the Midhills of Nepal*, ICIMOD Working Paper **2016/3**, (2016)

10. Penn State College of Agricultural Sciences Research and Extension Program, The Pennsylvania State University, *Spring Development and Protection*, **XH0024**, (2017)

MATEC Web of Conferences **138**, 06010 (2017) DOI: 10.1051/matecconf/201713806010
EACEF 2017

A Study on the Estimation of Flood Damage in Medan City

A. Perwira Mulia Tarigan[1,*], Asril Zevri[2], Rudi Iskandar[1], and Ivan Indrawan[1]

[1]Department of Civil Engineering, Engineering Faculty, University of Sumatera Utara (USU), Medan, Indonesia
[2]Department of Civil Engineering, Engineering Faculty, Muhammadyah University of Sumatera Utara (UMSU), Medan, Indonesia

Abstract. A large part of the Medan City, a city of more than 2 million people, experiences frequent floods which cost a lot of damage due to the inundation. The objective of this study is to estimate the area of inundation and the consequent damage cost with reference to the predicted discharges of the rivers crossing the city in a certain return period. Two main watersheds are involved here, i.e. the watersheds of Belawan and Deli.The relevant data for the flood estimation include rain fall, watersheds, as well as river profiles and cross-sections. The computed and mapped flood affected area is then stacked on top of the city maps in a GIS framework. By identifying the number of affected houses, the damage cost for a return period of flood can be estimated in a relative accurate manner. Also the affected city infrastructures can be shown and listed.

1 Introduction

There are two main watersheds, i.e. Belawan and Deli, which encompass Medan City within their river basins' areas of 417.63 km^2 and 459.86 km^2, respectively. A third watershed of Percut with a smaller basin of 194.88 km^2 includes merely the eastern fringe of Medan.Both watershed have their headwaters in the hilly district of Deli Serdang located south of the Medan City's administrative area. The Belawan River in the west side and and the Deli River in the east side cross the city in northerly courses before they reach the straits of Malacca (Figure 1).

In a study supported by JICA (1992), it is reported that the inundation area in Medan City during the annual flooding may reach up to 9000 ha. A suggestion from the study was to construct a flood canal that diverted a portion of the peak discharge of the Deli River to the Percut River before its water entered the center part of Medan. The canal was the then built and completed in 2008. However, the flood has still been occuring each year. Meanwhile inevitably land use changes and upstream deforestation due largely to a population increase of 0.26% per year would definitely cause more intense and frequent flooding in recent years [7]. More intense and frequent flooding would ensuingly mean more innundatian area with more affected houses and infrastructures in the city of more than 2 million people.

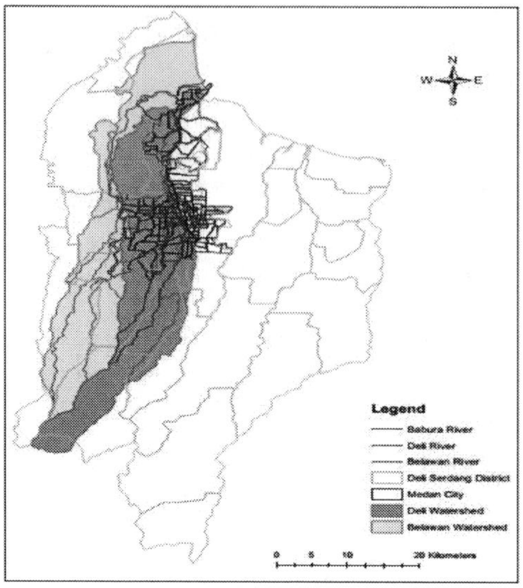

Fig. 1. Map of Medan City showing the watersheds of Belawan and Deli

Therefore it is of considerable importance to study the flood events and to estimate the resulting damage in the context of flood mitigation plan. The purpose of this study is to estimate inundation area and the consequent damage cost with reference to the predicted dischargesof the rivers in some certain return periods. The relevant data for the flood estimation include rain fall, watersheds, as well as river profile and cross-section.The resulting flood maps are to be stacked on top of the available city maps in GIS layers [9] [10] [12]. Hence by knowing the number of affected houses and infrastructures, the damage cost for a return period of flood is calculated in a relative accurate manner.

2. Basic Assumptions

Floods are an inevitable component of any river system, since runoff volumes vary and tend to increase substantially with time, especialy during the rainy season [1] [2] [7]. The precipitation within Belawan and Deli watershed in a year is tipically more than 2000 mm with the number of rainy days approximately the same as that of no rain days. The wet seasons in a year occur twice in a year, i.e. during March to May (120 – 150 mm of rainfall per month) and September to December (170 – 220 mm of rainfall per month) with the latter causing more intense and frequent flooding. Among others, [3] [4] [11] 14] have observed floods occurs in Medan City because high precipitation and large runoff during wet seasons in a year exceed the capacity of the river channels within the Belawan and Deli watersheds. This condition commonly take places especially in October until November when the the amount of rainfall likely exceed 200 mm in a month.

The capacity of the river channels vary across the city of Medan. The upper portions,which have not been normalized yet, have significantly smaller ones than the lower portion which has been normalized to protect more the city infratructures. Figure 2 shows a schematic picture of the capacity of the Deli River, whereas Table 1 tabulates the 10 to 100 years discharges at locations indicated in Figure 2 (JICA, 1992). It can bee seen that the capacity of the river can not accommodate even for 10 years flood of Q_1, Q_2, and Q_3. It can also be understood that the annual flood is commonplace in certain portions of the river channels.

It is a common practice to design a capacity of a primary drainage to sustain at least a 25 year flood flow. It is also a common practice to estimate flood events of a primary drainage like the river channel for at least 25 years recurrence interval. Hence in this study we focus our flood and damage cost estimation on the 25-years flood flow, although the results for 50 and 100 years flood are mentioned parallelly.

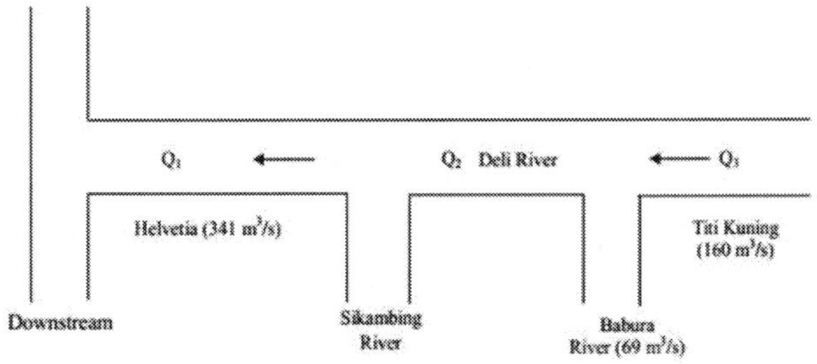

Fig. 2. Sketch of the capacity of the Deli River (JICA, 1992)

Table 1. Computed discharges for Deli River (JICA, 1992)

Flood flow (m^3/s)	Return period (years)				
	10	20	30	50	100
Q_1	460	530	570	620	690
Q_2	420	490	520	570	640
Q_3	260	300	320	340	380

3. Methodology

The watersheds under study are the Belawan River basin and the Deli River basin as shown in Figure 1. Both rivers are important components of the well being of Medan City since they flow through the middle of the city. The main inputs in the process of this study are the rainfall data, the river profiles , and the watersheds. The rainfall data consist of dayly maximum rainfall for 10 years from 2004 until 2013 obtained from 3 rainfall stations, one located in the upstream, one in the middle of the city, and the other in the downstream. Based on these rain fall data,the maximum dayly precipitation of certain return periods can be calculated.

From the maximum dayly precipitation and the characteristics of the watersheds, the floods of certain return periods can be computed. Next using HEC-RAS with the inputs of recurrence flood discharge, the cross and the longitudinal river profiles, and the watershed characters represented by the runoff coefficient , the water profile can be simulated to gain the spatial coverage of the inundation area for certain return periods of floods.

The resulting flood maps are then to be stacked on top of maps of Medan City in Gis layers. However, prior to the overlaying process, all layers of information should be preparedon the same coordinate system. Within the GIS framework, the houses and infrastructures affected in the inundation area can be identified. Afterward, the damage cost due to certain periods of floods can be computed using a simple formula. Figure 3 describes the scope and steps of the study in a schematic way. To simplify the process, we will demonstrate the results computed for the Babura River which is a tributary of the Deli River with frequent flooding [8]. The process could be continued for the whole rivers involved using the same procedure, and the total estimated damage cost is given.

4. Results and Analysis

4.1 River Hydrology

Based on the maximum daily rainfall data for 10 years, hourly rainfall intesity, and the runoff coefficient representing the landuse of the watersheds, the maximum discharge of certain return periods were predicted using the method of the Nakayasu synthetic unit hydrograph method. The resulting discharges for 2, 3, 5, 10, 25, 50 and 100 years recurrence intervals are tabulated in Table 2. It can be seen from the table that the flood magnitude of the Deli River is almost twice as large as that of the Belawan river for a same return period. Also, the Deli River's flood values given in Table 2 are relatively larger than that given in Table 1, as the watersheds have experienced landuse change and deforestation over the years. Note that the Babura River, a tributary of the Deli River, is included in Table 2 because the examples shown in this study are given from the calculation of the Babura River.

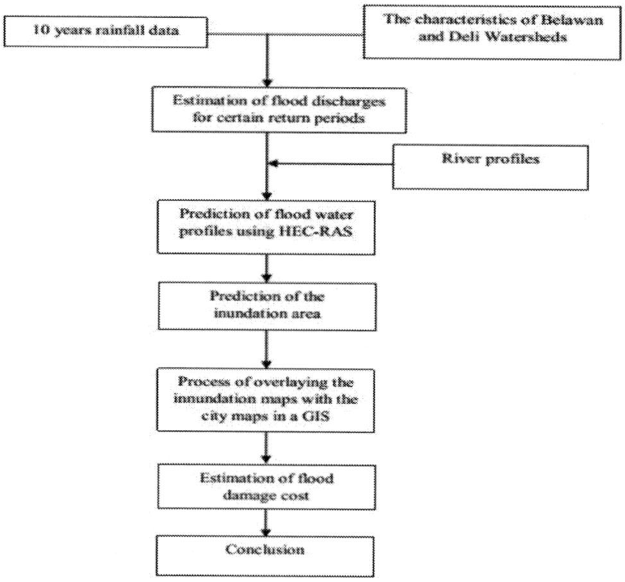

Fig. 3. The scope and the steps of the study

Table 2: Estimation of flood discharge for certain return periods

Deli River	maximum flood flow	return period (years)						
	Q maximum	2	3	5	10	25	50	100
	(m³/s)	401	453	510	582	674	741	809
Babura River	maximum flood flow	return period (years)						
	Q maximum	2	3	5	10	25	50	100
	(m³/s)	100	118	141	174	221	262	307
Belawan River	maximum flood flow	return period (years)						
	Q maximum	2	3	5	10	25	50	100
	(m³/s)	236	258	283	314	353	382	411

4.2 Surface Water Profiles

The flood values in Table 2 are used in HEC-RAS 4.0 to estimate the surface water profiles with reference to the associated return periods [13]. The larger is the return period, the deeper is the water depth at the river bank as well as the farther is the flood distance from the river bank. For example,

Figure 4 shows the 25-year water profile at a cross section in the Babura River. It can be seen from the figure that the water depth reaches about 2 m at the river bank and the flood distance reaches about 100 m to the left and 100 m to the right from the center of the river. The figures like that shown in Figure 5 were obtained using HEC-RAS 4.0 for all available data of the cross profiles of theBelawan and Deli river channels. Figure 5 shows the resulting flood map of the Babura River for the 25-years return period.

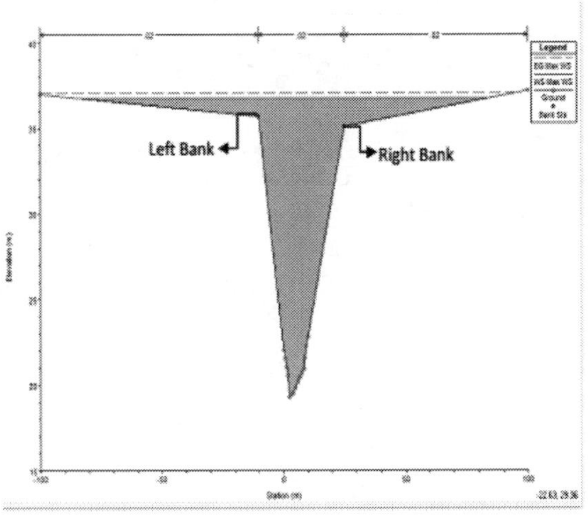

Fig. 4. 25 years water profile at a cross section in the Babura River, a tributary of the Deli River

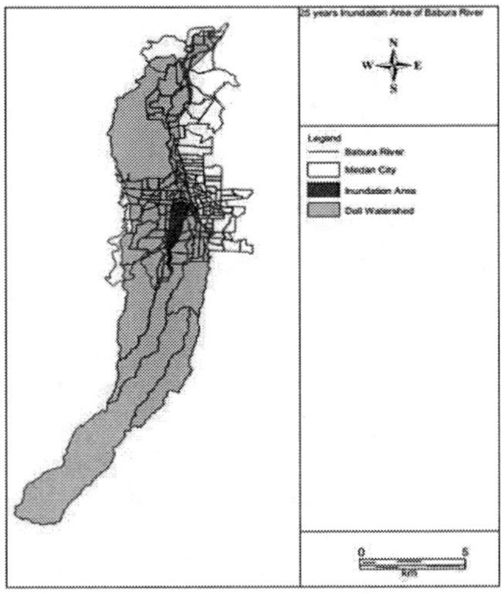

Fig. 5. The resulting 25-years flood map computed for the Babura River

4.3 Inundation Areas in a GIS Framework

To evaluate the extent of the inundation area, the resulting map of Figure 5 was stacked on top of city maps in GIS layers. Figure 6 shows the inundation area of the Babura River for the 25 years flood in GIS layers. The layer of the administrative areas of Medan City's subdistricts is included in the figure as a basis for the damage estimation. There are 7 affected subdistricts within 3 districts including, written consecutively from the upper to the lower stream, 1) Padang Bulan and 2) Darat subdistricts in the Medan Baru district, 3) Madras Hulu, 4) Polonia and 5) Angrung in the Medan Polonia district, and 6) Petisah Hulu and 7) Petisah Tengah in the Medan Petisah district. The figures like Figure 6 could be obtained using the same method for the whole Belawan and the Deli Rivers.

4.4 Estimation of Flood Induced Damage Cost

To estimate the number of people affected by the flood in a subdistrict, a simple proportionality formula is used as follows.

$$\text{Number of people affected} = \frac{\text{Inundation area}}{\text{Total area}} \text{ x Total population}$$

Along the Babura River there are a total of 13287 people affected by the 25-years flood. The calculation for each subdistrict is given in Table 3. To estimate the number of houses affected by the flood in a subdistrict, a similar formula is used as follows

$$\text{Number of houses affected} = \frac{\text{Inundationarea}}{\text{Totalarea}} x\text{Totalhouses}$$

Along the Babura River there are a total of 3128 houses affected by the 25-years flood. The calculation for each subdistrict is given in Table 4.

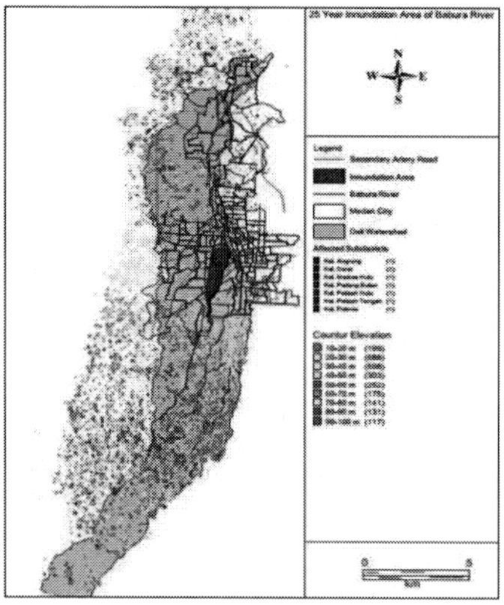

Fig. 6. The 25 years inundation area of the Babura River in a GIS format

Table 3: Calculation of the affected people by the 25-years flood for the Babura River

Districts	Subdistricts	Area (km²)	Inundation area (km²)	Population	Number of people affected
Medan Baru	Padang Bulan	0.66	0.06	9123	829
Medan Baru	Darat	0.16	0.04	1911	478
Medan Polonia	Madras Hulu	0.31	0.09	2785	809
Medan Polonia	Polonia	0.52	0.18	17392	6020
Medan Polonia	Angrung	0.12	0.03	1729	432
Medan Petisah	Petisah Hulu	0.27	0.08	4711	1396
Medan Petisah	Petisah Tengah	0.55	0.2	9137	3323
Total		2.59	0.68	46788	13287

Table 4: Calculation of the affected houses by the 25-years flood for the Babura River

Districts	Subdistricts	Area (km^2)	Inundation area (km^2)	Houses (unit)	Number of houses affected (unit)
Medan Baru	Padang Bulan	0.66	0.06	4457	405
Medan Baru	Darat	0.16	0.04	606	152
Medan Polonia	Madras Hulu	0.31	0.09	961	279
Medan Polonia	Polonia	0.52	0.18	2498	865
Medan Polonia	Angrung	0.12	0.03	428	107
Medan Petisah	Petisah Hulu	0.27	0.08	788	233
Medan Petisah	Petisah Tengah	0.55	0.2	2990	1087
Total		2.59	0.68	12728	3128

To estimate the flood damage cost for houses, a standard suggested by The National Development Planning Agency, Republic of Indonesiawas used as a reference [5] Table 5 lists the standard which is modified to account for inflation. For simplicity, the standard cost for moderate damage was used to calculate the damage cost using the following simple formula.

It is found that the damage cost for the 25 year flood of the Babura River was Rp 62,562,566,693. The calculation for each subdistrict is given in Table 6. Using the same procedure, the people and houses affected, as well as the damage cost can be estimated for the whole segments of the Belawan and Deli Rivers.

The damage cost = the number of houses affected x the standard cost

Table 5: A standard to estimate the flood damage cost

No	Name of main roads	Level of inundation
1	Jl. Djamin Ginting	Moderate
2	Jl.Dr Mansyur	Severe
3	Jl. Pattimura	Severe

Table 6: Calculation of the damage cost caused by the 25 years flood of the Babura River

Districts	Subdistricts	Area (km^2)	Inundation area (km^2)	Houses (unit)	Number of houses affected (unit)	Estimated flood damage cost (Rp)
Medan Baru	Padang Bulan	0.66	0.06	4457	405	8,103,636,364.00
Medan Baru	Darat	0.16	0.04	606	152	3,030,000,000.00
Medan Polonia	Madras Hulu	0.31	0.09	961	279	5,580,000,000.00
Medan Polonia	Polonia	0.52	0.18	2498	865	17,293,846,154.00
Medan Polonia	Angrung	0.12	0.03	428	107	2,140,000,000.00
Medan Petisah	Petisah Hulu	0.27	0.08	788	233	4,669,629,630.00
Medan Petisah	Petisah Tengah	0.55	0.2	2990	1087	21,745,454,545.00
Total		2.59	0.68	12728	3128	62,562,566,693.00

Figure 7.a shows the flood map layer of the Babura River stacked on top of the important public buildings layer. Based on these layers of GIS, the number and type of important building saffected could be indentified as shown in Table 7.

Table 7: The important buildings affected by the 25-years flood of the Babura River

Districts	Subdistricts	Hotel	Restaurant	Gas Station	Mall	Hospital	Junior High School	Senior High School	Primary High School
Medan Baru	Padang Bulan	0	0	0	0	1	0	0	3
Medan Baru	Darat	0	0	0	0	0	0	1	0
Medan Polonia	Madras Hulu	0	2	0	0	0	1	0	0
Medan Polonia	Polonia	0	8	0	0	1	2	2	7
Medan Polonia	Angrung	0	4	0	0	0	1	1	4
Medan Petisah	Petisah Hulu	0	0	0	0	0	0	0	2
Medan Petisah	Petisah Tengah	0	0	0	0	0	0	0	0
Total		0	14	0	0	2	4	4	16

Figure 7.b shows the flood map layer of the Babura River stacked on top of the road layer. Based on these layers of GIS, the main roads affected could be indentified as shown in Table 8 including the area of Polonia Airport.

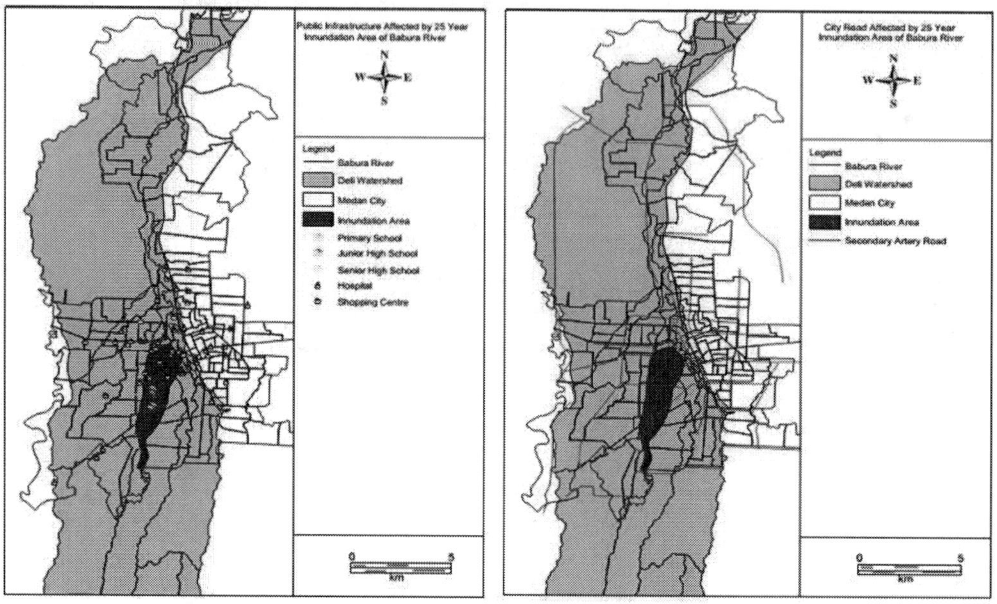

a. Flood map and public infrastructure b. Flood map versus the city roads

Fig. 7. The GIS layers of flood map

Table 8: The main roads affected by the 25-years flood of the Babura River

No	Name of main roads	Level of inundation
1	Jl. Djamin Ginting	Moderate
2	Jl.Dr Mansyur	Severe
3	Jl. Pattimura	Severe

Using the same procedure, the public infrastructures affected, including the roads and the transportation facilities, can be estimated for the whole rivers in the watersheds of Belawan and Deli. Figure 8 depicts the flood map for the whole rivers in the basins. Meanwhile, Table 9 tabulates the damage costs estimated for the Belawan and the Deli Rivers, which includes the Babura River, for 25, 50, and 100 years flood. It can be seen from Table 10 that the damage caused by the Belawan River is the least compared with that by the other rivers, whereas the Deli River causes the most damage. This is due to the fact that Deli River discharges comparatively a larger flood magnitude for the same return period and it flows right toward the center of the city where the population is mostly dense.

Table 9: Damage costs estimated for the Belawan River and Deli River for different return periods

River	Return period (Years)	Number of people affected (People)	Number of houses affected (Unit)	Estimated flood damage cost (Rp)
Deli River	25	28755	7803	156,055,718,288.00
	50	50434	13110	262,194,839,180.00
	100	219658	42336	846,750,183,208.00
Babura River	25	13287	3128	62,562,566,693.00
	50	32856	7064	141,279,006,406.00
	100	60711	13508	270,150,698,007.00
Belawan River	25	2751	528	10,556,879,257.00
	50	3009	1102	21,400,235,294.00
	100	64569	8909	41,940,528,190.00

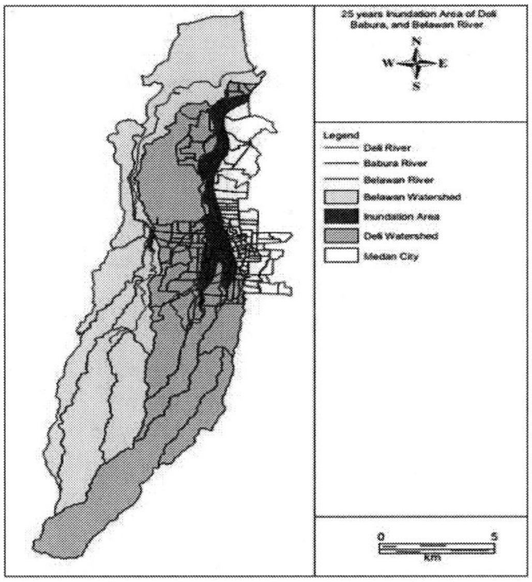

Fig. 8. Flood map of Belawan and Deli watersheds for 25 years flood

5. Conclusion

Based on the rainfall and the watersheds data, we have attempted to estimate the flood discharges of the main rivers in Medan City for certain return periods. It is found that the discharges estimated in this study are comparatively larger than that reported in a study supported by JICA (1992). Using HEC-RAS the flood discharges together with the river profiles and the runoff coefficient are inputted to obtain the flood maps. By stacking the resulting flood maps on top of city map and infrastructure layers, we are able to identify the extent of the inundation and estimate the affected people and houses as well as the ensuing damage cost. Also, the affected public infrastructures can be identified and categorized. Note that we demonstrated the results for 25 year flood of the Babura River which is a tributary of the Deli River. The same procedures could be conducted for all floods of the rivers within the basins for different return periods.

Acknowledgment

This study was supported by The Ministry of Reasearch, Technology, and Higher Education of Indonesia through the Research Scheme of "*Hibah Bersaing*" in 2015-2016.

References

1. T.V Cech, *Principles of Water Resources History, Development, Management, and Policy.* Second Edition. Wiley. USA (2005)
2. Freier. *Principle of Water Resources: History, Development, Management and Policy in GIS Flooding*, New Jersey (2005)
3. M Ginting, *Studi Potensi dan Mitigasi Banjir Kota Medan*. Prosiding Seminar Nasional-1 BMPTTSSI USU, Medan. (2012)
4. G.M. Hasibuan, *Model koordinasi kelembagaan pengelolaan banjir perkotaan terpadu*. Disertasi Perencanaan Wilayah USU. Medan (2004)
5. *http//www.scribd.com/doc/Bappena, Laporan Penilaian Kerusakan Kerugian Jabodetabek. (2007)*
6. JICA, Main Report,. *The Study on Belawan Padang Integrated River Basin Development*, Medan (1992)
7. J. Kodoatie, *Tata Ruang Air, On Integrated Water Resource Management in Indonesia*. Penerbit Andi (2005)
8. Kurniawan, A.. *Analisis Debit Banjir Rancangan Sungai Babura di Hilir Kawasan Kampus USU*. Bidang Studi Teknik Sumber Daya Air USU, Medan (2012)
9. Lyon, *Gis for Water Resources and Watershed Management*. London (2003)
10. N. Muhadi, and Abdullah, A.F. *Flood Damage Assessment in Agricultural Area in Selangor River Basin*. Jurnal Teknologi (Sciences & Engineering) **76:15 (2015) 111–117.**
11. D. Pasaribu,. *Konsep Pengelolaan Drainase Kota Medan*. Tesis Master, Program S2 Teknik Sipil USU (2007)
12. K. Trihono, *Penerapan Sistem Informasi Geografis dalam untuk Mereduksi Kerugian Akibat Banjir*. Seminar Nasional Aplikasi Teknologi Informasi (SNATI), Yogyakarta (2007)
13. U.S Army Corps of Engineers – Hydrologic Engineering Center (HEC). *Hydraulic Reference Manual HEC-RAS 3.1.3*. California: U.S. Army Corps of Engineers (2001)
14. A. Zevri, *Analisis Potensi Resiko Banjir pada DAS yang Mencakup Kota Medan dengan Sistem Informasi Geografis*. Tesis Master Prodi S2 Teknik Sipil USU (2014).

Motorcycles as public transport service based on smart phone android applications

Ansusanto Dwijoko[1,*], *Hangganararas* Sterya[1]

[1]Department of Civil Engineering, Universitas Atma Jaya Yogyakarta, Indonesia

Abstract. Growth of traffic in the city of Yogyakarta is dominated by private vehicle, while the urban public transport service is not satisfactory for consumer. In November 2015 the transport service based application called Go-Jek began operation in Yogyakarta. This service offers a transportation service that is safe, convenient, fast, cheap and more attractive the interest of consumers to use. The purpose of this study is to evaluate the operating system Go-Jek services, the public interest to this service, the feasibility of Go-Jek services as a mode of transportation in Yogyakarta. This research was conducted in the city of Yogyakarta and Sleman. The research method is done by using a questionnaire to the respondents: the driver and the consumer of Go-Jek. The analysis was conducted to assess the level of satisfaction and expectations of respondents using Importance-Performance analysis methods. Operating system Go-Jek is currently quite good, but needs to be further improved. Ranges of Income from 843.300 3.243.300 Indonesian Rupiah per month be perceived by the driver is good enough, while minimum income of workers on average in 2017 amounted 1.337.645 Indonesian rupiah per month in Yogyakarta Special District. Recommendations to the Go-Jek company in order to implement transparency in setting policies and systems to avoids the problems to the driver Go-Jek

1 Background

Congestion occurs almost everyday in Yogyakarta as a medium city, especially during office hours and on weekends. The condition of Yogyakarta City increasingly jammed and increasingly crowded by private vehicles, making the city of Yogyakarta is now considered to require adequate public transportation, considering the number and variety of public transport in Yogyakarta somewhat compared with other provincial capital.

On November 16, 2015, Go-Jek's transportation service began operations in Yogyakarta by offering safe, convenient, fast, practical, and inexpensive transportation services. PT.Go-Jek Indonesia is the first Indonesian technology company to provide application-based transportation services. In carrying out this business, PT.Go-Jek Indonesia cooperates in partnership with motorcycle and car drivers in major cities of Indonesia. Since application-based transportation services are currently in demand from the public, further study on the

* Corresponding author: dwiyoko@gmail.com

feasibility of application-based transportation services, especially Go-Jek services, as a mode of transportation in Yogyakarta.

1.1 Online Transportation in Indonesia

Online transport is a transportation based on a particular application, where consumers order a means of transportation through the application system in the smartphone. When customers place an order using an app, order details such as mileage, price, driver's identity, the length of time the driver arrives to the customer's location, and the data of the managing company is directly presented on the consumer smartphone screen. The entire identity of the driver is known for certain because the management company has done the verification process first before doing partnership cooperation with the driver.

With the online transport, passengers now no longer need to approach the motorcycle taxi station or no longer need to wait on the roadside to get a taxi. In addition, the passengers also do not have to be involved in the bargaining process because the tariffs that have been determined based on mileage. In addition to saving time, online transportation can also save money because of the many promos are offered. Simply by downloading the application on the smartphone, then register on the application, then include the location of pick-up and delivery.

1.2 Legalization of Online Transportation in Indonesia

The Ministry of Transportation Republic of Indonesia (Kemenhub RI) formally issues rules for application-based transportation services. The rules are summarized in Ministerial Decree (Permen) of Transportation no. 32 year 2016 on the Implementation of Transport of People with Public Vehicles Not in the Route. One of them mentioned that transport service companies are not in the route, for example taxis, are allowed to use the application. Provision of applications can be done alone or in cooperation with an application company already incorporated Indonesian law. The transport payment system may also be embedded at once in the application, provided that it remains subject to the provisions in the field of information and electronic transactions.

Company applications that provide passenger transport services using motor vehicles are required to follow the provisions of public transport conveyance contained in Articles 21, 22 and 23 of Permenhub. 32 of 2016. The provisions include requesting an application company to establish an Indonesian legal entity. The form of a recognized legal entity is a state-owned enterprise, a regional-owned enterprise, a limited liability company, or a cooperative. The application company is also required to carry out the transport permit of persons not on the route, with the requirement, among others, to have at least five vehicles on behalf of the company, the vehicle must pass the periodic test, have the pool and the workshop, and the driver must have a general driver license.

However, the legalization of online transportation is only valid for four-wheeled vehicles only, while two-wheeled vehicles can not be legalized as public transportation because they are not listed as public transportation in UU no. 22 of 2009. The Ministry of Transportation (Kemenhub) has adopted a policy of allowing motorcycles based on applications to operate as long as public transportation facilities and infrastructure in Indonesia are considered not fully feasible.

1.3 Quality of Service

Service quality is a long-term cognitive evaluation of consumers towards the delivery of a company's services [2]. The quality of service consist of five dimensions [3]:

a. Tangibles is the ability of a company in showing its existence to external parties.
b. Reliability is the ability of the company to provide services in accordance with the promised accurately and reliably.
c. Responsiveness is a policy to help and provide a fast service (responsive) and appropriate to consumers with clear information delivery.
d. Assurance is ability of company to increasingly consumer confidence to the company.
e. Empathy is a sincere and individualized or personal giving to the consumer by trying to understand consumer desires.

1.4 Go-Jek Transportation Service

The first step before using Go-Jek service is to register by entering e-mail address, name, phone number, and password. Once registration is successful, users can directly select the services available within the Go-Jek app. After selecting the Go-Jek service, the user then inputs the address of the place of origin and destination address, then the Go-Jek app will display information about the driver's existence, predictions of pick-up time, driver's identity (name, photo, phone number), and price. After the service use process is complete, the Go-Jek app provides an opportunity for users to provide an assessment of the services provided by Go-Jek drivers [1].

2 Operational System

This section will explain more about Go-Jek's operational system, which includes the tariff system, bonus, performance, suspend, guarantee, and Go-Jek driver application system.

2.1 Tariff system

During the Go-Jek operation in Yogyakarta, there are two tariff systems that have been used:

a. The tariff system is from November 2015 to July 2016
   For Go-Ride services, the rate per kilometer is Rp. 2.500 with a minimum payment of Rp. 15.000. In this period, Go-Car service has not been present in Yogyakarta city. Go-Car service began to be present in Yogyakarta in August 2016.
b. The tariff system for the period of August 2016 to date
   Go-Ride: per kilometer Rp. 2.000 with minimum payment of Rp. 4.000.
   Go-Car: per kilometer Rp. 4.000 with minimum payment of Rp. 10.000.

2.2 Performance System

Since August 2016, many system changes made by PT.Go-Jek Indonesia. In addition to tariff changes, there is a new system that PT.Go-Jek Indonesia began to implement. The system is a performance system, where the driver's performance is displayed in an updated and continuous manner by this system. Acceptance rate is the level of acceptance of incoming orders to each driver. This acceptance rate shows the comparison between incoming orders automatically to the driver's application and the order the driver takes. If the more orders are taken, the acceptance rate will be higher. Completion rate is the rate of completion of orders taken by the driver. The calculation of driver performance level is not only measured from the acceptance rate, but also from the order completion level.

2.3 Suspend system

The Go-Jek suspend system is a system in which a Go-Jek driver's account is temporarily closed by Go-Jek operational because the driver has committed an offense. If the Go-Jek driver's account is closed, the driver can not automatically accept the order until his account is reopened by the Go-Jek operational. This suspend system consists of two types, namely auto suspend and manual suspend. Automatic Suspend is a suspend that occurs based on the detection of fraud from Go-Jek system towards Go-Jek driver. Manual suspension is a suspend that occurs based on a complaint report from consumers Go-Jek or other parties to the behavior of the driver Go-Jek.

2.4 Application system

In the Go-Jek driver application system, there are some important things related to the process of receiving orders to the way Go-Jek driver in choosing the order in accordance with his wishes. The following will be explained more about the application system.

When there are consumers who make a reservation service Go-Jek, then the consumer order will automatically go into the application of one of the drivers who are located around the consumer location. If the driver within 10 seconds does not take or cancel the incoming order, then the offer of the order will automatically transfer to another driver. Although ignoring incoming orders can decrease the driver's performance level, but drivers prefer this way to get consumers with the desired mileage.

In the Go-Jek driver application there is one button that is the main reason of the many people who choose to work as a Go-Jek driver. The button is the on/off button. When the on button is turned on, then the driver is in active status and ready to receive orders from consumers. If the off button is turned on, then the driver is in the off status so there will be no order coming into the driver's application. With this button, then the driver can freely choose the time to work in accordance with his wishes.

3 Analyses

This section will explain the data from questionnaires distributed to Go-Jek drivers and consumer. These data include driver characteristics, driver operating costs, order characteristics received by the driver, as well as analysis of the satisfaction level and expectations of the driver by using the importance-performance analysis.

3.1 Driver Satisfaction and Expectation

A Go-Jek driver needs an operational cost in his work. Operating expenses consist of vehicle tax, fuel costs, and telephone call charges. In this study, taken the 43 drivers of respondents drivers Go-Ride (motorcycle), and 20 drivers Go-Car (car). The majority of the number of orders a driver can reach per day is about 10-15 orders, and the trip distance 5-10 km per order then the driver can earn 1.5 points per order. So only with 12 orders a driver can collect points as much as 18 points. Thus, the majority of Go-Jek drivers in Yogyakarta can achieve a daily bonus of Rp. 100,000. About 47% of respondents say that the time they take to complete each order is about 15-25 minutes, and 31% of respondents said that they have travel time per order for 10-15 minutes.

When under normal conditions, ie when it is not in the holiday season and also not in a rain condition, then the time required by the driver to get an order that is about 10-15 minutes. However, if it is in the holiday season then the driver only takes about 1-2 minutes to get an order, while if it is raining, especially for drivers Go-Ride must be willing to wait for orders entered up to over 1 hour. For the average value of expectations, performed with the same calculation with the average value of satisfaction.

Table 1. Order Trip Distance

Trip Distance (km)	% respondent
<5	22
5 – 10	47
10 – 12	11
12 – 15	11
>15	9

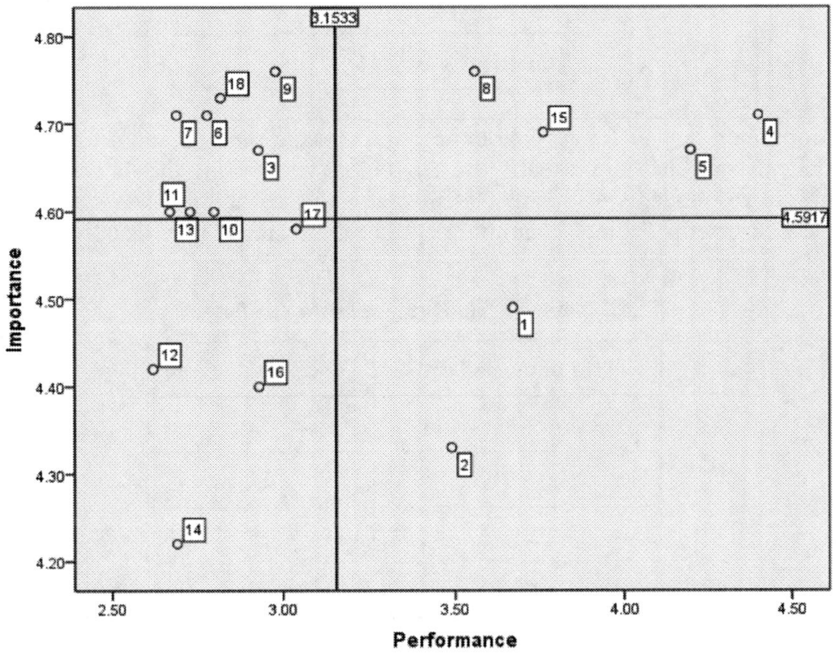

Figure 1. Importance-Performance of Go-Jek Driver

Table 2. Driver Satisfaction and Expectation

Quality	Satisfaction	Expectation	Gap
Tangibles	3.36	4.50	-1.13
Reliability	3.44	4.72	-1.29
Responsiveness	2.72	4.54	-1.82
Assurance	3.01	4.48	-1.47
Empathy	2.93	4.66	-1.72

3.2 Consumer characteristics

The majority of respondents Go-Jek user are students, about 71% of undergraduate students and 13% graduate students. About 82% of respondents are 20-25 years old. This is related to the majority of the Go-Jek consumer profession of students, who are generally aged between 20-25 years.

Table 3. Frequent Using Go-Jek

Trip Frequent	%
1 – 3 times	51
1 time/week	15
2 – 3 times/week	11
3 – 5 times/week	11
> 5 times/week	11

The consumer use Go-Jek service more often in the afternoon and afternoon. There are 26 respondents said that they usually use Go-Jek service during the day at 10:00 to 14:00 pm, while 20 respondents also said that they use the service Go-Jek in the afternoon at 14:00 to 18:00 pm. In Table 4 it can be seen that the distance they usually travel by using Go-Jek is 5-10 km.

Table 4. Trip Distance Using Go-Jek

Trip Distance (km)	%
<5	31
5 – 10	51
10 – 12	13
12 – 15	4

The majority of Go-Jek consumers have an income of 1-2 million rupiah per month, where the income comes from money that provided by their parents. PT.Go-Jek Indonesia has been quite successful in attracting the public by offering convenience in ordering transportation services.

Table 5. Waiting Time for Arival of Go-Jek Driver

Waiting time (minutes)	N
<3	3
3 – 5	19
5 – 10	14
10 – 15	9

DOI: 10.1051/matecconf/201713807001

The length of waiting time for the arrival of Go-Jek drivers based on Table 5 is between 3-5 minutes. This arrival time is usually influenced by the distance between the pick-up location and the spreading area of the Go-Jek driver, as well as the pick-up time. The location of the most frequent pickup of consumers is the area of residential, while the location of delivery is most often done is the university. In contrast, the second order pickup location that consumers often do is the university, while the second order delivery location is residential.

3.3 Consumer Satisfaction and Expectation

For the average value of expectations, performed with the same calculation with the average value of satisfaction, so that the results obtained as in Table 6.

Table 6. Consumer Satisfaction and Expectation

Quality	Satisfaction	Expectation	Gap
Tangibles	3.91	4.37	-0.46
Reliability	4.10	4.39	-0.29
Responsiveness	4.15	4.39	-0.24
Assurance	3.58	4.12	-0.54
Empathy	3.94	4.39	-0.45

In figure 2, Quadrant I shows variables that are considered important by Go-Jek's customers and are well implemented by Go-Jek companies. The highest value of satisfaction and expectation in this quadrant is about ease of getting or downloading application, with value of satisfaction equal to 4,58 and expectation equal to 4.73.

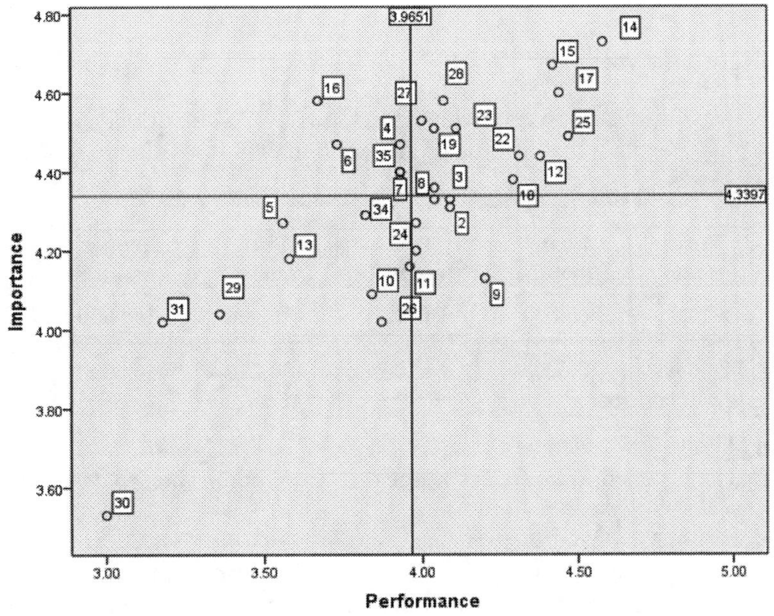

Figure 2. Importance-Performance Go-Jek Consumer

Quadrant II shows variables that are considered important by Go-Jek consumers but have not been well implemented by Go-Jek companies. There are 3 variables: timing of pick-up, the attitude of the driver when greeting the consumer, and customer complaints service. All three variables have a satisfaction level of 3.93 and expectations of 4.40.

Quadrant III shows variables that have not been implemented well by Go-Jek companies but are not very important for Go-Jek consumers. In this quadrant there is one variable that intersect with quadrant intersection line, that is variable about vehicle comfort with satisfaction level equal to 3.96 and expectation equal to 4.16.

Quadrant IV shows the variables that have been implemented well by Go-Jek company but not too important for consumers Go-Jek. In this quadrant there is one variable with positive gap value, that is variable about driver's travel speed.

4 Discussion

In this section will be discussed further on the problems occurring with the Go-Jek operational system, the level of Yogyakarta's interest in Go-Jek services, an analysis of the Go-Jek service feasibility level, and the comparison of transportation costs between Go-Jek and private vehicles.

4.1 Go-Jek Consumer

The 23 respondents as a new consumer of Go-Jek service just use 1-3 times, while only 10 respondents use Go-Jek routine every day. Therefore, it can be said that the people of Yogyakarta have not used Go-Jek service regularly every day. Usually Go-Jek service is used at certain times only, for example when going to the station or shopping center.

Go-Jek driver population in August 2016 is about 1.500 people. Although in August 2016 the presence of Go-Jek in Yogyakarta was only 9 months old, but the people of Yogyakarta already quite a lot who worked as a driver of Go-Jek. This proves that the presence of Go-Jek in Yogyakarta received a positive appreciation from the people of Yogyakarta.

4.2 Feasibility analysis

In this study, feasibility analysis is reviewed in terms of Go-Jek driver's revenue. Based on the data in Table 7 explain that income of driver per day is Rp. 100.000 to Rp. 200.000.

Table 7. Go-Jek Driver Nett Income, in Rupiahs (Rp.)

Income (/day)	Work Time	Loan (/month)	Tax (/month)	Fuel (/month)	Phone (/month)	Nett Income (/month)
100.000	24 days	700.000	16.700	480.000	360.000	843.300
200.000	24 days	700.000	16.700	480.000	360.000	3.243.300

Description of table 7.:

a. Work time is assumed to be 6 days a week, so in 1 month the driver's working time is 6 days x 4 weeks = 24 days.
b. The majority of Go-Jek driver's vehicle credit costs are 500-700 thousand rupiah per month, so the maximum value of Rp. 700.000 per month is obtained.
c. Vehicle tax about 100 to 200 thousand rupiah per year, (Rp. 16.700 per month).

d. The majority of Go-Jek driver's fuel cost is 10-20 thousand rupiah per day, so the maximum value of Rp. 20.000 per day (Rp. 480.000 per month) is calculated.
e. Go-Jek driver's phone call charges Rp.10-15 thousand/per day (Rp. 360.000 /month).

If the driver earns 100-200 thousand per day, then the driver will earn net income of about Rp. 843,300 to Rp. 3.243.300 per month. If the value is compared with the value of Minimum Salary of In Yogyakarta Province 2017 amounting to Rp. 1.337.645.

4.3 Comparison of transportation costs between Go-Jek and private motorcycle

Based on 30 respondents statements that have a private motorcycle it is necessary to calculate the cost of transportation between Go-Jek with a private motorcycle. About 51% of consumer respondents say that the distance they usually travel by using Go-Jek is about 5-10 km. If the tariff of Go-Ride service in Yogyakarta is Rp. 2.000 per km, then consumers have to spend transportation costs of Rp. 20.000.

Table 8. Transport Cost Using Go-Jek

Distance (km)	Respondents	%	Fuel Cost (Rp/km)	Total Cost (Rp)
<5	14	31,11	2.000	10.000
5 – 10	**23**	**51,11**	**2.000**	**20.000**
10 – 12	6	13,33	2.000	24.000
12 – 15	2	4,44	2.000	30.000

If the consumer traveled a distance of 10 km, then the consumer must spend fuel costs of Rp. 2.180. Although the cost of fuel for private motorcycles is much cheaper than Go-Jek transportation costs, but if using private vehicles there are still many other costs that must be incurred such as vehicle taxes, maintenance fees, and depreciation value of vehicle.

Table 9. Transport Cost Using Motorcycle

Distance (km)	Fuel Consume (km/lt)	Fuel Fare (Rp/lt)	Fuel Cost (Rp/km)	Total Cost (Rp)
<5	30	6.550	218	1.090
5 – 10	**30**	**6.550**	**218**	**2.180**
10 – 12	30	6.550	218	2.616
12 – 15	30	6.550	218	3.270

5 Conclusions and suggestions

5.1 Conclusions

The conclusions for this research are:

1. Go-Jek service is interesting for many people although those has not been used routinely by the people of Yogyakarta

2. Net revenues of the Go-Jek is approximately Rp. 843,300 to Rp. 3.243.300 per month, while the value of Minimum Salary in Yogyakarta Province 2017 is Rp. 1.337.645 quite feasible for the driver.
3. Although Go-Jek fare is Rp. 2,000/km and transport costs using private motorcycles is only Rp. 218/km, but by using Go-Jek services there are no other operational costs.
4. Go-Jek transport services are more suitable for non-routine travel (temporary travel) such as shopping, go to/from the station/airport/terminal.
5. Go-Jek's services can reduce dependence on private vehicles.

5.2 Suggestions

Some suggestions about operational systems and Go-Jek services are:
1. Go-Jek's company is expected to implement transparency in establishing policies and systems, so that when there is a change of policy/system does not cause turmoil to Go-Jek drivers and policy changes can be received by Go-Jek driver.
2. Should the problem of error/error that often occurs in the Go-Jek driver application immediately addressed well, and also preferably in the GPS applications Go-Jek more enhanced accuracy to make it easier for consumers and drivers of Go-Jek.
3. We recommend that the problem of fictitious order immediately addressed thoroughly.

References

1. F. D. Amajida, J. Kaj. Kom. **46**, 1, (2016)
2. C. H. Lovelock, L. K. Wright, Pen. Indeks, (2007)
3. Z. Parasuraman, and Berry, J. Retail, **1**, 64, (1988)

DOI: 10.1051/matecconf/201713807002

Numerical investigation on the performance of ducted propeller

Aldias Bahatmaka[1,3*], *Dong-Joon* Kim[2], *Deddy* Chrismianto[3], *Joga Dharma* Setiawan[4], and *Aditya Rio* Prabowo[1]

[1]Pukyong National University, Interdisciplinary Program of Marine Design Convergence, 48513 Busan, South Korea
[2]Pukyong National University, Department of Naval Architecture and Marine Systems Engineering, 48513 Busan, South Korea
[3]Diponegoro University, Department of Naval Architecture, 50269 Semarang, Indonesia
[4]Diponegoro University, Department of Mechanical Engineering, 50269 Semarang, Indonesia

Abstract. As a part of the important thing in ship design is propeller design. Propeller its self has many kinds of the design. This study investigates the effect on the performance of ducted propeller. The investigation was carried out by solving the Navier-Stokes equations with the Computational Fluid Dynamics (CFD) method. These simulations were compared in accordance SHUSKIN nozzle that has 3 types of design. Each type of thruster model indicates different force and torque. Thus, for the analysis will be generated by each model and it can be discovered that which model has the most optimal for the thrust. The analysis by using CFD indicates the change of fluid flow around the ducted propeller. The results showed that the nozzle on the propeller will give the extra of thrust for the performance of propeller.

1 Introduction

Many remote operated vehicle (ROV) have been developed for the maintenance and inspection of ocean structures. Most of them in practical use are self-propulsive with umbilical cable. To avoid the effects of noise force from the cable a free swim power would be ideal, but most are tethered, then the electrical power can be supplied and control signals and data obtained can be transmitted. A self-propulsive ROV can be guided to target more accurately than towed one. Consequently, a tethered self-propulsive which is usually used to inspect and maintain ocean structures [1]. In ship hydrodynamics, fixed propellers, also named screw propellers, have an important place among the propulsion systems to propel a ship. Propeller design is to obtain the optimum propeller which applies to minimum power requirements and against maximum efficiency conditions at an adequate revolution number. Usually two methods are used in the propeller design. The first is use diagrams obtained from the open water propeller experiments for systematic propeller series. The second is to use mathematical method (lifting line, lifting surface, vortex-lattice, boundary element method)

* Corresponding author: aldias@pukyong.ac.kr

based on circulation theory [2]. The characteristics of flow around propeller ducted with decelerating nozzle have been analysed [3]. Form from the previous research, the nozzle produces a decrease in the propulsive efficiency. Thus, the nozzle is important to use in propeller as ROV thruster.

2 Design model

2.1 ROV design

In this research, ROV project in Fig.1 has been designed [4] and the dimension is presented in Table 1.

Table 1. Principle dimensions of ROV

Item	mm
Length	602
Beam	409
Height	290

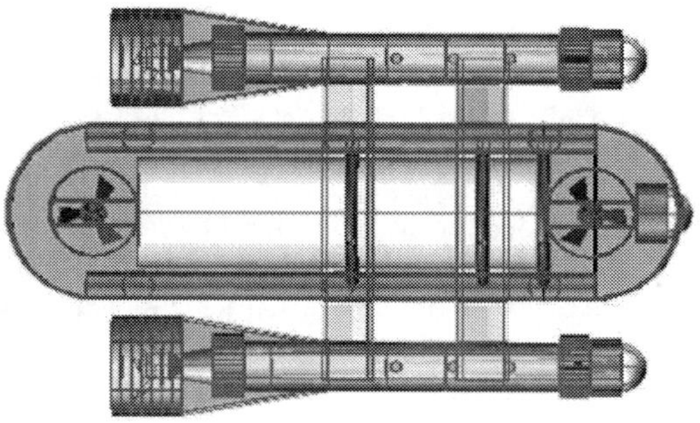

Fig. 1. ROV project design.

2.2 Propeller design

The propeller dimension and geometry design are listed in Table 2.

Table 2. Principle dimensions of propeller

Item	Unit
Diameter	130 mm
Pitch	78 mm
Amount of blade	5 blades
Expanded BAR	0.75
Rake of angle (B5-75)	10.00 degree
Rake of angle (Ka5-75)	15.00 degree

In this case, the propeller dimension was chosen by the product sold in the market. And by this dimension, propeller was designing in 3D model. For the design used 2 types of design, B-series (B5-75) and Kaplan-series (Ka5-75) as shown in Fig.2.

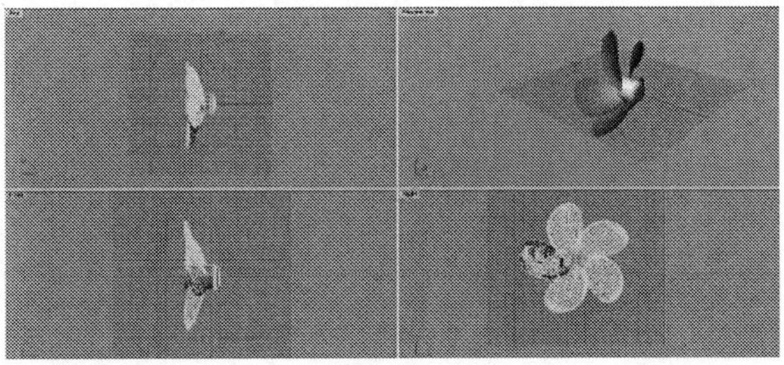

(a)

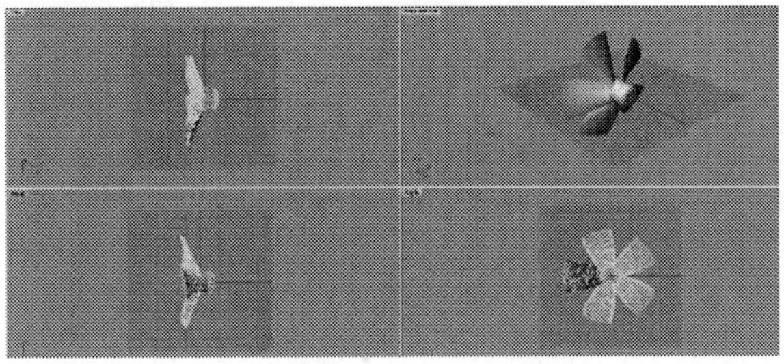

(b)

Fig. 2. Geometry models for propeller: (a) B-series (B5-75), (b) Kaplan-series (Ka5-75).

2.3 Nozzle design

For the calculation of nozzle, there are three types of nozzle as shown in the Fig.3 and the dimension is listed in the Table 3.

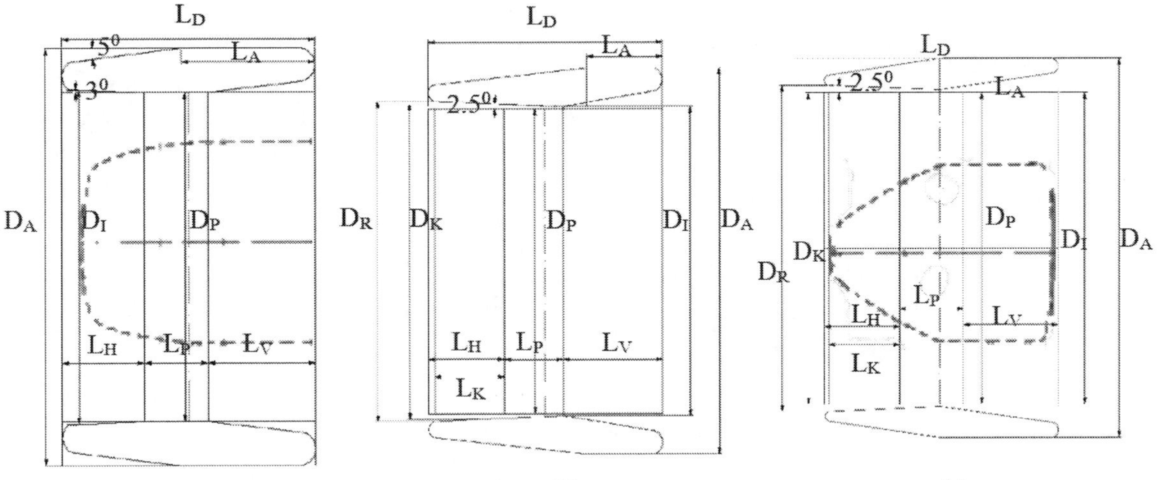

(a) (b) (c)

Fig. 3. Geometry models for nozzle: (a) type A, (b) type B, (c) type C.

 DOI: 10.1051/matecconf/201713807002

Table 3. Principle dimensions of nozzle

Item	Type A	Type B	Type C
LD/DP	0.75	0.75	0.75
DI /DP	1.015	1.015	1.015
Limits	20 mm < (DI –DP) < 60 mm	20 mm < (DI –D) < 60 mm	20 mm < (DI –DP) < 60 mm
LA/LD	1.25	0.32	0.50
LP/LD	0.53	0.25	0.50
LV/LD	0.27	0.425	0.40
LH/LD	0.40	0.325	0.35

3 Methodology

For the numerical on analysis was using computational fluid dynamics method (CFD). The CFD is one of method which is using numerical and algorithm for predicting and solving the engineering problems related fluid flow. Application of CFD to the completion of the flow problem. Reynold Averaged Navier-Stokes equations (RANS) have been developed based in the concept that a velocity and length scale are sufficient to described the effect of turbulence in flow. In a statically steady flow, every variable can be written as Eq.1 the sum of a time averaged value ($\overline{U}$) and fluctuation (u').

$$U = \overline{U} + u' \quad (1)$$

Where the time -averaging velocity component is defined as Eq.2

$$\overline{U} = \frac{1}{T}\int_0^T U dt \quad (2)$$

Where T is the averaging interval. This interval must be large compared to the typical time scale of the fluctuations. Thus, if T is large enough, ($\overline{U}$) does not depend on the time at which the averaging is started. Several steps for calculation such as geometry, mesh, setup, solution and result. Boundary condition was calculating for the open water propeller was presented [5]. The boundary condition can be shown in Fig.4.

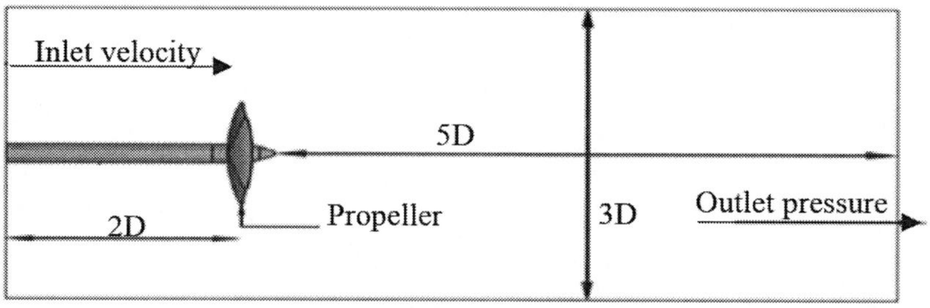

Fig. 4. Boundary condition for open water propeller.

4 Validation of thruster

In this research to validate the result of the test model, used the software test result that already existed and was conducted [6]. Validation of underwater thruster is used to determine the exact boundary condition. ROV thruster using CFD-based software. Reference is taken from the validated model for the testing thruster. Used propeller type of Ka5-75 series and based on the theory of Wageningen [7]. The open water propeller characteristics conventionally were presented in form of the thrust and torque coefficient K_T and K_Q in term of the advanced coefficient J which is given as Eq.3, 4 and 5.

$$K_T = \frac{T}{\rho n^2 D^4} \quad (3)$$

$$K_Q = \frac{T}{\rho n^2 D^5} \quad (4)$$

$$J = \frac{V_A}{nD} \quad (5)$$

5 Meshing process

For this process, there were the same input of parameters and produced different results for the mesh size. Fig.5 is represented as sample of meshing of the models. All of the nozzle's meshing size as shown in the Table 4.

Table 4. Formatting sections, subsections and subsubsections.

		Nodes	Elements
Type A	B-Series	151053	845769
	Kaplan	125480	702840
Type B	B-Series	204080	1146817
	Kaplan	179164	1007934
Type C	B-Series	239314	1343789
	Kaplan	214573	1206703

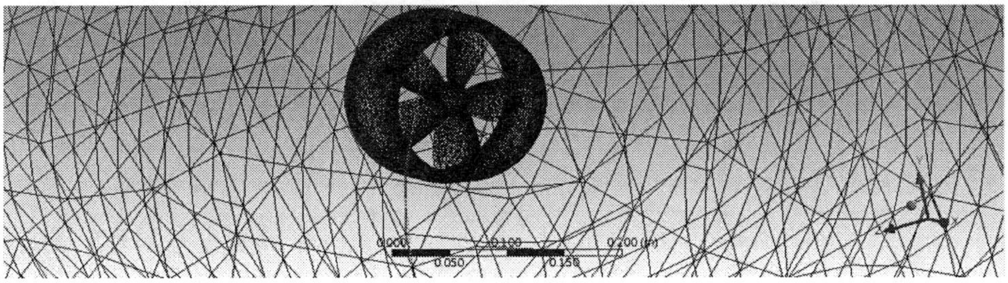

Fig. 5. Meshing of model.

DOI: 10.1051/matecconf/201713807002

6 Result and discussion

6.1 Streamline simulation

For the simulation of streamline, the input for all types were same. The rotational speed is 300 rpm. The results of streamline is presented in Fig.6.

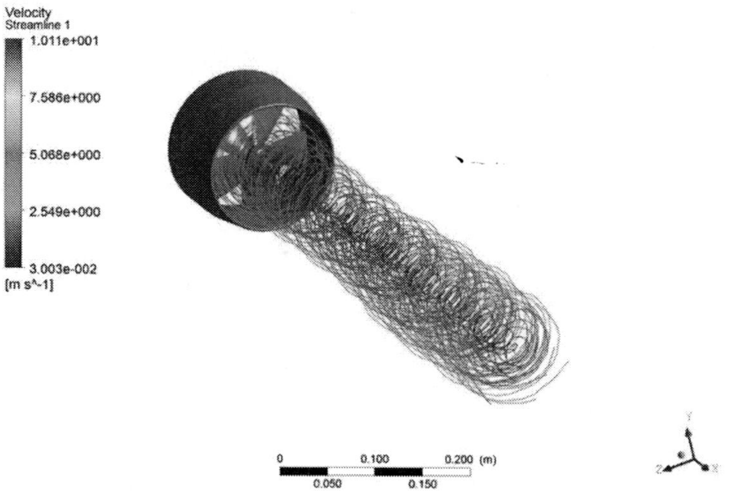

Fig. 6. Streamline of model.

6.2 Pressure Contour simulation

This process was indicating the pressure distribution on the blade area. The results of pressure contour as shown in Fig.7.

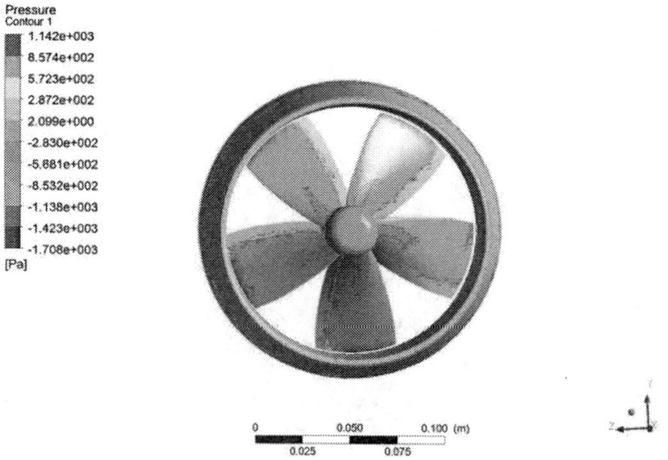

Fig. 7. Pressure contour of model.

DOI: 10.1051/matecconf/201713807002

6.3 Discussion

From the analysis of CFD method, it can be seen on the Table 5. For the B-series and Table 6. For Kaplan series. The changes of the force as clearly shown in Fig.8 and the change of the torque shown in Fig.9.

Table 5. Results of force and torque for B-series.

Output	Propeller Only (P)	P+NA	P+NB	P+NC
Force (N)	1.67	2.06	2.12	2.26
Torque (Nm)	0.016	0.020	0.021	0.023

Table 6. Results of force and torque for Kaplan-series.

Output	Propeller Only (P)	P+NA	P+NB	P+NC
Force (N)	2.02	2.34	2.50	2.53
Torque (Nm)	0.020	0.022	0.023	0.024

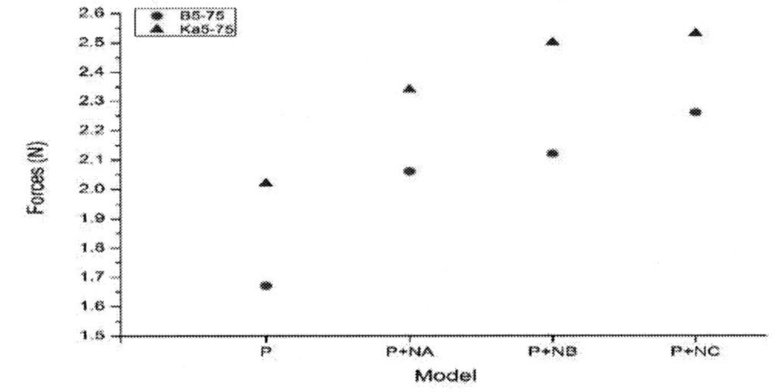

Fig. 8. Chart of force (T) on 300 rpm.

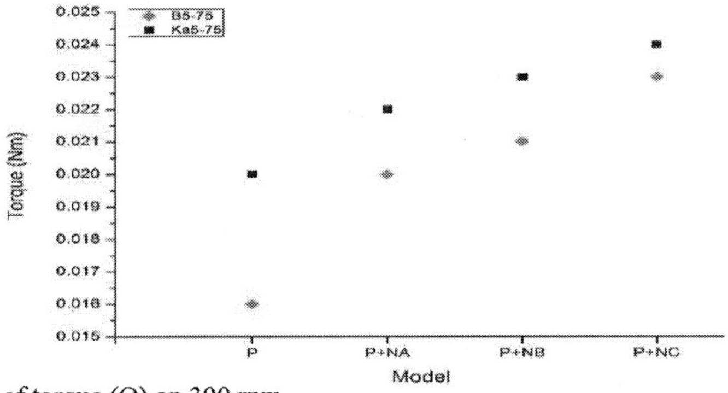

Fig. 9. Chart of torque (Q) on 300 rpm.

In this paper, we have described the specification and the design concepts of ducted propeller as the thruster of ROV. From 6 models, B-series (B5-75) and Kaplan-series (Ka5-75) with nozzle type A, B, and C. Based on the simulations which were conducted, can be concluded that the highest of force (T) was model Kaplan-series (Ka5-75) with type C of nozzle. The model could produce 2.53 N or 25.24% of extra thrust. The lowest of torque (Q) was model of B-series (B5-75) with type A of nozzle. Thus, the best model will be used for ROV thruster is model of Kaplan-series (Ka5-75) with type C of nozzle.

7 Conclusions

The present work described that the numerical investigation by CFD method has been demonstrated to be more effective as the problem solver for determining the ducted propeller performances. Nozzle can produce the extra thrust for the propeller and Kaplan series with type C nozzle is the best model for ROV thruster.

Acknowledgement

The authors would like to thank for the support of the BK21 plus MADEC human resource development group in 2016, Pukyong National University for providing the facilities in this research.

References

1. N. Masahiko, H. Kajiwara, W. Koterayama, Ocean Eng. **28**, 1-43 (2000)
2. S.Ekinci, Brodogradnja **62.2**, 123:129 (2011)
3. R. Bontempo, M. Cardone, M. Manna, Applied Ocean Research **58**, 322-330 (2016)
4. Adriono et al, *DIPO ROV design for exploration robot in Indonesia* (Diponegoro University, Semarang, 2013)
5. C. Parra, *Numerical investigation hydrodynamic performances of marine propeller* (University of Galati, Galati, 2013)
6. Muljowidodo, Indian Journal of Marine Science **38**, 338-345 (2009)
7. M. M. Barnitsas, D. Ray, P. Kinley, *Kt, Kq, and efficiency curves for the Wageningen B-series propellers* (1981)

Optimization of thrust propeller design for an ROV (Remotely Operated Vehicle) consideration by Genetic Algorithms

Aldias Bahatmaka[1,3*], *Dong-Joon* Kim[2], *Deddy* Chrismianto[3], *Nguyen* Hai[1], and *Aditya Rio* Prabowo[1]

[1]Pukyong National University, Interdisciplinary Program of Marine Design Convergence, 48513 Busan, South Korea
[2]Pukyong National University, Department of Naval Architecture and Marine Systems Engineering, 48513 Busan, South Korea
[3]Diponegoro University, Department of Naval Architecture, 50269 Semarang, Indonesia

Abstract. Remotely Operated Vehicle (ROV) is one of the technology that can facilitate for observing and exploring the underwater condition (sea). The design and manufacturing process of underwater robots needs many support and increase the performance of the underwater robot to do some missions. Thruster is one of main component which has function as an actuator during the operation. In the present study, propeller design the most important for solving the problem of an ROV. For the optimization of thrust, Genetic Algorithms (GA) can powerfully search for parameters in large multidimensional design space. Thus, the principle can be applied for determining the initial dimension of the propeller for producing optimum thrust of ROV. GA was successfully shown able to obtain an optimal set of parameters leading to propeller characteristics and best performances.

Nomenclature

A_E	propeller expanded area
A_0	propeller disk area
C_{Tn}	regression coefficient of thrust coefficient
C_{Qn}	regression coefficient of torque coefficient
D	diameter propeller
J	advance coefficient
K_T	thrust coefficient
K_Q	torque coefficient
n	propeller rotating speed
P	propeller blade pitch
P/D	pitch diameter ratio
Q	torque (Nm)
S_n	exponent of J
T	thrust
t_n	exponent of P/D
U_n	exponent of A_E/A_0
V_n	exponent of Z
Z	number of blades
ρ	density of water

* Corresponding author: aldias@pukyong.ac.kr

1 Introduction

Remotely Operated Vehicle (ROV) is an instrument formed mini-sized submission vehicle. ROV is usually to explore underwater object, such as underwater photography, military operation and underwater pipeline repairing. ROV is used for activities in human-unreached area like an underwater small cave. ROV is designed to have abilities to rescue operation and repairing of object from the surface [1].

Underwater robot design and manufacturing absolutely requires many supporting components to improve the underwater robot operation to perform a variety of underwater missions. Thruster is one of main components as locomotor of underwater robot to maneuver horizontally when it moves forward and backward and vertically to moves up and down [2].

The first proposal to use screw propeller appears to have been made in England by Hooke in 1680 [3]. The genetic algorithms are one of method that has tool for optimization for many difficult optimization problems with multi objectives. GA methods were shown to be able to replace the traditional computation method and design charts. In this case for determining the initial propeller that optimum for ROV thruster is difficult. For selecting the initial propeller, used optim-tools in Genetic Algorithms. However, the traditional design charts method and previous GA methods were limited in considering or maximum of hydrodynamic aspect, such as efficiency, thrust coefficient, etc. hence, the goals of present research to determine the optimum value of thrust for ROV by 3 parameters, pitch diameter ratio, expanded blade area ratio, and rotational speed of propeller. Applying GA methods to propeller design with consideration of thrust for reaching the best performance of propeller.

2 Propeller design

The preliminary propeller design problem is described in detail in the principles of Naval Architecture. Here is the basic design or initial design of propeller with the dimension for dependent variables as shown in Table 1. For independent variables as presented in Table 2.

Table 1. Dependent variables.

Items	Unit
Diameter	130.00 mm
Number of blades (z)	5 blades
Material	Mn.Bronze (2)
Rake ofangle	15.00 degree
Density of water	1.025 kg/m^3

Table 2. Independent variables.

	I	II	III
P/D	0.60	0.65	0.70
AE/A0	0.75	0.80	0.85
Rotational speed *(n)*	5 rps	10 *rps*	15 *rps*

2.1 Parameters of optimization

Selecting from a propeller series is a simple method to design a propeller. Among the series propeller is one of the most often used and studied [4]. And selection of the table was improved later.

3 Genetic algorithms (GA)

The basic theory has been described [5]. Genetic algorithms are search algorithms based on the mechanics of natural selection and natural genetics. They combine survival of the fittest among string structure with structured yet randomized information exchange to inform a structure algorithms with some of the innovative flair of human search.

Based on the theory, the research was conducted with set of parameters dependent and independent variables. For the first steps were identifying variables and functions the propeller of ROV. There are three variables that we could change randomly as like as P/D 9pitch ratio), AE/A0 (expanded blade area ratio), and n (rotational speed). From this process, we could get the value of K_T and K_Q. Thus, the thrust (T) amd torque (Q) can be expressed in Eq.1 and 2.

Function:

$$T = K_T\left(\rho n^2 D^4\right) \tag{1}$$

$$Q = K_Q\left(\rho n^2 D^5\right) \tag{2}$$

The fitness function we used as the functions of the blade number, blade area ratio, pitch ratio and advanced coefficient can be described in the Eq.6 and **7**.

Fitness function:

$$K_T = \sum_{n=1}^{39} C_{Tn} J^{Sn} \left(\frac{P}{D}\right)^{tn} \left(\frac{A_E}{A_0}\right)^{Un} Z^{vn} \tag{3}$$

$$K_Q = \sum_{n=1}^{47} C_{Qn} J^{Sn} \left(\frac{P}{D}\right)^{tn} \left(\frac{A_E}{A_0}\right)^{Un} Z^{vn} \quad (4)$$

Where C_{Tn} and C_{Qn} are the regression coefficients of the thrust and torque coefficients respectively, S_n, t_n, U_n and v_n are exponents of J, P/D, AE/A0, and Z respectively.

The constraint function can be created in Eq.5 and 6.

Constraints:

$$0.60 \leq \frac{P}{D} \leq 0.70 \quad (5)$$

$$5 \leq n \leq 15 \quad (6)$$

4 Optimization procedures

For the process of optimization shown in Fig.1. The optimization tools which is supported by MATLAB program for the solution of genetic algorithms method. This study, GA apply to know how is the optimum thrust for the initial propeller using the parameters and setup for GA is shown in Table 3.

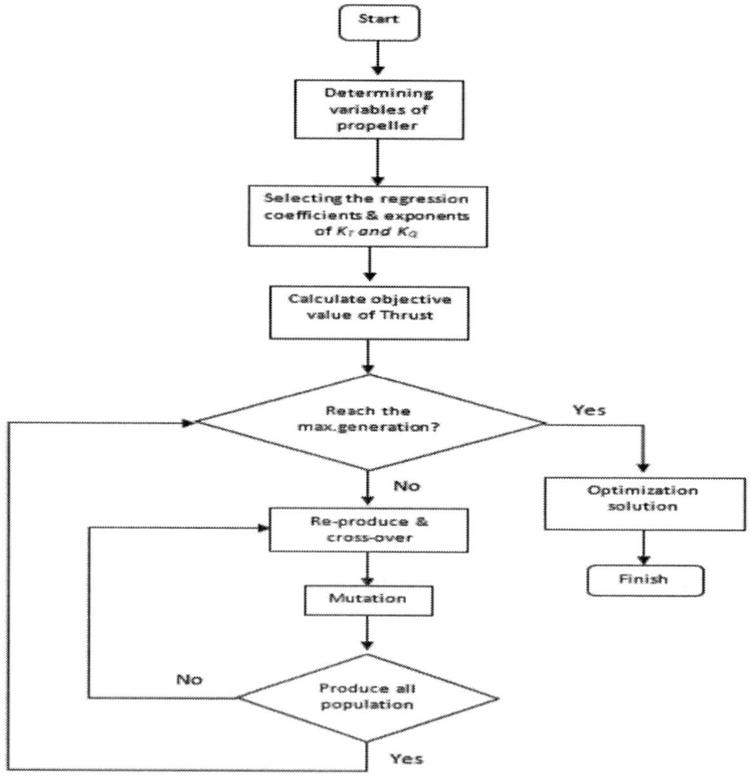

Fig. 1. Flowchart of genetic algorithms.

Table 3. Setting up the GA parameters.

Population size = 20	
Number of generations	80
Selection	Stochastic uniform
Crossover rate	0.8
Crossover function	Scattered
Mutation rate	0.2

5 Results and discussion

5.1 Optimization Output

The output of optimum thrust can be shown in Table 4. And results can be exported by genetic algorithms as shown in Fig.2.

Table 4. Output.

Iteration	**P/D**	**AE/A0**	**n (rps)**
51	0.60	0.79	5.00

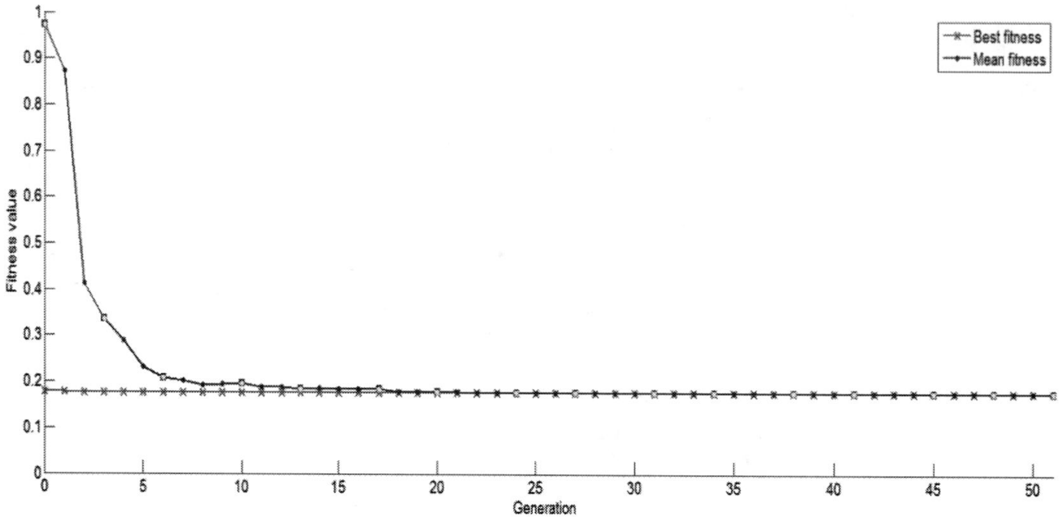

Fig. 2. Fitness value through the generations

At the 51 iterations using GA, the thrust will be optimum when the value of P/D (pitch ratio) is 0.60, AE/A0 (expanded area ratio) is 0.79 and the value of rotational speed is 5 rps.

5.2 Final Results

The difference of computational result can be seen in Table 5.

Table 5. Comparison the final results.

Results	Original	New	Differences
Force (N)	2.530	3.224	+27.43%

Based on the results, the thrust is increasing 27.43% from 2.530 N to 3.224 N. it means the optimization of the thrust for the propeller of ROV is reached and satisfied. And results of new design as shown in Fig.3 for the streamline simulation and Fig.4 for the pressure contour.

 DOI: 10.1051/matecconf/201713807003

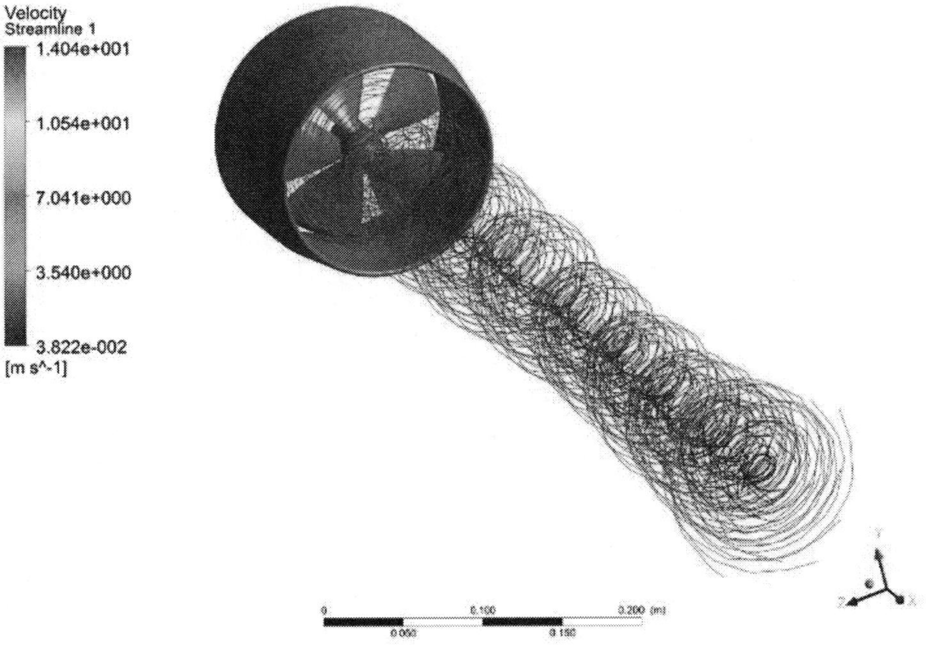

Fig. 3. Streamline simulation

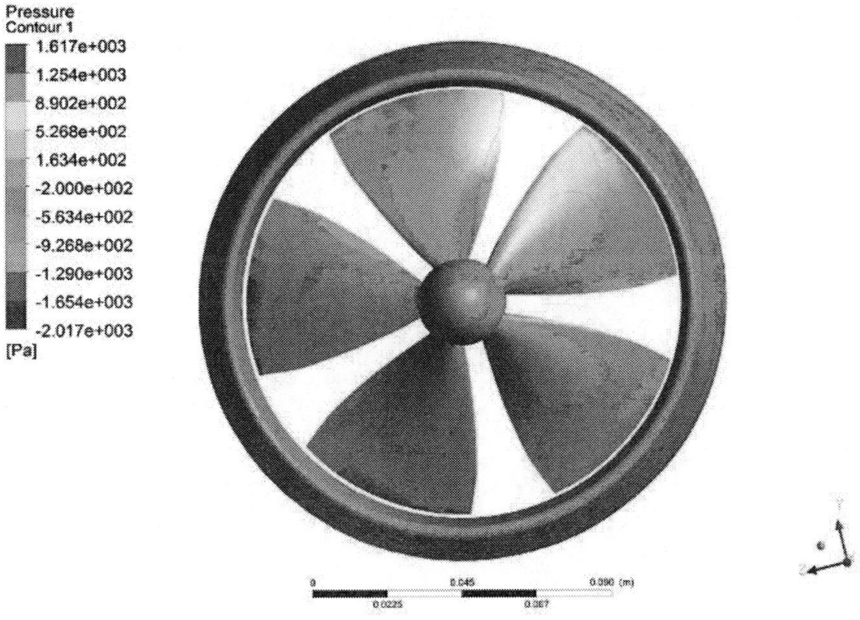

Fig. 4. Pressure contour

6 Conclusion

In this paper, study on development of the propulsion device for an ROV (Remotely Operated Vehicle). The GA is applied for optimization design of the initial propeller dimension and the results show that it is good for kind of the problems. The optimization of propeller for an ROV has been proposed and provides the solution to ROV thruster and GA optimization can be employed due to propeller performances.

Acknowledgement

This work successfully published with support from BK21 plus MADEC Human Research Development Group, Pukyong National University, South Korea. The corresponding author would like to thanks to his professor and the people who help and support for this research.

References

1. R. D. Roberto, Sr. L. Wernli, Robert, The ROV manual, Burlington, MA 01803, UK, (2007)
2. A. Sharkh, S., Harris, et al. *Design considerations for electric drives for the thrusters of unmanned underwater vehicles* (6th ECEA, Sevilla, pp.799-801, 1995)
3. J. D. V. Manen, P. Van Ossanen, PNA, **2** (1988)
4. G.Kuiper, *New developments around sheet and tip vortex cavitation on ships propellers* (Caltech, 2001)
5. D.E.Goldberg, *Genetic algorithms in search, optimization, and machine learning* (Addion Wesley, p.102, 1989)

Integrated sustainability for transportation infrastructure development in Indonesia: A case study of Karawang region

Ayomi Dita Rarasati[1,2,*], and *Tamara* R.F. Iskandar[1, 2]

[1]Civil Engineering Department, Faculty of Engineering, Universitas Indonesia, Indonesia
[2]Center for Sustainable Infrastructure Development, Faculty of Engineering, Universitas Indonesia, Indonesia

Abstract. Efficient infrastructure provides social and economic benefits to the society and there is a need to develop a modern infrastructure in order to ensure the well-being of the society to make a sustainable social, economic and environmental growth. However, costs always become the impeding factor for stakeholders when they contemplate sustainability initiatives. Therefore, it is crucial to balance the financial benefits with sustainability deliverables in transportation infrastructure development. This research aims to develop feasibility of implementation from an environmental, social, and economic sustainability perspective in comprehensive integrated infrastructure development. The results show that safety is the most influencing factors in selecting mode of transportation, more reliable service is needed to be improved to attract public transportation users and security is needed to enhance transportation facility to promote public transportation use.

1 Introduction

Efficient infrastructure provides social and economic benefits to the society. Furthermore, there is a need to develop a modern infrastructure in order to ensure the well-being of the society to make a sustainable social, economic and environmental growth. Sustainable development is about making societal investments that are sensitive to the natural environment and at the same time financially viable in the long term [1]. The choice to implement sustainable infrastructure, however, faces major obstructions for environmental sustainability. Sustainable infrastructure development requires drastic change in culture and behaviour of the society, but also needs to be incorporated in its underlying policies as well [2]. Additionally, although the application of sustainability in built assets is beneficial, it often involves major capital investment. Costs always become the impeding factor for stakeholders when they contemplate sustainability initiatives. Therefore, it is crucial to balance the financial benefits with sustainability deliverables in transportation infrastructure development.

* Corresponding author: ayomi@eng.ui.ac.id

DOI: 10.1051/matecconf/201713807004

The concern of current and future transportation development system originated from the lack of integration and interaction between the transportation system and the environment. Although infrastructure is crucial in modern economy, more infrastructures may not necessarily mean more growth. The issue with infrastructure expansion often comes at the expense of the local environment [3]. The realization that environmental quality has been affected tremendously by transportation activities, which continue to accumulate over time, aroused the concern of many people that the transportation development that has been carried out so far was a mistake in terms of policy making and planning [4]. Past and current practice of transportation infrastructure, both from the society and from the government, needs to shift forward and should be seen from a different perspective if improved quality of life wants to be achieved.

The success of regional development and the state is determined by the availability of natural resources, human resources and facilities and other supporting infrastructure. One of the contributing factors of successful development is the availability of infrastructure that aims to support the mobility of resources and facilitate the implementation of sustainable development. That is why infrastructure became one of the main priorities in the government's development plans because it generates a multiplier effect. A number of potential benefits hoped to be achieved from these development plans in construction of roads and the transport sector in general, among others: changes in the relative cost of new transportation means to transport long; increased availability of means of transportation; improved ride quality; increased economic activity that ultimately affect change people's consumption patterns and structures; the increase in per capita income of the people, and so forth. However, the existing conditions of the transportation infrastructure in Indonesia suffers issues particularly in having gaps between provinces, issues in developing the facilities, funding limitations in the transportation sector and many other issues regarding its development [5].

In accordance with the accelerated growth of the global community to eradicate poverty and tackle climate change, growth strategies still seem to fail to be proven sustainable. A common impetus that could provide the necessary initiative against the challenging obstruction is infrastructure development. A sustainable infrastructure will cut down unnecessary costs, lost production costs, as well as decrease energy consumption, waste and pollution.

The servicing and provision of Indonesia's infrastructure is not optimum and there are huge inconsistencies in achievement between different regencies/cities. This can be seen from the limited provision of infrastructure and less infrastructure network in areas outside Java Island. Java Island is a huge boost to Indonesia's economy, measuring up to 80% of its income. This integrated transportation system hopes to accelerate the rate of economic growth in Indonesia. In terms of infrastructure network, Java Island is the most developed where nearly all major provinces scored top 10 in the mapping of infrastructure in Indonesia [2]. However, effectiveness of sustainable infrastructure may also vary depending on the related party's ability to take advantage of the situation, in addition to other constraints such as poor managerial incentives, planning, externalities from missing components, as well as inefficient and unintegrated transportation system [3].

Therefore, this research aims to determine demands and conditions by Indonesian citizens, as well as to investigate financial constraints of sustainability measures and develop feasibility of implementation from an environmental, social, and economic perspective in comprehensive integrated infrastructure development in Karawang region in Indonesia, with a particular focus in transportation system. However, this paper will only discuss the critical demands and conditions according to Indonesian citizens in pursuing integrated urban transportation development based on transportation sustainability indicators in Karawang region in Indonesia.

2 Sustainable transportation indicators

There are three aspects of sustainable transport, namely sustainability in environmental, social, and economy. The goal of sustainable development is the creation of an equal balance between the economic, environmental, social, and governance aspects, in order to create more liveable and productive cities, towns, and villages by using methods that require the participation of decision makers and stakeholders and new tools to enable decision makers to respond properly [6]. Hence, sustainable development requires the effective participation and initiative of stakeholders, including governments, industry, academia, and society as the basis of its efforts to achieve a balanced development.

Sustainable infrastructure can be defined as an infrastructure which integrates these three key components in accordance with governance aspects into a project's planning, building and operating phases [7]. A sustainable infrastructure should also ensure resilience towards climate changes or shocks in order to improve the attractiveness of infrastructure investments by mitigating risks, creating tangible benefits and opportunities as well as reducing emissions and climate risks [8].

Socially sustainable infrastructure means that it inclusive and respects human rights. This includes meeting the needs of the poor by increasing infrastructure access, supporting general poverty and inequality reduction, and reducing vulnerability to climate change risks. Socially sustainable transport is a transport that can minimize the extent of accidents, and can enhance social justice and the level of health in the community. Therefore, in the social aspect, sustainable transport is defined as a transport system which is capable of creating equality towards every aspect against the use of transport itself, with high transportation safety levels, where the institutional system is capable of supporting the creation of a sustainable transportation system [9].

Economically sustainable infrastructure means that it positively impacts Gross Domestic Product (GDP) per capita and job outcomes. It does not burden governments with debt they cannot repay, or end-users—especially the poor—with tariffs they cannot afford. Furthermore, sustainable transportation in economic aspect is transportation that is affordable, operates efficiently, can provide alternative modes of transport, where it improves accessibility and supports the economic growth rate. This relates to the purpose of the transport sector which is to support the economic development of the region [9].

Environmentally sustainable infrastructure means that it establishes the foundation for a transition to a low-carbon economy, leaves a low-carbon footprint, and is also resilient to climate change. Environmentally sustainable transport among others is a transportation which minimizes air pollution, noise level, and the level of use of resources that are not renewable, in order to minimize public health impacts related to transport activities. A sustainable infrastructure can also employ fundamentally different ways of meeting infrastructure service needs, including the implementation of more responsive and integrated information systems that complement its hard infrastructure [9].

Sustainability indicators are selected, targeted, and compressed variables that reflect public concerns and are of use for the decision maker [10]. Indicators of sustainable development had been discussed at an international level, but what is at issue is how to apply it at the national level and even further at an urban level (city/county), as there is difference in focus between the national and local levels as well as the differences between the regions. In terms of sustainability indicators, many institutions have had defined their own standards and measures for this particular matter. Sustainability is generally evaluated using indicators, which are specific variables suitable for quantification or measurement. Such indicators are useful for identifying trends, predicting problems, setting targets, evaluating solutions and measuring progress. Which indicators are selected can significantly influence analysis results. A particular policy may seem beneficial and desirable if evaluated using one set of indicators but harmful and undesirable according to others. It is therefore important that people involved in sustainability planning understand the assumptions and perspectives of the performance indicators they apply.

3 Research methodologies

The research is positioned as mixed methods research that uses complementary primary data collection and secondary data collection that are computed using quantitative method. The research began with a quantitative phase, which is a survey as a primary source of data for the research. A review of the literature was undertaken to help establish a rationale for the research questions and to establish the extent and depth of existing knowledge on sustainable transportation indicators where it can be further assessed according to the development of current transportation system in Indonesia, as well as to determine the critical demands of pursuing sustainability in transportation in a certain area that serves as a benchmark for the research.

This research used questionnaire-based surveys as the method to identify the critical demands of the Indonesian community that emphasizes in pursuing a sustainable transportation system by bench-marking on a village in Indonesia, as well as to analyze the feasibility of the demands in a benefit cost ratio analysis towards the sustainable transportation indicators that are proposed. The questionnaire surveys were selected for the reason that questionnaire surveys are effective in gathering information about the characteristics, actions, or opinions of a large group of people [8]. In the questionnaire survey, Indonesian citizens (namely those who are 15 to 64 years of age) were asked to rate their demands and conditions in transportation development referring to the proposed sustainable transportation indicators, and were also asked about their availability in participating in the transportation development process itself.

Prior to the distribution of questionnaires to the respondents, a survey was initially piloted to a small sample of respondents. In addition, the pilot survey was also tested for its validity, using the Pearson-Product Moment Correlation, and reliability, with the Spearman-Brown Prophecy Formula and Cronbach's Alpha. Data collected from the questionnaire surveys are recorded and analyzed using a software program. Following the data collection, data analysis is done through a quantitative approach by quantification of the qualitative data into statistical analysis. Furthermore, data analysis is best used through the aid of software programs [12].

4 Results

A total of 500 questionnaires were prepared for the respondents with a cover letter from the Civil Engineering Department, Universitas Indonesia. The survey was administered throughout March – April 2017. Respondents that participated in this research come from various gender, age, education and occupational background to achieve heterogeneity of the data collected. Furthermore, respondents are also acquired in different locations in the Karawang region (i.e. north/beach area, south/mountain area, and centre/city area of Karawang). Out of the 500 questionnaires, the researchers managed to obtain a total of 344 questionnaires through questionnaire distribution and structured interview.

Following the employed mixed methods research of questionnaire survey and structured interview, the results were processed using the validity and reliability test. For the validity test, a significance level of 0.05 (5%), confidence level of 95%, with the degree of freedom (df) of 342 was determined from the data collection. Furthermore, a one-way tailed test with an r-table value of 0.089 was employed as the hypothesis used in the research consists of only one possibility. The r-table values of the validity test were compared with the r value computed from the research where all statements are proven to be 100% valid as all r values exceed the r-table values from the validity test.

Moreover, the type of reliability test that was used is the internal consistency reliability type, with the split-halves method using the Spearman-Brown Prophecy formula and Cronbach's Alpha. After computation of the data, the reliability test value for the split-halves method of the research is 0.94, which means according to the Spearman-Brown Prophecy Formula, the test would imply that it is approximately 94% reliable as it is true and yields consistent results of the same measure. Additionally, if calculated using the Cronbach's Alpha method, the reliability test value would yield 0.895 or essentially 0.9 if rounded up. This would suggest that if compared to the Internal Consistency Cronbach's Alpha table ($\alpha \geq 0.9$), the test is considered within the 'excellent' grade. Thereupon, as the outcome of the survey passed both the validity and reliability test, the research can then be further processed for data computation.

The administered mixed methods of questionnaire survey and structured interviews consisted of questions, each with a specific purpose of answering the research questions. Therefore, the findings of the data are framed around the components that will satisfy the aims and objectives of this study while synthesizing the need to address the research questions. The mixed methods focused on the basis of the identification of critical demands and needs of transportation in Karawang related to sustainable measures. The respondents were asked to indicate the extent of importance, agreements, and willingness of statements specified in the questionnaire survey on a four-point Likert rating scale.

Furthermore, the results are set out where the highest and lowest earning rates are highlighted for ease of data illustration. The highest earning rates are highlighted in light grey, whereas the lowest earning rates are highlighted in dark grey. As the survey was conducted at three locations—the city, mountain area, and beach area—the data are illustrated according to their locations as well, aside from the overall score of all locations.

Table 1 summarizes data for question on 'how certain factors may affect the respondents' choice of public transportation'. Overall, the respondents agree that 'safety' is the most crucial factor that affects their transportation choice with 8.83%. This is also true for respondents that are from the city and beach area. However, respondents from the mountain area imply that the most crucial factor is 'security', where 'safety' is in fact third in place after 'health benefits'. 'Weather' seems to be the most undecided factor that affects transportation choice, with the overall score of the respondents, as well as respondents residing in mountains and city area agreeing to the statement. In contrast, respondents residing in the beach area suggest that 'environmental concerns' as well as 'social stigma' as the most undecided factor for transportation choice, where 'weather' ranks third out of the thirteen statements.

Table 2 summarizes results for question on 'what factors should be improved in transportation to attract more public transportation users'. Overall, the respondents agree that a 'more reliable service' should be improved to attract more transportation users with 11.06%, where the lowest factor would be 'public commercial' with 8.86%. If divided to each location, city respondents suggest that they agree with the overall consensus for both highest and lowest factor in attracting more transportation users. However, data from mountain respondents imply that a 'cheaper fare' would attract more transportation users with 10.96% compared with 10.80% for a 'more reliable service', where the lowest factor is 'integration with other modes' and 'less environmental impact'. Additionally, respondents residing in and around the beach area suggest that 'better transportation facilities' as the most crucial factor with 11.16%, where the lowest factor is 'less environmental impact' with 8.37%.

DOI: 10.1051/matecconf/201713807004

Table 1. Factors influence in selecting mode of transportation.

Does the following factor affect your choice of transportation?	Rate in Percentage			
	Overall	City	Mountain	Beach
Affordability	7.13%	7.02%	7.46%	8.16%
Weather	6.88%	6.80%	7.15%	7.58%
Frequency of service	7.38%	7.29%	7.61%	8.16%
Environmental concerns	7.18%	6.82%	7.31%	6.41%
Convenience	8.07%	8.09%	7.91%	8.16%
Personal comfort	8.48%	8.56%	7.91%	8.45%
Safety	8.83%	8.98%	7.91%	8.45%
Social stigma	7.44%	7.51%	7.46%	6.41%
Distance to nearest stop/station	7.33%	7.25%	7.76%	7.58%
Travel time	7.77%	7.77%	7.61%	8.16%
Health benefits	7.76%	7.79%	8.07%	6.71%
Reliability	7.51%	7.51%	7.46%	7.58%
Security	8.57%	8.62%	8.37%	8.16%

Table 2. The need of transportation improvement to attract public transportation users.

Transportation improvement in the following factors will attract more public transportation users	Rate in Percentage			
	Overall	City	Mountain	Beach
More public transportation routes	9.93%	9.84%	9.97%	10.36%
Increased frequency of armada	10.12%	10.09%	10.30%	9.16%
Better path/road condition	10.87%	10.83%	10.63%	11.16%
Better transportation facilities	10.95%	10.95%	10.47%	11.16%
Cheaper fares	10.85%	10.78%	10.96%	10.76%
More reliable service	11.06%	11.13%	10.80%	9.56%
Integration with other transportation modes	8.96%	8.92%	9.14%	9.56%
Less environmental impacts	9.17%	9.15%	9.14%	8.37%
Access to transportation mode	9.32%	9.41%	9.30%	10.36%
Public commercial	8.86%	8.90%	9.30%	9.56%

Table 3. The need of transportation facility improvement to promote public transportation use.

Improvement in the following particular components of a transportation facility will promote public transportation use	Rate in Percentage			
	Overall	City	Mountain	Beach
Cleanliness	23.92%	23.84%	26.00%	23.53%
Increased Service	24.54%	24.51%	26.00%	24.51%
Security	26.00%	26.07%	24.00%	26.47%
Comfort	25.53%	25.57%	24.00%	25.49%

Table 3 summarizes results for question on 'what components should be improved in a transportation facility to promote public transportation use'. Overall, the respondents agree that improvement in 'security' of a transportation facility will promote transportation use, as did city and beach respondents. The trend seems to follow where 'cleanliness' is the lowest factor to promote transportation use from the looks of the overall data, city respondents, and beach respondents. A noticeable difference can also be seen for mountain respondents where the data seem to more or less agree for all factors as the percentages are quite similar to each other. 'Cleanliness' and 'increased service' is regarded as the most important component to improve in promoting transportation use with 26% each, whereas 'security' and 'comfort' is the opposite with 24% each.

5 Conclusion

This paper examines critical demands and conditions according to Indonesian citizens in pursuing integrated urban transportation development based on transportation sustainability indicators in Karawang region in Indonesia. The results show that safety is the most influencing factors in selecting mode of transportation, more reliable service is needed to be improved to attract public transportation users and security is needed to enhance transportation facility to promote public transportation use. Even though there are some low earning rates, this does not indicate that factors (i.e. weather, environmental concerns, and social stigma) which affect alternative decision of transportation are not important. This indicates that Karawang people prefer to have better safety, security and personal comfort for their preference.

The authors would like to acknowledge funding support from Universitas Indonesia through PITTA Grant No. 743/UN2.R3.1/HKP.05.00/2017 for conducting research and disseminating research results. Authors also gratitude the Center for Sustainable Infrastructure Development for the in kind contribution in supporting this research.

References

1. K. C. Goh, J. Yang, Importance of Sustainability-Related Cost Components in Highway Infrastructure: Perspective of Stakeholders in Australia, J. Infrastruct. Syst. **20**, 1 (2014)
2. Y. Latief, M. A. Berawi, A. D. Rarasati, L. S. Supriadi, A. R. B. Berawi, I. S. Hayuningtiyas, Mapping Priorities for The Development of The Transportation

Infrastructure in The Provincial Capitals of Indonesia, International Journal of Technology **7**, 4, pp. 544–552 (2016)

3. M. Fay, M. Toman, D. Benitez, S. Csordas, Infrastructure and Sustainable Development, *Postcrisis Growth and Development*, The World Bank, pp. 329-372 (2011)
4. A. Sjafruddin, *Pembangunan Infrastruktur Transportasi untuk Menunjang Pembangunan Berkelanjutan Berbasis Ilmu Pengetahuan*, pp. 1–11 (2011)
5. National Development Planning Agency (Bappenas), *Buku I Agenda Pembangunan Nasional* (2014)
6. I. G. Malkina-Pykh, Integrated assessment models and response function models: pros and cons for sustainable development indices design, Ecological Indicators **2**, 1-2, pp. 93-108 (2002)
7. K. N. Lee, Urban Sustainability and the Limits of Classical Environmentalism, *Environment and Urbanization* **18**, 1, pp. 9-22 (2016)
8. H. P. Egler, R. Frazao, *Sustainable Infrastructure and Finance - How to Contribute to a Sustainable Future*, pp. 1-41 (2016)
9. A. Bhattacharya, J. Oppenheim, N. Stern, *Driving Sustainable Development Through Better Infrastructure: Key Elements of a Transformation Program* **91**, pp. 1-49 (2015)
10. H. Gudmundsson, Indicators and performance measures for Transportation, Environment and Sustainability in North America, *Research Notes* **148** (2001)
11. J. W. Creswell, *Research Design* **4** (2014)
12. U. Sekaran, *Research Methods for Business* **4** (2003)

Analysis of The Time Value for Public Transport Passenger by Using Random Regret Minimization

Medis Surbakti[1] *and Fransiscus Pinem*[1]
[1]Department of Civil Engineering, University of Sumatera Utara, Indonesia

Abstract. This study aims to analyze the value of time a public transport passengers in Medan which is the amount of money issued by a person who was willing to save one unit of time traveling. An investment in transport projects should consider the high or low travel speeds for road users, whenever he or she is the driver, passengers or goods that were transported. Savings or reduction in the value of time is one of the advantages in transportation projects. Random Regret Minimization is a new method that can be used to calculate the value of time. The value of the travel time of passenger public transportation Medan with Random Regret Minimization method is Rp.6687.52/hour/person. Regret Minimization showed significantly different results with the Mode Choice Approach method and the Income Approach. This shows the tendency of public transport passengers in Medan to avoid the regret that will arise when the monorail will operate later.

1 Introduction

Transport Is an attempt to move people, animals or goods from one place to another by using a vehicle driven by human or machine. Along with the economic growth of a region, the demand for transportation will be even greater. So in general, the increase in the economic is reflected in the raise in the number and quality of mode of transportation. Medan is one of the largest cities in Indonesia, which has a lot of transportation problems that are quite complicated. To solve the problem, the city government in collaboration with the private sector (Public Private Partnership) is planning the construction and development of the monorail transportation modes.

Planning and construction of the Monorail Medan are still in the early stages of the study, but it is important to estimate the impact of the emergence of new modes of transport. In addition to the calculation of the negative impact, positive impact on the operation of the monorail also is analyzed. The value of travel time is a form of analysis that is needed in knowing the economic feasibility of a new mode of transportation. Passenger travel time value of different types of public transport modes that operate in the city of Medan will be one of the references of the time value.

The purpose of this study is to determine the value of public transport users in Medan, as a component for the calculation of the feasibility of operation modes of transport monorail, as a new mode of transportation in the city of Medan.

Transport is an attempt to move people, animals or goods from one place to another by using a vehicle driven by human or machine. Along with the economic growth of a region, the demand for transportation will be even greater. So in general, the increase in the

economic is reflected in the raise in the number and quality of mode of transportation. Medan is one of the largest cities in Indonesia, which has a lot of transportation problems that are quite complicated. To solve the problem, the city government in collaboration with the private sector (Public Private Partnership) is planning the construction and development of the monorail transportation modes.
Planning and construction of the Monorail Medan is still in the early stages of the study, but it is important to estimate the impact of the emergence of new modes of transport. In addition to the calculation of the negative impact, positive impact on the operation of the monorail also is analyzed. The value of travel time is a form of analysis that is needed in knowing the economic feasibility of a new mode of transportation. Passenger travel time value of different types of public transport modes that operate in the city of Medan will be one of the references of the time value.
The purpose of this study is to determine the value of public transport users in Medan, as a component for the calculation of the feasibility of operation modes of transport monorail, as a new mode of transportation in the city of Medan.

2 Literature Review

Time Value is the amount of money a person will pay to save one unit of travel time [5]. As is well known, investment in a transport project should take into account the high speeds of road users traveling. Therefore, the procurement of transport facilities should provide an opportunity for road users to save travel time so that the time saved can be used to perform other activities.

The existence of the monorail will be expected to facilitate the high mobility of the people who will use the monorail. In addition to being a mass public transport, the monorail is expected to serve the needs of passengers better than existing public transportation in terms of comfort and timeliness. The monorail is a metro vehicle that has its track, so its movement is not influenced by other vehicles. This causes the monorail to move more quickly and on time. With the efficiency of time and reliability of this monorail, it is expected to be worth the rate to be applied. Therefore it is important to know the value of travel time from passengers who have the potential to ride this monorail.

Some factors that affect travel time value [5]:
1. Income Level
2. Travel Length
3. Travel Mode
4. Time Period
5. The Purpose of Travel
6. Environmental Conditions

Estimating the value of travel time is to try to place the value of money on the travel time savings due to the emergence of a new transport facility. There are several methods to calculate travel time value. Among them is to use the preferred approach between the existing mode and the new mode to be implemented and the second is the income level approach.

2.1 Random Regret Minimization

There are two theories that explain the preferred approach between the existing mode and the mode that will operate, the regret minimizing theory and the theory of utility

maximization. In 2008, Chorus explained Regret's minimizing theory (RRM) that in choosing a choice, one would further minimize regrets that would arise rather than maximize the value of the utility [3]. The method of Random Regret Minimization is based on the intuition of an individual's behavior that has been based on the paradigm of choosing utility improvement (in this case the individual psychological approach is more prominent)[4].
Similar to the utility maximization method, the time value can also be calculated by the Random Regret Minimization method. The time value with the Random Regret Minimization Method can be calculated by the following formula:

$$\text{Value Of Time}_{\text{RRM}} = \frac{\sum_{i\neq j}\left(\frac{-\beta_{TT}}{1+\frac{1}{exp[\beta_{TT}(TTj-TTi)]}}\right)}{\sum_{i\neq j}\left(\frac{-\beta TC}{1+\frac{1}{exp[\beta_{TC}(TCj-TCi)]}}\right)} \quad (1)$$

Where:
TT = Travel Time (minute)
TC = Travel Cost (rupiah)
β_{TT} = Travel Time Attribute coefficient
β_{TC} = Travel Cost Attribute coefficient
i, j = alternative Mode (moda *i,j*)

2.2 Mode Choice Approach/Maximum Utility Approach

This method attempts to determine the time value of the model, to estimate the choice ratio of a mode of transport. In this method, the comparison of options is assumed to be a function of two variables: operating cost and time cost. The time value is defined as the ratio between the travel time coefficient and the travel cost coefficient.

$$Pi = a_0 + a_1(Ci - Cj) + a_2(Ti - Tj) \quad (2)$$

Where:
Pi = Ratio to choose mode i
C,T = Cost (C) and Time (T)
i, j = mode Alternative
$a_0, a_1, a_2,...$ = Coefficient
The value of travel time obtained from the method is the comparison between the coefficient of travel time and travel cost coefficient.

$$\text{Value of Time} = \frac{B\_TIME}{B\_COST.} = \frac{\beta_{TT}}{\beta_{TC}.} \quad (3)$$

2.3 Income Approach

The time value is usually proportional to per capita income and is a fixed ratio to the income level. But this is a rather bold assumption because there is little or no empirical data to support it [5]. Time value calculation taking into account per capita income (GRDP) can be formulated as follows:

$$Value\ of\ Time = \frac{PDRB/JP}{WKT} \quad (4)$$

Where :
PDRB = Gross Regional Domestic Revenue (per capita/Rp)
JP = Number of population (orang)
WKT = Total Work Time, Annual (hour)

3 Method

The stages of research activity in this study as follows:

1. Identify the travel characteristic of modes that will be operating. In this type of mode of transportation that will be the object of research is a monorail service that will be substitute another type of public transport even private transport.
2. Designing survey forms. To design the survey form, the technique used is the stated preference. By using this technique, the question can be posed to the respondents are changes in both the increase or decrease in the value of similar attributes in both modes and relevance to the modal choice preferences.
3. The data collection process. Survey forms that have been designed and is disseminated to users of the public transport and private transport which makes a start or end their travel from monorail service area corridor. Figure 1 bellow describes the plan of monorail route in Medan.

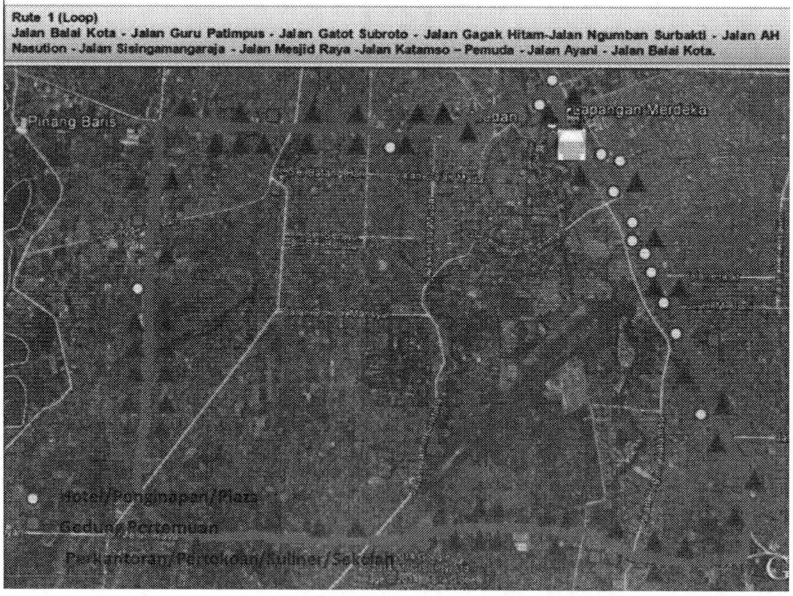

Figure 1. Route Plan to be passed by the monorail of Medan City

4. Mode Choice Modelling. The data have been collected and processed to obtain a modal choice models. The next step is a test of the model, to determine their validity
5. Analysis the Model. Models are declared valid by statistical analysis, and then analyzed the coefficient of travel time and travel cost, for random regret modelling and modeling choice approach (maximum utility approach).

4 Result and Discussion

To analyze the parameters needed to calculate the time value, in this study used Biogeme 2.2 software. (BIerlaire Optimization toolbox for GEv Model Estimation) [1]. For Random Regrete Minimization, output from Biogeme can be seen bellow:

Table 1. Parameters from Biogeme software result.

NO	PARAMETER	*VALUE*
1	Number of observations	1417
2	Number of individuals	1417
3	Null log-likelihood	-982.19
4	Cte log-likelihood	-965.19
5	Init log-likelihood	-982.19
6	Final log-likelihood	-678.04
7	Likelihood ratio test	608.299
8	Rho-square	0.31
9	Adjusted rho-square	0.306
10	B_TIME	-0.218
11	B_COST	-0.343
12	B_WT	-0.203

Table 2. Value of Time with RRM approach

NO	PARAMETER	*VALUE*
1	B_TIME	-0.218
2	B_COST	-0.343
$\sum_{i\neq j}\left(\frac{-\beta_{TT}}{1+\frac{1}{exp[\beta_{TT}(TTj-TTi)]}}\right)$		286.72
$\sum_{i\neq j}\left(\frac{-\beta TC}{1+\frac{1}{exp[\beta_{Tc}(TCj-TCi)]}}\right)$		257.25
Value Of Time		Rp 6.687

For maximization utility using discrete choice model can be seen bellow;

Table 3. Value of Time with Maximum Utility approach

NO	PARAMETER	*VALUE*
1	B_TIME	-0.218
2	B_COST	-0.343
	Value of Time (Rupiah/jam)	Rp 3.813

For the value of Time based on the Income approach, can be seen in Table 4.

Table 4. Value of Time based on Income approach

GRDP based on market price [2]	Rp 138.020.000.000.000
GRDP based on fixed price	Rp 108.222.000.000.000

Number of population	13.326.307
Income per capita	Rp 8.120.897,26
Work Hours/week	40
Work hours annually	2080
Value Of Time	Rp 3.904

From Table 1 to 4 above it can be concluded that there are differences in the results of the calculations of the three approaches, RRM has the greatest value of time compared to the other two approaches. This shows that "regret approach" will give the greatest preference value to "value of time" compared to an assessment based on "maximizing utility."
Value of time for different type of occupation can be shown bellow;

Table 5. Value of Time based on occupation

Occupation	Approach	VOT
Gov. Employees	VOT/HOUR (BASED ON RRM)	7809,4588
	VOT/HOUR (Maximum Utility APPROACH)	4441,55844
Private Employee	VOT/HOUR (BASED ON RRM)	10017,7027
	VOT/HOUR (Maximum Utility APPROACH)	3412,04819
Entrepreneur	VOT/HOUR (BASED ON RRM)	11334,4739
	VOT/HOUR (Maximum Utility APPROACH)	6428,57143
Students	VOT/HOUR (BASED ON RRM)	2623,54672
	VOT/HOUR (Maximum Utility APPROACH)	990,723562
Others	VOT/HOUR (BASED ON RRM)	8750,31801
	VOT/HOUR (Maximum Utility APPROACH)	4923,07692

From Table 5 above it is shown that for each type of work, the RRM approach will provide a higher VOT value than the maximum utility approach. The biggest VOT is the Private Employee, while the smallest is the student.

References

1. M. Bierlaire. Estimation of discrete choice models with Biogeme 1.8. Ecole Polytechnique Federale De Lausane. Transport and Mobility Laboratory. Switzerland. (2009)
2. Central Bureau of Statistics of North Sumatra. Area, Number of Population, and Population Density by Regency. [accesed 15 Agustus 2015] http://sumut.bps.go.id/frontend/linkTabelStatis/view/id/138. (2015)
3. C. Chorus. Random Regret-based discrete choice modeling: A tutorial. Springer, Hieldelberg, Germany. (2012)
4. M. Surbakti, A. F.Sadullah. Comparative Analysis Using RUM and RRM in the Mode Choice Behavior. Proceeding, Malaysian Universities Transport Research Forum Conference. (2013)
5. O. Tamin. Transportation Modelling and Planning: problems example and applications. ITB, Bandung. (2008)

MATEC Web of Conferences **138**, 07006 (2017) DOI: 10.1051/matecconf/201713807006
EACEF 2017

Effect of sedimentation volume on productivity and dwelling time of ships at bom baru port

Putranto Dinar DA[1*], and *Buchari* Erika[2]

[1]Civil Engineering,Faculty of Engineering, University of Sriwijaya, Km. 32,Inderalaya Highway Street, South Sumatera, Indonesia.
[2] Post Graduate Program, Faculty of Engineering, University of Sriwijaya, Bukit Besar, Palembang, Indonesia

Abstract. Many ports in Indonesia who have problems with limited dock land, yard and warehouse and equipment. There is a dock that has good equipment, but Yard Occupancy and Berth Occupancy ratio are high ratio. Boom Baru port in the city of Palembang, in Indonesia, has good hinterland and the number of ship visits is quite high, but can not increase productivity or capacity of the port due to the rapid sedimentation of the port basin and land constraints to accumulation of goods. In this study, the method used to determine the condition of the real problems of sedimentation conducted through analysis of sediment concentration and sediment discharge a total of either scenario sediment from upstream (river) as well as the source of sediments estuary (port basin)[1]. By knowing the pattern of flow and sediment caracteristic process of sedimentation and erosion in the estuary, can be calculated by the method of Yang[2].The calculation and analysis show that the sedimentation on the Musi river is high enough where the sludge reaches 40 cm/month. While the volume of sedimentation in the pools of Boom Baru port calculated using the Yang method, amounting to 13,081.42 m3/year.

1 Introduction

The length of river port services can be grouped into two categories, namely Waiting Time and Dwelling Time. Problems, dwelling time of record in 4 (four) national ports in the case of mal administration examined Ombudsman (2014) as follows: (1) delay is protracted, namely the length of the necessary permits prohibitions and restrictions (Lartas) from the relevant authorities, the lengthy process of publishing Identification Number Customs Office (NIK), the uncertainty of the time service physical examination of the examination process until a response from the Directorate General of Customs and Excise, uncertainty container classification and customs valuation which causes delay in clearance of containers; (2) Deviation procedures, including services at the Port was not optimal 24/7 and quarantine inspection is carried out outside the port area or container that is already out of the customs territory despite still waiting for test results, laboratories but entered the warehouse owner; (3) Any competent examiner's performance including the container on

*Corresponding author: dwianugerah@yahoo.co.id

DOI: 10.1051/matecconf/201713807006

the red line (behandle) and quarantine inspectors were not optimal, and not all human resources (HR) master of regulation; (4) Abuse of authority committed by some individuals, including the issuance of the Memorandum of Correction (NOTUL), the persons who facilitate or complicate expenditure containers; (5) Charges not authorized by the person, including service during the ascending lower (lift on - lift off) container terminal, forklift operator, opening the container in behandle, the withdrawal process container into behandle, physical examination process up to the issuance of SPPB.

Compounding the problem, the waiting time (waiting time) long in port caused by several factors including the weather factor, sedimentation, tides, loading and unloading equipment damage cranes and others. There were 9 (nine) types of waiting time (waiting time) at the Port, namely: (1) Approach Time (AT), is the time it takes ships from anchoring locations to tie the rope on the dock; (2) BerthTime (BT), is a period in the dock since ships first fastening strap (first line) until the ship off her leash (last line), consisting of: (a) Berth Working Time (BWT), namely working hours are available for loading and unloading boats at the dock; (b) Effective Time (ET)/Operating Time (OT), which is the amount of time used for loading and unloading activities at the dock; (c) Idle Time (IT), the time wasted during the ship is in dock due to the effects of weather and loading and unloading equipment damaged; (d) Not Operation Time (NOT), that time does not work planned during the ship at dock (preparation B/M and the rest of work); and (e) Berth Occupancy Ratio (BOR) or the level of use Pier, is comparison between the use of the dock to the time available in a specific time period that is expressed in percentage; (3) Turn Around Time (TRT) is the amount of time for the ship was in port, counted since the ship arrived at the location of anchorage outside the harbor waters, then dry dock and load and unload until the ship departs back. Turn Around Time (TRT) is called the Dwelling Time. Dwelling Time is the waiting time of ships in port or rather the time spent in port ship in order the process of unloading or loading of goods; (4) Postpone Time (PT) is a waiting time caused by administrative procedures in ports are calculated from the anchor to ship finished unloading at the dock. According to UNCTAD (United Nations Conference On Trade and Development), the level of use of the pier (BOR) can not exceed the value (%) that has been set.

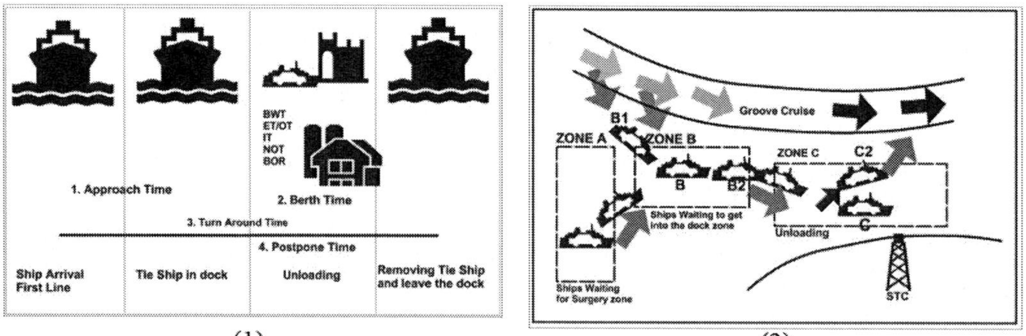

(1) (2)

Fig.1. Schematically wait time ships in the port (1) waiting time at pier; (2) zone ships in port

On this basis, the problem is: (1) How big is the volume and distribution of sedimentation occurred in waters around Boom Baru port basin, causing water levels are the lowest that can be passed vessel to be reduced; (2) How can flow management port waters of rivers and basin in order to obtain the waiting time shorter ships to load and unload.

The purpose of this research is: (1) to analyze the distribution and volume of sedimentation in the waters around the port basin, and the river channel; (2) Managing the

DOI: 10.1051/matecconf/201713807006

flow of cruise ships around the port basin to obtain the shortest time to load and unload due to tidal influence there. It required the making of port planning model in improving productivity and stevedoring services at the port so it can reduce the dwelling time of service multimodal transportation system that is integrated, effective and efficient.

2 Methodology

2.1 Research location

Research carried out at the port of Boom Baru, Palembang on Indonesia with geographical location coordinates between 104° 46' 30"- 104° 47' 0" East and between 2° 58' 48"- 2° 59' 05" Southern hemisphare. Boom Baru port, is river port built since the days before independence, which serves as the loading and unloading of goods on a large scale that goes to the South Sumatra province. Initial Condition Boom Baru port development, earmarked for ships weighing over 10,000 DWT with the port basin depth is 15 m (PT. Pelabuhan Indonesia II).

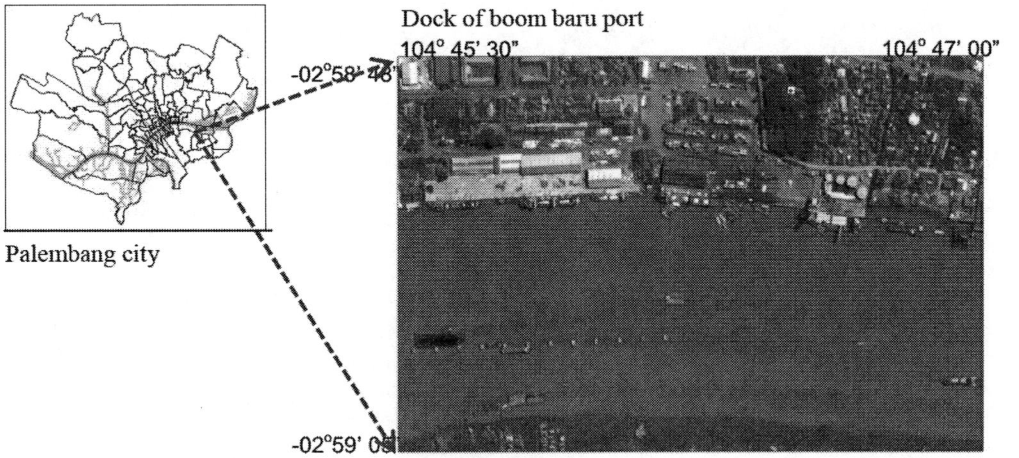

Fig. 2. Research sites of bom baru port, Palembang Cities

2.2 Material and methods

The study was conducted by using the following steps: (1) collecting field data in the form of measurement bathymetry using the tool Echosounder conducted transects along and width of port basin (500 x 500) m, with distance measurements every 25 mx 25 m, sampling sediment drift and sediment riverbed at each cross-section is done once and then conducted laboratory tests to determine the grain size, flow measurement using current meter on each cross-section and low tides conducted over 15 days; (2) Analysis of hydro-oceanography, sediment characteristics analysis and flow modeling using DIVAST-2D models; (3) Calibration models to determine the parameters of empirical field, which is obtained from laboratory analysis of the results of sampling field.

Divast 2-D model of a hydrodynamic modeling (waves and currents), sediment transport, temperature and pollutants. DIVAST 2-D model is the average depth of the horizontally oriented in the x and y directions. DIVAST models also use finite difference methods that are solved by using the Fortran language[3].

The equation used is as follows,

a) The equation of hydrodynamics wave number vector equation:

$$\frac{\partial K_i}{\partial t}+\left[C_g\frac{K_j}{k}+U_j\right]\frac{\partial K_i}{\partial x_j}+\frac{\sigma G}{2h}\frac{\partial h}{\partial x_i}+\frac{C_g}{2kA}\left[\frac{1}{A}\frac{\partial A}{\partial x_i}\frac{\partial^2 A}{\partial x_j^2}-\frac{\partial^3 A}{\partial x_i \partial x_j^2}\right]+K_j\frac{\partial U_j}{\partial x_i}=0 \quad (1)$$

The amplitude of the wave equation :

$$\frac{\partial A}{\partial t}+\frac{1}{2A}\frac{\partial}{\partial x_i}\left\{A^2\left[\frac{C_g}{k}K_i+U_i\right]\right\}+\frac{S_{ij}}{\rho g A}\frac{\partial U_j}{\partial x_i}+C^a A^2=0 \quad (2)$$

Continuity equation :

$$\frac{\partial \varsigma}{\partial t}+\frac{\partial (HU)}{\partial x}+\frac{\partial (HV)}{\partial y}=0 \quad (3)$$

$$H=h+\varsigma \quad (4)$$

Momentum equation :

$$\frac{\partial (HU)}{\partial t}+\frac{\partial (\beta HU)}{\partial x}+\frac{\partial (\beta HUV)}{\partial y}=fHV-gH\frac{\partial \varsigma}{\partial x}-\frac{gH^2}{2\rho\partial}\frac{\partial \rho}{\partial x}-\frac{\tau_{wx}-\tau_{bx}}{\rho}+$$

$$\frac{}{\upsilon_h H}\left[\frac{\partial^2 U}{\partial x^2}+\frac{\partial^2 u}{\partial y^2}\right]+\frac{1}{\rho}\left[\frac{\partial S_{xx}}{\partial x}+\frac{\partial S_{xy}}{\partial y}\right] \quad (5)$$

$$\frac{\partial (HV)}{\partial t}+\frac{\partial (\beta HUV)}{\partial x}+\frac{\partial (\beta HV^2)}{\partial y}=-fHU-gH\frac{\partial \varsigma}{\partial y}-\frac{gH^2}{2\rho}\frac{\partial \rho}{\partial y}-\frac{\tau_{wy}-\tau_{by}}{\rho}+$$

$$\frac{}{\upsilon_h H}\left[\frac{\partial^2 V}{\partial x^2}+\frac{\partial^2 V}{\partial y^2}\right]+\frac{1}{\rho}\left[\frac{\partial S_{xy}}{\partial x}+\frac{\partial S_{yy}}{\partial y}\right] \quad (6)$$

Where : ζ = water level; U, V = the average speed of the depth of x and y; H = the total depth of water; h = the depth of still water, A = wave amplitude, Cg = wave velocity in groups, β = momentum coefficient; τ_{wx},, τ_{wy} = shear stress components on the surface due to wind in the direction of x and y; τ_{bx}, τ_{by} = shear stress components at the base in the direction of x and y; υ = eddy viscosity; S_{xx}, S_{xy}, S_{yy} = stress caused by radiation wave field.

b) Sediment transport equations and basic elevation changes

This model explains patterns of sediment movement, who moderated the equation in the form of convection-diffusion equation that already included the erosion and deposition rate. The equation is written,

$$\frac{\partial (HC)}{\partial t}+\frac{\partial (\alpha HUC)}{\partial x}+\frac{\partial (\alpha HVC)}{\partial y}=\frac{\partial}{\partial x}(H\varepsilon\frac{\partial C}{\partial y})+\frac{\partial}{\partial y}(H\varepsilon\frac{\partial C}{\partial y})+HS_c \quad (7)$$

By giving the effect of concentration as in Van Rijn[4], the formulation of basic topographical changes based on the completion of the sediment continuity equation can be written,

 DOI: 10.1051/matecconf/201713807006

$$\frac{\partial z_b}{\partial t}+\frac{1}{\rho_s(1-n_p)}\left[\frac{\partial(DS_t)}{\partial t}+\frac{\partial q_{tx}}{\partial x}+\frac{\partial q_{ty}}{\partial y}\right]=0 \quad (8)$$

where :

$$q_{tx}=q_{s,x}+q_{b,x}$$

$$q_{ty}=q_{s,y}+q_{b,y}$$

where :

Z_b	= bottom elevation, n_p = porrossities and
q_{tx} dan q_{ty}	= the rate of total sediment transport in the x and y directions.
S_c	= source terms in connection with erosion and deposition.
	= D - E

c) Temperature equation

$$\frac{\partial(HT)}{\partial t}+\frac{\partial(HUT)}{\partial x}+\frac{\partial(HVT)}{\partial y}=\frac{\partial}{\partial x}(H\varepsilon\frac{\partial T}{\partial y})+\frac{\partial}{\partial y}(H\varepsilon\frac{\partial T}{\partial y})+HT_c \quad (9)$$

Where : H = total depth of flow, U, V = the average speed of flow depht, C = Concentration of sediment, D = deposition, E = erosion, T = average depth of water temperature, ε = depth- averaged eddy diffusity, T_c = source termsin respect of external influences on temperature feedback.

According to Yang[5], in the calculation of sediment transport need to be seen sediment load tpe to be counted. In this case the need to classify the type of sediment, namely: (1) Bed Load, particles are transported by means of rolling, sliding and jumping; (2) Suspended Load, is transported by floating particles. Yang[5]discusses the basic assumptions used in the equations derived sediment transport. It was concluded that the rate of sediment transport can be determined from the discharge of water, the flow rate of the average - average, energy slope or shear stress. The equation used :

$$Log\ Ct = 5{,}435 - \ {,}268\ log\frac{\omega d}{v} - \ {,}457 log\frac{U*}{\omega} + (1{,}799 - \ {,}409\ log\frac{\omega d}{v} -$$

$${,}314\ log\frac{U*}{\omega}) log\frac{vs}{\omega} - \frac{vcrS}{\omega}) \quad (10)$$

$$U* = (g\times D\times S)^{1/2} \quad (11)$$

Where shear rate (U*) is Reynolds shear speed figures are :

$$Re=\frac{U*\times d_{50}}{\upsilon} \quad (12)$$

for $1.2<\frac{U*\times d}{\upsilon}<70$, then $\frac{2{,}5}{\log(\frac{U*xd_{50}}{\upsilon})-0{,}06}+0{,}06$ (13)

$$\text{for} \quad 70 < \frac{U^* \times d}{\upsilon}, \text{ then} \qquad \frac{V_{cr}}{\omega} = 2.05 \tag{14}$$

where :

Ct= Total of sediment concentration	V	= speed of flow (ft/s)
Ω= speed of sediment falls (ft/s)	S	= slope drainage (%)
D= middle diameter of particle (ft)	Vcr	= critical speed
υ= Viscosity Kinematic (ft^2/s)	U*	= sheare rate (ft/s)

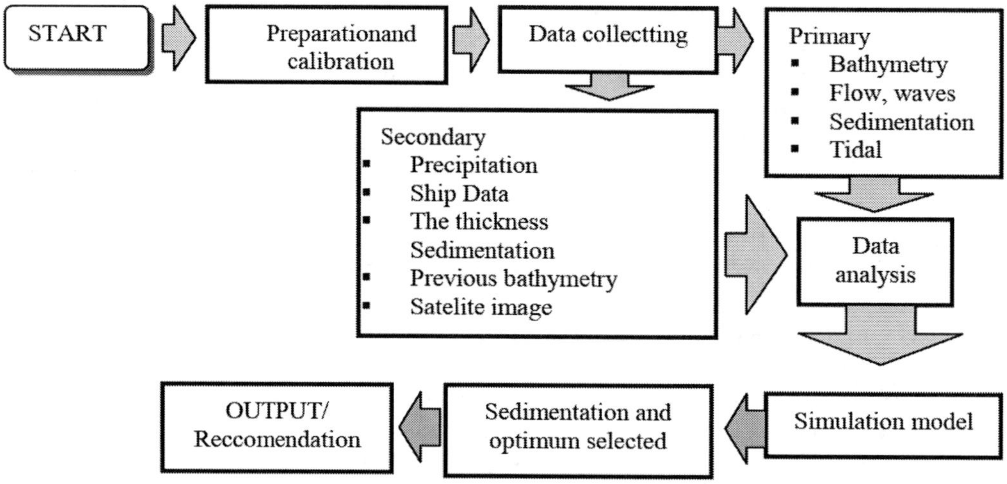

Fig. 3. Interaction diagram process assessment model sedimentation at port basin

3 Results and discussion

3.1 Depht, tide and currents

Conditions Musi river around the port of Boom Baru, generally shallow and very gentle and the general condition of the area around the port is under the influence of sedimentation due to the presence of several rivers that carry sediments and slopes of bottom waters. The depth of the waters around the port of Boom Baru measurement results show that the depth range (5.5 to 15.0) m, with an average depth of 9.91 m (spreads horizontally depth can be seen in Figure 4. The most shallow waters found around the docks (5.2 m), while the deepest waters found in the middle of the river flow (15.6 m).

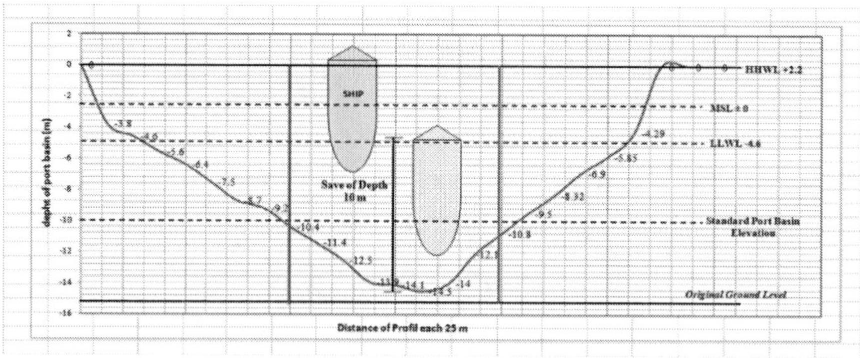

Fig.4. Depth and cross section of port basin

DOI: 10.1051/matecconf/201713807006

Port basin cross-sectional shape, of conditions such as bathymetry data above, can be described , that during low tide so the water level only reached (4-5) m from the mean sea levels. While during high tide, the water level can reach a height (15-16) m from the mean sea levels. Based on observations of tides, suggesting that this type of tidal influence is of a single type, meaning that in one day occurred one high tide and one low tide. The highest tide happened at about 20:00 to 5:00 pm, average as high as 11.87 to 15.56 m. While the lowest tide occurred at 7:00 to 14:00 pm, approximately 4.6 - 6.6 m.

Based on the above data, it's the height of the water surface in accordance traversed ships would load and unload at the port basin to adjust the distance of the hull weighing 10,000 DWT, the time available is only about 10 hours. This condition causes the adjustment time of arrival of the vessel into dry dock to be disrupted, because the waiting time of the tide.

Based on data from BMG from 2006-2016 year, average wind speed during the month of October is 3.27 Km/h = 0.907 m/s with an average direction of wind movement towards the southeast. Current measurement port basin waters around Boom Baru port obtained current velocity ranges from 0.02 to 0.306 m/s, with an average speed of flow during the observation is 0.159 m/s.

3.2 Total sediment load

To know the total load of sediment transport, then the calculation is based on parameters obtained in the port basin through sampling and testing of large grain sediment load carried out in laboratories[6]. Properties Index sediment testing includes testing sieve analysis which aims to get the size of a grain escaped percent 50% (D_{50}), and a specific gravity of sediment contained in the port basin[7].Based on the results of data analysis sieve (sieve analysis) of sediment samples in Boom Baru port basin, can be illustrated in the graph of particle diameter (mm) which passes 50% (Fig. 5), it is known that the grain size of the sediment downstream portion D_{50} is 0.04 mm and the upstream side was 0.06 mm. Thus, the average grain size of the sediment D_{50} in Boom Baru port is 0.05 mm. The result of the calculation of density sediment testing for upstream and downstream obtained an average value of 2.53 grams.

Table 1. Volume of sedimentation boom baru port basin (2005-2014)

Year	Volume of Sedimentation (Cm3/year)
2005	34,609.13
2006	98,430.13
2007	62,104.01
2009	56,739.94
2010	62,505.72
2012	75,544.92
2013	86,512.71

Sedimentation, can cause several problems including the activity disrupted navigation/maneuvering the vessel either at[8]: (1) Shipping-In the arrival of a vessel (dock to dock); (2) The activities of loading and unloading of goods; and (3) Shipping-Out is when the ship set sail again (the ship's departure). The depth of each point varies, from one another. The spread of sediment and depth of an uneven harbor that, due to changes in current direction due to the influence of waves and turbulence effects caused by the rotation of the blades (Propeller) Ships in the harbor bottom of the port.

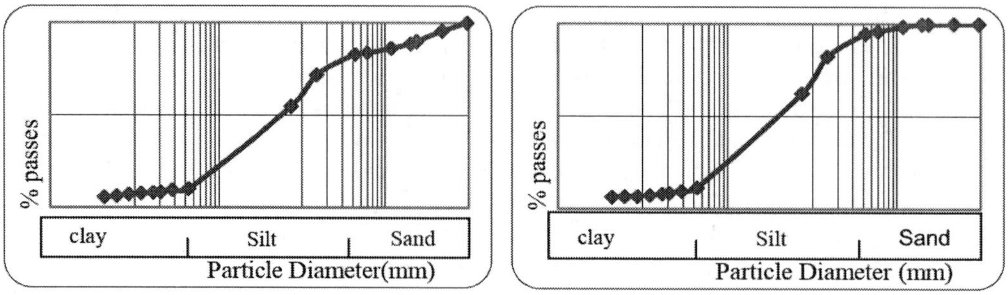

(b) upstream (a) downstream

Fig. 5. Graph distillate passes 50% (D_{50}) sediment samples

Figure 6, shows that the depth of the Musi river different between one area to another area with an average depth of 10.88 m. Based on data from Table 1, the conditions shipping channel depth due to sedimentation in the Boom Baru port basin, especially in front of the pier to change every year, which caused fluctuations in the depth of the water level up and down due to sedimentation.The condition of the water surface elevation of the Musi River for five (5) years from 2008 to 2014 and especially in 2014, the average distance between the bottom of the boat (keel) the maximum draft of 7.24 m when the surface of the bed river less than 10 meters to weight of 15,000 DWT ships. Based on Figure 6, Zone 1-3 at the time of the ship arrived (Shipping-In), Musi river water level at that post or position of the highest water level (HHWL) with a minimum requirement of 7.24 meter draft vessels, the depth (h) port basin at that position is 12.2 meters.

		1	2	3	4	5	6	7	8	9	10	11	12	13	14	15	16	17	18	19	20	21	22
A	1	0	0	0	0	0	0	0	0	0	0	0	0	0	0	0	0	0	0	0	0	0	0
B	2	-3.8	-4	-4.1	-3.85	-3.9	-4.1	-5.1	-5.4	-5.38	-5.43	-5.45	-5.41	-5.42	-5.36	-5.4	-4.95	-3.8	-3.65	-3.7	-3.65	-3.6	-3.55
C	3	-4,6	-4.5	-4.53	-4.6	-4.64	-4.75	-5.23	-5.34	-5.41	-5.46	-5.4	-9.5	-5.44	-5.4	-5.3	-5.12	-4.9	-4.83	-4.8	-4.7	-4.6	-4.64
D	4	-5,56	-5.55	-5.55	-5.53	-5.59	-5.6	-4.32	-5.2	-5.35	-5.37	-4.95	-4.2	-5.82	-5.3	-5.29	-5.03	-5.53	-5.57	-5.56	-5.55	-5.54	-5.55
E	5	-6.4	-6.5	-6.6	-6.65	-6.7	-6.3	-4.75	-4.9	-4.91	-4.66	-4.85	-4.78	-4.7	[illegible]	-4.83	-4.75	-6.95	-5.9	-6.35	-5.9	-7	-6.3
F	6	-7.5	-7.62	-7.7	-7.75	-7.7	-7.8	-6.9	-5.3	-6.5	-6.33	-6.74	-6.63	-6.9	-6.35	6.74	-6.9	-7.9	-7.9	-7.9	-7.6	-7.7	-7.6
G	7	-8.7	-8.75	-8.8	-8.79	-8.8	[illegible]	-8.57	-8.88	[illegible]	-8.33	-9.1	-9.2	-9.21	-9.21	[illegible]	-9.1	-9	-8.9	-8.95	-8.66	-8.76	-8.66
H	8	-9.2	-9.3	-9.35	-9.4	-9.35	-9.27	-9.29	-9.3	-9.4	-9.5	-9.6	-9.72	-9.63	-9.57	-9.6	-9.7	-9.69	-9.6	-9.5	-9.4	-9.32	-9.21
I	9	-10.4	-10.3	-10.5	-10.5	-10.4	-10.5	-10.5	-10.6	-10.6	-10.5	-10.5	-10.4	-10.2	-10.6	-10.4	-10.4	-10.2	-10.1	-10.1	-10.2	-9.85	-9.9
J	10	-11.4	-11.6	-11.5	-11.6	-11.7	-11.7	-11.7	-11.7	-11.7	-11.9	-11.4	-11.4	-11.3	-11.9	-11.3	-11.7	-11.1	-11.7	-11.5	-11.5	-11.5	-11.5
K	11	-12.5	-12.4	-12.5	-12.5	-12.5	-12.7	-12.6	-12.7	-12.6	-12.7	-12.6	-12.8	-12.7	-12.8	-12.7	-12.6	-12.5	-12.4	-12.4	-12.2	-12.5	-12.2
L	12	-13.9	-13.6	-13.7	-10.2	-13.6	-13.6	-13.5	-13.4	-13.4	-13.4	-13.3	-13.4	-13.5	-13.4	-13.2	-13.3	-13.4	-13.3	-13.2	-13.1	-13.2	-13
M	13	-14.2	-14.4	-14.2	-13.8	-13.7	-14	-14.1	-14.2	-14.1	-14	-14.1	-13.8	-13.7	-12.7	-13.6	-13.5	-13.5	-13.4	-12.9	-13.4	-13.3	-13.3
N	14	-14.5	-14.2	-13.8	-14	-13.9	-13.8	-13.6	-14	-14.1	-14.1	-14.2	-13.9	-13.8	-13.7	-13.9	-13.7	[illegible]	-13.5	-13.61	-13.5	-13.8	-13.3
O	15	-13.9	-13.1	-13.1	-13.2	-13.2	-13	-13.2	-13.3	-18.4	-13.3	-13	-12.9	-12.8	-12.9	-12.7	-12.8	-12.8	-12.5	-12.4	-12.6	-11.39	-12.3
P	16	-12.1	-12	-12.1	-12.4	-12.5	[illegible]	-12.6	-12.7	[illegible]	-12.1	-12	[illegible]	-11.6	-11.5	-12	-12	[illegible]	-11.3	-11.6	-10.9	-11.7	-11.5
Q	17	-10.8	-11.2	-11.2	-11.3	-11.2	-11.1	-11.4	-11.5	-11.4	-11.5	-11.3	-11	-10.8	-10.9	-11	-10.8	-10.7	-10.6	-10.5	-10.4	-10.3	-10
R	18	-9.5	-9.3	-10.1	-9.9	-9.3	-9.7	-10	-10.3	-10.2	-10.1	-9.3	-9.9	-9.3	-9.35	-9.7	-9.75	-9.5	-9.5	-9.4	-9.45	-9.3	-9.25
S	19	-8.32	-8.5	-8.6	-8.64	-8.7	-8.8	-8.8	-8.82	-8.7	-8.8	-8.8	-8.7	-8.7	-8.69	-8.58	-8.62	-8.6	-8.5	-8.4	-8.36	-8.2	-8.25
T	20	-6.9	-7.1	-7.2	-7.3	-7.2	-7.3	-7.5	-7.6	-7.5	-7.5	-7.63	-7.6	-7.5	-7.4	-7.3	-7.32	-7.25	-7.4	-7.44	-7.5	-7.4	-7.35
U	21	-5.85	-5.33	-5.93	-6.12	-6.2	-6.22	-6.25	-6.15	-6.2	-6.25	-6.23	-6.2	-6.2	-6.1	-5.36	-5.32	-5.3	-5.9	-5.8	-5.9	-5.35	-5.6
V	22	-4.29	-4.28	-4.34	-4.35	-4.37	-4.42	-4.43	-4.42	-4.45	-4.5	-4.44	-4.42	-4.39	-4.38	-4.39	-4.35	-4.25	-4.23	-4.25	-4.2	-4.25	-4.24
W	23	0	0	0	0	0	0	0	0	-3.8	-3.7	-3.6	-3.5	-3.7	-3.75	-3.6	-3.5	-3.4	-3.45	-3.5	-3.4	-3.42	3.25
X	24	0	0	0	0	0	0	0	0	0	0	0	0	0	-2.6	-2.5	-2.65	-2.7	-2.3	2.65	-2.6	-2.4	-2.2
Y	25	0	0	0	0	0	0	0	0	0	0	0	0	0	0	0	0	0	0	0	0	0	0

Fig. 6. Ship maneuver in boom baru port basin

In that position the ship is still safe depth, they are ideal for navigation or maneuvering the ship, because it is still in the position of the middle course of the Musi river. When the ship began to move closer towards the dock, the ship's position is at a depth (h) of 8.66 meters (2014). Conditions such as these, would make the vessel must carefully maneuver, because the keel (the base) boat almost touch the bottom of the pool and could cause the ship run aground or sunk. In Figure 6 could also note that when the arrival of the vessel

(Shipping-In), Musi river water level is at its lowest water level position (LLWL) and draft of the ship in a state of at least 7.24 meters. Seeing the condition of the depth (h) port basin (+ 7 meters) and a maximum draft of vessels (7.24 meters) is certainly the vessel would not be able to sail safely and smoothly.

Activities maneuvering boat at the dock are also affected by the tidal conditions, which usually ships usually go or load and unload when the water level started to put the Musi river. To see the relationship between conditions of high surface activity with the activities of the Musi river ship maneuvering in port navigation channel in the pool can be explained as follows, the time of arrival at the dock ships usually morning from 06.00 - 12.00, where the water depth reached + 6 meters. When departure from the dock usually late afternoon towards evening ranged between the hours 18:00 to 00:00 pm, when the depth of the Musi river ranged from 6 meters up to 7 meters and at that time made directly loading and unloading of goods. But sometimes there are conducted in the morning, around 06.00 pm when the water level of the Musi river at 6 meters - 7 meters and was completed during the day, at around 12:00 pm, and some even finished the afternoon at about 18:00 pm the water level Musi river reaches 6-7 meters. Based on observations of the activities of loading and unloading of ships at dock of Boom Baru port, the average time it takes each vessel at Terminal for the Interest Alone (TUKS), obtained an average as follows: (1) Turn Around Time (TRT) takes 123 hours and 21minutes , or about 5 days; (2) While the effective time (ET) and the Operating Time (OT), takes 53 hours and 25 minutes, or about two and a half days; and (3) Bert Time (BT), the time it takes the ship from mooring rope (first line) until the loose rope (last line) takes 117 hours and 38 minutes, or about 5 days. Based on the three data, the level of efficiency when unloading the ship was approximately 77.55%.

The volume of sediment that formed in Boom Baru port basin is + 13081.42 m3/year. Based on available data, the depth of the port at low tide basin about 5-8 meters and + 10 meters at high tide, so that when a ship with the highest draft or DWT (dead weight tons) ships up the process of navigation/maneuvering in the port basin have limitations , To cope with high sedimentation in the port basin and reduce duwelling time, because waiting for the tide, it is necessary to once a year dredging the port basin with a thickness of approximately 3m. The waiting time between the tide with the port basin depth in accordance with the needs of ship maneuvering average is 8 hours.

4 Conclusion

1.The rate of sedimentation on the Musi river about + 13,081.42 m3/year
2. Average Bert Time for each vessel is 117 hours 38 minutes
3. Time of arrival and departure is always carried out every morning until noon, when the water level around 7:24 am

References

1. L. Pinto, C.Munoz, T. Nalpaset all. Role of sedimentation during basin inversion in analogue modelling., Journal of Structural Geology Vol. **32,** 554-565 (2010)

2. Yang,Chih Ted, *Sediment Transport, Theory and Practice*, Mc.Graw Hill, London (1996)

3. Z. Easton, D. Fuka, E. White et all, A multi basin SWAT model analysis of runoff and sedimentation in the Blue Nile, Ethiopia, Journal Hydrology and Earth System Sciences, Vol. **14,** 1827-1841 (2010)

4. S. Yang, Formula for Sediment Transport in Rivers, Estuaries, and Coastal Waters, Journal of Hydraulic Engineering Vol. **131,** 968-979 (2005)

5. Yang Bai, JG. Duan, Simulating unsteady flow and sediment transport in vegetated channel network, Journal of Hydrology Vol. **515,** 90-102 (2014)

6. C. Ribers, C. Kergaravat, C. Bonn et all, Fluvial sedimentation in a salt-controlled mini-basin: Stratal patterns and facies assemblages, Journal Sedimentology (2015)

7. M.Lazzari , D. Gioia, M. Piccarreta et all, Sediment yield and erosion rate estimation in the mountain catchments of the Camastra artificial reservoir (Southern Italy): A comparison between different empirical methods, Journal Catena Vol. **127,** 323-339 (2015)

8. G. Figueredo De Olieviera, P. Cariou, The impact of competition on container port (in)efficiency, Transportation Research Part A: Policy and Practice (2015)

DOI: 10.1051/matecconf/201713807007

Strategy toward sustainable local road network infrastructure

Fajar Sri Handayani[1,*], *Ary* Setyawan[1], *F. Pungky* Pramesti[1] , and *M. Agung* Wibowo[2]

[1]Civil Engineering Department, Universitas Sebelas Maret, Surakarta, Indonesia
[2]Civil Engineering Department, Diponegoro University, Semarang, Indonesia

Abstract. With massive explosion of population, the urban inhabitants will rise up to 64% in 2050. This condition increases the pressure on the existing urban road infrastructure and requires substantial maintenance and often expansion of the inter/rural-urban as well as local road as mobility of people and goods upsurge extensively. The construction of these new network and maintenance of existing road infrastructures consumes huge natural resources which further might increase volume of waste and greenhouse gas emission. Currently, approximately 28% provincial road and 41% local (urban) road network in Indonesia are in unsatisfactory condition. There is obvious demand to improve regional road with minimum ecological impact despite limited amount of resources and budget. In achieving sustainable road network, local authorities require a strategy in tackling this complex, large-scale and interconnected problems. This study aims to formulate relevant strategy for local authorities – urban or inter rural-urban level within a province – in operation, maintenance and development of road network by applying strategic management principles. Rational decision model was used in strategy formulation in three consecutive steps: input stage, Matching stage, and Decision Stage. Rational decision model was also used in strategy establishment taking into account the Environmental Minister regulation and some important issues in the Department of public works employing sustainability concept. The strategy proposed is believed to be ecologically sound (environmentally friendly) while improving local road network quality.

1 Introduction

Civil infrastructure, especially road way, is a vital and essential for nation-development and economic growth. With massive explosion of population, the urban inhabitants will rise up to 64% in 2050. This condition puts the existing urban road infrastructure under pressure and requires substantial improvement of inter-urban as well as rural-urban road as the extensive mobility of people and goods upsurge.

The Directorate of Transportation of the Ministry of National Development Planning (BAPPENAS) reports that 28% of provincial road network and 41% of local road network are in unsatisfactory conditions. Further, the short design service life (10 years) even

* Corresponding author: fajarhani.uns@gmail.com

increased annualized life cycle coat. At the same time the road quality became deteriorated prematurely during its service life.

At the national level, however, the current Indonesian national road condition are quite well with 94% of it are in satisfactory.

Still, these are not the case for neither provincial nor local road network. Based on the 2014 transportation statistics, 171.295 km of 417.793 km of local road networks are in unsatisfactory conditions. It consumes a number of natural resources, increase waste volume, and require a massive human resources [1].

Therefore, a sound and sustainable planning, design, and construction approach is needed in road infrastructure development [2]. The sustainable concept is the crucial issue for the whole development activities, and even there is a growing demand to measure its implementation [3-6]. Applying sustainable construction via efficient contruction methods is appropriate strategy for project performance improvement [7]. In the strategic planning of the Ministry of Public Work and Housing, the sustainable development principle should be implemented in every building and infrastructure constructions. This principle harmonizes infrastructure and building constructions regarding environmental, climate, natural resources, economic, social and cultural aspects. Hence, such principles should be put into action in every building and infrastructure constructions including local road development.

2 Formulation of strategy

In general, the strategy is a method or plan chosen to bring about a desired future, such as achievement of a goal or solution to a problem [8]. Policy strategy is what government wants to change — its agenda, and the ways in which the agency will move to help it achieves this agenda. Strategy seems a good investment, provided it is done openly, and within acknowledged structures of accountability [9]. Strategic planning frameworks provide direction and justification for the flow of regularity and investment activity, which leads to changes in the location and form of development [10].

Rational Decision Model procedure as one of the comprehensive methods in formulating strategy has been suggested by David [11]. There are three stages to develop the procedure including input, matching, and decision. Each stage is shown in matrices which is consistent with the size and type of the organisation. Therefore, decision makers will be able to have an improvement in identifying, evaluating, and selecting the appropriate strategy.

Each of those three stages implies some necessary points to underline as stated in the following details. First, at the input stage, there are two kinds of matrix as Internal Factor Evaluation (IFE) and External Factor Evaluation (EFE) matrix. The IFE matrix is applied to acknowledge the internal factors of a company which are associated with the strengths and weaknesses that are considered necessary. Meanwhile, the EFE matrix is applied to evaluate the external factors of the company. We collect external data in analysing matters which relates to economic, social, cultural, demographic, environmental, political, governmental, law-related and technological issues and competition in markets where the company is located, among other relevant external data. Similarly, at the matching stage, we develop Strength Weakness Opportunities Threats (SWOT) matrix and Internal-External matrix (IE). The first step is to complete the SWOT items in the existing cells and then to develop alternative strategies with respect to the adjustment of internal to external factors.

It is imperative use alternative strategies logically consistent with various internal aspects of the company, according to the external trends, which are realistic and practicable (feasible). IE matrix is used to position Strategic Business Unit (SBU) of the company into a matrix which consists of nine cells, as shown in Figure 1. The IE matrix comprises two dimensions (i.e., a total score of the IFE matrix on the X-axis and total score of the EFE matrix on Y-axis). Therefore, by looking at the company's position in the nine of these cells, a conclusion can be drawn to formulate the strategy as follows: SBU which is located on cell I, II, or IV can be described as Grow and Build. A suitable strategy would be applied as an intensive strategy, such as market penetration, market development, and product development. We can select another strategy, such as integration strategy, or backward, forward and horizontal integration. SBU which is located in cell III, V or VII can be described as Hold or Maintain. A suitable strategy is market perpetration or product development. For SBU which is located on cell VI, VIII or IX, the right strategy to choose would be Harvest or Divesture.

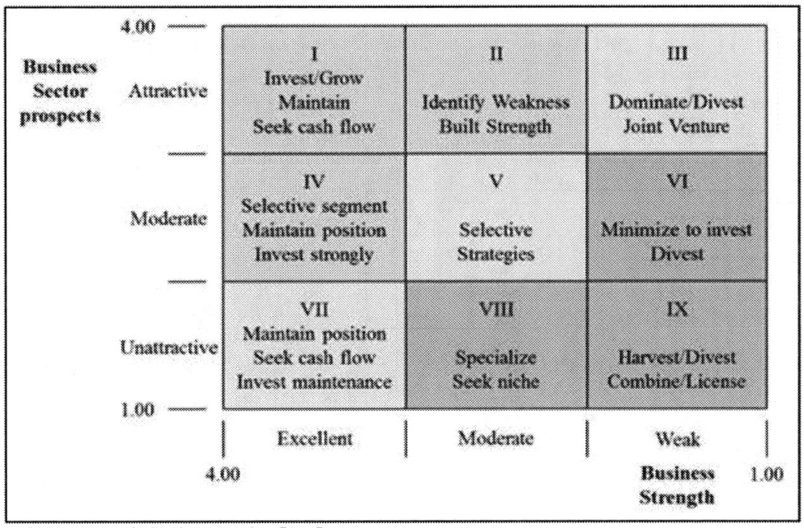

Fig. 1. Internal-External (IE) matrix [10]

The company which is able to control business in cell I will be the most successful company. Finally, at the decision stage, of the several alternatives available, Quantitative Strategic Planning Matrix (QSPM) is applied to select the best strategy. This technique provides an assessment of which strategy is the most appropriate. QSPM applies a matrix input from an external matrix, internal matrix, SWOT matrix and IE matrix. The highest Total Attractiveness Score (TAS) determines the strategy the contractors should apply.

In this study aims to formulate relevant strategy for local authorities-urban or inter rural-urban level within a province – in operation, maintenance and expansion of network by applying strategic management principles. Data were obtained through questionnaires and interviews. The participants of the research were road management, road planner, and road contractor in Central Java, Indonesia.

3 Result and discussion

3.1 SWOT factors

Prior to formulating a strategy toward sustainable local road network infrastructure, it is necessary to examine their current situation, condition, and SWOT (strength, weakness, opportunity and threats). This was carried out through interviews and questionnaires representative respondents. The following section describes the result of the survey and the process of formulating strategies. Table 1 recapitulates the result of SWOT factors inventory.

Table 1. SWOT factors

Strength	Weakness
1. Road has a significant role in the spatial structure and formation. 2. Road is an important infrastructure in advancing the economic growth. 3. An alternative funding concept has continuously developed to overcome the lack of adequate funding. 4. The human resources quantity of Directorate General of Highways is relatively large. 5. Government's commitment to build and to develop road	1. The condition of local road network quality is relatively low. 2. The sustainability aspects in the life cycle of local road design, construction, operational, and maintenance is not put into consideration yet 3. The road prescribed design life is still short-term oriented, which is approximately 7-10 years. 4. The management of environmentally friendly road construction is not yet optimal. 5. The low quality of human resources of local road management.
Opportunity	**Threats**
1. The development in the road innovation and technology concerns the aspects of sustainability, safety, and quality improvement. 2. Provincial Road Improvement and Maintenance (PRIM) funding schemes for local road. 3. The road preservation funding scheme is implemented in accordance with Act Number 22 Year 2009 on Traffic and Road Transportation. 4. The existence of international cooperation in road construction such as IMG-GT, BIMP-EAGA, Asian Highway, and ASEAN Highway. 5. There is an existence of bureaucracy reform planned by government in embodying the good governance.	1. The population growth followed by the massive growth of motor vehicle. 2. The geologic geographic condition (volcanoes, faults, monsoon, etc) poses threats of natural disasters and high rainfall. 3. The delays in the arrangement of regulatory framework needed for innovative funding and road preservation. 4. The institutional unit related to local road development is not yet integrated in supporting the local target. 5. Government (Directorate General of Highways) has not yet owned an accurate and accountable database of local road as a basis in the appraisal for Local Government proposed program.

3.2 SWOT matrix

SWOT factor analyses were then followed by developing of SWOT matrix. After filling SWOT items in the existing cells, the process of alternative strategy development took into account the appropriateness of internal factors (strength and weakness) and external factors (opportunity and threat). The S-O analysis was conducted in this part. This was to analyse how the strengths can be used to benefit the advantage provided by the opportunities. While W-O analysis was conducted to how the weakness can be coped with the existing opportunities. The S-T analysis was performed to study how the strengths can be used to overcome or minimalize the existing threats. Finally, the W-T analysis was carried out to analyse how the weakness can be solved to minimalize the threats. The alternative strategy

DOI: 10.1051/matecconf/201713807007

produced should be realistic and implementable. The result of analysis is presented in Table 2.

Table 2. SWOT matrix

	STRENGTH	**WEAKNESS**
	S1. Government's commitment to build and to develop road S2. An alternative funding concept has continuously developed to overcome the lack of adequate funding S3. Road is an important infrastructure in advancing the economic growth. S4. The human resources quantity of Directorate General of Highways is relatively large. S5. Road has a significant role in the spatial structure and formation	W1. The condition of local road network quality is relatively low. W2. The sustainability aspects in the life cycle of local road design, construction, operational, and maintenance is not put into consideration yet W3. The road prescribed design life is still short-term oriented, which is approximately 7-10 years. W4. The management of environmentally friendly road construction is not yet optimal. W5. The low quality of human resources of local road management.
OPPORTUNITY O1. The development in the road innovation and technology concerns the aspects of sustainability, safety, and quality improvement. O2. Provincial Road Improvement and Maintenance (PRIM) funding schemes towards local road. O3. The road preservation funding scheme is implemented in accordance with Act Number 22 Year 2009 on Traffic and Road Transportation. O4. The existence of international cooperation in road construction such as IMG-GT, BIMP-EAGA, Asian Highway, and ASEAN Highway. O5. There is an existence of bureaucracy reform planned by government in embodying the good governance.	**S-O Strategy** 1. Building a stable transportation connectivity concerning aspects of sustainability, safety, and quality improvement in order to support and encourage the local economic development. (S1,S3,O1) 2. Establishing international cooperation in the road construction in order to improve the quality of road. (S1,S2, O4) 3. Improving the human resources competence and conducting capacity building to embody good governance.(S4,O5) 4. Maximizing the PRIM funding schemes to build and to develop local road. (S2, O2)	**W-O Strategy** 1. Improving the quality of local road network by applying PRIM funding schemes and preservation funding schemes. (W1,O2,O3) 2. Implementing performance based pavement design taking into account sustainability, safety, and quality. (W3,O1) 3. Implementing environmentally friendly roads management by using innovative and sustainable technology. (W4,O1)
THREATS T1. The population growth followed by the massive growth of motor vehicle.	**S-T Strategy** 1. Developing concept and regulatory framework towards innovative funding	**W-T Strategy** 1. Improving the quality of local road network supported by related

T2. The geologic geographic condition (volcanoes, faults, monsoon, etc.) poses threats of natural disasters and high rainfall. T3. The delays in the arrangement of regulatory framework needed for innovative funding and road preservation. T4. The institutional unit related to local road development is not yet integrated in supporting the local target. T5. Government (Directorate General of Highways) has not yet owned an accurate and accountable database of local road as a measurement in the appraisal of Local Government proposed program.	and road preservation. (S2, T3) 2. Optimizing human resources quantity in the Directorate General of Highways in strenghtening accurate and accountable local road database government's commitment in constructing and developing roads. (S4,T5) 3. Building road network connectivity supported by the support of government's commitment and institutional unit of road development in order to attain local development target and to advance the economic growth. (S1,T4)	institutional unit in order to attain the local development target. (W1,T4) 2. Managing environmentally friendly road by integrating sustainability in the whole life cycle and local road maintenance aiming at minimizing from natural disaster and high rainfall. (W3,T2) 3. Creating local road accurate and accountable database and human resources capacity by conducting trainings. (W5,T5)

3.3 QSP matrix

Following the SWOT matrix analysis, several alternative strategies were found. These alternative strategies were formulated. These alternative strategies were then reformulated in Table 3. Alternative 1, 2, 3, and 4 - is S-O, W-O, S-T, and W-T strategy, respectively.

Table 3. Alternative strategies

Alternative	Strategy
1	- Creating a stable transportation connectivity concerning aspects of sustainability, safety, and quality improvement and maximizing the PRIM funding schemes in order to support and encourage the local economic development. - Establishing international cooperation in the road construction and improving the human resources competence by conducting several trainings in order to embody good governance.
2	- Managing and improving the quality of local road network by sustainability in the whole life cycle supported by PRIM funding schemes. - Implementing performance based design and managing environmentally friendly and safer roads by using and sustainable technology.
3	- Optimizing human resources quantity in the Directorate General of Highways in strengthening accurate and accountable local road database supported by the government's commitment in constructing and developing roads. - Building road networks connectivity supported by the government's commitment and institutional unit of road development to attain local development target and to advance economic growth.
4	- Managing environmentally friendly road by sustainability in the whole life cycle and minimizing the natural disaster and high rainfall. - Creating accurate and accountable local road database and building human resources by conducting trainings

Of some alternative strategies, the best strategy can be chosen by using quantitative strategic planning matrix (QSPM). This technique assesses the best strategy out of several alternatives. Based on the existing key factors to identify previous factors. The best strategy formulated by means of the QSPM is presented in Table 4.

DOI: 10.1051/matecconf/201713807007

Table 4. QSP matrix

Key factors	weight	Alternative 1		Alternative 2		Alternative 3		Alternative 4	
		AS	TAS	AS	TAS	AS	TAS	AS	TAS
S1	0,20	2	0.40	1	0,20	2	0,40	1	0,20
S2	0,15	3	0.45	3	0,45	3	0,45	1	0,15
S3	0,02	3	0.06	1	0,02	3	0,06	1	0,02
S4	0,01	3	0.03	1	0,01	3	0,03	1	0,01
S5	0,01	2	0.02	2	0,02	1	0,01	1	0,01
W1	0,27	2	0.54	4	1,08	1	0,27	4	1,08
W2	0,20	1	0.20	4	0,80	1	0,20	4	0,80
W3	0,10	1	0.10	3	0,30	1	0,10	3	0,30
W4	0,02	2	0.04	3	0,06	1	0,02	3	0,06
W5	0,02	1	0.02	1	0,02	1	0,02	1	0,02
O1	0,26	4	1.04	3	0,78	1	0,26	1	0,26
O2	0,20	4	0.80	4	0,80	1	0,20	1	0,20
O3	0,12	3	0.36	2	0,24	1	0,12	1	0,12
O4	0,11	3	0.33	4	0,44	1	0,11	1	0,11
O5	0,06	3	0.18	2	0,12	1	0,06	1	0,06
T1	0,17	1	0.17	1	0,17	2	0,34	2	0,34
T2	0,03	1	0.06	1	0,03	2	0,06	3	0,09
T3	0,02	2	0.04	1	0,02	4	0,08	2	0,04
T4	0,01	2	0.02	1	0,01	4	0,04	3	0,03
T5	0,02	1	0.02	1	0,02	3	0,06	4	0,08
Total score			4,85		5,59		2,89		3,98

From QSPM matrix calculation, it is obvious that alternative 2 is the most attractive strategy. However, by looking at a relatively small score difference, alternative 1 namely S-O strategy is also a quite interesting strategy. Hence, combining these two alternative strategies is promising. It may be deployed to increase local road quality by maximising PRIM funding scheme and to manage environmentally friendly and safer-local road by integrating sustainability aspects in design, construction, maintenance, and rehabilitation (whole life cycle). These strategies are in accordance with the policies and targets of local road development of Directorate General for Road Network, Ministry of Public Works and Housing Strategic Planning 2015-2019, which among others are the increasing of basic infrastructure availability; the expansion of PRIM pilot project; the satisfactory condition of provincial roads by 75%, and the satisfactory condition of local roads by 65%.

4 Conclusion

The appropriate strategy toward sustainable local road network infrastructure were the management of local road and improving the quality of local road network by concerning aspects of sustainability in the stages of design, construction, maintenance by applying PRIM funding schemes, the management of environmentally friendly roads with the long-term pavement design and with road innovation and technology.

References

1. J. Morrissey, U. Iyer-Raniga, P. McLaughlin, and A. Mills, *Environ. Impact Assess. Rev.*, vol. **33**, no. 1, pp. 55–65, (2012)
2. L. Shen and J. Zhou, *Habitat Int.*, vol. **44**, pp. 111–120, (2014).

3. G. Marsden, M. Kimble, J. Nellthorp, and C. Kelly, Int. J. Sustain. Transp, vol. **4**, pp. 189–211, (2010)
4. J. L. Anderson, *Measuring sustainability in civil engineering,* University of Washington, (2012)
5. L. O. Mills and N. Attoh-Okine, *International conference on sustainable development of critical infrastructure*, pp. 79–90 (2014)
6. A. Umer, K. Hewage, H. Haider, and R. Sadiq, Int. J. Sustain. Built Environ., vol. **5**, no. 2, pp. 604–619 (2016)
7. F.S. Handayani, *Procedia Engineering*, vol **171**, 387-395 (2017)
8. H. A. Rahardjo, D. Dinariana, and F. Suryani, *Procedia Eng.*, vol. **125**, pp. 541–546 (2015)
9. J. Stewart, Aust. J. Public Adm., vol. **63**, no. 4, pp. 16–21 (2004)
10. M. A. U. Rahman, *J. Urban Manag.*, vol. **5**, no. 1, pp. 16–22 (2016)
11. F. R. David, *Concepts in Strategic Management*. Prestice Hall, New Jersey (2000)

MATEC Web of Conferences **138**, 07008 (2017) DOI: 10.1051/matecconf/201713807008
EACEF 2017

Engineering model of traffic and transportation safety with pattern of cooperation between sustainable region in Bogor

Syaiful[1,*]

[1]Ibn Khaldun University, Civil Engineering Departement, 16162 Bogor, Indonesia

Abstract. Bogor and Bogor district is the buffer zone of the capital Jakarta Special Capital Region. Bogor city as the city has also become a pilot implementation of the one way system (SSA) and the Contra Flow pilot system, the system open and closed area of Puncak, Bogor. If examined more deeply traffic flow and patterns of population movement is very diverse city. The diversity of travel of the city resulted in the number of people who pass through the region is also increasingly diverse. Population mobility infrastructure also requires adequate transportation. Adequate facilities and infrastructure that meet the criteria as it is safe, convenient and affordable for the community. If explored further that the more the number of vehicles crossing the city of Bogor and the types of vehicles also increased. This increase raises the issue in the field of transport. The problem varies according to the level of speed, volume and density of vehicles resulted in a noise due to exhaust gas. Highway safety is very important to address reducing the risk of death and permanent disability is experienced riders of motor vehicle and highway users.

1 Background of the problem

Bogor city and Bogor district is the buffer area of Jakarta Capital City Special Capital Region. Bogor City and Bogor regency are not directly adjacent to Jakarta, but the city of Bogor is a transit area and the residents of Jakarta and surrounding areas to rest area. Bogor City is also a pilot as a city of application of one-way system (SSA) and pilot Contra Flow system, open close system in the area of Puncak Bogor. If studied more deeply the flow of traffic and patterns of urban population movement is very diverse. The diversity of the city's journeys has resulted in a growing number of people passing through the area [1]. Population mobility also requires adequate transportation facilities and infrastructure. Sufficient facilities and infrastructure that meet the criteria such as safe, convenient and affordable for the community [2]. If further examined that the more days the number of vehicles that cross the city of Bogor and the type of vehicle is also increasing. This increase poses problems in the field of transportation. The problem varies according to the speed,

* Corresponding author: syaiful@ft.uika-bogor.ac.id

volume and density of the vehicle resulting in noise caused by the flue gas. Noise is generated by the sound of motor vehicle engine and the remaining sound due to the reflection of objects around it [3].

While the relationship between Speed-Volume-Density is the relationship between volume (q) and density (k) is called the fundamental diagram for traffic, or the q-k curve. Empirical studies show that the rising density of the stream first increases and reaches its maximum and then turns down, but there is some conflicting evidence regarding the exact shape of the curve. The shape and calibration of such relationships is important, as they provide the basis for the selection of measures of effectiveness and definition of the range-of-service range for the expressway segment [4].

Influence other than those associated with the vehicle such as wind speed, temperature, humidity is very small then ignored the noise [5]. Noise generated by influential vehicles from long-term residents in a housing complex and deeply disturbed if they stay more than a year [6]. The noise acceleration depends on the three basic elements of the traffic flow, ie, the driver, the road, and the traffic conditions [7, 8]. Noise in question is the sound of motor vehicles simultaneously on the highway, generally this noise is related to the inconvenience of listeners nearby. Noise can be regarded as a vehicle speed interference from a uniform speed and can be identified as a measurement of the smooth flow of traffic. The predicted number of motor vehicles based on the noise level is influenced by the average speed, the passing vehicle type and the noise damping factor, the method used is fuzzy prediction system with 8% error rate [9 - 15].

2 Purpose of the study

The purpose of this research are:
Obtain road safety level Pajajaran Bogor city on the path of SSA (One Way System) based on the matrix of destination.

3 The frame of thought

The modeling of this research approach is to examine how much the effect of motor vehicle traffic condition has impacted the activity of road users either on straight road, turn road, road with smooth surface (paved road), road with rough surface (concrete road/rigid pavement) to safety riding due to motor vehicle noise is based on speed [2, 15]. In this research is presented in the form of a frame of thought as below (Figure 1);

DOI: 10.1051/matecconf/201713807008

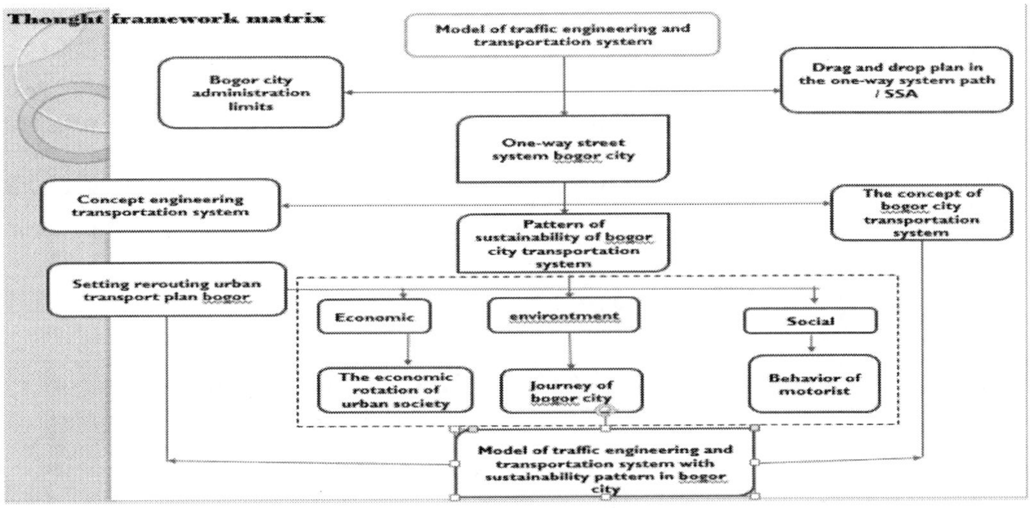

Fig. 1. Thougt framework matrix.

4 Problem Formulation

The formulation of the problem in the draft proposed research proposal of this dissertation plan is considered very complex, important and needs to be investigated. The existence of motor vehicles in the city of Bogor and Bogor district increasingly increasing and increasing [1]. The increase in the number of motor vehicles can no longer be prevented, due to various social factors that follow, including the easy to get motor vehicles such as motor credits and cars [2]. So also supported the existence of Mall/Supermarket growing in every corner of the city. The development of shopping facilities, public spaces and pedestrians, schools and places of worship and housing is a factor that triggers the increase of people's activities in daily life [3]. Preparation of research on traffic engineering model and transportation system with sustainable inter-regional cooperation pattern in Bogor city is to produce congestion model, transportation safety and noise level. The passing vehicles are classified as non-motorized vehicles (KTB), motorcycles (SPM), private car transport (AMP), freight cars (MAB) and passenger cars (MAP), as well as heavy transport vehicles (KAB). Several studies have shown that problems with congestion, transport safety and sustained traffic noise levels have not been satisfactorily answered or split.

Observing the facts and phenomenon above writer attempt to arrange research about. This is what makes researchers want to examine more about the model of noise levels caused by motor vehicles, the influence of motor vehicles on traffic congestion, noise and safety, especially in the city of Bogor. How much influence the public per capita opinion and performance patterns of urban communities. The influence of the behavior of passengers on public transport and road users and how motor vehicle users perform. And how big is the level of user annoyance that reflects the level of public awareness in addressing the circumstances surrounding the presence of highway traffic. Public disturbance to the noise caused by motor vehicle traffic in connection with daily activities how big.

The results of the equations obtained in this study are expected to be quantitative and qualitative in nature with field data taking and referring to the correct statistical methods. The result forecast is a model of traffic engineering and transportation system with sustainable inter-regional cooperation in Bogor city.

5 Hypothesis

Hypothesis testing is proposed is traffic congestion, driving safety and influence of motor vehicle noise level.
Highway traffic congestion level against noise generated by motor vehicles. The level of transport safety to noise affected by the number of vehicles.

6 Data Collection Method

In this research conducted various data collection methods in the field such as data retrieval by interview survey method, secondary data retrieval method at SSA location.

7 Methods of data analysis

Data processing is carried out in several stages, namely for data taken directly in the field such as speed, volume and traffic density of motor vehicles using fuzzy Logic Method [8] and expert system model of ISM Professional [9].
In the data transport safety data taken secondary data available in the police and data in hospitals with secondary data processing.
For sampling method using the number of respondents according to Slovin is by the formula:

$$ni = \frac{Ni}{N} \times n \tag{1}$$

$$n = \frac{N}{1 + N(e)^2} \tag{2}$$

N = sample size
N = population size
E = standard error (5% - 8%)
ni = sample size strata i
Ni = population size strata i
In noise model taking.

$$Leq = 10 \log (\sum ni \times 10^{Li/10}) / N \tag{3}$$

That is:
N = number of observations
ni = number of observations that cost L [11].

8 Research Design

The following submitted research has been and will be implemented for 12 months in the city of Bogor (Figure 2);

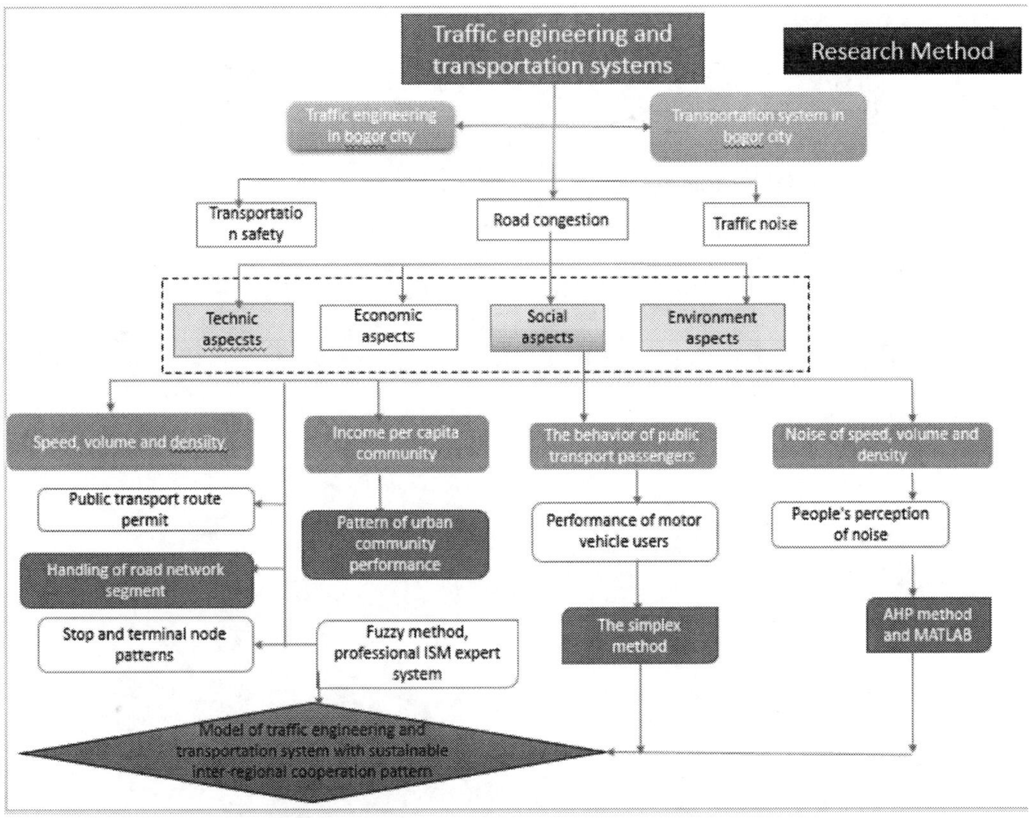

Fig. 2. Research method matrix

9 Place and time of research

This research takes place in Pajajaran street with One Way System of Bogor city and Bogor district. The study was conducted for 12 months ie from January 2017 to December 2017. Currently the research is still ongoing. However, as an illustration has been able to image data used to be processed as a preliminary material to plan something about the matrix of destination included secondary data prepared by the traffic service traffic and Bogor city transportation (Figure 3, Figure 4, Figure 5, Figure 6, Figure 7 and Figure 8);

Fig. 3. Junction crossing monument of kujang

Fig.4. Clutter around traffic light monument kujang

DOI: 10.1051/matecconf/201713807008

Fig. 5. Monument of kujang bogor city

Fig. 6. The faded zebra cross

Fig. 7. The New symbol of the monument kujang city bogor

Fig. 8. Pajajaran street in one way system in city bogor

10 Results of calculation of SSA path movement in Bogor city

The model of the movement generation is intended to produce a model of the relationship between the number of movements with the socio-economic characteristics of the perpetrators of the movement leading to and to a zone. This modeling is very necessary to know how big the movement that occurred in the future.

On modeling the rise of movement of motor vehicles and citizens using zone based movements. For Pajajaran road including subdistrict zone Bogor East urban village bogor based on urban village zone. The consideration of using the zone based on the village is based on the things that affect it, among others:

1) Knowing how much the movement generates (production and attraction of movement) for each village.
2) Analyze the social and economic level characteristics of the research area based on the data input from the village.

Based on the rise and based on the matrix of origin of the movement of the community in urban Bogor District can be observed more thoroughly. Similarly, the level of socio-economic characteristics of perpetrators of the movement in the region does affect the magnitude of the movement of the urban population. Getting the characteristics of the rate of urban population movement is obtained based on the survey, as for the survey is the survey of individual income, family dependents and ownership of homes and motor vehicles:

1) Individual income (a1)
2) How many family dependents (a2)
3) House ownership status (a3)
4) Status of ownership of motor vehicles (a4)
5) Travel duration at SSA location (a5)

For free variable is the amount of movement of urban community in research location that is SSA for every perpetrator of movement in time unit that is (y).

The process of modeling the rise of movement (production and pull) begins by looking at the correlation of the independent variable and the independent variable of the movement. The greater the correlation between these variables indicates that socio-economic characteristics greatly influence the number of movements. And vice versa, the smaller the correlation then the relationship between these variables do not affect each other. The correlation between independent and non-free variable of Bogor city movement can be seen in (Table 1 and Table 2);

Table 1. The relationship of independent variables with free variable pattern of urban community movement

Relationship	y	a1	a2	a3	a4	a5
y	1					
a1	0,24329	1				
a2	0,32547	0,04365	1			
a3	0,04421	0,05492	0,00758	1		
a4	0,04342	-0,07611	0,00421	0,00163	1	
a5	0,03121	0,006542	-0,00987	0,00729	0,00321	1

Source: Processing using expert system of Professional ISM [9]

Table 2. Relationship of independent variables with free variable pattern of urban community movement

Relationship	y	a1	a2	a3	a4	a5
y	1					
a1	0,03242	1				
a2	0,03215	0,64321	1			
a3	-0,06432	0,06576	-0,09643	1		
a4	0,03238	-0,04333	0,076512	0,02198	1	
a5	0,07432	0,04365	0,003265	0,04365	0,032347	1

Source: Processing using expert system of Professional ISM [9]

Table 1 means that all positive-valued variables have an effect on the pattern of community movement of Bogor city such as individual income of 0.24329, family dependent of 0.32547, home ownership of 0.04421, ownership of motor vehicle 0,04342 and length of time in passing SSA equal to 0.03121. Table 2 shows that the variables that influence the pull of urban movement are each of the family dependents of 0.04365, the home ownership is 0.00758, the motor vehicle ownership is 0.00163 and the length of time in the passing of the SSA is 0.00321.

For the movement of urban community due to tension with free variable is negative value of each negative value that is home ownership of **-0.07611** and the length of time passing on the road SSA **-0.00987**. This means that there is a negative value relationship is there is no relationship of mutual influence between positive values on the pattern of movement of the city of Bogor today.

11 Conclusions

The tentative conclusions of this study are:

1) There is a very close relationship between the number of tugs of the study area around the SSA line with safety in transport.
2) There is no correlation between the number of Bogor city community awakening to home ownership and time taken when passing on SSA road.

References

1. Bappeda K.B, *Bogor dalam angka,* **2**, hal. 45-46 (2014) (Indonesian)

2. *Directorate General of Highways*, **1**, hal. 32-33, Jakarta (1997) (Indonesian)

3. E. R. Purnomo, J.M.T, **4**, hal. 90-92, Yogyakarta (1997) (Indonesian)

4. L.E. Haefner, M-S. Li, L.A. Porello, CTRE, *Semisequentennial transportation conference proceedings*, **1**, pp 11-12, Ames, Iowa (1996)

5. Kep. M.N.L.H, **48**, pp. 7-9, Jakarta (1996) (Indonesian)

6. L.P. Bangun, I.M. Kamil, I.B.A. Putra, *Kebisingan lalu lintas dan hubungannya dengan tingkat ketergantungan masyarakat,* **1**, F.T.S.L, hal. EH-2-EH-11, Bandung (2009) (Indonesian)

7. L.E. Haener, *Preliminary data collection and analysis for traffic flow management on a freeway corridor*, D. Civil Engineering, Washington University, St. Louis (1996)

8. H. Martono, Wijayadi, *The health effects of environmental noise other than hearing loss*, P.A. Number **3311,** JN 7845, Australia (2004)

9. N.M.F. Saputra, T.Pynkyawati, J.O.I.T.N, **4**, hal. Reka karsa-1- 10, Bandung (2014) (Indonesian)

10. Resmana, J.T.E, **1**, hal. 23-24, Universitas Kristen Petra, Surabaya, (1999) (Indonesian)

11. R. Setiawan, T.D. Arief, N. Handayani, P. Sawitri, J.Teknik Sipil, **1**, hal. 1-8, Universitas Kriten Petra, Surabaya (2001) (Indonesian)

12. T.P.M. Panjaitan, B. Pramudya, Manuwoto, I.F.P. Poerwo, J. Sabuga, **1**, hal. 1-8, PWK, Universitas Sam Ratulangi, Manado (2011) (Indonesian)

13. Syaiful, Syaiful, *Analisis kebisingan arus lalu lintas dan geometri jalan di kawasan simpang lima kota Semarang*. Masters thesis, program Pascasarjana Universitas Diponegoro, Diponegoro University, Semarang: INSTITUTIONAL REPOSITORY Available : http://eprints.undip.ac.id/13253/ (2005) (Indonesian)

14. Syaiful, I.J.A.R, **4**, pp. 772-779, Available: http://www.journalijar.com/article/7687/ (2016)

15. Syaiful, Syaiful, Y. Elvira, I.J.T.I, **1**, pp. 34 - 40, Available: http://jurnal.narotama.ac.id/index.php/ijti (2017)

 DOI: 10.1051/matecconf/201713807009

Comparison of the behavior of motorists in traffic safety in Yogyakarta Special Region by gender

Poei Purnamasari[1], and *Dwijoko* Ansusanto[1]

[1]Civil Engineering Department, Atma Jaya Yogyakarta University, Yogyakarta, Indonesia

Abstract. Yogyakarta as a student city, a cultural city and tourist destination after Bali changing rapidly year after year. In the 70s, transportation mode in this small town were bicycles, tricycles, buggy and only few motor vehicles. Because of the development of the technology and the number of newcomers grew larger, Yogyakarta is now dominated by two-wheeled motor vehicles and cars. That facts resulted the higher risk of danger, if the riders are not well behaved and disregard the safety instructions on the road. Therefore this study had been conducted in Yogyakarta Special Region (DIY) with the objection to compare and observe the behavior of the road users based on the gender. The study used explanatory research with the reference to the Law of the Republic of Indonesia No. 22 of 2009 about the safety instructions for the riders on the road, with common people, high school and university students as the respondents.

1 Introduction

Yogyakarta as a student city, a cultural city and tourist destinations after Bali changing rapidly year after year. Almost 20% of the productive population in Yogyakarta are students. Many young people from Sabang to Merauke are studying in this small city, which covers only 0.017% of the land area of Indonesia. In the 70s, transportation mode in this small town were bicycles, tricycles, buggy and only few motor vehicles. The growth of the number of newcomers in town (either permanently or temporarily) as well as the development of the technology that made the vehicle go faster than the previous mode, causing the domination of two-wheeled motor vehicles and cars in Yogyakarta. With the need of easier and faster mobility as the main reason, having a motor vehicle nowadays seems to have become a necessity. It is easier than before, because there are motorcycles, that relatively affordable for everyone. Using only 500,000.00 rupiah as a down payment, anybody can take a new motorcycle home. Consequently, the number of vehicles in Yogyakarta is now dominated by two-wheeled motor vehicles and cars. As a dynamic city, Yogyakarta has a fast pace change that is also affecting road users to be eager to move sooner to the destination, hence the traffic become more chaotic. In this case, riders without ethical behavior and disregard for safety on the road would endanger the safety of both fellow motorists, pedestrians and other road users. Factors such as gender, age, educational background and physical condition, cognitive and psychomotor and knowledge in the traffic the factors can influence the behavior of the riders. In reference to the Law of the Republic of Indonesia No. 22 of 2009 [1], article 106 paragraph (2): every person who

DOI: 10.1051/matecconf/201713807009

drives a motor vehicle on the road shall prioritize the safety of pedestrians and cyclist s. Article 284 consist of : Any person driving a motor vehicle without priority to pedestrian or cyclists safety as referred to in Article 106 paragraph (2) shall be imprisoned for a maximum of two months or a fine of not more than Rp 500,000.00 (five hundred thousand rupiahs) Article 131 paragraph (1): Pedestrians are entitled to the availability of supporting facilities in the form of sidewalks, crossings and other facilities. (2) Pedestrians are entitled to priority when crossing the road at the crossing. (3) In the absence of facilities as referred to in paragraph (1) Pedestrians shall have the right to cross in the chosen place with due regard to their safety.

2 Subjects and Methodology

Heinrich [2] said that there exists an important distinction between overt and covert behaviour. Only overt behaviour can be observed and recorded. Covert behaviour is never open to direct observation, it can only be deduced from external behaviour. By covert behaviour we mean cognition, emotions, and the building up of attitudes or habits. This covert behaviour is usually considered to be the logical, temporal and structural precondition for overt behaviour, and for this reason it seems logical to direct attempts at modifying behaviour to the covert level. Heinrich [2] also mentioned that In the feeling of responsibility and in efforts to make road traffic safer it is frequently the first -mentioned meaning of the term change which dominates: we have to make people behave more safely and then traffic in general will become safer! But education which ignores the fact that the influence of situational factors on the behaviour of people may be just as strong as that of dispositional factors, is doomed to failure. data also came from the first year of Competitive grant research [4] this study was conducted in Yogyakarta Special Region (DIY), which consists of 4 districts and 1 city (Sleman; Bantul; Kulonprogo; Gunungkidul and city of Yogyakarta). There are total 1010 Questionnaire with high school students, university students and common people as the respondents. The comparison of the motorists' behavior based on gender in traffic safety in Yogyakarta Special Region from each districts would be conducted using explanatory research. Like in Fishbein [3], the question types in the questionnaire using Likert scale related to attitudes and behavior of road users are shown in the Table 1 below.

Table 1. Questions related to attitudes and behavior of road users.

No	Code	Questions
1	DF1	Road users are obligated to comply with traffic signs / road markings
2	DF2	Pedestrians crossing on the zebra-crossinging are protected by the Traffic Act
3	DF3	All road users must respect and prioritize road crossers
4	DF4	Zebra-crossinging makes it easy for pedestrians to cross the road
5	DF5	I feel safer when I cross in zebra Crossing, that's why I always cross the road using it.
6	DF6	Crossing on the zebra Crossing guarantees the safety of the crossers
7	DF7	Drivers should slowing down their speed as they approach the zebra Crossing
8	DF8	Zebra Crossing users are appreciated by Drivers in DIY
9	DF9	It is difficult to cross the street without the officer's help, even though it's in zebra Crossing.
10	DF10	Pedestrians feel free to cross anywhere along the way
11	DF11	When I need to cross the street and the zebra Crossing is far away from
12	DF12	There are differences in difficulty level between crossing in zebra-
13	DF13	In my opinion, many of the traffic signs in DIY blocked by trees / pole

 DOI: 10.1051/matecconf/201713807009

14	DF14	I always try to cross the intersection even though the traffic lights are
15	DF15	I always carry a complete driving documents when I drive car / motor
16	DF16	I always overtaking another vehicle either from the left or right side.
17	DF17	I overtaking another vehicle either from the left or right side.
18	DF18	When I want to turn or change lanes, I always turn on the turn signal
19	DF19	While waiting for the train to passed the train crossing, I always stay at
20	DF20	After I turned or moved to another lane, I always turn off the turn light
21	DF21	I always check the condition of the vehicle such as wind pressure,
22	DF22	In my opinion, many of the traffic lights in DIY blocked by trees / pole
23	DF23	When driving I always be aggressive and do not want to lose if passed
24	DF24	I always receive / call / see / send SMS while driving on the highway.
25	DF25	I will quickly replace the brake lining before it's worn out
26	DF26	Dysfunction brakes light will always be replaced quickly
27	DF27	I don't use a helmet if I drive in a short distance (about 1 km)
28	DF28	I always pay attention to the safety of the people around while driving
29	DF29	I'm impatient when I have to stop because of the red light is on and
30	DF30	I could be distracted by the sound surround me (for example loud
31	DF31	I always look around and give a sign when I want to overtake / turn /
32	DF32	I often rode with three or more people while driving on a motorcycle.
33	DF33	I do not always prepare helmets for the people I cared for, especially
34	DF34	I think the application of traffic rules in DIY is very good
35	DF35	In my opinion, the implementation of sanctions against traffic violators
36	DF36	I immediately stop the vehicle when the red light is almost on
37	DF37	During bad weather / rain / misty / smoky, I reduce the speed of the
38	DF38	I always use a helmet while driving motorcycle
39	DF39	I will cross the street marking/ move to another lane when I turn
40	DF40	I will leap over to another lane/track in uphill road
41	DF41	In my opinion, the number of traffic signs in DIY is sufficient
42	DF42	I always use seat belt while driving.
43	DF43	I will watch out for another vehicle before I open the car door.
44	DF44	When I pull the brakes, I will always use the front brake and rear brake
45	DF45	I will give way to the public bus that will pull over /out of the bus stop.

3 Analysis

From all of the 45 questions above can be divided into 5 groups of respondents' opinions: 1) about signs in DIY, 2) about traffic signaling tools (APILL), 3) about the completeness of driving documents, 4) attitudes and behavior of respondents about zebra Crossing, 5) the behavior of respondents while driving.

3.1 About signs in DIY

From Table 2. and Table 3., it appears that female and male respondents in 4 districts and 1 city have the same opinion in terms of priority sequences related to signs. They prioritize DF1 "Road users are obligated to comply with traffic signs / road markings" and assume DF13 "In my opinion, many of the traffic signs in DIY are blocked by trees / pole / banners/ not located in the appropriate location" less important.

Table 2. Female respondents opinion about signs in DIY

Bantul		Gunungkidul		Kulonprogo		Sleman		Yogyakarta	
Code	Mean (F)	Code	Mean (F)	Code	Mean (F)	Code	Mean (F)	Code	Mean (F)
DF1	3,66	DF1	3,69	DF1	3,67	DF1	3,74	DF1	3,76
DF41	2,73	DF41	2,66	DF41	2,64	DF41	2,64	DF41	3,07

DF13	1,97	DF13	2,10	DF13	1,95	DF13	2,03	DF13	1,99

Table 3. Male respondents opinion about signs in DIY

Bantul		Gunungkidul		Kulonprogo		Sleman		Yogyakarta	
Code	Mean (M)	Code	Mean (M)	Code	Mean (M)	Code	Mean (M)	Code	Mean (M)
DF1	3,53	DF1	3,61	DF1	3,55	DF1	3,79	DF1	3,65
DF41	2,71	DF41	2,82	DF41	2,76	DF41	2,74	DF41	2,96
DF13	2,19	DF13	2,11	DF13	1,97	DF13	2,10	DF13	2,02

3.2 Traffic Signaling Tools (APILL)

From Table 4. and Table 5., can be seen that female and male respondents in 4 districts and 1 city have the same opinion in terms of priority sequence regarding traffic signaling tools. They prioritize DF36 "I immediately stop the vehicle when the red light is almost on". Although the mean value is different. They consider DF22 "In my opinion, many of the traffic lights in DIY are blocked by trees / pole / banners / not located in appropriate location" less important.

Table 4. Female respondents opinion about traffic signing tools in 4 districts and 1 city

Bantul		Gunungkidul		Kulonprogo		Sleman		Yogyakarta	
Code	Mean (F)	Code	Mean (F)	Code	Mean (F)	Code	Mean (F)	Code	Mean (F)
DF36	3,03	DF36	3,28	DF36	2,99	DF36	3,51	DF36	3,08
DF14	2,81	DF14	2,86	DF14	2,64	DF14	3,25	DF14	2,83
DF22	1,95	DF22	2,23	DF22	2,08	DF22	2,26	DF22	2,12

Table 5. Female respondents opinion about traffic signing tools in 4 districts and 1 city

Bantul		Gunungkidul		Kulonprogo		Sleman		Yogyakarta	
Code	Mean (M)	Code	Mean (M)	Code	Mean (M)	Code	Mean (M)	Code	Mean (M)
DF36	3,15	DF36	3,19	DF36	3,11	DF36	3,04	DF36	3,12
DF14	2,78	DF14	2,61	DF14	2,76	DF14	2,88	DF14	2,53
DF22	2,11	DF22	2,34	DF22	2,13	DF22	2,32	DF22	2,19

3.3 Completeness of respondents while driving

Tables 6 and 7 show that women and men in 4 districts & 1 city prioritize DF15 "I always carry a complete driving documents when I drive car / motor cycle" except for women from Gunungkidul who prefered DF26 "Dysfunction brakes light will always be replaced Quickly ". The least important or lowest ranking according to Bantul women; Sleman and the city of Yogyakarta is DF34 "I think the application of traffic rules in DIY is very good", while the women of Gunung Kidul concerned less with DF35 "In my opinion, the implementation of sanctions against traffic violators in DIY is in accordance with the level and type of violation ", Kulonprogo women are less concerned with being associated with DF33 helmets. For Bantul men; Kulonprogo and Sleman are less prioritizing DF35, while Gunungkidul and Yogyakarta men are less prioritizing DF33 with regard to helmets.

 DOI: 10.1051/matecconf/201713807009

Table 6. Completeness of female respondents while driving in 4 districts and 1 city

Bantul		Gunungkidul		Kulonprogo		Sleman		Yogyakarta	
Code	Mean (F)	Code	Mean (F)	Code	Mean (F)	Code	Mean (F)	Code	Mean (F)
DF15	3,28	DF26	3,30	DF15	3,37	DF15	3,29	DF15	3,17
DF25	3,23	DF 25	3,25	DF25	3,00	DF26	3,17	DF26	3,15
DF26	3,15	DF15	3,16	DF26	2,98	DF25	3,12	DF25	2,99
DF21	3,07	DF33	3,14	DF21	2,94	DF21	2,83	DF21	2,77
DF33	2,70	DF21	3,04	DF34	2,81	DF33	2,72	DF33	2,60
DF35	2,61	DF34	2,73	DF35	2,71	DF35	2,61	DF35	2,57
DF34	2,58	DF35	2,39	DF33	2,65	DF34	2,54	DF34	2,57

Table 7. Completeness of male respondents while driving in 4 districts and 1 city

Bantul		Gunungkidul		Kulonprogo		Sleman		Yogyakarta	
Code	Mean (M)	Code	Mean (M)	Code	Mean (M)	Code	Mean (M)	Code	Mean (M)
DF15	3,38	DF15	3,23	DF15	3,35	DF15	3,26	DF15	3,27
DF25	3,15	DF26	3,31	DF25	3,10	DF26	3,10	DF25	3,08
DF26	3,14	DF25	3,23	DF21	3,09	DF25	3,04	DF26	3,02
DF21	2,87	DF21	3,12	DF26	3,09	DF33	2,87	DF21	2,90
DF33	2,77	DF34	2,87	DF34	2,95	DF21	2,75	DF35	2,80
DF34	2,54	DF35	2,83	DF33	2,69	DF34	2,75	DF34	2,72
DF35	2,50	DF33	2,74	DF35	2,63	DF35	2,70	DF33	2,66

3.4 Attitudes and behavior of respondents about zebra Crossing

Tables 8 and 9 show that female respondents from all areas of DIY have the same understanding and awareness of DF2 "Pedestrians crossing on the zebra-cross are protected by the Traffic Act". The same thing happened to male respondents. About DF10 which read "Pedestrians feel free to cross anywhere along the way" female respondents of Gunungkidul and Sleman have a high mean, for Bantul and Yogyakarta city close to 3 indicates that, although they are aware and understand the danger and risks, but they dare to do it. This is also the case for male respondents in Sleman with mean result greater than 3. No wonder DF12 " There are differences in difficulty level between crossing in zebra- crossing and not in zebra-crossing" Bantul, Gunungkidul and Sleman are considered by female respondents. Meaningless with the lowest score. For male respondents, DF12 in low mean values is applicable in Gunungkidul and Kulonprogo. There is only a few differences between Women in Kulonprogo and Yogyakarta, they both ignored DF9 "It is difficult to cross the street without the officer's help, even though it's in zebra Crossing". Similarly, DF9 with a low mean value occurs for male in Bantul, Sleman and Yogyakarta.

Table 8. Female attitudes behavior related with zebra Crossing in 4 districts and 1 city

	Bantul	Gunungkidul		Kulonprogo		Sleman		Yogyakarta	
Code	Mean (F)	Code	Mean (F)	Code	Mean (F)	Code	Mean (F)	Code	Mean (F)
DF2	3,63	DF2	3,63	DF2	3,56	DF2	4,20	DF2	3,60
DF4	3,57	DF3	3,54	DF3	3,51	DF4	4,13	DF3	3,56
DF3	3,49	DF4	3,51	DF4	3,36	DF3	4,04	DF4	3,51
DF5	3,30	DF7	3,31	DF7	3,01	DF5	3,82	DF5	3,13
DF6	3,15	DF10	3,18	DF5	3,00	DF6	3,64	DF6	3,01
DF7	3,10	DF8	3,16	DF6	2,92	DF7	3,59	DF10	2,96
DF10	2,94	DF6	3,09	DF8	2,83	DF10	3,41	DF7	2,87
DF8	2,89	DF5	3,06	DF10	2,70	DF8	3,34	DF8	2,72
DF11	2,44	DF11	2,58	DF11	2,18	DF11	2,83	DF11	2,40
DF9	1,99	DF9	2,29	DF12	2,12	DF9	2,30	DF12	2,12
DF12	1,97	DF12	1,89	DF9	1,95	DF12	2,28	DF9	1,92

Table 9. Male attitudes and behavior related with zebra Crossing in 4 districts and 1 city

Bantul		Gunungkidul		Kulonprogo		Sleman		Yogyakarta	
Code	Mean (M)	Code	Mean (M)	Code	Mean (M)	Code	Mean (M)	Code	Mean (M)
DF2	3,54	DF2	3,64	DF2	3,60	DF2	3,69	DF2	3,49
DF4	3,46	DF3	3,63	DF3	3,57	DF3	3,56	DF3	3,48
DF3	3,45	DF4	3,51	DF4	3,30	DF4	3,46	DF4	3,33
DF5	3,32	DF5	3,20	DF5	2,80	DF10	3,08	DF5	2,99
DF6	3,14	DF6	3,14	DF7	2,78	DF5	2,98	DF6	2,80
DF7	2,94	DF7	3,13	DF6	2,74	DF6	2,87	DF7	2,78
DF10	2,84	DF8	3,09	DF10	2,61	DF7	2,85	DF10	2,75
DF8	2,64	DF10	2,81	DF8	2,54	DF8	2,62	DF8	2,68
DF11	2,46	DF11	2,29	DF11	2,27	DF11	2,35	DF11	2,23
DF12	2,21	DF9	2,26	DF9	2,06	DF12	2,18	DF12	2,06
DF9	1,90	DF12	1,83	DF12	1,95	DF9	2,05	DF9	2,02

3.5 The behavior of respondents while driving.

For Tables 10 and 11, the highest mean of DF18 is dominated by female respondents from Gunungkidul, Kulonprogo, Sleman and Yogyakarta. The same thing happened also on the part of the male respondent. Both women and men in Bantul prioritize DF37 "During bad weather / rain / misty / smoky, I reduce the speed of the vehicle". For the lowest mean female respondents are Bantul; Kulonprogo and Yogyakarta with DF42 "I always use seat belt while driving.", Gunungkidul with DF43 and DF17 is Sleman. For male respondents, the lowest mean of Bantul and Sleman is DF17 "I overtaking another vehicle either from the left or right side. Whichever easy during traffic jam ", For Gunungkidul; Kulonprogo and Yogyakarta with DF42.

 DOI: 10.1051/matecconf/201713807009

Table 10. The behavior of female respondents while driving in 4 districts and 1 city

Bantul		Gunungkidul		Kulonprogo		Sleman		Yogyakarta	
Code	Mean (F)	Code	Mean (F)	Code	Mean (F)	Code	Mean (F)	Code	Mean (F)
DF37	3,47	DF18	3,90	DF18	3,43	DF18	3,58	DF18	3,41
DF20	3,41	DF38	3,48	DF20	3,33	DF37	3,51	DF45	3,24
DF18	3,39	DF20	3,40	DF37	3,32	DF32	3,45	DF38	3,23
DF38	3,32	DF37	3,38	DF28	3,27	DF20	3,42	DF37	3,23
DF31	3,31	DF28	3,36	DF38	3,27	DF38	3,41	DF31	3,19
DF28	3,30	DF31	3,33	DF31	3,17	DF43	3,36	DF20	3,16
DF45	3,23	DF45	3,33	DF45	3,14	DF23	3,33	DF28	3,12
DF24	3,16	DF24	3,31	DF44	3,14	DF24	3,33	DF32	3,12
DF44	3,14	DF32	3,18	DF24	3,20	DF28	3,32	DF44	3,07
DF32	3,10	DF23	3,10	DF32	3,12	DF31	3,32	DF24	2,97
DF19	2,98	DF40	3,09	DF40	2,96	DF29	3,29	DF29	2,97
DF23	2,98	DF29	3,03	DF23	2,92	DF45	3,26	DF23	2,85
DF40	2,98	DF19	3,00	DF29	2,89	DF19	3,16	DF27	2,85
DF29	2,81	DF27	2,90	DF39	2,74	DF39	3,11	DF19	2,79
DF39	2,78	DF44	2,86	DF19	2,67	DF40	3,11	DF39	2,77
DF16	2,58	DF16	2,79	DF16	2,55	DF44	3,08	DF40	2,77
DF27	2,58	DF39	2,79	DF27	2,44	DF42	3,03	DF16	2,73
DF17	2,51	DF17	2,66	DF30	2,40	DF27	2,74	DF17	2,33
DF43	2,43	DF30	2,49	DF17	2,38	DF16	2,70	DF30	2,24
DF30	2,42	DF42	1,76	DF43	0,81	DF30	2,47	DF43	2,19
DF42	2,30	DF43	1,75	DF42	0,80	DF17	2,38	DF42	2,04

Table 11. The behavior of male respondents while driving in 4 districts and 1 city

Bantul		Gunungkidul		Kulonprogo		Sleman		Yogyakarta	
Code	Mean (M)	Code	Mean (M)	Code	Mean (M)	Code	Mean (M)	Code	Mean (M)
DF37	3,38	DF18	3,48	DF18	3,54	DF18	3,53	DF18	3,38
DF28	3,35	DF37	3,39	DF20	3,50	DF37	3,38	DF38	3,37
DF38	3,32	DF38	3,35	DF38	3,34	DF20	3,35	DF20	3,33
DF18	3,31	DF28	3,31	DF37	3,33	DF28	3,35	DF37	3,31
DF20	3,29	DF31	3,28	DF28	3,30	DF32	3,32	DF28	3,26
DF31	3,29	DF24	3,27	DF44	3,30	DF31	3,30	DF31	3,22
DF44	3,29	DF20	3,26	DF31	3,29	DF24	3,29	DF45	3,21
DF24	3,27	DF29	3,18	DF45	3,24	DF38	3,25	DF44	3,15
DF45	3,26	DF45	3,17	DF32	2,94	DF45	3,20	DF32	2,96
DF32	3,20	DF44	3,10	DF23	2,91	DF29	3,18	DF24	2,94
DF19	3,04	DF32	3,08	DF24	2,84	DF44	3,15	DF23	2,79
DF43	3,01	DF39	3,02	DF40	2,84	DF23	3,14	DF19	2,67
DF39	2,98	DF23	3,00	DF29	2,71	DF40	3,04	DF29	2,65
DF23	2,90	DF40	2,99	DF39	2,60	DF43	3,04	DF40	2,63
DF40	2,85	DF19	2,89	DF27	2,33	DF19	2,96	DF39	2,57
DF29	2,77	DF27	2,70	DF19	2,27	DF39	2,88	DF27	2,56
DF42	2,77	DF16	2,65	DF30	2,27	DF42	2,80	DF30	2,35
DF27	2,76	DF30	2,43	DF16	2,25	DF27	2,69	DF43	2,31

DF30	2,55	DF17	2,33	DF17	2,09	DF30	2,50	DF17	2,23
DF16	2,47	DF43	0,99	DF43	2,07	DF16	2,46	DF16	2,10
DF17	2,43	DF42	0,95	DF42	1,91	DF17	2,27	DF42	2,07

4 Conclusion

The results of the analysis can be summarized as follows: a) For female and male respondents in 4 districts and 1 city have the same understanding about traffic sign and also APILL, b) men and women in all areas of Yogyakarta except women in Gunungkidul understand and realize the importance of carrying complete documents while driving, c) Although people in DIY, both men and women, understand and realize the importance of APILL and that crossing the road through the zebra Crossing protected by law, they still assume that people can cross anywhere anyway. d) all men and women in DIY (except male and female respondents from Bantul) give priority to the sign when they want to turn and move to another lane, while Bantul respondents prefer to reduce the speed of vehicle during bad weather / rain / fog. e) the highest mean of DF18 "When I want to turn or change lanes, I always turn on the turn signal lights" is dominated by female respondents from Gunungkidul, Kulonprogo, Sleman and Yogyakarta, while both women and men in Bantul prioritize DF37 "During bad weather / rain / misty / smoky, I reduce the speed of the vehicle".

References

1. Anonim, *Undang-undang Republik Indonesia No. 22 tentang Lalu lintas dan Angkutan jalan* (2009)
2. H. Ch. Heinrich, IATSS, **14 No 1**, 85 (Tokyo, 1990)
3. M. Fishbein, I. Ajzen, *Predicting And Changing Behavior,* 87 (Psychology Press, Routledge, 2015)
4. P. E. Purnamasari, D. Ansusanto, E-Journal. Uajy (2016)

MATEC Web of Conferences **138**, 07010 (2017) DOI: 10.1051/matecconf/201713807010
EACEF 2017

The Design of Road Evaluation and Monitoring System based on Geographical Information System

Ary Setyawan[1,*], *Muhammad* Farid *Al* Alimi[2] , *Dewi* Handayani[2], *Agus* Sumarsono[1]

[1]Roadmate Research Group, Universitas Sebelas Maret, Surakarta 57126, Indonesia
[2]Civil Engineering Department, Universitas Sebelas Maret Surakarta 57126, Indonesia

Abstract. Road infrastructure that are suffered by high traffic volume and repetitive loading will lead to a decline in the quality and capacity. As an indicator can be seen from the road surface conditions, both structural and functional condition. Damage that occurs varies greatly along the road and when left in the long term, it will worsen the condition of the road. The method used in the evaluation of damage to roads in Indonesia is PCI (Pavement Condition Index) method. PCI method is a method of road pavement condition assessment based on the type of damage, severity level and ethe density of damages compare to the wide of road surfacing. Application of PCI method tend to be less desirable because of the classification of the type of damage and the level of severity meausurement and density calculation are performed by manual observations at each segment and further calculation are still needed to get the index value. Road Evaluation and Monitoring System is a proposed system that lead to the adoption of Geographical Information Sistem based device to reduce the costs and labor for pavement evaluation and monitoring.

1 Introduction

Networks of road infrastructure in Indonesia are widely scattered. The whole road network requires evaluation and monitoring on a regular basis to determine the level of damage. Road manual evaluation and monitoring in all parts of Indonesian road require substantial resources, time consuming and expensive. For this reason, the tool is expected to facilitate this process of evaluation and monitoring that can be caried out faster, with less effort and cost. This study is proposed to design a tool named Road Evaluation and Monitoring System (REMS) to assist evaluation and monitoring road damage automatically. This design involves the use of various types of sensors and programs to assist in the detection of damage to the road as well as to input the data into Geographical Information System. Designing the REMS aims to perform evaluation and monitoring of road damage and expected to be significantly cheaper, faster and easier than the existing manual application [1].

* Corresponding author: arysetyawan @staff.uns.ac.id

2 Basic Technology for Development

Standardized evaluation of the severity level on the road in this research is divided into two types, namely, IRI and PCI. IRI standards obtained by noting the level of roughness of the road along the road and the process to get the value of IRI. Standard PCI damage is obtained by detecting the damage and measure their density of damage against the dimensions, followed by processing the data to get their PCI values [2].

The level of road roughness can be accounted for by a variety of methods. The method used in this research is the accelerometer sensor. The accelerometer sensor can be used to measure the acceleration of gravity and changes in real time. Initially, the beginning of the study the testing of accelerometer sensors technology were conducted to understand their performances and limitations of these sensors by using Android smartphones. Android smartphone is now generally implanted accelerometer sensor for multi-purpose applications. The sensor works internally and follow the user behavior. To test the accelerometer sensor on the Android smartphone, a simple application is carried out to detect changes that occur in the value of existing sensors on smartphones.

3 Hardware development and data flow

Road Evaluation and Monitoring System technology consists of two major systems: hardware and software. The REMS hardware system is designed based on phase 1, consists of a variety of sensors that are used to support the evaluation process of road damage. Accelerometer sensors mounted on a vehicle survey consists of 2 pieces each of which is connected to the microcontroller. When the system is running, the gravitational acceleration data change at any time are recorded and sent via microcontroller to the Central Processing Unit (CPU). Figure 1 shows the data flow process. Odometer mounted on the vehicle serves as registrar distance that has been taken during the survey. This distance changes in real time and the distance data transmitted via microcontroller to CPU.

On the CPU, accelerometer data are processed based on data from the odometer distance to get a profile of the road. The IRI value obtained approximately 100 meters each survey. The GPS device mounted on a vehicle survey serves to capture the satellite signals of GPS and GLONASS constellation. The signal is then converted into position data and then transmitted via USB to the CPU.

The CPU itself is in the form of high-performance laptop computer capable of receiving input from all sensors at once and be able to quickly processed the existing data into the road damage data and send some data to the server via the internet. Data stored in the CP, is stored in the Data Logger in the form of NAS (Network Attached Storage). The NAS serves as the storage of data during the survey. NAS is connected to the CP through Gigabyte Ethernet network that allows fast access up to 1 Gbps. NAT allows a survey carried out for a long time and a large capacity file backup. CP connected with the HSPA modem capable of GSM mobile network connected to the Internet. This connection can then be used by the CP to send data to the server so that the monitoring process of the survey can be done [3].

CPU is connected to an LCD or Android Smartphone which serves as a tool to show the display of survey process, the condition of survey equipment and the data flow connection conditions as well as to control of the entire equipment survey.

The entire existing hardware requires electrical energy supplied from the battery. The batteries to supply the energy to the inverter and then from the inverter, the electrical energy adapted to the respective hardware so that the output can be directly connected to the associated hardware securely. Figure 2 shows the Road Evaluation and Monitoring System (REMS) initial design car.

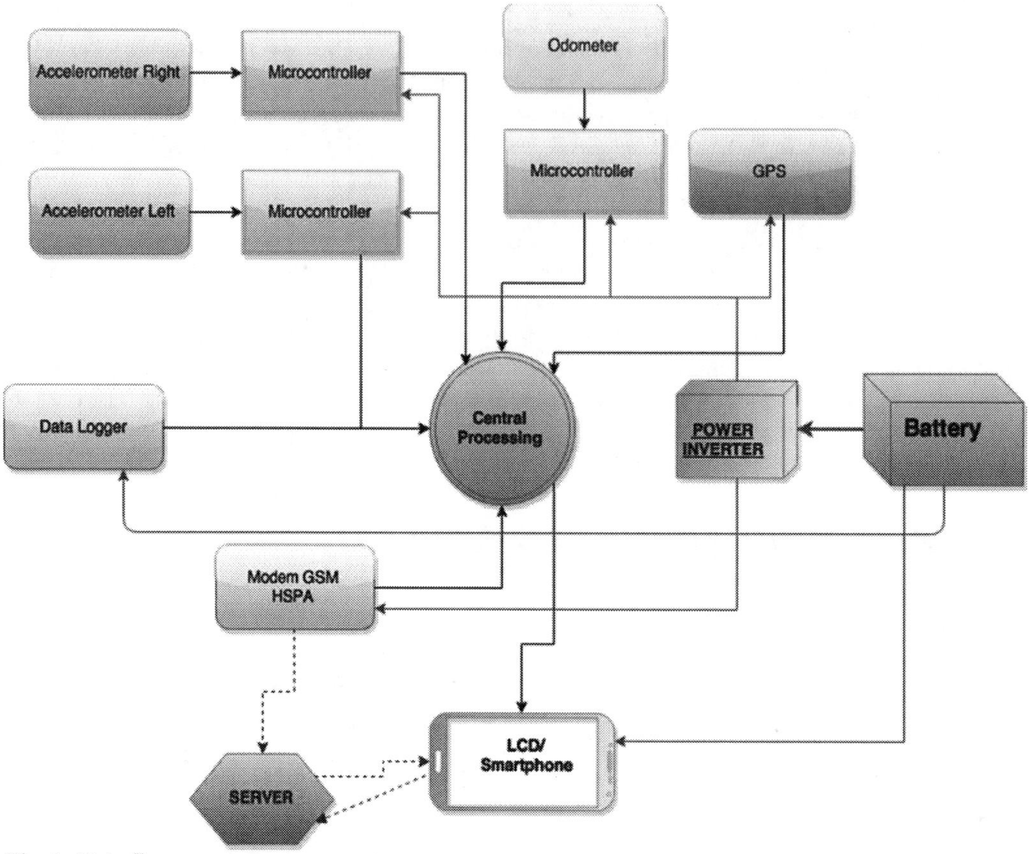

Fig. 1. Data flow process

3.1. Sensor Design

To detect the roughness of the road this study using the accelerometer sensor. Accelerometer sensor used is a 3- axis sensor. Sensor accelerometer has many types and specification according to the data requirements.

3.2. Design of Syncronitation Method Based on Distance

The output data from the accelerometer sensor must be associated with a distance that can be processed accurately. There are many methods for measuring the distances, such as GPS and odometer standard car, but the two are not able to provide the data range with accurate and fast output. Existing GPS has an error rate of about 3 meters. Car odometer measured with the smallest unit of 0.1 km. So that both are not feasible to measure distances with high precision.

Therefore, in this study designed custom odometer mounted on the gearbox with high detail capability. This odometer uses laser technology and filter are rotated with one axle gearbox car. As car gearbox ratios angular velocity with the car so that the wheel angular velocity filters can be processed into the actual distance traveled by car survey. In theory odometer is able to detect the distance with great detail, but the accuracy is limited research to a range between 3-5 cm.

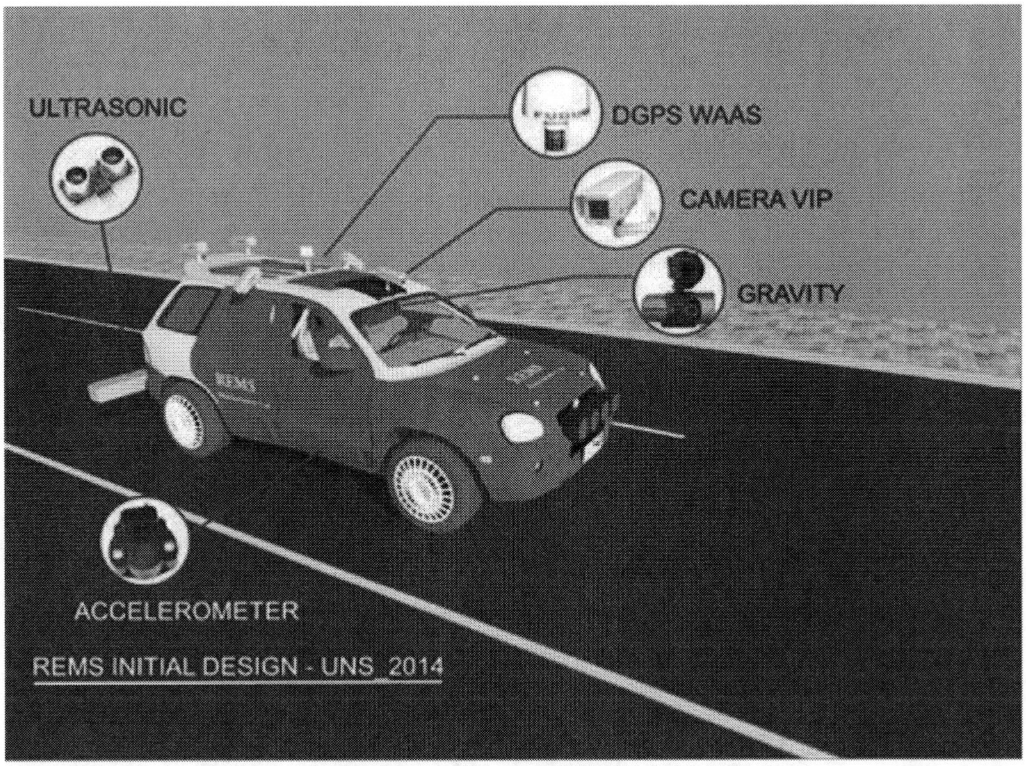

Fig. 2. Road Evaluation and Monitoring System (REMS) Initial Design

3.3. Design of GIS Integration and Connectivity

For connectivity and integration of GIS, GPS device uses the external antenna. Output GPS itself depends on the several factors, but in general the maximum output is 10 Hz GPS. The GPS output range is 0.1 Hz to 10 Hz. In addition, the GPS can be set to output also influenced by the change in position, it means that if the vehicle position is not changed then the GPS survey does not provide output.

The data output from the GPS is stored into the database on the CPU and collected along with the other sensor data so that the other data can be integrated geographically. GPS data are also sent to the server along with most other data for remote monitoring survey vehicle. The data are sent to the server using GSM HSPA modem with updates within 0.1 Hz to 1 Hz.

The position data is also transmitted from the CPU database to LCD or Android smartphone on the dashboard via a Bluetooth connection or the Internet. The position data can also be processed internally on a smartphone to facilitate the application by using the internal GPS sensor smartphone. The data required by the smartphone position to indicate the position, location, surveys progress and other data required by surveyors in the field.

3.4. Profilometer Design

Profilometer sensor consists of 15 pieces of high- speed laser distance to scan the road profile in the field. Lasers are used in the design is a Class 1 laser with the ability to scan up to 50 Hz [4,5]. The relationship between the car speed laser scan survey and can be seen in Table 1.

Table 1. Relationship between the car speed, laser scan survey and output profilometer

No.	Speed (km/hr)	Speed (m/s)	Scan (Hz)	Interval (m)	Interval (cm)
1	30	8.333333333	50	0.166666667	16.66666667
2	40	11.11111111	50	0.222222222	22.22222222
3	50	13.88888889	50	0.277777778	27.77777778
4	60	16.66666667	50	0.333333333	33.33333333

From Table 1, it can be seen that the lower the vehicle speed survey, the lower the interval between scans so the more accurate the results of the survey, however it need the longer time to conduct the survey. This problem could be solved by an interval of 20 cm to 27 cm and considered to be an ideal interval so that the car can run the surveys with an ideal speed and not exceed than 50 km/hour.

3.5. Design of Visual Damage Detection Sensor

The visual impairment in the form of visible damage to the eye is processed using computer vision technology with the sensor input of the camera. Because, it mounted on the running vehicle, then used a high-resolution camera sensor with a framerate of 30-60 fps [6,7]. Field of View is processed in the camera and set so that only the processing area with the transverse distance of 3.5 meters at a distance of 3 meters extending roads. The camera sensor mounted on balancer to reduce shacking or vibration that occurs when the survey is carryng out.

Summary

1. The schematic hardware of this REMS technology can be made and resolved
2. Design, implementation of road roughness using an accelerometer sensor, proximity sensor using a laser odometer, GPS sensor, camera sensor and profilometer sensor have been completed and ready to be implemented
3. Profilometer has a range of about 50 Hz and computer vision technology also has limitations in graphics processing so that the ideal speed survey is approximately 40 Km h
4. The accelerometer has a high output levels of 550 Hz so as to survey the damage only based IRI survey vehicle can drive up to 80 km/h

References

1. Direktorat Jendral Bina Marga, *Indonesian Public Work Department* (1998)
2. Ary Setyawan, Jolis Nainggolan, and Arif Budiarto, *Procedia Engineering* **125** :417-423 (2015)
3. Wix Richard, and Roland Leschinski, *International Public Works Conference, Darwin, Northern Territory, Australia* (2013).
4. Bennett, Christopher R., and Hanqing Wang, *Harmonising automated rut depth measurements*. No. **242** (2003).
5. Huidrom, Lokeshwor, Lalit Kumar Das, and S. K. Sud, *Procedia-Social and Behavioral Sciences* : 312-321. **104** (2013)

6. Ksaibati, Khaled, McNamara, Ronald and Armaghani, Jamshid. *Department of Transportation State of Florida – Report* (1999)
7. Ogonda, Godfrey Onyango, *Recovery of the geometric road design elements using low-cost sensors: a feasibility study* (2009).

MATEC Web of Conferences **138**, 07011 (2017) DOI: 10.1051/matecconf/201713807011
EACEF 2017

Analysis and design of road and bridge infrastructure database using online system

Joni Arliansyah[1,*], *Yadi* Utama[2], *Maureen* Arlini Wijayanti[3], *Rachmat* Gusti[3], and *Arifianto*[3]

[1]Department of Civil Engineering of Faculty of Engineering of Sriwijaya University, Palembang, Indonesia
[2]Department of Information System of Faculty of Computer Science of Sriwijaya University, Palembang, Indonesia
[3]The Office of Public Works and Spatial Planning of Palembang City, Indonesia

Abstract. A road database is an important component for observing a general picture of a road network, determination of treatment needs, road development and planning so that the roads in the road network in a city can receive treatment at the right time and action in accordance with its needs. The development of the road database system by the Office of Public Works of Palembang, Indonesia using Linux open source database platform, that is MySQL servers and application using the API (Application Protocol Interface) services provided by Google in order to represent the database as information in the form of a Google map. The developed menu was tailored to the needs of road maintenance management in Palembang including the menu input data, data view,minimum service standards, reporting and integration with national databases. Multi-level access into the developed system was applied to the users ranging from decision makers to field implementers. Since the development in 2014, the number of the recorded roads in Palembang City has been getting complete ranging from 952 roads with the road length of 618.9 km in 2014 to 1204 roads with the road length of 681.8 km in 2016. The number of recorded bridges increased from 125 bridges to 135 bridges in 2016. With the existing systems and data, the needs analysis of road maintenance based on the road conditions can be determined. In addition to the analysis of the road minimum service standards and integration with the national database system, the advantages andimprovement necessity in the implementation of the system are discussed in this paper.

1 Introduction

Palembang City as a city of trade and services increasingly grows rapidly from time to time. Infrastructure construction to support the movement necessity keeps going on along with the city development. Palembang City faces the increasingly complex problems that must be addressed, particularly with regard to the provision of city transport infrastructure.

Corresponding author: joniarliansyah@yahoo.com

DOI: 10.1051/matecconf/201713807011

Road database is an important component to provide an overview of road network, determine handling requirements, development and road planning so that the roads in the road network in a city will receive a handle at appropriate time and action according to the road necessity.

In 2014 a road data update was conducted in Palembang City. It was the update of road data in the road network and the road and bridge database system development using online system. The road and bridge database system is capable of providing an overview and details of the existing road and bridge network. It is an important device to support to determine the handling necessity, development and planning of roads and bridges in Palembang City.

The developed online database system refers to the road management needs of Palembang City and literature on the existing database management. Database development or studies on road and bridge database and road management system have been reported [1, 2, 3, 4].

The study aimed to analyze: 1) the development of online-based database system of roads and bridges and 2) the development of the advantages and improvement necessity of the developed database system.

2 Development of online database system

The online system for road and bridge infrastructure database was developed using information technology facilities such as the Internet and mobile devices. For the internet network, the system was built with the procurement of an online web server accommodating a database and a web-based application program. The database installed into a web server is a relational database holding all the data related to road and bridge infrastructures designed structurally, integratively, and easily developed. Web-based application program installed in a single location of a database server was used as a store of functions related to the entire business of data management process of road and bridge infrastructure and as a provider of web-based graphical interface to all system users. The mobile device was also used in this system that served as an additional medium for the interface user on the system. The developed online system network involved some hardwares connected through the internet in the form of a computer/PC client, mobile client devices, web servers and database servers. Architecture of online system network is given in Figure 1 below.

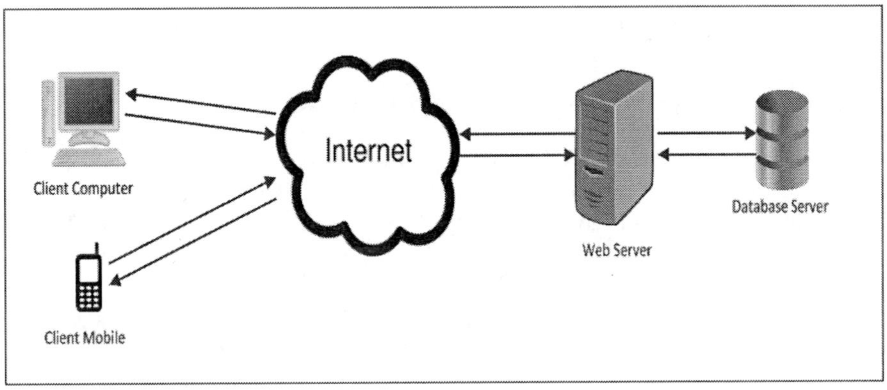

Fig. 1. Architecture of online system network

Some software aids that can be used in the online system are among other things:

a. Web Server:
   - Linux Operating System
   - Apache web server

 DOI: 10.1051/matecconf/201713807011

- PHP 5

b. Database Server with MySQL 5
c. Cloud Web Services: Google Maps API Services
d. Client Computer: Internet Browser
e. Client Mobile: OS Android dan GPS Tracking System

The online database system was designed using relational and integrated structure with the relational diagram given in Figure 2 as follows:

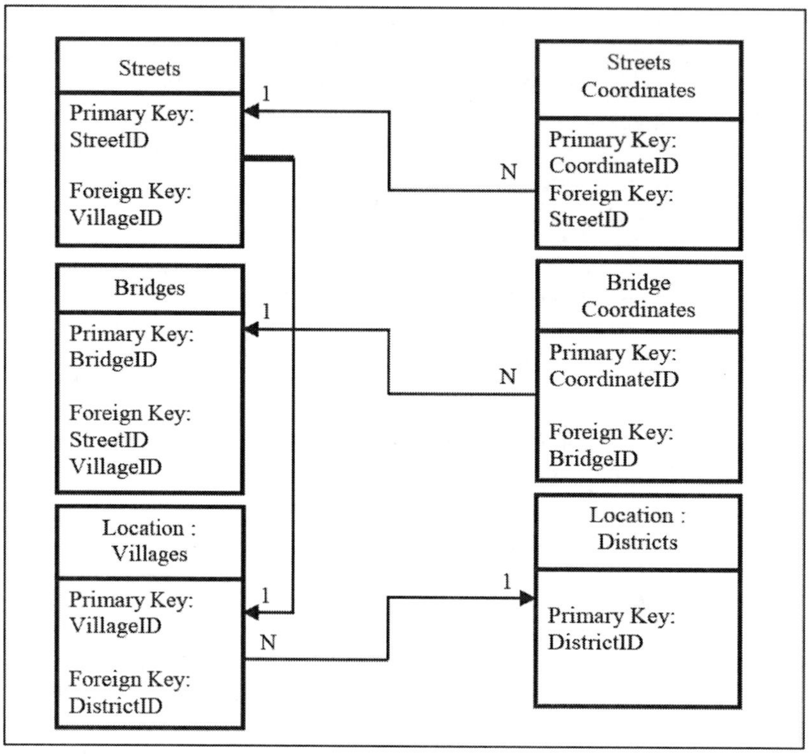

Fig. 2. Structure of online system database

3 Data management and implementation analysis

The concept of developed database systemdescribed in the menus contained in the online-based software of road and bridge database of Palembang City. They are input data menu, see the data, analysis of minimum service standard, report, handling history, integrated with national database and library. Database period is made annually in order to be able to update and monitor the development of road and bridge network data each year.In database development, the collected data of roads and bridges were, among others, data of inventory, coordinate, material types, and construction types as well as their conditions. Other collected data included data of roadsides, sidewalks, culverts, and other road necessities. All data were input into the system developed through the data input menu.

Menu of viewing the data is given in Figure 3 consisting of menus of viewing the data of roads, bridges, road networks, road conditions, networks of pavement-drainage-road shoulder, and culverts.

On the menu of viewing the data, one can see both the network and detailed data for each road such as the data of inventory, position, type and condition of the road. Figure 4 shows the examples of program output of a road segment.

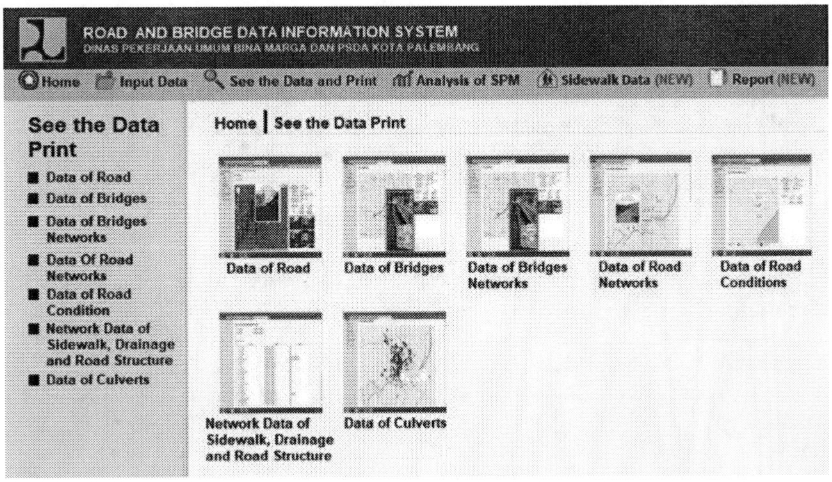

Fig. 3. Menu of viewing the data

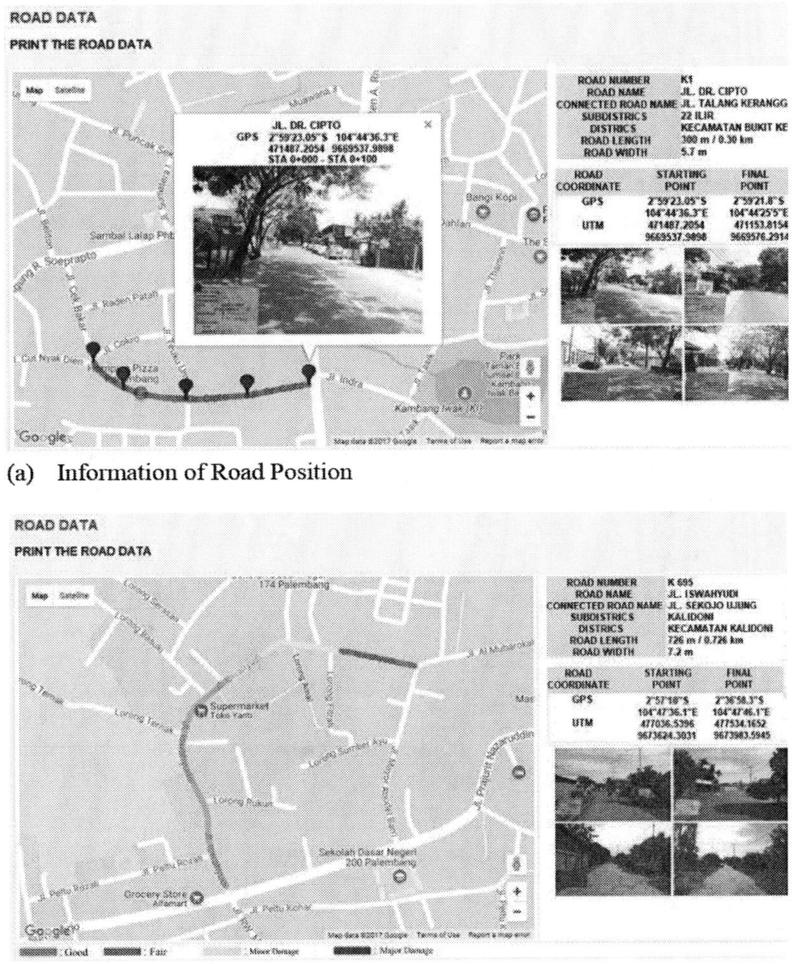

(a) Information of Road Position

(b) Information of Road Condition

Fig. 4. Output for a road segment

DOI: 10.1051/matecconf/201713807011

Analysis of data on the annual development of the road network can be done, among others, to see developments of network conditions and types of pavement. The development of pavement types and network conditions in the Palembang City in 2014 and 2016 are shown in Figures 5 and 6. Figures 5 and 6 show that:

- Despite an increase in the number and length of the road segments of the recorded data from 952 roads of 618.9 km in 2014 to 1204 roads of 681.8 km in 2016, the percentage distribution order of pavement types did not change, that is asphalt concrete, concrete with asphalt surface, concrete, dirt road, and gravel. The existence of dirt and gravel roads resulted from the fact that the government of Palembang City kept constructing pioneer roads, particularly those in newly developing suburban regions.

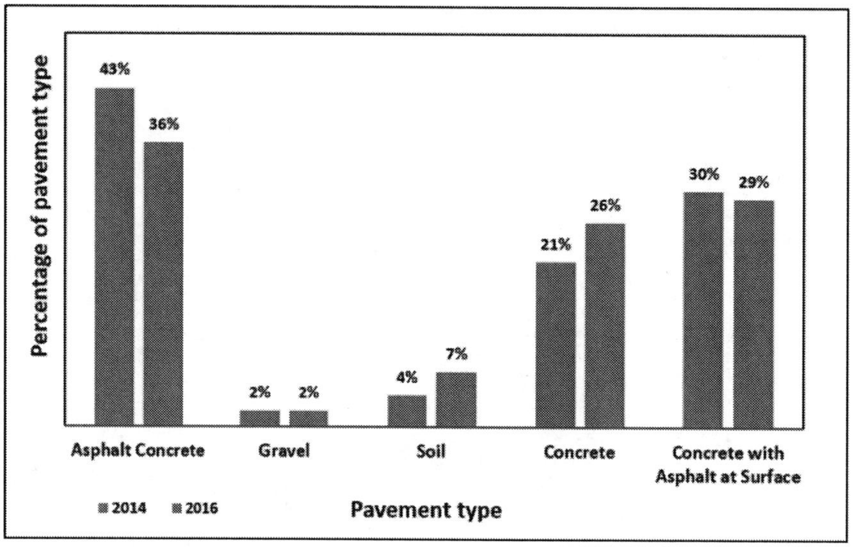

Fig. 5. Development of road pavement type

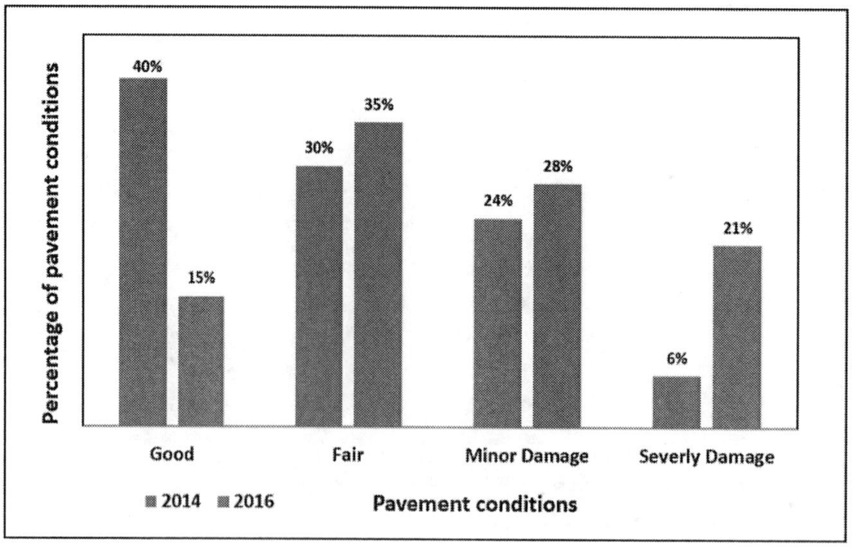

Fig. 6. Development of road network condition

- In 2016 the stable road condition (good, fair and minor damaged) was 78%, decreasing 16% compared with that of 2014. It indicates that there needs an immediate action to heavily damaged roads, although some of them occur in the pioneer roads constructed from dirt and gravel roads.

Until 2016 the number of bridges existing in Palembang City has been 156 bridges. The existing database is capable of showing the distribution of bridge locations, inventory data, bridge condition both in network and individually. The distribution of bridge locations in Palembang City on a road segment, bridge upper structure type and bridge condition is shown from Figures 7 to 10. Figures 9 and 10 show that:

- Most bridges in Palembang City have a concrete construction structure.
- Generally in 2016 there was a decline in the condition of bridges. Consequently, it is very critical to have the upkeep of the bridges whose bad conditions decline.

The results and analysis of road and bridge conditions both in network and individually can help the Office of Public Works in the upkeep program of roads and bridges in Palembang City.

The analysis of minimum service standard in the system developed is in accordance with the Regulation of Minister of Public Works of Republic of Indonesia No. 14/DRT/M/ 2010 on the minimum service standard of public works and spatial planning for achieving the minimum service standard [5]. The result of score output of the minimum service standard for accessibility, mobility, safety, road condition and speed is given in Table 1.

To improve the minimum service standard, the database system produces a list of road segments to be repaired in Palembang City. In 2016 as shown in Table 1, Palembang city will still have to increase the minimum service criteria of road network for mobility, safety, road conditions and speed.

Report menu in the database system is designed to produce a report according to the report format needed by the builder of Palembang City roads for both regional and national reporting.

Handling history menu is used to record the action of repair of road segments annually so that it can be recorded year after year. In the management of road maintenance, the data can be used for analysis of handling priority.

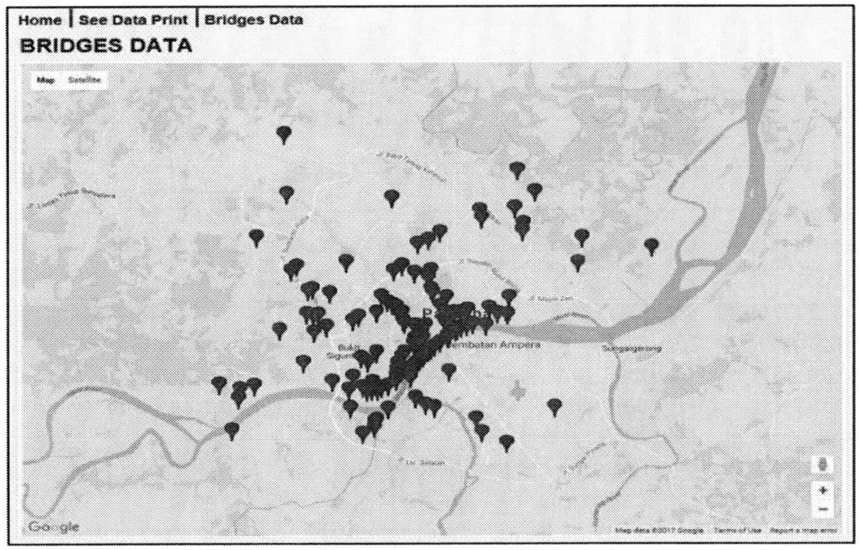

Fig. 7. Distribution of bridge locations in Palembang

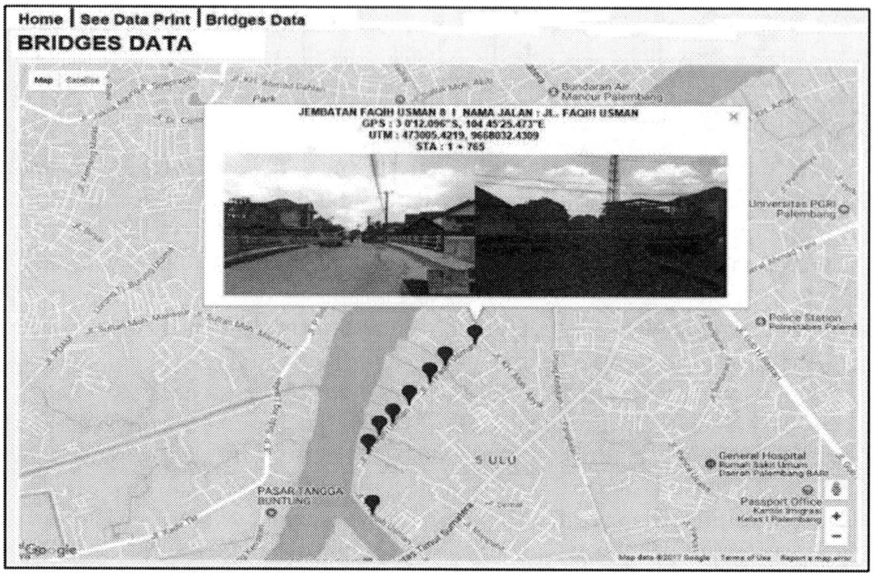

Fig. 8. Distribution of bridges on road segments

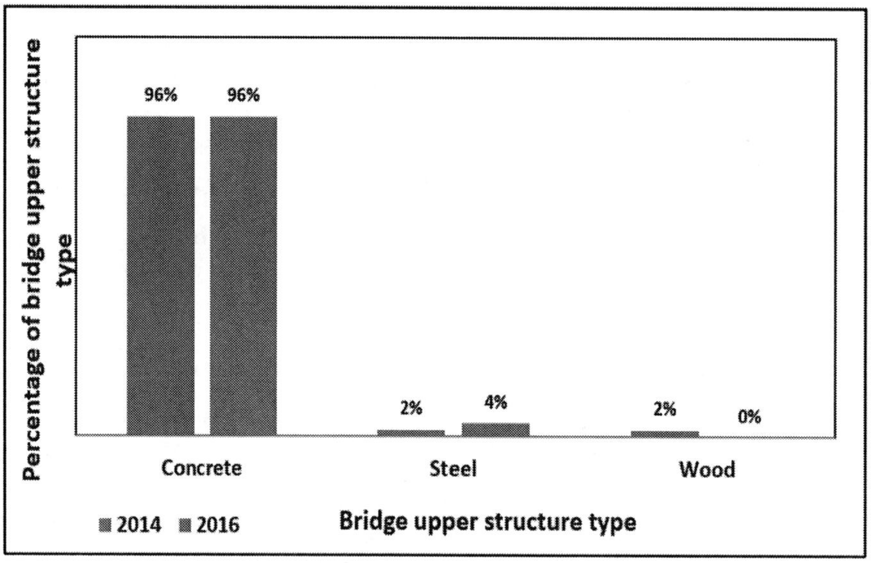

Fig. 9. Distribution of bridge upper structure type

Integration menu with national road database is used to prepare and upload road network data of Palembang City to national road database system. A complete road database of Palembang City at national level can be used by the central government to prepare a policy of road development in the regions.

After three years of running, the handling of the roads using the developed road database system has advantages, among other things:

- The road builder in Palembang City can easily have an overview of the network condition of roads and bridges.

Table 1. Standard score of minimum service

No	Minimum Service Criteria	Score	Standard	Remarks
1	Accessibility	1.62	1.5	High
2	Mobility	4.22	5	Need to improve
3	Safety	25.11	-	Need to improve
4	Road Condition	53.14	-	Need to improve
5	Speed	53.14	-	Need to improve

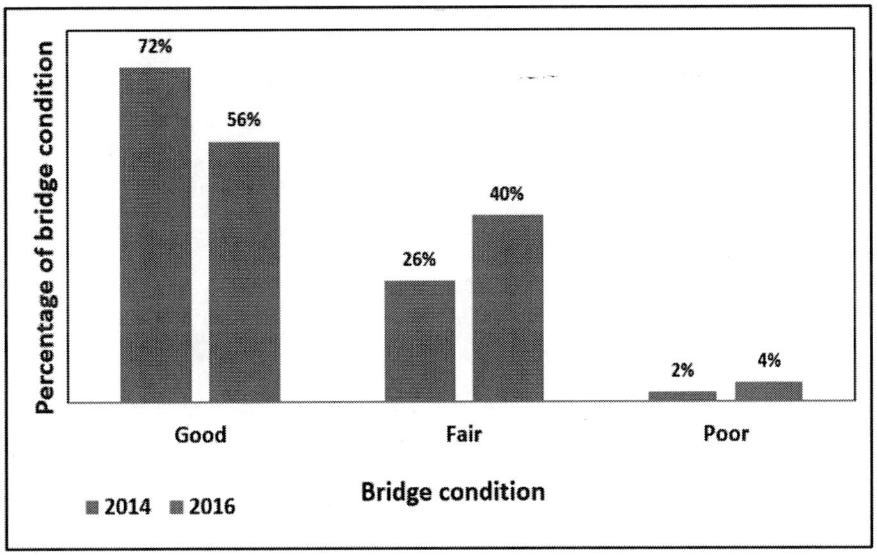

Fig. 10. Distribution of bridge condition

- The system can help road managers to view in detail the inventory data and conditions of each road and bridge so that an action can be executed fast and accurately.
- The above two mentioned can be conducted on other road components such as pavements and culverts.
- Due to the online-based system, the database system can help field technicians identify field problems by using cellular phone gadgets.
- The reporting system has been in accordance with the needed reporting format.

In a improvement of the existing database systems it showed that improvements in both data collection instruments for inventory data collection and the road condition used the practical tools that had a high degree of accuracy, given the road network of Palembang City is broad and complex.

4 Conclusions

The implementation of online-based road and bridge database system developed in 2014 indicates that the developed system is very helpful for road manager in looking at the overview of road and bridge network in Palembang City as well as helpful in preparing the planning of road and bridge handling more accurately and effectively. In the future, considering broad and complex road condition of Palembang City, the act of perfecting the output of the existing database system will be on the data collection instruments both for inventory data and road condition by using highly accurate practical.

Acknowledgment: The data used in this study were provided by the Public Works and Spatial Planning Department of Palembang City. Their cooperation is gratefully acknowledged.

References

1. B.M. Zulkipli, Y. Mohd, H. Hamidah, Geo. Inf. Sys. J. E **2**, 215-219 (2010)
2. P.I. Nyoman, H.E. Savitri, W.N.K. Ayu, Com. Sci. Iss. J. V **12**, 95-100 (2015)
3. T. Karthigaipriya, D. Srividya, Int. J. Innovations Sci. Eng. Res. V **1**, 18-23 (2014)
4. B. Andras, A.S. Kornelia, Acta Polytechnica Hungarica,V **2**, 91-102 (2005)
5. Regulation of Minister of Public Works of Republic of Indonesia No. 14/DRT/M/ 2010, *Minimum service standard of public works and spatial planning*, (2010)

DOI: 10.1051/matecconf/201713807012

Analysis of Model Loading and Unloading Time of Ships at Boom Baru Port, Palembang, Indonesia

Erika Buchari[1,*], *Dinar Dwi Anugerah* Putranto[2] , and *Aulia* Rahman[1]
[1]Centre of Excellence Multimodal Transportation of Sriwijaya University, Jl. Padang Selasa 524, Palembang 30139, Indonesia
[2]Department of Civil Engineering, Engineering Faculty Sriwijaya University, Palembang, Indonesia

Abstract. Logistic transport passing through road transport in Palembang city has been very high. To support green Transport, it is recommended to convert some of land logistic transports to river transport. However, the river transport has not been ready in terms of dredging and maintaining water level of Musi River. Furthermore, Boombaru Port, which has been shallow, and has high tidal level difference, can cause the process of loading and unloading become longer. High dwelling time is one of many problems that happen in port's container terminal in Indonesia. This is commonly caused by longer loading and unloading time than those ports with high navigability and good handling equipment. Loading and unloading time depends on some factors, i.e ship size, limitation of equipment and stacking yard, a shallow port basin, tidal period and tidal magnitude. This paper aims to model the loading and unloading Time at Container Terminal of Boombaru Port, and analyze the effect of some factors on loading and unloading time. The model derived is Y = 0.501 X1 – 0.021 X2 – 2.63 X3 – 29.745, which X1 is Length Over All (m), X2 is Low Tide Elevation (cm) and X3 is duration of tidal period (hour).

1 Introduction

Sometimes, it was found the fact that transport problem is not derived from transport sector itself, but it is mostly come from other independent sectors, which is not connected one from another. For example the problems from unconnected supply chain of goods transport. Logistic transport passing through road transport in Palembang city has been very high. To support green Transport, it is recommended to convert some of land logistic transports to river transport. However, the river transport has not been ready in terms of dredging and maintaining water level of Musi River.

Furthermore, Boombaru Port, which has been shallow, and has high tidal level difference, can cause the process of loading and unloading become longer. High dwelling time is one of many problems that happen in port's container terminal in Indonesia. This is commonly caused by longer loading and unloading time than those ports with high navigability and good handling equipment. Loading and unloading time depends on some

* Corresponding author: eribas17@gmail.com

factors, i.e. ship size, limitation of equipment and stacking yard, and shallow port basin, tidal period and tidal magnitude.

Dwelling time is accumulated time from unloading and loading containers, until the containers leave the terminal trough the main gate [1]. There are many factors that affect dwelling time; among other things is the speed of handling process. In terms of berthing handling, the process of handling is mostly unloading time.

The speed of unloading process is affected by many factors, such as ship size, performance of the equipment, number of handling equipment, qualifications from the operator, the depth of post basin. In the case of river port, tidal level is also the main but neglected factor for time of handling process.

Since the time of handling process is very important factors, therefore this research will discuss about the factors which affect unloading time which include tidal level difference from sedimentation.

1.1 Problem Formulation

Problem Formulated are

1. How to model the loading and unloading time at Container Terminal of Boombaru Port, and
2. How to analyze the effect of some factors on loading and unloading time

1.2 The aims of the study

This paper is aimed to

1) Model the unloading Time at Container Terminal of Boombaru Port, and
2) Analyze the effect of some factors on unloading time

2 Literature Study

2.1 Dwelling Time

Dwelling time is accumulated time from unloading and loading containers, until the containers leave the terminal trough the main gate [1].

At Boombaru Port, Berth Occupancy Ratio is about 51.03%, which is above standard required services, has caused long time delay for ships berthing [2]. As a result the flow of containers at Container Ports Boombaru has been less than maximum required capacity, which gives impacts to lower productivity.

Further research has been done to get the model of productivity at container terminal, Boombaru. By the means of Multiple Linear Regression model with Stepwise method, the model derived is $Y=1,543\ X_1 + 47,253\ X_2 + 0,195\ X_3 - 162.964$. Y is containers flow in TEU per day; X1 Average length of All in m' (LOA); X2 is density of containers in ship in TEU's/m' and X3 is Berth troughput in TEU's per day **[3].** The constant -162.964 has been annoying results, which reflect unknown parameter. Therefore this study will see what other parameters that also affect berthing time. It is realized that some problems could come from Port Basin in front of Berth and sedimentation along the tidal river/canal approach. Further research has been set up to find out what other factors that affect the productivity and dwelling time in port terminal.

2.2 Tidal Basin

Tides are the rise and the fall of sea levels caused by the combined effects of the gravitational forces exerted by the moon, the sun and the rotation of earth**.** (Widyantoro,

 DOI: 10.1051/matecconf/201713807012

2014). Tidal level are grouped to three main components, namely low tide, high tide, and tidal range. The tidal range is derived from the difference measure of high tide and succeeding low tide.

3 Research Method

The research Method are delineated as the following figure 1.

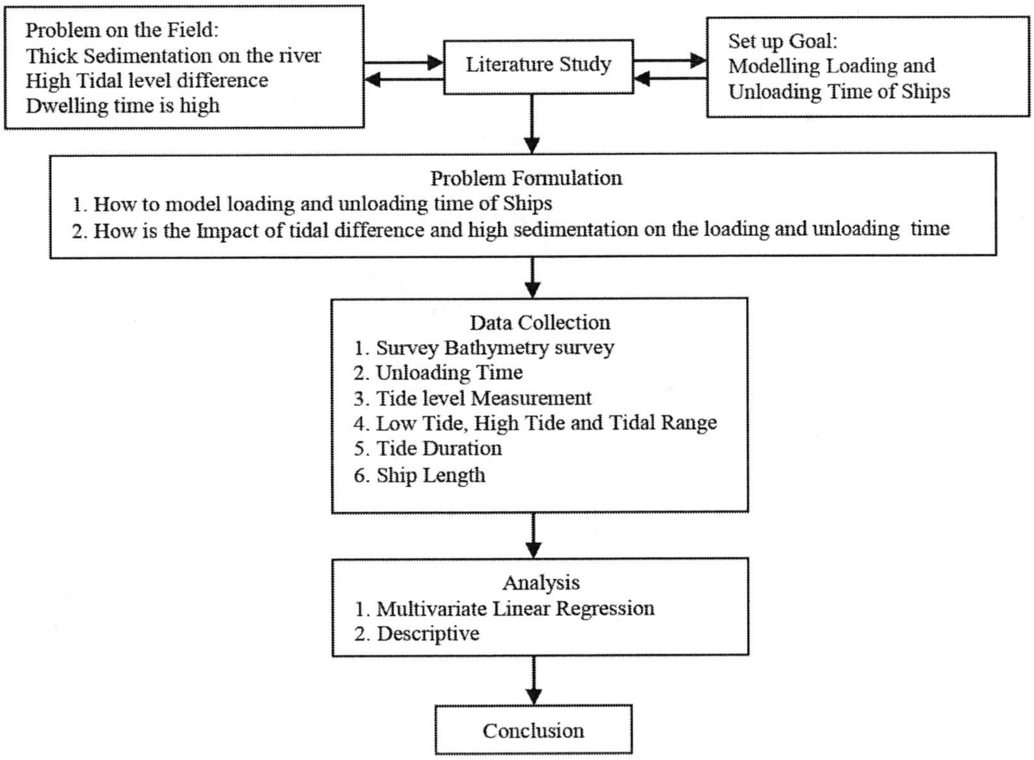

Figure 1 Flow chart of research

4. SURVEY RESULTS AND ANALYSIS

4.1 Survey Results

Data Variables

There are six variables that are predicted in finding productivity model, namely:

a. Average length of all (AvLOA)
b. Unloading Time
c. Low Tide
d. High Tide
e. Tidal Range
f. Duration of Tide

In order to find loading and unloading time model, some data collection had been done. The following **Table1** is presenting average length of all of ships for each month.

DOI: 10.1051/matecconf/201713807012

Table 1. Average Length of All Ships

No	Month	Average Length of All (m)
1	January	101.3
2	February	101.37
3	March	100.04
4	April	101.0
5	May	99.91
6	June	99.44
7	July	101.11
8	August	101.64
9	September	105.51
10	October	106.23
11	November	100.1
12	December	100.9
13	January	101.12
14	February	99.07
15	March	99.65

The following **table 2** is presenting average Loading and unloading time each month.

Table 2. Average Loading and unloading time each month

No	Month	Loading Unloading time each month (hour)
1	January	13.66
2	February	14.76
3	March	16.71
4	April	16.67
5	May	15.57
6	June	16.59
7	July	17.34
8	August	15.95
9	September	16.04
10	October	16.02

No	Month	Loading Unloading time each month (hour)
11	November	14.55
12	December	12.77
13	January	11.65
14	February	12.58
15	March	11.79

The following **Table 3** is presenting data of average Low Tide Elevation

Table 3. Average Low of Tide Elevation

No	Month	Low Tide Elevation (cm)
1	January	44.0
2	February	49.0
3	March	43.0
4	April	56.0
5	May	40.0
6	June	30.0
7	July	40.0
8	August	60.0
9	September	49.0
10	October	45.0
11	November	45.0
12	December	49.0
13	January	58.0
14	February	58.0
15	March	65.0

The following **Table 4** is presenting data of Average High Tide Elevation.

No	Month	High Tide Elevation (cm)
1	January	190.0
2	February	180.0
3	March	155.0

No	Month	High Tide Elevation (cm)
4	April	166.0
5	May	173.0
6	June	191.0
7	July	230.0
8	August	210.0
9	September	215.0
10	October	217.0
11	November	215.0
12	December	221.0
13	January	222.0
14	February	228.0
15	March	215.0

The following **table 5** is presenting data of Range of Tidal difference of fluctuation.

Table 5. Range of Tidal level

No	Month	Fluctuation (cm)
1	January	18.0
2	February	18.0
3	March	14.0
4	April	13.0
5	May	12.0
6	June	11.0
7	July	14.0
8	August	24.0
9	September	24.0
10	October	24.0
11	November	24.0
12	December	24.0
13	January	24.0
14	February	24.0
15	March	24.0

The following **Table 6** is presenting data of Average Duration of Tide.

Table 6. Average Duration of Tide

No	**Month**	**Average duration**
1	January	18.0
2	February	18.0
3	March	14.0
4	April	13.0
5	May	12.0
6	June	11.0
7	July	14.0
8	August	24.0
9	September	24.0
10	October	24.0
11	November	24.0
12	December	24.0
13	January	24.0
14	February	24.0
15	March	24.0

4.2 Analysis of Model of Unloading Ships

Before doing regression process, data were tested and had to pass assumption test. Classical Assumption Method is used in order to verify research data, such as Normality Test, Multicollinearity Index, Heteroscedasticity Test, auto correlation Durbin-Watson, and Linearity Test.

After it is processed with MLR and Step Wise Method, summary model is derived as it is presented in the following table:

Table 7. Summary Model Analysis

Model Summary				
Model	R	R Square	Adjusted R Square	Std. Error of the Estimate
1		.665	.479	1.37916
2		.664	.530	1.30988
3		.663	.571	1.25106
4		.657	.600	1.20801
a. Predictors: (Constant), Tide duration, SC, Tide difference, LOWT, LOA				
b. Predictors: (Constant), Tide duration, SC, LOWT, LOA				
c. Predictors: (Constant), Tide duration, LOWT, LOA				
d. Predictors: (Constant), Tide duration, LOA				

While the result of FTest is presented in the following table 7.

Table 7. Table F test

Model		Sum of Squares	df	Mean Square	F	Sig.
1	Regression	33.991	5	6.798	3.574	.047[a]
	Residual	17.119	9	1.902		
	Total	51.110	14			
2	Regression	33.952	4	8.488	4.947	.018[b]
	Residual	17.158	10	1.716		
	Total	51.110	14			
3	Regression	33.893	3	11.298	7.218	.006[c]
	Residual	17.217	11	1.565		
	Total	51.110	14			
4	Regression	33.598	2	16.799	11.512	.002[d]
	Residual	17.511	12	1.459		
	Total	51.110	14			
a. Predictors: (Constant), Tide duration, SC, Tide Dif, LOWT, LOA						
b. Predictors: (Constant), Tide duration, SC, LOWT, LOA						
c. Predictors: (Constant), Tide duration, LOWT, LOA						
d. Predictors: (Constant), Tide duration, LOA						
e. Dependent Variable: Unloading Time						

After analysis having been done, some equation models which represent unloading time of ships in Container Port terminal Boom Baru, namely:

1) Y = -0.41 Ship Arrival +0.444 LOA -0.14 LOWT +0.003 Tide Dif -0.270 Tide duration -23.540
2) Y = -0.35 Ship Arrival +0.448 LOA -0.21 LOWT -0.257 Tide duration -23.617
3) Y = 0.501 LOA -0.21 LOWT -0.263 Tide duration -29.745

From those three models derived, Model three is more accurate, because all independent variables, such as LOA, LOWT and Tide duration has significant impact (Sig. < 0,05), so that the third model is considered best-fit model. This model then can be used to predict unloading time in Boombaru Port.

5 Conclusion

From the above analysis and discussion, it can be concluded as the following:

1) Model the unloading Time at Container Terminal of Boombaru Port is Y = 0.501 LOA -0.21 LOWT -0.263 Tide duration -29.745
2) Loading and unloading time are affected by length of all, Low Tide level, and tide duration

References

1. UNCTAD Annual Report (2011)
2. A. Maulana Situmorang, E. Buchari. Proceedings of The 18th FSTPT, Lampung University, (2015)
3. A. Rachman, E. Buchari, Proceedings of The 19th FSTPT, UII (2016)
4. E. Buchari, H. Basri, Procedia Engineering **125**, 519-525 (2015)
5. E. Buchari, Proceedings of EASTS Conference (2015)
6. H. Basri, Port Engineering Module for Inland Waterways and Ferries Port Management Course, Mariana (2001)
7. E. Buchari, Port Design Module for Inland Waterways and Ferries Port Management Course, Mariana (2001)
8. M. Zilzaludin P.N., E. Buchari, Proceedings of The 19th FSTPT, UII (2016)

Providing facilities to improve accessibilities to railway station for all

Wanrawee Nawagawong[1] *and Wasaporn* Techapeeraparnich[*,2]

[1] Civil and Environmental Engineering Department, 25/25 Phutthamonthol Sai 4, Phutthamonthol, Salaya, Nakhonpathom, Thailand, 73170.

Abstract. Thailand is now entering to an aging society, as the number of elderly will increase to 25.2 million people in the year 2030 [1]. From the statistics of National Statistical office 2007, the number of people with disabilities (PwDs) is approximately 2.9% or 1.7 million people of the total population. The Thai government has recently enacted many policies and acts specifically to equal the right of these people to be able to live like a normal person. One of the aspects of these acts is enabling them to access transportation services without difficulties. As Thailand has recently invested large amount budget on developing many railway projects, thus this is an important opportunity to enhance the design standard of transportation facilities to meet the demands of all types of users. This research is aimed to compare the laws and guidelines of railway station and facilities designed for all at international level and to explore the principles of railway station design that used for railway station design in order to make the recommendation for Thai railway station design to improve accessibility in terms of designs and additional facilities.

1 Introduction

It is reported that 1 in 8 U.S. adults aged 35 to 64 is a disability and 1 in 3 adults aged 65 or older has a disability, meanwhile the population is approximately 309 million [2]. Similar to the European Union, the number of elderly and disabled people is currently estimated between 60 – 80 million people and disability are highest in the age group 65 over [3].Thailand is not different to United State of America. We are entering into aging society and people with disabilities (PwDs) caused from aging keep increasing. According to a statistic from the Office of National Economic and Social Development Board (2015), Thailand will have the elderly at approximately 21.2% in year 2025 and will increase to 25.2% in year 2030, at which Thailand will become a complete aging society. WHO's reported that disability is usually defined in terms of restrictions on the ability to perform activities of daily living (ADL) or the inability to function independent [4]. The limited of mobility is common to unhealthy elderly and 1 of 5 of elderly will have difficulty in mobility according to the age and occurring since middle age (50-64 years old).

[2]Certified from Mahidol University Institutional Review Board Committee project no. 2016-206.0912 dated 6 March 2017

[*]Corresponding author: wasaporn.tac@mahidol.ac.th

The main casual symptoms cited were "pain in the legs or foot" then "shortness of breath" [5]. Besides elderly problems, the record of Department for Empowerment of Persons with Disabilities as of October 2016 reported the statistic of registration of persons with disabilities approximate 1.7 million people across the country.

Recently, The Thai government has invested huge budget in the development of railway transportation development. Many projects are now progressing on feasibility study, design or even though under construction. Economic Intelligence Center reported that Thailand's Transport Infrastructure Investment Plan (2013-2020) has a total investment of 2 Trillion Baht including strategic development and improvement of transport system to raise mobility approximate 6 billion (60% of 2 Trillion Baht), such as Rapid Transit System Development Project for Urban Passenger Transport System and transport development project to linking core economic areas within the country[6]. Thus this is an important opportunity to enhance the standard of transportation facilities to meet the demands of all types of users. This research is aimed to compare the laws and guidelines of railway station and facilities designed for

all at international level and to explore the principles of railway station design that used for railway station design in order to make the recommendation for improving Thai standard to improve accessibility of these people to the railway station.

2 Literature review

To improve the accessibility for all, it is necessary to understand the meaning of "for all". The British Standards Institute (2005) defined "for all" as "the design of mainstream products and/or services that is accessible to, and usable by, as many people as reasonably possible without the need for special adaptation or specialized design." EIDD Design for All-Europe also defined that design for all is a design for human diversities, social inclusion and equality and aims to enable all people to have equal opportunities to participate in every aspect of society [7]. Therefore, the meaning term of "for all" means "designed for people of all ages in the community without restrictions on age, gender, children, the sick people, the elderly, pregnant women and PwDs". The Thai government has enacted Building Control Act 2005 (No.5) which effective from 2005 that the design and calculation of buildings must provide all facilities for PwDs and elderly as required[8]. The types of building that must provide the facilities to PwDs and the elderly are:

1) Hospital, health center, government building, State-owned enterprises buildings, educational institutions, libraries, museum, massed transit stations such as airports, train stations, bus stations, wharf which area of the building purpose is using for the public over 300 square meters.

2) Office, theater, hotel, auditorium, stadium, shopping center, department store which area of the building purposes is using for the public over 2,000 square meters.

Likewise, the Ministerial Regulations of Designing or Equipment Facilities of the Ministry of Transportation to enable people with disabilities to access and utilize (2013) states that railway station must provide facilities for PwDs and elderly as follows [9]. 1) Door for PwDs 2) Seating or wheelchair parking for PwDs 3) Ramp 4) Handrail 5) Surface floor for visual disability 6) Toilet with PwDs assistive 7) Lift 8) Telephone with PwDs assistive 9) Parking for PwDs and 10) PwDs facilities signage Additionally, The Thai government offers incentives to building owner to whom who provide facilities to accommodate to PwDs and elderly people according to the Act of Promote and Improve the Quality of Life of the Disabled (2008) [10]. There are many laws already enforced in Thailand However, the details of types, numbers, dimension of the assistive infrastructure and facilities are still undeveloped.

Many types of people are living in a society such as child, adult, elderly or even PwDs. All of those people have the right to access any public infrastructure as equality policy. To do this, PwDs may need the assistive facilities more than ordinary people. According to the International Classification of Functioning, Disability, and Health (ICF) had set up a framework using in inter-communication collaborative practice and person-centered care and it is the disability classification by physical model. The two major parts of ICF framework are 1) Functioning & disability: It means the impairments of body functions such as movement disabilities, learning disabilities and it effects to daily activities of individual PwDs. 2) Contextual Factors: this part means the participation or involvement of people participation in society and environment[11].

Similar to Disability Rehabilitation Act 1991 [12] issued by Ministry of Health, disabilities can be classified into 6 types. The details as shown in Table 1. below.

Table 1. Type of PwDs in Thailand [12]

Item	Type of PwDs	Characteristics
1	Vision disability	The individual who has limitation in participating activities in daily life or limited to participate in social activities, which is a result of vision impairments.
2	Hearing disability	The individual who has the limited in the participation of activities in daily life or limited to participate in social activities and loses the ability of hearing, which is a result of hearing. In this case, it is included the impairment of interpretation.
3	Mobility disability	The individual ability by prosthesis implantation or their physical aid this case including a wheelchair.
4	Mental or behavioral disability	The individual who has the limited in participation activities in daily life or to participate in social activities, which is a result of deficiency or abnormality of mind.
5	Intellectual disability	The individual who has the limited in participation activities in daily life or to participate in social activities, which is a result of cognitive impairments and the development slower than usual.

Remark: This research is going to study for 3 types of disabilities including vision, hearing and mobility disabilities only.

3 Design principle for public transportation facilities

The widely accepted design principle of building or facility including public transportation facilities and station design to enable the use of all are 1) Universal Design 2) Barrier Free Design and 3) Service Design.

Universal Design: Universal design means the design of composition and environment that allow everyone to access regardless of age, size, ability or disability matters. The concept of Universal Design is trying to eliminate the certain barriers that limit the usability of environment for people with disabilities. Universal Design was developed from "Barrier Free" and "Assistive Technology" (Peloquin, 1994). The following are main ideas of universal design. 1) EquiTable use 2) Flexible in use 3) Simple and intuitive 4) Perceptible Information 5) Tolerance for error 6) Low physical effort 7) Size and space for approach and use

Barrier Free Design: United Nation (UN) defined that barrier-free could fulfill the needs of all people equally. Barrier free design considers holistically allowing the people with disabilities (people with wheelchair or less mobility) to access any places such as buildings, restrooms, transportation stations, etc. or allow them to easily use the facilities[14].

Service Design: Service design is multi-disciplinary and integrative field. It helps to both innovate and improve services to make them more useful, usable, desirable for clients, as well as more efficient and effective for organizations. Service design consists of five concepts as following 1) User centered 2) Co-creative 3) Sequencing 4) Evidencing 5) Holistic

Therefore if people with reduced mobility is able to access to public facilities, it will be a benefit to all such as families with small children or prams or push-chair or people having a lot of luggage. People, carrying a suitcase in each hand, have difficulty entering a station if the door does not open automatically. Thus, the requirements of those people need to be considered for railway facilities design. Besides this issue, personal safety while waiting or departing at railway stations where criminals were tended to hide around the area [18]. Both accessibility

 DOI: 10.1051/matecconf/201713807013

and criminal's problems became the policy or legislation of EU countries which will be enforced to all licensed train and station operators to strictly implement.

4 Methodology

This research was firstly conducted by reviewing documents. The documents are standards and guidelines for railway stations and facilities design from different countries. Then the list is summarized and compared. List of standard and guidelines gathered and reviewed in this study are as shown in Table 2.

Table 2. List of standards and guidelines reviewed and considered in this study.

Item	Standard Name	Issued by	Country
1.	2006 Americans with Disabilities Act (ADA) Standards for Transportation Facilities[15]	United State Access board	The United State of America
2.	COMMISSION REGULATION No 1300/2014 of 18 November 2014 on the technical specifications for interoperability relating to accessibility of the Union's rail system for persons with disabilities and persons with reduced mobility (EU) [19]	European Union Agency for Railway	The European Union
3.	Accessibility for the Disabled A Design Manual for a Barrier Free Environment [14]	United Nation (UN)	-
4.	Disability Standards for Accessible Public Transport 2002 (DSAPT) [20]	Australian Government	Australia
5.	2007 Facility Accessibility Design Standard [21]	City of London	Canada
6.	Ministerial Regulations of Designing or Equipment Facilities of the Ministry of Transportation to enable people with disabilities to access and utilize [9]	Thai government	Thailand

Standards listed above, some of them for example 2006 Americans with Disabilities Act (ADA) was based on "Barrier Free Design" similar to Accessibility for the Disabled A Design Manual for a Barrier Free Environment which was developed by United Nation (UN). While European Union Agency for Railway had developed guideline with a broader consideration of both Universal and Barrier Free Design. The study of is the evidence that even by reducing the gap between the train and platform at underground stations and widening the doors of trains according to Universal Design and Barrier Free Design can have a significant impact in saving the train dwell time[21]. The results of research found that these are able to increase the benefit–cost ratio of a project and also specified that improving passenger mobility and network efficiency research indicates a cost benefit to operators.

5 Questionnaire and sampling

After reviewing the guidelines and standards, the researcher developed the questionnaire to ask respondents to assess the importance of facilities which are developed from the comparison of standards and guidelines. Data is collected from sampling group which composed of all types of prospective passengers. The result of this part will be preliminary analyzed by using descriptive statistic. The research sampling groups were identified as shown in Table 3. below.

Table 3. Research sampling groups

Type	Description	Quantity
1	Visual disability	5
2	Hearing or meaningful disability	5
3	Mobility disability	5
4	General passenger	5
5	Elderly passenger	5
6	Passenger with luggage	5
7	Passenger with pram	5
8	Passenger who travel with children (age under 10 years old)	5
	Total	40

 DOI: 10.1051/matecconf/201713807013

6 Findings

After reviewing standard guidelines from different countries, the researcher then made a comparison of the components of public transportation station facilities as shown in Table 4. It is found that these standards were moderately similar; by way of it concentrates on accessibility for all and sustainability of 2 groups those are 1) Compulsory facilities that are necessary for stations function design and safety reason regardless of station size. 2) Optional facilities that will be able to assist passengers to travel safer, more accessible and infrastructure usage.
The facilities indicated by the above standards able to be compared as shown in Table 4.

Table 4. Thailand Compulsory of public transportation and comparison to international standards

Item	Facility name	ADA[1]	EU[2]	UN[3]	CA[4]	DSAPT[5]	Thai[6]
1	Parking	✓	✓	✓	✓		✓
2	Street furniture			✓	✓	✓	
3	Areas of refuge Emergency exit and fire evacuation and area of rescue assistance	✓			✓		✓
4	Passenger loading zone	✓			✓		
5	Bus boarding and alighting areas	✓				✓	
6	Level track crossing at station or Overpass for PwDs		✓				✓
7	Curb ramp	✓		✓	✓		✓
8	Ramp	✓		✓		✓	✓
9	Boarding ramp					✓	✓
10	Accessible routes, Pedestrian for PwDs or Obstacle free routes	✓	✓	✓	✓		✓
11	Entrance	✓					
12	Doors, doorway and gate	✓	✓	✓			✓
13	Stair	✓		✓	✓	✓	✓
14	Stairway	✓					
15	Directional sign	✓		✓	✓	✓	✓
16	Floor or ground surfaces	✓		✓	✓	✓	✓
17	Highlighting of transparent obstacle	✓		✓			
18	Handrail	✓		✓		✓	✓
19	Lighting	✓		✓		✓	
20	Pathway			✓		✓	
21	Pedestrian crossing			✓			✓
22	Public address system	✓	✓		✓		
23	Toilet or restroom	✓	✓	✓	✓	✓	✓
24	Platform	✓	✓				
25	DetecTable warning surface or Tactile ground surface indicators	✓			✓	✓	
26	Turnstile				✓		
27	Protecting of protruding and overhead objects	✓			✓		
28	Maneuver area for wheelchair	✓				✓	
29	Resting point every 60 meters with seat				✓	✓	
30	Waiting area with 2 seats available to PwDs	✓				✓	✓
31	Braille or Visual information	✓	✓			✓	
32	automatic door to protect between rail / train and platform						✓
	Total	**26**	**9**	**16**	**15**	**16**	**16**

As the results of comparing standard in Table 4, among other ADA provided the highest number of compulsory facilities at 26 facilities. Even though other standards such as UN, CA and DSAPT are providing almost the same number of compulsory facilities but they indication of compulsory facilities are somewhat different. For the second group "optional facilities" is indicated by different standards can be compared as shown in Table 5. below.

 DOI: 10.1051/matecconf/201713807013

Table 5. Comparison of optional railway station facilities.

Item	Facility name	ADA[1]	EU[2]	UN[3]	CA[4]	DSAPT[5]	Thai [6]
1	Telephone	✓			✓		✓
2	Ticketing, information desks and customer assistance points		✓			✓	✓
3	Ticketing and automatic fare vending	✓			✓	✓	
4	Elevator or lift	✓			✓	✓	✓
5	Escalator	✓			✓		✓
6	Fire alarm system	✓			✓	✓	
7	Platform lift	✓			✓		
8	Drinking fountain	✓			✓		
9	Control operating mechanism				✓		
10	Landscaping and planting				✓		
13	Food and drink service					✓	
14	Belonging service					✓	
15	Automatic or power assist door					✓	
16	Elevated platform				✓		
17	Automatic teller machines	✓					
18	Two way communication system	✓					
19	Change machine	✓					
	Total	**10**	**1**	**0**	**10**	**7**	**3**

Note:

1. Americans with Disabilities Act (ADA)
2. European Union Agency for Railway
3. Accessibility for the Disabled - A Design Manual for a Barrier Free Environment
4. Disability Standards for Accessible Public Transport 2002
5. 2007 Facility Accessibility Design Standard (Canada)
6. The ministerial regulations for facilitation of disabled or handicapped and elderly facilities 2013 (Thailand).

Some facilities are additionally recommended for the convenience of the passengrers and for improving accessibility. The additional facilities are; 1) Automatic ticket gate 2) Cycle lanes on the road leading up to station access 3) Transit Hot Lines with human operators 4) Traveler's aid 5) Car rentals and car sharing 6) Bike sharing 7) Wireless internet access 8) Public lockers 9) Retail

It was found that the compulsory, optional and additional facilities for all are gathered as 56 facilities as shown in Table 4, 5 and additional requirement.

The researcher has designed the simple question to find out the approaches of 1) Compulsory 2) Optional 3) not required. These 3 categorized facilities are to be assessed the importance by respondents. The data was analyzed by using Frequency Index (FI) technique in order to classify facility into 3 categories. Also, FI technique will be employed to rank the importance of each facility. The respondents are required to prioritize the importance of facilities ranging from 1-3. FI technique was categorized on three-point Likert's scale as compulsory, optional and not required and will be calculated according to the following formula Frequency Index (F.I.) (%)

$$\sum a \left(\frac{n}{N}\right) * 100/3 \tag{1}$$

Where;

a =	Constant weighing that given to response (range from 1 Not required, 2 Optional and 3 Required)
n =	Frequency of the responses
N =	Total number of respondent

The assessment citeria of facility categories are shown in the below table

Table 6. Criteria of facility assessment

Category	Frequency Index Percentage
Compulsory	> 72.5%
Optional	< 55% ≤ 72.5%
Not required	< 55%

The follwing table is the results analysis of facilities importance

Table 7. Result of railway station facilities importance assessment

Item	Facility name	Percent of responses (%)			Conclude	F.I. (%)	Rank
		Compulsory	Optional	Not required			
1	Parking	60.00	32.50	7.50	Compulsory	84.17	20
2	Street furniture	30.00	60.00	10.00	Compulsory	73.33	40
3	Areas of refuge Emergency exit and fire evacuation and area of rescue assistance	37.50	45.00	17.50	Compulsory	73.33	43
4	Passenger loading zone	72.50	25.00	2.50	Compulsory	90.00	9
5	Level track crossing at station or Overpass for PwDs	62.50	35.00	2.50	Compulsory	86.67	14
6	Curb ramp	60.00	37.50	2.50	Compulsory	85.83	17
7	Ramp	60.00	40.00	-	Compulsory	86.67	14
8	Boarding ramp	55.00	42.50	2.50	Compulsory	84.17	20
9	Accessible routes, Pedestrian for PwDs or Obstacle free routes	70.00	30.00	-	Compulsory	90.00	9
10	Main entrance	60.00	40.00	-	Compulsory	86.67	14
11	Stairway	57.50	37.50	5.00	Compulsory	84.17	20
12	Directional sign	77.50	17.50	5.00	Compulsory	90.83	8
13	Floor or ground surfaces	65.00	32.50	2.50	Compulsory	87.50	13
14	Highlighting of transparent obstacle	60.00	32.50	7.50	Compulsory	84.17	20
15	Handrail	82.50	17.50	0.00	Compulsory	94.17	4
16	Lighting	87.50	12.50	0.00	Compulsory	95.83	1
17	Barrier free pathway	77.50	22.50	0.00	Compulsory	92.50	5
18	Pedestrian crossing	60.00	25.00	15.00	Compulsory	81.67	32
19	Public address system	70.00	27.50	2.50	Compulsory	89.17	12
20	Toilet or restroom	77.50	22.50	0.00	Compulsory	92.50	5
21	Platform	85.00	15.00	0.00	Compulsory	95.00	3
22	DetecTable warning surface or Tactile ground surface indicators	57.50	42.50	0.00	Compulsory	85.83	17
23	Turnstile	17.50	22.50	60.00	Not required	52.50	56
24	Protecting of protruding and overhead objects	50.00	47.50	2.50	Compulsory	82.50	28
25	Maneuver area for wheelchair	55.00	37.50	7.50	Compulsory	82.50	28
26	Resting point every 60 meters with seat	45.00	47.50	7.50	Compulsory	79.17	35
27	Waiting area with 2 seats available to PwDs	32.50	60.00	7.50	Compulsory	75.00	39
28	Braille or Visual information	55.00	37.50	7.50	Compulsory	82.50	28
29	Automatic door to protect between rail / train and platform	60.00	32.50	7.50	Compulsory	84.17	20

Item	Facility name	Percent of responses (%)			Conclude	F.I. (%)	Rank
		Compulsory	Optional	Not required			
30	Telephone	40.00	30.00	30.00	Optional	70.00	49
31	Ticketing, information desks and customer assistance points	70.00	30.00	0.00	Compulsory	90.00	9
32	Ticketing and automoatic fare vending	52.50	35.00	12.50	Compulsory	80.00	34
33	Elevator or lift	62.50	30.00	7.50	Compulsory	85.00	19
34	Escalator	52.50	40.00	7.50	Compulsory	81.67	31
35	Fire alarm system with light signalling	87.50	12.50	0.00	Compulsory	95.83	1
36	Platform lift	50.00	42.50	7.50	Compulsory	80.83	33
37	Drinking fountain	35.00	42.50	22.50	Optional	70.83	47
38	Emergency help point	75.00	25.00	0.00	Compulsory	91.67	7
39	Landscaping and planting	57.50	37.50	5.00	Compulsory	84.17	20
40	Food and drink service	37.50	42.50	20.00	Compulsory	72.50	44
41	Belonging service	17.50	70.00	12.50	Optional	68.33	50
42	Automatic or power assist door	45.00	27.50	27.50	Compulsory	72.50	44
43	Elevated platform	40.00	40.00	20.00	Compulsory	73.33	40
44	Automatic teller machines	55.00	40.00	5.00	Compulsory	83.33	27
45	Two way communication system	40.00	32.50	27.50	Optional	70.83	47
46	Change machine	45.00	40.00	15.00	Compulsory	76.67	36
47	Mobile garbage bin	35.00	30.00	35.00	Optional	66.67	51
48	Automatic ticket gage	45.00	37.50	17.50	Compulsory	75.83	38
49	Cycle lanes on the road leading up to station access	25.00	45.00	30.00	Optional	65.00	52
50	Transit Hot Lines with human operators	42.50	45.00	12.50	Compulsory	76.67	37
51	Traveler's aid	57.50	37.50	5.00	Compulsory	84.17	20
52	Car rentals and car sharing	15.00	57.50	27.50	Optional	62.50	53
53	Bike sharing	17.50	30.00	52.50	Not required	55.00	55
54	Free WiFi	37.50	40.00	22.50	Optional	71.67	46
55	Public lockers	15.00	45.00	40.00	Optional	58.33	54
56	Retail	32.50	55.00	12.50	Compulsory	73.33	40

Remark:

1. This research had prioritized only required facilities.
2. The conclusion of each facility is considering from the percentage of Frequency Index (F.I.) should more than 40%.

According to F.I. calculation, the first fifth compulsory facilities can be concluded as Table 8.

Table 8. Priority importance of facilities

Priority	Facilities	Frequency Index (F.I.)
1	Firm alarm with siren signal and Lighting	95.83%
2	Platform	95.00%
3	Lift	94.17%
4	Directional signage, Barrier-free pathway and Toilet	92.85%
5	Emergency help point	92.83%

The first fifth of importance facilities are 1) Firm alarm with siren signal and Lighting 2) Platform 3) Lift 4) Directional signage, Barrier free pathway and Toilet 5) Emergency help point.

The results from prioritizing the important of faciliities also shown that there are 45 compulsory facilities , 9 optional facilities whereas 2 facilities were specified that not required. Regarding to facilities specified that "Not required" from the respondents and they had explained about those facilities as follows

1) Turnstile especially manually-operated type can be seen as an obstacle for accessibility and made it more difficult to enter the place for passengers.
2) Bike Sharing; as Thailand can be considered as a motor-driven country, eventhough biking is encouraged, it supporting facilities, such as extensive bike lane network still by far lacking behind.

Besides, some of the respondents gave the opinion to other facilities such as drinking fountain may not be necessary since the quality of water from drinking fountain may not be fully hygiene. They also gave the opinion to the public telephone may also not necessary because nowadays nearly everyone has a personal mobile phone.

7 Conclusion

According to standards and guidelines reviewing in this research, it made the acknowledgment that the facilities are able to improve the accessibility to public transportation. However, those facilities will be work in full efficiency when they were built according to standards and guidelines. The Thai's legislation and standards need to be revised by providing more facilities in details or even provide the facilities list according to a specific type of public transportation station. Accessibility should be improved by providing facilities suit for each type of passengers and the simulation models should be provided for verification.

From the analysis, the researcher has found that there are 56 facilities need to provide for all and using the F.I. method found the 5 items compulsory are priority importance for accessibility to the railway station. Those 5 items are 1) Firm alarm with siren signal and Lighting 2) Platform 3) Lift 4) Directional signage, Barrier free pathway and Toilet 5) Emergency help point as shown in Table 7. Those facilities almost related to accessibility of public transportation. Notwithstanding, the improvement or providing facilities will spend a big budget and may not be able to improve or provided those facilities to all station at the same time. Therefore, this ranking would help to prioritize for improvement or providing those facilities.

Thailand is facing lacking of facilities and some of the existing facilities are not ready to use. The facilities providing or improvement should be considered in order to improve accessibilities to railway station which able to encourage the number of ridership increase, it will be a benefit to sustainable development for rail transportation system in Thailand.

From eight types of respondent shown in Table 3, the researcher found that interviewing with hearing disabilities are most difficult. This is because it required the help of hand language translator at all times. This research would like to express sincere thanks to Ratchasuda College (College for disabled), Mahidol University for facilitating with PwDs during data surveying. The researcher hopes that this research would be useful for improving public transportation facilities to be accessible by all in the near future.

8 Discussion

Due to this research is relied on the answer of respondents hence the results shown in Table.6 might be questionabilities to classified them into "Compulsory", "Optional" or "Not required". So these questionabilities might need to survey the data futher.

9 References

1. National statistic office, *Population and Housing Statistics Report Thailand,* (2015)
2. National Center on Elder Abuse, *Abuse of Adults with a Disability,* (to be published)
3. Eurostat, *Disability statistics - prevalence and demographics*, available online: http://ec.europa.eu/eurostat/statistics-explained/index.php/Disability_statistics_-_prevalence_and_demographics

4. Neena L. Chappell and Cooke H.A, International Encyclopedia of Rehabilitation, *Age Related Disabilities – Aging and Quality of Life*, Available online: http://cirrie.buffalo.edu/encyclopedia/en/article/189/
5. Elizabeth A Gardener, Felicia A Huppert, Jack M Guralnik and David Melzer, *Middle-Aged and Mobility-Limited: Prevalence of Disability and Symptom Attributions in a National Survey,* **21(10)** 1091–1096 (2006)
6. Economic Intelligence Center, *Outlook Quarter 3/2014* Available online: https://www.scbeic.com/th/detail/product/131
7. EIDD Design For All Europe, *Inclusive Design* Available online http://dfaeurope.eu/
8. Ministry of Interior, *Building Control Act 2005 (No.5)*, **132** 32-38 (2005)
9. Ministry of Transportation, The Government Gazzette, *Ministerial Regulations of Designing or Equipment Facilities of the Ministry of Transportation to enable people with disabilities to access and utilize*, **130** 4 (2013)
10. Ministry of Social and Human Security, *Promote and Improve the Quality of Life of the Disabled Act* (2008)
11. World Heath Organization, *International Classification of Functioning, Disability, and Health* (2007)
12. Ministry of Health, *Disability Rehabilitation Act*, **108** 205 (1991)
13. Edwin H.W. Chan, Grace K.L. Lee and Antony T.S. Chan, (2009) "Universal design for people with disabilities: A study of access provisions in public housing estates", Property Management, **27**(2), 138-146 (2009)
14. United Nation, *Accessibility for the Disabled A Design Manual for a Barrier Free Environment,* available online: http://www.un.org/esa/socdev/enable/designm/.
15. U.S. Department of Transportation, *Americans with Disabilities Act (ADA) Standards for Transportation Facilities* (2010)
16. National Council on Disability, *The Current State of Transportation for People with Disabilities in the United States. Washington* (2005)
17. Marc Stickdorn and Jakob Schneider, *This is service design thinking*, (to be published)
18. Kennedy D.M., Land Transport New Zealand Research Report 344, *Personal security in public transport travel in New Zealand: problems, issues & solutions* (2008)
19. European Union, Official Journal of the European Union, *COMMISSION REGULATION (EU) No 1300/2014 of 18 November 2014 on the technical specifications for interoperability relating to accessibility of the Union's rail system for persons with disabilities and persons with reduced mobility*, **L 356/110** (2014)
20. Office of legislative drafting and publishing A.-G. s. D, *Disability Standards for Accessible Public Transport 2002* (2011)
21. City of London Canada, *Facility Accessibility Design Standard* (2007)
22. Xenia Karekla, Taku Fujiyama and NickTyler, *Evaluating accessibility enhancements to public transport including indirect as well as direct benefits*, **2** 92-100 (2011)

A comparative study of performance measurement standards of railway operator

Siripong Pongjirawut[1], *Wasaporn* Techapeeraparnich[1,*], and *Waranon* Dilokkhunanan[2]

[1] Department of Civil and Environmental Engineering, Faculty of Engineering, Mahidol University, 999 Phuttamonthon 4 Road, Salaya, Nakorn Pathom 73170, Thailand
[2] Chief of Project Analysis Section, Acting Director of Analysis and Development Division, Mass Rapid Transit Authority of Thailand, 175 Rama IX Road, Huai Khwang, Bangkok 10320, Thailand

Abstract. The European standard (EN 13816), is one of the widely accepted standards for measuring the quality of public passenger transport (PPT) service. EN 13816 indicates 8 measurement criteria, 29 sub-criteria and 193 Key Performance Indicators (KPIs) to be used to measure the performance of railway operators. Nowadays, there are other addition criteria beyond EN13816, developed by various organisations. This research firstly aims to explore the service performance measurement of railway operators used by actual railway operators at international level and in Thailand. After an intensive review of performance measurement standards, 9 standards are compiled and compared in terms of criteria, sub-criteria and KPIs using a cluster analysis methodology. The result found additional performance measurement aspects at 2 sub-criteria and 91 KPIs in addition to EN 13816. This research summarized and compared different performance measurement standards to measure service quality of metro rail line.

Key Words

Performance Measurement Standards, Railway Operator, EN 13816

1 Introduction

The first section of the Chaloem Ratchamongkhon Line (Blue Line) running between Hua Lamphong Station – Bang Sue Station was officially opened in 2004 as Bangkok's second public transit system. The Metropolitan Rapid Transit (MRT) is the first underground rail line in Bangkok Metropolitan Region, with the total distance of 20 kilometers, totaling 18 stations. The second section, the MRT Purple Line Project running from Khlong Bang Phai Station - Tao Poon Station, with the total distance of 22 kilometers, totaling 16 stations, was later opened in 2016 [1]. The MRT is operated by the Bangkok Expressway and Metro Public Company Limited (BEM) under a concession granted by the Mass Rapid Transit Authority of Thailand (MRTA).

* Corresponding author: wasaporn.tac@mahidol.ac.th

In 2015, Thailand had planned to expand the metro-railway network. The MRT Blue Line extension will compose of 1) Hua Lumphong – Bangsue section and 2) Bangsue – Tha Phra section, and will connect with the MRT Purple Line Project, Tao Poon - Rat Burana Section [1]. Construction of the Red Line section from Bang Sue to Rangsit was started in May 2013 and will use some of the concrete infrastructures from the Hopewell project. In addition to the main route of the Red Line between Bang Sue and Rangsit, there are several extensions planned. These extensions include Bang Sue – Taling Chan, Bang Sue – Hua Lamphong, Rangsit – Thammasat University, Taling Chan – Salaya and Don Mueang – Phaya Thai (Improvement of Railway Passenger Services, 2014). In 2017, MRT Pink Line and Yellow Line projects are planning to be built. MRT Pink Line Project will start from Khae Rai Station to Min Buri Station, with the total distance of 35.5 kilometers, totaling 30 stations. MRT Yellow Line Project from Ladprao Station – Samrong Station, with the total distance of 30.4 kilometers, totaling 20 stations is planned to be built as a monorail line [2]. Along with the MRT, the Bangkok Mass Transit System (BTS), and the Suvarnabhumi Airport Rail Link (SARL), are also a part of Bangkok's railway transportation network. Figure 1 shows the Bangkok Metro-railway map.

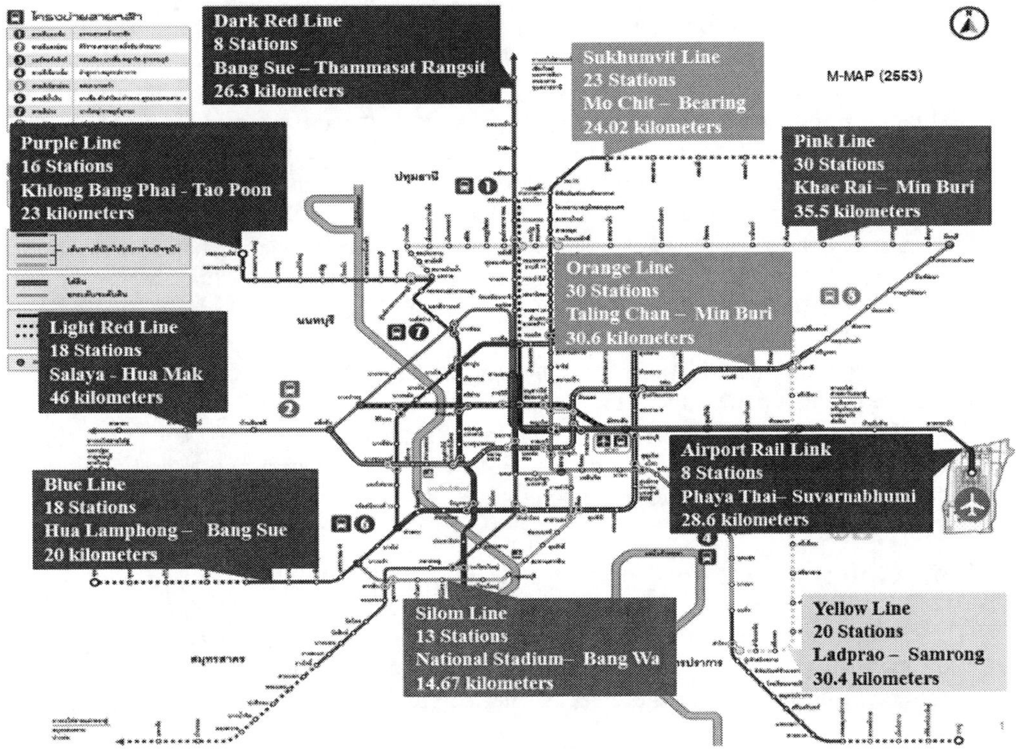

Fig. 1. MRT, BTS and SARL Systems Map [1].

Due to the fact that the 3 systems, namely MRT, BTS and SARL are operated by different companies, therefore performance measurement standards of these railways operators are somewhat different. Some of the operators develop their own performance measurements of service quality prioritising on market-driven and customer-oriented satisfaction. However, some of these operators' performance measurements criteria are still based on standards, such as European Standard (EN 13816), CoMET and Nova.

Other international measurements standards might have different viewpoints of measuring service quality. For example, Sweden developed a robust performance monitoring tool for Performance-based Railway Infrastructure Maintenance (PBRIM). Service quality for rail operators in Northern Italy consists of seven main characteristics of a railway service and 33 service quality attribute. In order to improve public transport and increase the use of public transportation, Boston Consulting Group (BCG) developed the Railway Performance Index (RPI) to provide a tool for measuring railway performance. The RPI is the most comprehensive benchmarking widely accepted by European railway operations.

The purpose of this research is to develop and compare performance criteria measuring service quality of metro rail lines operators at international level and Thailand.

2 Related Literature

Clustering is the unsupervised classification of data into groups of objects that share common characteristics [3]. Clustering can produce either disjointed or overlapping partition. In an overlapping partition, it is possible for a document to appear in multiple clusters whereas in disjointed clustering, each document appears in exactly one cluster [4]. Clusters are mainly categorized into partitioning and hierarchical methods. A partitional cluster is a method for grouping the data objects into non-overlapping subsets. A hierarchical clustering method works by grouping data objects into a tree of clusters. These methods can further be classified into agglomerative and divisive hierarchical clustering depending on whether the hierarchical decomposition is formed in a bottom-up or top-down fashion [5]. In this paper, we only focus on hierarchical clustering.

The performance measurement standards for measure quality of service that are internationally accepted are for example; European Standard EN 13816: 2002 [6], CoMET and Nova benchmarking groups [7]. Thailand also established a performance measurement standards for measure the quality of service but they are varied according to each operator such as; Suvarnabhumi Airport Rail Link (SARL), Mass Rapid Transit Authority of Thailand (MRTA). Most of performance measurement standard, used by Thai Railway Operators standards often adapted from international standards. Their main focuses of the performance are punctuality and reliability, by choosing the appropriate. Currently, some of the Thai Railway Operators are aiming at International arena since they are registered members of The CoMET and Nova group. However, the measuring system should be of a high standard and comparable among operators [8].

3 Data Collection

The researchers reviewed other performance measurement standards of railway service used by international railway operators and Thai's operators, then compared performance measurement criteria. Standards for measuring service quality of railway operators are as following;

3.1 European Standard EN 13816: 2002

European Standard (EN 13816) is the most comprehensive performance measurement aspects of service quality in transportation sector. It is also the international standards accepted worldwide. It is composed of 8 main criteria which are Availability, Accessibility, Information, Time, Customer Care, Comfort, Safety, Environmental Impact, 30 sub-criteria and 193 KPIs [9].

One result of Quality Approach in Tendering Urban Public Transport Operation (QUATTRO) is the quality loop of public transport system [10]. The diagram in Figure 2

DOI: 10.1051/matecconf/201713807014

illustrates the Quality Loop set out in EN 13816, which defines a clear process to ensure that the service provider can meet the needs of existing and potential users most effectively, and therefore be as convenient as possible [11].

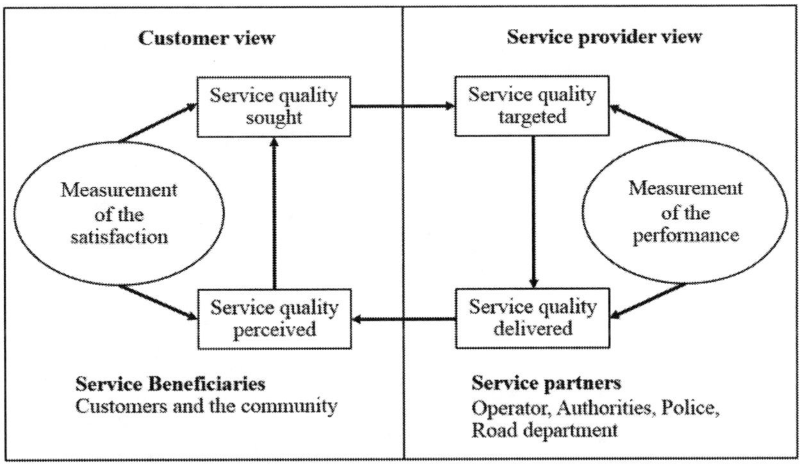

Fig. 2. EN13816 Service Quality Loop European [6].

3.2 CoMET and Nova Benchmarking Groups

The Community of Metros (CoMET) was established in 1994. The original members of the CoMET were Hong Kong's Mass Transit Railway Corp, New York City Transit Authority, Berlin U-Bahn, London Underground and Paris Metro. CoMET is a system of international railway benchmarking and consists of large metro systems from around the world. Following the success of the CoMET, the Nova group was later set up in 1998 as another benchmarking association for the more medium and small-scale metro systems [12]. Nowadays, CoMET and Nova comprised of 34 large and medium-sized metro systems from 32 cities around the world [7].

CoMET and Nova provider a confidential forum to share experiences, compare performance, and identify best practices to improve performance. It consists of 3 indicators, 6 criteria and 32 KPIs [12]. The data set is collected according to the standard definition set by all the metros of CoMET and Nova, on an annual basis [13].

3.3 Customer Satisfaction Surveys at Istanbul Railway Systems (IRS)

The Istanbul Transportation Corporation, an enterprise of the Istanbul Metropolitan Municipality was established in 1988 for operating the public rail transportation systems and for managing all other public transportation.

To assess the user satisfaction level, Istanbul Ulasim (IU) has been conducting annual passenger satisfaction surveys since 2005. The passenger-oriented performance index (POPIX) can be determined through the Passenger Satisfaction Surveys by classification that measures the indicators from the questionnaire, determination of each measure indicator satisfaction. The performance collected from the 6 criteria in the surveys are fares, reliability, cleanliness, comfort, security and security and 24 indicators [14].

3.4 The European Railway Performance Index

Boston Consulting Group (BCG) developed the Railway Performance Index (RPI) to provide a tool for measuring railway performance. The Railway Performance Index (RPI) measures railway systems' performance in three dimensions for both passenger and freight traffic. The three performance dimensions are Intensity of Use, Quality of Service and Safety. Each dimension comprises at least two sub-dimensions, and all were given equal weight percentage for each criterion because it's easier to understand and analyse as shown in Figure 3 [15].

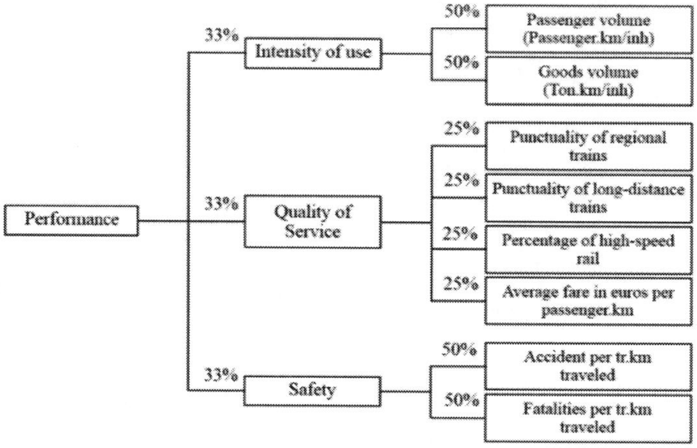

Fig. 3. The RPI Comprises Weighted Measures Across Critical Dimensions [15].

3.5 Maintenance in Swedish Transport Administration (STA)

The Swedish railway network system has approximately 20 companies that are now using the Swedish state's rail infrastructure. The maintenance of the railway network (rolling stock and infrastructure) is also managed by many different companies. Morant refers to Espling (2007) studied the maintenance strategy for a railway infrastructure in a regulated environment by implementing benchmarking techniques to compare different case studies from the Swedish railway network. The result of the analysis identified 4 risk areas, which are the risk of losing control over maintenance costs, asset condition, safety demands, core competence and asset knowledge [16].

Swedish Transport Administration (STA) is the standard developed to monitor the outcome of Performance-Based Railway Infrastructure Maintenance (PBRIM). It is composed of 4 objectives, 5 strategic and 19 measures. The outcome measure could be divided into strategic areas representing the basic maintenance objectives. [17].

3.6 Railway Service Operating in the North of Italy (RSI)

The number of passengers in the Northern Italian railway system (comprises of 32 regional lines, 9 suburban lines, 2 express lines) are increasing due to its benefits of alleviating the problems of traffic jams especially in metropolitan areas. In order to improve public transport, railway operators developed the appropriate tools for measuring and monitoring service quality. RSI developed service quality comprising 7 main characteristics of a railway service and other 33 Service quality attributes to measure customer satisfaction and service quality [18].

DOI: 10.1051/matecconf/201713807014

3.7 The National Rail Passenger Survey (NRPS)

The National Rail Passenger Survey (NRPS) provides a network-wide picture of customers' satisfaction with rail travel. Passenger opinions of train services are collected twice a year from a representative sample of passenger journeys. It composed of 2 indicators from station facilities, train facilities, and satisfaction with 33 specific aspects of service [19].

3.8 Suvarnabhumi Airport Rail Link (SARL)

The Suvarnabhumi Airport Rail Link (SARL) is a special mass rapid transit rail system project, linking the center of Bangkok to Suvarnabhumi international airport operated by State Railway of Thailand (SRT). The structure of the Suvarnabhumi Airport Rail Link is both underground and elevated, with the alignment accommodating journeys from the eastern and northern outskirts, covering a total of 8 stations, with a distance of 28.6 kilometers [1].

SARL moderately adopted from the guidelines of the European Standard EN 13816: 2002. This service quality guideline considers the needs and expectations of the customer in a comprehensive manner to ensure that remedial action can be taken to improve the quality of service. It is composed of 8 criteria, 27 sub-criteria and 97 KPI [20].

3.9 Mass Rapid Transit Authority of Thailand (MRTA)

The purpose of MRTA standard is to establish a framework for service performance targets that are intended to set the service quality levels to be achieved, particularly as regards the provision of train services and the operation besides maintenance to achieve continuous practicable improvements in the quality.

MRTA will use KPIs for measuring performances of the Concessionaire. Methods of measuring the performance are generally divided into 3 criteria, 5 sub-criteria, and 35 KPI. Firstly, direct performance measurement method of monitoring the actual performance of the service from operational records or physical observation. Secondly, MRTA'S mystery shopper measurement of each of the MRTA Survey KPIs shall be carried out by MRTA personnel undertaking a physical inspection of stations, trains, and related buildings. Thirdly, MRTA will carry out the customer satisfaction surveys during the operation services period and use the results of the customer satisfaction surveys to be attributes for making decisions to adjust the performance targets of service quality [8].

4. Data Integration

To find the suitable performance evaluation, the researcher reviewed the secondary data of the performance measurement standard at international level and Thailand from various sources; such as the database, documents, books, websites, researches, international journals, user manuals, etc. The researchers intensively studied the details of performance measurement criteria indicators, and then compared and analysed using cluster analysis method.

Cluster analysis is a method of evaluation or grouping things of the same characteristics or condition. Researchers would receive new criteria and indicators from integrations of EN 13816, CoMET and Nova Metros, SARL, MRTA and other international performance measurement standards. Researchers will then grouped the standards by clustering and comparing the key performance criteria.

EN 13816 is the main system for integration (criteria, sub-criteria and KPIs) because it has most comprehensive performance measurement aspects of service quality in the transportation sector. It is also the international standard used to measure service quality worldwide. The results of clustering analysis still display the same main criteria and condition as EN 13816 standard, which consists of 8 main criteria as shown in Figure 4.

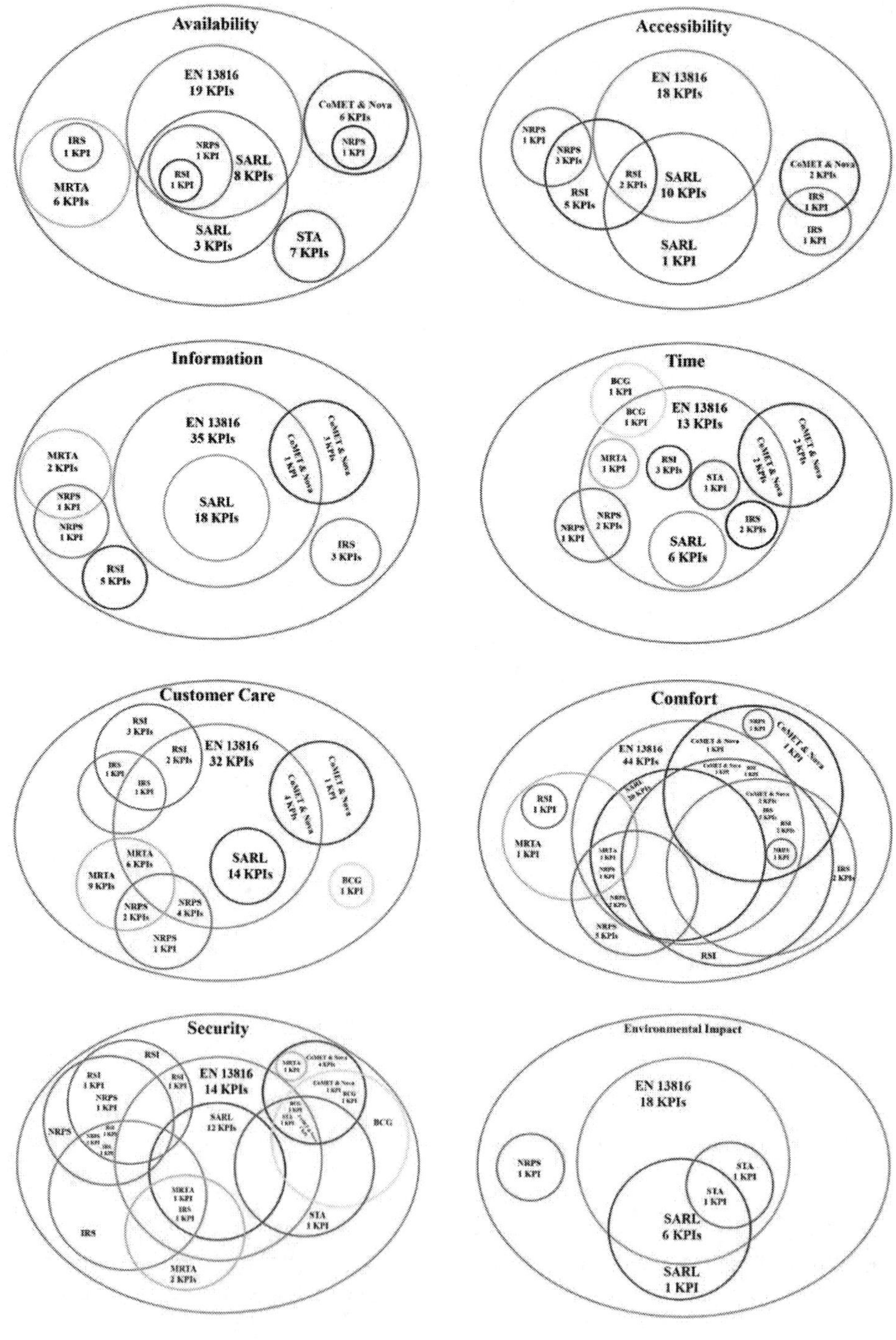

Fig. 4. The results of clustering analysis

5. Implementation and results

After reviewing both international and Thai performance measurement standards, a comprehensive list of performance measurement criteria will be obtained after cluster analysis method.

The result of clustering analysis found 2 additional sub-criteria of performance measurement aspects in term of Availability, Customer Care and new 91 KPIs in addition to EN 13816 which are 22 KPIs in Availability, 10 KPIs in Accessibility, 14 KPIs in Information, 4 KPIs in Time, 17 KPIs in Customer Care, 9 KPIs in Comfort, 13 KPIs in Security and 2 KPIs in Environmental Impact. The details of clustering analysis results and additional KPI to EN 13816 are shown in Table 1 respectively.

From the 8 main criteria of EN 13816, Environmental Impact criterion have not yet been seriously taken into consideration for measuring operator performance. This is because 5 standards do not indicate this criterion.

In total, most performance measurement standards have 4-12 KPI in compliance with EN 13816, except for SARL which has the highest number of KPI in accordance with EN 13816 at 91 out of 193 KPIs.

According to EN 13816, the highest numbers of KPIs are in the criteria of Comfort (44 KPIs), Information (35 KPIs) and Customer Care (32 KPIs) respectively. The big association like CoMET and Nova tried to set the minimum boundary by using only 13 KPIs of EN 13816 and created 19 more KPIs in addition to EN 13816. CoMET and Nova set more number of KPIs in Availability, Security, Comfort and Customer Care respectively.

The newly creative KPIs, are concentrated in the criteria of Availability (22 KPIs), Customer Care (17 KPIs), Information (14 KPIs) and Security (13 KPIs). The total number of new KPIs is 91 and is spreader to every criterion.

Thailand as a newcomer in Metro Railway system, has tried to comply with EN 13816 standard. This is especially for SARL who has the highest number of KPI at 91 KPIs complied with EN 13816 standard. On the contrary, MRTA might have only 8 complied KPIs but is in the process of intensively developing their own performance measurement model and guideline.

Table 1. Main performance measurement criteria after clustering analysis.

Criteria	International standards							Thai standards	
	EN 13816	CoMET & Nova	IRS	BCG	STA	RSI	NRPS	SARL	MRTA
Main Criteria	8	7	7	3	4	7	8	8	6
Additional Sub-Criteria	-	-	-	-	-	-	-	-	2
EN 13816 KPI	193	-	-	-	-	-	-	-	-
Additional KPI	-	19	5	21	9	6	8	13	10
Total KPI	284 KPIs								

6. Conclusions

This research introduces a performance measurement standard and service quality of railway operator that is developed on the cluster analysis methodology by reviewing several international practices and Thailand's practices. Starting from the development an integrated performance measurement model from EN 13816, CoMET and Nova, SARL, MRTA and other international performance measurement standards in terms of service quality developed to match with a case study performance. On the part of clustering analysis is evaluated with a total 8 measurement criteria, 31 sub-criteria and 284 KPIs. The additional performance measurement criteria and KPIs are the effort to make the system effective replenishment and to help solving problems that have occurred in the past. The development of current performance measurement standard however, needs supporting technology and regulation of government in order to achieve the defined quality. The outputs of this study will be an early phase of producing a comprehensive list of performance measurement criteria which will be the beginning of the next phase, to gather opinions of the expert panel and develop suitable performance measurement criteria for Thai's railway operators.

Acknowledgement

The authors would like to express our gratitude thanks to Mass Rapid Transit Authority of Thailand who grants research funding for this study. Sincere thanks are also extended to all representatives of relevant stakeholders appearing in this study as well as all research associates in the MRTA and Mahidol University. This study according to the protocol COA No. MU-CIRB 2017/016.0320 approved by Mahidol University Central Institutional Review Board (MU-CIRB) as of February 3, 2017.

References

1. Bangkok Expressway and Metro Public Company Limited, *A strong foundation for the limitless potential of the future integrated transport system*, 22-62 (2015)
2. N. Juneja, *Bangkok Metro (MRT)*, 3-4 (2015)
3. M.K. Rafsanjani, Z.A. Varzaneh, N.E. Chukanlo, *A survey of hierarchical clustering algorithms*, 229-240 (2012)
4. N. Shah, S. Mahajan, *Document Clustering: A Detailed Review*, (2012)
5. A. Jaiswal, N. Janwe, *Hierarchical Document Clustering: A Review*, (2011)
6. European Standard (EN 13816 : 2002), *Transportation - Logistics and Service - Public Passenger Transport - Service Quality Definition, Targeting and Measurement*, (2002)
7. CoMET (the Community of Metros) and Nova Metro Benchmarking Groups, Available Online: http://cometandnova.org/
8. Mass Rapid Transit Authority of Thailand, *MRTA's Requirements (MRT Purple Line Project Bang Yai to Rat Burana Bang Yai to Bang Sue Section) Appendix A Service Level Agreement*, (2017)
9. P. Fáskerty, G. Horváth, A. Rixer, J. Turi, *The Public Transport Services to Measure the Quality of Standard Bases*, 211 (2012)
10. I. Pticina, *The Methodology of Data Collection About Public Transport Service Quality*, (2011)

11. R. Anderson, B. Condry, N. Findlay, R.B. Ardao, H. Li, *Measuring and Valuing Convenience and Service Quality*, 12 (2013)
12. Metro Benchmarking Report, *CoMET and Nova deliver tangible benefits*, 16 (2003)
13. CoMET and Nova International Metro Benchmarking Groups, *CoMET and Nova Newsletter*, (2006)
14. A.S. Kesten, K.S. ÖĞÜT, *A New Passenger-Oriented Performance Measurement Framework for Public Rail Transportation Systems*, (2014)
15. S. Duranton, A. Audier, J. Hazan, V. Gauche, *The 2015 European Railway Performance Index*, (2015)
16. A. Morant, M. Westerberg, P.O. Larsson Kråik, *Knowledge management in a railway network: The case of signalling systems*, 274 (2014)
17. S.M. Famurewa, U. Juntti, U. Kumar, *Performance Based Railway Infrastructure Maintenance: Towards Achieving Maintenance Objectives*, (2011)
18. L. Eboli, G. Mazzulla, *Relationships between rail passengers' satisfaction and service quality: a framework for identifying key service factors*, 185-201 (2014)
19. National Rail Passenger Survey, *Spring 2016 Main Report*, (2016)
20. S.R.T. Electrified Train Company Limited, *Service Performance Target Appendix C to the Operations and Maintenance Management Services Conditions of Contract*, (2015)

MATEC Web of Conferences **138**, 07015 (2017)
EACEF 2017
DOI: 10.1051/matecconf/201713807015

Internet of Things (IoT) as Green City Economic Development Smart Transportation System

Winardi Slamet[1], *Kamisutara* Made[2], *Purworusmiardi* Tubagus[2], *Sukoco* Agus[3], and *Mudjanarko* Sri Wiwoho[4,*]

[1] Computer Science, Computer System Department, Narotama University, Arief Rachman Hakim No. 51, Surabaya, 60117,Indonesia
[2] Computer Science, Informatic System Department, Narotama University, Arief Rachman Hakim No. 51, Surabaya, 60117,Indonesia
[3] Economic & Business, Management Department, Narotama University, Arief Rachman Hakim No. 51, Surabaya, 60117,Indonesia
[4] Civil Engineering, Civil Department, Narotama University, Arief Rachman Hakim No.51,Surabaya, 60117,Indonesia

Abstract. The number of vehicles in Indonesia based on BPS data in 1987 amounted to 7.98148 million units to 104 118 969 units in 2013. An increase in the number of vehicles by 1304.51% is dominated by a motorcycle which was originally 5,554,305 units to 84,732,652 units. A large number of vehicles adds to problems of traffic regulation on the road. Internet of Things (IoT) can be used as the vehicle detection control device so as to create solutions Green Economic Development City Smart Transportation System that is good in handling motor vehicle. IoT sensor devices produced can identify the vehicles through model of information system connected with the detection system of motor vehicle license plate identification by means of sensors and data are stored in a digital chip which is mounted on any motor vehicle. With the chip of these vehicles will be produced a Green Economic Development City Smart Transportation System for the development of cities, especially in Narotama.

1 Introduction

Indonesia is a member of ASEAN which has the largest area, the most populous and has a metropolitan city with the largest number of vehicles. [1] In Java, the most populous island in Indonesia, more than 140 million people live in the region with an area of approximately 130,000 km2, with a density approaching 1,100 inhabitants per km2. The high number of demographic and economic growth that followed the increase in urbanization. Population four major cities in Java today has far exceeded millions of lives: Jakarta has nearly 10 million inhabitants, more than 3 million Surabaya, Bandung and Semarang about 2.5 million to 1.5 million residents. This rapid urbanization will continue to increase due to an estimated 60% of Indonesia's population will live in cities before 2025. The expansion of urban areas and the rapid economic growth resulted in an increased need for movement and transport demand is significant. In fact, urban transport has an

* Corresponding author: sri.wiwoho@narotama.ac.id

important economic impact. This condition requires Reduction Strategy Congestion City, which is intended to reduce the burden of congestion in urban areas by strengthening the Management Needs Transportation or a Transportation Demand Management (TDM) with Electronic Road Pricing (ERP), the parking system and the construction of the BRT system, the MRT and improve Traffic Supply Management (TSM) with the development of Intelligent Transportation Systems (ITS). Intelligent Transport Systems (ITS) to regulate the management capacity and priorities so as to create an environmentally friendly city. [10] Indonesia has a population of 248 818 100 people, an area of 1,910,931.32 km ^{2} and the number of motor vehicles as 104.118.969unit. Statistics Indonesia (2017),
the growth of motor vehicles in July 2016 reached 124 348 224 units consist of private cars, motorcycles, buses, and trucks. Growth potential vehicle traffic congestion and the potential for traffic violations are also getting bigger. Limited number of traffic officers will have difficulty in monitoring and vehicle settings. The composition of the number of traffic officers and the number of vehicles is not impartial. Solving these problems can be reduced using components such as digital chip attached on the vehicle license plate. Identification of any motor vehicle digitally facilitate supervision and regulation by traffic officers. Digital chip contains the identity of the motor vehicle and emitted continuously while in a highway through a wifi signal. The test was conducted on a digital chip limited area of Narotama, Surabaya, Indonesia.

2 System Description

2.1 Internet of Things (IOT)

Internet of Things (IOT) means "an open network of devices or things that are uniquely addressed based on standard communication protocols", the which is a novel prototype that is rapidly gaining ground in the scenario of modern wireless communications. The basic idea here is that the Omnipresence of a multifarious things or objects such as RF tags, sensors, actuators, mobile devices / things, etc are Able to interaction with each other to Obtain common goals through unique schemes of inscribing. The flairs offered by the IOT help in flourishing un-hackneyed protocols in varied domains (1) Traffic transportation and Logistics domain. (2) Healthcare domain (3) Smart environment (home, office, plant) domains. (4) Private and Social domains.

2.2 Positioning

A GPS consists of three distinct parts as it is embellished in the following Figure 1. Reviews These three parts are: the satellites in orbit, the ground control stations, and the users (satellite receivers found in land, air, sea). For the part of satellites in orbit, out of the twenty four, twenty-one are in operation. 21 Among These four satellites are visible at any time from any station on earth. The latitude and longitudinal position for each specific station is feasible to be Obtained in the form of 'x', 'y' and 'z' coordinates (position vector). The information concerning the speed (dx / dt, dy / dt, dz / dt) of a vehicle IS ALSO available all over the world, at any time, and under all weather conditions. Ground control station contains a master control station, monitoring station and injecting station. The master manipulates the satellite ephemeris and correction criterion of satellite clock, and injection of the data into the Reviews These satellites. Moreover it controls the satellite and issues Ordinance to it as well as associates in monitoring. Monitoring station perceives the reception of satellite signals; superintending the working status of satellites; injecting station effectuates the injection of satellite ephemeris and clock correction parameters of satellite.

DOI: 10.1051/matecconf/201713807015

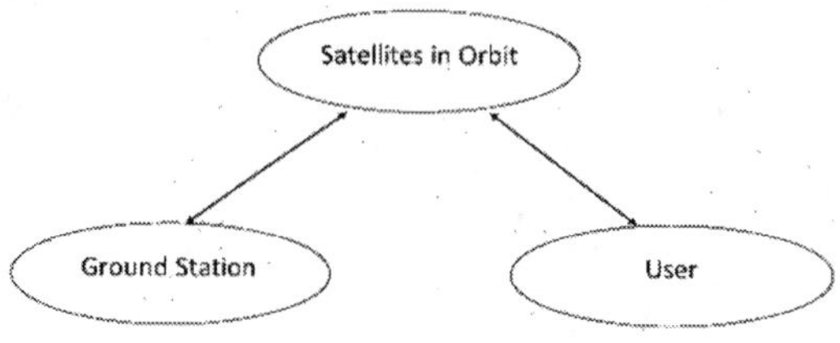

Fig. 1. Parts of GPS

2.3 Green Transport

[2] Green transport is sustainable transport concept is the use of environmentally friendly technology in an effort to meet the transportation needs of the community. In the context of urban planning, this concept is translated as an effort to improve community facilities for bikes, pedestrians, communication facilities, as well as mass public transport provision inexpensive and environmentally friendly such as electric trains and other public transportation to reduce private vehicle. [6] They are people, ideas, networks, knowledge, planning and visions. Later in the paper they will be illustrated, showing how they connect and interaction in the proposed draft strategy for Cluj-Napoca Romania. A smart city is a connected city. Just like the cases Analyzed in the previous pages, the which both located in strategic locations and poses an extremely well developed and advanced Infrastructures, representing around mobility hubs in their area. That is also available for Cluj-Napoca, a city which is said to be located in a advantaging position.

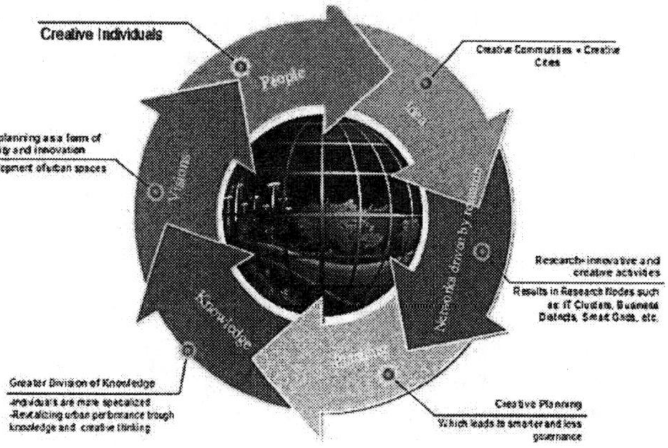

Fig.2. Conceptual Elements of Smart City. Own Sources [8]

[11] Center is a state of not moving a motor vehicle or motor can be the start of a journey to a specific period in accordance with the circumstances and needs require an area as stops were organized by the government and other parties who may be individuals or business entity. Various patterns are applied management in several cities in Indonesia and even in other countries, including in Malaysia. Parking problems do occur in various countries around the World. Parking management are applied starting with the conventional management and use of the latest management in the form of digital infrastructure. In the Klang Valley, vehicle ownership is an exceptional factor and plays a significant role in the Influencing parking demand as there are around 500,000 new vehicle registrations added to the region every year [5]. Also Malaysia tops the list as the country with the third highest car ownership in 2014, 88% of Malaysian car Also owners intend to upgrade Reviews their

 DOI: 10.1051/matecconf/201713807015

vehicle when they are financially-able [8]. With car ownership in the growing trend, the parking demand in the Klang Valley, especially at rail stations, is expected to increase is.

2.4 Management Needs Transport

[12] The elements of the transportation demand management (Transportation Demand Management (TDM)) consists of:

1. Restricting access to cars by improving public transport services and provide an incentive for users commuters.
2. Reduce car mobility reallocation of road space and limited traffic zones by improving public transport services, improving infrastructure buses, bicycles, pedestrians and improving mobility options
3. Land use planning are integrated with non-motorized transport planning
4. Repression / law enforcement through awareness-raising community awareness of traffic

2.5 Vehicle Identification Numbers

In the Law of the Republic of Indonesia Number 22 Year 2009 regarding traffic and road transport regulates TNKB as vehicle registration and vehicle identity to be registered. In article 64 to article 70 and article 78 regulates the process of registration and identification of vehicles and issued BPKB, vehicle registration, and TNKB as proof of identity of the motor vehicle. The shape of the motor vehicle identity issued by the police are as follows:

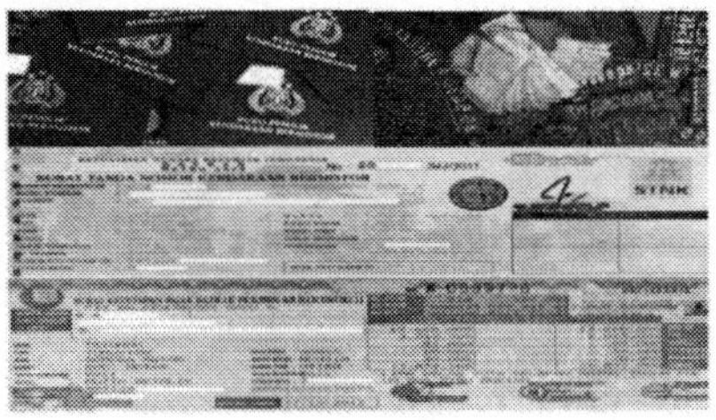

Fig.3 Form BPKB, vehicle registration, and TNKB

Such deficiencies should be owned by every owner of a motor vehicle, when driving a motor vehicle shall be equipped TNKB attached to the motor vehicle at the front and rear of a motor vehicle and a driver must carry vehicle registration as proof of driving a motor vehicle in accordance with Government Regulation No. 80 of 2012 and a driver's license as the identity of a driver who has been authorized by the police for driving a motor vehicle on the road in accordance with the type of vehicle. Installation TNKB vehicles conforming to the standards is as follows:

Fig.4. Installation Number Plate (TNKB)

3 Methodology

3.1. Vehicle Identification Numbers

Hardware Design of Digital Vehicle Identity
Identification of the vehicles were built using a chip ESP8266 provides vehicle data and the owners are always transmitted continuously via a wifi signal.

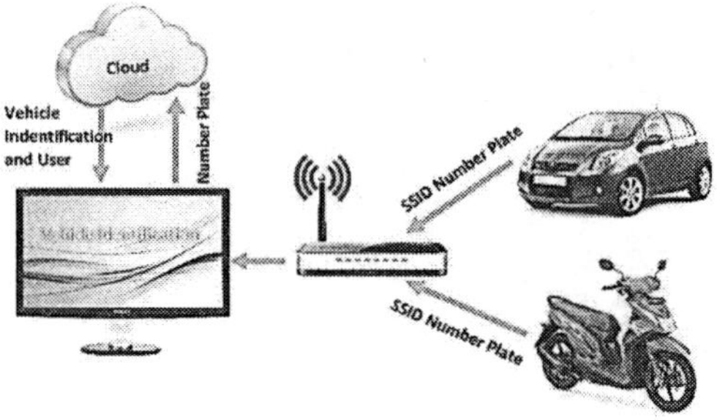

Fig.5. System Block Diagram

Every motor vehicle both cars and motorcycles are all given a chip that contains the identity of the vehicle as listed in the vehicle registration date. The chips are used to store the identity of a motor vehicle using the ESP8266. This chip will be given identity (SSID) in the form of any motor vehicle license plate number because it is unique (different every motor vehicle).

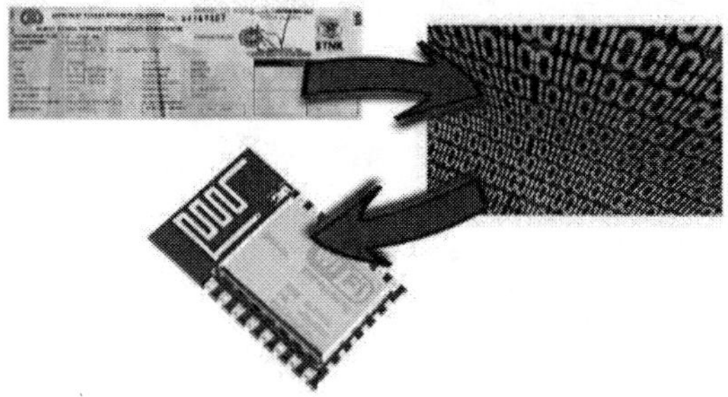

Fig.6. Embedded Chip Identification of Vehicles to ESP8266

Once all the data entered ESP8266 paired subsequent chip in a motor vehicle and serves as a client that transmits data. If found access point then this chip will connect and provide access point SSID to subsequently forwarded to the server to be recorded in the vehicle identification data base. Each client will have an IP that is obtained from the configuration after connecting with the wifi signal at the access point. Computer and chip ESP8266 must be connected to the same access point. Once the chip is filled ESP8266 next vehicle identification chip is mounted on a motor vehicle in accordance with the license plate number.

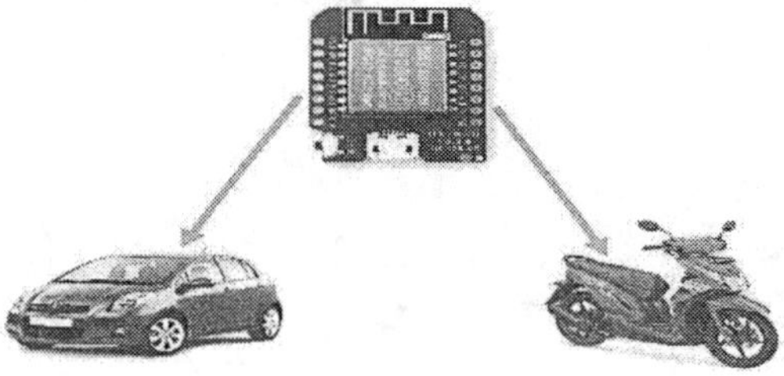

Fig.7. Vehicle Identification Digitally

DOI: 10.1051/matecconf/201713807015

3.2 Data Flow Analysis

Design Software

To realize this motor vehicle identification required two software namely Arduino IDE and PHP. ESP8266 chip is programmed using the Arduino IDE and Server using PHP. Arduino software to design hardware for the client and the router while PHP for the purpose of designing the web service.

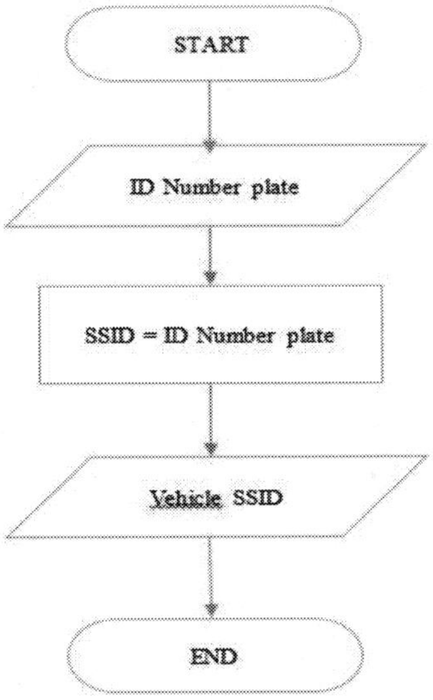

Fig.8. Flowchart Client

For client hardware mounted on a motorcycle or car in the form of chips ESP8266 with a program filled in the form of police numbers are used as SSID *(Service Set Identifier)* using the Arduino IDE and emitted continuously by a wifi signal.

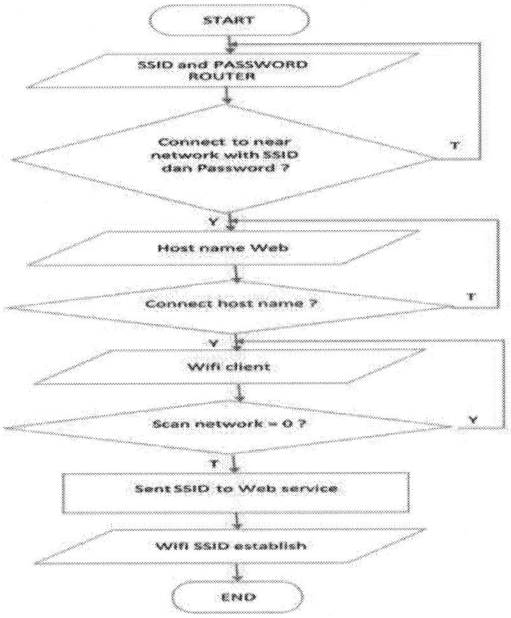

Fig.9. Flowchart Router

In this router also uses the chip ESP8266 filled program to detect the presence of a motor vehicle by capturing SSID motor vehicle emitted through a wifi signal. This router must first be connected to the Internet network via the access point nearby. The domain name system, or commonly abbreviated as DNS is a system that serves to translate IP addresses into domain names and vice versa, from domain names to IP addresses. Thus, the host computer sends queries in the form of a computer name and domain name server which is then mapped into an IP address by the DNS. After getting the IP and the next router will scan the wifi signal from the client (SSID motor vehicles) whether there are vehicles which are located around the router, If any SSID motor vehicle router will send the SSID to the server (web service) then the web server will bring the SSID the.

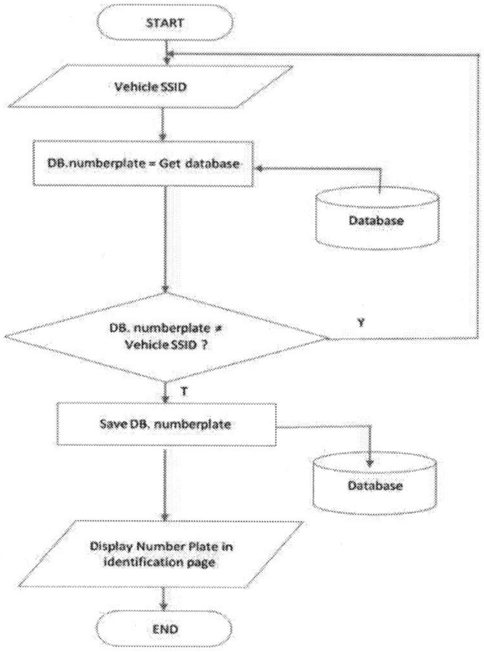

Fig.10. Flowchart Web Service

Once the router sends the SSID of the motor vehicle to the web service next system will check the database if the SSID is already registered or not, if not it will display the license plate number of the vehicle only, while when it is registered with the web service will show complete data motor vehicle besarta the identity of the vehicle owner motorized. The existence of the motor vehicle will be stored in the database and displayed on the monitor screen.

4 Results Analysis and Discussion

After designing the hardware client and the router using Wemos ESP8266 and identity database servers to accommodate motor vehicles. Routers are placed around the parking area off parking the motorcycle and the car park for the wifi signal range of about 10 meters. This router will capture client wifi signal in the form of motor vehicle setup SSID form plate number (license plate number). Then the router will forward the SSID to the data server and then be displayed on the monitor screen in the form of the identity of the motor vehicle.

Fig.11. Router Installation

ESP8266 chips that function as a client filled by the identity of the motor vehicle in the form of police numbers, then the chip would emit wifi signal continuously until it was discovered by a router. In this case the identity of the motor vehicle license number L 6125 QZ police, license plate number is used as the SSID name of the client. Every motor vehicle shall have a different SSID for using the identification number issued by the authorities, namely the police, and indeed any motor vehicle license number will be different.

Fig.12. Identification Motorcycles

As with motorcycles, cars were given chip ESP8266 who has filled the car's identity .. Identification of complete vehicles and vehicle owners are stored in the database server. In places parker Narotama University campus area in pairs of a router to identify vehicles that are around the router. The router will always scan wifi signals are scattered around, but the only format number plate only to be captured by the router and forwarded to the server to display the identity of a motor vehicle on the screen. This router will automatically send data to the server when the client has a wifi signal formats such as vehicle license plate. In this case use xenia car with police number L 1676 HB in parking near university mosque Narotama.

 DOI: 10.1051/matecconf/201713807015

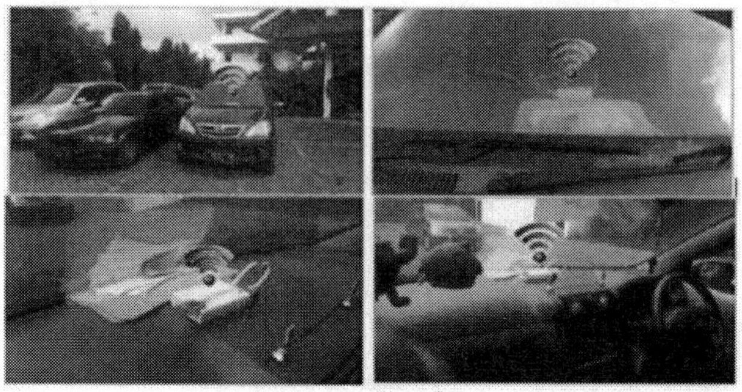

Fig.13. Identification Car

Regular users of motor vehicles to the campus of the University Narotama obliged to register the identity of the vehicle to a server computer to be recorded in the database of vehicle owners. Entitled to enter the motor vehicle identity data is the admin of this system. If you are already registered then the system will display the data of the motor vehicle owner, if a motor vehicle license plate will be displayed any time and date and hours of admission to college.

Fig.14. Database Own Vehicle

Vehicle data has been embedded in the chip ESP8266 in the motor vehicle will be recognized by a router that has been installed at certain places that can reach the existence of a motor vehicle. The signals emitted by the client will be caught by the next router and client SSID will be forwarded to the data server to retrieve the database that was inputted in its own server and will be displayed in a monitor to detect vehicles which are located around the router. For the time being displayed in this system is the last number of police and he is connected with the router. Data can be developed so that is displayed in the system more complete.

Fig.15. List Vehicle Identified

5. Conclusion

The conclusion of the identification system in the form of a motor vehicle with wifi technology

1. The chip is able to identify license plates with the code specified by the police of the Republic of Indonesia.
2. The chip detection can be developed in other countries based on their vehicle license and transportation system standards.
3. The chip can detect the presence of vehicles not more than 100 meters in distance by using wifi signal emitted by those vehicles.
4. This digital identification system can help identifying vehicles in case of crucial problems.
5. This digital identification system can help government programs in transportation, for examples restriction program on vehicles uses on highways using the rules of odd and even number plate and the restriction program on fuel purchases

References

1. AE. Andersson, C. Mellander, *Handbook of Creative Cities,* Edwar Elgar Publishing Limited, UK, (2011)
2. B. Ganesh, Harisudha, *Efficient Transportation System Using the Internet of Things*, the International Journal of Applied Engineering Research, ISSN 0973-4562 **5** 10 pp. 3979-3982, (2015)
3. D. Gusnita, *Green Transport: Transport and Sustainable Contribution to Reduce Pollution Air, Research Center of Atmospheric Sciences and Climate*, LAPAN, Aerospace News **11,** pp. 66-71, 6 (2010)
4. F. Kawsar, D. Fitton, and V. Sundramoorthy, *Smart objects as building blocks for the internet of things,* Internet Computing, IEEE **14**, pp. 44-51, (2010)
5. Government Transformation Program, *Improving Urban Public Transport, National Key Results Areas, GTP Annual Report 2014*, Performance Management and DeliveryUnit (GUIDE), (2014)
6. H. Phooi, *Future Parking Demand at Rail Stations in the Klang Valley*, ISCEE pp. 103.09001 (2017)
7. J. Lee, J. Jeong, T. Oh, J. Jun and SH. Son, *DCMAC: Data-Oriented Cluster-Based Media Access Control Protocol for Vehicular Networks,* 30th Int Conf on Advanced Information Networking and Applications workshops (WAINA), Crans-Montana, pp. 258-261 (2016)
8. J. Supriono , SW. Mudjanarko, *Performance Evaluation Parking In RSU Haji Surabaya,*. E-Journal Spirit Pro Patria, (2016)
9. Ministry PUPR, *Statistical Information Infrastructure, Public Works and Public* Housing (BIS-PUPR), Statistics Indonesia, pp : 34 (2015)
10. R. Nielsen, *Middle Class Will Drive Demand in the Global Automotive ComingTwo Years*, (2014)
11. T. Rye, *Parking Management: A Contribution to the city livable, Deutsche Gesellschaft Fur Internationale Zusammernabeit (GIZ) GmbH, Federal Ministry For Economic Cooperation and Development (BMZ)*, (2010)
12. The Agence Française de Développement, *Finding a solution to urban transport: the challenge of financing and network integration* **6** (2014)

The Assessment of Natural Rock Asphalt for Thin Surfacing Mixture Design

Ary Setyawan[1,*], *Arif* Budiarto[1] , *Djumari* Djumari[1] and *Salvatore* Sukmana[2]

[1]Roadmate Research Group, Universitas Sebelas Maret, Surakarta 57126, Indonesia
[2]Civil Engineering Department, Universitas Sebelas Maret Surakarta 57126, Indonesia

Abstract. Thin surfacing hot mixed asphalt is one of the alternatives to anticipate the problem of thickness on overlay that are being developed as environment-friendly pavement, but the use of thin surfacing was still not maximum with some weakness. One way to improve the qualities of bitumen is by the use of the modification which has been available in the market. A new materials available in market is retona. Retona (Refined Buton Asphalt) is produced by the extraction of the natural asphalt from Buton island, Indonesia that were developed with a distillation process and extraction. The study has been carried out to define optimum bitumen content and characteristic in hot mix thin surfacing by using retona blend 55 and asphalt penetration 60 / 70 with the experimental method using the Marshall to define the stability, flow, density, percentage number of pores, Marshal Quotient (MQ). Result of Marshall test thin sufacing hot mix asplhat for optimum bitumen content retona belnd 55 asphalt is 5,87%, using the same gradations achieved optimum bitumen content of asphalt penetration 60/70 of 5.80%. In each of the asphalt on the conditions of optimum bitumen content obtained difference for characteristic and properties. Stability value retona blend 55 asphalt 525,61 kg, and asphalt penetration of 60/70 is 651,16 kg. The density value of retona blend 55 asphalt 2,06%, and asphalt penetration 60/70 is 2,11%. The percentage of the pores value of retona blend 55 asphalt is 18,45%, and asphalt penetration 60/70 is 15,8%. Flow value retona blend 55 asphalt is 2,11mm, and asphalt penetration 60/70 is 2,23mm. Thus asphalt penetration have resistance to traffic loads which is stronger, with a more solid density value, the value of smaller pores and higher flow values than the natural rock asphalt blend retona 55 asphalt.

1 Introduction

Thin Surfacing Hot Mix Asphalt is one technology that is being developed as a preventive and resurfacing for pavement. Thin Surfacing HMA is a very thin surface layer such as surface dressings and slurries, this thin layer of coating has a thickness from 30 mm to 40 mm [1].

Gilbert et al. (2004) states that the main purpose of using Thermal Surfacing Hot Mix Asphalt is for the treatment of road pavement surfaces. The thin layer of HMA can extend

* Corresponding author: arysetyawan@staff.uns.ac.id

the service life and improve pavement performance such as smoothness, comfort, agitation, noise reduction [2]. The advantages of thin asphalt overlays are long service life, smooth surfaces, able to withstand heavy traffic and large shear stress, high resistance skid, and easy maintenance [3].

Asphaltic characteristics affect the performance of asphalt mixtures. Therefore, asphalt with good quality will produce a pavement mixture with good performance. The use of a thin layer of asphalt mixture is very much required as an alternative or environmentally friendly innovation on the manufacture or maintenance of roads in Indonesia.

Retona (Refined Buton Asphalt) is the product of natural asphalt extraction from Buton Island. An asphalt of 55 blend retona product, from a variety of research blend 55 blend can serve high traffic. Increased stability, fatigue resistance and adhesion due to temperature (fatigue life ratio 4.73-32.62 times greater according to Tu Delft), high adhesion and cohesion strength, water resistance due to retona nitrogen base 5.61 (± 400%), Longer service life (at least twice), so the maintenance cost is cheap, easy to use as normal asphalt, Marshal stability up to 30%, dynamic stability rises so that 400% (average above 3000 trajectory / min), stability Dynamic for heavy loaded and heavy traffic road is minimum 3000 trajectory / min. In this research will be tested the effect of asphalt use for thin layer using asphalt which have been available in Indonesia that is: retona blend 55 asphalt and normal penetration asphalt (Pen.60 / 70).

Several research have been conducted to assess the performance of thin surfacing in application for overlay design as reported by Setyawan et al [4,5]. The application of overlay as the conventional method to maintain the surface quality of pavement need a large number of fresh natural aggregate and bitumen due to the relatively thick layer and consequently the road geometric must be corrected, so that regarded as not environmentally friendly. The alternative solution is Ultra-Thin Surfacing Hot Mix Asphalt (UTSHMA) which has a thickness of 12-15mm. UTSHMA gradation is designed based on the highest Marshall Quotient value tested by performing Marshall Test of five variations gradation from Standard Specification Construction of Transport Systems. The Optimum Bitumen Content (OBC) obtained by Marshall Test. The bitumen used for this is 60/70 penetration. The several structural tests were performed on the best gradation at OBC consist of Marshall tests, Indirect Tensile Strength (ITS), Unconfined compressive strength (UCS) tests and water permeability test. The UTSHMA mixture with coarse gradation has the greatest Marshall Quotient at 6.35% OBC accompanied by the higher ITS and UCS properties compare to conventional asphalt concrete. The coefficient permeability of UTSHMA is lower than conventional AC, indicated that UTSHMA is more resistant to water penetration. The research concluded that UTSHMA has compromised the standard specification and has sufficient structural properties compare to conventional AC and can be regarded as structural layer of pavement for overlay and road preservation design [6].

Meanwhile the application of thinner layer such Hot Mixture Asphalt Concrete Overlays (THMACO's) are comprised of a thin layer less than one inch hot mix asphalt concrete layer and a binder material/tack coat. The uses of thin layer are necessary to compromise with the environment issue especially to protect the natural resources and save the energy used during asphalt construction. THMACO's can be used as preventative maintenance on pavement preservation or as a new surface on a pavement construction. This layeris usually regarded as non-structural layer at pavement design. However, this paper presents the structural assessment of this materials and their comparison to the conventional asphalt concrete mixtures. This research was conducted based on indirect tensile strength (ITS), unconfined compressive strength (UCS) and indirect tensile stiffness modulus (ITSM) of thin hot mixture asphalt. The results show that the thin hot mixture asphalt has performed slightly different on the Marshall and structural properties compare to conventional asphalt concrete [7,8]

2 Research Method

2.1 Aggregate and Gradation

Aggregates are mineral particles in the form of granules which are one of use in combination with various types ranging from as materials in cement to forming concrete, road layer, filler material, etc. [9]

Mixed gradation planning is based on the National Asphalt Pavement Association (NAPA), North Carolina. The gradation used is presented in Table 1.

Table 1. The Gradation of Thin Surfacing Hot Mixture Asphalt

Sieve Sizes (mm)	Specification	Median	Selected Gradation
3/4" (19 mm)	100		100
1/2" (12,7 mm)	85 - 100	92.5	92.65
3/8" (9,51 mm)	60 - 80	70	69.3
No.4 (4,76 mm)	28 - 38	33	33.62
No.8 (2,38 mm)	19 - 32	25.5	25.16
No.50 (0,297 mm)	8 - 13	10	10.6
No.200 (0,074 mm)	4 - 7	5.5	5.68

2.2 The optimum bitumen content

The optimum bitumen content can be determined by Marshall Test or often called Asphalt Institute method. Marshall testing is done based on the estimated asphalt content of 4.5%, 5%, 5,5%, 6%, 6,5%. Before the Marshall test is done, Volumetric test firstly includes measurement of diameter, thickness and weight in the air, then calculated to get density value, Si_x, and porosity. Then just done Marshall testing and got the value of stability, flow, and Marshall Quotient. From these values can be determined the best mixed properties or optimum asphalt content which is then used as the basis in the manufacture of the next test object.

From the Marshall test results obtained parameters in the form of flow value, stability dam Marsahll Quotient, from the data will get the value of the optimum asphalt content of asphalt used.

3 Results and Discussion

3.1 Marshall Stability

From the results in Table 2, it can be seen that the use of 6% retona asphalt obtained the highest Marshall stability of 568.38kg; 573.42kg; 662.43kg with an average of 601.41 kg, while for 60/70 asphalt penetration with 6% asphalt content obtained the Marshall stability rate of 743.86kg; 760.59kg; 684.74kg; With an average of 729.73kg

Table 2. The Results of Marshall Test

Bitumen contents (%)	Stability (kg) Retona	Stability (kg) 60/70pen	Flow (mm) Retona	Flow (mm) 60/70pen	Marshall Quotient (kg/mm) Retona	Marshall Quotient (kg/mm) 60/70pen
5	552.76	460.69	2.3	3.25	240.33	141.75
5	462.64	592.98	2.1	2.1	220.31	282.37
5	319.18	487.43	2.75	1.65	116.06	295.41
	444.86	513.7	2.38	2.33	192.23	220.16
5.5	395.53	707.66	1.9	2.1	208.17	336.98
5.5	514.43	672.01	2.3	2.95	223.66	227.8
5.5	429.52	490.87	2.5	1.7	171.81	288.75
	446.49	623.51	2.23	2.25	201.22	277.12
6	568.38	743.86	2.1	2.1	270.66	354.22
6	573.42	760.59	1.9	2.8	301.8	271.64
6	662.43	684.74	2.3	1.65	288.01	414.99
	601.41	729.73	2.1	2.18	286.82	334.23
6.5	413.95	480.91	2	2.2	206.97	218.59
6.5	452.21	427.64	1.9	2.1	238.01	203.64
6.5	427.44	446.64	1.7	2.2	251.44	203.02
	431.2	451.73	1.87	2.17	232.14	208.49
7	361.66	334.5	1.8	2.1	200.92	159.28
7	431.52	454.92	1.7	2.3	253.84	197.79
7	311.15	440.05	1.9	2.1	163.76	209.55
	368.11	409.82	1.8	2.17	206.17	189.15

3.2 Density

Density is the ratio between weight and volume (indicating the density of a mixture). The results of the study show that the density decreases with increasing the asphalt content, but will increase after reaching asphalt level ± 6%. This is because the mixture has not reached the maximum density at ± 6% asphalt level, or not saturated, so that as the asphalt level increases its density will rise.

The optimum level of retona asphalt content, the value of density 2,06 from the density value still increases at 6%, 6,5% and 7%, it shows that the optimum Retona asphalt content has not obtained maximum density, therefore Will decrease the asphalt stability value. For asphalt penetration the highest density value is at the optimum asphalt level of 5.80% by 2.11%.

3.3 Marshall Flow

Flow is a state of change in the shape of a mixture due to a load to the limit of collapse. The flow value indicates the degree of flexibility or stiffness of the mixture. High flow values indicate a high degree of flexibility, and vice versa. Increased mixed asphalt levels will also increase the amount of asphalt that envelop aggregate, fatigue time increases in length so that at the time of loading will be better able to follow the change of shape. The results of the research presented in the figure show that the flow value tends to decrease along with the increase of bitumen content.

The flow value of the optimum Retona asphalt level was 5.87% by 2.11 mm. While the asphalt penetration of 2.23 mm at the optimum level. In the research, the value of flow tends to decrease with increasing the bitumen content. However, the decrease of the flow value is not too large only ranges from ± 0.15mm from 0.5% of the difference of asphalt content, while the penetration asphalt is between ± 0.10mm.

3.4 Marshall Quotien

Marshall Quotient is the result between stability and flow. The results of the research shown in the figure can be seen that the MQ value increases with the amount of asphalt content, but will decrease slightly when reaching asphalt level ± 6%. Directly the value of Marshall Quotient relates to the value of stability and fatigue.

The results of the research presented in the picture of the relationship between asphalt content with stability, it can be determined the optimum bitumen content of the mixture. The optimum bitumen content is the bitumen content which will yield the best characteristic properties of an asphalt mixture. To find the value of optimum asphalt content, calculation of the regression equation relation of asphalt level with stability.

Conclusion

1). Characteristics of Marshall of Thin Surfacing Hot Mix Asphalt mixture obtained an optimum Retona rate of 5.87%. The results showed that by using the same gradation obtained optimum asphalt binder content Retona greater than 60/70 penetration bitumen, asphalt That's because it still contains filler Retona 2%, so that the filler in asphalt retona used more than asphalt penetration, nature Filler which tend to absorb the asphalt cause the use of asphalt retona cinderung more than asphalt penetration which its optimum bitumen content is 5,80%.
2). From the Marshall test the properties of retona asphalt cold be concluded that Figures stability of asphalt retona lower than 60/70 penetration bitumen, asphalt stability retona value at the optimum bitumen content 5.87% by 525,61Kg, while at the optimum level of penetration bitumen gained 5.80% niali setabilitas 651,16Kg. This illustrates that the retona cinderung asphalt is weaker accepting the load when compared to the asphalt concentration.
3). At the optimum bitumen content retona 5.87% 2.06 density values obtained, the results of research still increased density value at a level of 6%, 6.5% and 7%, this shows the optimum bitumen content retona maximum density is not obtained, so that Will decrease the asphalt stability value. For asphalt penetration the highest density value is at the optimum asphalt level of 5.80% by 2.11%.
4). The flow value at the optimum retona asphalt level was 5.87% by 2.11 mm. While the asphalt penetration of 2.23 mm at the optimum level. In the research, the value of flow that cinderung decreased with increasing the bitumen content. It should increase the level of asphalt, the mixture will be more flexible due to the increasing number of asphalt that

envelop aggregate. But the decline in flow value is not too large only between ± 0.15mm of each binder content at 0.5% difference, while the decrease of penetration bitumen ranges between ± 0,10mm.

References

1. J.C. Nicholls, I. Carswell and J.T. Williams. *Durability of Thin Asphalt Surfacing Systems: Part 1 Initial Findings United Kingdom* (2002).
2. T.M. Gilbert, P.A. Olivier and N.E. Gale, *Conference on Asphalt Pavements for Southern Africa* (2004)
3. D. E. Newcomb, and K. R. Hansen, *International Conference on Perpetual Pavements. Columbus, Ohio*, (2009)
4. B. Yulianto and A. Setyawan, In *IOP Conference Series: Materials Science and Engineering*, **176(1)**, 012034 (2017).
5. M.I. Matar, A. Setyawan, and K.A.Sambowo, *Jurnal Teknik Sipil*, **2(1)**. (2016)
6. O.M Kabbash, A. Setyawan, and D. Handayani, *Jurnal Teknik Sipil*, **2(2)**. (2016)
7. A. Budiarto, H. R. Purnomo, and A. Setyawan. In *App.Mech and Mat*, **754**, 828-832 (2015)
8. I. N. Iskandar, A Setyawan, and S. J. Legowo. *Matriks Teknik Sipil* **4(2)** (2016).
9. D. Woodward and A. Woodside, *The International Journal of Pavement Engineering and Asphalt Technology,* **12(2)**.63-77 (2002)

Can the design of space-transport development strategies influence on noise pollution?

Lasmini Ambarwati[1*], *Amelia* K. Indriastuti[2], *Yatnanta* P.Devia[1] and *Deputri* N.Sari[3]

[1]Department of Civil Engineering, University of Brawijaya, 65142 Malang, Indonesia
[2]Department of Civil Engineering, Diponegoro University, 50275 Semarang, Indonesia
[3]Alumnus of Department of Civil Engineering, , University of Brawijaya, 65142 Malang, Indonesia

Abstract. The rapid growth of transportation in developing countries brings advantages and disadvantages. One of the disadvantages gives negative impacts, such as traffic jam, which directly decreases a highway performance as well as increases air and noise pollution. A suitable policy has not been formulated to overcome the traffic jam in most highways in big cities. However, traffic demand management regard to noise pollution is required in most cities of developing countries, particularly Surabaya City, as a case study area. This study aims to understand the influence of saturated flow of highway on noise pollution for several types of road corridors. The results reveal that saturated flow of highway has a significant effect on noise pollution. By examining the alternatives of space-transport development strategies, the intention is to reduce noise pollution together with decreasing traffic jam due to decreasing the number of light vehicles. This research proposes the improvement of public transport (PT) by planning a Mass Rapid Transit ((MRT). In addition, a structure of compact zones is designed as urban spatial strategy alternatives. Its substantially reduction is achieved with the design of improvement of PT systems and compact zone. The best option for reducing noise pollution level, the implementation of an integrated space-transport development strategy, is then recommended.

1 Introduction

Rapid expansion of city areas is occurring world-wide. The phenomenon of urban sprawl has an important role in the build-up of urban areas, influencing living conditions and mobility.

Urban sprawl has been identified in most Indonesian cities. It reflects a pattern of urban development with increasing settlement growth in the suburbs since the beginning of the 20th century. The phenomenon of urban sprawl also occurs in the city of Surabaya as case study area, are characterized by an estimated 38% of the population lives in the suburbs [1]; most of them commute every day to work in the central urban area.

Surabaya, as a case study area, is the capital city of East Java Province with the second dense population after the state capital of Jakarta. Surabaya has an area of 333,063 km^2.

* Corresponding author: lasmi68@yahoo.com; labarwati@ub.ac.id

Surabaya is divided into five areas namely Central Surabaya, North Surabaya, East Surabaya, South Surabaya, and West Surabaya.

Traffic congestion in the city of Surabaya has strongly increased since the late 1990s; predominantly due to increasing numbers of motorcycles and private cars. Private car (30%), motorcycle (62%), and other vehicle types (8%, mostly minibuses) are the existing transport mode chosen by residents. This situation results in significant increase of time, cost and productivity losses, and an increase in air and noise pollutions. A major consideration for reducing pollution and improving mobility is undertaken by developing strategies for sustainable transportation linked to spatial planning. Such a strategy is expected to change modal split from private vehicles to public transport, to reduce noise from motorized vehicles, and to decrease travel time and distance. Private vehicle is currently a substantial source of noise pollution and traffic congestion in urban areas.

Integrated a road traffic noise model and air pollution model was conducted by employing Geographic Information System (GIS) [2]. The model was developed in order to investigate the existing four urban forms on the Macao Peninsula influence on vehicle transport and street environment. The urban forms in historical areas with narrower roads, complex road networks and a higher density of intersections lead to lower traffic volumes and consequently to lower noise pollution. The Calculation of Road Traffic Noise (CRTN) prediction model was employed to predict noise levels at 28 locations in the City of Amman. The model was successful in predicting noise levels at most of the locations and more accurate predictions for night-time measurements [3].

Some researchers reveal that the mitigatory measures to overcome such type of traffic noise pollution through a design of noise barrier along the road. This effort is useful for motivating people towards the use of public transport system [4]. The light motor vehicles are the main source of noise pollution in the city and have higher correlation coefficient values in generating noise pollution under interrupted traffic flow conditions in urban areas of Indian cities [5].

Previous research on decreasing noise from traffic, has focused insufficiently on understanding change of the number of private vehicles on highway, urban form and land use in the city. Linkage between spatial and transport development strategies in minimizing noise pollution is insufficient due to the fact that the existing efforts only consider on road corridor activities.

The aim of study is to determine the road performance and the degree of noise pollution by improving public transport and reducing the use of private vehicles linked to management of the urban form.

The research results are useful to understand the change of noise level by developing space-transport development strategies and to assess the effects of transport policies addressing to reduce noise.

This paper is organized as follows: Section 2 explains research method. Section 3 explains space-transport development strategies by considering improvement of public transport and design of urban development scenarios. Section 4 identifies the change of modal split for each alternative of space-transport development strategies. Comparison noise level on differences of road corridors is stated in Section 5. Section 6 shows the change of noise level for each alternative. Finally, Section 7 discusses conclusions and recommendations for further analysis.

DOI: 10.1051/matecconf/201713807017

2 Research Method

The city of Surabaya, a case study area, is the second biggest city in Indonesia, as seen in Fig. 1. The city has a population of 2,765,908 [1]. The population density of the city was estimated more than 11,000 people per km² in 2011. This city has a major problem related to settlement development for its residents. Since the beginning of the 20th century, urban sprawl has been identified in most Indonesian cities. It reflects a pattern of urban development with increasing settlement growth in the suburbs. The phenomenon of urban sprawl occurs in the city of Surabaya, are characterized by an estimated 45% of built-up area in the central urban areas and 38% of the population lives in the suburbs [1]; most of them commute every day to work in the central urban area. This extensive growth in economic and residential development has significant consequences for high mobility, as well increasing pollution.

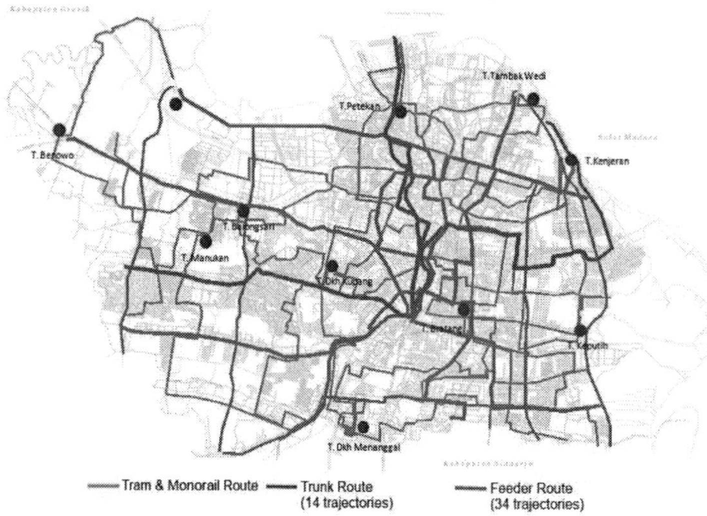

Fig. 1. Study Area, Surabaya City and monorail routes.

The research framework consists of the background condition of public transport system and spatial urban development, further plan of space-transport development. The result is expected to reduce the use of private vehicles and noise pollution, to increase the beneficial received from design of space transport development, as shown in Fig. 2.

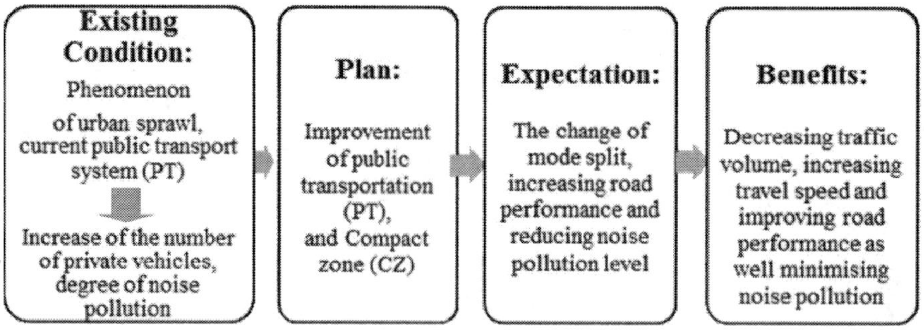

Fig. 2. The research framework.

This study is empirical research supporting by transport data, collected by an extensive survey. The questionnaire was distributed to 554 respondents spread out over the 163 zones (neighborhoods). The questions were divided into three groups: socio economic background, trip characteristic and transport mode choice, and residents' satisfaction regarding public transport. Other data, i.e. traffic counting, speed study and road geometric inventory were investigated on 31 road-sections of Surabaya City. Traffic volume, speed and road corridors are analyzed to assess noise level. Five alternatives of space-transport development strategies are proposed to analyze the differences of noise levels.

3 Space-Transport Development Strategies

A set of simulations, consisting a spatial settlement strategy and a transport system were carried out. The impacts from this set of simulation cases were compared and formed as input for the formulation of strategic options for integrated development of the city. A total of three spatial settlement scenarios and two alternative (public) transport situations are combined into five simulation alternatives (cases). The cases are: (1) Existing (2010) settlement, (2) Settlement 2030 based on a continuation of current trends, (3) Settlement 2030 with the consideration of a so called compact zone in the western part of the city, (4) Situation 2030 with improvement of MRT + LRT, and (5) Situation 2030 currently planned (compact zone)+ PT systems (MRT and LRT).

Five alternatives of space transport development are simulated in order to reduce the large number of private vehicles. Alternatives which promote a modal shift towards public transport and will positively contribute to increased accessibility and minimized noise level.

3.1 Improvement of Public Transport

The existing Surabaya public transport network comprises minibus/paratransit, busses and trains. There are 68 available paratransit routes, and 22 bus routes (minibus accommodates 8-12 passengers, while bus has 50-55 seats). Most bus routes use national/provincial roads and toll roads in the middle area of city, other areas are served by paratransit. The paratransit routes have no fixed time schedule or frequency. Bus and paratransit are driven more during the day than the night. The frequency of minibusses and busses are approximately 20-25 vehicles/hours and 5-6 vehicles/hours respectively.

The improvement of the public transport service is regarded as an alternative to reduce the number of private vehicles. Several needed improvements of public transport have been identified. The set up and modeling of those public transport systems, including model calibration based on an extensive survey, have been elaborated in previous study [6].

The Local Government has planned to improve transport infrastructures such as a MRT and LRT system, as explained in Fig. 3. Monorail (MRT) accommodates the passengers in the western and eastern regions. Improvement with this system is expected to increase the public transport passengers. Thus, noise level will reduce because of declining the number of private vehicles and increasing travel speed.

The properties of the proposed MRT and LRT are: a ticket price of approximately \$0.82, stop spacing of 0.5-2 km, seat capacity of 177 seats for the MRT and 100 seats for the LRT, frequency of 12 vehicles/hour, and an average speed of 55 km/h for the MRT and 35 km/h for the LRT [7,8]. The improvements to public transport are expected to encourage the users of cars and motorcycles to shift their transport mode to the public transport systems.

DOI: 10.1051/matecconf/201713807017

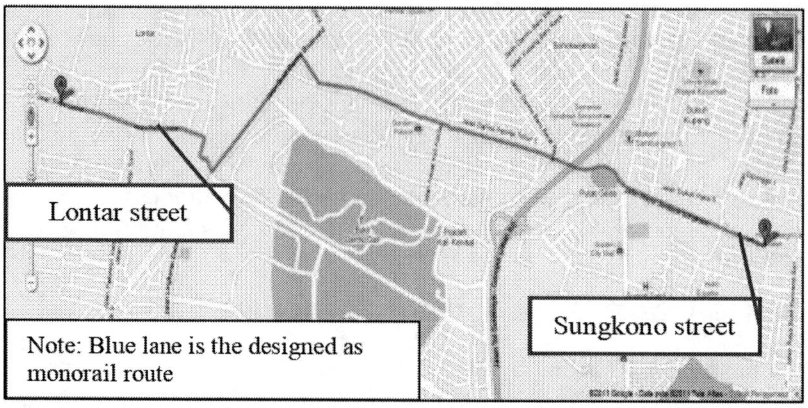

Fig. 3. Monorail (MRT) routes passing from eastern to western side of city, through two road sections with different road corridors.

3.2 Spatial Development Scenarios

Simulation of spatial development strategies is based on the existing spatial panning and transport network, the design of a compact zone in particular regions related to the Spatial Masterplan of the city of Surabaya. Three alternative city structures are considered with distinct spatial planning. They are elaborated below.

3.2.1 Current situation and Current trend

The current situation 2010 describes the current settlement for the city of Surabaya. The spatial schematization comprises 163 administrative zones ('*desas*") of Surabaya City; 5 regions are considered (central, north, south, west, and east), as shown in Fig. 4.

The total area of the city is approximately 33,637 hectares, about 65% of which can be found in the west and east regions. These regions have the built-up area less than 15%, meaning that there is more potential for development in these regions than in other regions of the city.

Settlement development has extended to the outskirts of the urban area, particularly on the western and eastern sides of the city since the 1980s. The 135,000 households live in the suburbs. The highest job density due to new industrial enterprises has developed on some zones of the western regions. These regions create a significant number of jobs for the city, approximately 20% of total employment. On the other hand, a low population density occurs in the regions, furthermore resulting in a high employment-population ratio. The transport infrastructures in these regions are inadequate, and as a consequence it takes more than 90 min by motorized vehicles and more than 110 min by PT to reach the city center.

The current trend 2030 comprises the projected development of Surabaya in 2030, based on a continuation of the current situation and implementation of current planned transport interventions; further an annual growth of 5% GDP and 1% employment has been adopted over a 20 year period (2010-2030). In this scenario, the population will reach approximately 4.5 million residents, with the lowest population in the western regions. Employment growth is more equally divided over the city with a total number of almost 1.6 million jobs [8].

 DOI: 10.1051/matecconf/201713807017

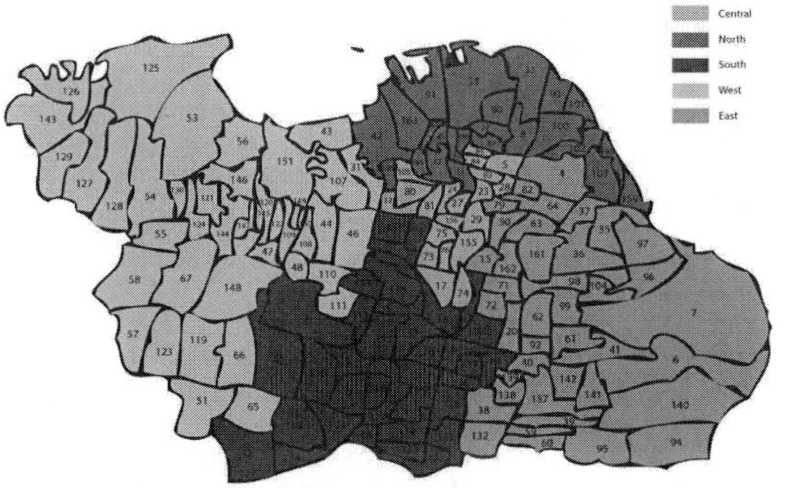

Fig. 4. The "*Desa*" level and regions of the city of Surabaya.

3.2.2 Compact zone balanced with distribution employment

A compact zone was designed at a high density with mixed land-uses and complementary functions such as housing, shopping and offices. The compact zone is considered a strategy to reduce energy use in households and the extent of everyday travel in densely populated area [9]. Based on this study, the compact zone strategy is considered in the Spatial Masterplan of the city of Surabaya. Observation on the distribution of employment and residential density in the city indicates relatively high employment in the western part of the city attracting workers from other parts of the city. In this part, the local government has set up new factory centers, and this has led to a higher employment density in this area of the city. Currently, the residential areas are located away from industrial areas and other business areas [8].

The compact zone strategy in the present analysis has been designed by using a similar overall current trend 2030 with GDP growth 5% and employment rate of 1%. However for this zone (the western part of the city) the ratio of employment to population increases 3% from the current trend 2030. The compact zone strategy in the present analysis has been designed in previous research [8].

4 The Modal Split for Each Alternative

This section presents the assessment of change of modal split and road performance in order to assess the noise level for each alternative. Five alternatives of space-transport development are simulated in order to reduce the large number of private vehicles. Alternatives which promote a modal shift towards public transport and will positively contribute to increased accessibility and to minimize noise level.

Each alternative has an effect on the change of modal split for each vehicle class. Table 1 presents the results on modal split for different alternative. Improvement of public transport systems are intended to increase the use of public transport in each alternative [8].

The following results in Table 1 can be made:

- The design of combination of MRT and LRT systems (A4) encourages the use of public transport up to 37%.

- Spatial strategy with the design of a compact zone in the western regions (A3) also does not have an influence on the increase of public transport use either.
- A fifty one percent increase in the use of PT indicates that public transport passengers are more interested in alternative with a design incorporating a space-transport development strategy than other alternatives. This finding means that the alternative of the design of a space-transport development strategy will reduce significantly the load of private motored vehicles and noise level on the road corridors.

Table 1. Modal split for each alternative (%).

Transport mode	Spatial Development Strategies			Transport Options	
	2010	2030			
	(A1) Current Base	(A2) Current trend	(A3) Compact zone	(A4) MRT and LRT	(A5) MRT + LRT and compact zone
Public transport	5.82	5.62	5.66	7.72	8.45
Motorcycle	58.73	63.39	64.3	58.7	59.34
Car	34.88	30.23	29.2	28.7	27.11
Bicycle	0.57	0.75	0.82	5	5.09

Provision of sufficient PT is necessary to take over from other modes but an adequate urban structure is needed to make a new PT system successful and is expected to minimize noise level.

5 Comparison Noise Level on Highway Corridor

The existing traffic flow on two road corridors, i.e. business activities and residential area is analysed to compare noise level. This assessment is to understand the extensive use of motorized vehicles influencing noise level.

Analysis of noise level is based on basic noise level with some corrections for mean speed and the number of heavy vehicles, gradient, pavement surface, condition of noise source to recipient, building condition, and view angle.

The values were analysed using the procedure for determining noise level for motorized vehicles [10]. The basic noise level ((dBA) is based on the assumption for normal traffic flow, no heavy vehicles on the traffic, and average speed 75 km/h, explained as follows:

$$L_{10} = 42.2 + 10 \log Q \quad (1)$$

which Q is traffic volume.

5.1 The Differences Noise Level in the Existing Condition

The existing road geometry for two road sections, i.e. Mayjen. Sungkono Street (road corridor as business area) and Lontar Street (as residential areas) are described six-lane two-way and two-lane two-way local highways respectively. The existing road performance in term of degree of saturation (DS) are 1.3 (Mayjen. Sungkono street) and 0.6 (Lontar street).

The noise level on road corridor as business areas is amounted to 80.881 dBA, exceed than the Noise Standard Level (70 dBA). Other road corridor as residential area is estimated of 77.963 dBA which is much greater than the Noise Standard Level (55 dBA). Thus, the values of both road corridors were exceeds the noise threshold stated in the Noise Standard Level. The values mean that traffic flow should be decreased by designing improvement public transport and spatial planning in order to reduce the number of private vehicles and noise level. Table 2 explains the average values of noise level in the existing condition for both road corridors.

Table 2. The Average values of noise level in the existing condition.

Street	Road corridor	Average Value of Noise Pollution (dBA)			Average (dBA)
		Morning	Noon	Afternoon	
Mayjen. Sungkono	Business activities	81.518	80.161	80.964	80.88
Lontar	Residential area	78.561	77.284	77.954	77.96

Degrees of saturations are assessed in the morning and noon in order to understand the correlation with noise level for each peak hour and off-peak hour of traffic flow. Fig. 5 and Fig. 6 show the relationship between the degree of traffic saturation with noise level. From the results (R^2) of fitting the regression linear, it means that degree of saturation of both traffic flow on those road corridors have an effect to the noise level.

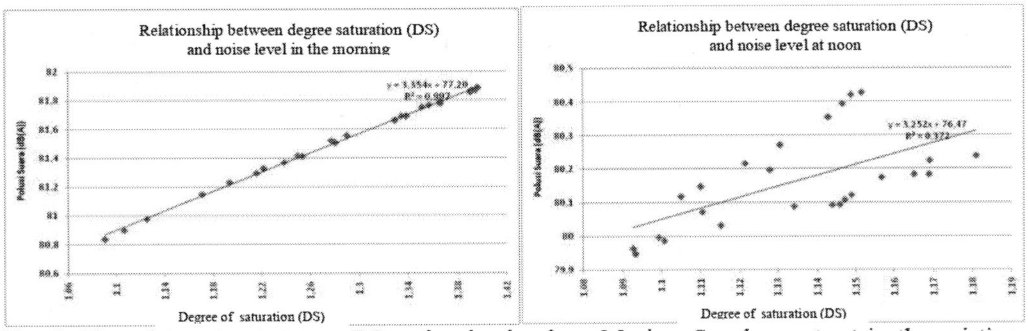

Fig. 5. The relationship between DS and noise level on Mayjen. Sungkono street in the existing condition.

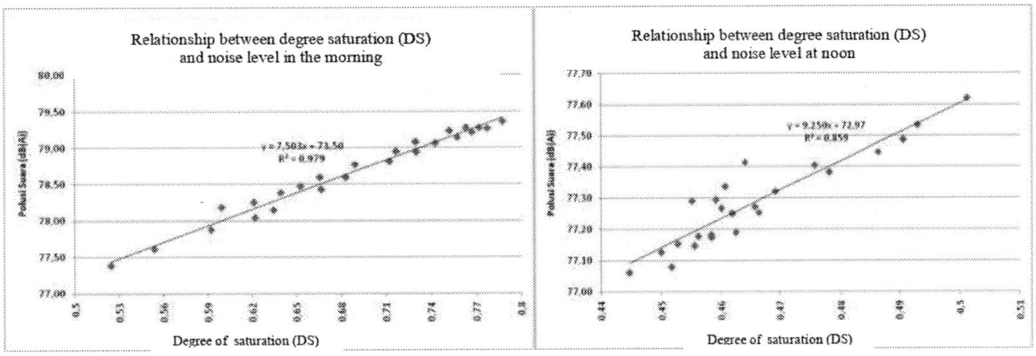

Fig. 6. The relationship between DS and noise level on Lontar street in the existing condition.

5.2 The Difference Noise Level by Reducing 10% of Light Vehicle (LV) and Heavy Vehicle (HV)

This section explains the reducing 10% of light vehicle (LV) and heavy vehicle (HV) influencing noise level in both road corridors. Minimizing noise level on both road corridors is expected below the Noise Standard Level and to be accepted by human health.

Table 3 explains the decreasing noise level by reducing the number of LV and HV in traffic flow on both road corridors (business activity and residential area). Table 3 shows that decreasing the number of LV is significantly reducing noise level on both road corridors, estimated 0.5%. The reduction occurs due to the number of light vehicles much greater than decreasing heavy vehicle volume.

Table 3. The change of noise level due to reducing 10% LV and HV.

Street	Road corridor	Existing condition	Reducing 10% LV	Decreasing (%)	Reducing 10% HV	Decreasing (%)
Mayjen. Sungkono	Business activities	80.881	80.448	0.54	80.858	0.03
Lontar	Residential area	77.963	77.556	0.52	77.936	0.04

6 The Change of Noise Level for Each Alternative

Concerning the mode choice, the combination of improved PT systems (i.e. MRT + LRT) and compact zone (A5) encourages more passengers to use public transport. This alternative is expected to decrease noise level. The results of noise level in each alternative are presented in Table 4.

Table 4. The noise level for all cases (dBA).

Street	Road corridor	2010	2030			
		Spatial scenarios			PT system options	
		(A1)	(A2)	(A3)	(A4)	(A5)
Mayjen. Sungkono	Business activities	80.881	87.3	82.04	74.9	74.12
Lontar	Residential area	77.963	84.15	79.08	72.19	71.45

The following results in Table 4 can be made:

- The design of combination of MRT and LRT systems (A4) has a significantly effect on noise level
- Spatial strategy with the design of a compact zone in the western regions (A3) does not have an influence on reducing noise level.
- A significant change of noise level is identified by a design incorporating a space-transport development strategy (A5). Thus, noise level on road corridor as business activities is approximately the National Noise Level Standard.
- Briefly, noise level is under level stated in the National Noise Level Standard, in which the reduction of light and heavy vehicle volume is approximately 14% of traffic volume on Mayjen. Sungkono Street (trade/business area) and 30% on Lontar Street (residential area) respectively.

7 Conclusions and Recommendations

This study demonstrated an integrated approach, comprised of interventions in urban structure and public transport, is necessary to address the mobility and noise level problems in fast growing urban areas and the associated effects of urban sprawl. An important factor in urban transport planning is an assessment of noise level. The change of noise level was analysed for five alternatives.

The alternative with improved PT systems such as MRT and LRT combined with design of compact zone (A5) was characterized by the smallest noise level, while the spatial development strategy with compact zone (A3) demonstrated an insignificant decrease of noise level.

The compact zone strategy does not provide better noise level. Other strategy in urban planning should be designed, such as examining closest job housing, in order to assess the effect on reducing noise level due to expectation of decreasing travel length and the number of use of private vehicles. Integration of spatial-transport strategies as well as an environmental assessment should provide more sophisticated and long-term advantages, especially improvement of environmental quality when designing improvements of public transport systems related to housing planning.

Acknowledgement

The authors thank to Libertina Widyamurti Ambari for her valuable assistance in writing the manuscript of draft paper.

References

1. Statistic Bureau of Surabaya, *The Figures of Surabaya City* (in Indonesian) (2010)
2. U.W. Tang, Z.S. Wang, Environ. Model. Softw. **22,** 1750-1764 (2007)
3. A. Jamrah, A. Al-Omari, R. Sharabi, Environ. Monit. Assess. **120,** 499–525 (2006)
4. R.K. Mishra, M. Parida, S. Rangnekar, Int. J. Environ. Sci. Technol. **7 (4)**, 737-750 (2010)
5. S. Agarwal, B.L. Swami, Environ. Monit. Assess. **172,** 113–120 (2011)
6. L. Ambarwati, R. Verhaeghe, A.J. Pel, B. van Arem, WIT Trans. Ecol. Environ. **138,** 653-668 (2014b)
7. Department of Transport of Surabaya City, *Surabaya MRT: Surabaya Mass Rapid Transportation (SMART).* Technical Report. Surabaya, Indonesia (2013)
8. L. Ambarwati, R. Verhaeghe, A.J. Pel, B. van Arem, Transp. Res. D Trans. Environ. **44,** 134–146 (2016)
9. E. Holden, I.T. Norland, Urban Stud. **42**, 2145–2166 (2005)
10. C.S. Papacostas, P.D. Prevedouros, *Transportation Engineering Planning.* Hawai (1993)

Riverside sustainability – Overcome problems and reliable approaches

Irma Novrianty Nasution[1]*, *Syahreza* Alvan[2]

[1]State University of Medan, Civil Engineering Department, Medan, North Sumatera, Indonesia
[2]State University of Medan, Civil Engineering Department, Medan, North Sumatera, Indonesia

Abstract. The growth of population is a feature behind the urbanization. The city's facilities are an attraction for migrants to work and live. It has led to the emergence of urban problems in terms of providing employment, legal settlements, and livable formal housing. The inability of the low-income community economically will lead it to live in area on riverside, railroads, and Chinese cemetery. Then, it will be turned into a slum area. Kampong Aur is part of a slum area of Medan, also have the same problem. Therefore, it is the importance of research. The data collected by an observations, previous studies, journals, planning documents, and aerial photographs. The data obtained were analyzed and elaborated using qualitative descriptive method. The results show that Kampong Aur can be observed from the land tenure, physical environment, livelihood, public facilities, and socio-cultural conditions. It is a unique type of settlement with consists of formal and informal housing built side-by-side. Informal settlements are usually located along the river borderline. And it is categorized into slum settlements. Originally, slums need to be managed integratedly and well organized, improvement or relocation, with a humane and sustainable approach through the participation of all parties. In this way, participatory and cooperative designs are the reliable approaches that can be applied to the settlement conditions in Kampong Aur.

Key Words
riverside, environment, sustainable, settlement

1 Introduction

Rapid urbanization is happening across Asia, with more and more people in need of housing. Providing adequate housing to everyone in the cities is not an impossible goal. It's possible to solve the serious housing problems, if it can begin to see urban poor settlements not as problems, but as sources of energy and important contributions to the production of housing. And it's possible if it can look at the poor not as beneficiaries of someone else's ideas, but as the primary actors at the centre of their own development.[1]

* Corresponding author: irma_nst11@yahoo.com

Over one billion people live in 'squatter', 'slum', or 'informal' housing settlements globally. This population is expected to double by 2030, making such settlements the major form of urban housing.[2] And more than a half of them live in Asia.[3] These settlements often characterised by unclear tenure status, poor basic services, inadequate construction an over crowded living conditions, are an urban emerging due to various social, economic, and political factors.[4] The most obvious place to choose is unauthorized public land, such as: river banks, fire access lanes, railway yards, and even Chinese cemeteries.[5]

According to Kamil (2017), in 2025 nearly 60 % of people will live in urban areas and human civilization in the city.[6] The city will grow and develop according to the dynamics of the economy, population, and social culture. Urban problems are increasingly complex and increasing. Problems are not only in physical form, but encompassing people and society. Kamil declares that Indonesia in the future can only be saved by human beings. Where urban improvement and implementation activities are implemented taking into account the sustainability of life and the future of human civilization in the city.

Medan is the third largest city in Indonesia by population, having the same problem. Rapid urban growth is characterized by population growth, availability of adequate facilities and infrastructure, wide and diverse employment, and rapid development of social, economic and cultural life. And this condition has an impact on the growth of urban settlements. Settlements in Indonesia developed formally and informally in locations directly related to the downtown area or other supporting areas.[7] Formal settlements are generally owned and occupied by middle and high-income communities. Meanwhile, informal settlements are inhabited by low-income communities. Low-income communities actually have limitations in getting a decent board. These communities will look for illegal lands to live and live in society. Because the settlement for the marginal is not just a place for shelter and gathering but also a place of business, a social event, and various multi-dimensional activities.[8]

Kampong Aur as a study area is a mix of formal and informal settlements along the Deli River, Medan. Deli River which divides the city center into a place that is not interesting to visit or be seen. Because of the river basin area has been polluted by a pile of rubbish and a slum of informal settlements. Most of the riverside in Medan City does not geographically from the city's features because the river is used as the back of the city that is not utilized optimally.

The problem of urban settlements in almost all developing countries is not an easy thing to solve. Regulations, policies, and evaluations related to economic, social and environmental sustainability should be carefully considered and thoroughly devoid of the community as a benchmark for planning and design, and an active role in decision-making.

Therefore, this paper was conducted as a preliminary study aimed at understanding, identifying, and looking at settlement problems in Kampong Aur, as well as finding the right solution, with a sustainable and humane approach. The main issue is having a reliable approach that can improve the quality of life of the riverside settlements. At next level of study, it becomes a baseline to planning and design with considering the existing condition of settlement – poverty, health problems, and poor living condition. The study was conducted by collecting data from various secondary sources, such as planning documents, an official site from the local government, previous studies, and aerial photographs. The primary data collection derived from observation for the formal and informal riverside settlements condition. The data obtained were elaborated and analyzed using a qualitative descriptive method.

2 An overview of riverside settlement

2.1 Cities and settlements

The city is a center of activity and community life. The city is identical with social, economic, and environmental problems. These three are the basic pillars of the city that should grow and develop in a balanced way with an integrated approach. The developments of the city in general only focus on economic aspects. Social and environmental aspects, which are related to residence, are often underpaid.

The construction of housing for the masses in Indonesia is massively passive and static. So far the role of government only as a provider is not an enabler. The houses and settlements available have not been planned through a humane approach and according to the needs of the community.

Community involvement in the planning and design of urban settlement environments is a new alternative in bridging between government, planners and the community itself. With the active role of the community will create a sense of belonging so that the motivation to maintain and improve the environment will increase.[9]

Rapid population growth without equitable housing and infrastructure development is occuring in almost all developing countries. This condition is obviously in low-income communities living in urban areas with inadequate infrastructure, service, and environmental facilities. These people choose to live in legal and illegal urban lands, with formal or informal environments. And often the existing settlements ignore aspects of good health, comfort, and environmental quality. Meanwhile, formal settlements that provide better housing facilities are economically inaccessible to low-income communities due to price mechanisms that do not meet the standards of community life at this level.

The problems of urban settlements are inseparable from the context of slums, informal and low-income communities. The problem of informal settlements can be seen as a result of many influencing factors, and in many urban scenarios, informality is regarded as illegal. A consequence of their commonly perceived illegality, informal settlements, exist outside the formal frameworks – both socially and economically slum dwellers often lack rights to their homes, political recognition, and services from their government.[10] Rosmarin & Connor also quoted that informality should be seen not solely a symptom of poverty – but rather as an alternative and potentially superior method of creating our urban environments.

2.2 Informality of riverside settlements

Riverside settlements are characterized by houses built on river borderline with simple materials and construction. Settlements are often not equipped with good sanitation, water supply, drainage and electricity systems. People living by the river and use it as a waste disposal, bathing, and washing, and household disposal systems that are not equipped septic tank. The distance between buildings is close so the occupancy does not get good light and the wind all day, so the house is not healthy and comfortable to live in. Riverside settlements are identical to disaster-accessible locations, such as the risk of floods, fires, and earthquakes.

Riverside settlements are chosen by the community as the nearest alternative land to work and socio-cultural activities that are comfortable for low-income people. In principle, people living on the riverside have an understanding and a desire to maintain their environment. However, cooperation between the community and local government needs to be improved in order to facilitate the needs and problems of the community regarding the quality of the urban settlement environment. In addition, the level of understanding and community life, as well as the flow of urbanization supported by loose regulation resulted in these settlements continuing to grow along the riverside.

2.3 Riverside sustainability in the future

Development of urban settlements, especially on the riverside, has been done in many places in the world. Activities are undertaken from eviction, providing low-cost housing, relocation, slum upgrading, land sharing, and lending to low-income communities to obtain a proper dwelling. Development activities are conducted between government and non-government organizations (NGOs), Community Based Organizations (CBOs), and local communities. Repair of slums consists of improvements in infrastructure including improvements to toilets and walkways, street lighting, drainage and water supplies.[11]

The most prominent settlement improvement in Indonesia is the Kampung Improvement Program (KIP).[12] The objective of the project is to provide access to physical infrastructure such as roads, paths, drainage, waste systems, water supply and social infrastructure such as schools and health centers for the poor living in informal settlements.

3 The Deli riverside of Medan

3.1 Medan at a glance

Medan City is the third largest city in Indonesia, located in North Sumatra Province. Medan City is bordered by Aceh Province and West Sumatra Province. The city of Medan, as one of the most important regional economic centers on the island of Sumatra, has a strategic position, function, and role as the main gateway for regional and international trade and financial services activities in western Indonesia. The strategic city position is supported by the availability of airports and seaports and other urban infrastructure and utilities.

The characteristics of the region are 265.10 km2 or 3.6 % of the total area of North Sumatra Province. Administratively, Medan City is bordered by the Malacca Strait in the north, and borders on Deli Serdang Regency in the east, west, and south. The area of Medan City is accessible to flooding and it causes the geographical position of the city which is located at an altitude of 2.5 - 5.0 M above sea level with relatively flat topography with a slope of 0 - 3 %.

Medan City is hydrologically influenced and surrounded by several major rivers and branch such as Percut River, Deli River, Babura River, Belawan River and other rivers. The rivers passing in Medan City not only affect the physical form, space, and environment but also affect the development pattern of Medan City. The rivers are still used by the community as one source of water that lives in the area around the river. In addition, these rivers also serve as the main drainage in the framework of flood control, as well as rainwater disposal. As a result, river functions tend to be more limited due to silting and environmental degradation. And this becomes a problem for the city of Medan.

The population of Medan City until 2009 is estimated to reach 2.1 million people with an average population growth of 1.1 % per year. The projection of population growth rate of Medan City until 2015 is about 1.07 % per year.[13] The highest population growth is caused by the rate of urbanization. The high rate of urbanization due to the city of Medan is an economic attraction for people in its hinterland. As population increases, the population density increases to 8.001 life / Km^2. The density of the population is relatively high and can cause an imbalance between the carrying capacity and urban environment less than optimal.

The attempt to provide housing for low-income communities in the city of Medan has not been comparable with the number of people who still need a house that is habitable, resulting in the growth of housing/slums. Currently, slum areas in Medan City are estimated to reach 22.5% of the total area of Medan City which consists of 88.166 housing units or 16.87% of the total existing houses in Medan City. Slum areas are spread in 145 villages and are generally located on the riverside and railroads in the city center.

3.2 Riverside settlements along Deli River

In early days of the Medan City development, the Deli River perform as a pulse of trade to other areas, namely as a transportation route in trading activities. At that time, the Deli River became a benchmark for the development of Medan City. Medan city grew and developed by the crossing of Deli River and Babura River. These two rivers became a busy traffic line and developed into a transit port of traders, in the past. And the river affects the historical, cultural and social development of the people of Medan City until now. However, the low appreciation of urban communities to rivers and the surrounding environment makes the river decrease water quality and the environment.

The Deli River is located in Medan Maimoon sub-districts of Medan City. The Medan Maimoon sub-district is one of the 21 sub-districts of Medan City that is located in the city center, near the center of Medan City Government. It covers a 3.342 Km^2, with borders of; Medan Polonia sub-district in the West, Medan Kota sub-district in the East, Medan Johor sub-district in the South, and Medan Barat sub-district in the North. There are 66 cluster associations and 6 neighborhood associations. According to 2012 census data, the Medan Maimoon sub-district has 39,581 inhabitants belonging to 137 households.[14] The study area is one part of 6 neighborhood association in Medan Maimoon sub-district which called Kampong Aur.

Fig. 1. Riverside settlements along Deli River

Source: http://google.co.id, accessed on 24th April 2017

3.3. The study area – Kampong Aur

Kampong Aur is one of the urban villages classified as slum settlement.[15] Kampong Aur consists of four neighborhoods that are part of Medan Maimoon sub-district, with an area of 34.2 Ha, and a population of 715 people with a density number of 21.

The total number of buildings in Kampong Aur is 320 units and 26 % of it is well ordered. Kampong Aur is known as a settlement with a high occupancy density of 45 units / Ha. And only 80 % of housings have the condition of the roof, floors, walls, according to technical requirements.

The existing infrastructure of Kampong Aur is quite good. Environmental accessibility has been complemented by a proper road as well as technical requirements. In addition to the high density of the building, Kampong Aur is also known as a flood-accessible area due to the Deli River which overflowed in the rainy season, where almost all neighborhoods are flooded and drainage systems are not able to anticipate the condition. The availability of environmental drainage is sufficient, but it is necessary to review the appropriate design for Kampong Aur's environmental conditions. Electricity and water infrastructure is well available and accommodated by the local government. More than 90 % of people in

Kampong Aur have been served by clean water and electricity. Meanwhile, waste management and fire protection still need special attention and handling.

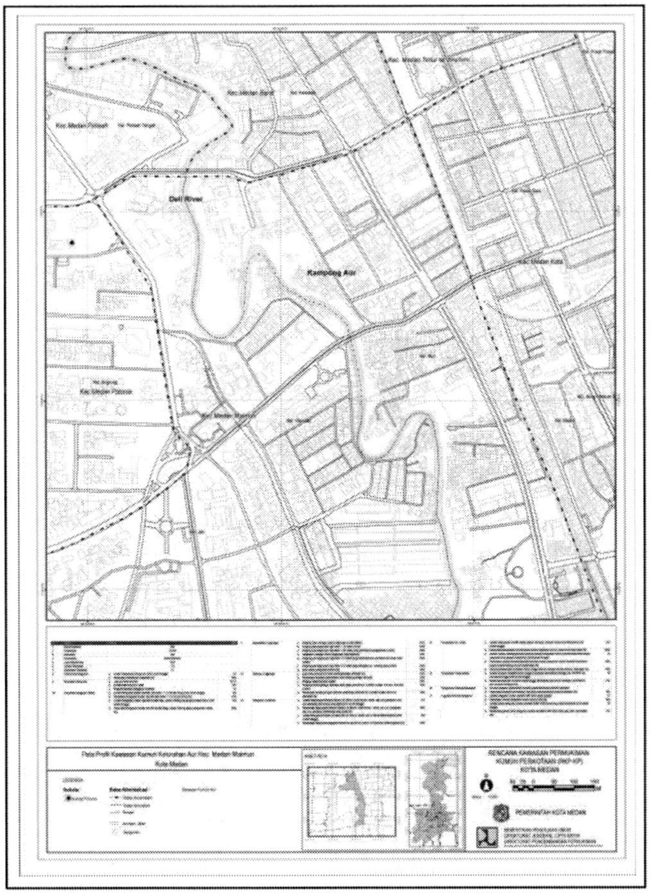

Fig. 2. Map of study area

Source: RKP-KP Pemerintah Kota Medan, 2016

4 Problem: cause and approach

The problems of urban settlement in the city center, such as Kampong Aur, have their own character that can be observed in the physical environment, land tenure, public facilities, economic features and livelihoods, and social and culture conditions.

4.1 Land tenure

From the total of 320 occupations, only 11 % have legal land tenure, and the rest do not have its certificates. Based on information in the field, the land tenure of Kampong Aur in the past is owned by the Deli Sultanate and one of the largest entrepreneurs in Medan City, namely Chong A Fie. Possibly, the condition becomes one of the reasons behind the arrangement of Kampong Aur settlement area has not been done properly.

According to land tenure, it is necessary to establish communication between various parties such as academics, community, local or regional authorities, non-governmental organizations (NGO's), which are integrated through the efforts of land consolidation. Finally, it will ensure the security and comfort of living in the community, development activities, and to improve the quality of the environment and preservation of natural resources.

4.2 Physical environment

The physical condition of the Kampong Aur environment has diverse characteristics. The physical form of the settlement is a permanent house which is one-storeyed and two. While the physical form at the fringe, which is on the river borderline, is identified as slum area that used materials construction such as wood, zinc, and plywood. Physically, the characteristics of the settlement area contoured with the lowest elevation orientation on the riverside. The density of buiding, irregular, and contoured arrangements produce a slum settlement. The approach of land consolidation can be improved the physical condition of a better environment for the sustainability of the community.

Housing in Kampong Aur can not be said a proper dwelling because of the settlement area is not well ordered. In addition, there is a decline in the quality capability of environmental facilities and infrastructure caused by the low maintenance and development of infrastructure after it was built. This is concerned with the performance of the community is low. This problem is a great task for the local government because of there is no specific reference to addressing the environmental issues of the settlements in Kampong Aur.

The improvement of the physical environment, both settlements, and housing, in Kampong Aur, can be done through cooperative design approach or known as co-design approach. It is an approach to design attempting to actively involve all parties (e.g. settlers, academics, NGO's, local authorities, investors, private sectors) in the design process to help ensure the result meets their needs and is usable.

Fig. 3. Physical condition of settlements and environment

4.3 Public facilities

The establishment of public facilities in Kampong Aur are mosques and religious schools or *madrasah*. Open space as a performance of people's social interaction is determined in the roads, porch, riverside, and gathering hall of resident house. Public utilities such as water, electricity, drainage systems, and waste disposal mechanisms are lack. For example, not every house has a septic tank. Septic tank also has the same sewage system as the environmental drainage system. The existing of the drainage system accommodates all housing and environmental disposal systems, such as rainwater disposal and solid waste. The locked structure of drainage system causes an overflow of water when flooding which is not possible to accommodate trash and flooded in the drain. It causes poor living conditions for local residents.

Public facilities in the settlement need to be renewal and reconstruction. A passed narrow road path by various vehicles causes uncomfortable track road to be through. For example, pavement systems can be renewal into porous pavements that can absorb water into the ground so that when a flooding through the road, it can be absorbed by the ground well. In addition, the road with a different level of height should be made up a ladder so that the road path comfortable to pass. Other efforts to prevent overflow from flooding into residential areas are should be greening along the riverside. And slum settlements along the river borderline can be suggested to be relocated to an appropriate place. By the relocation, planning and design activity should not be neglected the nature of habit of community in the previous settlements. And it should be done by considerate and humane.

Fig. 4. Public facilities of Kampong Aur

4.4 Economic features and livelihoods

Kampong Aur is a strategic urban village located in the inner city and it is a wider riverside settlement than other. This condition allows the community to be able to move from and to the location of work. In addition, the location is very close to city government center, traditional markets, and commercial activity centers. These provide opportunities for different sources of livelihoods. The main sources of livelihood are street vendors, carrying market stuff for the consumers, parking keepers, casual laborers, sewing laborers, *warung* or small stalls, construction workers, and government employees.

Kampong Aur community is very active in various social activities related to the development of family welfare. They have an active community group. Through these community groups, all parties such as private companies, academics, and government agencies have a strong interest in providing training, counseling, and physical actions to improve community's economic lives. The activities were conducted together with local women's groups with various types of activities, including knitting training, inorganic fertilizer training, cooking, gardening of vegetable plants, and others. With the efforts made by all parties are expected the welfare of the community and insight into the source of livelihood is increasing.

Home economic, such as small stalls, should be trained and facilitated to obtain a proper workplace for an attractive and comfortable environment. Another effort that can be done is to build the people into a tourism-conscious community. The activity, both economics, social, and cultural, encouraged income opportunity to invite local tourists and non-local to visit the settlement.

Fig. 5. Economic activities around the environment

4.5 Socio cultural conditions

In general, the community in informal settlements lives with threat conditions that affect the environment and itself. River as a natural resource used for bathing, clothing, and even toilet. Nevertheless, the social bonds between peoples are well established with the physical intimacy, common problems of life, and their social character.

However, Kampong Aur does not have the facility to engage in social interaction, especially for children, ceremonies and religious activity. Kampong Aur communities are the low-income people with a poor living condition. There are children of school age who can not continue their education to the next level because of a cost. So, it becomes a vandalism and drugs activity among it. People, in general, are apathetic to their life. However, some of the people in Kampong Aur neighborhood are engaged in positive activities with the external.

To maintain a positive social life can be done through counseling continuously, open an economic opportunities, creating interest activities for children, young, and woman. Activities need to be sustainably maintained so that it becomes a culture and profitable for the communities. For example, organize kampong festivals, mural festivals, as well as being environmental agents around. All activities are the participation of all levels of community and outside agent roles as facilitators and supported by the establishment of kampong councils as respected community agents.

Fig. 6. Social interaction among people around the environment

5 Conclusions

The Deli riverside problem in the study area is low-income community and migrants from rural. This community lives in a forbidden location in the middle of the city because of the urge of the rapid development of the city. The need for a place to live close to work sites, public transport, and activity centers has also become the main reason. To obtain a livable house in a formal settlement is not a priority because of an inability economically of the communities.

Poor living conditions and poverty is a common feature of informal settlements with chain reactions among different socio economic, political, and environmental factors. And evictions are not the acceptable method to solve informal settlement problems. By evicting it means violating human rights and the poor are the most affected victims.

A possible approach is a combination of integral improvement and relocation activities for the sustainability of the riverside. The relocation, especially in the borderline, can be moved to a new location around the settlements. And the settlement improvement can be done through renewal by the participatory approach in planning and design. Through this approach, it is expected that communication gaps between the communities, local authorities, NGO's, private sectors, or investors, to determine policy, and financing can be well organized during the planning process. The proposed approach aims and focuses on the informal community interests so that they do not return to a poor living. Finally, increased quality of settlements will affect city life more humane and sustainable in the future.

References

1. UN-Habitat. *Housing the poor in Asian Cities: Quick guide 1*, Published by UNESCAP, pp. 1-29 (2008)
2. UNCHS (United Nations Centre for Human Settlements). *Challenge of slum, Global report on human settlements 2003*, UN Habitat, Earthscan, London and Sterling, pp. 195-228 (2003)
3. et al, (2003)
4. T. Rosmarin, J. Connor. *Lesson of Informality, Part I: 'A global phenomenon urban informality'*, Regular Spread, University of Melbourne, Melbourne School of Design, pp. 12-15 (2016)
5. Dick, 2002 quoted by S., A.A. Sari. The emergence of informal riverside settlements and challenge for planning: The case of Kali Mas in Surabaya, Indonesia, Proceeding: *The 3rd International Seminar on Tropical Eco Settlements; Urban Deprivation: A Challenge to Sustainable Urban Settlements,* A-14, pp. 137-148 (2012)
6. http://www.youtube.com/watch?v=hDw-Tq1anKU, TEDxJakarta : Ridwan Kamil-Creativity and design for social change in cities, accessed on 6 March (2017)
7. N. Hamidah. "Kampung sebagai model permukiman berkelanjutan di Indonesia", INERSIA. J. V **12** N 2: 24-34 (2017)
8. E. Budiharjo. *Percikan masalah : Arsitektur, perumahan, perkotaan*. Penerbit : Gadjah Mada University Press, Yogyakarta, pp. 67-68 (2006)
9. et al, pp. 54 (2006)
10. Mc Kean, 2009 quoted by T. Rosmarin, J. Connor. *Lesson of informality, Part I: 'A global phenomenon urban informality'*, Regular Spread, University of Melbourne, Melbourne School of Design, pp. 12-15 (2016)
11. UNESCAP, 2003 quoted by S., A.A. Sari. The emergence of informal riverside settlements and challenge for planning: The case of Kali Mas in Surabaya, Indonesia, Proceeding: *The 3rd International Seminar on Tropical Eco Settlements; Urban Deprivation: A Challenge to Sustainable Urban Settlements*, A-14, pp. 137-148 (2012)
12. S., A.A. Sari. The emergence of informal riverside settlements and challenge for planning: The case of Kali Mas in Surabaya, Indonesia, Proceeding: *The 3rd International Seminar on Tropical Eco Settlements; Urban Deprivation: A Challenge to Sustainable Urban Settlements*, A-14, pp. 137-148 (2012)
13. ___________Peraturan Pemerintah Kota Medan Nomor 14 Tahun 2011 Tentang Rencana Pembangunan Jangka Menengah Daerah (RPJMD) Kota Medan Tahun 2011-2015, Pemko Medan (2011)
14. http://pemkomedan.go.id/hal-medan-maimun.html, accessed on 30th March (2017)
15. ___________Rencana Kawasan Permukiman Kumuh Kota Medan (RKP-KP), Kementrian PU Direktorat Jenderal Cipta Karya ; Direktorat Pengembangan Permukiman (2016)

Leachate of Demolition Waste

Yatnanta Padma Devia[1,*], *Eko Andi* Suryo[1]

[1]Department of Civil Engineering, Brawijaya University, Jl Mayjen Haryono 167 Malang, Indonesia

Abstract. Demolition waste contains not only non-hazardous and uncontaminated material but also consists a small fraction of residual hazardous materials that need to be separated, processed or disposed of. The demolition wastes are utilized as aggregates in new concrete and road construction application, initial resources in reuse construction and disposal materials of landfill. When demolition wastes are disposed in unprotected landfills can cause environmental and health problems as leaching that may have consequence in groundwater and soil. The majority of demolition wastes researches in Indonesia addressed its classification, quantification and causes, while environmental effect do not explore yet. Therefore, it is necessary to investigate demolition waste in Indonesia especially its leaching effects. This preliminary research is showing result of leaching test of demolition waste using toxicity characteristic leaching procedure on copper, zinc, lead, sulphate, chloride and fluoride. A concentration of lead showed in higher level (0.65 mg/L) than Indonesian standard of toxicity characteristic leaching procedure (0.50 mg/L). Some elements were detected reach near and over the limit value of drinking water standard, such as sulphate and fluoride. Sorting is better to do to minimize the risk of hazardous waste from demolition waste.

1 Introduction

Demolition waste defined as waste debris from destruction of a construction [1] as a part of construction and demolition waste (CDW), which the waste arises from construction, renovation and demolition activities [2]. In Indonesia, demolition waste is classified into two categories, called valuable waste and non-valuable waste [3]. Generally, valuable waste is reuse or re-sale, such as wood, roof tile, steel or other metals. A non-valuable waste, such as inert waste (concrete, brick masonry, sand, gravel) utilize as aggregate in new concrete, road construction application (Fig. 1), a virgin resources in reuse construction (Fig. 2) and disposal material of landfill [4]. A large quantities of demolition waste that produced from buildings and civil engineering works demolition are landfilled without pre-treatment after separation of valuable waste for reuse and recycling [5]. Furthermore, demolition waste contains not only non-hazardous and uncontaminated material but also consists a small fraction of residual hazardous materials that need to be separated, processed or disposed of [6]. When demolition waste is disposed in un-protected

* Corresponding author: ypadmad@gmail.com

landfills can cause health and environmental problems as leaching that may result in waterborne contaminant potentially affecting subsoil and groundwater [4, 7]. Mostly the previous studies, especially in Indonesia, focussed on classification, quantification and cause of demolition waste [3, 8], but in the contrary its environmental effects still not explore yet. This study objective is to investigate leachate of demolition waste.

Fig. 1. Demolition waste for road construction

Fig. 2. Demolition waste for reuse construction

2 Methods

2.1 Quantification of demolition waste used for test

A complexity of demolition waste in type, age, activity, site and properties made uneasy quantification of sample obtained [7, 9]. In developing countries, such as Indonesia, information of generation of demolition waste is less than construction waste due to manual method applied more than blasting methods [10]. Quantification construction and demolition waste (CDW), according to Wu *et al.* [10], were site visit method, waste generation rate method, lifetime analysis method, classification system accumulation method, variables modeling method [11] and other particular methods. Ten samples were taken from demolition building site, which half of them from building in the age of less than 10 years and in other half its more than 10 years. However, a suitable portion of sample for leaching test that represent demolition waste in a site were not available on previous studies especially in Indonesia. A simple method of sampling that referred to Trankler [9], was used in this experiment. Each sample during demolition days was taken around 1% of day's waste. This sample was mixed divided to obtain sample that meet standard agitation test of toxicity characteristic leaching procedure. The particle size range of 0 – 8 mm or categorized as fine fraction was chosen because higher size will less endanger environment by its leachate [9].

2.2 Toxicity Characteristic Leaching Procedure for demolition waste

Samples of demolition waste from houses demolition in Surabaya City, Indonesia were collected and the Figure as shown in Fig. 3. Samples were categorized in their construction year or age of the house which less and more than 10 years. Samples were taken for the determination of the standard agitation test by using rotary agitator. The leachate was analysed according to toxicity characteristic leaching procedure (TCLP). The copper (Cu), lead (Pb), zinc (Zn), sulphate (SO_4), chloride (Cl^-) and fluoride (F) content were determined by referring to the standard methods [12] using a atomic absorption spectroscopy for Cu, Pb and Zn; a spectrophotometry for SO_4 and F; and an argentometry method for Cl.

Fig. 3. Sample of demolition waste

3 Results and Discussion

Quantification of demolition waste sample summarizes in weight percentage that consist of inert waste and source of hazardous waste, such as asbestos and paint. Table 1 summarizes demolition waste in weight percentage to the total quantity of demolition waste.

Table 1. Weight percentage of demolition waste sample

Demolition waste	%mass
Brick masonry	54
Concrete	35
Cement	7
Asbestos	2
Ceramic	3
Paint	0.2

Six elements were analyzed: copper (Cu), lead (Pb), zinc (Zn), sulphate (SO_4), chloride (Cl) and fluoride (F). In a comparison, Table 2 shows leachate concentration of samples at minimum/maximum values and leaching limit values for the acceptance of waste of the Indonesia TCLP standard by regulation PP RI no 101/2014 Appendix IV Standard TCLP Test. Copper, zinc, sulphate, chloride and fluoride were detected in considerable levels on both samples of demolition wastes which represent waste of less than (samples N1 – N5)

DOI: 10.1051/matecconf/201713808002

and more than 10 years (samples O1 – O5). Similarly, a limit value of copper, zinc, sulphate, chloride and fluoride of European Council Decision (2003/33/CE) that used by Roussat *et al.* [7] on their leaching concentration from the percolation test were 0.6, 1.2, 1500, 460 and 2.5 mg/L, respectively. Based on their values, copper, zinc, sulphate, and chloride still were much lower than the limits.

Table 2. TCLP result

	Demolition waste <10 years (mg/L)		Demolition waste >10 years (mg/L)		Indonesia TCLP Standard (mg/L)*
	Min	Max	Min	Max	
Copper	0.12	0.20	0.20	0.35	10.00
Lead	0.52	0.65	0.42	0.60	0.50
Zinc	0.47	0.55	0.8	0.87	50.00
Sulphate	191.60	240.00	150.40	220.30	**)
Chloride	24.00	30.00	20.50	55.00	12,500.00
Fluoride	1.10	1.50	2.40	2.75	75.00

*) PP RI no 101/2014 Appendix IV about Standard TCLP Test
**) No limited value

Low concentration of cooper (Fig. 4) and zinc (Fig. 5) in both sample, less and more than 10 years, can be due to reduction by adsorption phenomena within the porous structure, which makes diffusion barrier between water and these elements [7]. In the site, an affinity of heavy metal with sub-surface geology, which consists of clay can cause absence of heavy metal in groundwater [13]. These kinds of heavy metals may constitute environmental problems if the leachates migrate into the groundwater. If copper in the undesirable limit on groundwater will give the effects such as astringent taste, discoloration and corrosion of pipe, fitting and utensils [14].

DOI: 10.1051/matecconf/201713808002

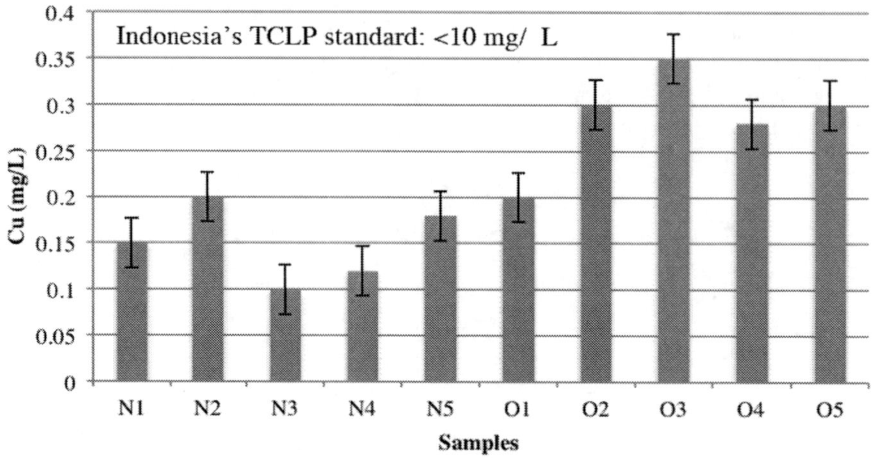

N1 – N5 samples from building demolition in the age < 10 years
O1 – O5 samples from building demolition in the age > 10 years

Fig. 4. TCLP result of copper

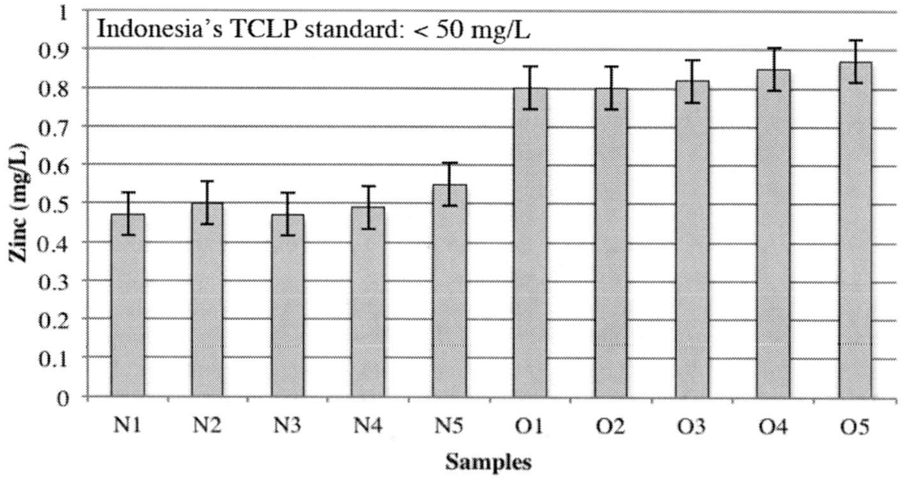

N1 – N5 samples from building demolition in the age < 10 years
O1 – O5 samples from building demolition in the age > 10 years

Fig. 5. TCLP result of zinc

A concentration of lead (Fig. 6) on leachate sample of demolition waste in the age less than 10 years is higher than the limit value (0.65 mg/L instead of 0.50 mg/L for the limit value). Lead is categorized as hazardous waste. It contain was probably came from chip plaster coated with lead-base paint. There were no significant difference values between two kinds samples, because the age of samples of more than 10 years were derived from building demolition in the age around less than 20 years. Lead was a common additive to paint prior to 1980, even though still be found in household paint produced by industries in countries that have not banned the use of lead, recently. This metal is characterized as toxic for drinking water. The permissible limit of lead in drinking water is 0.05 mg/L [15], which leachate sample concentration was higher beyond the limit value. The high impact of lead made this heavy metal represent as environmental impact parameter units (kg Pbeq/m^2) within a Life Cycle Analysis of CDW [16].

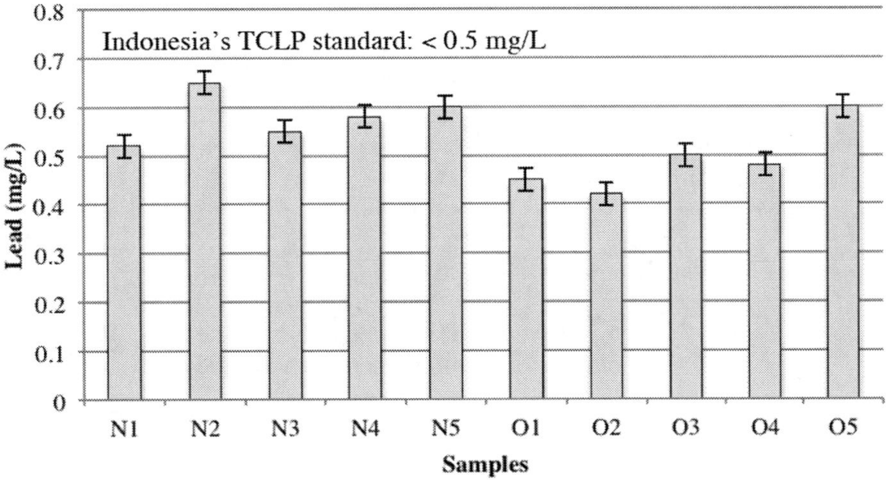

N1 – N5 samples from building demolition in the age < 10 years
O1 – O5 samples from building demolition in the age > 10 years

Fig. 6. TCLP result of lead

Sulphates has no limited value in Indonesia TCLP standard, but the sulphate contain have a significant leachability [7] and can be dragged away to the environment and indicate an environmental problem. Sulphate concentration is come from gypsum plaster that contained calcium sulphate, which contribute sulphate concentration in leachate. Sulphate is not considered as hazardous waste. A leachate sample of demolition waste on sulphate (Fig. 7) in the age less than 10 years and more than 10 years were 240 and 220 mg/L, respectively, These values are detected higher than sulphate limit value of drinking water on 200 mg/L [15]. High concentration of sulphate in drinking water can have a laxative impact when combined with hardness elements, calcium and magnesium [14].

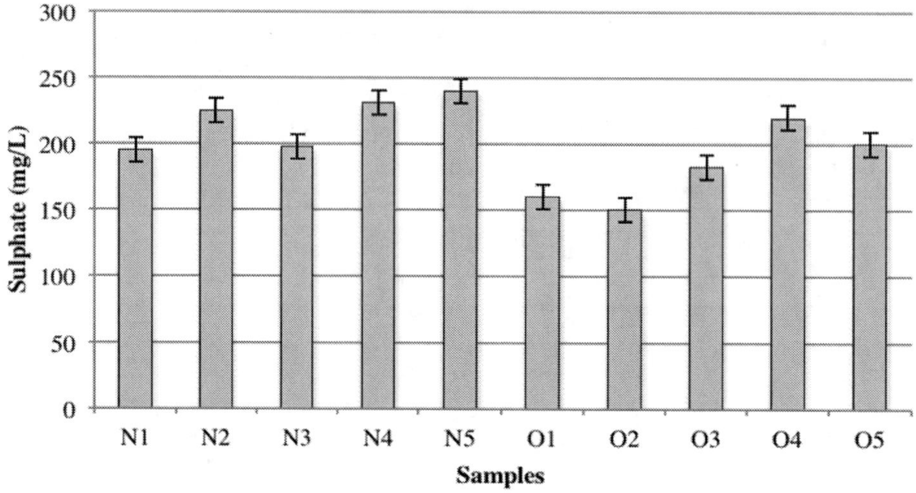

N1 – N5 samples from building demolition in the age < 10 years
O1 – O5 samples from building demolition in the age > 10 years

Fig. 7. TCLP result of sulphate

The maximum chlorides (Fig. 8) in leaching were 30 mg/L and 55 mg/L for demolition waste in the age less and more than 10 years, respectively. These values are indicating that these waste do not exposed to sea water or salts nor affected by calcium chloride that some occasion be added as hardening accelerator for concrete [17]. Chloride concentration on leaching samples did not exceed allowable limit of Indonesia TCLP (12,500 mg/L) and drinking water standards (600 mg/L).

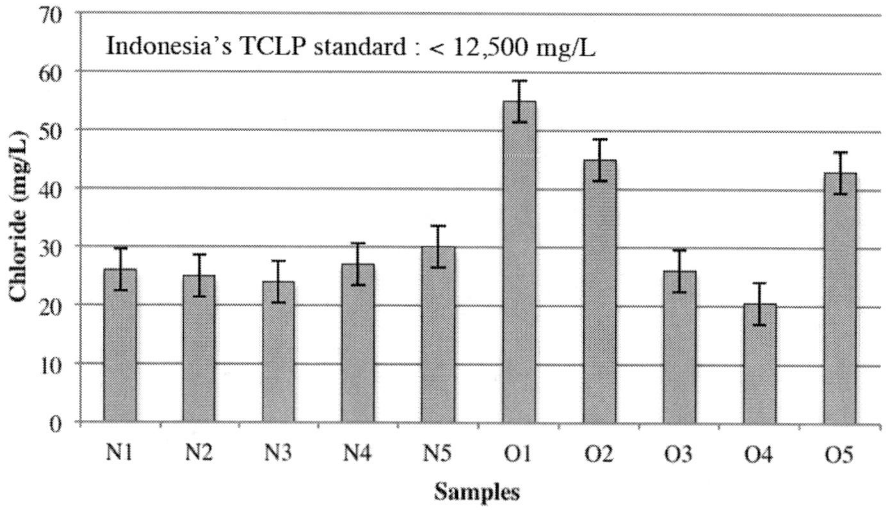

N1 – N5 samples from building demolition in the age < 10 years
O1 – O5 samples from building demolition in the age > 10 years

Fig. 8. TCLP result of chloride

Flouride (Fig. 9), in the case compared with limit value of European Council Decision percolation test from previous research [7], had concentration 1.5 mg/L and 2.75 mg/L for demolition waste less and more than 10 years, respectively, that were over the limit value (2.5 mg/L). The Indonesia TCLP standard for fluoride is less strict leaching limit value on 75 mg/L.

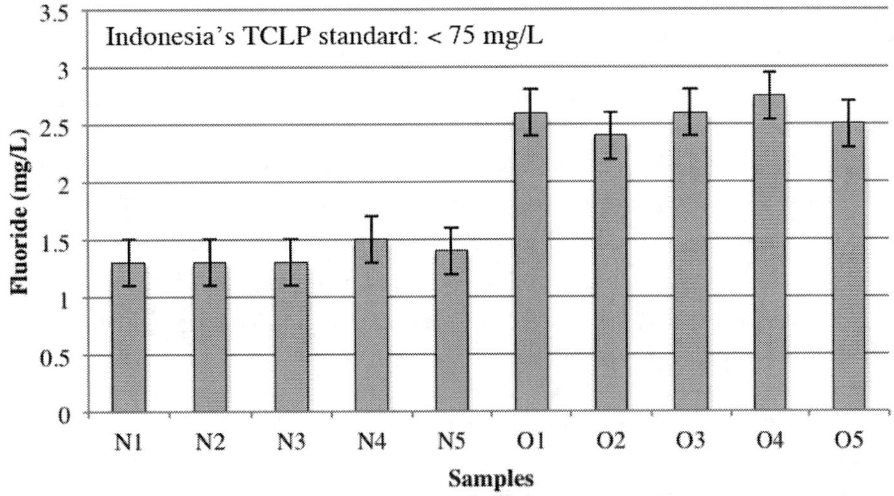

N1 – N5 samples from building demolition in the age < 10 years
O1 – O5 samples from building demolition in the age > 10 years

Fig. 9. TCLP result of fluoride

This preliminary result shows that leachate of demolition waste in Surabaya - Indonesia especially for cooper, zinc, sulphate, chloride and fluoride are detected on considerable level compared to Indonesia TCLP standard, excluded lead concentration. This opportune condition might be caused by kinds of material and manual sorting and crushing processes of valuable waste that can reuse or re-sale, separated from a non-valuable waste. However, in the future, the main problem for recycling demolition waste is the cost of sorting, crushing and disposal due to the manual processes [18]. A machinery process will replace manual process and implicate difficulties on sorting waste. Furthermore, a less sorting process will give more complexity on environmental impacts.

4 Conclusions

This preliminary research used TCLP test for demolition waste to knowing its leachate concentration. Analyses of the leachate shows that lead, as a hazardous waste, exceeded the limit value, which can cause health and environmental impacts, in particular if it is deposited in an unprotected landfill mixed with inert demolition waste. A chip plaster coated with lead-base paint, which probably lead's source, is better sorting to minimize the risk of hazardous waste in environment. A drinking water standard might be considered to realize that the impact of leachate of demolition waste could reach on soil and groundwater in environment.

References

[1] M. R. Ponnada, Kameswari. Int. J. Adv Sci and Tech **84**, 19-46 (2015)

[2] H. Yuan, L. Shen. Waste Manage **31**, 670-679 (2011)

[3] A. Nugroho, T. Tongthong, T. Shin-ei. Int. J. of Civil & Env. Eng. **13**, 35-41 (2013)

[4] S. Butera, T.H. Christensen, T.F Astrup. Waste Manage **44**, 196-205 (2015)

[5] M.D. Bovea, J.C. Powell. Waste Manage **50**, 151-172 (2016)

[6] M. Bernardo, M.C. Gomes, J.de Brito. Waste Manage **49**, 156-169 (2016)

[7] N. Roussat, J. Mehu, M. Abdelghafour, P. Brula. Waste Manage **28**, 2032-2040 (2008)

[8] S. Alwi, K. Hampson, S. Mohamed. *Proc. of the 1st International Conference of CIB W107 – Creating a Sustainable Construction Industry in Developing Countries* (2002)

[9] J.O.V. Trankler, I. Walker, M, Dohmann. Waste Manage **16**, 21-26 (1996)

[10] Z. Wu, A.T.W Yu, L. Shen, G. Liu. Waste Manage **34**, 1683-1692 (2014)

[11] A.F. Masudi, C.R. Che Hasan, N.Z. Mahmood, S.N Mokhtar, N.M. Sulaiman. Int. J. Global. Environ. **12** (2-4), 269-281 (2012)

[12] American Publich Health Association (APHA), American Water Works Association (AWWA) and Water Environment Federation (WEF) *Standard method for the examination of water and wastewater.* 20th edition, American Public Health Association, Washington DC (1998)

[13] A. Adeolu O, A. Gbenga A, O. Adebayo A. African J. of Env. Sci. and Tech **5** (11), 933-940 (2011)

[14] N. Raman, D. Sathiya Narayanan. Rasayan J. Chem. **1** (4), 826-836 (2008)

[15] WHO: Guideline for Drinking Water Quality. 3rd edition. Geneva: World Health Organization; 940–949 (2008)

[16] A. Coelho, J.de Brito. Waste Manage **32**, 532-541 (2012)

[17] S. Butera, T.H. Christensen, T.F Astrup. J. of Hazardous Material **276**, 302-311 (2014)

[18] N. Kartam, N. Al-Mutairi, I. Al-Ghusain, J. Al-Humoud. Waste Manage **24**, 1049-1059 (2004)

Removals of colour and turbidity from stabilized leachate by using alum and glutinous rice flour dual coagulants

Shaylinda Mohd Zin Nur[1,*], and *Azizan* Mohd Omar[1]

[1] Faculty of Civil and Environmental Engineering, Universiti Tun Hussein Onn Malaysia, 86400 Parit Raja, Batu Pahat, Johor, Malaysia

Abstract. The effects of alum and glutinous rice flour as dual coagulant on the removals of colour and suspended solid from stabilized leachate were investigated. A series of jar test were conducted by optimizing the dose and pH of single coagulant (alum and glutinous rice flour) and its combination. It was observed that low removals (colour: 46%, suspended solid: 16%) were recorded by glutinous rice as single coagulant at pH 7 and dose 1.2 g/L. However, addition of glutinous rice flour (0.12 g/L) as coagulant aid for alum at pH 6 and dose of alum 2.5g/L, show better removals of colour (86%) and suspended solid (92%) from alum. At same pH and dose of alum as single coagulant, the removals achieved only 79% and 83% of colour and suspended solid, respectively. Thus, addition of glutinous rice flour as coagulant aid improved the colour and suspended solid removals from the stabilized leachate

1 Introduction

Due to increment of population expansion and economic development it will contribute to solid waste generation. According to Diaz and Savage [1], improper solid waste management will poses a hazard to human health and the environment. Uncontrolled dumping area and inconsistent waste handling causes a variety of problems such as contamination of leachate into water bodies. Leachate contain large amount of organic contaminants and hazardous compound such as chemical oxygen demand (COD) and biological oxygen demand (BOD), suspended solids, significant concentration of heavy metals, ammonia, and inorganic salts. If leachate not treat properly and safely disposed it could be a potential source of threat to quality of water. Regarding to Marttinen and Wang [2], usually the treatment of landfill leachate often involves a combination of various method such as aerobic and anaerobic biological process, photo-oxidation and membrane process, chemical oxidation and precipitation, activated carbon and absorption and coagulation-flocculation.

Nowadays, coagulation and flocculation process has been given full priority by many researches in improving the sustainability to the environment [3]. Normally the coagulant

* Corresponding author: nursha@uthm.edu.my

use in the treatment of leachate is chemical coagulant such as Alum, Ferric Chloride, Aluminium Sulphate and Polyaluminium chloride (PAC). Even though the uses of chemical coagulant show the good performance and efficiency, some of them causes environmental effect and promote secondary pollution [4]. Therefore, to reduce this problem various study on the efficiency of dual coagulant which consist of chemical coagulant and natural coagulant were carried out. Normally, coagulation just utilized single coagulant, for example, Alum, Ferric Chloride, Aluminium Sulfate and other synthetic coagulant. Hence, the use of dual coagulant consist of chemical and natural coagulant is utilized to enhance the productivity of flocs, and removal colour and turbidity from leachate.

In this research, alum was used as the primary coagulant while glutinous rice flour as the coagulant aids. Alum is one of the conventional chemical coagulant agent used in Malaysia. However, the application of alum produced after used issues such as high volume sludge production and high aluminium content in treated waste water. Application of coagulant aid able to reduced volume of sludge and dose of primary coagulant. However, most of coagulant used is mainly from synthethic polymer material. Which is believed had influence on health issues. Application of natural coagulant is organic and contain no toxic elements. Natural coagulant can be harvest from plant or animal based. Starch is a potential coagulant aid material. The polysaccharides element in starch able to increase the coagulation and flocculation efficiency. Starch is a carbohydrate polymer derived from various sources of food plant structures such as corn, rice, wheat, potato, tapioca, nuts, mung bean, green pea and sago [5]. Unlike most of the existing of natural coagulant, the use of glutinous rice flour as natural coagulant for leachate treatment has not widely been investigate. Hence, this study investigates the effectiveness of dual coagulant consist of alum and glutinous rice flour for the treatment of stabilized leachate by evaluating the colour and turbidity removal.

2 Methodology

Samples of stabilized leachate were collected from Simpang Renggam Landfill site (SRL) located in Simpang Renggam, Johor, Malaysia. The sampling procedures for leachate were conducted according to the method of collection and preservation of the samples. All the collected samples were immediately transported to the laboratory and stored in a cold room at $4^{0}C$ to minimize biological and chemical reaction.

Table 1. Optimum coagulation factors determined from preliminary experiment.

Coagulation factors	Unit	Value
Rapid mixing speed	RPM	200
Rapid mixing time	minutes	4
Slow mixing speed	RPM	30
Slow mixing time	minutes	15
Settling time	minutes	30

Jar test was conducted by using a jar test apparatus. The jar test was performed considering the factors listed in Table 1. NaOH (0.1 mol) and HCL (0.1 mol) were used to

DOI: 10.1051/matecconf/201713808003

adjust leachate (500 mL) pH. Percentage removals of colour and turbidity were calculated by measuring concentration of aforesaid parameters before and after the jar test.

3 Results and discussion

The range of coagulant dose tested was shown in Figure 1. Alum as single coagulant achieved higher removals at dose 5 g/L with 99% and 98% of SS and colour removals respectively. While application of glutinous rice (GR) coagulant as single coagulant shown higher removals at 1.2 g/L with 16% and 38% SS and colour removals. Thus, GR as single coagulant is less effective compare to Alum. Combination of Alum and GR as dual coagulant efficiency are shown in Figure 1 (c). At dose of 3.5 g/L alum and 0.12 g/L of GR, the best removals were recorded (97% for SS, 89% for colour). Slightly lower removal of SS (94%) and colour (88%) was recorded by single alum at 3 g/L. At 2.5 g/L dose of single alum 83% and 79% of SS and colour were removed respectively. At the same dose with addition of 0.12 g/L GR, 92% and 86% SS and colour removals were recorded. Addition of GR into <2.5 g/L alum able to increase the removal rate of single alum. Thus at lower dosage of alum, GR able to increase the coagulation efficiency. At higher or near optimum value of alum in dual coagulation system, the increment of removal is not that significant. This probably resulted on the dominant removal mechanism during the reaction. At optimum dose of alum in dual coagulation, the action of alum is much more dominant compare to GR. While at point where the alum dose is not optimum, GR reaction able to work together with alum in improving the coagulation efficiency.

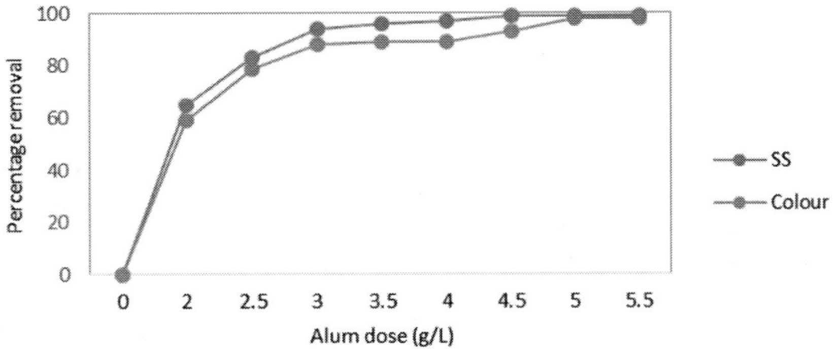

(a)

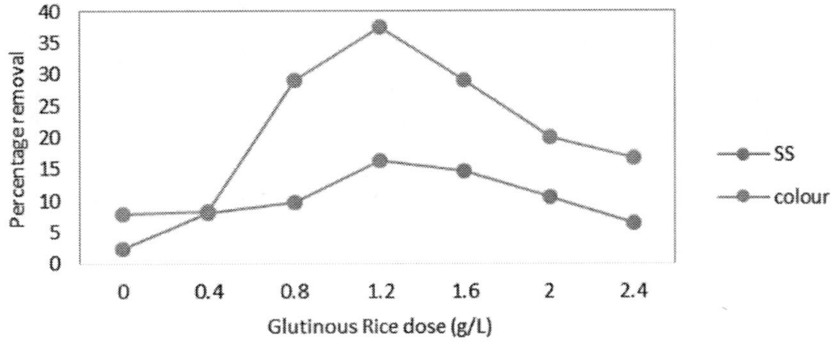

(b)

DOI: 10.1051/matecconf/201713808003

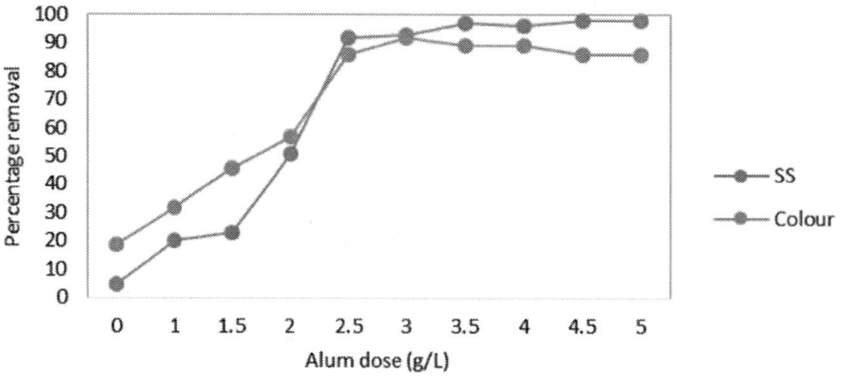

(c)

Fig. 1. Percentage removal of suspended solid and colour under the influence of dose (a) Alum as single coagulant at pH 6(b) Glutinous rice as single coagulant at pH 7 (c) Dual coagulant (Alum + 1.2 g/L Glutinous rice) at pH .

4 Conclusion

The effectiveness of dual coagulant made from alum and GR on stabilized SRL leachates was determined. Dual coagulant optimum removal of SS was recorded at pH 6, alum (4.5 g/L) and GR (0.12 g/L) for 98% removal. While, for optimum colour removal was at pH 6, alum (3 g/L) GR (0.12 g/L) with 92% removal. At lower dose of alum (2.5 g/L) with 0.12 g/L of GR, the removals of colour and turbidity is better than single alum (2.5 g/L) and single GR (1.2 g/L).

Acknowledgement

This work was supported by Universiti Tun Hussein Onn Malaysia under FRGS grant scheme FRGS Vot 1570. The authors also wish to acknowledge Majlis Perbandaran Kluang for their assistance during the sampling process.

References

1. L.F. Diaz, G.M. Savage, L.L. Waste Mgmt, **25(6)**, 626-637 (2005)
2. S.K. Marttinen, R.H. Kettunen, K.M. Sormunen, R.M. Soimasuo, J.A. Rintala, Chem., **46**, 851-858 (2002)
3. K.S. Amit, NR. Mandre, AB. Panda, P. Sagar, Carbo. Pol., **95**, 753–759 (2013)
4. Y.B. Song, J. Zhang, WP. Gan, JP. Zhou, LN. Zhang, Ind. Eng. Chem. Res., **49(3)**, 1242 1246 (2010)
5. A.J.F. Carvalho, *Biopolymers: Biomedical and environmental applications* (John Wiley & Sons, Inc. & Scrivener Publishing LLC, (2011)

Analysis of Microbial Air Quality in the Surrounding Hospital's Wastewater Treatment Plants in Jakarta, Indonesia

Gabriel Andari Kristanto[1,*], and *Fatma Nur Rosana*[1]

[1]Department of Civil Engineering, Faculty of Engineering, Universitas Indonesia, Depok 16424, Indonesia

Abstract. As hospitals require the usage of various hazardous drugs and chemicals in its daily operation, the often contaminated wastewater of health facilities needs to be processed to comply with the quality standards specified in Regulation of the Governor of DKI Jakarta No. 65/2013. However, wastewater treatment plants (WWTP) can also pollute not only water but also the ambient air of facilities. The objective of this research is to identify airborne bacteria and the concentration of fungi due to the presence of WTP and to analyse the types of bacteria that exist through Gram Staining. Microbial air samples taken using EMS E6 showed that the average concentration of bioaerosol at hospital's WWTP was about 810-1,915 CFU/m^3 for bacteria and 1,450-1,955 CFU/m^3 for fungi. Furthermore, the highest concentration of bacteria and fungi found in the surrounding equalization and aeration tanks was 1,915 CFU/m^3 and 1,955 CFU/m^3, respectively, which far exceeded the background value of 80 CFU/m^3 for bacteria and 440 CFU/m^3 for fungi. Gram staining of the bacteria found in the air surrounding WWTP showed 94% Gram-negative with the bacteria being pathogenic. Therefore, it is pertinent that further action is taken to prevent bioaerosol from spreading around WWTP by building better ventilation surrounding the plants, indicating the technology of the treatment plant to be used, and enforcing personal safety measures on WWTP's employees.

1 Introduction

Hospital wastewater contains pathogenic, toxic, degradable, and radioactive contaminants which have harmful effect on health [1]; thus, it requires treatment prior to the surface water discharge. The function of wastewater treatment plant (WWTP) is to eliminate the contaminants and pathogens in wastewater in order to prevent it from polluting the environment and harming people's health. Biological treatment process generally utilizes microorganism for degrading pollutant compound. However, it should be noted that the process in WWTP can also produce air pollution, known as bioaerosol. Bioaerosol is one type of air pollutant which contains microorganism and has the potential for affecting

* Corresponding author: gakristanto@gmail.com; andari@eng.ui.ac.id

 DOI: 10.1051/matecconf/201713808004

human health in the forms of infectious diseases, allergies, and cancer [2]. Bioaerosol can be naturally produced in decomposition of organic matter by microbes dispersed to atmosphere. Furthermore, bioaerosol emission can be produced through human activities, and one of them is waste disposal facility including biological treatment plant, MBT (mechanical biological treatment) unit process, material recovery facilities and solid waste transfer station, landfill, and WWTP [3, 4, 5, 6, 7]. The microorganism contained in bioaerosol such as bacteria and fungi are commonly found in the surrounding WWTP [8, 9]. Bioaerosol in WWTP comes from bubbles generated from aeration process or oxygen from degradation process [10] which can be pathogenic and can cause allergy if inhaled thus making this bioaerosol existence highly influential in the quality of the ambient air [11]. The magnitude of this bioaerosol emission is influenced by factors such as WWTP capacity, applied treatment plant technology, applied treatment plant machines, applied treatment method, and content of contaminants in wastewater [10]. In addition, factors such as nutrients, humidity, temperature, and oxygen can affect the growth of microorganism [12]. The purpose of this research is to analyze: 1) the concentration of bacteria and fungi in WWT's ambient air; 2) the influence of temperature, humidity, and wind on WWTP's ambient air; and 3) the existence of pathogenic bacteria in air sample.

2 Methodology

2.1 Sampling Location and Time

Sampling took place at waste water treatment plant (WWTP) in Dr. Suyoto Rehabilitation Center of the Ministry of Defense and conducted in January and April 2016 in wet/rainy and dry seasons. The sampling points were on WWTP site as shown in Fig. 1. The distance of sampling from WWTP was 0, 10, 20, 30, 40, and 50 meter(s), whereas background concentrations were obtained from the Faculty of Engineering, Universitas Indonesia, Depok and used as a comparison to identify the effect of WWTP's existence on air quality.

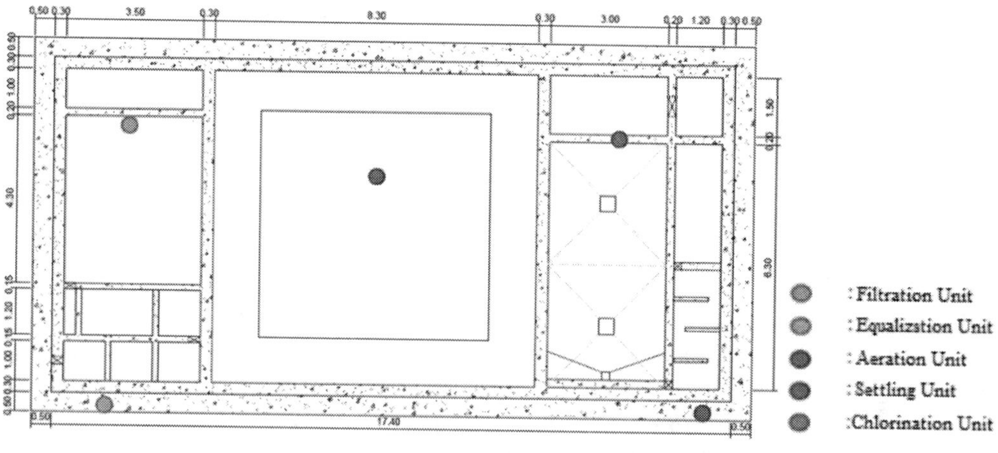

Fig. 1. Sampling Points

DOI: 10.1051/matecconf/201713808004

2.2 Parameters & Method

The sampling of bacteria and fungi was done using Single-Stage Multi Orifice Bioaerosol with a capacity of 25 L/minute vacuum pump. The measurement of air temperature, humidity, and wind velocity parameters used anemometer instrument of Lutron type AH-4223. The agar medium used was Tryptic Soy Agar (Oxoid) as the medium for the growth of bacteria at incubation temperature of 37°C for 24 hours and Malt Extract Agar (Oxoid) as the medium for the growth of fungi at incubation temperature of 29 °C for 72 hours. The counting of bacteria and fungi growing in the media was done using Total Plate Count (TPC) method. Gram staining was used to find out the Gram type of the bacteria which involved 4 dye solutions such as crystal violet, Lugol's iodine, alcohol, acetone, and safranin. The observation of bacteria was conducted using Olympus IX73 Research Inverted microscope with a magnification of 100 and Olympus DP73 Color-Monochrome Dual CCD 17.28 MP.

2.3 Analysis of Data

The average concentration of bacteria and fungi in CFU/m^3 (Colony Forming Unit/cubic meter) was calculated as follows:

$$Microorganism\ Concentration = {}^{CFU}/_{Plate} \times \frac{1}{0.0250\ {}^{m^3}/_{minute}} \times \frac{60\ {}^{second}/_{minute}}{t\ second} \quad (1)$$

The impact of physical parameter on the concentration of bacteria and fungi was obtained statistically by employing correlation test with SPSS 16.0.

3 Results and Discussion

3.1 Concentration of Bioaerosol in the Surrounding WWTP

The sampling of bioaerosol in WWTP was conducted in wet season when it was drizzling and raining lightly. The data measurement of bioaerosol in WWTP showed varying results (Fig. 2.):

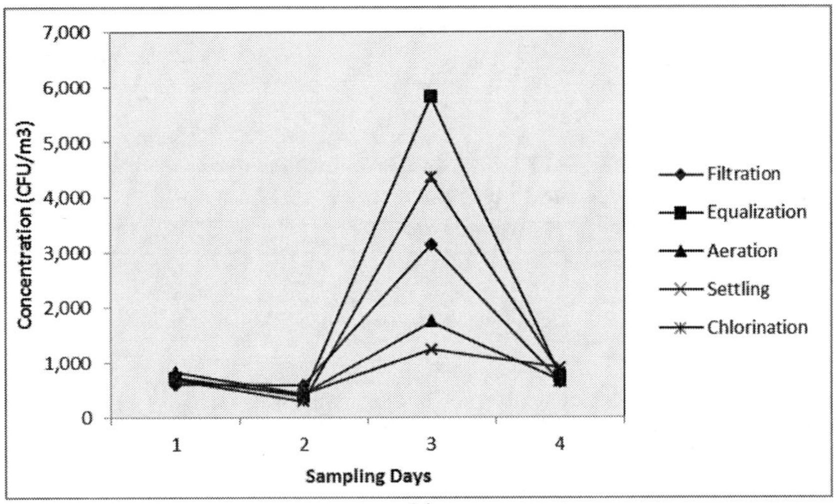

(a)

DOI: 10.1051/matecconf/201713808004

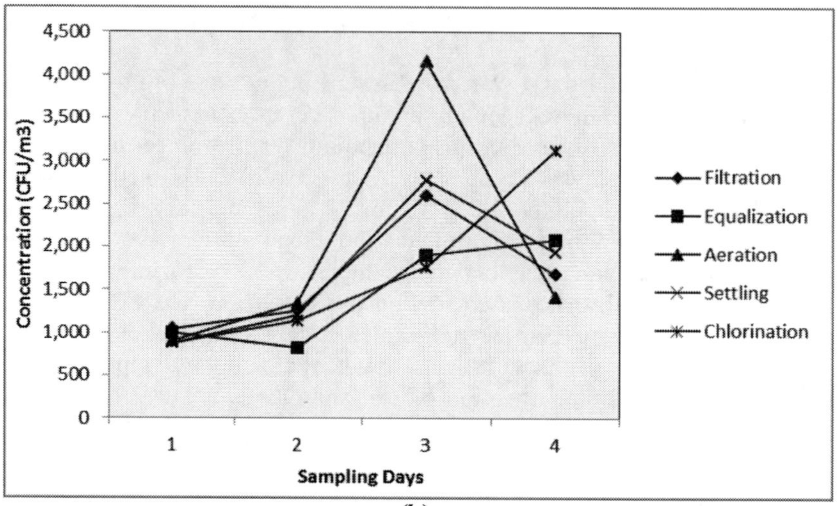

(b)

Fig. 2. (a) Bacteria and (b) Fungi Concentration in WWTP

It was found that the concentration of bacteria and fungi was higher during rainy days due to higher air humidity. According to a research, the concentration of bioaerosol in bacteria and fungi during rainy season will be three to seven times higher than that of in normal dry (not raining) condition [13]. As for this research, the average concentration of bioaerosol in bacteria increased three to fifteen times during the rainy season and two to five times in fungi during the dry season. Therefore, seasonal factor can also affect the concentration of bioaerosol in bacteria and fungi.

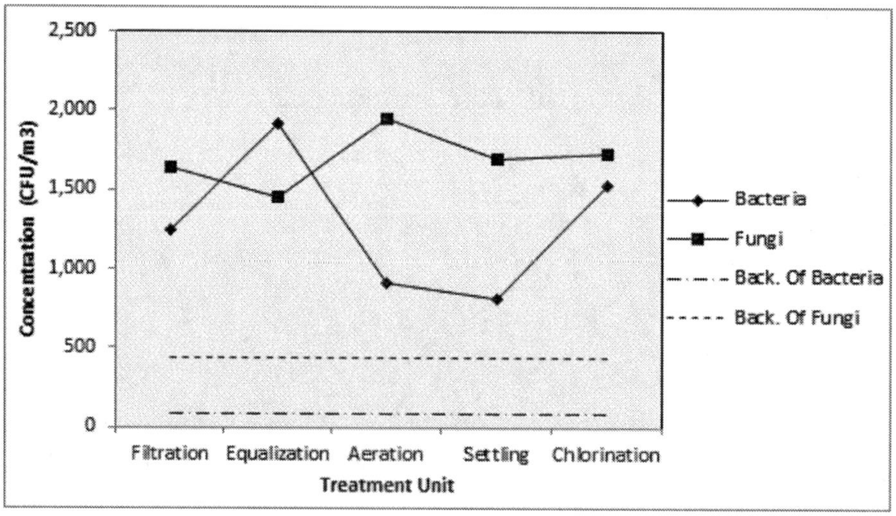

Fig. 3. Average Concentration of Bioaerosol in WWTP

As illustrated in Fig. 3, the concentration of bacteria and fungi in WWTP have exceeded the background concentration which means WWTP can be a factor in the increased concentration of bioaerosol in the surrounding area. The highest concentration of bacteria in equalization unit was 1,915 CFU/m^3, while the highest concentration of fungi in aeration unit was 1,955 CFU/m^3. This high concentration in equalization unit indicated the

DOI: 10.1051/matecconf/201713808004

utilization of microorganism used in wastewater treatment plant to reduce organic pollutants in wastewater. This finding was supported by the fact that equalization unit functions as biological treatment involving microorganism. A research showed that oxidation ditch produced the highest microbe aerosol in domestic WWTP in China [14]. Like the previous research, shows that the highest mesophilic bacteria are produced in trickling filter unit in domestic WWTP in Italy [15]. Similarly, one hospital in Jakarta, Indonesia, shows that hospital's WWTP can also be the source of bioaerosol pollution, and the highest concentration of bacteria is found in aeration unit (Integrated WWTP 1) and equalization unit (Integrated WWTP 2), while the highest concentration of fungi is found in equalization unit (Integrated WWTP 1) and in sedimentation unit (Integrated WWTP 2) [16]. The aeration using diffusors can increase the concentration of bioaerosol in the air through the bursting of water bubbles that releases microorganism. The bubbles formed from aeration process contain microorganism in high concentration, and through the research, aeration unit become the main source of bioaerosol [17].

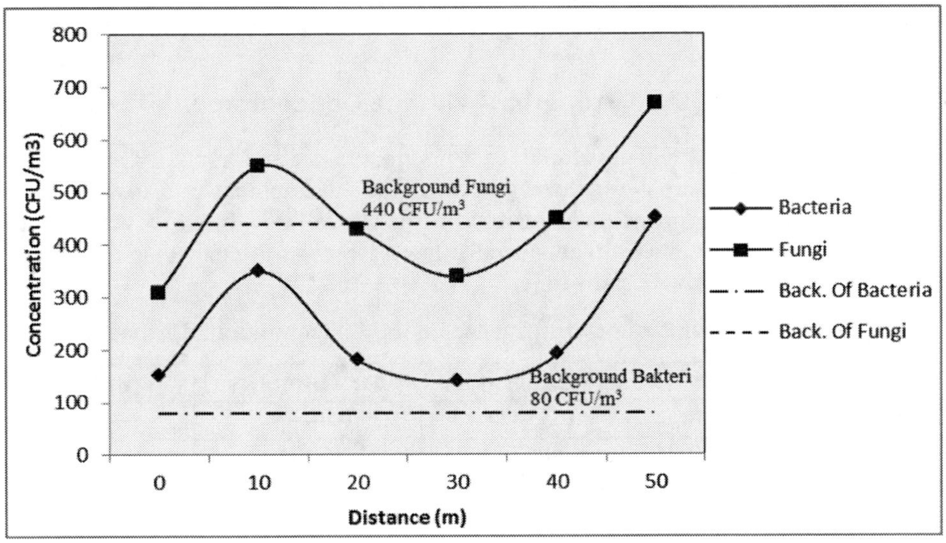

Fig. 4. Bioaerosol according to distance variance

Apart from that, environmental factor can also promote the growth and distribution of microorganism in air. The distribution of bioaerosol in WWTP is shown in the concentration of bioaerosol to distance as illustrated in Fig. 4. The samples were taken in dry season, when it was sunny. The concentration of bioaerosol varied at a distance of every 0 – 50 meter(s), and this concentration of bacteria exceeded the threshold limit allowed by the Ministry. On the other hand, the concentration of fungi at a distance of 0, 20, and 30 meter(s) although it was still below the limit. When sampling was conducted, the aeration unit was not in operation hence the concentration of bioaerosol at 0 meter from WWTP was lower than the 10 meter(s) sampling point. Apart from that, the high temperature and low humidity can lower bioaerosol concentration. At a2 distance of 10 and 50 meter(s) the concentration of both bacteria and fungi showed an increase, which could be caused by the factors of sampling location and environment such as temperature and humidity. Another increase was shown at a distance of 40 and 50 meter(s), which may be caused by other sources of bioaerosol in the physiotherapy room. There was a decrease in the concentration of bioaerosol at a distance of 20 and 30 meter(s) influenced by factors such as the distance to WWTP, wind velocity, and even the altitude of the location may also cause a faster distribution of bioaerosol in the air.

In terms of variance in seasonal condition between the rainy and the dry season, the concentrations in rainy season are higher than that of in dry season. To prevent this difference, the observed point was the aeration point where the sampling location for both seasons took place at a distance of 0 meter. The concentrations of bacteria during the rainy season are 400 – 2,800 CFU/m^3 and the fungi concentrations are 840 – 4,520 CFU/m^3, while during the dry season the concentration of bacteria and fungi was around 100 – 200 CFU/m^3 and 240 – 380 CFU/m^3 respectively. These results showed that the concentrations of bioaerosol in dry season are lower than in rainy season. Statistical correlation shows that there is a significant relation between temperature and humidity in the air toward the concentration of bioaerosol. The moderately high temperature can affect the growth of microorganism due to the tendency of bacteria to die in high temperature, while the moderately high humidity will support the growth of microorganism. Moreover, WWTP's condition can also have effect on the concentration of bioaerosol when bioaerosol may be from the bubbles produced by aeration process or the oxygen produced by degradation process [10], may be pathogenic and develop allergy if inhaled [11]. When the samplings took place in dry season, no aeration occurred, and this caused lower bioaerosol emission.

3.2 Correlation of Physical Properties with the Concentration of Bioaerosol

Microorganisms can grow and develop in certain temperature and humidity; in other words, there are necessary conditions for them to live. If the environment is below the optimum level, it will hamper their growth; conversely, if it is above the optimum level, it will make them lethal. Furthermore, microorganisms also have a certain life expectancy and are only likely to survive in a relatively short time.

Table 1. Physical Parameter Influence on the Concentration of Bioaerosol

	Air Temperature	Air Humidity	Wind Velocity
Bacteria concentration	-0.314*	0.401**	-0.303*
Fungi concentration	0.063	0.570**	0.140

Signification: **P value*<0,05 and ***P value* <0,01
N: 64

Table 1 shows the correlation between physical parameters, such as air temperature, air humidity, and wind velocity and the concentration of bacteria and fungi. There is a significant correlation between these parameters and bacteria concentration (*P value* = 0.012; *P value* = 0.001; *P value* = 0.015 respectively) and as for fungi concentration, there is only significant correlation with air humidity (*P value* = 0.000). Air humidity has the highest correlation with both the concentration of bacteria and fungi indicating that air humidity can influence bioaerosol concentration (the higher the humidity, the higher the concentration of bioaerosol). On the other hand, it does not have any significant correlation with fungi concentration. The correlation of air temperature and humidity with that of bioaerosol can be overlooked in the measurement results in both rainy and dry season. It is shown by the existence of bioaerosol in a lower concentration when the temperature is high in dry season and higher bioaerosol concentration will occur if humidity in wet season is high. Air temperature and wind velocity have a moderately significant correlation with fungi concentration [18].

3.3 Identification of Gram-positive and Gram-negative Bacteria

The majority of bacteria in WWTP are Gram-negative bacteria making up about 94% of the population, whereas the remain 6% are Gram-positive bacteria. The shapes of the observed

 DOI: 10.1051/matecconf/201713808004

bacteria are rod and coccus, yet the most observed bacteria are the rod-shaped ones which vary in size from 0.73 to 11.04 μm. These bioaerosols have the ability to give impacts on living things in the forms of infectivity, allergy, toxicity, pharmacology, or other processes [19]. Fig. 5 shows the result of Gram staining for Gram-negative and Gram-positive bacteria.

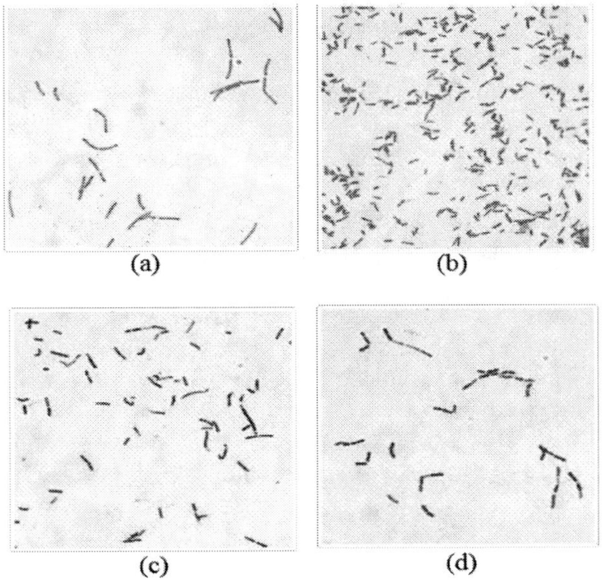

Fig. 5. (a) and (b) Gram Negative Bacteria; (c) and (d) Gram Positive Bacteria

4 Conclusion

The average concentration of bioaerosol surrounding the WWTP was around 810 – 1,915 CFU/m^3 for bacteria and 1,450 – 1,955 CFU/m^3 for fungi. The highest concentration of bacteria and fungi was 1,915 CFU/m^3 in equalization unit, whereas the second highest was 1,955 CFU/m^3 in the aeration unit. There was a moderately strong correlation between air humidity and the concentration of bacteria and fungi. This correlation can be overlooked in research result showing high air humidity and an increased concentration of bacteria or fungi. The factor of aeration system used in wastewater treatment was also able to increase the concentration of bioaerosol in the air. The bacteria produced in WWTP showed that 94% of them were gram negative and predominantly rod-shaped in highly varying size from 0.73 to 11.04 μm.

References

1. Prayitno, Z. Kusuma, B. Yanuwiadi, R.W. Laksmono, Res. Inv.: Int. J. of Eng. and Sc. **2**, 13-16 (2013)
2. J. Douwes, P. Thorne, N. Pearce, D. Heederik, Ann. Occup. Hyg. **47**, 187-200 (2003)
3. V. Kummer, W.R. Thiel, Int. J. of Hyg. and Env. Health, **35**, 299-307 (2008)
4. G.A. Kristanto, E. Novita, Yudithia, World App. Sc. J. **33**, 453-459 (2015)
5. L. Pascual, S. Pérez-Luz, M.A. Yáñez, A. Santamaría, K. Gibert, M. Salgot, D. Apraiz, V. Catalán, Aerobiologia. **19**, 261-270 (2003)
6. M.A. Sánchez-Monedero, M.I. Aguilar, R. Fenoll, A. Roig, Water Res. **42**. 3739-3744 (2008)

7. K. Kruczalak, K. Olańczuk-Neyman, Pol. J. of Env. Stu. **13**. 537-542 (2004)
8. V.M. Cavinatto. *Influency of Environmental Factors in the Dispersion of Particles Originated in a System of Biological Treatment of Sewage in Vals of Oxidation* (Universidade de São Paulo, São Paulo, 1991)
9. M.D. Azevedo, M.A.Couto, S.G. Leite, B.R. Azevedo, P.A. Almeida, A.S. Procópio, et al., Int. J. of Eng. & Tech. IJET-IJENS. **14**, 48-55 (2014)
10. M. Vítězová, T. Vítěz, H. Mlejnková, T. LošákActa Univ. Agr. et Silviculturae Mendelianae Brunensis, **LX**, 233-240 (2012)
11. N. Patentalakis, A. Pantidou, N. Kalogerakis, Water, Air, & Soil Pollution: Focus. **8**, 107-115 (2008)
12. L.M. Prescott, J.P. Harley, D.A. Klein, *Microbiology, Seventh Ed.* (McGraw-Hill, New York, 2008)
13. K.J. Heo, H.B. Kim, B.U. Lee, J. of Aerosol Sc. **77**, 31-37 (2014)
14. Y. Li, H. Zhang, X. Qiu, Y. Zhang, H. Wang, Aerosol and Air Quality Research, **13**, 1807-1814 (2013)
15. L. Fracchia, S. Pietronave, M. Rinaldi, M. Giovanna Martinotti, Water Res. **40**, 1985-1994 (2006)
16. P.N. Sari, Effect of The Availability of Wastewater Treatment Plant on The Microbial Air Quality (A Case Study*: Dr. Cipto Mangunkusumo Hospital)* (Universitas Indonesia, Depok, 2013)
17. Z. Filipkowska, W. Janczukowic, M. Krzemieniewski, J. Pesta, Pol. J. of Env. Stud. **9**, 273-280 (2000) (5)
18. S. Niazi, M.S. Hassanvand, A.H. Mahvi, R. Nabizadeh, M. Alimohammadi, S. Nabavi, et al., Env. Sc. and Poll. Research, **22**, 16014-16021 (2015) (10)
19. C.S. Cox, C.M. Wathes, *Bioaerosols Handbook* (CRC Press, United States of America, 1995)

 DOI: 10.1051/matecconf/201713809001

Deaggregation of probabilistic ground motions in the Kota Kinabalu and Lahad Datu towns of Sabah, Malaysia

Harith Noor Sheena Herayani[1,*], *Adnan* Azlan[2] , and *Shoushtari* Abdollah Vaez[3]

[1]Civil Engineering Program, Faculty of Engineering, Universiti Malaysia Sabah, Jalan UMS, 88400 Kota Kinabalu, Sabah, Malaysia
[2]Department of Structure and Materials, Faculty of Civil Engineering, Universiti Teknologi Malaysia, Skudai, 81310, Johor Bahru, Malaysia
[3]Department of Civil Engineering, Islamic Azad University, Khorasgan Branch, University Blvd, Arqavanieh Jey Street, Isfahan, Iran 81595-158

Abstract. Sabah has witnessed an increase in low to moderate seismic activities due to a few active fault lines. According to the historical records ranging between 1900 and 2017, the region already experienced a devastating earthquake with a size Mw 5.8 in Lahad Datu and just last two years with a size Mw 6.0 in Ranau. Over the years, more than 400 with size ranging from Mw 2.9 to 6.0 are known to have occurred. There are three tectonic features that affected Sabah namely; subduction zone, strike-slip earthquake and local background seismicity zone. The effects of the earthquakes should be anticipated in order to mitigate the catastrophic failure of structures. Therefore, seismic hazard analysis, such as probabilistic seismic hazard analysis (PSHA) is required. This study deaggregated the PSHA results of 2% and 10% probability of exceedance in 50 years. The deaggreagation of selected cities in Sabah such as Kota Kinabalu and Lahad Datu to help understand the relative control of local fault sources in terms of distances and magnitudes. The contribution to hazard indicated that the distance from the earthquake which contributes most to the hazard at each city is mainly controlled by shaking by near seismic area sources.

1 Introduction

Sabah is one of the Malaysian states bordered internationally with Brunei Darussalam and Kalimantan, Indonesia located on the island of Borneo. Sabah is considered as a relatively stable continental shield region with moderate seismicity [1, 2]. Sabah is situated in the eastern part of East Malaysia and is bounded by three seas; South China Sea, Sulu Sea and the Celebes Sea. The total land area of Sabah is nearly 72,500 km^2. By referring to the 2015 Malaysian Census, the population of Sabah being the third most populous state in Malaysia with a total of 3,543,500 including non-citizens population at 870,400. The population is

* Corresponding author: sheena@ums.edu.my

particularly concentrated in the coastal areas. Sabah's economy is mainly based on agriculture, forestry, petroleum, tourism and services.

Sabah is prone to earthquake activities if compared to other parts of Malaysia. The region already experienced a devastating earthquake with a magnitude of M_W 5.8 on 26th July 1976 centered in Lahad Datu area. The largest ever recorded with M_W 6.0 on 05th June 2015 had occurred in Ranau area. A mild earthquake measuring 3.4 occurred near Lahad Datu on 29th March 2017. Even though the magnitude is small, Sabah has records of more than 400 earthquake events occurred between the years 1900 to 2017 (Figure 1) with size ranging from M_W 2.9 to 6.0 are known to have occurred. The condition of the tectonic setting seems to be very complex, with southeasterly subduction in Northwest Borneo and extension in the southeast in the Celebes Sea and Makassar Strait, with evidence of several geological events found in the references [3, 4, 5, 6].

The seismic activity across the region concerns low-to-moderate earthquakes, as presented earlier in Leyu [7] and recent earthquake databases monitored by Alexander [1, 2]. However, it has been recently recognized that this region, even at rather remote distances, is at significant seismic risk from two main active earthquake sources i.e. Indonesia and Philippines. Recently, a large earthquake measuring 7.3 occurred in the Celebes Sea on 10th Jan 2017 with distance 466 km from east of Sabah. All signs indicate that the same problem will continue in the future. The highest intensity of these earthquakes reached VIII degrees in MMI scale, causing serious economic loss and social unrest.

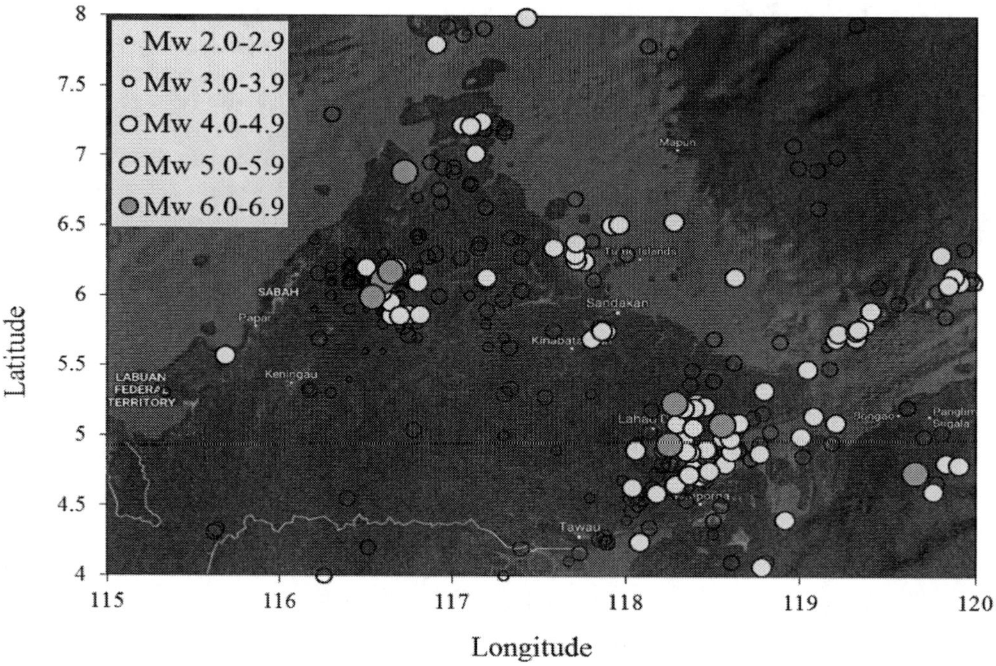

Figure 1. Earthquake events distribution of Sabah province.

Probabilistic Seismic Hazard Analysis (PSHA) approach was adopted to investigate seismic hazard distribution across Sabah. Potential sources of seismic activities in the region were identified, and their earthquake recurrence relationships were developed from instrumental and historical data. The probabilistic seismic hazard assessment (PSHA) is performed in 2% and 10% probability of exceedance (PE) in a design time period of 50 years or the corresponding to return period of approximately 500 and 2,500 years, respectively.

The methodology of this study is based on five main components; (i) data input based on historical earthquake data and the seismotectonic map, (ii) characterization of seismic source with distance up to 600 km, (iii) selection of suitable ground motion prediction equation, (iv) calculation of the seismic hazards and (v) deaggregation analysis. Deaggregation plots can provide useful information on the distance and magnitude of predominant sources, which can be used to generate scenario earthquakes and select corresponding time histories for seismic design.

1.1 Methodology

1.1.1 Earthquake Database

The geometries of the seismotectonic model were constructed based on earthquake spatial distributions and regional tectonic setting of Sabah. The national earthquake catalog was compiled from historical and instrumental recordings from 1900 to 2017 for a magnitude of 2.0 and above in the vicinity of Sabah region. A large dataset provided by various national and international agencies was collected. The international agencies included the United States Geological Survey/National Earthquake Information Centre (USGS/NEIC), Advanced National Seismic System (ANSS), International Seismological Centre (ISC), National Geophysical Data Centre (NGDC), Harvard Centroid Moment Tensor (CMT), the national agency of the Malaysian Meteorological Department (MMD), and various literature [7, 8].

There are two types of earthquake catalogs covered in this study, namely historical and instrumental databases. The historical parts of the catalog are compiled from various literatures, such as Leyu [7] and Alexander [1, 2]. The available catalogue was carefully reviewed for uncertainties and completeness. The magnitudes were unified in terms of moment magnitude (M_W). The combined catalogues cover areas from 115^{0}E to 120^{0}E longitude and from 4^{0}S to 8^{0}N latitude and include more than 400 earthquake events occurred between the years 1900 to 2017 (Figure 2).

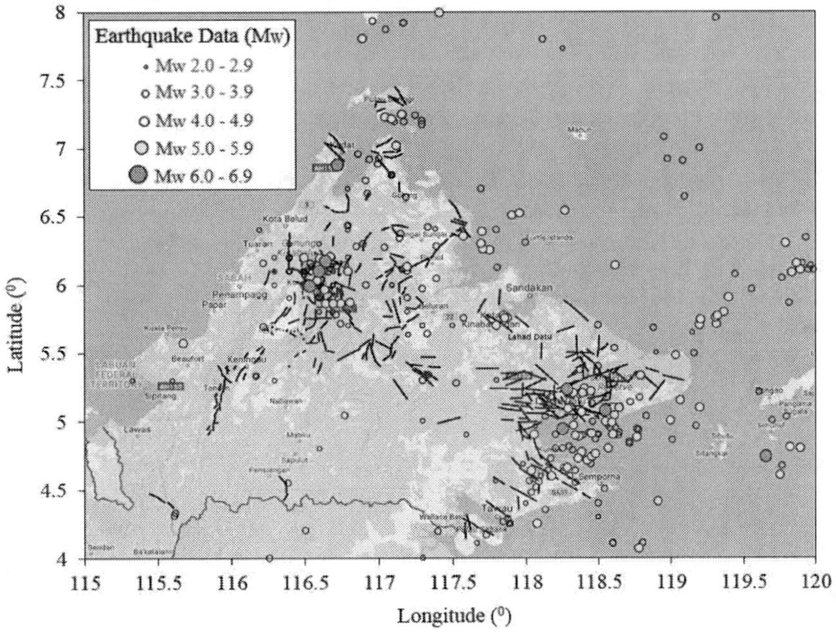

Figure 2. Earthquake events distribution and tectonic map of Sabah province.

1.1.2 Seismic Characterization

In this study, the model of Gardner and Knopoff [9] is being considered in the analysis. The model provides a specific space-time distance as a function of the mainshock magnitude to identify aftershocks. This technique is one of the simplest forms of identifying aftershocks. The historical earthquake data that occurred between 1900 and 2017 is analyzed for completeness. Seismic hazard analysis requires a complete possible history of earthquakes in or near the region of interest. It is necessary to have a comprehensive knowledge of historical recording coverage in order to assess the completeness level of an earthquake catalog.

The historical earthquake data from the earthquake catalogs were then combined with seismotectonic conditions to develop the seismic source model. The seismic source model was characterized by seismic hazard parameters such as a-b values, rates, and maximum magnitudes. The occurrence of earthquakes in a seismic source is assumed as the Poisson distribution. The seismicity parameters are summarized depending on the minimum and maximum magnitude of each zone used in this analysis, which were calculated in the recurrence relationship in Table 1.

Table 1. Recurrence rate parameters for source zones

Earthquake Source	Recurrence Rate		
	β	λ	α
Zone 1	0.921	0.152	1.852
Zone 2	0.824	0.043	1.270
Zone 3	0.921	0.201	2.341
Zone 4	0.824	0.043	1.270
Zone 5	1.051	0.175	2.709

1.1.3 Ground Motion Prediction Equation (GMPE)

The prediction of ground motion parameters using probabilistic approach is calculated for the design parameters of earthquake ground motion such as the acceleration at bedrock. The two main uncertainty parameters, ground motion prediction equation, GMPE and maximum magnitude, M_{max} incorporated in the hazard estimate used the logic tree to model variable alternatives to avoid being too conservative by over- or underestimating the PGA values. The Next Generation Attenuation (NGA) of the Abrahamson and Silva [10] model was used to analyze ground motions of an earthquake occurring within the active shallow crustal seismogenic zones including Fukushima and Tanaka [11] and Sadigh [12]. In addition, the equation by Zhao [13] was aimed at predicting accelerations for a subduction region (Table 2).

Table 2. Ground motion Prediction Equations (GMPE) adopted in this study

Reference	Ground motion Prediction Equation (GMPE)
Fukushima and Tanaka [11]	$\log(PGA) = 0.42M_W - \log(R + 0.025\times 10^{0.42M_W}$ $- 0.0033R + 1.22$
Sadigh [12]	$M_W \leq 6.5;$ $\ln(PGA) = -0.624 + 1.0M_W + (-2.1\ln(R +$ $\exp(1.29649 + 0.25M_W)))$ $M_W > 6.5;$ $\ln(PGA) = -1.274 + 1.1M_W + (-2.1\ln(R +$ $\exp(-0.48451 + 0.524M_W)))$
Abrahamson and Silva [10]	$\ln(PGA) = 0.804 + 0.231(M_W - 6.75) + (-0.0372$ $(8.5 - M_W)^2 - 0.9679 + 0.265(M_W - 6.75))\ln(R)$
Zhao [13] - Megathrust	$\ln(PGA) = 1.101M_W + (-0.0564R) - \ln(R)$ $+ 0.01412 + S_I + C_1 + Q_I(M_W - M^2) + W_I$
Zhao [13] - Benioff	$\ln(PGA) = 1.101M_W + (-0.0564R) - \ln(R)$ $+ 0.01412 + S_S + S_{SI}[\ln(R_{hypo})] + C_1 + P_S(M_W - M)$ $+ Q_S(M_W - M)^2 + W_S$

1.1.4 Seismic Hazard Assessment

The probability distribution is defined in terms of the annual rate of exceeding the ground motion level at the site under consideration, due to all possible pairs of magnitude and epicentral distance (M, R) of the earthquake event expected around the site. The seismicity parameters for each source zone were calculated using the method by Gutenberg-Richter (G-R). Table 3 shows the PGA at major towns and villages in Sabah at bedrock level. PGA value at more than 16% of g can be seen in Lahad Datu, for 500 years return period, whereas in Kota Kinabalu shows less than 5% of g. At 2,500 year return period, Lahad Datu gives 27.43% of g and Kota Kinabalu at 5.6% of g.

Table 3. PGA values for town areas in Sabah

No	Town Name	Latitude (°)	Longitude (°)	PGA (10% PE @ 500-yr) %g	PGA (2% PE @ 2500-yr) %g
1	Kota Kinabalu	5.99	116.07	3.34	5.60
2	Lahad Datu	5.02	118.33	16.36	27.43

The probabilistic seismic hazard assessment (PSHA) procedures allow computation of the mean annual rate of exceedance at a particular site based on the aggregate risk from potential earthquakes of many different magnitudes (M) occurring at many different sources to site distances (R) [14]. The rate of exceedance computed in a PSHA, therefore, is not associated with any particular earthquake magnitude or sources to site distance. In some cases, however, it may be useful to estimate the most likely earthquake magnitude and/or the most likely sources to site distance. These quantities may be used, for example, to select existing ground motion records (recorded in earthquakes of similar magnitude at similar source-site distance) for response analysis. Moreover, the results of PSHA can be exploited to determine predominant sources of seismic hazard and can provide deterministic design magnitudes and distances through a process known as seismic hazard deaggregation processes [15, 16]. The mean annual rate of exceedance as a function of magnitude is shown in equation 1.

$$\lambda_{y*}(m_j) \approx P[M = m_j]\sum_{i=1}^{N_S}\sum_{k=1}^{N_R} v_i P[Y > y^* \mid m_j, r_k]P[R = r_k] \quad (1)$$

The mean annual rate of exceedance as a function of source-site distance is shown in equation 2.

$$\lambda_{y*}(r_k) \approx P[R = r_k]\sum_{i=1}^{N_S}\sum_{j=1}^{N_M} v_i P[Y > y^* \mid m_j, r_k]P[M = m_j] \quad (2)$$

Finally, it is possible to compute the mean annual rate of exceedance as functions of both earthquake magnitude and source-site distance is shown in equation 3.

$$\lambda_{y*}(m_j, r_k) \approx P[M = m_j]P[R = r_k]\sum_{i=1}^{N_S} v_i P[Y > y^* \mid m_j, r_k] \quad (3)$$

Performing the PSHA in order to compute the ground-motion parameters (e.g., PGA) at the specific site with the defined percentage of probability of exceedance in a considered exposure time (e.g., 10% and 2% probability of exceedance (PE) in 50 years). Based on the above selections the corresponding total mean annual rate of exceedance (λ_y*(Total)) due to the summation of the calculated seismic hazard curves of the considered seismic sources can be obtained. Then, using the equation above, the mean annual rate of exceedance corresponding to each pair of M and R, for the obtained y* should be calculated for each of the seismic sources.

2 Results

In order to have more insight into the probabilistic seismic hazard analysis, an attempt is made to deaggregate these ground motions for the two selected cities (i.e. Kota Kinabalu and Lahad Datu). The latitude and longitude for the downtowns of these cities are used to deaggregate their ground motions in terms of bin pairs of magnitude and distance (M, R).

 DOI: 10.1051/matecconf/201713809001

Deaggregation was carried out based on a magnitude interval of 0.4 and a distance interval of 25 km. Here we account for both aleatory and epistemic uncertainties in the ground motion selection through PSHA deaggregation. The results of the hazard analysis represent the mean of the log-normal distribution ground motion outputs obtained from computations using the above attenuation relationships on bedrock conditions (Vs=1500 m/s). Figure 3 through Figure 6 reflect the deaggregation results for 5% damping with respect to 10% and 2% probability of exceedance in 50 years, respectively.

The mean distance from the controlling earthquake in the Kota Kinabalu city is ranging between 26-75 km at both 2% and 10% probabilities. Contrarily, the more the control is expected from distant very near and ranges between 0-50 km for both 2% and 10% probabilities. The seismic sources are dominated by the source located at a distance less than 100 km. Therefore, at a given probability level, it is clear that for larger and more distant both cities show the less influence to the cities. The controlling earthquake for 10% and a 2% probability at Kota Kinabalu city is dominated by an earthquake located at a mean distance of 50 km and its mean magnitude equals 5.8. In Lahad Datu, the controlling earthquake for 10% and 2% probability is dominated by an earthquake located at a mean distance of 25 km and its mean magnitude equals 5.8.

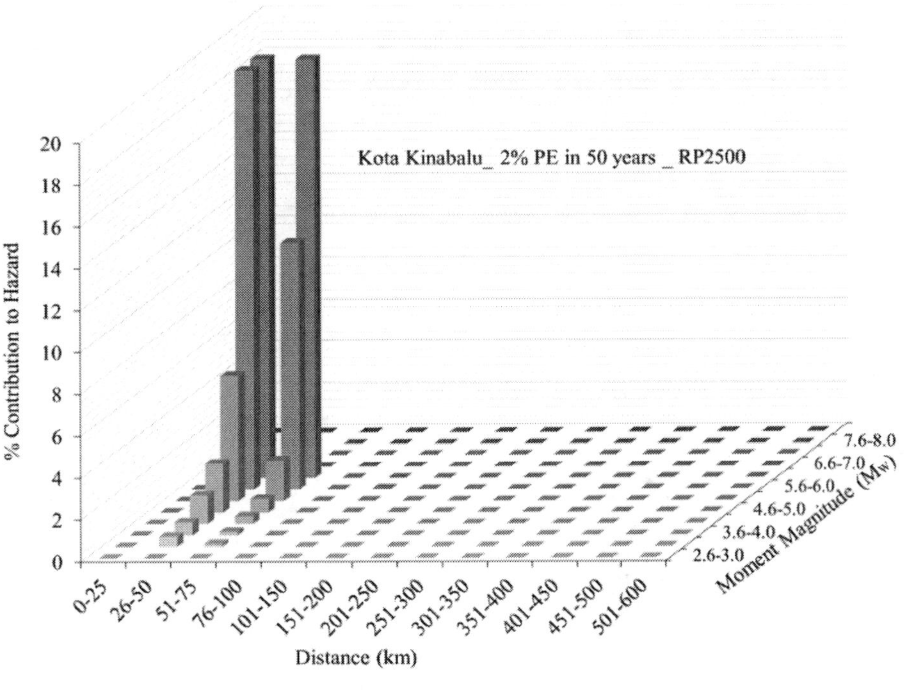

Figure 3. Deaggregation of PSHA results of Kota Kinabalu city for PGA at bedrock at 10% probability of exceedance.

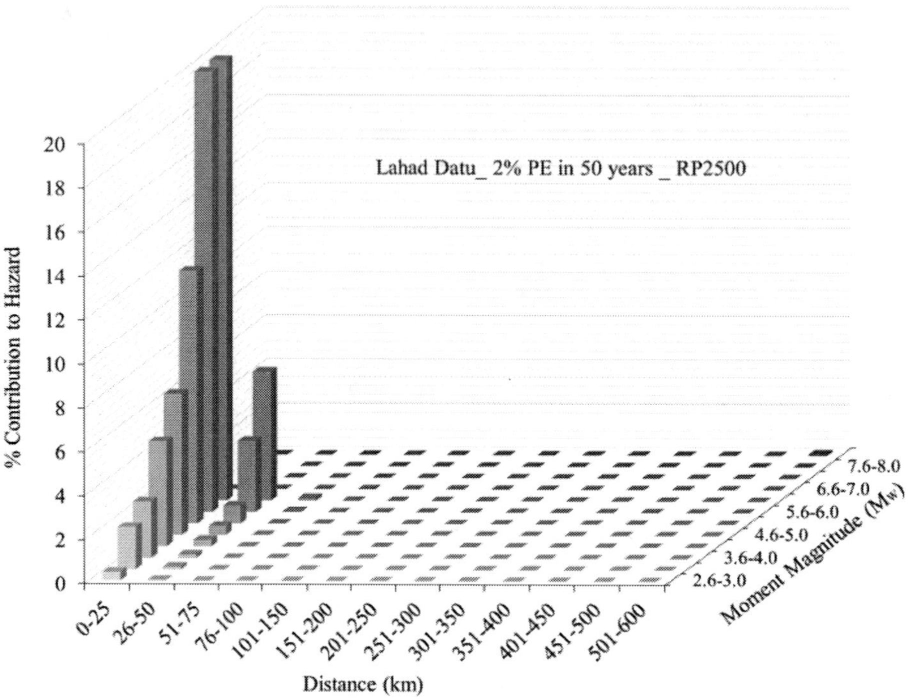

Figure 4. Deaggregation of PSHA results of Lahad Datu city for PGA at bedrock at 10% probability of exceedance.

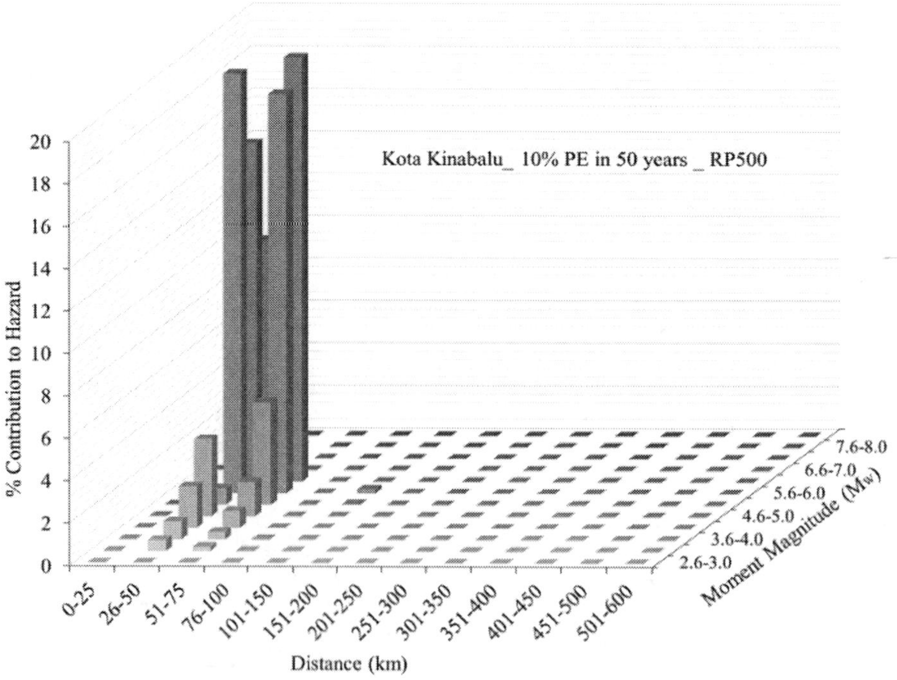

Figure 5. Deaggregation of PSHA results of Kota Kinabalu city for PGA at bedrock at 2% probability of exceedance.

DOI: 10.1051/matecconf/201713809001

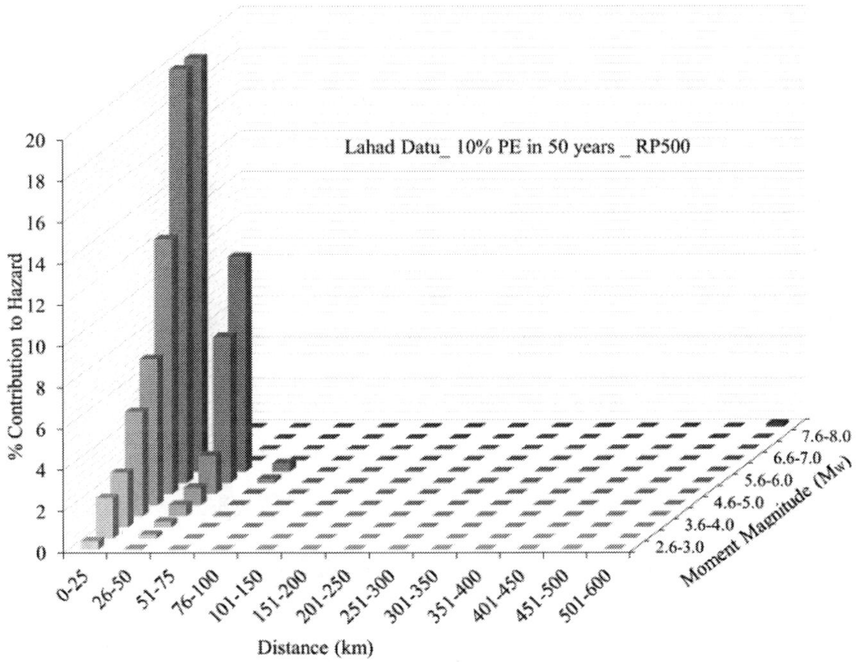

Figure 6. Deaggregation of PSHA results of Lahad Datu city for PGA at bedrock at 2% probability of exceedance.

3 Conclusion

In the present study, the deaggregation of the seismic hazard is performed in order to identify the sources with specific faults that have a major contribution to the hazard of Sabah cities. This study has deaggregated the hazard model of a probability of exceedance for 2% and 10% exceedance probability in 50 years. The use of such parameters would better describe the hazard scenarios and allow civil engineers, geologists, seismologists and municipal policy makers to identify either the local and distant sources, or predominant hazardous earthquakes in their design, planning and risk mitigation. The results indicate that very near seismic sources of relative higher magnitudes are the dominating sources of hazard for the selected Sabah cities, where the most probable earthquake is equal to 26-75 km from Kota Kinabalu city and 0-50 km to Lahad Datu city for a probability of exceedance in 50 years. The contributions from distant sources, at more than 400 km are relatively low.

References

1. Y. Alexander, S. Suratman, A.Liau, M. Hamzah, M. Y. Ramli, H. Ariffin, M. Abd. Manap, M. B. Mat Taib, A. Ali, H. D. Tjia. *Report on the Geological and Seismotectonic Information of Malaysia*. (Ministry of Natural Resources and Environment, 2006)
2. Y. Alexander, A. Liau, M. Hamzah, M. Y. Ramli, M. B. Mat Taib, A. Ali, H. Ariffin, B. Ismail, H. D. Tjia. *Report on Assessment of the Seismic Threats to Malaysia from Major Earthquake in the Southeast Asian Region*. (Ministry of Natural Resources and Environment, 2008)

3. C. S Hutchison. *Geology of North West Borneo: Sarawak, Brunei and Sabah*. (Elsevier, Boston, 2005)
4. C.S. Hutchison. Marine Geology **271**, 32-43 (2010)
5. S. Hesse, S. Back, D. Franke. GSA Bulletin **121**, 939-953 (2009)
6. A. B. Cullen. Petroleum Geoscience **16**, 3-29 (2010)
7. C. H. Leyu, C. F. Chong, E. P. Arnold, S. L Kho, Y. T. Lim, M. Subramaniam, T. C. Ong, C. K. Tan, K. S. Yap, Y. K. Shu, H. L. Goh. *Series on Seismology Malaysia*. (E. P. Arnold Southeast Asia Association of Seismology and Earthquake Engineering (SEASEE), 1985)
8. J. F. Pacheco, L. R. Sykes. Bulletin of the Seismological Society of America **82**, 1306-1349 (1992)
9. J. K. Gardner, L. Knopoff. Bulletin of the Seismological Society of America **64**, 1363-1367 (1974)
10. N. Abrahamson, W. Silva. Earthquake Spectra **24**, 67-97 (2008)
11. Y. Fukushima, O. Köse, T. Yürür, P. Volant, E. Cushing, R. Guillande. Journal of Seismology **6**, 379-396 (2002)
12. K. Sadigh, C. Y. Chang, J. A. Egan, F. Makdisi, R. R. Youngs. Seismological Research Letters **68**, 180-189 (1997)
13. J. X. Zhao, J. Zhang, A. Asano, Y. Ohno, T. Oouchi, T. Takahashi, H. Ogawa, K. Irikura, H. K. Thio, P. G. Somerville. Bulletin of the Seismological Society of America **96**, 898-913 (2006)
14. S. L. Kramer. *Geotechnical Earthquake Engineering*. (Prentice Hall PTR, 1996)
15. P. Bazzurro, C. Allin Cornell. Bulletin of the Seismological Society of America **89**, 501-520 (1999)
16. S. Harmsen, D. Perkins, A. Frankel. Bulletin of the Seismological Society of America, 89, 1-13 (1999)

Specifying risk management standard for flood risk assessment: a framework for resources allocation

Anastasia Yunika[1, 2*], *Matthijs* Kok[1], and *Jacob Gerrit (Jarit)*de Gijt[1]

[1]Hydraulic Structures and Flood Risk Section, Delft University of Technology, Delft, the Netherlands
[2]Civil Engineering Department, Universitas Atma Jaya Yogyakarta, Yogyakarta, Indonesia

Abstract. General risk management standard, e.g. ISO 31000:2009, approaches risk as a coin with a pair of two sides, i.e. the threat and the opportunity. However, it is hardly the case of flood events which mainly come as threats. Despite the contrary, this study explores the potential applicability of the available risk management standards specifically for flood. It then also synthesizes the components to result a framework for allocating resources among various strategies to result the optimum flood risk reduction. In order to review its applicability, the framework is then reviewed using several historic flood risk reduction cases. Its results are qualitatively discussed and summarized including the possible improvement of the framework for further applications.

1 Introduction

In the daily life terminology, "risk" is commonly perceived as something with harmful outcome or is simply related to more possibility of negative result, shortly, a situation of threat. In general, there is no standard available for risk management in such condition. Different persons faced to a particular circumstance with a certain type and level of risk will make decision or take action regarding that situation differently one from another based on personal preferences, priorities, experiences, and many other aspects. On the other hand, in the economy-related fields, e.g. finance, business, and investment, "risk" is defined more neutrally by considering both the possibility of gain (positive yield) and loss (negative yield). The basic understanding is that there is the probability of difference between the actual and expected returns [1]. In this field, a risk management standard has been developed and at the international level, ISO 31000:2009 is the latest one, in which "risk" is defined as the "effect of uncertainty on objectives" [2].

More specific definition of "risk" used in different areas including flood risk management is the product of occurrence probability and its consequence of hazardous event or phenomenon [3, 4]. The second frequently used formulation of "risk" includes three components, i.e. hazard, exposure, and vulnerability (a.k.a. "Risk Triangle" [5]). Referring to those two definitions, despite its essence as a natural phenomenon, flood is

* Corresponding author: a.yunika@tudelft.nl

perceived as more of a disturbance, problem, or even disaster at its higher scale due to the impacts of its interaction with humans. This perspective tends to abandon the positive side of flood's essential existence as "an integral part of the hydrologic cycle" [6], e.g. environmental and ecological balancing process, agricultural soil nourishment, water sources recharge, river flushing, as well as topographical formation through erosion and deposition. Consequently, the flood risk reduction measures as part of flood risk assessment and management will less take into account the possibility of gaining the benefit from the flood existence due to the focus merely on minimizing the damage and loss.

This study tries to adapt the concept of general risk management standard, i.e. ISO 31000:2009 specifically for flood risk reduction so that both sides of the coin, i.e. the threat ("loss and detriment") and opportunity ("gain and benefit") can be more proportionally accommodated. By doing so, it is expected that innovative measures in the flood risk reduction can be identified so that resources will be optimally allocated among strategies.

2 RISK MANAGEMENT

Despite the semantically self-explaining terminology, ISO 31000:2009 [2] describes "risk management" as shown in Figure 1a in which "risk assessment" is the core part. In addition to it, there are three other components, i.e. "establishing the context", "communication and consultation", "monitoring and review". Similarly, a simplified process described by Jonkman [7] is shown in Figure 1b. On the other hand, a different concept of risk management was presented by Gerrard and Petts [8] as a cyclic process shown in Figure 2 which oppositely considered the "risk perception and communication" as the core component and located it at the center of the cycle while "risk assessment" as one of the outer components. Taking into account the formulation of the process as a well-documented standard, this study opts to adopt the model presented in ISO 31000:2009 in developing the proposed framework that will be discussed further.

3 RISK ASSESSMENT

As shown in Figure 1a and 1b, risk assessment is recognized as the core of the risk management process. However, it is well understood that in terms of decision making, there are many different risk assessment approaches. Taking two of them to mention are "quantitative risk assessment" and "comparative risk assessment". The first focuses on the probability quantification of "adverse impact due to exposure" while the second pays attention on "ranking risk issues by their severity for prioritization and justification of resources allocation" [8]. Both approaches will be incorporated in the proposed framework of this study.

4 PROPOSED FRAMEWORK AND ITS SYNTHESIZATION

This study emphasizes on the optimum resources allocation among alternatives for optimum flood risk reduction strategy. However, since flood risk reduction as the last phase is inseparable from the preceding phases of risk assessment (Figure 1a), a simplification inappropriately be made more than the four-phase iterative process shown in Figure 3. The first phase of "flood hazard analysis" also covers two subcomponents equal to "establishing the context" and "risk identification" in Figure 1a. They are "mapping of area and its flood risk system" and "flood hazard analysis" itself which are logically and chronologically required. Other proceeding phases of the framework are corresponding to the ones of ISO 31000:2009. The two types risk assessment mentioned earlier in Section 3 are adopted in

 DOI: 10.1051/matecconf/201713809002

the second and last phases of the framework. The "quantitative risk assessment" is applied to the "flood consequence analysis" while the "comparative risk assessment" is used in the "flood risk reduction".

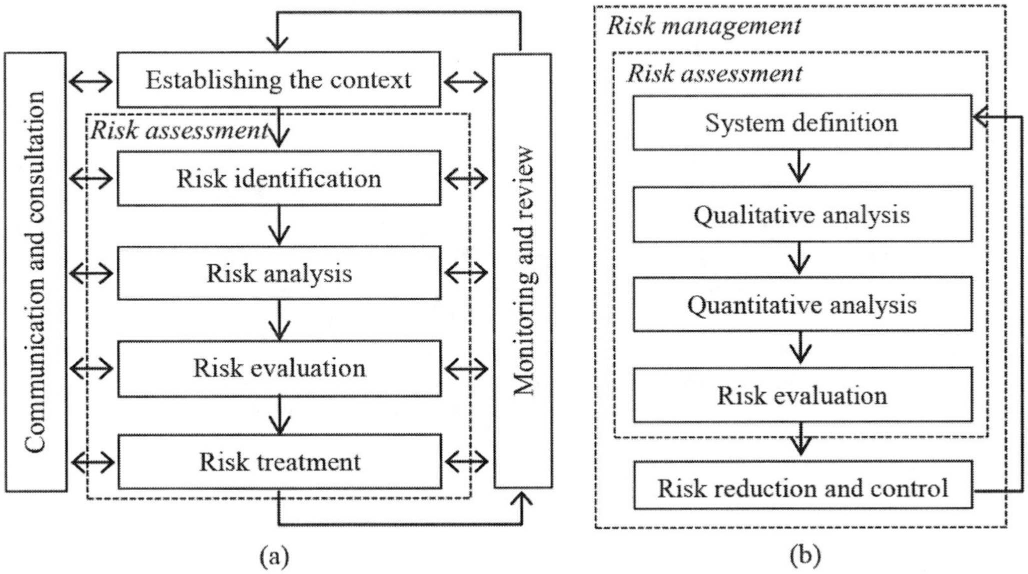

Fig. 1. Risk management process according to (a) ISO 31000:2009 [2] and (b) Jonkman [7]

Fig. 2. Risk management cycle according to Gerrard and Petts [8]

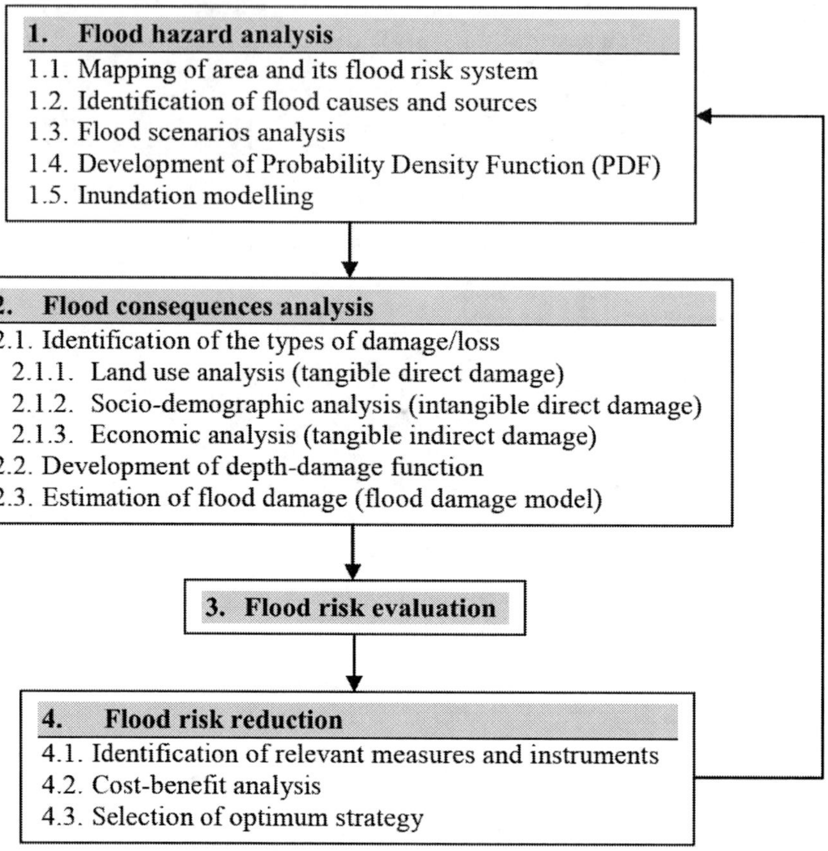

Fig. 3. Proposed framework of this study

In spite of well-recognized concept that risk assessment is a part of risk management, the study by Gerrard and Petts presented an intriguing question of "whether risk assessment as a scientific process can and should be separated from risk management". The idea of either isolation or integration was initiated by the difference between two perspectives in looking at risk assessment, i.e. as "one of the many decision making tools or the decision making process itself". Furthermore, historically, it is also supported by the distinct believes between "logical positivism" and "cultural relativism" upon the "wholly objective" nature of science and scientific process. The earlier supports the exclusion of social and political aspects of decision making since risk assessment is and must be kept objective. The latter argues oppositely since it is impossible. In between the two lies the "scientific proceduralism" which focuses on the process of conducting scientific activities so that "subjective value judgements within technical risk assessment can be acknowledged and dealt with in an appropriate manner". The concept presented in [8] that "science might somehow be conducted isolation and occasionally deliver objective information" is adopted in this study so that the components of risk management in Figure 1a other than the "establishing the context" and risk assessment are excluded from the proposed framework as the first synthesization.

The second synthesization of the framework is conducted upon last phase of Figure 1a. In ISO 31000:2009, "risk treatment" has several possible types, i.e. risk avoidance (i.e. by cancelling the activity, e.g. development, and/or by modifying the condition under which activity is conducted), risk removal (i.e. removing the source of risk), risk modification (i.e. changing the likelihood and/or consequence), risk sharing (i.e. insurance and/or outsourcing), and/or risk retention (when the risk level meets the criteria of acceptance). Those types of risk treatment will be used in identification of relevant alternatives of flood risk reduction towards optimum strategy. The criteria of "optimum" here is based on the cost-benefit consideration.

5 RESOURCES ALLOCATION

Resources are never unlimited. As mentioned earlier in Section 1, this study tries to develop a framework which result a more balanced strategy in flood risk reduction by accommodating idea of positive perspective upon flood's potential impacts. It is expected that the strategy might consist of a single or combination among various measures, both structural and non-structural ones. Consequently, the limited available resources need to be optimally allocated among the measures in order to result the minimum flood risk. Another study will specifically deal with this resources allocation which applying the concept of "comparative risk assessment" mentioned in Section 3. Furthermore, it is also a challenge to evaluate the effectiveness of implementation of certain flood risk reduction strategy [3].

6 HISTORIC CASES ON FLOOD RISK REDUCTION

The past studies aiming for flood risk reduction are categorized into several points as elaborated for review below.

6.1 Balance vision between the losses and the benefits

A study by Penning-Rowsell and Pardoe [9] demonstrated the benefit of flood by showing that flood risk reduction did not only bring benefits, but also losses for different parties. It disclosed that in addition to benefited parties ("gainers") such as owners of property with flood risk well as insurance companies and their shareholders, there were also the "losers", i.e. other taxpayer who are not related to any flood risk and people who rely on the work of flood damage repair or replacement. Further, Ferrier and Hague [10] stated that "flood risk management would provide a framework for balancing the multiple complementary and competing factors that affect risk. At the same time, a carefully crafted flood risk management strategy must also consider associated risks and opportunities, such as protecting natural floodplain functions from the detrimental impacts of human use."

6.2 Balance management between the flood and its related aspects

Ferrier and Hague [10] argued that flood risk management should be a more comprehensive approach then the old flood management of controlling the flood itself at various times and places. It must also cover the management of the

1. building and other development taking place in flood prone areas,
2. land area considered to be susceptible to flooding,
3. flood damage (with relief measures, insurance, and recovery assistance),
4. floodplain functions and resources (with regulatory controls or land management), and even
5. vulnerability of development (by applying site-specific mitigation measures).

A symposium conducted by the ASFPM Foundation [11] tried to identify the key benefits of a natural flooding process and the key resources provided by natural floodplains as well as the availability of models to prioritize or otherwise describe them.

6.3 Improved flood risk analysis

Out of several definition upon "risk", this study opts to define flood risk as the product between probability of occurrence and its consequence of a flood event. Hence, the two main components in the analysis is the hazard and consequences.

6.3.1 More accurate flood hazard analysis

Hazard is not risk although they are interrelated and one is part of the other [10, 12]. Flood hazard analysis is the first part of flood risk analysis. Ferrier and Haque [10] studied the environmental hazard identification using the current major tools, i.e. research, screening, monitoring, and diagnosis. Other than the frequency (or probability) of occurrence of a given flood, a number of issues must also be explored, e.g. the time of last event occurred and the corresponding mitigation measures. A number of different sources to address the issues are required such as experts (scientists and researchers), published materials (textbooks, journal articles, databases), historical records in the library and/or municipality, newspapers, long-time residents, all with different qualities to consider and differently treated.

6.3.2 More comprehensive flood consequence analysis

In addition to flood hazard analysis, flood consequences analysis is the second part to be conducted in the flood risk analysis. The "quantitative risk assessment" mentioned earlier in Section 3 will be the main method used in this part. The benefits of flooding are included in this component of the flood risk equation, such as agricultural soil nourishment and water sources recharge. On the other hand, the ASFPM [11] revealed that individual and household behaviors in dealing with a particular threat are not related to their understanding of the actual risk of such events and/or their consequences. Hence, it is not completely true that by having a better knowledge on flood risk, people would take action to avoid or reduce it and be safer. The further question then be how to promote compulsory collective behavior which certainly cannot be conducted individually. This kind of education can also be considered as another benefit of flooding in improving the people's quality of life by living in harmony with the surrounding both physical and non-physical, including social environments.

6.4 Complementary soft engineering to hard engineering for flood risk reduction measures

Having the fact that flood defense cannot be against all floods [13] or in other words recognizing the limits of structural flood defense [14–16], and witnessing that the extensive investment in flood control works does not decrease the flood occurrence nor damages [17], solution for flood risk reduction has moved from hard to soft engineering solution and non-structural approaches [18]. It is more promising to meet the "demands of sustainable development, a more strategic, holistic and long-term approach" of flood risk management [19]. Hooijer et al. [3] concluded their study that "the most effective and sustainable reduction of flood risks could be achieved by reducing the potential damage (vulnerability) in flood-prone areas through adapted land use and spatial planning".

The introduction of more public and private involvement to the government's role and responsibility in "modern flood risk management" in Japan has become more common practice in some other countries as well [16], e.g. Germany [20] and Vietnam [21] despite a higher level of heterogeneity in terms of Willingness to Pay (WTP) preferences of the latter as a developing country.

7 CHALLENGES FOR FRAMEWORK IMPLEMENTATION: FROM VISION TO MISSION

Taking into account most of the lesson learnt elaborated in Section 6, limiting the scope of this study to only flood risk assessment technical aspect shown in Figure 1a, potentially challenge the implementation of the proposed framework in terms of the following points. First, the United Kingdom with its *Making Space for Water (MSW)* undergoes the question of how the vision can be turned into mission (i.e. "policy prescription" [18]). In order to achieve successful implementation, the incorporation of the corresponding "legislative, financial, cultural, organizational, and social changes" aspects is significant. The second challenge is related to the stakeholders' (e.g. lay people's) perception [22] and "preferences for risk assessment indicators and assessment deliverables" [23], public participation [24] in decision making process [25] instead of "technocratic approach" [26], as well as policy of differing conceptual approaches to risk [27, 28]. Those two points imply the necessity to see to what extent the flood risk assessment framework can be implemented in a scope isolation of this study.

8 CONCLUSION

Sharing a common understanding upon terminologies and having a same perspective in discussing the flood risk and its assessment are important to "address definitional and conceptual ambiguities". From that departure point, some countries have shifted their paradigm from "keeping flood water out", i.e. "flood fighting", to "flood defense", and finally "flood risk management", e.g. the United Kingdom, the Netherlands, and Germany. The latest concept strives for balance inclusion between managing the loss and benefit of flood event. Taking the defined "flood risk" as the function of the probability occurrence of a flood event and its consequences, those two sides of the coin are accommodated in the proposed framework of this study, specifically in their quantifications of the consequences analysis phase. For the reason of scope limitation, the framework is isolated only in the flood risk assessment as the core part of the whole flood risk management process. However, it is well understood that the non-technical aspects of "communication and consultation" as well as "monitoring and review" cannot be fully separated. It implies the necessity of anticipating its impact on the implementation results. In addition, the effectiveness of the resources allocation as the purpose of the framework needs to be measured, e.g. in a way represented by the effectiveness of the selected flood risk reduction strategy as the output of the framework implementation.

This study is a part of the doctoral research conducted by the corresponding author and sponsored by the Directorate General of Resources of Science, Technology, and Higher Education, Ministry of Research, Technology, and Higher Education.

References

1. P. Kungwani, IOSR J. Bus. Manag. Ver. III **16**, 83 (2014)
2. ISO31000, *Risk Management - Principles and Guidelines*, First Edit (International Standard Organization, Geneva, Switzerland, 2009)
3. A. Hooijer, F. Klijn, G.B.M. Pedroli, and A.G. van Os, River Res. Appl. **20**, 343 (2004)

4. Business Dictionary.com (2017)
5. D. Crichton, Nat. Disaster Manag., 102 (1999)
6. P. Samuels, F. Klijn, and J. Dijkman, Irrig. Drain. **55**, S141 (2006)
7. S.N. Jonkman, *Loss of Life Estimation in Flood Risk Assessment - Theory and Applications* (2007)
8. S. Gerrard and J. Petts, in *Risk Assess. Risk Manag.*, edited by R.E. Hester and R.M. Harrison, 1–20 (The Royal Society of Chemistry, Cambridge, UK, 1998)
9. E.C. Penning-Rowsell and J. Pardoe, Environ. Plan. C Gov. Policy **30**, 448 (2012)
10. N. Ferrier and C.E. Haque, Nat. Hazards **28**, 271 (2003)
11. ASFPM, 1 (2009)
12. B. D. W. Davis and D. F. Munger, 1 (2014)
13. J. Bourn, *Inland Flood Defence* (London, 2001)
14. C. Johnson, E.C. Penning-Rowsell, and D. Parker, Geogr. J. **173**, 374 (2007)
15. C. Johnson, E.C. Penning-Rowsell, and S. Tapsell, Area **39**, 214 (2007)
16. G. Zhai, T. Sato, T. Fukuzono, S. Ikeda, and K. Yoshida, J. Am. Water Resour. Assoc. **42**, 927 (2006)
17. Z.W. Kundzewicz and K. Takeuchi, Hydrol. Sci. J. **44**, 417 (1999)
18. C.L. Johnson and S.J. Priest, Water Resour. Dev. **24**, 513 (2008)
19. DEFRA, *Making Space for Water - Developing a New Government Strategy for Flood and Coastal Erosion Risk Management in England* (London, 2004)
20. D. Osberghaus, *The Determinants of Private Flood Mitigation Measures in Germany - Evidence from a Nationwide Survey* (Mannheim, 2014)
21. A. Reynaud and M.H. Nguyen, Environ. Model. Assess. **21**, 603 (2016)
22. R. Raaijmakers, J. Krywkow, and A. van der Veen, Nat. Hazards **46**, 307 (2008)
23. S. Scheuer, D. Haase, and V. Meyer, Comput. Environ. Urban Syst. **37**, 82 (2013)
24. L.S. Matott, J.E. Babendreier, and S.T. Purucker, Water Resour. Res. **45** (2009)
25. W. Kenyon, Ecol. Econ. **64**, 70 (2007)
26. J.D. Brown and S.L. Damery, Trans. Inst. Br. Geogr. **27**, 412 (2002)
27. C. Butler and N. Pidgeon, Environ. Plan. C Gov. Policy **29**, 533 (2011)
28. J.A. Bradbury, Sci. Technol. Hum. Values **14**, 380 (1989)

Assessing the impact of climate change on the water intake of the headworks on the Red River Basin in Viet Nam

Lan Pham Thi Huong[1,*], *Nguyet* Le Minh[1,**]
[1]Faculty of Hydrology and Water Resources, Thuy Loi University, Hanoi, Vietnam

Abstract. The irrigation systems for the Red River Delta usually get water by pumping and gravity.The results determine the possible minimum water level line in the dry-season months under climate change scenarios (B2, SC5 and SC6 and current scenario is B2, SC4) on the river system of the Red River – Thai Binh River in study areas of project as below: Water levels on the Red River in the dry season under climate change scenarios significantly reduced compared with 2010. At Ha Noi station on Red River, the water levels reduce in 2050 about 6.5cm (B2) and 7.5cm (A2) and 6cm (B1). At Nam Dinh station on Dao River, the water levels reduce in 2050 about 4cm (B2) and 5cm (A2) and 3cm (B1). At Cat Khe on Thai Binh River, the water levels reduce in 2050 about 2cm (B2) and 3cm (A2) and 1.5cm (B1). At An Bai on Kinh Thay River, the water levels reduce in 2050 about 5cm (B2) and 6cm (A2) and 4cm (B1). At Truc Phuong on Ninh Co River, the water levels reduce in 2050 about 6.5cm (B2) and 7.2cm (A2) and 5.8cm (B1). In the scenario in which 50% of the socio-economic development plans are achieved (SC6), the water demand is lower than the scenario of full development (SC5). The decrease in the average water level in the whole Red River system is about 1.3-1.8 cm less than in SC5, and in Thai Binh river this figure is about 1.4-2.1cm. The lowering of the water level under climate change in the river has caused problems for irrigation during the dry season, in order to reach the target level in Hanoi.

Key Words

Climate change, water intake, Red River Basin

1 Introduction

The Red River Basin is a trans-boundary river basin flow through three countries Vietnam, China and Laos with the total area of approximately 170.000km2 in which around 88.000 km2 in the territory of Vietnam, accounting for 51% of the entire basin area. The river ends in a delta area, located entirely within the territory of Vietnam, which

* Corresponding author: lanpth@wru.vn
** nguyetlm410@wru.vn

 DOI: 10.1051/matecconf/201713809003

covers about 17.000km2.

The length of the Red river in the territory of Vietnam is about 328km. The area located in China is 81.000 km2, accounting for 48% of the entire basin area. The part of the catchment located in Laos is 1,000km2, accounting for 0.7% of the entire basin.

The Red river delta consists of 11 provinces: Hanoi, Hai Phong, Hai Duong, Hung Yen, Bac Ninh, Vinh Phuc, Quang Ninh, Thai Binh, Ha Nam, Nam Dinh, Ninh Binh. This study focuses on four of them: Hai Duong, Nam Dinh, Hai Phong and Thai Binh.

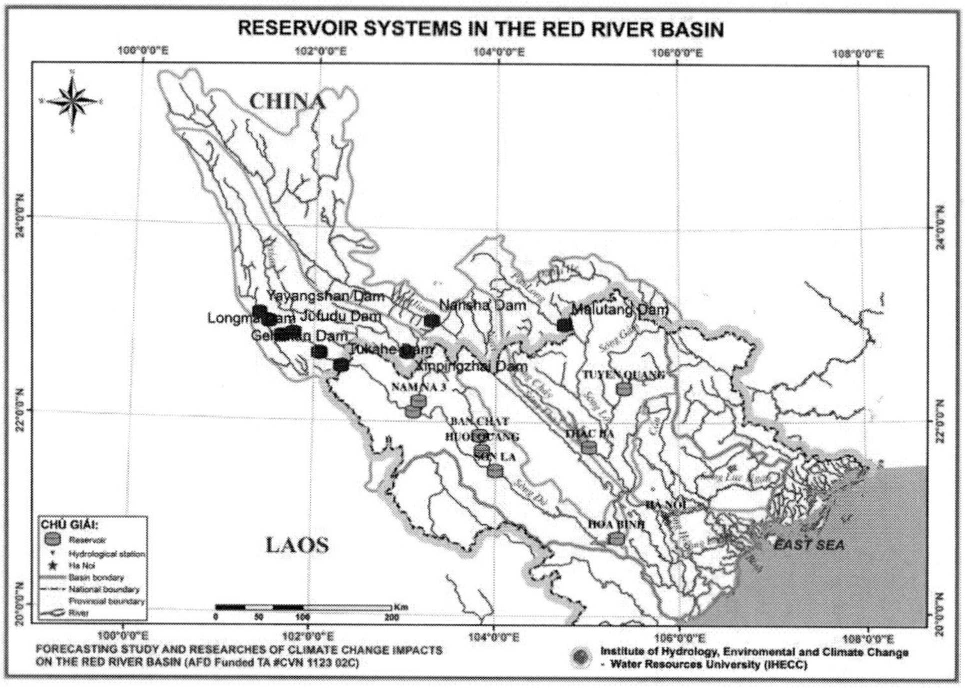

Fig 1. The Red River basin

The climate of the Red River Basin is characterised by high humidity and temperature typical of the tropic climate with a clear seasonal division. At the higher latitudes of the catchment, however, the climate has some similarities with the subtropical area.

The flow in the Red River – Thai Binh River basin is generally abundant. The mean annual total amount of the flows is about 118 billion m3 in Son Tay corresponding to 3743 m3/s. When the Thai Binh river, Day river and the delta area are considered, the total flow reaches to 135 billion m3 per year, of which 82.54 billion m3 (approximately 61.1%) produced in Vietnam and 52.46 billion m3 (approximately 38.9%) produced in the territory of China. However, due to terrain and uneven distribution of rainfall, the flow in different parts of the basin is also very different. As a major branch of the Red River, the Da River has the largest flow that accounts for 42% of the total flow measured in Son Tay. The Thao River covers a drainage area approximately equal to that of the Da River but has the smallest flow, which accounts for 19%. The Lo River has the smallest drainage area but has a significant flow after the Da River, accounting for 25.4%.

The main dams: In China have been build 9 reservoirs in Đà river and 2 reservoirs in Thao river, mainly built in 2007 (Đà river), 2005 (Lô Gâm river), 2010 (Thao river) with a capacity of about 4 billion in which total capacity on Da River about 2,251 billion billion m3. The reservoirs in China are mainly used for hydroelectric power generation, not downstream flood protection; In Viet Nam: In the Red Thai Binh system in Vietnam there are now 4 large multi-purpose reservoirs namely Thac Ba, Hoa Binh, Tuyen Quang and Son La is responsible for downstream flood control and Ban Chat, Huoi Quang, Lai Chau has no task downstream flood prevention, only power production tasks. The total active storage capacity is about 18,53 billion m3, in which the capacity to flood control about 8,45 billion billion m3, and the installed hydropower capacity of 6722 MW. Systems for integrated use reservoirs for water supply, downstream flood control, power generation, combined transport, aquaculture and eco-environmental protection on the main line of the Red River systeminclude:

- Thac Ba Reservoir on the Chay river with flood control capacity is 0.45 billion m3.
- Hoa Binh and Son La Reservoir on the Da River with flood control capacity is 7 billion m3.
- Tuyen Quang Reservoir on the Gam River with flood control capacity is 1 billion m3.

The total flood control capacity for RRD's existing reservoir systemis 8.45 billion m3.

In addition to the large reservoir above the Red River basin, there are some other large reservoirs but only electricity production tasks, no task Flood control, including:

- Reservoir Lai Chau, a capacity is 1,215 million m3 full, useful capacity is 711 million m3
- Reservoir Ban Chat, a capacity is 2137.7 million m3, usefulcapacity is 1720 million m3.

The total area required for irrigation is 883 578 ha, the irrigation designed area is 857 735 ha, actively irrigated area is 609 877 ha, accounting for 69% of requirement (mainly for paddy-rice areas, while the perennial is no irrigation). The irrigated area is shown in the following figure 2:

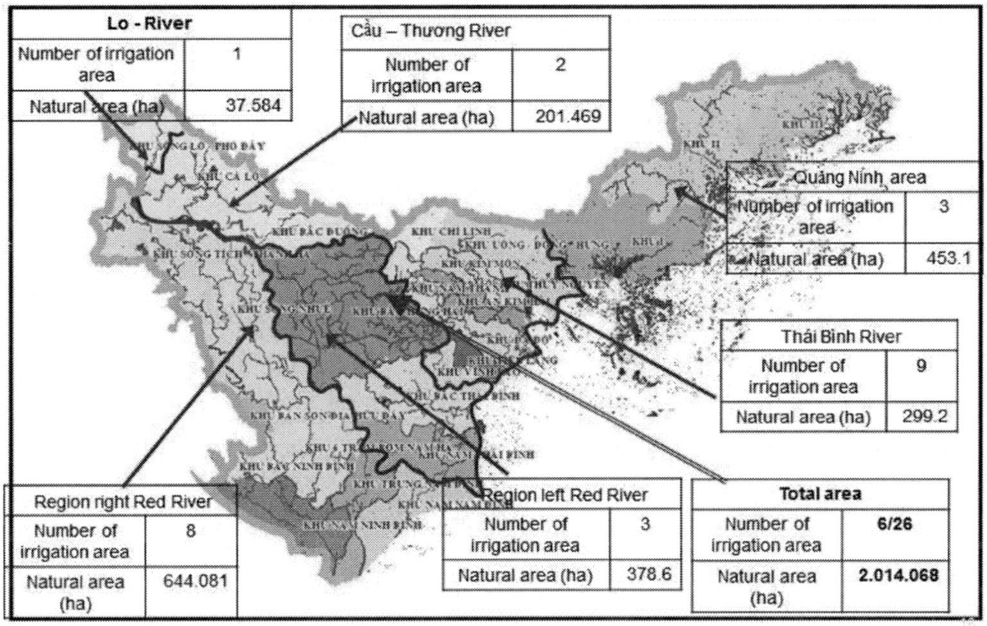

Fig 2. The irrigated area

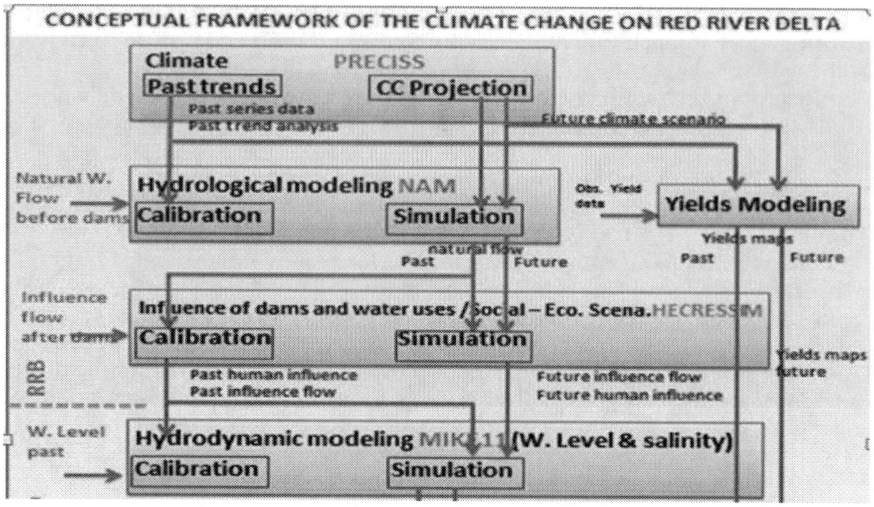

Fig 3. Methodology

2 Methodology

We used three models—Rainfall-Runoff, Hydrodynamic, and Salinity models—to capture the effects of all the factors that affect salinity in the downstream of Red River basin (fig. 3)

3 The results

The irrigation systems for the Red – Thai Binh River Delta usually get water by pumping and gravity:

- The systems with the high terrain, lying inland (inland system) are usually provided with water by pump stations in the pumping method. The pump stations get water from the rivers through the water intakes below the dykes and directly carry water into the irrigational canal system(floating canal).
- The irrigational systems with the low terrain and lying on the seaside often get water in the gravity method through water intakes by sluice.
- The water is carried from the canals into the buried canal system, and then it runs or is pumped into the irrigational canals (floating canal).

Some systems combine gravity method and pumping.

The intake operates as follows: - When water level in river lower than in the field: intake is closed. - When water level in river higher than in the field and salinity level less than or equal to 0.1%: intake is opened. (Crops can tolerate salinity of less than 0,1%, hence when river salinity is greater than 0.1%, the intake has to be closed).

The results determine the possible minimum water level line in the dry-season months under climate change scenarios (B2, SC5 and SC6 and current scenario is B2, SC4) on the river systemof the Red River – Thai Binh River in study areas of project as below:

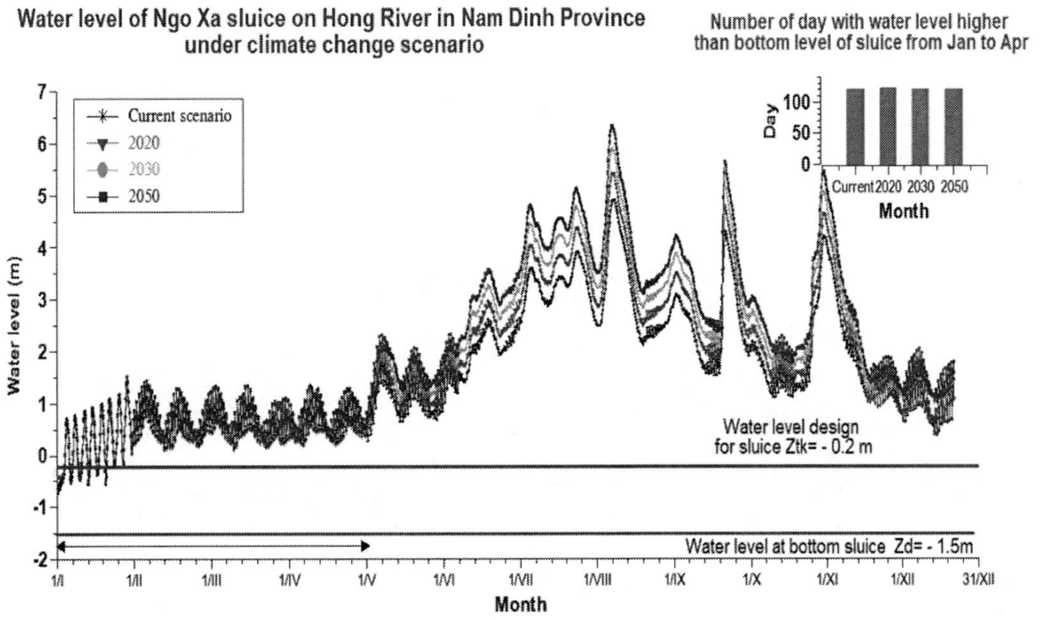

Fig 4. Water level of Ngo Xa sluice on Hong River in Nam Dinh Province under climate change scenario

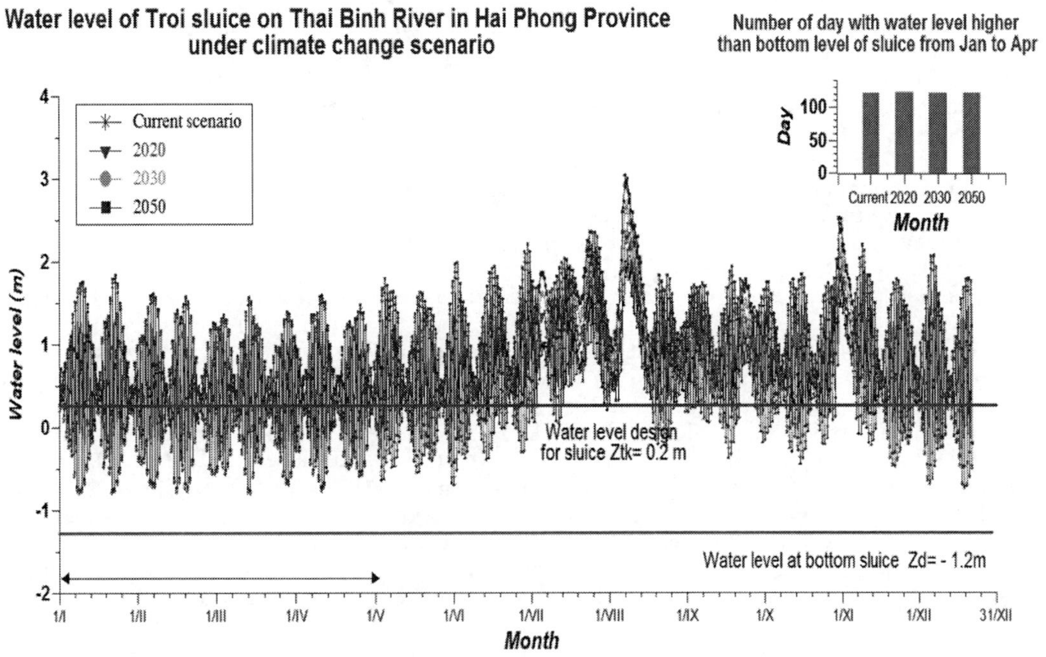

Fig 5. Water level of Troi sluice on Thai Binh River in Hai Phong Province under climate change scenario

 DOI: 10.1051/matecconf/201713809003

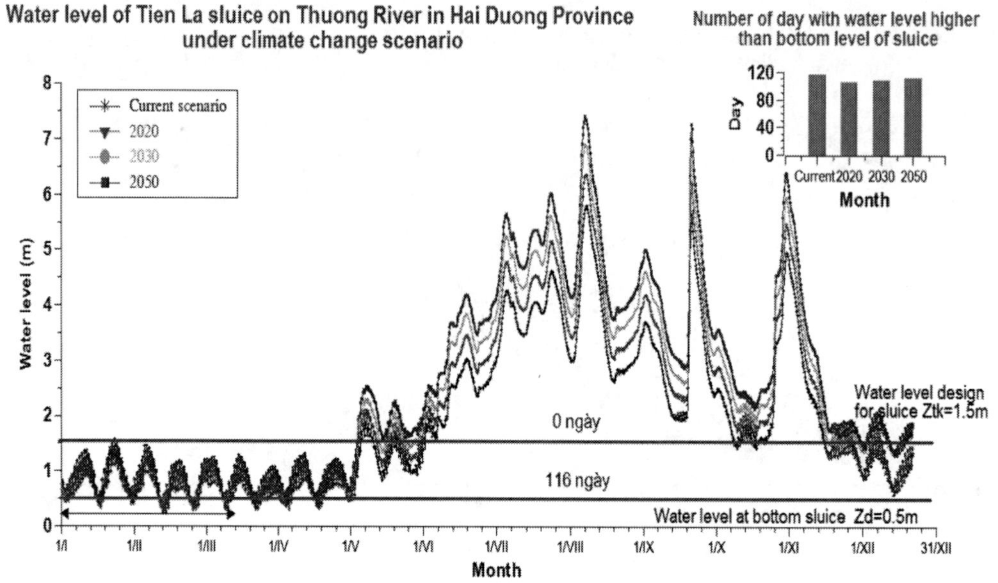

Fig 6. Water level of Tien La sluice on Thuong River in Hai Duong Province under climate change scenario

Some commented as follows:
- Water levels on the Red River in the dry season under climate change scenarios significantly reduced compared with 2010. At Ha Noi station on Red River, the water levels reduce in 2050 about 6.5cm (B2) and 7.5cm (A2) and 6cm (B1). At Nam Dinh station on Dao River, the water levels reduce in 2050 about 4cm (B2) and 5cm (A2) and 3cm (B1). At Cat Khe on Thai Binh River, the water levels reduce in 2050 about 2cm (B2) and 3cm (A2) and 1.5cm (B1). At An Bai on Kinh Thay River, the water levels reduce in 2050 about 5cm (B2) and 6cm (A2) and 4cm (B1). At Truc Phuong on Ninh Co River, the water levels reduce in 2050 about 6.5cm (B2) and 7.2cm (A2) and 5.8cm (B1).
- In the scenario in which 50% of the socio-economic development plans are achieved (SC6), the water demand is lower than the scenario of full development (SC5). The decrease in the average water level in the whole Red River system is about 1.3-1.8 cm less than in SC5, and in Thai Binh river this figure is about 1.4-2.1cm.
- In January and February is the time to use the most water for agriculture, the water level at some sluice is smaller than culvert high, thus leading to a number of days in January and February did not take enough water. Water levels on the Red River in the dry season under climate change scenarios significantly reduced compared with 2010. At Ha Noi station on Red River, the water levels reduce in 2050 about 6.5cm (B2) and 7.5cm (A2) and 6cm (B1). At Nam Dinh station on Dao River, the water levels reduce in 2050 about 4cm (B2) and 5cm (A2) and 3cm (B1). At Cat Khe on Thai Binh River, the water levels reduce in 2050 about 2cm (B2) and 3cm (A2) and 1.5cm (B1). At An Bai on Kinh Thay River, the water levels reduce in 2050 about 5cm (B2) and 6cm (A2) and 4cm (B1). At Truc Phuong on Ninh Co River, the water levels reduce in 2050 about 6.5cm (B2) and 7.2cm (A2) and 5.8cm (B1).

The lowering of the water level under climate change in the river has caused problems for irrigation during the dry season, in order to reach the target level in Hanoi.

4 Discussion and conclusions

The outputs from the MIKE 11 model steady state runs used to derive rating curves at each gauge location. It was typically found that the MIKE 11 model results matched well with gauged estimates of streamflow. The MIKE 11 model produced good calibrations and verification to all hydraulic stations on the Red River.

The model was used to model the impacts of climate change on flow regime in the Red River basin. In order to operate the MIKE 11 model, several data inputs are required, including the river network alignment, channel and floodplain cross sections, boundary data, and roughness coefficients. In hydraulic modeling, boundary conditions are required to provide the model input at the edges of the domain. Upper boundary is hydrographs formulated by MIKE NAM model from the RRB in the following 11 sub-catchment and the lateral discharge of the model is the process lines of flow Q = f(t) entering the middle sub catchment calculated by hydrological model in currents condition and the future climate change.

References

1. P.T.H. Lan. *Report on the results: forecasting sudy and reseaches of climate change impacts on the Red River Basin*. AFD Funded TA # CVN 1123 02C (2015).

2. T. Thuc, N.V. Thang, H.T.L. Huong, M.V. Khiem, N.X. Hien, D.H. Phong. *Climate change scenario of Viet Nam 2012*. (MONRE, Vietnam, 2012).

DOI: 10.1051/matecconf/201713809004

Mapping of Biomass Production of Land Damage Assessment to Reduce Environmental Changes In East Java Probolinggo

Ali Mahrus[1], *Hariyadi* Bambang Wicaksono[2], *Nurlina*[3], *Hasyim* Cholil[4] , *and Mudjanarko* Sri Wiwoho[5],*

[1,2,3] Agriculture, Agrotechnology Department, Ketintang VII/2, Merdeka University, Indonesia,60232
[4] Civil Engineering, Civil Department, Darul Ulum University, Gus Dur No.29A, Mojopahit, Jombang 61419 Indonesia
[5] Civil Engineering, Civil Department, Narotama University,Arief Rachman Hakim No. 51, Surabaya, 60117,Indonesia

Abstract. The increasing population in Probolinggo, East Java, Indonesia require residential land. Excessive use of housing land reduces agricultural lands will impact environmental changes surrounding area. Limitations of use of agricultural land due to the construction of housing in need of good quality soil in increasing food production. This study aims to determine the value of the damage of Land For Biomass Production to reduce the environmental changes surrounding area. The method used in the form of land damage mapping biomass production. Variable ground damage assessment status of each parameter is done by weighting and scoring methods. The results showed that the status of land degradation in the District Paiton is the status of damaged land with an area of 3922.74 hectares of land and/or land for biomass production with an area of 147,234.60 hectares, or approximately 2.66% of the total area of the District Paiton. The area of a degraded land category is at the Village Bhinor, Jabung comb, middle Alas, Kalijajar wetan and Plampang.

1 Introduction

The total area of ProbolinggoSub districtis approximately 1696.16km²consisting 147,74km² of settlements, 373.13 km² of Rice Field, 513.80 km² of moor, 32,81km² of plantation, 426.46 km² of forest, 13.99 km² of pond/pool, and 188.23 km² of other utilization [1]. Terrestrial environment is a stretch of the earth's surface which is the unity of space with energy, materials of various types of biomass and living creatures and man with all his behavior. Terrestrial environment with its high environmental or natural resources leads a lot of human activity. Some areas of land are used as residential, industry, mining, agriculture, livestock farming, aquaculture fresh water and other areas. Many areas of land

* Corresponding author: sri.wiwoho@narotama.ac.id

have experienceddegradation; one of which is due to soil erosion as a result of high rainfall, low vegetation covered-land, conversion of land from forests to agricultural crops, steepland utilization and misused of land can cause erosion, sedimentation and siltation of river. As a result, river water flow decrease and cause flooding and degradation of soil fertility which in turn decrease agricultural production and eventually farmers' income. Control of terrestrial environment damage can be grouped with preventive and reactive approaches. Preventive approach is performed through consistent spatial planning and appropriate space management and the use of environmentally friendly technologies. Reactiveapproachincludes conducting outreach to the community in order to manage the land damaged by erosion and pollution as well as to avoid the occurrence of land conversion. Preserving environment is not merely the responsibility of government or heads of state, but every human being on earth. Everyone should make an effort to save the environment in accordance with the capacity of each;the involvement of individual beneficial for the realization of habitableearth for the next generation. Measurement procedure for the of standard criteria of damaged land for biomass production is structured to explain steps to be performed by the regent/mayor related to monitoring and surveillance.

2 Research Methods

The research was conducted at critical lands inside and outside forest areas in five villages in PaitonSub district, Probolinggo in 2014. Case study method, with land based on a survey by scale detail in the ratio of 1: 10,000 or 1: 25,000 was used as an object. Data was collected and then proceededto describeobjective conditions of critical land in Probolinggo as a basis for policy making ofdeveloping the damaged land into productive one. Soil sampling was carried out by using *overlay* or *superimpose* on some thematic maps to obtain an overview of the areas that can potentially damage the land/soil and from the result of coordination with relevant agencies.

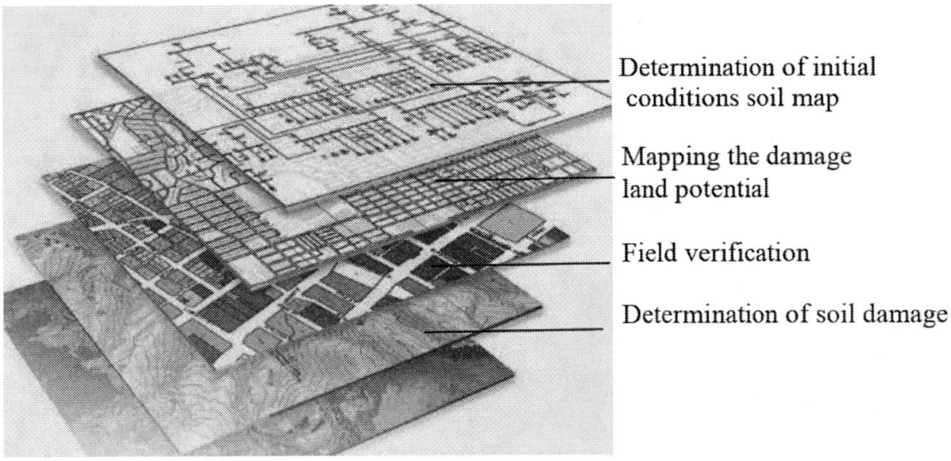

Fig. 1. Illustration of the superimpose / overlay thematic map for the determination of land analysis unit.
Source: Data Analysis Year 2014. [1]

3 Results And Discussion

Based on the analysis of land use map overlay (for biomass production), slope maps, soil maps and rainfall maps in Probolinggo Sub district, 8 land units for four study location villages inPaiton Sub districts. Theresult of the observation and interpretation is explained as follow: [2-9]

2.1 Paiton Subdistrict

Paiton sub district includesBhinor, Jabung Sisir, Alas Tengah, Kalijajar Wetan dan Plampangvillages. The soil characteristics ineach village vary as follow:

2.1.1 The Bhinor Village (Plantation)

Table 1. Results of Analysis and Interpretation of Land Damage in Bhinor (P) Village, Paiton Sub district,Probolinggo.

No.	Parameter	Critical threshold (PP150 / 2000)	Observations/ Analysis	Exceeding/Not exceeding
1.	Solum thickness	<20 cm	> 90 cm	not Critical
2.	Surface rocks	40%	<3%	not Critical
3.	fraction composition	<18% colloidal 80% sand kuarsitik	6.26%	not Critical
4.	weight content	1.4 g / cm ^{3}	1.18 g / cm ^{3}	not Critical
5.	porosity Total	<30%; > 70%	50.52%	not Critical
6.	The degree of release of water	<0.7 cm / hour; 8.0 cm / hour	1.16 cm / hour	not Critical
7.	pH (H $_2$ O) 1: 2,5	<4.5; > 8.5	6.6	not Critical
8.	Electrical Conductivity	4.0 mS / cm	0.259 mS / cm	not Critical
9.	redox	<200 mV	59.0 mV	Critical
10.	number of Microbes	<10 ^{2}cfu / g soil	Bacteria> 300 x 10 ^{9} Mushrooms 16 x 10 ^{9}	not Critical

Note: Exceeding = Critical
No = Not Critical

Based on the land conditionobtained from field, the redox was low by 59.0 mV less than 200 mV indicating that the land was water saturated and low oxygen presented in the soil that fe and mn elements are conditioned in redox state, rooting faces obstacles in its development.

The solution to restore land is by manufacturing irrigation channels and drainage to enable oxygen penetrates into the soil pores, continuousorganic fertilizer givenin ech soil cultivation is preferable to increase the pore space of soil and organic matter availability.

2.1.2 Bhinor Village (Rain fed rice field)

Table 2. Results of Analysis and Interpretation of BhinorVillageLand Damage (STH), Paiton Sub district,Probolinggo.

No.	Parameter	Critical threshold (PP150 / 2000)	Observations / Analysis	Exceeding / Not exceeding
1.	Solum thickness	<20 cm	> 90 cm	not Critical
2.	Surface rocks	40%	<3%	not Critical
3.	fraction composition	<18% colloidal 80% sand kuarsitik	3.24%	not Critical
4.	weight content	1.4 g / cm 3	1.26 g / cm ^{3}	not Critical
5.	porosity Total	<30%; > 70%	46.13%	not Critical
6.	The degree of release of water	<0.7 cm / hour; 8.0 cm / hour	1.51 cm / hour	not Critical
7.	pH (H_2O) 1: 2,5	<4.5; > 8.5	6.6	not Critical
8.	Electrical Conductivity	4.0 mS / cm	0.176 mS / cm	not Critical
9.	Redox	<200 mV	55.2 mV	Critical
10.	number of Microbes	<10 2cfu / g soil	Bacteria 2 X 10 ^{9} Mushrooms 2 X 10 ^{9}	not Critical

Note: Exceeding = Critical
No = Not Critical
A constraint faced in Bhinor village is similartoBhinor village plantation with constraints on redox by 55.2 mv.Intensive processing and long period ofland inundationhampers oxygen fromentering soil. This causes toxic elements highlyinfluenced the development of plant roots.
The solution to be taken to overcome critical land is creating irrigation channels and drainage as well as giving organic fertilizer and manure or hay or remnants of harvests. The soil should be also processed as minimally toincrease pore space with minimum inundation.

2.1.3 Bhinor Village (Production Forest)

Table 3. Results of Analysis and Interpretation of the BhinorVillage Land Damage (HP), PaitonSub district, Probolinggo.

No.	Parameter	Critical threshold (PP150 / 2000)	Observations / Analysis	Exceeded / Not
1.	Solumn thickness	<20 cm	> 90 cm	not Critical
2.	Surface rocks	40%	<5%	not Critical
3.	fraction composition	<18% colloidal 80% sand kuarsitik	3,07%	not Critical
4.	weight content	1.4 g / cm ^{3}	1.16 g / cm ^{3}	not Critical
5.	porosity Total	<30%; > 70%	51.48%	not Critical
6.	The degree of release of	<0.7 cm / hour;	1.32 cm / hour	not Critical

	water	8.0 cm / hour		
7.	pH (H_2O) 1: 2,5	<4.5; > 8.5	6.8	not Critical
8.	Electrical Conductivity	4.0 mS / cm	0.090 mS / cm	not Critical
9.	redox	<200 mV	42.9 mV	Critical
10.	number of Microbes	$<10^2$cfu / g soil	bacteria - mushrooms -	Critical

Note: Exceeding = Critical
No = Not Critical

Bhinor village land, Paiton subdistrict is categorizedas critical condition with problems in the redox and the number of microbes. However, there are many alternative solutions available. The characteristic of production forest is similar to other Bhinor land utilization. The difference is on production forest land withzero number of microbes. This could be caused by redox reactions and organicpesticide use that hampers bacteria growth microbial life in the soil.
Soil should not be cultivated intensivelyas the more intensive soil cultivation the higher the compaction and surface water frequently floods impermeable surfaceslayer. Critical threshold for BhinorVillage area is needs attentionto prevent it from happening again that soil physical and chemicalpropertiesdo not degrade the land quality. Land management should be maintained by creating irrigation channels and drainage as well as minimum tillage for annual or seasonal crops is adapted to the suitability of land.Organic pesticidesis better used for eradication of pests and diseases to avoid bacterial decomposers death due to poisoning elements of inorganic pesticides. Organic materials in addition to the manure and dry leaves are more advisable.

2.1.4 Bhinor Village (Moor)

Table 4. Results of Analysis and Interpretation of the BhinorVillage Land Damage (T), PaitonSub district, Probolinggo.

No.	Parameter	Critical threshold (PP150 / 2000)	Observations/Analysis	Exceeded / Not
1.	Solumn thickness	<20 cm	> 90 cm	not Critical
2.	Surface rocks	40%	<5%	not Critical
3.	fraction composition	<18% colloidal 80% sand kuarsitik	14.80%	not Critical
4.	weight content	1.4 g / cm^3	1.31 g / cm^3	not Critical
5.	porosity Total	<30%; > 70%	43.89%	not Critical
6.	The degree of release of water	<0.7 cm / hour; 8.0 cm / hour	0.20 cm / hour	Critical
7.	pH (H_2O) 1: 2,5	<4.5; > 8.5	6.6	not Critical
8.	Electrical Conductivity	4.0 mS / cm	0.228 mS / cm	not Critical
9.	redox	<200 mV	64.1 mV	Critical
10.	number of Microbes	$<10^2$cfu / g soil	Bacteria> 300 X 10^9 Mushrooms 3 X 10^9	not Critical

Note: Exceeding = Critical
No = Not Critical

Moor inBhinorvillage, Paiton subdistrict indicatedcritical land with constraints on the parameters of water discharge and redox degrees. The degree of water discharge is 0.20 cm/

hour passing critical limit and indicated that there is problem on the rate of water entering the ground, as a result, nutrients are swept away by the water following the flow of the surface that carry fertilizer in it. Waterlogged or flooded land tends to experience flooding due to slowflow ofwater into the ground. The redox is lowthat soil is frequently poisoned byFe and Mn that affect plant growth.

The solution for this problem is to reduce intensive tillage with minimum processing and making drainage channels. To improve the soil, organic fertilizer from cow manure can be usedto create nutrients and ongoing harvest. The use of organic pesticides is more advisable to increase soil porosity.

2.1.5 Jabung Village (Rice Irrigation)

Table 5. Results of Analysis and Interpretation of JabungSisirVillage Land Damage (SI), Paiton Sub district,Probolinggo.

No.	Parameter	Critical threshold (PP150 / 2000)	Observations / Analysis	Exceeded / Not
1.	Solumn thickness	<20 cm	> 90 cm	not Critical
2.	Surface rocks	40%	<5%	not Critical
3.	fraction composition	<18% colloidal 80% sand kuarsitik	4.46%	not Critical
4.	weight content	$1.4\ g / cm^3$	$1.38\ g / cm^3$	not Critical
5.	porosity Total	<30%; > 70%	41.43%	not Critical
6.	The degree of release of water	<0.7 cm / hour; 8.0 cm / hour	0.04 cm / hour	Critical
7.	pH (H_2O) 1: 2,5	<4.5; > 8.5	6.4	not Critical
8.	Electrical Conductivity	4.0 mS / cm	0.154 mS / cm	not Critical
9.	redox	<200 mV	48.9 mV	Critical
10.	number of Microbes	$<10^2$ cfu / g soil	Bacteria> 300 X 10^9 Fungi 6 X 10^9	not Critical

Note: Exceeding = Critical
No = Not Critical

Land inJabungSisir village, Paiton subdistrict indicated critical land with problem on degree of water discharge and redox. Based on the landcondition obtained by field research, the problems is identical to that ofBhinorvillage. The degree of water discharge is very low that a lot of water stagnate the surface of the land because of land compaction. The redox availability is low due to less supportive irrigationchannels, causing that plants cannot grow and develop well.

The solution to preserve the land is to create of irrigation channels and use organic fertilizers continuously in every tillage. Other land conservation efforts is also needs to be implemented to plant annual crops intercropped with seasonal crops and farmers should reduce the use of inorganic pesticides for land sustainability.

2.1.6 Alas Tengah

Table 6. Results of Analysis and Interpretation of Land Damage in Alas Tengah Village, Paiton Sub district,Probolinggo

No.	Parameter	Critical threshold (PP150 / 2000)	Observations / Analysis	Exceeded / Not
1.	Solumn thickness	<20 cm	> 90 cm	not Critical
2.	Surface rocks	40%	<5%	not Critical
3.	fraction composition	<18% colloidal 80% sand kuarsitik	12.07%	not Critical
4.	weight content	1.4 g / cm ^{3}	1.28 g / cm ^{3}	not Critical
5.	porosity Total	<30%; > 70%	44.36%	not Critical
6.	The degree of release of water	<0.7 cm / hour; 8.0 cm / hour	0.32 cm / hour	Critical
7.	pH (H_2O) 1: 2,5	<4.5; > 8.5	6.9	not Critical
8.	Electrical Conductivity	4.0 mS / cm	0.043 mS / cm	not Critical
9.	redox	<200 mV	56.4 mV	Critical
10.	number of Microbes	<10 ^{2}cfu / g soil	Bacteria> 300 X 10 ^{9} Mushrooms 2 X 10 ^{9}	not Critical

Note: Exceeding = Critical
No = Not Critical

Soil in Alas Tengah Village, Paiton sub districtindicated problem on the degree of the water discharge and redox. The land problem in this area is similar to that of JabungSisisr village with the degree of water discharge is higher in the latter village yet with change in redox due to frequent cultivation that the land lost organic matter and the soil became more compact.

The existing problem can be addressed with irrigation channels and drainage, organic fertilizer with manure or harvest residue of hay or remnants of harvest along with minimal processingto improve redox. The use of organic pesticides is preferable because in addition to increase the number of bacteria and fungi also help the availability of organic matter in the soil.

2.1.7 Kalijajar Wetan Village

Table 7. Results of Analysis and Interpretation of Soil Damage in KalijajarWetan Village, PaitonSub district, Probolinggo.

No.	Parameter	Critical threshold (PP150 / 2000)	Observations / Analysis	Exceeded / Not
1.	Solumn thickness	<20 cm	> 90 cm	not Critical
2.	Surface rocks	40%	<3%	not Critical
3.	fraction composition	<18% colloidal 80% sand kuarsitik	5.66%	not Critical
4.	weight content	1.4 g / cm ^{3}	1.46 g / cm ^{3}	not Critical

5.	porosity Total	<30%; > 70%	33.22%	not Critical
6.	The degree of release of water	<0.7 cm / hour; 8.0 cm / hour	0.40 cm / hour	Critical
7.	pH (H_2O) 1: 2,5	<4.5; > 8.5	6.2	not Critical
8.	Electrical Conductivity	4.0 mS / cm	0.247 mS / cm	not Critical
9.	redox	<200 mV	86.6 mV	Critical
10.	number of Microbes	<10^2cfu / g soil	Bacteria> 300 X 10^9 Mushrooms 11 X 10^9	not Critical

Note: Exceeding = Critical
No = Not Critical

The land in KalijajarWetan Village was in critical condition with problems in water discharge and redox degree, while there are vary alternative solutions available. The characteristic of land in KalijajarWetan village is also marked with problem of water movement patterns. The degree of water discharge was low that water movement is slow, this makes nutrients quickly be swept away by surface runoff so that plant roots grow poorly.Annual crop of this village grows better such as star fruitand mango tresswith improvements that need the participation of the whole community. The critical threshold for KalijajarWetan village area needs attentionto avoid critical threshold and reduction of physical and chemical properties of the soil.
The existing problem can be addressed by making irrigation and land management. This can be carried out by growing annual crop according to land suitability and reduce the degree of small water discharge with the use organic matter from cow dung and dry leaves.

2.1.8 Plampang village (Rice Irrigation)

Table 8. Results of Analysis and Interpretation of PlampangVillage Land Damage (SI), PaitonSub district, Probolinggo.

No.	Parameter	Parameter	Critical threshold (PP150 / 2000)	Observations / Analysis	Exceeded / Not
1.	Solumn thickness	Solum thickness	<20 cm	> 90 cm	not Critical
2.	Surface rocks	Surface kebatuan	40%	<5%	not Critical
3.	fraction composition	composition fraction	<18% colloidal 80% sand kuarsitik	7,82%	not Critical
4.	weight content	weight content	1.4 g / cm^3	1.39 g / cm^3	not Critical
5.	porosity Total	porosity Total	<30%;> 70%	38.32%	not Critical
6.	The degree of release of water	The degree of release of water	<0.7cm/hour; 8.0 cm / hour	3.50 cm / hour	not Critical
7.	pH (H_2O) 1: 2,5	pH (H_2O) 1: 2,5	<4.5; > 8.5	6.5	not Critical

8.	Electrical Conductivity	Electrical Conductivity	4.0 mS / cm	0.242 mS / cm	not Critical
9.	redox	redox	<200 mV	58.7 mV	Critical
10.	number of Microbes	number of Microbes	<10 ^{2}cfu / g soil	Bacteria 43 x 10 $_7$ mushrooms -	Critical

Note: Exceeding = Critical
No = Not Critical

Irrigation rice fieldindicated degraded land with problem initiated on redox and number of microbes. Water conductivity was low by 3.50 cm/hour or at critical limits indicated in nutrients lost to runoff andlow availability of organic material. The problems are indicated by firm ground structure and the soil is difficult to be destroyed due to bonding clay and sedimentary limestone.High rainfall and intensive land management makes upper soil layer lost. Land improvement by planting cover crop that soil is not easily lost and by giving crops remnants or organic fertilizer to improve soil organic matter content in the soil.
The number of redox was low by 58.7 mV thatthe soil was dominated by Fe and Mn which can be poisonous and therefore contributes to a very few number of microbes. This makes bacteria cannot live optimally function as decomposers to provide availability of nutrientto plants and provide organic matter into the soil.Intensive cultivation of land cause land to loss its main nutrient of organic material.
The solution is to reduce intensive tillage or Minimum Tillage and planting cover crops. Plant intercropping of annual and seasonal crop can be also implemented that have benefits for maintaining water that it is not easily lost to prevent nutrient erosion on the surface layer of top soil. Soil maintenance can be also carried out according to the land conditions and post-harvest treatment with the returned harvested crops. The use of organic fertilizer from cow manure is also goodto maintain nutrient availability fromongoing harvest. The use of organic pesticides is recommended. Redox can be increasedwith seedbed that allows oxygen to enterthe soil, allowing oxidation process can run.

4 Conclusion

The following conclusion is drawn from the research:

1. Land that qualify as critical in accordance observations indicated low quality of land as a result of intensive land cultivation, low attention onpost-harvest handling regarding soil fertility level and continuousland use.
2. Villages in the Paitondistrictindicated problems that vary according to the condition of bulk density factor, the degree of water dischargeand influencing redox and microbes.
3. Establishing irrigationand drainage channels, Organic fertilizer usage by utilizing the manure from crop, crop rotation patterns and conservation of the land by planting vegetation and in accordance with the conditions are preferable to be carried out for each region.
4. Participation supportfrom the community for sustainable land.

Reference

1. BPS. Probolinggo (2014), the Central Statistics Agency Figures Probolinggo. Probolinggo, East Java.

2. KLH. Repulic Indonesia. (2009), Preparation of Technical Guidelines Map Status Land Damage Biomass Production, Jakarta, 31Halaman.
3. K. Minister of Forestry, **Nomor** 20 / Kpts- Ii / (2001), About PU and Criteria and Standards of Forest and Land Rehabilitation, Jakarta.
4. P. Government **Nomor** 150 of the Year (2000), Land Degradation Control For Biomass Production, Jakarta.
5. PMLH. **Nomor** 07 Year (2006), Procedures Criteria, Damage Measurement of Soil Materials for Biomass Production, Jakarta.
6. Y.A. Butler, K. Wikantika, and F. Hadi, (2005), Spatial Analysis Critical Areas In North Bandung City Grass Using Open Source, Department of Geodesy, Faculty of Civil and Environmental Engineering, Institut Teknologi Bandung - ITB, Jl. Ganesha Nomor 10, Bandung, Indonesia.
7. KH. Tan (1991), Soil Chemistry Basics, UGM, Yogyakarta
8. WH. Utomo (1985), Soil Physics, Brawijaya University Press, Department of Soil Science, Faculty of Agriculture, University of Brawijaya, Poor.
9. Zhiddiq (2005), *Mapping of Critical Areas* Atlas, Your Community, Napier South Celebes Indonesia.

AUTHOR INDEX

AUTHOR INDEX

AUTHOR INDEX